ANIMMA 2021 - Advancements in Nuclear Instrumentation Measurement Methods and their Applications

AA003715

EPJ Web of Conferences Volume 253 (2021)

Prague, Czech Republic
21 - 25 June 2021

Editors:

Abdallah Lyoussi
Michel Carette
Rastislav Hodak
Igor Jencic
Patrick Le Du
Stanislav Pospisil
Christelle Reynard-Carette
Luka Snoj
Ivan Stekl
Ludo Vermeeren

ISBN: 978-1-7138-3952-1

Printed from e-media with permission by:

Curran Associates, Inc.
57 Morehouse Lane
Red Hook, NY 12571

Some format issues inherent in the e-media version may also appear in this print version.

No changes have been made to the content of these proceedings. There may be changes to pagination and minor adjustments for aesthetics.

Printed with permission by Curran Associates, Inc. (2021)

For additional information, please contact EDP Sciences – Web of Conferences at the address below.

EDP Sciences – Web of Conferences
17, Avenue du Hoggar
Parc d'Activité de Courtabœuf
BP 112
F-91944 Les Ulis Cedex A
France

Phone: +33 (0) 1 69 18 75 75
Fax: +33 (0) 1 69 28 84 91

contact-edps@webofconferences.org

Additional copies of this publication are available from:

Curran Associates, Inc.
57 Morehouse Lane
Red Hook, NY 12571 USA
Phone: 845-758-0400
Fax: 845-758-2633
Email: curran@proceedings.com
Web: www.proceedings.com

TABLE OF CONTENTS

https://doi.org/10.1051/epjconf/202125300001

June 21-25, 2021, Prague, Czech Republic

Dear colleagues and friends,
The conference on Advancements in Nuclear Instrumentation Measurement Methods and their Applications ANIMMA has always strived to create a special meeting place shared by all those working in nuclear instrumentation and its applications, as we strongly believe that cross-border exchanges between scientists, engineers and industrialists can only lead to the most developed ideas, the best solutions and the most efficient collaborations and partnerships. ANIMMA continues to maintain a high level of scientific and technical quality by presenting not only the latest advances but also the state of the art in each field through the participation of international specialists and experts.
Since the beginning of 2020, travelling around the world to attend meetings, congresses, symposia, and conferences is highly handicapped by the Covid-19 pandemic whose development differs from place to place and time to time. Teleworking, virtual meetings, distance courses are used as substitutes for the traditional face to face meetings, with an inevitable loss of quality of the human contacts which are believed to be of paramount importance to build enthusiasm for a project, an idea, a collaboration, a grouping of individuals.
In such a context, the ANIMMA2021 committees (Program, Organizing and Steering committees), by following the sanitary rules as well as the Czech government recommendations, dedicated best and huge efforts in designing the Conference in hybrid layout to make it highly attractive and very efficient for both the on-site participants as well as the remote ones.

Therefore the best possible communication means gave the opportunity not only to remotely follow or deliver presentations, but also to interact online in the best possible way with the authors and contributors as well as the exhibitors, with the other connected experts, and to access recorded materials in delayed mode to cope with the time-zone differences. This required also a fully committed and flexible conference management team to create a conference that is rewarding for both presenters and attendees.

ANIMMA-2021 was held from June 21st-25th in hybrid mode at Clarion Congress Hotel in Prague, Czech Republic. The conference was locally organized by the Czech Technical University represented by the institute of experimental and applied physics (IEAP) in a strong partnership with CEA (France), SCK-CEN (Belgium), Aix-Marseille University (France), Josef Stefan Institute (Slovenia). The ANIMMA2021 was technically sponsored by IEEE/NPSS

https://doi.org/10.1051/epjconf/202125300001

ANIMMA 2021

Prague, Czech Republic **Hybrid mode** **June 21-25, 2021**

The conference program was dealing with:

- 2 full-day short-courses: The classical lectures have been partly given on-site, partly via remote connection. They have been streamed to all students simultaneously via a Zoom meeting, allowing for interaction between students and lecturers. The practical sessions have been performed by IEAP and CAEN. A part of exercises was in remote mode, simultaneously for the students on-site and those remotely connected and the other part as a real hands-on course for the students on-site. Separately a remote interactive demonstration session has been organized for the remotely connected students. Participants to the Short Courses had the opportunity to perform immediately after the courses a multiple choice test (35 multiple-choice questions on all courses).
- 3 international workshops: The topics of the workshops addressed in this 7th ANIMMA edition focused on instrumentation for major research facilities in fusion and fission fields, crystalline materials for radiation detection in harsh media and prospective technologies for the future of nuclear medicine. 19 speakers and 71 people participated from 22 countries. Each workshop included time slots for discussion and exchange between speakers, experts and participants. These periods were useful to share opinions, cross-fertilize skills and expertise and identify collaborative research axes to go further by solving scientific and technological challenges.
- 3 prestigious keynote lecturers :
  - Dr. William D. MAGWOOD, IV, Director-General, OECD Nuclear Energy Agency (NEA) *on "Nuclear Solutions: Innovation for an Energy Transition"*
  - Dr. Alan OWENS, European Space Agency, ESTEC and IEAP on *"Recent developments in sensor materials"*.
  - Prof. Dr. Katia PARODI, Ludwig-Maximilians-Universität München, Department of Experimental Medical Physics on *"Getting to the point in particle therapy"*

- 4 plenary sessions in a new transversal topic on *"current trends in development of radiation detectors"*,
- *A second* new topic on "Space sciences and technology".
- Over 270 contributions presented in 31 parallel oral sessions
- A permanent poster session

ANIMMA-2021 welcomed 320 attendees from 29 nationalities including 65 graduated students. 30% of the attendees participated physically and 70% attended remotely.
The ANIMMA2021 attendees deserve our warmest thanks and consideration for making this seventh edition an exceptional success in extraordinary situation.
All the committee members are looking forward to welcoming everyone again at the ANIMMA2023 in Pisa and Viareggio in Tuscany, Italy on June 12-16, 2023.

Prof. Dr. Abdallah LYOUSSI
ANIMMA General Chairman and Program Committee Chair

(*) Advancements in Nuclear Instrumentation Measurement Methods and their Applications

Statement of Peer review

In submitting conference proceedings to EPJ Web of Conferences, I certify to the Publisher that I adhere to the **Policy on Publishing Integrity** of the journal in order to safeguard good scientific practice in publishing.

1. All articles have been subjected to peer review administered by the proceedings editors.
2. Reviews have been conducted by expert referees, who have been requested to provide unbiased and constructive comments aimed, whenever possible, at improving the work.
3. Proceedings editors have taken all reasonable steps to ensure the quality of the materials they publish and their decision to accept or reject a paper for publication has been based only on the merits of the work and the relevance to the journal.

Title, date and place of the conference

ANIMMA2021 : Advancements in Nuclear Instrumentation Measurement Methods and their Applications

Prague, June 21-25, 2021

Proceedings editor(s):

Abdallah Lyoussi, Michel Carette, Rastislav Hodák, Igor Jencic, Patrick Le Du, Stanislav Pospíšil, Christelle Reynard-Carette, Luka Snoj, Ivan Stekl, Ludo Vermeeren.

Date and editor's signature

LYOUSSI Abdallah 133404

Signature numérique de LYOUSSI Abdallah 133404
Date : 2021.10.22 09:08:37 +02'00'

Novel neutron detector assembly based on SiPM readout to be coupled with the Active Target for SPES

Felix Pino[1], Jessica Delgado[1], Giorgia Mantovani[1], Lucio Pancheri[2], Daniela Fabris[3], Cristiano L. Fontana[4], Vladimir Ruiz[1], Davide Brunelli[2] and Sandra Moretto[1,3]
[1]Department of Physics and Astronomy "Galileo Galilei", University of Padova, Italy
[2]Department of Industrial Engineering, University of Trento, Italy
[3]INFN Sezione Padova, Italy
[4]European Commission, Joint Research Centre, Geel, Belgium
felixeduardo.pinoandrades@unipd.it

***Abstract*— The Active Target ATS (Active Target for SPES) is a new time-projection chamber designed for reaction and decay studies with nuclei far from stability. The physics cases for the new-generation active target are related to the ongoing developments of facilities for radioactive ion beams. Thanks to its flexibility, this instrument will be capable of taking advantage of the most exotic beams which will become available at the SPES facility under construction at the Legnaro National Laboratories in Italy. Particular attention will be also paid to couple it with ancillary detectors, for both charged and neutral (gamma and neutrons) particles. In particular, in this work, we will focus the attention on the neutron ancillary detectors. The proposed prototype is a compact device able to discriminate, by performing pulse shape analysis, between neutrons and gamma. The device take advantage of recent improvements in silicon photomultiplier (SiPM) technology and the development of new plastic scintillators exhibiting neutron/gamma discrimination capability. Our work is focused on the read-out with silicon photomultipliers arrays of EJ-276 (and its old version EJ-299) and EJ-276G scintillators of several sizes (ranging from 20 mm to 50 mm diameter). Moreover, we will show the comparison of discrimination performances between SiPM and standard photomultiplier read-out configurations.**

***Keywords* —Neutron/gamma discrimination, SiPM, plastic scintillators, neutron detector.**

I. Introduction

THE SPES (Selective Production of Exotic Species) project is conceived as a multidisciplinary facility that is being installed at Legnaro National Laboratories (INFN, Italy) [1]. It has been designed as a second-generation Radioactive Ion Beam (RIB) facility using the Isotope Separation On-Line (ISOL) technique. A thick target (Uranium Carbide) is bombarded with a high intensity (~250 μA) primary proton beam (E_p ~ 30-70 MeV) to induce up to 10^{13} fissions/s. The project is organized in four phases: α, β, γ and δ. SPES-α spans the installation of the proton driver (cyclotron produced by the Best Cyclotron Systems Inc.). SPES-β covers the production and post-acceleration of RIBs. SPES-γ includes the production of radionuclides focused mainly for medical applications. And, SPES−δ covers the production of neutron fields for applied science studies (material science, biology, etc.).

In the framework of the SPES-β phase, which is focused on the use of high intensity and high-quality neutron-rich nuclei beams to perform forefront research in nuclear structure and reaction dynamics, there is an ongoing project aimed to design and build an ACtive TARget and Time Projection Chamber (ACTAR TPC) [2, 3]. This device is a novel gas-filled detector, and at the same time a gaseous thick target, that allows the tracking of charged particles in three dimensions and provides a precise reaction energy reconstruction from the vertex position. For many cases of interest [4, 5], coupling ancillary detectors (for neutrons and gamma) to the ACTAR TPC can help to optimize the data acquisition and to improve the precision of the results. In particular, in the present work, we will focus the attention on the neutron ancillary detectors. We propose a compact detector able to discriminate, by performing pulse shape analysis, between fast neutrons and gamma photons. The device take advantage of recent improvements in silicon photomultiplier (SiPM) technology and the development of new plastic scintillators exhibiting neutron/gamma discrimination capability.

Organic scintillators with discrimination capabilities exhibit a fast decay component (τ_f ~ tens of ns) and a slow decay component (τ_s ~ hundreds of ns). The last one is the most important to discriminate particles, because its intensity changes as a function of the linear energy transfer (dE/dx) of the particle that excites the scintillator. The pulse shape discrimination technique is based on the reconstruction of the time profile of the scintillation decay. With fast photomultipliers (rise time ~ units of ns) this can be done in a straightforward way, but when using SiPM there are several aspects to take into account: the number and recovery time of the microcells, and the dimension (active area) of the SiPM. In a SiPM the output signal is made of the sum of the individual signals coming from the fired microcells. Organic scintillators produce ~10000 photons/ 1 MeV e^-, so, in order to collect most information we need at least the same order of magnitude of microcells. On the other hand, the recovery time of the microcells goes from 10 to 100 ns depending on the equivalent RC circuit. Shorter recovery time could potentially help the performance of the SiPM in terms of particle discrimination with organic scintillators. Collection of the total scintillation light is important to get accurate reconstruction of time profiles, so, using SiPM arrays is necessary in order to increase the active area up to units of cm^2, and be able to read-out large-sized scintillators.

II. Experimental set-up

A. Plastic Scintillators

Three types of organic plastic scintillators with neutron/gamma discrimination capabilities, from Eljen Technology – Texas US, were used. The scintillators have cylindrical shapes and several dimensions, we used: two EJ-299 (20 mm dia. x 20 mm thick. and 50 mm x 50 mm) acquired in 2013 and 2015 respectively, one EJ-276 (25 mm dia. x 25 mm thick.) and one EJ-276G (25 mm dia. x 25 mm thick.), these last two acquired in 2019. Note that the EJ-276 model replaced the EJ-299 one. The producer declared that the improvements were mostly in terms of physical hardness and long-term stability of scintillation and optical characteristics. Essentially, features like density, scintillation efficiency and emission curves are very similar.

B. Silicon Photomultiplier Arrays

Two types of silicon photomultipliers arrays, from AdvanSiD – Trento IT, were tested. A near ultra-violet (NUV) and a red-green-blue sensitive (RGB) SiPM. The main characteristics of the SiPM can be seen on Table 1. In Fig. 1 the photon detection efficiency (PDE) of the SiPM and the emission curves of the plastic scintillators are reported. As can be seen, there is an excellent match between the EJ-276 emission and the photon detection efficiency peak of the NUV-SiPM (as happen with the standard photomultiplier). On the other hand, the EJ-276G emission is between the sensitivity peak of the NUV and RGB-SiPM. For comparison purposes, a 2" fast linear focused photomultiplier (PMT) Hamamatsu, model H1949-51, was used.

TABLE I
MAIN CHARACTERISTICS OF THE SILICON PHOTOMULTIPLIER ARRAYS USED

	NUV-SiPM	RGB-SiPM
Array size	16.88 mm x 17.43 mm	
Single SiPM size	4 mm x 4 mm	
Cell number	9340 / ch	
Cell size	40 µm x 40 µm	
Recharge time constant	70 ns	50 ns
Peak sensitivity wavelength	420 nm (43%)	550 nm (32.5%)
Breakdown voltage, typ	26 V	27 V
Dark count rate	< 100kHz/mm²@ 4V OV	<200kHz/mm² @ 4V OV

C. Read-out board for the SiPM 4x4 arrays

The read-out board for the SiPM arrays is divided in two parts: the connection of all 16 pixels of the SiPM array (to sum the single signals and to supply the bias voltage) and, the amplification of the anode signal, see Fig. 2. According to [6], connecting in parallel four sets made of 4 single SiPM connected in series is the optimal configuration to get a good sum signal, in terms of its amplitude and timing characteristics. Also, in this way the supply of the required bias voltage to each pixel can be managed easily. By the other hand, the other part of the read-out board consists in a transimpedance preamplifier based on a Ultra-Low Noise, High-Speed OpAmp (LMH6629). The circuit was simulated theoretically and tested illuminating the active area of the SiPM array with an attenuated pulsed-laser (width ~ 4 ns). An excellent agreement was found between the theoretical predictions and the experimental signals. These exhibit a fast leading edge (< 10 ns) and a decay time dominated for the recovery time of the SiPM (~100 ns).

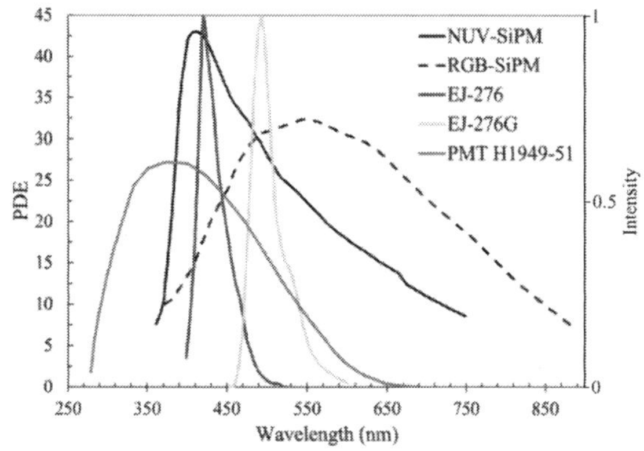

Fig. 1. Photon detection efficiency (PDE) corresponding to the NUV and RGB SiPM (left axis) compared with the emission curves of the plastic scintillators (right axis). Quantum efficiency of the H1949-51 PMT is also given (left axis).

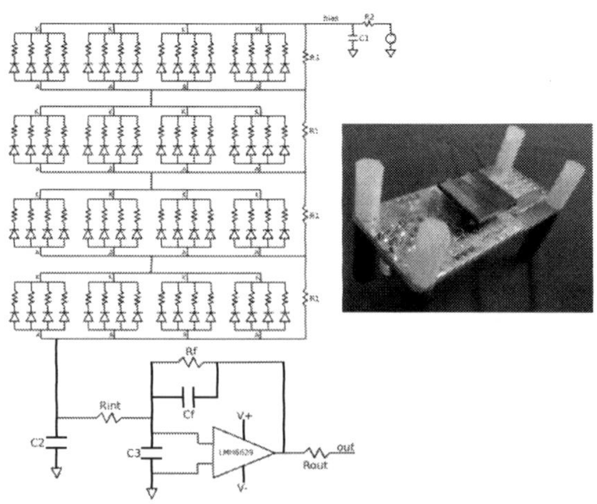

Fig. 2. Scheme (left) and photo (right) of the transimpedance preamplifier used to read-out the anode signals coming from the SiPM arrays.

D. DAQ and data processing

Output signals (coming from the PMT or the transimpedance amplifier) were acquired using a CAEN DT5725 fast digitizer (250 MSamples/s, 14-bit ADC resolution). The digitizer allows to record the signals as digitized waveforms, to perform subsequent off-line analysis. Also, the installed firmware (Digital Pulse Processing for Charge Integration and Pulse Shape Discrimination, DPP-PSD) can pre-process the data, determining the pulses timestamps and integrating them over two gates, in order to perform particle discrimination using the so-called double integration method [7].

Bias voltage was applied using a CAEN V6533 VME HV Power Supply Module. A CAEN V1718 VME-USB 2.0 Bridge was used to control the HV power supply unit. A CAEN A3818 PCI Express CONET2 Controller installed on a PC was used to control the digitizer (by means of optical fiber connections using the CAEN CONET2 protocol). All parameters of the nuclear electronics, and also of the data acquisition, were managed using the ABCD (Acquisition and Broadcast of Collected Data) software [8-10], released as an open-source project (https://github.com/ec-jrc/abcd). The OpAmp was powered using the a power supply (Oltronix) model B703DT.

A ^{22}Na calibration gamma source (~300kBq) and a AmBe source (neutron emission rate ~2x10^{5} s^{-1}) were used for the measurements. In Fig. 3 a typical experimental set-up is presented. As can be seen, the plastic scintillator was wrapped with Teflon tape (PTFE) to improve the light collection. The assembly scintillator/SiPM/read-out board was located inside a black box to isolate it from external light (not shown in Fig. 3).

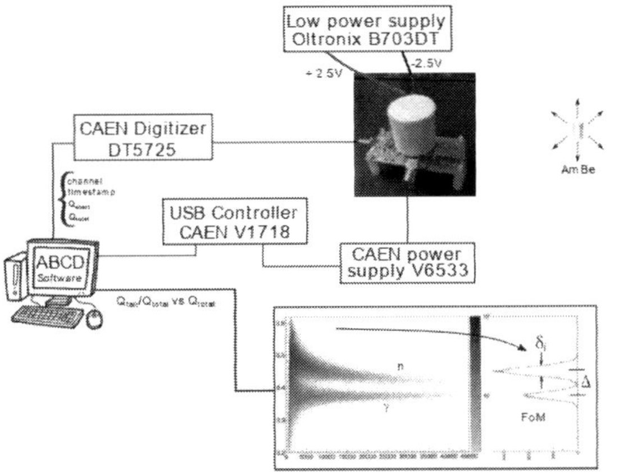

Fig. 3. Experimental set-up. 1" x 1" EJ-276 scintillator coupled to the NUV-SiPM array is shown.

The double gate integration method was applied to perform the pulse shape discrimination analysis. The pulse shape parameter (PSP) is computed as: PSP = Q_{tail} / Q_{total}, where Q_{tail} is the integral over the tail of the signal (computed as: (Q_{total} - Q_{short}). Q_{short} is the integral over a short gate, including the fast rise time region and part of the fastest decay component of the pulse, and Q_{total} is the integral over a long gate, encompassing the majority of the pulse.

In order to characterize the discrimination capability of the scintillator/read-out device assembly, the Figure of Merit (FoM), defined as FoM = $\Delta/(\delta_n + \delta_\gamma)$, was computed by analyzing the PSP distribution of events with energies between 1 MeVee $\pm$ 20%. Δ is the difference between the two mean values of the neutron and gamma distributions and ($\delta_n + \delta_\gamma$) is the sum of the gamma and neutron Full Width at Half Maximum (FWHM).

III. Results and Analysis

A. EJ-299, 20 mm dia. x 20 mm thick.

The first thing that we want to compare is the shape of the signals coming from the PMT and from the transimpedance preamplifier output (SiPM). For this purpose we used the smaller plastic scintillator. In Fig. 4 average neutron and gamma induced signals coming from the PMT and the NUV-SiPM array can be seen. PMT signals decay is very fast, it is dominated by the fast decay component of the scintillator (~units of ns). By the other hand, the decay of the SiPM signals is much slower (~hundreds of ns), in this case it is dominated for the recovery time of the microcells of the SiPM, however, good differences between neutron and gamma induced signals can be observed.

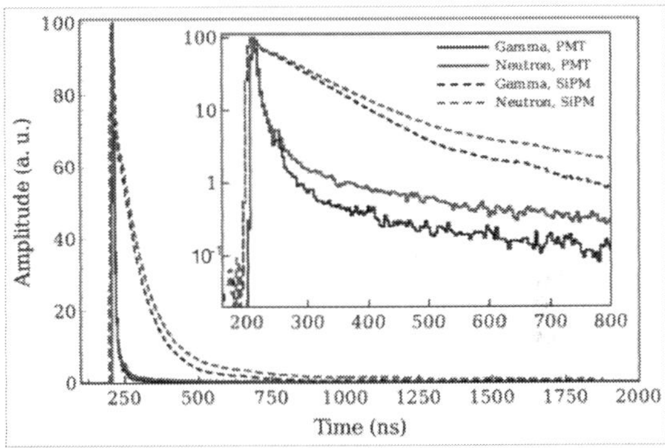

Fig. 4. Neutron and gamma average signals coming from the PMT anode and preamplifier output (NUV-SiPM) using an EJ-299 (20 mm dia. x 20 mm thick.).

In order to cover as much as possible the scintillator face, our first tests were conducted using a small plastic scintillator. In this specific case we were able to cover ~87% of the area. In Fig. 5 the ^{22}Na energy spectra acquired with the small EJ-299 coupled to the PMT and to the NUV-SiPM array are shown. As can be seen the energy resolution obtained with the SiPM array is very good. Qualitatively it can be evaluated using the magnitude of the slope of the Compton edges, larger slopes mean better energy resolution. By following this approach, the small EJ-299 exhibits slightly better energy resolution when it is coupled to the NUV-SiPM array. This result is mainly due to the combination of covering a large fraction of the scintillator face and the high PDE of the SiPM (see Fig. 1).

To assess the neutron/gamma discrimination performance of

the small EJ-299 coupled to the NUV-SiPM array we did some measurements using the Am-Be source (shielded with 5 cm of Pb to attenuate the 59.5 keV gamma intensity). In Fig. 6 2D plots, PSP versus Q_{total} (proportional to the light output), corresponding to the measurements when the scintillator was coupled to the PMT (a) and to the NUV-SiPM array (b) are reported. SiPM bias voltage was set in 120 V, i.e. overvoltage (OV) equal to 4 V. As can be seen excellent n/γ discrimination is found in both cases.

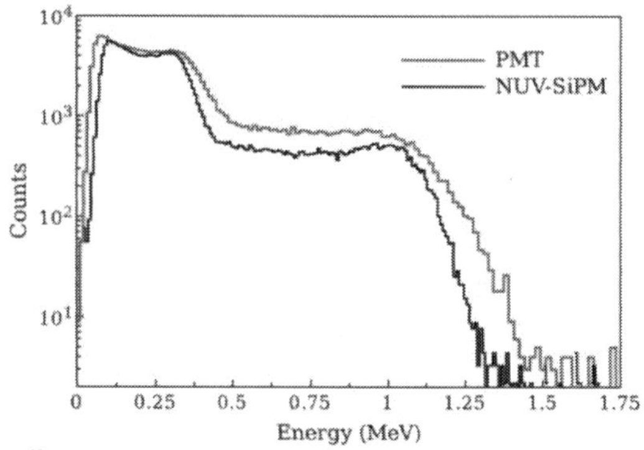

Fig. 5. ^{22}Na spectra using the PMT and the NUV-SiPM array coupled to the EJ-299 (20 mm dia. x 20 mm thick.).

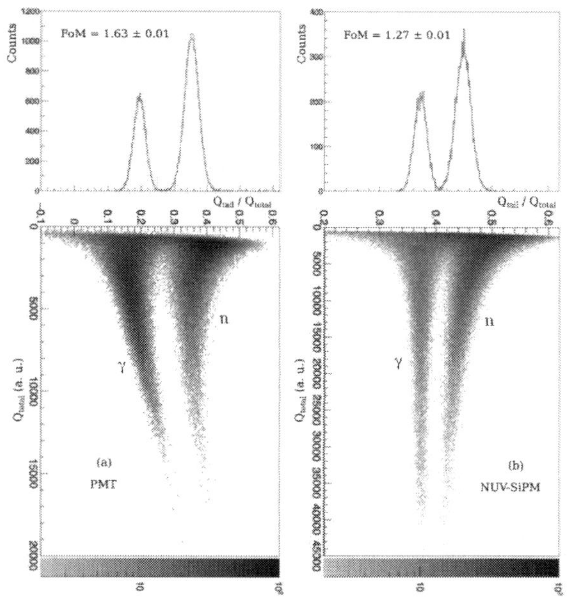

Fig. 6. 2D plots, PSP versus Q_{total}, corresponding to the measurements using the Am-Be source with the EJ-299 (20 mm dia. x 20 mm thick.) coupled to the PMT (a) and to the NUV-SiPM array (b). PSP distributions and FoM values (at ~1 MeVee) are reported on top.

The short and long integration gates in the case of the NUV-SiPM were optimized and found to be 100 ns and 720 ns (measured from the leading edge of the signal) respectively. FoM value (at ~1 MeVee) obtained with the NUV-SiPM array is ~80% respect to the one we got when using the PMT. The results exceed the performance reported by [11], using a SensL SiPM array, and are consistent with [12], using a cubic (6 mm x 6 mm x 6 mm) plastic scintillators and SensL C- and J-series SiPM arrays, and [13], using Hamamatsu SiPM arrays.

B. EJ-276 and EJ-276G, 25 mm dia. x 25 mm thick.

Our next step was to study the n/γ discrimination capability of the assembly composed of a 25 mm dia x 25 mm thick. EJ-276 plastic scintillator coupled to the NUV-SiPM array. In this case the covered area of the scintillator face is ~ 60%, even so, similar results, in terms of energy resolution and n/γ discrimination, were obtained compared to the smaller EJ-299 scintillator. In Fig. 7 the 2D plots, PSP versus Q_{total}, corresponding to the measurements (with the Am-Be source) when the scintillator was coupled to the PMT (a) and to the NUV-SiPM array (b) are given. Here a good FoM value (~1 at ~ 1 MeVee) was found using the NUV-SiPM, that is ~80% the one obtained with the standard PMT. This means that the lack of light collection was compensated by the high PDE of the SiPM, letting a good separation between neutrons and gamma events. These results are consistent with data reported using a larger SiPM (8x8 array from SensL), in which slightly better results have been obtained [14].

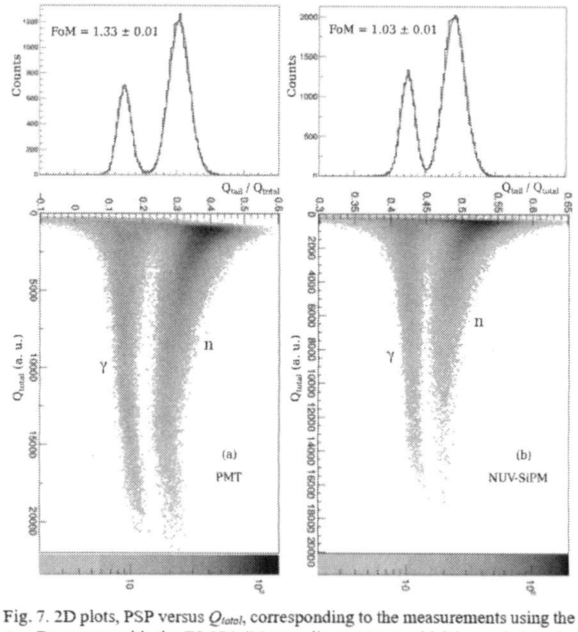

Fig. 7. 2D plots, PSP versus Q_{total}, corresponding to the measurements using the Am-Be source with the EJ-276 (25 mm dia. x 25 mm thick.) coupled to the PMT (a) and to the NUV-SiPM array (b). PSP distributions and FoM values (at ~1 MeVee) are reported on top.

Another interesting study was carried out with the EJ-276G scintillator (25 mm dia. x 25 mm thick.). It is known as the green emitting EJ-276 scintillator, however, as can be seen in Fig. 1 its emission is balanced between blue and green wavelengths. For this reason, NUV and RGB types of SiPM arrays were tested in order to find the optimum assembly to get better n/γ discrimination.

In this case, when the scintillator was read-out with the PMT the FoM value (at ~1 MeVee) was (1.16 ± 0.01). In Fig. 8 the 2D plots, PSP versus Q_{total}, corresponding to the measurements (with the Am-Be source) when the scintillator was coupled to the NUV-SiPM (a) and to the RGB-SiPM (b) arrays are presented. RGB-SiPM bias voltage was set in 124 V in order to compare the results at same overvoltage conditions. Better results were obtained with the NUV-SiPM array, and taking into account that the covered area is the same, and that the PDE in the emission range of the EJ-276G is very similar for both types of SiPM, the difference can be ascribed to the dark count rate of each type of SiPM (at OV = 4V), which is much lower in the case of NUV-SiPM (see Table 1). The assembly EJ-276G/NUV-SiPM shows a FoM value ~75% respect to the one we obtained with the standard PMT, which is consistent with the small EJ-229 and EJ-276 results because the PDE is larger for those scintillators, so better performance is expected.

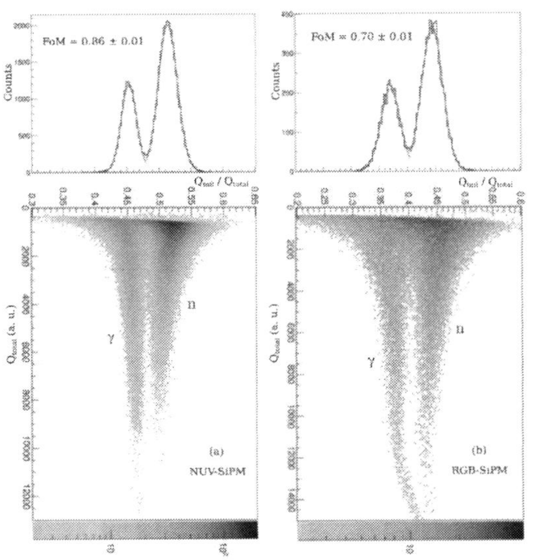

Fig. 8. 2D plots, PSP versus Q_{total}, corresponding to the measurements using the Am-Be source with the EJ-276G (25 mm dia. x 25 mm thick.) coupled to the NUV-SiPM (a) and to the RGB-SiPM array (b). PSP distributions and FoM values (at ~1 MeVee) are reported on top.

C. *EJ-299, 50 mm dia. x 50 mm thick.*

Finally, an incipient study of the n/γ discrimination using a large-sized plastic scintillator coupled to a SiPM is shown in this section. A 2" right cylinder EJ-299 scintillator was coupled to a NUV-SiPM array, i.e. covering only ~15% of the scintillator face. The results are reported in Fig. 9. In this case, when the large scintillator is read-out with the NUV-SiPM the n/γ separation is poor (FoM ~ 0.43 at ~1 MeVee), meaning that despite the high PDE exhibited by the SiPM, the light collection is very little to be able to get accurate signals. Following this idea we tried to use a conical light guide (D1 = 50 mm, D2 = 20 mm, H = 55 mm) to improve the light collection.

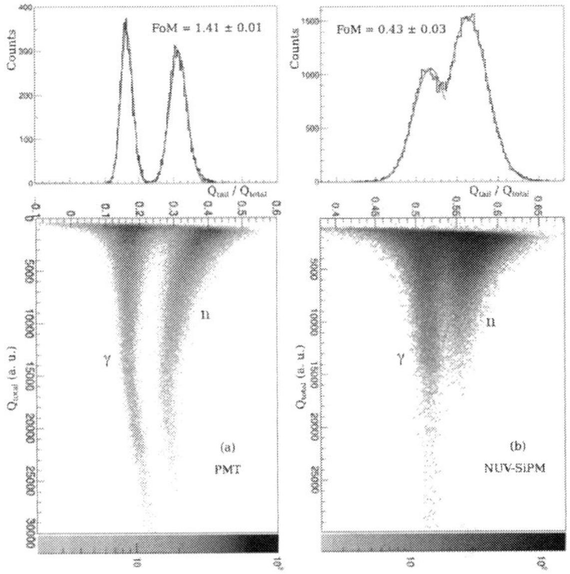

Fig. 9. 2D plots, PSP versus Q_{total}, corresponding to the measurements using the Am-Be source with the EJ-299 (50 mm dia. x 50 mm thick.) coupled to the PMT (a) and to the NUV-SiPM array (b). PSP distributions and FoM values (at ~1 MeVee) are reported on top.

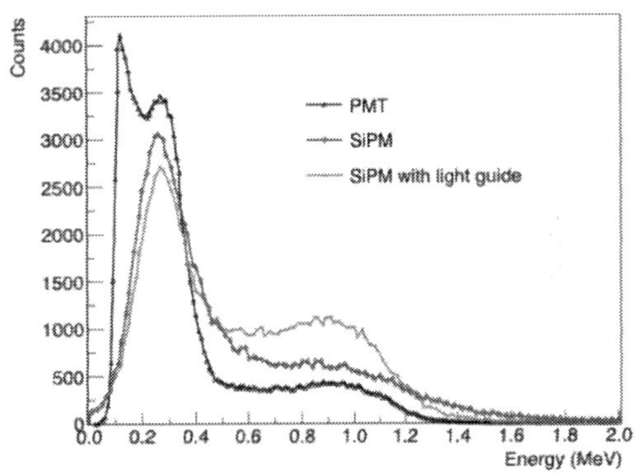

Fig. 10. [22]Na energy spectra corresponding to the 2" right cylinder of EJ-299 coupled to the PMT (H1949-51)and to the NUV-SiPM with and without light guide.

In Fig. 10 we show [22]Na spectra acquired with the large scintillator coupled to the PMT (black), to the NUV-SiPM without light guide (blue) and to the NUV-SiPM with light guide (red). Better energy resolution (at ~ 1 MeV) was found using the light guide, however, neither with nor without light guide is possible to recognize the Compton edge of the 511 keV photon (i.e. at ~ 300 keV), meaning that the light collection is still too little. Furthermore, same FoM values were obtained with and without the light guide. These results suggest that in order to improve the n/γ discrimination capability of the assembly large-sized plastic scintillator/SiPM array, the covered area should be increased. For example, four NUV-SiPM array would cover 60% of the scintillator face, so, we could expect a performance similar to the 1" right cylinder of EJ-276 coupled to one NUV-SiPM array.

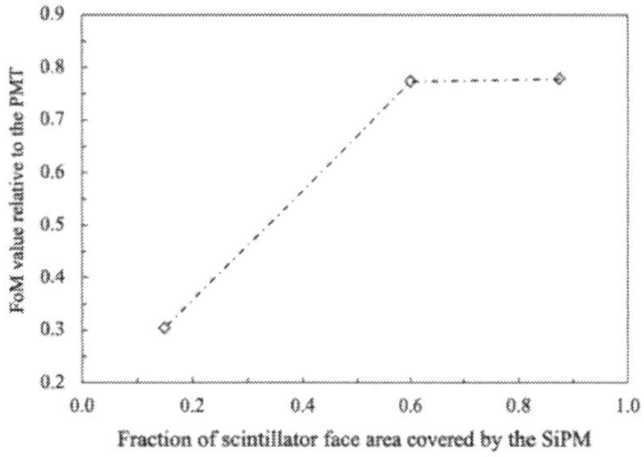

Fig. 11. Relationship between the performance of the detector assembly, in terms of n/γ discrimination, as a function of the fraction of the scintillator face area covered by the SiPM.

IV. Conclusions

The main purpose of this work was to develop a neutron/gamma detector prototype, based on a plastic scintillator read-out with a SiPM array, to be used as a secondary detector of the Active Target for SPES (at Legnaro National Laboratories, Italy). In order to achieve that, we have presented an innovative preamplifier designed to manage the sum of the individual signals and the bias supply of each pixel of 4x4 SiPM arrays, produced by AdvanSiD (Trento, Italy). We have studied the response of small, medium and large plastic scintillators with neutron/gamma discrimination capabilities coupled to a 4x4 SiPM array and to a standard 2" PMT. The small (EJ-299) and medium (EJ-276 and EJ-276G) plastic scintillators exhibit excellent n/γ discrimination (FoM values > 1) reaching almost 80% of the performance of a standard PMT. Besides, good light collection combined with the high PDE of the SiPM produced excellent results in terms of energy resolution. Furthermore, we have found that even if the EJ-267G scintillator is known as the green emitting EJ-276, its performance is better when it is coupled to the NUV-SiPM (because it has lower dark count rate). By the other hand, we have obtained poor n/γ discrimination with the large-sized EJ-299 scintillator (2" right cylinder) when it is coupled to the NUV-SiPM. The main reason is the lack of light collection. As a summary, the relationship between the performance of the detector assembly, in terms of n/γ discrimination, as a function of the fraction of the scintillator face area covered by the SiPM is presented in Fig. 11. It is clear that small scintillators are not useful for most of applications, and for that reason, the next steps will be adding SiPM arrays in order to increase the light collection and at the same time the discrimination performance (at least 3 or 4 arrays to cover more than 50% of the scintillator face). The other solution could be the use of brightest neutron detectors such as stilbene crystals or liquid scintillators.

References

[1] G. Prete, A. Andrighetto, J. Esposito, P. Mastinu, J. Wyss, "The SPES project: a second generation ISOL facility," Physics Procedia, vol 26, pp. 274-283, 2012, 10.1016/j.phpro.2012.03.035.

[2] T. Roger, J. Pancin, G.F. Grinyer, B. Mauss, A.T. Laffoley, P. Rosier, et al., "Demonstrator Detection System for the Active Target and Time Projection Chamber (ACTAR TPC) Project," Nuclear Inst. and Methods in Physics Research, A, vol 895, pp. 126-134, 2018, 10.1016/j.nima.2018.04.003.

[3] B. Mauss, P. Morfouace, T. Roger, J. Pancin, G.F. Grinyer, J. Giovinazzo, et al., "Commissioning of the ACtive TARget and Time Projection Chamber (ACTAR TPC)," Nuclear Inst. and Methods in Physics Research, A, vol 940, pp. 498–504, 2019, 10.1016/j.nima.2019.06.067.

[4] M. Cinausero, V. L. Kravchuk, T. Marchi, F. Gramegna, G. Collazuol, G. de Angelis, et al., "Angular Distribution and Cross Section Measurement of the ^{6}Li(^{3}He,n)^{8}B Reaction at 5.8 MeV", EPJ Web of Conferences 66, 03048, 2014, 10.1051/epjconf/20146603048.

[5] J. J. Valiente-Dobón, A. Poves, A. Gadea, B. Fernández-Domínguez, "Broken mirror symmetry in ^{36}S and ^{36}Ca", Physical Review C, vol. 98, 011302(R), 2018, 10.1103/PhysRevC.98.011302.

[6] T. Cervi, M. Babicz, M. Bonesini, A. Falcone, A. Menegolli, G. L. Raselli, et al., "Characterization of SiPM arrays in different series and parallel configurations," Nuclear Inst. and Methods in Physics Research, A, vol 912, pp. 209–212, 2018, 10.1016/j.nima.2017.11.038.

[7] F. Pino, L. Stevanato, D. Cester, G. Nebbia, L. Sajo-Bohus, G. Viesti, "Detecting fast and thermal neutrons with a boron loaded liquid scintillator, EJ-339A," Applied Radiation and Isotopes, vol 92 pp. 6–11, 2014, 10.1016/j.apradiso.2014.05.025.

[8] C. L. Fontana, et al., "A Distributed Data Acquisition System for Signal Digitizers with On-Line Analysis Capabilities," in *Proc. 2017 IEEE NSS/MIC*, Atlanta, USA. 2017, pp. 1-5. 10.1109/NSSMIC.2017.8533063

[9] C. L. Fontana, et al., "Resource sharing in nuclear physics laboratory classes: A distributed data acquisition system for experiments with shared resources and data management," in Proc. 2017 CAARI, Texas, USA, 2019. p. 050024. 10.1063/1.5127716.

[10] C. L. Fontana, et al., " A Distributed Data Acquisition System for Nuclear Detectors," in Proc. CRETE17, Crete, Greece. 2018. 10.1142/S2010194518601187.

[11] Can Liao, Haori Yang, "Pulse shape discrimination using EJ-299-33 plastic scintillator coupled with a Silicon Photomultiplier array," Nuclear Inst. and Methods in Physics Research, A, vol 789, pp. 150-157, 2015, 10.1016/j.nima.2015.04.016.

[12] M. P. Taggart, P. J. Sellin, "Comparison of the pulse shape discrimination performance of plastic scintillators coupled to a SiPM," Nuclear Inst. and Methods in Physics Research, A, vol 908, pp. 148-154, 2018, 10.1016/j.nima.2018.08.054.

[13] M. Grodzicka-Kobylka, T. Szczesniak, M. Moszyński, L. Swiderski, D. Wolski, J. Baszak, et al., "Study of *n*-γ discrimination by zero-crossing method with SiPM based scintillation detectors," Nuclear Inst. and Methods in Physics Research, A, vol 883, pp. 159-165, 2018, 10.1016/j.nima.2017.11.045.

[14] F. Ferrulli, N. Dinar, L. G. Manzano, M. Labalme, M. Silari, "Characterization of stilbene and EJ-276 scintillators coupled with a large area SiPM array for a fast neutron dose rate detector," Nuclear Inst. and Methods in Physics Research, A, vol 1010, 165566, 2021, 10.1016/j.nima.2021.165566.

The Data Acquisition System for the ATLAS Phase-II Tile Calorimeter Demonstrator

Fernando Carrió Argos[1], on behalf of the ATLAS Tile Calorimeter System
[1]Instituto de Física Corpuscular (CSIC-UV), Spain
fernando.carrio@cern.ch

***Abstract*—The Tile Calorimeter (TileCal) is the central hadronic calorimeter of the ATLAS experiment at Large Hadron Collider (LHC). The TileCal readout system consists of about 10,000 channels. In 2025, the LHC will be upgraded leading into the High Luminosity LHC (HL-LHC). The HL-LHC will be capable to deliver an instantaneous luminosity up to seven times compared to the LHC nominal luminosity. The TileCal Phase-II upgrade will replace the majority of the on-detector and off-detector electronics using a new readout schema for the HL-LHC era. The on-detector electronics will digitize and transmit calorimeter signals to the off-detector electronics at the bunch crossing frequency. In the counting rooms, the off-detector electronics will store the digitized samples in pipelined buffers and compute reconstructed trigger objects for the first level of trigger.**

The TileCal Phase-II upgrade project has undertaken an extensive R&D program which includes the development of a Demonstrator module to evaluate the performance of the new clock and readout architecture for the HL-LHC. A Demonstrator module containing the upgrade on-detector readout electronics was built, tested during several test beam campaigns, and inserted into the ATLAS experiment. The Demonstrator module is operated and read out using a Tile PreProcessor (TilePPr) Demonstrator which enables backward compatibility with the present ATLAS Trigger and Data AcQuisition and the Timing, Trigger and Command systems.

This contribution describes the components of the clock distribution and data acquisition system for the Demonstrator module, and its implementation in the ATLAS experiment.

***Keywords* —Data Acquisition Systems, Calorimetry, High Energy Physics.**

I. INTRODUCTION

THE Tile Calorimeter (TileCal) [1] is the main hadronic calorimeter of the ATLAS experiment [2] at the Large Hadron Collider (LHC). It covers the central part of the experiment up to a pseudorapidity of $|\eta|<1.7$. This subdetector contributes to the measurement of hadrons, jets, taus, and missing transverse energy.

TileCal is divided into four cylindrical readout sections along the beam axis: one central long barrel (LBA, LBC) and two extended barrels (EBA, EBC). Each barrel is segmented azimuthally into 64 wedge-shaped TileCal modules.

As illustrated in Fig. 1, TileCal modules are composed of alternating layers of steel plates and plastic scintillator tiles, which are placed perpendicularly to the beam direction and radially divided into eleven sections.

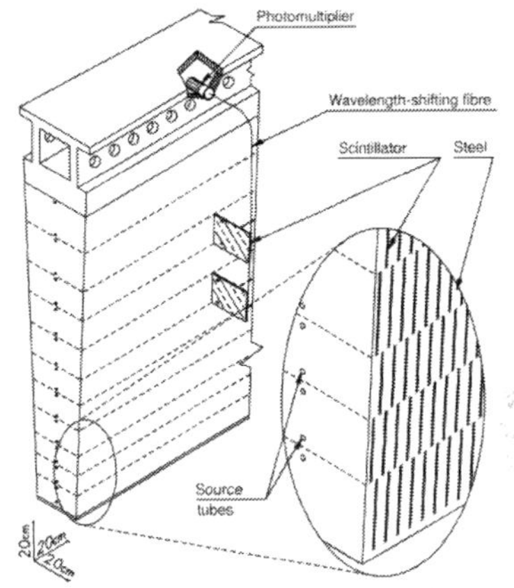

Fig. 1. Detailed sketch of a TileCal module showing its components and structure [1].

The light produced by a charged particle passing through a plastic scintillating tile is collected by Wave-Length Shifting (WLS) fibers and transported to photomultiplier tubes (PMTs) in the outermost part of the modules.

The WLS fibers from individual tiles are grouped in cells, with a dimension of $\Delta\eta \times \Delta\Phi = 0.1 \times 0.1$ in the first two radial layers (A and BC) and $\Delta\eta \times \Delta\Phi = 0.2 \times 0.1$ in the last layer (D). Each cell is read out by two PMTs, where Long Barrel modules are equipped with 45 PMTs and Extended Barrel modules with 32 PMTs. The complete readout of the detector is done using approximately 10,000 PMTs.

In the current data acquisition system, the analog PMT signals are amplified with two gains (1:64 ratio), digitized, and stored in pipeline memories in the front-end electronics at 40 MHz. Upon the reception of an accept trigger signal from the Level-1 trigger system, the selected events are transmitted via optical fibers to the Read-Out Drivers (RODs) [3] in the counting rooms at a maximum trigger rate of 100 kHz. Fig. 2 presents a block diagram of the present readout architecture in TileCal.

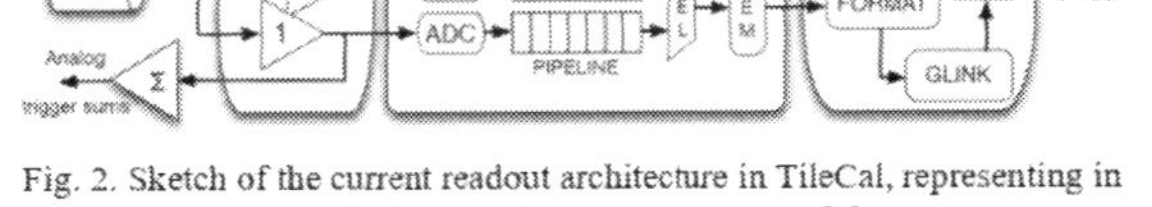

Fig. 2. Sketch of the current readout architecture in TileCal, representing in detail all the on-detector components [4].

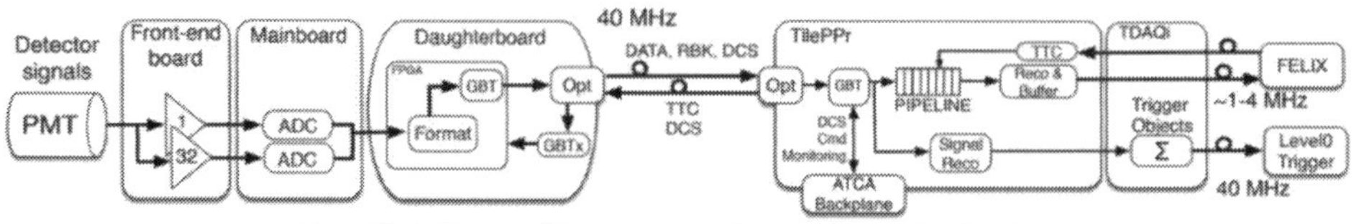

Fig.3. Block diagram of the HL-LHC readout architecture in TileCal.

II. ATLAS PHASE-II UPGRADE

The LHC will undergo a series of upgrades during the Long Shutdown 3 (2025-2027) leading to the High-Luminosity LHC (HL-LHC). The HL-LHC is aimed to deliver an instantaneous peak luminosity up to $7.5 \cdot 10^{34} cm^{-2}s^{-1}$. It is expected that HL-LHC provides a total integrated luminosity of 4000 fb^{-1} by the end of the data-taking.

The ATLAS Phase-II Upgrade [4] will accommodate all the subdetectors and its electronics to operate in the stringent radiation environment of the HL-LHC and with the elevated pile-up levels, which can reach up to 200 simultaneous collisions per bunch crossing.

As part of the ATLAS Phase-II Upgrade, TileCal will completely replace the on-detector and off-detector electronics implementing a new readout strategy with radiation tolerant electronics. The new readout architecture will be capable of providing full-digital information per bunch crossing to the ATLAS Trigger and Data Acquisition (TDAQ) system for the computation of complex trigger algorithms.

Fig. 3 presents a block diagram of the TileCal upgraded readout architecture. The upgraded on-detector readout electronics will transmit digitized signals from the PMTs for every bunch crossing (~25 ns) to the off-detector electronics located in the counting rooms, about 100 meters away from the detector.

In the off-detector electronics, 32 Tile PreProcessor (TilePPr) modules will handle and store the detector data in pipeline memories until the reception of a trigger acceptance signal, activating data transmission to the ATLAS Front End LInk eXchange (FELIX) [5] system. The upgraded readout system will comprise 6,000 optical fibers and a total bandwidth of 40 Tbps.

In parallel to the detector data acquisition, the TilePPr modules will compute the cell energies per bunch crossing and will transmit the results to the Trigger and Data Acquisition Interface (TDAQi) boards, which will calculate and transfer pre-processed trigger objects to the first level trigger.

III. ATLAS TILE DEMONSTRATOR MODULE

As part of the Phase-II Upgrade Demonstrator program, a full-size Demonstrator module containing all the upgraded readout electronics was constructed in 2014 as a Long Barrel module. The Demonstrator module was tested during several test beam campaigns between 2015 and 2018 in the H8 beam line of the SPS accelerator at CERN. The test beam setup included other TileCal modules instrumented with legacy electronics in order to study the performance of the new electronics.

In July 2019, the Demonstrator module was installed into the ATLAS experiment to complete the validation of the new readout electronics for the HL-LHC. The performance of the upgrade electronics has been studied with Charge Injection, Laser, and Cosmic runs, showing excellent performance in terms of low noise, signal quality, and timing.

A. Demonstrator on-detector electronics

The Demonstrator module consists of four independent mini-drawers hosting 12 PMT blocks, all the services, and electronics required for its operation (Fig. 4). Each mini-drawer is composed of a mechanical aluminum substructure which supports one Mainboard [6], one Daughterboard v4 [7], one high voltage regulation board, and up to 12 PMT blocks equipped with upgraded 3-in-1 front-end boards (FEBs).

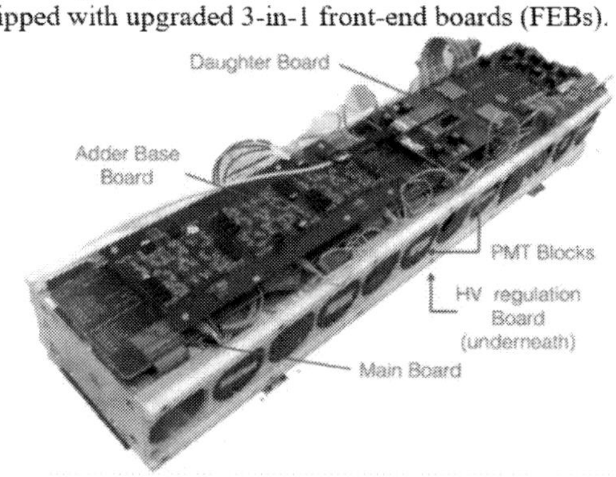

Fig.4. Picture of one assembled mini-drawer with all the on-detector electronics [4].

For improved reliability and robustness, all readout electronics and power distribution are separated into independent halves each of them hosting 6 FEBs.

The upgraded 3-in-1 FEBs provide shaping and amplification with two gains (1:32 ratio) to the PMT analog signals, which are transmitted to the MainBoard for digitization. These FEBs also provide calibration capabilities and include an integrator readout chain for Cesium calibration and luminosity measurements. The upgraded 3-in-1 FEB provides better linearity and lower noise levels than the previous version.

One MainBoard digitizes the signals from 12 PMT blocks using dual 12-bits ADCs at 40 MSps for the two gain signals and 16-bit ADCs for the slow integrator signal. This board hosts 4 Altera Cyclone FPGAs to control and configure the FEBs.

The digitized samples are transferred for every bunch crossing to the DaughterBoard via LVDS lines. The DaughterBoard packs and transfers the digitized data to the off-detector electronics using four redundant GigaBit Transceiver (GBT) [8] links operating at 9.6 Gbps. The timing, control and configuration commands for the operation of the front-end electronics are received via a 4.8 Gbps GBT link from the off-detector electronics. Each DaughterBoard v4 is equipped with two Xilinx Kintex 7 FPGAs, two GBTx [9] ASICs, and two QSFP modules. The GBTx ASIC recovers the LHC clock with fixed and deterministic latency. The recovered clock is distributed for the PMT sampling and to drive the FPGA high-speed transceivers.

Finally, in order to keep backward compatibility with the current ATLAS trigger system, the adder cards sum the low gain analog signals of the FEBs to produce trigger towers, which are transmitted to the Level-1 Calorimeter trigger system.

B. Tile PreProcessor Demonstrator

The Tile PreProcessor Demonstrator (TilePPr) [10] is responsible for operating and handling the data coming from the Demonstrator module, acting as a bridge between the upgraded on-detector electronics and the current ATLAS TDAQ and Trigger, Timing, and Control (TTC) systems.

The TilePPr Demo (Fig. 5) is a double mid-size Advanced Mezzanine Card (AMC) form factor, which includes one Xilinx Virtex 7 (Main FPGA) and one Kintex 7 FPGAs (Trigger FPGA). The Main FPGA interfaces with four QSFP modules providing high-speed communication with the on-detector electronics through 16 GBT links operating at 4.8/9.6 Gbps. It also communicates with the ROD and FELIX systems for event data transmission. On the other hand, the Trigger FPGA is intended for the development and testing of complex pre-processing trigger algorithms for the HL-LHC.

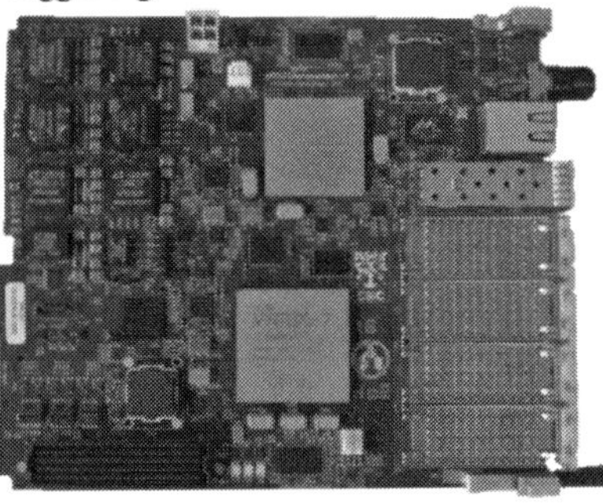

Fig. 5. Picture of the double AMC Tile PreProcessor Demonstrator [10].

IV. DATA ACQUISITION SYSTEM

The data acquisition system for the Demonstrator module is implemented according to the readout strategy envisaged for the HL-LHC and presented in Section II. In addition, the Demonstrator module provides analog trigger tower signals to the current Level-1 calorimeter trigger system as the rest of the legacy modules.

As explained in the following sections the Demonstrator module keeps backward compatibility with the ATLAS TDAQ system via the TilePPr Demo, which enables the translation of commands from the TTC system and transmits triggered data to the ATLAS TDAQ system.

A. Configuration and data readout path

The configuration for the front-end electronics is received in the TilePPr Demo from the TTC system via an optical fiber at 80 Mbps. The TilePPr Demo extracts the timing and configuration commands and translates them into upgraded commands before transmitting them via 3 downlinks to each DaughterBoard. Commands are decoded in the DaughterBoard FPGAs and then executed or forwarded to the MainBoard FPGAs for the configuration of the FEBs. Fig. 6 presents a complete block diagram with all the firmware blocks involved configuration and data readout of the Demonstrator module.

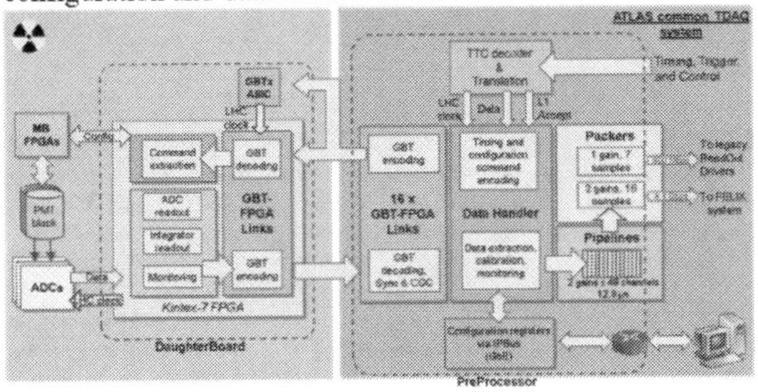

Fig. 6. Block diagram describing the main blocks used for the implementation of the clock and readout architecture for the HL-LHC.

The downlink communication uses the GBT protocol with Forward Error Correction (FEC), where the FPGA transceivers are configured to operate with fixed and deterministic latency at a data rate of 4.8 Gbps.

Regarding the data readout path, the TilePPr module receives continuously event data for every bunch crossing via 4 uplinks per DaughterBoard (16 uplinks in total), each pair sending redundant data from one mini-drawer side. Uplinks are implemented using a modified version of the GBT protocol with fixed and deterministic latency. The modified version doubles the original data rate to 9.6 Gbps and solves the Clock Domain Crossing between the on-detector and off-detector clock domains using Blind Oversampling Clock and Data Recovery units [11].

The Main FPGA extracts the integrator, monitoring and data samples routing them to the proper interfaces after tagging the data with the corresponding Bunch Crossing Identifier (BCID).

In the case of the data samples, they are stored in 96 pipeline memories (one per gain and channel) with a depth of 12.8 μs. Once a Level-1 trigger acceptance signal is received via the TTC interface, the selected events are extracted from the pipeline memories, packed, and transmitted to the legacy RODs (7 samples, 10-bit precision, 1 gain) and to the FELIX system (16 samples, 12-bit precision, 2 gains) using the proper data formats. The integrator and monitoring data are read out via IPbus [12] protocol upon request.

The Demonstrator module is fully integrated with the ATLAS TDAQ software, enabling the acquisition of calibration and physics runs together with the rest of the TileCal modules. From the point of view of the ATLAS TDAQ system, the Demonstrator module is operated and read out as any other legacy TileCal module.

Fig 7. compares the average noise levels between the Demonstrator module and a legacy module obtained with dedicated runs through the ATLAS TDAQ software and the RODs. As observed, the Demonstrator module (LBA14) shows lower noise levels compared with the legacy module (LBA15).

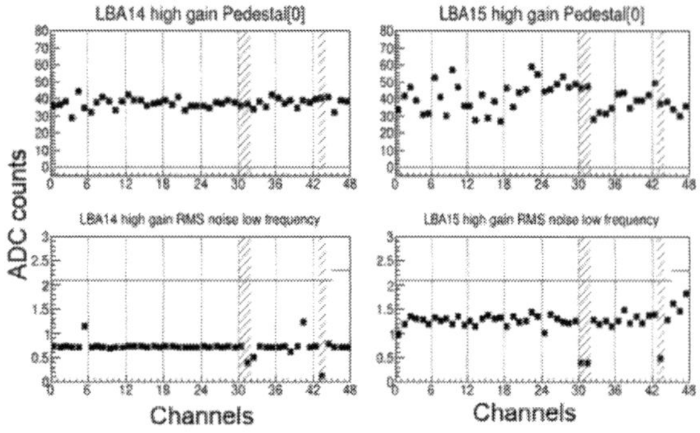

Fig. 7. Average pedestal noise for the Demonstrator module (left) and for a contiguous legacy module (right). The Demonstrator presents lower noise levels compared with the legacy module.

B. LHC clock distribution

An adequate distribution of the LHC clock is required for the synchronization of the on-detector electronics with the accelerator. This is achieved through a combination of high-speed links with fixed and deterministic latency and GBTx ASICs.

First, the TilePPr Demo recovers the LHC clock from the legacy TTC system using an Analog Devices ADN2814 chip. The recovered clock is externally jitter-cleaned using a Texas Instruments CDCE62005 chip to obtain a high-quality clock to drive the FPGA transceivers. Then, the TilePPr Demo distributes the LHC clock to the on-detector electronics embedded with configuration commands via the downlinks.

In the on-detector electronics, one GBTx ASIC per DaughterBoard recovers the LHC clock from the incoming GBT link at 4.8 Gbps. The FPGAs of DaughterBoard receives the LHC clock and buffers it to the MainBoard for the digitization of the PMT signals. The LHC clock is also driven to the FPGA transceivers in the DaughterBoard for proper communication with the off-detector electronics.

V. CONCLUSIONS

The new HL-LHC conditions imply the complete redesign of the readout electronics of the Tile Calorimeter. The new readout electronics will operate in a high radiation environment using a new readout strategy able to provide full-digital input for the ATLAS trigger system with improved granularity.

A Demonstrator module was constructed in 2014 with the upgraded on-detector readout electronics. The Demonstrator module was tested in several test campaigns with particle beams between 2015 and 2018 to study the performance of the new readout electronics. The upgraded clock distribution and readout strategy for the HL-LHC was implemented for the operation of the Demonstrator module. The on-detector electronics transmit digitized data to the TilePPr Demo at the LHC frequency, which also distributes the LHC clock for the sampling of the PMT signals embedded in the downlinks.

The TilePPr Demo interfaces the Demonstrator module with the current ATLAS TDAQ and TTC systems, enabling backward compatibility. One of the goals of the Demonstrator program is to keep the Demonstrator module in ATLAS during Run-3 (2022-2025) to continue studying its performance in real operation conditions.

REFERENCES

[1] ATLAS Collaboration, "*Readiness of the ATLAS Tile Calorimeter for LHC collisions,*" in European Physics Journal C, Vol. 70, Issue 4, pp. 1193-1236, December 2010.

[2] ATLAS Collaboration, "*The ATLAS Experiment at the CERN Large Hadron Collider,*" 2008 JINST 3 S08003.

[3] A. Valero et al., "*ATLAS TileCal ReadOut Driver production,*" 2007 JINST 2 P05003.

[4] ATLAS Collaboration, "*Technical Design Report for the Phase-II Upgrade of the ATLAS Tile Calorimeter,*" CERN-LHCC-2017-019, 2017.

[5] J. Anderson et al., "*FELIX: a High-Throughput Network Approach for Interfacing to Front End Electronics for ATLAS Upgrades,*" in Journal of Physics: Conference Series 664 (2015) No.8, 082050.

[6] F. Tang et al., "*Design of the front-end readout electronics for the ATLAS tile calorimeter at the sLHC,*" in IEEE Transactions on Nuclear Science, Vol. 60, No. 2, pp. 1255-1259, April 2013.

[7] E. Valdes Santurio, S. Silverstein, and C. Bohm. "*Readiness of the ATLAS Tile Calorimeter link daughterboard for the High Luminosity LHC era*". PoS, TWEPP2019:087, 2020.

[8] M. Barros et al., "*The GBT-FPGA core: features and challenges,*" 2015 JINST 10 C03021.

[9] P. Moreira, J. Christiansen and K. Wyllie, "*GBTx Manual – GBT project,*" Manual v0.15, 2016.

[10] F. Carrió, P. Moreno and A. Valero, "*Performance of the Tile PreProcessor Demonstrator for the ATLAS Tile Calorimeter Phase II Upgrade,*" 2016 JINST 11 C03047.

[11] S. I. Ahmed and T. A. Kwasniewski, "*Overview of oversampling clock and data recovery circuits,*" Canadian Conference on Electrical and Computer Engineering, 2005., 2005, pp. 1876-1881.

[12] C. Ghabrous Larrea, K. Harder, D. Newbold, D. Sankey, A. Rose, A. Thea and T. Williams, "*IPbus: a flexible Ethernet-based control system for xTCA hardware*", JINST 10 (2015) no.02, C02019

The MULTI spectrometer for measurement of β-decay process in exotic nuclei

Ivan Siváček[1,2], Sergey Stukalov[1], Yuri Sobolev[1], Dmitry Testov[1], Vladimir Smirnov[1], Yuri Penionzhkevich[1,3] and Zhasulan Zeinulla[1,4]
[1]Flerov Laboratory of Nuclear Reactions, Joint Institute for Nuclear Research, 141980 Dubna, Russia
[2]Nuclear Physics Institute of the Czech Academy of Sciences, 25068 Řež, Czech Republic
[3]National Research Nuclear University, 115409 Moscow, Russia
[4]Institute of Nuclear Physics, Almaty, Kazakhstan
sivacek@ujf.cas.cz

Abstract—**The construction of the portable spectrometer, MULTI, is designed for the measurement of β-γ-neutron coincidences on beams of radioactive exotic nuclei. It will be used in measurements of the β-decay process of neutron-rich nuclei with large Q_β values accessible in radioactive beams facilities. The experimental technique includes the measurement of the ratios between gamma-ray emission and multi-neutron emission, following beta-decay of radioactive nuclei implanted into an active target. The construction of a gamma-ray spectrometer based on high-volume $CeBr_3$+NaI(Tl) phoswich scintillation detectors is introduced. Results of the measurements of its fundamental characteristics – total detection efficiency, peak efficiency and Compton suppression effect with a ^{60}Co source are presented. Design of the neutron counter array, consisting of 40 proportional counters filled with ^{3}He gas, was optimized with Monte Carlo simulations. Results were validated with a ^{252}Cf spectroscopic source, leading to expected detection efficiency of ~35% for neutrons with energy < 1MeV. Further development plans are discussed.**

Keywords — **Monte Carlo simulations, Phoswich detector, gamma-ray spectrometer, ^{3}He neutron counters.**

I. Introduction

The MULTI spectrometer was developed as a setup for measurement of the total reaction cross section with radioactive beams [1]. Measurements were carried out on the ACCULINNA fragment separator beam in the Flerov Laboratory of Nuclear Reactions, Joint Institute for Nuclear Research in Dubna (FLNR JINR) [2]. The Nuclear Physics Institute of the Czech Academy of Sciences (NPI CAS) is involved in the collaboration since 2012 [3]. Fundamental properties of the spectrometer, important for the evaluation of results, include detection efficiency of the gamma-ray spectrometer and registration efficiency for nuclear reaction events as a function of gamma-ray energy and multiplicity were earlier evaluated by Monte Carlo simulations in Geant3 [1] and later Geant4 [4]. The spectrometer consisted of an in-beam telescope for measurement of beam characteristics and a gamma-ray spectrometer for identification of nuclear reaction events by prompt gamma-rays. A lot of effort was put into development and optimization of the active collimators [5] and the method of measurement with an active target. Energy dependence of the total reaction cross section for light neutron-rich nuclei ^{6,8}He, ^{9,11}Li was measured [6,7], results were successfully interpreted theoretically by the time-dependent Schrödinger equation for ^{11}Li+^{28}Si [8]. The measurement technique includes detection of prompt gamma-rays accompanying the nuclear reactions, which provide a tag for a reaction event. The MULTI spectrometer was upgraded to enhance registration efficiency for nuclear reaction events and lower its dependency on multiplicity of prompt gamma-rays [9]. Monte Carlo results were verified using measurements with a ^{60}Co spectroscopic source for gamma-ray multiplicity up to $M_\gamma \leq 5$.

Possibilities of upgrading the MULTI spectrometer with novel high-resolution detectors were analyzed by Monte Carlo method, aimed to improve detection and registration efficiency with possible gamma-ray spectroscopy measurements for identification of the reaction channels [10]. The spectrometer was upgraded with $CeBr_3$+NaI(Tl) phoswich scintillation detectors and the electronics migrated from CAMAC to VME digital acquisition standard. Measurements with an active ^{28}Si target were already carried out in 2019, and unfortunately, they can be completed only after the upgrade of the U400M cyclotron in FLNR JINR in 2022.

The recently commissioned fragment separator ACCULINNA-2 offers an opportunity to make measurements on secondary beams of higher purity with lower background [11]. It offers isotopes of light, neutron-rich nuclei, including isotopes with weakly bound neutrons in the neutron halo and neutron skin. Pygmy dipole resonance (PDR) is defined as collective oscillations of the neutron halo against the inert core of the nucleus and is directly related to these phenomena, which can also be studied with gamma-ray spectroscopy methods. Re-configuration of the detectors for gamma-ray spectroscopy measurements was the subject of Monte Carlo simulations. These exotic nuclei are β^- emitters, often with large Q_β values, emitting multiple neutrons and gamma-rays following β-decay. Based on our collaboration with Institut Physique Nucleaire (IPN) in Orsay, France on neutron counter TETRA [12], the neutron counter assembly with proportional counters filled with ^{3}He gas was introduced into simulations. We are preparing the

MULTI spectrometer for β-gated gamma-ray spectroscopy and multi-neutron emission measurements, similar to our measurements on $^{82,83,84}Ga$ isotopes in collaboration with IPN Orsay [13]. We will follow in search of properties of weakly bound neutrons in the region of light neutron-rich nuclei, accessible on the ACCULINNA-2 fragment separator. For this purpose, the correctness of Monte Carlo simulations must be verified with measurements. Initial measurements of the discussed assembly were carried out to validate Monte Carlo simulations of gamma-ray spectrometer and neutron counter array. Results of the measurements are compared to the simulations and discussed.

II. The MULTI Spectrometer

Two methods are widely used for the PDR studies: the investigated nuclei can be excited with an external source of radiation, or can be studied in the decay process leading to the excited states of daughter nucleus. We propose the second mentioned method. The process of de-excitation of an implanted exotic nucleus after β-decay can go through multi-neutron emission, or trough emission of gamma rays. As proposed for ^{134}Sn [14], energy of the β-particle determines the mode of de-excitation. Similar studies including measurement of β-decay strength function are being prepared. The new spectrometer, MULTI, is now developed for measurement of the β-decay process in neutron-rich nuclei with large Q_β values. It will be able to measure coincidences between emission of β-particles, gamma-rays, and multiple neutrons from the decay of the selected radioactive nucleus implanted into the active target.

The MULTI spectrometer consists of an in-beam multi-detector telescope with active target, 4π β-detector, a high-efficiency gamma-ray spectrometer, and a 4π assembly of ^3He neutron counters. Current geometry of the setup was optimized with the Monte Carlo method using the software toolkit Geant4 [15-17]. Scheme of the spectrometer, consisting of a gamma-ray detector array and a neutron counter assembly, is shown in figure 1. Photo of the neutron counters assembly is shown on figure 2. The detector for β-particles, which will be located in the center of the setup around the active target, is currently being developed.

The in-beam multi-detector telescope is designed for beam transmission measurement and identification of secondary beam particles. It consists of two active collimators, a dE_0 silicon detector, and an active target. Active collimators with plastic scintillator define the irradiated area of the target and provide information about beam transmission and time of flight (TOF) for secondary beam projectiles. Particles are identified by a dE-TOF method in a 243 μm thick silicon dE_0 detector. A detailed description is given in [9]. The active target is designed for the implantation of secondary beam nuclei and for measurements of β-particles from decay with pulse-shape analyzing electronics. The energy of the secondary beam below 10 MeV/u will be adjusted with polyethylene absorbers.

The active target will be surrounded by a β-particle detector covering a large solid angle, made of optically connected fast (plastic) and slow (crystalline) scintillator phoswich. It will be designed to distinguish between high and low energy β-particles, using pulse-shape discrimination techniques.

Validation of the Geant4 results for neutrons and gamma-rays is necessary for finalization of the detector design. This work is dedicated to the measurement of fundamental properties of gamma-ray spectrometer and validation of the Geant4 model of the setup for gamma-rays and neutrons. It will help with optimization of the spectrometer geometry for measurements on secondary beams of different facilities as well as with development of the 4π β-detector.

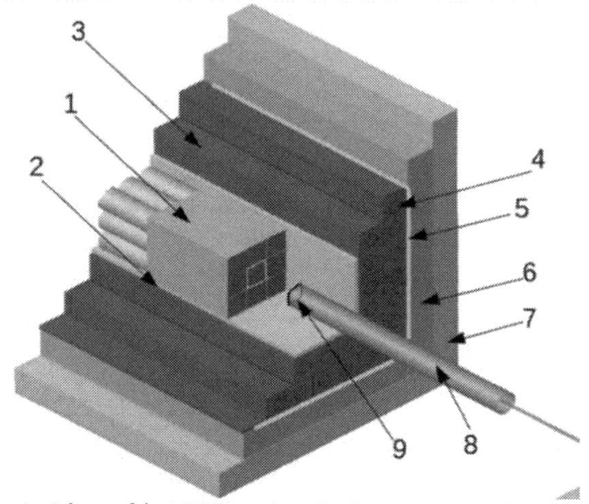

Fig. 1. Scheme of the MULTI spectrometer: 1- gamma-ray spectrometer, 2- spectrometer frame, 3- polyethylene blocks, 4- 3He neutron counters, 5- Al frame, 6- high-density polyethylene, 7- borated polyethylene 8%, 8- beam pipe, 9- place for active target and β-detector.

Fig. 2. Photo of neutron counters assembly with central channel for gamma-ray spectrometer. Nine neutron counters are placed in the inner layer of the polyethylene blocks for detection efficiency measurement.

III. Gamma-ray Spectrometer

The gamma-rays carrying information about nuclear resonances from excited nuclei following beta decay will be registered by high-volume $CeBr_3$ + NaI(Tl) phoswich scintillation detectors. The detectors are of rectangular shape with a 528 cm^3 active volume (fig. 3). The energy resolution of this detector is 3,9 % (661 keV). Its rectangular shape allows for the construction of clusters of detectors of the same type with enhanced characteristics. The PARIS collaboration

(Photon Array for studies with Radioactive Ion and Stable beams) [18] is connecting different laboratories to build large clusters of phoswich detectors with $CeBr_3$ or $LaBr_3$:Ce crystals. We use a 3x3 cluster of the detectors for our measurements.

The signal from the detector is divided into a fast component ($CeBr_3$ crystal) and slow component (NaI(Tl) crystal) with pulse-shape analyzing electronics. Slow signal can be used for Compton suppression of the gamma-ray spectrum recorded by a high-resolution $CeBr_3$ crystal. Moreover, neighboring detectors in the cluster can serve as Compton suppression detectors as well. It is important in the search for Pygmy resonance states within the background induced by high-energy gamma rays.

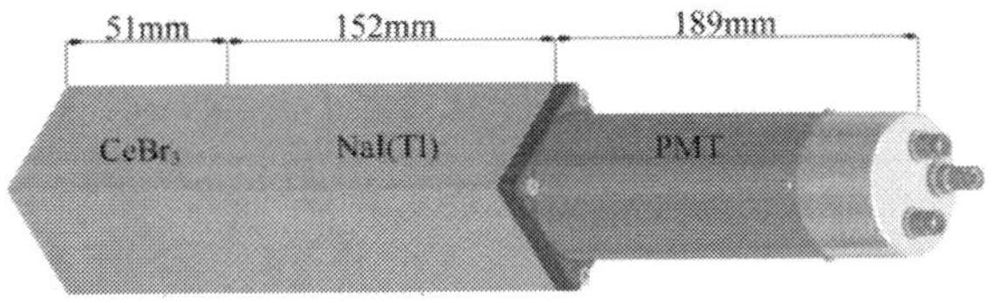

Fig. 3. $CeBr_3$ + NaI(Tl) phoswich scintillation detector. Dimensions of the crystals are 51x51x51 mm ($CeBr_3$) and 51x51x152 mm (NaI(Tl)).

Fundamental characteristics of the spectrometer (peak efficiency, overall registration efficiency over the threshold, and Compton suppression effect) were measured with a ^{60}Co spectroscopic source. The method of measurement is based on tagged gamma-rays [9]. The ^{60}Co source emits two gamma rays in coincidence – 1173,2 keV and 1332,5 keV. An external detector was employed to register the 1332,5 keV gamma-ray peak. It initiates the start of the acquisition, while the other 1173,2 keV gamma-ray was registered by the spectrometer. The external detector was a single crystal $CeBr_3$ of the same dimensions placed at 10 cm distance from the source. Multiple measurements were carried out with the distance of gamma-ray spectrometer at 10 cm and 40 cm from source.

The differentiation of the fast and slow components of the pulse is achieved by setting two gates for one pulse by the VME Modular Digital Pulse Processor. The first gate for acquisition of the fast signal had 25 ns, followed by a 600 ns gate for the slow component of detector impulse. The two-dimensional fast-slow signal spectrum of tagged 1173,2 keV gamma-rays is shown on fig. 4. This allows for the use of a NaI(Tl) crystal of phoswich detector for Compton rejection. The Compton suppression effect is defined as a ratio of mean height of the Compton part of the spectrum with and without Compton rejection. Measurement with tagged gamma-rays at a 40 cm distance (quasi-parallel rays) showed a 58% suppression effect for the central detector of the 3x3 cluster. Geant4 simulations showed a 66% effect of Compton suppression under the same conditions. Measurements at 10 cm distance (fig. 5) showed 63% suppression effect, while simulations resulted into 72%. Results of the simulations and measurements are compared in Table I.

TABLE I

FUNDAMENTAL CHARACTERISTICS OF THE GAMMA-RAY SPECTROMETER. COMPARISON OF SIMULATION AND MEASUREMENT AT 10 CM AND 40 CM DISTANCE.

	^{60}Co measurement	Geant4 calculation
Source distance 10 cm:		
Total efficiency >400 keV	8,6 %	7,7 %
Peak efficiency 1173,2 keV	2,5 %	3,2 %
Compton suppression	58 %	66 %
Source distance 40 cm:		
Total efficiency >400 keV	0,6 %	1,2 %
Peak efficiency 1173,2 keV	0,3 %	0,4 %
Compton suppression	63 %	72 %

The difference between simulations and measurements is mainly due to uncertainty of the source position (x,y), which was set very roughly. In the simulations, the source is point and has defined coordinates. The distance along the z-axis was measured with uncertainty ± 2 cm. Simulations of Compton suppression are more complex task. In the first approximation, rejection can be simulated as an energy deposition in the NaI(Tl) crystal with some threshold. However, in the measurements, there is always a small part of the slow signal in the fast gate and vice versa. To correctly reproduce measured values with simulations, these "mixture coefficients" should be determined to fit the slope of the measured 2-dimensional spectrum (fig. 4). Previous results with tagged gamma rays achieved excellent agreement with the simulations [9].

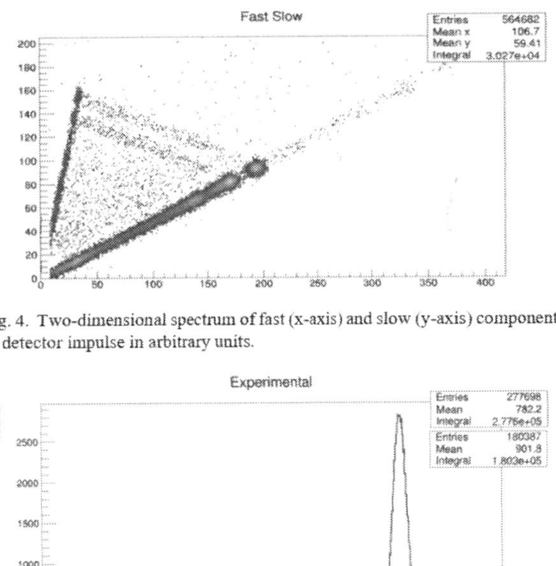

Fig. 4. Two-dimensional spectrum of fast (x-axis) and slow (y-axis) component of detector impulse in arbitrary units.

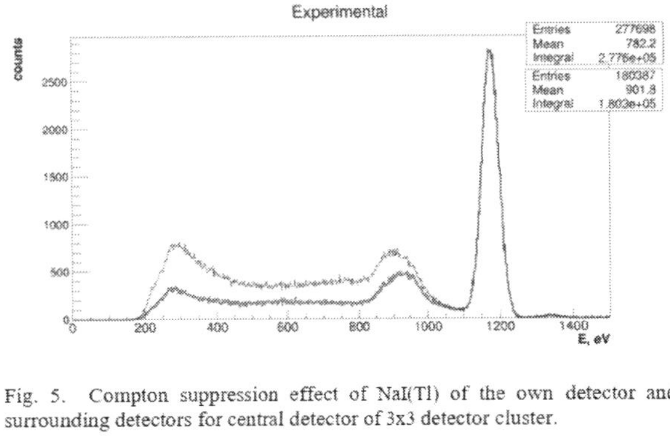

Fig. 5. Compton suppression effect of NaI(Tl) of the own detector and surrounding detectors for central detector of 3x3 detector cluster.

IV. Neutron Counters Assembly

Neutrons are measured with proportional counters filled with ^3He gas placed in high-density polyethylene blocks. They were chosen due to their large volume, high absorption cross-section for thermal neutrons, and zero cross-talk, important for multiplicity measurements. Their modularity allows one to easily adjust geometry, and they show low sensitivity to gamma rays. Neutrons are registered indirectly through the reaction $n + {}^3He \rightarrow p + {}^3H + 764$ keV.

Materials and design of the setup optimized were with Monte Carlo method with the aim of maximizing the detection efficiency. It is built from polyethylene blocks with dimensions 130x70x600 mm. Each block has two circular apertures, Ø 33 mm, localized symmetrically with center in the middle of block height, at ¼ and ¾ of the block width. A total of 20 blocks for 40 neutron counters are situated around the gamma-ray spectrometer in two layers, with 16 and 24 neutron counters respectively (fig. 2). The polyethylene blocks are fixed in a 5 mm thick aluminum case. The case is surrounded with 50 mm thick polyethylene moderator layer and a 75 mm thick 8% borated polyethylene layer for neutron background suppression. The back panel contains 33 mm wide circular apertures through all of the layers for inserting neutron counters and cables.

Neutron counters are in the form of 600 mm long tubes with a 32 mm diameter. The effective length of the detector is 528 mm, and the pressure of the ^3He gas in the mixture is 7 atm. Pre-amplifiers were removed; detectors are directly connected to the neutron-counter electronics module. The module was developed in FLNR, and is similar to the one used in the ALTO facility for the TETRA neutron counter [12]. It uses one high-voltage supply for feeding all counters and a common threshold for pulse-height discrimination. It counts every registered neutron with a 2 μs cycle and provides multiplicity measurements in a 128 μs time frame. It uses TTL logic to communicate with a computer and has an analogue output for every single channel for single spectra acquisition.

Simulation results were verified with measurements with a ^{252}Cf spectroscopic neutron source. The main parameter used for the verification was the detection efficiency of the neutrons in the setup. The aim of the measurements was to validate the results of the simulations, used for optimization of the setup geometry and materials. For this purpose, measurements were performed without a gamma-ray spectrometer inside the setup. A total of nine neutron counters were used for the measurements. The subject of the comparison was detection efficiency in different neutron-counter configurations and measurements with the neutron source placed in center of the setup and at the edges of the polyethylene blocks containing the neutron counters.

Two methods of efficiency evaluation were used. One result is obtained as a ratio of net number of neutrons registered in the setup and the total of the number of emitted neutrons from the ^{252}Cf activity, evaluated for the day of the measurement (fig. 6 yellow triangles). Another method to evaluate detection efficiency is based on the well-known neutron multiplicity distribution (N_1, N_2, N_3, ...) of ^{252}Cf. The neutron multiplicity ratios (N_1/N_2, N_1/N_3, N_2/N_3, ...) depend on the detection efficiency of the setup. With known multiplicity ratios for the source, efficiency of the setup can be extracted from measured ratios (fig. 6 red diamonds). The method is described in [19]. This method is generally more precise, because it does not depend on the activity of the source. However, it is more suitable for large arrays with high registration efficiency.

The threshold on the discriminator, common for all counters, was set to relatively high level of 1,6 V (maximum amplitude was ~ 5 V) due to electronic noise on the stand. Analysis of spectra from neutron counters showed that part of the spectrum related to neutrons lost is due to threshold in some of the detectors. It is still suitable for the test measurements and verification of the Geant4 model of the setup; however, improvements must be made before on-beam measurements.

Efficiency was evaluated from the single spectra analysis with variable thresholds, taking into account the full neutron-related spectrum from the proportional counter (fig. 6 blue triangle). The acquired single spectra also allowed us to introduce the equivalent of a 1,6V common threshold into the Monte Carlo calculations for each detector on the corresponding position separately (fig. 6 violet triangle). It resulted in the simulated detection efficiency being artificially lowered, taking into account the electronics threshold. Results of the measurements and simulations are shown on fig. 6.

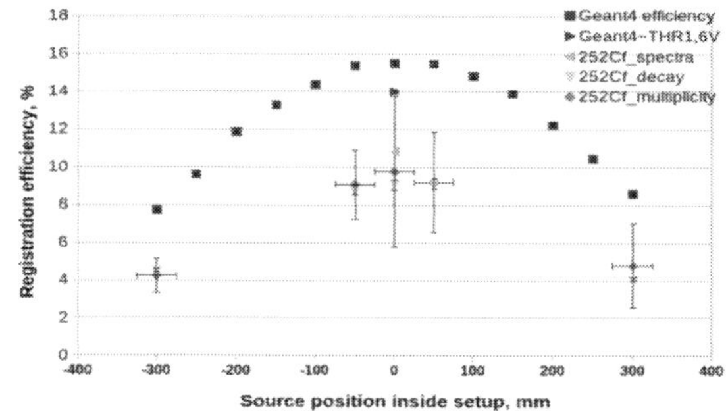

Fig. 6. Comparison of neutron detection efficiency obtained by measurement with ^{252}Cf and Geant4 calculation. Nine neutron counters were placed in the inner layer moderator blocks (fig. 2).

Detection efficiency results obtained by multiplicity and integral measurements are in good agreement. A source of large uncertainties in the multiplicity measurements are from the variance of the evaluated efficiencies from different neutron multiplicity ratios. This is partially due to the effect of a high common threshold and a low number of neutron counters used for detection of multiple neutrons. Results obtained with single spectra analysis shows that only a relatively small portion of detection efficiency is lost due to the high common threshold.

Comparison of measured values and calculations shows a relatively large discrepancy. Ratios between the detection efficiency measured in the central region of the setup and borders are ~0,5 for measurements and simulations both. A slightly lower efficiency is observed in the position -300 mm. It is due to apertures in the back panel of the setup. Detection efficiency obtained with common threshold measurements (integral and multiplicity) compared to artificially lowered Geant4 result gives measurement-to-simulation ratio of 0,67.

The same ratio is obtained in comparison of single spectra analysis and Geant4 simulation results. As a conclusion, Monte Carlo simulations are succeeding in reproducing measurements with a large systematic error of ~ 33%. Variations of simulation parameters showed that the most important parameter for sensitivity to neutrons is the amount of ^{3}He in the counters, represented by the gas density. Further investigation is required to understand this systematic error and successfully quantitatively reproduce measured values.

In the result a correction factor of 0,67 was introduced to Monte Carlo calculations. Expected energy dependence of the detection efficiency for single neutron in the full assembly of 40 neutron counters is shown on fig. 7. The detection efficiency is ~35% and it is relatively flat up to 1 MeV. Further investigation of the spectrometer characteristics will be carried out with the aim of evaluating the detection efficiency for neutrons in the gamma-ray spectrometer and gamma-rays in the neutron counters with pulse-shape analyzing electronics.

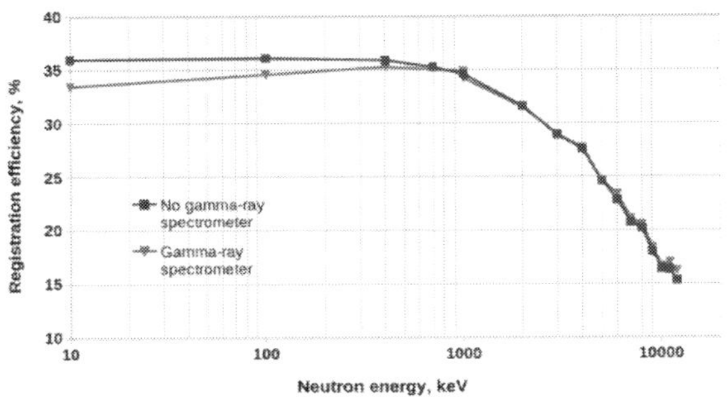

Fig. 7. Expected efficiency for 40 neutron counters based on Monte Carlo simulations corrected by measurements with ^{252}Cf.

V. Conclusions

The MULTI spectrometer is designed for measurement of β-γ-neutron coincidences from the decay of exotic nuclei implanted into and active target. It consists of in-beam multi-detector telescope for identification of secondary beam particles by the dE-TOF method, a gamma-ray spectrometer, and a 4π neutron counter assembly. Radioactive nuclei are implanted into an active target, which provide particle identification and possibly information about β-particle energy. The active target will be surrounded by a 4π scintillator phoswich detector for β-particles, capable of identification of high and low energy β-particles from the decay of implanted nuclei.

The gamma-ray spectrometer is assembled from nine $CeBr_3$+NaI(Tl) phoswich scintillation detectors into a 3x3 cluster. The fundamental characteristics of the assembly were measured with a ^{60}Co spectroscopic source by a tagged gamma-rays method. It showed 2,5 % detection efficiency for 1173,2 keV at a distance of 10 cm and are in relatively good agreement with simulations. Pulse-shape analyzing electronics can enhance performance of the spectrometer, when the NaI(Tl) crystal can be used as a Compton suppression detector. Measurements showed a 58% effect at a 40 cm distance from source, when mean height of the Compton part of the spectrum was lowered by factor 2,4. The effect is even more significant at a 10 cm distance, achieving suppression of 63%. This is important for gamma-ray spectroscopy measurements on background induced by high-energy gamma rays.

A 4π phoswich scintillation detector for measurement of β-particles is currently being developed with help of Geant4 simulations. It will be able to distinguish between high and low energy beta particles, using pulse-shape analyzing electronics.

Multi-neutron emission following beta decay of exotic nuclei with large Q_β values will be measured with a neutron counter assembly located around the gamma-ray spectrometer. It consists of 40 proportional counters filled with ^{3}He gas. It covers a large solid angle, close to a 4π geometry. The geometry was optimized with Geant4 simulations with respect to maximum detection efficiency. Results compared to measurements with a ^{252}Cf neutron source showed systematic error, resulting in a measurement-to-simulation factor of 0,67. Further improvements must take place before measurements on the ACCULINNA-2 fragment separator beam. According to Monte Carlo simulations corrected with experimental data, expected detection efficiency achieves value of 35% for a single neutron with energy up to 1 MeV.

The MULTI spectrometer will take place in measurements on ACCULINNA and ACCULINNA-2 fragment separators in the Flerov Laboratory of Nuclear Reactions of Joint Institute for Nuclear Research in Dubna, Russia. Moreover, the MULTI spectrometer is constructed to be portable. It can be optimized with validated Geant4 simulations and deployed on beams at different laboratories.

Acknowledgment

This work was supported by Program of cooperation JINR – Czech Republic (3+3) "Development of the MULTI-purpose spectrometer" (2020, 2021) and Grant of the Plenipotentiary of the Government of the Czech Republic in JINR "Total cross section study in reaction with exotic nuclei by MULTI setup".

References

[1] Yu. G. Sobolev et al., "A setup for measuring total cross sections of nuclear reactions", Instruments and Experimental Techniques 55 (2012) pp. 618.
[2] A. S. Fomichev et al., "Studies of light exotic nuclei in the vicinity of neutron and proton drip lines at FLNR JINR", Phys.-Usp. 59 (2016) pp. 321.
[3] Yu. G. Sobolev et al., "Studying the excitation function of the full cross section of a reaction using a modified transmission technique: Initial results", Bulletin of the Russian Academy of Sciences: Physics 76 (2012) pp. 952.
[4] I. Sivacek et al., "Response function investigation of 4π γ-spectrometer "MULTI" by Geant4 Monte Carlo code", Proceedings of the International symposium on Exotic Nuclei: EXON 2016, Kazan, Russia, World Scientific, ISBN 9789813226531, pp. 356.
[5] Yu. G. Sobolev et al., "Active collimators in experiments with exotic nuclear beams", Instruments and Experimental Techniques 54 (2011) pp. 449.
[6] Yu. G. Sobolev et al., "Measuring the Total Cross Sections for Reactions in Collisions of $^{6,8}He$ + ^{28}Si and ^{9}Li+ ^{28}Si", Bulletin of the Russian Academy of Sciences. Physics 83 (2019) pp. 402.
[7] Yu. G. Sobolev et al., "Total Reaction Cross Sections for $^{6,8}He$ and ^{9}Li Nuclei on ^{28}Si, ^{59}Co, and ^{181}Ta Targets", Bulletin of the Russian Academy of Sciences. Physics 84 (2020) pp. 948.

[8] Yu. E. Penionzhkevich et al., "Energy dependence of the total cross section for the $^{11}Li + ^{28}Si$ Reaction", Phys. Rev. C 99 (2019) 014609.
[9] I. Sivacek et al., "MULTI-2, a 4π spectrometer for total reaction cross section measurements", Nucl. Instr. Meth. A 976 (2020) 164255.
[10] I. Sivacek et al., "Upgrade possibilities of the spectrometer MULTI", Proceedings of the International Symposium on Exotic Nuclei EXON 2018, World Scientific, ISBN 9811209448, pp. 444.
[11] G. Kamynski et al., "Status of the new fragment separator ACCULINNA-2 and first experiments", Nucl. Instr. Meth. B 463 (2020) pp. 507.
[12] D. Testov et al., "The ^{3}He long-counter TETRA at the ALTO ISOL facility", Nucl. Instr. Meth. A 815 (2016) pp. 96.
[13] D. Testov et al., "Pygmy Gamow-Teller resonance in the N=50 region: New evidence from staggering of β-delayed neutron-emission probabilities", Phys. Rev. C. 95 (2017) 054320.
[14] A. P. Severyukhin et al., "Pygmy dipole resonance of ^{134}Sn", Proceedings of the International Symposium on Exotic Nuclei EXON 2018, World Scientific, ISBN 9811209448, pp. 341.
[15] S. Agostinelli, et al., "Geant4 – a simulation toolkit", Nucl. Instrum. Methods A 506 (2003) pp. 250.
[16] J. Allison, et al., Geant4 developments and applications, IEEE Trans. Nucl. Sci. 53 (2006) pp. 270–278.
[17] J. Allison, et al., Recent developments in Geant4, Nucl. Instrum. Methods A 835 (2016) pp. 186–225.
[18] F. Camera, A. Maj, PARIS White Book, Published by the Henryk Niewodniczański Institute of Nuclear Physics, Polish Academy of Sciences, Kraków 2021, ISBN 978-83-63542-22-1
[19] M. Dakowski et al., "Reconstruction of particle multiplicity distributions using the method of statistical regularization", Nucl. Instr. Meth. 113 (1973) pp. 195.

Measuring the delayed neutrons multiplicity and kinetic parameters for the thermal induced fission of ^{235}U, ^{239}Pu and ^{233}U

Benoit Geslot[1], Alix Sardet[2], Pierre Casoli[1], Pierre Leconte[3], Grégoire De Izarra[1], Abdelhazize Chebboubi[3], Grégoire Kessedjian[3], Olivier Méplan[5], Diane Doré[4], Torsten Soldner[6] and Paolo Mutti[6]
[1]CEA, DES, IRESNE, DER, SPESI, Cadarache, F-13108 Saint Paul Lez Durance, France
[2]CEA, DES, IRESNE, DTN, SMTA, Cadarache, F-13108 Saint Paul Lez Durance, France
[3]CEA, DES, IRESNE, DER, SPRC, Cadarache, F-13108 Saint Paul Lez Durance, France
[4]CEA DRF, IRFU, DPhN/LEARN, Saclay, France
[5]CNRS IN2P3 LPSC, Grenoble, France
[6]Institut Laue-Langevin, Grenoble, France
benoit.geslot@cea.fr

Abstract— **In the normal operation of nuclear reactors, the kinetic behavior of the neutron population in the core is driven by the so-called delayed neutrons (DN). The DN yield per fission, their average lifetime and their groups' abundances are the main physical parameters used to predict the kinetic behavior of the reactor and its reactivity. The overall uncertainty associated to reactivity predictions, which is linked directly to the nuclear reactor safety margins, is thus closely dependent on a few parameters associated with DN.**

Depending on the nuclear data library, DN kinetic parameters present significant discrepancies, even for major fissile isotopes such as ^{235}U or ^{239}Pu. In this context, CEA has long been working for improving DN nuclear data. In 2018, CEA launched the ALDEN project (Average Lifetime of DElayed Neutrons) in the framework of a collaboration between CEA/DES, CEA/DRF, CNRS/IN2P3 (LPSC, CENBG, LPC), ENSICAEN and Caen University. This program aims at measuring the kinetics of the delayed neutrons to derive the DN yield, their average lifetime and abundances.

Two experimental campaigns focusing on the thermal fission of ^{235}U were conducted in 2018 and 2019. They demonstrated the concept feasibility and gave high quality estimations of the DN multiplicity (1.631 ± 0.014 %) and average lifetime (8.82 ± 0.6 s) for ^{235}U. More recently in March 2021, a more ambitious irradiation campaign was conducted on ^{239}Pu and ^{233}U. This paper details the new experimental setup, which was upgraded to meet ILL safety requirements for handling plutonium. The data analysis process is presented, with a focus on the problem of dead time correction. Finally, some preliminary delayed neutron decay curves are showed and discussed.

Keywords—**Delayed neutrons, fission chambers, ^{3}He proportional counters, fissile isotopes, uranium 235, uranium 233, plutonium 239**

I. Introduction

In nuclear reactor physics, some fission products are referred to as "delayed neutron precursors" because they undergo a (β+n) radioactive decay and thus produce neutrons long after the fission (from milliseconds to several minutes). In a thermal fission nuclear reactor, the DN population is roughly 0.7 % of the total number of neutrons, but it still drives the kinetic behavior of the nuclear reactor. In normal operations, the prediction of the reactor's reactivity is strongly dependent on the kinetic parameters of the DN precursors. In nuclear data libraries, such as ENDF-B/VII or JEFF3.1.1, inconsistencies on some major parameters describing DN precursors can still be found [1]. This is especially true for the average lifetime of DN precursors, as well as for the average number of DN per fission (also called DN multiplicity or DN yield).

In this context, CEA has long been working on improving nuclear data evaluation of DN precursors. Among other studies, it is paramount to produce new and well documented data sets of DN kinetic parameters. This is the reason why a new experiment was designed with the aim of studying the thermal fission of several actinides of interest, among them: ^{235}U, ^{239}Pu, ^{233}U and ^{241}Pu.

The ALDEN project was created for that purpose within a collaboration regrouping laboratories involved in nuclear data (CEA DES and DRF, CNRS/LPSC, CNRS/CENBG, CNRS/LPC Caen, ENSICAEN and University of Caen). The principle of the experiment is to induce fissions in a miniature fission chambers loaded with a small amount of actinide (in the range of 100 µg to 200 µg) and then detect DN emitted right after stopping the irradiation. An efficient fast neutrons detection system is used to monitor the decay of the DN precursors in the target. A fast beam shutter was also designed to quickly cut the beam, which allows measuring the short lived delayed neutrons.

Designed to take advantage of the PF1b cold neutron beam at ILL (Institut Laue Langevin, Grenoble, France), the experimental setup was first tested in September 2018 [1-3]. A second experimental campaign dedicated to ^{235}U was conducted in June 2019. The major outputs of these campaigns are the estimations of the DN multiplicity (1.631 ± 0.014 %) and average lifetime (8.82 ± 0.6 s) for ^{235}U thermal fission. These results are in very good agreement with the literature and have reduced associated uncertainties.

The present paper focuses on the last experimental campaign [4] performed in March 2021 on the PF1b facility [5]. Due to safety requirements for ^{239}Pu, the setup was modified to add a vacuum chamber around the target. Because the DN multiplicity of ^{239}Pu is 2.5 times lower than the one of ^{235}U, a key point for ensuring good quality results was to improve the signal to background ratio. This was done by reducing the detection background as much as possible.

After presenting the experimental setup, the last measurement campaign is detailed. The data analysis procedure is presented with a focus on the problem of dead time correction. Then, preliminary results on ^{239}Pu are shown and discussed.

II. Experimental Setup

A. Fissile targets

The fissile targets used in this work were parallel plate miniature fission chambers (MFC) manufactured by CEA at Cadarache [8]. The fissile material (with a mass of pure isotopic actinide between 100 and 200 µg of pure isotopic actinide depending on the target) was electroplated on the inner electrode of the detector (Fig. 1). The chamber's outer diameter is 12 mm. On the miniature connector side (4 mm in diameter), a feedthrough ensures the airtightness of the detector (which encapsulate pure argon at a pressure of 12 bars).

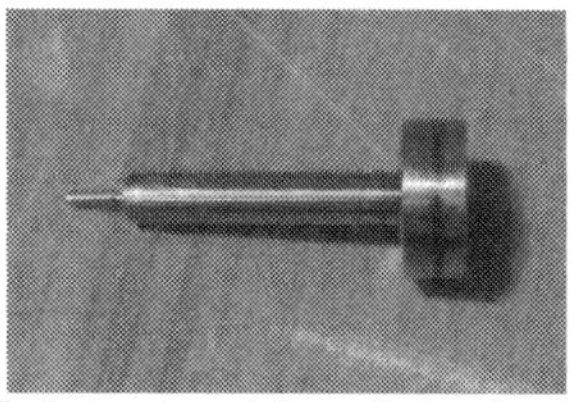

Fig. 1. View of a fissile target: a miniature fission chamber with a flat fissile deposit (8 mm in diameter) coated on its inner electrode.

B. Detection system

The LOENIEv2 detection system (named after the "long counter" LOENIE designed for the LOHENGRIN fission fragment separator at ILL) is made of high density polyethylene (PEHD) in which sixteen ^{3}He proportional counters (PC) are embedded. Neutrons are thermalized in the PEHD and then induce $^{3}He(n,t)$ reactions in the PCs. The detectors are arranged in two rings: the outer ring detectors are more efficient with fast neutrons, whereas the inner ring detectors are more efficient with lower energy neutrons. This way, the total efficiency of the detection system, summed up over the 16 counters, is nearly flat in the range 0.1 to 1 MeV of initial neutron energy (around 2 % variation at maximum). This way, the total count rate induced by DN is very little dependent on their energy spectrum). A central cavity accommodates the fissile target, placed inside an airtight tubing device. Boron rubber shields cover the outer and inner faces of the PEHD blocks to cut the background thermal neutrons from entering the detection system.

First, LOENIEv2 was installed in the irradiation area, in front of the neutron guide and aligned with the beam using a laser pointer (Fig. 2a). Then, the MFC coaxial cable was placed in the central tube, with the beam stop attached to it. Its role is to absorb neutrons from the beam at the output of the detection system before they undergo scattering reactions (Fig. 2b).

Then, the ^{3}He PCs were put in place and connected to the preamplifiers by means of coaxial cables. Last, the vacuum chamber, in which the fissile target had been installed inside a glove box, was inserted in LOENIEv2 from the rear side (Fig. 2c and 2d).

C. Beam shutters

The beam shutters allow to stop the irradiation of the target by the cold neutrons. Their neutron capture efficiency is crucial as it drives the signal to background ratio of the measurement.

CEA developed in 2018 a fast, rotating shutter that cuts the beam in less than 10 ms. It is made of 2 screens of boron carbide (B_4C) and cadmium.

Because of concerns about a small part of alpha particles that could be converted into fast neutrons by $^{12}C(\alpha,n)$ reactions, it was decided to use a second beam shutter to cut these fast neutrons. Therefore, ILL developed a transversally motioned borated polyethylene block (10 cm), which was installed in line with the first (Fig. 3). Indeed, the background was reduced, but not as much as what was expected (see section III.D for more details).

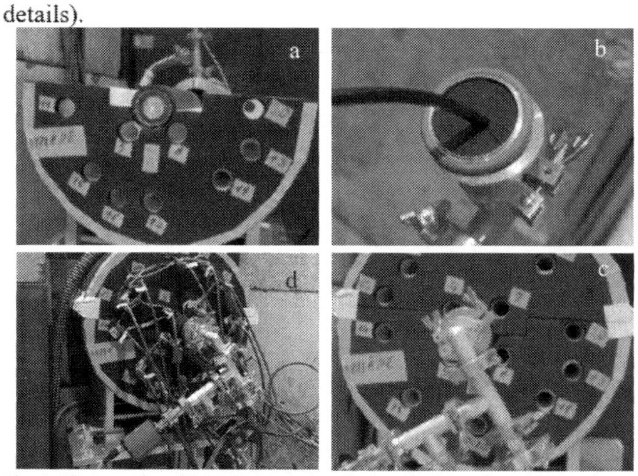

Fig. 2. Photographs of LOENIEv2 during the installation: when aligning the system with the neutron beam (a); placing the beam stop at the end of the airtight tube (b), installation of the target chamber inside the detection setup (c); after connecting all the detectors (d).

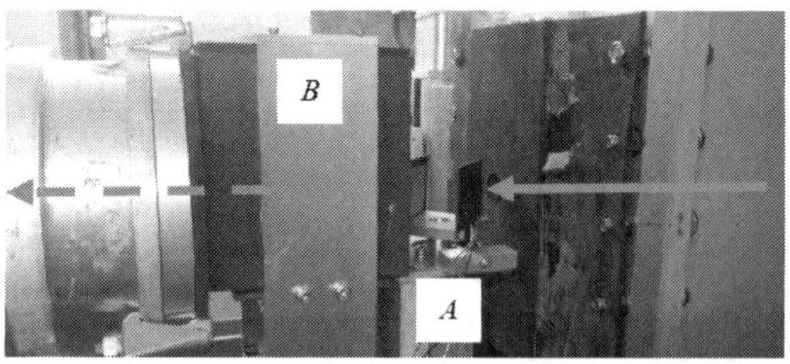

Fig. 3. Fast shutter (A) and polyethylene shutter (B) installed downstream the neutron guide line. The impinging neutron beam is featured by a solid arrow before and by a discontinuous arrow after being shaped by the shutters.

D. Collimation and shielding

LOENIEv2 was designed to efficiently detect neutrons in the 0.1 MeV-1 MeV energy range (its global efficiency is 22 %). If it was not for the boron rubber shielding, it would also be very efficient to detect scattered neutrons in the epithermal and thermal energy ranges. Even a small gap in the shield can lead to spurious detections, and this would dramatically degrade the signal to noise ratio (SNR).

On the front side, a wall of boron polyethylene bricks was installed as extra shielding and a central collimator shaped the beam to a diameter of 12 mm. On the rear side, we had difficulties to efficiently shield the system, because of the signal cables. Nonetheless, boron rubber flanges were placed around the detectors' connectors and a tube of boron rubber covered the external part of the central device (Fig. 4).

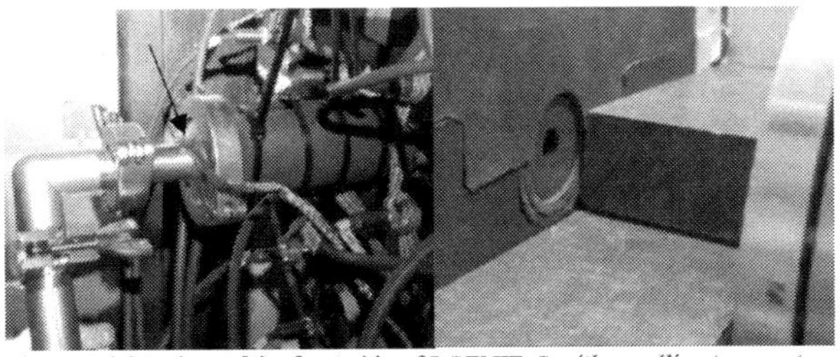

Fig. 4. Right: view of the front side of LOENIEv2 with a collimator aperture of 12 mm in diameter. One can see a part of the shielding wall made of borated polyethylene bricks. Left: view of the rear side of LOENIEv2 with a cylindrical shielding (5 mm of borated rubber) added for the ^{233}U campaign around the central tubing. The beam stop is placed in the iso-KF flange (see the arrow).

E. Data Acquisition System

The ILL instrument control group put together the acquisition system in 2018 for the purpose of the first experimental campaign.

A 16 channels charge sensitive preamplifier card (developed by ILL) shapes the signals of the16 ^{3}He PCs. They are then fed to two synchronized digitizers (CAEN V1724, 8 channels, 14 bits 100 MS/s Flash ADC) housed in a VME crate. The digitizers run a real time digital pulse processing embedded program (CAEN DPP-PHA) to process the data flow, detect events and output their arrival time and energy (i.e. the pulses' amplitudes). Raw data buffers are stored into binary files in "list mode" for further data processing. The crate houses a third CAEN acquisition card (V1751) to record monitoring signals from the beam shutters. An optical fiber links a monitoring computer to the VME crate. This allows handling a large data rate without any loss of data buffers.

Given the large counting rate delivered by the MFC under irradiation (around 2 10^5 s^{-1}), it was too costly to process its signal like it was done for the ^{3}He PCs, and a signal reduction had to be implemented first. This is the reason why an analog amplifying stage based on a Canberra ADS7820 as broadband amplifier-discriminator was used. It shapes the current pulses and issues TTL pulses (5V, 100 ns) thanks to its integrated discriminator stage.

III. The experimental campaign of March 2021

A. Overview of the experiments

Acquiring high quality data on ^{239}Pu was a major objective of the last experimental campaign, conducted in March 2021. As reported in JEFF3.1, the DN yield in the ^{239}Pu thermal fission is about 0.0065(3), which is much smaller than the yield for ^{235}U 0.0162(5). So, the signal to noise ratio was expected to be 2.5 times lower with ^{239}Pu than with ^{235}U, should the detection background be the same. Since the background has to be subtracted to the DN decay curve, it degrades in particular the estimation of the first DN group, which has the longest decay time.

Our target uncertainty associated to the estimation of ^{239}Pu 1st DN group was about 2 to 3 %. This implied to optimize the estimation of the background's level. First, we had to reduce its level as much as possible, by using a second beam shutter for instance. Second, we optimized the irradiation sequence for allowing a larger measurement window for estimating and monitoring the background level.

The campaign spread over March 2021 and was focused on irradiating three fissile targets (^{235}U, ^{239}Pu and ^{233}U) and a dummy one (see Table I).

The use of a dummy target was required for measuring the extra counting rate due to scattered neutrons (coming directly from the beam and scattered before undergoing fission, or coming from outside LOENIEv2). This was the only way to account for a bias that amounts to a few percent of the total counting rate during irradiation.

Most of beam time was used to acquire DN signals in various irradiation scenarios of interest (see table 1). The model used to describe DN precursors decay depend on whether the irradiation is large compared to the average lifetime. If so, the DN emission is saturated and close to its maximum level. We studied four main scenarios:

a. Short cycles (3s/0.25s)
b. Long irradiation (typically 100s/500s)
c. Short irradiation (5s/395s)
d. Short irradiation with even irradiation and cool down (typically 10s/10s)

In the case of 235U, only scenarios a. and b. were performed, for the sake of comparison with previous experiments and to seek for a better estimation of its 1st DN group's abundance.

TABLE I
SHORT LIST OF EXPERIMENTS CONDUCTED WITH ^{235}U, ^{239}PU, ^{233}U AND DUMMY TARGETS.

Day	Target	TYPE	Type : Objective
5	^{235}U	Short (3s/0.25s)	DN multiplicity
7	^{235}U	Long (50s/450s)	Groups abundances
8	^{235}U	Long (100s/500s)	Groups abundances
8	^{235}U	Long irradiation	Metal foil activation
9	Dummy	Long irradiation	Background
10	^{239}Pu	Short (3s/0.25s)	DN multiplicity
12	^{239}Pu	Long (150s/750s)	Groups abundances
17	^{239}Pu	Long (50s/450s)	Groups abundances
19	^{239}Pu	Short (10s/10s)	Groups abundances
20	^{239}Pu	Short (5s/5s)	Groups abundances
20	^{239}Pu	Short (3s/3s)	Groups abundances
21	^{239}Pu	Long (5s/395s)	Groups abundances
22	Dummy	Long irradiation	Background
23	^{233}U	Irradiation at low flux	Dead time correction
24	^{233}U	Short (3s/0.25s)	DN multiplicity
25	^{233}U	Long (100s/500s)	Groups abundances
26	^{233}U	Long (50s/450s)	Groups abundances
26	^{233}U	Long (5s/395s)	Groups abundances
28	^{233}U	Short (10s/10s)	Groups abundances
29	^{233}U	Short (5s/5s)	Groups abundances
29	^{233}U	Short (3s/3s)	Groups abundances

B. Filtering parameters of the ^{3}He proportional counters

After the preamplifier stage, signals issued by the ^{3}He PCs were digitized, processed and filtered by the DPP-PHA (which stands for Digital Pulse Processing, Pulse Height Analysis) algorithm implemented in the CAEN V1724 digitizers. The overall quality of the measurements is directly linked to the parameters of the filtering algorithm (see Tab. II). The parameters used for data taking are listed in Tab II. The signal threshold, triggering pulse processing, was tested individually for each channel. This trigger threshold (TT) has a role of discriminating the pulses from the fluctuations of the baseline (statistical and/or parasitic signals).

If the TT is too low, many non-physical events are processed and recorded: that generates additional pile up and dead time. On the opposite, if the TT value is too high, then valid neutron events are not recorded. This is especially the case at high counting rate, when the energy spectrum tends to shift a bit towards the lower energies (illustration on Fig.5).

Because of the high level of electromagnetic perturbations that was observed in the PF1b environment, we had to set the TT values as high as possible, without damaging the energy spectrum of the signals. This was checked at low and high neutron fluxes. The optimum TT value was found to be 450 LSB (i.e. Least Significant Bit: unit associated to the signals after digitization).

Several channels showed noisier signals: in the energy histogram, the low energy region (under 1000 LSB) was more populated that for other channels. For those channel the TT value was increased until the energy spectrum exhibited a normal behavior.

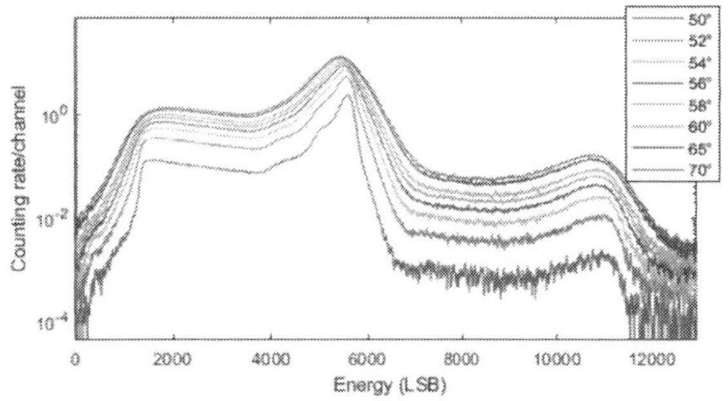

Fig. 5. Energy distributions of pulses in channel 0 (inner ring detector) for various shutter angles. Increasing the counting rate changes the shape of the energy spectrum: the peak becomes larger and slightly shifts towards low energies.

TABLE II
PSD FILTERING PARAMETERS FOR 3HE PROPORTIONAL COUNTERS

Parameter	Value
TT (LSB)	450 (except for #2:550, #5 and #6:500)
Polarity	Positive
DC Offset	40
Smoothing	32
Trigger hold off (µs)	1.0
Rise Time (µs)	0.35
Decay Time (µs)	1.65
Trapezoid Rise (µs)	2.0
Trapezoid Flat top (µs)	1.5
Baseline mean	256
Trapezoid Gain	1
Peaking delay (µs)	1.20
Peak mean	4
Baseline hold off (µs)	1.0
Peak Hold off (µs)	1.0

C. *Flux dosimetry and targets' fission rates*

Metal foil activation is a widespread technique for measuring the intensity of neutrons fluxes. In short, by measuring the activity of a particular radionuclide after the irradiation, it is possible to measure the neutron capture reaction rate R.

To perform metal foil activation on gold, we used an aluminum foil containing $c = 0.1$ % of gold in mass. Because the amount of gold is low enough, any self shielding can be neglected. The foil diameter was 8 mm, its thickness was 0.1 mm and its mass was $m = 13.946(1)$ mg. It was stuck on the front face of the ^{235}U target. Then, the target was placed in the detection system the same way as usual. The metal foil was irradiated during T = 15000 s at the maximum flux possible (the shutters wide open).

The CNRS/LPSC (Laboratory of Subatomic Physics and Cosmology, Grenoble, France) performed the gamma spectrometry measurement. The measured activity per mass unit was $A_m = 12480(340)$ Bq/g, which gave us a reaction rate of 3985(108) s^{-1}. Most of the uncertainty (2.7 %, at 1 sigma) is due to the efficiency calibration of the gamma spectrometry chain.

The thermal point of gold microscopic cross section is a dosimetry standard, reported at 98.7 b in the literature. Using this value, one can estimate an "equivalent neutron flux", i.e. the intensity of the neutron flux that would have given the same reaction rate if the neutrons were all at the thermal energy of 25 meV. The equivalent neutron flux Φ_{eq} is equal to 9.47(26) 10^8 n/cm²/s. Using this value, one can calculate a rough estimate of the fission rate F in a target, by using the microscopic cross section of the fissile isotope at 25 meV (see Table III).

TABLE III
FISSION AND CAPTURE RATES ASSOCIATED TO THE EQUIVALENT NEUTRON FLUX (AT 25 MEV).

Target	Cross section JEFF3.1 (b)	Mass (µg)	Expected reaction rate (s^{-1})
^{197}Au	98.7 b	13.946	3985
^{235}U	580 b	204.3 [a]	285 600
^{239}Pu	648 b	113.7 [a]	174 500
^{233}U	504 b	204 [b]	250 100

[a]estimated based on activity measured by gamma spectrometry.
[b]not measured but taken equal to ^{235}U target's mass.

D. *Detection background*

In the context of this study, the detection background is defined as the total counting rate of ^3He PC (that fall within the energy region of interest) due to particles other than prompt or delayed neutrons emitted by the target. During the measurements, the neutron beam can be either on (shutters open) or off (shutters closed). Thus, we are dealing with two types of background:

1. When the shutters were closed (b_{off}), spurious detection of particles from the environment surrounding the detectors (cosmic rays, natural radioactivity, neutrons from surrounding experiments or coming through the beam shutter);
2. When the shutters was open (b_{on}), particles coming from interactions other than fission between the beam

and the surrounding materials, that is mainly thermal neutrons from the beam scattered inside or outside the detection system.

The b_{off} background is the one that limits the SNR associated mainly with the long lived DN (groups 1 and 2). If the irradiation cycle is long enough, b_{off} can be easily estimated at the end of each cycle.

The b_{on} background can introduce a bias of a few percent on the prompt neutron counting rate and thus, on the normalization of the DN activity curves.

IV. Preliminary Results and Discussion

A. Data analysis

Because each detection event in the ^{3}He PC was recorded by the acquisition system, a very large amount of binary data was generated during the campaign (around 2 TB). Each irradiation cycle, called one "run" was recorded into a single binary file.

The first step was obviously to convert the binary files for better handling of the lists of events, for instance in the Matlab environment. Then, several steps were performed to convert the raw data into curves to be fitted by a mathematical model:

1. Energy scaling: the energy histogram of each channel is rescaled, so that the same energy region of interest (ROI) can be applied to all channels.
2. Background discrimination: events outside the region of interest (typically 1000 to 3 10^4 LSB) are discarded.
3. Time synchronization: the fast shutter closing time is used as a reference time to synchronize the cycle.
4. Time histogramming: time histograms of counting rates are created. The background $b_{off}(i)$ is measured for each detector i by averaging the counting rate at the end of the run (typically over the last 50 s).
5. Dead time correction: histograms are dead time corrected by applying a non extendable dead time model with a parameter τ that is yet to be evaluated with precision. We know that τ is roughly equal to the veto time of the DPP-PHA algorithm: 4.7 µs, based on the parameters given in table 1.
6. Summation: if no outliers are detected among the 16 histograms, they are summed up to get the overall neutron emission detection rate.

B. Dead time corrections

The so-called "dead time" phenomenon refers to the lost detections when dealing with random events, like in the case of radiations emitted by a radioactive source. In short, the measurement chain (from the detector itself to the data recording) is not always in capacity to process an incoming event. This is the case, if the waiting time between two events is shorter that a duration τ, which corresponds to the standard time required for processing a detection signal.

Models can be applied to correct *a posteriori* the observed counting rate. Models are derived based on a few solid assumptions: events must be random and all follow the same probability distribution over time (but not necessarily Poisson) and they must be independent and not correlated.

When dealing with neutrons from fission, the last two assumptions are not perfectly true. One fission event can lead to multiple detection events, thanks to the prompt neutron multiplicity. Moreover, the time behavior of the signals is affected by correlation: this is because of the random time needed for each neutrons they reach a detector. The probability distribution of time between two events is not constant in this case. These phenomena are well known in reactor physics as they are exploited by the so-called "pile noise techniques" to determine kinetics parameters of a nuclear reactor.

The amplitude of the correlated part of the signal is proportional to the square of the detector's efficiency. This is the reason why it is usually negligible compared to the random Poisson type signal. This is not the case in LOENIEv2 because the detectors' efficiencies are large (roughly 3.5 % for inner detectors and 1 % for outer detectors).

This behavior was brought to light when analyzing signals at various flux levels. By putting the shutter at intermediate angles (between 50° and 70°), it was possible to progressively increase the intensity of the flux on the target. The correlated signals could be observed on the waiting time histogram, i.e. the time probability between two consecutive events. If events follow a Poisson type distribution, the distribution of waiting times features an exponential curve. At low flux especially (Fig. 6, blue curve), the behavior observed is a mixed case in which the correlated signal adds up to the Poisson-type signal.

Because of this phenomenon, the observed dead time is greater than the one expected from standard dead time models.

Analysis are ongoing to implement a method to estimate the proportion of dead time due to correlated events. At high flux fortunately, it seems that the correlated signal fades out and the standard dead time correction can be applied (Fig. 6, yellow curve).

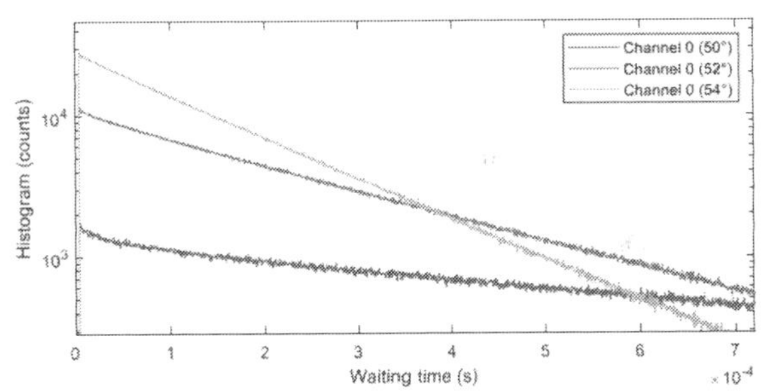

Fig. 6. Histograms of waiting times for 3 angles of the shutter (channel 0, inner ring).

C. Preliminary results for ^{239}Pu

The ^{239}Pu target was irradiated for 12 days in a row, from the middle of the 2nd week until the end of the 3rd week of the beam time. Short irradiation cycles (3 s of irradiation) were first performed to obtain the counting rate just after the irradiation is stopped (Fig. 7). This is required to obtain an experimental estimate of the ratio ν_p/ν_d. This factor is then used to rescale the DN emission curves.

Most of the beam time was then used to accumulate statistics in scenarios with a long decay time, to fully deplete the target from DN precursors. Three main cases were tested with various irradiation times: 150 s (448 runs of 900 s), 50 s (354 runs of 500 s) and 5 s (398 runs of 400 s). The neutron emission as measured by the ^{3}He PC is shown on Fig. 7 and Fig. 8. Data processing was performed as detailed in section IV.A, except for the dead time correction. The irradiation time has a clear effect on the DN activity, but also on the shape of the curve.

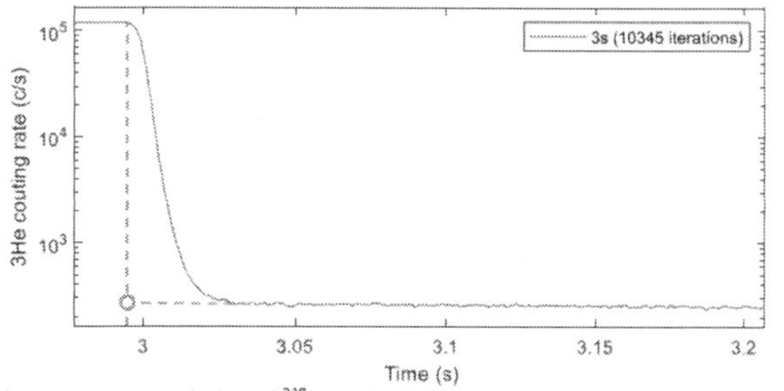

Fig. 7. Neutron emission of ^{239}Pu in the case of short irradiation cycles (3 s of irradiation).

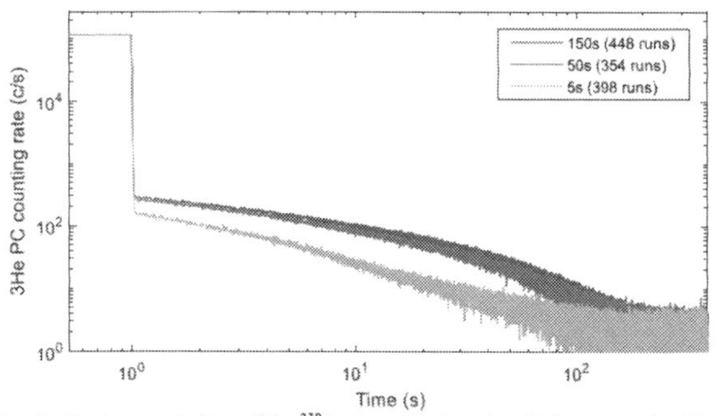

Fig. 8. Neutron emission of the ^{239}Pu target in three irradiation scenarios (5 s, 50 s and 150 s of irradiation). No dead time correction was taken into account.

V. Conclusions

CEA is actively involved in producing new measurements of DN kinetic parameters. This is the goal of the ALDEN experiments that have taken place since 2018 at ILL on the cold neutron beam PF1b. The experimental setup was designed to measure the neutron emission of a miniature fission chamber accommodated at the center of very efficient neutron detection setup. By fitting the neutron emission, it is possible to estimate the DN yield per fission, their average lifetime or the DN groups' abundances.

After two experiments dedicated to ^{235}U, a more ambitious campaign performed in March 2021 focused on ^{239}Pu and ^{233}U. Thirty days of beam time were allocated for this campaign and half of it was controlled remotely using the remote control tools implemented by ILL.

The campaign was a success, on nearly all technical aspects: miniature fission chambers worked perfectly, as well as the data acquisition system. The SNR was improved thanks to a careful work on the detection background (reduced down to 2 c/s). In the future, the irradiation conditions should be reproducible, notably the alignment of the detection setup with the beam.

The comprehensive data analysis is still to be performed, and final results will be published elsewhere, but preliminary results show that the measurements' quality is very satisfactory.

Acknowledgment

This work was partly funded by I3P Institute gathering CEA, EdF and Framatome, and by the NEEDS project gathering CEA and CNRS. The authors would like to thank warmly the ILL staff members who were actively involved in the preparation of ALDEN experiments.

References

[1] D. Foligno, "New evaluation of delayed-neutron data and associated covariances", PhD thesis, N°2017AIXM0001/001ED62. Aix Marseille Université, France, Oct. 2019.

[2] P. Leconte, A. Billebaud, G. De Izarra, D. Doré, D. Foligno, B. Geslot, G. Kessedjian, U. Köster, F.R. Lecolley, J-L. Lecouey, X. Ledoux, L. Mathieu, O. Serot, N. Marie-Nourry, T. Soldner, (2019) "Absolute yield and time dependence of delayed neutron emission in 235U(n,f) and 239Pu(n,f)", Institut Laue-Langevin. (ILL) doi:10.5291/ILL-DATA.3-07-380

[3] B. Geslot, D. Bernard, A. Billebaud, P. Casoli, G. De Izarra, M. Diakaki, D. Doré, D. Foligno, D. Garnier, G. Kessedjian, U. Koester, F-R. Lecolley, P. Leconte, J-L. Lecouey, X. Ledoux, L. Matthieu, O. Méplan, A. SardetO. Serot, T. Soldner (2019). "Absolute yield and kinetic parameters of delayed neutrons from the neutron-induced fission of Pu-239" Institut Laue-Langevin (ILL) doi:10.5291/ILL-DATA.3-07-385

[4] P. Leconte, D. Bernard, A. Billebaud, P. Casoli, A. Chebbo ubi , G. De Izarra, D. Doré, D. Garnier, B. Geslot, G. KessedjianU. Koester, F-R. Lecolley, J-L. Lecouey, X. Ledoux, N. Marie, L. Mathieu Ludovic, O. Méplabn, A. Sardet; O. Serot, T. Soldner, (2021). Absolute yield and kinetic parameters of delayed neutrons from the neutron-induced fission of 233U. Institut Laue-Langevin (ILL) doi:10.5291/ILL-DATA.3-07-396

[5] H. Abele et al., Nucl. Instr. Meth. A 562 (2006) 407.

[6] M. A. Lone et al, Nucl. Instr. Meth. 174 (1980) 521.

[7] B. Geslot, P. Leconte, G. de Izarra, D. Foligno, O. Serot, A. Chebboubi, A. Billebaud, L. Mathieu, N. Marie, O. Méplan, F.-R. Lecolley, J.-L. Lecouey, U. Koester and T. Soldner. "A new experiment at ILL for measuring the kinetic behavior of delayed neutron precursors from thermal induced fission of 235U". IEEE. Trans. Nucl. Sci. To be published.

[8] B. Geslot, et al., "Development and manufacturing of special fission chambers for in-core measurement requirements in nuclear reactors," 2009 1st Int. Conf. on Advancements in Nuclear Instrumentation, Measurement Methods and their Applications. ANIMMA, 7-10 June 2009, Marseille, France.

EPJ Web of Conferences **253**, 01005 (2021) https://doi.org/10.1051/epjconf/202125301005
ANIMMA 2021

Production and Monitoring of Neutron Flux by Activation Detectors

Ivan Haysak[1], Vasyl Martishichkin[1], Yevgen Harapko[1], Robert Holomb[1,2], and Karel Katovsky[2]
[1]Uzhgorod National University, Ukraine
[2]Brno University of Technology, Czech Republic
Corresponding author: ivan.haysak@uzhnu.edu.ua

Abstract—**The neutron generation technique was tested on the microtron M-10 with an output electron beam of 8.7 MeV. Given the low energy that the microtron can provide to electrons, the bremsstrahlung induced photonuclear reaction ^{9}Be (γ, n), which has a low threshold, was chosen for neutron generation. Cobalt and indium targets were tested as activation detectors to estimate the neutron flux density. In the cobalt target, the isomeric state of ^{60m}Co with an energy of 58.6 keV and a half-life of 10.5 minutes is well activated. Two well-known additional gamma lines of standard cobalt source permit to clarify the absolute value of the neutron flux. The activated indium target has four gamma lines bound to the ^{116m}In isomer β⁻ decaying with the half-life of 54.4 minutes, what is convenient for measurement of gamma spectrum. Despite the low energy of the output electron beam, at a beam intensity of 5 μA it is possible to obtain an almost isotropic neutron flux of 10^7 n/(s·cm^2).**

Keywords — **Neutron production, microtron, gamma spectroscopy, neutron activation detectors.**

I. Introduction

On the microtron M-10 of Uzhgorod National University there are carried out experimental investigations of photonuclear reactions [1] as well as applied studies of irradiation influence on the properties of new technological materials and electronical components [2]. For irradiation it is possible to use electrons of energies 4-10 MeV and bremsstrahlung. But it was demonstrated [3] that low energy electron accelerator microtron MT-25 with 22 MeV output electron beam can serve as neutron source, which is interesting to compare with the generation of neutrons by a proton accelerator at the same beam energies [4].

Neutrons are unique particles that are of interest from both a fundamental and an applied point of view. Of the unsolved experimental fundamental problems, we note the problem of neutron lifetime [5], the question of the presence of the electric dipole moment of the neutron [6], and the search for neutron clusters [7]. In addition, neutron registration still represents an uneasy problem.

Among the current applied research, we note the study of different types of structures in the physics of condensed matter, the study of the effect of neutron fluxes on living tissues and organisms, neutron diffraction analysis [8], neutron activation analysis. Of great practical importance for materials science is the availability of accurate data on the cross-sections of the interaction of neutrons with nuclei. For energy needs, namely for reactors controlled by accelerators, experimental data on neutron generation processes are required.

In this paper, we present the results of neutron generation on the M-10 microtron of Uzhgorod National University. The experiment consist of few stages – 1) output electron beam generate bremsstrahlung spectrum in reaction $e^- + Z \rightarrow \gamma + e^- + Z$, 2) neutrons creation in photonuclear reaction (γ,n), 3) activation reaction (n,γ) and 4) measuring and analysis of gamma spectra of activated sample.

II. Experimental set up

A. Schematic view

The Fig. 1 shows a schematic of the experiment. The output electron beam falls on the braking tungsten target after which the bremsstrahlung beam hits the beryllium target. Neutrons are created on a beryllium target in reaction

$$\gamma + {}^{9}Be \rightarrow n + 2\alpha, \qquad (1)$$

starting from the threshold E_γ=1.57 MeV. The intensity of generated neutrons is registered by the activation sample (target) due to (n, γ) reaction. Analysis of gamma spectra of the activated sample (detector) gives the value of the neutron flux.

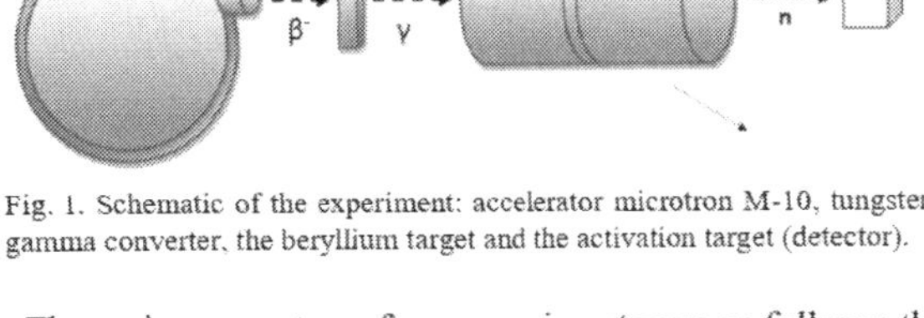

Fig. 1. Schematic of the experiment: accelerator microtron M-10, tungsten gamma converter, the beryllium target and the activation target (detector).

The main parameters of our experiment were as follows: the energy of the induced electron beam was 8.7 MeV and a plate of natural tungsten of size 93 x 55 x 2 mm (thickness 2 mm) served as a braking target. The size of the beryllium target is Ø10 x 14 cm, and its weight is 2 kg.

To measure neutrons flux we use ^{59}Co as activation detectors. The cobalt detector (boxed powder $CoCO_3$*$Co(OH)_2$*nH_2O, mass 31 g) was irradiated during 10 min by neutrons Cobalt detector is activated by neutrons to both ^{60}Co ground state and isomeric level ^{60m}Co with half-life 10.47 minutes and following isomeric transition $E_\gamma = 58.6$ keV to the ground state. After β-decay (half-time 5.27 years) the ground state gives well-known two lines $E_\gamma = 1173$ keV and $E_\gamma = 1332$ keV [9]. As a neutron indicator, we also used an aluminum plate coated with indium powder.

B. Gamma Spectrometer

A NaI(Tl) crystal scintillation detector with size Ø63 x 63 mm looking through by photomultiplier 19-M and with the SBS-40 amplitude analyzer board served as a gamma spectrometer (Fig. 8). The accumulation of spectra occurs with the help of a specialized computer program AkWin. The number of channels of the analyzer is variable and can take values from 256 to 8192.

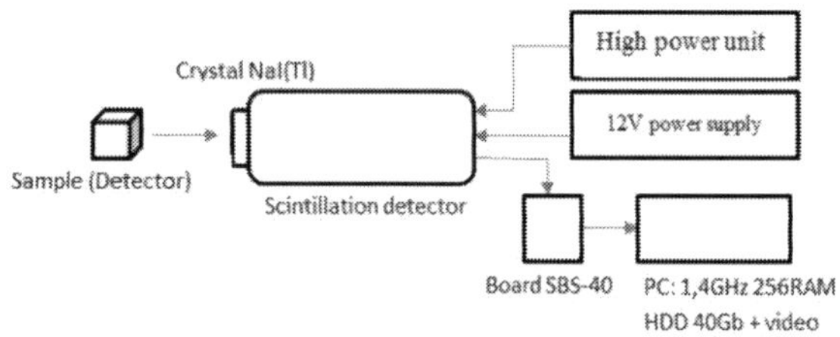

Fig. 2. Schematic of the scintillation spectrometer.

The measured energy resolution (FWHM) of the spectrometer for calibration lines and the simulated total peak efficiency ε for the detected gamma lines of the neutron-activated sample 59Co are shown in Table I.

TABLE I
THE EFFICIENCY AND THE ENERGY RESOLUTION OF THE GAMMA SPECTROMETER

E_γ, keV	58.6	59.5	661.7	1173	1332
$\varepsilon \times 10^2$	23.2	-	-	4.71	4.16
FWHM keV	-	16.2(2)	71.2(3)	82.0(5)	88.8(9)

III. BREMSSTRAHLUNG SPECTRUM

The gamma-ray spectrum created on a 2 mm thick tungsten target in the forward direction by a 5 MeV electron beam is showed at the Fig. 2. The spectrum is simulated by the FLUKA code [10]. The bremsstrahlung spectrum is a wide peak in the interval of energy $E\gamma$=0.1-5 MeV. In the energy region, less than 0.1 MeV the characteristic X-ray lines of tungsten atom are present, also the annihilation line of 0.511 MeV is evident. That is, in a tungsten conversion target with a thickness of 2 mm, there is a significant probability of a three-stage process, namely, 1) the braking process on the nucleus $e^- + Z \rightarrow \gamma + e^- + Z$; 2) formation in the field of the nucleus of electron-positron pairs $\gamma + Z \rightarrow e^+ + e^- + Z$; 3) the formation of the positronium atom $e^+ + e^- \rightarrow Ps$; 4) annihilation of positronium $Ps \rightarrow \gamma + \gamma$. The first two processes take place with the participation of the atomic nucleus, and the last two are the atomic processes.

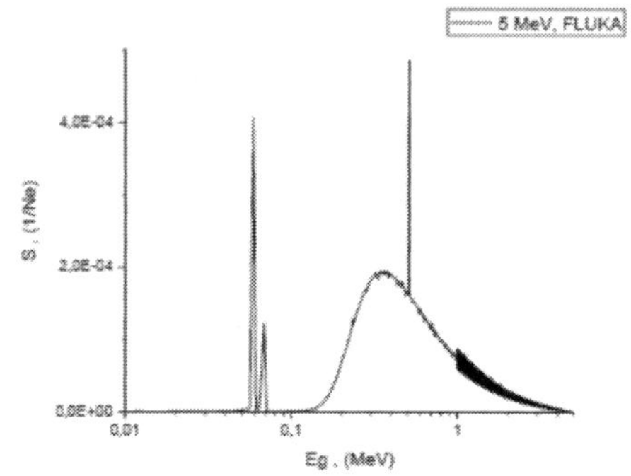

Fig. 3. The simulated braked gamma spectrum of 5 MeV electron beam in 2 mm tungsten converter.

In Fig. 3 is shown a comparison of the bremsstrahlung spectra simulated for electron beam of energies 4 MeV, 5 MeV, 8 MeV, and 9 MeV [11]. It is seen that the shape of the peak is preserved, and with increasing electrons energy the yield of gamma quanta is increased. In the study of nuclear reactions, only a part of the gamma spectrum with energy greater than the reaction threshold "works". In our case of neutrons generation on a beryllium target, it is gamma quanta with energies greater than 1.6 MeV. But at a set of the absorbed dose, it is necessary to consider the whole bremsstrahlung spectrum.

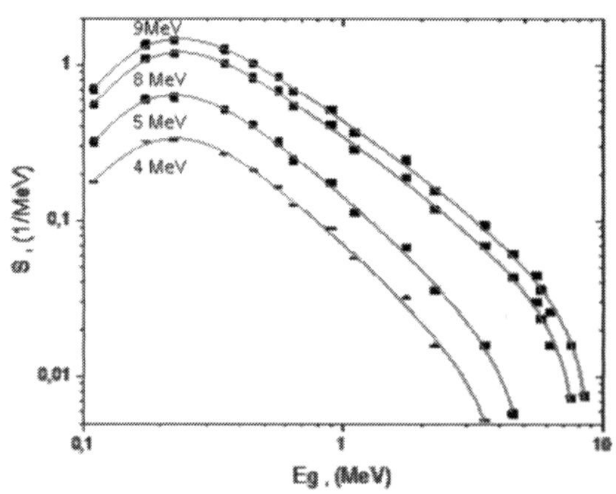

Fig. 4. The simulated bremsstrahlung for different electron beam energies 4, 5, 8 and 9 MeV.

It should be noted that the bremsstrahlung spectrum depends on both the electron energy and the thickness of the braking target and its atomic composition. The penetration depth of a monoenergetic electron beam with energies $3 < E_e < 20$ MeV in aluminum is well described by phenomenological formula

$$R = 0.53 \cdot E_e - 0.106\,, \qquad (2)$$

where R is the depth of penetration in g/cm², E_e is the electron energy in MeV [12]. For other matter the penetration depth can be estimate by correction

$$R_x = R_{Al}\frac{(Z/A)_{Al}}{(Z/A)_x}, \quad (3)$$

where $(Z/A)_x$ is the charge to mass ratio of the corresponding element. The estimated penetration depth for an electron beam of 8.7 MeV in tungsten is 2.8 mm.

IV. Neutron production

The measured and evaluated data of the energy dependence of the photonuclear reaction cross-section $\sigma(\gamma, abs)$ for beryllium, is as follows Fig. 4. For electron energy of 8.7 MeV(and maximal energy of gamma), the cross-section is about 1.5 mb. Up to threshold energy of channel (γ, p) 16.89 MeV there are open channels (γ, nx) where x are 2α or ^{8}Be

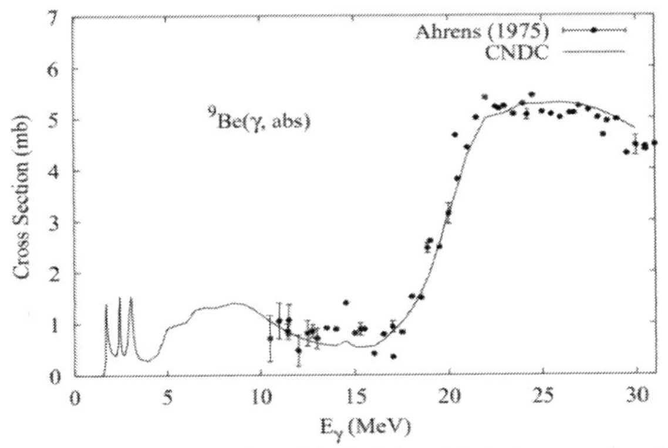

Fig. 5. The measured and the model calculation of the total cross-section of photoabsorption reaction ^{9}Be(γ,abs) [13].

in the ground or excited states (the three lowest well-separated levels of ^{8}Be are 3.04, 11.4, and 16.6 MeV). The thresholds of these channels are indicated in Table I. So for electron beam 8.7 MeV the three channels of reaction ^{9}Be(γ,n) are opened.

TABLE II
THE THRESHOLD ENERGIES FOR DIFFERENT REACTION CHANNELS

Threshold energies (MeV)			
γ, n+2α	γ, n+^{8}Be	γ, n+^{8}Be*(3.04)	γ, n+^{8}Be*(11.4)
1.57	1.67	4.71	13.1

Analysis of the kinematics of the ^{9}Be(γ, n)^{8}Be reaction shows that the kinetic energy of a neutron is weakly dependent on its emission angle. Fig. 5 given the dependence of the neutron kinetic energy distribution versus the angle of the outgoing neutron relative to the gamma beam direction for energies of gamma quanta 8, 6, 4, and 2 MeV. However, the shape of the angular distribution can be deformed by the dynamics (matrix element) of the nuclear reaction. Fig. 3 shows that the intensities of the corresponding energies of gamma quanta differ by an order of magnitude. That is, one can conclude that the reaction (γ, n) forming an almost isotropic distribution of neutrons in the whole energy region from zero up to 6 MeV. But in the beryllium target itself, neutrons are moderated by elastic multiple scattering on beryllium nuclei. Therefore, the spectrum of neutrons outside the beryllium target will be much softer.

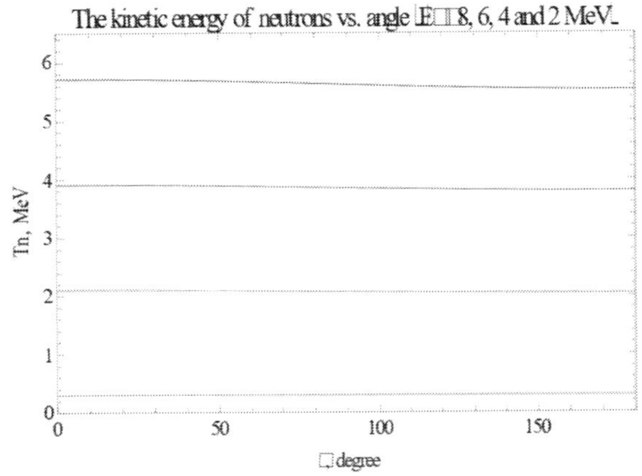

Fig. 6. The angle dependence of kinetic energy of outgoing neutrons for reaction γ + ^{9}Be → n + ^{8}Be at energy E_γ=8, 6, 4 and 2 MeV (violet, red, yellow, and green line respectively).

V. Activation detectors

We chose samples of ^{59}Co and natural indium as activation detectors which have satisfactory rate activation and disintegration constants. The activation sample was located close to the side surface of the beryllium cylinder.

A. Cobalt sample

The effective cross-section of the reaction ^{59}Co(n,γ) is shown in Fig. 6. In the region from thermals up to 6 MeV neutrons, the cross-section belongs to interval 2-40 barns [14].

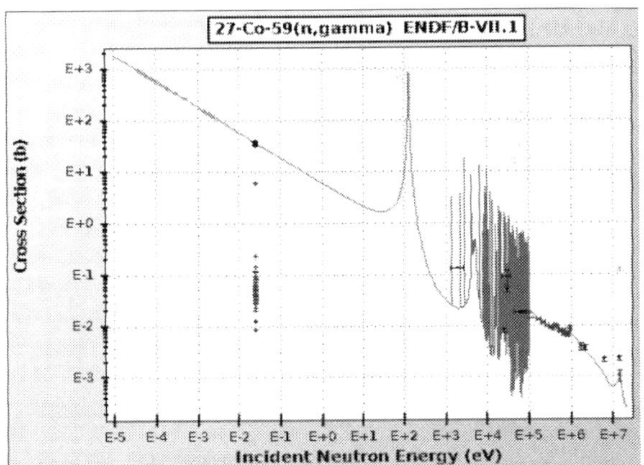

Fig. 7. The energy dependence of cross-section of reaction n + ^{59}Co → γ + ^{60g,m}Co [14].

Neutron activation of ^{59}Co leads to the formation of ^{60g}Co in the ground state and in the isomeric state ^{60m}Co. The decay schemes of cobalt from the ground ($T_{1/2}$=5.27 years) and isomer ($T_{1/2}$=10.47 minutes) states [9] are shown in Fig. 7. Isomer state has two decay modes – 99.76% isomeric gamma transition with E_γ=58.59 keV and 0.24% β^- decay. The ground state is a well-known standard cobalt-source which after β^- decay gives two gamma lines 1333 keV and 1173 keV.

B. Indium sample

Natural indium consist of ^{113}In(4.29%) and ^{115}In(95.7%) isotopes [15]. After activation by neutrons in reaction (n,γ) isomer ^{116m}In is created which after β^- decay with half-life 54.29 m transforms into ^{116}Sn in excited states with gamma lines $E\gamma$=417, 1097, 1293 and 2112 keV. The cross-section of reaction ^{115}In(n,γ) is near 200 barn [14].

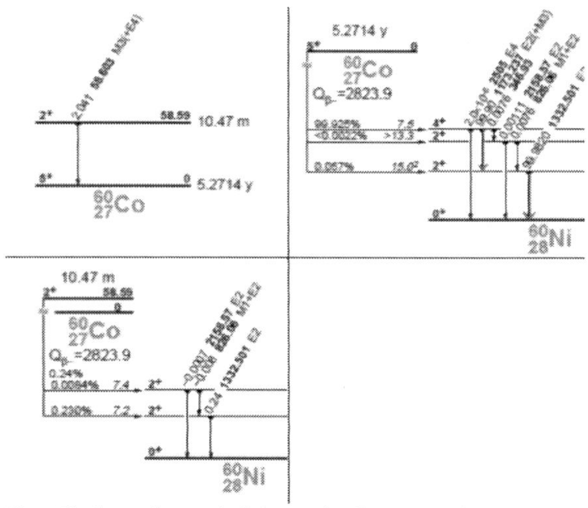

Fig. 8. The decay schemes of cobalt ground and isomer states [8].

VI. Results and conclusions

A beryllium target and activation samples were irradiated for 10 minutes with a braking beam of gamma quanta obtained on a tungsten conversion target by an electron beam with an energy of 8.7 MeV and an intensity of 5 μA. Fig. 9 shows the gamma spectra from samples of indium, cobalt, and background. The isomeric cobalt line is not shown because it is located near channel 20 where the background is very high.

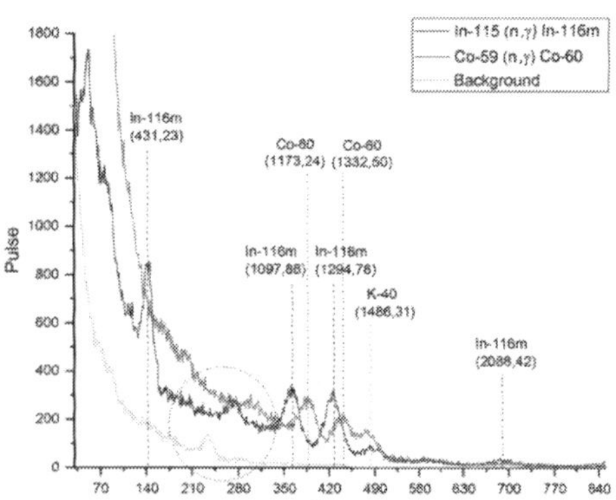

Fig. 9. The gamma spectra from activated ^{59}Co and ^{115}In samples and background spectrum.

The preliminary result of processing the gamma spectra of indium is given in [16], and the processing of cobalt spectra confirmed it, namely, at an electron beam current of 5 μA at an energy of 8.7 MeV and selected bremsstrahlung and neutron producing targets on the M-10 microtron we obtain a neutron flux 2×10^7 n/(cm^2·s).

The questions of neutron energy distribution and the ratio of neutron and gamma fields remain open [17,18].

Acknowledgment

This work was supported by the grant of the Ministry of Education and Science of Ukraine No. 0115U001098. V. Martishichkin thanks professor O.M. Parlag for a fruitful discussion about neutron detectors.

References

[1] V.I. Zhaba, I.I. Haysak, A.M. Parlag, V.S. Bokhinyuk, and M.M. Lazorka, "Energy dependence of cross section of photonuclear reactions on indium isotopes", *Problems of Atomic Science and Technology*, vol. 115, 2018, pp.155-158.

[2] A.A. Molnar, H.V. Vasylyeva, I.I. Haysak, M.T. Sabolchiy, I.M. Stoika, and A.A. Grabar, "Influence of the electron and gamma-irradiation on dielectric Properties of Sn2P2S6", *Uzh. Univ. Sci. Her. Ser. Phys.*, iss. 40, 2016, pp.13-17, 10.24144/2415-8038.2016.40.13-17

[3] M. Kralik, J. Solc, D. Chvatil, P. Krist, K. Turek, and C. Granja, "Microtron MT 25 as a source of neutrons", *Rev. Sci. Instrum., vol. 83, 083502*, Aug. 2012, 10.1063/1.4739404.

[4] D De Luca, L. Campajola, and P. Casolaro, "Neutron production with particle accelerators", *J. Phys.: Conf. Ser.*, vol. 1226, 012021, May 2019.

[5] A. P. Serebrov, M. E. Chaikovskii, G. N. Klyushnikov, O. M. Zherebtsov, and A. V. Chechkin, "Search for explanation of the neutron lifetime anomaly", *Phys. Rev. D 103, 074010*, Apr. 2021.

[6] C. Abel et al., "Measurement of the permanent electric dipole moment of the neutron", *Phys. Rev. Lett. 124, 081803*, Febr. 2020.

[7] M. D. Higgins, C. H. Greene, A. Kievsky, and M. Viviani, "Comprehensive study of the three- and four-neutron systems at low energies", *Phys. Rev. C 103, 024004*, Febr. 2021.

[8] A. Nagornyi, V. I. Petrenko, M. Rajnak, I. V. Gapon, M. V. Avdeev, B. Dolnik, L. A. Bulavin, P. Kopcansky, M. Timko, "Particle assembling induced by non-homogeneous magnetic field at transformer oil-based ferrofluid/silicon crystal interface by neutron reflectometry", *Appl. Surf. Sci.*, vol. 473, 2019, pp. 912–917, 10.1016/j.apsusc.2018.12.197.

[9] R.B. Firestone, "Table of Isotopes". Eight edition // CD ROM Edition, Wiley-Interscience, 1996.

[10] A. Ferrari, P.R. Sala, J. Ranft, U. Siegen, "FLUKA: A multi-particle transport code". SLAC-R-773. (2015).

[11] I.I. Haysak, O.V. Takhtasyev, R.R. Holomb, K. Katovsky, A. Tanchak, J. Khushvaktov, A.A. Solnyshkin, "Monte Carlo simulation of bremsstrahlung spectra for low energy electron accelerators", in *2020 21st International Scientific Conference on Electric Power Engineering, EPE 2020*, Prague, Czech Republic, 2020, pp. 1-4, 10.1109/EPE51172.2020.9269252.

[12] V.P. Rudakov, "Passage of Ionizing Radiation Through Matter", in *Physical quantities: Handbook*, I. S. Grigoriev and E. Z. Meilikhova eds., Energoatomizdat, Moscow, 1991, p. 1170.

[13] "Handbook on photonuclear data for applications cross-sections and spectra", *IAEA*, Vienna, Austria, IAEA-TECDOC-1178, Oct. 2000.

[14] National Nuclear Data Center https://www.nndc.bnl.gov/sigma/

[15] J.K. Tuli, "Nuclear Wallet cards", *Brookhaven Nat. Lab.*, Upton, N-Y, USA, Jan. 2000.

[16] V. Martyshychkin *et al.*, "Generation of neutrons on Microtron M-10", in *2020 21st International Scientific Conference on Electric Power Engineering, EPE 2020*, Prague, Czech Republic, 2020, pp. 1-6, 10.1109/EPE51172.2020.9269218.

[17] X. Mao, K.R. Kase, and W.R. Nelson, "Giant dipole resonance neutron yields produced by electrons as a function of target material and thickness", *Health Phys.*, Feb. 1996, vol. 70, no. 2, p.207-214. 10.1097/00004032-199602000-00008.

[18] T.G. Soto-Bernal, A. Baltazar-Raigosa, D. Medina-Catro, and H.R. Vaga-Carrillo, "Neutron production during the interaction of monoenergetic electrons with tungsten foil in the radiotherapeutic energy range", *NIM Phys. Res. A*, vol. 868, Oct. 2017, P.27-38.

A High speed data link optimization for digitalized transfer to processing FPGA

J. Collado[1,2], V. Gonzalez[2], and A. Gadea[1]
[1]ETSE Universidad de Valencia, Spain
[2]IFIC-CSIC, Spain

javier.collado@uv.es

***Abstract*—State-of-the-art arrays of detectors, that require digital processing, may have a sizeable number of digitalized signal links. This is the case in several experimental nuclear physics instruments. Moreover, the data rate of the sampled signals, defined primary by the signal bandwidth of the individual detectors, may not exhaust the capabilities of a single FPGA transceiver input.**

The preprocessing is usually carried out in a modern FPGA with transceiver data rate capabilities over 10Gbps. Moreover, cost effective FPGA have a limited number of transceivers for given FPGA processing capabilities. The investigation of a cost-effective and efficient solution to the mismatch between both data rates, optimizing simultaneously the use of the FPGA resources, is the topic of the present work.

We have developed a solution based on the Time Domain Multiplexing link aggregation, in the form of a Mezzanine board. This mezzanine combines four channels from an optical or copper input up to 2.5 Gbps to one up to 10Gbps, and serves them to the FPGA via the mezzanine connector. The board itself is controlled by a small FPGA by the Two Wire Interface (TWI) protocol as a standalone intelligent device, so minimum slow control is needed. The solution has been also developed for a motherboard housing a SoM module and FMC connector as an alternative implementation. An associated firmware has been developed to de-aggregate the data in the FPGA and recover the original sampled data, based on JESD204 communication protocol, inside the FPGA. The method has been validated and applications, beyond the development of the AGATA electronics, may be envisioned.

***Keywords* — FPGA optimization, Time Domain Multiplexing, Communication serial link, Electronic instrumentation.**

I. Introduction

The main motivation for this work is the development of the new Phase 2 electronics for the AGATA experiment[1]. The AGATA detector is an array of segmented high purity germanium crystals for gamma spectroscopy. The segmentation is required to grant position sensitive capabilities to each crystal, and in sum with all the array, being capable to perform tracking algorithms. This is a huge advance in nuclear science gamma spectroscopy because it allows for searching of gamma multiplicities and angular correlations, in addition to an unprecedented energy resolution. These crystals are segmented in 36 segment contacts plus a core contact. To perform the positioning of an event on the crystal, a processing of the pulse shape on each segment is needed.[2] This implies a digitalization of each contact signal. In the case of AGATA, the sampling is performed with 14 bit (for an ENOB of 12) at 100MHz, in free-running operation, due to the bandwidth of the detector and preamplifier response, limited to about 30 MHz. Each ADC has a link to the preprocessing system at 2Gbps using the JESD204 protocol. [3]

As in many instrumentation electronics for these kind of experiments we just need the information when an event arrives to the detector, therefore a triggering is performed on the pre-processing electronics.[4] There are also metrics needed for each event, as the energy or the timing for each event. The triggering, energy extraction, packaging of traces and readout is performed in real time using an FPGA. On AGATA each FPGA is capable of perform this preprocessing for one entire crystal, which implies that 38 transceivers are needed to receive serialized data from the ADCs. There are detector arrays with even a higher amount of ADCs to be processed in the FPGA. All these issues lead to the main problem solved in this work.

II. Problem

The problem presented in this electronics development was the high bandwidth mismatch between the ADC boards and the processing FPGA. As already mentioned, the detector preamplifier response is 30MHz and the sampling frequency of the digitizer to acquire this bandwidth is 100MHz, in order to obtain the data precision required to carry out positioning algorithms. A serialized data output from 8 to 16 bit requires about 1-2Gbps. In the case of AGATA there is a 2Gbps JESD204 encoded differential line per ADC. As a pair of ADC are integrated in each IC it is possible to send data as fast as 4 Gbps in each link in pair. Being AGATA case an example, there are many configurations with the same high speed side data link bandwidth, depending on the ADC sample rate, the number of bits and the amount of digitalizes per integrated circuits.

All cases require a certain amount of links between the ADCs and the destination FPGA for real time processing. In the segmented array detectors, or similar instrumentations, this becomes a critical element. While FPGA processing is possible in parallel on the same device for huge amount of channels, the number of input transceiver is limited in the device and has a direct impact on cost. Nevertheless, the FPGA transceivers have an input data rate maximum much higher than 1-5 Gbps, in modern FPGA is about 10-40Gbps depending of the technology used[5], [6].

An optimization of the aggregated bandwidth has benefits on two parts of the developments. The first benefit is the direct reduction of FPGA cost and design, allowing the use of lower performance FPGA. In the case of AGATA only one FPGA is capable to preprocess the 38 channels of a segmented crystal in one mid-low performance FPGA as a Xilinx Zynq Ultrascale+ XZU15[7]. The second benefit is to open the possibility to use SOM, COTS boards. Such SOM modules (using high speed connectors or FMC) include a reduced number of high speed channels depending on the model and connector.

III. Solution

A logical solution for the described problem is to multiplex incoming channels. The simplest option for this solution is to perform a Time Domain Multiplexing (TDM). In this case we can combine several lines from 1-5Gbps to a higher speed link of about 5-20Gbps.

The solution could be implemented in several ways. The first possibility is to implement the TDM in low-cost FPGAs, with a limited number of transceiver . The second possibility is to search for specific commercial ASICs used in other applications to perform the TDM.

This second solution led us to explore the possibility to use Ethernet ASICs for communication systems. In particular, there is a TDM family from Texas Instruments that performs a 2:1, 4:1 or 8:1 multiplexing for 1-5 Gb serial links. Their primary function is to aggregate Ethernet links to reduce the connection number. For our implementations TLK10002 and TLK10022[8] are selected. They perform a 4:1 aggregation of up to 1-2.5 Gbps to 10Gbps maximum per link.

For the development, there is a huge gap to cover due to the nature of these ASICs as they are meant to work in pairs with one device aggregating and sending data through a long transmission lines and another device at the receiver to recover the original links. Each of them has a Tx and Rx part to operate in full duplex mode. Our development includes the FPGA firmware reception side development to disaggregate the data inside the FPGA, in order to obtain the advantage of the transceiver number reduction.

As mentioned before, the use of our development for different instruments and applications has been considered. In order to do that, two implementations, allowing to extend the use for several configurations of data rates for input and output, have been developed.

IV. Implementation

We have implemented two options for delivering data to the FPGA. One is an FMC VITA 57.1 standard[9] mezzanine to connect to a custom motherboard with the selected FPGA or to an evaluation board in case it's used in the development. The second implementation is a motherboard with a SOM socket to host the selected FPGA, including also an FMC connector to perform the readout from the FPGA, as in the case of AGATA.

The data input is done by either optical or copper links from the ADC boards. To maintain compatibility, both boards have been designed to cope with these two possibilities, optical and copper connection.

Both implementations have been developed looking to comply as much as possible to the standards. The have been designed and simulated to work with very low jitter and synchronized clocking and stable data links.

A. FMC version

The FMC version is called IDM (Input Data Mezzanine)[10]. This board is built to minimize the form factor. Nevertheless, is not a standard FMC form factor but slightly wider. This issue is foreseen to be corrected in future versions of the board.

This mezzanine implements a megArray 9x9[11] connector for input data and are fully compatible with miniPod[12] for optical transceiver. This miniPod can be easily converted to 12 line MTP optical fibers or multiple LC if needed. Each of the 4 input connectors is capable to receive 12 lines. Although, only five TLK10022 are used, each of them is capable of perform 2 aggregations (channel A and B). Therefore, we can process up to 40 lines and aggregate them into 10. These 10 high speed serial lines are connected to a High Pin Count (HPC) FMC connector as is the VITA57.1.

The board also includes a small form factor, low power and low cost FPGA to perform slow control and autonomous initialization. The board is seen from the FMC I2C lines as one device with all the element memory mapped on this FPGA. It has also alerts for temperature and voltage anomalies.

On the AGATA environment the board has been tested aggregating from 40 differential lines of 2Gbps to 10 differential serial lines of 8Gbps.[13]

Fig. 1. The IDM mezzanine with the aggregation concept. It is the original demonstrator for AGATA Phase 2 and a modular solution for other instrumentation systems.

B. Motherboard version

The second implementation is built as a motherboard (called PACE-CAP) with a socket for a TE0808 SOM from Trenz Electronic GmbH[14], demonstrating the possibility of integrating aggregation and processing on the same board. In our case, this board is the final design for AGATA Phase 2 electronics and implements an FMC connector to send the processed data through Ethernet using a custom board called STARE[15].

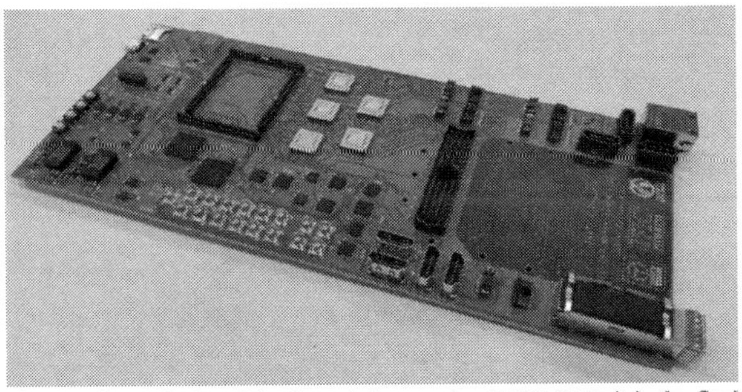

Fig. 2. The motherboard implementation version PACE-CAP, it is the final solution for AGATA Phase 2 electronics in combination with STARE board and the FPGA SOM.

C. Firmware

One of the main developments is the firmware to recover the data on the FPGA, on this case is done for the JESD204B[16] protocol, but the IP from the disaggregation is standard to any 8b/10b based communication protocol.

The firmware is capable also to perform alignment of the 38 ADC channels for the AGATA digitalizes.

V. Results

The IDM board is built using a High Density Interconnect process to ensure the minimal size to cope with FMC size. Both boards are designed carefully with previous simulations on the high speed signals parts.

There is a test bench designed for each of the boards with a setup of AGATA digitizers. On the IDM case, the mezzanine is connected to an evaluation board, Trenz TEBF0808, with a TE0808 SOM board as FPGA. The data is extracted by internal memory reading.

The testbench for the PACE-CAP is done with a set of AGATA Phase 2 digitizers in final configuration. This board is also capable to perform the slow control of the ADC and the general clocking.

Both boards were tested with stable links over periods longer than 24h. The IDM has been tested with optical connections and the PACE-CAP is tested using copper links. Future tests may include copper connections for the IDM and optical ones for the PACE-CAP.

VI. Conclusions

Our implementation of the Time Domain Multiplexing has probe to be a valid method to aggregate data from a large amount of ADCs (30-40) optimizing the FPGA transceiver occupancy. The 40 serial lines from 2Gbps links are aggregated up to 10 links and data is recovered and synchronized. The two possible boards are valid and stablish a solution that cope with other complex instrumentation involving many channels of sampling electronics

Future developments are on the way to optimize the IDM to fit in a standard FMC form factor and updating the input connections to extent compatibility. On the PACE-CAP, there are some implementations on study to extent its capacity to new configurations of line aggregation beyond 2:1 and opening the use for different applications.

References

[1] AGATA Colaboration, "AGATA Colaboration Web." https://www.agata.org/about (accessed Sep. 01, 2019).

[2] S. Akkoyun *et al.*, "AGATA - Advanced GAmma Tracking Array," *Nucl. Instruments Methods Phys. Res. Sect. A Accel. Spectrometers, Detect. Assoc. Equip.*, vol. 668, pp. 26–58, Mar. 2012, doi: 10.1016/j.nima.2011.11.081.

[3] A. Pullia, D. Barrientos, D. Bazzacco, M. Bellato, D. Bortolato, and R. Isocrate, "Digi-opt12: 12-channel 14/16-bit 100/125-MS/s digitizer with optical output for GALILEO/AGATA," 2012.

[4] M. Bellato *et al.*, "Sub-nanosecond clock synchronization and trigger management in the nuclear physics experiment AGATA," *J. Instrum.*, vol. 8, no. 7, 2013, doi: 10.1088/1748-0221/8/07/P07003.

[5] Intel, "Intel Programable." https://www.intel.es/content/www/es/es/products/programmable.html.

[6] Xilinx, "Xilinx." www.xilinx.com.

[7] Xilinx Inc., "Zynq UltraScale + MPSoC Data Sheet : Overview - DS891 v1.7," 2018.

[8] Texas Intruments, "TLK10022 datasheet." http://www.ti.com/product/TLK10022 (accessed Dec. 07, 2016).

[9] Vita, "ANSI-Vita 57.1 Standard," 2019.

[10] J. Collado, V. González, S. Member, J. M. Blasco, E. Sanchis, and S. Member, "Design of a mezzanine card with bandwidth aggregation for HPGe gamma spectroscopy," in *2016 IEEE Nuclear Science Symposium, Medical Imaging Conference and Room-Temperature Semiconductor Detector Workshop (NSS/MIC/RTSD), Strasbourg, 2016*, 2016, pp. 1–2, doi: 10.1109/NSSMIC.2016.8069736.

[11] Amphenol ICC, "MEG-Array. Proven BGA connector for reliabilityand high data rates.," 2019. [Online]. Available: https://cdn.amphenol-icc.com/media/wysiwyg/files/documentation/datasheet/mezzanine/mezz_megarray.pdf.

[12] Avago Technolgies, "MiniPOD™ AFBR-812VxyZ,

AFBR-822VxyZ Product Brief," 2013.
[13] J. Collado, "Contributions to Phase Two of AGATA electronics," Universidad de Valencia, 2020.
[14] Trenz Electronic, "TE0808 Technical Reference Manual v.32," 2019.
[15] N. Karkour *et al.*, "40 Gbps Readout interface STARE for the AGATA Project." 2020.
[16] JEDEC STANDARD, "Serial Interface for Data Converters - JESD204B," South Arlington, 2012. [Online]. Available: papers2://publication/uuid/E1A55D26-05F8-4FE4-8D5E-8713EDC575D1.

EPJ Web of Conferences **253**, 01007 (2021) https://doi.org/10.1051/epjconf/202125301007
ANIMMA 2021

Commissioning the ATLAS Liquid Argon Calorimeter Phase-I Upgrade

Ellis Kay, on behalf of the ATLAS Liquid Argon Calorimeter Group
Department of Physics and Astronomy, University of Victoria, Victoria, BC, V8W 2Y2, Canada
E-mail: ellis.kay@cern.ch

Abstract

Liquid argon (LAr) sampling calorimeters are employed by ATLAS for all electromagnetic calorimetry in the pseudo-rapidity region $|\eta| < 3.2$, and for hadronic and forward calorimetry in the region from $|\eta| = 1.5$ to $|\eta| = 4.9$. In the first LHC run (Run-1), a total luminosity of around 26 fb^{-1} was collected at centre-of-mass energies of 7-8 TeV. After detector consolidation during a long shut-down, Run-2 started in 2015 and around 150 fb^{-1} of data at a centre-of-mass energy of 13 TeV was recorded. With the end of Run-2 in 2018, a multi-year shutdown for the Phase-I detector upgrades began. As part of the Phase-I upgrade, new trigger readout electronics of the ATLAS LAr Calorimeter have been developed. Installation be-gan at the start of the LHC shut down in 2019 and is expected to be completed in 2020. A commissioning cam-paign is underway in order to realise the capabilities of the new, higher granularity and higher precision level-1 trigger hardware in Run-3 data taking. This contribution will give an overview of the new trigger readout system and the ongoing commissioning and installation efforts.

I. The ATLAS Liquid Argon Calorimeter

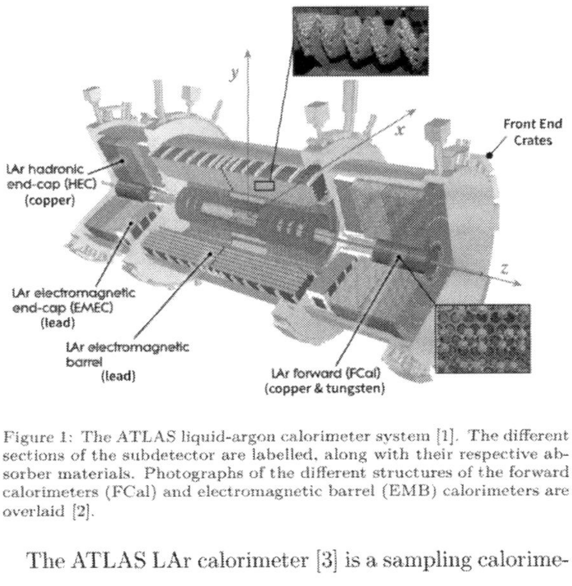

Figure 1: The ATLAS liquid-argon calorimeter system [1]. The different sections of the subdetector are labelled, along with their respective absorber materials. Photographs of the different structures of the forward calorimeters (FCal) and electromagnetic barrel (EMB) calorimeters are overlaid [2].

The ATLAS LAr calorimeter [3] is a sampling calorimeter with full coverage in azimuthal angle[1]. It uses liquid argon as an active medium, with copper electrodes and various absorber materials (see Figure 1). Incident particles shower in the absorber material and subsequently ionise the LAr. An electric field applied across the LAr gap causes the ionisation electrons to drift across it, leading to a triangular pulse which is collected by the electrodes (see Figure 2). The subdetector is housed in one 'barrel' and two 'endcap' cryostats kept at -184°C. A high granularity electromagnetic (EM) calorimeter with an accordion geometry covers the barrel (EMB) region of $|\eta| < 1.475$ and the end-cap (EMEC) region of $1.375 < |\eta| < 3.2$. Two parallel-plate hadronic endcap (HEC) calorimeters cover the region $1.5 < |\eta| < 3.2$, while forward calorimeters (FCal), which use cylindrical electrodes running parallel to the beam axis within a honeycomb structure of metal, cover the region $3.1 < |\eta| < 3.9$. Each section of the calorimeter is segmented in depth in up to three layers, referred to as front, middle and back. For $|\eta| < 1.8$, an additional 'presampler' layer provides measurements of energy lost upstream of the subdetector. The LAr calorimeter, with each of these technologies mirrored in the A and C sides of the detector, consists of 8 subsystems, referred to as 'partitions', namely EMBA, EMBC, EMECA, EMECC, HECA, HECC, FCalA and FCalC.

II. Readout Electronics

The ionisation signals from the 182 486 LAr readout channels leave the cryostats via 114 feedthroughs. Each feedthrough is connected to one of the Half-Front-End-Crates (HFECs), which sit radially around the detector and house several kinds of electronic boards connected to a baseplane. Front-End Boards (FEBs) [5] are responsible for receiving the analogue signals from up to 128 channels and sampling and digitising them before transmitting them, via optical links, to the back-end electronics which reside in an underground service cavern separate from the

[1]ATLAS uses a right-handed coordinate system with its origin at the nominal interaction point (IP) in the centre of the detector and the z-axis along the beam pipe. The x-axis points from the IP to the centre of the LHC ring, and the y-axis points upward. Cylindrical coordinates (r, ϕ) are used in the transverse plane, with ϕ being the azimuthal angle around the z-axis. The pseudorapidity is defined in terms of the polar angle θ as $\eta = -\ln\tan(\theta/2)$. In ATLAS, the ends of the detector with positive and negative z coordinates are referred to as 'A-side' and 'C-side', respectively.

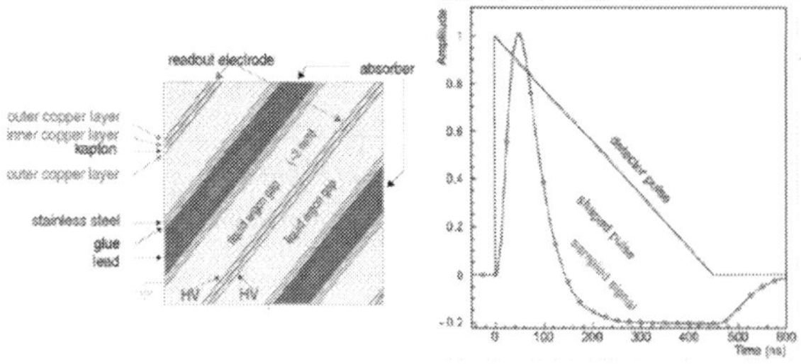

Figure 2: Diagrams of the structure of a sampling layer of the ATLAS EM calorimeter (left) and a pulse shape read out from the detector before and after shaping and sampling (right) [4].

experiment. The signal from each channel is split into one of three gains (low, medium and high) and, for each gain, the triangular ionisation pulse is shaped with a bipolar (CR-(RC)^2) analogue filter. The shaped signals are sampled at the LHC bunch-crossing frequency of 40 MHz and the samples are stored in a Switched Capacitor Array (SCA) analogue memory buffer while awaiting a Level-1 (L1) trigger accept. In addition to reading out the individual calorimeter channels, the FEB performs the first stages of summing in preparation of analogue sums which are used by the L1 trigger system [6] to inform L1 accepts. Each of the shaper chips in the FEB outputs a partial sum of input channels, which are routed to two mezzanine boards known as Layer Sum Boards (LSBs) which perform further summing, depending on the number of layers in the relevant detector region. These sums are sent through the HFEC's baseplane to Tower Builder Boards (TBBs) or Tower Driver Boards (TDBs), which sum data from calorimeter cells in all layers in so-called Trigger Towers (TTs), which correspond to slices of the detector of approximately $\Delta\eta \times \Delta\phi = 0.1 \times 0.1$. The HFECs also contain calibration boards, which are used to inject known triangular pulses into the cryostat in order to simulate the LAr ionisation signal and calibrate the electronics.

III. Phase-I Upgrade

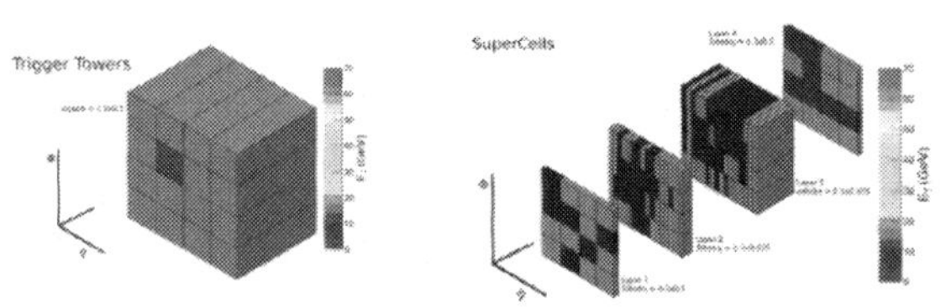

Figure 3: Diagrams showing the energy deposited by an electron with $E_T = 70$ GeV as seen by a trigger tower (left) and by the equivalent SCs for a typical region in the EM calorimeter [7].

The instantaneous luminosity and pileup will increase for Run-3 of the LHC, yet the current ATLAS trigger system does not allow for a proportionate increase of the L1 trigger rate. In order to compensate for this while maintaining similar triggering performance, an improvement of the trigger system itself is required. Upgraded L1 algorithms will have better discrimination between photons, electrons, taus and jets, and new Feature EXtractor (FEX) modules will use topological shower shape variables to discriminate between these different objects [8]. Such improvements require higher granularity information from the LAr calorimeter than what is provided in the Run-2 trigger readout path. Currently, signals from cells in different calorimeter sampling layers are grouped into ~ 5 400 trigger towers, as described in the previous section. As a result of this summing, information about the development of shower shapes is lost, and cannot be used as a discriminating variable in the trigger decision. In the Phase-I upgrade, cells will instead be grouped into ~ 34 000 'Super Cells' (SCs), which provide up to four-layer information and a tenfold increase in granularity. Figure 3 shows a depiction of an electron energy deposit as seen through a TT compared to the same deposit measured by SCs for a typical EM η, ϕ slice. In this case, the SCs provide information for $\Delta\eta \times \Delta\phi = 0.1 \times 0.1$ for the presampler and back layers and $\Delta\eta \times \Delta\phi = 0.025 \times 0.1$ for the front and middle layers. Figure 4 shows the expected improvement in performance from the Phase-I LAr upgrade.

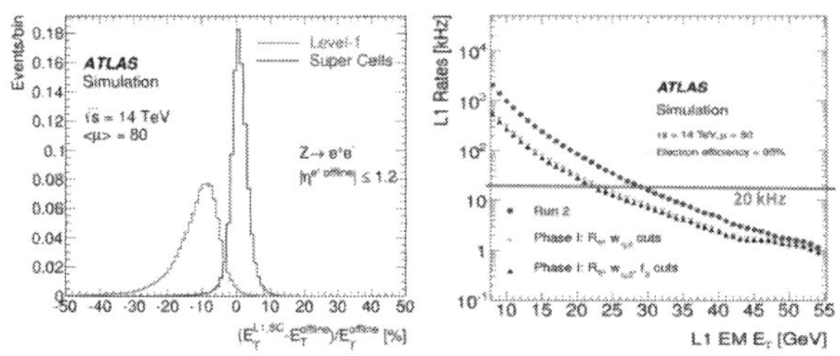

Figure 4: Expected performance of the Phase-I trigger upgrade [7]. In the plot on the left, the percent difference between the E_T of electrons reconstructed offline (E_T^{offline}) and using the Run-2 L1 is shown in red, while the difference between this offline reconstruction and the energy from SCs is shown in blue, where the E_T has been simulated using $Z \to e^+e^-$ events in the EMB calorimeter. The plot on the right shows expected L1 trigger rates for $\mathcal{L} = 3 \times 10^{34}$ cm^{-2}s^{-1} as a function of the EM E_T threshold for the case of the Run-2 trigger (blue) and the Phase-I trigger scheme with different selections of shower shape variables (black and green). A red line indicates the L1 limit of 20 kHz for electromagnetic objects, which will remain in Run-3.

In order to transmit this higher granularity information from SCs to the L1 Calorimeter (L1Calo) system, the LAr front-end and back-end electronics must be upgraded. Figure 5 shows a schematic of the readout electronics, with these new changes shown in red. The most notable change to the front-end is the introduction of 124 new LAr Trigger Digitiser Boards (LTDBs), which serve to digitise the analogue signal of supercells, processing up to 320 SC signals each. There are seven different 'flavours' of these boards depending on their location, with one flavour for the barrel and six for the end-caps. In order to accommodate this new addition, the baseplanes in each FEC must be replaced with new ones, which have the necessary number of slots. These new baseplanes also deal with changes to the signal path which are introduced in this upgrade, including the routing of legacy trigger tower sums for some layers through the LTDB to the TBB, and

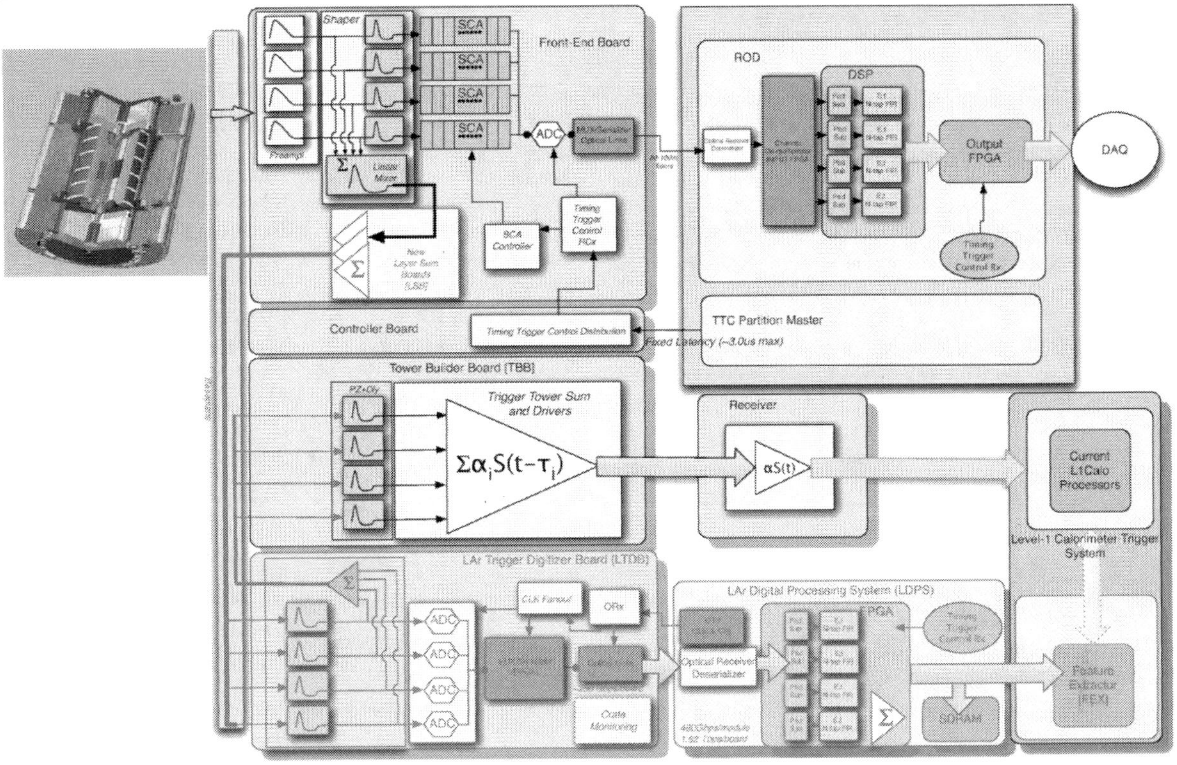

Figure 5: Schematic of the Phase-I upgrade architecture [7]. The new components are indicated by the red outlines and arrows. The main readout path (from FEBs to DAQ) and legacy trigger path (from FEBs to L1 via TBB) are also shown.

the routing of SC sums through the LSBs to the LTDB. The latter path also requires an exchange of the LSBs on all FEBs. The reception of this digital SC data and its transmission to the new FEXes is handled by new back-end hardware, collectively referred to as the LAr Digital Processing System (LDPS). This system, connected to the LTDBs through a total of 7 000 optical fibres, serves to receive ADC data and align timing, compute E_T for individual SCs in each bunch crossing, sum E_T for SCs over defined areas of the detector and send the result to the L1 trigger modules. The LDPS consists of advanced mezzanine cards, named LAr Trigger prOcessing MEzzanines (LATOMEs), and Advanced-TCA (ATCA) LAr Carrier (LArC) blades with Rear Transition Module (RTM) cards and Intelligent Platform Management Controller (IPMC) mezzanine cards [9]. Each set of LArC and four LATOMEs is referred to as a LAr Digital Processing Blade (LDPB). In total, the LDPS consists of 116 LATOMEs and 30 LArCs distributed across three ATCA shelves.

The main readout path, as described in section II, remains unchanged by this upgrade, though data must pass through the new baseplanes and refurbished FEBs. The legacy trigger path will be maintained during Run-3, running concurrently with the new digital trigger, and should also be unchanged, though trigger tower sums for some layers will be routed through the new LTDBs.

IV. Installation Status

Over the course of the ongoing Long Shutdown (LS2) of the LHC in preparation for Run-3, the installation of the new front-end and back-end LAr electronics has been progressing. The removal and replacement of all baseplanes on the detector has been completed. This required, as a first step, the removal of all boards in each HFEC; a challenging tasks in a restricted space. In order to exchange the LSB on each FEB, every one of these boards needed to be removed from the cavern and refurbished. This has been achieved for all FEBs on the detector. Simultaneously, the cooling plates which surround each board were replaced with newly manufactured ones, as well as the ageing hoses that supply cold water to each board.

The installation of LTDBs continues at the time of this publication. All LTDBs were manufactured, tested and delivered to CERN, with just six barrel (EMBC) and four end-cap (EMECC) boards awaiting installation due to a lack of access to the relevant regions of the detector.

The back-end installation of LDPBs is complete for the equivalent LTDBs, with data fibres fully connected between the LTDBs and LATOMEs for all of the ATCA crates set up in the ATLAS service cavern.

V. Main Readout Path Validation

Although the main readout path is unchanged by the Phase-I upgrade, much of the contributing electronics has undergone refurbishment. Signals pass through new baseplanes to FEBs whose LSBs have been exchanged, and some of these FEBs have been returned to different positions within the HFECs, or to different HFECs entirely. Checking data from the main readout path is therefore an important first step in confirming that the system is performing as it was before any disturbances due to installation. Basic low-level checks such as connectivity scans are used to ensure that all FEBs and channels are connected properly, while full calibration sets with comparisons to reference runs from the end of Run-2 provide a detailed check that the pulses read from the detector look the same as before, confirming that the electronics are stable. Three different types of calibration runs, namely, pedestal, delay and ramp runs, are used in order to calibrate the system and validate the readout electronics. Pedestal runs involve reading the detector output with no input signal, providing pedestal information from the average read out signal (ADC), noise from the RMS and autocorrelation from the timing correlation of the samples. Delay runs consist of injecting pulses with a single input signal current (DAC) with an increasing time shift in steps of 1.04 ns in order to reconstruct the full pulse shape. Ramp runs, where signals of increasing DAC are injected, provide the gain slope, which is extracted from a fit of the DAC versus ADC curve. Figure 6 shows, as an example, the comparison between pedestal RMS values measured before and after the installation process for the middle layer of EMBA. The data are in clear agreement, confirming that the system is stable and the hardware upgrades have not impacted this readout path. Such stability has been seen for all calibration run types in all of the refurbished crates.

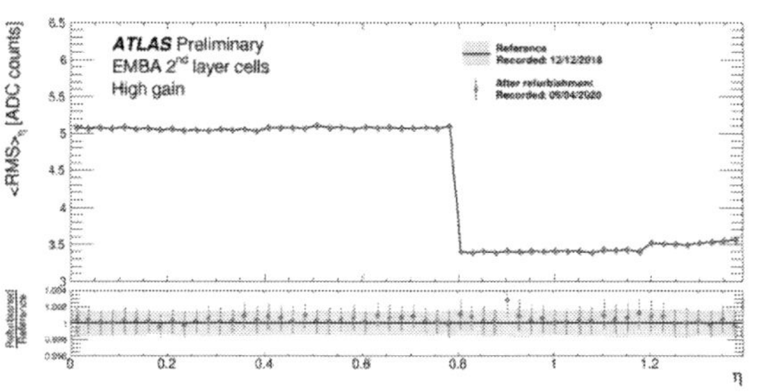

Figure 6: The mean value of the RMS of the pedestal for a given η range measured during a reference run at the end of Run-2 (black line) compared to values measured after the front-end refurbishment (purple dots) during LS2 [10]. The data shown is only taken from the second layer of EMBA. The displayed uncertainties are computed as $\sigma_{rms,cells}/\sqrt{N_{cells}}$, where $\sigma_{rms,cells}$ is the RMS of the RMS values over the cells in the given η range and N_{cells} is the number of cells in the given η range. The step at η=0.8 can be attributed to a change of the electrode geometrical structure in this region.

VI. Legacy Trigger Path Validation

As described in section III, the analogue legacy sums for some layers will be routed through the LTDBs before reaching the TT builders in the upgraded electronics scheme. This increased path through the system will introduce a delay in the L1 system for reception of the signal between different layers, which must be measured and accounted for in the trigger decision. Measuring these delays is an important step in validating the new hardware, since the introduced timing difference must reflect the expected performance of the system. Gain and timing scans performed in collaboration with the L1 trigger group are taken upon the refurbishment of each crate in order to check connectivity and measure this delay between the sums from different layers. Figure 7 shows the delays obtained from these timing scans for each layer in EMBA. A timing difference of up to 10 ns is observed between some layers, which is consistent with the expected change due to the increased path through the LTDB.

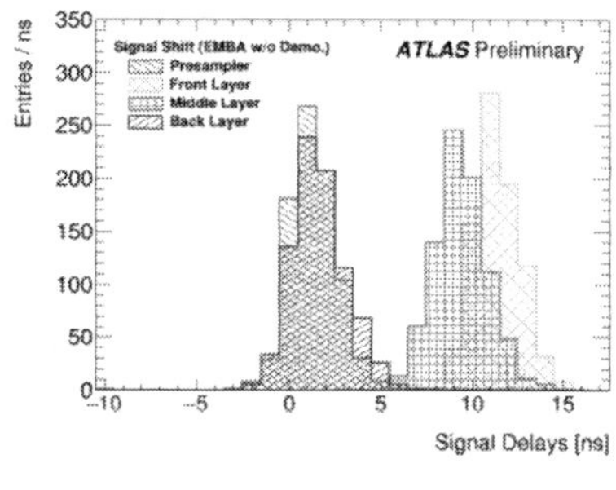

Figure 7: The signal delays introduced by the Phase-I front-end electronics for L1Calo trigger signals, collected from all trigger towers in EMBA [10]. Signals from the region $0 < \eta < 1.5$ and $1.8 < \eta < 2.2$ are excluded, since these were covered by Phase-I test electronics which were installed during Run-2.

VII. Digital Trigger Path Validation

In order to validate the full digital trigger front-end and back-end path, the different calibration runs, as described in section V, are used, with ADCs for full SCs read from the LATOME output, as opposed to the individual LAr cell data from the main readout path. This validation requires the full chain of the Phase-I system, starting with configuration of the LTDBs and LDPS to take data. Special calibration patterns must also be developed, in order to pulse all of the cells in each tested SC. Overnight pedestal or pulsed runs are taken in order to monitor the stability of the system. These runs are also used to extract information such as the pedestal value and peak position for the pulses from each LATOME, which is obtained using data processing software which is automatically applied to the raw data following each run. Full pedestal, delay and ramp calibration sets are taken for each fully refurbished and upgraded crate. Figure 8 shows examples of processed calibration data from a gain and ramp run for a SC in EMBA. Here we see a pulse shape from the delay run, which has the expected bipolar form, and an ADC versus DAC curve from the ramp run. In the latter plot, the ADC increases linearly with deposited transverse energy up to around 800 GeV, where saturation of the SC pulse occurs. Such saturation is expected to occur in some channels for very high energies. Unlike the main readout, no reference data exists for this newly installed system, therefore validation of the digital trigger requires checking that the calibration data behaves as expected based on the predicted performance. In many cases, select variables from the processed data (such as pulse position or gain slope) are compared to expected thresholds, which are determined through an understanding of the expected SC output. From these runs, calibration constants used for energy reconstruction are obtained, and the offline E_T can be compared to what is computed by the LATOME. For the LATOMEs tested so far, this E_T is found to be in good agreement.

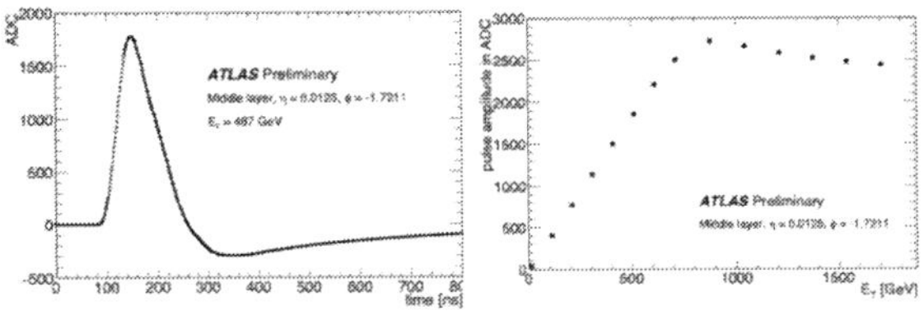

Figure 8: Calibration data read from the Phase-I trigger path [11]. The plot on the left shows the measured pulse shape of a SC in EMBA, obtained from delay run with an injected charge of DAC = 4400 (corresponding to a transverse energy of 487 GeV). The plot on the right shows the measured amplitude (ADC) of calibration pulses as a function of injected charge (DAC) from a ramp run for this same SC.

VIII. Conclusion

In this contribution, an overview of the installation and commissioning of the ATLAS LAr Calorimeter Phase-I upgrade has been presented. Excellent progress has been made in the installation, despite additional challenges brought on by recent global events, with almost the full front-end and back-end electronics installed, pending access to the remaining sections of the detector. The commissioning of crates involves checks of the main readout, legacy trigger and digital trigger path. The main readout checks have shown that the system is stable, with no deviations from the performance seen at the end of Run-2, and the legacy trigger checks show the expected timing shifts introduced by LTDBs. Connectivity and mapping checks, as well as full calibration sets are being used to check newly connected Phase-I hardware, with automatic processing and comparisons to newly defined thresholds used to validate each crate. These tools will be used to sign-off all of the upgraded HFECs and to prepare for processing real data in upcoming pilot runs and Run-3, which starts in March of 2022.

Acknowledgments

I would like to thank the organisers of ANIMMA 2021, for inviting me to present these results and for organising a rich programme of diverse content, even during this difficult time. I would also like to thank my colleagues in the ATLAS Liquid Argon Calorimeter group for affording me the opportunity to present this content on their behalf.

References

[1] ATLAS Collaboration, "The ATLAS Experiment at the CERN Large Hadron Collider," JINST, vol. 3, S08003, 2008. doi: 10.1088/1748-0221/3/08/S08003.

[2] P. Krieger, "The atlas liquid argon calorimeter: Construction, integration, commissioning and performance from selected particle beam test results," in IEEE Nuclear Science Symposium Conference Record, 2005, vol. 2, 2005, pp. 1029–1033. doi: 10.1109/NSSMIC.2005.1596428.

[3] ATLAS Collaboration, "ATLAS Liquid Argon Calorimeter: Technical Design Report," ATLAS-TDR-2; CERN-LHCC-96-041, 1996. [Online]. Available: https://cds.cern.ch/record/331061.

[4] ATLAS Collaboration, "Drift Time Measurement in the ATLAS Liquid Argon Electromagnetic Calorimeter using Cosmic Muons," Eur. Phys. J. C, vol. 70, p. 755, 2010. doi: 10.1140/epjc/s10052-010-1403-6. arXiv: 1002.4189 [hep-ex].

[5] N. J. Buchanan et al., "ATLAS liquid argon calorimeter front end electronics," Journal of Instrumentation, vol. 3, no. 09, P09003–P09003, Sep. 2008. doi: 10.1088/1748-0221/3/09/p09003. [Online]. Available: https://doi.org/10.1088/1748-0221/3/09/p09003.

[6] ATLAS Collaboration, "ATLAS Level-1 Trigger: Technical Design Report," ATLAS-TDR-12; CERN-LHCC-98-014, 1998. [Online]. Available: https://cds.cern.ch/record/381429.

[7] ATLAS Collaboration, "ATLAS Liquid Argon Calorimeter Phase-I Upgrade: Technical Design Report," ATLAS-TDR-022; CERN-LHCC-2013-017, 2013. [Online]. Available: https://cds.cern.ch/record/1602230.

[8] ATLAS Collaboration, "ATLAS TDAQ System Phase-I Upgrade: Technical Design Report," ATLAS-TDR-023; CERN-LHCC-2013-018, 2013. [Online]. Available: https://cds.cern.ch/record/1602235.

[9] CERN-IPMC: Technical Overview. [Online]. Available: https://cern-ipmc.web.cern.ch/doc.

[10] Level-1 Calorimeter Trigger Public Results. [Online]. Available: https://twiki.cern.ch/twiki/bin/view/AtlasPublic/L1CaloTriggerPublicResults.

[11] LAr Calorimeter LS2 Public Results. [Online]. Available: https://twiki.cern.ch/twiki/bin/view/AtlasPublic/LArCaloPublicResultsLS2.

Fast timing detectors with applications in cosmic ray physics and medical science

Christophe Royon[1]
[1]Department of Physics and Astronomy, The University of Kansas, Lawrence, USA
Christophe.royon@cern.ch

***Abstract*—We describe two applications concerning fast timing detectors and fast signal sampling in cosmic ray measurements in collaboration with NASA and in dose measurements for cancer treatment.**

***Keywords* —cosmic ray measurement, radiation measurement, space physics, dose measurement in medicine.**

I. Introduction: Signal Amplification and Measurement

THE idea is to discuss some applications to measure radiations using Fast Silicon detectors that are used in a standard way nowadays in high energy physics at CERN [1,2,3]. We will first describe the principle and methods of the technics that we use and describe in detail two applications related to measurements of cosmic ray in space in collaboration with NASA and of doses received by patients in cancer treatment especially in flash proton therapy.

The principle behind the measurements is shown in Fig. 1. We first use a Silicon detector in order to detect particles that go through the detector. In particular, we use fast detectors so that the signal emitted by the particle has a short time length, typically a few nanoseconds. This has the main advantage of reducing the probability of overlaps between the signal produced by many particles.

The signal originating from the Silicon detector (or the diamond detector) is first amplified with very limited shaping (so without modifying the good timing properties of the signal such as its duration). This is performed using front end electronics designed at the University of Kansas using standard components [4]. The following step consists in performing a fast sampling and digitization of the signal as shown in Fig. 1. This presents two advantages. First, it allows obtaining very precisely a time stamp when the signal (or in other words the particle) arrives in the detector (this is used to measure time-of-flight of particles in high energy physics with a precision of 15-20 picoseconds). Second, it allows reconstructing the full amplitude and shape of the signal by performing an interpolation between the sampled points (there can be 64 points in a few nanoseconds) if the signal is short in time.

We installed a full test stand at the University of Kansas allowing to test the performance of the Si detectors and their electronics and preliminary measurements showed resolutions between 15 and 20 picoseconds for ultra fast Silicon detectors as an example.

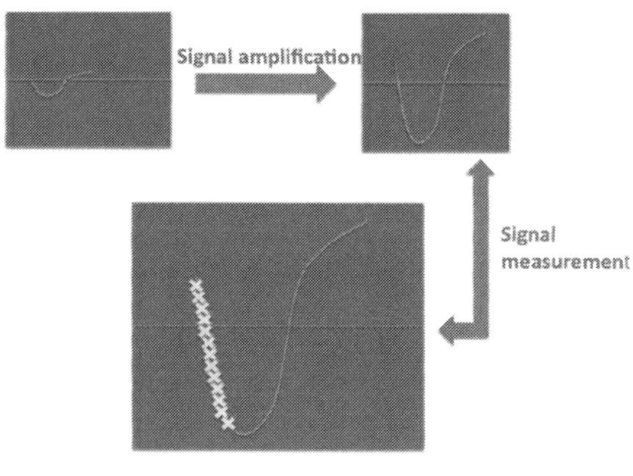

Fig. 1. Principle of signal analysis. A particle entering a fast Silicon detector induces a signal that is first amplified and then analyzed by performing a fast sampling of the signal (yellow crosses) that allow to measure the arrival time of the signal, its amplitude, shape and duration.

Fig. 2.View of the Silicon detector to be sent into space in collaboration with NASA (AGILE project) together with the readout electronics built in Kansas.

II. MEASURING COSMIC RAY IN SPACE: THE AGILE PROJECT

The goal of the AGILE project in collaboration with NASA [5] is to measure cosmic ray particles in space in the MeV-GeV range. The goal is to identify the type of particle (electron, proton, He, Pb, Au…) and at the same time measure their energy in space using a compact and affordable detector that can be part of a cube satellite. The basic idea is to use different layers of fast Silicon detector and to measure and reconstruct the signal where the particle stops in a given layer, the so called Bragg peak, using the fast sampling technique. A view of one layer of the detector to be sent into space together with the readout electronics is shown in Fig. 2.

The results discussed in this section come from a direct simulation of the signal using GEANT (particle simulation and interaction with the detector), WEIGHTFIELD (simulation of the silicon detector) and LTSPICE (simulation of electronics) as shown in Fig. 3. The output of this simulation is to obtain a database of signal characteristics that lead to particle identification and energy measurement.

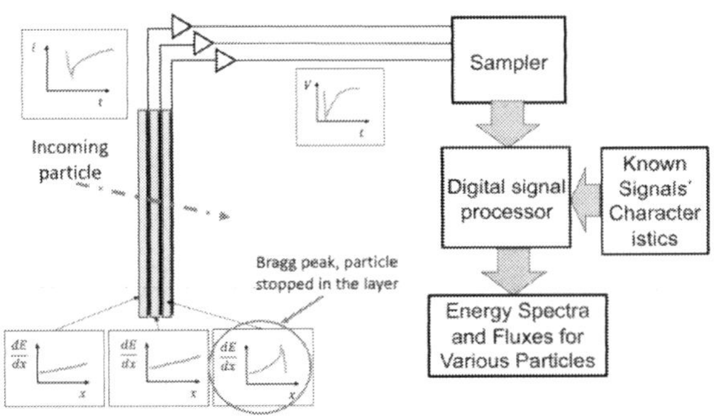

Fig. 3. Schematic of the simulation of the events measured by AGILE. We first use te signal collected by three layers of the fast Silicon detector. We use a fast sampling method to sample that signal, which is then digitized. It leads to a database of energy spectra and fluxes for different particles that allow to define the best characteristics for particle identification and energy measurement.

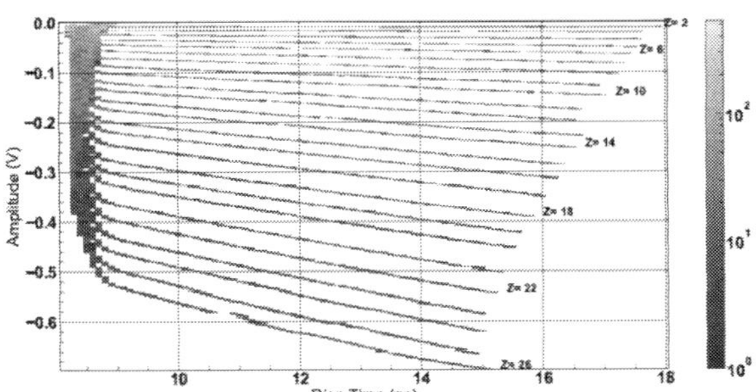

Fig. 4. Amplitude versus rise time for different particles (the value of Z is indicated on the figure). Above a rise time of about 6 ns, the curves are well separated which shows that it is possible to distinguish between different particle species and to measure their energy.

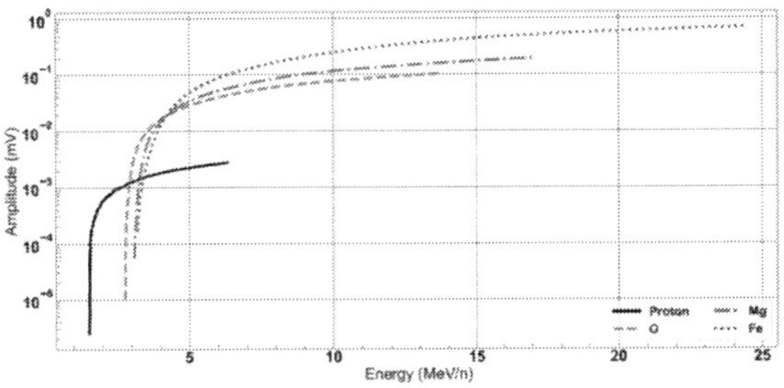

Fig. 5. Amplitude versus energy for different kinds of particles (proton, oxygen, magnesium and iron). If we assume that the kind of particle is known from the rise time measurement (see Fig. 2), the amplitude allows measuring the energy of the particle on a wide range above 5 MeV.

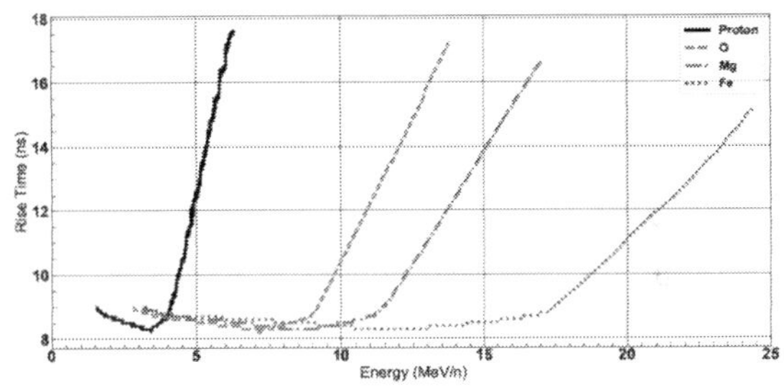

Fig. 6. Rise time versus energy for different kinds of particles (proton, oxygen, magnesium and iron). We see that the type of particle can be easily identified over a wide domain in energy by measuring the rise time of the signal.

In order to identify the type of particle and to measure their energy, we use two estimators. The first one is the rise time and the second one the amplitude of the signal, both measured in the layer where the particle stops (from the trigger point of view, we will require no signal in the next layer).

The amplitude as a function of rise time in the stopping layer, as put in the reference database, is shown in Fig. 4 for different particles with Z between 2 and 28. We see that all curves (corresponding to different values of Z) do not overlap for a rise time above 5 ns. It is thus clear that particle identification or in other words, the measurement of the Z value, can be performed using the rise time and amplitude of the signal. Once the type of the particle is known, it is possible to measure its energy using the amplitude of the signal as shown in Fig. 5 for proton, oxygen, magnesium and iron. Energy measurements can clearly be performed for all kinds of heavy ions in space above about 5 MeV with good resolution. The rise time versus energy of the particle is shown on Fig. 6 for the same particles. We clearly see the reason why the rise time has to be above 5 ns due to the shape of the curves.

The two measurements of the signal amplitude and rise time can be performed online in the cube satellite using a processor or offline on earth after data have been transmitted. The launch of the satellite is foreseen for Spring 2022 with three layers of Silicon detectors only as a prototype. The goal is to have a detector fully approved by NASA which is the case if the full chain is successful, namely if we are able to send back data to earth. If this is the case, the next step is to send a larger detector into space with more layers (up to 30) with two objectives : a mapping of cosmic rays emitted by the sun can be performed using a network of satellites, and a precise mapping of the radiation between the earth and Mars can be performed before sending astronauts to Mars. Obviously, this is high priority for NASA.

III. MEASURING RADIATION IN CANCER TREATMENT

The second application that we are going to discuss is related to the measurement of dose delivered to patients when treated for cancer especially for flash beam treatments (using high intensity proton beams as an example) [5]. The same technique applies for beam monitoring in hospitals or also for any commercial or research beam infrastructure.

Our silicon detectors and readout system were put in an electron beam used in the past for radiotherapy at St Luke Hospital, Dublin, Ireland. A schematic of this test is shown in Fig. 7 where we can compare the information from our ultra-fast Silicon detector (UFSD) with the usual ion chambers commonly used in hospitals to measure doses delivered to patients. As we will see, our system can provide a precise and instantaneous measurement of dose without the need of calibration and with high spatial resolution (typically mm). The new idea is that we have a fast detector that is able to discriminate and count single particles (electrons or protons) benefiting from the fact that the UFSD detector is fast enough.

The first step was to compare the information from our UFSD with traditional measurement of doses using ion

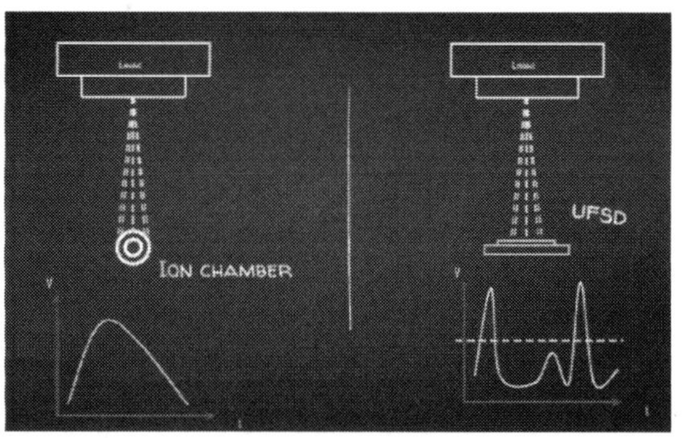

Fig. 7. Schematic of the tests performed at the St Luke hospital at the University of Dublin, Ireland, using an ion chamber or a fast Silicon detector. The signal from the ion chamber is integrated over a few seconds in time whereas the UFSD information can be instantaneous.

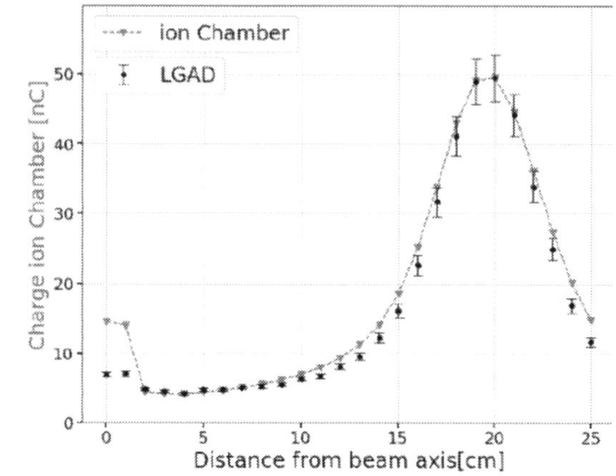

Fig. 8. Comparison between the measurement of the charged ions by the ion chamber (red points) and the measurement using our fast Silicon detector. We note a good correlation between both measurements.

chambers as performed in hospitals. Results are shown in Fig. 8 as a function of distance from the beam axis, downstream a permanent magnet (to create a spectrometer). Results are in very good agreements except at low deviations, that can be explained by the fact that the electron beam is highly contaminated by photons (not affected by the magnetic field) in that region.

The second step was to measure the instantaneous dose delivered to the patient by counting the number of electrons going through our detector. The principle is described in Fig. 9. Each spike in the UFSD is considered as a particle (here an electron) crossing our detector. In the bottom of Fig. 9, we clearly see the signal above the noise that trigger our detector (red triangles) as a function of time. We notice that having a short signal (a few nanoseconds), as well as a very fast sampling method, is crucial to be able to measure each particle crossing the detector. Counting the number of triggered spikes

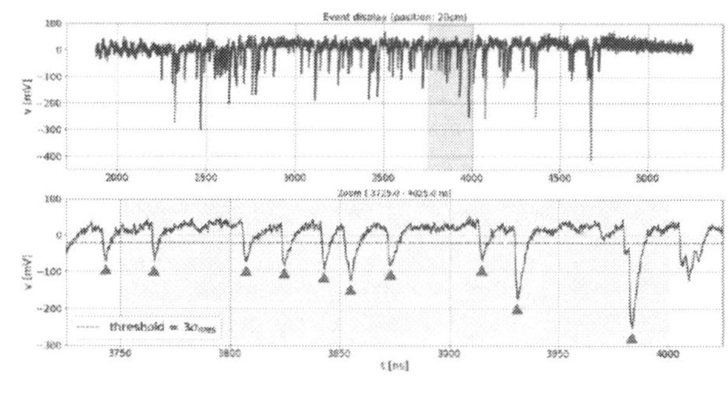

Fig. 9. Signal measured as a function of time. Each spike shows the beam delivered to a potential patient. The red area is zoomed in the bottom plot and the red triangles show the spikes that can be triggered using the Silicon detector.

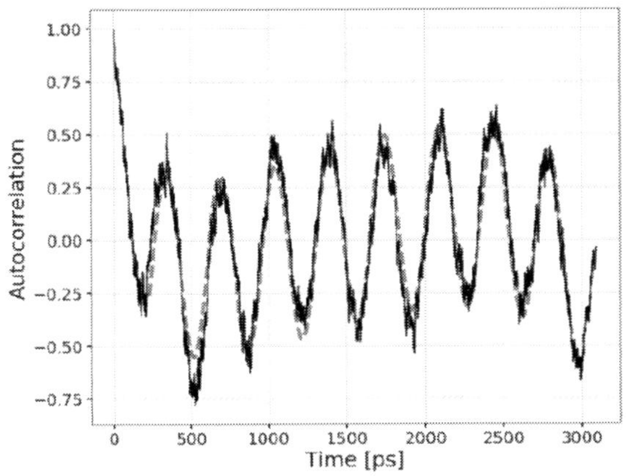

Fig. 10. Bunch structure as measured by the fast Silicon detector. For the first time, we were able to measure the structure in timescales of about 50 ps that allow to measure and control instantaneous doses delivered to patients.

is directly proportional to the dose (with a small efficiency when two consecutive spikes overlap). Fig. 9 clearly shows that this inefficiency is very small and our detector effectively counts all delivered electrons or protons, without the need of temperature and aging calibration, that allow to measure precisely and instantaneously the doses that are delivered to patients during cancer treatment.

Results coming from our UFSD are shown in Fig. 10. We see for the first time the bunch structure delivered by the machine at the St Luke hospital, Dublin. The periodicity of the beam is about 330 ps and measurements using UFSDs can be performed every 50 ps. In comparison, the usual ion chamber integrates measurements over times of the order of few seconds, making impossible to use for flash proton therapy. Everyday monitoring of proton beams used for flash proton therapy is also possible using our system.

IV. CONCLUSION

In this short article, we described two applications using fast Silicon detectors that are used nowadays in a standard way at the LHC.

The first application developed in collaboration with NASA aims at measuring cosmic rays originating from the sun (or other galactic objects) with high precision. Our technique of measuring both the rise time and the amplitude of the signal in the Silicon layer where the particle stops allows identifying the type of particle and measuring its energy using a compact detector that can be sent into space using a cube satellite.

The second application is to measure instantaneously the dose delivered to patients during cancer treatment (especially during flash proton beam therapy). Our technique allows counting the number of particles that are delivered to patients that it is directly proportional to the dose. Another directly related application is everyday monitoring of the beam quality before treatment

Many other applications are also possible using the same technique. It is for instance possible to measure precisely th process of catalysis in chemistry by measuring and analyzing for instance the interface between two liquids or a gas and a liquid and its time variation with a 50 ps precision using interferometry technics. Other applications include beam monitoring for private or public accelerators as well as any instantaneous dose measurements that might be required.

ACKNOWLEDGMENT

These works come from a collaboration with NASA and the University of Dublin, Ireland, namely F. Gautier, A. Greeley, S.G. Kanekal, T. Isidori, G. Legras, P. McCavana, B. McClean, R. McNulty, N. Minafra, and I would like to thank them for this nice collaboration.

REFERENCES

[1] A. Apresyan et al., "Studies of uniformity of 50 mum low-gain avalanche detectors at the Fermilab test beam", Nucl. Instr. Meth. A 895 (2018) 158.

[2] N. Minafra et al., "Test of ultra-fast silicon detectors for picosecond time measurements with a new multipurpose read-out board", Nucl. Instr. Meth. A 867 (2017) 88.

[3] D. Breton et al., "Measurements of timing resolution of ultra-fast Silicon detectors with the SAMPIC waveform digitizer", Nucl. Instr. Meth. A 835 (2016) 51.

[4] F. Gautier, A. Greeley, S. G. Kanekal, T. Isidori, G. Legras, N. Minafra, A.Novikov, C. Royon, Q. Schiller, "A novel technique for real-time ion identification and energy measurement for in situ space instrumentation", ArXiv:2103.00613,

[5] T. Isidori, P. McCavana, B. McClean, R. McNulty, N. Minafra, C. Royon, "Performance of a low gain avalanche detector in a medical characterization of the beam profile", Phys. Med. Biol. 66 135002.

Metrology of Acquisition Chains and Signal Processing of LMJ Experiments

Vincent Trauchessec[1]
[1]CEA, DAM, DIF, F-91297 Arpajon, France
vincent.trauchessec@cea.fr

Abstract **— Since the first experiment in 2014, more and more plasma diagnostics are being deployed on the Laser MegaJoule (LMJ) facility manufactured by C.E.A/D.A.M. These diagnostics aim at measuring radiations or particles emitted during laser experiments to study high-energy physics, especially inertial confinement fusion (ICF). Different types of sensors surround the LMJ target chamber and realize the conversion of the quantities of interest to an electric signal. The signal is then transmitted via coaxial cables, acquired by a broadband oscilloscope, and digitally post-processed. Each step of this typical acquisition chain adds measurement errors and increases the global uncertainty. First, a numerical model of the digitizer alongside a specific hardware system designed to perform its metrology in situ will be presented. It computes errors sources such as offset, gain and skew, and provides a measurement of the effective number of bits (ENOB) of the digitizer. The experimental characterization of the electrical chain via its transfer function measurement will also be detailed. Finally, the numerical methods deployed to handle the inverse problem, based on deconvolution processes, will be introduced, including future developments exploiting Bayesian inferences and statistical approaches.**

Keywords **— Metrology, LMJ, Digitizer, Deconvolution**

I. Introduction

The Laser MegaJoule (LMJ) is a major component of the French "Simulation Program" led by the Military Applications Division of the Atomic Energy Commission (CEA-DAM). This program aims at validating theoretical models by performing numerical simulations and experiments, especially in high-energy physics domains. To reach these extreme conditions, LMJ will deliver shaped pulses from 0.7 ns to 25 ns with a maximum energy of 1.8 MJ of UV light on the target. In 2019, the first successful ICF campaign has been carried out and the associated nuclear diagnostic will serve as example for the following sections. The aim was to measure the primary neutron yield, the ion temperature and the neutron bang time in deuterium (D_2) implosion by using a neutron time-of-flight (nTOF) diagnostic. This diagnostic is composed of several detectors made of fast plastic scintillators coupled to gated photomultiplier tubes (GPMT). These detectors are placed all around the target chamber, which is a 10-meter diameter aluminum sphere, covered with a neutron shielding made of 40 cm thick of borated concrete. The electric signal coming out of the GPMT is then transmitted through ultra-low-loss coaxial cables to a digital oscilloscope, which is located in a Faraday room next to the target chamber. Many parameters have to be estimated precisely (integral value, full-width at mid-height, instant of maximum), however the cable length can be up to 80 meters and attenuators of tens of dB are used along the lines, which make the output signal strongly filtered and attenuated. This requires determining accurately the transfer function of the chain so that deconvolution algorithms can be implemented to retrieve the initial signal measured by the detector. Moreover, error sources due to the digitizing process itself decrease the signal to noise ratio. This has motivated the development of a customized system to perform different metrology procedures with the digitizers associated to this diagnostic. Based on these results, uncertainties associated to the acquisition would be estimated more accurately.

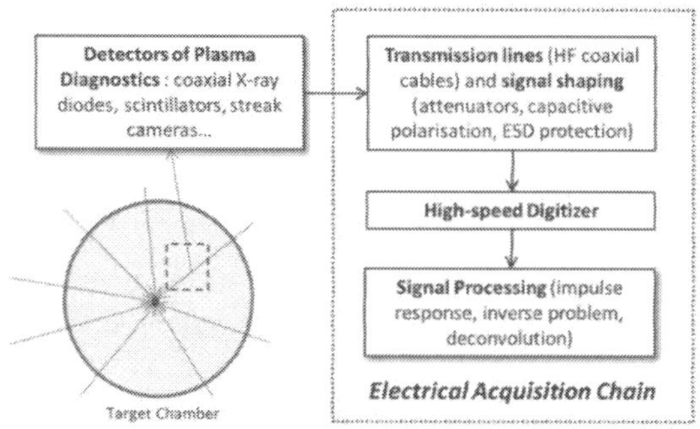

Fig. 1. Synoptic of a diagnostic acquisition chain at LMJ

II. In Situ Metrology

A. Digitizer model

A typical model of a digital oscilloscope is shown on Fig. 2. The digitization process adds different types of random errors:

- Electronic noise σ_{therm} generated by the thermal agitation of the charge carriers inside an electrical conductor.
- Phase noise σ_{phase} representing random fluctuations of the sampling clock of the analog-to-digital converter (ADC).

- Quantization noise σ_{quant} due to the replacement of continuous input values by an approximation from a finite set of 2^N discrete values, where N is the theoretical number of bits of the ADC.
- The vertical non-linear behavior of the ADC σ_{nl}, proportional to the signal amplitude, and characterized by a coefficient δ_{nl}

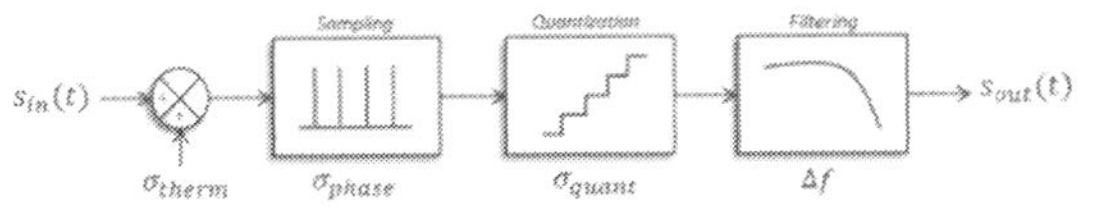

Fig. 2. Simplified model of a digitizer showing error sources during the two steps of the digitization: sampling and quantization [3].

As indicated in Table 1, the influence of these noise sources increases with both the frequency and the amplitude of the input signal. The bandwidth of the digital oscilloscopes used for the ICF campaign was set to 4 GHz, which implies an absolute noise floor value σ_{therm} of 57 μV_{RMS}. Then, since the electric signal transmitted by the detector has a frequency signature up to 100 MHz, the phase noise σ_{phase} is estimated around 885 μV_{RMS} for a specified intrinsic jitter value of 1 ps_{RMS}. Finally, the input signal is splitted before being sent to the 4 channels of the digitizer, while each of them is set to a different vertical sensitivity value. This configuration allows to measure the signal of interest with the highest and most appropriate full-scale ratio and it can reduce the quantization noise σ_{quant} by a factor of 10. Indeed, since the vertical resolution of the ADC is 8 bits and the full-scale is varying between 0.4 V and 4 V, σ_{quant} can be bounded between 450 μV_{RMS} and 4.5 mV_{RMS}.

TABLE 1 : NOISE SOURCES

Symbol	Quantity	Expression
σ_{therm}	Thermal noise	$\sqrt{4.k_B.T.R.\Delta f}$
σ_{phase}	Phase noise	$\sqrt{2}\pi.f.J.FSR.FS$
σ_{quant}	Quantization noise	$FS/(2^N.\sqrt{12})$
σ_{nl}	Non-linearity	$FSR.FS.\delta_{nl}/\sqrt{2}$

k_B = Boltzmann's constant (J/K), T = temperature (K), R = input impedance (Ω), Δf = bandwidth of the digital oscilloscope (Hz), f = frequency of the input signal (Hz), J = jitter (s), FSR = full-scale ratio (%), FS = full-scale (V), N = number of bits of the ADC, δ_{nl} = non-linearity coefficient (%).

B. *Effective Number of Bits (ENOB)*

An easy way to get access to the effective noise due to the random errors described above is to measure the effective number of bits (ENOB) of the digital oscilloscope. The procedure is the following:

- A pure sinusoidal signal is sent to the input of the oscilloscope
- The signal is fitted using a least-square optimization
- The standard deviation of the difference between the measurement and the fit is computed to obtain the effective noise σ_{eff}
- ENOB is computed by using the following formula:

$$ENOB = N - log_2\left(\frac{\sigma_{eff}}{\sigma_{quant}}\right) = log_2\left(\frac{FS}{\sigma_{eff}.\sqrt{12}}\right)$$

This value quantifies the quality of the ADC and is given for a specific full-scale ratio (FSR). Fig. 3 shows the ENOB values for each settings of vertical sensitivity of the digitizer (from 5 mV/div to 1 V/div) and for a sinusoidal input signal which frequency is varying between 1 kHz and 20 MHz (upper image). The simulations based on the noise sources identified in Table 1 can be extended to 4 GHz (lower image). It can be seen that the phase noise starts being the main limiting factor of the ENOB above 1 GHz, while the quantization noise and thermal agitation limit the ENOB in the lower frequency range. The nominal value is around 6.3 ENOB.

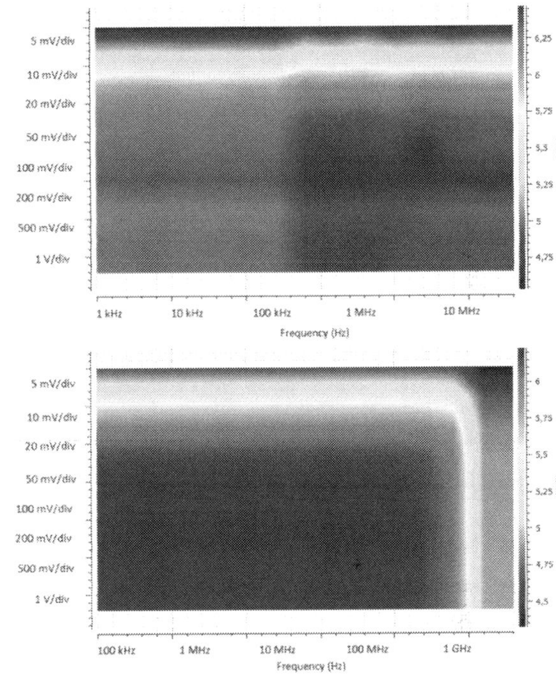

Fig. 3. Effective Number of Bits (ENOB) for every vertical sensitivity measured on a Keysight DSO91204 up to 20 MHz (up) and computed up to 4 GHz according to the previous model (down).

C. *Customized system for in situ metrology*

During the ENOB measurement, the sinusoidal signal is fitted to compute the residual noise. This operation cancels the influence of other error sources due to the oscilloscope such as:

- Zero Volt error
- Offset error
- Gain error
- Timebase error
- Impedance mismatch
- Skew error

To estimate these errors, each of them requires a specific metrology procedure that could only be carried out by the manufacturer. Moreover, the constraints of the LMJ facility impose to place the acquisition devices inside Faraday rooms and to remove these instruments periodically for metrological purposes would drastically reduce the operational time for experiments. To prevent this, a customized system has been designed so that each digital oscilloscope of every diagnostic can be tested in situ at all times. The system depicted on Fig. 4 is composed of a digital multimeter, a waveform generator, a high-voltage fast pulse generator, and a pure sinusoidal crystal generator. A metrology test can be launch by the operator in the control room and a full report containing all the errors for every channel measured at each vertical sensitivity is generated, in order to help quantifying the uncertainties to the acquisition system.

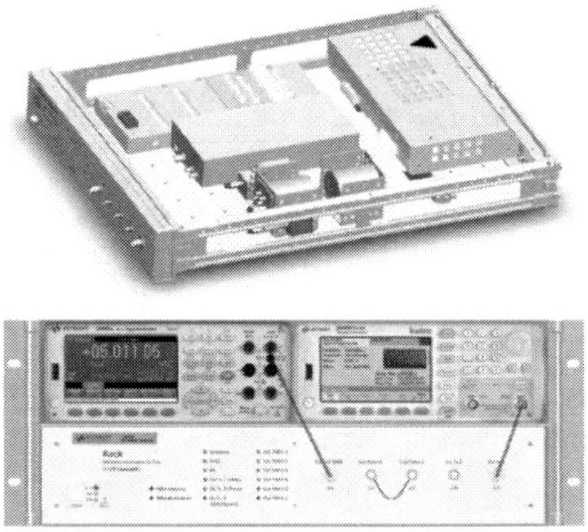

Fig. 4. Customized rack designed to realize in situ metrological measurements

III. Signal Processing

A. Transfer function

After the complete characterization of the oscilloscope, the effect of the other elements of the acquisition chain (coaxial cables, attenuators, filters…) has to be accurately quantified. The measurement chains are considered as linear and time-invariant systems. A simple convolution model relates the output signal *s(t)* to the unknown input signal of interest *e(t)* :

$$s(\tau) = \int h(t-\tau)\, e(t) dt$$

where *h(t)* is the impulse response of the chain.

In order to measure *h(t)* for the nuclear diagnostic, an extremely fast step generator featuring a rise time *tr* of 45 ps was used. The step is first sent directly to the digitizer as the reference signal, and then it is sent in the chain at the detector level. The time derivative of the step response gives the impulse response. However, the reference signal is not a perfect Dirac i.e. an infinitely high and fast signal of unit energy. It is fast enough to cover the whole bandwidth of the experiment (BW $\approx 0.35/t_{rise} \approx 7.7$ GHz), but its amplitude has to be calibrated by dividing the step response by the integral value of the reference signal. The Fourier transform of the impulse response gives the transfer function in the frequency domain *H(f)*, which is used to check the behavior of the chain in terms of attenuation and delay in the bandwidth of interest (cf. Fig. 5). The low-pass filtering effect clearly appears on the amplitude curve (red) and it will be corrected by the deconvolution algorithm.

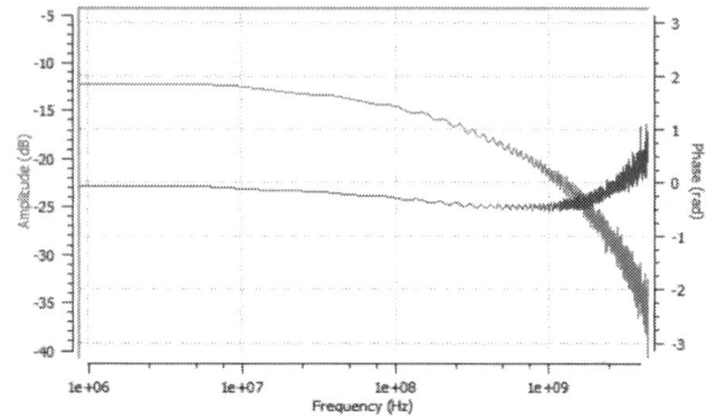

Fig. 5. Amplitude (red) and phase (blue) of the impulse response in the frequency domain *H(f)* of one channel of the nuclear diagnostic.

B. Deconvolution

One of the simplest approaches to deconvolution is to set up an inverse filter. Using Fourier transform, the convolution equation becomes a simple product, and the estimated input signal *e(t)* is obtained by dividing the output by the transfer function:

$$e(t) = FFT^{-1}\left(\frac{FFT(s(t))}{H(f)}\right)$$

Fig. 6 shows an example of inverse filtering. The blue trace is a typical output of the nTOF detector. The main spectral components of this signal are located around 50 MHz. According to the amplitude plotted on Fig. 5 the attenuation due to the chain is -13,5 dB (i.e. a factor of 4,73). A simple gain correction leads to the red curve, while the green curve shows the estimated input by inverse filtering. In the case of the nuclear diagnostic, several information must be extracted from the parameters of these signals: the area under the curve corresponding to the neutron yield, the full-width at half-height (FWHM) corresponding to the ionic temperature, and the instant corresponding to the maximum of emission i.e. the bang time.

Table 2 details how the deconvolution correction is much more efficient and how it improves the accuracy of the measurement. Taking into account the behavior of the chain in terms of amplitude and delay for the whole bandwidth allows to straighten the curve and to obtain a rise and fall time in good agreement with the predictions.

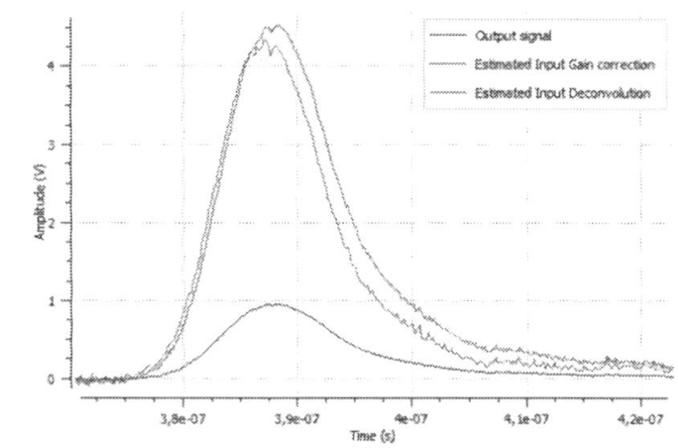

Fig. 6. Output signal (blue) and estimated inputs using a simple gain correction (red) and an inverse filter deconvolution (green).

TABLE 2 : COMPARISON OF PARAMETERS VALUE DEPENDING ON THE CORRECTION (GAIN OR DECONVOLUTION)

Parameters	Gain Correction	Deconvolution Correction	Error
Integral (yield (a.u))	6.90	6.25	10.4%
FWMH (ionic temperature (a.u))	11.14	10.47	6.4%
Maximum instant (bang-time (ns))	388.0	387.2	0.8 ns

In the case of the nuclear diagnostic, the signal to noise ratio is high enough so that the deconvolution by inverse filtering is efficient. However, this process may not give relevant results in many other cases. To illustrate the limits of this method, the convolution equation can be re-written by constructing the matrix $\boldsymbol{H}$ with the elements h_i of the impulse response:

$$s = \boldsymbol{H}.e \quad \text{where} \quad \boldsymbol{H} = \begin{pmatrix} h_0 & 0 & \cdots & \cdots & \cdots & 0 \\ h_1 & h_0 & \ddots & \ddots & \ddots & \vdots \\ \vdots & \ddots & \ddots & \ddots & \ddots & \vdots \\ h_n & \ddots & \ddots & \ddots & \ddots & \vdots \\ 0 & \ddots & \ddots & \ddots & \ddots & 0 \\ 0 & 0 & h_n & \cdots & h_1 & h_0 \end{pmatrix}$$

Solving the inverse problem then requires satisfying the three Hadamard's conditions [2]: existence, uniqueness, and stability. Indeed:

- The solution may not exist: the operator $\boldsymbol{H}$ may not be invertible ($\boldsymbol{H}^{-1}$ may not exist). In many systems, the first term h_0 of the impulse response is equal to zero $(h(t) \sim t.e^{-t})$ in which case the matrix $\boldsymbol{H}$ is singular. It can be avoided by taking h_1 as the first term and to consider that the impulse is displaced one unit of time from its true position.

- The solution may not be unique: $\boldsymbol{H}$ admits more than one inverse $(\exists G1, G2 \,/\, G1(\boldsymbol{H}) = G2(\boldsymbol{H}) = I)$. To make sure the solution is unique one has to narrow the solutions to the space of functions whose spectrum is included into the system bandwidth.

- The solution may not be stable, a small variation *ds* of the output generates a large deviation of the estimate ($\lim_{ds \to 0} \|\boldsymbol{H}^{-1}(s) - \boldsymbol{H}^{-1}(s + ds)\| \neq 0$). The stability can be estimated by computing the conditioning factor of $\boldsymbol{H}$, which is the ratio between its highest and its lowest eigenvalues. If this factor is much larger than 1, then a small noisy perturbation of the output will result in a large variation of the estimated input.

These inherent limitations of inverse filtering lead to the consideration of other deconvolution methods, that could take into account *a priori* information about the input to be recovered. This is one of the main advantages of probabilistic methods.

C. *Probabilistic approaches*

Since deterministic methods lack tools to handle previous knowledge about statistical properties of unknown parameters and noise errors, probabilistic approaches can be implemented to solve the inverse problem. One of them is based on the Bayes theorem [1]:

$$P(e_j/s_i) = \frac{P(s_i/e_j).P(e_j)}{\sum_{k=1}^{m} P(s_i/e_k).P(e_k)}$$

Where $P(e_j)$ is the *a priori* probability of the cause e_j and $P(s_i/e_j)$ the probability that the output s_i was caused by the input e_j. This term $P(s_i/e_j)$ represents one of the element of the matrix $\boldsymbol{H}$ containing the impulse response that has been measured on the diagnostic.
Using this Bayesian inference approach will make it straightforward to include uncertainties on the data and on the computed solution of the inverse problem (e_j) via probability laws.

IV. Conclusion

The paper describes how the acquisition chains of plasma diagnostics installed on the LMJ facility are controlled. First by designing a customized instrument to perform the metrology procedures used for the digital oscilloscopes, especially the ENOB measurement, that can be launched before every campaign. Then the way the behavior of the entire chain is characterized by the measurement of its impulse response is detailed, and the final results obtained with the deconvolution algorithm based on inverse filtering are illustrated. New methods based on Bayesian approach will be used to be able to infer more precisely the uncertainty of the computed solutions.

References

[1] G. D'Agostini, "Improved iterative Bayesian unfolding" ArXiv:1010.0632v1, Oct 2010.

[2] A. Mohammad-Djafari, "Bayesian inference for inverse problems" ArXiv:physics/0110093v1, Oct 2001

[3] P.Semécurbe, V.Allouche, M.Prat, B.Villette, "New method to define and optimize the diagnostic electrical acquisition channels application to the LMJ DMX plasma diagnostic" in Proc. ANIMMA, Marseille, France, 2009

EPJ Web of Conferences **253**, 03002 (2021) https://doi.org/10.1051/epjconf/202125303002
ANIMMA 2021

Development of a real-time signal processing unit for diamond detectors of ITER Vertical Neutron Camera

Michael Zhuravlev[1], Grigorii Nemtcev[1], Nikita Nagornyi[1], Sergey Meshchaninov[1], Roman Rodionov[1], Andrey Mironov[1], Anzhela Zvonareva[1], Ekaterina Mironova[1], Sergey Portone[1]

[1]Institution "Project Center ITER", Russian Federation
Corresponding author: M.Zhuravlev@iterrf.ru

***Abstract* - CVD Diamond Detectors are going to be used in a number of neutron diagnostics of the International Thermonuclear Experimental Reactor (ITER) during its power operation phase [1]. One of such diagnostics is the Vertical Neutron Camera (VNC), which is being developed by the ITER Russian Domestic Agency (Institution "Project Center ITER").**

In VNC Diamond detectors are used for real-time neutron flux and neutron spectrum measurements. In this paper we present current advancements in the development of a digital signal processing unit for the VNC diamond detector measurement channel. We give an overview of the signal properties, hardware requirements and firmware/software solutions used in the current version of the signal processing unit. We also present preliminary test results for this measurement system and explain the discovered problems and possible solutions.

***Keywords*—ITER, Vertical Neutron Camera, CVD Diamond Detector, Digital Signal Processing.**

I. Introduction

VERTICAL Neutron Camera (VNC) is a multichannel neutron collimator intended to characterize ITER plasma as a fusion neutron source. It measures the time resolved neutron emission profile for both Deuterium-Deuterium (DD) and Deuterium-Tritium (DT) plasmas, providing the evaluation of the neutron and α-source emissivity profile, as well as fusion power density, fusion power and total neutron flux [5].

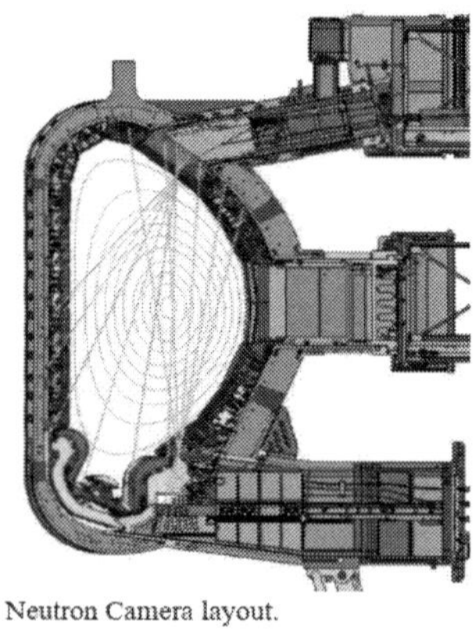

Fig. 1. Vertical Neutron Camera layout.

VNC consists of two subsystems: upper and lower, as illustrated in figure 1. Each of them has six Detector Units, and each detector unit consists of four detectors: two Fission Chambers and two CVD Diamond Detectors, as shown in figure 2. This amounts to 24 Fission Chamber and 24 Diamond Detector measurement channels.

CVD Diamond Detectors are radiation-hardened semiconductor particle detectors. In VNC their primary purpose is neutron flux and neutron spectrum measurements, based on which VNC measurement parameters are calculated. In this article we describe the techniques and the hardware, used to process the diamond detector signal in real-time and to provide the required signal parameters: *pulse count rate* and *pulse amplitude spectrum*, along with a number of auxiliary parameters like pulse rejection rate and the number of registered pulses.

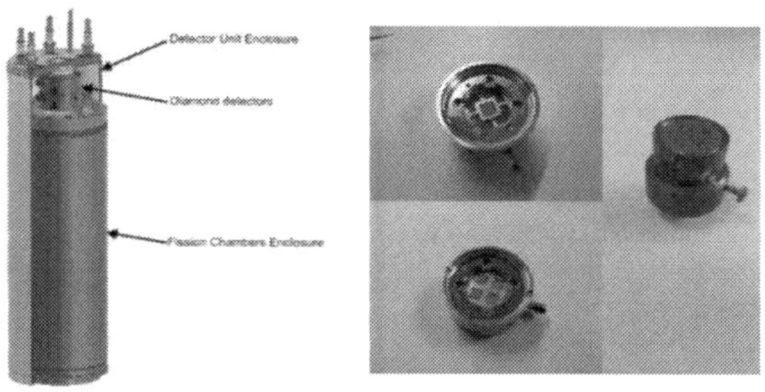

(a) (b)

Fig. 2. VNC Detector Unit layout (a) and CVD Diamond Detectors (b).

II. Diamond detector Signal Properties

CVD Diamond Detectors can register various types of radiation. They are responsive to α, β and γ radiation, as well as neutrons.

Upon an interaction with an ionizing particle the diamond detector produces a pulse of electrical current. Its duration depends on the voltage applied to the crystal, and the amplitude depends both on the voltage and on the energy of the incident particle.

For diamond detectors of VNC, the duration of the initial pulse from the detector is 10 ns, and the amplitude is 0 to 10 μA [1]. Such a pulse is too weak to be detectable by conventional ADCs, and thus preamplification of the signal is required.

For this purpose, we currently use a matched charge preamplifier CAEN A1426. The amplitude of pulses after preamplification is 0 to 2 V, and their duration-at-base goes up to 50 ns. This duration is the same for all pulses, and this property of the signal can be used to distinguish single pulses from pile-ups.

When diamond detectors are used for neutron detection, carbon acts as a neutron converter. This means that ionization in the body of the detector is induced not directly by neutrons, but by products of a range of nuclear reactions happening between carbon atom nuclei and neutrons of different energies. Though there is a variety of such reactions [2], there are three of them, which are most prominent:

1. $^{12}C(n, \alpha)^{9}Be$, which is induced by high-energy neutrons (above 8 MeV), and specifically by 14 MeV neutrons from the DT reaction;
2. $^{12}C(n, 3\alpha)n'$, induced by medium-energy neutrons;
3. $^{12}C(n, n')^{12}C$, elastic scattering of low-energy (scattered) neutrons, which leads to excitation of a carbon atom nucleus with further deexcitation and emission of high energy γ.

The shape of a typical pulse after preamplification is shown on figure 3. As demonstrated in this illustration, the duration of the pulse is measured as the time interval between threshold crossings.

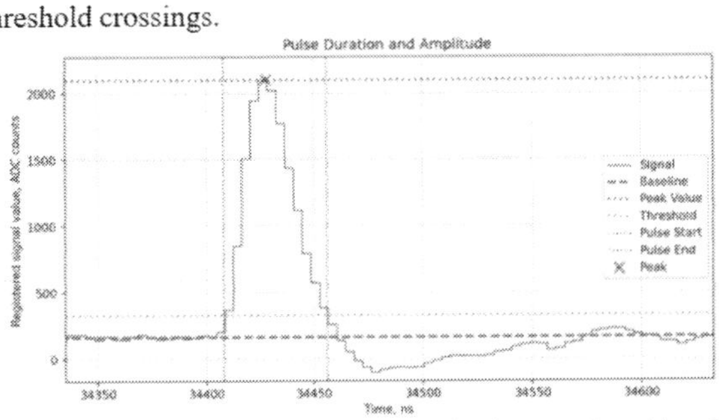

Fig. 3. The shape of the diamond detector signal pulse, as registered by a 250 MS/s ADC, and its parameters.

Because the amplitude of a diamond detector pulse is proportional to the energy of the particle that produced it, a diamond detector placed in a field of fusion neutrons, produces a very distinct range of pulse amplitudes, forming an *amplitude spectrum*. An example of a CVD Diamond Detector pulse amplitude spectrum, acquired from a DT neutron generator NG24M in JSC SRC RF TRINITI, Troitsk, Russia is shown in figure 4. Peaks corresponding to aforementioned nuclear reactions are prominent.

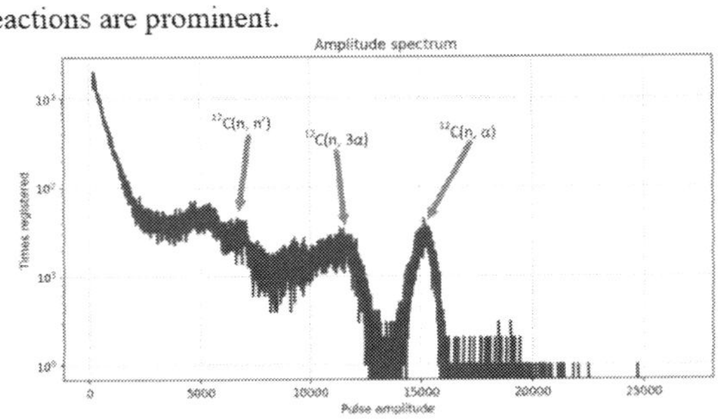

Fig. 4. CVD Diamond Detector pulse amplitude spectrum from an experiment at NG24M (x-axis values are in ADC counts).

III. Hardware

As already mentioned, diamond detector pulses are very short in duration (50 ns between threshold crossings, or less), and thus for their detection a fast ADC is required.

The rise time of a pulse is close to 20 ns, time at peak is less than 5 ns, and the fall time is 25 ns. If we assume, that we want to register at least 3 points at the peak of the pulse, this means we need a sampling frequency of

$$\frac{3}{5\ \mathrm{ns}} = 0.6\ \mathrm{GS/s},$$

or 600 MS/s. There is a number of ADCs on the market, that loosely fit this requirement. A number of manufacturers offer 500 MS/s ADC, and some provide solutions with data acquisition speeds of 800 MS/s and higher.

Another important factor, which affects the choice of hardware, is that by ITER standards all high-speed data acquisition and control is performed by so called Fast Controllers (Plant Controller Fast, PCF), which is a name for a compound device, consisting of an industrial computer and I/O chassis. In our case, PCI eXtention for Instrumentation (PXIe) platform was chosen for VNC Instrumentation and Control (I&C) system. It is an industrial standard for modular measurement equipment, developed by National Instruments (NI) and supported by a range of measurement hardware manufacturers worldwide.

The final factor is that the amount of raw data gathered by the system is very big. If 500 MS/s ADCs are used, the total data acquisition rate just from diamond detector measurement channels reaches 12 GS/s, which means, depending on the data representation, up to 24 GB/s of raw data feed. And due to ITER requirements, streaming of raw data (ADC codes) is obligatory for archiving and offline analysis. Thus, hardware should support such data transmission rates. At the same time, to calculate the required parameters, the signal has to be preprocessed. Preprocessing allows us to extract only relevant signal information (pulse peak values, their duration and pulse rejection statistics), which can reduce the data stream by as much as two orders of magnitude, down to hundreds of MB/s.

At this stage of system development, it was decided, that we should not compromise measurement quality, and thus a 12 bit 800 MS/s 2 channel ADC NI-5772 from National Instruments was chosen. It is an ADC adapter for NI FlexRIO reconfigurable digital input/output PXIe boards, which can be used for signal preprocessing and data rate reduction. The board used in our system is equipped with a Xilinx Kintex-7 FPGA, and supports up to 3.2 GB/s data streaming rates through Direct Memory Access (DMA). The hardware assembly is demonstrated in figure 5.

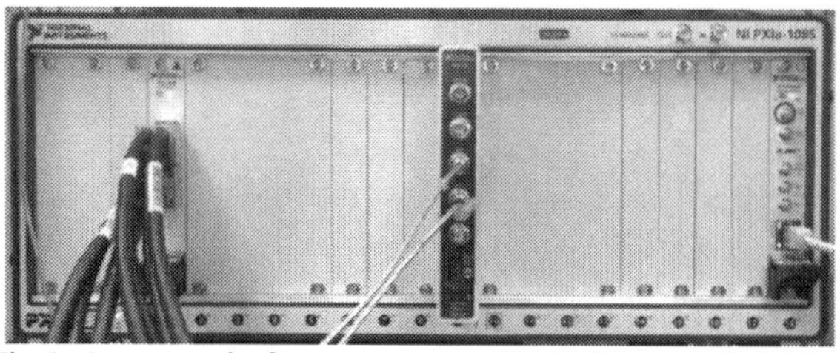

Fig. 5. Measurement hardware

IV. Signal processing algorithm

The diamond detector signal has to pass through several processing stages before the required neutron source parameters are going to be available to the plant I&C system. Namely, these parameters are:

- baseline correction;
- pulse detection and peak calculation;
- pile-up rejection;
- count rate and amplitude spectrum calculation.

We will go through them one-by-one.

First, the analogue signal is acquired by the ADC. When a signal is digitized, its *baseline* (zero-level, background), most of the time does not match the zero count of the ADC. Besides, this level tends to drift both due to processes in the ADC electronics and due to external factors like power grid interference, static charges, electromagnetic radiation from other devices in the system, etc. For this reason, the first stage of signal processing is *baseline correction.*

There are a number of approaches to calculating the signal baseline. They vary between different fields of science and between specific applications. In our case, when the signal is a continuous stream of ADC values (16-bit integers), the most straightforward approach would be measuring the distribution of signal values over a certain period of time, and then calculating the *mode* of this distribution. This approach was used by us when processing pre-recorded data to calculate the amplitude spectrum, shown previously in figure 4.

In real-time conditions and on FPGA-based computational devices this approach requires a lot of hardware resources and introduces significant latency. At the same time, this method is robust, and as we will later discuss, may have its benefits over the currently implemented solution.

Another alternative solution worth mentioning is the utilization of a digital low-pass filter applied to a downsampled version of the input signal. This solution alone is not reliable, and two potential problems with this approach have been identified. First, due to the fact that the useful signal is represented by short pulses, its frequencies are difficult to isolate, especially in a downsampled signal, and a short pulse can cause a low-frequency ripple in the proceeding data. The other significant problem is that on FPGAs we should try to avoid using floating-point calculations, instead resorting to fixed-point numbers, which makes an efficient digital filter implementation much harder, especially if there are no constraints on the numerical range of input data.

This brings us to the solution which is currently implemented in our system. The signal, acquired at 800 MS/s, is split into two streams, one of which gets downsampled to 40 MHz. This downsampled signal is used to calculate the baseline.

The calculation consists of two stages. First, the signal is passed through a median filter. It is a non-linear digital filter, and its most useful property is that it suppresses any sudden spikes in data (spatially or temporarily localized deviations from the range of most common values), while retaining other features of the signal, including sharp edges and long-term trends in data. An example of a median filter suppressing a data spike is shown in figure 6.

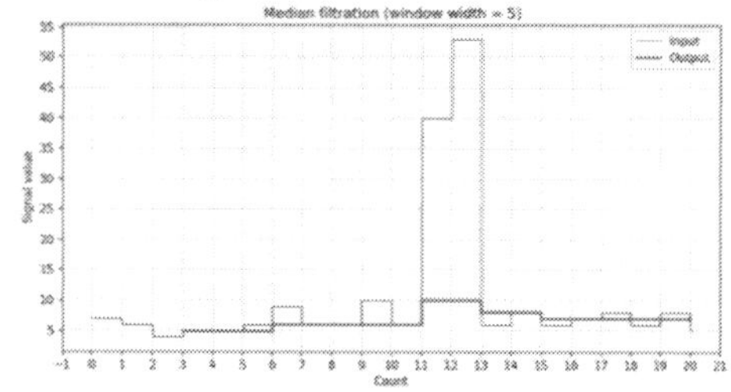

Fig. 6. Suppression of a data spike by a median filter (window width 5)

By using a median filter, we can eliminate pulses from the data stream. At 40 MHz a single pulse takes up:

$$50\text{ ns} * 40\text{ MHz} = \frac{50\text{ ns}}{25\text{ ns/tick}} = 2\text{ ticks.}$$

A typical two pulse pileup lasts up to two times as much, that is 4 clock ticks or less. A spike of any magnitude can be fully suppressed by a median filter, as long as it takes up less than half the width of the median filter window. Thus, a window width of 7 has been chosen. This is likely to change due to reasons described further.

It should be noted, that median filtration is a known and tried method, which has been successfully applied in other similar cases before, for example in [3].

On the second stage of baseline calculation, a signal devoid of any spikes is passed through a low-pass filter (Butterworth) with a cutoff frequency of 2.5 MHz. This has to be done due to the fact that the unfiltered baseline signal remains noisy, which can cause unnecessary distortion of registered pulse amplitudes. Figure 7 shows the calculated baseline compared to the input signal.

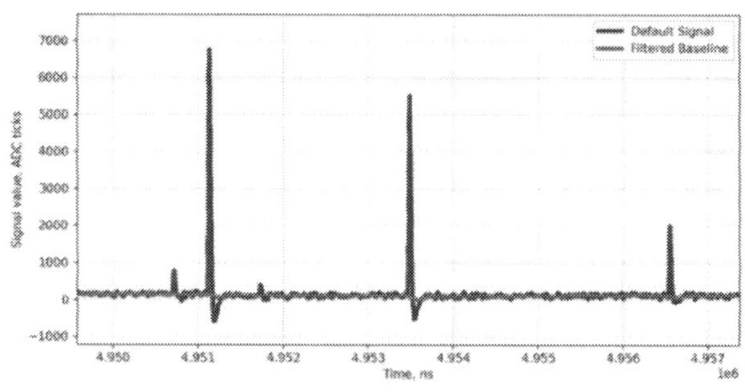

Fig. 7. Input signal (blue) and the calculated baseline (red).

Figure 8 shows the frequency spectrum of the calculated

baseline before and after filtering.

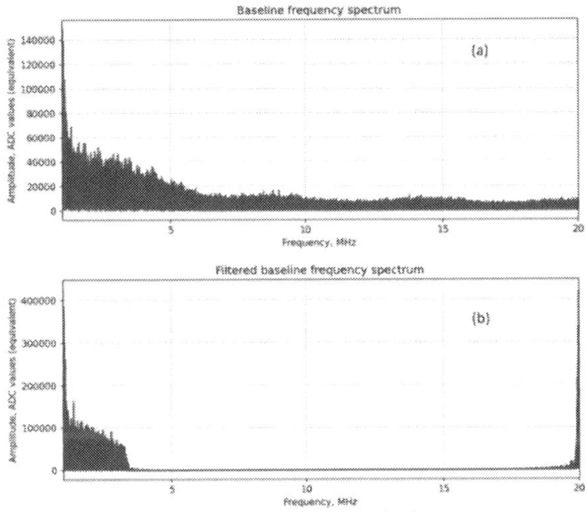

Fig. 8. Baseline frequency spectrum before and after filtering.

After the baseline is calculated, it is subtracted from the signal. Next, a manually set amplitude threshold is used to distinguish pulses from noise. Simultaneously, the peak of the pulse is calculated, and pile-up rejection based on pulse duration is performed. The algorithm can be represented as follows:

1. If the threshold has been exceeded, start acquiring data.
2. Set value of *maximum* to the initially acquired value.
3. Acquire next value. If this value is greater than *maximum*, set *maximum* equal to the newly acquired value.
4. If the pulse has not ended and the *duration threshold* has not been exceeded, go back to step 3. Otherwise, end.

All signal processing stages discussed above are performed on the FPGA. *Pulse counting* is also performed by the FPGA code (both for single pulses, as well as pulses together with pile-ups), and thus the *count rate* parameter is provided as well. The signal at this stage of processing is a stream of pulse peak values. This stream is sent to the host computer of the system, where a histogram of pulse amplitudes, the *amplitude spectrum*, is calculated, as well as the *pulse rejection rate* (the ratio of the number of registered pile-ups to the total number of times the threshold is exceeded).

As already mentioned, to calculate the baseline we could use two approaches: a reliable statistical method (calculation of the mode of the signal value distribution), and the approach we currently use - the median filter.

The implementation of the first one requires a lot of hardware resources, and specifically memory, to store and constantly update statistical data about the signal. Besides, memory transfers introduce latency into the system, and in some cases this latency is not fully predictable. Also, there is the problem of potential mode ambiguity.

The median filter solution is much more deterministic, and requires much less FPGA resources. At the same time, with a window width of 7 its performance with pile-ups can be poor, which may cause unwanted 'bumps' in the baseline, as illustrated in figure 9.

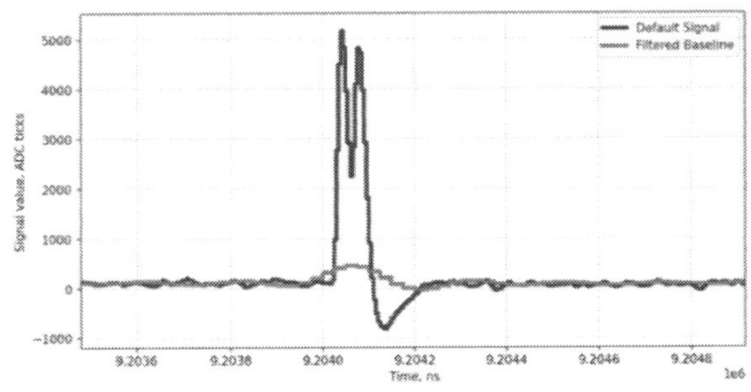

Fig. 9. Median filter response to a pileup of equal pulses (window width 7).

This can be mitigated by a wider median filter window, for example, 15 or 17 values. But such a filter loses its low-latency advantage over the statistical method. Besides, at high loads, the number and the width of pile-ups will increase significantly, which will require even wider windows. The statistical method is expected to be more stable across a wider range of neutron loads, while being insensitive to the ratio of pile-ups and single pulses. Currently the diamond detector measurement channel is being developed to support only pulse counting, and the use of median filters may potentially limit the maximum possible count rate. For this reason, the two solutions have to be tested side-by-side to determine the one, which is best-suited for the application. There is also the possibility that the two approaches may be combined.

V. System testing

The two main parameters that are measured by the signal processing unit are:

1. pulse amplitude spectrum;
2. pulse count rate.

To get an estimate of the quality of measurement of these parameters, two separate tests were devised.

It should be noted first, that the following experiments were not performed in accordance with any metrological standards, yet. Thus, the presented characteristics should be considered merely as estimates. Later, a dedicated and detailed plan for a test campaign is going to be developed to properly test the system's performance.

First, to test the accuracy of pulse amplitude measurements, the full diamond detector measurement channel was assembled:

1. CVD Diamond Detector;
2. CAEN A1426 matched charge preamplifier;
3. FlexRIO device with ADC adapter, mounted in PXIe-1085 chassis;
4. Host computer.

During the experiment, a calibrated source of α particles (americium-241) was used, which was placed in the immediate vicinity of the diamond detector. The stream of pulses created by the source was, on average, 30 s^{-1}. The exposure time was 30 minutes. The measurement setup, as well as the resulting amplitude spectrum are shown in figure 10.

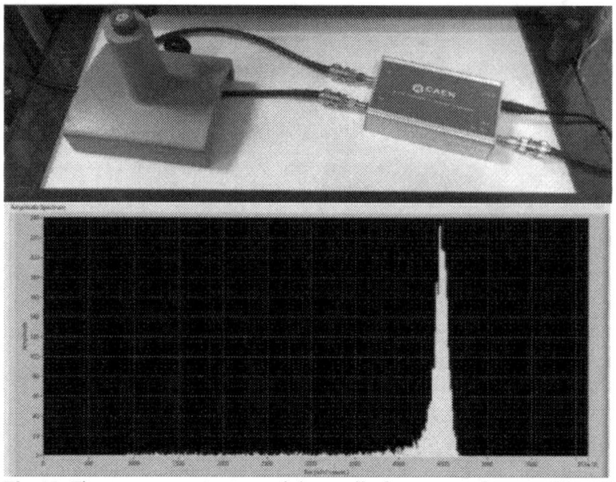

Fig. 10. The measurement setup and the amplitude spectrum from a calibrated americium-241 α-particle source.

The relative amplitude measurement error, calculated in this experiment, was 4%, but figures for the noise to signal ratio in the channel are yet to be acquired.

Next, to estimate the count rate measurement accuracy, an arbitrary waveform generator has been connected directly to the ADC. This waveform generator was configured to play back a single pre-recorded diamond detector pulse at a frequency of 1 MHz. The shape of the pulse, and the registered count rate are shown in figure 10.

As can be seen, the system correctly counts 1 million pulses per second. On the other hand, a clear deviation from exactly 10^6 is prominent. The reason for this is that the generator we use is not a calibrated instrument. Verified test equipment will be used for such measurements at the next stage, and on the side of the measurement system various hardware synchronization techniques should be implemented to mitigate this effect.

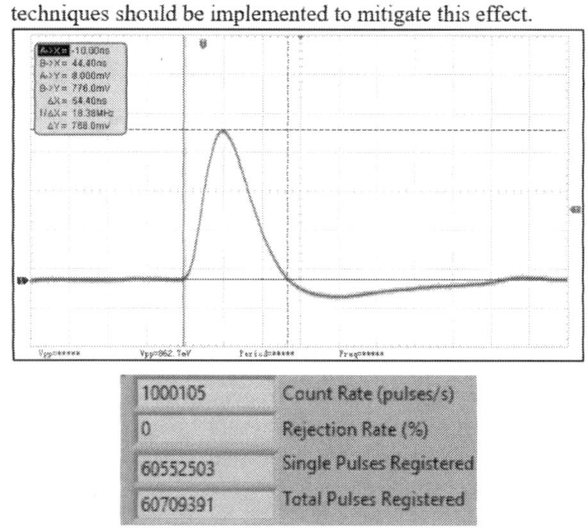

Fig. 11. The test signal and the registered count rate.

VI. Conclusions

Research presented in this paper is a crucial part of the VNC diagnostic I&C development process. Its main result is a functioning proof-of-concept digital signal processing unit, which after further development and testing will become an integral part of the final measurement system. Besides being a test-bed for concepts and solutions, which are going to be used in the VNC diagnostic, our digital signal processing unit and its infrastructure can serve as a platform for development of other ITER diagnostics I&C.

Acknowledgment

This work has been carried out in accordance with the state contract dated May 11, 2021 No. H.4a.241.19.21.1038 "Development, pilot production, testing and preparation for the supply of special equipment to ensure the fulfillment of Russia's obligations under the ITER project in 2021".

References

[1] H. Pernegger, S. Roe, P. Weilhammer, V. Eremin, H. Frais-Kölbl *et al.* "Charge-carrier properties in synthetic single-crystal diamond measured with the transient-current technique". J. Appl. Phys. 97, 073704 (2005); DOI: 10.1063/1.1863417.

[2] V. Amosov, S. Meshaninov, N. Rodionov, R. Rodionov. "Fast neutron diamond spectrometer. Diamond and Related Materials" 20(8) (2011), 1239–1242. DOI: 10.1016/j.diamond.2011.07.004.

[3] A. W. Moore and J. W. Jorgenson. "Median filtering for removal of low-frequency background drift". Analytical Chemistry 1993 65 (2), 188-191; DOI: 10.1021/ac00050a018.

[4] A. Fernandes, N. Cruz, B. Santos, P. F. Carvalho, J. Sousa, *et al.* "FPGA code for the data acquisition and real-time processing prototype of the ITER Radial Neutron Camera". IEEE TNS 66:7 (2019), 1318-1323; DOI: 10.1109/TNS.2019.2903646

[5] L. Bertalot, A.V. Krasilnikov, L.Core, A. Saxena, N. Yukhnov, R. Barnsley, M. Walsh. "Present Status of ITER Neutron Diagnostics Development". Journal of Fusion Energy; DOI: 10.1007/s10894-019-00220-w

Data Acquisition System Prototype of the ITER Diagnostic Divertor Neutron Flux Monitor Testing at Research Nuclear Facilities

E.S. Martazov[1], Yu.A. Paryshkin[1], N.A. Selyaev[1], V.A. Fedorov[1], V.A. Vorobev[2], A.S. Dzhurik[2], Yu.A. Kashchuk[2], S.Yu. Obudovsky[2], and M.V. Bulavin[2, 3]
[1]National Research Nuclear University "MEPhI", Russian Federation
[2]Institution "Project Center ITER", Russian Federation
[3]Joint Institute for Nuclear Research, Russian Federation
ESMartazov@mephi.ru

***Abstract*—The Divertor nuclear flux monitor (DNFM) is one of the ITER neutron diagnostics. This diagnostic consists of the three same subsystems. Each subsystem concludes the detector module with fission chambers (FCs) and the data acquisition system (DAQ). To solve the task of the neutron flux measurements in the range of 7 orders of magnitude and 1 ms of time resolution the multidetector module is used. To confirm the possibility of the neutron flux measurements in a wide range using such a detector module and to evaluate the characteristics of the DAQ prototype a number of tests were carried out under conditions of the intense neutron radiation. The detector module and the DAQ, which are the prototype of the equipment planned for use on site were used for the tests. The tests were carried out at the plasma neutron diagnostic stand based on the NG-24M neutron generator and at the IBR-2 pulsed reactor of the Joint Institute for Nuclear Research. During the tests at the plasma neutron diagnostic stand the data for the calibration of the DNFM DAQ measuring channels were collected. During the tests at the IBR-2 pulsed reactor the signals from the measuring channels of the DNFM subsystem were obtained while the neutron flux was changed. This report shows the test results and the subsystem calibration techniques.**

***Keywords* —Fusion reactors, Tokamak devices, Neutrons, Reactor instrumentation.**

I. Introduction

The neutron diagnostic "Divertor neutron flux monitor" (DNFM) is one of the instruments to determine the neutron yield and the thermonuclear power of the International thermonuclear experimental reactor ITER. According to the project requirements this diagnostic should provide the measurements of these parameters in the range of 7 orders of magnitude and 1 ms time resolution.

The DNFM diagnostic consists of the three identical subsystems. Each subsystem comprises the detector module and the data acquisition system (DAQ). The detector modules will be placed on the inner part of the ITER vacuum vessel under the divertor cassette. The expected neutron flux at the detector modules locations is from 10^6 to 10^{13} n×cm^{-2}×s^{-1}. The neutron flux measurements should be provided with 1 ms of time resolution and 10 % of uncertainty. To meet these requirements the multidetector module consists of six FCs with different sensitivities and the DAQ system processed FCs signals with different methods is used.

At the first stage the DAQ system prototype was manufactured. With this prototype the software for measuring and diagnostic procedures was checked and validated [1].

The next stage of the DNFM neutron diagnostic research and developing was field testing of the DAQ system prototype with the detector module mockup. The first aim of tests was to check the operation parameters of the DNFM subsystem under the conditions of neutron and gamma radiation. The second aim was to obtain the data to validate the methods and techniques of the calibration and combination of the measurement chains of each measurement channel taking into account the FC parameters.

II. Structure of the DNFM subsystem

Figure 1 shows the structure of the DNFM subsystem for the single detector module. The detector module concludes two detector units (DU). Each DU is the combination of three FCs. DU U5 consists of 235U FCs and DU U8 consists of 238U FCs. The expected sensitivity of the FCs in the single DU has a ratio 1:10:100. Using such a detector module allows one to provide the dynamic range of the DNFM diagnostic.

The DNFM data acquisition system has a spatially distributed structure. The Preamplifier Unit (PAU) consists of six Preliminary amplifier modules (PAM) and the module rack. The main function of PAM is to separate the pulse and DC components of the FC signal and amplify the pulse signal. To reduce the communication lines length from the FCs and the influence of electromagnetic interference PAU will be placed in a shielded protective box in the Port Cell as close as possible to the detector module. The circuitry of the PAM modules has been optimized to minimize the effect of electromagnetic fields and radiation and to reduce the shielded volume.

The DNFM Signal Processing Unit (SPU) conclude six Control and preprocessing modules (CPPM) and six remote

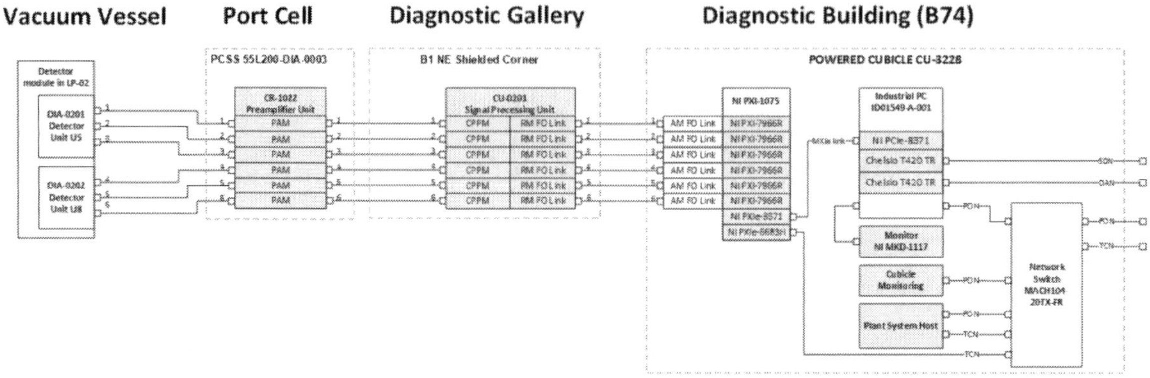

Fig. 1. The structure of the DNFM subsystem for single detector module.

fiber optic modules (RM FO). SPU will be placed in the gallery of the Tokamak building and will be connected to the PAU via communication lines of at least 30 m of length. The main functions of the SPU are the PAM signals conditioning and preliminary analog processing, converting it to digital form and transfering to the PXIe-system. The conversion of signals into digital form is carried out using an analog to digital convertor (ADC) with 14 bits resolution and 120 MS/s sample rate. The data from the ADC outputs are converted from parallel to serial form and transferred to the PXIe-system at 2.5 Gb/s ratio. The modular design of the PAU and SPU allows, if necessary, to rapid replace the modules without dismantling the rest of the equipment.

The PXIe-system receives digital data from the SPU, processes them via the NI PXIe-7966R module software, and transfers the results for further processing and calculations to the industrial personal computer (PC). The industrial PC publishes "raw" and calculated data on the ITER plant-wide network. Herein and after the "raw" data means signals from the output of the analog part of the measuring chains converted into a digital code with a maximum sampling rate without any digital processing. The Industrial PC and PXIe-system will be placed in the diagnostic cubicle along with additional equipment in the diagnostic building.

Signals from the detector module to the BPU unit and from the BPU to the SPU unit are transmitted via triaxial cables. The SPU unit is connected to the PXI system via optical lines.

III. Combination of the data of the DNFM subsystem measurement chains

To increase the dynamic range of the neutron flux measurements using one FC the data acquisition system processes the signal from each chamber with four methods: pulse counting, fluctuation (MSV, Campbelling), current, and pulse-fluctuation. The pulse-fluctuation chain overlaps the operating range of the pulse counting and MSV chains. It had been included in the system processing channel to increase interference immunity and implement the principle of diversity. The main feature of this chain is the use of a charge-sensitive amplifier with a linear selector [2]. The selector is implemented in digital form in the FPGA software. The subsequent processing is provided by the PC software.

To take to a single physical value dissimilar data of all measuring chains of single FC channel in the areas of their combinations the system uses the concept of a "normalized" pulse count rate. For a pulse counting chain the "normalized" count rate is the count rate of impulses at a conditionally zero threshold, i.e. the number of all neutrons detected by the FC for a specified time frame. To calculate the "normalized" count rate of pulse counting chain the measured count rate should be multiplied by a dimensionless calibration factor (greater than one). Calibration factor depends on the magnitude of the threshold in the chain. For the rest chains it is also necessary to determine these coefficients (have the appropriate dimensions). For example for the current chain the calibration factor is inversely proportional to the average charge of the FC current pulse. To obtain the values of the neutron flux one should divide the "normalized" pulse count rates for each measuring chain by the pulse sensitivity coefficient of the FC. Ideally, the obtained neutron flux should be the same for all chains of all channels with the FC used in the DNFM diagnostic.

The following methods can be used to determine the calibration coefficients in the measuring chains of single channel of DAQ with the FC:

- Experimental method;
- Computational method;
- Mixed experimental and computational method.

A. Experimental method

The experimental method is the best in terms of the accuracy of determining the calibration coefficients. The method consists in exposing the FC to the sufficient neutron radiation to measuring the signals of the chains in the regions of their combination. The use of powerful neutron sources such as research reactors leads to high activation of the detector module and the impossibility of its further use as a result. Therefore, this method can be used at the ITER only after the first launches or during tests as a reference method for assessing the uncertainty of the remaining two methods.

B. Computational method

The computational method requires mathematical modeling of both physical processes in the FC taking into account the characteristics of the electrodes, gas pressure, tracks of fission fragments, and the characteristics of the analog part and processing algorithms of the I&C digital parts. Calculations of FCs processes are complex and the results have significant errors.

C. Mixed experimental and computational method

The mixed experimental and computational method is based on the functions of the averaged pulse shape and the pulse-height spectrum, as well as the distribution of pulses by electric charges. These data are obtained experimentally for each FC. The pulse shape and pulse-height spectrum are recorded from the outputs of the pulse counting chains. The distribution of pulses by electric charges is recorded from the outputs of the pulse-fluctuation chains. The DAQ system implements the function of recording impulses (sampling frequency of signals 120 MS/s) and spectra (up to 4096 channels). These data are the basic for subsequent mathematical or semi-natural modeling in order to calculate the calibration coefficients.

In mathematical modeling it is possible to use software-implemented FC models, in which the corresponding normalized pulse shape and pulse-height spectrum are set, and the models of the measuring chains signal responses.

In semi-natural modeling it is advisable to use a full-scale sample of a data acquisition system (a prototype, for example) and an I&C complex (fission chamber signal simulator [3,4]), which implements the mathematical model of forming the FC signal with set parameters and provides an output of an electric pulse-current signal.

The last mentioned method of determining the calibration coefficients was verified during the tests.

IV. DNFM PROTOTYPE TESTING AT THE NEUTRON STAND

The DNFM DAQ prototype tests with mockups of detector units were carried out at the neutron diagnostics stand of plasma of the JSC "SSC RF TRINITY". The stand is build on the basis of the NG-24M neutron generator [5]. Figure 2 shows photos of the equipment during tests at the neutron diagnostic stand of plasma.

During tests at the neutron stand the average pulse shapes and pulse-height spectra were recorded for six FCs of the detector module prototype. The data for calibrating the DAQ measuring chains were obtained and the relative sensitivity of the FCs of the detector units was estimated.

Some experiments were carried out to confirm the possibility of calculating the calibration coefficients by the semi-natural modeling method. To determine the characteristics the high sensitivity FC with a 235U was used. To increase the thermal neutron flux the FC was placed in the paraffin neutron moderator. This made it possible to obtain a counting rate of up to 5×10^{6} s^{-1}. The averaged over 1000 counts pulse shape and pulse-height spectrum were recorded. The average electrical charge per pulse has been calculated. The obtained data were set to the FC signals simulator built on the basis of NI equipment (NI PXIe-5451 generator and NI PXIe-7966R FPGA module for FlexRIO). To restore the current pulse shape which should have been set from the FC signal simulator to the PAM input the obtained pulse shape was processed using a dynamic model corresponding to the inverse transfer function of the pulse chain analog part. This allows one to simulate the FC output signal taking into account its parameters and characteristics of the communication line between the FC and the input of the preamplifier. The pulse-height spectrum was also normalized to set into the FC signal simulator. Figures 3 and 4 show the normalized averaged pulse shape and the pulse-height spectrum. They were set to the FC signal simulator during calibration of the measuring chains.

Figure 4 also shows the threshold set in the pulse counting chain and reflects the area with the total number of pulses that were registered in the chain (above the threshold) and the area of unregistered pulses (below the threshold level). For the current distribution the amount of registered pulses was 83% and unregistered pulses – 17%.

Using the FC signal simulator the signals with count rate in the range from 1E3 to 1E8 1/s were sequentially set to the input of the measurement channel of the DAQ prototype. The measuring chains data were recorded. Using the obtained data the calibration coefficients of the measuring chains were

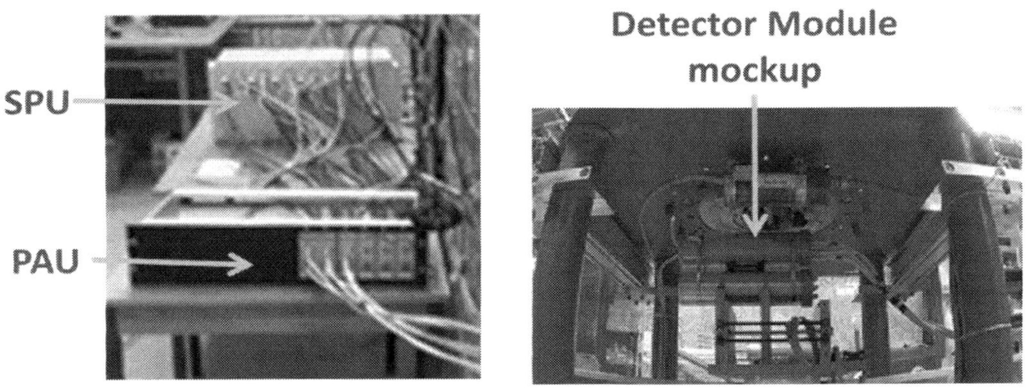

Fig. 2. Images of the equipment during tests at the plasma neutron diagnostics stand.

calculated. The calibration coefficients were set into the prototype configuration file. Further, the data of the normalized count rate of the measuring chains were recorded during the process of changing of the neutron generator power. Figure 5 shows the measuring data of the normalized count rates of the DAQ measuring chains.

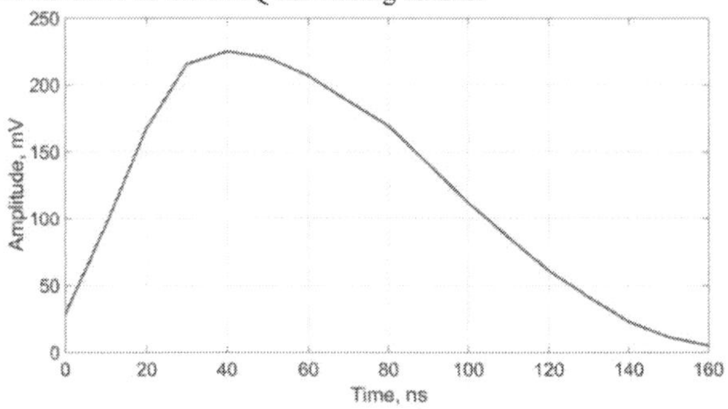

Fig. 3. Normalized averaged pulse shape of the high sensitivity FC with a 235U to set into the FC signal simulator.

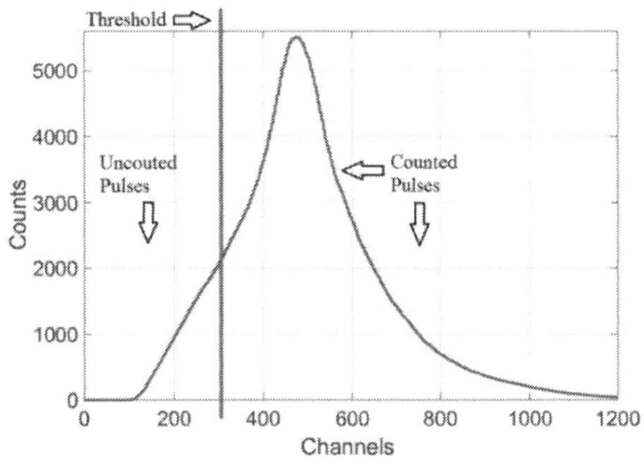

Fig.4 . Normalized pulse-height spectrum of the high sensitivity FC with a 235U to set into the FC signal simulator.

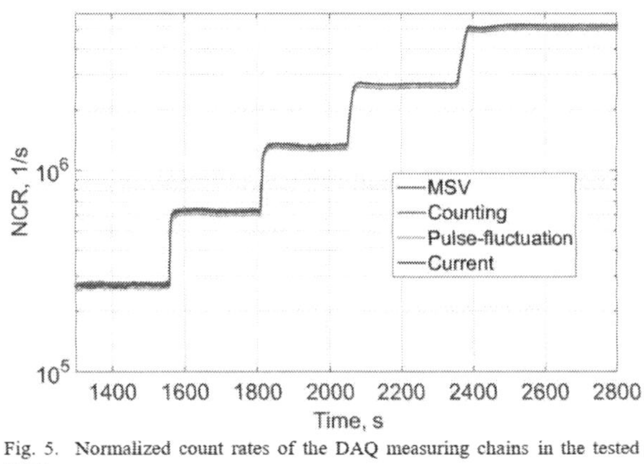

Fig. 5. Normalized count rates of the DAQ measuring chains in the tested range.

The deviations in the measuring data of the normalized count rate from the average value of all four measuring chains in the tested range did not exceed 2%. This result shows the possibility of determining the calibration coefficients set of the DAQ for calculating the normalized count rates in the chains by the mixed experimental and computational method.

V. DNFM PROTOTYPE TESTING AT THE IBR-2M PULSED REACTOR

The tests with the FC under conditions of intense neutron and gamma radiation were carried out at the IBR-2M pulsed reactor at JINR [6, 7]. During the tests the signals of the DAQ measuring chains with connected detector unit mockups were obtained at constant power levels and in the process of its change (ramp up and ramp down) of the reactor facility. A specific feature of the IBR-2M pulsed reactor is the repetitive power increase pulses with a frequency of 5 Hz. This means that every 200 ms the neutron flux increases by several orders of magnitude in a time of the order of 240 μs. This feature of the reactor makes it possible to investigate the time resolution of the measurement channels and study the operation of the DAQ during short-term overloads of the measuring chains. But this had to be taken into account when processing the results. Figure 6 shows the experimental channel of the IBR-2 reactor where the detector module mockup was placed. Figures 7 and 8 show the normalized count rate of the MSV chain while the reactor power rump-up and at a constant power of the plant of 450 kW over a short time interval.

Fig. 6. Experimental channel of the IBR-2 reactor, where the detector module mockup was placed during the tests.

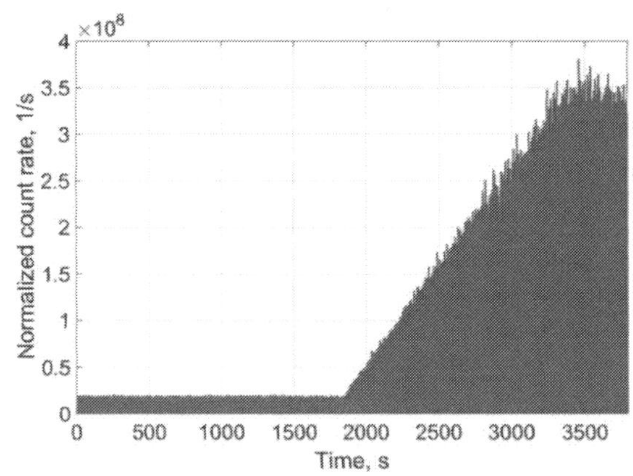

Fig. 7. Normalized count rate of the MSV chain while the reactor power rump-up

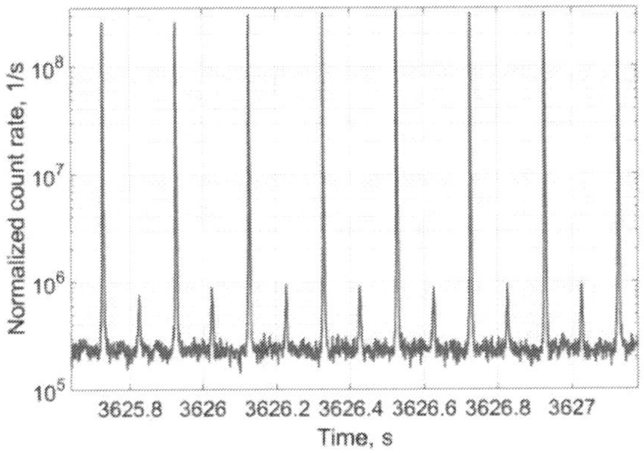

Fig. 8. Normalized count rate of the MSV chain at a constant power of the plant of 450 kW over a short time interval.

Despite the pulsed operation of the reactor, during the tests it was possible to obtain the averaged pulse shapes and pulse-height spectra for the four most sensitive to the neutron flux FCs and to estimate the boundaries of the linear operation regions of the measuring chains.

VI. Conclusions

The tests of the DNFM data acquisition system prototype at research nuclear facilities within the FC multidetector units and communication lines, similar to those that are supposed to be used in the diagnostics "Divertor neutron flux monitor", confirmed the possibility of measuring the neutron flux in a wide range with specified measurements uncertainty and dynamic characteristics of measurement channels. During the tests methods of measurements and data processing were checked out.

An important result of the tests is the approbation of the mixed computational and experimental method for calibrating the measuring chains of the DAQ according to the data on the average pulse shape and the pulse-height spectra. The method makes it possible to determine the values of the calibration coefficients of the measuring chains during the in-situ calibration of the DNFM diagnostics using the neutron source at the ITER facility for subsequent set it into the DAQ configuration file. In case of a change in the pulse chains of the thresholds values (determined by noise, electromagnetic interference and the expected level of gamma) their calibration coefficients are easily corrected according to the pulse-height spectra. To automate this process, a service program is used that contains a database of spectra of all 18 fission chambers included in the DNFM diagnostics.

Acknowledgment

The authors would like to acknowledgment to the staff of the neutron diagnostics stand of plasma of the JSC "SSC RF TRINITY" and the IBR-2M pulsed reactor at JINR for their help in preparing and conducting the tests.

This work was carried out with the financial support of the Institution "Project Center ITER" in the framework of the state contract with State Corporation "ROSATOM".

References

[1] V.A. Fedorov, *et al* "Study of the Data Acquisition System for ITER Divertor Neutron Flux Monitor Diagnostic," IEEE Transactions on Nuclear Science, vol. 67, no. 4, pp. 688-693, April 2020.

[2] V.P.Alferov, *et al* "Applicaton of Modern Tools for NPP Neutron Monitoring Systems Upgrading," in *Proc. International Topical Meeting on Nuclear Plant Instrumentation, Controls, and Human-Machine Interface Technologies (NPIC&HMIT 2000)*, Washington, D.C., USA, November 2000, pp. 393-401.

[3] A.V. Batyunin, *et al* "Fission Chamber Simulator for Data Acquisition Performance Tests," in *Proc. Conference on Advancements in Nuclear Instrumentation, Measurements Methods and their Application ANIMMA-2013*, Marceille, France, June 2013, pp. 1-4.

[4] I. V. Alyaev, *et al* "Wide-range fission chambers signal simulator," in *Proc. 10th International Topical Meeting on Nuclear Plant Instrumentation, Control, and Human-Machine Interface Technologies (NPIC&HMIT 2017)*, San Francisco, CA, USA, June 2017, vol. 3, pp. 1991-1998.

[5] [5]. S.V. Syromukov, *et al* "NG-24 Neutron Generator for Nuclear Medicine and Thermonuclear Research," Atomic Energy, vol. 119, pp. 58-60, October 2015.

[6] IBR-2 Reactor Parameters. Available: http://flnph.jinr.ru/en/facilities/ibr-2/parameters. Accessed on: 08 12, 2021.

[7] Yu.G. Dragunov, *et al* "Modernization of the IBR-2 pulsed research reactor," Atomic Energy, vol. 113, pp. 29-38, November 2012.

In-situ gamma irradiation testing of radiation hardened chips till 1 MGy

David Geys[1], Ying Cao[1], Marco Van Uffelen[2], Laura Mont Casellas[2], Ludo Vermeeren[3], and Andrei Gusarov[3]
[1]Magics Instruments, Belgium
[2]Fusion for Energy, Spain
[3]SCK CEN, Belgium
David.geys@magics.tech

***Abstract*—Ten samples of a custom tailored Mega-Gray hardened resolver/LVDT-to-digital converter, a resistive base sensor-to-digital converter and a RS485 communication application specific integrated circuit (ASIC) were combined in 1 irradiation campaign for Fusion for Energy (Barcelona, Spain). Radiation resistance of these ASICs, developed by Magics Instruments (Geel, Belgium) for Fusion for Energy, was assessed for a total ionizing dose (TID) above 1 MGy using the Co-60 gamma underwater irradiation test facility at SCK CEN (Mol, Belgium).**

The 3 different ASICs were irradiated at an average dose rate of 484 Gy/h and their performance was continuously measured (in-situ) during 97 irradiation days. A fully autonomous and modular test setup was developed to perform these measurements and ensure continuous operation by implementing a recovery and warning system in case of failure to restrict measurement data loss to minimum.

An in-situ post-irradiation assessment was performed afterwards to observe recovery from the irradiation in a so-called annealing phase. Annealing was done for seven days at room temperature followed by another 7 days of high temperature annealing at 100 °C to accelerate the recovery effect.

During the full test campaign all data was saved in a database, post-processed with Python into readable plots to deduct possible performance shifts due to the irradiation and afterwards during recovery.

During the complete testing campaign of these ASICs the ESCC22900 (Total Dose Steady-State Irradiation Test Method) standard was followed.

***Keywords* —Rad hard ASIC, 1MGy TID, In-situ measurements**

I. Introduction

Most of today's commercial off-the-shelf (COTS) electronics are not specified to meet the demanding requirements of advanced nuclear applications requiring MGy-level Total Ionizing Dose (TID) tolerance. Examples are maintenance and diagnostics tasks in future burning plasma fusion reactors, for example the International Thermonuclear Experimental Reactor (ITER). Applications such as interventions during nuclear accidents, dismantling of old nuclear power plants and disposal of radioactive waste also call for rad-hard electronics. Therefore, first, the development and, secondly, the testing of custom tailored MGy hardened integrated solutions become necessary for use in these environments. It will not only reduce the shielding requirements but will make it possible to place electronics closer to front-end sensor transducers and actuators in a radiation environment.

Three ASIC designs of such custom tailored MGy hardened integrated solutions, developed by Magics Instruments (Geel, Belgium) for Fusion for Energy, were tested in-situ to monitor their performance during irradiation and prove their MGy radiation hardness. Of each ASIC design, 10 samples were measured. Average dose rate during the irradiation experiment was 484 Gy/h to reach at least a TID-level of 1 MGy after 97 days of irradiation for all 30 samples in the test. During these 97 days of irradiation, specifications of all samples were measured continuously in a repeated fashion. Therefore, a fully automated modular test setup was developed to perform these measurements.

The irradiation assessment was performed at the Co-60 gamma underwater facility (RITA) at SCK CEN (Mol, Belgium). The irradiation assessment was followed by a post-irradiation assessment split up in an annealing at room temperature for 7 days followed by high temperature annealing at 100 °C for another 7 days. This post-irradiation assessment was performed in the temperature chamber (CLARA) also located at SCK CEN.

II. Background

The 3 ASIC designs tested in this test campaign are called: ASIC1, ASIC2 and ASIC5 and are part of a set of 5 radiation hardened ASICs developed by Magics Instruments:

- ASIC1: resolver/LVDT-to-digital converter ASIC
- ASIC2: resistive bridge sensor signal conditioner ASIC
- ASIC3: 10-channel limit switch readout ASIC
- ASIC4: 10-channel relay driver ASIC
- ASIC5: BiSS (RS-485) interface communication ASIC

With these 5 ASICs it is possible to build an actuator and sensor read-out network in a harsh environment up to 1 MGy TID. Such networks could be used for maintenance and diagnostic tasks in fusion reactors (like the ITER). A typical sensor/actuator network is depicted in Fig. 1. The presented network exists out of 3 RS-485 slave nodes. In each of these slave nodes a combination of up to 3 ASICs is used to read out different kind of sensors or drive relays. The first 4 ASICs out of the set of 5 listed above can be used for this purpose. The measured sensor data are combined into packages and sent over the long bus cable via instances of ASIC5 in the bus. Communication in the bus to/from the RS-485 master of sensor/actuator data from/to the RS-485 slave nodes is possible over a long CAT7 cable of up to 215 meters long. All ASICs in this network need to be supplied by the rad-hard DC-DC power module. This module is outside the scope of this work presented here.

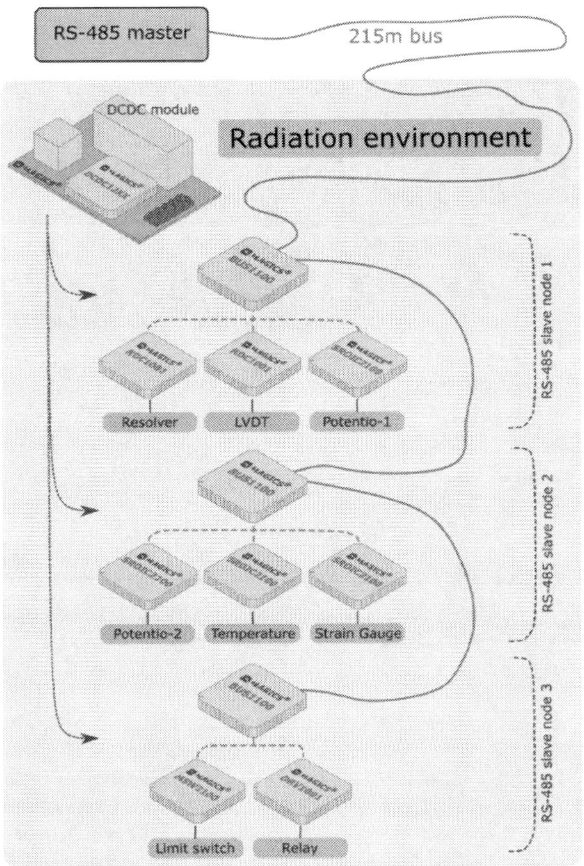

Fig. 1. Schematic representation of a network inside a radiation environment built up by custom tailored MGy hardened ASICs to read out sensors and drive actuators.

With the network presented in Fig. 1 it is possible to build closed loop motor control networks as they are needed for remote handling.

The big advantage for placing these electronics in the radiation environment itself is dual. (a) Only a standard single data communication cable is needed from the radiation environment to the safe zone and therefore significant cost savings could be achieved on multiple expensive cables in case these electronics were placed in the safe zone. (b) Additionally, the need for thick shielding is not stringent anymore. Consequently, also here a significant cost saving exists and furthermore, the system could be built smaller.

III. Gamma irradiation testing

The goal of gamma irradiation testing in this project is to prove the functionality and measure specifications of the ASIC designs when irradiated up to an accumulated dose of at least 1 MGy.

The characteristics of the ASICs change with the received TID. This shift occurs in 2 directions, as a result of the well-known recovery effect, typically observed at 1-10 kGy. Therefore, it is important to monitor characteristics of the DUTs *in-situ*, during irradiation.

In-situ measurements can only be performed when the ASICs are accessible during irradiation. Therefore, numerous interconnection cables are needed between ASICs and lab equipment measuring the specifications.

During this gamma irradiation assessment samples of ASIC1, ASIC2 and ASIC5 were irradiated in the Co-60 test facility at SCK CEN at an average dose of 484 Gy/h. To reach a TID of 1 MGy the ASICs were exposed for 97 days. The applied dose rate is in-line with the dose rate of 300 Gy/h expected in the ITER during maintenance mode.

The irradiation was interrupted twice to investigate the dynamic balance between radiation-induced changes and recovery. These interruptions were held at an average TID of 10 kGy and 100 kGy each time for a duration of 1 hour.

IV. Post-irradiation testing

More measurements were performed after irradiation in the so-called annealing phase. Goal of these measurements is to monitor possible recovery effects of the performance of all ASICs after irradiation.

Following the standard ESCC22900 [1] the post-irradiation testing was split up into 2 main parts. (a) 7 days annealing at room temperature. This room temperature value may not differ more than 5 °C compared to the temperature during irradiation with a maximum of 30 °C. (b) 7 days annealing at high temperature. This accelerated ageing phase is carried out at a temperature of 100 °C ±5 °C.

The post-irradiation testing has been performed in the CLARA temperature chamber at SCK CEN. This temperature chamber is located in another building than the RITA irradiation facility on the same site of SCK CEN. The first 24h of room temperature annealing have been performed in the RITA facility. In this period the largest recovery is expected. When moving the setup immediately after stopping of the irradiation from the RITA facility to the CLARA facility, precious time would be lost in disconnecting all cables, perform contamination checks, moving the full setup, re-connect all cables and perform a test run. After 24h the test setup was moved to the CLARA facility for 6 more days of room temperature annealing and 7 days at 100°C.

In-situ measurements were performed during the full post-irradiation assessment.

V. Test setup

During the full irradiation and post-irradiation assessment in-situ measurements were required to monitor current consumption, accuracy and resolution of readout chains, verify communication and many more. In total more than 30 specifications of the ASICs were continuously measured during the 97 days of irradiation and 14 days of annealing. To ensure a long-term stable measurement setup a modular automated test setup was developed by Magics Instruments. With this setup it is possible to perform fully automated parallel measurements repeated in a sequential fashion of many ASICs placed close to each other in a remotely controlled setup with a minimum set of lab equipment. Furthermore, a data storage redundancy and failure warning mechanism with the possibility to self-restore was built in. A schematic representation of this setup is depicted in Fig. 2.

All ASICs in the test setup are mounted individually on a small substrate PCB which is plugged in, together with 3 other substrate PCBs holding other DUTs, on a test tower PCB slice. These PCB slices are stacked on top of each other constructing a test tower. This test tower, containing all DUTs in the setup, is placed in the RITA container for irradiation or in the CLARA temperature chamber for annealing.

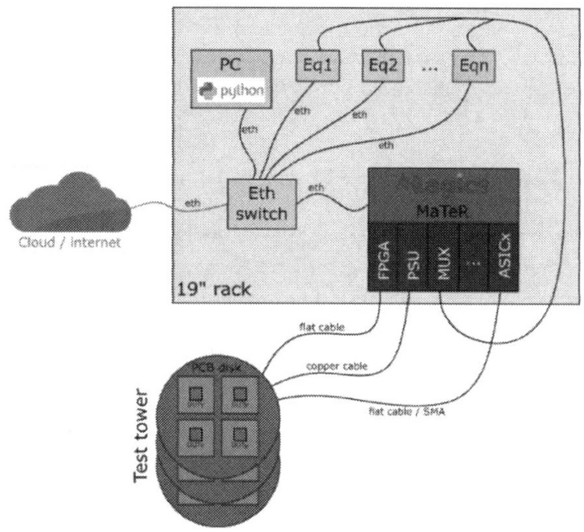

Fig. 2. Schematic representation of the developed in-situ automated test setup for (post-)irradiation testing.

Measurements are performed in-situ and therefore many connection cables are needed to route all signals and power from/to the test tower to/from the lab equipment in the safe test zone. Different types of interconnection cables are used depending on the signals they carry: flat cables for digital signaling, copper cables for power routing and SMA cables for highly sensitive analog signals. All these cables are routed from the test tower in 4 flexible tubes to the 19 inch rack. These cables have a length of 10 meters making it challenging to ensure high speed connections.

All equipment in this setup is mounted in a 19 inch rack placed in a safe zone. A rack PC is used to run all scripts and to drive the lab equipment and read from it. Multiple custom test rack computers (MaTeR) are also included householding multiple functionalities like there are FPGA, MUX, PSU and others. An ethernet switch connects all devices in this 19 inch rack and ensure a connection towards the cloud.

A. MaTeR

The MaTeR (Magics' Test Rack) is an in-house development of a 19 inch test rack PC that can be used to include many DUTs in a test setup with a minimal set of lab equipment. A photo of 1 such a MaTeR is depicted in Fig. 3.

Fig. 3. Modular Magics' Test Rack (MaTeR) filled with 1 test card.

It is a modular design that can be adapted depending on the amount and type of DUTs in the test. Each MaTeR contains a back panel PCB in which up to 10 slide-in PCBs with different functionalities can be slit in. Currently the following types of slide-in PCBs are available:

- FPGA for digital signaling via flat cable and/or UTP
- PSU for power routing towards DUTs
- MUX for multiplexing SMA and flat cables
- CLK for generating different kinds of clock signals
- TC for reading out type-T class 1 thermocouples
- ASICx as ASIC specific test cards like f.e. a resolver simulator card to test ASIC1

In this project 4 FPGA cards were used to route in total 170 digital signals and 6 UTP cables. Six different PSU cards routed 80 supply lines towards the DUTs. Eighty four analog sensitive signals were multiplexed via 11 MUX cards to the limited set of lab equipment in the 19 inch rack. One resolver simulator card was integrated to test the LVDT and resolver functionality of all samples of ASIC1 in the test. Besides this also 20 clock signals were needed generated by the 2 CLK cards in the test. To monitor the temperature dependency of the measured specifications 1 TC card was used.

These MaTeRs are mounted in a 19 inch rack of which 2 of them exist in this project. Fig. 4 shows the smallest rack in which 3 of the 4 MaTeRs are located in. All cables needed to make connection to the DUTs are plugged/screwed in.

Fig. 4. Front view of a 19 inch rack with 3 MaTeRs fully connected.

B. *Test Tower*

The test tower is specifically designed to fit in the container of the irradiation facility. DUTs are arranged in a way it is possible to allow many interconnection cables to each DUT in the test tower and include many DUTs very close to each other. In this project the following DUTs are included:

- ASIC1:
  6 (in-situ) + 2 (in-situ @ worst-case bias) + 2 (0 V bias)
- ASIC2:
  6 (in-situ) + 2 (in-situ @ worst-case bias) + 2 (0 V bias)
- ASIC5:
  6 (in-situ) + 2 (in-situ @ worst-case bias) + 2 (0 V bias)

Fig. 5 shows the test tower fully connected before the start of the irradiation experiment.

It is very important to locate all the DUTs close to each other because the dose rate varies along the vertical axis of the container. Before the actual start of the irradiation experiment the dose rate (dosimetry) was determined by SCK CEN. The dose rate was determined at specific heights (250, 300, 350, 400 and 450 mm). These data were used to determine the dose rate experienced at each level of the DUTs in the test tower by defining a 4th order polynomial fit through the data from the dosimetry. An average dose rate for all DUTs in the test was calculated at 484 Gy/h. Along the height of the test tower, the dose rate varied from 438 Gy/h (-9.5%) at the lowest located DUT to 516 Gy/h (+6.6%) at the highest located DUT in the test tower. After 97 days of irradiation all samples received at least 1.014 MGy of TID. DUTs located at the top of the test tower almost received 1.2 MGy. The height of each DUT in the test tower with the corresponding dose rate and TID at the end of the irradiation experiment are depicted in Fig. 6.

Fig. 5. Fully connected test tower before the start of the irradiation experiment.

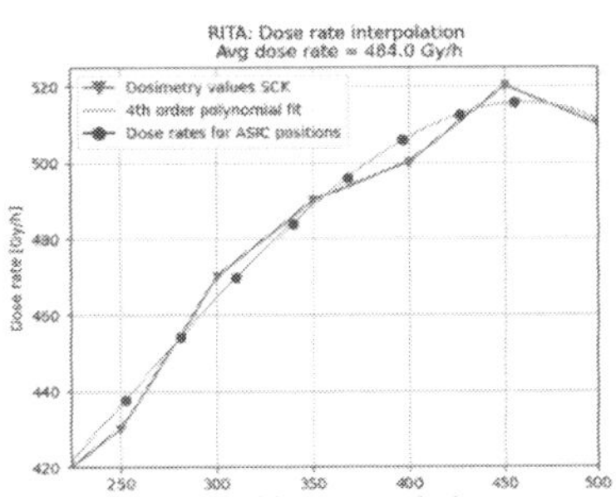

Level in container [cm]	TID @ end [MGy]
25.2	1.014
28.1	1.052
31.0	1.088
33.9	1.121
36.8	1.149
39.7	1.172
42.6	1.187
45.5	1.194

Fig. 6. Dose rate versus height in the test tower and TID at the end of the irradiation experiment.

C. *Software*

The rack PC integrated in the 19 inch test rack of the setup (see Fig. 2) runs several scripts, all written in Python, in parallel:

- 3 measurement scripts: 1 script for each ASIC design in the test was running. This script is a combination of writing to or reading from the lab equipment in the test to measure all specifications of the DUTs. This script also handles the saving of the measured data in a database.
- 1 temperature script: This script monitors the

temperature via the 3 type-T class 1 thermocouples in the test tower. Temperature is measured each minute.

- 1 data replica script: Data locally stored on the hard disk of the rack PC is continuously copied to an on-line database at a cloud server for reasons of redundancy.
- 1 warning script: In case the setup comes to a halt a warning email and test message is sent to the engineers of Magics.

The above described scripts run all in parallel in different Python processes. This lowers the risk for the full setup to stop due to a failure in a single script. Each of the scripts runs sequentially for the duration of the experiment. This has the advantage that the test cycle time of 1 ASIC design could be kept small enough to observe shifts in performance of the DUTs due to irradiation. The test cycle time is defined as the total time needed to measure all specifications of all samples of 1 ASIC design in the test setup. Because dose rates are high in this test, each DUT must be measured fast enough to measure it again after the other 9 samples are measured to capture possible shifts in the performance due to irradiation. Target value for the test cycle time is to be smaller than 30 minutes.

If 1 or multiple scripts crash, the event is logged in a text file on the test rack PC. Thereafter, the warning script will send out a warning email and a text message to communicate to the engineers of Magics Instruments that the setup needs to be checked. Sending a text message was added to the methodology of this warning mechanism to cover the event the internet connection is down at the irradiation facility. Some possibility is built in the test setup to self-restore in the event of a script crashing. If that action is not successful, the setup could still be controlled remotely via the secured connection to restart scripts or power cycle lab equipment before going to the facility itself.

VI. Facilities

Facilities used in this project were all located at the Research Centre for Nuclear Energy (SCK CEN) in Mol, Belgium. RITA is the irradiation facility and CLARA is the temperature chamber used for the annealing.

A. RITA – Radio Isotope Test Arrangement

RITA is the Co-60 underwater irradiation facility of SCK CEN located at their site in Mol, Belgium. This facility allows in-situ measurements. An overview photo of this facility together with the test setup of this project is shown in Fig. 7. At the top, the RITA container is visible. In this container the test tower can be loaded. After sealing of the container with its lid, it can be lowered in the pool where the Co-60 sources are located at the bottom. From the moment the container is in between these sources, the irradiation is ongoing. The scheduled interruptions were performed by lifting the container out of the irradiation flux. It was chosen to not lift the container fully to the top of the 6 meter deep water basin to lower the risk of failure of cables, connecting the test setup to the test tower. With the TID accumulation the risk of the failure increases because cable insulation material becomes brittle and short-circuiting may occur during displacement of the container.

Fig. 7. RITA – Co-60 irradiation facility at SCK CEN (Mol, Belgium) with the test setup of this project before the start of the irradiation experiment.

The photo in Fig. 7 is taken just before the start of the irradiation experiment. The test tower was still located on a small stand next to the open RITA container to connect all interconnection cables. This is a very delicate job as a single wrong connection could damage the DUTs in the setup. All cables were connected wearing proper ESD protection to limit possible ESD event damaging the DUTs.

The test tower is connected at the bottom side of the lid of the RITA container. Four flexible tubes guide all interconnection cables from the lid-test tower combination to both 19 inch racks of the setup located next to the water basin. These 19 inch racks give place to the rack PC, MaTeRs and all lab equipment needed to perform the testing of all ASICs in the test.

The parameter control panel of the RITA facility itself is visible at the top left of the photo. Besides temperature monitoring by Magics, another temperature monitoring was performed by SCK CEN via thermocouples fixed in the RITA container.

Fig. 8 shows the RITA container during irradiation. The container is located at the bottom of the 6 meter deep water basin in between the Co-60 radio-active sources. The sources can be identified by the well-known blue Cherenkov light. The 4 flexible tubes make it possible to perform in-situ measurements.

One day after the irradiation was finished, the test tower was disconnected from all its cables, contamination check was performed, and it was moved together with the test racks to the CLARA facility.

Fig. 8. RITA at SCK CEN (Mol, Belgium) when the container is in between the Co-60 sources.

B. CLARA – CLimate chamber for Automated temperature and humidity Resistance Assessment

The post irradiation assessment was performed in the temperature chamber CLARA at SCK CEN in Mol, Belgium. This climate chamber offers a useful capacity of 2 m^3. Holes are present to make the DUTs inside the chamber accessible for in-situ measurements.

An overview photo of the setup during annealing is shown in Fig. 9. All interconnection cables were routed through the hole at the side of the temperature chamber. Both 19 inch racks were placed, face-to-face, next to the temperature chamber to keep interconnection cable length low. Another set of cables than the ones used for the irradiation experiment was used. Re-use of the cables of the irradiation experiment is not possible because unscrewing these resulted in irreparable damage as they were brittle due to the outgassing by the irradiation. Cables used for post-irradiation are also shorter in length. An interconnection cable of 3 meters long is sufficient to make connection between the 19 inch test racks and the DUTs in the temperature chamber.

For the annealing, the test tower of the irradiation assessment was split up into 2 parts for ease of accessibility during connecting the cables. Before the start of the measurements during this annealing everything was verified for correctness and the photo in Fig. 10 was taken.

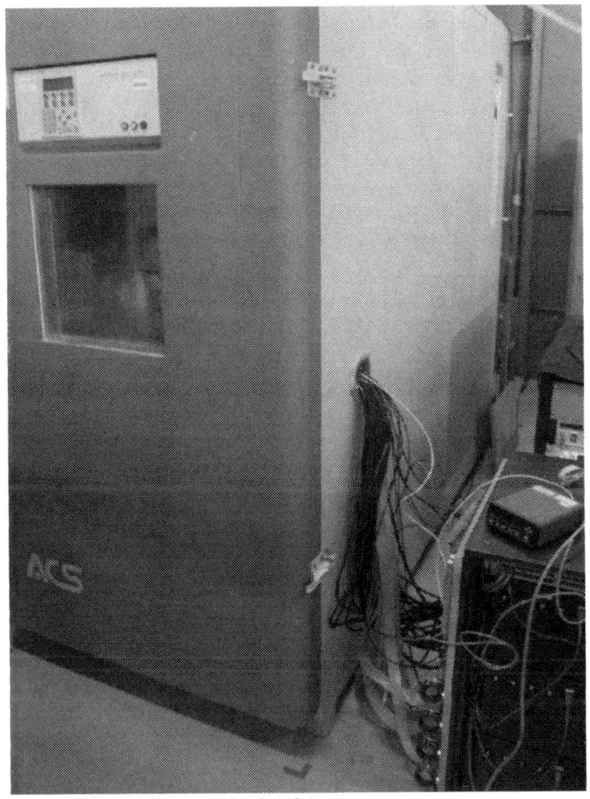

Fig. 9. CLARA at SCK CEN (Mol, Belgium).

Fig. 10. Setup fully connected in CLARA before start of in-situ measurements during annealing.

VII. OUTCOMES

All data measured during the irradiation and post-irradiation assessment was saved in a MySQL database locally stored on the test rack PC. Continuously a replica of this database was saved on the cloud database server for reasons of redundancy. A lot of measurement data was generated during this more than 100 days lasting experiment. Python was used to perform post-processing on this data. In total more than 48k lines of code were written divided over 262 Python files to generate more then 10k of graphs. These graphs were all studied to detect possible shifts in performance due to the irradiation and possible recovery during annealing to prove functionality of all ASICs in the test during irradiation up to 1 MGy TID and beyond.

Thanks to the designed automated setup of Magics some drifts of the more than 30 measured specifications along the received TID could be observed. Radiation induced drifts up to 10% are monitored but all were still acceptable. All ASICs in the test were always fully functional and therefore can be concluded that ASIC1, ASIC2 and ASIC5 are 1 MGy radiation tolerant.

VIII. Conclusions

The project to prove the 1 Mega-Gray TID radiation tolerance of 3 ASICs types, earlier developed by Magics Instruments for Fusion for Energy, was performed in co-operation with SCK CEN.

In total, 30 ASICs were measured in-situ during an irradiation campaign for almost 100 days to reach a TID of at least 1 MGy for all samples under the test. Additionally, samples were in-situ measured during a 14 day post-irradiation assessment split up in 7 days annealing at room temperature and another 7 days accelerated ageing at 100 °C. To perform these measurements non-stop in an automated fashion, Magics Instruments developed a modular test setup to measure the performance of all ASICs in the test.

After almost four months of continuous online measurements, less than 3 hours of cumulative down-time of the automated test setup were reported. Thanks to automatic detection of down-time and notifications to the team members, multiple issues were quickly identified and all of them, except one, were fixed in less than 15 minutes.

Magics test platform has proven to be a very stable and robust solution for automated bench test and long term irradiation measurements. This test platform supported to successfully confirm the 1 MGy radiation tolerance of all three ASICs under test.

The full assessment followed the total dose steady-state irradiation test method described in the ESCC22900 standard.

IX. Disclaimer

The work leading to this publication has been funded by Fusion for Energy under Specific Contract F4E-OFC-358-2-01-03. This publication reflects the views only of the author, and Fusion for Energy cannot be held responsible for any use which may be made of the information contained therein.

References

[1] ESCC22900, Total Dose Steady-State Irradiation test, Version: Issue 5 June 2016 [Online]. Available: http://escies.org/escc-specs/published/22900.pdf

Long Term Neutron Activation in JET DD Operation

Andrej Žohar[1,2], Igor Lengar[1], Paola Batistoni[3], Sean Conroy[4], Aljaž Čufar[1], Renata Kierepko[5], Bor Kos[1], Stefano Loreti[3], Jerzy W. Mietelski[5], Chantal R. Nobs[6], Lee W. Packer[6], Mario Pillon[3], Vladimir Radulović[1], Marilia I. Savva[7], Luka Snoj[1,2], Ion E. Stamatelatos[7], Žiga Štancar[1], Theodora Vasilopoulou[7], Anna Wójcik-Gargula[5] and JET contributors*

[1] Reactor Physics Department, Jožef Stefan Institute, Jamova cesta 39, 1000 Ljubljana, Slovenia
[2] University of Ljubljana, Faculty of Mathematics and Physics, Jamova cesta 19, 1000 Ljubljana, Slovenia
[3]Department of Fusion and Technologies of Nuclear Safety and Security, ENEA C. R. Frascati, via E. Fermi 45, 00044 Frascati (Roma), Italy
[4]Department of Physics and Astronomy, Uppsala University, Box 516, SE-75120 Uppsala, Sweden
[5]The Henryk Niewodniczanski Institute of Nuclear Physics (IFJ PAN), Polish Academy of Sciences, 31-342 Kraków, Poland
[6]United Kingdom Atomic Energy Authority, Culham Centre for Fusion Energy, Culham Science Centre, Abingdon, Oxon OX14 3DB, UK
[7]INRASTES, NCSR 'Demokritos' 153 10, Agia Paraskevi Attikis, Greece
*See the author list of 'Overview of JET results for optimising ITER operation' by J. Mailloux et al. to be published in Nuclear Fusion Special issue: Overview and Summary Papers from the 28th Fusion Energy Conference (Nice, France, 10-15 May 2021)"

andrej.zohar@ijs.si

***Abstract*— In the 2019 C38 Deuterium-Deuterium campaign at JET several different ITER-relevant materials and dosimetry foils were irradiated in a specially designed long-term irradiation station located inside the vacuum vessel with the purpose of testing the activation of ITER materials by fusion neutrons. The samples were exposed to a neutron fluence of 1.9E14 n/cm² during JET discharges performed in the experimental campaign over a period of 5 months. Gamma ray spectroscopy measurements were performed on irradiated samples to determine the activation of different long-lived isotopes in the samples. Monte Carlo computational analysis was performed to support the experiment by using the measured neutron yield and irradiation time. In this paper we focus on the computational analysis of the dosimetry foils that are used in order to measure the local neutron energy spectrum and flux. The foils were chosen to cover different neutron energies: thus Yttrium and some of the Nickel and Cobalt reactions were used to determine the Deuterium-Tritium fusion fraction, while Scandium and Iron and some of the Nickel and Cobalt reactions were used for comparison of the computed activity with the experimental measurements. The obtained C/E values show a reasonably good agreement between calculated and measured activity, thus validating the computational methodology and providing the basis for the analysis of the ITER-relevant materials and future experiments performed at JET in the Deuterium-Tritium campaign.**

***Keywords* — JET, Long Term Neutron Irradiation, MCNP**

I. Introduction

The neutron irradiation impact on materials is an important research field for the development of fusion technologies. The Joint European Torus (JET) tokamak is currently the only tokamak capable of performing fusion plasma experiments with Deuterium-Deuterium (DD), Tritium-Tritium (TT) and also Deuterium-Tritium (DT) fusion reactions, thus producing high 2.45 MeV and 14.1 MeV neutron fluence. The high neutron fluence can be used to perform neutron irradiation experiments of different materials, of which the most important are materials that will be used in components of the ITER tokamak.

To support ITER and study the impact of neutron irradiation on materials, especially for DT plasmas, several experiments are planned to be performed in the upcoming TT and DT experiential campaigns at JET planned to commence in 2021/22. In preparation for the DT campaign at JET, several important experiments have already been performed [1], such as 14 MeV neutron calibration [2,3], neutronics benchmark experiments [4-8] and activation measurement experiments [9-11]. For the preparation of experiments on the neutron induced activation on materials, samples of the main components of the ITER tokamak and dosimetry foils were irradiated during the 2019 C38 Deuterium experimental campaign at JET [12,13]. Samples were irradiated in a long-term irradiation station (LTIS) located close to the plasma boundary for several months. The goal of the experiment was to study the DD neutron activation of ITER materials and preparation of the experimental methodology for the upcoming TT and DT experimental campaigns. The developed methodology consists of the execution of irradiation experiments, measurements of sample activation and computational support to the experiment.

To properly evaluate the activation of different isotopes in the ITER samples and validate the computational methodology, several different dosimetry foils (Cobalt, Nickel, Iron, Yttrium, and Scandium) were irradiated together with the ITER samples in the LTIS. The dosimetry foils were chosen to cover different neutron energy regions, from thermal neutrons (energies up to 1 eV) up to fast neutrons (from 0.1 MeV to 14 MeV). In this paper, the comparison between experimental measurements of irradiated dosimetry foil activities and calculated activities will be presented.

The paper is organized as follows. In the first section, the irradiation experiment is presented. Key aspects concerning the experiment are: the long-term irradiation station together with different ITER samples and dosimetry foils, the conditions to which samples in the LTIS were exposed during tokamak operation and the measurements of sample activity performed after irradiation. In the second section, the computational methodology is presented. The methodology consists of a detailed and validated JET Monte Carlo N-Particle transport code (MCNP [14]) model and the procedure to calculate the isotope activities in dosimetry foils at the end of the irradiation. In the last section the results of activity measurements for dosimetry foils will be compared with calculated activities for validation of the computation methodology for upcoming TT and DT irradiation experiments at JET.

II. Experimental measurements

A. Long-term irradiation station

For the neutron irradiation impact on materials experiment in the DD, TT and DT campaigns at JET, a new LTIS was made. The LTIS was designed to hold several different samples of different sizes and be positioned in one of the irradiation positions located close to the plasma boundary but not within the vacuum vessel. The chosen position allows the samples to be exposed to high neutron fluence and be removed between experimental campaigns at JET without breaching the vacuum barrier. The LTIS position inside the JET vacuum vessel is presented in Fig. 1. For LTIS to hold samples of different size, two holders were made caller 'RADA' holder and 'ACT' holder. The 'ACT' holder has 26 equal size positions in which many samples can be inserted simultaneously and photo of the 'ACT' holder is presented in Fig. 2. The 'RADA' holder has 28 positions of different sizes.

For the experiment performed in the C38 deuterium campaign, the 'ACT' holder in the LTIS was loaded with 27 ITER material samples and 70 samples of different dosimetry foils. The thicknesses of the samples varied from 0.1 mm to 1 mm with the majority of samples being 0.5 mm thick. Fig. 2 shows the 'ACT' holder loaded with different samples for the irradiation experiment. The irradiated ITER material samples include material samples of poloidal field coil jacket, toroidal field coil radial closure plate steels, EUROFER 97-2 steel, Tungsten and CuCrZr materials from the divertor, Inconel 718, CuCrZr and 316L stainless steel for the blanket modules and vacuum vessel forging samples [13].

Fig. 1 CAD model of the LTIS position in the JET tokamak. The LTIS is located in the 7th octant below the middle plane of the tokamak.

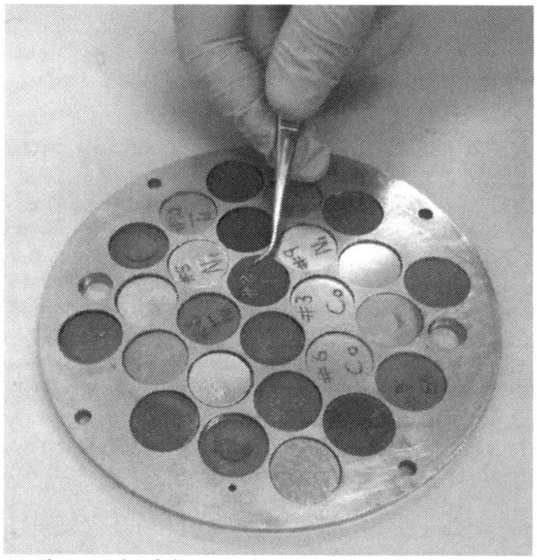

Fig. 2 Photograph of the outer long-term irradiation station sample holder loaded with ITER samples and dosimetry foils. The samples were irradiated in the C38 deuterium campaign at JET.

The dosimetry foils in LTIS were chosen to cover different neutron energy regions, from thermal neutrons to fast neutrons. The dosimetry foils used in the irradiation experiment were made from Cobalt, Nickel, Iron, Scandium and Yttrium.

B. Experimental condition and gamma ray spectroscopy measurements

The LTIS was installed on 27th July 2019. The samples were irradiated in the C38 campaign from 29th July 2019 till 20th December 2019. During the campaign, there were 101 days of irradiation and 46 days of no experimental irradiation. The neutron irradiation during plasma discharges was short (10 s to

20 s per plasma discharge) with cooling time in between discharges. The total neutron yield of the C38 experimental campaign at JET was $3.15 \cdot 10^{19}$ neutrons. The neutron fluence the samples in LTIS were exposed to during the C38 experimental campaign was $1.9 \cdot 10^{14}$ n/cm^2.

The samples from LTIS were retrieved on 3rd January 2020, approximately 13 days after the end of irradiation. The ITER samples and dosimetry foils were sent to different European laboratories for analysis: CCFE, ENEA, IFJ-PAN and NCSRD. All laboratories performed high resolution gamma ray measurements of samples with similar methodologies for calibration, analysis and reporting of the sample activity results. The reported measured activity of all isotopes in samples were decay-corrected to the end of the JET irradiation period – 20th December 2019 at 21:30:00 GMT. Besides decay-corrected activities, all laboratories also reported sample mass and measurement uncertainty. A detailed description of the measurement process and obtained results for ITER samples are presented in a separate paper [13].

III. Sample activation calculation methodology

The computational support for the neutron activation of materials in the LTIS was performed with the MCNP (version 6.2 [14]) code together with the nuclear data library FENDL-3.1d [15]. The computations were performed with an already existent MCNP model of the tokamak JET, which has been validated on several experiments in the past at JET [2,3,4,8]. The model contains all larger components inside the vacuum vessel with slight simplifications with the aim to omit the use of toroidal surfaces in the description of components. The reason is that in MCNP models, containing toroidal surfaces, the computations take on average longer than in models without torus surfaces for the same number of simulated particles. The outside of the tokamak structure is partly defined with a low density mixture of Inconel with additions, representing the equipment surrounding the torus (homogenized surrounding region) and some of the ex-vessel diagnostic equipment. The drawing of the MCNP model is presented in Fig. 3.

The neutron emissivity profile used for calculations was taken from a pre-computed TRANSP plasma transport simulation for a NBI heated JET plasma [16,17]. As in the DD plasma some DT neutrons are produced due to the burnup of tritium generated in the DD reactions, therefore the neutron source used for computational support was a combination of DD and DT neutron sources [18]. The neutron spectrum for DD and DT plasma used in the analysis was assumed to be Maxwellian, computed for a plasma with average ion temperatures of T_{th} = 20 keV. All of the neutrons were presumed to be born isotropically in direction.

A. LTIS MCNP model

To properly support the neutron activation experiment a detailed model of the LTIS was added to the validated MCNP model of the JET tokamak. The MCNP model was constructed from the CATIA model with the help of GRASP (Grasshopper-Rhino analyzing surfaces for MCNP) which was developed at the Reactor physics department of the Jožef Stefan Institute [19]. In the model of the LTIS all details were preserved and all irradiation position were filled with specified samples. Fig. 4 represents the MCNP model of LTIS used in the analysis. The bulky part of the LTIS is facing the plasma and the RADA holder is positioned closer to the plasma compared to the ACT holder.

B. Activity computation methodology

For comparison with measurements the activity of different isotopes in the dosimetry foils can be calculated using the equation [20,21]:

$$A = R\left(1 - e^{-\lambda t_{irr}}\right)e^{-\lambda t_{cool}}, \qquad (1)$$

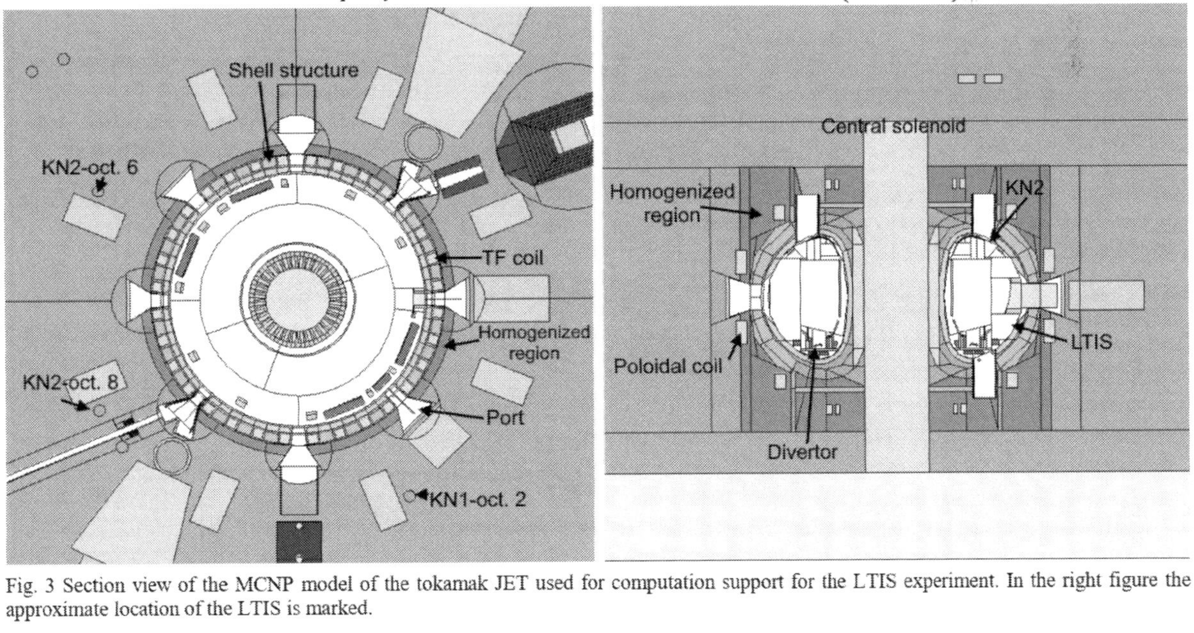

Fig. 3 Section view of the MCNP model of the tokamak JET used for computation support for the LTIS experiment. In the right figure the approximate location of the LTIS is marked.

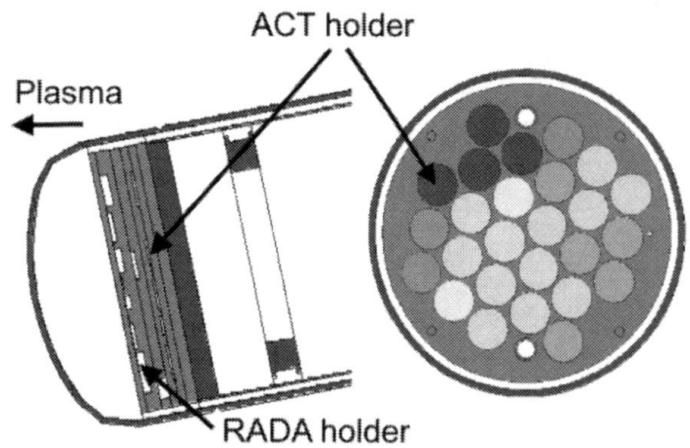

Fig. 4 MCNP model of the LTIS with marked ACT and RADA holder and direction of plasma. The ACT holder is filled with different ITER samples and dosimetry foils.

where R is the calculated reaction rate for the studied isotope in the dosimetry foil, λ is the decay constant of the activated isotope, t_{irr} is the irradiation time of the samples and t_{cool} is the cooling time. As isotopes Co-58 (half-life 70.86 days) and Sc-46 (half-life 83.79 days) have a relatively short half-life the experimental campaign cannot be represented in calculations as a continuous irradiation of 101 days followed by 46 days of cooling. To properly calculate the activity of isotopes at the end of C38 deuterium campaign the irradiation and cooling days were merged on weekly basis – 22 weeks of irradiation. The final activity of samples is calculated using the equation (1) plus the decay of activated isotopes produced in previous weeks. The equation used can be thus written as:

$$A_n = R\left(1 - e^{-\lambda t_{irr}}\right)e^{-\lambda t_{cool}} + A_{n-1}e^{-\lambda t_w}, \quad (2)$$

where n is the number of the week and t_w is time in one week. For a more accurate calculation the last day was simulated without any merger of pulses.

The reaction rates were calculated using MCNP in the modelled samples in the LTIS. The analysis presented in this paper focused on reactions presented in Table I. The cross sections for calculation of reaction rates were taken from the IRDFF-II [22] nuclear data library and JEFF-3.3 [23] nuclear data library for reactions missing in the IRDFF-II nuclear data library (reaction Ni-58(n,np)Co-57).

Table I
Studied reactions in the dosimetry foils from LTIS in the C38 deuterium campaign at JET.

Dosimetry foil material	Reaction
Cobalt	Co-59(n,γ)Co-60
	Co-59(n,2n)Co-58
Nickel	Ni-58(n,p)Co-58
	Ni-60(n,p)Co-60
	Ni-58(n,np)Co-57
Iron	Fe-54(n,p)Mn-54
Scandium	Sc-45(n,γ)Sc-46
Yttrium	Y-89(n,2n)Y-88

Despite the LTIS position being close to the neutron source and due to the very small volume of tally regions, variance reduction method was used to reduce the Monte Carlo statistical uncertainty in the thermal region of the neutron spectra for activation reactions produced predominantly by thermal neutrons (Co-59(n,γ)Co-60). This was achieved using weight windows generated by the code ADVANTG [24].

IV. Comparison of calculated and measured activity

A. Fast DT neutron reactions

As mentioned, the dosimetry reactions were chosen to cover different regions of the neutron spectra. The reactions Co-59(n,2n)Co-58, Ni-60(n,p)Co-60, Ni-58(n,np)Co-57 and Y-89(n,2n)Y-88 have a threshold energy above the energy of DD neutrons (2.45 MeV). Due to this, they can only be a result of interaction with fast DT neutrons (14.1 MeV) that are produced in the DD plasma due to tritium production in DD reactions. Thus, the analysis of these reactions represent an opportunity to determine the DT neutron fraction in the DD plasma. The cross sections for the studied reactions from the IRDFF-II and JEFF-3.3 nuclear data libraries are presented in Fig. 5.

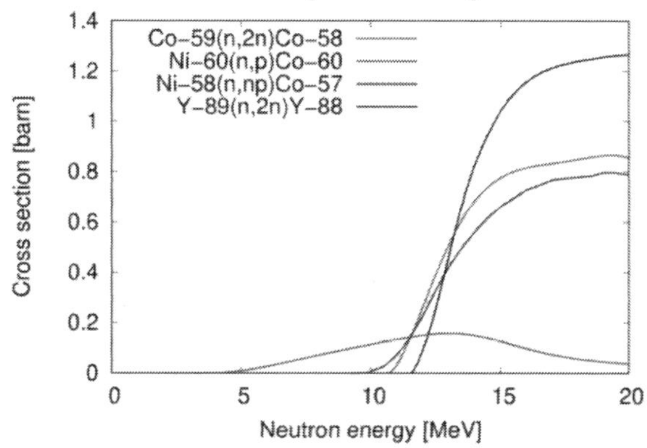

Fig. 5 Cross sections for neutron reactions with high threshold energies. The reaction Ni-58(n,np)Co-57 cross section is taken from the JEFF-3.3 nuclear data library while other cross sections are taken from the IRDFF-II nuclear data library.

In general a 1 % DT neutron fraction in a DD plasma is assumed. However, the C/E values for the activities of isotopes produced by fast reactions were below 1 in all cases. Due to this, the DT fraction was varied to obtain a better agreement between measurements and computations. The analysis showed that a 1.4 % DT neutron fraction produced the calculated-to-measured activity ratios closest to 1. The C/E values for 1.4 % DT neutron fraction for isotopes produced by fast DT neutrons are presented in Fig. 6.

The uncertainties in the C/E results presented in Fig. 6 were computed from the 2 sigma statistical uncertainties of MCNP calculations, uncertainties in measured sample activity and uncertainty of the neutron yield. The statistical uncertainty for all calculated reaction rates with MCNP was below 2 %. The major part of the uncertainty of the C/E values are thus due to the measurement uncertainties and neutron yield uncertainty.

The C/E values presented in Fig. 6 vary from 0.9 to 1.6 for reactions sensitive to fast DT neutrons, assuming the total neutron yield includes 1.4 % DT neutrons. The most significant difference between calculations and measurements is observed for cobalt samples measured by IFJ. The dispersion in the C/E

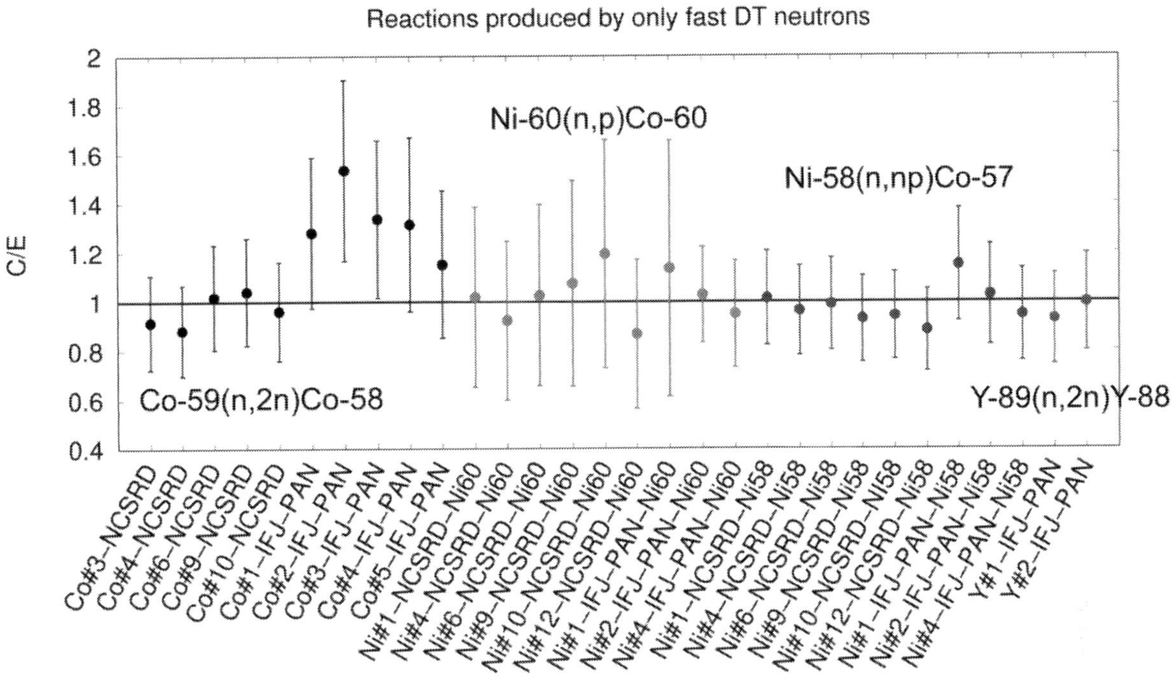

Fig. 6 Comparison of the calculated activity with measured for isotopes in dosimetry foils produced by fast neutrons - Co-59(n,2n)Co-58, Ni-60(n,p)Co-60, Ni-58(n,np)Co-57 and Y-89(n,2n)Y-88. The DT neutron fraction used in the analysis was 1.4 %. The error bars represent 2 sigma statistical uncertainty of MCNP calculations and uncertainty in measured sample activity. The number next to the material name represents the ID of the sample.

values presented in Fig. 6 can be attributed to the differences in the measurement methodology at different institutions and different size of samples measured by institutions. Despite the variations in the C/E values the computed activities are in reasonably good agreement with the measured activity by different institutions. The determined 1.4 % DT fraction was used in further analyses of reactions produced by DD and DT neutrons.

B. DD and DT neutron reactions

The reactions studied on isotopes, activated by both DD and DT neutrons were Co-59(n,γ)Co-60, Ni-58(n,p)Co-58, Fe-54(n,p)Mn-54 and Sc-45(n,γ)Sc-46. The cross sections for the reactions from the IRDFF-II are depicted in Fig. 7. Reactions Ni-58(n,p)Co-58 and Fe-54(n,p)Mn-54 are threshold reactions with thresholds at 0.8 MeV and 0.7 MeV while the Co-59(n,γ)Co-60 and Sc-45(n,γ)Sc-46 reactions are produced predominantly by thermal neutrons.

The C/E values for these reactions are presented in Fig. 8. As in the case of solely DT neutron reactions the uncertainties in the C/E values were computed from the 2 sigma statistical uncertainties of MCNP calculations, uncertainties in measured sample activity and uncertainty of the neutron yield. The MCNP statistical uncertainty for all calculated reaction rates with MCNP was below 2 %. The major part of the uncertainty of the C/E values are thus due to the measurement uncertainty and neutron yield uncertainty.

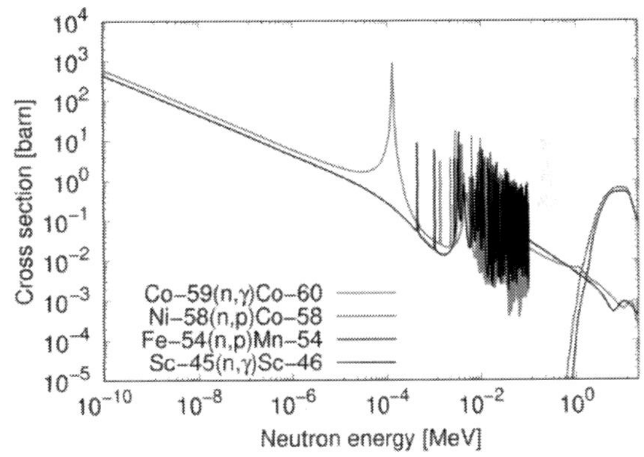

Fig. 7 Cross sections for reactions produced by DD and DT neutrons. All cross sections are taken from the IRDFF-II nuclear data library.

For the reactions, predominantly produced by thermal neutrons, the C/E values vary for cobalt samples measured by two different institutions. For the samples measured by the NCSRD show C/E values around 1, while the cobalt samples measured by IFJ show higher C/E values of around 1.5 similar to results presented in Fig. 6. Sc samples exhibit C/E values of around 1.3. As in the case of DT neutron activation reactions the Co samples measured NCSRD and IFJ were of different size.

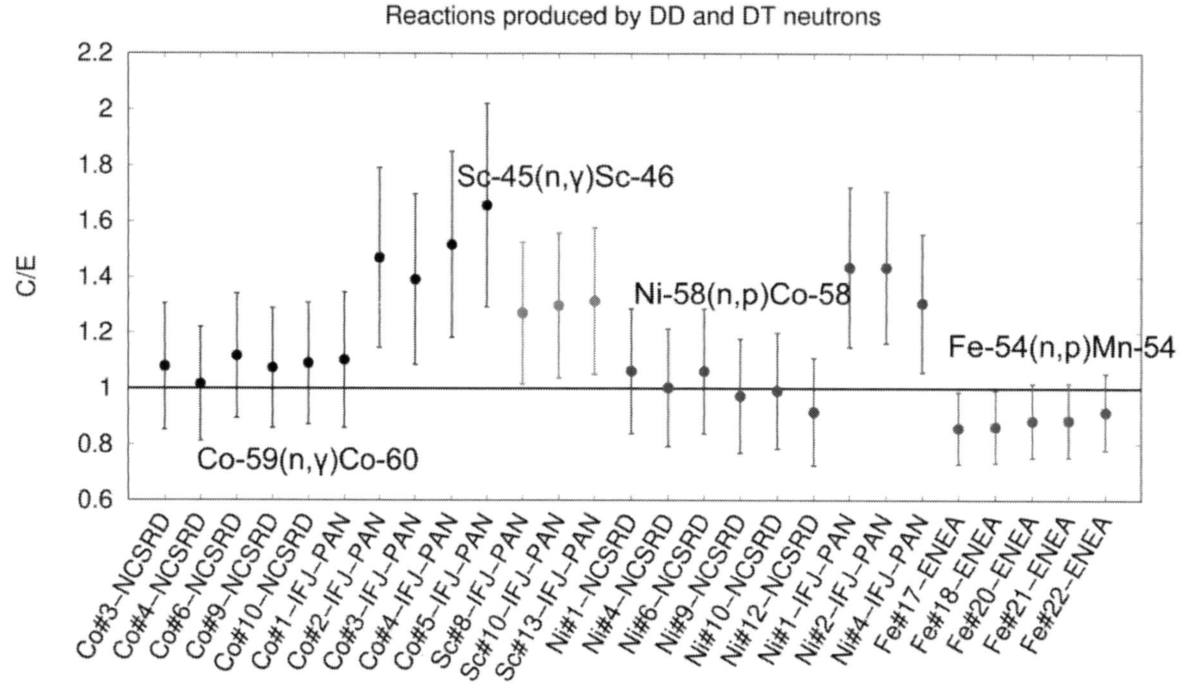

Fig. 8 Comparison of the calculated activity with measured isotopes in dosimetry foils produced by DD and DT neutrons - Co-59(n,γ)Co-60, Sc-45(n,γ)Sc-46, Ni-58(n,p)Co-58 and Fe-54(n,p)Mn-54. The DT neutron fraction used in the analysis was 1.4 %. The error bars represent 2 sigma statistical uncertainty of MCNP calculations and uncertainty in measured sample activity. The number next to the material name represents the ID of the sample.

For the Ni-58(n,p)Co-58 reaction measured by NCSRD the C/E values are around 1. The C/E values for nickel samples measured by IFJ are around 1.4. The C/E values for iron samples measured by ENEA are around 0.9. An analysis is currently under way with the aim to explain the differences in the C/E values between different institutions.

Despite the variations in the C/E values the computed activities are in reasonably good agreement with the measured activity by different institutions.

V. Conclusions

The JET tokamak currently offers one of the most suitable environments to study the effects of fusion neutron irradiation in materials. In the 2019 C38 deuterium experimental campaign several different ITER material samples and dosimetry foils were irradiated in the long-term irradiation station. The samples were exposed to a neutron fluence of $1.9 \cdot 10^{14}$ n/cm^2 during JET discharges performed in the experimental campaign over a period of 5 months.

Gamma ray spectroscopy measurements were performed on irradiated samples by several different European laboratories to determine the activation of different long-lived isotopes in the samples.

The computational methodology for calculation of activation of isotopes in different samples was performed with the MCNP code. To properly simulate activation, a detailed model of the LTIS was made. All dosimetry foils in the ACT holder were modelled and for each sample reaction rates in the sample were calculated. The activities for the different isotopes at the time the holder was removed from JET were calculated taking into account the measured neutron yield and irradiation time.

The calculated values were compared to the measured. The C/E values vary from 0.9 to 1.6 for reactions sensitive to only fast DT neutrons, assuming the total neutron yield includes 1.4 % DT neutrons. For reactions produced by DD and DT neutrons the C/E values vary from 0.9 to 1.6.

Despite the variations in the C/E values, the computed activities of isotopes in samples are in a reasonably good agreement with the measured activities by different institutions within the total uncertainties, thus validating the computational methodology and providing the basis for analysis of the ITER relevant materials and future experiments performed at JET in the Tritium-Tritium and Deuterium-Tritium campaign. Additional analysis is being performed with the aim to explain the differences in the C/E values between different institutions.

Acknowledgment

This work has been carried out within the framework of the EUROfusion Consortium and has received funding from the Euratom research and training programme 2014-2018 and 2019-2020 under grant agreement No 633053. The views and opinions expressed herein do not necessarily reflect those of the European Commission.

The work was supported by the Slovenian Ministry of Education, Science and Sport (projects codes: P2-0073 Reactor Physics; P2-0405 Fusion technologies; 1000-17-0106-6 - Training of young researchers).

References

[1] E. Joffrin, et al., "Overview of the JET preparation for deuterium-tritium operation with the ITER like-wall", Nuclear Fusion, vol. 59, 112021, 2019.
[2] A. Ćufar, et al., "Calculations to support In situ neutron yield calibrations at the joint European torus", Fusion science and technology, vol. 74, 370-314, 2018.
[3] P. Batistoni, "On the absolute calibration of neutron measurements in fusion reactors", Fusion Engineering and Design, vol, 105, 58-69, 2016.
[4] P. Batistoni, et al., "Benchmark experiments on neutron streaming through JET Torus Hall penetrations", Nuclear Fusion, vol. 55, 042017, 2015.
[5] R. Villari, et al., "ITER oriented neutronics benchmark experiments on neutron streaming and shutdown dose rate at JET", Fusion Engineering and Design, vol 123, 171-176, 2017.
[6] T. Vasilopoulou, et al., "Neutron streaming along ducts and labyrinths at the JET biological shielding: Effect of concrete composition", Radiation Physics and Chemistry, vol. 116, 359-364, 2015.
[7] J. Naish, et al., "Comparison of neutron flux streaming calculations to the 2019-2020 JET experimental deuterium-deuterium results", Fusion Engineering and Design, vol. 170, 112538, 2021.
[8] B. Kos, et al., "Application of ADVANTG to the JET3 - NEXP streaming benchmark experiment", Fusion Engineering and Design, vol.147, 1-10, 2019.
[9] I. Lengar, et al., "Characterization of JET neutron field in irradiation locations for DD, DT and TT plasmas", Fusion Engineering and Design, vol. 146, 1967-1970, 2019.
[10] L.W. Packer, et al., "Status of ITER material activation experiments at JET", Fusion Engineering and Design, vol. 124, 1150-1155, 2017.
[11] G. Stankunas, et al., "Analysis of activation and damage of ITER material samples expected from DD/DT campaign at JET", Fusion Engineering and Design, vol. 125, 307-313, 2017.
[12] L.W. Packer, et al., "Activation of ITER materials in JET: nuclear characterization experiments for the long-term irradiation station", Nuclear fusion, vol. 58, 096013, 2018.
[13] L.W. Packer, et al., "Technological exploitation of the JET neutron environment: progress in ITER materials irradiation and nuclear analysis", to be published in Nuclear Fusion.
[14] C.J. Werner, "MCNP User Manual – Code Version 6.2", Los Alamos National Laboratory, report La-UR-17-29981, 2017.
[15] IAEA Fusion Evaluated Nuclear Data Library - FENDL-3.1d, International Atomic Energy Agency, https://www-nds.iaea.org/fendl/
[16] J. Breslau, et al. "TRANSP", 2018
[17] Ž. Štancar, et al. "Multiphysics approach to plasma neutron source modelling at the JET tokamak", Nuclear Fusion, vol 59, 096020, 2019.
[18] S. Conroy, et al., "Time resolved measurements of triton burnup in JET plasmas", Nuclear Fusion, vol. 28, 2127, 1988.
[19] B. Kos and L. Snoj, "On using Grasshopper add-on for CAD to MCNP conversion", in PHYSOR, Sun Valley, USA, 2016.
[20] A. Žohar and L. Snoj, "On the dose fields due to activated cooling water in nuclear facilities", Progress in Nuclear energy, vol. 177, 103042, 2019.
[21] A. Žohar, et al., "Water activation experiments and calculations at JSI TRIGA research reactor", Proceedings: NENE2020: 29th International Conference Nuclear Energy for New Europe, 2020.
[22] A. Trkov, et al. "IRDFF-II: A New Neutron Metrology Library", Nuclear Data Sheets, vol. 163, 1-108, 2020.
[23] A.J.M. Plompen, et al., "The joint evaluated fission and fusion nuclear data library, JEFF-3.3", The European Physics Jurnal, vol. 56, 181, 2020.
[24] S.W. Mosher, "ADVANTG-An Autometed Variance Reduction Parameter Generator," ORNL/TM-2013/416 Rev.1, Oak Ridge National Laboratory, 2015.

Experimental study of ISHTAR thermostatic irradiation device for the MARIA research reactor

Maciej Lipka*, Anna Talarowska, Grzegorz Wojtania, Marek Migdal
National Centre for Nuclear Research, Poland
*corresponding author: Maciej.lipka@ncbj.gov.pl

***Abstract*—Materials and core components for the next generation power reactors technologies require testing that can be performed in existing research reactors. Such experiments employ devices dedicated to reflect the relevant thermal and neutron parameters simulating conditions present in, for example, but not limited to, HTGR reactors. A novel thermostatic irradiation device named ISHTAR (Irradiation System for High-Temperature Reactors) has been designed and constructed in the MARIA research reactor. Its mission is to enable irradiation of samples in controlled, homogeneous temperature field reaching 1000°C and inert gas atmosphere. The high temperature is achieved by a combination of electric and gamma heating, together with carefully designed thermal insulation. Additionally, samples holder made of graphite with high thermal conductivity enables the temperature homogenization in all directions. Device will be placed inside the Beryllium matrix of MARIA core and cooled with forced circulation of water from the reactor pool loop. This paper presents the outcome of experiments conducted with the rig prototype in external hydraulic mock-up of the MARIA reactor irradiation channel. The results have proved that the desired conditions for irradiation of the samples were achieved and their comparison against computational data has shown the adequacy of the design process. Finally, the loss of flow scenario was tested in protected and unprotected conditions (meaning with and without the safety system based on temperature feedback), proving the operational safety of the ISHTAR design. Experimental results will be used in the future to validate the numerical models (two and three dimensional) of the irradiation rig, providing an improved understanding of free convection and radiation phenomena modeling.**

***Keywords* —Research reactor, materials testing, Irradiation, High-temperature Reactor, thermostatic device, MARIA, HTGR**

I. Introduction

MARIA is a channels-in-pool type, water-cooled research reactor. It has a nominal power of 30 MWth with a core containing 20+ individually cooled fuel elements located in pressure tubes. Beryllium blocks and light water provide neutrons moderation. The reactor core is surrounded by the graphite reflector [1]. MARIA's core is highly flexible – fuel channels and beryllium (moderator) blocks might be freely shuffled between the reactor cycles to meet the requirements of the broad scope of irradiation programmes and experimental studies for each operation cycle. Irradiation positions with neutron flux and spectrum optimized for the thermostatic devices programme are presented in Figure 1.

As the High-Temperature Reactors (HTGR or HTR) seem to be considered by the Polish government as a meaningful branch of scientific studies in the field of nuclear energy [2][3] (however separate from the Polish Nuclear Power Programme [4]), the need for the development of thermostatic rig that will enable irradiations in the HTGR conditions emerged. As a result, a state-funded research programme "Gospostrateg-HTR" was initiated. One of its tasks was designing, building, and testing the irradiation rig in the research reactor core. The rig enables graphite samples irradiation in thermal conditions as close as possible to HTGR operating conditions, namely 1000°C and helium gas atmosphere. The initial rig design was carefully optimized to minimize heat flux in axial direction and homogenize the temperature field within the holder filled with materials samples. The optimization process, which leads to the final axial insulation arrangement, required tight cooperation between analysts and design engineers. Test of the prototype was performed to measure thermal and hydraulic conditions in the external mock-up of the vertical irradiation channel of the MARIA reactor – coolant flow correlated with the known pressure drop. When the hydraulic conditions were known (and derived from them – heat transfer coefficient), thermal measurements became possible: the temperature at various points inside and outside the rig was measured to map the temperatures and compare them with calculation results of the measurements to prove the adequacy of the design.

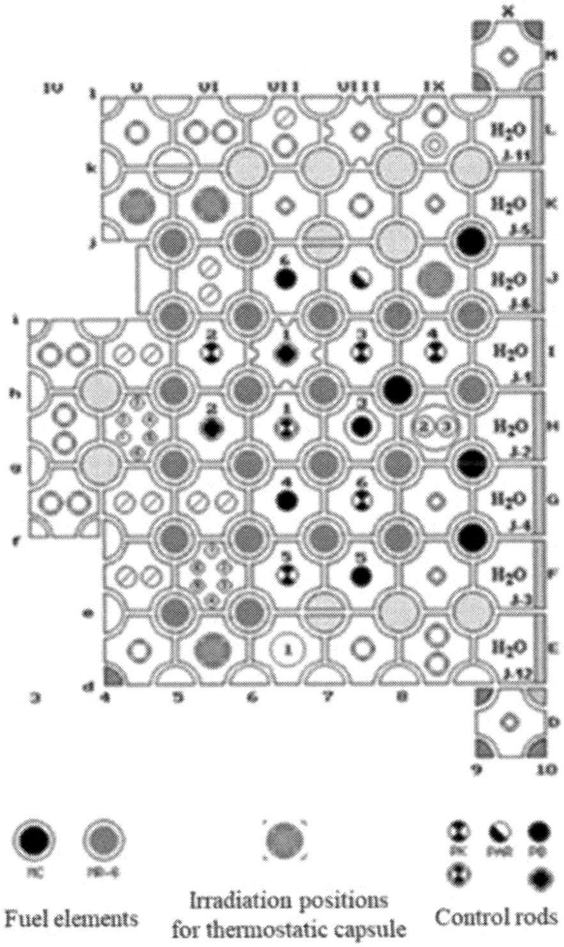

Fig. 1: Irradiation positions in MARIA reactor core

II. ISHTAR DESIGN

The ISHTAR (Irradiation System for High-Temperature Reactors) thermostatic irradiation rig ensure homogenous temperature both by nuclear and electrical heating during in-core irradiation. The prototype used for the mock-up tests has an elevated electrical heating to fill in the nuclear heating during out of the core safety testing procedures.

The main feature of the rig is the homogeneous irradiation temperature of 40 cm specimen's holder length which is quite impressive comparing to the 1 m of active reactor core height. The homogenous temperature of 1000°C was achieved by helium gas insulation layer. The outer diameter of the rig d = 48.3 mm can be easily fitted to the vertical channel which internal diameter equals 54 mm. The rig is cooled by the forced convection of water in reactor pool. The main goal of ISHTAR thermostatic device was to demonstrate the possibility of graphite specimen irradiation in high temperatures. The thermostatic device irradiates four graphite specimen, each 100 mm long. Samples from this irradiation campaign are intended for the pull tests. The samples are fitted with six thermocouples as part of the temperature monitoring inside the thermostatic device. The specimen is implemented inside the rig's tray, with cylindrical loading space of diameter 28 mm and 400 mm long. The tray encapsulating the loading space is wrapped in 7 high-temperature heaters. The heaters are placed in spiral grooves and interlaced at the bottom, allowing both cold ends of each heater directly upwards. The heater interfaces stainless steel sleeve tube's outer diameter allows maintaining 3.15 mm isolating gap between thermostatic device tray and thermostatic device outer tube. The gap is filled with helium with a pressure of 1.5 bar, providing isolation necessary to reach a temperature of 1000°C inside the tray. Thermal expansion shrinks the gap to about 2.85 mm, which was taken into account in the calculations. The upper and lower ends of the tray in the sleeve are additionally isolated by rings manufactured from Zirconium-Yttrium ceramic with low thermal conductivity of 2 $Wm^{-1}K^{-1}$ combined with the helium volumes. The whole assembly is locked within the stainless steel (AISI316) cylinder, providing structural rigidity to the assembly. The construction is presented in the Figure 2.

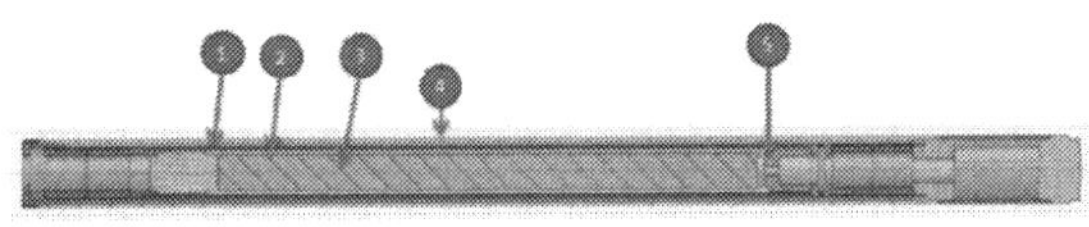

Fig. 2: Geometrical model of the in-core part of the thermostatic device.
1 – inner shell, 2 – electrical heater, 3 – samples with holder, 4 – outer shell, 5 – "crown" (lower insulation).

III. EXPERIMENTAL SETUP

Due to several major uncertainities (such as coolant velocity in the channel) as well as due to the complex geometry of the heater it is difficult to simulate the temperature field numerically within the rig. Therefore, it has been decided to perform an additional experimental investigation in order to determine cooling parameters and verify technical solutions by performing out-of-core experiments in reactor channel mock-up. It allowed to adequately test selected insulators, connectors, etc. A purpose-designed mock-up of the irradiation channel with full simulation of its cooling capacity has been therefore constructed. The mock-up called FLOW-30 is presented in Figure 3. The transparent structure of the mock-up made out of the plexiglass tubes enables observation of the flow type: laminar or turbulent after the addition of the reagent to water, also boiling and its regimes can be investigated in this manner if they occur.

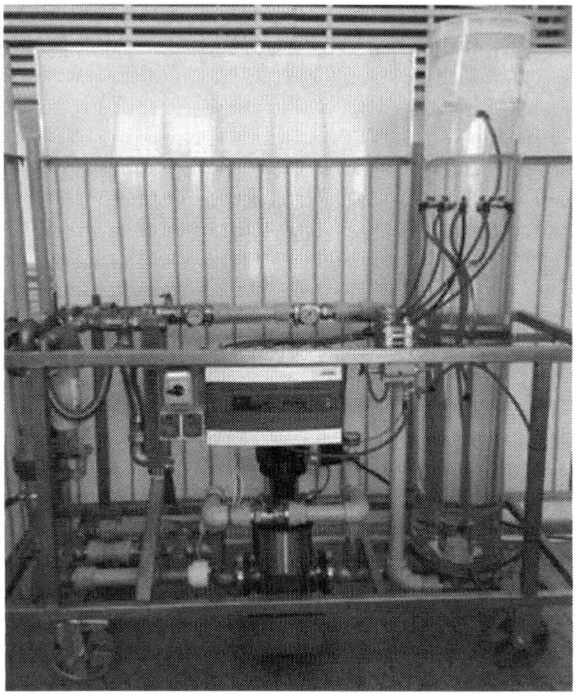

Fig. 3: FLOW-30 Experimental hydraulic stand

The experimental stand FLOW-30 allows performing thermal and hydraulic measurements in 1:1 scale, both in heat transfer and geometrical terms. The stand is equipped with water temperature and flow rate measurement and control systems. A number of pressure sensors allow to determine pressure drop. It is possible to correlate the flow rate with pressure drop and water gap thickness. The heat transfer coefficient [$W \cdot m^{-2} \cdot K^{-1}$] can be therefore determined.

Parameters of stand FLOW-30:

- Flow rate: 0÷30 $m^3 \cdot h^{-1}$
- Measurements of pressure drop: 0÷250 kPa
- Heating power: 0÷5000 W
- Cooling power: 0÷7000 W
- Measurements of vacuum on the end of the channel: -100÷0 kPa
- Nominal working fluid temperature: 0-60°C

Temperature inside of the irradiation device and the water temperature were measured with the k-type thermocouples.

IV. Methodology

Initially a set of measurements was performed in order to determine the pressure drop characteristics of the irradiation channel with thermostatic rig inside. Linear pressure drop was consistent with the smooth-pipe regime theory, however local pressure drop at the outlet of the channel due to the complexity of the geometry behaves somewhat differently and is slightly higher than in the simple 1-D theory. The channel outlet was printed in 3D and put inside of the measurement stand. This part is presented in Figure 4, as it can be seen, the geometry contains multiple outlets with different diameters, so the pressure drop is a combination of linear and local ones. Pressure drops of the whole outlet part were measured with cylinders of different diameters attached to the outlet, representing the linear part of the ISHTAR irradiation device. The results of the measurements have been presented in Figure 5. Those results, combined with the pressure drop measurements in the linear part, enabled hydraulic calibration of the irradiation channel. They were used to create a mathematical model for calculating the flow in a channel with variable geometry.

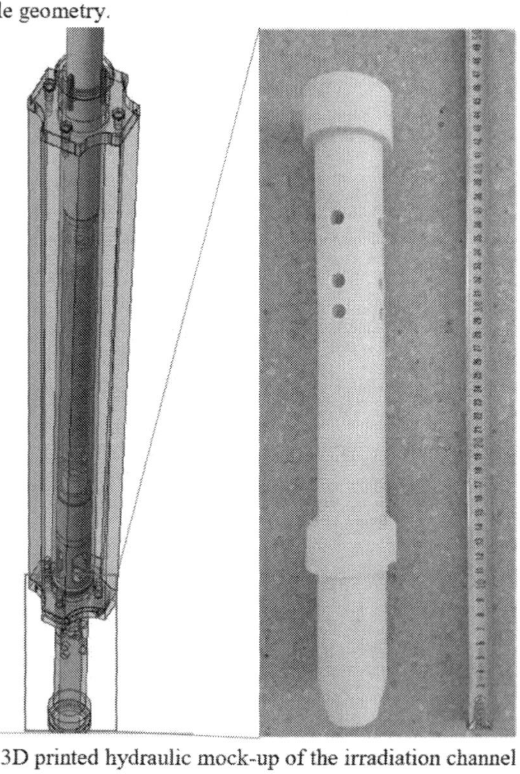

Fig 4: 3D printed hydraulic mock-up of the irradiation channel outlet. Multiple outflows visible. On the left the whole irradiation channel, filled with the ISHTAR device is presented.

In the second part of the measurement, the coolant flow (7 m^3/h) in the channel was set to be consistent with the known pressure drop in the channel (1.4 kPa). ISHTAR irradiation device was placed in the channel and the electric power was gradually increased to 7.2 kW at which the desired temperature >1000°C was achieved in the sample's region. The time series (temperatures and the electric heaters power) of the experiment are presented in Figure 6. During the measurement, thermocouples T1, T2 and T3 were put inside the graphite filler, Twall was welded in a special groove inside the wall.

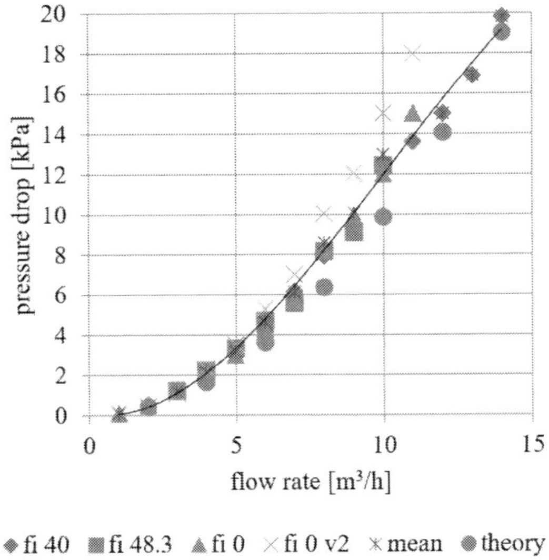

Fig 5: Pressure drop measurements of the irradiation channel outlet.

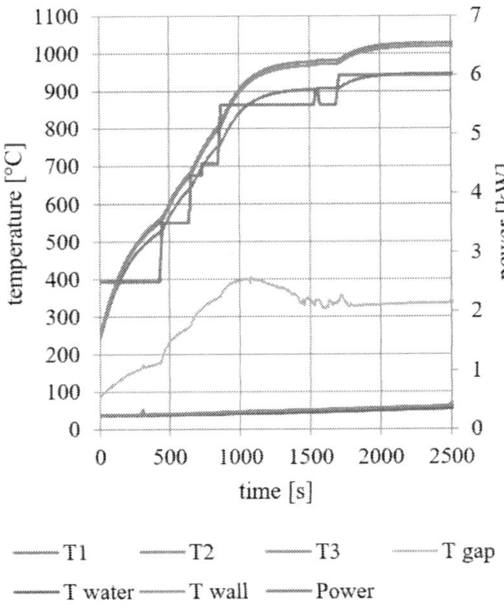

Fig. 6: Time series of the experiment.
T1, T2, T3 – temperatures in the samples region, T gap – temperature in the insulation helium gap, T wall – outer wall temperature, T water – cooling water temperature, power – electric power

V. Results discussion

When steady state was achieved, comparison of the computations with the measurements results became possible. As it was presented in figures 7 and 8, both 1D and 3D calculations proved to be in-line with the measurements. As expected, temperature inside the insulation gap was rising alongside the electrical heaters power increase. Temperature of the helium gap was initially rising and then stabilized at the level 330°C, probably due to the stabilization of the natural circulation cell within the gas. Heat transfer coefficient was consistent with the theoretical one, calculated from the pressure drop measurements with Gnielinski formula [5] .

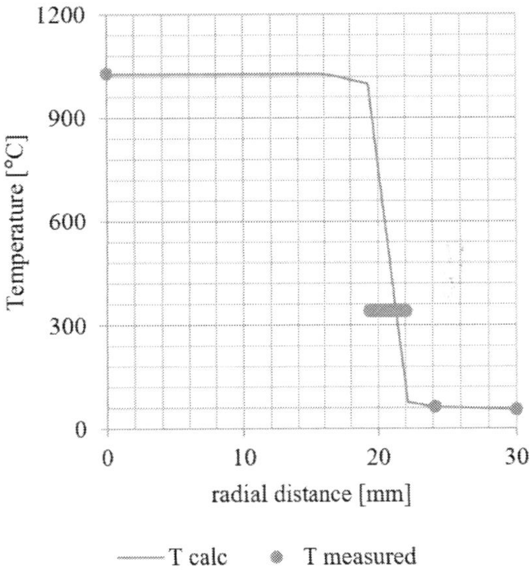

Fig. 7: 1D calculation comparison with temperature measurements. One of the measurement points in the gas gap not well defined.

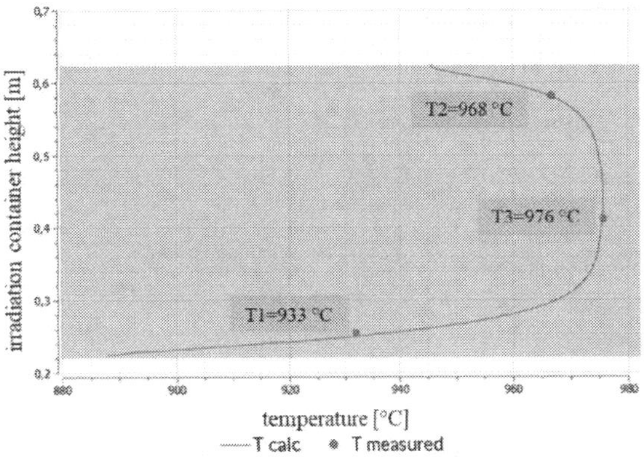

Fig. 8: 3D calculation comparison with temperature measurements

VI. Conclusions

Measurements of the pressure drop in the irradiation channel filled with the thermostatic device were performed., followed by thermal measurements of the temperature field inside and outside the ISHTAR capsule. The temperature measurements results proved the adequacy of the computational methods and the design. ISHTAR thermostatic irradiation device is able to achieve desired parameters, that is, temperature >1000°C. The ISHTAR device will be tested inside the core of the MARIA reactor in the fourth quarter of 2021.

Acknowledgment

This work was supported by the National Centre for Research and Development (NCBR) as a part of the GOSPOSTRATEG-HTR programme.

References

[1] Prokopowicz, R., Pytel, K. (2016). Determination of nuclear fuel burn-up axial profile by neutron emission measurement. Nuclear Instruments and Methods in Physics Research Section A: Accelerators, Spectrometers, Detectors and Associated Equipment, 838, 18–23. DOI: 10.1016/j.nima.2016.09.021

[2] https://www.gov.pl/web/klimat/realizacja-prac-projektowych-reaktora-htgr--umowa-podpisanaProgram polskiej energetyki jądrowej

[3] Talarowska A., Lipka M., Wojtania G.,(2021), Preliminary computational and experimental design studies of the ISHTAR thermostatic rig for the High-Temperature Reactors materials irradiation., Nukleonika, (Reviewed)

[4] Official Gazette of the Government of the Republic of Poland, M.P. poz. 946, 6.10.2020. Polish Nuclear Power Programme

[5] Y.A. Cengel, "Internal Forced Convection", in *Heat Transfer: A Practical Approach 2nd ed. USA: McGraw-Hill, 2002. Pp 419-458*

Characterization of the X-ray spectrum of a linear accelerator

Maëva Maulin[1], Pierre Guy Allinei[1], Daniel Eck[1], Nicolas Estre[1], Emmanuel Payan[1], David Tisseur[2], Grégoire Kessedjian[2]
[1]CEA, DES, IRESNE, DTN, Cadarache F-13108 Saint-Paul-lez-Durance, France.
[2]CEA, DES, IRESNE, DER, Cadarache F-13108 Saint-Paul-lez-Durance, France.
Maeva.maulin@cea.fr

***Abstract*—The Nuclear Measurement Laboratory (LMN) at CEA Cadarache in France uses a high-energy electron linear accelerator, LINAC (9-21 MeV), to characterize nuclear waste drums. It enables to explore new examination modalities, such as active photon interrogation or dual-energy CT to scan large concrete objects with diameters up to 140 cm. These techniques require precise awareness of the photon spectrum emitted by the LINAC. However, direct measure of this photon energy spectrum cannot be achieved because of the accelerator pulses causing detector saturation. During the last few years, a large number of indirect methods has been developed. From an experimental point of view, the simplest indirect method for spectrum estimation method is ransmission measurements. Because it can be set up easily and accurately using an ionization chamber as well as an appropriate screen. The obtained transmission curve depends on the photon energy spectrum, which can be estimated using inverse models.**

In this paper, we present the development of a numerical model to determine the energy spectrum from an attenuation curve via transmission measurements which combines two types of inverse models: a continue model and a discrete model.

We validate this tool using a test spectrum and its transmission curve obtained via Monte-Carlo simulation. This qualification allowed us to determine its sensitivity (signal-to-noise ratio, SNR) in order to have a good convergence. We show that if the SNR is less than 4%, we have a good estimation of the photon energy spectrum. Then, it was experimentally tested with a transmission curve obtained at the laboratory.

***Keywords*—Linear electron accelerator, High energy computed tomography, Spectrum characterization, LINAC**

I. INTRODUCTION

FOR the last few years, the characterization of energy spectra from linear accelerators becomes a major issue for medical sector (radiotherapy) or industry (to develop new technics of non-destructive testing). The Nuclear Measurement Laboratory (LMN) at CEA Cadarache has an imaging system currently implemented in the CHICADE experimental platform allowing radiographic and tomographic analyses on large objects (Fig. 1). A major upgrade of this system was launched involving the provision of a new high-energy accelerator called SATURNE, with photon energy up to about 15 MeV [1]. The latter is made of new elements and elements from an older accelerator (such as accelerating waveguide), consequently its characteristics, like the electron spectrum or the photon maximal energy, are not accurately know.

This high-energy accelerator allows exploring new examination modalities, such as active photon interrogation (IPA) [2, 3] or dual-energy CT [4]. These techniques require precise awareness of the photon spectrum emitted by the LINAC.

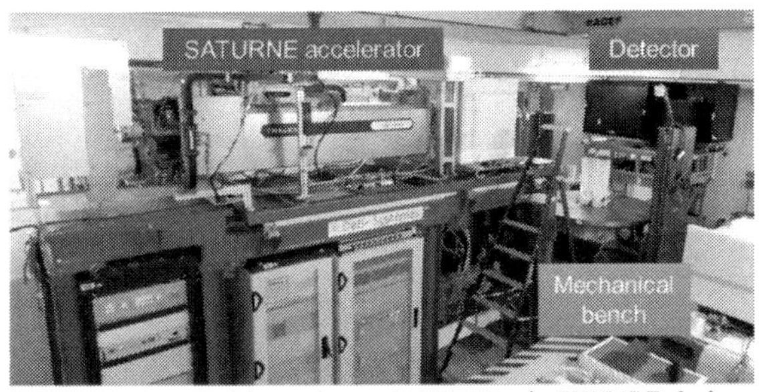

Fig. 1. Panoramic view of the accelerator SATURNE in the CHICADE platform (July 2019)

However, direct measurement of high-energy linear accelerators photon energy spectrum (for example, with a hyper-pure germanium spectrometer) is difficult to implement. The SATURNE accelerator used by the laboratory is a pulsed accelerator with 5 μs pulses during which the photon flux produced is very high. Spectrometers placed directly in the output beam saturate during the pulse, making impossible the direct measurement.

Consequently, to characterize the photon spectrum of high-energy accelerators, a large number of indirect methods have been developed.

One indirect method employs Monte-Carlo simulation to determine without experimental protocol the accelerator photon energy spectrum [5]. Nevertheless, this method requires a precise knowledge of the accelerator characteristics including electron beam energy spectrum, which is difficult to assess accurately experimentally or numerically. A second indirect method is based on an experimental protocol using energy spectrum from the spectroscopy of Compton-scattered photons [6]. Photons emerging from the accelerator interact by Compton scattering in a screen. The Compton spectrum measurement at a specific scattering angle can be connected directly to the photon spectrum by a linear correlation between the incident photon energy and the recorded photon position. The main drawback of this method is that it requires a complex experimental set-up. Indeed, a massive shielding is necessary in order to avoid detector saturation and the consequent loss of resolution. Another experimental method, based on photo-activation, is based on the photonuclear reaction cross-section of different materials according to energy and radioisotopes produced by these reactions [7]. Screens of different materials, with well-known photonuclear cross-section and nuclear reaction, are irradiated. The type and quantity of radioactivity produced by this irradiation for the different screens depends on the photon energy spectrum. This method requires a large number of screens of known composition, not always available at the laboratory.

The simplest indirect method from an experimental point of view is the spectrum estimation method from transmission measurements. It consists on dose rate measurements through more or less attenuating known materials, leading to the construction of a transmission curve [8, 9]. With this curve and attenuation coefficients depending on the energy for the material used, the energy spectrum can be estimated using inverse models. Attenuation measurements can be set up easily and accurately using an ionization chamber of small volume as well as an appropriate screen. Moreover, spectrum estimation from transmission measurements requires only the use of an adapted mathematical inverse model. It is for these reasons that the spectrum estimation from transmission measurements was chosen in our study.

II. Numerical Models

Relative transmission $T(x_i)$, at attenuator thickness x_i can be expressed by (1).

$$T(x_i) = \frac{S(x_i)}{S(x_0)} \quad (1)$$

With $S(x_0)$ the detector signal without an attenuating screen. If the energy is partitioned into m discrete energy channels E_j, then it may be possible to trace the energy spectrum of the photons, $\varphi(E_j)$, for example via an inverse process, from the transmission measurements, with the equation (2).

$$T(x_i) = \frac{\sum_{j=1}^{m} R(E_j).\varphi(E_j).\exp(-\mu(E_j).x_i)}{\sum_{j=1}^{m} R(E_j).\varphi(E_j)} \quad (2)$$

With $\varphi(E_j)$ the value of energy spectrum for energy E_j, $R(E_j)$ the detector response and $\mu(Ej)$ the linear attenuation coefficient of the attenuator screen at energy E_j [10].

Two methods for energy spectrum estimation from the transmission measurements have been selected:

- discrete numerical model,
- continuous numerical model.

Discrete models are the most widely used in the literature [8]. The searched spectrum is initialized via a given function (often triangular shape) which is then corrected by an iterative process in order to minimize the cost function summing the quadratic deviations between the experimental transmission curve and the estimated spectrum transmission. The final spectrum will then be estimated by solving the linear transmission equations from (2).

Continuous models [5] are based on mathematical equations and the fact that different photon spectra obtained from linear electron accelerators always have a similar shape (Bremsstrahlung spectrum), where only the energy of the peak flux and the width of the spectrum vary. The photon flux initially increases monotonically with the energy to a maximum (peak flux), and then decreases monotonically to the maximum photon energy (Fig. 2) equal to the maximum electron energy.

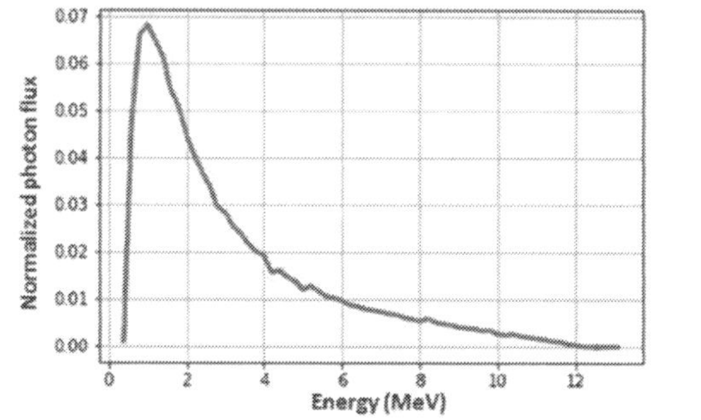

Fig. 2. Energy spectrum obtained by Monte-Carlo simulation of the SATURNE accelerator, for monoenergetic electrons of 12 MeV.

Some work has proposed the use of a continuous model where these changes can be represented by a lognormal function [5]. Then, the energy spectrum can be estimated from the lognormal function equation in discrete form for the different energy channels via (3).

$$\varphi(E_j) = \exp\left[-\frac{(\ln(E_j)-\mu_{logn})^2}{2.\sigma_{logn}^2}\right] \quad (3)$$

With μ_{logn} the expectation and σ_{logn} the standard deviation of the lognormal function which will be updated at each iteration. The estimated energy spectrum is used to calculate the corresponding transmission curve through the attenuating screen thicknesses and compare it with the experimental transmission curve.

For both models, the estimated spectrum is obtained by minimizing the cost function defined as the sum of the squared deviations between the reference and the estimated spectrum transmission curve for each attenuator screen thickness.

A. Advantages and drawbacks of the two models

The discrete model requires a hypothesis on the initial shape of the spectrum, based on the knowledge of the peak flux and the maximum energy of the photons, which can strongly influence the result. Indeed, the accelerators used in the literature (especially in the medical field) are standard and very well characterized. The maximum energy of the photons as well as the energy of the flux peak are known. For example, in Orman Zaid Assatel's thesis [8], the initialization is done with a triangular distribution which requires to fix the peak energy of this function corresponding to the peak flux energy of the searched energy spectrum. Since in this case the energy of the flux peak is known, the energy of the peak of the triangular function can be fixed and a monotonicity condition can be added to constrain the spectrum to be monotonically increasing before this peak and monotonically decreasing after this peak. In this way, the peak flux energy and the maximum photon energy are fixed when the spectrum is initialized and will not change during the estimation. The model used without the monotonicity condition then gives too much freedom in the estimation of the spectrum and therefore provides spectra whose shape does not make physical sense. However, when initialized with a function close to the spectrum to be estimated, this model gives very good results thanks to a channel by channel adjustment.

With the continuous model, initializing the spectrum with a lognormal function avoids hypotheses about the peak flux energy and allows realistic spectra to be obtained that are close to the simulated reference spectra. However, it imposes a lognormal distribution shape. The continuous model therefore gives an initial idea of the general shape of the spectrum without any prior hypothesis. It gives good results at low energies and provides the energy of the peak flux of the photon spectrum.

B. Combination of both models

The challenge of our study is to characterize the energy spectrum of an accelerator without first knowing the shape of the spectrum to be estimated. In view of the disadvantages and advantages of the two models, a combination of the two would allow this characterization to be carried out. The continuous model is run first in order to have a first spectrum shape and the energy of its flux peak. The discrete model is then initialized with the spectrum estimated via the continuous model (instead of the triangular function). Then, the condition on monotonicity before and after the flux peak can be added.

The thickness of the Bremsstrahlung target being very well characterized (5 mm), it allows a first estimation of the maximum energy of the photons as a function of the measured dose from Monte-Carlo calibration simulations where the electrons are monoenergetic. As this energy is not accuratly known, it cannot be fixed but will vary randomly during the different iterations within a limited range compared to the estimate (approximately ± 2 MeV, i.e., 8 channels of 0.43 MeV). Furthermore, the peak energy found with the continuous model will also vary within a small range around the initial value (about ± 0.5 MeV or 2 channels of 0.43 MeV).

C. Selection of the attenuating screen

Spectrum evaluation method based on transmission curve requires the use of different thicknesses of attenuating screen. The choice of the attenuator screen can be made on the basis of its total linear attenuation coefficients [8]. The variation of the linear attenuation coefficient of the chosen material must be monotonous over the energy range of the spectrum of interest in order to avoid two energies having the same linear attenuation coefficient value. In our case, the attenuating screen chosen is aluminum 5754 alloy. Indeed, its linear attenuation coefficient is monotonous over the energy range of the accelerator (up to 17 MeV) and it is inexpensive so easily procurable in quantity. Moreover, its density of 2.68 g.cm^{-3} limits the total thickness required. It takes about 60 cm to obtain 1% transmission at 15 MeV instead of 150 cm of PMMA (Polymethyl methacrylate) for example.

III. Validation of the model by simulation

A. Model parameters

In order to validate the method with conditions close to the experimental ones, a modelling of an operating point of the SATURNE accelerator is carried out with the MCNP6.1 Monte-Carlo calculation code [11] allowing to obtain a known photon spectrum. The electrons are monoenergetic with an energy of 12 MeV corresponding to the maximum energy of the photon spectrum. From this simulated spectrum, a transmission curve is generated via (2) through 30 plates of 2 cm thick aluminum 5754 alloy, i.e., 60 cm. As the resulting curve has 30 measurement points, the spectrum must have a maximum of 30 energy channels.

The spectrum is estimated over a range from 0.2 MeV to 13.1 MeV with channels of 0.43 MeV (30 channels). Indeed, the lower limit of 0.2 MeV is justified by the fact that below 0.4 MeV, the photons have a contribution to the spectrum of less than 4.10^{-3} %, which is 40 times smaller than the average uncertainty of our experimental dose measurements of 0.26 % (cf paragraph 4). The final channel will vary randomly between 9.66 MeV and 13.1 MeV (i.e. 8 channels of 0.43 MeV).

One of the parameters in equation (2) of the transmission is the ionization chamber response, $R(E_j)$. Experimentally, the detector used for the dose measurements is a small volume (0.125 cm^3) ion chamber of the PTW brand (type 31010) [12]. The response of this ionization chamber is determined by Monte Carlo simulations (MCNP6.1) where, for each incident photon energy, the average deposited energy is calculated. This response of the ionization chamber, shown in Fig. 3, is used in our algorithms.

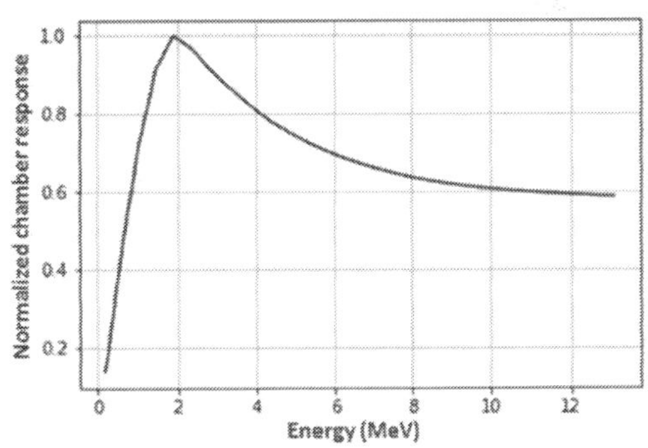

Fig. 3. Response curve $R(E_j)$ of the ionization chamber as a function of energy.

However, we have checked that the convergence of the method depends only slightly on the response used, but the response must be consistent with the transmission measurements. The convergence is the same within 10% even we use a monotonic increasing or flat response.

B. Results and validation by simulation

The method described in section 2.2 is launched 5000 times to obtain 5000 estimated spectra. For each of these spectra, 2000 iterations are performed during the channel by channel modification. The number of iterations has been fixed so that convergence is always obtained. Indeed, the convergence of the method always happens before 2000 iterations as can be seen on the example in Fig. 4.

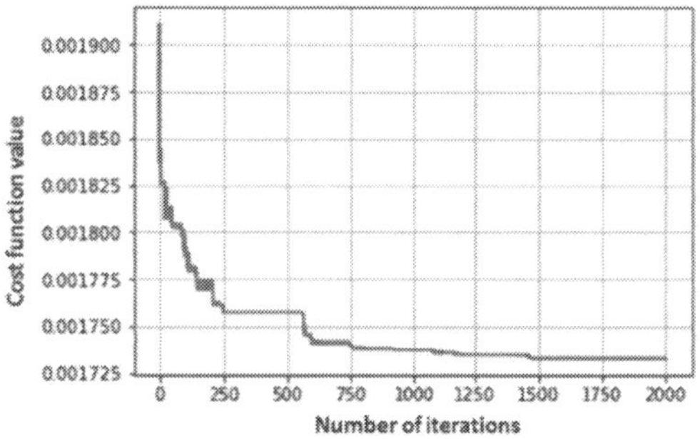

Fig. 4. Variation of the cost function as a function of iterations on a given spectrum.

The energy of the final channel is set randomly at the beginning of each spectrum estimation. Among all spectra obtained, 10 % (i.e. 500 spectra) with the lowest cost function, and therefore the best convergence, are kept (the choice to keep only 10 % of the spectra is detailed in paragraph 5). The energy of the end channel of the estimated final spectrum, E_{max}, is determined via (4) with N equal to 10 % of the 5000 estimated spectra, i.e. 500 here, and $E_{end\ chan}$ (i) the energy of the end channel of spectrum i. The end channel of the estimated spectra is defined as the mean last non-zero channel.

$$E_{max} = \frac{1}{N} \cdot \sum_{i=1}^{N} E_{end\ chane}\ (i) \tag{4}$$

The average energy of the final channel is here 12.65 ± 0.02 MeV. The energy of the end channel of the reference spectrum is 12 MeV so the average energy found with our method is correct within 2 channels (0.65 MeV).

The final spectrum is estimated by averaging the spectra whose final channel corresponds to the channel of E_{max} or the channel closest to this energy. In the end, 213 spectra corresponding to this condition are averaged to obtain the estimated final spectrum, shown in blue in Fig. 5 (in linear Fig. 5. (a) and logarithmic Fig. 5. (b) scales). The red spectrum is the reference spectrum from Monte-Carlo simulations.

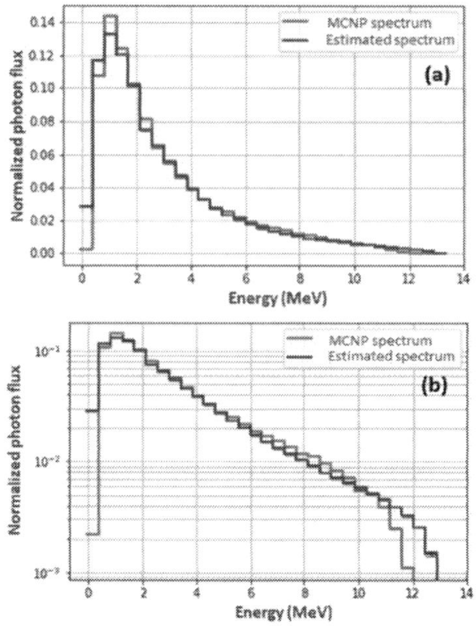

Fig. 5. Estimated average energy spectrum in linear (a) and logarithmic (b) scales with the green error bars representing the uncertainty of each channel.

For each channel the uncertainty is represented by the green error bars in Fig. 5. The relative average uncertainty per channel with respect to the average spectrum is $1.3.10^{-4}$. This is defined by (5), where m is the number of channels in the energy of the spectra, x_{ji} is the flux value of channel j for spectrum i, $\bar{x}_j$ is the average flux of channel j for the n number of retained spectra, here n=213.

$$\sigma_{spectrum\ average} = \frac{\sum_{j=1}^{m} \sqrt{\frac{1}{n_j} \sum_{i=1}^{n} (x_{ji} - \bar{x}_j)^2}}{m.\sqrt{n}} \tag{5}$$

The spectra are therefore relatively close to each other. Moreover, the estimation has converged well since graphically the reference transmission curve from the models and that of the estimated average spectrum (Fig. 6. (a)) are very close: the relative deviations between these two curves (defined by (6), with T the transmission) are between $-1.6.10^{-4}$ and $2.5.10^{-4}$ and the average relative deviation per point is $-2.1.10^{-5}$ (Fig. 6. (b)).

$$Relative\ deviation = \frac{T_{estimated\ spectrum} - T_{reference\ spectrum}}{T_{reference\ spectrum}} \tag{6}$$

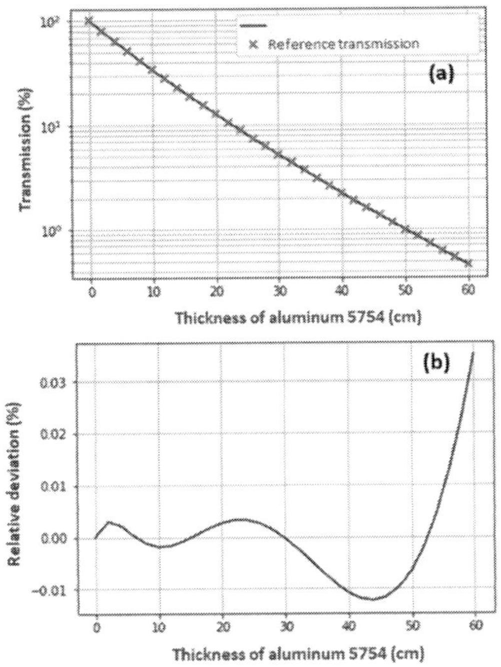

Fig. 6. (a): Transmission curve from modelling (red dots) and transmission curve from the estimated spectrum (blue) - (b): Relative differences between the two transmission curves as a function of energy.

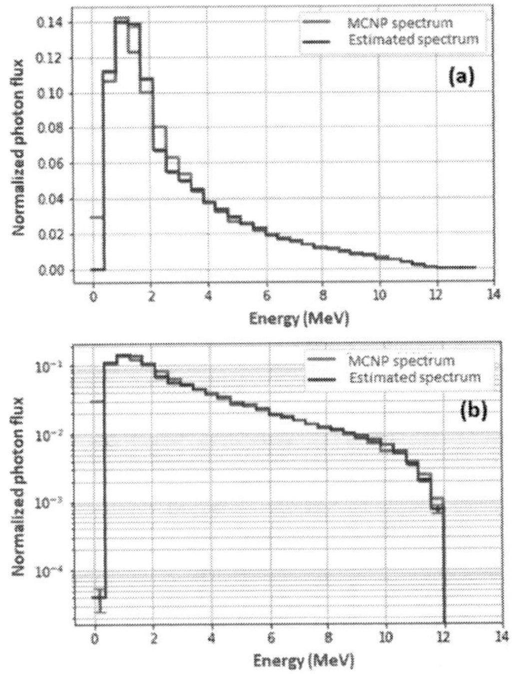

Fig. 7. Estimated average energy spectrum in linear (a) and logarithmic (b) scales after noise suppression of the reference transmission, with the green error bars representing the uncertainty of each channel

C. Adding Gaussian noise

In order to finalize the validation of our method, a Gaussian noise is added to the transmission curve resulting from the modelling. The standard deviation of this normal distribution is fixed at $\sigma = 2$ % in order to be higher than the maximum uncertainty on the experimental measurements which is 1.35 % with a mean uncertainty of 0.26 % (cf. paragraph 4).

The method described in section 2.2 is performed 5000 times. Among all spectra obtained, 10 % with the lowest cost function are chosen (i.e. 500 spectra). The average energy of the end channel is 11.75 ± 0.04 MeV (defined by (4)). The estimated average final spectrum is shown in Fig. 7 (in linear Fig. 7. (a) and logarithmic Fig. 7. (b) scales). The red spectrum is the reference spectrum from Monte-Carlo simulations.

For each channel the uncertainty is represented by the green error bars in Fig. 7. The average uncertainty per channel with respect to the average spectrum is $3.5.10^{-4}$ (defined by (5)), which is twice as large as for the noiseless estimate of the transmission curve. The noisy reference transmission curve and the estimated average spectrum curve are plotted in Fig. 8. (a).

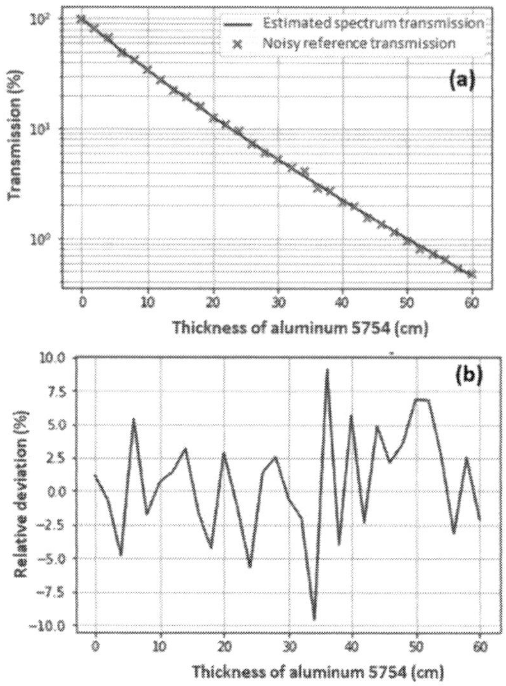

Fig. 8. (a): Transmission curve from modelling with noise (red dots) and transmission curve from the estimated spectrum (blue) - (b): Relative differences between the two transmission curves as a function of energy.

The relative deviations of the two transmission curves (defined by (6)) are between $-9.6.10^{-2}$ and $9.1.10^{-2}$, with an average relative deviation per point of $3.4.10^{-2}$ (Fig. 8. (b)). This deviation, which is much greater than for the previous part without noise (which was $6.1.10^{-5}$), is explained by the noise of the reference transmission curve, which is therefore no longer monotonous.

The Gaussian noise added to the reference transmission curve with a standard deviation of 2 % encompasses the uncertainty in the experimental measurements (which is at most 1.35 %). Nevertheless, the estimated spectrum has a shape close to the expected spectrum. The estimation method developed is therefore validated for a spectrum derived from Monte-Carlo simulations and can be used to estimate the spectrum of the SATURNE accelerator.

IV. EXPERIMENTAL SET-UP

The transmission measurements are performed with the SATURNE accelerator through 30 aluminum 5754 alloy plates of 2 cm thickness and 10 cm side. The experimental setup is shown in Fig. 9.

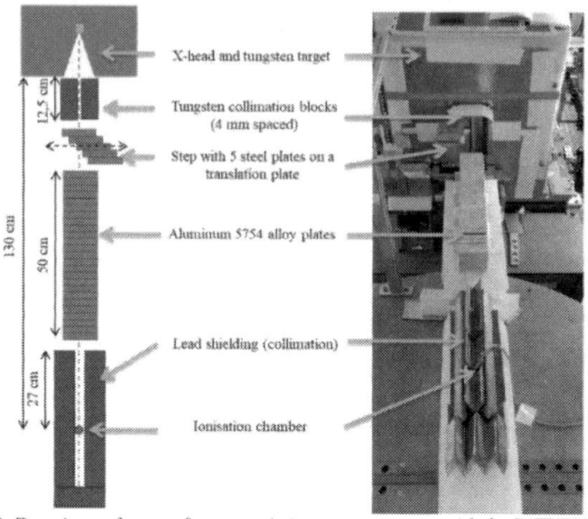

Fig. 9. Experimental set-up for transmission measurements with the SATURNE accelerator.

Dose measurements are performed with a small volume (0.125 cm^3, type 31010) PTW ionization chamber [12] surrounded by its 3 mm PMMA cap, located after the attenuator screen plates. In order to correct for any variation in dose rate during the measurements, a CdTe probe is integrated next to the target. The measured signal is used to normalize the signal obtained with the ionization chamber (beam monitoring). At the output of the accelerator, two tungsten blocks spaced 4 mm apart allow the beam collimation. The ionization chamber is also collimated with lead, including two 20 cm blocks separated by 1 cm in the direction of beam incidence.

For each aluminum thickness, 120 dose measurements are made and averaged. The average uncertainty of the 3600 measurements made is 0.26 % with a maximum of 1.35 %. In order to make the measurement faster by reducing the number of trips to remove each aluminum plate, a step, made up of 5 plates staggered by one centimeter, is placed on a translation plate that can be remotely controlled. Thus, moving the step by one centimeter simulates the removal of a plate. Every measurements, 5 aluminum plates are removed. The experimentally obtained transmission curve is shown in Fig. 10 (with the measurement uncertainty for each thickness in light blue).

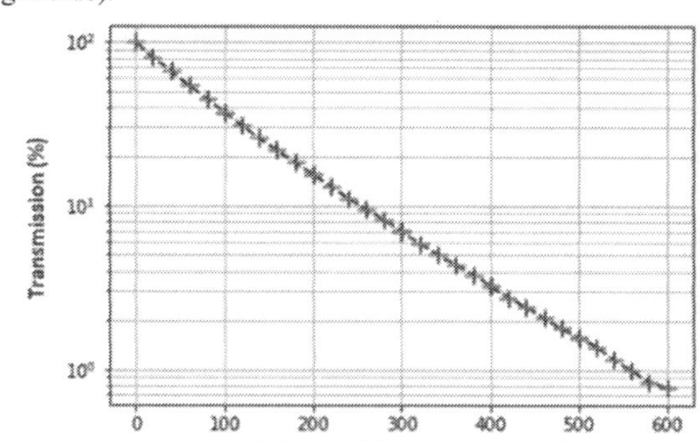

Fig. 10. Experimental transmission curve obtained with the SATURNE accelerator and 30 plates of 2 cm aluminum 5754 alloy in linear scale.

 https://doi.org/10.1051/epjconf/202125304002

The method described in section 2.2 is executed and 5000 spectra are obtained. They are sorted according to their increasing cost function in order to select only the 10 % (i.e. 500 spectra) with the best convergence. The end channel energy of the estimated final spectrum is taken to be equal to the average end channel energy of these 500 spectra.

The average energy of the final channel is 11.61 ± 0.4 MeV (defined by (4)). The spectrum is estimated by averaging the spectra whose final channel corresponds to the channel of the mean energy or to the channel closest to this energy. In the end, 77 spectra match this criterion and are averaged to obtain the estimated final spectrum shown in Fig. 11.

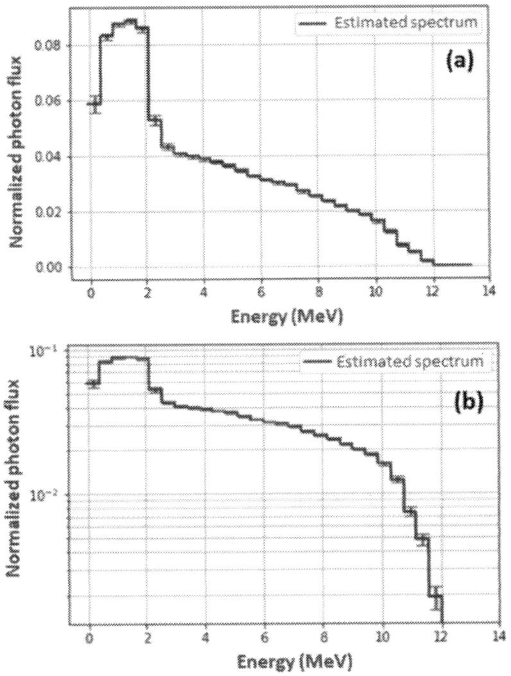

Fig. 11. Estimated mean energy spectrum, in linear (a) and logarithmic (b) scales, from the experimental transmission curve obtained with the SATURNE accelerator, with the green error bars representing the uncertainty of each channel.

For each channel, the uncertainty is represented by the green error bars in Fig. 11. The average uncertainty per channel with respect to the average spectrum is $7.1.10^{-4}$ (defined by (5)).

This value is close to the average uncertainty found when estimating the spectrum from the noisy reference transmission curve (which was $3.5.10^{-4}$). The experimental transmission curve and the estimated average spectrum curve are plotted in Fig. 12. (a).

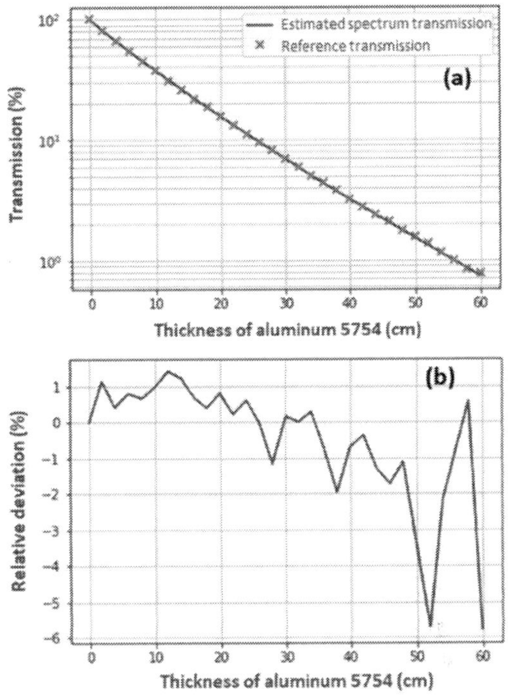

Fig. 12. (a): Experimental transmission values (red dots) and estimated spectrum transmission curve (blue) - (b): Relative differences between experimental and estimated values as a function of energy.

The relative deviations of the two transmission curves (defined by (6)) are between $-5.4.10^{-2}$ and $1.4.10^{-2}$, with an average relative deviation per point of $-4.9.10^{-3}$ (Fig. 12. (b)). This deviation is also close to that found with the noisy reference transmission curve (which was $3.4.10^{-2}$).

V. Discussion

Fig. 13 provides a visual comparison of the estimated spectrum from the experimental measurements and the spectrum from Monte-Carlo simulations. The estimated spectrum looks different from the simulated spectrum with a much slower decay after the flux peak. However, as the SATURNE accelerator is composed of an assembly of elements of different origins, its characteristics are not precisely known, so the models probably do not reflect reality. Moreover, it is unlikely that the electrons are monoenergetic.

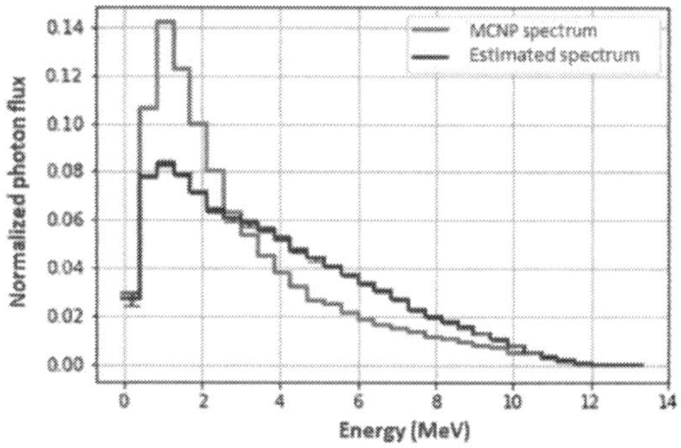

Fig. 13. Comparison of the average energy spectrum estimated from the experimental transmission curve obtained with the SATURNE accelerator (in blue) and the reference spectrum from the MCNP6.1 simulations (in red).

For this study, 10% of the spectra are selected for the estimation of the energy spectrum shape. This value is the result of an optimization between the execution time of the method to have a good convergence and the accuracy of the results. For example, for the spectrum estimated from the transmission curve from the simulations:

- By keeping 5 % of the spectra instead of 10 %, there are 102 spectra kept for the estimation of the spectrum shape instead of 213, i.e. 70 % less spectrum for the same execution time. But the average value of the cost function of the estimated spectrum decreases by 50 % (better accuracy).
- By retaining 15 % of the spectra instead of 10%, there are 321 spectra retained for spectrum shape estimation instead of 213, i.e. 40 % more spectrum for the same execution time. But the average value of the cost function of the estimated spectrum increases by 34 % (worse accuracy).

Finally, the discrete model used in our method has also been implemented with the simulated annealing method [13] which gives equivalent results with a 5 times higher convergence time. For this reason, this method has not been retained in our study.

VI. Conclusions

In this paper, the results of the development of an experimentally simple and robust tool for characterizing the photon spectrum of a high-energy accelerator via transmission measurements are presented. The combination of a discrete and a continuous numerical model enables the estimation of the energy spectrum as well as the maximum photon energy (to the nearest 0.65 MeV, compared to the reference spectrum derived from Monte Carlo simulations) of an accelerator whose characteristics are not precisely known. The method has been developed and tested using Monte-Carlo simulations. The shape of the estimated spectrum is physically realistic and very close to the reference spectrum used for comparison. The addition of noise, corresponding to the uncertainty of the experimental measurements, on the transmission curve resulting from the simulations allows to validate our method under conditions close to the experimental ones. This method was then tested on a transmission curve obtained with the SATURNE accelerator. The energy spectrum of the accelerator was estimated as well as the maximum energy of the photons, which is 11.6 ± 0.4 MeV.

References

[1] N. Estre, D. Eck, J.-L. Pettier, E. Payan, C. Roure, and E. Simon, "High-energy X-ray imaging applied to non destructive characterization of large nuclear waste drums," in *2013 3rd International Conference on Advancements in Nuclear Instrumentation, Measurement Methods and their Applications (ANIMMA)*, 2013, pp. 1–6. DOI: 10.1109/ANIMMA.2013.6727987.

[2] F. Carrel, M. Agelou, M. Gmar, F. Laine, B. Poumarede, and B. Rattoni, *"Measurement of Plutonium in Large Concrete Radioactive Waste Packages by Photon Activation Analysis"*, Nucl. Sci. IEEE Trans. On, vol 57 p. 3687-3693, Jan. 2011, DOI: 10.1109/TNS.2010.2087356.

[3] E. Simon, F. Jallu, B. Perot, and S. Plumeri, *"Feasibility study of fissile mass quantification by photofission delayed gamma rays in radioactive waste packages using MCNPX"*, Nucl. Instrum. Methods Phys. Res. Sect. Accel. Spectrometers Detect. Assoc. Equip., vol 840, p. 28-35, Dec 2016, DOI: 10.1016/j.nima.2016.09.047.

[4] L. Tamagno, C. Eleon, N. Estre, S. Plumeri, and D. Tisseur, *"Atomic number reconstruction using multi-MeV X-ray DECT in concrete drums"*, IEEE Nucclear Science Symposium and Medical Imaging Conference Proceedings, Sydney, Australia, p. 1-3, Nov. 2018, DOI: 10.1109/NSSMIC.2018.8824646.

[5] G. Li, H. Zheng, G. Sun, and Y. Wu, *"Photon Energy Spectrum Reconstruction Based on Monte Carlo and Measured Percentage Depth Dose in Accurate Radiotherapy"*, Prog. Nucl. Sci. Technol., vol. 2, p. 160-164, Oct. 2011, DOI: 10.15669/pnst.2.160.

[6] Texin Li *et al.*, *"Re-evaluation of Compton scattering method in characterization of radiation therapy beams"*, in Proceedings of the 22nd Annual International Conference of the IEEE Engineering in Medicine and Biology Society (Cat. No.00CH37143), Chicago, IL, USA, 2000, vol. 2, p. 1082-1084, DOI: 10.1109/IEMBS.2000.897915.

[7] R. Nath and R. J. Schulz, " *Determination of high-energy x-ray spectra by photoactivation*", Med. Phys., vol. 3, n° 3, p. 133-141, Jun. 1976, DOI: 10.1118/1.594217.

[8] O. Assatel, *"Characterisation of photon and neutron spectra in medical linear accelerators using theoretical and experimental techniques"*, PhD, University of Surrey, 1996

[9] M. Krmar *et al.*, *"A simple method for bremsstrahlung spectra reconstruction from transmission measurements"*, Med. Phys., vol. 29, p. 932-940, Jul. 2002, DOI: 10.1118/1.1469632.

[10] NIST, *"Xcom Photon Cross Sections Database"*. [Online] Available : https://www.physics.nist.gov/PhysRefData/Xcom/html/xcom1.html. Accessed on : Jun. 18 2020.

[11] B. Pelowitz, *"MCNP6 Users Manual - Code Version 1.0 - LA-CP-13-00634"*, 2013.

[12] *"Semiflex Ionization Chamber 31010"*. [Online], Available : https://www.ptwdosimetry.com/en/products/semiflex-ionization-chamber-31010/ . Accessed on : Jan. 10 2020.

[13] I. Charon, O. Hudry, andA. Germa, "METHODES D'OPTIMISATION COMBINATOIRE", Elsevier Masson, 1996.

Characterization of neutron emission during pulse mode of low output electronic neutron generator

Tomas Bily [1], Ondrej Huml [1]
[1]Czech Technical University in Prague, Faculty of Nuclear Sciences and Physical Engineering, Department of Nuclear Reactors, V Holesovickach 2, Prague 8
tomas.bily@fjfi.cvut.cz

Abstract— The paper deals with the characterization of the neutron output during operation the neutron generator in the pulsing mode. The characterization is performed by a cadmium covered He-3 detector. First, the adequacy of a cadmium covered He-3 detector to study the pulsed mode operation of neutron generator is assessed using Monte Carlo simulations in MCNP6 code. Second, the pulsing mode operation of the D-D P-385 Thermo Fisher Scientific neutron generator with a maximal neutron output of 7 x 106 n/s is studied. The study shows the dependence of neutron emission and pulse shapes on the frequency, duty factor, high voltage and beam current.

***Keywords* — portable neutron generators; pulse characterisation.**

I. Introduction

PORTABLE electronic neutron generators are an established technology that can be used for many lab and field applications. Offering significant regulatory and operational benefits over radionuclide neutron sources, they start to be more widely used at universities and research centres. Beside the established industrial applications (e.g., in logging or mining industry), applications are being developed for fields including security, safeguards, or waste characterization. They are becoming more used for E&T in nuclear engineering, nuclear analytical techniques, or nuclear related fields. They can be utilized for many analytical techniques (including NAA, PGNAA, DNC, DDA). In many cases, the pulsed operation of the generator is required or can be used to improve the efficiency of the applied method. Consequently, the knowledge on the actual production of neutron and its dependence on operational parameters is important.

The paper provides such characterization performed by a cadmium covered He-3 detector. First, the adequacy of cadmium covered He-3 detector to study the pulsed mode operation of neutron generator is assessed utilizing Monte Carlo simulations in MCNP6 code. Second, the pulsing mode operation of the D-D P-385 Thermo Fisher Scientific neutron generator is analyzed.

II. Materials and Methods

A. P-385 neutron generator

The P-385 is a D-D portable neutron generator produced by Thermo Scientific [1]. It has maximum emission rate of 7 x 10^6 n/s [2]. It can be operated in continuous or pulsed mode with frequency up to 20 kHz with adjustable duty cycle between 5 % and 100 % (5 μs pulse width minimum). The neutron output of the generator can be adjusted by set-up of high voltage (up to 130 kV) and beam current (up to 70 μA).

The generator can provide information on neutron production via a TTL signal that can be used for triggering the data acquisition system. Within this paper the SOURCE (synchronized) signal was used. The SOURCE pulse drives the ion source pulse of the generator. As stated by the producer, there is a slight delay due to the source capacitance charging before a neutron pulse is generated. This delay varies based on the beam current and pulse width settings [1]. Besides the SOURCE signal, three other DELAY signals are available [1].

B. Detector and data acquisition

Measurements were performed with He-3 chamber DEXTRAY 65NH45 (length/diameter 650/25 mm) and TEMA EMK-310 analyzer. A system based on a programmable FPGA was developed to collect data from the neutron detector, which are time-synchronized with the pulse mode of D-D neutron generator operation (i.e. synchronized with the SOURCE signal). The system is based on the RedPitaya STEMLAB 125-14 platform containing the Xilinx Zynq 7010 FPGA. This platform contains up to 16 digital input/output ports. These are used for data collection purposes as inputs for fast independent counters implemented inside the FPGA. The first counter monitors the signal on start of neutron production in each cycle of the generator. It assures the synchronization of the measurement cycles of data collection with the pulse cycles of the generator. The developed software allows to divide the neutron response in each cycle into 1000 time-bins. At a D-D generator frequency of the order of hundreds Hz, it is then possible to achieve a time resolution of the order of tens of μs. Good statistical response can then be achieved by repeating a large number of cycles and accumulating results.
For the measurements the detector was placed inside a coverage made by cadmium to cut the contribution of thermal neutrons

and thus to approach the time dependency of the detector response to the time dependency of neutron emission.

C. Experimental set-up

The experimental set-up is shown in the Fig. 1 and 2. It is used for experiments with radionuclide neutron sources and electronic neutron generators [3]. The neutron generator was placed inside a central channel of the polyethylene box of 50 x 50 cm x 100 cm, the cadmium covered detector was placed in one of the dry experimental channels with diameter of 7 cm, the centre of the detectors was located 10.5 cm from the centre of the tube. The measurements were performed for duty factors ranging from 10 % to 50 % and with pulsing frequencies from 100 Hz to 500 Hz, i.e. for rather lower part of the frequency range, starting even below the minimum stated frequency of 250 Hz. Such low frequencies can be used e.g. for delayed neutron counting applications (see e.g. [4]) Measurements with various high voltage and beam current have been performed to show the behaviour at maximum tube parameters (130 kV and 70 μA) and at lower ones.

Fig. 1. Neutron generator control unit, generator tube, cadmium coverage and He-3 detector used in the study on top of polyethylene block

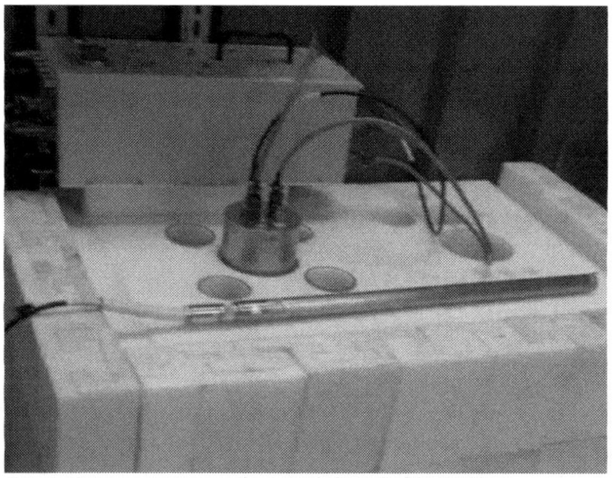

Fig. 2. Neutron generator in the central channel of the PE block

D. Calculations

Calculations in Monte Carlo code MCNP6 [5] have been performed to prove the adequacy and to assess the limitations of cadmium covered thermal neutron detector to study time-dependent behavior of neutron generator pulsed mode operation. The neutron source was simulated using SDEF card of the MCNP as a point isotropic source with the energy of 2.45 MeV. All neutrons have been set to start at time equal to zero. The calculated quantities included the time-dependent neutron flux and detector response. For comparison both the case of cadmium covered and bare He-3 detectors were included. The tallies were binned to understand the contribution of neutrons below 1 eV, 1eV – 1 MeV, and above 1 MeV. The 1eV boundary was chosen due to its closeness to the cadmium edge. The quantity corresponding to detector response was calculated as He-3(n,p)H-3 reaction rate. Two time-binning schemes were used to observe fast and slower part of the response. The former calculated the response in the first 50 μs with the time step of 0.1 μs, the latter in first 5 ms with the time step of 10 μs. Such calculations can describe the time scales between the emission of neutron by generator and reaction occurring in the detector. The further delays caused by charge collection time in detector or those caused by electronic are neglected here. All calculations were performed with the ENDF/B-VIII nuclear data library [6]. Ca 2 x 10^8 initial neutrons were used for each simulation.

The geometry is shown in the Fig. 3. The time dependent response of neutron detectors has been calculated both for bare He-3 detector, and cadmium covered He-3 detector.

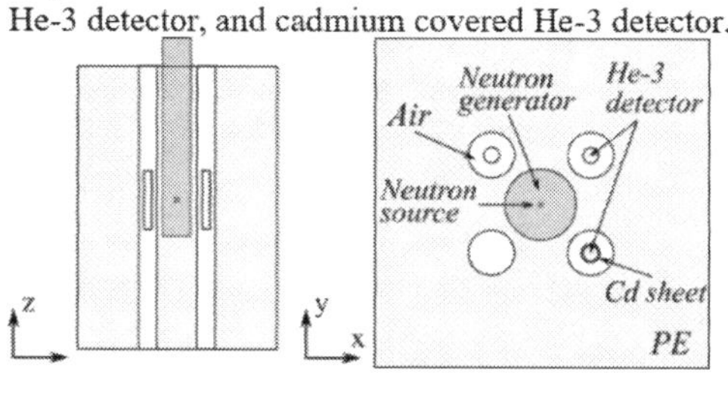

Fig. 3. Geometry used in calculations

The neutron generator was placed to the centre of the geometry. The generator internals were modelled in detail based on the input provided by the tube producer. The detectors assumed in calculations were He-3 tubes with active length of 20 cm, diameter of 2.5 cm, and with helium pressure in the tube of 2 bars. The centre of the detectors was placed 10.5 cm from the centre of the tube (neutron source position); vertically the centre of the detector corresponds to the source plane. Only active volumes of the detectors were simulated. In case A (moderated geometry), the generator was placed inside PE block of 50 x 50 x 100 cm; the detectors were placed in dry channels with diameter of 7 cm. One detector was simulated as bare, another was covered by 1 mm thick cadmium sheet with density of 8.65 g/cm^3.

III. Results of calculations

The results are shown in Figs. 4-6. Fig. 4 shows the time-evolution of neutron flux. In the first 5 μs, the flux is dominated by neutrons above 1 eV, after that, the neutrons below 1 eV became dominant. Initially, the response of bare He-3 detector (see Fig. 5) drops sharply reaching minimum around 5 μs, then it starts to rise due to onset of neutrons below 1 eV, and their high reaction cross-section. The maximum response is achieved around 30 μs after neutron emission from the generator. Then, the response starts to decrease slowly. For detector covered in cadmium (see Fig. 6) the low energy neutrons are efficiently cut and the response drops in less than 15 μs to 1% of its value at time 0 (for response evaluated with time step of 0.1 μs).

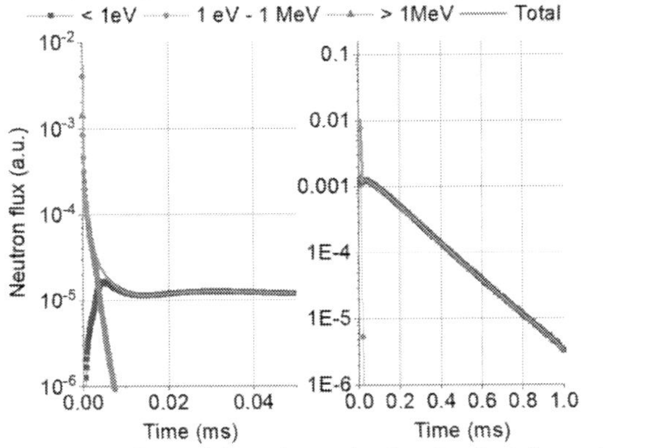

Fig. 4. Neutron flux evolution at detector location of moderated geometry. Left: short term behavior (time step of 0.1 μs); right: long term behavior (time step of 10 μs)

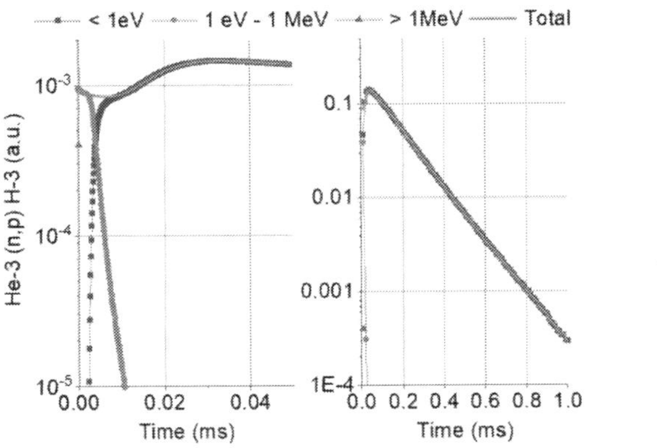

Fig. 5. Relative detector response of bare detector in moderated geometry. Left: short term behavior (time step of 0.1 μs); right: long term behavior (time step of 10 μs)

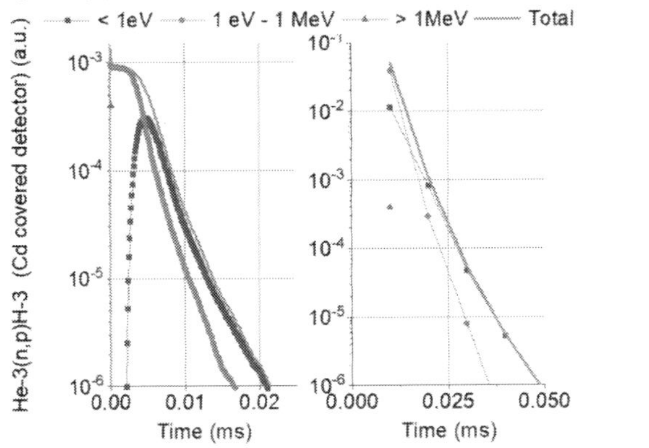

Fig. 6. Relative detector response of Cd covered detector in moderated geometry. Left: short term behavior (time step of 0.1 μs); right: long term behavior (time step of 10 μs)

Based on the calculations, one can conclude that characterisation of neutron generator pulse mode behaviour by Cd-covered He-3 detector in moderated geometry is reasonable, and especially for lower range of frequencies (say below 1 kHz) it can provide precise information on pulse shape. For higher frequencies the applicability of the approach would be limited as the response delay after the neutron emission from the generator might form a significant part of the cycle.

IV. Results of Measurements

The following section summarizes the measurements performed with the Cd-covered He-3 detector and P-385 neutron generator to characterize its pulse mode operation. The characteristics includes the generator start-up, pulse stability, total neutron output dependence on pulsing set-up, and neutron emission pulse shape dependence on pulsing set-up.

A. Generator start-up & pulse stability

The start-up of neutron generator for various set-ups is illustrated in the Fig. 7. In the figure, the time evolution starts when the first pulses have been detected. One may observe the generator reaches its nominal output ca. 5 to 8 seconds after the start-up, first 3 to 5 seconds producing only negligible number of neutrons. The rise time seems to be more dependent on duty factor set-up than on high voltage and beam current set-up.

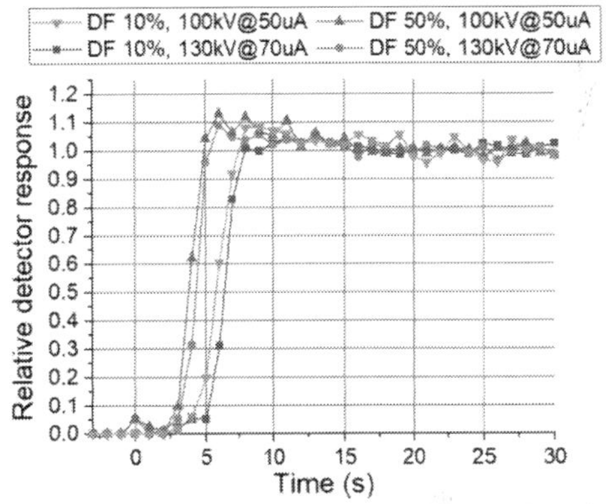

Fig. 7. Relative detector response of Cd-covered detector in moderated geometry to generator start-up for four set-ups at 250 Hz.

The pulse stability has been observed for five-minute operation of the tube with various set-ups of pulsing frequencies, duty factors, high voltages, and beam currents. The detector response was summed per five seconds, thus creating 60 batches per each five-minute-long run. The results are given in the figure 8. One can observe typically one or two batches with lower neutron output that correspond to the generator start-up. Then, the pulse shapes remained stable throughout the operation.

B. Neutron output dependence on pulsing set-up

The generator pulsing set-up may affect the neutron emission rate. The dependence on pulsing frequency and beam current, measured as number of neutron counts detected over five minute operation, is illustrated in the figure 9 for three frequencies and several values of duty factor of generator operating at nominal high voltage and beam current (130 kV, 70 μA). Results were normalized to the value obtained at 250 Hz with duty factor of 10 %. Figure 10 shows the neutron output dependence on high voltage and beam current at two fixed values of duty factor at frequency of 250 Hz. Obviously, the

https://doi.org/10.1051/epjconf/202125304003

neutron emission rises linearly with beam current with the slope given by high voltage and duty factor set-up.

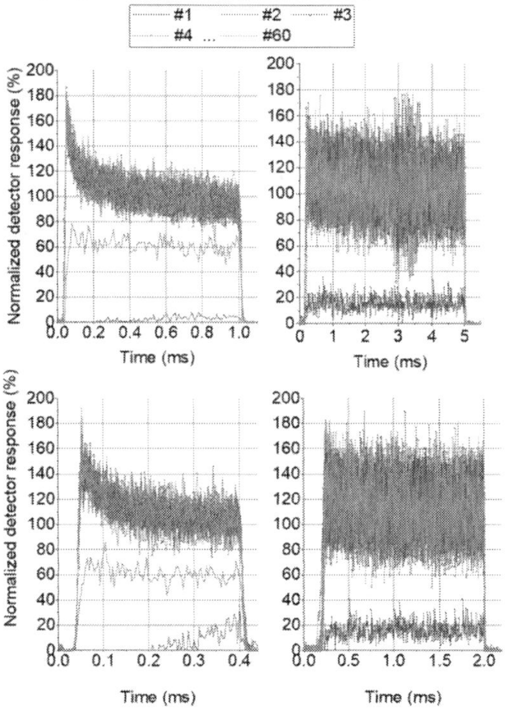

Fig. 8 Cadmium-covered detector response to P-385 D-D neutron generator operated in pulsed mode. Counts integrated per 5 s for 5 minutes of operation at 130 kV @70 μA (i.e. 60 batches denoted as #1 to #60. For 100 Hz, 1 batch sums over 500 cycles, for 250 Hz one batch sums over 1250 cycles). Top left: 100 Hz, DF 10%. Top right: 100 Hz, DF 50%. Bottom left: 250 Hz, DF 10%. Bottom right: 250 Hz, DF 50%.

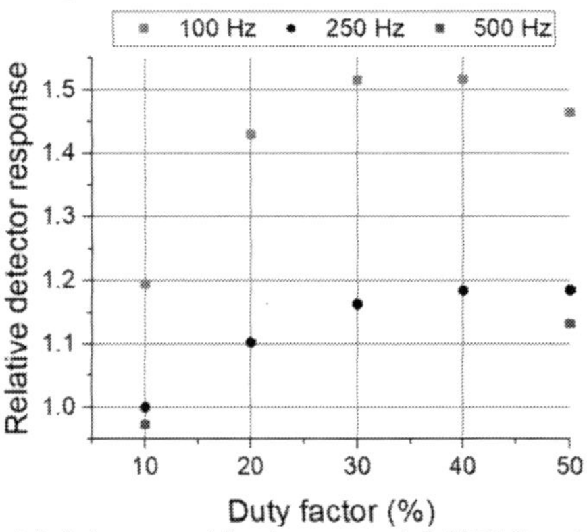

Fig. 9 Cadmium-covered detector response to P-385 D-D neutron generator operated in pulsed mode demonstrating the neutron output dependence on pulsing frequency and duty factor at constant high-voltage and beam current.

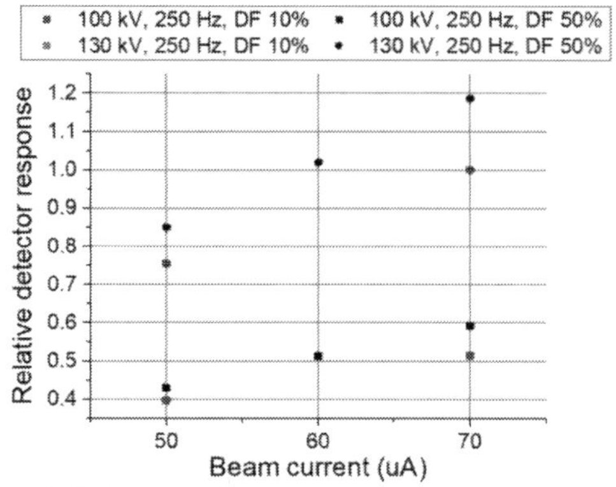

Fig. 10 Cadmium-covered detector response to P-385 D-D neutron generator operated in pulsed mode demonstrating the neutron output dependence on beam current

C. Neutron emission pulse shape dependence on pulsing set-up

The neutron emission pulse shapes for two frequencies were observed at nominal high voltage and beam current parameters (130 kV, 70 μA) for two frequencies and five values of duty factor (see Fig. 11). For all studied cases deviations from idealized rectangular shape can be observed. At these high-voltage and beam current set-ups, the neutron emission was higher at the pulse start and decreased to a steady value. This effect was stronger for lower values of duty factors. On the other hand, the higher was the duty factor, the longer took the pulse rise-time, reaching more than 0.2 ms for duty factor of 50 % and frequency of 250 Hz.

Fig. 12 shows the comparison of pulse shapes at 250 Hz for two fixed values of duty factors for varying high voltage and beam current. One can see that the emission rise time differ slightly for various set-ups, being ruled mostly by the beam current. Contrary, at the end of the pulse, the emission drops rapidly and practically identically for all studied set-ups. Further, one can observe that for high values of duty factor (here DF = 50 %), the pulse is nearly rectangular whereas for low values of duty factor (here DF = 10 %), the pulse shape depends strongly on applied high voltage. Finally, Fig. 13 provides the comparison of neutron emission pulse shapes for three frequencies and two values of duty factor. The time scale corresponds to percent of a cycle. One can see the neutron emission delays after the indicated pulse start (SOURCE signal driving the ion source). This is more relevant for higher values of duty factor and higher frequencies, e.g., for the frequency of 500 Hz at duty factor of 50 %, the pulse started after 10 % of the cycle time (i.e. after 0.2 ms). At 250 Hz, the delay was similar (ca. 0.2 ms) but related to cycle length it accounts to only 5% of the cycle length.

https://doi.org/10.1051/epjconf/202125304003

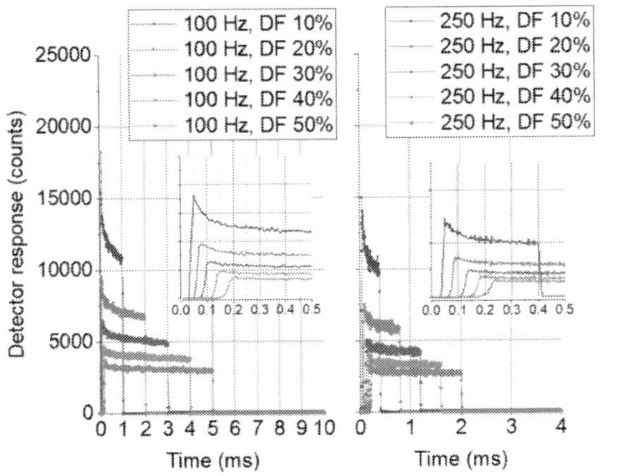

Fig. 11 Cadmium-covered detector response to P-385 D-D neutron generator operated at 130 kV and 70 μA for pulsed operation at 100 Hz and 250 Hz with duty factor ranging from 10 % to 50 %.

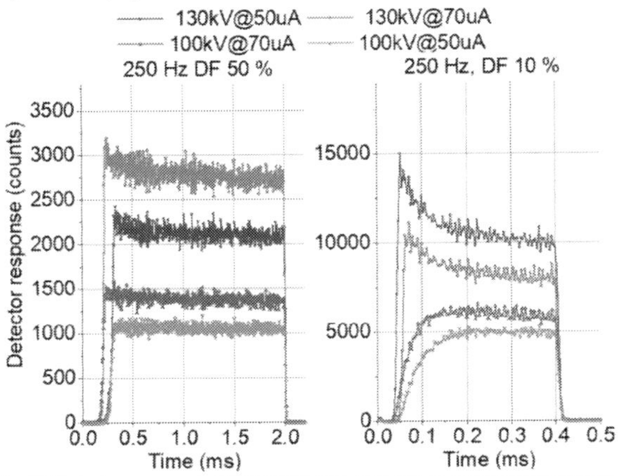

Fig. 12 Cadmium-covered detector response to P-385 D-D neutron generator operated in pulsed mode at 250 Hz with duty factor ranging from 10 % to 50 % with varying high voltage and beam current setup.

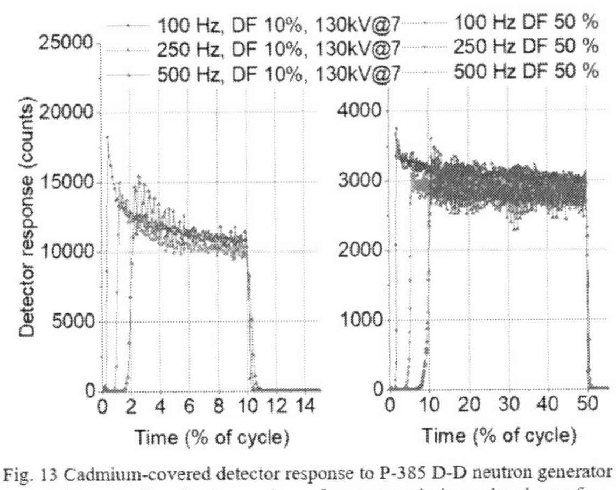

Fig. 13 Cadmium-covered detector response to P-385 D-D neutron generator operated in pulsed mode. Comparison of neutron emission pulse shapes for three frequencies and two values of duty factor at 130 kV @ 70μA.

V. Conclusions

The paper deals with the characterization of low-output neutron generator pulsed mode operation. First, the calculations performed in the MCNP6 code proved the viability to use He-3 thermal neutron detector covered by cadmium for the purpose, especially for lower frequencies (say below 1 kHz). Further, various aspects of neutron production in pulsed mode were demonstrated for P-385 neutron generator operated between 100 Hz and 500 Hz, duty factors between 10 % and 50 % and for various set-ups of high voltage and beam current. This might help to improve the utilization of the generator in applications utilizing the generator pulsed mode.

Acknowledgment

This work was supported by the project "Utilisation of radiation-based methods for detection and identification of CBRNE materials" (project nr. VI20172022171) supported by Ministry of the Interior of the Czech Republic.

References

[1] Thermo Fisher Scientific Inc., "P 385 Neutron Generator Operation Manual". Tech. Rep. Manual P/N 120006-A 062907, 2010.

[2] J. Rataj, T. Bílý, O. Huml, et al., "Verification of the MCNP model of Manganese Bath," Prague, *2020 21st International Scientific Conference on Electric Power Engineering (EPE)*, Czech Republic 2020, doi: 10.1109/EPE51172.2020.9269226

[3] J. Rataj, P. Suk, T. Bílý, et al., "Characterisation of neutron field in the polyethylene neutron irradiator", *Applied Radiation and Isotopes*, vol. 168, 2021. DOI. 10.1016/j.apradiso.2020.109529

[4] T.-H. Lee, H.-S. Shin, and H.-D. Kim, "Delayed Neutron Measurement in High Frequency Mode and Its Application to the Determination of Fissile Content in Spent Fuel Assembly", Valencia, *2011 IEEE Nuclear Science Symposium Conference Record N53-7*, Spain, 2011

[5] C.J. Werner, et al., "MCNP6.2 Release Notes". Los Alamos National Laboratory, report LA-UR-18-20808, 2018.

[6] D.A.Brown, M.B.Chadwick, R.Capote, et al., "ENDF/B-VIII.0: The 8th major release of the nuclear reaction data library with CIELO-project cross sections, new standards and thermal scattering data", Nucl. Data Sheets 148(2018)1, DOI. 10.1016/j.nds.2018.02.001

Jules Horowitz Reactor Irradiation Devices: Inspection Methods proposal

Seppo Hillberg[1], François Baque[2], and Stéphane Gaillot[2]
[1] VTT Technical Research Centre of Finland Ltd, Finland
[2] CEA, France
Seppo.Hillberg@vtt.fi

Abstract — **Jules Horowitz Reactor (JHR) irradiation test devices (so called ADELINE and MADISON) must undergo a periodic inspection every 40 months (French ESP(N) Regulation). The first step of inspection proposal was performed from October 2019 to March 2020 and concerns application of non-destructive methods for electron beam welds, while examining the possible methods and locations that could be utilized.**

This study has been performed in collaboration between VTT Technical Research Centre of Finland Ltd and French CEA with VTT employee secondment at CEA Cadarache.

The initial conclusions of this work are the following:

- **Two inspection methods are likely needed.**
- **Eddy current is likely suitable for surface examination while volumetric ultrasonic inspection can be used for tube bodies.**
- **Inspection under-water in a storage pool (EPI) is likely the best option for location in the JHR.**
- **Some automation and remote controls will be needed during the inspection process. The inspection area is quite large as the whole tube bodies will be inspected.**
- **Internal surfaces located between the nested tubes likely cannot be inspected for corrosion and therefore, sealing the space from oxygen and water will be likely be the best option.**
- **Experimental reference is a critical next step of the work to assess and qualify the inspection methods.**

As the next step, CEA will continue discussions on requirements of the mock-ups: for eddy current inspection, a Zircaloy-4 planar specimen with artificial reflectors is needed. For ultrasonic reference, a representative mock-up of the cylinder with a weld and artificial defect(s) is needed. The mock-up testing will provide an experimental reference to CIVA calculations as simulations of this work have been performed with CIVA software, which is extensively used in the industry and research organizations in simulation of non-destructive control.

This study will allow further steps in the future, up to the qualification of inspection methods for JHR test devices.

Keywords — **JHR, irradiation devices, NDT, NDE, periodical inspection, Non Destructive Examination, Ultrasounds, Eddy Currents, CIVA software**

I. Introduction

JHR irradiation test devices (so called ADELINE and MADISON) must undergo a periodic inspection every 40 months (French ESP(N) Regulation). The first step of inspection proposal concerns application of non-destructive methods for electron beam welds while examining the possible methods and locations that could be utilized. This initial study has been performed in collaboration between VTT Technical Research Centre of Finland Ltd and CEA. It was performed as VTT employee secondment at CEA Cadarache.

As the process of inspection planning has only just been started, it is pointed out that the methods chosen to studied here are not restrictive and are not presented as the only option. Lack of an aspect or a method should not be interpreted as having been considered and ruled out as insignificant or not suitable.

Further steps beyond, up to the qualification of inspection methods for JHR test devices, will include tests and simulations with CIVA software [1], which is extensively used in the industry and research organizations in simulation of non-destructive control.

II. JHR test devices

In use, irradiation test devices are inserted into the vicinity of the reactor core and contain material or fuel samples. This report focuses mostly on the ADELINE device, which is shown in Fig 1. Orange and green weld markers show the approximate positions of the inner and outer tube welds.

The main body of the device consists of two vertical nested tubes – the main interest of the work described in this report. The thicker inner tube acts as a pressure vessel while the outer tube and the gas gap offer thermal insulation.

Once in use, due to deformations of thermal and radiation origins, these tubes are likely impossible to be separated (removal and re-assembly in a hot cell during inspection phase).

ADELINE material is Zircaloy-4 and MADISON device may have a stainless steel inner tube. The length of the inner tube is 3726 mm, with inner tube diameter ranging from 38.8 mm to 66.0 mm and thickness ranging from 7.9 to 10.3 mm. Gap width between the tubes ranges from 0.4 to 1.0 mm. Outer tube outer diameter ranges from 64.6 mm to 108 mm and thickness from 4.0 mm to 9.7 mm. ADELINE inner tube has three narrow electron beam welds and the outer tube has two [3,4]. Additionally, the tubes may be welded together from the top.

Machining and welding test information have been developed and qualified [5,6].

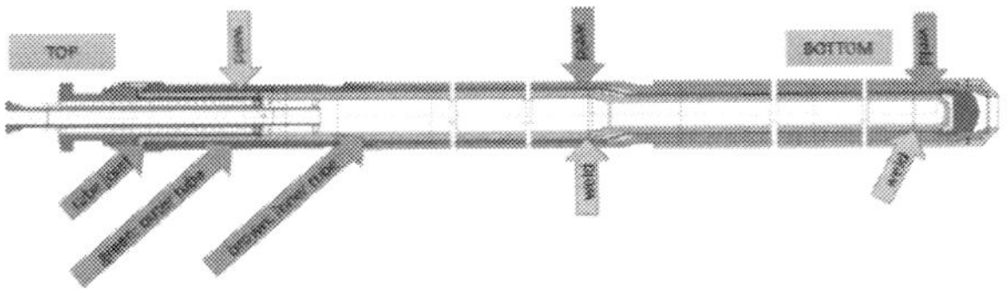

Fig. 1. ADELINE irradiation test device [2].

III. Periodical inspections

JHR irradiation test devices fall into N2 class of French nuclear pressure equipment regulation [7]. This criterion is for the equipment not falling into N1 criteria but activity release exceeding 370 GBq in case of failure. Inspection interval of the devices is 40 months.

Only the inner and outer tubes of ADELINE device have to be inspected. Internal apparatus will have been dismantled in the hot cell before inspection operation.

At this stage, it is known that the inspection must respect the requirements of RCC-MRx [8]. As RCC-MRx is designed as guide for construction phase, it does not explicitly specify the requirements of periodical inspections.

According to radiographic inspection criteria (Table I), no crack type of defects are allowed. This definition does not specify the required accuracy of the inspection.

During this initial planning phase, it has been assumed (as best estimate) that the smallest defect that must be detected is 0.2 mm large.

As the RCC-MRx is not intended for inspection phase, it does not cover corrosion. For the purposes of this study, it has been assumed that the corrosion layer would be 0.05 mm thick.

TABLE I
RCC-MRX 7714.32: ACCEPTANCE CRITERIA FOR RADIOGRAPHIC EXAMINATION ALUMINUM AND ZIRCONIUM ALLOYS [8]

N°	Imperfection designation	EN ISO N°6520-1	Remarks	Within thickness	Class N_2Rx
1	Crack	100		t ≥ 0.5 mm	Unacceptable
2	Crater crack	104		t ≥ 0.5 mm	Unacceptable
3	Porosity	2011	Maximum dimensions of the blowhole.	t ≥ 0.5 mm	d < 0.3s or 0.3 a max 5mm
4	uniformly distributed spherical blisters	2012	The defects are assessed with respect to the projected surface. See the examples given in standard EN ISO 10042	0.5 mm ≤ t ≤ 3 mm	A ≤ 2 % L = 100 mm
				3 mm < t ≤ 12 mm	A ≤ 4 % L = 100 mm
				12 mm < t ≤ 30 mm	A ≤ 6 % L = 100 mm
				30 mm < t	A ≤ 8 % L = 100 mm
5	Clustered porosity	2013	The cluster is represented by a circle of diameter DA including all the porosities	t ≥ 0.5 mm	DA < 20 mm
6	Worm holes	2016		t ≥ 0.5 mm	l < 0.3s or 0.3 a max 4mm
7	Oxyd inclusions	303	If many inclusions,, l1, l2...+ l= l1+l2 +...	t ≥ 0.5 mm	l < 0.5s or 0.5 a max 5mm
8	Tungsten inclusions	3041		t ≥ 0.5 mm	l < 0.3s or 0.3 a max 4mm

A. Possible methods

During the different working discussions, the following methods below were discussed in some capacities. Underlined are the ones that were ultimately chosen for the first part of the study:

- Ultrasonic:
  - o Volumetric,
  - o Guided waves;
- Eddy currents:
  - o Surface C-scan (perpendicular to the object),
  - o Bobbin probe (cylindrical);
- X-rays;
- Visual inspection;
  - o Dye penetrant.

As the internal apparatus of the device will be removed in the hot cell, it may be beneficial to perform a preliminary visual inspection before transporting the device to the next inspection location. This possibility will need to be discussed later as the inspection planning progresses.

Originally, also ultrasonic guided waves were intended to be studied but after receiving feedback from CEA Saclay experts, this was discarded as too lengthy study to be performed within the available time frame. A study of this kind would need to be performed in stages, starting from a highly simplified geometry and progressing towards real geometry.

From the very early stage of the study, it has been the impression that at least two methods are needed for the inspection. While ultrasonic volumetric inspection can be used for inspecting the tube bodies, it is not well suited for corrosion inspection. Eddy current examination, however, is very efficient in finding surface defects. Whether this extends to a small corrosion layer is discussed later in this report. In the present conditions, eddy currents do not penetrate more than a couple of millimeters into the tube so this method is devoted to surface-only examination.

B. Inspection location and execution

Once the test device is removed from the reactor core, sample holder of the device can be removed already in the reactor pool. From there, the device will be transported to the hot cell for disassembly of the internal parts. Possible inspection locations are the hot cell and the storage pool near the hot cells (Fig. 2). Both locations are accessible with existing transportation means.

Inspection in a hot cell would mean non-submerged inspection in a place where space is premium, whereas inspection in a storage pool would take place underwater and in an area where a dedicated inspection rig can be constructed. In ultrasonic inspection, a liquid film needs to be introduced between the probe and the inspected surface. This would not rule out hot cell inspection, but is a benefit of the in-pool inspection.

Due to these considerations, it is likely that inspection in a storage pool is the better option.

It is also a possibility that an existing bench could be adapted and used for these inspections. This approach has been recognized in the discussions but has not been studied.

Inspection operations will need to be performed for both the inner and the outer tube. How the device and the probe would be handled has not been studied in detail but one possibility is:

- Outer tube inspection: Device remains vertical and it will be rotated. The probe is introduced to the device radially from the side. The probe is capable of being moved vertically and radially in relation to the device.
- Inner tube inspection: A manipulator arm is inserted into the device which remains vertical. It is capable of moving vertically and radially in relation to the device. Either the manipulator arm or the device is rotated.

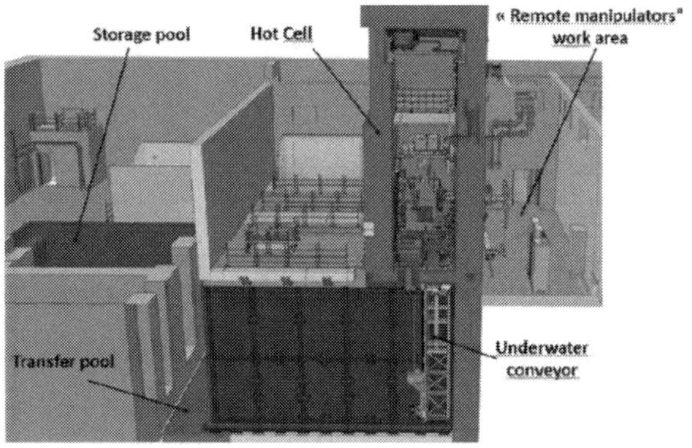

Fig. 2. JHR Storage Pool and Hot Cell.

Note: It is currently undecided what the final connection method of the inner and outer tubes is.
The inner and outer tubes can be welded together with similar means as other welds. For ADELINE, this weld (Fig. 3) should be relatively similar to inspect than the other welds. For MADISON, the connection would be between stainless steel and Zircaloy-4. The specifics of this inspection are not covered in this work but irradiation aspects may be significant [9].

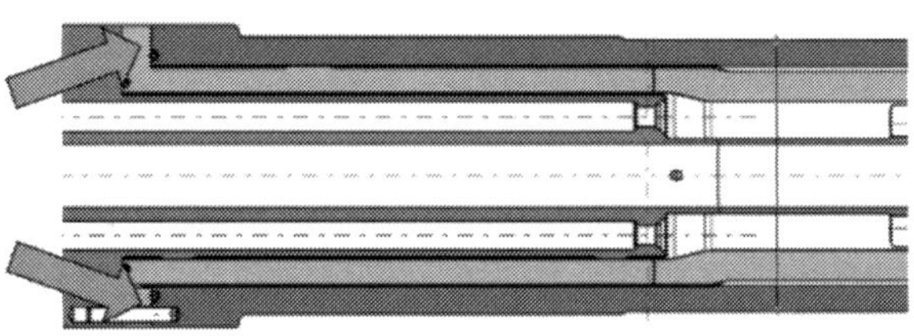

Fig. 3. Location of the optional end-weld.

IV. Simulations

CIVA software version used in the simulations was CIVA_TECNA_2020-02-04.

A. Geometry and used materials

The simulations were done in a cylindrical geometry, within the cross section of each tube (Fig. 4). Diameter and wall thickness were taken from ADELINE device dimensions. Smallest inner tube diameter and largest wall thickness were used. Material in ultrasonic simulations was CIVA Zircaloy ready-made material with density changed to correspond RCC-MRx.
The following values were then used:

- bulk density : 6500 kg/m^3,
- transversal ultrasonic wave velocity: 2360 m/s and
- longitudinal ultrasonic wave velocity: 4720 m/s.

In the eddy current simulations, Zircaloy-2 ready-made material was used:

- conductivity: $1.39 \cdot 10^6$ S/m and
- relative permeability: 1.0

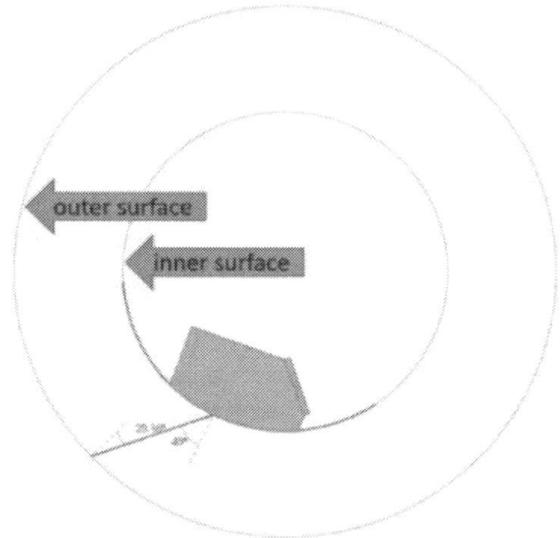

Fig. 4. Geometry used in the simulations.

B. Ultrasonic simulations

Ultrasonic testing is a family of non-destructive testing techniques based on the propagation of ultrasonic waves in the object or material tested. In most common applications, very short ultrasonic pulse-waves with center frequencies ranging from 0.1-15 MHz are transmitted into materials to detect internal flaws or to characterize materials.

Such ultrasonic simulation can be performed using CIVA module which simulates the beam-defect interaction and predicts the amplitude and the time of flight of various echoes: direct echo, corner effect, tip diffraction echo, etc. It can also calculate echoes scattered back by the geometry (backwall, entry surface and interior specular interfaces echoes) and takes into account mode conversions.

Ultrasonic sensors chosen for this study are 2 MHz frequency, 7x7 mm and 10 MHz , 8x8 mm size. The 2 MHz sensor was chosen as it represents a very common inspection frequency.

The sensor size was chosen as it produces a homogenous field into the inspected body when a flat sensor is used. It is, however, possible that this frequency will not allow detection of the smallest defects. The 10 MHz was chosen since a similar probe is listed in a manufacturer catalog [10].

Both probes were used as contact probes. The used wedge length was 15 mm, width 10 mm and height 10 mm. Beam incidence angle to the tube body was 45 degrees.

In actual inspection, array sensors will be used. They were not used in this study since the use of single transducers was suitable for this first stage of study. Array sensors will be able to direct and focus the ultrasonic beam without mechanical means and this process is very fast. This will be highly beneficial for the inspection since not all type of defects cannot be detected from all angles (beam directing). Beam focusing will allow moving the strong focal spot throughout the inspected body.

https://doi.org/10.1051/epjconf/202125304004

Beam verification is performed with a side drilled hole (SDH) as a reference reflector. SDH is a simple cylindrical shape (typically 1 mm or 2 mm diameter) with a flat base. It can be located at any distance from the sensor, within the depth of the piece to be controlled, perpendicular to the acoustic beam incidence to get an echo which can be used as a 'reference' state.

In practice, the characteristics of the different beams that will be used for the inspection are measured with side drilled holes. These measurements, simulated with CIVA and displayed on the next figures, allow evaluation of the beam characteristics like the refracted angle, the exit point, the beam focal point depth and the beam diameter at focal point. For example, for the 2 MHz single planar probe, the amplitude echodynamic curve shows the weak amplitude variation of the transversal radiated beam with the depth (the variation is less that 2 dB from 2 mm depth (maximum amplitude) to 10 mm depth). Inspection geometry, a B-scan and an amplitude echodynamic curve are shown in Fig. 5 to 7.

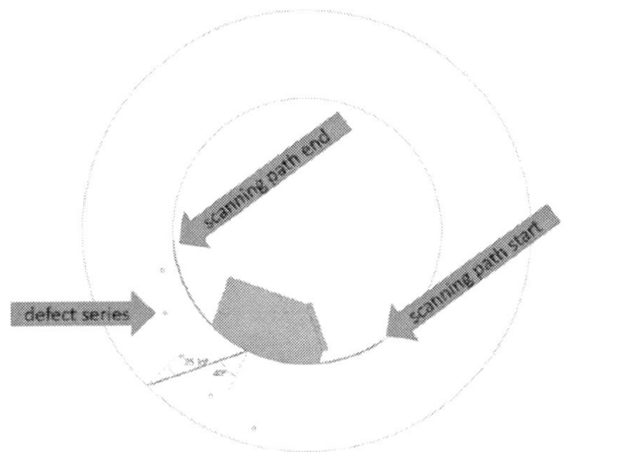

Fig. 5. Geometry in the inspection simulation and defect positions.

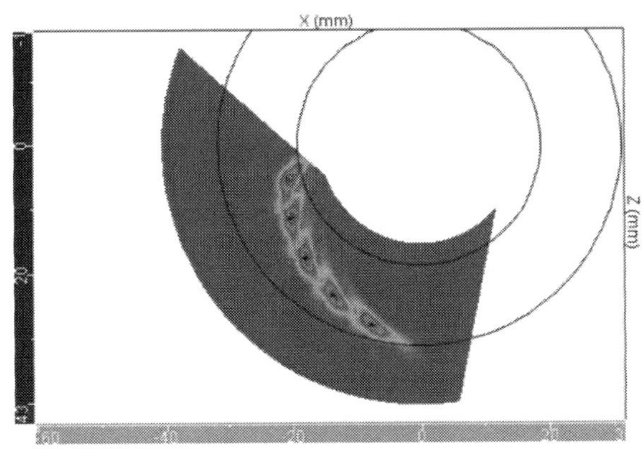

Fig. 6. B-scan of a SDH defect series.

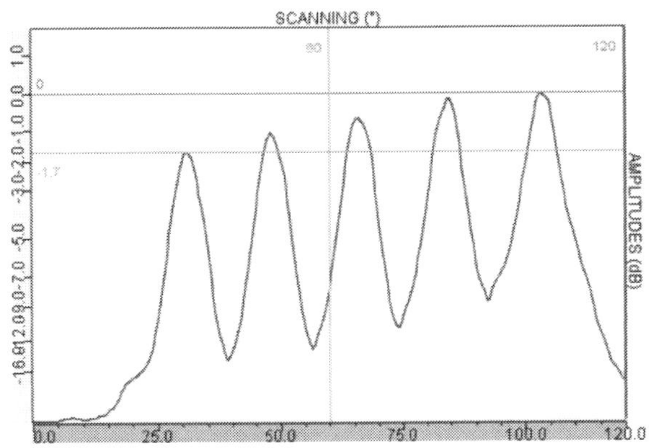

Fig. 7. Amplitude echodynamic curve.

Ultrasonic echoes identification is made considering that, even if only transversal direct waves are radiated into the specimen, mode conversion will occur at the back wall or defects leading to a variety of echoes (see Fig. 8). This is especially relevant in this case as the back-wall is curved. It will likely not be possible to direct longitudinal echo out of the tube body. CIVA simulations can help to predict the expected inspection echoes. Two examples are given in Fig. 9 and in Fig. 10.

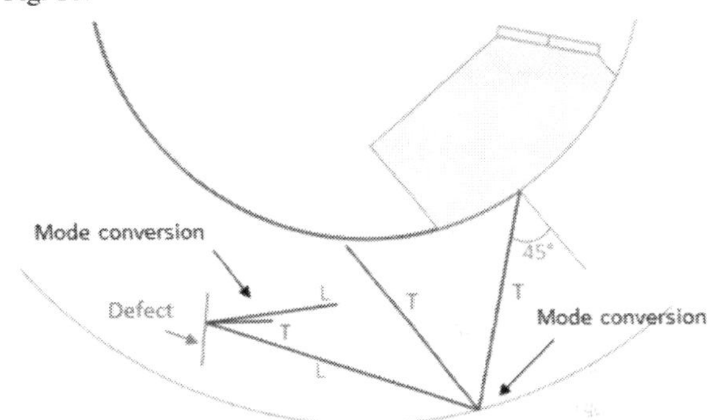

Fig. 8. T45° radiated beam, mode conversion on the back-wall and on the defect.

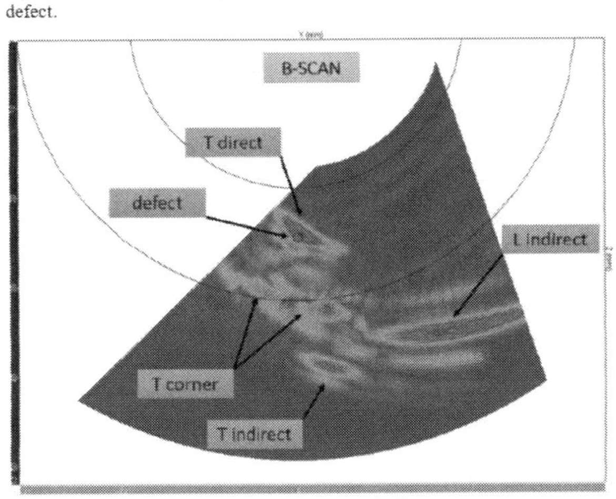

Fig. 9. Identified echoes: 1 mm SDH at 5 mm depth.

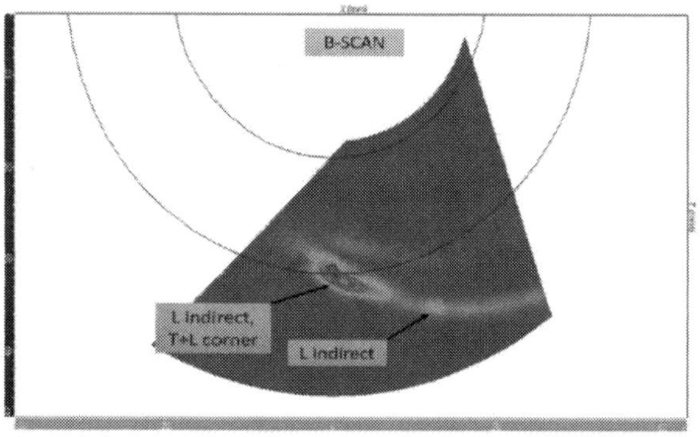

Fig. 10. Identified echoes: 1 mm SDH at back wall.

For beam focusing, 2 MHz (7x7 mm rectangular) and 10 MHz (8x8 rectangular) spherical sensor beam profiles were calculated in order to see the range of possible focus. For the 2 MHz probe, 10 mm spherical radius produces a 1.6 mm x 4.9 mm focal spot (-3 dB) in the inner surface. Increasing the spherical radius will not move the focal point (strongest point of the beam) to the outer surface but instead the beam turns effectively into a flat sensor beam. All of the tube body is, however, in the focal spot area from 50 mm radius onward. Spherical 200 mm radius produces a 4.0 mm times 17.5 mm focal spot. The beam profiles are shown in Fig. 11.

For the 10 MHz probe, 10 mm spherical radius places a 0.3 mm times 1.0 mm focal spot into the inner surface. Using 25 mm spherical radius, the focal spot is at the other surface and its size is 0.7 mm times 5.7 mm. 10 MHz beam profiles are shown in Fig. 12.

These simulated beams show that, for a given probe and frequency, it will be possible to compute the radiated beam with CIVA in order to check the insonification of the specimen thickness and the focal spot dimension and adjust the probe's parameters and frequency if necessary.

With the use of array sensors focusing will be much easier, as they provide wide possibility of focus.

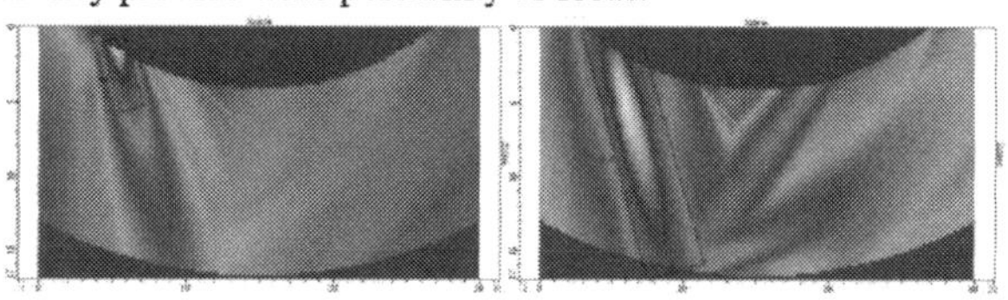

Fig. 11. Beam profiles: 2 MHz, 7x7mm rectangular probe, spherical focus r=10 mm (left) and 200 mm (right).

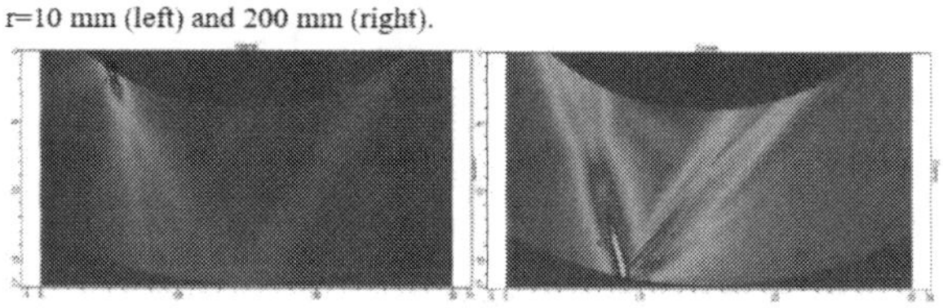

Fig. 12. Beam profiles: 10 MHz, 8x8mm rectangular probe, spherical focus r=10 mm (left) and 25 mm (right).

C. Eddy current simulation

Eddy-current testing is one of many electromagnetic testing methods used in non-destructive testing making use of electromagnetic induction to detect and characterize surface and sub-surface flaws in conductive materials. It is why eddy current control simulation has been done in order to evaluate the detection of corrosion and surface defects. Conventional coil sensor has been chosen.

As not all needed geometrical details were available in manufacturer catalogues, a reasonable assumption for a sensor was identified.

A representation of a single eddy current simulation coil is shown in Fig. 13.

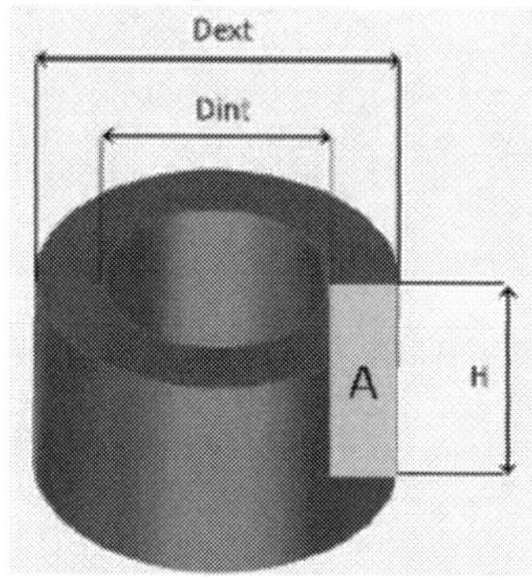

Fig. 13. Eddy current coil dimensions.

In actual examination, many of these coils would be used in array setup. Unlike ultrasonic array probes, in evaluation of defect response, eddy current probes gain only larger inspection area from array use and a simulation of a single probe is sufficient for determining the response.

It is assumed here that half of the coil area is conductive material (wire). Area of a single coil wire is then calculated as follows:

$$A_{s.\,coil} = \frac{0.5 \cdot H \cdot (Dext - Dint)/2}{n_{coil}} , \tag{1}$$

where $A_{s.coil}$ is the area of a single coil wire, H is the height of the coil, D_{ext} is the coil external diameter, D_{int} is the coil internal diameter and n_{coil} is the number of turns in one coil.

Usually, 5 amperes per square millimetre can be put through a single coil wire. Therefore maximum current for the simulation is:

$$I_{max} = 5 \cdot A_{s.\,coil} , \tag{2}$$

where I_{max} is the maximum usable current in the simulation. Using values shown in Table II, we end up with maximum usable current of 28 mA. Value 25 mA was chosen for simulations.

TABLE II
EDDY CURRENT SIMULATION COIL VALUES

Acronym	Value	Unit	Note
H	1.5	mm	Height of the coil
D_{ext}	2.0	mm	Coil external diameter
D_{int}	0.8	mm	Coil internal diameter
n_{coil}	80	-	Number of turns in one coil
A	0.9	mm²	Area of the coil bundle
$A_{s.coil}$	0.005625	mm²	Area of a single coil wire
I_{max}	0.028125 (25 mA chosen)	A	Maximum current in eddy current simulation

Another type of eddy current sensors could be used (bobbin probe), having a much larger diameter, approximately the diameter of the tube to be controlled. It would be moved along (for external inspection) or within (for internal inspection) the tube, giving a more global information on the specimen state.

For simulation, parametric variations of lift-off were performed at 1 MHz frequency. In these simulations, a scanning pattern shown in Fig. 14 was used. As result a C-scan and a maximum amplitude of each permutation were obtained. Lift-off values ranged from 0 mm to 1.8 mm in 0.2 mm increments.

Defects in these simulations were vertical and rectangular. The defect sizes and used mesh points are shown and explained in Table III and Fig. 14. Fig. 15 visualizes the loss of resolution when a higher lift-off is used. The results of the parametric variations are shown in Fig. 16.

As the noise of close to sensor electronics is in the range of 1 nV, it is difficult to detect the signals with amplitudes of the same value. That is why the defects detection limit is defined as 1 nV in amplitude. All of the simulation results are over this value.

In this regard, these results are promising even for corrosion detection (0.05 mm thickness) but no conclusions should be drawn at this early state. Experimental testing with a mock-up will give a reference for future simulations.

As for the preferred value of lift-off, a smaller value is better. How small this value can be will likely depend on robotic and mechanical aspects. If contact probes cannot be used, handling accuracy will define a limit for the lift-off. In order to have the possibility to move the sensor inside the tube without it damaging the tube, lift-off range should be 0.1-0.2 mm.

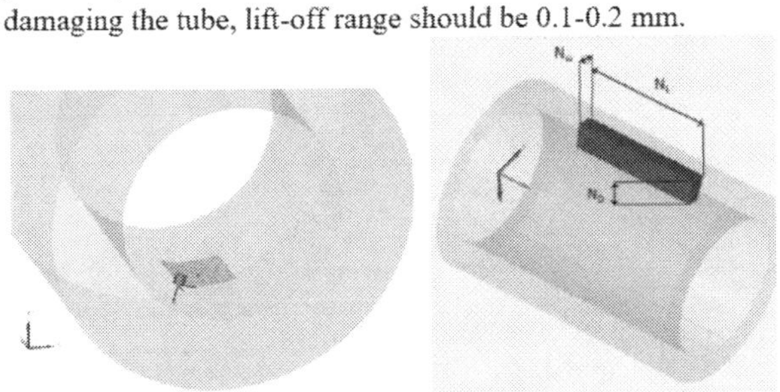

Fig. 14. Scanning pattern in eddy current simulations with geometrical explanation of Table III variables.

TABLE III
DEFECT MESH POINTS IN THE EDDY CURRENT LIFT-OFF VARIATION

	NL x ND x NW 0.5 x 0.5 x 0.1 mm	NL x ND x NW 0.2 x 0.2 x 0.1 mm	NL x ND x NW 0.1 x 0.1 x 0.1 mm	NL x ND x NW 0.5 x 0.05 x 0.1 mm
Mesh points	10 x 10 x 2	4 x 4 x 2	2 x 2 x 2	10 x 1 x 2

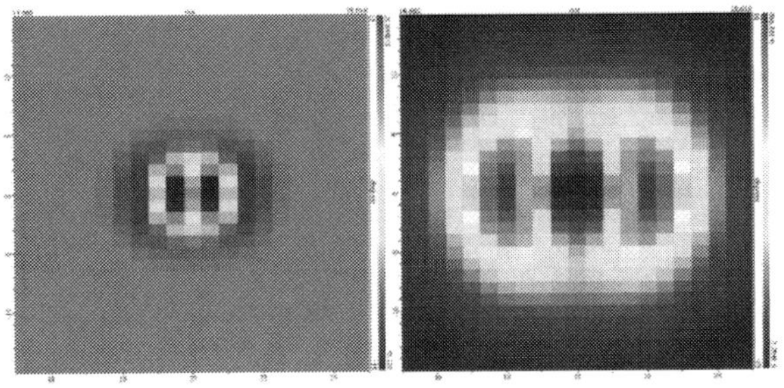

Fig. 15. Eddy current C-scan, 0.5 mm x 0.5 mm x 0.1 mm defect, real part : 0.2 mm lift-off (left C-scan), 1.8 mm lift-off (right C-scan).

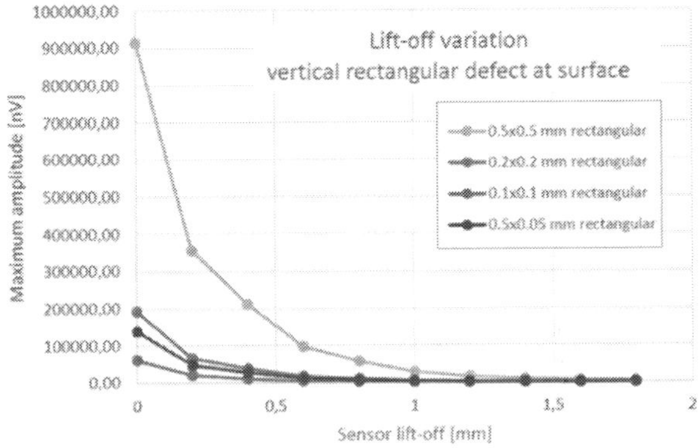

Fig. 16. Eddy current maximum amplitudes in lift-off variation simulations (defect width=0.1 mm).

V. Proceeding to mockup testing

Proceeding to experimental phase will be mandatory for establishing the NDT methods qualification. CIVA simulations are performed in two steps: a first step before experimental phase for guiding the preliminary choice of NDT methods and associated sensors, and a second step after experimental phase for ranging various relevant parameters and confirming methods and sensors, as not all aspects of inspection could be taken into account in CIVA simulations. The final sensor or sensors will need to be determined.

For Eddy current inspection, a Zircaloy-4 calibration block is needed. Reference defects will need to be introduced and their sizes to be decided.

As of now, the Eddy current inspection is suspended and the joint effort is focusing on the volumetric ultrasonic inspection: two representative mock-ups of the cylinders, each of them including a weld and artificial defects (reflectors) have been manufactured (Fig. 17). Some side drilled holes and slots will be machined within the molten area of the weld, and also in the Heat Affected Zone, as reflectors for inspection tests.

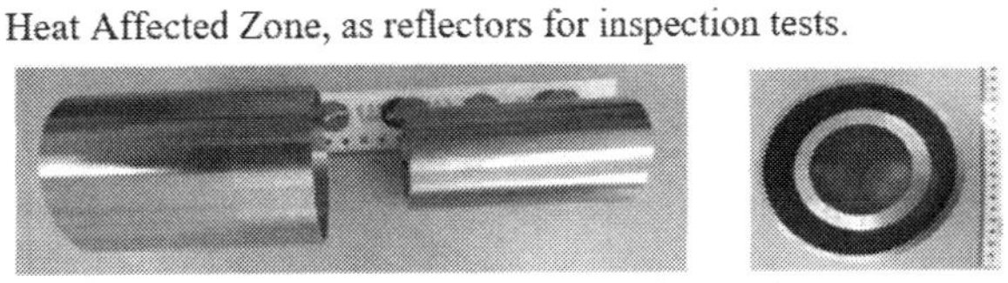

Φe68mm e4mm L110mm and Φe58mm e8mm L108mm

Fig. 17. Representative mock-ups with Electron Beam weld join

CEA will perform in water tests with a linear phased array (128 elements at 15MHz) and VTT will make the corresponding CIVA simulations.

VI. Conclusions

Conclusions of this work can be summarized as follows (2020 position):
- Two inspection methods are likely needed:
  - o Eddy current is likely suitable for surface examination
  - o volumetric ultrasonic inspection can be used for tube bodies.
- -Under-water storage pool is likely the best inspection location.
- Since it is assumed that all of the tube body has to be inspected with the same resolution as the welds, area to be inspected becomes rather large. Therefore, it is likely that some level of automation in the scanning process is needed.
- It is very likely that the surface between the inner and outer tubes cannot be inspected for corrosion issues. If tubes were to be welded to each other and the space between would be filled with inert gas, probability of corrosion occurrence would be low.
- Experimental reference as the next step is critically important.

As the next step (in 2021), CEA and VTT are continuing the inspection work with two representative mock-ups (with Electron Beam welds). Artificial defects have been specified (side drilled holes and slots).

Under water tests will provide an experimental reference to CIVA calculations.

Proceeding to experimental phase is mandatory for establishing the NDT methods qualification. CIVA simulations are performed in two steps: a first step before experimental phase for guiding the preliminary choice of NDT methods and associated sensors (linear phased array at about 15MHz), and a second step after experimental phase for confirming methods and sensors, as not all aspects of inspection could be taken into account in simulations.

References

[1] CIVA NDT simulation software. Available: http://www.extende.com/civa-in-a-few-words
[2] F. Delassalle, "Minimal and maximal values for diameter, thickness and gas/water gap for each device section latest device", March 2020, unpublished
[3] RJH - ADELINE - Mecanics – Assembling internal tube device, A3 Drawing, December 2018, unpublished
[4] RJH - ADELINE - Mecanics – Assembling internal tube device, A3 Drawing, December 2018, unpublished
[5] Test program for the machining and assembly of Zircaloy, September 2019, TESTE society, unpublished
[6] Electron beam weld test on a Zircaloy mockup, October 2019, BODYCOTE-TECHMETA society, unpublished
[7] French Nuclear Pressure Equipment Regulation, 2005
[8] Design and Construction Rules for Mechanical Components in high-temperature structures, experimental reactors and fusion reactors, RCC-MRx. Available: AFCEN, 2018
[9] S. Gaillot, "RJH Preliminary estimation of radioactive doses within irradiation device tubes during mandatory periodical inspection", 2017, unpublished
[10] Olympus NDT Solutions, Small-Footprint Probes. Available: https://www.olympus-ims.com/en/probes/small-footprint/

Some Considerations on the Energy Deposition During a RIA Transient Based On Monte Carlo Simulations

Julia Bartos[1], Adrien Gruel[1], Claire Vaglio-Gaudard[2] and Christine Coquelet_Pascal[1]
[1]CEA, DES,IRESNE,DER,SPESI,LP2E, Cadarache, 13108 Saint Paul lez Durance, France
[2]CEA, DES,IRESNE,DER,SESI, Cadarache, 13108 Saint Paul lez Durance, France
Corresponding author: julia.bartos@cea.fr

***Abstract* — Specific research reactors are capable of reproducing reactivity injection accidents in order to study the behavior of the nuclear fuel pins in accidental situations. In the CABRI research reactor, the fuel pin to be examined (test pin) is placed in the center of the core in a dedicated test loop. It is then subjected to a power transient, obtained by the fast depressurization of the ^{3}He neutron absorber gas from the transient rods located in the core. One of the central parameters of the experiment is the energy deposition in the test pin, which is currently not measured during a transient. Instead, it is assumed that the relative energy distribution between the core and the test pin is constant regardless the operational state of the reactor. Currently, this correlation is measured in steady state. As such, the impact of the variations in the neutron flux, fuel and moderator temperatures during the transient is assumed equivalent on the energy deposition in the core and in the test pin. The goal of this work is to improve our knowledge on the mechanisms involved in the transient energy deposition. The aim of this paper is to present a methodological approach for the energy deposition estimation during a CABRI transient, based on static Monte Carlo calculations. The results suggest that the transient energy deposition rate is mainly dependent on the helium pressure and the Doppler feedback, and the relative energy distribution between the core and test pin changes during the transient.**

***Keywords* — Energy deposition, CABRI, RIA, Monte Carlo calculations, TRIPOLI4**

I. Introduction

The term Reactivity Initiated Accident (RIA) refers to nuclear accidents involving a sudden and unexpected reactivity increase in the core. It is characterized by an extremely rapid power excursion with a significant energy deposition in the fuel assemblies surrounding the perturbation. In order to minimize the possibility of fuel failure and its consequences (such as the fuel dispersal in the primary circuit), the fuel safety criterion associated with this kind of accident is defined as a limit on the energy deposited in fuel pins [1].

Specific research reactors are capable of reproducing reactivity injection accidents in order to study the behavior of the nuclear fuel pins in accidental situations. In the CABRI research reactor [2], funded by the French Institute for Radioprotection and Nuclear Safety (IRSN) and located in the Cadarache research center of the French Alternative Energies and Atomic Energy Commission (CEA), the fuel pin to be examined (test pin) is placed at the center of the core, in a dedicated test loop, and is then subjected to a power transient obtained by the fast depressurization of the neutron absorber ^{3}He gas from the transient rods located in the core. The main parameters of the power pulse such as the peak power, the pulse length and the total energy deposition, depend on the operational characteristics of the reactor and are adjustable by experimental design.

The energy deposited in the test fuel pin during the transient is currently not measured. It is usually derived from the core characteristics at steady state, assuming a constant relative power distribution in the core and the test pin during all operational states of the reactor. The aim of this work is to improve our knowledge of the mechanisms involved in the energy deposition and relative energy distribution during a transient with static Monte Carlo calculations.

The transient is simulated by running several static calculations with different core configurations representative of the different stages of the transient. In order to define these core configurations, the different parameters (physical, experimental or computational) that could influence the relative energy distribution between the core and the test pin are first identified. This step is followed by a sensitivity study to determine the main parameters that the energy deposition depends on.

The methodologies currently used for estimating the energy deposition along with the summary of the parameters of interest is presented in section II. The results of the sensitivity studies are discussed in section III.

II. Calculation Methodology

The methodology presented in this section is based on the use of the TRIPOLI-4® 3D continuous energy Monte Carlo neutronics code developed by CEA for particle transport calculations [3]. The required nuclear data, such as cross sections and fission spectra come from the JEFF-3.1.1. evaluation file [4].

The reactor core is accurately modeled with a 3D native TRIPOLI-4 geometry (see Fig. 1), and the initial core configuration corresponds to the critical state of the CABRI reactor at 100 kW with a helium pressure of 11.4 bar in the transient rods. The test pin is a fresh 6% enriched UO_2 pin, identical to those found in the core. The initial moderator parameters in the pressurized water loop are considered identical to the core: water at ambient pressure and 20°C.

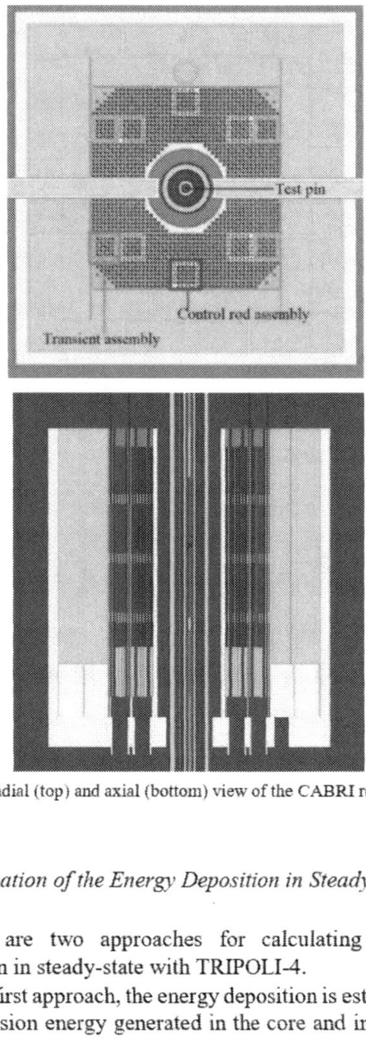

Fig. 1. Radial (top) and axial (bottom) view of the CABRI reactor core

A. Estimation of the Energy Deposition in Steady-State

There are two approaches for calculating the energy deposition in steady-state with TRIPOLI-4.

In the first approach, the energy deposition is estimated based on the fission energy generated in the core and in the test pin. First, the fission rates ($\tau_{fission,i}$) are computed with TRIPOLI-4® for each fissile isotope i both in the core and in the test pin. The power generated is then calculated by taking into account the individual contribution of each fissile isotope based on their reaction rate and their fission energy ($E_{fission,i}$)[1], as shown in (1). Finally, the energy deposition can be calculated by integrating the core and the pin power over the defined duration.

$$P = \sum_i \tau_{fission,i} \times E_{fission,i} \qquad (1)$$

In this approach, however, it is assumed that the energy generated by fission, which appears as the kinetic energy of fission products, neutrons, γ and β particles is deposited entirely where the fission took place. Although the largest part of the generated energy is transmitted to the fission products, which convey their energy to the surrounding medium, that is to say the fuel pin, without being transported, the energy carried away by the prompt and delayed neutrons, γ and β particles can be deposited in the surrounding fuel pins, moderator or structural materials.

This spatial repartition of the energy generated during fission can be taken into account in the nuclear heating scheme. Nuclear heating is expressed in terms of energy deposition in a given volume.

The total nuclear heating is composed of:

- prompt heating which comes from the kinetic energy of fission fragments, kinetic energy of neutrons, energy of prompt photons emitted by neutron interactions (mainly fission, radiative capture and inelastic scattering);
- delayed heating which comes from the energy deposition of delayed particles (mainly photons) emitted by the decay of fission and activation products.

The prompt and delayed heating can be estimated separately. The transport of 4 (neutrons, photons, electrons and positrons) or 3 particles (photons, electrons and positrons) is simulated in 3D TRIPOLI-4® calculations, depending whether the prompt or delayed heating is calculated. The electromagnetic cascade shower is modelled with both electrons and positrons in the TRIPOLI-4® calculations. The photon transport is made on the basis of the EPDL97 photo-atomic library developed at Lawrence Livermore National Laboratory [4].

The prompt heating is obtained by running a criticality calculation in TRIPOLI-4®. In this approach, the energy allocated to the charged particles and recoil nucleus is calculated by subtracting the energy carried away by transported particles from the initial energy released from fission. The energy deposition can be directly computed in TRIPOLI-4®.

Prior to calculating the delayed nuclear heating, the delayed source has to be defined. Given that TRIPOLI-4® does not model the decay of fission and activation products, the delayed γ and β sources associated with the decay of fission and activation products are determined with the DARWIN [5] depletion code. This source is then propagated with a fixed-source TRIPOLI-4® calculation, in which the delayed contribution to the nuclear heating is computed.

Table I presents the instantaneous energy deposition in the test pin (fuel and cladding) and in the core, normalized to 1 MW core power and calculated with the two methodologies presented in this section. When estimated with the fission energy scheme, this value represents the energy generated by all fissions in the test pin and assumed to be deposited

[1] 193,72 MeV for ^{235}U and 197,31 MeV for ^{238}U

entirely within the pin. While when estimated with nuclear heating it represents the energy deposited in the test pin by taking into account the contribution to the energy deposition of the particles coming from the core and the energy 'loss' due to the particles that migrate outside the pin, and thus deposit their energy elsewhere.

TABLE I
ENERGY DEPOSITION VALUES IN THE TEST PIN, NORMALIZED TO 1MW CORE POWER, CALCULATED WITH THE FISSION ENERGY AND THE NUCLEAR HEATING SCHEME

Fission energy scheme		**Nuclear heating scheme**	
Power deposition (W/MW)	Uncertainty[2] (%)	Power deposition (W/MW)	Uncertainty (%)
2113	0,1	1948	0,3

Table II presents the decomposition of the energy deposition. The majority of the energy deposition is due to the fissions taking place in the pin ('pin to pin'), however, as suggested from prior calculations, 13% of the heating generated in the test pin is deposited outside of the test pin ('pin to outside'), while 1% of the deposited energy is due to the particles coming from the core. The delayed heating contribution in the test pin is around 3% of the total heating. In a power reactor in equilibrium cycle condition, this value is about 7%. The discrepancy is due to the short irradiation duration considered for these calculations (5 minutes).

TABLE II
DECOMPOSITION OF ENERGY DEPOSITION BY SPATIAL DISTRIBUTION AND BY PROMPT AND DELAYED COMPONENTS, NORMALIZED TO 1MW CORE POWER

	Prompt energy deposition (W/MW)	**Delayed energy deposition (W/MW)**	
		Gamma (W/MW)	Beta (W/MW)
Pin to pin	1878 ± 0,3 %	9,75 ± 0,03 %	43,75 ± 0,02 %
Core to pin	15 ± 0,7 %	1,57 ± 3 %	0,13 ± 4 %
Pin to outside	250 ± 1,4%	28,22 ± 0,03 %	4,69 ± 0,02 %

Even though there are different approaches for estimating the energy deposition, all methodologies are derived from the same physical quantities, calculated in each case with TRIPOLI-4®, the neutron flux and reaction rates. Since the stationary transport equation is solved, the impact of the variation of the neutron flux and spectra during the transient is not taken into account.

Further approximations are introduced by the fact that the kinetic parameters and Doppler coefficient are considered constant, whereas these parameters may vary during the transient. Moreover, neither calculation scheme takes into account the thermal-hydraulic feedbacks and the thermo-mechanical behavior of the fuel during the transient.

[2] The uncertainty values presented in this paper refer to the Monte Carlo statistical uncertainty, at one standard deviation

The possible impacting parameters on the energy deposition are summarized in the following section.

B. Identification of the Impacting Factors on the Transient Energy Deposition

The impacting parameters on the transient energy deposition can be classified as experimental, physical or calculation parameters.

In the case of the CABRI reactor, the main parameters characterizing the power pulse such as the peak power, the pulse length and the total energy deposition, depend on the operational characteristics of the reactor and are adjustable by experiment design. The initial pressure of the ^3He gas and the kinetics of the depressurization determine the maximum core power attained during the test and the duration of the transient while the total energy is finely tuned with the timing of the control rod drop. Similarly, the choice in the test pin, i.e. UOX or MOX pin, its burnup or enrichment can also influence the energy deposition during a transient.

The helium depressurization leads to a rapid power excursion (length of the power pulse between 8 to 100 ms) and instantaneous temperature increase in the core. The resulting fuel temperature increase (in the order of hundreds of K) terminates the power surge due to the Doppler feedback. The impact of the increase of the moderator temperature and the subsequent density decrease in both the core and the test loop, although slower and less important in amplitude, needs also to be considered.

The reactor core, including the test loop, is accurately modelled with a 3D native TRIPOLI-4® geometry. However, there are several approximations made in the modelling of the fuel pins. The isotopic composition of the fuel, as well as the radial temperature profile within the fuel pins, is considered homogeneous. Yet, for instance in the case of a test fuel with a high burnup, the initial uniform fissile material distribution is changed due to the enhanced neutron resonance capture on the pellet periphery, leading to a local buildup of transuranium isotopes at the pellet outer ring, called the rim region. The radial temperature profile also varies during a transient as discussed in [6]. In addition, local geometry changes in the pellet and/or clad during the transient may affect the heat exchange between the different media, that could lead to rapidly varying temperatures.

In conclusion, the main impacting parameters on the energy deposition and its calculation are the following:

- Experimental parameters:
  - initial helium pressure and depressurization kinetics (short or long pulse)
  - test pin characteristics (UOX or MOX fuel, burnup and enrichment)
- Physical parameters:
  - reactivity, thermomechanical and thermal-hydraulics feedbacks

- Simulation parameters:
  - modelling approximations (concerning the temperature and isotopic concentration discretization for the fuel pin), nuclear data library
  - calculation methodology: estimation based on fission energy or based on nuclear heating
  - computational scheme: static or kinetic Monte Carlo calculation

The sensitivity analysis presented in this paper is focused on the variations in both the physical and experimental parameters, more precisely the helium pressure and fuel and moderator temperature, and their impact on the energy deposition during a transient, presented in the next section.

III. Sensitivity studies on the transient energy deposition

The dynamic aspect of the transient is approximated with static TRIPOLI-4® calculations by running several calculations with different core configurations, i.e. simulation points, representative of the different stages of the transient. At each of these simulation points the helium pressure, the fuel and moderator temperatures are modified. The aim of this approach is to account for the reactivity variation (due to both the helium pressure and the reactivity feedbacks) associated with the studied transient, in order to determine the energy deposition during the transient. Given that this is a sensitivity analysis, it was assumed that the simplified approach relying on static calculations is relevant for the identification of the parameters influencing the energy deposition. Then at each simulation point, an instantaneous energy deposition is estimated based on the fission energies, as shown in (1).

Fig. 2 presents the characteristics of the hypothetical transient used for this study. The transient characteristics, such as the helium depressurization, the average fuel temperature and core water temperature variations are obtained with the SPARTE code. SPARTE [2] (Simulation Prediction and Analysis for RIA Transients an Excursions) is a recently developed code dedicated to R&D studies on the physical characteristics of the CABRI transients, such as the power peak, transient length and power trace.

The moderator temperature and consequently the moderator density variation is considered identical during the transient. SPARTE computes one average transient temperature profile for all the fuel pins in the core. Given that the test pin is identical to the core fuel pins, it is assumed that the temperature profile of the test pin can be reproduced based on the temperature profile of the pin with the highest radial form factor.

A. Variations in Energy Generation Induced by the ^3He Depressurization

As described in chapter II, the energy deposition is estimated based on the reaction rates and fission energies. In order to understand the behavior of the energy deposition in the test pin during the depressurization, we have to study the normalized fission rate distribution in the core.

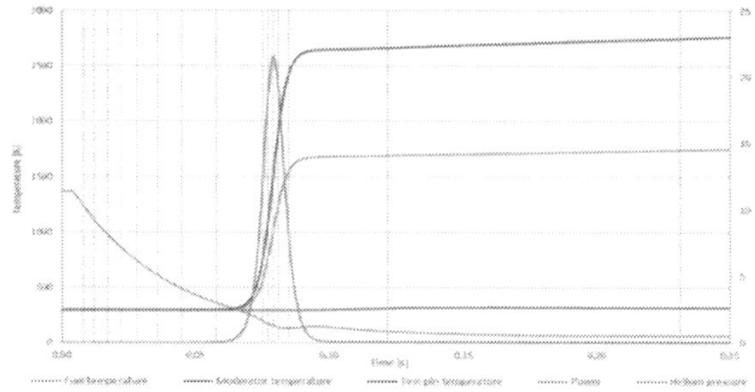

Fig. 2. Pressure (blue) and temperature (red for the test pin and orange for the pins in the core) during the studied transient and the chosen simulation points. Results obtained with the SPARTE code

At each simulation point, the reaction rates are normalized to the average fission rate. This normalization is applied on all fission distribution results presented in this paper.

Fig. 3 (left) shows that in steady state, prior to the depressurization, the reaction rate is peaked in the center of the core, with the lowest values in the transient rods and in the periphery due to the increased absorption and neutron leakage in those regions. On the right-hand side figure, this peak is observable even after the depressurization, however it is less important, given that the fission rate in the transient rods increases by ~ 60%, while there is no significant change in the fission rate in the rods close to the center. Concerning the test pin, the fission rate decreases by 4% between the two states.

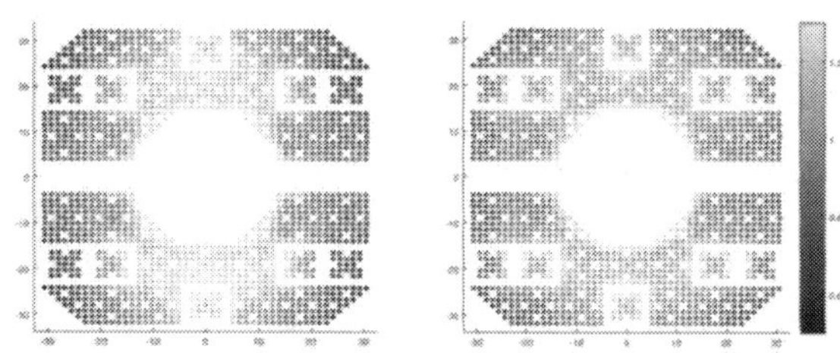

Fig. 3. Fission rate distribution in the core at steady state (left) and after the helium depressurization (right). Values normalized to the k_{eff} and the average fission rate

Fig. 4 depicts the energy deposition in the core and test pin estimated from fission energies according to the helium pressure. As suggested by the fission rate variations, the energy deposition in the core is monotonically increasing during the helium depressurization, due to the increased fission rate in the transient rods. The deposition rate in the test pin however, is decreasing, as suggested by the flux variation. The statistical relative uncertainties are within the marker size, around 0.2%.

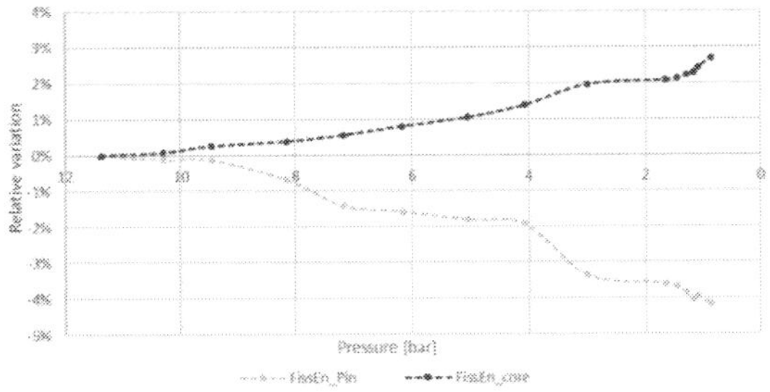

Fig. 4. Relative variation of the energy deposition in the core and test pin according to the helium pressure

B. Variation in Energy Generation Induced by Doppler Feedback

The helium depressurization leads to an extremely rapid power excursion and fuel temperature increase. With the increasing temperatures the neutron flux decreases as a result of the Doppler broadening of the resonance capture cross section. The individual impact of the Doppler feedback can be seen on Fig. 5. For these simulations, the helium pressure was kept constant while the temperature increased according to Fig. 3.

Contrary to the helium depressurization, the Doppler feedback has the same impact on the energy deposition rate in the core and in the test pin. While the depressurization has local effects in the reaction rates in the core (see Fig. 3), the Doppler feedback has a global and instantaneous impact.

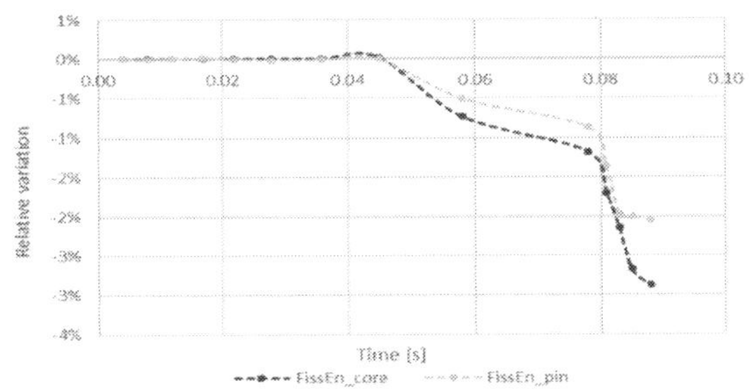

Fig. 5. Relative variation of the energy deposition in the core and test pin due to the Doppler feedback during the transient

As Fig. 6 shows, the fission rate distribution itself does not change with the increasing temperature. It is to be noted, that in the current calculation scheme, each fuel pin in the core is assumed to have the same temperature. This approximation needs to be improved in further studies.

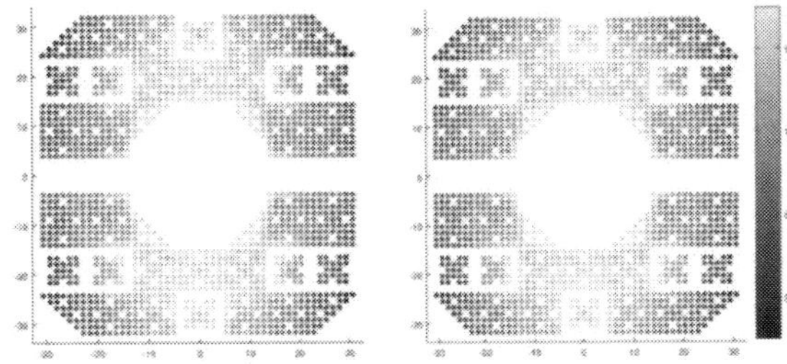

Fig. 6. Fission rate distribution at steady state (left) and after the transient (right), normalized to the k_{eff} and average fission rate

C. Variations in Energy Generation Induced by Moderator Feedback

The moderator temperature increase, and consequently its density decrease appears with a delay compared to the Doppler feedback. As discussed in [7] it takes around 150 ms for heat transfer to take place between the fuel and the surrounding moderator (Fig. 8). Given the duration of the studied transient, 9 ms, the moderator effect is not relevant for the energy generation during the transient. As Fig. 7 shows, the relative variation in the energy deposition is under 0.3%, which is within the convergence uncertainty of the calculations.

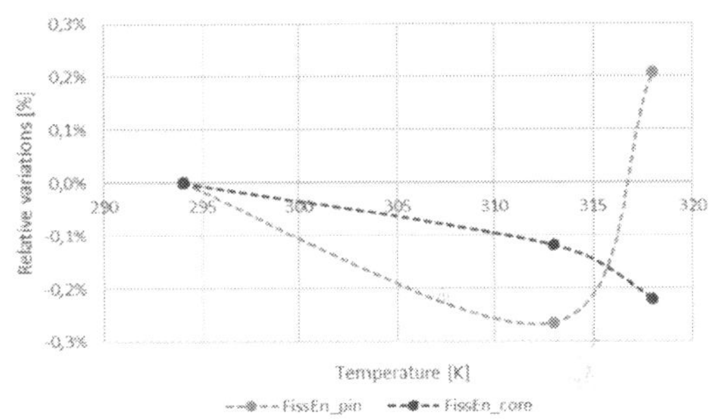

Fig. 7. Relative variation in the energy deposited in the core and in the test pin versus the moderator temperature increase

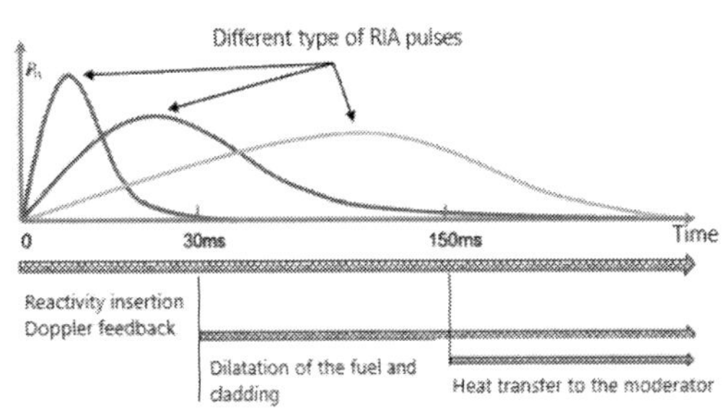

Fig. 8. Chronology of the reactivity feedbacks depending on the length of the transient [7]

D. Variations in Energy Generation During a Transient

Fig. 9 presents the relative variation of the energy deposition in the test pin and in the core according to the time. Given that these are static calculations, the time and the simulation instants actually represent a set of values for the initial parameters (helium pressure, fuel and moderator temperature) representative of different stages of the hypothetical transient chosen to study.

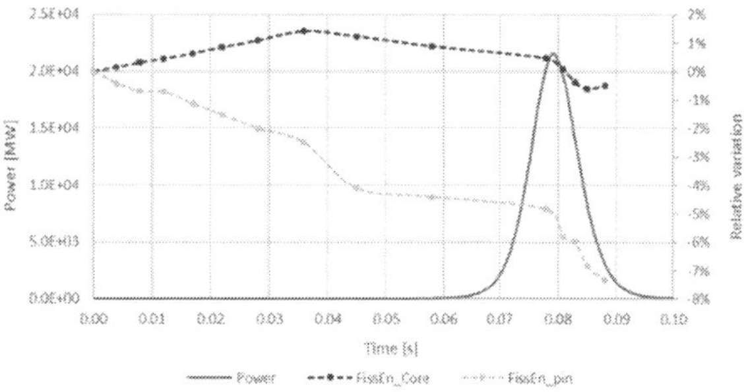

Fig. 9. Relative variation in the energy deposition in the core and test pin during the transient

The general trend in the test pin is the decrease of the energy deposition, all throughout the transient. Special attention needs to be given to the variations within the power peak itself, since the major part of the energy deposition happens during the power peak. When considering the power peak, current results suggest that the most impacting parameter on the energy deposition is the Doppler feedback. The energy deposition rate during the peak decreases by 0.5% when only the helium pressure varies, while it decreases by 2% when the fuel temperature increases. The corresponding relative uncertainties are around 0.2% (within the marker size).

Whereas, on the core level, the helium depressurization and the Doppler feedback have opposite effect on the energy deposition. During the early stages of the transient the energy deposition rate is increasing due to the lower helium pressure, however, as soon as the fuel starts to heat up, the Doppler feedback and the resulting increased absorption rate lead to a steady decrease in the energy deposition rate. This decrease is more important during the power peak, according to the jump in the temperature increase.

IV. Conclusions and Perspectives

This paper presents methodological elements, based on impact studies, aimed at improving our knowledge of the mechanisms involved in the energy deposition during a transient with static Monte Carlo calculations. The transient is modeled by running several static TRIPOLI-4® calculations with different core configurations representative of the different stages of the transient.

The steady state energy deposition is computed both based on fission energies and nuclear heating for the test case studied in order to study the limitation of the fission energy scheme concerning the spatial repartition of the deposited energy. In the first approach, the energy generated by the fissions in the test pin is assumed to be deposited locally. Computing the energy deposition with the nuclear heating scheme, which takes into account the spatial distribution of the energy, shows that the former approach overestimates the energy distribution in the pin by 11%. When looking only at the energy distribution following a fission in the test pin, the calculations show that 13% of the energy generated is deposited outside the pin, while 1% is of the energy deposited in the test pin is due to the particles coming from the core. This is in accordance with the aforementioned overestimation.

Furthermore, in order to evaluate the most impacting physical parameters on the energy deposition during a transient, a sensitivity study is carried out. The results illustrate that the energy deposition is most dependent on the helium pressure variations and the Doppler feedback. A future objective of this study is to study the possible impact of other experimental parameters (control rod position, initial helium pressure, depressurization kinetics,..) as well as the impact of current modelling approximations on the energy deposition.

Although this approach does take into account the impact of the helium depressurization and of the resulting reactivity feedbacks, the dynamic behavior of the system can not be modelled with static calculations. The goal of the future calculations is to define the time-dependent energy deposition in the core and in the test pin during the transient with kinetic Monte Carlo simulations.

References

[1] OECD, "Nuclear fuel behaviour under reactivity-initiated conditions, state-of-the-art report," NEA/CSNI/R(2010)1, 2010.

[2] O. Clamens, P. Blaise, J.-P. Hudelot, J. Lecerf, B. Duc and B. Biard, "Coupled Experimental and Computational Approach for CABRI Power Transients Analysis," *IEEE Transactions on Nuclear Science,* vol. 65, no. 9, pp. 2434-42, 2018.

[3] E. Brun, F. Damian, C. Diop, F. Malvagi, O. Petit and A. Zoia, "TRIPOLI-4, CEA, EDF and AREVA reference Monte Carlo code," *Annals of Nuclear Energy,* vol. 82, pp. 151-160, 2015.

[4] A. Santamarina, B. Bernard and Y. Rugama, "The JEFF-3.1.1 Nuclear Data Library, JEFF Report 22," 2009.

[5] A. Tsilanizara, C. M. Diop, B. Nimal and et al., "DARWIN: An EvolutionCode System for a Large Range of Applications," *Journal of Nuclear Science and Technology*, vol. 37, pp. 845-849, 2014.

[6] I. Guenot-Delahaie, J. Sercombe, T. Helfer, P. Goldbronn, E. Frederici , T. Le Jolu, A. Parrot, C. Delafoy and C. Bernaudat, "Simulation of RIA transients on UO2-M5 fuel rods with ALCYONE V1.4 fuel performance code," *Water Reactor Fuel Performance Meeting Proceedings,* 2017.

[7] J. Labit, N. Marie, J-P. Hudelot and E. Merle, "An advanced validation experimental methodology of multiphysics calculation tools on CABRI transients," in *Proceedings of International Conference M&C*, Portland, USA, 2019.

Numerical and experimental characterization of the reaction rates in the core of the CNESTEN's TRIGA Mark II research reactor

H. Ghninou[1] (hamza.ghninou@cea.fr), A. Gruel[1], A. Lyoussi[1], C. Reynard-Carette[2], C. El Younoussi[3], B. El Bakkari[3], B. Nacir[3], Y. Boulaich[3], H. Bounouira[3]
[1]CEA, DES, IRESNE, DER (SPESI/LP2E), Cadarache F-13108 Saint-Paul-Lez-Durance, France
[2]Aix Marseille Univ, Université de Toulon, CNRS, IM2NP, Marseille, France
[3]CNESTEN/CENM, POB 1382, Rabat, Morocco

Abstract— **Education, training and isotopes production are the most important uses of the Moroccan 2 MW TRIGA Mark II reactor situated at the National Center for Energy Sciences and Nuclear Techniques (CNESTEN, Morocco). To develop new R&D projects in research reactors, the particular and advanced knowledge of neutron and photon flux distribution, within and around the reactor core, is crucial. In order to precisely preparing the experiments in the CNESTEN's TRIGA reactor, a detailed model was developed using the 3D continuous energy Monte Carlo code TRIPOLI-4 and the continuous energy cross-section data from the JEFF3.1.1 nuclear data library. This new model was used to carry out preliminary neutron and photon calculations to estimate flux levels in the irradiation channels as well as to calculate kinetic parameters of the reactor, core excess reactivity, integral control rods worth and power peaking factors. As a first step of the validation of the model, the obtained results were compared with the experimental ones available in the Final Safety Analysis Report (FSAR) of the TRIGA reactor. A study is being carried out at the end of which the results will be published as an evaluated benchmark. Furthermore, this work aims at experimentally characterize the reaction rates in various irradiation channels inside and outside the reactor core. The measurements are carried out using the neutron activation technique. To set up the experimental design for the activation experiments a series of preliminary calculations were performed using the TRIPOLI-4 model to calculate the expected gamma flux/intensity levels of various materials after irradiations in different positions in the irradiation facilities. Different activation foils with known characteristics are then irradiated and the activity of several isotopes is measured with the Gamma Spectrometry Method. The measured relative reaction rates are then compared with the calculated ones evaluated through the new TRIPOLI-4 reactor model. Fairly good agreement was found, which indicates that the new computational model is accurate enough to reproduce experiments.**

Keywords **—TRIGA reactor, neutron activation analysis, TRIPOLI-4 simulation, γ spectrometry.**

I. Introduction

The precise and advanced knowledge of the neutron flux is crucial to develop new R&D projects in research reactor and to benchmark the computational models of the reactor. The Moroccan TRIGA MARK-II research reactor, operated by the CNESTEN in Rabat, is a pool type light-water reactor, with a maximum steady state power of 2 MW.

A relevant effort has been made by the CNESTEN in the past few years to accurately model the reactor to support planning, designing and implementing new experiments in various fields of nuclear research as well as to interpret and analyze the corresponding experimental results [1] [2] [3] [4].

The present paper describes the preliminary characterization of reaction rates in various irradiation and experimental channels of the CNESTEN's TRIGA Mark II research reactor. For this purpose, different metallic foils were irradiated and the resulted activities were analyzed *via* γ spectrometry method. Through comparisons of measured and calculated reaction rates, the computational scheme of the reactor can be experimentally validated.

This study was conducted in the framework of the bilateral collaboration between the CEA-Cadarache and the CNESTEN.

II. The CNESTEN's TRIGA Mark II reactor

The 2 MW TRIGA (Training Research and Isotope production General Atomics) Mark II is a research reactor designed and manufactured by General Atomics. It is a pool-type light water moderated and cooled reactor. The reactor core is immerged into an 8 m high and 2.52 m diameter tank filled with water, it is a hexagonal array type consisting of seven concentric rings where the cylindrical fuel elements (FEs) are arranged, and two aluminum grid plates which provide accurate spacing between the FEs.

The reactor core assembly is composed of 101 FEs, including 5 fuel-follower control rods, 17 graphite elements, a central channel and a pneumatic transfer system (Fig. 1). The FEs of the TRIGA reactor are contained within cylindrical rods made of stainless steel. The height of the active fuel region of the FE is 38.1 cm, the upper and lower endcaps are designed to enable safe positioning in the core. Two cylindrical graphite elements, serving as axial neutron reflectors, are stoppered on the top and bottom end of the fuel region (Fig. 1 A).

The fuel material consists of a uniform mixture of uranium (8.5 %wt, enriched to 19.7% of ^{235}U) and zirconium hydride leading to a neutron moderation inside the fuel, which makes the reactor inherently safe and suitable for manpower training, nuclear research and radioisotopes production. The reactor is also equipped with a thermal column, a rotary groove and 4 beam ports channels that are clad with aluminum and filled with air. Beam port NB1 is a tangential port, whereas beam ports NB2, 3 and 4 are radially oriented ports (Fig. 2).

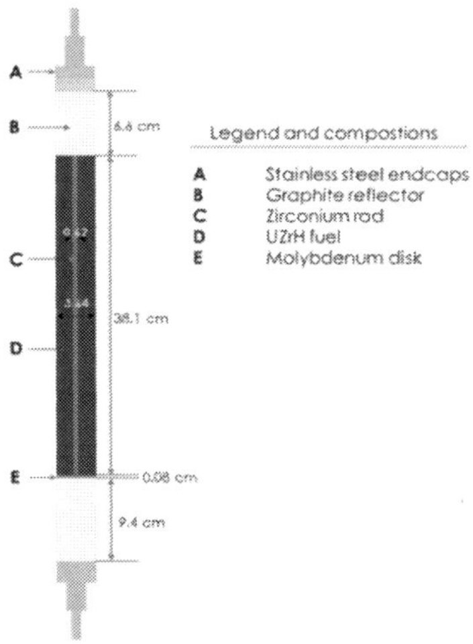

(a)

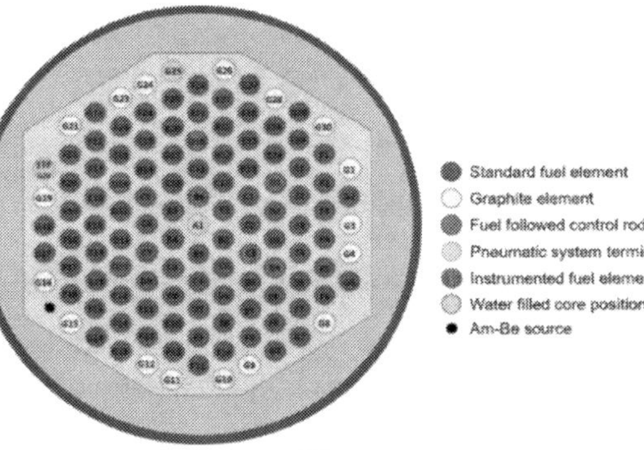

(b)

Fig. 1. (a) TRIGA standard fuel element, (b) the actual core configuration of the TRIGA reactor.

The reactor core reactivity is controlled by means of five absorbing rods, labeled SHIM I, SHIM II, SHIM III, SHIM IV and REGULATING (located in D4, D10, D13, D1 and D16 respectively). They are contained within stainless steel cylindrical rods. The lower part of the rod is filled with fuel, whereas the upper part is filled with B_4C neutron absorber. When the control rods are fully inserted, the upper part with neutron absorber takes place inside the reactor at the same level of the active fueled section of fuel elements.

A. Computational model

A detailed new computational model of the TRIGA Mark II reactor was developed using the 3D continuous energy Monte Carlo code TRIPOLI-4.11 [3] and the continuous energy cross section from the JEFF3.1.1 nuclear data library [4]. In this new model, FEs, graphite elements and control rods are modelled with full available precision, meaning that the zirconium rod, stainless steel cladding and endcaps, air gaps and molybdenum supporting disc are explicitly described. Furthermore, the supporting grids, graphite reflector, annular lead bloc, thermal column, rotary groove and beam ports are also taken into account in this model. In order to expand its utilization and to verify its validation against the MCNP reference computation model as well as against experiment [1], this new model was used to carry out preliminary neutron calculations to estimate flux levels in the irradiation channels, kinetic parameters of the reactor, control rods worth and power peaking factors. The preliminary obtained results have shown a good agreement compared to MCNP and experimental ones, and will be soon published as an evaluated criticality benchmark.

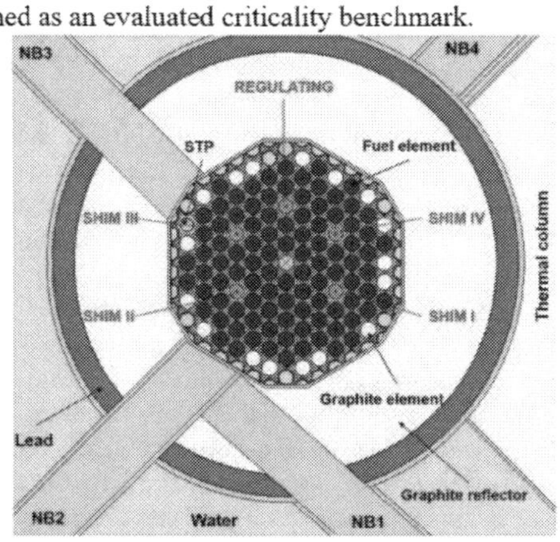

(a)

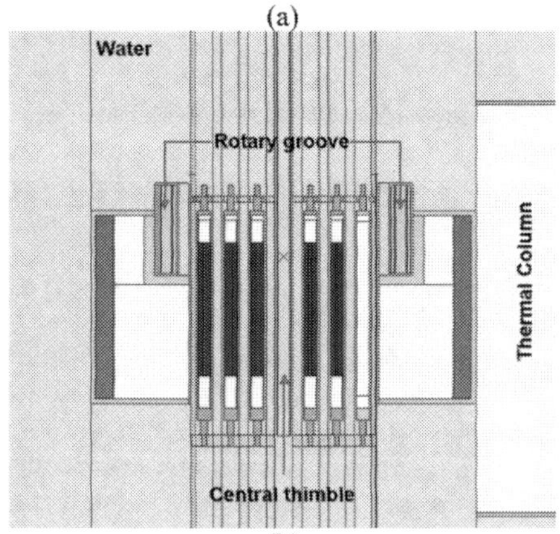

(b)

Fig. 2. (a) Side and (b) top view of the TRIGA reactor core as described in the TRIPOLI-4 model.

B. Experimental design and preliminary calculations

Preliminary simulations with the TRIPOLI-4 model were carried out in order to precisely prepare and design the experiments in the irradiation channels of the reactor. Through these simulations we can choose dosimeters that respond to a desired specific region of the neutron spectra, we can also set the optimal irradiation conditions (reactor power level and irradiations duration) in order to optimize the measurement

https://doi.org/10.1051/epjconf/202125304006

time in HP-Ge detector so that we can sequence the dosimeters to be measured considering their half-lives.

Six dosimeters were selected to cover different regions in the energy range of 0 – 10 MeV of the neutron spectra: ^{197}Au(n, γ), ^{59}Co(n, γ), ^{58}Ni(n, p), ^{56}Fe(n, p), ^{46}Ti(n, p) and ^{27}Al(n, α). The selection of these dosimeters is based on different criteria such as: the cross section of the neutron reaction of interest, energy range of response, energy and intensity of decay γ rays' emission.

The reaction rates calculated in the core reactor center are shown in Fig. 3 and Fig. 4. The cumulated reaction rates are calculated by the following formula:

$$C = \frac{\sum_{1}^{j} R_j}{\sum_{1}^{N} R_j}$$

Where N is the total number of energy groups, R_j is a reaction rate of an energy group "j". The cumulated reaction rate for resonant dosimeters, ^{197}Au and ^{59}Co, shows a large contribution of thermal neutrons. Furthermore, one might notice that the resonant dosimeters present peaks that are consistent with resonant zones of the (n, γ) cross section Fig. 5. Therefore, the dimensions of resonant dosimeters should be small enough so as to reduce the self-shielding and to avoid the neutron flux disturbance in the irradiation channels. Consequently, the gold foils have been chosen with a 0.1 mm thickness, and 1 mm in the case of cobalt wires.

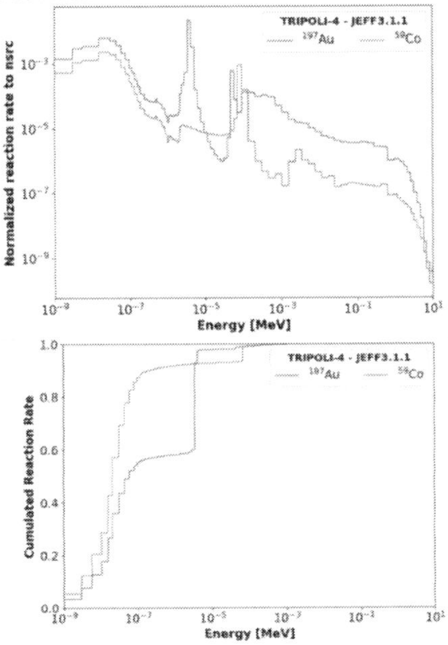

Fig. 3. Distribution of reaction rates and cumulated reaction rates for resonant dosimeters.

For threshold dosimeters, the cumulated reaction rate shows that the latter are very sensitive to the fast region of the neutron energy spectra. Threshold activation reactions, in particular (n, p) and (n, α) reactions, generally have a threshold energy around 1 MeV [5]. The thickness of the threshold irradiated disk-form dosimeters was selected as follows: 0.128 mm for Ni and Fe, 0.25 mm for Al and 0.5 mm for Ti.

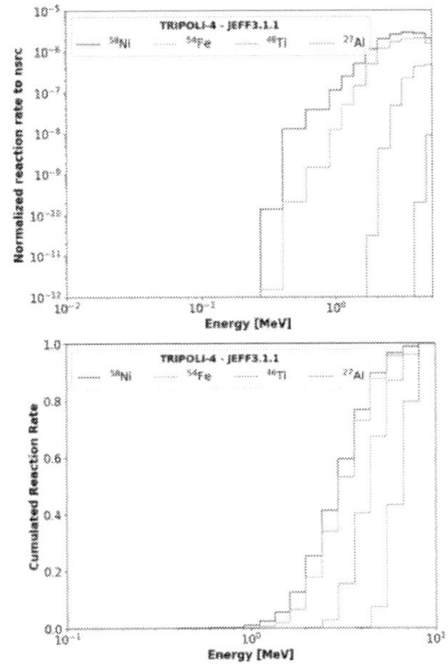

Fig. 4. Distribution of reaction rates and cumulated reaction rates for threshold dosimeters.

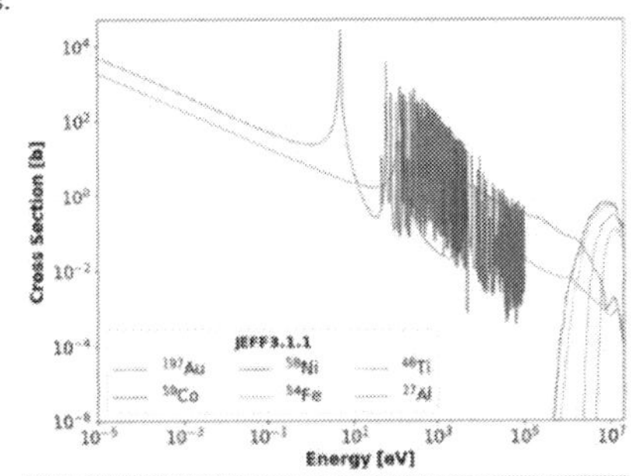

Fig. 5. Cross-sections for resonant and threshold dosimeters from JEFF3.1.1 nuclear data library [8].

C. *Experimental set-up*

The above-mentioned dosimeters were irradiated in 4 different irradiation facilities:

- Core central channel (CC) – gold and nickel dosimeters were irradiated in different axial positions in order to characterize the axial fast and thermal neutron flux profiles. For the purpose of this irradiation, a new specific device was manufactured (Fig. 6) in order to precisely position the dosimeters in this irradiation channel.
- Rotary groove (RG) – gold dosimeters were irradiated in 8 symmetrical positions among the 40 circularly distributed channels, with an inner diameter of 3.8 cm, in that irradiation facility. All six dosimeters were then irradiated in IC#10 to characterize the neutron spectra in that position.

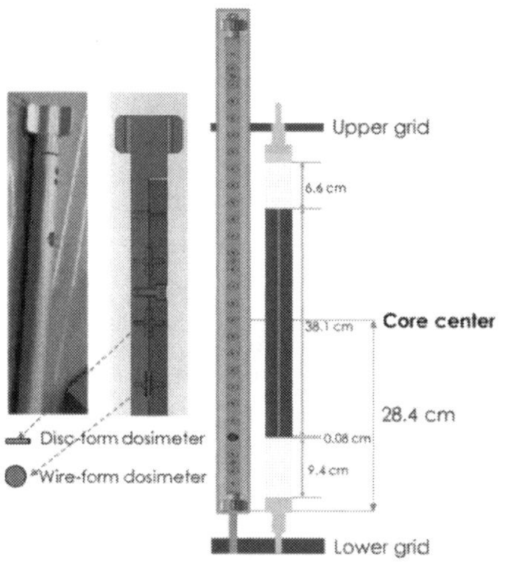

Fig. 6. experimental irradiation device in the core central channel.

- Pneumatic transfer system (PTS) – the PTS is used to irradiate and transport samples between the radiochemistry lab and an in-core location. As for the case of RG, all 6 dosimeters were irradiated in PTS in order to characterize spectra in that core location. The dosimeters. Before irradiation in the RG and PTS, dosimeters were prepared and fixed in different capsules depending on the irradiation facility (Fig. 7).
- NB1 tangential channel – gold and nickel dosimeters were irradiated in the closest position to the reactor core.

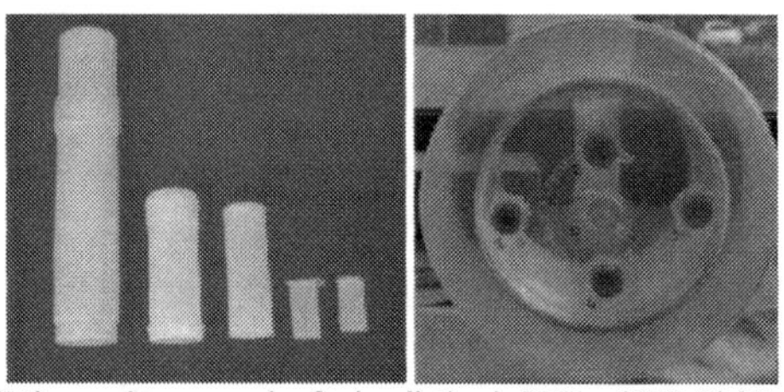

Fig. 7. (left) the specimen capsules for irradiation in PTS and RG, (right) device used for irradiation of dosimeters in the NB1 channel.

An illustration is given in Fig. 8, which shows the neutron spectra, for 315 energy groups, in the central channel, pneumatic tube and in the NB1 channel. A normalization factor was applied in order to scale the calculated flux scores to the actual reactor power level [8] :

$$C_{MC} = \frac{P \times \bar{\nu}}{w_f \times k_{eff}} \left[\frac{neutrons}{s} \right]$$

With P is the reactor power in [W], $\bar{\nu}$ is the average number of neutrons per fission, w_{eff} is the energy released per fission [MeV/fission] and k_{eff} is the effective multiplication factor of the reactor.

The simulations were performed in cells that have the same thicknesses as the actual dosimeters and then normalized to the full reactor power; 2 MW. The 1σ relative statistical uncertainty, shaded area in the graph, is within 0.4% to 5% in the energy range of 10^{-10} – 10 MeV.

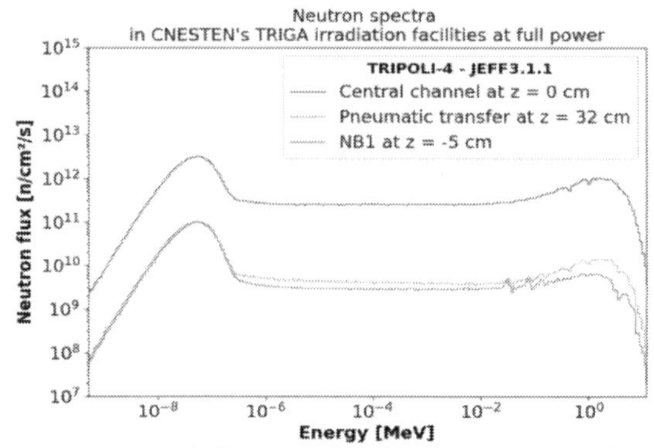

Fig. 8. Neutron spectra in the Central Channel (CC), Pneumatic Transfer System (PTS) and the tangential channel NB1.

III. Results and discussions

A. Irradiations and measurements

Four irradiations were performed, between December 2020 and March 2021, in different position within and beyond the reactor core. The irradiations lasted different times (t_{irr}) depending on the flux intensity in the irradiation facilities. The reactor power level was chosen to be 20 kW for irradiations of resonant dosimeters in both CC and RG (Table I). For threshold dosimeters the reactor power level was 100 kW and 1 MW in the case of ^{46}Ti dosimeter irradiation.

TABLE I
THE IRRADIATIONS PERFORMED IN DIFFERENT REACTOR FACILITIES.

Date of irradiation	Facility	Power level	Irradiation time	Irradiated samples
08/12/2020	Center CC and NB1	100 kW	30 min	Co, Fe and Au
09/12/2020	Axial profile CC	20 kW	15 min	Ni and Au
26/02/2021	Axial profile CC and RG	20 kW	15 min	Au
29/03/2021	Center CC and PTS	1 MW	7min 30 s	Co, Ni, Al, Fe, Ti, Au

The post-irradiation analysis of the irradiated dosimeters was performed by means of the wide-spread γ spectrometry technique. In addition to the factors determined by the conditions of measurement, the activity of the radioactive products depends on neutron flux at the position of irradiation as well as the corresponding cross section of the targeted material. The relationship between the reaction rate and the activity is evinced by the following basic formula:

$$A = N\sigma\phi(1 - e^{-\lambda t_i})$$

Where A is the activity produced in a dosimeter containing N numbers of nuclei when irradiated for a t_i irradiation time at a ϕ [n.cm^{-2}.s^{-1}] neutron flux. σ [b] and λ [s] refer to the microscopic reaction cross section and the decay constant, respectively. At the end of irradiation, the activity A_j for the jth peak of an isotope, in the irradiated sample, may be written as:

$$A_j = \frac{N(E_j)}{m \times I_\gamma(E_j) \times R^P(E_j)} \times \frac{\lambda \times t}{(1 - e^{-\lambda t}) \times e^{-\lambda t_d}} \sum_i C_i$$

Where $N(E_j)$ is the net area under the total absorption peak at energy E_j ; $I_\gamma(E_j)$ is the emission probability of the photon with energy E_j of the material in quest; t is the counting time, $I_\gamma(E_j)$ data is taken from [10] ; $R^p(E_j)$ is the total absorption yield for the photon with energy E_j; t_d : time between the end of irradiation and the beginning of the measurement; m is the mass of the dosimeter; $\sum_i C_i$ is the sum of the different correction factors of geometry, self-absorption of the photon of energy E_j in the foil and summing effects due to cascade emission of photons.

The yield calibration of the CNESTEN's HP-Ge detector was carried out with a certified europium multi-gamma radioactive source of small size (considered as a punctual source) for different geometries. Depending on the dosimeter's activity, the spectrometry γ measurements were performed at two source-detector distances: 21.53 and 6.93 cm.

B. Normalized values

In this section, we present the comparison of the calculated and the measured specific activities represented as "relative quantities" e.g. normalized to a certain reference value according to the following equations:

- Profiles in central channel (CC)

$$A_{i,norm}^{CC} = \frac{A_i^{CC}}{A_{\max}^{CC}}$$

- Rotary groove (RG)

$$A_{i,norm}^{RG} = \frac{A_i^{RG}}{\frac{1}{i_{\max}}\sum_{i}^{i_{\max}} A_i^{RG}}$$

- Pneumatic transfer system (PTS)

$$A_{i,norm}^{PTS} = \frac{A_i^{PTS}}{\frac{1}{i_{\max}}\sum_{i}^{i_{\max}} A_i^{PTS}}$$

Where the A_i is the measured and calculated specific activities (in Bq/mg).

Figs. 9-12 refer to the comparisons of the normalized measured and calculated values. The blue area in the middle side of the graphs denotes the active fuel region of the standard FE of the TRIGA reactor and the error bars represent the 1σ statistical and measurement uncertainties. The relative differences (C/E-1) between experimental and calculated results are presented below each graph.

The simulations were performed in criticality mode using the detailed computational TRIPOLI-4 model of TRIGA Mark II reactor (see section II.A) and JEFF3.1.1 nuclear data library. To ensure consistency between the measured and calculated values, the dosimeters and the irradiation devices were modelled explicitly inside the irradiation facilities. Furthermore, the control rod positions during the experiments were taken into account in calculations. It should be noted that the control rods travel vertically a distance of 38.1 cm between their fully withdrawn and inserted positions. The control rod scale is divided into 1000 steps – 0 fully inserted and 1000 fully withdrawn.

As shown in Figs. 9-10 the calculated normalized activity profiles for gold (thermal and epithermal neutron flux) and nickel (fast neutron flux) foils are in good agreement with the experimental results. For the axial distribution of the $^{197}Au(n, \gamma)^{198}Au$ reaction rate, the relative differences (C/E-1) are around 3% in the active fuel region and around 7% in the reflector area of the FE. In the case of the $^{58}Ni(n, p)^{58}Co$ reaction rate, the relative differences (C/E-1) are around 2 to 10 % in the active fuel region and around 20% in the reflector parts of the FE. The preliminary analysis of the axial gold and nickel reaction rate profiles has shown a slight systematic shift in the axial direction, which was corrected by simulations. This shift may be related to a slight positioning error in the central channel.

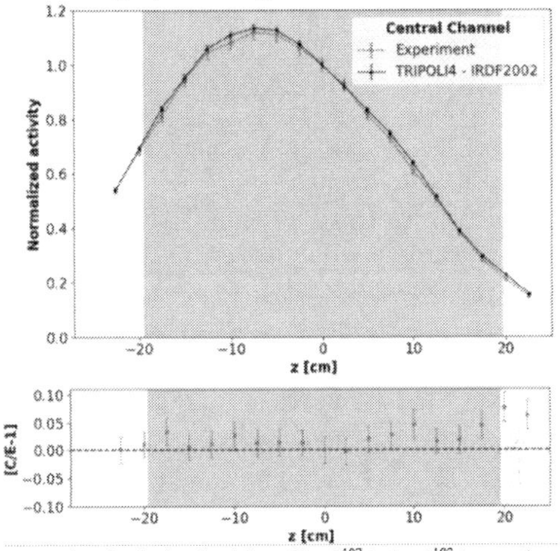

Fig. 9. Measured and calculated axial profile of $^{197}Au(n, \gamma)^{198}Au$ reaction rates in the reactor core central channel normalized to the maximal value. The error bars represent the 1σ uncertainty. Area marked in blue denotes the active fuel region of the standard fuel element.

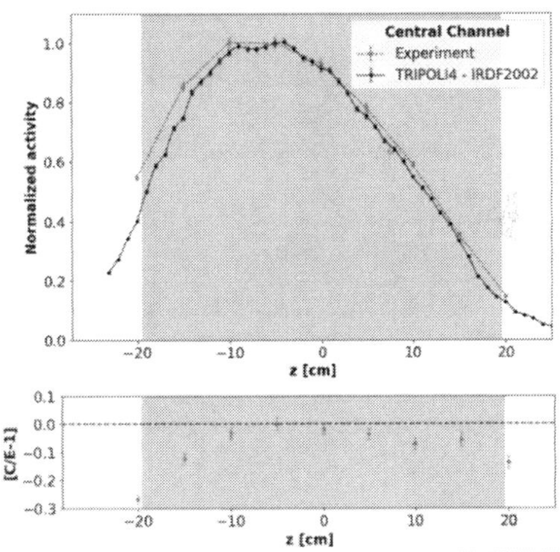

Fig. 10. Measured and calculated axial profile of $^{58}Ni(n, p)^{58}Co$ reaction rates in the reactor core central channel normalized to the maximal value.

Fig. 11 shows a good agreement between the measured and calculated normalized gold reaction rates in 8 positions among the 40 in the rotary groove facility. The relative differences between the calculated and experimental values are in average around 6% (1σ). However, the 1σ statistical uncertainty in the calculated reaction rates have varied between 1% to 5% (in IC#20).

Fig. 12 displays the calculated and measured reaction rates for 5 types of dosimeters; ^{197}Au(n, γ), ^{27}Al(n, α), ^{46}Ti(n, p), ^{58}Ni(n, p) and ^{59}Co(n, γ). In order to qualitatively compare the calculated and measured values, the reaction rates were normalized to the average reaction rates of all dosimeters irradiated in the PTS (see section III.B).

A good agreement among the results was found for all dosimeters irradiated in the PTS. The C/E-1 relative differences are within 1% to 4% (maximum difference for ^{46}Ti dosimeter) which indicates that the TRIPOLI-4 model can accurately reproduce the experimental results.

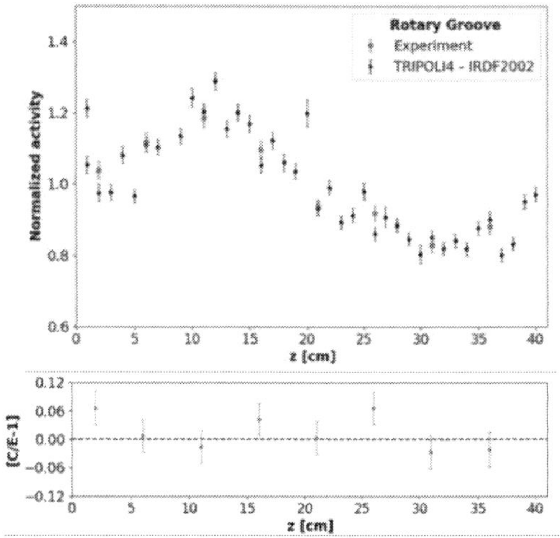

Fig. 11. Measured and calculated ^{197}Au(n, γ)^{198}Au reaction rates in the RG facility normalized to the average value.

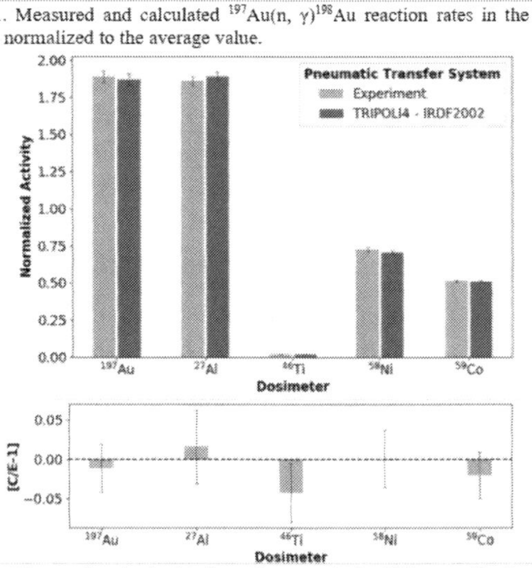

Fig. 12. Measured and calculated reaction rates in the PTS normalized to the ^{197}Au(n, γ)^{198}Au reaction rates in the core center.

IV. Conclusion and perspectives

This paper sums up the results of the preliminary analysis of the measurements carried out during the first experimental campaign which was implemented in the TRIGA Mark II reactor of the CNESTEN. Different dosimeters were selected to cover different regions in the energy range of 0 – 10 MeV of the neutron spectra. The experimental design was accomplished with Monte Carlo simulations using a new detailed computational TRIPOLI-4 model of the reactor TRIGA. Comparisons, in terms of C/E-1, have been made between the normalized experimental and calculated reaction rates of the irradiated foils. As a good agreement among the results was deduced, we can assume that the TRIPOLI-4 reactor model is able to correctly reproduce the experimental values.

As a sequel to this work, the spectrum unfolding is going to be achieved by the use of the CALMAR dosimetry adjustment code for neutron spectrum in order to validate the neutron flux and spectra in the irradiation facilities. Moreover, the comparison between absolute values of reaction rates will soon be performed.

References

[1] B. El Bakkari et al., «Monte Carlo modelling of TRIGA research reactor,» *Radiation Physics and Chemistry*, vol. 79, pp. 1022-1030, 2010.

[2] Y. Boulaich et al., «Steady-state thermal-hydraulic analysis of the Moroccan TRIGA MARK II reactor by using PARET/ANL and COOLOD-N2 codes,» *Nuclear Engineering and Design*, vol. 241, pp. 270-273, 2011.

[3] E. Chham et al., «Neutronic and thermal–hydraulic analysis of new irradiation channels inside the Moroccan TRIGA Mark II research reactor core,» *Applied Radiation and Isotopes*, vol. 116, p. 178–184, 2016.

[4] H. El Yaakoubi et at., «Neutronic study of the 2-MW TRIGA MARK-II research reactor by the deterministic codes DRAGON5 & DONJON5,» *Applied Radiation and Isotopes*, vol. 157, 2020.

[5] E. Brun, F. Damian, C. Diop, E. Dumonteil, F. Hugot, C. Jouanne, Y. Lee, F. Malvagi, A. Mazzolo, O. Petit, J. Trama, T. Visonneau et A. Zoia, «Tripoli-4®, CEA, EDF and AREVA reference Monte Carlo code,» *Annals of Nuclear Energy* , vol. 82, pp. 151-160 , 2015.

[6] A. Santamarina, D. Bernard , P. Blaise, M. Coste, A. Courcelle, T. Hyunh, C. Jouane, P. Leconte, O. Litaize, S. Mengelle, G. Noguère, J.-M. Ruggiéri, O. Sérot, J. Tommasi, C. Vaglio et J.-F. Vidal, «The JEFF-3.1.1 Nuclear Data Library,» *Nuclear Energy Agency Organisation for Economic Co-operation and Development, ISBN 978-92-64-99074-6, OECD* , 2009.

[7] N. Soppera, M. Bossant et E. Dupont, «JANIS 4: An improved version of the NEA java-based nuclear data information system,» *Nucl. Data Sheets*, vol. 120, pp. 294-296, 2014.

[8] N. Soppera, M. Bossant et E. Dupont, «JANIS 4: An Improved Version of the NEA Java-based Nuclear Data information System,» *Nuclear Data Sheets*, vol. 120, pp. 194-196, 2014.

[9] Zerovnik et al., «On normalization of fluxes and reaction rates in MCNP criticality calculations,» *Annals of Nuclear Energy*, vol. 63, pp. 126-128, 2014.

[10] «Library for gamma and alpha emissions - CEA/LNHB http://www.lnhb.fr/donnees-nucleaires/module-lara/».

PISTIL, a reactivity modulation device to probe the transfer function of the nuclear reactor CROCUS

Yifeng Jiang[1,2], Benoit Geslot[1], Vincent Lamirand[2], Pierre Leconte[3], Daniel Godat[2], Laurent Braun[2], Pavel Frajtag[2], Christine Coquelet-Pascal[1] and Andreas Pautz[2]
1 CEA, DES, IRESNE, DER, SPESI, LP2E, Cadarache, F-13108 Saint-Paul-Lez-Durance, France
2 Laboratory of Reactor Physics and Systems Behaviour, Ecole Polytechique Fédérale de Lausanne, CH-1015 Lausanne, Switzerland
3 CEA, DES, IRESNE, DER, SPRC, LEPh, Cadarache, F-13108 Saint-Paul-Lez-Durance, France

Corresponding author: yifeng.jiang@cea.fr

***Abstract*—The present article summarizes the development and testing of a reactivity modulation device developed by the French Atomic and Alternative Energies Commission (CEA). It was installed in the CROCUS reactor of the Swiss Federal Institute of Technology in Lausanne (EPFL). Experimental tests were performed in the framework of a collaboration between CEA and EPFL.**

The so-called PISTIL device aims at measuring the nuclear reactor transfer function in the frequency range of interest between 1 mHz and 200 Hz, in order to probe the in-core kinetic behavior of prompt and delayed neutrons. The reactivity modulation is obtained from the rotation of cadmium foils.

The design of the system was driven with the objective of installing PISTIL at the center of the CROCUS reactor. Neutronic simulations with TRIPOLI-4 Monte Carlo code were performed to select the suitable design parameters and meet the safety requirements of the reactor operation.

The total reactivity worth of the device, as estimated by TRIPOLI-4 Monte Carlo calculation, was approximately 0.16 \$ and the maximum amplitude of the reactivity modulation was about 0.013 \$. In-core reactivity calibration was then performed and were consistent as compared to TRIPOLI-4 estimations.

***Keywords* —In-core integral experiment, reactivity modulation, zero-power reactor transfer function, reactor kinetics**

I. Introduction

The pile oscillator method is well-known as an experimental approach to evaluate the model consistency of transient calculations in nuclear reactors with respect to its temporal behavior in reactor physics [1], as the corresponding knowledge are crucial to the design, operation and safety analysis of nuclear reactors. Instead of studying the temporal response of the in-core neutron population subject to variations in reactor configuration (e.g. control rod movement, temperature feedback effect, etc.), a periodic reactivity modulation is introduced to investigate the overall reactor kinetic behavior in the frequency domain through the measurement of the reactor transfer function (RTF) by Fourier analysis.

A particular feature of the RTF of low power reactors, namely the Zero-Power reactor Transfer Function (ZPTF), is that it is only dependent on the kinetic behavior of prompt and delayed neutron population produced in the fission chain reaction. While an extensive amount of computational work were conducted to improve the modeling of reactor kinetics, recent studies [2-4] have shown the potential of an alternative experimental approach, through in-core reactivity modulation, towards the better understanding of the underlying physical phenomena. In this context, the PISTIL device (Periodic reactivity Injection System Transients Induced Locally) was developed by CEA in the framework of a collaboration between CEA and EPFL. The device has two aluminum tubes holding cadmium foils. The rotation of the center tube, driven by a brushless motor, generates a controlled reactivity modulation through the variation of the radiative capture reaction rate of the cadmium. The height of the ensemble, as well as the relative position between rotary and static parts are adjustable, offering flexibility on reactivity worth and reactivity amplitude. The reactivity calibration test of the device was performed in the zero-power reactor CROCUS at EPFL for the experimental investigation of its reactor kinetic parameters by frequency domain analysis. In this work, we present the design methodology and first calibration results of this experimental device.

II. Context and background

A. Point kinetic reactor model and Zero-power transfer function

The point kinetic model in reactor physics describes the temporal behavior (integrated in space and energy) of the neutron population in the core of a nuclear reactor, with respect to a reactivity variation [1]:

$$\begin{cases} \dfrac{dN(t)}{dt} = \dfrac{\rho(t) - \sum_i^n \beta_i}{\Lambda} N(t) + \sum_i^n \lambda_i C_i(t) \\ \dfrac{dC_i(t)}{dt} = \dfrac{\beta_i}{\Lambda} N(t) - \lambda_i C_i(t) \quad i = 1,2,\dots,n \end{cases}$$

Where $N(t)$ represents the neutron population, $\rho(t)$ is the reactivity, $C_i(t)$ is the delayed neutron precursor concentration of group i in an n-group formulation. The kinetic parameters involved in this formulation are the prompt neutron generation time Λ (s), the effective delayed neutron fractions $(\beta_i)_{i=1,2,\ldots,n}$ and associated decay constants $(\lambda_i)_{i=1,2,\ldots,n}$ (s^{-1}) for the precursor group i. As mentioned earlier, the temperature feedback effects are considered negligible in the case of zero-power reactors.

When a small and periodic reactivity perturbation $\rho(t) = \rho_0 + \delta\rho(t)$ is generated in the reactor core, its kinetic behavior with respect to the neutron population $N(t) = N_0 + \delta N(t)$ can be modelled by a linear time invariant system represented in the frequency domain by the ZPTF [6]:

$$\frac{N(j\omega_0)}{\rho(j\omega_0)N_0} = H(j\omega_0) = \left(j\Lambda\omega_0 + \sum\nolimits_i^n \frac{j\omega_0\beta_i}{j\omega_0 + \lambda_i} - \rho_0\right)^{-1}$$

Where j is the unit imaginary number and ω_0 is the angular frequency of the perturbation (rad.s^{-1}). The function, as in the complex domain, is conventionally represented by its modulus (amplitude) and argument (phase).

The measurement of the ZPTF can be achieved by modulating periodically the reactivity at several given frequencies and measure its effect on the neutron flux. Thanks to standard Fourier analysis techniques, the kinetic parameters are inferred in the frequency domain: the reactivity variation profile and the neutron detector response. An example of simulated modulation experiment and resulting analysis for ZPTF, in a point reactor kinetic formulation manner, is shown in Figure 1. When a sine-like reactivity modulation is introduced to the reactor, a power (neutron flux) variation of similar profile can be measured. The analysis of the amplitude ratio and phase lag is measured at the characteristic fundamental frequency of the modulation.

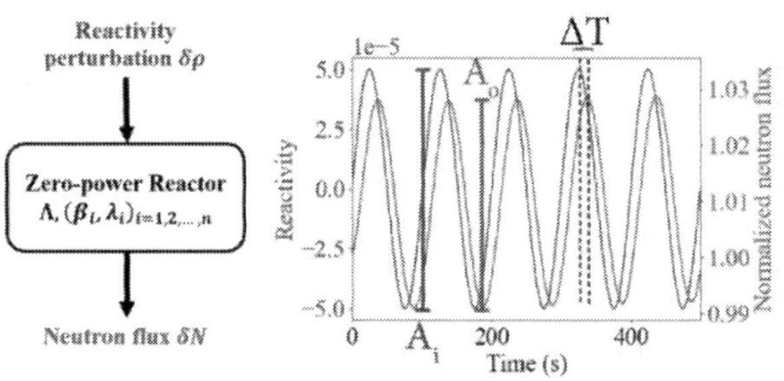

(a) Reactivity perturbation and simulated reactor power response

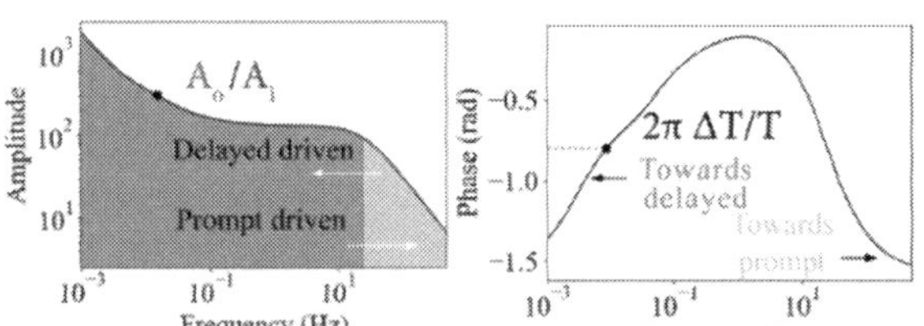

(b) Amplitude and phase representation of the ZPTF in the frequency domain

Figure 1. Simulated zero-power reactor behavior in a modulation experiment

B. *The CROCUS reactor*

CROCUS is a zero-power research reactor operated at EPFL. Figure 2 shows a cross sectional view of the reactor core and its associated structures. The maximum power of the reactor is limited to 100 W, corresponding to a neutron flux of 2.5×10^9 cm^{-2}.s^{-1} in the core center. The reactor is operated at atmospheric pressure and room temperature using a controlled water loop. The power is adjusted either by changing the in-core water level using a spillway, or by moving two B_4C control rods.

The reactor core is light-water moderated and uranium fueled. The active zone of the core is approximately a cylinder of 60 cm in diameter and 1 m in height. It is composed of two interlocked fuel zones: an inner zone of 336 UO_2 cylindrical rods enriched to 1.806% and an outer zone of 176 U_{metal} cylindrical rods enriched to 0.947%. Two aluminum grid plates spaced 1 m apart maintain the fuel rods vertically. PISTIL was designed to be held in a similar manner in an experimental channel available at the core-center of 10 mm diameter. In the following parts of the article, all mentioned axial positions are those of the reference height documented in the CROCUS benchmark [9].

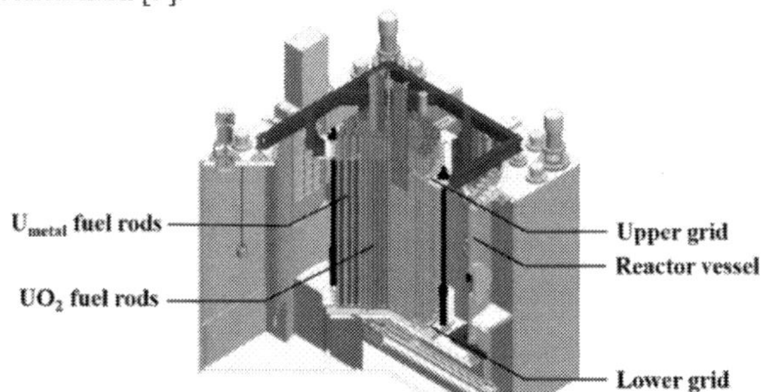

Figure 2. Cross-sectional view of the CROCUS reactor

III. Neutronic design

A. *Estimation of the differential reactivity worth*

The neutronic design of the PISTIL device was conducted using the transport Monte Carlo code Tripoli-4® [7]. A two-step approach was adopted for the study.

In the first step, criticality calculations using a reduced scale model, representative of the fuel lattice of the core center, as shown in Figure 3, were first performed with the objective of selecting the best isotopes to achieve the reactivity modulation. A qualitative concept of absorbers in rotation was produced. The reactivity effect are induced by the variation of the distance between two cadmium foils: when they are far from each other, the macroscopic cross sections is maximum. In contrast when they are close, the cross sections decreases due to self-shielding effects.

In the second step, calculations of the whole core of CROCUS, with PISTIL modeled at the core center, were conducted for the estimation of the overall reactivity worth and reactivity modulation amplitude.

Between two different reactor configurations, the differential reactivity worth was estimated as:

$$\Delta\rho_{1\to2} = \frac{1}{k_{eff,1}} - \frac{1}{k_{eff,2}}$$

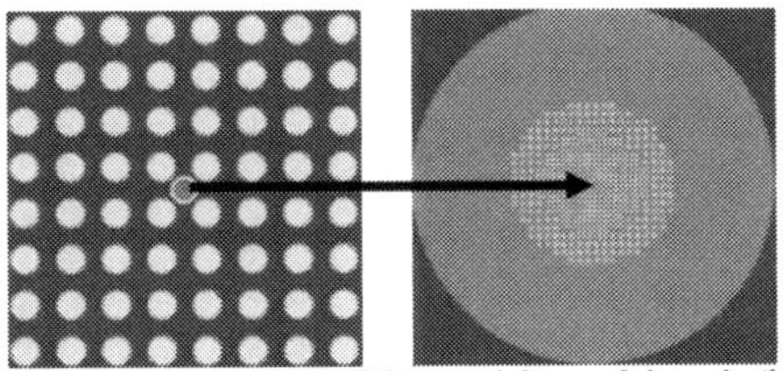

Figure 3. TRIPOLI-4 modeling of the reactivity modulator in the center of reduced scale fuel lattice (left) and in the full core of CROCUS

B. Conceptual design

The design consists in several cylindrical elements positioned vertically in the reactor core. The main components of the neutronic design of PISTIL, in a cross-sectional view, are shown in Figure 4. The system is composed of a watertight aluminum tube, a static subset (stator) of two cadmium foils and a rotary subset (rotor) composed of another two cadmium foils and a high-density polyethylene (HDPE) cylinder. The change in the relative angular position between the rotor and the stator generates variation in the (n,γ) absorption reaction rate [8]. The HDPE has the effect of enhancing the capture reaction rates by inducing a local thermalization of the neutron spectrum. Cadmium was selected as the absorber material for its high (n,γ) cross section in the thermal energy range (< 0.5 eV) compared to other conventional materials that are easy to machining to foils of less than 0.1 mm thickness. The foil geometry was chosen to minimize the strong self-shielding effect within the volume of the cadmium material, which reduces considerably the modulation effect.

A rotary motion drives the reactivity modulation induced by the cadmium foil on the rotor. As the two foils on the stator and the rotor cover each one a 90° section and separated 90°, the device has a 2-fold rotational symmetry. Such a design allows the doubling of the modulation frequency as compared to the mechanical frequency in concern of relieving safety considerations at frequencies as high as 100 Hz. The rotary motion was preferred than the linear one in this aspect, as it would introduce much less vibration at high frequency.

In the following sections, the 0° will refer to the angular position configuration of the rotor and stator shown in Figure 4. This corresponds to the minimum reactivity worth of PISTIL. In contrast, the maximum reactivity corresponds to the configuration of a 90° rotation from the 0° position.

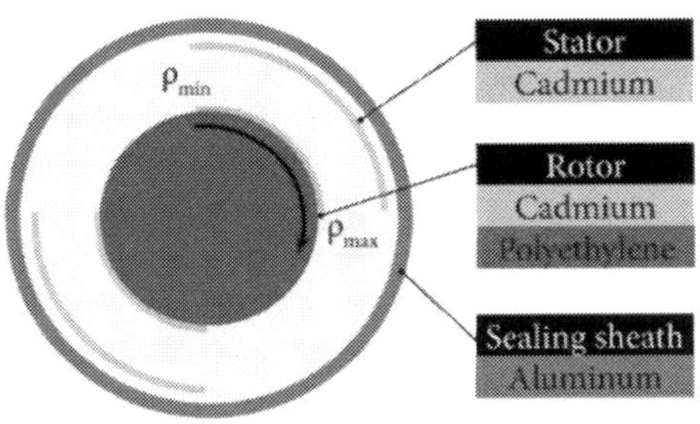

Figure 4. Main components of the neutronic design of PISTIL

C. Reactivity worth and modulation amplitude

Full core calculations were performed for several configurations of the reactor. All these calculations use the JEFF 3.3 library [11] for nuclear data, and the in-core water level was considered at 96.5 cm.

The cadmium elements and the HDPE were modelled at the same axial position centered at 550 cm. This configuration will also be considered for the experimental calibration discussed in section V. The calculation-based estimation showed that the expected maximum anti-reactivity insertion is about 0.16 $ while the modulation amplitude is 1.3 ¢. The amplitude of the modulation is sufficiently large for the ZPTF measurement in the frequency range of interest.

The choice of the maximum reactivity worth was driven by the limit of the additional reactivity reserve of 0.2 $ available in the current configuration of CROCUS [14] to bring the reactor to critical, while optimizing the precision of the measurements by maximizing the modulation amplitude.

TABLE I
CALCULATED MULTIPLICATION FACTOR IN VARIOUS REACTOR CONFIGURATIONS

Configuration	k_{eff}	Relative uncertainty (%)	Differential worth (pcm)
Reference without PISTIL	1.002858	2.881890 10^{-3}	-
PISTIL 0°	1.001729	2.48895 10^{-3}	-112.4 ± 3.8
PISTIL 45°	1.001688	2.48910 10^{-3}	-116.4 ± 3.8
PISTIL 90°	1.001628	2.49976 10^{-3}	-122.4 ± 3.8

IV. Mechanical Design

The available in-core space in the center of CROCUS, with a outer diameter of 10 mm due to the presence of the two grids, led to the design of only including the elements mentioned in the neutronic design in the active core. The other structures of PISTIL and the motor are placed above the fuel zones. Other positioning options of the device less restrictive with respect to the geometry, such as the periphery regions, were discarded due to the insufficient reactivity modulation amplitude.

A. Mechanical components

The dimensions of the elements present in the fuel zone are shown in Table II. Figure 5 shows a mechanical representation of the whole device.

TABLE II
DIMENSIONS OF ELEMENTS IN THE NEUTRONIC DESIGN OF PISTIL

Element	Axial length (mm)	Outer diameter (mm)	Inner diameter (mm)
Sealing sheath	1000.0	10.0	9.0
Cadmium on stator	100.0	8.63	8.4
Cadmium on rotor	100.0	5.70	5.46
HDPE	100.0	5.50	0

The bottom of the device was designed to be in contact with the base plate of CROCUS, while its upper edges are fixed to the reactor's structural support above the core.

 https://doi.org/10.1051/epjconf/202125304007

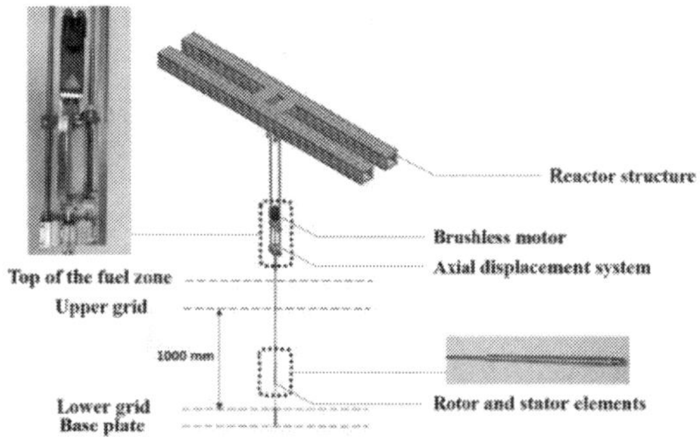

Figure 5. Schematic representation of the mechanical elements of PISTIL

The mechanical concept of PISTIL included the possibility to modify the axial position of the rotor with respect to the stator over 100 mm range, while the rotor/stator ensemble can also be adjusted from 550 mm to 950 mm. This can be achieved when the device is installed inside the core.

Figure 6. PISTIL installed in the core center of CROCUS (left), its motor and position adjustment system (right)

B. *Motion command and data communication system*

A TGS1 brushless motor with a nominal speed of 6000 rpm (100 Hz) drives the mechanical motion of the rotor of PISTIL. The motion profile is controlled by a 230 V AC INFRANOR XtrapulsePac servo-controller. The motion commands are communicated through the numerical control TrioMotion Flex-6-Nano P600, using a program developed with National Instrument LabView [12] via the ActiveX TRIO PC Motion [10].

A number of motion parameters, such as the angular position and the servo current of the motor are acquired in parallel to the rotation by the numerical control at a maximum sampling rate of 1 kHz. These data are received by buffer in the aforementioned LabView program through a User Datagram Protocol (UDP) transmission.

Two types of instructions are currently implemented for the motion of PISTIL. The motor can rotate either continuously at a given frequency, or following a stepwise predefined profile in angular position. In Figure 7, the two modes are illustrated: a continuous rotation (left) and a pseudo-random sequence of quarter flip motions (right). The angular positions are shown in degree (°), while the neutron flux are normalized to its mean value during the experiment.

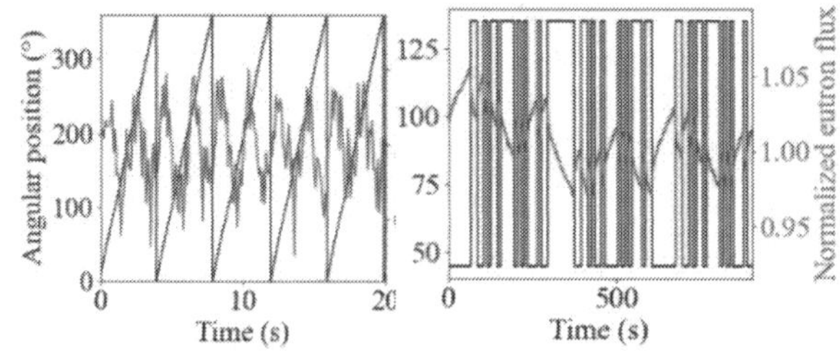

Figure 7. Measured motion profile of PISTIL in blue and the measured modulation amplitude in red for the continuous rotation (left) and the stepwise movement (right). Time is normalized to the period of the motion

V. Preliminary in-core calibration

The calibration aimed at establishing PISTIL's reactivity worth function vs. angle.

Two methods were tested and compared: modulation experiments during which PISTIL was operated at a constant speed and a step reactivity insertion experiments. During the experiments, the reactor was brought to criticality at a power of approximately 1 W while PISTIL remained static. The calibrations were then performed through either a reactivity modulation or a step reactivity variation.

A. *Experimental setup*

The experiments were conducted in two different manners. While the B_4C control rods were fully extracted, the criticality was maintained by adjusting the core water level. Alternatively the water level remains constant when operating with the rods. The neutron instrumentation consisted in 4 Photonis CFUL01 fission chambers located symmetrically with respect to the center of the core as detailed in the reference configuration [9]. Their output currents were sampled at 1 kHz. Two additional Photonis CFUM21 detectors were also used for operation monitoring purposes.

The criticality water level reached in the experiments were compared to the operation histories of CROCUS without PISTIL installed. The overall reactivity effect introduced by PISTIL is approximately -0.13 \$. A precise estimation of the reactivity worth was be made due to differences in detector set-ups.

B. *Differential reactivity worth*

The response of the reactor with respect to reactivity modulation were measured during a constant frequency rotation at 1.1 10^{-2} Hz. The resulting reactivity profile were estimated by inverse kinetic analysis [1] using values of adjoint-weighted kinetic parameters of CROCUS calculated in Tripoli4 [13] with JEFF 3.3 library. The reactivity as a function of the angular position of PISTIL obtained from the 4 signals are shown in Figure 8. The solid lines represents the reactivity effects with a angular resolution of 1° and the shaded area are the associated estimated uncertainties. The measured profiles were consistent between detectors. A maximum reactivity modulation amplitude of 1.35 ± 0.26 ¢ (10.24 ± 1.97 pcm) was observed, while the angular position between the maxima and the minima was 90 ± 1 °. The results are consistent with the simulation predictions. However, a shift of approximately 30° was observed between the expected position of the extrema (at 0 and 90°). This could be due to errors during the assembling process of the device that will be examined by out-of-core verifications.

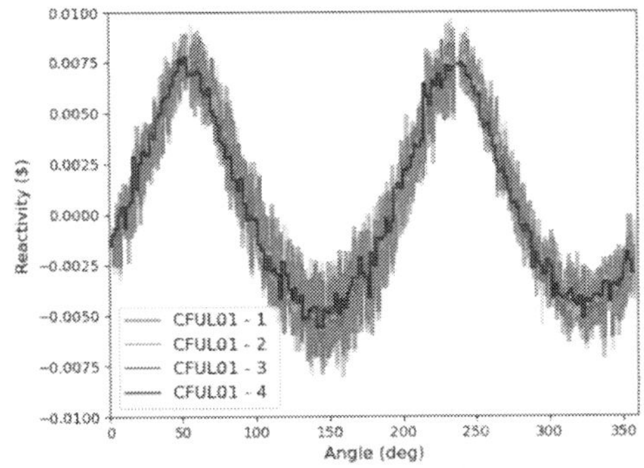

Figure 8. Measured reactivity profile of PISTIL with respect to its angular position from the dynamic calibration method

C. *Asymptotic period measurement*

Verification of the measured reactivity profile was also conducted using the asymptotic period method. The critical state water levels of PISTIL at 45° and 135° were first measured. Positive and negative reactor period were then measured during 30 minutes through a step reactivity variation from the critical state by the position switch of PISTIL (i.e. from 45° to 135° and *vice versa*)..The uncertainties are evaluated with respect to the determination of least-square fitted reactor periods. It was observed that for both positive and negative periods are of the same order of magnitude. The differences in the measurements could be related to the mechanical precision of the reactor operation and thus the difficulty in reaching the identical initial critical state. The resulting differential reactivities are approximately 8 pcm when evaluated using the inhour equation. The values are consistent with the dynamic measurements.

TABLE III
ASYMPTOTIC PERIOD MEASUREMENT CONFIGURATIONS AND PRELIMINARY RESULTS

Core water level (mm)	B_4C rod position (mm)	PISTIL position (°)	Period (s)	Reactivity (pcm)
984.0	1000.0	45	1044.0 ± 0.8	8.68
981.9		135	-1096.1 ± 0.9	-8.81
990.0	715.0	45	1052.4 ± 1.1	8.58
990.0	678.0	135	-1129.7 ± 1.1	-8.54

VI. CONCLUSION AND OUTLOOKS

The PISTIL device was developed for the in-core experimental study of reactor kinetics through reactivity modulation. Its design makes it possible to generate modulations in a large frequency range (from 1 mHz to 200 Hz). Its design was achieved by Monte Carlo calculations and adapted to a mechanical concept to be installed in the core center of the CROCUS reactor. The preliminary analysis of reactivity calibration tests in the CROCUS core showed a good consistency between Monte Carlo estimations and measurements. This suggests the validity of the developed design methodology with respect to the estimation of small reactivity variations through differential calculations. Further development will be performed towards better characterization of the device and the optimization of future experiments. The first experimental campaign for the measurement of the zero-power transfer function was carried out in June 2021 in CROCUS. Modulations covering the whole frequency range of the design were performed. A second campaign is under preparation in CROCUS and is expected in October 2021.

ACKNOWLEDGMENT

This work was partly funded by I3P (CEA, EdF, Framatome) and by the CNRS/Défi NEEDS project.

REFERENCES

[1] K.O. OTT, R.J. NEUHOLD, "INTRODUCTORY NUCLEAR REACTOR DYNAMICS", AMERICAN NUCLEAR SOCIETY, 1985

[2] Y. YEDVAB, I. REISS, "DETERMINATION OF DELAYED NEUTRONS SOURCE IN THE FREQUENCY DOMAIN BASED ON IN-PILE OSCILLATION MEASUREMENTS", IN *PROCEEDING OF PHYSOR-2006*, 2006, PP.1–9

[3] E. GILAD ET AL., "EXPERIMENTAL ESTIMATION OF THE DELAYED NEUTRON FRACTION β_{eff} OF THE MAESTRO CORE IN THE MINERVE ZERO POWER REACTOR", *J. NUCL. SCI. TECHNOL. 52*, 2015, PP.1026–1033

[4] P. LECONTE ET AL., "NEW DELAYED NEUTRON GROUP CONSTANTS AND COVARIANCES FOR LWR CORE APPLICATIONS, COMBINING SUMMATION CALCULATIONS AND INTEGRAL EXPERIMENTS", *ANN. NUCL. ENERGY*, VOL.139, 2020

[5] D. FOLIGNO, "NEW EVALUATION OF DELAYED-NEUTRON DATA AND ASSOCIATED COVARIANCES", PH.D. DISSERTATION, AIX-MARSEILLE UNIVERSITY, 2019

[6] Y. JIANG ET AL., "REVIEW OF KINETIC MODULATION EXPERIMENTS IN LOW POWER NUCLEAR REACTORS", *EPJ NUCLEAR SCI. TECHNOL.* VOL. 6, 2020

[7] E. BRUN ET AL., TRIPOLI-4®, "CEA, EDF AND AREVA REFERENCE MONTE CARLO CODE", *ANN. NUCL. ENERGY*, VOL.82, PP. 151-160. 2015

[8] A. JEBB,. "KINETIC STUDIES ON THE UNIVERSITY OF LONDON NUCLEAR REACTOR", PH.D. DISSERTATION, IMPERIAL COLLEGE OF SCIENCE AND TECHNOLOGY, 1973

[9] U. KASEMEYER ET AL., "BENCHMARK ON KINETICS PARAMETERS IN CROCUS", NEA/NSC/DOC. 2006

[10] TRIO MOTION TECHNOLOGY LTD, "USING THE PC MOTION ACTIVEX CONTROL", TRIO MOTION

TECHNOLOGY MOTION COORDINATOR – TECHNICAL REFERENCE MANUAL, 2008

[11] O. CABELLOS ET AL., "BENCHMARKING AND VALIDATION ACTIVITIES WITHIN JEFF PROJECT", *EPJ WEB CONF.* VOL.146, PP.4–9, 2017

[12] ELLIOTT C ET AL., "INSTRUMENTS LABVIEW: A PROGRAMMING ENVIRONMENT FOR LABORATORY AUTOMATION AND MEASUREMENT", *JOURNAL OF THE ASSOCIATION FOR LABORATORY AUTOMATION*, VOL.12. PP. 17-24, 2007

[13] GUILLAUME TRUCHET ET AL.. "COMPUTING ADJOINT-WEIGHTED KINETICS PARAMETERS IN TRIPOLI-4 BY THE ITERATED FISSION PROBABILITY METHOD", *ANN. NUCL. ENERGY*, VOL.85, 2015

[14] V. LAMIRAND ET AL., THE COLIBRI EXPERIMENTAL PROGRAM IN THE CROCUS REACTOR: CHARACTERIZATION OF THE FUEL RODS OSCILLATOR, *EPJ WEB CONF.* VOL.225, 2020

https://doi.org/10.1051/epjconf/202125304008

CAREDAS: a Comprehensive Architecture for a Redundant and Evolutive Data Acquisition System for JHR reactor

Fabrice Leroux[1], Lionel Ducobu[1] and Frédéric Milleville[1]
[1]CEA Cadarache, DSTG/STIC, France
Corresponding author: fabrice.leroux@cea.fr

***Abstract*—A new material testing reactor Jules Horowitz Reactor is under construction at CEA Cadarache. The materials to be irradiated will be placed into experimental devices around the reactor. Process and measurements of experimental devices will be carried out by command control. A data acquisition system having processing performances will be associated to the programmable logic controller. The challenge is to design and realize for twenty experiment devices a high availability data acquisition system architecture for 50 years of sustainability. The real time target will achieve 24/7 data acquisition and real time processing. This scalable architecture could be use as well for JHR experimental devices with high availability as for testbed. This architecture could be run on a standalone station for a measuring bench or deployed on cluster for high availability. CAREDAS's design is modular and use proven widely used open source solutions. All parts are independent from each other and can be replaced with another technology with the same functionalities. This ensures sustainability and control of software sources.**

***Keywords* —CAREDAS, MQTT, InfluxDB, Grafana, data acquisition system, DAQ, ExprTk, TSN, EtherCAT, OPC UA.**

I. Introduction

A new material testing reactor Jules Horowitz Reactor (JHR) [1] is under construction at CEA Cadarache. The materials to be irradiated will be placed into experimental devices around the reactor. Process and measurements of experimental devices will be carried out by command control (Fig. 1).

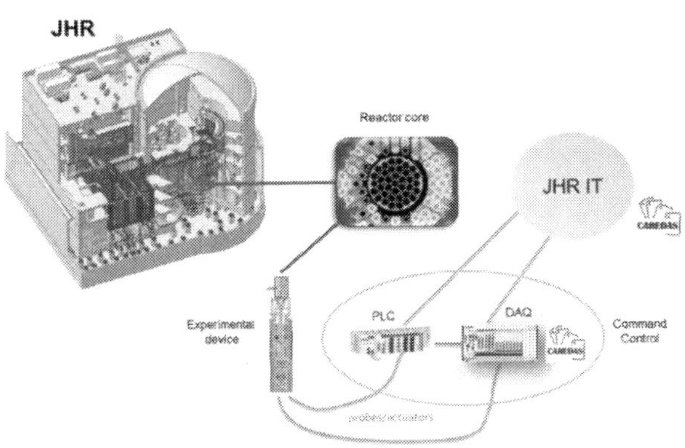

Fig. 1. Control command of JHR experimental devices.

Programmable Logic Controller (PLC) dedicated on each experimental device will realize this function. Some experiment devices will need to achieve complex real time processing that PLC cannot reach. A data acquisition system (DAQ) having these processing performances will be associated to the PLC. This system will realize measurements and processing. The challenge is to design and realize for twenty experiment devices a high availability data acquisition system architecture for 50 years of sustainability. The real time target will achieve 24/7 data acquisition and real time processing. The other components will be in charge of data storage, online and offline data visualizations, experiment setting and processing without modifying software core.

This scalable architecture could be used as well for JHR experimental devices with high availability as for testbed. This architecture could be run on a standalone station for a measuring bench or deployed on cluster with redundant data acquisition system for high availability. CAREDAS's design is modular and use proven widely used open source solutions of IIoT world (Industrial Internet of Things). All parts are independent from each other and can be replaced with another technology with the same functionalities. This ensures sustainability and control of software sources.

The real-time DAQ is modular according to the user's needs with various inputs/outputs such as analog inputs, digital inputs, temperature sensor. A DAQ management software, which can be used by a non-IT specialist, combines the functions of parameters setting, supervision and visualization of measurements. Real-time processing is defined in this software with a scientific programming language such as Scilab.

This architecture is modular and scalable according to the needs. It offers high availability on demand at all levels.

II. CAREDAS architecture

CAREDAS architecture (Fig. 2) was designed to ensure high availability for 20 independent acquisition systems and adaptable according to the needs of the plant. It ensures the archiving of a large volume of data during experiment. The acquisition chain has a sustainability of at least 50 years. Solutions selected are proven and if possible open source. All the functions are synchronized by Network Time Protocol (NTP) that distributes a clock. DAQ are synchronized by Time Sensitive Network (TSN).

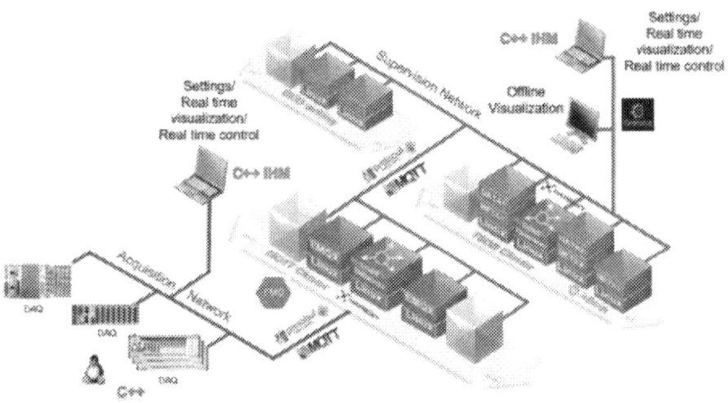

Fig. 2. CAREDAS architecture.

At the bottom of the chain, there are real-time DAQ allowing continuous acquisition of up to 300 sensors while performing 200 real-time processing for 50ms cycle time. No data loss is provided by redundant DAQ. DAQ software is independent of the hardware on which it runs. Linux operating system with real-time functionality is used.

The user can, through a configuration tool, configure his experience, add or remove measurement channels, hardware without software programming. He will define his real-time processing in a structured scientific programming language without programming DAQ. A DAQ supervision and a real-time visualization of measurements allow to check settings as close as possible to the hardware.

At the top of the chain, data and system settings archiving ensure high availability of storage and access to data. The database selected is InfluxDB [2], a Time series database (TSDB) which is able to absorb and store a large amount of data. For example, a DAQ with the maximum number of channels and 50ms of cycle time represents an amount of 14 GB/day.

The real-time acquisition chain is made up of two networks: the supervision network and the acquisition network. The supervision network and the acquisition network are common to all DAQ and are based on separate Ethernet networks. This division isolates and secures DAQ. The acquisition network must ensure no data loss and its isolation allows to not be disturbed by traffic that the connected systems are not receiver.

Communication between all these functions is ensured by messaging using the open source MQTT (Message Queuing Telemetry Transport) protocol [3] that is widely used and proven.

Offline data visualization is achieved with Grafana to compose dashboards that query directly InfluxDB databases.

Only three software in C++ language are homemade to ensure sustainability: real time data acquisition, writing data into InfluxDB, tools for experiment setting and supervision.

All software run on station or cluster nodes on the Linux operating system (Red Hat distribution).

III. Communication between functions: MQTT

MQTT messaging is the heart of CAREDAS. It manages the dialogue between the different functions of the acquisition chain without loss of messages. MQTT is an open source messaging protocol widely used in the world of Internet of Things (IoT) and in Web as in monitoring servers of data center for example. MQTT is based on publish/subscribe principle. A server named "broker" is responsible for collecting messages published by clients and distributes them to subscribed clients. A client has the possibility to define a level of quality of service for message transmitted. The level of quality tell on broker if it must or not ensure that the message have been received by clients.

In view of the number of measurements transmitted by DAQ in each cycle, it is not reasonable to transmit one MQTT message per measurement. Measurements are grouped to optimize message transmission. A data serialization, Google Protocol buffers [4], is using to make a MQTT message (Fig. 3).

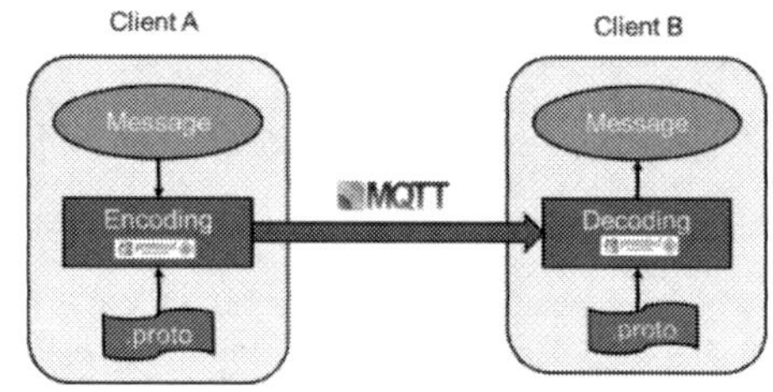

Fig. 3. Google Protocol buffers.

It was developed by Google and available under an Open Source License. Protocol buffers is a language for describing a data structure that allows a program to produce or read a message.

MQTT CAREDAS architecture (Fig. 4) is composed of a cluster of three brokers and two load balancer. The cluster ensures high availability of MQTT messaging. In case of a node is lost in the cluster or a network failure, other nodes take over. The selected open source MQTT broker is EMQ X from EMQ [5].

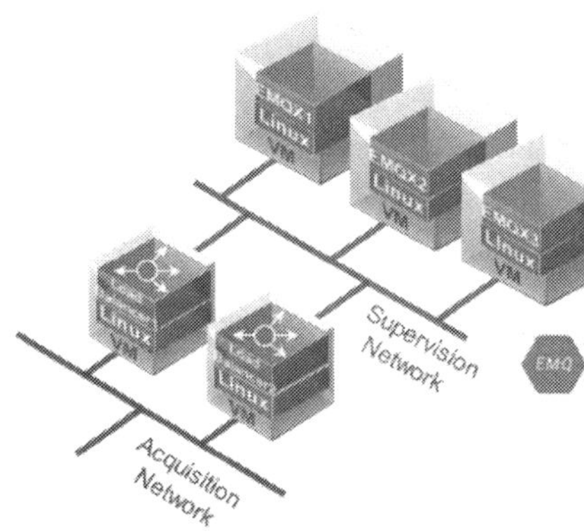

Fig. 4. CAREDAS MQTT architecture.

The load balancer is the single point of entry between clients and brokers. It distributes the connection requests from clients to a broker with the least load. Brokers are only connected with the load balancer. It separates the supervision network from the acquisition network.

IV. Data archiving

InfluxDB is the database to store DAQ measurements. Data storage is based on a cluster architecture that is a collection of nodes or servers (Fig. 5). A cluster enables high availability and load balancing across multiple nodes. Like MQTT cluster, a load balancer distributes requests from clients to InfluxDB nodes following nodes load. Data are automatically replicated on all cluster nodes.

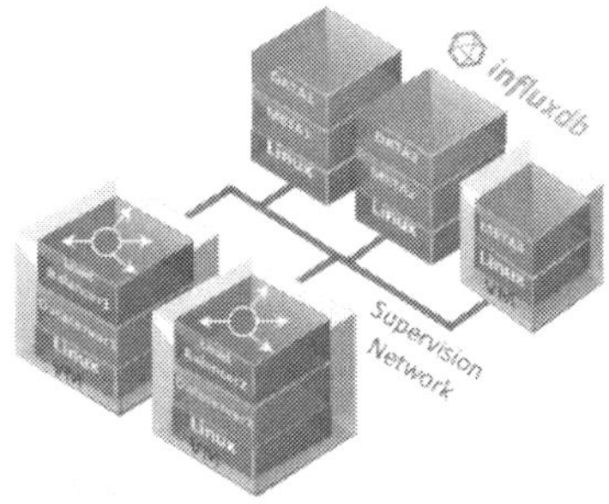

Fig. 5. CAREDAS data archiving.

The software in charge of writing data into InfluxDB database is homemade. It receives data from DAQ by MQTT message and write it in a dedicated DAQ database. This software makes the arbitration and consistency test of data received from redundant DAQ before storing it.

V. Data visualization

Two types of data visualization are possible: online and offline.

Online data visualization displays data directly from DAQ in subscribing to MQTT messages (Fig. 6). It is reachable in the configuration tool. The user can tune DAQ parameters and displays the result with the same tools near the DAQ.

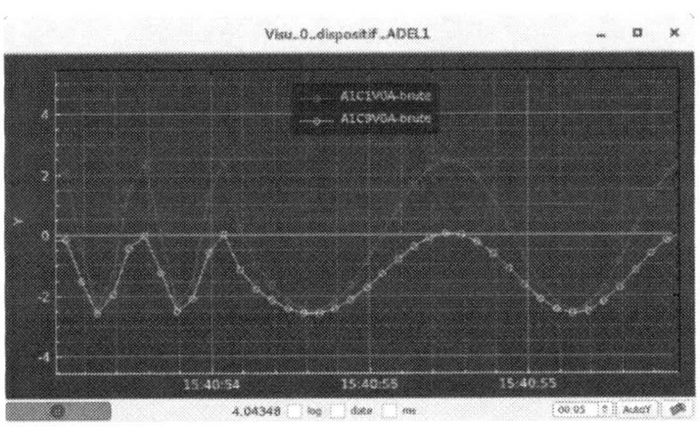

Fig. 6. Online data visualization.

Offline visualization displays data from InfluxDB databases with Grafana [6]. Grafana is an open source software to display data into dashboards. User makes dashboard (Fig. 7) with various panel like graph display, gauge, table of values. Dashboard are designed and reachable in a web browser. Alarms on threshold or data transformation can be performed in dashboard.

Fig. 7. Grafana dashboard.

A Grafana plugin allows to display diagrams on dashboard with objects animated by data value like SCADA for industrial process survey (Fig.8). Objects appearance can be change on data threshold like color change in a rule.

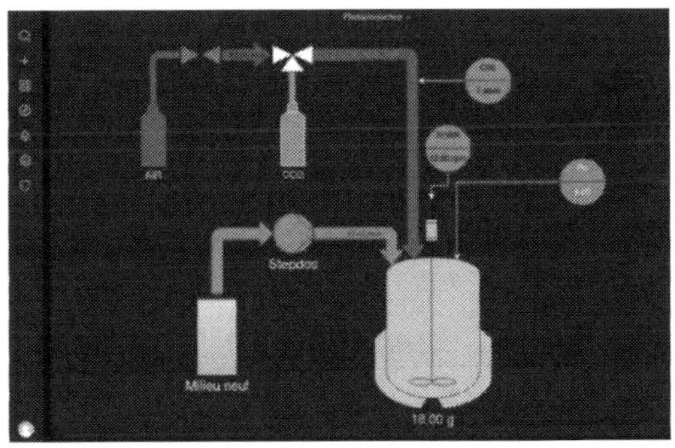

Fig. 8. SCADA on Grafana dashboard.

Grafana run as service on a Linux server. A high availability is getting by running many Grafana services on different nodes.

VI. Data acquisition system

Real time data acquisition software is modular and technology independent (Fig. 9). It can compose of many chassis with different technologies following the user needs. Proven industrial system technologies have been selected to ensure sustainability on 10 years. Technologies implemented are PXIe bus, CompactRIO and CompactDaq from National Instruments, EtherCAT protocol and various measurement units over Ethernet. The operating system is Linux with PREEMPT_RT real time functionality. CAREDAS data acquisition software can also run on a Raspberry PI. To ensure the high availability and no data loss many mechanisms are implemented. First, data are stored locally on a disk. MQTT client of the DAQ stores messages in memory while few minutes during a network loss. Messages stored in memory will be transmitted if the link is restored before this period. If this deadline is exceeded then data stored on disk will be transmitted. DAQ can also be redundant with three systems.

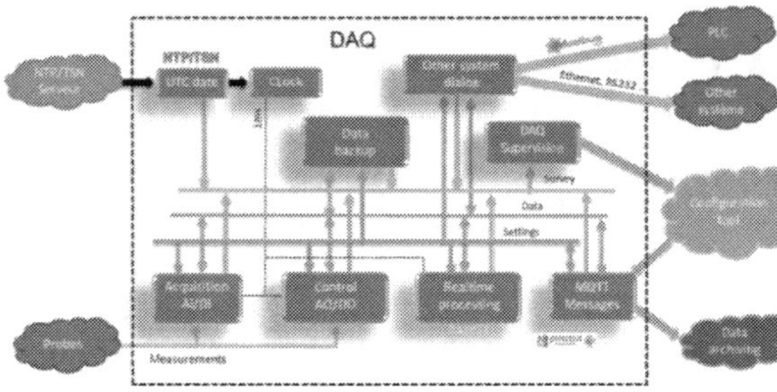

Fig. 9. DAQ architecture.

DAQ provides data timestamp in UTC (Universal Time Coordinated) with a 1ms, 1µs or 1ns precision. Electronic board synchronization is realized by distributed hardware clocks. These clocks are generated from system clock that is synchronized by TSN protocol [7]. Thus, all DAQ are synchronized at 1µs.

DAQ software is homemade in C++ language. It is modular, technology, operating system and hardware independent. This architecture allows adding new hardware or replace a library without impact on all DAQ software.

Real time processing are describing in configuration tool, thus user do not need to modify DAQ software. An open source mathematical expression parser ExprTk [7] performs real time processing. On the start of data acquisition, ExprTk produces an instance with an AST (Abstract Syntax Tree) of the real time processing and use it in run time to evaluate ExprTk expression. The ExprTk library has capabilities of mathematical functions, operators, control and loop structure and user function that permit to realize complex processing like digital processing or machine learning.

VII. Data acquisition setting

The configuration tool (Fig. 10) allows user to define intuitively all necessary information for DAQ. It surveys the DAQ operation and visualizes online data in real time directly from DAQ MQTT messages. The user can tune DAQ parameters near the DAQ with a laptop.

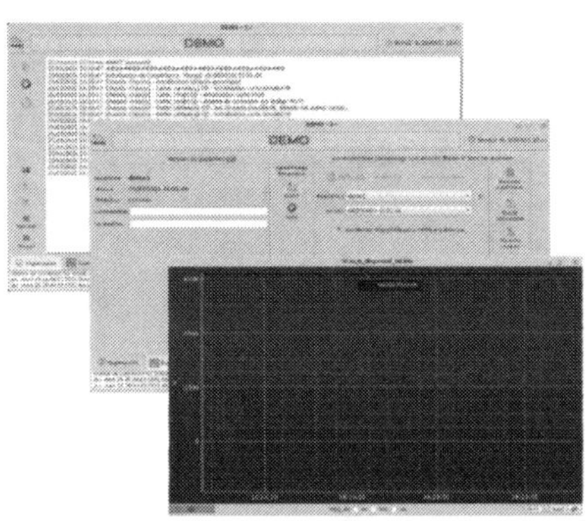

Fig. 10. CAREDAS configuration tool.

He can define an unlimited number of configuration stored in InfluxDB database. Every parameter backup generates a new configuration version. The tool is able to generate the CAREDAS configuration for all the acquisition chain when adding a new DAQ (MQTT topics, create database, DAQ system configuration). The configuration tool is a homemade software in C++ language usable on any operating system.

Multiple tabs allow to simply define chassis, electronic boards and channels. Hardware channels are defined in channel tab and associates the hardware channel to the measurement name in database. Real time processing are written in this tab with the possibility to create user functions in a library and share variables between real time processing or user functions (Fig. 11).

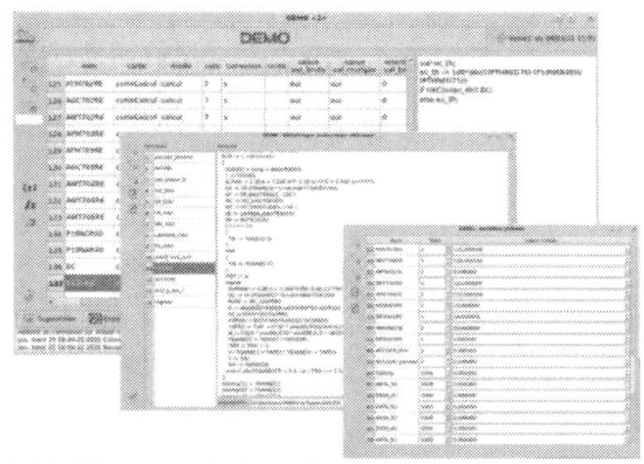

Fig. 11. Real time processing in configuration tool.

VIII. CAREDAS as single station

All the CAREDAS architecture is able to run on single station like personal computer (PC) without software modification (Fig. 12). On this PC run EMQ X MQTT broker, InfluxDB database, offline data visualization with Grafana and CAREDAS configuration tool. CAREDAS data acquisition software can also run on this station when the hardware is not open like a black box. CAREDAS is then adapted to a standalone measuring bench.

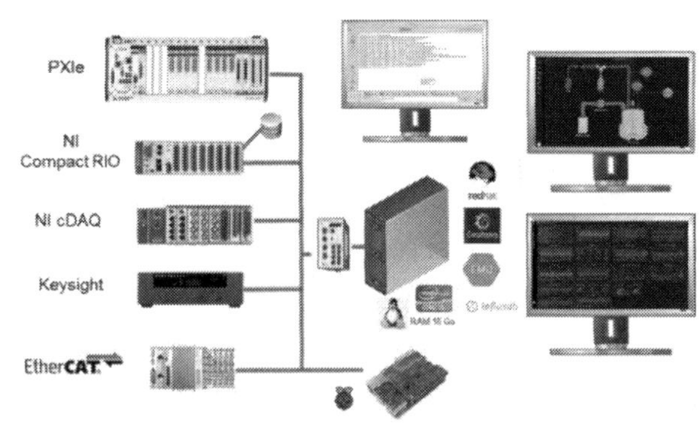

Fig. 12. CAREDAS single station architecture.

IX. Conclusions

A proof of concept has been realized by the end of 2019. Data flow of 20 DAQ experimental devices were simulated 24/7 for two months without data loss (Fig. 13). Experimental devices had 800 channels each for a cycle time of 50ms. Two real DAQ with CompactRIO and PXIe DAQ are used and Raspberry PI 3 simulates the 18 others. This represents 16000 measurement channels every 50ms ether a flow of 320000 measurements by second and 28 billion measurements by day. MQTT cluster is made of three DELL R630 dual processor server with 64 GB of RAM to store messages in case of client network failure. InfluxDB cluster is composed of two data nodes and three meta nodes with SSD disk for databases. Computers are DELL R630 dual processor server with 16 GB of RAM.

Fig. 13. DAQ experimental devices for CAREDAS POC

Now, CAREDAS is under development and a first stable version for single station is planning for September 2021. EtherCAT [9] and OPC UA [10] protocol are being integrated into CAREDAS to have a wide choice of technologies. Meanwhile, a beta version was deployed on five measuring bench on some CEA Cadarache plant. This strategy allowed us to have feedback of CAREDAS use and improves software along developments. CAREDAS version for JHR is planned for March 2022.

References

[1] JHR Web site http://www-rjh.cea.fr
[2] Influx DB from influxdata https://www.influxdata.com/
[3] MQTT https://mqtt.org/
[4] Google Protocol Buffers https://developers.google.com/protocol-buffers/
[5] EMQ X from EMQ https://www.emqx.io/
[6] Grafana https://grafana.com/
[7] TSN https://en.wikipedia.org/wiki/Time-Sensitive_Networking
[8] ExprTk http://www.partow.net/programming/exprtk/
[9] EtherCAT https://www.ethercat.org/
[10] OPC UA https://opcfoundation.org/about/opc-technologies/opc-ua/

3-D thermal and radiation-matter interaction simulations of a SiC solid-state detector for neutron flux measurements in JSI TRIGA Mark II research reactor

V. Valero[1], L. Ottaviani[1], A. Lyoussi[2], H. Ghninou[2], V. Radulović[3], L. Snoj[3], A. Pungerčič[3], A. Volte[1], M. Carette[1], C. Reynard-Carette[1]
[1]Aix Marseille Univ, Université de Toulon, CNRS, IM2NP, Marseille, France
[2]CEA/DES/IRESNE/DER, Section of Experimental Physics, Safety Tests and Instrumentation, Cadarache, F-13108, Saint Paul-lez-Durance, France
[3]Reactor Physics Division, Jožef Stefan Institute, Ljubljana, Slovenia
valentin.valero@im2np.fr

***Abstract*—Neutron detection is a relevant topic in the field of nuclear instrumentation. It is at the heart of the concerns for fusion applications (neutron diagnostics, measurements inside the Test Blanket Modules TBM) as well as for fission applications (in-core and ex-core monitoring, neutron mapping or safety applications in research reactors). Moreover, due to the even more harsh conditions of the future experimental reactors such as the Jules Horowitz Reactor (JHR) or International Thermonuclear Experimental Reactor (ITER), neutron detectors need to be adapted to high neutron and γ fluxes, high nuclear heating rates and high temperatures. Consequently, radiation and temperature hardened sensors with fast response, high energy resolution and stability in a mixed neutron and γ environment are required. All these requirements make wide-bandgap semiconductors and, more precisely, Silicon Carbide (SiC) serious candidates due to their intrinsic characteristics in such extreme environments. Thus, since the last decades, SiC-based detectors are developed and studied for neutron detection in various nuclear facilities. In this paper, a SiC-based neutron detector is 3-D designed and studied through thermal and radiation-matter interaction numerical simulations for a future irradiation campaign at the Jožef Stefan Institute TRIGA Mark II research reactor in Slovenia. Firstly, this paper presents the scientific background and issues of our SiC-based detectors. In a second part the 3-D geometry is shown. Thereafter, the 3-D numerical thermal simulation results are reported. Finally, the 3-D numerical radiation/matter interaction simulations results are presented.**

***Keywords* —Neutron detection, silicon carbide, irradiation campaign, thermal, radiation-matter interaction, 3-D simulations.**

I. Introduction and Issues

In a nuclear reactor, either for fission or fusion applications, it is crucial to measure key parameters such as neutron and γ fluxes or nuclear heating rate for a better understanding of the behavior of nuclear fuels or materials subjected to nuclear radiation and for the monitoring of the nuclear facilities. All this places neutron detection and measurement at the heart of these scientific objectives. As the conditions in fusion and fission research reactors become more and more harsh, it is crucial to develop sensors adapted to such characteristics. For this reason, wide-bandgap semiconductors and more precisely 4H-SiC are interesting. Their main advantages are their fast response, their stability versus radiations, their high energy resolution, their ability to discriminate between neutrons and γ, their high thermal conductivity and their low semiconductor defect concentration. In addition, thermal and fast neutrons are able to be detected by adding Neutron Converter Layer (NCL) and /or by considering threshold energy reactions for this type of detectors [1-5].

At present, such SiC-based neutron detectors are studied within the framework of the LIMMEX laboratory (Laboratory of Instrumentation and Measurement Methods in EXtreme media, a joint laboratory between Aix-Marseille University and the CEA). This topic started, more precisely, with a previous joint European project, called I_SMART (Innovative Sensor for Material Ageing and Radiation Testing). The greatest achievement of this project was the development of SiC-based p^+n junction diodes with a thermal neutron converter layer with Boron-10 implantation in order to measure thermal and fast neutron fluxes.

Thermal neutron measurements with these SiC-based diodes have been achieved at MINERVE Zero Power Reactor at CEA Cadarache in France for a maximal neutron flux of around 9.4×10^8 n·cm^{-2}·s^{-1} including 31 % of thermal neutrons [2]. A fluence study in order to demonstrate the stability under radiation has been performed at the BR1 research reactor at Moll in Belgium for a maximum fluence of around 2.0×10^{13} n·cm^{-2} [3]. Afterwards, these detectors have been tested to detect 14.1 MeV fast neutrons at the Deuterium-Tritium generator of Technical University of Dresden in Germany. As a first step, the SiC-based diode signal stability has been studied to room temperature up to 500 °C with fast neutron fluxes of around 3.1×10^7 n·cm^{-2}·s^{-1} [4]. As a second step, the response and the energy resolution of the SiC-based diode were compared to those of a diamond-based detector purchased from CIVIDEC Instrumentation company for a neutron flux of 9.4×10^6 n·cm^{-2}·s^{-1} [5].

Consequently, there are three main objectives. The first aim corresponds to the increase of the measured total neutron flux: 5.5×10^{14} n·cm^{-2}·s^{-1} for an energy > 1 MeV expected in the JHR core and from 10^9 n·cm^{-2}·s^{-1} to 10^{14} n·cm^{-2}·s^{-1} for an energy > 14.1 MeV expected in the ITER tokamak. The second aim is the increase in the irradiation duration as well as the adaptation of the housing of the detector for in-core measurements in a research reactor. The latter can be achieved by means of 3-D numerical thermal simulations and 3-D numerical simulations of radiation/matter interactions. The first kind of simulation can be used in order to evaluate the maximum temperature and the temperature fields. The second kind of simulation can provide neutron and γ fluxes and attenuation, nuclear heating rate values and other information such as the activation.

However, intermediate steps are necessary for the SiC-based detectors in order to reach these aims. A qualification of the SiC-based detector for measurements of increased neutron fluxes (especially higher than 10^{10} n·cm^{-2}·s^{-1}) with a new specific acquisition chain and nuclear-hardened instrumentation cables will be carried out during an irradiation campaign at the Jožef Stefan Institute TRIGA Mark II research reactor in Slovenia. The main characteristics of this reactor are its maximum thermal power (at steady state) i.e. 250 kW and its maximum neutron and γ fluxes at the center of the core i.e. 2.0×10^{13} n·cm^{-2}·s^{-1} and 2.0×10^{13} γ·cm^{-2}·s^{-1} respectively. The measurements will be carried out in a Triangular Irradiation Channel (TIC) which is an in-core dry air channel with natural convection (low heat transfer coefficient) and a maximum ambient temperature of around 40 °C. Another main feature of this channel is its size, as it measures almost 6 cm in diameter. The maximum total neutron flux (thermal and fast) is about 1.2×10^{13} n·cm^{-2}·s^{-1}, the maximum γ flux is about 1.2×10^{13} γ·cm^{-2}·s^{-1} and an estimated nuclear heating rate through neutron and photon KERMA rate (Kinetic Energy Released per MAss unit) in Silicon lower than 0.055 W·g^{-1} [6]. Consequently, this paper focuses on 3-D numerical simulation studies dedicated to the definition of the most optimized device for this campaign.

In a first part, the 3-D detector geometry is shown and detailed. The latter consists of all the detector components (housing, base, diode placed on its support, connectors, screws and instrumentation cables).

In a second part, the 3-D thermal simulations realized by means of COMSOL Multiphysics finite element code are presented. The aim of these thermal simulations is to determine the maximum temperature and the temperature field inside the system and in particular in the SiC-based diode for various boundary conditions and the expected nuclear heating rate range.

In the last part, 3-D numerical simulations of interactions between nuclear radiations and matter are realized with the SiC detector inserted inside the MCNP computational model of the JSI TRIGA Mark II reactor. In order to reduce the calculation time, a simplified system as that implemented for 3-D numerical thermal simulations is used. The goal of these simulations is to estimate the neutron and γ fluxes expected inside the SiC-based diode and the KERMA rate in each part of the studied system at the vertical central position of the Triangular Irradiation Channel.

II. Detector 3-D Geometry

The studied SiC-based detector (cf. Fig. 1) is composed of three main parts. The first one is a machined housing made of Duralumin. Moreover, four fillets caused by the drill during the machining are positioned at the four inner corners of the housing. The latter induces an Air-gap located inside this housing. Finally, four Stainless Steel screws are placed on the top of the housing for the hood closing.

The second part is composed of the SiC-based diode placed on its Alumina support. Then, the latter is fixed on a base made of Duralumin and the whole assembly is maintained to the housing by means of two Stainless Steel screws in and two Alumina washers.

The last part consists in a connector made of Stainless Steel screwed on the housing by means of an inner part. Inside this connector, two sheaths pass through it corresponding to a k-type thermocouple and the two power and signal recovery wires of the SiC-based diode respectively.

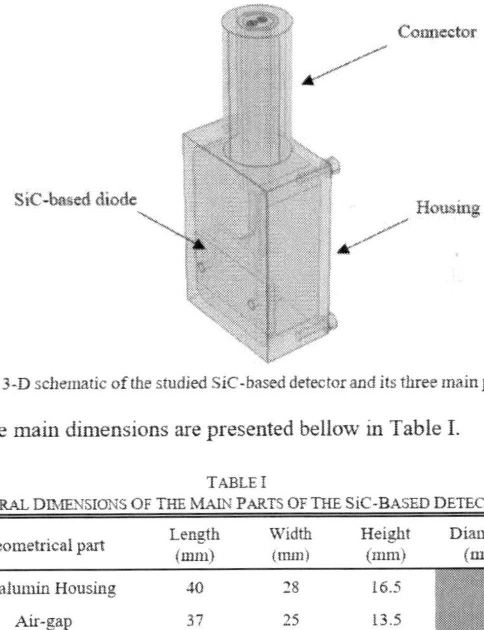

Fig. 1. 3-D schematic of the studied SiC-based detector and its three main parts.

The main dimensions are presented bellow in Table I.

TABLE I
GENERAL DIMENSIONS OF THE MAIN PARTS OF THE SiC-BASED DETECTOR

Geometrical part	Length (mm)	Width (mm)	Height (mm)	Diameter (mm)
Duralumin Housing	40	28	16.5	
Air-gap	37	25	13.5	
SiC-based diode	7	5	0.371	
Alumina support	16	14	1.0	
Duralumin base	12	25	1.5	
Stainless Steel connector (external part)	28			14
Stainless Steel connector (inner part)	36			8

https://doi.org/10.1051/epjconf/202125304009

The 3-D geometry given in Figure 1 is used for the 3-D thermal simulations. By against to reduce the calculation time in the case of the 3-D numerical radiation/matter interaction simulations, this 3-D geometry is simplified. The screws, washers, k-type thermocouple, SiC-based diode wires and the inner part of the connector are neglected.

III. 3-D Numerical Thermal Simulations

COMSOL Multiphysics finite element code coupled with a heat transfer module was used for this study. Regarding the considered 3-D thermal model, the heat equation is solved only by considering conductive heat transfers with thermal conductivity depending on temperature (convective and radiative exchanges are neglected inside the detector) and for the stationary state. Moreover, volumetric heat sources (Q) are applied in all parts of the detector. They depend on the material density (ρ_i) and the nuclear heating rate (E_n) value ($Q = \rho_i \cdot E_n$). The nuclear heating rate imposed is independent of the material nature. The nuclear heating rate range is considered from the literature, the range is from 0 $W \cdot g^{-1}$ to 0.055 $W \cdot g^{-1}$ [6] by steps of 0.0025 $W \cdot g^{-1}$. An extra-fine mesh was used with element sizes from 102 μm to 2380 μm. The applied boundary conditions correspond to external natural convection of dry air. The imposed temperatures of the coolant fluid (T_{fluid}) are 35 °C, 40 °C and 45 °C. The external heat transfer coefficient (h) is calculated and imposed by COMSOL Multiphysics finite element code by taking the geometry into account. Thus, four heat transfer coefficients are applied for the vertical housing walls, the lower housing wall, the upper housing wall, and the vertical cylinder of the connector. Finally, a constant heat transfer coefficient is applied on the upper surface of the connector and fixed at 2 $W \cdot m^{-2} \cdot K^{-1}$.

A. Boundary conditions influence

The nuclear heating rate influence on the external heat transfer coefficient has been studied for a dry air temperature equal to 40 °C for the complete 3-D detector (cf. Fig. 2). As shown, there is an increase in the heat transfer coefficient value with the nuclear heating rate. In fact, the increase in nuclear heating rate leads to higher wall temperatures which induce an increase of the thermal exchanges and thus the heat transfer coefficient values.

Moreover, as shown in Figure 2, the heat transfer coefficient values calculated by COMSOL Multiphysics code are coherent with those obtained by using correlations of the literature (from 5 $W \cdot m^{-2} \cdot K^{-1}$ to 25 $W \cdot m^{-2} \cdot K^{-1}$).

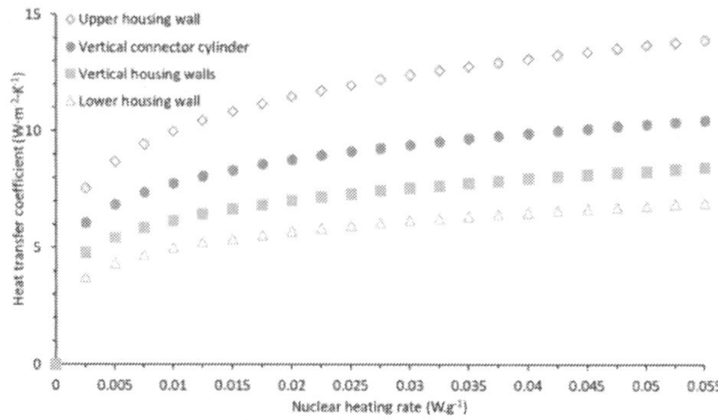

Fig. 2. Various heat transfer coefficient values calculated by COMSOL as a function of the nuclear heating rate for a fluid temperature of 40°C.

B. Nuclear heating rate influence on the maximum temperature

The nuclear heating rate influence on the maximum temperature and the temperature field has been studied in order to observe the thermal behavior of the complete SiC-based detector.

As expected, the maximum temperature is obtained under the highest extreme conditions: a nuclear heating rate value of 0.055 $W \cdot g^{-1}$ and a fluid temperature of 45 °C, giving rise to the maximal value of around 110 °C. For an expected fluid temperature of 40 °C and a maximum nuclear heating rate value, the maximum reached temperature is around 105 °C. These values are much lower than the Duralumin melting point i.e. around 600 °C. Concerning the SiC-based diode temperature, the obtained values are about 99 °C and 104 °C for fluid temperatures of 40 °C and 45 °C respectively. Thus, these obtained temperatures allow the use of standard Tin alloys welding for the SiC-based diode connection and wiring (melting point between 180 °C and 200 °C).

Regarding the thermal behavior of the SiC-based detector and as shown in Figure 3, the highest temperature is reached inside the connector. This behavior is confirmed whatever the imposed nuclear heating rate (cf. Fig. 3). In fact, the latter is induced by the low heat transfer coefficient value applied on the upper surface of the connector, and by a high energy deposited inside this part due to the high density of the used material i.e. Stainless Steel (ρ = 7850 $kg \cdot m^{-3}$).

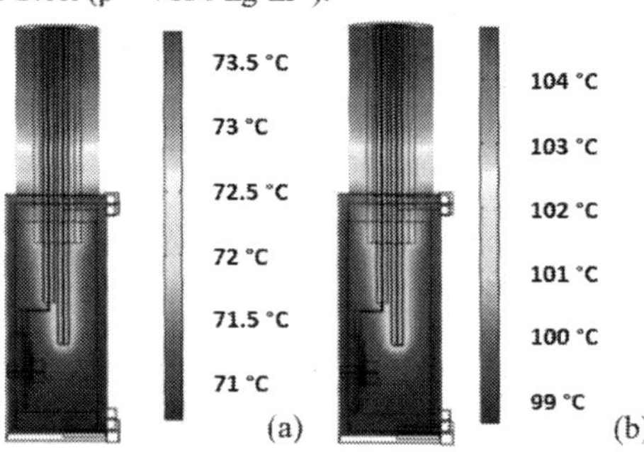

Fig. 3. Vertical cross section of the complete 3-D detector for nuclear heating rate values of 0.025 $W \cdot g^{-1}$ (a) and 0.055 $W \cdot g^{-1}$ (b) and a fluid temperature of 40 °C.

C. Comparison of the thermal behavior of the two geometries

As explained above, in order to gain time for the 3-D numerical radiation/matter interaction simulations, a simplified geometry has been designed. In order to check if any thermal differences are induced, a comparison with the same parameters than those used for the complete detector geometry was performed. The same thermal behavior was observed (cf. Fig. 4) for the two detectors and the most part of the heat is deposited in the connector. The geometry simplification leads to a very low difference of maximum temperature: -2.3 °C for a nuclear heating rate value of 0.055 $W \cdot g^{-1}$ and a fluid temperature of 40 °C. Moreover, the lower the condition values are, the lower the maximum temperature difference is. In fact, the geometry reduction leads to a lower mass of the SiC-based detector which induces a lower energy deposition and thus a lower temperature. As a conclusion, this simplified geometry can be used for the 3-D numerical radiation/matter interaction simulations since no significant thermal differences have been found.

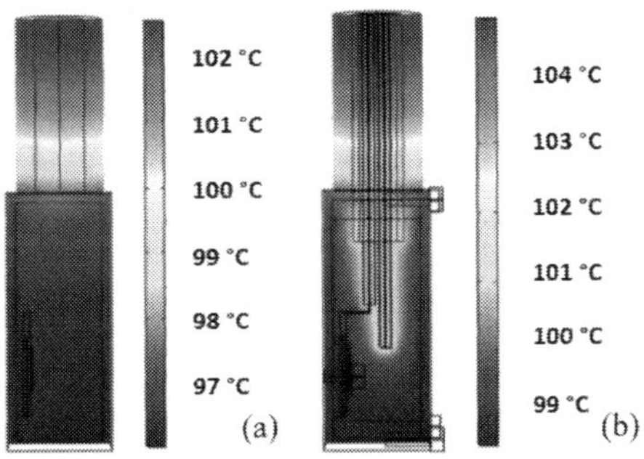

Fig. 4. Vertical cross section of the simplified 3-D detector (a) and the complete 3-D detector (b) for a nuclear heating rate value of 0.055 $W \cdot g^{-1}$ and a fluid temperature of 40°C.

IV. 3-D Numerical Radiation/Matter Interaction Simulations

Monte-Carlo N-Particle transport (MCNP) code [7] was used in order to estimate the prompt neutron and γ deposited energy with KERMA in all cells of the simplified geometry of the SiC-based detector (F6 tally in $MeV \cdot g^{-1}$) and the prompt neutron and γ fluxes in the SiC-based diode (F4 tally in $1 \cdot cm^{-2}$) at different power level. Thus, this SiC-based detector geometry was built using MCNP6 code and then inserted inside the MCNP computational model of the JSI TRIGA Mark II reactor. More precisely, the detector was simulated inside the TIC at the center of the channel i.e. the middle axial position. For these simulations, the ENDF/B-VII.0 cross-section nuclear data library at 297.15 K was also used.

A. Prompt neutron and γ KERMA values

Nuclear heating rate can be estimated by the KERMA which is defined as the sum of the initial kinetic energies of all the charged particles liberated by the uncharged ionizing radiation i.e. the neutrons and photons. The track length F6 tally with a neutron–photon coupled mode (NP mode) was also used in order to obtain the sum of the KERMA induced by prompt neutrons and the one induced by prompt photons (γ). As projected, the maximal KERMA values are obtained at the maximal reactor thermal power i.e. 250 kW (cf. Table II). Moreover, the obtained values are coherent with the literature i.e. 0.055 $W \cdot g^{-1}$ [6]. The statistical uncertainties are lower than 0.3 %, except the one for SiC-based diode slightly higher and equal to 0.54 %. The small size of the latter and the low calculation time (1.51×10^{12} particles are considered) can explain this value which is not so high for this first study.

These values will be used for future more accurate 3-D thermal simulations by considering a nuclear heating rate value depending on the detector part.

TABLE II
CALCULATED KERMA VALUES FOR VARIOUS REACTOR THERMAL POWERS

	Prompt Neutron and Photon KERMA Values ($W \cdot g^{-1}$)		
Geometrical part/Reactor power level	50 kW	150 kW	250 kW
Duralumin Housing	0.008	0.023	0.039
SiC-based diode	0.009	0.028	0.046
Alumina support	0.009	0.027	0.045
Duralumin base	0.008	0.025	0.042
Stainless Steel connector	0.010	0.029	0.048

B. Prompt neutron and γ fluxes in the SiC-based diode

In order to prepare our irradiation campaign, prompt neutron and γ fluxes in the SiC-based diode have been calculated in order to work on the writing of a first operating protocol. Indeed, as explain on the first section, the SiC-based diode has to be qualified for the measurement of increasing neutron fluxes, especially fluxes higher than $10^{10} n \cdot cm^{-2} \cdot s^{-1}$. Thus, these results can provide interested information concerning the necessary reactor thermal power needed to measure a precise neutron flux or to calculate the measurement time in order to reach a targeted neutron fluence. As shown in Table III, the increase in the prompt particle flux varies constantly when the reactor thermal power increase. As expected, the maximal fluxes are obtained at full reactor thermal power i.e. 250 kW. Thus, the obtained maximum prompt neutron flux is $9.94 \times 10^{12} n \cdot cm^{-2} \cdot s^{-1}$ and the obtained maximum prompt γ flux is $1.06 \times 10^{13} \gamma \cdot cm^{-2} \cdot s^{-1}$. Concerning the uncertainties, they are equal to 0.5 % for the prompt photons and 0.31 % for the prompt neutrons.

TABLE III
CALCULATED NEUTRON AND γ FLUXES FOR VARIOUS REACTOR THERMAL POWERS INSIDE THE SIC-BASED DIODE

	Prompt Neutron and Photon Fluxes (particle$\cdot cm^{-2} \cdot s^{-1}$)		
Kind of particle/Reactor power level	10 W	1 kW	100 kW
Prompt neutrons	3.98E+08	3.98E+10	3.98E+12
Prompt γ	4.25E+08	4.25E+10	4.25E+12

V. Conclusions and Outlooks

Thanks to the 3-D thermal simulations and 3-D radiation/matter interaction simulations, various conclusions were obtained for a future qualification of the SiC-based detector in the JSI TRIGA Mark II reactor. For the 3-D numerical thermal simulations, firstly the heat transfer coefficient values calculated by COMSOL are coherent with those estimated by using correlations of the literature. Moreover, the maximum SiC-based detector temperature is around 110 °C under the highest extreme boundary conditions (maximum nuclear heating rate and coolant fluid temperature). This value is much lower than Duralumin melting point even with a safety margin and permits the use of standard Tin alloys welding for the SiC-based diode connection (melting point between 180 °C and 200 °C). Afterwards, the most part of the heat is deposited in the connector and in order to reduce its temperature a material with a lower density could be used or a smaller connector could be designed. Finally, the low temperature difference of the SiC-based diode obtained between the two geometries has permitted the use of the simplified geometry for MCNP code simulations. Thus, the first preliminary results of the 3-D numerical simulations of radiation/matter interactions confirm the right nuclear heating rate range used for the 3-D thermal simulations. In spite of a short calculation time (due to a low number of particles considered), the maximum uncertainty is 0.54 % and majority are below 0.3 % and are satisfying for first results. Lastly, the estimation of the neutron flux for several reactor powers give us an idea of the future irradiation campaign conditions targeted to test our SiC-based detector.

Regarding the outlooks, about the 3-D numerical simulations of radiation/matter interactions, the key values will be calculated through the complete 3-D geometry of the detector for different axial positions inside the TIC, for various reactor powers or in another irradiation channel. Moreover, the integration of another estimator for nuclear heating rate calculations could be made. In fact, the electron energy deposition in matter (*F8:E pulse height tally with MCNP code) could be used in order to estimate the nuclear heating rate and not only the KERMA as performed in this paper. The neutron activation of the materials which compose the detector is another interesting parameter in particular for long time experiments. Afterwards, the use of the real detector geometry and assembly with the screws, washers, thermocouple, wires and maybe a new connector can be also interesting to study when considering a longer calculation time. For the 3-D numerical thermal simulations, new heat sources can be applied by considering the MCNP code results leading to a maximum temperature and temperature field comparison. The integration of radiative and convective heat transfers can be also a great opportunity in order to consider a more accurate model.
Furthermore, the optimized detector will be tested during an irradiation campaign at the JSI TRIGA Mark II reactor in the last quarter of 2021.

To conclude, within the framework of the LIMMEX laboratory, the great results previously obtained have led to study a new measurement device by coupling a non-adiabatic calorimeter to a SiC-based neutron detector for measuring nuclear heating rate and neutron fluxes (thermal and/or fast) in a research reactor core simultaneously [8-9].

References

[1] F. Ruddy, A. Dulloo, J. Seidel, M. Das, S. Ryu and A. Agarwal, "The Fast Neutron Response of 4H Silicon Carbide Semiconductor Radiation Detectors," *IEEE Trans. Nucl. Sci.*, vol. 53, no. 3, pp. 1666-1670, June 2006.

[2] O. Obraztsova, L. Ottaviani, B. Geslot, G. de Izarra, O. Palais, A. Lyoussi and W. Vervisch, "Comparison between Silicon-Carbide and Diamond for Thermal neutron detection at Room Temperature," *IEEE Trans. Nucl. Sci.*, vol. 67, no. 5, pp. 863-871, May 2020.

[3] F. Issa, L.Ottaviani, V. Vervisch, D.Szalkai, L.Vermeeren, A. Lyoussi, A. Kuznetsov, M. Lazar, A. Klix, O.Palais, R. Ferone and A. Hallén, "Study of the stability of 4H-SiC detectors by thermal neutron irradiation," *Materials Science Forum*, vol. 821-823, pp. 875-878, June 2015.

[4] O. Obraztsova, L.Ottaviani, A.Klix, T. Döring, O. Palais and A. Lyoussi, "Comparing the Response of a SiC and a sCVD Diamond Detectors to 14-MeV Neutron Radiation," *IEEE Trans. Nucl. Sci.*, vol. 65, no. 9, pp. 2380-2384, Sep. 2018.

[5] D. Szalkai, R. Ferone, F. Issa, A. Klix, M. Lazar, A. Lyoussi, L. Ottaviani, P. Tütto, and V. Vervisch, "Fast Neutron Detection With 4H-SiC Based Diode Detector up to 500 °C Ambient Temperature," *IEEE Trans. Nucl. Sci.*, vol. 63, no. 3, pp. 1491-1498, June 2016.

[6] K. Ambrožič, G. Žerovnik and L. Snoj, "Computational analysis of the dose rates at JSI TRIGA reactor irradiation facilities," *Applied Radiation and Isotopes*, vol. 130, pp. 140–152, Dec. 2017.

[7] J. T. Goorley, M. R. James, T. E. Booth, F. B. Brown, J. S. Bull, L. J. Cox, J. W. J. Durkee, J. S. Elson, M. L. Fensin, R. A. I. Forster and et al., "Initial MCNP6 Release Overview-MCNP6 version 1.0", Los Alamos National Laboratory (LANL), June 2013.

[8] V. Valero, L. Ottaviani, A. Lyoussi, V. Radulović, L. Snoj, A. Volte, M. Carette and C. Reynard-Carette, "Thermal simulations of a new SiC detector design for neutron measurements in JSI TRIGA Mark II research reactor," in *Virtual Issue on SiC 2020 - ECSCRM 2020/2021*, Virtual conference, to be published, 2021.

[9] V. Valero, A. Volte, L. Ottaviani, A. Lyoussi, M. Carette, J. Brun and C. Reynard-Carette, "Parametrical study for the coupling of a silicon carbide (SiC) detector and a calorimeter to measure simultaneously neutron flux and radiation absorbed dose rate in research reactors," in *RRFM 2020*, Virtual conference, 2020.

Assessment of Irradiation Performance in the Jules Horowitz Reactor (JHR) using the CARMEN Measuring Device

David Blanchet [1], Muriel Antony [2], Hubert Carcreff [3], Sébastien François [4], Philippe Guimbal [5], Bernard Pouchin [1]

[1] CEA Cadarache, DES/IRESNE/DER, F-13108 Saint-Paul-lez-Durance, France
[2] CEA Cadarache, DES/DIMP/DCET, F-13108 Saint-Paul lez-Durance, France
[3] CEA Saclay, DES/ISAS/DM2S, F-91190 Gif-sur-Yvette, France
[4] CEA Cadarache, DES/DIMP/DPED, F-13108 Saint-Paul-lez-Durance, France
[5] CEA Saclay, DES/ISAS/DMN, F-91190 Gif-sur-Yvette, France

david.blanchet@cea.fr

***Abstract*—The development of the JHR experimental devices rely on the operational feedback from previous French material testing reactors (i.e. SILOE and OSIRIS). The experimental devices used for the irradiation of structural material were already facing technological limitations, in particular regarding the control of irradiation temperature and of the thermal gradients in the experimental samples, which is essential to ensure the quality of the experiments. Obtaining satisfactory thermal fields (in compliance with the setpoint and the homogeneity) is all the more difficult as the level of nuclear heating is higher in the JHR. This paper attempts to characterize the irradiation conditions in different experimental positions of the JHR and to compare them with the conditions and the empirical criteria of maximum acceptable temperature measured in OSIRIS. The study shows that the irradiation conditions obtained inside the experimental devices can sometimes be significantly different from the measured conditions using instrumentation devices. The interpretation of the experimental results and their transposition to other situations will always require a calculation versus measurement adjustment and the intensive use of computer simulation. However, despite all simulation and transposition efforts, the control of temperature conditions is not yet fully demonstrated and nothing will ultimately replace experimental validation.**

***Keywords* —Material Test Reactor (MTR), differential calorimeter, experimental devices, Reactor Pressure Vessel (RPV).**

I. Introduction

The Jules Horowitz Reactor (JHR) is a Material Test Reactor currently under construction at Cadarache. It will serve to conduct experiments for the qualification of materials under irradiation and to produce radioisotopes for medical applications [1][2]. The material testing experiments require accurate monitoring of the irradiation conditions in terms of neutron flux (level and spectrum) and nuclear heating. A key aspect is to regulate the temperature level and to get uniform temperature distributions (absence of gradients) in the material samples during the irradiation. These requirements ensure the quality of the experiments and therefore contribute to the value of the irradiation service to future JHR clients.

CARMEN is a multi-detector measuring device that will provide a comprehensive characterization of the radiation field in the experimental channels of the JHR. The measurements performed with this device can serve as a preliminary assessment of unperturbed irradiation conditions, before the introduction of the real experimental device at that location. The present study aims at reproducing the CARMEN measurements by simulation and connecting them to the irradiation conditions expected inside the MICA and the OCCITANE test devices - which are respectively devoted to in-core and out-core experiments for the irradiation of steel materials.

Section 2 presents the MICA and the OCCITANE test devices for structural material irradiation in the JHR. Section 3 presents the CARMEN measuring device. Section 4 examines modelling aspects for nuclear heating measurements with a differential calorimeter. In section 5, the study evaluates the transfer function between the CARMEN measurements (unperturbed conditions) and the actual nuclear heating inside test devices (perturbed conditions). Finally, section 6 attempts to integrate the feedback from the IRMA test device operation in the OSIRIS reactor.

II. The experimental devices for the irradiation of structural materials in the JHR reactor

A. The MICA test device

The MICA test device provides irradiation capabilities to study the structural material aging at high dpa-rates. There are ten in-core positions available to host this device: seven small channels located inside the fuel elements, and three large channels that could receive either a large irradiation device or an additional fuel element (see Fig. 1). Design studies and development of these devices rely on accurate nuclear heating calculations thanks to the experimental validation programs conducted in zero power reactors [3].

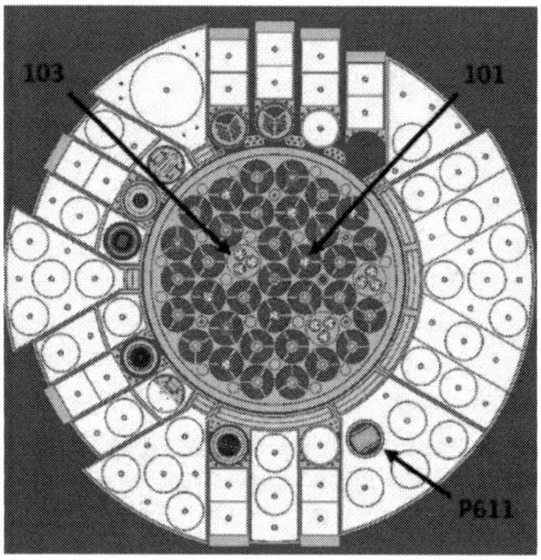

Fig. 1. Radial layout of the JHR reactor core model with TRIPOLI-4®.

As illustrated in Fig. 2, the MICA test device is inserted in the central axis of a fuel element. The material samples are loaded in a sample holder in the centre of the MICA device. The inner stainless steel tube of 24 mm in diameter limits the available space for the samples, which are immersed in static NaK eutectic. The good thermal conductivity of this liquid metal ensures optimal cooling and homogeneous thermal distribution in the samples. Surrounding the inner tube, some electric heating elements are embedded in order to reach and to regulate the setpoint temperature in the experimental zone. They allow flattening of the axial temperature gradients imposed by the axial nuclear heating profile. Several thermocouples are monitoring online temperatures while specific dosimeters are dedicated to evaluate, after irradiation, the integrated fast neutron fluence.

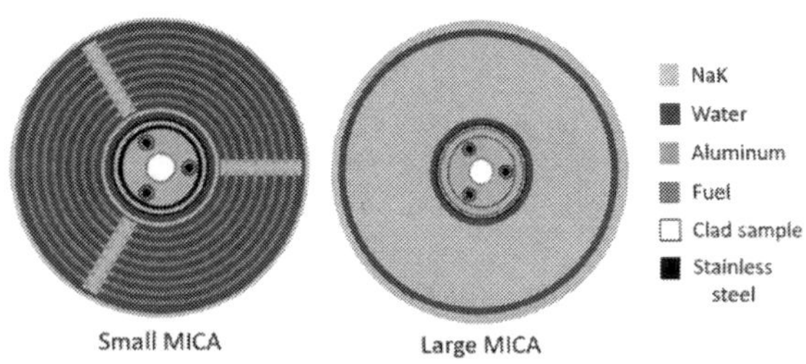

Fig. 2. The MICA test device.

B. The OCCITANE test device

The OCCITANE is a capsule-type experimental device used to study the accelerated aging and embrittlement of structural materials under irradiation, with the aim of anticipating the service life of steel vessels in Light Water Reactors. The RPV (Reactor Pressure Vessel) steel samples should be irradiated at low damage rate of about 1 mdpa per EFPD (Equivalent Full Power Day) to reach a total of 0.1 or 0.2 dpa (i.e. it takes about one year to reproduce the damage cumulated during 40 years in a standard NPP). Hence, the OCCITANE will be located outside the core, at position P611 (see Fig. 1), in the first row of JHR reflector. Independently of the neutron spectrum and nuclear heating imposed by the JHR, the test device must make it possible, if needed, to modify the irradiation conditions in the studied material samples. An important requirement is to be able to adapt the neutron spectrum and to regulate the internal temperature to cover the widest experimental field. Thus, the OCCITANE test device must meet the following specifications:

- A fast neutron flux in the range of $3 \cdot 10^{12} \leq \varphi_{>1\,MeV} \leq 5 \cdot 10^{12}$ $n \cdot cm^{-2} \cdot s^{-1}$;
- A homogeneous temperature distribution regulated between 260°C and 330°C;
- A neutron spectrum ratio, denoted as Rs (see the forthcoming definition below), lying between $2.3 \leq Rs \leq 2.5$.

The neutron spectrum transitions observed between internal and external walls of a RPV are often characterized by a physical quantity defined as the spectrum ratio: $Rs = \Phi\,(E > 0.1\ MeV) / \Phi\,(E > 1\ MeV)$, varying in the range of 2.4 at the inner wall, up to 8 at the outer wall [4]. In connection with experimental data from the irradiation programs previously carried out in the OSIRIS reactor, the test device must be able to reproduce a subset of irradiation conditions identical to those of the IRMA capsule in OSIRIS reflector, offering a spectrum ratio close to 2.4. However, it is also desirable to propose extended spectral tailoring capabilities, to be able to increase the spectrum ratio up to Rs = 8, if needed.

The experimental zone, covering a surface area of 30 x 60 mm^2, can host an arrangement of various type of samples (creep, tensile, Charpy and microstructure specimens) stacked on top of each other, on 60 cm height corresponding to the fissile column of the core. The axial damage gradient follows the fast neutron flux profile (see Fig. 5). Between each cycle, the sample holder is rotated by an angle of 180° to azimuthally homogenise the neutron fluence in the experimental samples.

In order to control the irradiation temperature with homogeneous distribution, it is necessary to regulate with electric heating elements and to provide thermal insulation of the furnace, meanwhile reducing the nuclear heating in the experimental zone. Therefore, two screens are embedded in the device and optimized for that purpose (see Fig. 3):

- A flat tungsten screen of variable thickness, placed against the sample holder, to attenuate the gamma rays coming directly from the core;
- A cylindrical neutron screen of variable thickness attached to the outer refrigeration tube, made of boron carbide to absorb thermal neutrons.

Thermal neutrons are indeed responsible for reactions of radiative capture in steel material, emitting gamma rays that contribute locally to nearly 30% of total nuclear heating in the experimental zone.

The inert (helium) gas gap provides a thermal insulation of the furnace. The heating elements are distributed along the experimental zone to compensate for the thermal gradients created by the axial and radial nuclear heating profile. In addition, the instrumentation includes typically 35 dosimeters installed as close as possible to the specimens to integrate the fast neutron fluence, and 18 thermocouples to provide in-line control of the temperatures inside the sample holder, and in-line regulation of the multi-zone heating system.

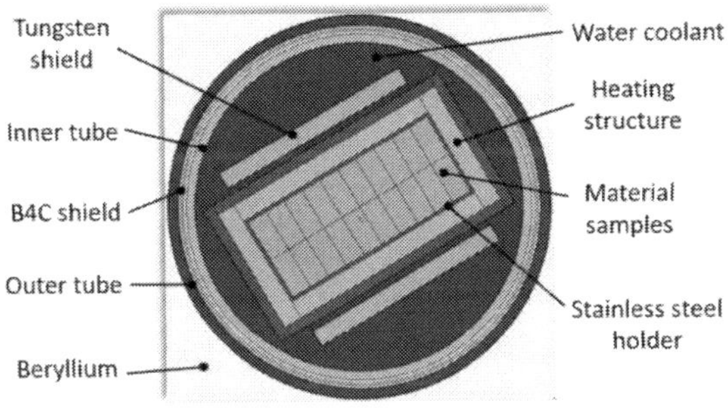

Fig. 3. The OCCITANE test device.

III. The CARMEN device

The CARMEN is a multi-probe measuring device ensuring a high-level metrology service and providing a comprehensive nuclear radiation characterization in the experimental channels of the JHR. It will serve at first during reactor start-up commissioning tests, to verify core physics and safety parameters, and then, later, in support of reactor operation, to measure neutron flux and irradiation conditions. Hence, it will contribute to improve modelling aspects and better predict irradiation conditions for the different devices. It will also provide useful information for designing new equipment (in terms of performances or safety parameters).

The design of CARMEN integrates significant feedback from the experimental campaigns carried out in the OSIRIS reactor to qualify the instrumentation (CALMOS and CARMEN1) [5][6][7]. The qualification beyond 70 MW up to the power level of the JHR (100 MW) remains to be investigated. The system is equipped with state-of-the-art instrumentation, embedding 4 sensors (see Fig. 4):

(1) An ionization chamber (low sensitivity to neutrons) operating in current mode to measure the gamma flux;
(2) Two fission chambers, as part of the FNDS® system developed by the CEA [8], operating in fluctuation mode (practically insensitive to the gamma field) to measure the neutron flux: the first one equipped with a U235 deposit, providing high sensitivity to thermal neutrons, and the second one with a Pu242 deposit, being more selective in the fast energy range;
(3) A two cells differential calorimeter with an aluminum-based sample and a self-calibration system to measure the total heating rate.

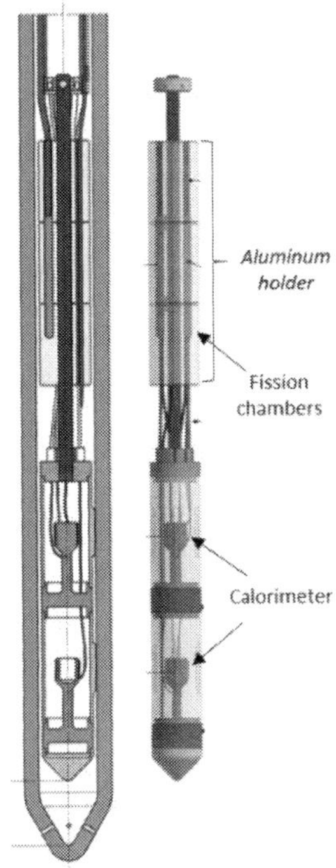

Fig. 4. The CARMEN device.

The use of various type of sensors having selective spectral responses to either neutron or gamma fields ensures better reliability and offers more possibilities of correlations for the interpretation and the transposition of the experimental results. CARMEN is well suited to access various experimental channels: on a fixed position in the core, and either on a fixed position or on a radial displacement system in the reflector. It has a motorized vertical displacement system with an incremental precision of 1 mm, moving at a variable speed from 0.1 to 10 mm/s, between -650 mm and +1000 mm with respect to the axial core mid-plane (knowing that the fissile zone lies between -300 mm and +300 mm). This wide axial displacement range allows the measurement cells to move up in a nearly zero-flux park position, which is of outmost importance for maintaining performance over time and avoiding drift problems, de-calibration, aging under flux (especially for fission chambers), or even damage of the sensors.

Fig. 5 depicts the axial profile of thermal ($E \leq 0.625$ eV) and fast ($E \geq 1$ MeV) neutron flux obtained at the MICA (in-core) and OCCITANE (reflector) irradiation positions.

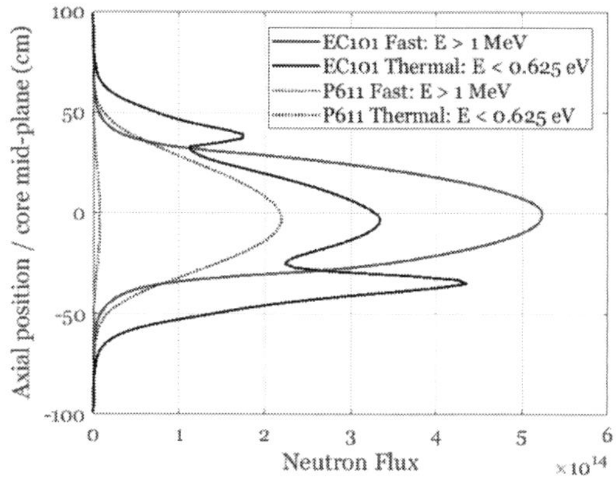

Fig. 5. Axial profile of thermal and fast neutron flux at the core (for MICA) and reflector (for OCCITANE) irradiation positions.

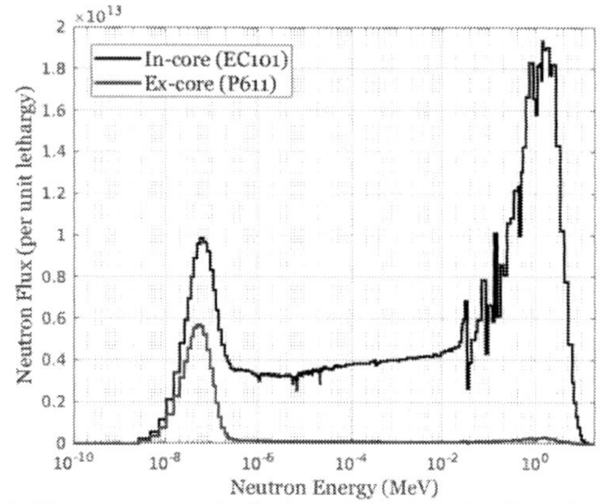

Fig. 6. Neutron spectrum distributions at the core and reflector irradiation positions.

IV. Modelling of nuclear heating measured by CARMEN

The JHR core neutronics is simulated in critical mode using the continuous-energy Monte-Carlo particle transport code TRIPOLI-4® [9], relying on the JEFF3.1 nuclear data library. Fig. 7 depicts the TRIPOLI-4® geometry of the aluminum-based differential calorimeter and, in Fig. 8, the axial profile of nuclear heating in aluminum obtained at the MICA and the OCCITANE irradiation positions. The advanced modelling of nuclear heating and energy deposition in the calorimeter requires the use of four-particle simulations involving neutron, photon, electron and positon. According to the results summarized in Table I and Table II, the impact of taking into account the transport of secondary charged particles produced by gamma interactions appears negligible. The absorbed dose obtained in (n,γ) simulations is actually consistent with the absorbed dose obtained in (n,γ,e,p) simulations, which indicates that electronic equilibrium conditions are satisfied and that gamma heating is well-estimated by the use of a KERMA gamma response function.

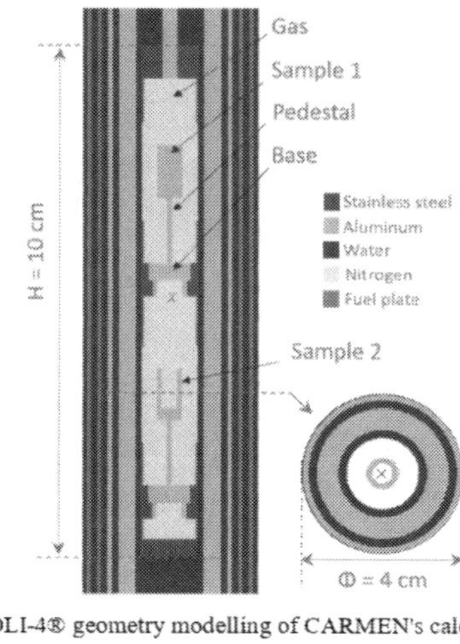

Fig. 7. TRIPOLI-4® geometry modelling of CARMEN's calorimeter.

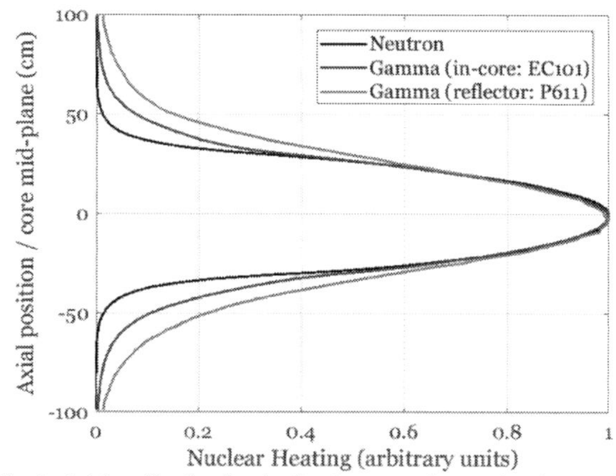

Fig. 8. Axial profile of nuclear heating (neutron and gamma) at the core (for MICA) and at the reflector (for OCCITANE) irradiation positions.

The operating principle of a differential calorimeter is to measure a temperature difference related to a difference in nuclear heating between an empty cell and a cell containing a material sample (see Fig. 7). Each cell is equipped with thermocouples (K-type) measuring a temperature difference proportional to the heat transfer by conduction through the pedestal, from the top ("hot side") down to the bottom ("cold side") part of the pedestal. Indeed, assuming a purely conductive heat transfer from the sample to the pedestal, neglecting the emissivity of inner surfaces (heat loss by radiation) and the heat conduction through the filling inert gas, and considering an optimal heat sink, the temperature difference is directly proportional to the difference in nuclear heating. The target temperature of the aluminum sample is around 300°C for a maximum nuclear heating of ~20 W/g expected in the JHR.

https://doi.org/10.1051/epjconf/202125304010

TABLE I

TRIPOLI-4® (N,γ) AND (N,γ,P,E) SIMULATIONS OF NUCLEAR HEATING MEASURED BY CARMEN'S CALORIMETER INSERTED IN JHR FUEL ELEMENT EC101, AT THE AXIAL MID-PLANE OF THE CORE.

	Mass g	Total Heating (n,γ) ± 2σ	Total Heating (n,γ,p,e) ± 2σ	Difference % ± 2σ
Cell 1	1.601	28.91 W ±0.2	28.21 W ± 0.3	2.5% ± 1.3%
Cell 2	1.124	20.71 W ±0.1	20.18 W ± 0.3	2.6% ± 1.5%
Δ = (1-2)	0.477	8.20 W ±0.2	8.04 W ± 0.4	2.1% ± 5.8%
Δ = (1-2)	0.477	17.19 W/g ±0.5	16.84 W/g ± 0.9	2.1% ± 5.8%

TABLE II

TRIPOLI-4® (N,γ) AND (N,γ,P,E) SIMULATIONS OF NUCLEAR HEATING MEASURED BY CARMEN'S CALORIMETER INSERTED IN P611 POSITION, AT THE AXIAL MID-PLANE OF THE REFLECTOR

	Mass g	Total Heating (n,γ) ± 2σ	Total Heating (n,γ,p,e) ± 2σ	Difference % ± 2σ
Cell 1	1.601	2.53 W ±0.05	2.37 W ± 0.08	6.8% ± 4.2%
Cell 2	1.124	1.82 W ±0.05	1.68 W ± 0.07	7.9% ± 4.8%
Δ = (1-2)	0.477	0.72 W ±0.07	0.69 W ± 0.11	4.1% ± 18.4%
Δ = (1-2)	0.477	1.50 W/g ±0.15	1.44 W/g ± 0.22	4.1% ± 18.4%

According to Table I and Table II, the expected nuclear heating rate measured by CARMEN is ~17 W/g in the core, at the hottest irradiation position of the MICA test device, and ~1.5 W/g in the reflector, at the irradiation position of the OCCITANE test device.

Future work might address heat transfer modelling within the calorimeter, using a multi-physics code to get temperature distributions. This might be useful for having better insight into the CARMEN measurements interpretation and for the investigation of possible non-linearity effects that could be observed at high nuclear heating levels [10].

V. DISCUSSION

The assessment of irradiation conditions in various experimental positions in the Jules Horowitz Reactor core and reflector will rely on measurements performed using the CARMEN device. However, there is always a difference between the flux measured by sensors and the actual flux inside the experimental devices inserted at the same position. On the one hand, the introduction of the detector system may induce some perturbation of the environment that one wishes to characterize, and, on the other hand, the experimental device is modifying the external conditions to adapt them to irradiation specifications. Regarding the possible perturbations induced by the detector itself, design efforts were made for instance to limit the amount of moderator (water) close to the fission chambers while preserving margins for cooling requirements.

By design, the OCCITANE test device proposes spectral tailoring capabilities in compliance with the irradiation conditions specifically required for RPV steels. These conditions are achieved by adding screens of different natures to adjust the spectrum ratio and to reduce it at the target value of 2.4, whereas it is naturally close to ~3 in the sample holder when the device is located in the beryllium reflector.

Due to the lack of space available for adding screens in a small MICA test device inserted in a fuel element (only 24 mm in diameter), it is a far more difficult task to modify efficiently the spectrum ratio, which is by design naturally equal to 2.3 in the core. In spite of that lack of space, it is possible to add small neutron filters to decrease the part of thermal neutrons while preserving the fast flux level, with possible applications for Gen-IV material irradiations for instance. However, there is much more latitude in the large MICA test device occupying a whole in-core location (without fuel element), since there is up to 80 mm in diameter of free space available that could allow for a spectrum ratio adjustment, at least in the range of 2.0 up to 3.5 (see Table III). Therefore, in these situations, one can expect a difference between the spectral responses of detectors in unperturbed conditions and the actual spectral conditions adjusted by design in the experimental devices.

TABLE III

TAILORING CAPABILITIES IN A MICA DEVICE OCCUPYING A LARGE POSITION.

Performance in Al Target		Neutron Screen			
Parameter	Units	H_2O	Al	Be	W
$\Phi_{E \leq 0.625\,eV}$	$n.cm^{-2}.s^{-1}$	$1.6 \cdot 10^{15}$	$2.8 \cdot 10^{14}$	$4.9 \cdot 10^{14}$	$1.3 \cdot 10^{14}$
$\Phi_{0.625eV \leq E \leq 0.1MeV}$	$n.cm^{-2}.s^{-1}$	$6.7 \cdot 10^{14}$	$8.2 \cdot 10^{14}$	$1.1 \cdot 10^{14}$	$4.7 \cdot 10^{14}$
$\Phi_{E \geq 0.1\,MeV}$	$n.cm^{-2}.s^{-1}$	$5.9 \cdot 10^{14}$	$9.2 \cdot 10^{14}$	$8.7 \cdot 10^{14}$	$7.7 \cdot 10^{14}$
$\Phi_{E \geq 1\,MeV}$	$n.cm^{-2}.s^{-1}$	$2.9 \cdot 10^{14}$	$3.9 \cdot 10^{14}$	$3.4 \cdot 10^{14}$	$2.2 \cdot 10^{14}$
$\Phi_{\geq 0.1MeV}/\Phi_{\geq 1\,MeV}$	-	2.0	2.3	2.5	3.5
Neutron Heating	W/g	0.4	0.6	0.5	0.4
Gamma Geating	W/g	17.0	11.7	13.4	2.8
Total Heating	W/g	17.4	12.3	13.9	3.2

CARMEN's calorimeter has a large dynamic measurement range going from lower than 0.5 W/g up to 20 W/g (in aluminum). Measurements in the inner hole of fuel elements using the differential calorimeter are perfectly matching the nuclear heating conditions in the experimental samples of the MICA device (the transfer function remains the identity function). On the contrary, there is a significant difference between the nuclear heating in the OCCITANE material samples (~ 0.5 W/g) and the measurements performed using CARMEN at that same location (~ 1.5 W/g). In any case, the interpretation of experimental results by the calorimeter and their transposition to other situations will always require a calculation versus measurement adjustment and the intensive use of simulation.

VI. MODELLING OF A GRAPHITE-BASED DOUBLE-DIFFERENTIAL CALORIMETER IN OSIRIS

The previous experimental program of RPV steel irradiations in OSIRIS was stressing the impossibility to ensure optimal temperature regulation in material samples (or specimens) when the measured nuclear heating by a graphite-based double-differential calorimeter was exceeding around 0.5 W/g. This empirical criterion was relying only on experimental characterizations performed with this calorimeter. The interpretation and the direct transposition of this criterion in JHR configuration is not an easy task. There are indeed several structural differences between OSIRIS and JHR core configurations, resulting in different irradiation conditions and finally each device has specific design features.

Fig. 9 and Fig. 10 present the radial layout of the OSIRIS reactor core and the IRMA capsule (containing the steel specimens for aging under irradiation) inserted in the water reflector. Fig. 11 describes the graphite-based calorimeter used to measure nuclear heating at the location of the IRMA capsule.

https://doi.org/10.1051/epjconf/202125304010

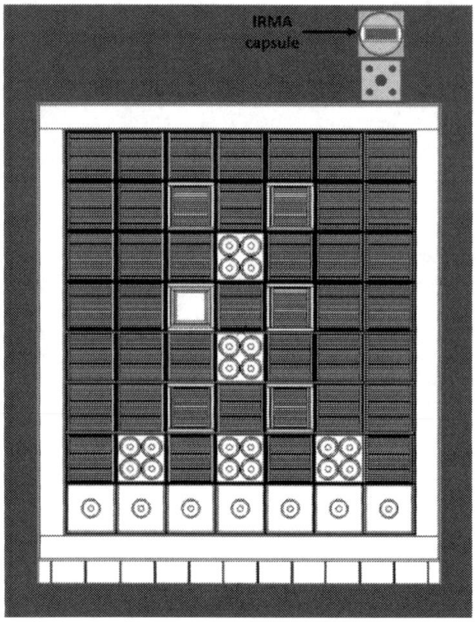

Fig. 9. Radial layout of the OSIRIS reactor core modelling with TRIPOLI-4®

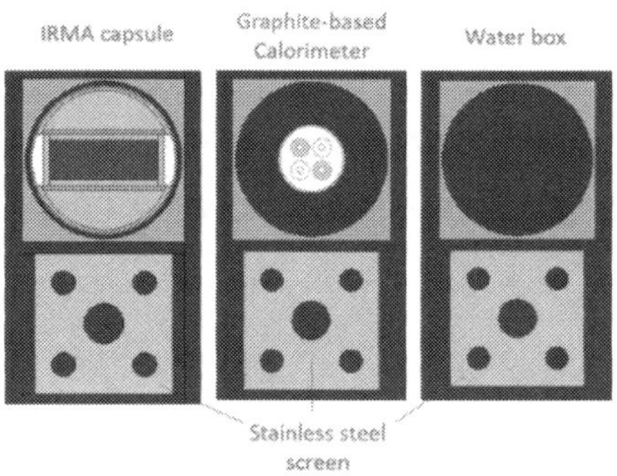

Fig. 10. TRIPOLI-4® modelling of the IRMA capsule and the graphite-based double differential dosimeter in the water reflector of the OSIRIS MTR.

Table IV presents results of energy deposition in the calorimeter, obtained by Monte-Carlo simulations with the TRIPOLI-4® code. It seems that there is no significant difference induced by the transport of secondary charged particles. Although these results neglect the contribution of delayed gamma emissions (produced by decay of fission products and activation products), they provide a first estimate of the nuclear heating measured by the calorimeter (~0.4 W/g), and a relative comparison with the nuclear heating in the experimental zone (sample holder) of the IRMA capsule, evaluated at ~0.3 W/g. Therefore, the previously mentioned criterion of 0.5 W/g of maximum nuclear heating measured by the calorimeter was actually corresponding to 0.375 W/g (=0.5*0.3/0.4) in the experimental zone of the IRMA capsule. A direct transposition of this criterion to JHR configuration is not obvious and requires further analysis.

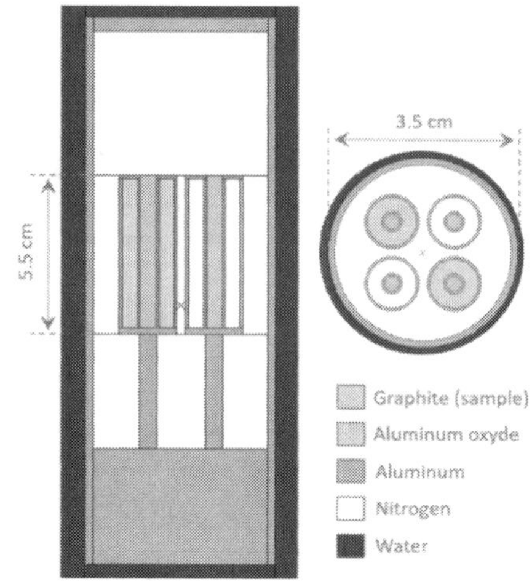

Fig. 11. Detailed geometry of the graphite-based double differential dosimeter.

TABLE IV
TRIPOLI-4® SIMULATIONS OF NUCLEAR HEATING (PROMPT GAMMA EMISSION ONLY) MEASURED BY A GRAPHITE-BASED DOUBLE-DIFFERENTIAL CALORIMETER INSERTED AT THE IRRADIATION POSITION OF IRMA TEST DEVICE IN OSIRIS.

	Mass g	Total Heating (n,γ) ± 2σ	Total Heating (n,γ,p,e) ± 2σ	Difference % ± 2σ
Cell 1	18.12	7.59 W ±0.11	7.03 W ± 0.08	8.0% ± 2.0%
Cell 2	8.24	3.53 W ±0.07	3.04 W ± 0.05	16.4% ± 2.9%
Δ = (1-2)	9.89	4.06 W ±0.12	3.99 W ± 0.10	1.6% ± 4.1%
Δ = (1-2)	9.89	0.41 W/g ±0.01	0.40 W/g ± 0.01	1.6% ± 4.1%

VII. Conclusion

CARMEN is a multi-probe measurement device ensuring a high-level metrology service for the characterization of nuclear radiations in the experimental channels of the JHR. The present study evaluates the irradiation conditions as measured by CARMEN, and compare them to the actual conditions in two experimental devices, namely MICA and OCCITANE, for the irradiation of structural materials, respectively at high dpa rate in the core, and at low dpa rate in the reflector.

Apart from neutron spectrum aspects, the study focused on nuclear heating since it plays a major role in temperature regulation inside these two devices, which is of primary importance for the quality of future irradiation experiments. Advanced modelling of the energy deposition in the aluminum sample of the CARMEN calorimeter is performed using continuous-energy Monte-Carlo particle transport simulations with the TRIPOLI-4® code. These simulations exhibited the absence of transport effects from the secondary charged particles. The choice of aluminum as material sample for the calorimeter appears to be more representative of the JHR environment (many structures of the core are made of aluminum). In addition, aluminum is less sensitive to fast neutrons, in comparison to graphite.

The study evaluated also a transfer function between the CARMEN measurements and the actual irradiation conditions for structural materials inside test devices (MICA and OCCITANE). The interpretation of the experimental results and their transposition to other situations will always require a calculation versus measurement adjustment and the intensive use of computer simulation. An attempt was made to integrate the feedback from the IRMA capsule, which was dedicated to RPV steel irradiations in the OSIRIS reactor, and to transpose it for JHR. In particular, there are on-going investigations to analyse an empirical criterion on nuclear heating measured by a graphite-based double-differential calorimeter in unperturbed conditions, which was required not to exceed 0.5 W/g in order to guarantee, when the IRMA capsule is in its location, the optimal temperature regulation in the capsule. The direct transposition of this criterion on the measured nuclear heating rate is not an easy task, taking into account all differences between core configurations, irradiation conditions and design of the devices. Preliminary comparisons do not yet demonstrate that temperature regulation is ensured for the OCCITANE device in the JHR. Therefore, it will be necessary to verify the thermal hydraulic characteristics beforehand, using dedicated model experiments for validation.

Future work will address heat transfer modelling within the calorimeter to get temperature distributions using a multi physics code. This might be useful for having better insight into the CARMEN measurements interpretation and for the investigation of possible non-linearity effects that could be observed at high nuclear heating levels.

References

[1] J.P. Coulon et al., "Radioisotopes Production for Medical Use: Preliminary Design of the Jules Horowitz Reactor Facilities", RRFM - IGORR, March 18-22 2012 Prague, Czech Republic.

[2] G. Bignan et al., "The Jules Horowitz Reactor: a new high performances Europeans MTR with modern experimental capacities: toward an international centre of excellence", RRFM - IGORR, March 18-22 2012 Prague, Czech Republic.

[3] C. Vaglio-Gaudard *et al.*, "Analysis of the AMMON experimental program in the EOLE facility supporting the qualification of the JHR neutron and photon tools", ANIMMA, 2013, pp. 1-9, DOI: 10.1109/ANIMMA.2013.6727887.

[4] Koji Fukuya, "Current understanding of radiation-induced degradation in light water reactor structural materials", Journal of Nuclear Science and Technology, 50:3, 213-254,DOI: 10.1080/00223131.2013.772448, (2013).

[5] H. Carcreff, L. Salmon, V. Lepeltier, Jm. Guyot, E. Bouard, "Last Improvements of the CALMOS Calorimeter Dedicated to Thermal Neutron Flux and Nuclear Heating Measurements inside the OSIRIS Reactor", ANIMMA, Jun 2017, Lieges, Belgium. ⟨hal-02417753⟩.

[6] Lyoussi et al., "Advanced methodology and instrumentation for accurate on line measurements of neutron, photon and nuclear heating parameters inside Jules Horowitz MTR reactor", in Proceedings of RRFM '12, Prague, Czech Republic, Mar. 2012.

[7] D. Fourmentel et al., "Nuclear Heating Measurements in Material Testing Reactor: A Comparison between a Differential Calorimeter and a Gamma Thermometer", IEEE Transactions on Nuclear Science, Vol. 60, N°1, February 2013.

[8] L. Barbot, D. Fourmentel, J.-F. Villard, C. Destouches, B. Geslot et al., In-pile qualification of the fast-neutron-detection-system. EPJ Web of Conferences, EDP Sciences, 2017, 170, pp.04025. ⟨10.1051/epjconf/201817004025⟩. ⟨cea-02437051⟩

[9] E. Brun, F. Damian, C.M. Diop, E. Dumonteil, F.X. Hugot, C. Jouanne, Y.K. Lee, F. Malvagi, A. Mazzolo, O. Petit, J.C. Trama, T. Visonneau, A. Zoia, *Tripoli-4®, CEA, EDF and AREVA reference Monte Carlo code*, Annals of Nuclear Energy 82, 151-160 (2015).

[10] H. Carcreff et al., "Recent developments in nuclear heating measurement methods inside the OSIRIS reactor", Nuclear Instruments and Methods in Physics Research, A 942 (2019) 162310.

Sensitivity Analysis of an Advanced Transmission Measurement Method for Thermal Neutrons Absorbers Detection in Irradiated Beryllium.

Małgorzata Wróblewska[a,b], David Blanchet[b], Abdallah Lyoussi[b], Patrick Blaise[c], Zuzanna Marcinkowska[a], Jacek Jagielski[a], Agnieszka Boettcher[a]
[a] NCBJ 05-400 Otwock-Świerk, Poland
[b] CEA Cadarache, DES/IRESNE/DER, F-13108 Saint Paul lez Durance, France
[c] CEA Saclay, DES/EC/DSE, F-91191 Gif-sur-Yvette , France
Malgorzata.wroblewska@ncbj.gov.pl

***Abstract*— This paper presents an advanced optimization analysis of the newly developed transmission measurement method conducted in the MARIA MTR reactor for thermal neutrons absorbers estimation in irradiated beryllium elements. Several neutron sources in combination with various thermal neutron detectors are investigated, along with the optimization of the moderating polyethylene layer to improve the signal to background ratio. It was concluded that the use of ^{239}PuBe or ^{241}AmB neutron source with polyethylene of 0.95 g/cm^3 density and ~4.5 cm thickness, as well as either ^{235}U lined fission chamber or BF$_3$ detectors are meeting the requirements for the use in the experiment.**

***Keywords* —Transmission measurements, Sensitivity analysis, neutron sources, thermal neutron detectors, beryllium**

I. INTRODUCTION

Experience in Material Testing Reactors (MTR) operating with beryllium shows important changes in this material properties, under the neutrons exposure [1-3]. Two main effects observed are mechanical damage and neutronic properties worsening [4,5]. The last effect is related to accumulation of thermal neutron absorbers (so-called poisons), induced by fast neutrons through (n,α) reaction on ^{9}Be [6,7].

A newly developed experimental method allows for inferring the actual build-up of absorbers in beryllium element from the transmission and absorption measurement of thermal neutrons in beryllium moderator block element of the MARIA pool type, material testing reactor, located in Poland [8,9].

The experimental setup is composed of a neutron transmission source, a neutron moderator and a neutron detector, mounted on a bench with a beryllium sample placed between the source and the detector. The moderator used to thermalize the neutron source is a High Density Polyethylene (HDPE). The conical shape of the beryllium blocks contributes to uneven axial neutron flux distribution in the core. In order to perform measurements, both neutron source and detector move simultaneously along the block. A nonirradiated beryllium block constitutes the reference point. Comparing measurement results of an irradiated beryllium block to the reference results allows evaluating thermal neutrons absorbed in the irradiated beryllium.

A parametric analysis has been performed to optimize the setup and ensure qualitative experiments. All the steps of the experiment were simulated with the SERPENT2 [10] Monte-Carlo code, using the JEFF3.1.1 [11] nuclear data library.

Several neutron sources were considered in this study: ^{241}AmBe, ^{241}AmB, ^{239}PuBe, SbBe, ^{252}Cf. The most commonly used detectors to measure thermal neutrons that were considered here are: fission chamber with ^{235}U neutron converter deposit, ^{3}He detector, ^{10}B lined-detector and ^{6}Li scintillator.

<u>Neutron Transmission Experiment</u>

The scheme of the experimental setup for neutron transmission is presented in Fig. 1. It consists of a neutron source placed in an aluminium cylinder. A polyethylene half bushing is placed between the neutron source and the examined beryllium block. In the centre of the beryllium block is a slot, in which another aluminium tube containing the detector is placed. Both the neutron source and the detector move simultaneously along the block, with the minimum possible air gap between each element. The entire experimental setup is placed in a hot cell, to which beryllium elements are delivered through a system of pools connected with the reactor pool.

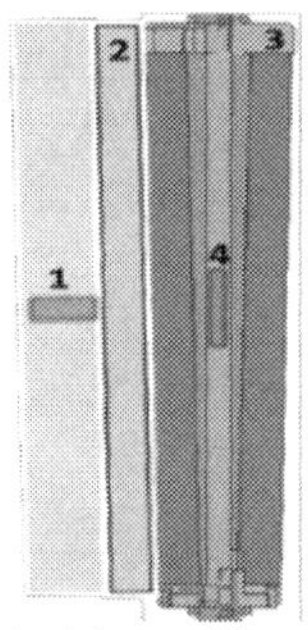

Fig. 1 - Scheme of the experimental setup

The experiment is based on the transmission method, where neutron beam from a chosen neutron source is transmitted through beryllium and induces reactions in the detector delivering counting rates. The ratio of neutrons passing through a nonirradiated beryllium block (I_1) to those passing through an irradiated beryllium block (I_2) and reaching the detector can be expressed with the transmission (T) function:

$$T = \frac{I_2}{I_1} \quad (1)$$

,where $I_1 = I_0 \cdot e^{-\Sigma_1 x}$, $I_2 = I_0 \cdot e^{-\Sigma_2 x}$, I_0 is the incident neutron beam, beryllium thickness x, I_1 neutron flux for nonirradiated and I_2 neutron flux for irradiated beryllium. A full description of the methodology can be found in [12].

To optimise and allow for the most accurate measurements, the most suitable neutron source for the experiment should emit thermal neutrons, have narrow energy spectrum be accompanied by low or no gamma emission. The detector should have maximal sensitivity to thermal neutrons, small size, high detection efficiency and low gamma sensitivity. The polyethylene moderator is intended to thermalize incident neutrons.

II. EXPERIMENTAL SETUP CONFIGURATION

A. *Neutron source*

Isotopic neutron sources are commonly employed for the calibration of nuclear instrumentation. They are very practical due to their size and portability. Each source has a particular neutron energy distribution and strength (the emission rate). Regarding the reaction yielding the neutron production, one can distinguish three main types of sources:

a. **(α,n)** sources – alpha emitter is mixed with a low Z material. The ^{9}Be is most frequently used because of its relatively high neutron yield due to its low neutron binding energy (≈1.67MeV). Other target materials such as ^{10}B, ^{19}F ^{13}C or ^{7}Li are also used [13]. The most common alpha emitters are ^{241}Am, ^{238}Pu and ^{239}Pu
b. **(γ,n)** sources – gamma emitter of specific gamma ray energy, higher than the Q-value (threshold energy) of (γ,n) reaction of the target material [14]. The only two target materials with threshold energy below 4 MeV are ^{9}Be (~1.67 MeV) and ^{2}H (~2.23 MeV) [15].
c. Spontaneous fission neutron sources – contain isotopes that undergo a spontaneous fission; most commonly used are ^{238}Pu, ^{242}Cm, ^{252}Cf.

Based on their characteristics given in Table 1, 5 neutron sources were chosen in the frame of the present study. The antimony source spectrum was calculated with iSourceC [16] code, whereas all the others are ISO calibration spectra [17]. All results presented in the following were obtained by Monte-Carlo simulations, using the SERPENT code. These simulations enabled to tally the detectors' responses. Preliminary calculations were performed for a setup with manufacturer reference values, with a layer of nonborated polyethylene of 0.93 g/cm³ density and 4.5 cm thickness, using the beryllium composition of a fresh MARIA reactor's element having density of 1.84 g/cm³ [8].

TABLE 1 - CHOSEN PROPERTIES OF SELECTED NEUTRON SOURCES [18,19].

SOURCE	$T_{1/2}$	$\bar{E}_n$	Emission Rate	Av. γ dose rate	$\bar{E}_\gamma$
	y	MeV	n/s	µGy/h at 1m/GBq	MeV
^{241}AmBe	433	~4.2-5.0	$8\cdot10^{7}$	0.68	4.4
^{241}AmB	433	~2.6	$8\cdot10^{7}$	1.75	2.8
^{239}PuBe	24000	~4.2-5.0	$2\cdot10^{7}$	0.027	
^{124}SbBe	60.4 h	~0.025	$2\cdot10^{7}$	270	1.7
^{252}Cf	2.65	~2.0-2.3	$5\cdot10^{7}$	140	0.2-1.8

B. *Polyethylene*

As no pure thermal neutron source exists, the use of a polyethylene layer to maximize the number or thermal neutrons reaching beryllium was necessary. Polyethylene, thanks to its high content of hydrogen atoms, is a very good moderator. One can observe good thermalization of the neutron energy in Fig. 2, when a 4.5 cm layer of polyethylene has been inserted in the setup of Fig. 1. The corresponding values of reaction rates in a fission chamber (with ^{235}U deposit) are given in the Table 2.

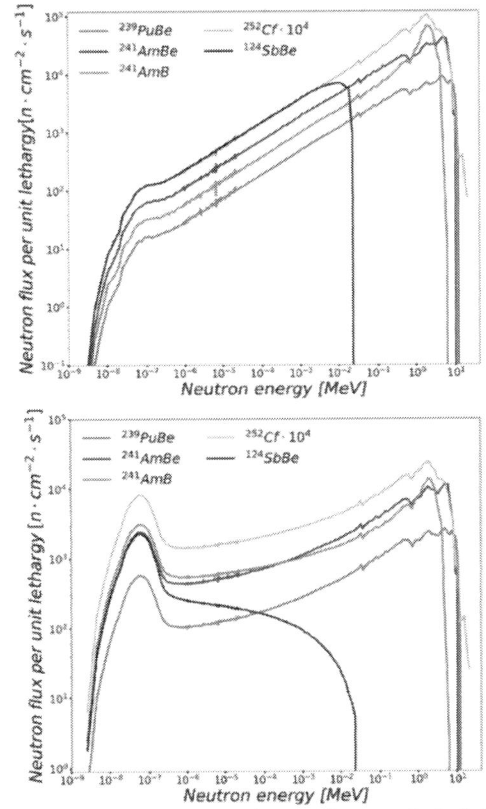

Fig. 2 – Un-moderated (up) and moderated with 4.5 cm of polyethylene (down) neutron source spectra at the detector's position.

TABLE 2 – REACTION RATES FOR DIFFERENT MODERATED (M - 4.5 CM) AND NONMODERATED (NM) NEUTRON SOURCES.

(MAXIMAL UNCERTAINTY 2 Σ = 0.3%).

	^{239}PuBe	^{241}AmBe	^{241}AmB	^{252}Cf	^{124}SbBe
			[s^{-1}]		
NM.	9.01E+04	3.54E+05	2.25E+05	7.46E+09	3.98E+05
M	5.68E+05	2.40E+06	3.01E+06	7.98E+10	2.07E+06

The most common densities of polyethylene range from 0.93 g/cm³ to 0.95 g/cm³ (HDPE – nonborated). Beryllium used for nuclear applications has ~99% ^{9}Be purity and a density of 1.85 g/cm³. Nonetheless, depending on the manufacturer, they can mainly differ in impurities content and beryllium grain size. One of the factors used to describe the amount of impurities in beryllium is the so-called danger factor denoted as D_f and defined as:

$$D_f = \frac{\widetilde{\Sigma_a}}{\Sigma_a} \quad (2)$$

Where $\widetilde{\Sigma_a}$ represents the effective cross section for thermal neutron absorption in beryllium with impurities, and Σ_a is the effective cross section of pure beryllium [20].

For the current MARIA reactor's beryllium elements, $D_f \approx 1.94$, whereas for the BR2 beryllium elements, this value was estimated to $D_f \leq 1.75$ [20]. The high-density polyethylene has less branching than low-density polyethylene, which allows the polymer chains to pack closely together, resulting in a dense, crystalline material, hence higher neutron thermalization [21] and stronger signal in the detector compared to the same polyethylene thickness of lower density.

For ^{239}PuBe neutron source, the impact of polyethylene density in reaction rates is slightly more important than for ^{241}AmBe. A similar trend is observed when the density of beryllium changes.

When choosing the neutron source, the optimal thickness of the polyethylene layer would slightly differ, depending on the neutron detector. An illustration of the signal in fission chamber for several polyethylene thicknesses and for selected neutron sources is presented in Fig. 3

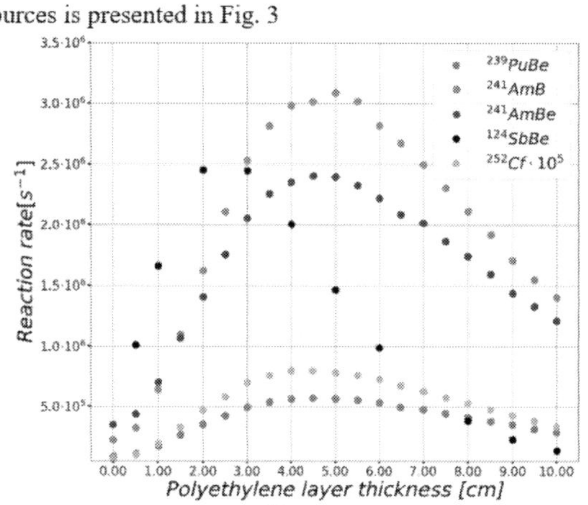

Fig. 3 – Fission reaction rates (^{235}U FC) induced by the neutron sources, depending on the polyethylene thickness (density 0.93 g/cm^3). (The maximum statistical uncertainty is $2\sigma = 3\%$), calculated in SERPENT2 [10].

C. *Neutron detector*

Four detector types having high enough reaction cross-section with thermal neutrons [22] have been tested: A detector filled with ^{3}He gas, relying on a (n,p) reaction; A detector filled with BF_3 gas, with (n,α) reaction on ^{10}B; A scintillator with ^{6}Li, based on the (n,α) reaction; a fission chamber lined with ^{235}U. Some of the most important aspects of these detectors can be found in the Table 3.

Fig. 4 presents the results obtained for a ^{239}PuBe neutron source only. One can observe that signal in the detector with ^{3}He is the strongest (blue dots), whatever the selected neutron source, whereas the fission chamber (orange dots) records the lowest signal. Although the ^{3}He filled tube provides the highest response, it is significantly more sensitive to gamma radiation than ^{10}B lined tube or fission chamber as it can be seen in the Table 3. This important criterion makes the choice of a fission chamber the most suitable.

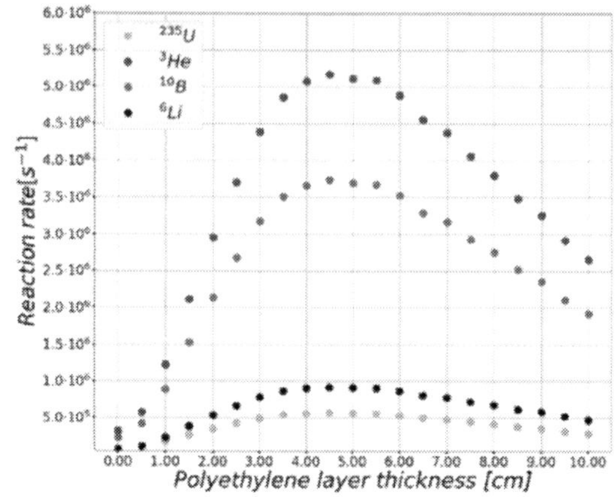

Fig. 4 – Microscopic reaction rates related to different detectors, using ^{239}PuBe neutron source, calculated in SERPENT2 [10].

TABLE 3 - PROPERTIES OF THE SELECTED THERMAL NEUTRON CONVERTER FOR NEUTRON DETECTION [18].

Target isotope	Reaction Type	Reaction cross-section n_{th} [barn]	Maximum equivalent rate* [Gy/hr]
^{10}B	n,α	3837	10^{-1}
^{3}He	n,p	5333	10^{-2}
^{6}Li	n,α	937	10^{-4}
^{235}U	n,fiss	582	10^4

*Approximate limit of γ dose for the efficiency of neutron detection.

D. *Sensitivity of the setup to poisons concentration*

Irradiation time of the beryllium element affects the accumulation of ^{6}Li and ^{3}He. The concentration of these isotopes determines the macroscopic absorption cross section of irradiated beryllium and thus the absorbed fraction of thermal neutrons incoming from the source.

As an example, Fig. 5 presents the difference between irradiated and nonirradiated beryllium, considering a fission chamber with ^{235}U deposit, a ^{239}PuBe neutron source and a variable polyethylene thickness.

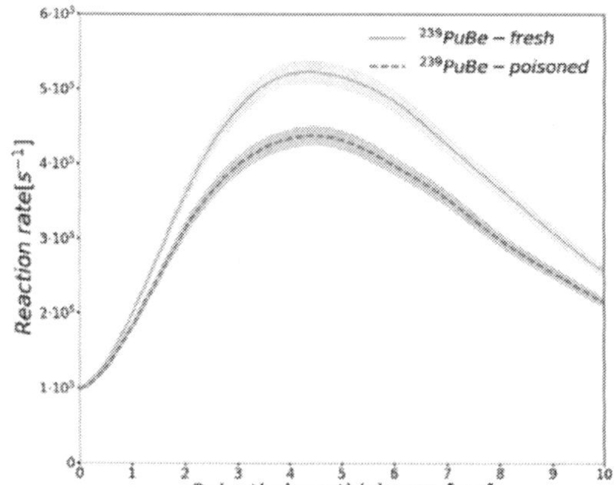

Fig. 5 – Comparison of reaction rates in fission chamber through irradiated and nonirradiated beryllium, using SbBe and PuBe neutron source with 3% uncertainty marked, calculated in SERPENT2 [10].

The shaded area corresponds to the 2σ uncertainty of about 3% resulting from the first campaign of experimental measurements performed in 2020 in the MARIA facility.

The accumulation of ^{6}Li and ^{3}He was calculated using an average neutron flux in beryllium element, calculated by a 3D full-core Monte-Carlo simulation of the MARIA reactor with the SERPENT2. Typical cycles of MARIA reactor have been considered, consisting of 100 h of reactor operation at nominal power, followed by an outage period of 68 h have been considered. The periods varied from 4 weeks to 14 years with 30 days of outage. The resulting poisons concentrations are presented in Table 4 and Fig. 6. Because these are analytically calculated values as described in [12], they do not contain uncertainties.

TABLE 4 - CONCENTRATION OF ^{6}LI AND ^{3}HE FOR SEVERAL IRRADIATION SCENARIOS.

Time [weeks]	^{6}Li	^{3}He
	at.barn^{-1}cm^{-2}	
4	2.2877E-07	1.9553E-10
8	4.0204E-07	7.8198E-10
12	5.3326E-07	1.6981E-09
24	7.6494E-07	5.7567E-09
36	8.6559E-07	1.0882E-08
48	9.0932E-07	1.6470E-08
72	9.3657E-07	2.8137E-08
96	9.4171E-07	3.9984E-08
144	9.4286E-07	6.3752E-08
192	9.4291E-07	8.7529E-08
240	9.4291E-07	1.1131E-07
336	9.4291E-07	1.5886E-07
384	9.4291E-07	1.8264E-07
480	9.4291E-07	2.3020E-07
720	9.4291E-07	3.4908E-07

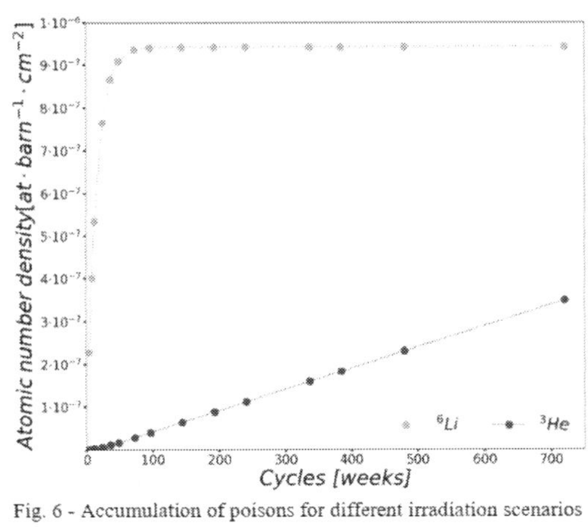

Fig. 6 - Accumulation of poisons for different irradiation scenarios .

Using SERPENT2 code, the detector's response was simulated for all neutron sources described in section A, depending on the poisons' concentration presented in the Table 4. Fig. 7 presents a percent difference between irradiated and nonirradiated beryllium element. One can observe that after the saturation of ^{6}Li, the major impact on reactions induced in the detector has accumulation of ^{3}He. The time needed to reach the minimum accumulation of poisons, will be thus very dependent on the overall outage time.

For all neutron sources, calculations were performed for the setup without polyethylene and the 4.5 cm thick polyethylene layer. In Fig. 7, the results of the simulations are presented with the 2σ statistical uncertainty is 0.3%.

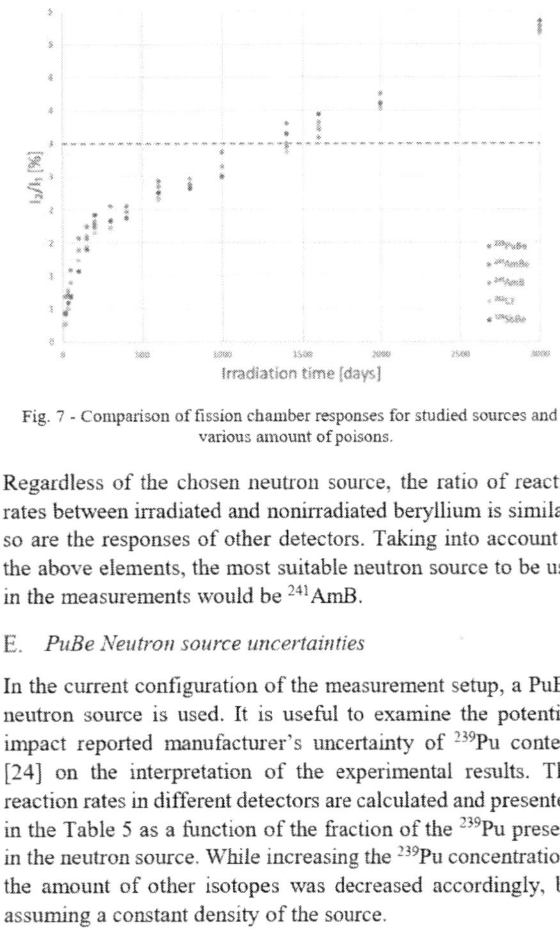

Fig. 7 - Comparison of fission chamber responses for studied sources and various amount of poisons.

Regardless of the chosen neutron source, the ratio of reaction rates between irradiated and nonirradiated beryllium is similar, so are the responses of other detectors. Taking into account all the above elements, the most suitable neutron source to be used in the measurements would be ^{241}AmB.

E. *PuBe Neutron source uncertainties*

In the current configuration of the measurement setup, a PuBe neutron source is used. It is useful to examine the potential impact reported manufacturer's uncertainty of ^{239}Pu content [24] on the interpretation of the experimental results. The reaction rates in different detectors are calculated and presented in the Table 5 as a function of the fraction of the ^{239}Pu present in the neutron source. While increasing the ^{239}Pu concentration, the amount of other isotopes was decreased accordingly, by assuming a constant density of the source.

TABLE 5 – REACTION RATES FOR DIFFERENT DETECTORS, DEPENDING ON THE ^{239}PU FRACTION IN THE ^{239}PUBE NEUTRON SOURCE. MAXIMUM UNCERTAINTY 2 Σ=0.3%.

^{239}Pu %	U	He	Li	B
		s^{-1}		
0.70	5.96E+05	5.44E+06	9.67E+05	3.93E+06
0.85	5.94E+05	5.42E+06	9.65E+05	3.92E+06
0.96	5.91E+05	5.39E+06	9.59E+05	3.89E+06

The impact of such variation of the ^{239}Pu content in the neutron source is restricted to the statistical uncertainty level. As can be seen in Fig. 8, the spectrum of PuBe neutron source simulated with iSourceC code differs in the intensity, thus with a properly measured emission rate of the used source, the impact of spectrum variation is negligible.

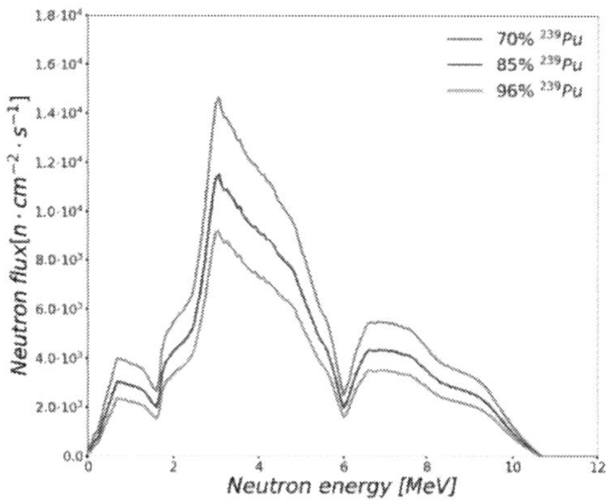

Fig. 8 – ^{239}PuBe neutron spectra depending on the ^{239}Pu content calculated with iSourceC program.

III. Conclusions and perspectives

The accumulation of thermal neutron absorbers in beryllium moderators and reflectors is an important problem in MTRs exploitation. An experimental method using neutron transmission has been developed to measure the concentration of these so-called poisons. This study presents a parametric analysis of the aforementioned setup, aiming to investigate experimental and numerical uncertainties associated with the method, as well as to optimize it.

Several neutron sources were considered, based on their characteristics. Two of them - ^{241}AmB and ^{239}PuBe are the most suitable for the experiment. These sources are characterized by low gamma emission and provide high counting rates. Depending on the reactor operation periods, a minimum of 1-3 years of irradiation time is needed to perform the measurements. Out of all investigated neutron detectors, fission chamber with ^{235}U deposit provides a safe choice for beryllium irradiated for a long period. A BF_3 type detector could be used alternatively if the gamma dose from beryllium activation is sufficiently low, because it provides higher sensitivity to thermal neutrons than the fission chamber.

The present study will be completed with gamma and uncertainty analysis whose results will be published in a companion paper. As the presence of several gamma sources in the measurement set affects the above choice, it is necessary to note that the two most important gamma sources would be the neutron source and activation of the impurities in beryllium element. The activity of the beryllium block used in the study [12] was measured at ~10^{13} Bq for the entire block (that is about 1GBq/g). The uncertainties related to nuclear data should have some impact on the numerical analysis of the experiment, however further work is needed to perform full sensitivity analysis.

References

[1] G.R. Longhurst, R.D. Rohe, Beryllium Use in Advanced Test Reactor. 8th IEA International Workshop on Beryllium Technology. INL/CON-07-12959. Portugal. 2007.

[2] A. Gołąb, Aging management of beryllium and graphite blocks in research rector MARIA. National Centre for Nuclear Research, Poland. 2013.

[3] H. J. Stander *et al.*, Replacement of the Core Beryllium Reflector in the SAFARI-1 Research Reactor. IAEA-TECDOC-1748. 2014. Web

[4] T.A. Tomberlin. Beryllium - A Unique Material in Nuclear Applications, article, November 1, 2004.

[5] A. Burkholtz, Irradiation damage in beryllium. Centre d'Etude de l'Energie Nucléaire. EUR3055.e. Belgium. 1966.

[6] S. Kalcheva, B. Ponsard, E. Koonen, Reactivity Effects due to Beryllium Poisoning of BR2. PHYSOR 2004. USA. 2004.

[7] V.V. Kalygin, A. P. Malkov, V.V. Pimenov, Effect of 3He and 6Li accumulation in beryllium blocks in the neutron-physical characteristics of the MIR reactor. Atomnaya Énergiya, Vol. 104, No. 2. Russian Federation. 2008.

[8] M. Wroblewska *et al.*, Beryllium Poisoning for research reactors. European Research Reactor Conference. Rotterdam, Netherlands. 2017.

[9] M. Wroblewska *et al.*, Qualification of Beryllium Poisoning Models for MARIA and Jules Horowitz Reactors. European Research Reactor Conference, Munich, Germany. 2018.

[10] J. Leppänen, *et al.* The Serpent Monte Carlo code: Status, development and applications in 2013. Ann. Nucl. Energy, 82 (2015) 142-150. 2015.

[11] A. Koning, R. Forrest, M. Kellett, R. Mills, H. Henriksson, Y. Rugama. The JEFF-3.1 Nuclear Data Library. Technical Report. JEFF Report 21, NEA – OECD. 2006.

[12] M. Wroblewska, D. Blanchet et.al. A review and analysis of the state of the art on beryllium poisoning in research reactors. Ann. of Nucl. Energy. (163)2021.

[13] H.R. Vega-Carrillo, S.A. Martinez-Ovalle. Neutron spectra and dosimetric features of Isotopic neutron sources: A review. ISSSD 2015 proceedings. Mexico. Sept. 2015.

[14] A. Wattenberg. Photo-Neutron sources. Physical Review. 71,497. 1947.

[15] IAEA. Neutron Monitoring for Radiological Protection. Technical Reports Series No: 252. IAEA. 1985.

[16] O.Bouland. Code iSourceC: Calcul de sources intrinsèques de neutrons. Private communication May 2021.

[17] IAEA. Compendium of Neutron Spectra and Detector Responses for Radiation Protection Purposes. Technical Reports Series No. 318. IAEA. 2001.

[18] D. Reilly, N. Ensslin, H. Jr. Smith, and S. Kreiner. Passive nondestructive assay of nuclear materials. United States. 1991. Web

[19] B.C. Anderson, K. E. Holbert, H. Bowler. Design, Construction, and Modeling of a 252Cf Neutron Irradiator.

[20] K.Pytel, J.Tombacher. Wymiana matrycy berylowej reaktora MARIA. IEA . Internal report [in polish] 1991.

[21] J.R. Granada, J. Dawidowski, R.E. Mayer, V.H. Gillette. Thermal neutron cross section and transport properties of polyethylene. Nuclear Instruments and Methods in Physics Research A261 Netherlands.1987.

[22] A. Oed. Detectors for thermal neutrons. Nuclear Instruments and Methods in Physics. Reas. Sec. A. Accelerators, Spectrometers, Detectors and Associated Equipment. 525(1). 2004.

[23] H.R. Vega-Carrillo, V.M. Hernández-Dávila, T. Rivera-Montalvo, A. Sánchez. Characterization of a 239PuBe Isotopic Neutron Source. Proccedings of the ISSSD 2012.

[24] J. Bagi, C.T. Nguyen, L. Lakosi, Assessment of the Pu content of Pu–Be neutron sources, Nucl. Instr. Meth. B222 (2004) 242-248.

EPJ Web of Conferences 253, 04012 (2021) https://doi.org/10.1051/epjconf/202125304012
ANIMMA 2021

New Neutron Imaging Facility development at the Penn State Breazeale Nuclear Reactor

Alibek Kenges, Kenan Unlu, Daniel Beck
Pennsylvania State University, USA
Corresponding author: azk63@psu.edu

***Abstract*— Preliminary results of characterization experiments for the New Neutron Imaging Facility (NIF) that is being developed at the Penn State Breazeale Nuclear reactor are presented in the following sections. The methodology of neutron beam characterization described in the American Society for Testing and Materials (ASTM) documents for the neutron imaging systems have been followed to improve the NIF at Penn State to a Category I facility by ASTM designation of quality. Preliminary results showed that our system is capable of producing images of high quality, corresponding to Category I; however, further experiments are needed for full declaration of our facility as such. Additionally, the effective collimation ratio (L/D ratio) of our current system is ~110 with the capability of improvement to ~150. The thermal flux at the exit surface of the neutron beam is equal to $5.4 \times 10^{6} n\, cm^{-2} s^{-1}$ at 1MWth reactor power, which corresponds to the industry comparable value.**

***Keywords* —Neutron imaging, neutron radiography, neutron filters, collimator components, beam characterization.**

I. Introduction

Neutron imaging can be described as one of the most useful techniques in non-destructive testing field that effectively uses the attenuation properties of neutrons to image the objects of interest. A neutron imaging facility has been available at Radiation Science and Engineering Center (RSEC) for the last several decades and with the newly installed core moderator assembly and the beam ports, it has been decided to rejuvenate the NIF at RSEC to its new levels. The new NIF has collimators installed over the past several months with the capability of utilizing variable apertures. It is also on the right track of using the state-of-the-art equipment and software for conventional neutron radiography and tomography. The potential of all RSEC facilities is highlighted by the Penn State Breazeale Reactor (PSBR), which first went critical in 1955, and is the nation's longest continuously operating university research reactor. The PSBR is a 1 MW, TRIGA with moveable core in a large pool and with pulsing capabilities [1]. The current layout for the RSEC NIF and the beam port dimensions are given in the Figure 1. Although the final design for the RSEC NIF is still under development, it has been decided to continuously characterize our system by the universal standards for the Neutron Radiography (NR). This would allow us to properly examine the capabilities of the facility at each point of the development process. While there are some other ways, most of the NR facilities around the world characterize their systems using American Society for Testing and Materials (ASTM) standards dedicated to neutron imaging techniques.

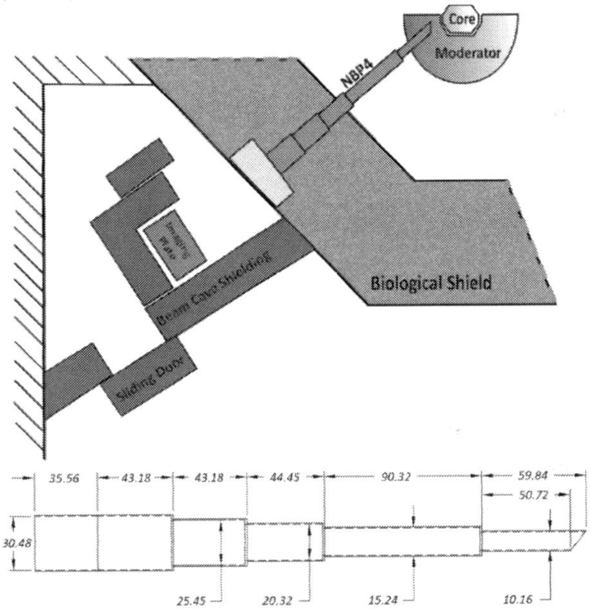

Fig. 1. RSEC NIF layout and the beam port (dimensions are in centimeters).

Therefore, it has been decided to follow the same methodology in categorization process of RSEC – NIF. The starting point of the system characterization was the assessment of Open Beam (OB) configuration, which is the beam port that does not have any collimator components installed [2]. Once the collimators components are installed the ASTM testing can be repeated to see the improvements in the system, and those steps are briefly summarized in the following section.

II. Characterization of the RSEC – NIF

Although we have the capability of implementing the film radiography, for these characterizations indirect method has been utilized, which assumes the conversion plate to be exposed without the film [3]. Instead, our system exploits the computed radiography (CR) technique that is made possible by the CRxFlex by Baker Hughes GE company and corresponding IPU phosphor imaging plates.

The OB configuration of the RSEC – NIF has been characterized by the end of the 2020. As a result, it has been determined that the images of medium level of quality can be produced corresponding to a facility of Category IV by ASTM designation of quality. The objective then was to improve the system in all NR relevant aspects to achieve the Category I via installation of collimator components shown in the figures below and discussed further.

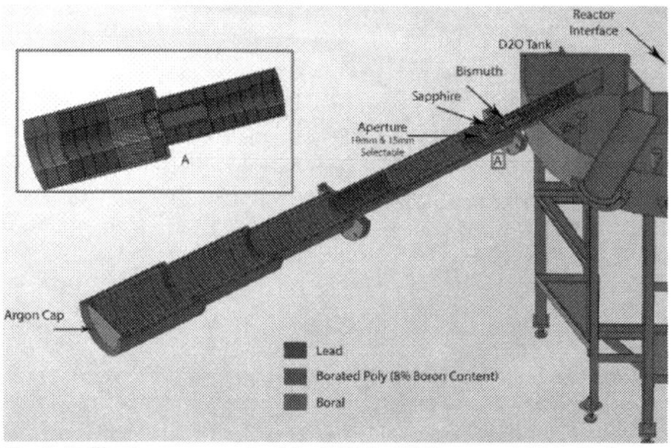

Fig. 2. Collimator components of the RSEC – NIF beam port. Note: Both, the Bismuth and Sapphire filters can be removed and placed to the beam port depending on the neutron spectrum needs.

To improve the collimation ratio, it has been decided to manufacture the aperture with the selectable diameters. Boral apertures of 20 mm and 30 mm in diameter have been selected as the preliminary suitable choice. There is the upper limit for the aperture diameter due to the diameter of the Bi-filter, which is 36 mm; however, the lower limit is only dependent on the value of the desired thermal neutron flux that will be reduced with the smaller diameter apertures. Optionally, the cadmium lining can also be added to the aperture to sharpen the edges.

Fig.3. Collimator components for the RSEC – NIF beam port: Bi single crystal in the aluminum encapsulation and two boral apertures (left), installation of the lead and poly steps using the in-house 3D printed delivery rails (right).

As the collimator beam walls (convergent and divergent parts), the lead and borated-poly material (8% Boron content) steps have been chosen with the filters for the gammas (Bismuth) and optional filter for the fast neutrons (Sapphire) can be installed pre and post aperture respectively. The choice for the mentioned collimator components were based on analysis of the existing NR facilities and runs of the appropriate MCNP calculations.

Fig.4. Collimator components for the RSEC – NIF beam port: insertion of the Bi-gamma-filter (left) and the boral aperture (right) in the corresponding section of the beam port.

The figure above shows of the installation process of the collimator components. It should be noted that this process can also be performed in the horizontal position once the components are installed in the beam port. The resulting view of the Collimated Beam (CB) port of the RSEC – NIF with all collimator components installed is shown in the figure below.

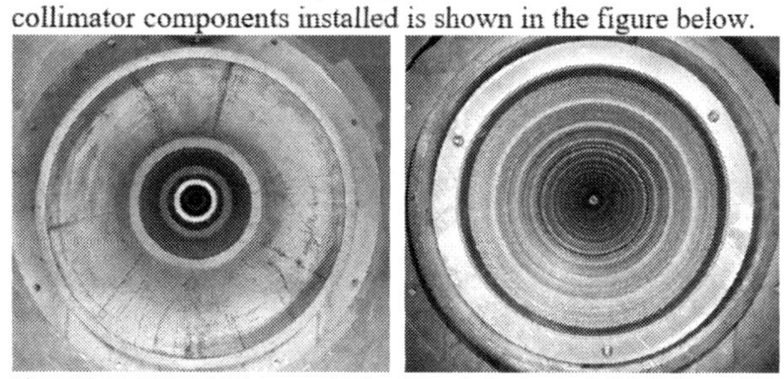

Fig.5. The RSEC – NIF beam port (view towards the reactor); OB (left) and CB (right).

The assessment of the system can be started by the examining the uniformity of the beam, which can be done by taking the image without any object placed within the beam. To consider the beam as uniform, the optical density values across the resulting radiograph should not deviate more than 5% as indicated in ASTM E545 [3]. In the figure below, although the intensities of the grey values for the OB and CB configurations are slightly differ, better uniformity in the beam profile can be seen for the latter.

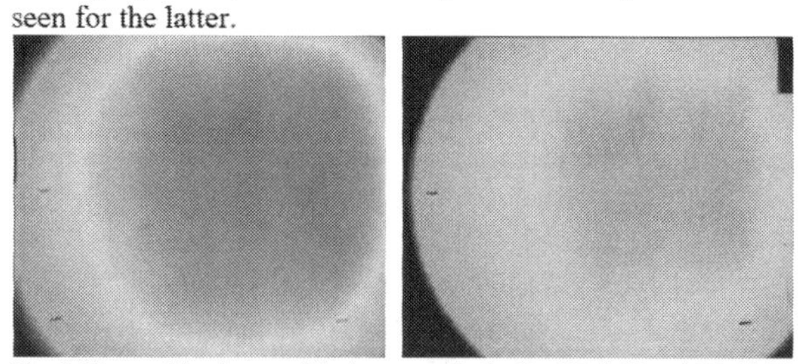

Fig. 6. The resulting image of the beam profile for the OB (left) and CB (right) configurations (a filter is used for visual illustration).

The difference in the grey value intensity is first due to the reduction in the thermal neutron flux at the surface of imaging plate. Gold foil activation technique was used for the experimental values of thermal neutron flux, that resulted in $3.7 \times 10^8 n\ cm^{-2} s^{-1}$ per 1 MW reactor power for the OB configuration. On the other hand, the value was reduced to $5.4 \times 10^6 n\ cm^{-2} s^{-1}$ per 1 MW power for the CB configuration. Deviation of the grey values across the entire image in Fig.3 changed from ~40% for the OB to ~24% for the CB. In the middle section of the image these numbers changed from ~13% to ~10%. Both values are more than 5% to declare the beam as uniform; however, this issue could be related to IPU phosphor imaging plate and its condition throughout the experiments. To make it right, the direct film radiography needs to be performed to access the beam uniformity for the CB configuration.

A. Effective Collimation Ratio (L/D)

Collimation ratio (L/D) of the imaging system defines the spatial resolution and hence can be treated as one of the most important features. The length of the beam port from the aperture the beam shutter end surface would be equal to 257 cm. Therefore, depending on the placement of the imaging plate post beam shutter, the value of (L) can be anywhere between 260 to around 600. On the other hand, the diameter of the aperture (D) is currently selectable between 20 cm and 30 cm. Simple division of two values would result in the geometric L/D but the effective collimation ratio, on the other hand, differs from the geometrical ratio due to the scattering events within the beam. ASTM E803 document provides the details of the experiment to find the effective L/D of the system using the no umbra (NU) device [4]. Two resulting images for this device are given in the figure below for the OB and CB configurations.

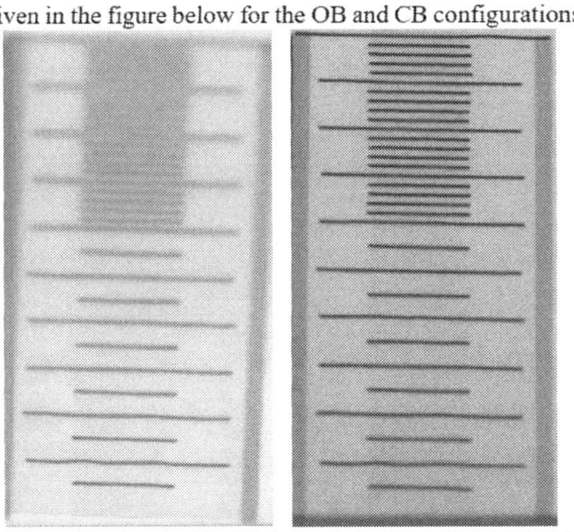

Fig. 7. The resulting image of the NU-device for the OB (left) and CB (right) configurations. Note: In general, sharp shadow lines (umbra shadow) of Cd-wires represent higher L/D value.

There are two different techniques of finding the effective L/D value by analyzing the image of NU-device that are given in ASTM E803 document. One of them is the visual analysis of the NU-device where the user should identify the wire for which the umbra shadow disappears. The second method is finding the umbra shadow disappearing wire with the optical densitometry measurements. For the computed radiography system, this can be done by the grey value histogram across the Cd-wire shadows. Subsequently, knowing the diameter of the Cd-wire and the distance between the wire and the imaging plate, one can calculate the effective collimation ratio. For the OB configuration the effective L/D was between 34.6 and 42.5. For the CB configuration, on the other hand, these values increased to between 107 and 115, which is more than 100 (initial goal value). For both cases the values are less than geometric L/D ratios (46 for OB and 135 for CB), which indicates to the existence of slight scattered neutron content.

B. Beam Quality: Beam Purity Indicator (BPI) & Sensitivity Indicator (SI)

There are two indicators that are described in the ASTM E545 and can be used to define the designation of the quality level of a NR facility. The analysis of resulting radiographs of Beam Purity Indicator (BPI) and Sensitivity Indicator (SI) leads to determination of two important factors, which are the thermal neutron content and sensitivity level [5]. These two factors define the Category of the system by the ASTM designation of quality.

According to the ASTM E545 document, the visual analysis of the BPI shown in Fig. 7, two areas are of special interest: the cadmium wires (vertical lines) and the lead disks (lower two). If one of the Cd wires is significantly sharper than the other, which is the case for the OB configuration, the L/D ratio of the system is considered to be low with the need to be improved. In contrast, for the CB configuration both Cd wires are comparably sharp. In addition to that, for the OB configuration, the lead disks (lower two) are distinguishable comparing to the polytetrafluoroethylene material of the BPI. That indicates to the high gamma content within the beam. Using the Bismuth single crystal in the CB configuration, we have filtered the beam from the gamma, which can be seen by the absence of lead disks in the BPI image.

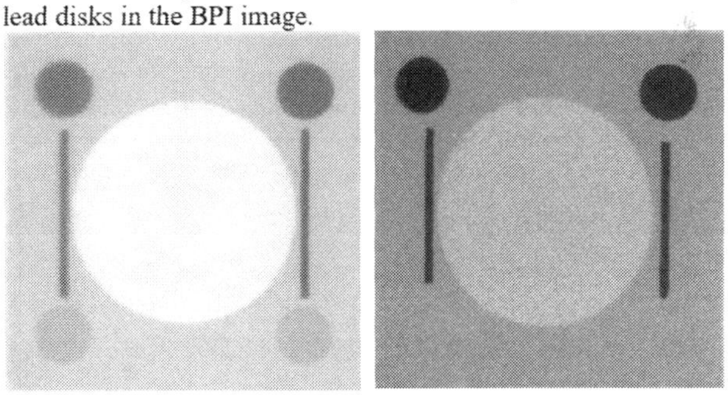

Fig. 8. The resulting image of the BPI for the OB (left) and CB (right) configurations. Note: the intensities of grey values are not comparable due to the lower neutron flux for the CB configuration.

The SI, on the other hand, is a step-wedge device containing gaps and holes of known dimensions that can be used to access the system based on the largest consecutive numbered visible hole (H) and the thinnest visible gap (G) between the thickness-changing absorber steps. Basically, the analysis of SI radiograph can give the general idea on NR system's spatial resolution.

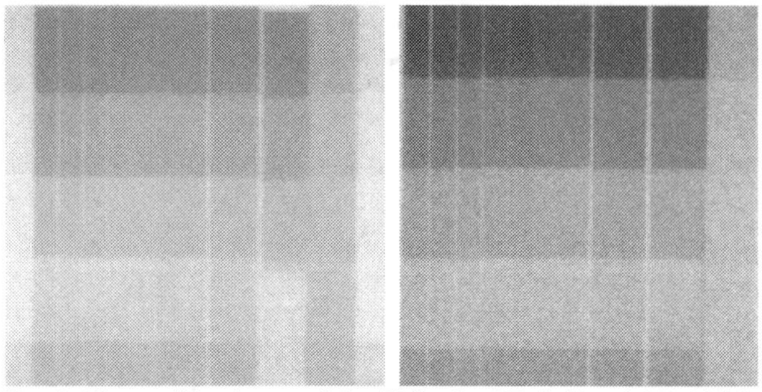

Fig. 9. The resulting image of the SI for the OB (left) and CB (right) configurations. Note: another SI has been used for the CB configuration that does not have the holes manufactured.

The optical densiometric analysis for different areas of the BPI and SI radiograph can be performed since the visual analysis can be somewhat challenging. To do so, the conversion of the gray values to the optical density values has been performed using the predicted calibration curve. The thermal neutron content (NC), as well as H and G values have been determined following the steps given in ASTM E545. For the OB configuration the NC was equal to 70.63 and the values for the H and G were equal to 4 and 6, respectively. Those results let to the conclusion that the ASTM designation of quality level for the OB configuration corresponded to Category IV. For the CB configuration, on the other hand, the initial calculations showed the promising results of NC value being more than 75 and the H and G values both being 6. Conventionally, more experiments including the film radiography are needed for the official and full declaration of the category; however, we are confident that all subsequent experiments will solidify the initial results that shows that the RSEC – NIF corresponds to the Category I facility by the ASTM designation of quality.

III Conclusions

Although additional characterization experiments are needed, it is fair to say that after the installation of collimator components and the gamma filter, the new Neutron Imaging Facility at the Penn State Breazeale Nuclear Reactor (or RSEC) can produce images of high standards that corresponds to facility of Category I by ASTM designation of quality. The current system has an effective L/D ratio between 107 and 115 and can be increased to around 150 deviating the distance to the imaging plate and selecting different apertures with different diameters. The neutron flux measurement performed with gold foil activation technique at the exit surface of the beam port resulted in the average thermal neutron flux that is equal to $5.4 \times 10^{6} n\ cm^{-2} s^{-1}$ for 1 MW reactor powe r with the standard deviation of 6.4%. Both mentioned values were in the scope of the initial objectives for the system and are successfully achieved. Further, the development of the neutron tomography capability is seen as one of the strategic next steps for the RSEC – NIF as well as the subsequent active application of our system in the field of NR.

Acknowledgment

The development process of the RSEC – NIF is financially supported by the internal funds of RSEC at Penn State. The authors would like to thank Alison Portanova, Jeffrey Geuther and other RSEC staff members for all their support.

References

[1] K. Ünlü, B. J. Heidrich, "The Penn State Breazeale Reactor", IAEA Technical Report Series No: 482- History, Development and Future of TRIGA Research Reactors, IAEA Vienna, 2016.

[2] A. Kenges, K. Ünlü, J. Geuther, D. Beck, "Development of a New Neutron Imaging Facility for Radiation Science and Engineering at the Penn State University", Transactions of the American Nuclear Society, Vol. 123, November 2020, pp. 414-416 [Online]. Available: https://www.ans.org/pubs/transactions/article-48728/

[3] ASTM International, E 748: Standard Guide for Thermal Neutron Radiography of Materials, 1995

[4] ASTM International, E803: Standard Test Method for Determining the L/D Ratio of Neutron Radiography Beams, 2017

[5] ASTM International, E545: Standard Test Method for Determining Image Quality in Direct Thermal Neutron Radiographic Examination, 2014

CABRI test events monitoring through three measurement systems

Quentin Grando[1], Léna Lebreton[1], Vincent Chevalier[1], Jacques Di Salvo[1], Stéphane Eymery[1], Claude Gaillard[1], Nathalie Monchalin[1], Jérôme Guillot[1]
[1]Nuclear Safety and Radiation Protection Institut - IRSN, France
quentin.grando@irsn.fr

***Abstract*— The CABRI experimental pulse reactor, located at the Cadarache nuclear research center, southern France, is devoted to the study of Reactivity Initiated Accidents (RIA). When the thermal-hydraulic conditions representative of a pressurized water nuclear reactor are reached (mostly temperature, pressure and flowrate conditions), a fuel test rod is submitted to a power excursion, triggered by the specific ^{3}He reactivity injection system, in order to simulate a control rod ejection accident. The experiment, managed by IRSN, aims at studying both the fuel and cladding behavior of the test rod placed into the center of the reactor during the power excursion.**

Several test rods pre-irradiated in pressurized nuclear power plants and with different characteristics (burn up, cladding material, fuel type, Zirconia thickness) are considered for the programs performed in the CABRI reactor.

Physical phenomena occurring during the power transient are monitored by various measuring systems designed or operated by IRSN. Each system provides information linked with the different phases of the experiment. Three main measuring systems will be considered in this paper:

- **The test devices, a sample holder that is also implemented with almost fifty sensors used to monitor the environmental parameters in the test channel such as temperature and pressure, and to control the rod behavior during the test sequence;**
- **The Hodoscope, an online fuel motion measurement system, which aims at analyzing the fuel motion deduced from the detection of fast neutrons emitted by the test rod, with a time step of 1ms during the transient;**
- **The IRIS facility, conceived to perform X-ray radiography and tomography imaging before and after a power transient thanks to a linear electron accelerator, as well as quantitative gamma scanning analyses.**

During the experimental sequence, different events are recorded. This paper focuses on a cladding failure that occurred during the transient and that can be revealed by the three systems mentioned above. The test device instrumentation allows to determine the timing of the failure and to analyze its consequences in the vicinity of the test rod from a thermal and hydraulic point of view, while the hodoscope measures fuel elongation and relocation during the power excursion. The IRIS facility, then, helps to confirm the failure, its location and its extent. These three systems are complementary and they allow the analysis of the same event from different perspectives. Their combination will ease the interpretation of the events in the next steps for the test results analysis. The study case taken into account in this paper concerns nuclear fuel after three irradiation cycles in a commercial PWR.

***Keywords* —CABRI, Reactivity Initiated Accident, Test Device, Hodoscope, IRIS.**

I. Introduction

FOR enhancing safety of nuclear power plants (NPP), the Nuclear Safety and Radiation Protection Institute (IRSN) carries out some experimental programs in order to improve the understanding of the fuel behavior under severe accident conditions. Experiments are then driven into the CABRI experimental reactor, to test fuel sample under PWR conditions and submitted to a RIA accident.

The postulated design-basis reactivity accident for PWR is initiated by the ejection of a control rod bundle following a control rod housing rupture. For simulating the subsequent neutronic transient, the key feature of the CABRI reactor is its helium-3 reactivity injection system [1], which generates the power transients starting from a max 100 kW initial power (HZP - Hot Zero Power conditions). It is composed of 96 tubes located within the core, replacing the same number of fuel rods, grouped in 4 so-called "transient rods". These transient rods can be pressurized at up to 15 bar of helium-3, a strong neutronic absorber gas. These rods are all linked and gather in a collector. The depressurization towards a 1000 L discharge tank is made through two discharge lines (low and high flow rates), each one being equipped with a fast-opening valve followed by a controlled valve. A specific control command application is operated from the remote-control room. The helium-3 ejection leads to a reactivity injection (up to 3.9 \$) in a short time (< 80 ms). The transient characteristics (Full Width at Half Maximum - FWHM, energy, power) depend on the sequence applied to the two fast-opening valves:

• Short FWHM (< 10 ms) and high power (up to 30 GW) transients are obtained by opening a unique fast valve (the high flowrate channel one);

• 20 to 80 ms FWHM high power transients (up to 10 GW) are obtained by successive and accurate opening of the two fast-opening valves.

Results shown in this article will concern the d-day experiment. The reactor reach 100kW in steady state condition with helium-3 into the transient rods, the experiment is triggered and the power excursion arises. The reactor is scrammed to end the chain reaction.

In order to analyse the test rod behaviour and collect data, the CABRI facility is equipped with various measurement systems. Among them, three systems are very specific and designed or operated by IRSN teams.

The first one is the *CABRI test device*. It holds the test rod and is also equipped with a variety of sensors that allow to record several thermal and hydraulic parameters of the system (such as temperature, pressure and flowrate) in the channel where the test rod is placed, as well as some other more exotic sensors used to measure more complex phenomena (like microphones, transient flowmeters, etc…).

The *hodoscope*, the second main measuring system coupled with the CABRI reactor, is a unique online fuel motion monitoring system, dedicated to the measurement of the fast neutrons emitted by the tested rod during the power pulse. This system is used to observe the degradation of the fissile column, and is able to measure fuel motion, potentially linked to clad failure, if it arises during the power excursion.

The third equipment is the *IRIS (Installation for radiography, imaging and spectrometry) facility* dedicated to pre and post non-destructive examinations of the rodlet. This measuring system is equipped by two main parts:

• A gamma spectrometry system which is used to measure gamma rays emitted by the fission products in the test rod.

• An X-ray facility which aims at performing X-ray radiography and tomography images.

The next section of this article gives a brief description of the three systems. Section four provides simplified test data showing how the three systems are complementary to describe the events that occur during the experimental phase.

II. Measurement systems overview

A. Test device description

1) General description of the system

To achieve the test objectives, a specific equipment was designed by IRSN. The test device acts as test rod holder. It allows to maintain it in place, at the right elevation at the center of the reactor core. It also allows water from the water loop at PWR conditions to run through the test channel where the test rod is placed. The primary water also runs around the test device in order to cool the outside walls of the device as well as the inner walls of the loop. Instrumentation is embedded in it. Mainly dedicated to measure thermal and hydraulical parameters such as flow rate, pressure and temperature into the test channel, it also includes various specific sensors much more dedicated to investigate the phenomenology of the CABRI tests.

In this paper, a focus will be made on the measurements of the temperature in the vicinity of the rod and the flowrate at the moment of the interaction between hot fuel and water. The test device is made of three specific parts, detailed from the bottom to the top in the following paragraphs.

The first part is the lower part. It is placed below the flange that allows to insert the test rod. This part especially holds the upstream transient flowmeter.

The next one is the core zone where the test rod is held. It is made of Zircaloy for neutrons transparency purpose. Only thermocouples are implemented in this area. The most important of them are the twelve sensors that are placed in the vicinity of the test rod into the channel, and held by a spacer system.

The last part is the upper part. At its upper end stands the water outlet. The downstream transient flowmeter and a rod elongation sensor are implemented there. Some more specific sensors are also implemented but not detailed in this article. Above the water outlet is the head of the test device. It allows to close tightly the pressurized water cell and also holds the connectors that route the signals to the conditioners and the acquisition systems.

The following figure, extracted from [1], shows an overview of the test device with a focus on the implemented sensors.

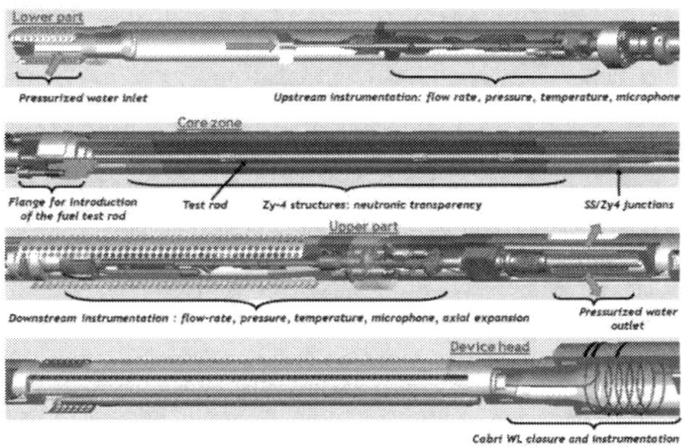

Fig. 1. Test device simplified description.

2) Instrumentation implemented in the device

The key sensors are located upstream and downstream the test rod to avoid being in the core zone, except for twelve K-type thermocouples that are placed in its vicinity. The rod is centered on the core axis and has a total length of around 1.1 m with a ~56 cm long fissile column.

The thermocouples are implemented on a structure called spacer system that allows the rod being centered and the thermocouples being placed at the right axial location along the rod length. They are distributed along the fissile column to measure the fluid temperature in the vicinity of the rod.

The flowmeters of interest in this paper are the two fast response flowmeters, based on the LVDT technology. They are located upward and downward the rod.

B. Hodoscope description

The CABRI reactor has been equipped with the hodoscope since 1977. This system measures in real time the behaviour of the test rod in the center of the CABRI core during the transient. The hodoscope is able to quantify the amount of fuel ejected in the milliseconds following failure, time-dependent axial fuel mass distributions and follows the fuel clusters after failure. In addition, the initial state of the test rod is precisely known thanks to the hodoscope measurement of the fissile length and the axial power profile.

The hodoscope centerpiece is a 3 m steel collimator (in green in Fig. 2) of 3 columns and 51 rows, making 153 channels. Behind each one of these line-of-sights, a ^{237}Np fission chamber detector (FC) and a proton recoil proportional detector (PR) measure the fast neutrons coming from the core and test rod. Two different technologies of detectors are used in CABRI in order to follow all the experiments from low core power (~50kW) up to 20 GW. Proton recoil counters are better suited for low power (up to ~1 GW which means 2.5 million impulses per second), thanks to a higher efficiency and no discrimination of the γ noise, whereas ^{237}Np fission chambers are used for higher power. The signal acquisition rate can be adjusted between 1 ms and 10 s according to the count rate. The associated electronics have been renewed taking benefit from the renovation period [2].

The collimator was built in order to pixelate the test rod placed in the center of the reactor. Each detector looks at a slice of the test rod. In steady state conditions, the detectors measure the axial power profile of the test rod, whereas in transient conditions (i.e. a RIA pulse) the signals vary according to a fuel displacement, a fuel densification (yielding to an increasing signal), or a fuel dilution or ejection (yielding to a decreasing signal). It should be noticed that two lines of sights, row 17 and 35, are filled with stainless steel in order to mechanically strengthen the collimator.

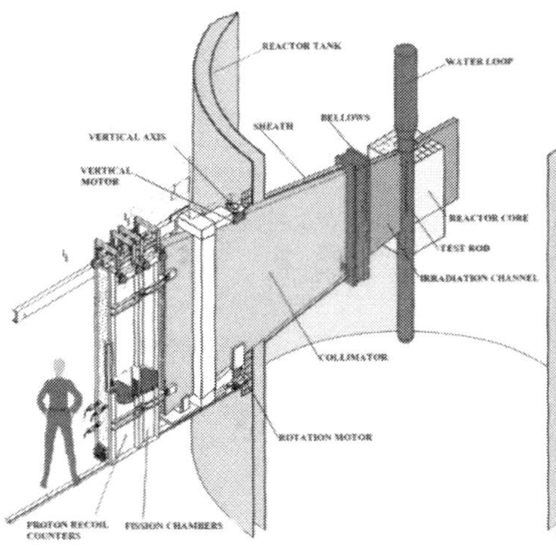

Fig. 2. Hodoscope detector system.

C. IRIS facility

The IRIS facility (Installation for Radiography, Imaging and Spectrometry, Fig. 3) is a non-destructive examination bench. IRIS can perform three types of examination:

- Quantitative gamma-scanning.
- X-ray transmission radiography,
- X-ray transmission tomography.

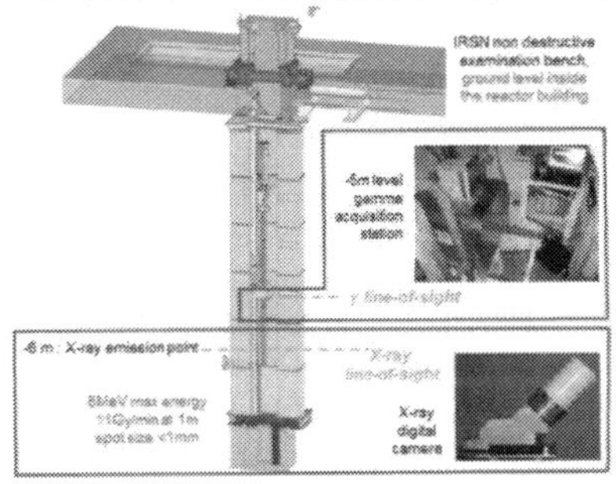

Fig. 3. IRIS facility.

Located within the CABRI reactor building, the mechanical bench is installed in vertical position, for avoiding modifying the post-test final state of the rod, inside a concrete and iron radiological shielding. The test device is placed within a tightness sheath that is motorized in vertical translation and in rotation in front of two examination line-of-sights.

Quantitative gamma-scanning is performed with vertical translation in front of a 1 mm height collimation. The detector is a high purity Germanium detector, coupled to a remote-controlled acquisition system. Before the test, the distribution of these radionuclides allows to verify the integrity of the fissile column. The results of gamma spectrometry will not be developed here.

X-ray radiographies and tomographies are carried out by means of a NMOS camera. The X-ray beam is generated by a high energy electrons linear accelerator (the maximum energy reaches 8 MeV, mean energy is around 2.5 MeV), producing a dose rate of more than 10 Gy/min at 1 m. Radiographies are acquired by vertical translation of the device in front of the camera, equipped with a ~0.5 mm entry slot collimator, while tomographies combine step-by-step rotation (0.4° azimuthal step) at a given elevation. The spatial resolution of the reconstructed images is close to 200 µm.

Before the test, from the radiography acquired at 0.5 mm step, the bottom and the top of the fissile column (BFC and TFC) are determined and compared to the ones measured by gamma spectrometry. The fissile column length is then confirmed. The location and integrity of the device sensors are verified also from radiographic examination.

The tomographies at positions chosen in relation to the zones of interest determined by radiography, allow to carry out a sectional view of the rod. When the density of the material crossed by the rays becomes greater, the image appears blacker. Before the test, these images make it possible to check the centering of the rod in the channel tube, the integrity of the spacer rings and the position of the sensors.

The tomographies also make it possible to measure the diameter of the rod before and after the test and thus to evaluate a possible rod swelling due to the transient. The dimensional measurements are made relative to the known values of the channel tube and the pressure tube diameters.

III. Results

Several noticeable events were observed during the transient. For this article, focus will be made on the event arising at the time TOP_{ONSET} + 379 ms (*Transient Over Power ONSET, which will be noted t0,* corresponds to the beginning of the neutron population rise), leading to a clad failure on the top part of the test rod.

Fig. 4 summarizes the main characteristics of the pulse and illustrates the time interval of interest in the test sequence: the three measurement systems will be further analyzed within this interval. It runs from t0 + 375 ms to t0 + 392 ms.

They all will be taken from the point of view of their own specific skills allowing them to monitor specific parameters evolution, but also from their ability to complement each other and widen the view of the monitored events.

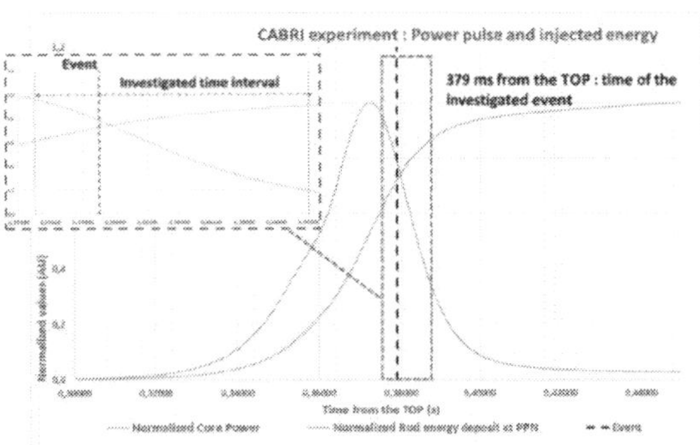

Fig. 4. Core power and rod energy deposit during the pulse.

A. *Measurements from the test device instrumentation point of view*

As already mentioned, a clad failure event occurs around 379 ms after TOP_{ONSET}.

The first sensor that detects the event is the DA, with a fast, but limited, decrease of the normalized elongation followed by a return to a different baseline (Fig. 5).

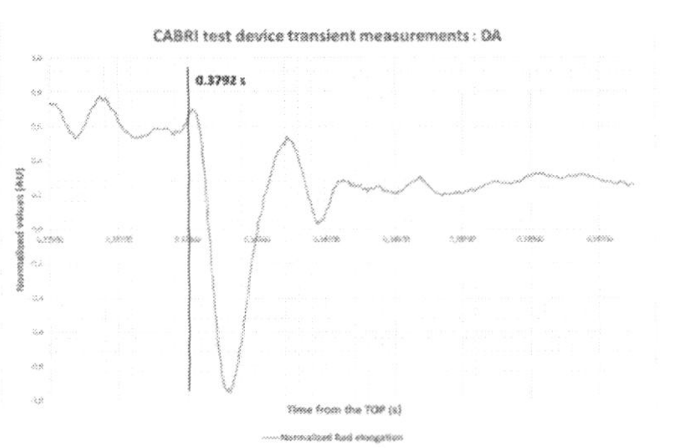

Fig. 5. Test device measurements of interest: elongation (DA).

The event can then be detected with the transient flow rates sensors QT1 & 2 (379.5 ms on Fig. 6). They show opposite variations that are a sign of water ejection in the two available directions. QT2 catches the event 0.7 ms before QT1 (this timing is obtained by the use of the moving variance technique, but it's difficult to associate a uncertainty to this value), indicating that this event occurs closer to the top of the rod.

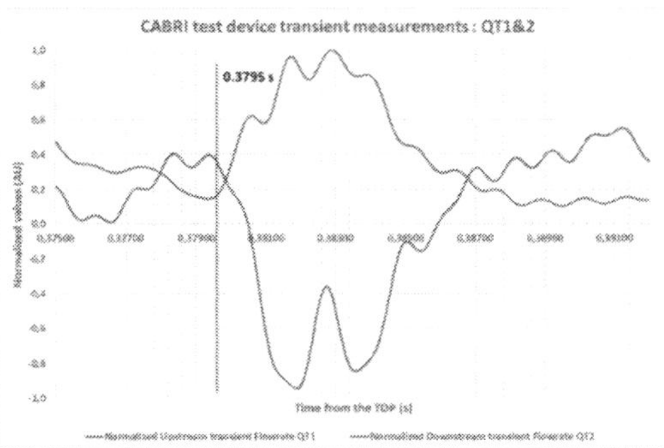

Fig. 6. Test device measurements of interest: transient flowrate (QT1&2).

The behavior of the sensors shown in this paragraph tends to confirm that a fuel-coolant interaction occurred. The time difference between downstream and upstream sensors tends to indicate that the event occurred in the upper part of the rod. A two phase mixture can also be guessed with one of the TCs placed on the rod cladding (not shown in this paper), that expresses a very high increase of the clad temperature from 372.5 ms after the top, marking a possible boiling. This information and the fact that only one TC reached a so high temperature tends also to indicate that both the water state and the temperature distribution are not homogeneous in the channel along the rod, which increases the difficulties to give a precise date and location of the event.

B. *Hodoscope and test device data collected during the transient*

1) *Accumulation of fuel detected by the hodoscope equipment*

One of the most important measurement performed by the hodoscope is the evolution of the axial fuel distribution during the power transient for each hodoscope row. The measurements are based on the count rates of the ^{237}Np fission chambers (fission threshold at 600 keV), better suited for the high level power reached during the transient. As already mentioned, for each row of detectors, only fast neutrons are detected by the hodoscope sensors. Thanks to the acquisition rate with a time step of 1 ms, it is possible to determine the moment and the characteristics of the degradation.

We define [4]:

- $S_i(t)$: signal created on row i by fast neutrons coming directly from the test rod at time t.
- $B_i(t)$: the so-called « noise background » created by fast neutrons coming from the core and fast neutrons scattered by the test loop and detected on row i at time t.
- $SNR_i(t)$: the Signal over Noise Ratio $SNR_i(t) = S_i(t)/B_i(t)$.

- $SNR_i(BT)$: the average SNR, recorded during the power plateau at 100kW just before the transient (Before TOP onset) .

As the variations of $SNR_i(t)$ are very small, the analysis system computes a normalized Signal over Noise Ratio, $SNR_i(t)/SNR_i(BT)$. This is the quantity to which the signal to fuel mass conversion abacuses can be applied [5].

The hodoscope system plots the evolution of $SNR_i(t)/SNR_{BT}$ during the power plateau and the power transient.

$SNR_i(t) > SNR_i(BT)$ or $SNR_i(t) < SNR_i(BT)$ means a fuel displacement during the power transient in the ith row.

In this article, the rows 14 to 30 corresponding to the top part of the fuel rod are particularly considered. To understand the Fig. 7, one have to know: a ratio $SNR_i(t)/SNR_i(BT) > 1$ will appear in green, corresponding to a fuel accumulation; on the contrary, a ratio $SNR_i(t)/ SNR_i(BT) < 1$ will be represented in red, meaning a lack of fuel. The blue curve overriding the different channels corresponds to the power level measured by the experimental chambers of the CABRI reactor. The grey dashed rectangle delimits the time interval [t0+375 ms; t0+392 ms] during which the failure occurred in the upper part of the fuel rod.

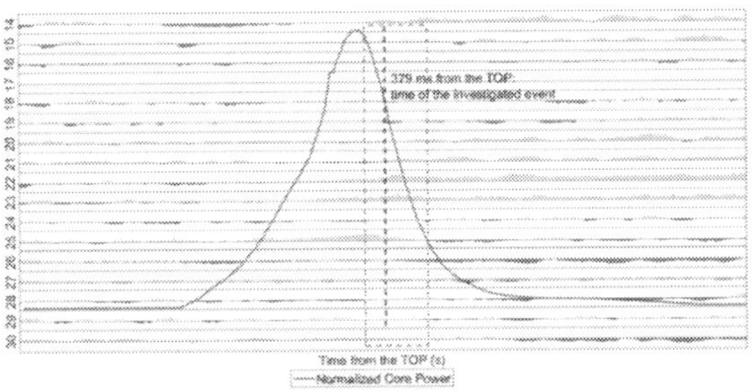

Fig. 7. Variation of the S/B before and during power excursion measured by the hodoscope's detectors for the upper part of the fuel rod (rows 14 to 30).

During and at the end of the transient, one can note some rows for which accumulation of fuel material is observed (row 14 linked with the ~top of the fuel rod, rows below 15, 20, 21, 22, 23) and others for which lack of fuel can be noticed (rows 25, 26, 28 corresponding to the ~middle of the fuel rod, and row 30). Accumulation of fuel for the rows 20 to 23 corresponds to an axial zone located between -120 to -200 mm below the TFC.

2) Clad failure detected by the test device instrumentation

Focusing on the time interval between [t0+375 ms; t0+392 ms] on Fig. 8, in the test channel, the large flowrate divergence is observed (transient flowmeters QT1 and QT2) and superposed to hodoscope results.

After the event, the hodoscope recorded a continuous fuel relocation in the upper part of the test rod.

Therefore, transient flowmeters signals combined with hodoscope measurements may reveal clad failure.

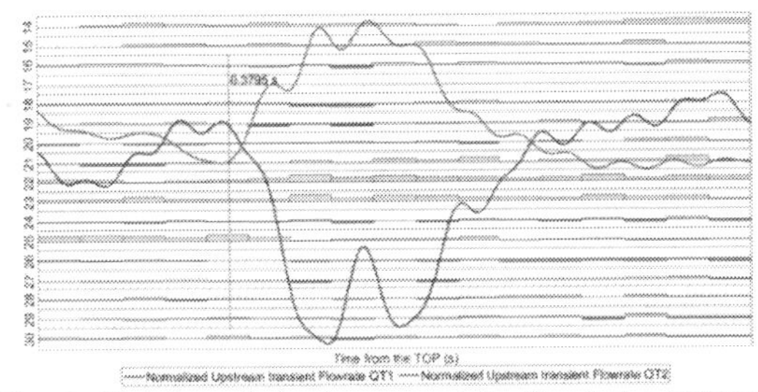

Fig. 8. Variation of signals measured by the device instrumentation (flowmeters) and S/B ratio evolution obtained with the hodoscope detectors for the upper part of the fuel rod, for a time range [t0+375ms; t0+392 ms].

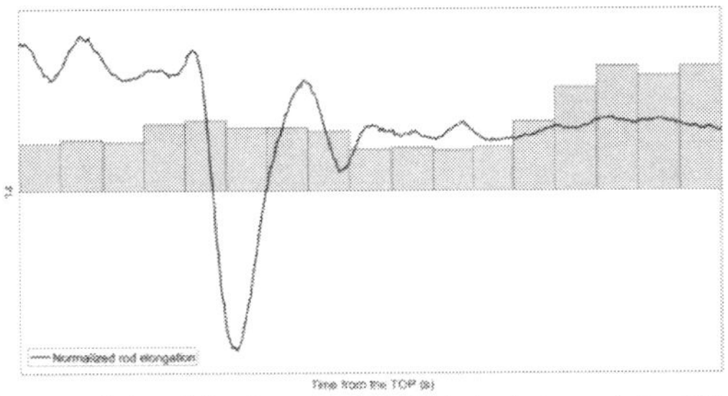

Fig. 9. Variation of signal measured by the device instrumentation (DA sensor) and S/B ratio evolution deduced from the hodoscope detectors for the row 14, for a time range [t0+375ms; t0+392 ms].

Consistent with the event detected by the axial expansion sensor DA, the hodoscope row 14 reveals the elongation of the fissile column (Fig. 9).

C. Hodoscope and IRIS data collected after the transient

1) Final state of the fuel rod detected by the hodoscope equipment

The final state of the rod after the transient is measured during a power plateau at 100 kW for 1 hour (in order to get enough statistics), starting ~ 75 minutes after the power transient. The fuel distribution is estimated as the ratio of the post-test signal to the pre-test signal.

The results obtained are presented for both PR and FC sensors. Fuel accumulation appears in green, while fuel loss is traced in red. The accumulation zone is evidenced in the upper part, and the lack of fuel is observed in the center part.

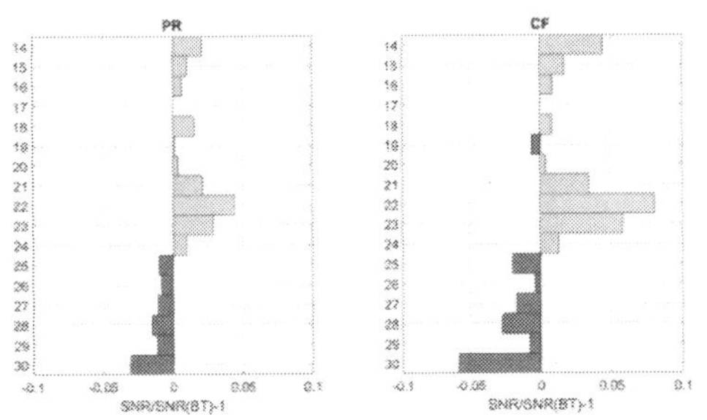

Fig. 10. Final state of the upper part of the fuel rod deduced from the hodoscope's detectors (rows 14 to 30).

2) Final state of the fuel rod obtained with IRIS X-ray measurements

The X-ray results are presented. Fig. 11 compares the radiography of the test rod performed before the transient (on the left, with only one angular position shown because the geometry was axisymmetric), to the three other ones performed after the test at three different angles: Θ°, Θ+60° and Θ+120° (on the right). The density and thickness of the materials crossed by the X-rays lead to blurring of the image (scattering of the rays). Concentric circles appear and define, from the center outwards, the pellet delimited by the clad, the space between the rod and the channel tube, the channel tube, the space between the channel tube and the pressure tube and finally the pressure tube. Small dots indicate the presence of sensors in the space between the rod and the channel tube and in the space between the channel tube and the pressure tube.

Ejection of fuel led to lack of fuel at some places and accumulation at other places. A zone of fuel accumulation is roughly located between -110mm and -180mm below the TFC, in consistency with hodoscope results.

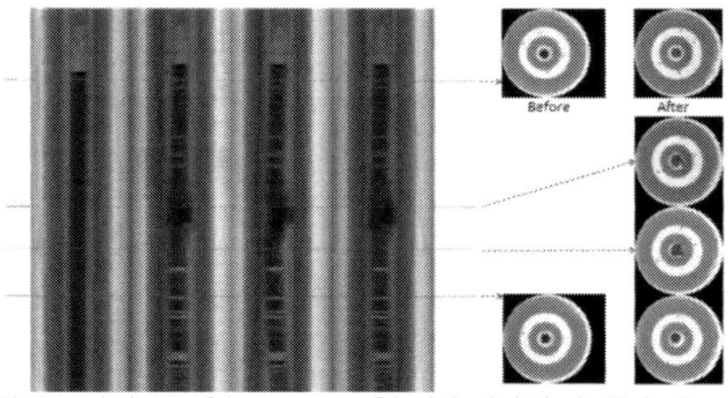

Fig. 11. Final state of the upper part of the fuel rod obtained with the X-ray measurements of IRIS equipment.

The rod is clearly not centered towards the vertical axis on the tomographies: a S-shaped deformation of the rod is observed. After the test, a residual elongation of the fissile column of at least 8mm is observed. Radiographies show a clear crushing of the spring, which length is reduced of ~8mm after the test. The top end of the upper plug did not move axially. The spacer rings are identified at the same location before and after the test.

On the tomographies after the test, the clad is clearly opened on the upper part of the rod. Fuel has been ejected, fuel fragments are located between the clad and the channel tube. The fuel remaining in the clad shows a fragmented structure.

The tomographies make it possible to locate the angular position of the clad opening. The combination of radiography and tomography allows to define the length of the opening.

3) Combination of the results

One has to be very careful when comparing the results of the different equipment in terms of axial localization, because the conditions of the measurements are quite different: hot (PWR) conditions for hodoscope measurements and cold (room temperature) for IRIS measurements.

However, we can observe on Fig. 12 the very good consistency obtained for qualitative results coming from these two equipments: the hodoscope PR measurements (on the left) and the X-radiography (on the right). At the location of the clad failure, the accumulation of fuel is clearly visible at the same level. One must keep in mind that for the hodoscope sensors, the results are expressed in relative terms, i.e. positive values means that, at the final state, more fuel is detected at the sight of view of the row than before the pulse. The results can be obtained very quickly after the transient and helps to the first interpretation of the results in combination with the other sensors.

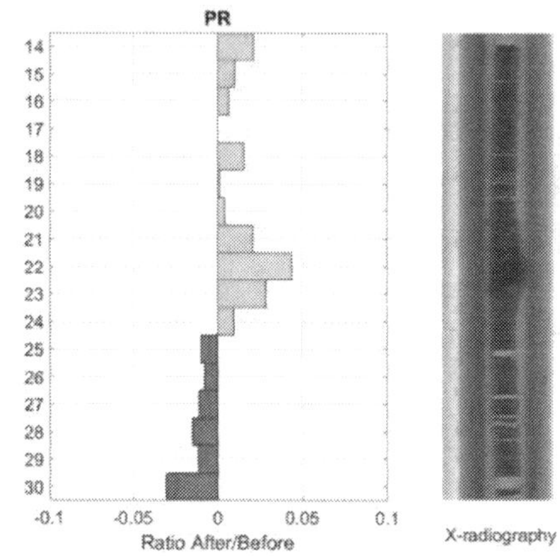

Fig. 12. Comparison of the final state of the upper part of the fuel rod obtained with the different devices.

IV. Conclusions

In this paper, we highlighted that a sequence of events can be very precisely monitored from various and complementary points of view thanks to the different devices implemented in CABRI.

Each one has its own skills and monitors specific quantities. First, the test device sensors give information from a thermo-hydraulic point of view, allowing a timing of the events and raw location. The hodoscope then monitors it from the fuel point of view and, when combined with the test device sensors, sharpen timing and location estimation. They both give a view of the events occurred during the test sequence and allow also deeper analysis later after the test sequence once the raw data are processed. Finally, the IRIS equipment provides the final state of the fuel rod, confirming and localizing more precisely the event. The possible damages to the cladding or fuel stack can be further analyzed. The precious information produced is very helpful for the definition of Post Irradiation Examinations of the tested fuel rod in a hot laboratory.

Acknowledgment

The author express their acknowledgments to all the contributors of this project, and especially the experimental physicist team, the engineering team and the operators of the CABRI reactor working at the CEA.

References

[1] B. Biard et al., "Reactivity Initiated Accident Transient on irradiated fuel rods in PWR conditions: the CABRI International Program CIP" Annals

of Nuclear. Energy , vol. 141, 2020, https://doi.org/10.1016/j.anucene.2019.107253.

[2] S. Mirotta et al., "Qualification and characterization of electronics of the fast neutrons Hodoscope detectors using neutrons from the CABRI core" presented at the fifth ANIMMA conference, Liege, Belgium, June. 19-23, 2017.

[3] V. Chevalier, S. Mirotta, J. Guillot: The CABRI fast neutron Hodoscope: renovation, qualification program and first results following the experimental reactor restart, "In ANIMMA2017 Conference Proceedings, Liege, June 2017.

[4] J. Di Salvo, S. Mirotta, V. Chevalier, "Analysis of the Signal over Noise Ratio of the hodoscope determined by Monte Carlo calculation", presented at the 7th International Conference ANIMMA, June 21-25, 2021

[5] Baumung et Augier. Quantitative fuel motion determination with the CABRI fast neutron Hodoscope: Evaluation Methods and Results, november 1990.

Measurement of prompt gamma field above the VR-1 water level

Tomáš Czakoj[1], Michal Košťál[1], Zdeněk Matěj[2], Evžen Losa[3], Jan Šimon[1], Filip Mravec[2], František Cvachovec[4]
[1]Research Centre Řež, Husinec-Řež, Czech Republic
[2]Masaryk University, Brno, Czech Republic
[3]Czech Technical University, Prague, Czech Republic
[4]University of Deference, Brno, Czech Republic
tomas.czakoj@cvrez.cz

***Abstract*— Pool-type reactors are an excellent source of high energy photons without neutrons because the water above fuel acts as an outstanding neutron shielding, whereas for high energy photons, it is nearly transparent. The absence of neutrons is a valuable property essential for testing semiconductor detectors because they are sensitive to neutrons, which might cause their damage. The high energy gammas are worth studying because they are responsible for radiation induced heating of reactor internal components in energy producing reactors, which might lead to their radiation induced damage by void swelling. At this effect, the helium produced in (n,α) reactions influenced by locally increased temperature clusters into bigger bubbles, which are increasing material volume, and therefore changing the geometry of internals. The referred method could be utilized e.g. for resource exploration by detecting prompt capture gammas. The prompt gamma field was measured at the VR-1 reactor with two different cores and water levels. The stilbene measurements showed relatively good agreement with the spectrum shape computed by MCNP6.2. An important result is a confirmation that the stilbene deconvolution matrix can be used with a rotated crystal. The testing measurement with HPGe detectors showed a possibility for measurement in such type of geometry.**

***Keywords* —Capture gammas, Scintillation detectors, HPGe detector, Prompt gamma**

I. Introduction

The pool-type reactors have a special benefit for scientific research – they can serve as an excellent source of high energy photons with absence of neutrons. The neutrons are absent due to the shielding by water present above fuel assemblies, whereas water is poor shielding material for high energy photons.

The proper knowledge of the gamma spectrum in the vicinity of the reactor core is essential in the characterization of the radiation field. It is important for ensuring radiation safety. The high energy gammas are responsible for radiation induced heating of reactor internal components, which might lead to their radiation induced damage. In this void swelling effect, the helium produced in (n,α) reactions influenced by locally increased temperature clusters into bigger bubbles, which are increasing material volume, and thus changing the geometry of internals [1].

On the other hand, prompt gamma radiation has several benefits - a well-described gamma field can be used for testing of radiation measuring devices, it can be used for the measurement of the reactor power [2]. Another application is the Earth crust exploration since the capture gammas are an important signature of the presence of materials present in Earth crust [3].

The field in the vicinity of the reactor core is interesting for testing because in contrary to commonly use gamma fields formed by common gamma sources, in the reactor fields, there are high energy gammas formed by neutron interactions with reactor structural components (prompt gamma capture reactions, inelastic scattering). Due to this mechanism, the gammas in the reactor field are mostly accompanied by the presence of neutrons.

An interesting situation may occur in deep penetration problems in water because a deep-water slab is an excellent fission neutron absorber but relatively weaker gamma absorber. Behind the deep-water slab, one can expect high energy capture gamma field with a negligible share of neutrons. This criterion is well filled above the water surface of the VR-1 research reactor.

The paper deals with the results of different measurements – both with nominal and reduced moderator height. The experiment with reduced water level was extended by additional measurement with a lower position of the detector.

II. Measurement

A. VR-1 reactor

The measurements were performed at the VR-1 reactor operated by Czech Technical University in Prague. It is a zero-power pool-type reactor. The reactor core consists of IRT-4M tubular fuel assemblies (19.7% ^{235}U in the form of UO_2-Al, active length approximately 68 cm) moderated and shielded by light water. The height of the water column above the assemblies is usually about 300 cm (see Fig. 1). The core is also equipped with several dry vertical channels, radial channels, and other experimental equipment.

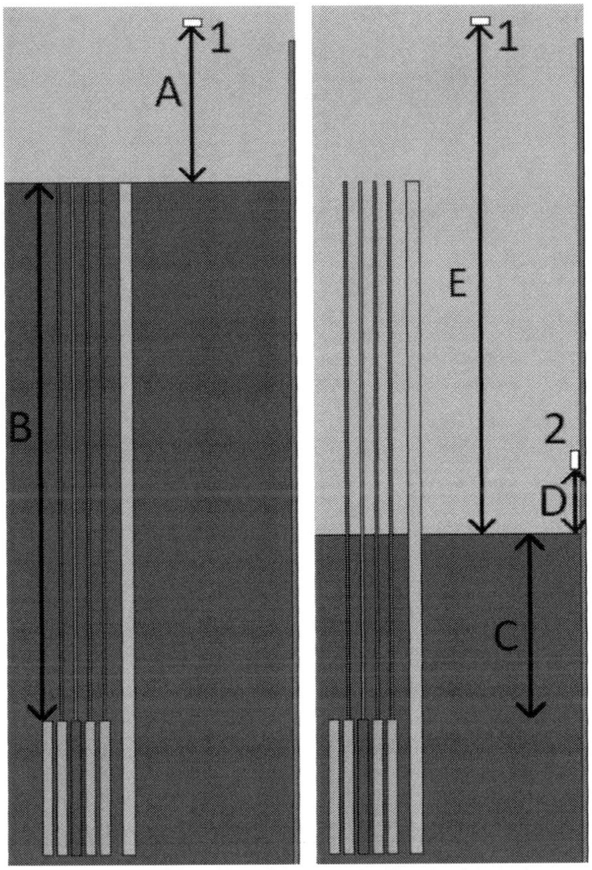

Fig. 1 Visualized axial position of detectors- Position 1 - detector in upper position, position 2 - detector in lower position, distance A is 86 cm, distance B is 306 cm, distance C is 95 cm, distance D is 34 cm, distance E is 291 cm

Fig. 2 Photography of measurement geometry

B. Stilbene measurement apparatus

The measurement was performed using the spectrometric system NGA-01 [4]. A 45 x 45 mm cylindrical stilbene organic scintillator was used as a detector. An innovative active high voltage divider is used for the measurement (Fig. 3). The divider was developed to compensate for the nonlinearity caused by the high frequency of pulses and their amplitude (Fig. 4). We use an internal negative high voltage source as a bias for the Hamamatsu R329-02 photomultiplier. The negative voltage allows fast pulse processing up to frequencies greater than 1E6. The NGA-01 system uses parallel processing of differently amplified inputs from several fast ADC converters at a resolution of 12 bits and a sampling frequency of individual ADC converters of 500 MS / s. The online data processing is taken care of by the field programmable gate array (Virtex 6), which communicates via a network interface located on the measuring card FD-17 (Fig. 5). A server is located inside the device. Subsequent processing and transfer of stored data are realized by remote connection to a server with the Linux operating system. The instrument spectra are stored on a RAID disk array inside the NGA-01 device for possible reprocessing or further evaluation. Using measured data and data from calibration sources (Cs-137, Co-60) we convert the apparatus spectrum by unfolding to flux density. We use response matrices calculated and validated in PTB Braunschweig for these calculations [5].

Fig. 3 Implementation of a linearized active divider for high pulse frequencies

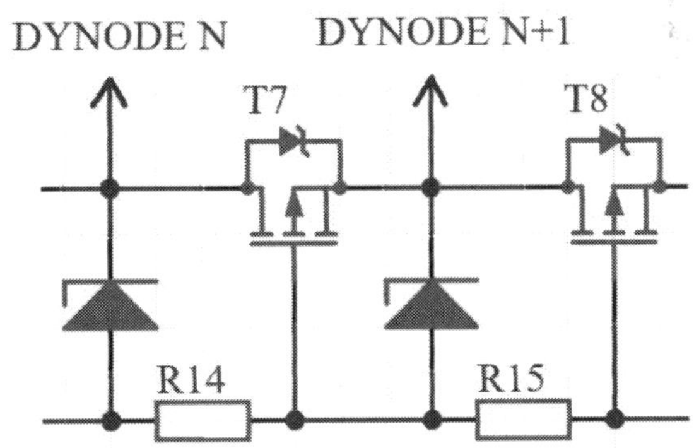

Fig. 4 Connection of a linearized active divider for high pulse frequencies

Fig. 5 Digitization card FD-17 with the FPGA (Virtex 6) and two ADC converters with a resolution of 12 bits and a sampling frequency of 500 MS/s containing four differential pair inputs and Ethernet output

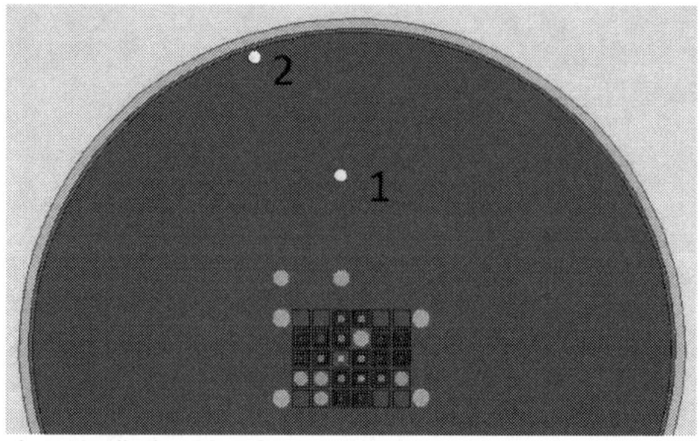

Fig. 6 Visualized position of detectors compared to the core in center of core horizontal plane. Position 1 - detector in upper position, Position 2 - detector in lower position

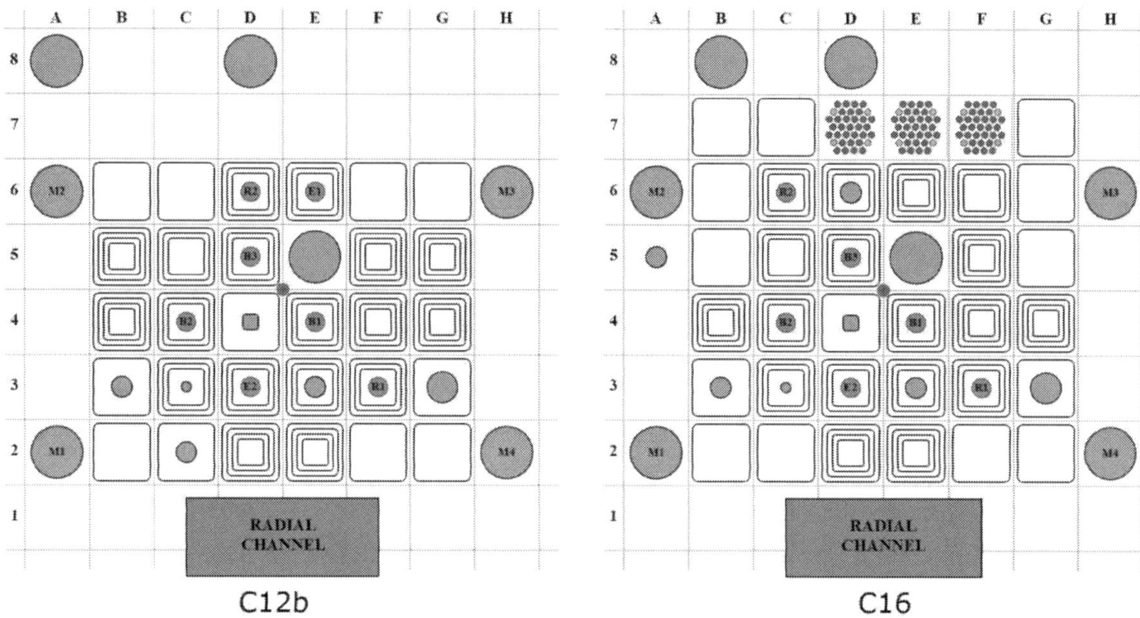

Fig. 7 Comparison of C12b and C16 core configurations. Stainless steel reflector is composed of three blue assemblies

C. Measurement geometry

The measurements were carried out at different times; therefore, two different cores were in the experiment. The experiment with nominal and reduced moderator height was performed with the core C12b. Later, the measurement with the nominal moderator level was repeated with the core C16. The main difference between the cores is a presence of stainless-steel reflector (see Fig. 7)

During experiments with the nominal and reduced moderator height, the detector was placed on the manipulation platform fixed in the reactor vessel above the water surface (same position in both cases). The detector center was 88.6 cm above water and its position was shifted to 48.3 cm from the nearest fuel assembly (see Fig. 1 and Fig. 6). The nominal and reduced water level is 306 cm and 95 cm, respectively, above the upper end of the active part of fuel.

The experiment with reduced water height was also performed with the stilbene detector center placed 36.5 cm above the water level. The detector was placed next to the reactor vessel and its position can see in Fig. 1 and Fig. 6.

During the measurement with nominal moderator level, the possibility of measurement with HPGe detector was tested. Portable HPGe coaxial detector Canberra Big MAC (GC2518 type with the 2002C preamplifier) was fixed next to the stilbene detector at the same height (see Fig. 2).

III. Calculations

Calculations were performed in MCNP6.2 Monte Carlo code [6] using ENDF/B-VIII.0 library [7] for neutrons and MCPLIB84 library for photons. Due to the deep transport problem, it was necessary two split the calculation to two parts. The first part was determination of photon source in critical mode. The photon spectrum was computed in the cell comprising the fuel assembly closest to the detector position. Since it was needed to obtain photon spectrum coming dominantly from fission, neutron capture, and inelastic scattering, the physics settings for photons and electrons were optimized to suppress bremsstrahlung and fluorescence.

It was also necessary to determine photon source distribution. This was determined based on the assumption that most of the photons are produced during nuclear fission and during reaction close to the fuel. Therefore, the tally F7 (fission energy deposition averaged over a cell) was scored in fuel assembly. Additionally, tally F7 was also scored in segments of each tube in the z-direction (per 2.5 cm).

Based on the scored photon spectrum, axial and radial distribution, the photon source was determined in a fixed source model with variance reduction techniques. Each part of fuel assembly has the same spectrum, but its photon emission probability (from tally F7) and own axial distribution (from segmented tally F7). Using the developed fixed source model, the photon spectrum in the detector volume was computed. The obtained spectrum was adjusted by Gaussian broadening with parameters for used stilbene detector.

This experiment was performed as an initial study; therefore, it was decided to focus on the spectrum shape. Thanks to that fact it was not necessary to deal with scaling the calculation to absolute values. Nevertheless, the experiment with focus on absolute value comparison is planned. The compared spectra from measurement and calculation were normalized in the 1 - 3 MeV range for a clear spectrum shape comparison.

IV. Results

Fig. 8 shows results for the stilbene measurement with different detector positions and different moderator height. The effect of water on photon flux attenuation is well visible in comparison of measurements with nominal and reduced water level. This additional 2.5 m of water decreases the flux by factor of 120 in 8 MeV region and of about 530 in 2 MeV region. The measurement uncertainties are within 7% in 1 – 5 MeV energy range and within 10 % in energies above 7 MeV.

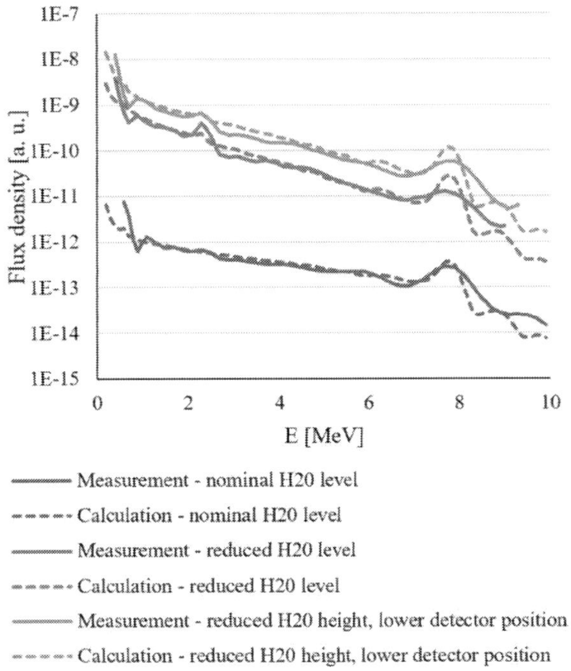

Fig. 8 Comparison of measured and normalized calculated gamma spectra for different moderator levels and detector positions

The comparison for measured gamma spectrum with the different core is visualized in Fig. 9. The C16 core configuration is equipped with the stainless-steel reflector. Its influence is clearly visible in higher energies. The main sources of high energy gammas in stainless-steel are ^{53}Cr (9718.8 keV and 8884 keV), ^{54}Fe (9297.8 keV), and ^{58}Ni (8998.6 keV and 8533.7 keV) peaks. In the 6-8 MeV region, the additional gammas compared to C12b configurations come from ^{56}Fe peaks (7645.6 keV, 7631.2 keV) [8]. On the other hand, the spectra in lower energy regions (bellow 3 MeV) are almost identical.

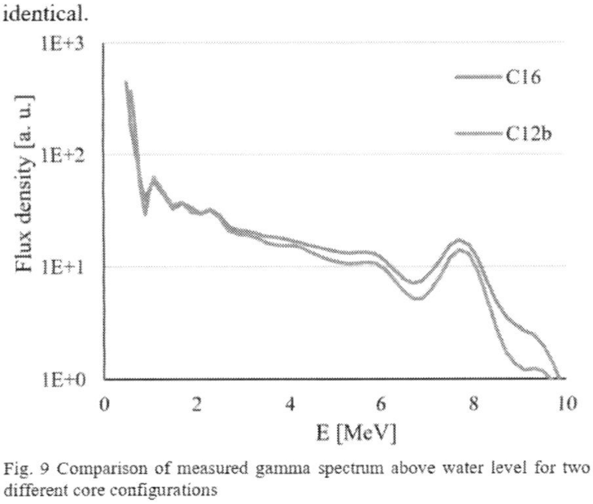

Fig. 9 Comparison of measured gamma spectrum above water level for two different core configurations

In another measurement, the influence of the measurement geometry was assessed. The stilbene detector is usually used in the optical axis parallel to the direction of the neutron field and the deconvolution matrix is derived for this measurement geometry. With advantage, the stilbene detector could be used in different applications if the evaluated photon spectrum would be independent of the crystal orientation. Therefore, the validity of measurement with the perpendicular optical axis was tested. During the test, the gamma spectrum was measured with the same geometry and neutron flux, only with a rotated stilbene axis.

The result of this experiment is visualized in Fig. 10. The presented results correspond to maximum reactor power during irradiation experiment. It is apparent that the results are almost identical and therefore the deconvolution matrix can be used with rotated stilbene.

The test measurement with HPGe detector was performed and the obtained spectrum is depicted in Fig. 11. The figure shows results from measurement in C16 core configuration. The spectrum is shown only in the most interesting energy range between 6 – 10 MeV. This energy range allows identification of important peak.

The most dominant peak in this energy range is 7724 keV ^{27}Al peak and its single and double escape peaks (single and double escape peaks were not identified in Fig. 11). One can also see peaks coming from stainless steel reflector and from other structural components. The effect of stainless steel reflector was also visible in stilbene core comparison (see Fig. 9).

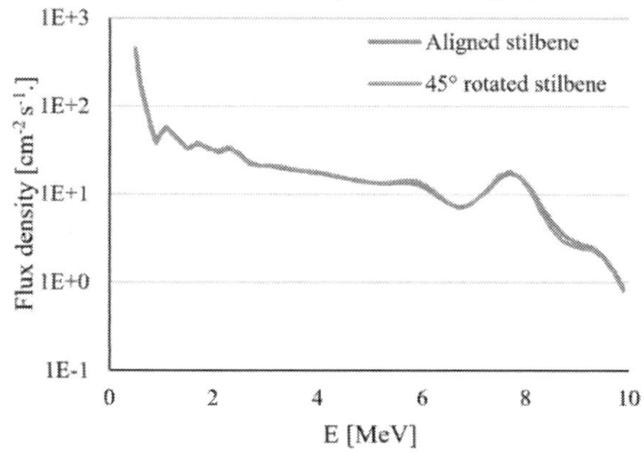

Fig. 10 Measured gamma spectrum with stilbene detector aligned with reactor axis compared to spectrum with rotated stilbene

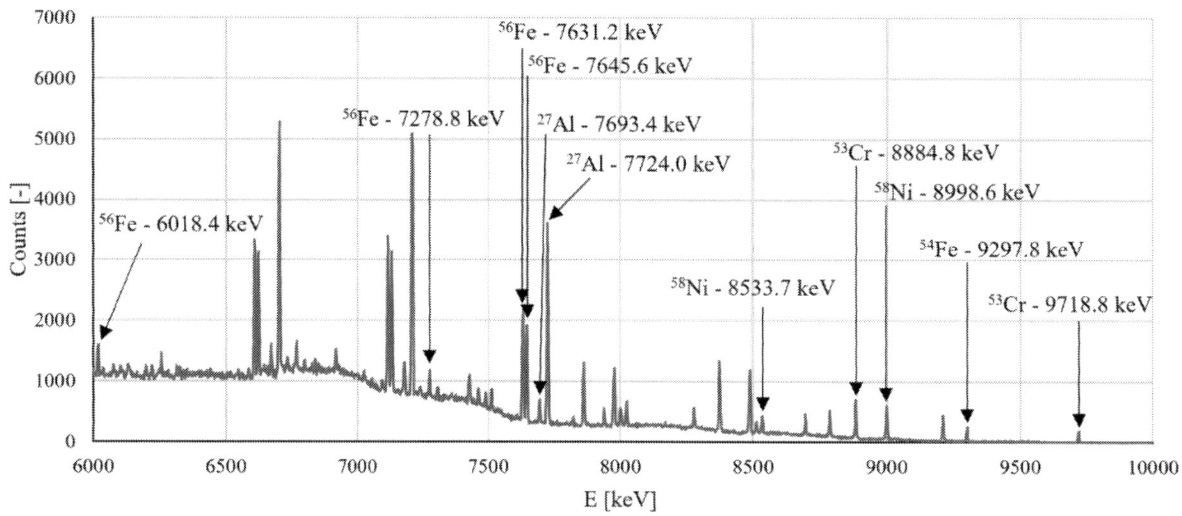

Fig. 11 Measured HPGe gamma spectrum with identification of important peaks, other visualized peaks are mostly single or double escape peaks

Taking into account the fact that the broad peak in scintillation detector measured spectra (see Fig. 8) in 7 – 9 MeV is formed by capture gammas from Al, Fe, Ni, Cr (see Fig. 11), it can be concluded that the magnitude of this broad peak is proportional to thermal neutron flux in the core, because the capture dominates in thermal region. Due to the fact that the VR-1 core is driven by thermal neutrons [9], it can be concluded the broad peak magnitude is proportional directly to reactor power, and such scintillation detector can be used as an independent power monitor.

V. Conclusions

The gamma field above the reactor pool is suitable for detector testing using high energy gammas. Thanks to the absence of neutrons, the methodology is simplified and allows testing of detectors also sensitive to damage by neutron field. This fact is important especially for semiconductor detectors, or detectors with sensitive volume which might be activated by neutrons.

The experiments prove the usability of organic scintillation spectrometry of high energy gamma rays for independent monitoring of zero power pool type reactors , which is based on the fact, that well distinguishable broad high energy peak magnitude is proportional to the number of thermal neutrons absorbed in structural components and the reactor power is also proportional to the thermal neutron flux.

The experiment proved the possibility of measurement of the prompt gamma field using the HPGe detector. The obtained spectrum shows well separated peaks. Thanks to that, the comparison of the measurement with a calculation for selected peaks is planned.

Acknowledgment

The presented work has been realized within Institutional Support by the Ministry of Industry and Trade and with the use of the infrastructure Reactors LVR-15 and LR-0, which is financially supported by the Ministry of Education, Youth and Sports – project LM2015074.
The authors would like to thank to VR-1 staff headed by F. Fejt for their effective help during the experiments and for the precise monitoring of the reactor power. The VR-1 operation was supported by LM2018118 project: VR-1 - Support for reactor operation for research activities, which was granted by The Ministry of Education, Youth and Sports of the Czech Republic

References

1. D. Harutyunyan, I. Mirzov, M. Košťál, et al, "Estimation of Void Swelling in VVER-1000 Baffle Using Benchmark in LR-0 Reactor". In: Reactor Dosimetry: 16th International Symposium. ASTM International, Santa Fe, NM, USA, 2018, pp 321–334

2. M. Jalali, M.R. Abdi, M.M. Davati, "Prompt gamma radiation as a new tool to measure reactor power". Radiat. Phys. Chem., vol.91, pp. 19–27, October 2013, 10.1016/j.radphyschem.2013.05.033

3. M.-L. Mauborgne, R.J. Radtke, C. Stoller, F. Haranger, "Impact of the ENDF/B-VIII.0 library on modeling nuclear tools for oil exploration". EPJ Web Conf., vol.239, pp. 20007, September 2020, 10.1051/epjconf/202023920007

4. M. Pavelek, Z. Matěj, O. Herman, et al, "Fast digital spectrometer for mixed radiation fields". In: 2017 IEEE SENSORS. IEEE, 2017, pp 1–3

5. F. Cvachovec, J. Cvachovec, P. Tajovsky, "Anisotropy of light output in response of stilbene detectors". Nucl. Instruments Methods Phys. Res. Sect. A Accel. Spectrometers, Detect. Assoc. Equip., vol.476, no.1–2, pp. 200–202, January 2002, 10.1016/S0168-9002(01)01431-0

6. C.J. Werner, J.S. Bull, C.J. Solomon, et al, "MCNP Version 6.2 Release Notes". February 2018, Los Alamos, NM (United States)

7. D.A. Brown, M.B. Chadwick, R. Capote, et al, "ENDF/B-VIII.0: The 8th Major Release of the Nuclear Reaction Data Library with CIELO-project Cross Sections, New Standards and Thermal Scattering Data". Nucl. Data Sheets, vol.148, pp. 1–142, February 2018, 10.1016/J.NDS.2018.02.001

8. H. Choi, R. Firestone, R. Lindstrom, et al, Database of Prompt Gamma Rays from Slow Neutron Capture for Elemental Analysis, STI/PUB/12. International Atomic Energy Agency, 2007, Vienna

9. M. Košťál, E. Losa, Z. Matěj, et al, "Characterization of mixed N/G beam of the VR-1 reactor". Ann. Nucl. Energy, vol.122, pp. 69–78, December 2018, 10.1016/j.anucene.2018.08.028

Development of a Wideband Current Amplifier dedicated to Fission Chamber Measurement

Danilo Bisiach[3], Loïc Barbot[1], Grégoire De Izarra[1], Christophe Destouches[2], Manuel Cargnelutti[3]and Sebastjan Zorzut[3]

[1]CEA, DES, IRESNE, DER, SPESI, Instrumentation Sensors and Dosimetry Laboratory, Cadarache, F-13108 Saint Paul-lez-Durance, France

[2]CEA, DES, IRESNE, DER, Experimental Physics, Safety experiment and Instrumentation Section, Cadarache, F-13108 Saint Paul-lez-Durance, France

[3]Instrumentation Technologies, d.o.o., Velika pot 22 - 5250 Solkan - Slovenija, Slovenia

danilo.bisiach@i-tech.si

Abstract— **Fission chambers are widely used in nuclear reactors, either occasionally to probe the neutron flux maps in nuclear power plants or as part of the nuclear instrumentation used in research reactors to characterize neutron flux levels in irradiation locations.**

Wide dynamic range acquisition systems are very limited in the nuclear instrumentation market. Thus, since 2012, CEA is developing the MONACO system, which allows measurement in pulse, fluctuation, and current mode running simultaneously on the same channel.

The simultaneous use of three measurement modes places high requirements on the amplification stage. Indeed, the current mode exploits a bandwidth between 0 and a few kHz, while the fluctuation mode is based on the frequency range of 100 kHz - 1MHz. Finally, for the pulse mode, it is necessary to identify current pulses of a few uA in amplitude and with a duration of about ten nanoseconds, which implies a bandwidth that extends up to 10 or 20 MHz.

In the frame of the industrialization of the MONACO system, a new preamplifier was designed to meet those requirements. With a first prototype, LDCI and I-TECH performed very promising bench tests in October 2020. Tests with synthetic signals corresponding to pulse and fluctuation mode showed very good performance of the electronics. The positive and negative current pulses are correctly amplified and the measurement noise was estimated at 0.1 µApp thanks to proper electronic shielding. The DC measurement is also satisfactory in the tested range with an accuracy of less than nA and a bandwidth of 1kHz. Preamplifier qualifications are planned within realistic experiments using a fission chamber placed in a neutron field.

Keywords — **Miniature Fission Chamber, Neutron Flux Measurement, Transimpedance amplifier, Low Current Measurement**

I. Introduction

The MONACO system [1][2] (Multichannel Online Neutron Acquisition in Campbell mOde) was developed at CEA during the past 10 years until the TRL7 level and demonstrated its efficiency as a functional prototype for Fission Chamber (FC) signal acquisition. The main feature of the system is to make possible to acquire simultaneously three different FC working mode: pulse, Campbelling and, current mode. Within the commissioning of the new Jules Horowitz nuclear reactor, this system will be integrated into the experimental part of the data acquisition system. The industrialization of the MONACO system is a result of a mutual efforts between CEA who developed the prototype and Instrumentation Technologies. The industrialized version of the MONACO system will be available on the market from 2023 and will be supported for at least 10 years aftermarket.

One of the key parts of the MONACO system is the frontend amplifier originally developed by CEA and based on analog components. Instrumentation Technologies redesigned the preamplifier to meet CEA performance requirements using new discrete and high-performance components.

The entire redevelopment process was divided into different phases, including a CEA specific PO for the first phase and a bilateral collaboration agreement for the remaining phases, to lower the risk connected to every step of industrialization:

- **Phase 1:** Redevelopment of the integrated preamplifier unit. This phase will consist of the redevelopment of the preamplifier frontend and of the modules that are critical for the MONACO system performance: the HV module, and the current measurement unit. This phase was successfully concluded and will be described in the next paragraphs of this paper.
- **Phase 2:** Integration of a 4-channel system based on the hardware prototype developed in phase 1. This phase will expand the 1-channel integrated frontend to a fully functional 4-channel system with the integration of a digitizer unit aimed to collect and process the digitalized data. The hardware developed within this phase will consist in the finalized version of the MONACO system First testing at a research reactor facility to confirm the hardware performance are planned in August 2021.
- **Phase 3:** Development of the Digital Signal Processing (DSP) interface on the digitizer module integrated into the 4-channel system.
- **Phase 4:** Development of the user Graphical User Interface (GUI) software and final testing of the device.

This paper will cover the first phase of the process where a functional pre-industrialization 1-channel version of the preamplifier was developed and tested.

https://doi.org/10.1051/epjconf/202125304015

II. Integrated Preamplifier Requirements

The MONACO system developed by CEA[MOL] is based on different building blocks that fulfill the final user application requirements. Each of the 4 channels is composed of a processing chain with the following blocks:

- Preamplifier frontend: aimed to collect the signal from the FC.
- High voltage source (independent for all the 4 channels) that polarizes the FC.
- Current measurement unit (independent for all the 4 channels) that measures the current provided to the MFC.
- Digital acquisition board and FPGA processing.
- Host PC with GUI.

The block system is well described in Fig.1.

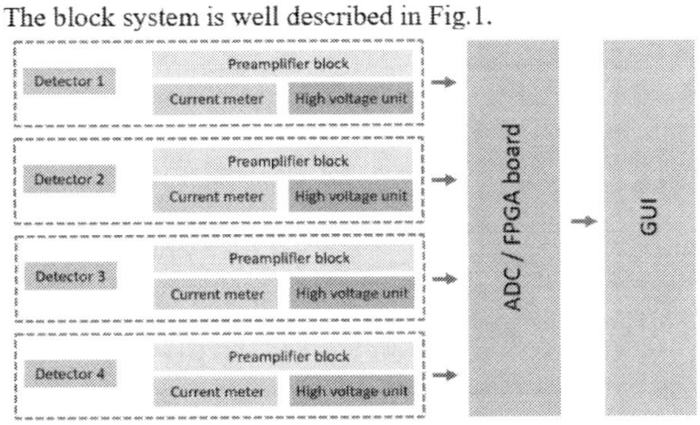

Fig. 1. The 4 channel MONACO system block schematics

In the following paragraphs, we will refer to the "preamplifier frontend" as the frontend unit which is converting and amplifying the signals from the FC, and to "integrated preamplifier" as the unit that was developed by Instrumentation Technologies that integrates the preamplifier frontend, the HV module, and the current meter.

Within Phase 1 , the goal is to provide a 1-channel prototype that can replicate the preamplifier frontend, high voltage source, and current meter unit integrated into a standalone system that can lead to the same or better results than the existing MONACO prototype developed by CEA.

The preamplifier development began with the review of the specifics of the actual MONACO system. As described in Fig. 1 the unit is composed of three separated modules, the preamplifier frontend, an HV unit, and a current measurement unit connected with wires in a common enclosure. All these blocks should work as an integrated subsystem, so the first choice was to integrate all the units on a common PCB to simplify the communication between the units and to improve EMC (noise immunity).

The preamplifier frontend converts the current signal retrieved from the FC detector into a voltage signal and amplifies it with some specific performance parameters in terms of gain and noise. The performance reference parameters were taken from the measurements of the existing CEA preamplifier.

The HV unit needs to deliver the polarization voltage to the detector with a ± 1 kV $\pm$ 1 % accuracy and a maximum of 1 mA output for each channel.

The current measurement unit provides an operative range from 10 nA to 1 mA with 1 nA of resolution at lower currents and integrates a digital sampling system with a sampling rate of 1 kS/s.

III. Integrated Preamplifier Development

In this paragraph, we will review the development of the preamplifier frontend, HV, and current meter unit. As reported before the starting point was the review of the performances of the existing CEA preamplifier. This unit seemed to be the most critical for the system performance since it is responsible for the HV decoupling, noise/disturbances immunity of the system, and overall performance in terms of bandwidth.

The block schematics of the integrated preamplifier is described in Fig.2:

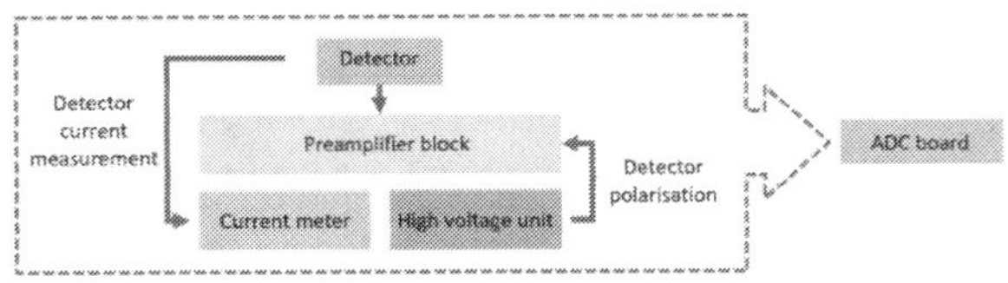

Fig. 2. Integrated preamplifier board concept

As mentioned before the choice to integrate all the modules on a common PCB will allow easier integration and interface debugging during the system development. Having a common interface for unit power supply and communication will also make easier the extension from the 1-channel to the 4-channel system.

A. Testing of the original CEA frontend preamplifier

The CEA preamplifier, delivered as a sealed box, consists of 1-channel electronics capable to collect the signal from the FC with given performance specifications and some additional features such as cable capacity immunity and the possibility to deliver also polarization voltage using an external high voltage source up to ± 1 kV. The CEA preamplifier parameters were measured at Instrumentation Technologies and are listed in Table I, these values will be the reference values for the I-Tech preamplifier development:

TABLE I
CEA Preamplifier Specifications

Feature	Specifications
Bandwidth	10 kHz – 18 MHz
Transimpedance gain	17500
Output noise	0.2 μA peak to peak
Cable capacitance immunity	~ 50 nF
Current limit	20 uA

The CEA preamplifier interfaces are composed of an HN input connector, two lemo output connectors, fine and coarse trimmers for gain adjustment and calibration, DC supply input. The electronics are packed and shielded in a high immunity enclosure.

The noise measured on the CEA preamplifier without any termination on the HN input was 40 mVpp. By attaching a 23 m long shielded high-immunity cable the total noise figure increased to 50 mVpp, demonstrating the strong cable capacity immunity of the preamplifier.

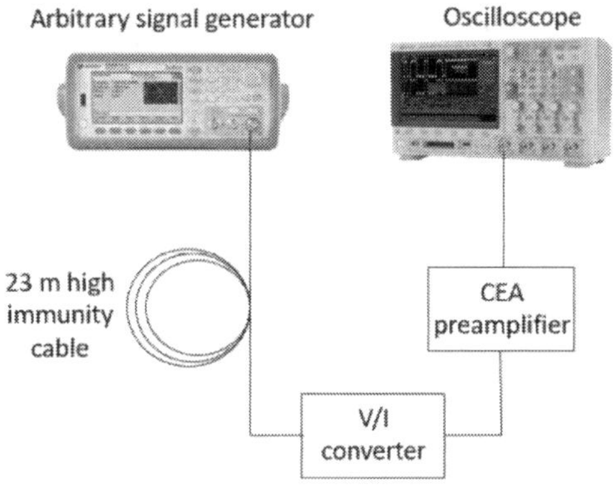

Fig. 3. The measurement setup of the CEA frontend preamplifier consisted of a Keysigth/Agilent arbitrary signal generator in combination with a Keysigth InfiniiVision DSOX3104T 1 GHz, 5 GSa/s oscilloscope

The next step was the evaluation of the effects of pulse signal reflection in the cable. Negative and positive pulses with different widths were provided to the preamplifier, from 20 ns to 200 ns width. The reflection time was influencing all the pulses depending on the pulse width. An example of a 20 ns pulse response is reported in Fig. 1:

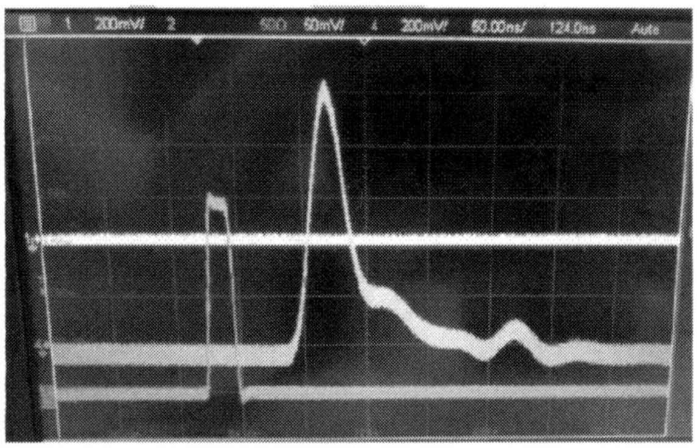

Fig. 4. A 20 ns positive pulse (blue) was delivered to the I/V converter through a 23 m high immunity long cable. The amplifier response (pink) presented a 10 ns rise time with a reflection occurring after the pulse bounced through twice the cable length.

The preamplifier S parameters were tested with an Agilent E5052B Signal Source Analyzer. The total bandpass was confirmed to be around 20 MHz as reported in the S21 parameter plot in Fig. 5.

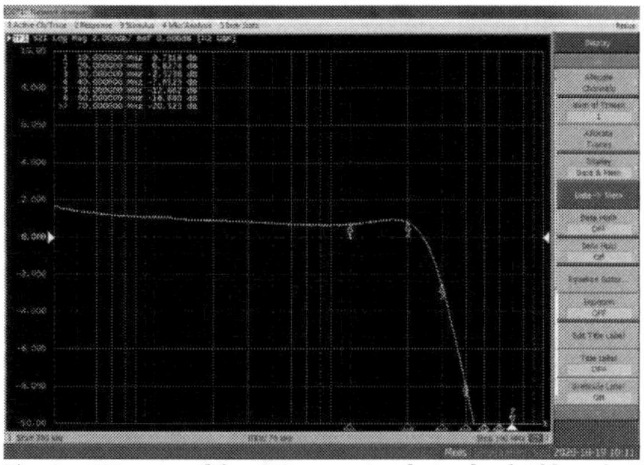

Fig. 5. *CEA preamplifier S21 parameter during bandwidth evaluation. Marker 2 is positioned at 20 MHz.*

B. I-Tech integrated preamplifier development

Based on the requirements listed in paragraph A, the Instrumentation Technologies preamplifier was redesigned by using discrete components. As described before, since this module needs to integrate all the key components of the final MONACO system (preamplifier, HV source, and current meter), special attention was given to noise immunity and DC-DC converter shielding. All these blocks will find their place on a dual-side PCB with a strong amplifier stage shielding as reported in Fig.6.
The preamplifier input is retrieved by a HN connector and a decoupling condensators allows the further stages to get only the AC component without the HV bias that is provided to the detector.

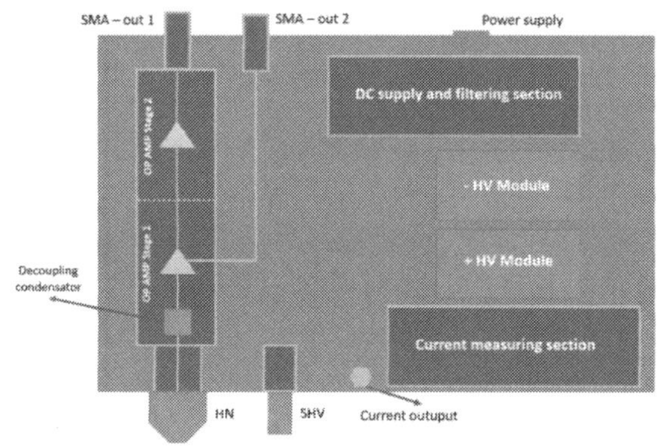

Fig. 6. Integrated preamplifier sections: the preamplifier chain is divided into two subchains and can provide the HV polarisation and the current measurement.

To investigate different amplification levels and to allow a quicker circuit debugging, the preamplifier frontend was split into two different gain outputs that were compared within the same pulse signal, both outputs are 50 Ohms matched. Different configurations between stages 1 and 2 lead to different pulse responses and noise figures. The HN connector shielding plays an important role in the prevention of external interferences. Additional connectors allow to retrieve the HV value (SHV connector) and the current that is provided from the HV modules.

C. I-Tech HV unit

The HV unit was designed using negative and positive HV modules that were combined in a common circuit separated by a relay . It allows the user to set the high voltage value within the +1kV to -1kV range. Strong filtering on the connector output was needed to prevent the ripple noise and to level the voltage level within the specifications. The addition of output leveling capacitors also leads to review the safety certification needs due to the amount of charge that can be released by the system.

D. I-Tech current meter unit

The current meter was designed using a shunt resistor system and by amplifying the voltage drop on this resistor with an operational amplifier chain. The main difficulties during the development of the unit came from the parasitic currents and the shielding of the unit. The measurement of few nA resolution within several hundred volts was possible by adding additional shielding to the unit and by isolating it within the PCB on a floating reference.

IV. Integrated System Prototype

All the units listed in the previous paragraphs (preamplifier frontend, HV module, current measurement module) were integrated on a single PCB board that also includes the power supply unit and part of the control logic. This proof of concept was also lately applied to the effective MONACO prototype by extending the setup to a 4-channel board and adding the ADC acquisition and control system.

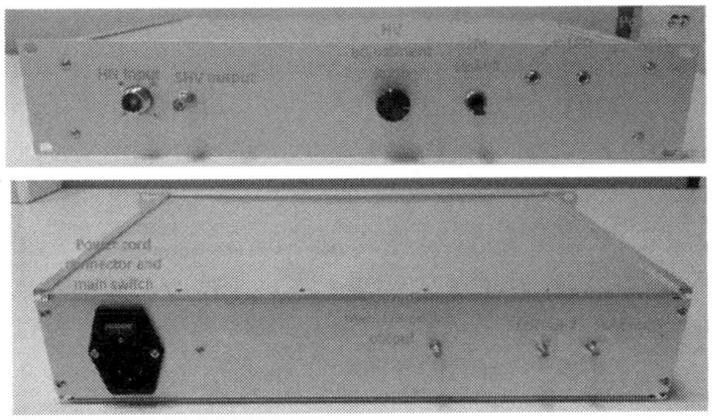

Fig. 7. Preamplifier integrated sections: the preamplifier chain is divided into two subchains and can provide the HV polarization and the current measurement.

The hardware enclosure allows the testing of the integrated preamplifier by shielding the inner components which are very sensitive to external perturbations. The power supply was also integrated in the final design. The input signal from the detector is delivered to the HN connector and the HV can be adjusted with an external knob. The outputs, as mentioned before, allow the user to evaluate the signal from the preamplifier stages (1 and 2) and from the current measurement section.

V. Performance Measurement

A. Preamplifier frontend evaluation

The system performance was evaluated by simulating the pulse train using an Agilent 33250A/ 80 MHz Arbitrary Waveform Generator and a Tektronix AWG5012 arbitrary 1 GHz signal generator. Within this setup it was possible to assess the bandwidth and the rise time of the preamplifier frontend and to confirm the absence of unexpected behaviors (e.g., noise, spikes, ringing…).

The pulse measurements were performed both for positive and negative pulses and the performance was comparable to the existing CEA preamplifier with an increased bandwidth so with a lower rising time of the measured pulses.

The preamplifier frontend bandwidth was later lowered to prevent the increase of the noise figure and of the cable ringing effect that was noticed lately during the preamplifier frontend comparisons.

Within the finalized prototype, the base noise level of the preamplifier was measured to be 40 mVpp but with some spikes coming from the DC/DC converters on the prototype board at approximately 100 kHz. With the use of ad-hoc low drop-out integrated into the DC supply module the total noise figure improved.

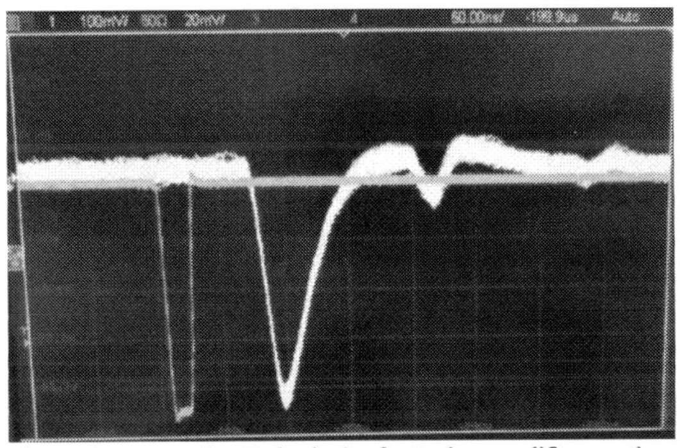

Fig. 8. Instrumentation Technologies frontend preamplifier negative pulse response. The oscillations after the pulse are due to the cable.

The S21 parameters showed a high increase in the amplifier bandpass:

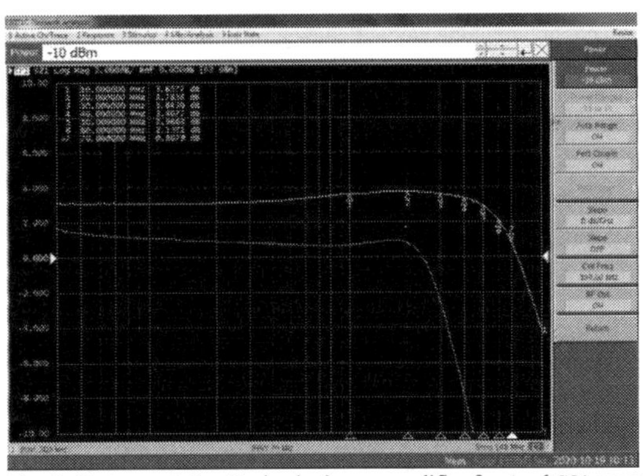

Fig. 9. Instrumentation Technologies preamplifier frontend S21 parameter (upper curve) vs CEA preamplifier frontend (lower curve)

By injecting white noise in the preamplifier frontend and recording the spectral response on an oscilloscope the bandwidth was confirmed. The addition of the 23m length high immunity mineral cable did not influence significatively the noise figure even if some reflections were noticeable. In any case, the cable reflections are not estimated as a showstopper for the implementation of the DSP algorithm that will be running on MONACO. With this test, the result confirmed the performance of the Instrumentation Technologies amplifier to be similar to the original CEA preamplifier frontend.

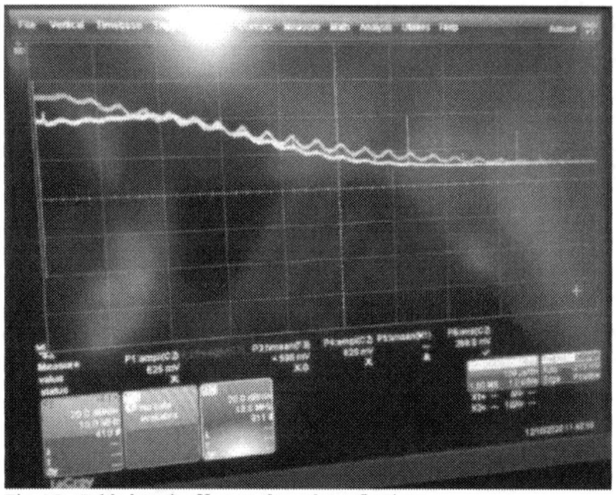

Fig. 10. Cable length effect on the pulse reflections.

B. High voltage module performance

The high voltage module was tested for high voltage stability and the presence of ripples. The overall performance was linked to the output capacitors' ability to smooth the incoming noise from the DC supply modules. The measurements showed a maximum of 1kV with a ripple below 60 mV.

C. Current meter module performance

The overall performance of the unit was measured to be in the range of 4nA and will be lowered in the next designs.

The operative range from 0 to 1kV was measured by comparing the current value measured by the current meter unit and the Keithley 6517B Electrometer/Resistance meter. The results are reported in Fig. 11 where strong linearity is seen between the two systems. A 100 MOhm resistor was connected to the HV output and the relative current was measured. The data relate very well in the entire span from 0 to -350 V with a final difference at the highest voltage and current value due to the prototype amplifier saturation (green box).

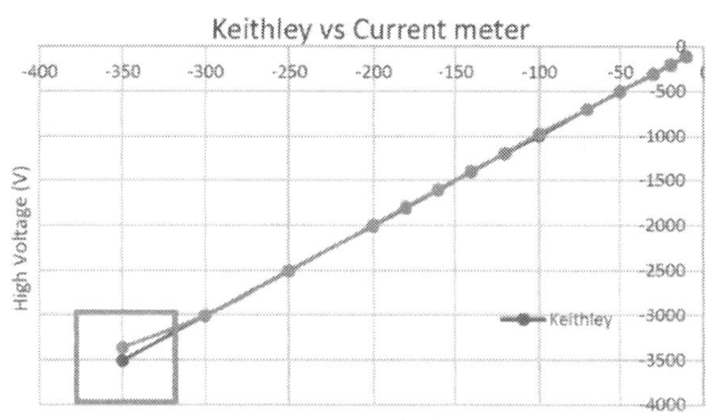

Fig. 11. I-Tech vs Keithley 6517 current measurement performance. A shunt resistor was polarized with the HV output and the relative current was measured from 0 to -350 V.

The bandpass of the current meter was measured using a sinusoidal signal from the Keithley current source, with a sweep from 50 Hz with measured amplitude 35.1 mV to 1kHz with measured amplitude 15.2 mV. This confirmed the bandpass point of -3dB of the current meter to be 1 kHz.

VI. Conclusions

The development of the integrated preamplifier prototype was the first milestone and the proof of capability for the redesign of the MONACO system's main building blocks. The results reported in the measurements allowed Instrumentation Technologies to evaluate all the critical points of the preamplifier, HV module, and current meter unit. Some of the main difficulties were due to the preamplifier design (cable capacitance…) and logistics due to the actual Covid19 pandemic (prototype testing, electronic components delivery).

The integrated preamplifier was also tested at the CEA facility with a neutron source, the results will be reported in the next publications.

The next steps of industrialization are already in progress with the foreseen delivery of the 4-channel integrated version of MONACO as a hardware prototype in July 2021. This unit will allow the confirmation of the 4-channel hardware architecture and the start of the CE certification procedure.

The MONACO industrialized version is planned to be available on the market in early 2023.

References

[1] L. Barbot, D. Fourmentel, G. De Izarra, C. Destouches, J-F. Villard, Y. Moline and H. Hamrita, "MONACO v2: Multipurpose and Integrated Data Acquisition System for On-line Neutron and Gamma Measurements", ANIMMA 2019, June 17-21, 2019, Portoroz, Slovenia.

[2] Y. Moline, A. Dabat-Blondeau, H. Hamrita, S. Klasa, G. Corre, J-M. Bourbotte, K. Boudergui, L. Barbot, C. Destouches, "MONACO: A Case Study of a New System on Chip and Stream Based Architecture for Nuclear instrumentation", ANIMMA 2019, June 17-21, 2019, Portoroz, Slovenia.

Study review of the CALORRE differential calorimeter: definition of designs for different nuclear environments

A. Volte[1], M. Carette[1], A. Lyoussi[2], G. Kohse[3], C. Reynard-Carette[1]

[1]*Aix Marseille Univ, Université de Toulon, CNRS, IM2NP, Marseille, France*
[2]*CEA/DES/IRESNE/DER, Section of Experimental Physics, Safety Tests and Instrumentation, Cadarache, F-13108, Saint Paul-lez-Durance, France*
[3]*Massachusetts Institute of Technology, Nuclear Reactor Laboratory, Cambridge, Massachusetts, USA*

Corresponding author: adrien.volte@univ-amu.fr

***Abstract*—This paper deals with the CALORRE differential calorimeter patented by Aix-Marseille University and the CEA in 2015. Firstly, the paper focuses on the presentation of the first prototype of CALORRE calorimeter qualified under real conditions during the MARIA irradiation campaign in 2015. Then, a review of the studies restricted to one CALORRE calorimetric cell realized thanks to experimental characterizations under laboratory conditions is detailed. Several configurations were studied to determine the influence of the cell height, its horizontal fin geometry and the nature of the material of its structure on its response for a calibration protocol: linearity, sensitivity, range, reproducibility, response time and absolute temperatures. Finally, within the framework of the new CALOR-I project, an optimization of the calorimeter assembly and its design were carried out in order to remove contact thermal resistances and provide a new configuration of CALORRE calorimeter suited for the in-core water loop of the MIT reactor (2 W.g^{-1}). The response of this new calorimeter is estimated thanks to thermal simulations.**

***Index Terms*—Calorimeter, Nuclear Absorbed Dose Rate, On-line Measurements, Calibration, Irradiation campaign.**

I. Introduction

The high neutron and gamma fluxes as well as strong displacements per atom characterize conditions of material testing reactors (MTRs). These research reactors constitute major support research facilities and allow experimental and real-condition studies of the behaviour of fuels and inert materials (vessel, reflector, cladding, etc.) in an extreme radiation environment. These studies are important as they lead to progress on the understanding of the accelerated ageing of materials and/or of phenomena for advanced scenarios up to accidental conditions and consequently they bring data for safety issues, the life span of existing nuclear power plants and their advancements with new concepts.

Therefore, new instrumentation is needed to measure online key parameters both before the experiments for the device design and during the experiments for the result interpretation. The construction of the Jules Horowitz Reactor, a new MTR of 100 MWth nominal power with unequalled performances in Europe (high fast neutron flux of 5.5×10^{14} n/(cm^2.s) from 1 MeV leading to a high accelerated ageing up to 16 dpa/year and a high nuclear absorbed dose rate up to 20 W.g^{-1} in aluminum), initiated new collaborative research work. Since 2009 Aix-Marseille University and the CEA (within the framework of the joint laboratory LIMMEX - Laboratory for Instrumentation and Measurement in Extreme Environments) and its IN-CORE program - Instrumentation for Nuclear Radiations and Calorimetry online in REactor) have been developing a new research topic. More precisely, they have been focusing on innovation in instrumentation and advanced measurement methods for the quantification of key nuclear parameters such as neutron and photon fluxes and nuclear absorbed dose rate, also called nuclear heating rate. The online measurement of this latter quantity requires specific sensors: non-adiabatic calorimeters. With regard to the state of the art, two distinct sensors are used in MTRs: French differential calorimeters (CALMOS, CARMEN or CALORRE type) and European single-cell calorimeters (such as gamma thermometers or KAROLINA-type calorimeters) [1-5].

These two types of calorimeter allow the quantification of the nuclear absorbed dose rate thanks to temperature measurements and preliminary calibration under non-irradiation conditions from steady thermal states in the case of integrated heating elements or from transient thermal states in other cases. One main objective targeted for French differential calorimeters within the framework of the IN-CORE program and more recently of the new research program, called CALOR-I, funded by Aix-Marseille University foundation (A*Midex) and involving the Nuclear Reactor Laboratory of the MIT and the CEA is the reduction of the sensor size with the CALORRE calorimeter.

The paper will present a review of recent work carried out on this calorimeter by AMU and the CEA in 2015. The work allows the design and the characterization of several CALORRE configurations over a wide range of nuclear absorbed dose rate (up to 20 W.g^{-1}).

The first part of the paper will describe the first design of CALORRE calorimeter fabricated for a range up to 1 W.g^{-1} and its first successful qualification under real conditions during an irradiation campaign inside the Polish MARIA reactor (in November 2015). The second part will be dedicated to the characterization of the response of 6 new configurations of CALORRE calorimetric cell by means of a comprehensive approach coupling experimental, theoretical and numerical work from laboratory conditions to nuclear environments. The experimental metrological characteristics of the 6

configurations, in terms of sensitivity, linearity, range, reproducibility and response time, will be detailed.
The last part will present new results obtained within the framework of the CALORI, program (2020-2022) which will allow the mapping of an in-core water loop of the MIT reactor in terms of nuclear absorbed dose rate by means of calorimeter for the first time. The criteria for the choice of a new configuration of CALORRE calorimeter will be detailed (mass, size, temperature, sensitivity). Then, the numerical responses under real conditions of a new CALORRE differential calorimeter assembly will be presented for three values of external heat transfer coefficient.

II. First CALORRE calorimeter prototype qualified under real conditions

One main objective of the irradiation campaign realized in 2015 inside the MARIA reactor was to qualify a first prototype of CALORRE calorimeter by comparing the obtained nuclear absorbed dose rate axial profile to those measured with other already qualified calorimeters such as CARMEN calorimeter.

A. The CALORRE calorimetric cell

Figure 1 shows schematics of the first type of CALORRE calorimetric cell which was developed tested under real conditions during the irradiation campaign in the MARIA reactor. This configuration had the same head as that of CARMEN calorimetric cell in terms of geometry, height (equal to 23.1 mm) and assembly with a heating element in order to compare the responses of these two sensor-designs under laboratory and/or real conditions.

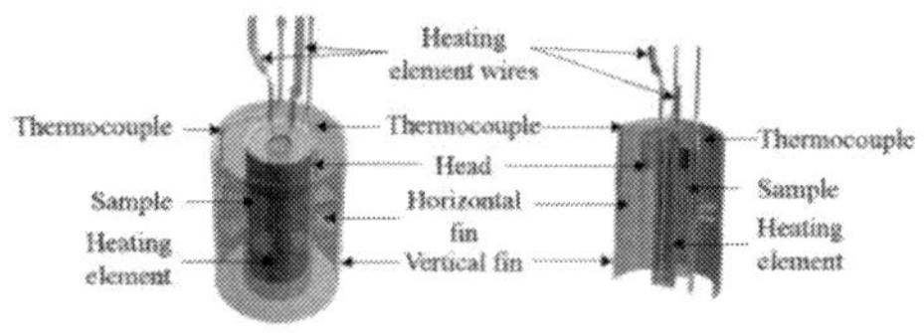

Fig. 1. A 3-D schemes of the first CALORRE calorimetric cell prototype (on the left-hand section) and of an internal view of this cell (on the right-hand section).

The heating element allows the simulation of the nuclear absorbed dose rate by Joule effect. The different parts of CALORRE calorimetric cell were already presented: a structure made of stainless steel AISI 316L which is composed of a head, a half horizontal-fin (8 metal sectors and 8 empty sectors with a same angle equal to 22.5 °) and a vertical fin , and an heating element with its holder, a sample in the case of the measurement cell and two K-type thermocouples to measure a temperature difference between two key points at hot and cold temperatures (T_{hot} and T_{cold}) respectively [1, 6]. This CALORRE calorimetric cell is called full-height and half horizontal-fin.

B. The CALORRE calorimeter

The CALORRE sensor is composed of two superimposed full-height and half horizontal-fin calorimetric cells. These identical cells host a graphite sample in the case of the measurement cell and gas for the reference cell. The reference cell is needed to remove the effect of the energy deposition on the structure of the measurement cell, the heating element, the heating element holder and the thermocouples.

Figure 2 gives a diagram of the complete CALORRE differential calorimeter. The two calorimetric cells are superimposed through the use of different thin spacers inside a jacket and filled with di-nitrogen. This external additional jacket is made of stainless steel (internal diameter equal to 17.0 mm and a thickness equal to 0.5 mm).

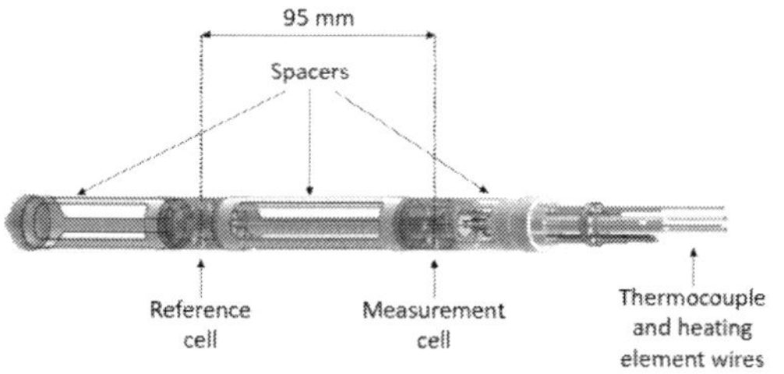

Fig. 2. 3-D scheme of the CALORRE differential calorimeter without its external additional jacket.

During the MARIA campaign, the distance between the two CALORRE cells was fixed to 71.9 mm (leading to an inter-sample space of 95 mm), as seen in the Figure 2, in order to reproduce the distance existing between the two heads of the usual CARMEN differential calorimeter.

C. Calibration of a differential calorimeter under laboratory conditions

For each calorimetric cell (reference and/or measurement), a preliminary calibration step under laboratory conditions (without nuclear rays) is crucial to obtain its calibration curve. This kind of curve represents the difference of the mean steady temperatures between the two key points (T_{hot} -T_{cold}) versus the electrical power generated by the heating element integrated inside the cell head.

In the case of the first prototype of the CALORRE differential calorimeter, the two calorimetric cells were calibrated simultaneously. Thus, the calorimeter was inserted inside a Polish bench with a fluid flow vein at fixed temperature (40 °C) and velocity (corresponding to a Reynolds number equal to 17340) [7]. Then, different successive electrical current values were injected to each heating element in order to generate a power range from 0 to 6 W with an increment of 1 W in each calorimetric cell.

The responses of the two cells of the differential calorimeter are different leading to two different calibration curves:

$$(T_{hot} - T_{cold})_{measurement} = \Delta T_S = -0.42P^2 + 21.09P \quad (1)$$

$$(T_{hot} - T_{cold})_{reference} = \Delta T_R = -0.41P^2 + 19.41P \quad (2)$$

with P the electrical power injected in W.

D. Operating protocol and MARIA irradiation campaign results

The calorimeter CALORRE was inserted into the experimental channel H-IV A thanks to a hollow cylindrical pipe (24.0 mm in internal diameter). The device was moved by applying an axial increment equal to 95.0 mm (space induced by the position between the heads of the two superimposed calorimetric cells) from -463.0 mm to 487.0 mm in order to obtain a mapping of the nuclear absorbed dose rate on ten axial positions. After each increment, 25 minutes were waited to reach a steady state inside the calorimeter. The nuclear absorbed dose rate at each axial position (z) is obtained by taking the two calibration curves into account, and by measuring the mean steady temperature difference for the measurement cell located at the z-position (step i) and the mean steady temperature difference for the reference cell moved at the same z-position (step i+1).

Figure 3 shows the axial profile of the nuclear absorbed dose rate obtained with the first CALORRE calorimeter prototype having full-height and half horizontal-fin calorimetric cell configuration. This profile and the profiles obtained with the other calorimeters were similar in term of shape once a normalization by the maximal value is performed.

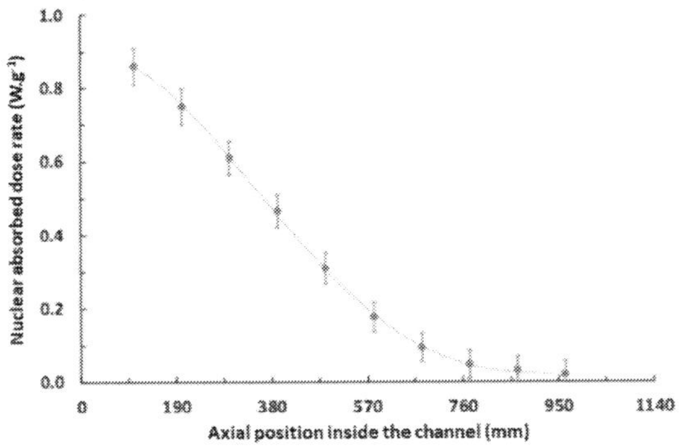

Fig. 3. Nuclear absorbed dose rate profile inside the H-IV A experimental channel of MARIA reactor.

In conclusion, the first CALORRE calorimeter prototype was well behaved under real conditions. Consequently, a parametrical study on new configurations of calorimetric cell were performed in order to show the influence of different parameters, such as the structure material nature, the height of the cell and the horizontal-fin design, on the cell response. These results are presented in the next part.

III. Parametrical study under laboratory

In this section, the responses and metrological characteristics of 7 configurations of calorimetric cell (including that used during the MARIA irradiation campaign, called configuration N°1) are presented in Table 1 by precising the height of the cell (H), the structure material nature (Material), the horizontal-fin geometry (Fin), the response time (t), the calibration curve coefficients A_1 and A_2 (of first and second orders respectively), the sensitivity at 6 W (S_{6W}), the total mass of the head structure (including shim, heating element and heating-element holder) and the sample (m), the absolute temperature located at the hot spot (T_{hot}) and the simulated nuclear absorbed dose rate (E_n) under laboratory conditions. There were obtained with the same set-up and operating protocol: only one calorimetric cell (the measurement cell) hosted by an external jacket [6] for same applied experimental conditions (T_{fluid}=33 °C and Re=1607) and with an electrical power range up to 6 W.

As expected, the absolute temperature, the non-linearity and the sensitivity decreased with the increase of the number of metal sectors within the horizontal-fin geometry and with a higher thermal conductivity of the structure material [6]. The change in cell height (11.55 mm instead of 23.1 mm) did not impact significantly the sensitivity (in °C.W^{-1}) under laboratory conditions but reduced the mass of the cell and so the mass of the sample as well as the reached absolute temperature and the response time. On the other hand, the reduction in sample mass led to a lower sensitivity (in °C.g.W^{-1}) under real conditions. This reduction in sensitivity under real conditions was confirmed by a validated predicted model based on a heat balance [8]. For example, the configuration N°1, had a sensitivity of 88.9 °C.g.W^{-1} as against 49.6 °C.g.W^{-1} for the configuration N°6 (same characteristics except for its reduced height) at 1 W.g^{-1}.

TABLE I Metrological characteristics for seven configurations of a CALORRE calorimetric cell

Configuration	N°1	N°2	N°3	N°4	N°5	N°6	N°7
H (mm)	23.1	23.1	23.1	23.1	23.1	11.55	11.55
Material (-)	AISI 316L	Al 5754	Al 5754	AISI 316L	AISI 316L	AISI 316L	TA6V
Fin (-)	1/2	1/4	1/2	1/4	1	1/2	1/4
t (s)	291	97	88	418	208	174	287
A_1 (°C.W^{-1})	22.65	5.04	2.25	30.03	11.05	18.70	48.15
A_2 (°C.W^{-2})	-0.54	-0.02	-0.01	-0.82	-0.17	-0.30	-1.78
S_{6W} (°C.W^{-1})	16.2	4.8	2.1	20.3	9.0	15.1	26.8
m (g)	5.22	3.20	3.20	5.22	5.22	2.92	2.16
T_{hot} (°C)	169	72	62	209	122	157	272
E_n (W.g^{-1})	1.15	1.90	1.90	1.15	1.15	2.05	2.80

This parametrical study under laboratory conditions showed a high modularity of the CALORRE calorimetric cells: sensitivity from 2.1 °C.W^{-1} to 26.8 °C.W^{-1} at 6 W leading to a wide measurement range. The aluminum configurations are suited to high nuclear absorbed dose rate measurement [8]. More precisely the full-height and half-surface aluminum configuration (configuration N°3) was qualified under laboratory conditions up to 60 W thanks to a new heating element system. This very large calibration range allowed the demonstration of the good running calorimetric cell for the maximal value of the nuclear absorbed dose rate inside the future JHR (20 W.g^{-1}). The titanium configuration (configuration N°7) was the more sensitive configuration but had the most non-linear response (A_2=-1.78 °C.W^{-2}) and the highest absolute temperatures (272 °C).

After this experimental parametrical study, a validated 3-D thermal model under real conditions allowed the estimation of the responses for a whole differential calorimeter assembly and its reduction in height by decreasing the important inter-cell space (13.2 mm instead of 71.9 mm) as well as the height of the cell (11.55 mm).

IV. New design of a specific MIT-R CALORRE configuration

In this part, work conducted for the new CALOR-I research program, will be detailed. The two main aims are the design of a new compact CALORRE calorimeter prototype and the mapping of an in-core water-loop in the MIT reactor in term of nuclear absorbed dose rate.

The core of the MIT reactor has a height of 56 cm and a diameter of 38 cm. The main maximal features are a thermal power of 6 MW, a thermal neutron flux of 3.6×10^{13} n.cm^{-2}.s^{-1}, a fast neutron (E > 1 MeV) flux of 1.2×10^{14} n.cm^{-2}.s^{-1} and an expected nuclear absorbed dose rate of 2 W.g^{-1} in titanium [9]. The temperature of the fluid flow as well as its velocity can be changed inside the in-core water-loop. By considering the constraints of the experimental set-up used for the calibration under laboratory conditions [6, 8], the in-core water-loop fluid flow temperature will be fixed around 50 °C but the required speed of the fluid flow will be studied to ensure energy evacuation.

A. The new CALORRE cell design

Regarding the previous irradiation campaign results, recent work [8], the results obtained for 6 other calorimetric cells and the expected nuclear absorbed dose rate inside the MITR [9], a new CALORRE cell design was defined and studied thanks to 3-D numerical thermal simulations. The aims were to remove the thermal contact resistances by means of a simplification of the assembly and the head and to reduce the total height of the calorimeter.

As for a configuration with a reduced-height calorimetric cell (11.55 mm) and a half horizontal-fin design made of stainless steel, the experimental results under laboratory conditions and the predicted response under real conditions showed low absolute temperatures, low response time, a linear response and a good sensitivity, the configuration N°6 was chosen. But its vertical fin was increased in order to remove and replace the additional jacket which induced thermal contact resistances. Moreover, the sample and the head structure were made from a single block to avoid heater holders and other thermal contact resistance. Finally, the vertical fins of the measurement and the reference cells will be welded in order to eliminate the spacers and to obtain the whole calorimeter (cf. Fig 4). An inter-cell space of 18 mm (leading to an inter-sample space of 29.55 mm) and a total height (excluding nose and cap) of 73.7 mm were defined.

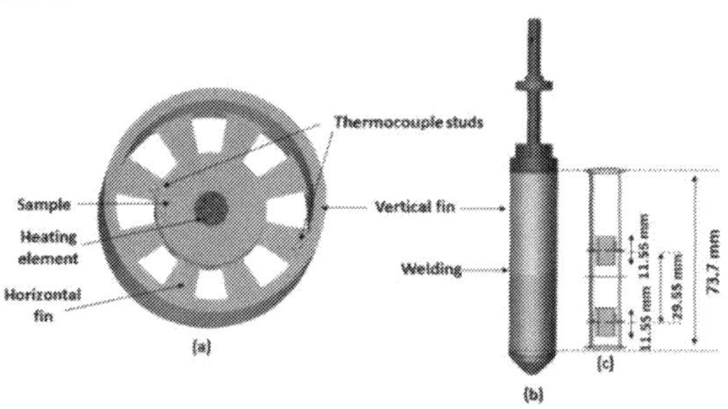

Fig. 4. Diagrams of the new CALORRE measurement calorimetric cell (a), the new assembly (b) and the geometry (c) considered for the 3-D numerical thermal simulations.

B. Study of the response under real conditions by 3-D numerical thermal simulations

The response of the new calorimeter design under real conditions (up to 2 W.g^{-1}) was studied by considering only the two cells and thanks to the validated 3-D thermal model [8] (cf. Fig 4). A study of the influence of the heat transfer coefficient on the response, maximal and wall temperature was carried out for a fluid flow temperature equal to 50 °C.

The Figure 5 shows the numerical responses of the calorimeter versus the nuclear absorbed dose rate for three values of the external heat transfer coefficient. There is a low influence of the heat transfer coefficient on the calorimeter response. As expected, a non-linear response is obtained $\Delta\Delta T=-2.51E_n^2+55.38E_n$ leading to a sensitivity of 45.3 °C.g.W^{-1} at 2 W.g^{-1} for a heat transfer coefficient of 10000 W.m^{-2}.K^{-1}. In contrast, the maximal temperature in the calorimeter and the maximal wall temperature decrease from 298 °C to 275 °C and from 85 °C to 56 °C respectively when the heat transfer coefficient varies from 1000 to 10000 W.m^{-2}.K^{-1}. The temperature field for a heat transfer coefficient of 10000 W.°C^{-1}.m^{-2} at 2 W.g^{-1} is provided in Figure 5.

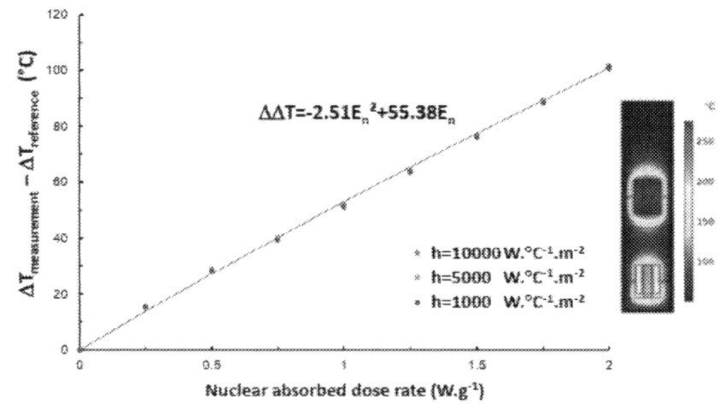

Fig. 5. 3-D numerical responses of the new design of CALORRE calorimeter for three values of the external heat transfer coefficient (on the left-hand section) and the temperature field for a nuclear absorbed dose rate equal to 2 W.g^{-1} (h=10000 W.°C^{-1}.m^{-2} and T_f=50 °C) (on the right-hand section).

V. Conclusions

A first prototype of the new CALORRE differential calorimeter was fabricated and qualified successfully under real conditions during an irradiation campaign in the MARIA reactor in 2015. Then, this irradiation campaign allowed the validation of predictive and 3-D models under real conditions. Then, 6 new configurations of CALORRE calorimetric cell were studied under laboratory conditions to show the influence of the structure material nature, the height of the cell and its horizontal-fin design on the cell response. One of these configurations suitable for JHR conditions up to 20 W.g^{-1} was qualified under laboratory conditions up to 60 W. A new calorimeter assembly dedicated to this maximal nuclear absorbed dose rate was designed and then optimized by reducing its total height. Finally, a new design of CALORRE calorimeter for future experiments in the MIT reactor was studied under real conditions thanks to the validated 3-D thermal model. The new calorimeter defined will have a reduced height and the thermal contact resistances previously observed will be removed thanks to new cells without heating-element holders in particular, and a new calorimeter assembly without spacers and without an additional jacket. A good behaviour and the influence of the heat transfer coefficient on the response and the maximal and wall temperatures were observed and determined numerically for this new calorimeter under real nuclear conditions.

The outlooks are on the one hand the simulations of the interactions between radiations and matter with the MCNP Monte-Carlo transport code and nuclear data library in order to define the local heat sources in each part of the new calorimeter for MIT-R conditions and on the other hand the experimental characterization of this new sensor under laboratory conditions and then under MIT-R conditions.

ACKNOWLEDGMENT

"The CALOR-I project leading to this publication has received funding from the Excellence Initiative of Aix-Marseille University - A*Midex, a French "Investissements d'Avenir" programme".

REFERENCES

[1] J. Brun, M. Tarchalski, C. Reynard-Carette, K. Pytel, A. Lyoussi, J. Jagielski, D. Fourmentel, J-F. Villard, M. Carette, "Responses of Single-Cell and Differential Calorimeters: From Out-of-Pile Calibration to Irradiation Campaigns," IEEE Trans. Nucl. Sci., vol.63, no. 3, pp. 1630 1639, Jun. 2016.

[2] A. Luks, K. Pytel, M.Tarchalski, N. Uzunow, T. Krok, "Modelling of thermal hydraulics in a KAROLINA calorimeter for its calibration methodology validation", NUKLEONIKA, vol 61, no. 3, pp 453-460, Oct. 2016.

[3] C. Reynard-Carette, G. Kohse, J. Brun, M. Carette, A. Volte, A. Lyoussi, "Review of nuclear heating measurement by calorimetry in France and USA", Proc. ANIMMA 2017, Jun. 2017, EPJ Web of Conferences 170, 04019 (2018).

[4] H. Carcreff, L. Salmon and F. Malouch, "Recent developments in nuclear heating measurement methods inside the OSIRIS reactor", Nucl. Instrum. Methods Phys. Res. A, Accel. Spectrom. Detect. Assoc. Equip., vol. 942, pp. 1-19, Oct. 2019.

[5] R. Van Nieuwenhove, L. Vermeeren, "Nuclear heating measurements by gamma and neutron thermometers", in Proc. ANIMMA, Jun. 2019, EPJ Web of Conferences 225, 04003 (2020).

[6] A.Volte, C. Reynard-Carette, A. Lyoussi, J. Brun, M. Carette, "Study of the Response of a New Compact Calorimetric Cell for Nuclear Heating Rate Measurements", IEEE Trans. Nucl. Sci., vol. 65, no. 9, pp. 2461-2470, Sept. 2018.

[7] M. Tarchalski, K. Pytel, P. Siréta, A. Lyoussi, J. Jagielski, C. Reynard-Carette, C. Gonnier, G. Bignan, "Principle of calibration of the simple calorimeter for nuclear heating measurements in MARIA reactor and transposition to the case of JHR reactor", Proc. ANIMMA 2013, Jun. 2013.

[8] A.Volte, J. Brun, A. Lyoussi, M. Carette, C. Reynard-Carette, "Qualification of a New Differential Calorimeter Configuration Dedicated to Nuclear Heating Rates up to 20 W.g-1", IEEE Trans. Nucl. Sci., vol. 67, no. 11, pp. 2405-2414, Nov. 2020.

[9] S. Kim, D. Carpenter, G. Kohse, L. Hu "Hydride fuel irradiation in MITR-II: Thermal design and validation results", Nucl. Eng. Des., 277 (2014), pp. 1-14.

Characterization of calorimeter responses under laboratory conditions thanks to an optimized transient thermal test bench

J. Rebaud[1], A. Volte[1], M. Carette[1], A. Lyoussi[2], G. Kohse[3], R. Prokopowicz[4], C. Reynard-Carette[1]

[1]Aix Marseille Univ, Université de Toulon, CNRS, IM2NP, Marseille, France
[2]CEA/DES/IRESNE/DER, Section of Experimental Physics, Safety Tests and Instrumentation, Cadarache, F-13108, Saint Paul-lez-Durance, France
[3]Massachusetts Institute of Technology, Nuclear Reactor Laboratory, Cambridge, Massachusetts, USA
[4]National Centre for Nuclear Research, Otwock, Poland

Corresponding author: jeremy.rebaud@etu.univ-amu.fr

Abstract - This paper presents an experimental study of the calibration of a non-adiabatic single-cell calorimeter called KAROLINA. This type of sensor is used to quantify the nuclear heating rate inside MTRs (Material Testing Reactors). The calibration of this sensor type is carried out in laboratory conditions without irradiation. The calibration corresponds to a preliminary step, which is crucial to use this sensor in real conditions. In the case of the studied single-cell calorimeter owning no heating element, the calibration is realized with a specific test bench, called BERTAN, based on thermal transient regimes. First of all, the paper describes the methodology of the determination of the calorimeter sensitivity and the dedicated experimental set-up developed recently. Then the paper is devoted to a parametric study carried out on the KAROLINA single-cell calorimeter to determine the influence of several parameters on its thermal time constant (temperature, velocity, phase, method).

Keywords — Nuclear Heating Rate, Calorimetry, Calibration

I. Introduction

The design of future nuclear power reactors requires the study and understanding of the behavior of advanced material and fuel expected to be used within. One of the means of study is the use of Material Testing Reactors. These reactors, specially dedicated to research, allow the study of the ageing of the materials, the nuclear fuel behavior but also the testing of innovative in-pile sensors. These sensors are required to quantify accurately online key quantities in order to lead to the best understanding of nuclear phenomena. The nuclear heating rate is one of these key quantities. It corresponds to an energy deposition rate per unit of mass (W.g^{-1}) and is measured with heat-flow calorimeters. This type of non-adiabatic sensors is based on the measurement of temperature increase due to the large amount of energy deposition induced by the radiation-matter interactions. There are two families of these sensors [1]: the differential calorimeter [2-4] and the single-cell calorimeter *[5]*. There are many similarities between these two families: a sample, an inert gas inside a jacket, and K-type thermocouple(s). However, there are two main differences: the first one is the number of cells that composes the sensor (one for the single-cell calorimeters and several for the differential calorimeters (sample cell(s) and reference(s))), and the second one, is the presence of heating elements inside the differential calorimeters. These differences lead to advantages and drawbacks of use for each family (cf. Table I).

TABLE I: ADVANTAGES IN GREEN AND DRAWBACKS IN RED OF TWO HEAT-FLOW CALORIMETER FAMILIES

	Differential calorimeter	Single-cell calorimeter
Size	≥ 200 mm	≤ 72 mm
Wire number	≥ 9	≤ 2
Measurement methods	3	1
Range	≤ 20 W.g^{-1}	≤ 20 W.g^{-1}
Measurement Time	Response time + Stabilization time + One displacement at least	Response time + Stabilization time
Sample height	≤ 20 mm	≥ 15 mm
Calibration methods	Steady	Transient

Because of these different designs, there are two different calibration methods (cf. Fig. 1) [6]. Indeed, to calibrate the differential calorimeter, an electrical current is injected into each heating element in order to generate a Joule effect, which will allow the simulation of the nuclear heating rate. The lack of heating elements inside the single-cell calorimeters requires another calibration method. This other calibration method is based on a thermal transient protocol for which the calorimeter passes rapidly from a medium 1 at a temperature 1 to a medium 2 at a temperature 2 (cf. Fig. 1).

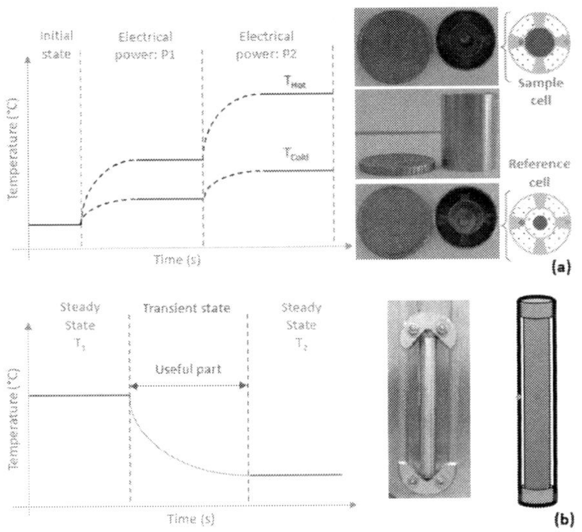

Fig. 1 Calibration protocols for: (a) Differential calorimeters; (b) Single-cell calorimeters.

The calibration protocol for the single-cell calorimeters under laboratory conditions was first implemented with a manual transfer of the calorimeter from an oven to a water bath. This method has three main drawbacks. Firstly, it does not allow reproducible tests. Secondly, it leads to a transfer time that can be long. Lastly due to the test protocol, only one type of transient state is exploitable (hot source to cold source) [7]. The automation and improvements of this calibration method led to the design of the BERTRAN test bench, which gave conclusive preliminary results [8]. Consequently, a new test campaign was then set up in order to study the calibration of a single-cell calorimeter called KAROLINA with various controlled experimental conditions.

The paper will focus on this parametrical study. More precisely this paper will first describe the single-cell calorimeter KAROLINA. Then it will present the methodology for the determination of the thermal time constant (k) and thus the sensitivity of the calorimeter. Afterwards, it will detail the BERTRAN bench used for the experimental tests, and the applied protocol. Finally, the parametric study will be shown and the analysis of the results for a test campaign will be discussed in order to define the influent factors on the thermal time constant value.

II. KAROLINA CALORIMETER

The KAROLINA calorimeter is a single-cell calorimeter developed by the National Center for Nuclear Research (NCBJ) to carry out experiments for the mapping of nuclear heating rate under real conditions in the MARIA research reactor [9].
The calorimeter shape is a cylinder with a height of 72 mm and an external diameter of 9.4 mm (cf. Fig. 2). A graphite sample is located in the center of this calorimeter, surrounded by a thin layer of inert gas in order to thermally isolate the sample from the outside environment and to impose a specific sensitivity. This sample is held in the center with two sample holders made of zirconium oxide which have a porous composition in order to minimize their impact by having a thermal conductivity close to the gas. The assembly is packaged in a stainless-steel jacket with an external diameter of 9.4 mm and a thickness of 0.5 mm. Regarding the instrumentation of this sensor, it has two K-type thermocouples of 0.5 mm diameter. The two thermocouples are located at the same height. The first one is in the middle of the sample (named sample thermocouple), and the second one is welded on the outer wall of the jacket (named fluid thermocouple).

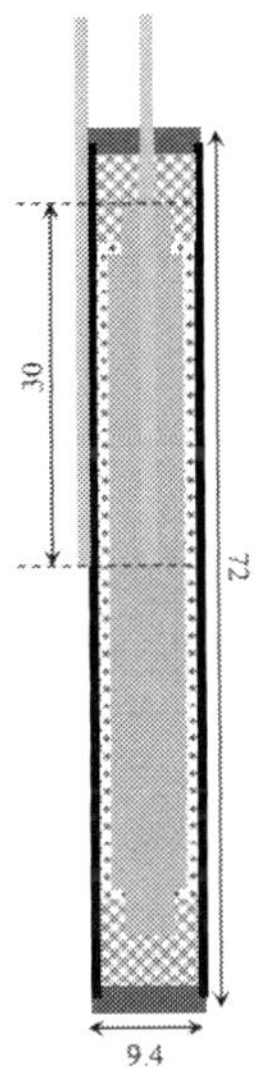

Caption	Name	Compostion
	Jacket	Stainless stell
	Cover	Stainless stell
	Sample holder	Oxide Zirconium
	Sample	Graphite
	K-type thermocouple	Chromel and Alumel wires + Mineral insulator + Stainless steel (jacket)
	Gas	Helium

Fig. 2 Scheme of KAROLINA single-cell calorimeter (dimensions expressed in millimeters).

III. METHODOLOGY FOR THE DETERMINATION OF THE SENSITIVITY OF A SINGLE-CELL CALORIMETER

A 1-D theoretical study of the KAROLINA calorimeter response, built thanks to a thermal balance taking into account some hypotheses, allows the determination of the temperature of the calorimeter sample versus time under irradiation conditions.
The hypotheses are:
- No thermal radiation,
- Heat exchange only on the radial axis,
- Isothermal sample,
- Isothermal wall,
- No contact thermal resistance,
- Energy deposition only into the sample.

The heat balance applied on the sample is given by this equation (1) (the parameters and their units are in Table II):

$$m_s\, E_n = \rho\, C_p\, V\, \frac{\partial T_s}{\partial t} + \left(\frac{T_s - T_f}{R}\right) \qquad (1)$$

The resolution of this equation leads to the temporal response of the calorimeter under irradiation conditions (2):

$$T_s(t) = T_f + E_n\, \frac{k}{C_p} + \left(T_{s_0} - T_f - E_n\, \frac{k}{C_p}\right) e^{-\frac{t}{k}} \qquad (2)$$

with $k = R\,\rho\,C_p\,V$ which corresponds to the calorimeter thermal time constant and T_{s_0} the initial sample temperature.
When the steady state is reached for a given nuclear heating rate ($t \to +\infty$) at a specific location inside the reactor, the expression (2) becomes:

$$E_n = \frac{(T_{s+\infty} - T_f)C_p}{k} \quad (3)$$

Consequently, the nuclear heating rate can be deduced from a measurement of steady temperatures and knowing its sensitivity which is equal to:

$$S = \frac{k}{C_p} \quad (4)$$

TABLE II
UNITS AND SYMBOLS FOR USED PARAMETERS

Symbol	Quantity	Unit
m_s	Sample mass	g
E_n	Nuclear Heating rate	$W.g^{-1}$
ρ	Sample density	$kg.m^{-3}$
C_p	Sample specific heat	$J.kg^{-1}.K^{-1}$
V	Sample volume	m^3
T_f	Fluid temperature	K
T_s	Sample temperature	K
R	Total thermal resistance between the sample and the fluid	$K.W^{-1}$
t	Time	s
k	Thermal time constant	s
S	Sensitivity	$kg.K.W^{-1}$

The value of the specific heat of the sample versus the temperature can be determined by using a Differential Scanning Calorimeter [10]. The thermal time constant can be measured under laboratory conditions thanks to a thermal transient protocol moving the calorimeter from a bath N°1 with a temperature T_1 to a bath N°2 with another temperature T_2.
Under laboratory conditions, out-of-pile, there is no nuclear heating rate in the sample ($E_n = 0$) and consequently the equation 1 becomes:

$$\rho\,C_p\,V\,\frac{\partial T_s}{\partial t} + \left(\frac{T_s - T_f}{R}\right) = 0 \quad (5)$$

At the beginning of the experiments when the calorimeter is immersed in the bath N°2 at $t=t_1$, the calorimeter sample temperature is equal to T_{s1}. The resolution of the equation (3) by taking this initial condition into account leads to this expression (6):

$$\frac{T_s - T_f}{T_{s_1} - T_f} = e^{-\frac{t}{k}+\frac{t_1}{k}} \quad (6)$$

The expression (6) can be transformed:

$$\ln\left(\frac{T_s - T_f}{T_{s_1} - T_f}\right) = -\frac{t}{k} + \frac{t_1}{k} \quad (7)$$

Two values of k can be deduced by plotting this expression (7) with the temporal results in temperature (cf. Fig. 3) and by using a linear fit to calculate the slope ($a = -\frac{1}{k}$) and the y-intercept ($b = \frac{t_1}{k}$).
Another method is to plot directly the sample temperature versus time. The thermal time constant corresponds to the response time of the sensor at 1 τ (63%) or one third of the response time at 3 τ (95%).

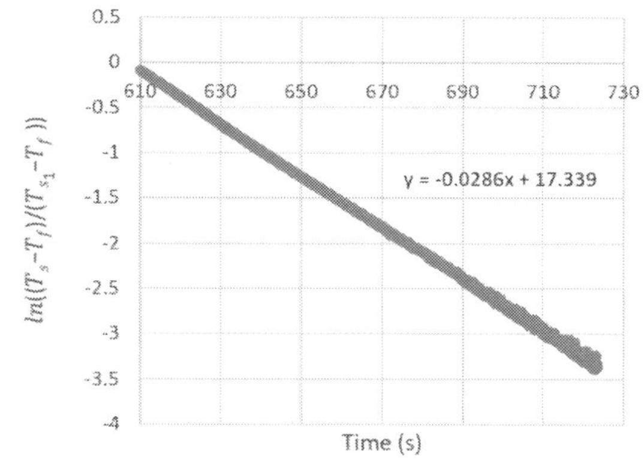

Fig. 3 Temporal curve representing the linearization of the transient state during the heating phase step with a propeller velocity of 1000 rpm and with hot and cold temperatures of 250 °C and of 31°C respectively.

IV. PRESENTATION OF THE TEST BENCH AND THE OPERATING PROCEDURE.

The calibration of the single-cell calorimeter KAROLINA has been performed with a test bench, named BERTRAN, specially designed to apply a thermal transient protocol in an automated way.

A. BEnch Regime TRANsient (BERTRAN)

The purpose of the BERTRAN test bench is to determine the transient response of the calorimeter when it changes rapidly from boundary conditions in temperature (from hot temperature to cold temperature or vice-versa). For this, it is necessary to have two media at two different temperatures and a mechanism to transfer the sensor automatically between these two media. These media are generated by means of two thermostatic baths filled with silicone oil. The silicone oil is chosen as heat transfer fluid because it as can be used up to 250 °C at atmospheric pressure. The two thermostatic baths are dedicated to having a cold environment (called cold bath) and a hot environment (hot bath) respectively.
The transfer mechanism is composed of a rail on which a pole is fixed and a mechanical drive system. At the end of the pole is put the KAROLINA calorimeter. An infrared sensor allows the control of the calorimeter position and its transfer. The transfer time of the calorimeter is around 1.5 s.
These main elements of the BERTRAN bench are given in Figure 4.

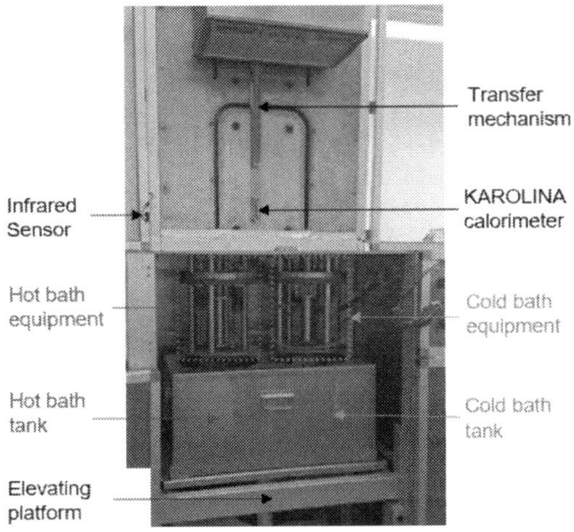

Fig. 4 Picture of calibration bench BERTRAN.

Each thermostatic bath has two parts (cf. Fig. 4 and Fig. 5). On the one hand, there is a silicon oil tank and on the other hand, there is an equipment, which allows the heating or the cooling of the silicone oil, the temperature stabilization and homogenization, and the generation of a downward forced convection.

Fig. 5 Picture of the two bath-systems.

The two bath-systems are quite similar. They differ only in the elements within their temperature range (3 heating cartridges for the hot bath against a heating cartridge and a coil connected to a cryostat for the cold bath). These elements associated to temperature sensors and regulators allow the regulation of the bath temperature to reach a steady state in each bath. Moreover, in the center of each bath is placed a cylindrical cavity (vein) which simulates an experimental channel of reactor by its inner diameter of 34 mm. At the bottom of each vein is located a propeller which causes a downward fluid flow in order to impose similar velocity and heat exchange conditions than those occurring in an experimental channel. In each vein there is a K-type thermocouple to control the temperature of the fluid flow in the vicinity of the sensor.

TABLE III: ACQUISITION CHANNELS OF BERTRAN BENCH

	Sensors	Quantity	Location
Channel 1	K-type thermocouple	Temperature (°C)	KAROLINA external jacket
Channel 2	K-type thermocouple	Temperature (°C)	KAROLINA sample
Channel 3	K-type thermocouple	Temperature (°C)	Cold bath vein
Channel 4	K-type thermocouple	Temperature (°C)	Hot bath vein
Channel 5	Infrared sensor	Voltage (V)	-

This test bench is connected to an Agilent data acquisition unit (34972A) with a 20 channels acquisition card (34901A) as multiplexer. Five channels are used for temperature and electrical voltage measurements (cf. Table III) with a sampling frequency of 10 Hz and a duration of 30 minutes (cf. section Operating protocol).

B. Operating protocol

To determine this thermal time constant (k), a transient thermal protocol is achieved (cf. Fig. 6).

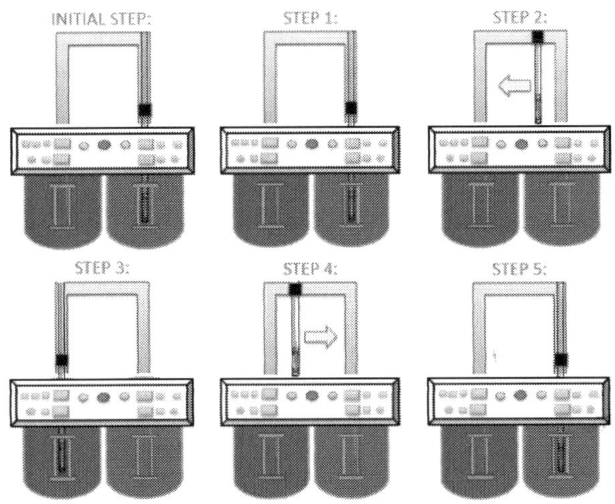

Fig. 6 Diagram of the thermal transient protocol.

Before starting the test, a specific temperature and velocity of the silicone oil in each thermostatic bath is imposed (initial step), and the calorimeter is immersed in the cold vein. The imposed temperatures in the cold bath and the hot bath can be from 25 °C to 37 °C and from 170 °C to 250 °C respectively. The rotation speed of the propeller is between 500 rpm to 1250 rpm. After reaching a steady thermal state, data acquisition is started. The calorimeter remains in the cold vein for 10 minutes (step 1) to acquire its initial steady response. Then the mechanical rail quickly moves the calorimeter to the hot vein: heating phase displacement (step 2). The calorimeter remains10 minutes in the hot vein (step 3) to observe its transient response for a heating phase and to reach a second steady thermal state inside the sensor. The reverse transfer of the calorimeter is then performed, from the hot vein to the cold vein: cooling phase displacement (step 4). Finally, the calorimeter remains for 10 minutes in the cold vein in order to observe its transient response for a cooling phase and to reach a final thermal equilibrium (step 5).

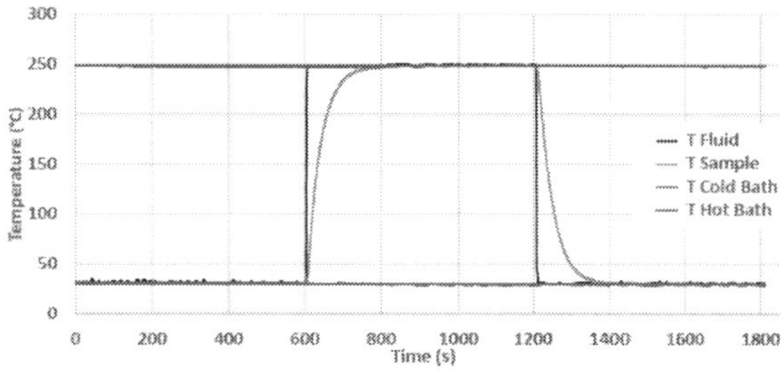

Fig. 7 Temperatures versus time for a test with a propeller velocity of 1000 rpm and with hot and cold temperatures of 250 °C and of 31 °C respectively.

Figure 7 gives the temperature measurements versus time for a test carried out with KAROLINA calorimeter by applying this operating protocol with a propeller velocity of 1000 rpm, a hot and cold temperatures of 250 °C and of 31 °C respectively. The different successive phases (steady and transient) can be observed.

V. Results

A test campaign took place from July 9 to July 23, 2020. The purpose of this campaign was to study the evolution of the value of the thermal time constant (k) according to the hot bath temperature, the cold bath temperature, the rotation speed of the propeller inside each bath and the cooling and heating phases. This section focuses first on a brief description of the test campaign. Then the repeatability of the results obtained for the tests is presented. Finally, the influence of each parameter is quantified with an Hadamard matrix.

A. *Test campaign*

In order to define the influence of the parameters, it was chosen the extreme temperatures of each range and a median value:
- 3 temperatures for the hot bath: 170, 210, 250 °C
- 3 temperatures for the cold bath: 25, 31, 37 °C

Regarding the speed of rotation of the propeller, it is necessary to have a sufficient high speed to be able to homogenize the baths, so the studied speed range is from 500 rpm to 1250 rpm and four values were tested:
- 4 speeds of the rotation of the propeller: 500, 750, 1000, 1250 rpm

By taking these conditions into account, the test campaign was composed of 35 tests. For each test, the values of the thermal time constant k are determined by applying the method corresponding to the equation (7). This presents a number of 140 values of the thermal time constant (k) (4 values per test: 2 values for the heating phase and 2 values for the cooling phase). Figure 8 gives the distribution of the values. This histogram shows a difference in results of k between the heating and cooling phases.

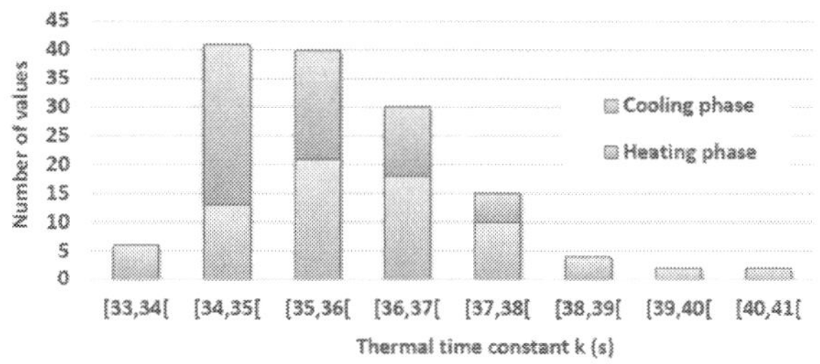

Fig. 8 Distribution of k values.

The difference between the lower value and the higher value is 7.38 s with $k_{min} = 33.11\ s$ and $k_{max} = 40.49\ s$. The average is $\bar{k} = 35.77\ s$, and the standard deviation is $\sigma_k = 1.35\ s$.

B. *Repeatability study*

A repeatability study was realized by fixing specific experimental conditions (hot bath at 210 °C, cold bath at 31 °C, propeller rotation speed at 1000 rpm) and for four successive experiments as described in Table IV.

TABLE IV: REPEATABILITY TESTS

	Start date of the test	Transient phase	Method	k value (s)
1	21/07/2020 13: 54	Heating	a	34.36
			b	34.25
		Cooling	a	35.59
			b	35.34
2	21/07/2020 14: 25	Heating	a	34.84
			b	34.72
		Cooling	a	35.09
			b	34.97
3	21/07/2020 14: 59	Heating	a	34.60
			b	34.48
		Cooling	a	35.71
			b	35.59
4	21/07/2020 15: 30	Heating	a	34.97
			b	34.84
		Cooling	a	35.97
			b	35.71

In order to have repeatability conditions, it is necessary to consider separately the results obtained during the heating phase and those obtained during the cooling phase.

A good repeatability is obtained for each phase: $\bar{k}_{heating} = 34.6\ s$, $\sigma_{k_{heating}} = 0.25\ s$ and $\bar{k}_{cooling} = 35.5\ s$, and $\sigma_{k_{heating}} = 0.34\ s$. The study confirms the discrepancy between the values of the thermal time constant obtained for a heating phase and those for a cooling phase. This behavior can be explained by a different thermal resistance in the two cases. The method of determination of the k-value (a thanks to the slope or b thanks to the y-intercept) does not affect the value of k significantly (for instance for the heating phase the difference of the two mean values of k is about 0.1 s with a similar standard deviation of 0.25 s). Consequently, a global study could be carried out by considering the k value whatever the applied method.

C. *Influence factors*

It is necessary to identify the factors that have an influence on the value of k. 5 factors were identified with more or less significant influence on the value of the thermal time constant (k). Three of these parameters are linked to the external boundary conditions: hot bath temperature, cold bath temperature, rotation speed of the propellers. Whereas the two last parameters are due to the phase and the data processing respectively: Transient phase (cooling or heating) and the determination method (slope or y-intercept).

The experimental results can be analyzed by using an Hadamard matrix which allows the identification of which factors have the most influence.

For the definition of the matrix, the experiments are considered with factors at 2 levels, the most extreme possible, in order to increase their influence. Table V gives the factors and their values for the two levels.

TABLE V: INFLUENCE FACTORS

	Factor	Low level	High level
U_1	Hot bath temperature	170 °C	250 °C
U_2	Cold bath temperature	25°C	37°C
U_3	Propeller rotation speed	500 rpm	1250 rpm
U_4	Transient state	Cooling	Heating
U_5	Determination method	Slope	y-intercept

The obtained Hadamard matrix is given in Table VI.

TABLE VI: HADAMARD MATRIX

	U_1	U_2	U_3	U_4	U_5	k values (s)
	1	-1	-1	1	-1	36.50
	1	1	-1	-1	1	40.16
	1	1	1	-1	-1	35.84
	-1	1	1	1	-1	35.44
	1	-1	1	1	1	34.12
	-1	1	-1	1	1	35.97
	-1	-1	1	-1	1	34.01
	-1	-1	-1	-1	-1	36.90
b_i	**6.30**	**3.88**	**-12.1**	**-6.87**	**1.59**	

By applying this matrix coupled with a Pareto analysis defined by equation (8) the influence in % for each factor is obtained (cf. Fig. 9).

$$P_i(\%) = 100 \times \frac{b_i^2}{\sum b_i^2} \quad (8)$$

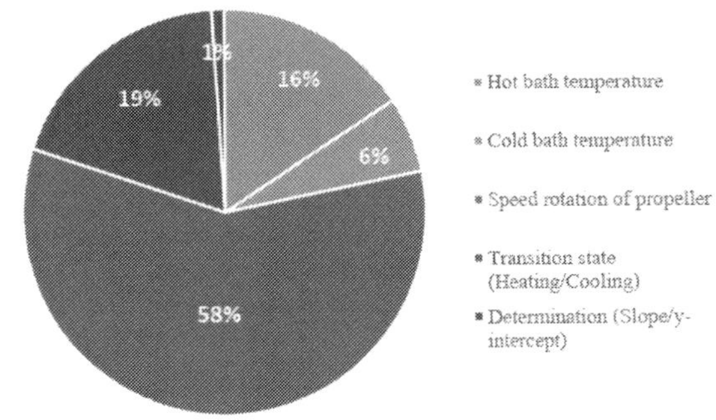

Fig. 9 Quantification of the influence of each factor.

The factor that mainly influences the value of k is the speed of rotation of the propeller, thus the velocity of the fluid around the sensor in the bath vein. Indeed the increase of the propeller speed leads to the increase of the thermal exchange coefficient (h; ($W.m^{-2}.K^{-1}$)) and consequently the decrease of the thermal resistance between the sample and the fluid. Two other factors also have a significant influence: the transient state (heating/cooling) and the hot bath temperature.

On the opposite, the determination method of k has no influence, thus this factor can be neglected. For the next campaign, an average between the two values obtained by the two methods could be calculated.

VI. CONCLUSIONS AND OUTLOOKS

A first experimental study was carried out to determine the thermal time constant of the Polish KAROLINA single-cell calorimeter thanks to a specific thermal transient protocol applied in the BERTRAN bench reproducing reactor conditions (temperature, heat exchange, geometry). This thermal time constant is a key value for the calibration of the sensor and thus for the determination of its sensitivity. The experimental campaign under laboratory conditions confirmed the repeatability of the calibration protocol and determined the influence of five parameters by using an analysis involving Hadamard matrix. The thermal time constant value is independent of the two calculation methods. But it depends on the cooling and heating phase. The temperature of the hot bath also impacts the thermal time constant value. These results and approach will be used to define new test campaign conditions and determine the sensor sensitivity and its uncertainties. Moreover, the instrumentation of the bench will be improved with the adding of a flowmeter in order to determine the fluid velocity ($m.s^{-1}$) near the calorimeter during the calibration. This quantity is crucial to interpret the results and to compare them with numeric results obtained thanks to a 1-D theoretical model and thanks to 3-D thermal simulations.

The optimized bench will be used for a next test campaign focusing on other calorimeters targeted for a future irradiation campaign within the framework of the CALOR-I project (such as a gamma thermometer or a CALORRE differential calorimeter).

ACKNOWLEDGMENT

"The CALOR-I project leading to this publication has received funding from the Excellence Initiative of Aix-Marseille University - A*Midex, a French "Investissements d'Avenir" programme".

REFERENCES

[1] C. Reynard-Carette, G. Kohse, J. Brun, M. Carette, A. Volte, A. Lyoussi, "Review of nuclear heating measurement by calorimetry in France and USA", Proc. ANIMMA 2017, Jun. 2017, EPJ Web of Conferences 170, 04019 (2018).

[2] J. Brun, C. Reynard-Carette, A. Lyoussi, C. De Vita, M. Carette, M. Muraglia, D. Fourmentel, P. Guimbal, J.-F. Villard, "Comparison of the thermal calibration response of two calorimetric cells dedicated to nuclear heating measurements", IEEE Trans. Nucl. Sci., vol. 61, no. 4, pp. 2254-2261, Aug. 2014.

[3] A.Volte, C. Reynard-Carette, A. Lyoussi, J. Brun, M. Carette, "Study of the Response of a New Compact Calorimetric Cell for Nuclear Heating Rate Measurements", IEEE Trans. Nucl. Sci., vol. 65, no. 9, pp. 2461-2470, Sept. 2018.

[4] A.Volte, J. Brun, A. Lyoussi, M. Carette, C. Reynard-Carette, "Qualification of a New Differential Calorimeter Configuration Dedicated to Nuclear Heating Rates up to 20 $W.g^{-1}$", IEEE Trans. Nucl. Sci., vol. 67, no. 11, pp. 2405-2414, Nov. 2020.

[5] R. Van Nieuwenhove, L. Vermeeren, "Nuclear heating measurements by gamma and neutron thermometers", Proc. ANIMMA 2019, Jun. 2019, EPJ Web of Conferences 225, 04003 (2020).

[6] J. Brun, M. Tarchalski, C. Reynard-Carette, K. Pytel, A. Lyoussi, J. Jagielski, D. Fourmentel, J-F. Villard, M. Carette, "Responses of Single-Cell and Differential Calorimeters: From Out-of-Pile Calibration to Irradiation Campaigns," IEEE Trans. Nucl. Sci., vol.63, no. 3, pp. 1630 1639, Jun. 2016.

[7] A. Luks, K. Pytel, M.Tarchalski, N. Uzunow, T. Krok, "Modelling of thermal hydraulics in a KAROLINA calorimeter for its calibration methodology validation", NUKLEONIKA, vol 61, no. 3, pp 453-460, Oct. 2016.

[8] J. Brun, C. Reynard-Carette, M. Carette, A. Lyoussi, A. Volte, M. Tarchalski, R. Prokopowicz, "Calibration of a Single-Cell Calorimeter in a New Transient state Test Bench", Proc. ANIMMA 2019, Jun. 2019, EPJ Web of Conferences 225, 04009 (2020).

[9] M. Tarchalski, K. Pytel, P. Siréta, A. Lyoussi, J. Jagielski, C. Reynard-Carette, C. Gonnier, G. Bignan, "Principle of calibration of the simple calorimeter for nuclear heating measurements in MARIA reactor and transposition to the case of JHR reactor", Proc. ANIMMA 2013, Jun. 2013.

[10] M. Tarchalski, "Nuclear heating measurements in the Maria reactor and implementation of neutron and photon calculation scheme", Ph.D manuscript, Aix-Marseille Université, Marseille, France, 2016.

CEA-JSI Experimental Benchmark for validation of the modeling of neutron and gamma-ray detection instrumentation used in the JSI TRIGA reactor

Clément Fausser[1], Nicolas Thiollay[1], Christophe Destouches[1], Loïc Barbot[1], Damien Fourmentel[1], Benoît Geslot[1], Grégoire De Izarra[1], Adrien Gruel[1], Gilles Grégoire[1], Christophe Domergue[1], Vladimir Radulović[2], Tanja Goričanec[2], Klemen Ambrožič[2], Gasper Žerovnik[2]. Igor Lengar[2], Andrej Trkov[2], Žiga Štancar[2], Anže Pungerčič[2], and Luka Snoj[2]
[1]CEA, DES, IRESNE, DER, SPESI, LDCI, Cadarache, F-13108 St-Paul-Lez-Durance, France
[2]Reactor Physics Department, Jožef Stefan Institute, Jamova cesta 19, SI-1000 Ljubljana, Slovenia

clement.fausser@cea.fr

***Abstract*—Constant improvements of the computational power and methods as well as demands of accurate and reliable measurements for reactor operation and safety require a continuous upgrade of the instrumentation. In particular, nuclear sensors used in nuclear fission reactors (research or power reactors) or in nuclear fusion facilities are operated under intense mixed neutron and gamma-ray fields, and need to be calibrated and modeled to provide selective and accurate neutron and gamma-ray measurements.**

The French Atomic Energy and Alternative Energies Commission (CEA) and the Jožef Stefan Institute (JSI) have started an experimental program dedicated to a detailed experimental benchmark with analysis using Monte Carlo particle transport calculations and a series of neutron and gamma-ray sensor types used in the JSI TRIGA Mark II reactor.

CEA has setup a simplified TRIPOLI-4® modeling scheme of the JSI TRIGA reactor based on the information available in the IRPhEP benchmark in order to facilitate analysis of future neutron and gamma-ray measurements. These allow the CEA to perform a TRIPOLI-4 instrumentation calculation scheme benchmarked with the JSI MCNP model.

This paper presents the main results of this CEA calculation scheme application and the analysis of their comparison to the JSI results obtained in 2012 with the MCNP5 & ENDF/B-VII.0 calculation scheme. This paper will conclude with some information about the new experimental program to be carried out in 2022 in the TRIGA reactor core.

***Keywords*—Monte Carlo, TRIGA, TRIPOLI-4, MCNP, JEFF, IRDF, benchmark, fission chamber, fission rate.**

I. Introduction

The collaboration on nuclear instrumentation between the CEA Experimental Physics, Safety experiment and Instrumentation Section (SPESI) and the Reactor Physics Division of the Jožef Stefan Institute (JSI) started in 2008 in the frame of the bilateral agreement signed in 2006 between the CEA and the Slovenian Ministry of Higher Education, Science and Technology [1]. The mutual interest relies on the needs for a detailed neutron and gamma-ray characterization of the 250 kW TRIGA reactor for JSI and a need to diversify the available irradiation facilities for instrumentation testing for CEA.

The Instrumentation, Sensors and Dosimetry Laboratory (LDCI), a subsection of CEA SPESI, has been designing, manufacturing and using nuclear sensors dedicated to R&D and power nuclear reactors for a long time.

The reference model of the reactor core using Monte Carlo N-particle transport code (MCNP) [2] was published in a detailed NEA report available in the International Reactor Physics Evaluation Project (IRPhEP) handbook [3]. It revealed that the detailed MCNP5 model of the TRIGA reactor provides a satisfactory agreement with the measurements: The average deviations within the active fuel height are indeed less than 5%, which is the approximate value of the combined uncertainty of measurements and modeling assumptions.

However, a set of experimental measurements could complete the characterization of the neutron flux and the photon flux of this benchmark. On the other hand, a comparison between the modeling schemes TRIPOLI-4 and MCNP should allow an improvement of the interpretation of the experimental results with in particular the recent improvement of the modelling of the prompt and delayed photons.

Thus, LDCI has proposed to develop a Monte Carlo model using the 4th generation of the tridimensional Monte Carlo code for polykinetic continuous-energy radiation transport [4], TRIPOLI-4® version 11.0 (2020). This paper presents the TRIPOLI calculation scheme and the first analysis of the comparison of the results with those of the JSI one.

II. Description of the measure: miniature fission chambers

The specificity of the measurements of the 2011-2012 benchmark lies on absolute fission rates measured by two miniature fission chambers (FC) provided by LDCI [5] (see Fig. 1), which assembled a ^{235}U fission chamber (FC5) and a ^{238}U fission chamber (FC8) in its fission chamber workshop at CEA Cadarache [6].

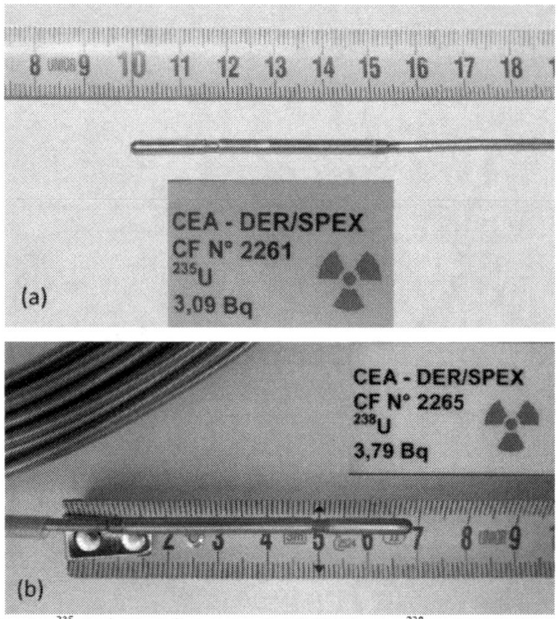

Fig. 1. ^{235}U miniature fission chamber "FC5" (a) and ^{238}U miniature fission chamber "FC8" (b).

Both FC5 and FC8 consist of a 4 mm long active part, which composition is detailed in NEA report [2]. The measurements were carried out in two different series of steady-state reactor power levels: one with FC5 at nominal power 103.5 W and the other with FC8 at 1034 W. One should notice that the nominal power is determined based on measurement from ex-core neutron detector without taking into account the uncertainty and possible variations (e.g. due to control rods movements) of the conversion factor between the nominal factor and the FC.

Measurements were performed in the axial (Z) direction by a pneumatic system from above the core upper grid to below the core lower grid. The FCs are guided inside an aluminum tube filled with water. 9 measurement positions (MP) are explored as shown in the Fig. 2: 6 having a 8 mm diameter (MP14 to 15 and 21 to 23) and 3 around the central channel having a larger diameter (10.5 mm, MP17, 20 and 25).

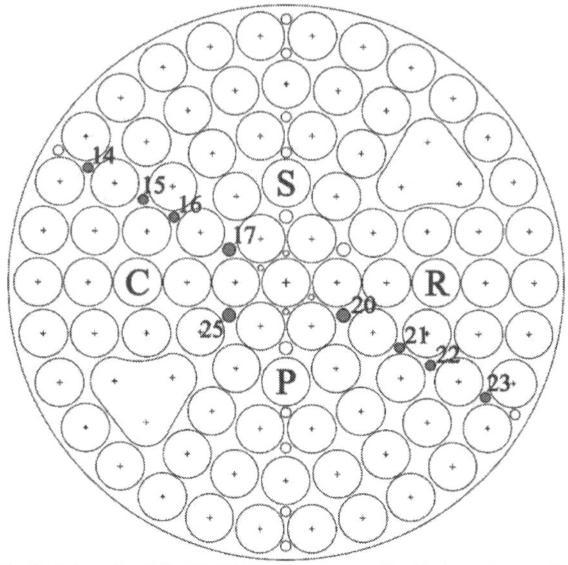

Fig. 2. Schematic of the JSI TRIGA core upper grid with denoted measuring positions.

III. Description of the TRIPOLI-4 model

LDCI built the TRIPOLI-4 model applying four general principles.

The first principle is to be as rigorously identical as possible to the JSI MCNP model published in the NEA report. The MCNP model is translated in the TRIPOLI-4 format, keeping the same geometric and composition values, as well as identical modeling assumptions. For example, the fuel in standard fuel elements and fuel-follower control rods are modeled fresh, FCs are not physically depicted into their guide tubes (the fission rates being collected into a 4 mm long water cylinder instead), etc.

The second principle is to use surface-based volumes only (no TRIPOLI-4 combinatory volume nor ROOT-based volume [7], as a reminder "volume" in TRIPOLI-4 is the exact equivalent of a MCNP "cell"). Not only do these surface-based volumes speed-up computation, but in addition they can eventually supply CAD / MCNP / TRIPOLI-4 automatic converters or viewers, such as McCAD [8] or MCAM [9].

The third principle is to use the point-wise nuclear cross section library supplied by default with TRIPOLI-4.11.0, CEAv5.1.2. CEAv5.1.2 is mainly based on JEFF-3.1.1 [10] for transport cross-sections and contains point-wise IRDF-2002 cross-sections [11], which are used to calculate ^{235}U and ^{238}U fission rates in FCs.

The fourth principle is to use the same control rods positions as the two MCNP models used to provide the MCNP results in the NEA report.

The main difference between the TRIPOLI-4 and MCNP models lies in the absence of transformation matrixes in TRIPOLI-4.11.0. As the JSI TRIGA core is composed of 6 rings (with an additional one for the rotary groove) at regular angle pitch, 161 transformation matrixes allow the MCNP code to automatically copy and move each of the different generic geometries (one for a standard fuel elements, one for a fuel-follower control rod, etc.). For the TRIPOLI-4 model, LDCI created a LibreOffice Calc file for each generic geometry, which automatically generates TRIPOLI-4 surface-based volumes from the values of the MCNP transformation matrixes. The TRIPOLI-4 volumes obtained are then copied and pasted into the TRIPOLI-4 data file, which leads to a longer data file only for the geometry section.

Fig. 3 and Fig 4. respectively illustrate a side view and a top view of the TRIPOLI-4 model obtained in FC5 configuration.

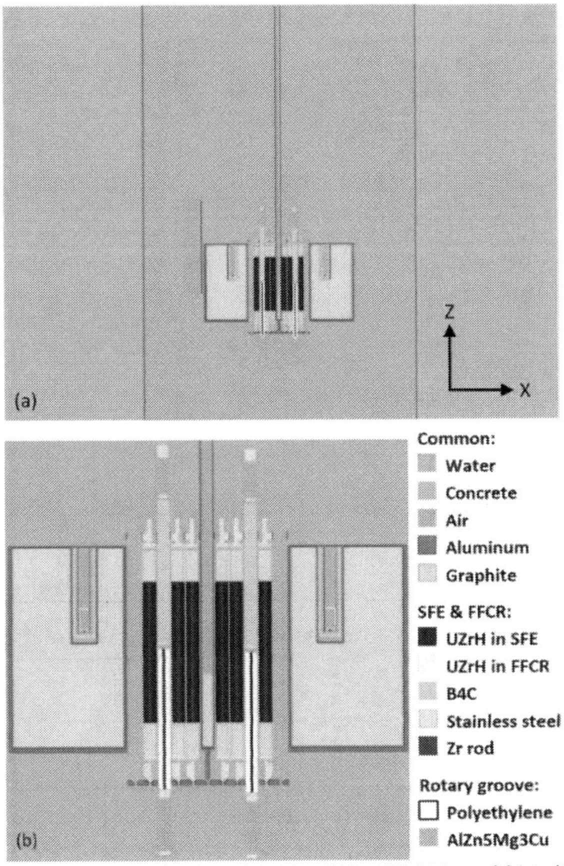

Fig. 3. Y = 0 cm XZ side-view of FC5 TRIPOLI-4 JSI TRIGA model (a) with zoom on the core (b).

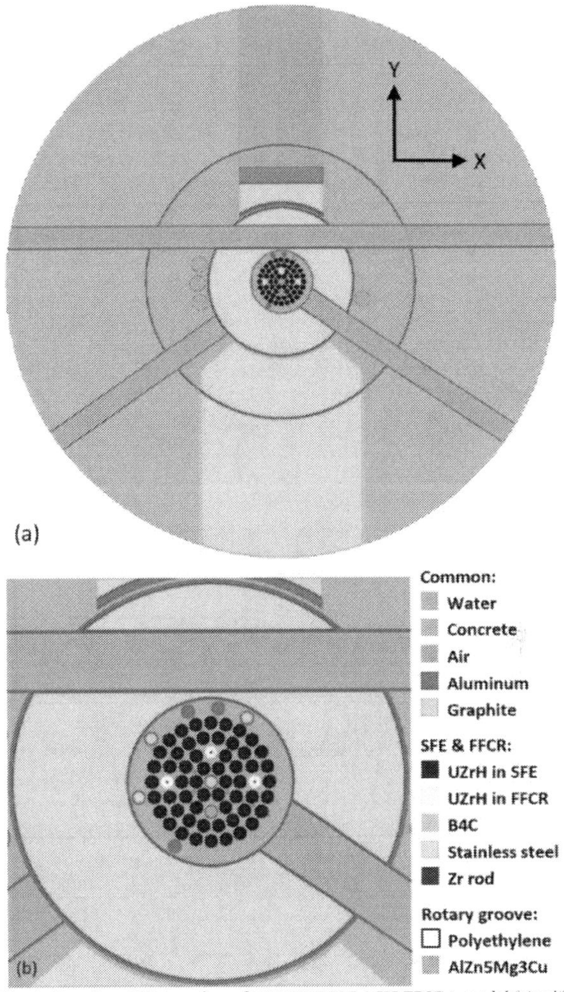

Fig. 4. Z = -8 cm XY top-view of FC5 TRIPOLI-4 JSI TRIGA model (a) with zoom on the core (b).

IV. From Monte Carlo raw results to absolute total fission rates

As with the MCNP models, the neutron source is normalized to one in the TRIPOLI-4 models. Thus, two normalization factors (one for FC5 and the other for FC8) must be processed in order to convert the fission rates obtained in the MCNP or TRIPOLI-4 results in order to compare to the FC absolute fission rates. Even if the irradiations are always performed with a critical core (k_{eff} is exactly 1), the use of fresh fuel in the models leads to k_{eff} values different from 1 in the criticality calculations. The normalization factor K to be applied to Monte Carlo microscopic fission rates to obtain the absolute macroscopic fission rates in FC5 or FC8 deposits is given in equation (1):

$$K = \frac{P\cdot\bar{\nu}}{w_f\cdot k_{eff}}\cdot N\ , \qquad (1)$$

With P the thermal power (unit: W), $\bar{\nu}$ the average number of neutrons released per fission (no unit), w_f the average deposited energy by fission event (unit: J) and N the number of uranium atoms in the FC deposit (unit: $\times10^{+24}$). The unit of K is then:

$\times 10^{+24}$/s. LDCI used for the present work the NEA report values for P (103.5 W for FC5 and 1034 W for FC8), $\bar{\nu}$ (2.439) and w_f (193.7 MeV converted to J).

LDCI calculates N values for FC5 and FC8 of respectively 2.270×10^{16} atoms and 2.319×10^{17} atoms and obtained k_{eff} values of 1.04829 for FC5 and 1.04857 for FC8 (with statistical uncertainties of ±2.0 pcm and ±1.4 pcm at k_{eff} = 1) with TRIPOLI-4 models. The k_{eff} values obtained with TRIPOLI-4.11.0 & CEAv5.1.2 (mainly based on JEFF-3.1.1) for FC5 and FC8 differ only by -106 pcm and -108 pcm from MCNP5 & ENDF/B-VII.0 k_{eff} results (i.e. -0.10%, MCNP statistical uncertainties being of ±2 pcm). The recalculated values of the K factors for FC5 and FC8 are respectively $1.76 \times 10^{+5}$ and $1.80 \times 10^{+7}$. These values differ by -1.2% and -1.4% from those published in the NEA report. In the present work, the K recalculated values are used not only for the TRIPOLI-4 results, but also for the MCNP ones (the MCNP results in the NEA report are multiplied by 0.9884 for FC5 and 0.9865 for FC8).

V. Absolute fission rates results: comparison between the modeling results and the benchmark-model measurements

JSI carried out thorough uncertainty estimation work on raw absolute fission rates measurements to provide benchmark-model measurements, which take into account some biases due to the modeling assumptions, such as fresh fuel or absence of modeling of the real FC geometries. Only one small diameter MP and far from the central channel (MP14) and one large diameter MP and close to the central channel (MP25) are illustrated in the present work. These illustrations depict the comparisons between TRIPOLI-4.11.0 & CEAv5.1.2 & IRDF-2002 (TRI4), MCNP5 & ENDF/B-VII.0 (MCNP), and benchmark-model measurements (Meas). Fig. 5 and Fig. 6 correspond respectively to FC5 and FC8 in MP14. Fig. 7 and Fig. 8 correspond respectively to FC5 and FC8 in MP25. A diagram of a standard fuel element at the same altitudes accompanies all figures to facilitate analysis. Similar to the corresponding figures in the NEA report for MCNP, a green line provides the results of a fine grid (4 mm window every mm, as for MCNP) with dedicated dots with error bars when the altitude corresponds to an experimental position. As for MCNP, the error bars associated with TRI4 quadratically combine the Monte Carlo statistical uncertainty and ±2.6% for FC5 or ±2.7% for FC8 (corresponding to the combined uncertainties of the K factor and the nuclear data). As published in the NEA report, the measurements error bars contain not only the uncertainty of the count rate of the real measurement, but also all the uncertainties of the modeling assumptions (e.g. fresh fuel). When TRI4 and benchmark-model measurements are compared in these figures, the present work uses the same convention as the NEA report for the error bars, i.e. a sum of the two uncertainties, σ_{SUM}, instead of a quadratic combination, σ_{QUAD}. Nevertheless, in these figures, the comparison between TRI4 and MCNP uses $\pm\sigma_{QUAD}$ error bars made by the quadratic combination of each of their statistical uncertainties and the nuclear data library uncertainties estimated in the NEA report as ±1.4% for FC5 and ±1.6% for FC8. K factors being theoretically identical for TRI4 and MCNP, their combined uncertainty does not considers it.

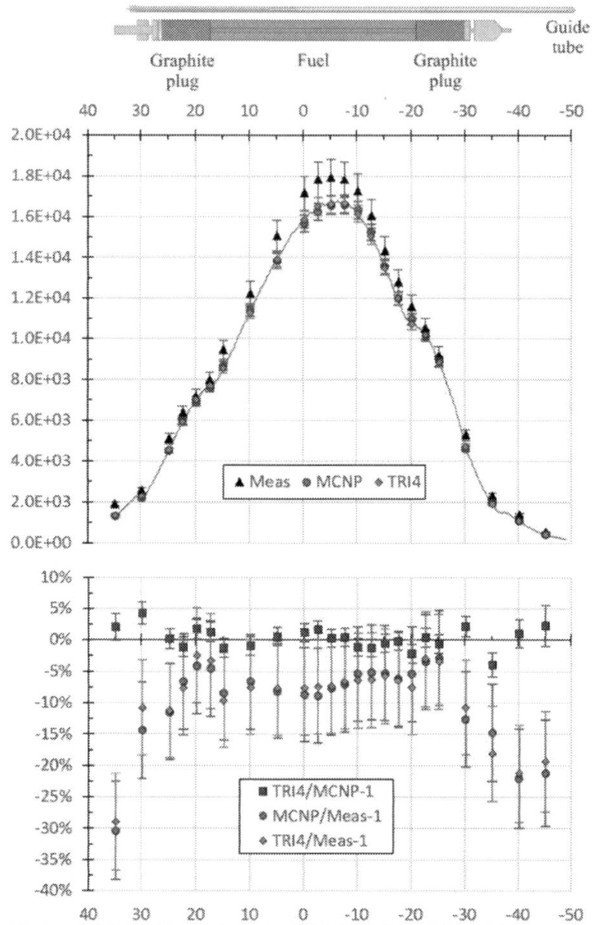

Fig. 5. Axial absolute fission rates (/s) in MP14 in FC5 model versus the altitude Z (cm) with a schematic of a standard fuel element at same altitudes.

As MCNP model absolute fission rates and benchmark-model measurements are accessible in the NEA report, only TRIPOLI-4 absolute fission rates are presented in the present work: FC5 ones in Table I and FC8 in Table II for every MPs, except MP20, which is in Table III for both FC5 and FC8. "Val." means the absolute fission rate and "σ_{Stat}" means only the statistical uncertainty (i.e. without ±2.6% for FC5 or ±2.7% for FC8 due to the combined uncertainty of the K factor and the nuclear data).

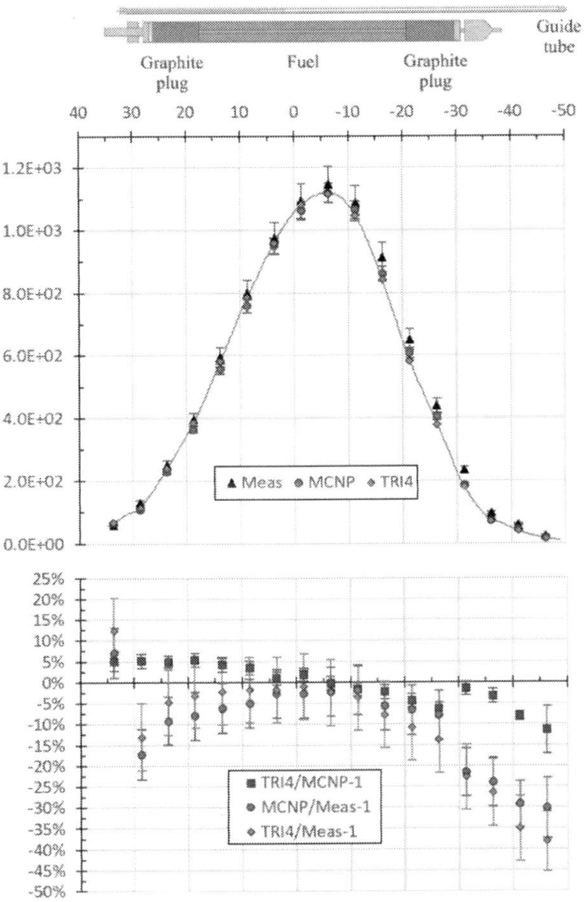

Fig. 6. Axial absolute fission rates (/s) in MP14 in FC8 model versus the altitude Z (cm) with a schematic of a standard fuel element at same altitudes.

Let consider the difference between TRI4 and MCNP relatively to their σ_{QUAD}: $\left(\frac{TRI4}{MCNP}\text{-}1\right)/\sigma_{QUAD}$. Let now consider the axial profiles obtained in the 9 MPs, which correspond to a benchmark-model measurement set of 220 values for FC5 and 159 for FC8 (the Z = +35.7 cm altitude in MP20 is outside modeling volume for the MP, so it is not taken into account). $\left(\frac{TRI4}{MCNP}\text{-}1\right)/\sigma_{QUAD}$ is on average -0.7 for FC5 and -0.2 for FC8, with respective standard deviation of ±1.6 for FC5 and ±3.1 for FC8. Thus, there is a light negative bias between TRI4 and MCNP (especially for FC5) as well as standard deviations greater than one σ_{QUAD} (especially for FC8). However, the high standard deviation for FC8 masks a tendency to obtain positive values of $\left(\frac{TRI4}{MCNP}\text{-}1\right)$ above Z = 0 and vice-versa below Z = 0, as in Fig. 6 and 8. At fuel altitudes in the standard fuel elements, $\left(\frac{TRI4}{MCNP}\text{-}1\right)$ is on average only -1.7% for FC5 and +0.4% for FC8 with standard deviation of ±1.7% for FC5 and ±3.0% for FC8. These values are lower than those of $\left(\frac{MCNP}{Meas}\text{-}1\right)$, which is on average -4.7% for FC5 and -3.6% for FC8 with standard deviation of ±2.1% for FC5 and ±3.3% for FC8. In addition, the standard deviations of $\left(\frac{TRI4}{MCNP}\text{-}1\right)$ are also much lower than the uncertainties associated with the benchmark-model measurements, which are ±5% for both FC5 and FC8. Therefore, TRI4 and MCNP give reasonably similar results with respect to the uncertainties of the benchmark-model measurements.

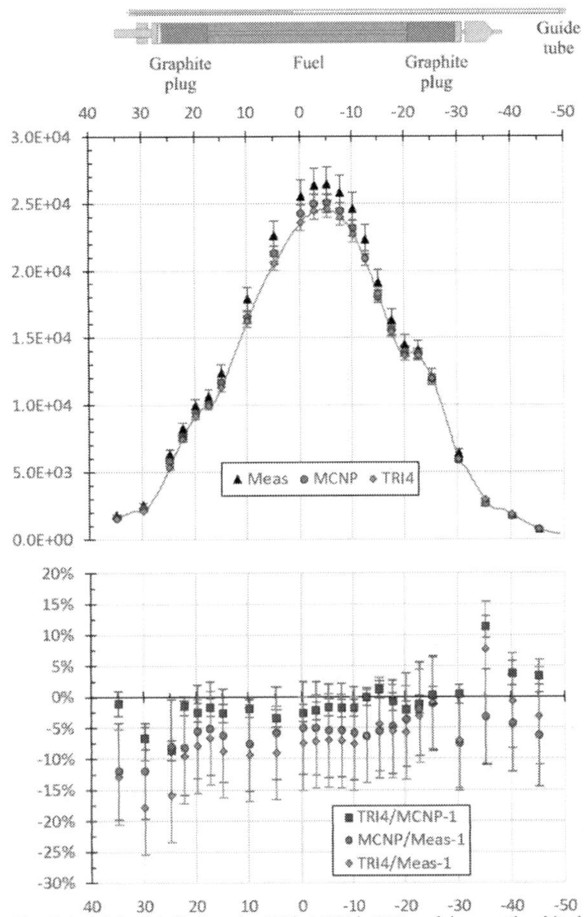

Fig. 7. Axial absolute fission rates (/s) in MP25 in FC5 model versus the altitude Z (cm) with a schematic of a standard fuel element at same altitudes.

Let now consider the difference between the absolute fission rates from TRI4 and from benchmark-model measurements with respect to their σ_{QUAD} in the 9 MPs at every altitude: $\left(\frac{TRI4}{Meas}\text{-}1\right)/\sigma_{QUAD}$. The average is -1.4 for FC5 and -1.8 for FC8, with respective standard deviations of ±1.0 for FC5 and ±1.8 for FC8. These values show a negative bias between TRI4 and the measurements and a standard deviation for FC8, which exceeds the expected ±1 value. Nevertheless, the study of $\left(\frac{TRI4}{Meas}\text{-}1\right)/\sigma_{QUAD}$ throughout the altitudes shows that, as encountered between MCNP and benchmark-model measurements, the Monte Carlo simulations are not satisfactory at extreme altitudes corresponding to the regions around the edge and outside of the active fuel height.

https://doi.org/10.1051/epjconf/202125304018

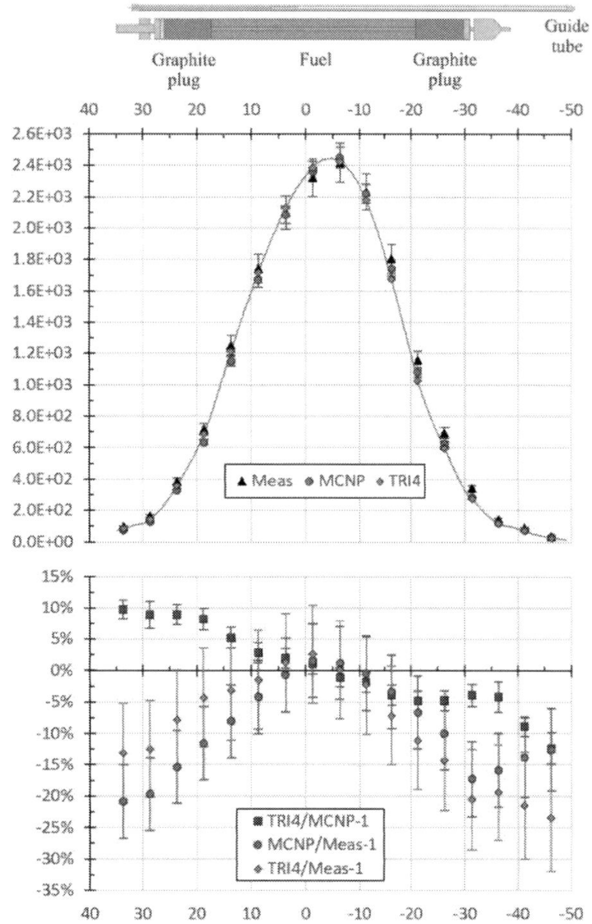

Fig. 8. Axial absolute fission rates (/s) in MP25 in FC8 model versus the altitude Z (cm) with a schematic of a standard fuel element at same altitudes.

However, $\left(\frac{TRI4}{Meas}\text{-}1\right)/\sigma_{QUAD}$ becomes much more satisfactory within fuel region of the standard fuel elements: The average is -1.1 for FC5 and -0.6 for FC8, with respective standard deviations of ±0.4 for FC5 and ±0.8 for FC8, with extreme values of [-2.0, +0.5] for FC5 and [-2.8, +0.8] for FC8. In terms of $\left(\frac{TRI4}{Meas}\text{-}1\right)$, this corresponds to averages of -6.3% for FC5 and -3.2% for FC8, with respective standard deviations of ±2.4% for FC5 and ±4.4% for FC8, with extreme values of [-11.1%, +3.0%] for FC5 and [-16.2%, +4.7%] for FC8. The further away upwards or downwards from the altitude of the maximum absolute fission rates (close to Z ≈ -4 cm), the more the relative difference between the simulated absolute fission rates (either TRI4 or MCNP) and the benchmark-model measurements, relatively to the latter, increase. This phenomenon has already been noted in the NEA report. It may come from the fuel modeling assumptions, which consider only fresh fuel. The NEA report showed that this might have led to a large underestimation of the actual fission rates at the extreme parts of the fuel. However, this is not a major drawback if the TRI4 model is used for pre-design of the instrumentation. New experimental and modelling benchmark should allow solving this discrepancy. Nevertheless, for verification and validation purposes, both of the TRI4 and MCNP models using fresh fuel are not sufficiently satisfactory.

TABLE I
TRIPOLI-4.11.0 & CEAV5.1.2 & IRDF-2002 FC5 ABSOLUTE FISSION RATES (UNIT: PER SECOND)

Z	MP14		MP15		MP16		MP17	
(cm)	Val.	σ_{Stat}	Val.	σ_{Stat}	Val.	σ_{Stat}	Val.	σ_{Stat}
-45.2	427	±7	569	±8	655	±9	727	±9
-40.2	1081	±12	1311	±13	1482	±14	1792	±15
-35.2	1896	±15	2006	±15	2089	±16	2839	±18
-30.2	4716	±23	4295	±22	4505	±22	5894	±26
-25.2	8848	±32	9364	±32	9879	±33	11910	±36
-22.7	10188	±34	10447	±34	11114	±35	13598	±39
-20.2	10730	±35	9714	±32	10386	±33	13721	±38
-17.7	11980	±37	10484	±33	11279	±34	15618	±40
-15.2	13500	±39	12431	±36	13472	±37	18561	±44
-12.7	15041	±41	14071	±38	15312	±39	21233	±47
-10.2	16144	±42	15351	±40	16755	±41	23321	±49
-7.7	16645	±43	16185	±41	17616	±42	24525	±51
-5.2	16612	±43	16468	±42	18042	±43	25081	±51
-2.7	16508	±43	16153	±41	17792	±42	25012	±51
-0.2	15854	±42	15456	±40	16883	±41	24152	±50
+4.8	13919	±39	13034	±37	14298	±38	21096	±47
+9.8	11304	±36	10167	±33	11133	±34	16652	±41
+14.8	8553	±31	7046	±27	7618	±28	11457	±35
+17.3	7692	±29	6703	±27	7109	±27	10043	±32
+19.8	7000	±28	6681	±27	7067	±27	9230	±32
+22.3	5902	±26	5644	±25	5932	±25	7480	±29
+24.8	4541	±23	4085	±21	4251	±21	5378	±25
+29.8	2282	±15	1539	±12	1403	±12	2113	±15
+34.8	1347	±13	1076	±11	1154	±12	1548	±13
Z	**MP21**		**MP22**		**MP23**		**MP25**	
(cm)	Val.	σ_{Stat}	Val.	σ_{Stat}	Val.	σ_{Stat}	Val.	σ_{Stat}
-45.2	635	±9	538	±8	393	±7	738	±9
-40.2	1461	±13	1265	±12	984	±11	1799	±15
-35.2	2023	±15	1935	±15	1712	±14	2900	±18
-30.2	4400	±22	4136	±21	4259	±22	5918	±25
-25.2	9633	±32	8934	±32	8672	±32	11950	±36
-22.7	10798	±34	9972	±33	10059	±34	13632	±38
-20.2	9992	±32	9368	±32	10486	±35	13659	±38
-17.7	10835	±33	9889	±32	11569	±36	15420	±40
-15.2	12742	±36	11685	±35	13087	±39	18265	±43
-12.7	14628	±39	13202	±37	14375	±40	20901	±47
-10.2	15948	±40	14476	±39	15496	±42	22741	±49
-7.7	16781	±42	15174	±40	16287	±43	23972	±50
-5.2	17059	±42	15475	±40	16700	±44	24580	±51
-2.7	16761	±41	15195	±40	16888	±44	24424	±50
-0.2	15919	±40	14514	±39	16688	±43	23603	±50
+4.8	13622	±37	12373	±36	15136	±41	20543	±47
+9.8	10624	±33	9642	±32	12204	±37	16213	±41
+14.8	7310	±27	6732	±27	8774	±32	11318	±34
+17.3	6915	±27	6448	±26	7757	±30	9910	±32
+19.8	6864	±27	6418	±27	6981	±28	9169	±32
+22.3	5758	±25	5466	±25	5737	±26	7450	±29
+24.8	4107	±21	3893	±21	4214	±22	5327	±24
+29.8	1366	±11	1430	±12	1700	±13	2097	±15
+34.8	1092	±12	981	±11	895	±10	1540	±13

TABLE II
TRIPOLI-4.11.0 & CEAV5.1.2 & IRDF-2002 FC8 ABSOLUTE FISSION RATES
(UNIT: PER SECOND)

Z	MP14		MP15		MP16		MP17	
(cm)	Val.	σStat	Val.	σStat	Val.	σStat	Val.	σStat
-46.3	15	±0	20	±0	23	±0	26	±0
-41.3	39	±0	50	±0	56	±0	67	±0
-36.3	71	±0	80	±0	90	±0	112	±1
-31.3	181	±1	196	±1	215	±1	270	±1
-26.3	379	±1	452	±1	501	±1	592	±1
-21.3	581	±1	757	±1	857	±1	1028	±2
-16.3	841	±1	1187	±2	1399	±2	1707	±2
-11.3	1047	±2	1532	±2	1822	±2	2238	±2
-6.3	1118	±2	1692	±2	2019	±2	2492	±3
-1.3	1082	±2	1634	±2	1955	±2	2453	±3
+3.7	958	±2	1413	±2	1688	±2	2182	±2
+8.7	785	±1	1125	±2	1343	±2	1762	±2
+13.7	581	±1	788	±1	937	±2	1240	±2
+18.7	385	±1	471	±1	533	±1	693	±1
+23.7	240	±1	258	±1	285	±1	359	±1
+28.7	113	±1	100	±0	104	±1	141	±1
+33.7	66	±0	56	±0	62	±0	83	±0
Z	MP21		MP22		MP23		MP25	
(cm)	Val.	σStat	Val.	σStat	Val.	σStat	Val.	σStat
-46.3	22	±0	20	±0	14	±0	27	±0
-41.3	56	±0	49	±0	37	±0	68	±0
-36.3	90	±0	78	±0	65	±0	115	±1
-31.3	209	±1	189	±1	164	±1	271	±1
-26.3	488	±1	434	±1	359	±1	595	±1
-21.3	833	±1	725	±1	555	±1	1027	±2
-16.3	1340	±2	1124	±2	788	±1	1675	±2
-11.3	1737	±2	1447	±2	974	±2	2182	±2
-6.3	1919	±2	1593	±2	1074	±2	2423	±3
-1.3	1848	±2	1540	±2	1078	±2	2385	±3
+3.7	1604	±2	1341	±2	986	±2	2125	±2
+8.7	1286	±2	1076	±2	807	±1	1718	±2
+13.7	904	±2	760	±1	586	±1	1212	±2
+18.7	520	±1	452	±1	378	±1	686	±1
+23.7	279	±1	248	±1	226	±1	358	±1
+28.7	102	±0	94	±0	94	±0	142	±1
+33.7	59	±0	52	±0	48	±0	83	±0

TABLE IIII
TRIPOLI-4.11.0 & CEAV5.1.2 & IRDF-2002 MP20 ABSOLUTE FISSION RATES
(UNIT: PER SECOND)

Z	FC5	
(cm)	Val.	σStat
-45.2	750	±9
-40.2	1779	±14
-35.2	2854	±18
-30.2	5907	±26
-27.7	9132	±32
-25.2	11900	±36
-22.7	13483	±38
-20.2	13588	±38
-17.7	15216	±40
-15.2	18086	±43
-12.7	20599	±46
-10.2	22435	±48
-7.7	23660	±49
-5.2	24302	±50
-2.7	24132	±50
-0.2	23297	±49
+2.3	22095	±48
+4.8	20415	±46
+7.3	18336	±44
+9.8	16066	±41
+12.3	13612	±38
+14.8	11157	±34
+17.3	9883	±32
+19.8	9127	±31
+22.3	7444	±29
+24.8	5291	±24
+29.8	2042	±14
+34.8	1535	±14

Z	FC8	
(cm)	Val.	σStat
-44.3	40	±0
-39.3	91	±0
-34.3	165	±1
-29.3	368	±1
-24.3	745	±1
-21.8	962	±2
-19.3	1279	±2
-16.8	1598	±2
-14.3	1886	±2
-11.8	2120	±2
-9.3	2285	±3
-6.8	2385	±3
-4.3	2422	±3
-1.8	2382	±3
+0.7	2281	±3
+5.7	1964	±2
+10.7	1520	±2
+15.7	988	±2
+18.2	724	±1
+20.7	536	±1
+23.2	384	±1
+25.7	251	±1
+30.7	111	±1

VI. CONCLUSION

Thanks to the CEA-JSI benchmark published in the IRPhE Project handbook, LDCI was able to set up Monte Carlo model of the neutron transport in the JSI TRIGA Mark II reactor using the CEA reference code: TRIPOLI-4. Associated to a JEFF-3.1.1 library for transport cross-sections and IRDF-2002 for fission cross-sections. Results are satisfactory, compared to either the JSI MCNP5 & ENDF/B-VII.0 model or the benchmark-model measurements results. Nevertheless, the use of fuel modeled fresh, without any absorbent fission product or transuranic elements, might be the reason why the simulations move away from the measurements outside the peak flux altitude and to a k_{eff} value well above 1, whatever Monte Carlo code is used. Even if this is not a major drawback for instrumentation predesign, this simplification of fuel modeling may become problematic for validation and verification purposes, especially if coupled neutron and gamma-ray transport has to be considered. The new experimental campaigns scheduled in 2022, involving not only miniature fission chambers, but also thorough activation dosimetry campaign, some self-powered detectors, thermoluminescent dosimeters, ionization chambers and calorimeters, are expected to drastically improve the TRIPOLI-4 and MCNP models used by CEA and JSI. These new calculation schemes should use the most up-to-date Monte Carlo codes and nuclear data libraries, such as TRIPOLI-4.11.1 with JEFF-3.3 [12] and IRDFF-II [13] or MCNP6.2 [14] with ENDF/B-VIII.0 [15].

ACKNOWLEDGMENT

This work is carried out as part of a CEA-JSI 2019-2022 collaborative project within the CEA-MTZV bilateral agreement [1].

References

[1] Christophe Destouches et al., "A Review of 10 years of JSI – CEA collaboration on nuclear instrumentation development," Proceedings of 27th International Conference Nuclear Energy for New Europe, NENE, 10-13 Sep. 2018, Portorož, Slovenia.
[2] X-5 Monte Carlo Team, "MCNP™ – A General Monte Carlo N-Particle Transport Code, Version 5," LA-UR-03-1987, 2004.
[3] Žiga Štancar, Luka Snoj, and Loïc Barbot, "Reaction Rate Distribution Experiments at the Slovenian JSI TRIGA Mark II Research Reactor," TRIGA-FUND-RESR-002, International Handbook of Evaluated Reactor Physics Benchmark Experiments. Paris: NEA/NSC/DOC(2006)1, OECD NEA, 2017.
[4] Emeric Brun et al., "TRIPOLI-4®, CEA, EDF and AREVA reference Monte Carlo code," Annals of Nuclear Energy, Vol. 82, 2015, pp. 151-160, ISSN 0306-4549, doi: 10.1016/j.anucene.2014.07.053.
[5] Loïc Barbot et al., "Neutron Field Characterization of Irradiation Locations Applied to the Slovenian TRIGA reactor," 2013 3rd International Conference on Advancements in Nuclear Instrumentation, Measurement Methods and their Applications (ANIMMA), 2013, pp. 1-5, doi: 10.1109/ANIMMA.2013.6727954.
[6] Benoît Geslot et al., "Development and manufacturing of special fission chambers for in-core measurement requirements in nuclear reactors," 2009 1st International Conference on Advancements in Nuclear Instrumentation, Measurement Methods and their Applications, 2009, pp. 1-4, doi: 10.1109/ANIMMA.2009.5503816.
[7] René Brun and Fons Rademakers, "ROOT - An Object Oriented Data Analysis Framework, Proceedings AIHENP'96 Workshop," Lausanne, Sep. 1996, Nucl. Inst. & Meth. in Phys. Res. A 389 (1997) 81-86.
[8] Lei Lu, Yuefeng Qiu, and Ulrich Fischer, "Improved solid decomposition algorithms for the CAD-to-MC conversion tool McCad," Fusion Eng. Des., 124 (2017), pp. 1269-1272.
[9] Lei Lu et al., "Application of MCAM in generating Monte Carlo model for ITER port limiter, Chinese Journal of Nuclear Science and Engineering," 27 (3) (2007), pp. 277-281.
[10] Arjan Koning et al., "Validation of the JEFF-3.1 Nuclear Data Library," Nuclear Energy Agency, OECD, 2013.
[11] "International Reactor Dosimetry File 2002," IAEA Technical Reports Series No. 452, STI/DOC/010/452 92-0-105106-9, 2006.
[12] Arjan J. M. Plompen et al. The joint evaluated fission and fusion nuclear data library, JEFF-3.3. Eur. Phys. J. A 56, 181 (2020), doi: 10.1140/epja/s10050-020-00141-9.
[13] Andrej Trkov et al., "IRDFF-II: A New Neutron Metrology Library," Nuclear Data Sheets, Vol. 163, 2020, pp. 1-108, ISSN 0090-3752, doi: 10.1016/j.nds.2019.12.001.
[14] Christopher J. Werner et al., "MCNP6.2 Release Notes", LA-UR-18-20808, 2018.
[15] David A. Brown et al., "ENDF/B-VIII.0: The 8th Major Release of the Nuclear Reaction Data Library with CIELO-project Cross Sections, New Standards and Thermal Scattering Data," Nuclear Data Sheets, Vol. 148, 2018, pp 1-142, ISSN 0090-3752, doi: 10.1016/j.nds.2018.02.001.

Reactor Pulse Operation for Nuclear Instrumentation Detector Testing – Preparation of a Dedicated Experimental Campaign at the JSI TRIGA Reactor

Vladimir Radulović[1], Loïc Barbot[2], Grégoire De Izarra[2], Julijan Peric[3] and Igor Lengar[1]
[1]Jožef Stefan Institute, Reactor Physics Division, Slovenia
[2] CEA, DES, IRESNE, DER, Instrumentation Sensors and Dosimetry Laboratory, Cadarache, F-13108 Saint Paul-lez-Durance, France
[3]University of Ljubljana, Faculty of Mathematics and Physics, Slovenia
vladimir.radulovic@ijs.si

Abstract— **The availability of neutron fields with a high neutron flux, suitable for irradiation testing of nuclear instrumentation detectors relevant for applications in nuclear facilities such as material testing reactors (MTRs), nuclear power reactors and future fusion reactors is becoming increasingly limited. Over the last several years there has been increased interest in the experimental capabilities of the 250 kW Jožef Stefan Institute (JSI) TRIGA research reactor for such applications, however, the maximal achievable neutron flux in steady-state operation mode falls short of MTR-relevant conditions. The JSI TRIGA reactor can also operate in pulse mode, with a maximal achievable peak power of approximately 1 GW, for a duration of a few ms. A collaboration project between the JSI and the French Atomic and Alternative Energy Commission (CEA) was initiated to investigate absolute neutron flux measurements at very high neutron flux levels in reactor pulse operation. Such measurements will be made possible by special CEA-developed miniature fission chambers and modern data acquisition systems, supported by the JSI TRIGA instrumentation and activation dosimetry. Additionally, measurements of the intensity of Cherenkov light are proposed and being investigated as an alternative experimental method. This paper presents the preparatory activities for an exhaustive experimental campaign, which were carried out in 2019-2020, consisting of test measurements with not fully appropriate fission chambers, activation dosimetry and silicon photomultipliers (SiPMs) The presented results provide useful and promising experimental indications relevant for the design of the experimental campaign.**

Keywords — **Reactor Pulse, Fission Chamber, Activation Dosimetry, Cherenkov Light, Silicon Photomultiplier.**

I. Introduction

THE availability of neutron fields with a high neutron flux, suitable for irradiation testing of nuclear instrumentation detectors (such as fission and ionization chambers and self-powered neutron detectors) relevant for applications in nuclear facilities such as material testing reactors (MTRs), nuclear power reactors and future fusion reactors is becoming increasingly limited. Over the last several years there has been increased interest in the experimental capabilities of the 250 kW Jožef Stefan Institute (JSI) TRIGA research reactor for such applications. This is thanks to extensive experimental and computational efforts in the past to characterize the experimental conditions in the reactor, in large part in collaboration with the Instrumentation, Sensors and Dosimetry Laboratory (LDCI) of the French Atomic and Alternative Energy Commission (CEA) – Cadarache. The maximal achievable neutron flux in the reactor in steady-state operation mode is approximately 2×10^{13} n cm^{-2} s^{-1}, in contrast to the MTR-relevant neutron flux range 10^{14} – 10^{15} n cm^{-2} s^{-1}. But the JSI TRIGA reactor can also operate in pulse mode, where one of the reactor control rods is ejected from the reactor core, thus introducing a sufficiently large reactivity to achieve prompt supercriticality. This operation mode is made possible by the TRIGA fuel, in the form of a dispersion of uranium and zirconium hydride, which gives rise to a large, prompt and negative temperature coefficient of reactivity. The resulting time dependence of the reactor power is a pulse, the peak power and duration depending on the initial inserted reactivity. In pulse mode, the maximal achievable peak power is approximately 1 GW, corresponding to a peak neutron flux level of the order of 10^{17} n cm^{-2} s^{-1}, for a duration of a few milliseconds.

Up to now pulse operation mode at the JSI TRIGA reactor has been investigated mainly in the academic context, for the validation and improvement of the Fuchs-Hansen model [1-6] describing the reactor behaviour during reactor pulses and for educational activities, e.g. in the framework of practical courses for students of the University of Ljubljana or dedicated courses, to be offered in the framework of the ENEEP platform [7], however its application for irradiation experiments was of limited scope. Recognizing the potential of reactor pulse mode for nuclear instrumentation detector testing, in particular the possibility of extending the useful neutron flux range up to MTR-relevant levels, in 2019, a bilateral collaboration project between the CEA and JSI was initiated. The aim of the project is the performance of absolute neutron flux measurements at very high neutron flux levels in reactor pulse operation. The measurements will be made possible by special CEA-developed miniature fission chambers (fissile deposit mass targeted at 50 ng) and by modern, validated, wide dynamic range data acquisition systems, in particular the CEA-developed MONACO system. In addition to the JSI TRIGA nuclear instrumentation, providing information on the peak power and released energy, activation dosimetry will be employed as a reference for the normalization of the recorded detector signals (neutron flux integral) for reactor pulses with differing characteristics. Finally, as an alternative experimental method enabling the measurement of the reactor power dependence during pulse operation, measurements of the intensity of Cherenkov light are proposed and being investigated.

This paper presents the 2019-2020 preparatory activities for an exhaustive experimental campaign to be carried out at the JSI TRIGA reactor jointly by CEA and JSI researchers in 2021. A series of test measurements using not fully appropriate fission chambers in reactor pulse operation was performed using a Keithley electrometer and the MONACO fission chamber acquisition system. Activation dosimetry measurements were performed for several thermal and fast neutron sensitive nuclear reactions. Photodiodes and silicon photomultipliers (SiPMs) have been tested in steady state and pulse operation modes. The presented results provide useful and promising experimental indications relevant for the design of the experimental campaign.

II. Fission Chamber Measurements

The LDCI Lab has been manufacturing specific miniature fission chambers with MI cable for more than two decades at the CEA Cadarache fission chamber (FC) workshop. The FCs are supported with some unique calibration methods [8] also developed at CEA, enabling absolute neutron flux measurements.

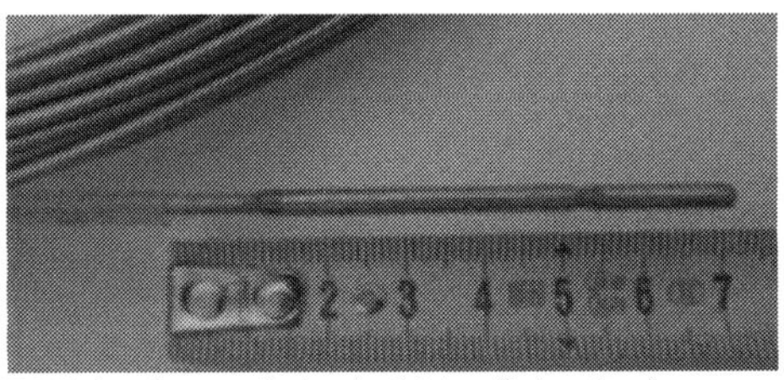

Fig. 1. CEA Cadarache manufactured miniature fission chamber.

In the last nine years, CEA developed a state-of-the art electronic acquisition system, named MONACO, which stands for 'Multichannel Online Acquisition in Campbell mOde'. In its current implementation, the system enables FC acquisition in pulse mode (i.e. counting pulses due to ionizations in the active gas, caused by energetic fission products), and in variance, or Campbell, mode (applicable when pulses are piling up, the signal variance being proportional to the incoming neutron flux) and current mode (pulse number become so high that their individual current are forming a measurable DC current proportional to the neutron and gamma environment). This implementation has been validated through numerous dedicated experimental campaigns, including several campaigns in the JSI TRIGA reactor [9, 10]. The current implementation (i.e. MONACOv2), now at TRL7 level, will be used in this project; the commercial version being currently industrialized in collaboration with the INSTRUMENTATION TECHNOLOGIES Company [11]. For this project, efforts will also be directed into increasing the acquisition rate of the MONACO2 system with a target of 1000 readings per second, enabling satisfactory discrimination of reactor pulses.

Fig. 2. CEA developed MONACOv2 acquisition system for fission chambers.

In preparation of the project experimental campaign, measurements in steady state and reactor pulse mode were performed in September 2019 using two existing CEA miniature fission chambers, with ~10 µg ^{235}U fissile deposit. The FCs had an outer diameter of 3 mm, and were positioned in an aluminium guide tube, located within Measurement Positions (MPs) in the lattice of the JSI TRIGA reactor core. A schematic view of the reactor core, a guide tube and a fission chamber is presented in Fig. 3.

Measurements were made using the MONACOv2 acquisition system [12]. A sequence of reactor pulses was carried out with varying inserted reactivity. Table 1 reports the reactor pulse operation parameters, as obtained from the JSI TRIGA reactor command control instrumentation: the pulse ID number, the prompt reactivity, the peak power, the full width at half maximum (FWHM) and the released energy.

TABLE I
PULSE PARAMETERS OBTAINED FROM THE JSI TRIGA INSTRUMENTATION

Pulse ID	ρ_{prompt} [$]	Peak P [MW]	FWHM [ms]	Rel. E. [MWs]
360	1,10	1,0	503*	0,1
361	1,15	1,3	6471*	0,1
362	1,20	1,9	3818*	0,2
363	1,25	3,3	2378*	0,3
364	1,30	6,0	267	0,5
365	1,35	10,3	166	0,9
366	1,40	16,7	119	1,4
367	1,40	20,0	106	1,6
368	1,40	19,9	107	1,6
369	2,00	240,2	27	7,5

*FWHM values obtained from the TRIGA instrumentation for low peak power pulses are not reliable.

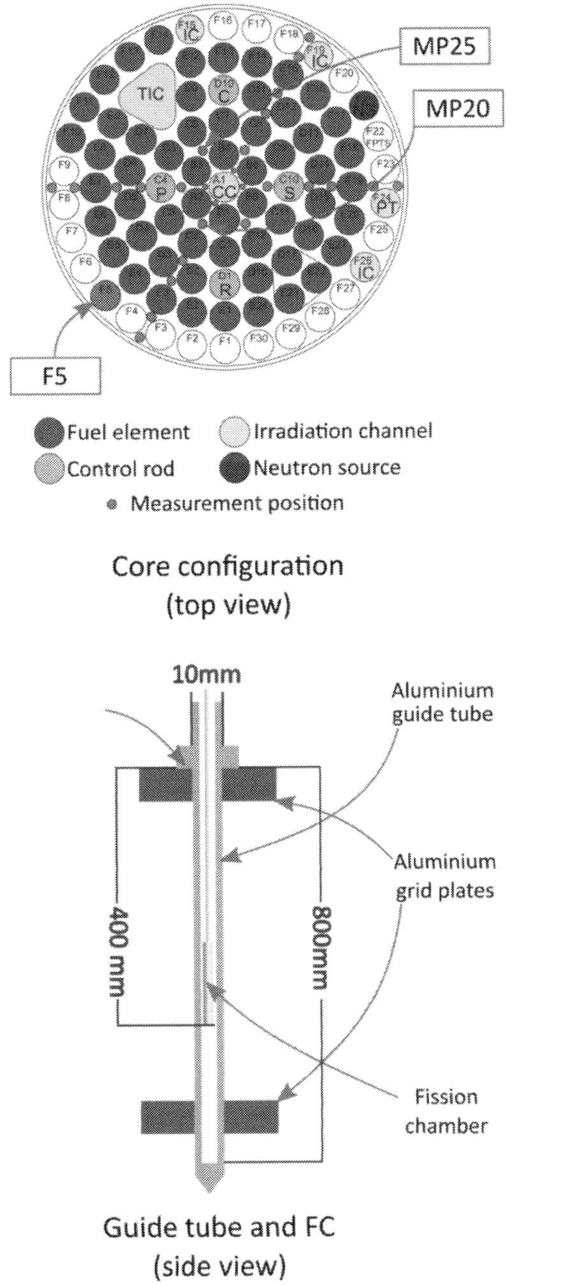

Fig. 3. Top: Schematic view of the JSI TRIGA reactor core configuration (top view). Bottom: Schematic view of a fission chamber and a guide tube located between the reactor support plates.

The signal acquisition rates for the MONACO system varied approximately from 10 to 1000 samples s^{-1}. The measured pulse signals are displayed in Figs. 4 and 5.

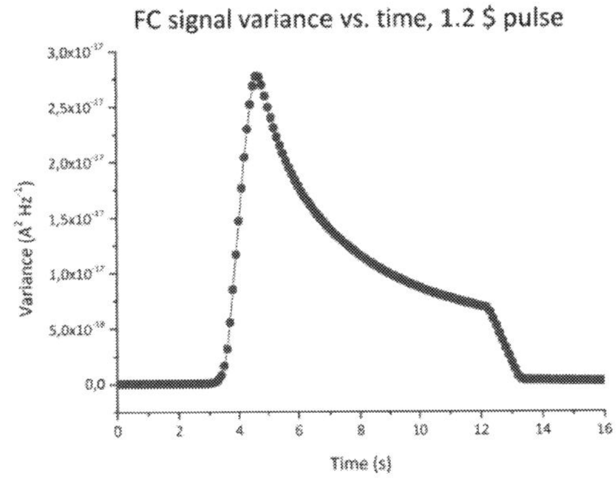

Fig. 4. Measured FC signals during a 1.2$ reactor pulse. ^{235}U FC inserted in the core at the fuel mid plane.

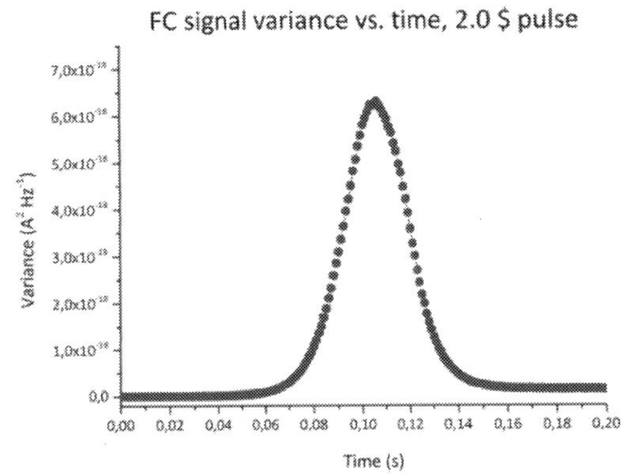

Fig. 5. Measured FC signals during a 2$ reactor pulse. ^{235}U FC placed ~200mm above the core.

Despite the fact that 10 μg ^{235}U fissile deposits were not optimal for such measurements, the preliminary tests show a very satisfactory linearity between the maximal FC signals and the measured pulse peak power, as shown in Fig. 6.

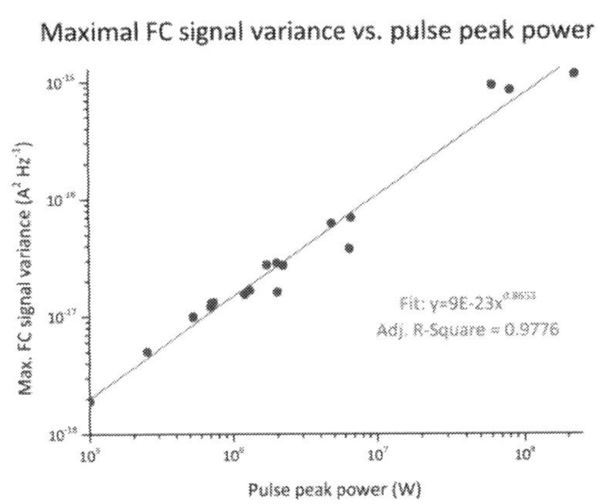

Fig. 6. Preliminary linearity of ^{238}U FC signals versus reactor peak power.

Strong efforts will also be devoted for the final experimental campaign to optimize FC fissile deposits to high reactor power to perform absolute neutron flux measurements. Thanks to a series of three fine chemical analysis, target fissile masses of ^{235}U and ^{238}U are planned down to <50ng.

III. Activation Dosimetry Measurements

Neutron activation dosimetry is the reference method in the determination of the neutron flux or fluence, and are included in the present work as a reliable experimental technique to obtain measurements of the total neutron fluence which can be delivered in pulse operation mode to samples of interest (e.g. neutron detectors, etc.). A series of test irradiations was performed in reactor pulse mode in which samples of certified reference materials Al-0.1%Au, Al-0.1%Co and pure Ni were irradiated within polyethylene capsules in the Central Channel (CC) of the JSI TRIGA reactor. The induced activities in the samples were subsequently measured using two absolutely calibrated High-Purity Germanium (HPGe) detectors.

Fig. 7 displays the measured specific end-of-irradiation (EOI) activities (in Bq per mg of target material – e.g. per mg of Co in Al-0.1%Co) vs. the pulse prompt reactivity. The induced activities are proportional to the total neutron fluence received during the irradiations, which is in turn proportional to the released energy in the pulse. According to the Fuchs-Hansen model [1] for reactor pulse operation, the energy released in a pulse is proportional to the prompt reactivity. The observed dependences clearly exhibit a linear behavior, taking into consideration the experimental uncertainties. In this work, these are relatively high (of the order of 3-4 %) in the context of reactor dosimetry measurements, on account of the low induced activities.

In the campaign, dosimetry measurements will serve as the reference for absolute normalization of the measured time-dependent signals. In order to be able to complete a relatively high number of activation measurements during the experimental campaign (of the order of 100), the choice of optimal sample composition will be considered, enabling higher induced sample activities.

IV. Cherenkov Light Measurements

As an alternative experimental method to record in detail the time dependence of the reactor power in real time, Cherenkov light measurements are considered. Cherenkov light is clearly visible in the reactor core during operation, especially in the outermost and mostly empty "F" ring, starting from approximately 10 kW. To investigate this possibility, a measurement setup was created consisting of an aluminium channel (inside diameter approximately 34 mm), closed on the bottom end and inserted into positon F4 in the reactor core. The channel had a slight curvature (approximately 0.5 m of side displacement between the top and bottom ends, 6 m apart) in order to limit neutron and gamma radiation streaming from the reactor core to the reactor platform. Approximately 1 liter of demineralized water was poured into the channel, serving as a

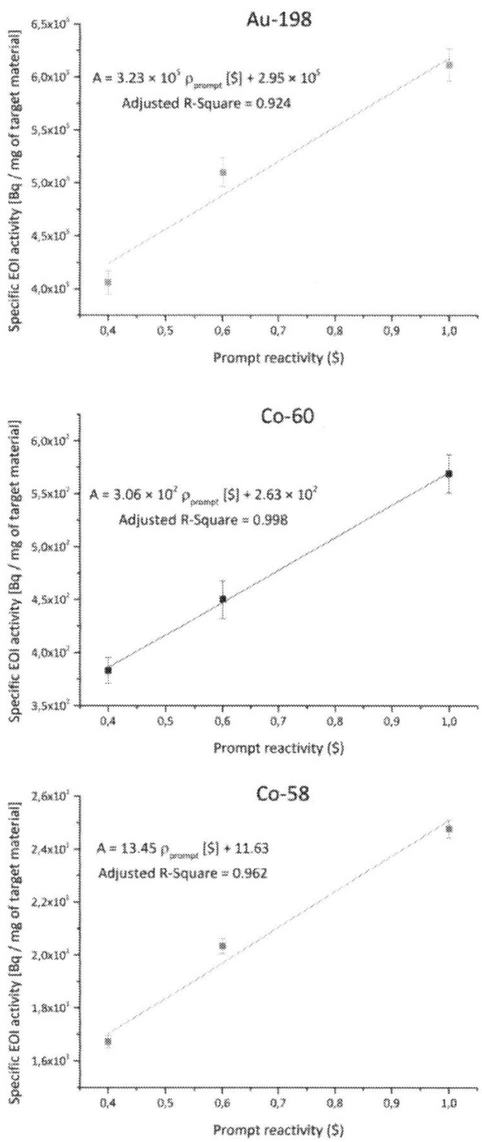

Fig. 7. Measured specific end-of-irradiation activities [Bq per mg of target material] for Au-198, Co-60 and Co-58, originating from the Au-197(n,γ), Co-60(n,γ) and Ni-58(n,p) reactions, as a function of the prompt reactivity. A linear dependence is expected.

Cherenkov light radiator. The inner surface of the channel was smooth, which enabled the Cherenkov light produced in the water to be reflected and to be visible at the reactor platform level Fig. 8 a photograph of the setup on the JSI TRIGA reactor platform and Fig. 9 shows a photograph of the Cherenkov light inside the channel during reactor operation.

https://doi.org/10.1051/epjconf/202125304019

Fig. 8. Photograph of the Cherenkov light measurement setup on the JSI TRIGA reactor platform.

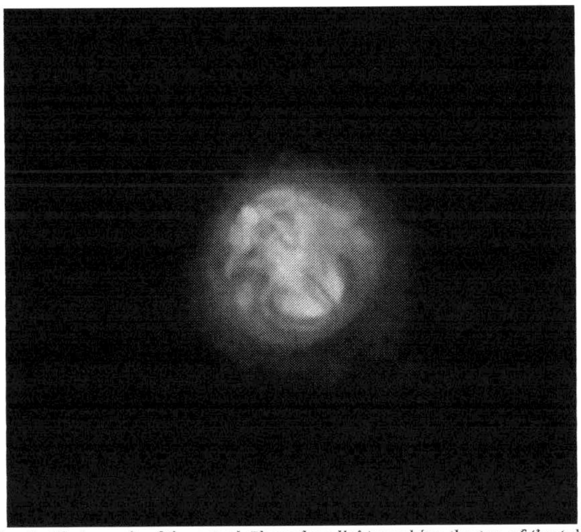

Fig. 9. Photograph of the actual Cherenkov light reaching the top of the tube during reactor operation.

The detector selected for light intensity measurements was a silicon photomultiplier (SiPM). Such detectors are nowadays inexpensive and are able to operate in single photon (pulse) mode as well as in current mode. The SiPM module used in the present work, along with the readout board, power supply were aquired from KETEK. A RedPitaya board [12] operated by a laptop computer was used to measure the voltage across the readout resistor in the SiPM readout board, proportional to the current flowing through the SiPM. A calibration of the SiPM detector was carried out in steady-state operation at several power steps from 2 kW to 250 kW. Fig. 10 displays the measured SiPM signal with the power calibration applied and the reactor power in steady state operation, as a function of time, during the calibration phase. The SiPM signal dependence is clearly directly proportional to the reactor power, suggesting the feasibility of Cherenkov light measurements to obtain the time dependence of the reactor power during reactor pulses.

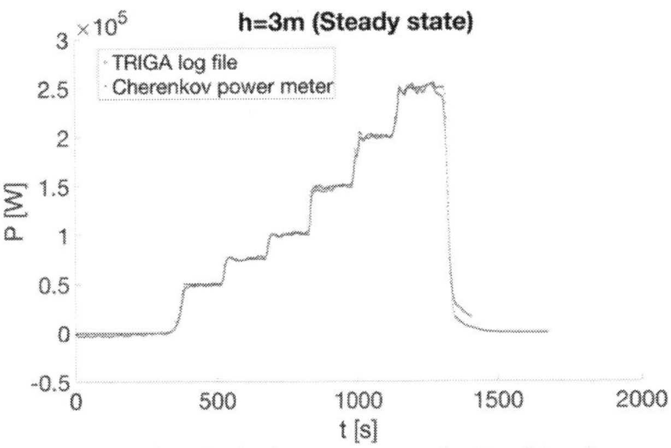

Fig. 10. Measured SiPM signal and reactor power as a function of time, for steady state reactor operation.

Subsequently, a series of pulses with increasing inserted reactivity was made and the SiPM signal was recorded. Fig. 11 displays the measured SiPM signals in reactor pulse mode, multiplied by the calibration factor determined in the calibration phase.

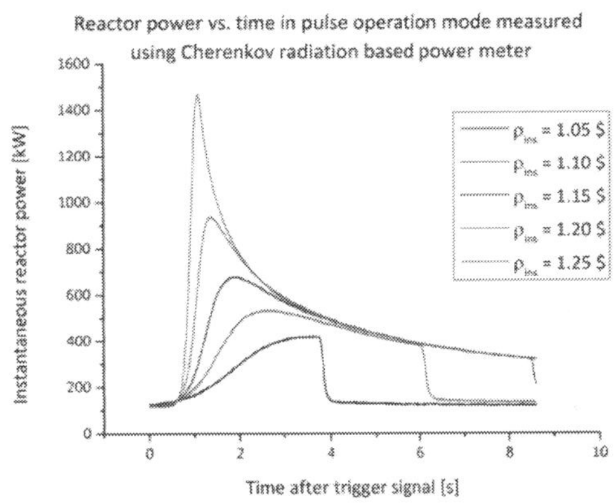

Fig. 11. Measured SiPM signal during reactor pulse operation, with power calibration factor applied.

A comparison of Figs. 4 and 11 (1.2 $ pulse) shows that the measured FC and SiPM signals are qualitatively very similar, additionally confirming the applicability of the use of Cherenkov light measurements in reactor pulse mode.

V. Discussion

The results obtained in the experimental activities presented in this work in preparation for a dedicated campaign demonstrate several important points. Miniature fission chambers in conjunction with the MONACO acquisition system perform well, in terms of their ability to measure a signal changing rapidly by several orders of magnitude. The measurements in the present work were performed in fission chamber variance mode. However, pulse acquisition mode (rapid counting of individual pulses, generated by fission events) is considered for the experimental campaign, as this operation mode enables more reliable absolute fission rate measurements and was used for benchmark fission rate measurements [13].

The obtained neutron dosimetry results demonstrate the feasibility of such measurements in reactor pulse mode for monitoring the total neutron fluence or the released energy. The EOI activities were relatively low (in the actual samples at most around 60 kBq for Au-198 in Al-0.1%Au samples and at minimum 2 Bq for Co-60 in Al-0.1%Co samples. In the future experimental campaign, the low induced activities may be compensated to some degree by longer counting times in the gamma spectrometry measurements - however this may not be achievable for a large number of monitored pulses – and larger sample masses. In the present work, the material Al-0.1%Co was used for the Co-59(n,γ) reaction rate. To optimize the Co-60 activities for subsequent measurements, in the experimental campaign use will be made of Al-1%Co, supplied by the CEA, as it is currently unavailable from the supplier.

The obtained results from the Cherenkov light intensity meter also clearly demonstrate its applicability for reactor pulse monitoring.

The experimental campaign will enable clear indications to support technological improvements in the JSI TRIGA pulse recorder, which was designed to cover a wide range of peak power levels, up to 2 GW, and consequently is not able to provide accurate information for low peak power pulses, which are deemed most interesting for instrumentation testing activities.

VI. Conclusions

This paper presents the results of preparatory activities for a dedicated experimental campaign at the JSI TRIGA reactor on pulse operation. Measurements with miniature fission chambers and neutron dosimetry are of primary importance, and good results have been observed for both techniques. The results will also support required adjustments to further optimize the measurements, e.g. using longer counting times and higher sample masses for neutron dosimetry and extremely low fissile deposit masses in FCs using very accurate chemical analysis techniques.

Measurements of Cherenkov light were tested as an alternative and inexpensive technique to obtain experimental data on the pulse shapes. A simple setup based on a SiPM detector was created, refined and tested with success. Its use is proposed for in the dedicated experimental campaign.

The campaign results will enhance the JSI TRIGA reactor experimental capabilities in the context of detector and instrumentation testing activities, research in radiation hardness and potentially enable the JSI TRIGA reactor to support research in new fields or applications.

Acknowledgment

The authors acknowledge the financial support from the Slovenian Research Agency (research core funding No. P2-0073 – Reactor Physics, research project No. NC-0011 - Absolute radiation measurements at very high neutron flux levels in reactor pulse mode).

References

[1] G. I. Bell, S. Glasstone, Nuclear reactor theory, Van Nostrand Reinhold Company, New York, 1970.

[2] M. Ravnik, "Experimental Verification of Adiabatic Fuchs-Hansen Pulse Model", 4th Regional Meeting Nuclear Energy in Central Europe, September 7-10, 1997, Bled, Slovenia.

[3] A. Pungerčič, I. Vavtar, L. Snoj, "Analysis of the JSI TRIGA pulse experiments", in *Proc. RRFM 2018*, Munich, Germany, 2018.

[4] I. Vavtar, A. Pungerčič, L. Snoj, "Utilisation of JSI TRIGA pulse experiments for testing of nuclear instrumentation and validation of transient models", in *Proc. ANIMMA 2019*, Portorož, Slovenia, 2019. (EPJ Web of Conferences, ISSN 2100-014X, Vol. 225, 2020).

[5] I. Vavtar, L. Snoj, "Evaluation of approximations in reactor pulse", in *Proc. PHYSOR 2020*, Cambridge, UK, 2020.

[6] I. Vavtar, D. Ćalić, A. Pungerčič, L. Snoj, "Evaluation of approximation and uncertainties in theoretical models describing pulse mode operation" in *Proc. RRFM* 2020.

[7] M. Cagnazzo, H. Boeck, Š. Čerba, "The European Nuclear Experimental Educational Platform (ENEEP) for education and training", Atw Internationale Zeitschrift fuer Kernenergie, 65(5), 2020, 251-256.

[8] B. Geslot, T.C. Unruh, P. Filliatre, C. Jammes, J. Di Salvo, S. Bréaud, and J-F. Villard, "Method to Calibrate Fission Chambers in Campbelling Mode", IEEE Transactions on Nuclear Science (Volume: 59, Issue: 4, Aug. 2012).

[9] L. Barbot et al., "MONACO v2: Multipurpose and Integrated Data Acquisition System for On-line Neutron and Gamma Measurements", in *Proc. ANIMMA 2019*, Portorož, Slovenia, 2019 (EPJ Web of Conferences, ISSN 2100-014X, Vol. 225, 2020).

[10] Y. Moline et al., "MONACO: a Case Study for a New System On Chip and Stream based Architecture for Nuclear Instrumentation", *Proc. ANIMMA 2019*, Portorož, Slovenia, 2019 (EPJ Web of Conferences, ISSN 2100-014X, Vol. 225, 2020).

[11] L. Barbot, G. De Izarra, C. Destouches, M. Cargnelutti, S. Zorzut and D. Bisiach, "Development of a wideband current amplifier dedicated to Fission Chamber measurement", to be presented at the 7th ANIMMA Conference, Prague, June 21-25, 2021.

[12] https://www.redpitaya.com/

[13] Ž. Štancar, L. Snoj, and L. Barbot, "Reaction Rate Distribution Experiments at the Slovenian JSI TRIGA Mark II Research Reactor, TRIGA-FUND-RESR-002," in International Hand-book of Evaluated Reactor Physics Benchmark Experiments. Paris:NEA/NSC/DOC(2006)1, OECD NEA, 2017.

Online optical refractive index measurement in research reactor core

Gary Fourneau[1], Marion Agoyan[1, 2], Guy Chemol[1], Ayoub Ladaci[1], Hicham Maskrot[1], Christophe Destouches[3], Damien Fourmentel[3], Sylvain Girard[2], Aziz Boukenter[2]

[1] CEA, DES/ISAS/DPC/SEARS/LISL, Saclay, France
[2]Université Jean Monnet, CNRS, Graduate School Optics Institute, Laboratoire Hubert Curien UMR 5516, Saint-Etienne, France
[3]CEA, DES/IRESNE/DER/SPESI, Cadarache, France
gary.fourneau@cea.fr

Abstract*— There is a growing interest in fiber optic measurements for applications in radiation environments. Optical fiber sensors and diagnostics can monitor many parameters of interest inside a research reactor core. For some applications, fiber optics are combined with an optical system that collects or focuses the light beam. The Radiation-Induced-Refractive-Index-Change (RIRIC) of the used glasses appears then as major phenomenon as it is a determining value for the sensor optical function. In the framework of the development of a radiation hardened confocal chromatic sensor, we implemented an on-line refractive index measuring device in order to test in a reactor core various glasses, candidates to be implemented into the sensor. The measurement relies on interferometry, which is a challenge because of the small volume available, the impossibility to make optical adjustments once installed, and the high temperature of operation. Precisely, the quantity retrieved is the optical path (product of the length L by the optical refractive index n), when L is well known, we can deduce n. But under high neutron fluence, some variation in density can be observed. The targeted online measurement of the refractive index therefore becomes an optical path measurement. We will present the device, the principle of measurement, and the first results of some index change measurement, produced by a temperature ramp from 20 °C and 350 °C. We have obtained original data for most of the candidate glasses used to design the optical system.

***Keywords* —Optical fiber sensor, Radiation-Induced-Attenuation, Interferometry, Radiation-Induced-Refractive-Index-Change, Interferometric measurement, Rad-hard Glasses, High neutron fluence.**

I. Introduction

THIS work is made in the framework of the Instrumentation for Nuclear Applications (INSNU) project, aiming to develop instrumentation and sensors for experiments in the Jules Horowitz Reactor (Cadarache, France). Among other sensors, we develop optical ones and particularly a confocal chromatic sensor [1]. While on some measurements on intrinsic optical fiber sensor, the fiber is the only passive component exposed to the harsh environment constraint, in the case of extrinsic sensors, such as the confocal chromatic [1], additional components are involved and need to be characterized. In our case, the sensor relies on the use of an optical system made of several glass components, thus leading to some issues caused by the influence of radiation and high temperatures on their performances. As for fiber optics in such reactor environment [2, 3], the two main radiation effects on bulk glasses are the Radiation Induced Attenuation (RIA) and the density change leading to dimensional change and Radiation Induced Refractive Index Change (RIRIC). Unfortunately, little data is available in the literature, a few studies are available but without the neutron contribution and at much lower gamma dose levels, they are then not representative of what is expected in an in-pile experiment [4].

In order to collect data on glass darkening and refractive index change we will design and perform in 2022 two specific experiments inside the TESCA (Test in Capsule) irradiation devoted to fiber optic sensor (FOS) -in Belgium Reactor 2. The two small measurement systems aim to provide online measurements, allowing to characterize both transient and permanent effects. First systems is designed to collect RIA data while the second is devoted to refractive index change measurements, more precisely for optical path measurement n × L. We focus here on the n×L measurement using a white light interferometry approach.

The specifications expected for the system are as follows:

- Mechanical diameter : 9 mm
- Relative accuracy for variations: $10^{-4} – 10^{-5}$
- Maximum temperature: 350°C
- Neutron fluence : 3.10^{19} n/cm^2
- Gamma dose : 1 GGy
- Duration of the experiment : 3 to 4 weeks
- Spectral range: 800 nm to 1200 nm [1]

Of course, it would have been preferable to measure separately n and L. But with such constraints, the usual techniques making it possible to measure n and L separately cannot be implemented inside a reactor core. We are also preparing additional samples for measurements to be done before and after irradiation on glass samples to try to decorrelate between the variations of n and L.

In order to collect as much information as possible, the n×L measuring device is expected be able to:

- Test various types of glass.
- Measure the optical path variation induced by

radiations at a constant temperature of irradiation.

- Operate at several wavelengths within the spectral domain of interest
- Also (out of pile experiment) measure the variation of the optical path induced by the temperature only (thermal dilatation and impact on the refractive index).

The optical fiber will carry the probe beam and will collect and carry back to the optical signal to the conditioner.

To perform such a measurement, we monitor the evolution of the channeled spectrum resulting from the interference between the two parallel faces of a cylindrical glass sample. In this paper, we present the architecture of the measurement device and its operation principle. We will present the preliminary results obtained in the laboratory varying the temperature only.

II. The optical path measurement device, signal conditioner and signal processing

A. Experimental setup

To respond to our needs, we developed an opto-mechanical device using a Single Mode (SM) optical fiber. Fig. 1 shows the developed setup.

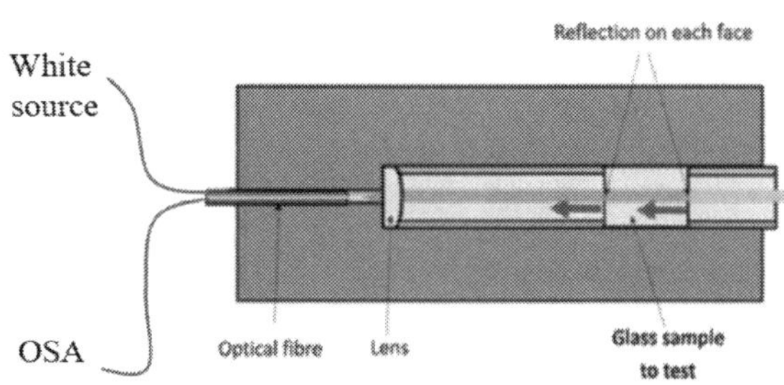

Figure 1 : Schematic diagram of the interferometric setup to measure refractive index or optical path change under radiative environment.

The light comes from a white supercontinuum source, which has a nearly constant intensity from 800 nm to 1500 nm (except a peak around the pump signal at 1064 nm); this allows us to use a single source to observe the wide spectral domain of interest. The light beam coming out of the fiber goes through a lens to collimate it. A part of the beam is then reflected on each face of the glass sample under test. These reflected rays interfere and are reinjected inside the fiber. Finally, the light is sent to an Optical Spectrum Analyzer (OSA) which allows observing the channeled spectrum (Fig. 2) resulting from the interference. The theoretical shape for such spectrum is given by the following formula, it is then needed to process the measured signal in order to extract the n×L value:

$$I(\lambda,T) = I_0(\lambda)\left[A + B\cos\left(2\pi\left(\frac{2\big(n(\lambda,T)\times L(T)\big)}{\lambda}\right) + \varphi_0\right)\right] \quad (1)$$

where $I_0(\lambda)$ is the source's intensity spectrum, A and B are constants, n is the refractive index of the glass sample and L its length. λ is the wavelength and T the temperature.

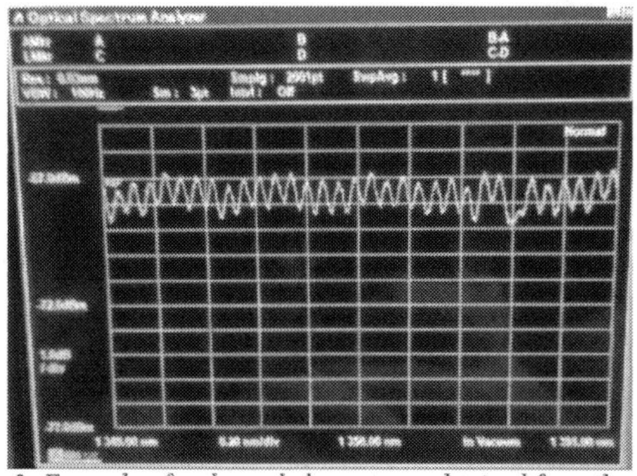

Figure 2: Example of a channeled spectrum obtained from the device.

B. Improved setup: the double interferometer

The Interfringe (If) of the channeled spectrum is :

$$If \sim \frac{\lambda^2}{2.n.L} \quad (2)$$

Our samples are approximately 10 mm long. With n~1.5 and λ= 1 µm, (2) gives: If~ 30 pm which is roughly the minimum spectral resolution available with our spectrometer. Working with an If in the same range than the spectral resolution affects the modulation, especially at lower wavelengths.

This led us to design a more complex setup than the one illustrated in Fig. 1, that consists in two interferometers in series. This architecture allows also increasing the accuracy of the measurement. The first interferometer is the previously described small measurement system while the second one is made of a cavity in the air with a length Lc close to n x L. The resulting spectrum corresponds to the product of the channeled spectrum of both cavities taken separately:

$$I_{tot}(\lambda,T) = I_1(\lambda,T)\times I_2(\lambda,T)$$

$$I(\lambda,T) \approx \cos\left(2\pi\left(\frac{2n(\lambda,T)\times L(T)}{\lambda}\right)\right)\times\cos\left(2\pi\left(\frac{2Lc}{\lambda}\right)\right) \quad (3)$$

With a little bit of trigonometry we can show that this signal is composed of four main frequencies:

- $2n \times L$
- $2Lc$
- $2(n \times L + Lc)$
- $2|n \times L - Lc|$

n x L and Lc both give high frequency signals, affected by the limited resolution of the OSA, which also does not permit to observe at all the signal with the $(n \times L + Lc)$ frequency.

However, since n x L and Lc are about the same length, the frequency last remaining in our signal is rather low and easy to measure accurately. This frequency only depends on the difference of length between the two cavities. If we know the exact length of the second cavity, we have access to the length of the optical path measured with a far better accuracy than with our first setup.

C. Signal processing principle

In order to have access to the optical length's value, we have to process the channeled spectrum. Firstly, let us consider that n is mainly independent of the wavelength. In that case, the spectrum is a periodic function of $\frac{1}{\lambda}$ with a frequency: $2n \times L$. To determine this frequency, we apply a Fast Fourier Transform (FFT) on the acquire signal and we search for a peak near the expected value of 2n×L, the abscissa of this peak gives us the exact value of the optical path measured.

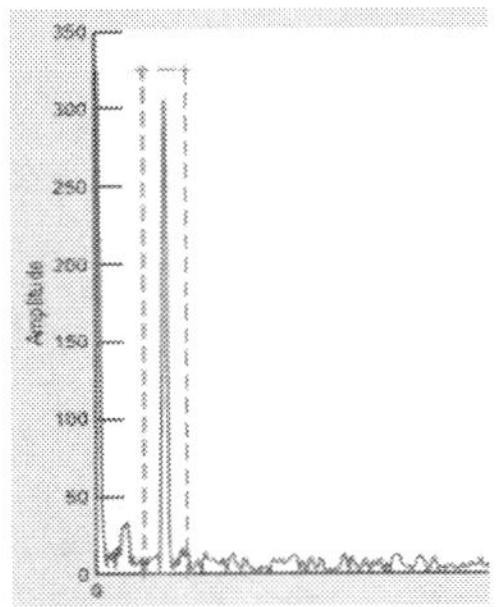

Figure 3: Example of a FFT applied to an experimental signal. The red rectangle corresponds to the domain where we expect to find a peak beforehand. The X-axis corresponds to the optical length measured

That is what happens with an extrinsic Fabry-Perot interferometer [5]. However, doing this way, experimental results are shifted by 150 µm to 250 µm. In fact, the glasses' refractive indexes depend on the wavelength. The chromaticity is responsible for the error observed. With this method, we do not have access to the refractive index but to the group index of the tested glass, the actual frequency retrieved by the FFT is not $n(\lambda_0) \times L$ as expected but rather :

$$n' \times L = \left(n(\lambda_0) - \frac{dn}{d\lambda}(\lambda_0).\lambda_0\right) \times L$$

It is then needed to add a correction to the initial value obtained by reading the peak's abscissa:

$$n(\lambda_0) = n' + \frac{dn}{d\lambda}(\lambda_0).\lambda_0 \qquad (4)$$

This mean that we need to know the value of $\frac{dn}{d\lambda}(\lambda_0)$ beforehand or we assume that this value does not vary with the temperature or radiations and we confirm it afterwards.

We have checked that using this signal processing and the correction on a simulated spectrum with a given glass and length, we get n × L with a relative error of 10^{-6} which may be caused by computing errors.

We also developed another method for processing the signal that consists of following the evolution of a single peak of the channeled spectrum to determine the variations of optical path [6]. While much more accurate in theory, this solution is actually not adapted to the TESCA experiment as it requires small variation between too acquisitions of the peak position to ensured that the peak observed has not been replaced by another one. We will test several samples in parallel, with therefore some delay between two acquisitions. Operation in a reactor core lacks stability to allow this measurement; which on the other hand permitted us to measure the stability of the compensating interferometer.

III. Results for variation of n×L with the temperature only

Obtaining and maintaining the optical adjustment which gives the interferometric signal is a difficult task, especially with the disturbances due to the increase in temperature. Once that was guaranteed, we measured the variation of the refractive index with the temperature of various glasses. Here, we know the value L(T), thanks to datasheets giving the thermal dilatation up to high temperatures, therefore by dividing the n(T)×L(T) product by L(T), we obtained n(T). Fig.6 display the result obtained for SF6G05 at three wavelengths.

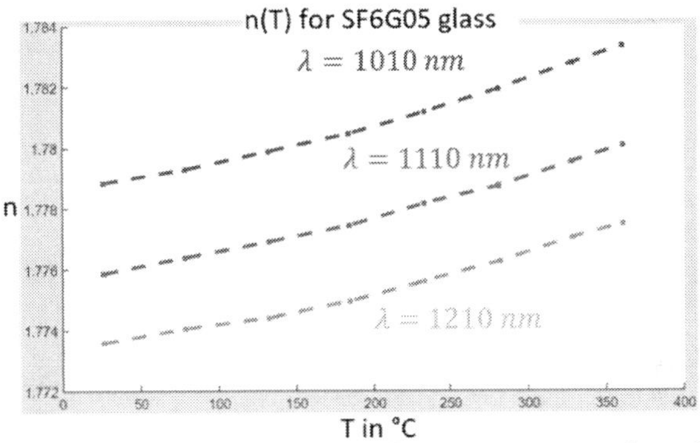

Figure 4 : n(T) obtained for SF6G05 glass using our setup and signal processing

We observe a difference of nearly 10^{-3} for absolute values of the refractive index at low temperature compared to the theoretical values provided by the glassmaker Schott. For other glasses, the offset is in the same range (few 10^{-3}). Possible reasons for these errors are some differences due to different glass batches, error in measuring the glass sample or the second cavity lengths. These errors will remain constant in the planned irradiation and should not affect our ability to measure n×L variations under irradiation.

IV. Conclusion

In order to develop a rad-hard confocal chromatic sensor, data on the behavior of glasses under radiation are expected to result from the TESCA irradiation (2022, in the BR2 research reactor). We have developed a smart miniature device in order to measure by interferometry the variations of the optical path n × L of some glass samples during in-core irradiation. This device will be soon set in the research reactor's core for online measurements.

Today, the robustness of the device has been tested by measuring the n×L variations with temperature, giving access to n(T) up to 350°C, data not currently available in the literature for most glasses. Work is still in progress to consolidate the accuracy of our setup and remains necessary to design an optical system able to operate at high temperature.

REFERENCES

[1] M. Agoyan et al, « Confocal chromatic sensor for displacement monitoring in research reactor », Animma 2021.

[2] S. Girard et al., "Overview of radiation induced point defects in silica-based optical fibers," Reviews in Physics, 2019, doi: https://doi.org/10.1016/j.revip.2019.100032

[3] L. Remy, G. Cheymol, A. Gusarov, A. Morana, E. Marin and S. Girard, "Compaction in Optical Fibres and Fibre Bragg Gratings Under Nuclear Reactor High Neutron and Gamma Fluence," in *IEEE Transactions on Nuclear Science*, vol. 63, no. 4, pp. 2317-2322, Aug. 2016, doi: 10.1109/TNS.2016.2570948.

[4] C. Muller, T. Lépine, T. Allanche, A. Boukenter, P. Paillet, S. Girard, and Y. Ouerdane "Potential performance loss and compensation techniques of a lens under ionizing radiations", Proc. SPIE 10690, Optical Design and Engineering VII, 106900T (5 June 2018)

[5] G. Cheymol, A. Verneuil, P. Grange, H. Maskrot and C. Destouches, "High-Temperature Measurements With a Fabry–Perot Extensometer," in *IEEE Transactions on Nuclear Science*, vol. 67, no. 4, pp. 552-558, April 2020, doi: 10.1109/TNS.2020.2975707.

[6] C.Z. Tan, J. Arndt, "Temperature dependence of refractive index of glassy SiO2 in the infrared wavelength range", *Journal of Physics and Chemistry of Solids*, vol. 61, pp. 1315–1320, Nov. 1999

Confocal chromatic sensor for displacement monitoring in research reactor

Marion Agoyan[1,2], Gary Fourneau[1], Guy Cheymol[1], Ayoub Ladaci[1], Hicham Maskrot[1], Christophe Destouches[3], Damien Fourmentel[3], Sébastien Gérand[4], Jérôme Gaillard-Groléas[4], Matthieu Desjacques[4], Sylvain Girard[2] and Aziz Boukenter[2].
[1]CEA, DES/ISAS/DPC/SEARS/LISL, Saclay, France
[2]Université Jean Monnet, CNRS, Graduate School Optics Institute, Laboratoire Hubert Curien UMR 5516, Saint-Etienne, France
[3]CEA, DES/IRESNE/DER/SPESI, Cadarache, France
[4]STIL Marposs, Aix-en-Provence, France
Corresponding author: marion.agoyan@cea.fr

Abstract—**Confocal chromatic microscopy is an optical technique allowing measuring displacement, thickness, and roughness with a sub-micrometric precision. Its operation principle is based on a wavelength encoding of the object position. Historically, the company STIL based in the south of France has first developed this class of sensors in the 90's. Of course, this sensor can only operate in a sufficiently transparent medium in the used spectral domain. It presents the advantage of being contactless, which is a crucial advantage for some applications such as the fuel rod displacement measurement in a nuclear research reactor core and in particular for cladding-swelling measurements. The extreme environmental conditions encountered in such experiments i.e. high temperature, high pressure, high radiations flux, strong vibrations, surrounding turbulent flow can affect the performances of this optical system. We then need to implement mitigation techniques to optimize the sensor performance in this specific environment. Another constraint concerns the small volume available in the irradiation rig next to the rod to monitor, implying the challenge to conceive a miniaturized sensor able to operate under these constraints.**

Keywords — **Confocal Chromatic Sensor, Pressurized Water Reactor, Radiation-Induced-Attenuation, Radiation-Induced-Refractive-Index-Change, Temperature, Optical design.**

I. Introduction

THIS work is made in the framework of the instrumentation for nuclear applications (INSNU) project, aiming to develop instrumentation and sensors for experiments in the Jules Horowitz reactor (Cadarache, France). Among other technologies are developed optical sensors: fiber Bragg gratings [1], fiber-optic distributed sensors to measure temperature and strain, pyrometric measurement [2], Fabry-Perot extensometer [3] and confocal chromatic sensor: a displacement sensor which is the object of this article.

After a certain operating time, fuel stresses the rod and leads to its deformation [4]. We develop the confocal chromatic sensor in order to monitor this deformation. This technology allows avoiding contact between the probe and the rod when performing the measurement and thus we will avoid risking scratching and initiating rod rupture.

As the sensor will be in an irradiation rig simulating the operation in a Pressurized Water Reactor (PWR) we have to consider its typical environment. This means temperature about 350°C, pressure of 150 bar. Very high radiation levels are expected: fluence of $10^{19}\ n, fast/cm^2$ and gamma dose about GGy. We will also have to consider turbulent flow (3 m/s) of corrosive water. Indeed, at such temperatures, water becomes corrosive for an optical window. The confocal chromatic sensor is set in front of a rod in the nuclear reactor. Then, three sensors located around the rod may be necessary to monitor its displacement due to possible buckling of the fuel rod. For this specific measurement of fuel rod swelling in a research reactor, all the interrogation part of the instrumentation will be deported by optical fiber outside the reactor core.

The overall principle of a confocal chromatic sensor is based on the physical concept of chromatism, with the change in optical refractive index with wavelength. Usually, chromatism is an optical aberration to avoid, but not in our case. When focusing a white light source with an optical system, not all the wavelengths are focusing at the same spot on the light propagation axis. In our system, a white source is focused through an optical chromatic system on the fuel rod. The rod's surface, made of an oxidized alloy of zirconium, is reflective enough to be considered as a diffusing surface. Therefore, the light hitting the rods goes back through the whole system. It is only partially reinjected in the fiber because of the spatial filtering due to the small fiber core diameter. This filtering depends on how the beam is focused on the surface in the optically conjugate plane. That produces a spectral selection, which can be visualized on an optical spectrum analyzer at the transport fiber output. Thus, we will detect a peak centered on a particular wavelength. If the fuel rod was swelling, we would then detect a change of the peak central wavelength. As we have an encoding between wavelengths and distance, we can then deduce of how many micrometers the rod swelled. The difference between the initial wavelength and the final one would give an equivalent value of distance's difference (Fig. 1).

Here are the targeted specifications for this sensor:

- Diameter of the sensor: 4 mm (mechanic), 3 mm (optics)
- Total length < 10 cm
- Accuracy : 1-10 µm
- Working distance (distance between the sensor output and the rod): 0.5 mm
- Measuring range (chromatism): 1.5 mm

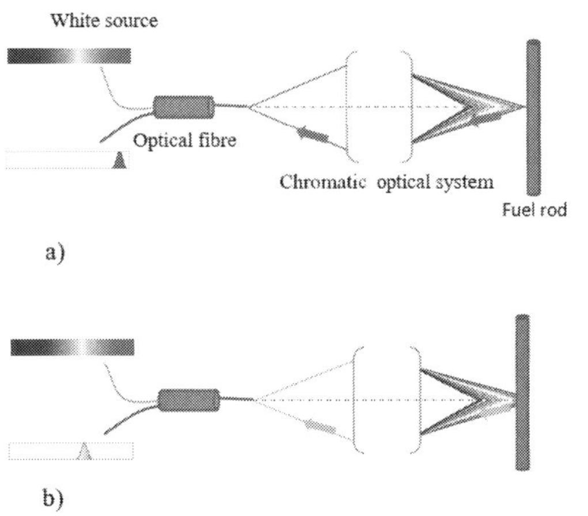

Figure 1: confocal chromatic sensor global principle. a) confocal chromatic in front of a fuel rod before swelling , b) after swelling

Some of the main challenges of this confocal chromatic sensor design are listed hereafter:

- The size of the sensor (both length and diameter) has to be very small to fit the reduced available space and to prevent perturbations on the cooling flow. However, as a wide measuring range (chromatism) is expected, this is more difficult to obtain with such small dimensions. The optical design software –Zemax- is exploited to conceive and optimize the sensor optical system.
- When a glass bulk is heated, its length is modified (*thermal dilatation*), as well as its refractive index, which will lead to displacement of the focusing points and therefore to measurement errors. Data relative to refractive index dependence to temperature from the literature are needed and if none is available, dedicated experiments will be mandatory.
- Radiation darkens the optical glasses [3] and optical fibers [5]. Moreover, large fast neutron fluence is also known to modify the glass refractive index, at least of the silica [6] [7]. Depending on the amplitudes of these effects, additional measurement errors can occur up to a loss of the signal. We then have to find radiation tolerant optical glasses and acquire data about their radiation responses.

In this paper we will present a brief state-of-the-art regarding the effects of radiation and thus explain the need to collect data during an irradiation campaign in BR2 (Belgium Reactor2) in 2022 (irradiation "TESCA"). We first present interferometric measurement technique to measure the changes of refractive index of a glass under irradiation. First, the setup is validated in laboratory conditions (variations with temperature only) before its future implementation in the 2022 irradiation campaign. In a second part, we focus on the optical design of the confocal chromatic sensor, describing the main challenges and the solutions to overcome them.

II. Change of refractive index and attenuation with radiations

A. State of the art

For many years [8], we have known that radiations influence optical fiber and optical glass properties. For our targeted environments, the two most impacting radiation effects on optical fiber and glasses are the Radiation Induced Attenuation (RIA) and the Radiation Induced Refractive Index Change (RIRIC).

RIA corresponds to the glass darkening under radiations related to an increased attenuation of the signal. RIA is caused by point defects created in the optical fiber core and cladding [5]. This is quite a complex phenomenon, depending on both the structural and optical properties of the point defects and on how the harsh environment constraints (*dose, fluence, dose rate, and temperature*) influence their generation mechanisms. Moreover, the RIA can recover after radiations through the thermal or photo bleaching of the point defects. For our confocal chromatic design, we need to choose glasses or optical fibers presenting lower RIA in the spectral domain of interest. For optical fibers, few data are available in the literature at high neutron fluence with a lack of information on the behavior at high temperature and/or a large spectral domain [9-16]. We will select a pure silica (or slightly F-doped silica) core optical fiber with a metallic coating to withstand the high temperature. About bulk glasses for the chromatic optical system, some cerium doped radiation-hardened glasses are considered, but no data at the very high targeted dose levels is available. We also plan to use pure-silica glass, which the behavior in a high radiation environment is better known. Finally, we plan to operate at wavelengths between 800 and 1200 nm where the optical fibers and glasses present the lowest RIA levels. This is also compatible with operation in water that is strongly absorbing above 1200 nm.

RIRIC is mostly observed under neutron exposure and corresponds to the radiation induced change of the refractive index. This is also a topic with little data available in the literature. Moreover, temperature will also induce a variation of refractive index [17].

For the development of the confocal chromatic sensor, we need to consider the glass darkening, if we want to perform the measurement with a better signal to noise ratio. We also have

 https://doi.org/10.1051/epjconf/202125304021

to consider the change of refractive index when glasses are exposed to high fast neutron fluence and/or high temperature in order to avoid measurement errors. Most of all, we want to reduce the degradation of the focusing quality of the chromatic optical system.

In order to consider glass darkening and refractive index change we have designed an experiment to collect data on candidate optical glasses. That will be one of the objectives of the TESCA experiment in Belgium Reactor 2 in 2022.

B. Experimental setups and preliminary results at high temperature

We developed two small measurement systems aiming to collect online data when performing TESCA irradiation. First one is designed to collect RIA data and the second one for refractive index change measurements, more precisely for optical path measurement $n(\lambda, radiations, T) \times L(radiations, T)$.

We have developed an opto-mechanical device bringing a light beam at the exact center of a mirror curvature. Light is fully reinjected in the fiber except when the RIA of a glass sample (placed in between) attenuates the beam (Fig. 2). We use an halogen lamp to illuminate the system and we collect the output signal on an Optical Spectrum Analyzer (OSA). RIA at a particular wavelength corresponds to the ratio of intensity going in the system and the one going out.

Before irradiation, we had to check if the setup is functioning well under high temperature i.e. that heating the device will not unsettle the optical alignment. The results of a preliminary test are presented in Fig. 3. Blue curve represents the time evolution of temperature (we aim 350-400 °C) and the orange one illustrates the power evolution at 970 nm. This illustrates the good stability of the setup through time and so with temperature. We conclude that the setup is suitable for radiation experiments at high temperature.

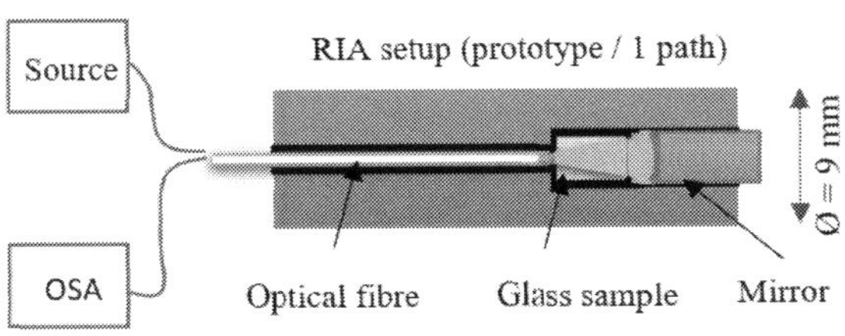

Figure 2: RIA setup

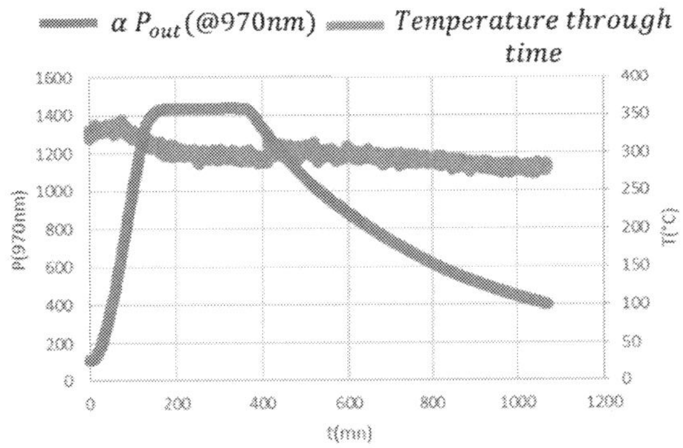

Figure 3: Stability of RIA setup with temperature

We also developed a setup in order to measure the variations of the refractive index with radiations. The first objective is indeed to measure the refractive index change. But since high neutron fluence is likely to produce density change leading to dimension change, the targeted online measurement of refractive index (n) therefore becomes an optical length (n × L) measurement. We are also preparing pre and post-irradiation measurement of glass samples to decorrelate variations of n and L. The setup is represented in the Fig.4. This is an interferometric measurement; we get a channeled spectrum from the interferences coming from both sides of the glass sample. From that spectrum and after signal processing, we get the value of the glass optical length which allows deducing the refractive index if the length is known. We have not yet tested the setup under radiative environment but the setup performances were estimated on different glasses, at different wavelengths and at different temperatures up to 350°C.

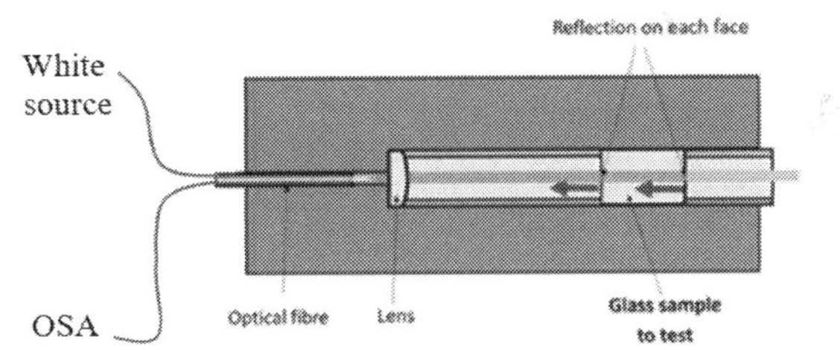

Figure 4: Schematic diagram of the interferometric setup to measure optical path or refractive index change under radiative environment.

Table 1 compares the theoretical values of refractive index and the data collected from our experimental setup.

λ [nm]	CO 79-40	SF6G05	LF5G19	BK7G18	K5G20
1010_{th}	1.4503	1.7789	1.5813	1.5102	1.5132
1010_{exp}	1.4524	1.7798	1.5769	1.5116	1.5176
1110_{th}	1.4491	1.7761	1.5796	1.5089	1.5119
1110_{exp}	1.4512	1.7768	1.5752	1.5104	1.5164
1210_{th}	1.4479	1.7737	1.5782	1.5076	1.5108
1210_{exp}	1.4500	1.7745	1.5738	1.5090	1.5151

Table 1: Refractive index at room temperature

We can observe a difference of about 10^{-3} for absolute values of refractive index at 1010, 1110 and 1210 nm for optical glasses CO 79-40, SF6G05, LF5G19, BK7G18 and K5G20. The theoretical refractive index values of our glasses at room temperature (not under radiation environment) are obtained from the glassmaker Corning and Schott datasheets. We might observe some differences due to different glass batches. Other possible sources for these differences might be a shift in the wavelength calibration of the OSA or an error when measuring the length of the sample. All these errors will remain constant during the irradiation campaign and should not affect the ability of the developed system to measure the refractive index changes.

This can be confirmed by measuring the evolution of refractive index for a particular glass, at a specific wavelength, with the temperature aiming 350°C. By doing so, we also tested the robustness of the device with temperature. When increasing it, we have been able to observe relative changes in the optical path in the range of few 10^{-3}. We have obtained data that is not yet available in the literature, and which is necessary to design the targeted optical system operating at high temperature. The setup is then almost ready to perform online measurement under a high radiative environment in 2022.

III. Optical Design

We discuss in this part, the optical design of the optical sensor for the fuel rod swelling monitoring in a research core reactor. At this design stage, the application requires a small size of the sensor. Temperature and radiation constraints are considered.

We require reaching 1.5 mm measuring range while keeping a good focus quality. Indeed, when the light reflects on the rod and goes back on the optical fiber, we need to read a wavelength with the best selectivity possible in order to give a swelling measurement with the ~1-10 µm accuracy. We then have to optimize our optical design with a compromise between chromatism and spot quality while fulfilling the specifications in terms of diameter and total track.

If we look at a common lens commercialized by Thorlabs, LB4879, f= 35.0 mm, Φ1″ UV Fused Silica Bi-Convex Lens, for instance- with an object numerical aperture of 0.12, we get the results described below.

The initial configuration for this bi-convex lens is near to a 2f-2f and we get:

- An image numerical aperture of 0.10 i-e. a paraxial magnification of about.1.15
- A focal shift (chromatism) of 1878 µm
- A very bad quality of spot due to a high spherical aberration. In order to improve this point we could consider the lens curvatures as variables and even aspherize one face of the lens.

Now, if we consider the same lens but with smaller diameter, the numerical aperture is reduced leading to reduced wavelength selectivity too. To avoid that we can change the configuration (object-image position). By doing so, we will not get better results in terms of chromatism and selectivity.

Table 2 illustrates the impossibility to design a confocal chromatic sensor with a single lens of the required size. This table shows the results obtained from the Thorlabs –LB4879 lens with an object numerical aperture of 0.12 and for wavelengths between 800 and 1200 nm. The Design referenced as "1" on this table refers the 2f-2f configuration spoken below and design "2" were obtained after optimization aiming a bigger image numerical aperture, a good spot quality with aspherization on the first surface of the lens and both curvatures set as variables –also aiming smaller diameters of the lens. On table 2, D1 refers to the distance between the source and the first surface of the lens and D2 corresponds to the distance between the second surface of the biconvex lens and the Image at 800 nm.

Design	1	2
D1 [mm]	63.749	10,41
D2 @ 800 nm [mm]	70.486	71,1185
Thickness [mm]	7.440	7.440
Curvature 1 [mm]	30.96	2,5
Curvature 2 [mm]	-30.96	-2,5
Conic of Curvature 1	x	-0.882
Clear Semi Diameter 1 [mm]	**13**	**1,319**
Clear Semi diameter 2 [mm]	**13**	**0.338**
Spot diagram	x	v
Airy Radius [µm]	5	2,755
Chromatism (Focal shift) [µm]	**1878**	**70**
Numerical Aperture (Image)	0,10	0,23

Table 2: Thorlabs-LB4879, f=35 mm, Φ1″ UV Fused Silica Bi-Convex Lens. Design 1: starting design. Design 2: constraint on diameter of optics with aspherization.

As we can see, with a single lens (even with a more chromatic glass), it is impossible to design a small diameter confocal chromatic sensor. We have to work on a more complex design, which finally will have to combine different radiation hardened glasses, especially the most dispersive ones.

IV. Conclusions

We are developing a confocal chromatic sensor in order to measure fuel rod's swelling in a research nuclear reactor. We have discussed the main scientific and technological challenges that have to be overcome: temperature, radiations and optical design under strict specifications. Temperature changes the refractive index of about 10^{-3}. Radiation also induces a change in refractive index and a glass darkening. Small dimensions of the sensor probe are limiting the measuring range resulting from chromatism.

New data on the behavior of bulk glasses under radiation are expected to result from the TESCA irradiation (2022, in the BR2 research reactor). We have developed miniature devices in order to measure RIA and the variations of the optical path of some glass samples under harsh environment. These devices will be set in the research reactor's core for online measurements; they have been tested and validated under high temperature. The optical path-measuring device gives data on the refractive index of glasses at high temperature. These data are necessary for the design of the confocal chromatic sensor, and are not yet all available in literature. Our next step will be the irradiation campaign in 2022 at BR2. While we will consolidate refractive index change setup and signal processing in order to get a better accuracy, confocal chromatic sensor design on Zemax with silica and a mix of silica and radiation-hardened (and more dispersive) glasses will be achieved.

REFERENCES

[1] S. Nehr et al., "On-line monitoring of multiplexed femtosecond Fiber Bragg Gratings exposed to high temperature and high neutron fluence", 26th International Conference on Optical Fiber Sensors, OFS-26, Poster presentation, Lausanne, Switzerland, 24-28 September, 2018

[2] B. Bouvry et al., "Multispectral pyrometry for surface temperature measurement of oxidized Zircaloy claddings," Infrared Physics and Technology, vol. 83, pp. 78-87, 2017

[3] G. Cheymol, A. Verneuil, P. Grange, H. Maskrot and C. Destouches, "High-Temperature Measurements With a Fabry–Perot Extensometer," in *IEEE Transactions on Nuclear Science*, vol. 67, no. 4, pp. 552-558, April 2020, doi: 10.1109/TNS.2020.2975707.

[4] P. Van Uffelen, G. Pastore, "Oxide Fuel Performance Modeling and Simulation", Comprehensive, Nuclear Materials, vo. 2, 2020, pp 363-416.

[5] S. Girard et al., "Overview of radiation induced point defects in silica-based optical fibers," Reviews in Physics, 2019, doi: https://doi.org/10.1016/j.revip.2019.100032

[6] Primak, "Fast-neutron-induced changes in quartz and vitreous silica", Phys.Rev.B 110 (6).1240-1254(1958).

[7] L. Remy, G. Cheymol, A. Gusarov, A. Morana, E. Marin and S. Girard, "Compaction in Optical Fibres and Fibre Bragg Gratings Under Nuclear Reactor High Neutron and Gamma Fluence," in *IEEE Transactions on Nuclear Science*, vol. 63, no. 4, pp. 2317-2322, Aug. 2016, doi: 10.1109/TNS.2016.2570948.

[8] David L. Griscom, "γ and fission-reactor radiation effects on the visible-range transparency of aluminium-jacked, all-silica optical fibers," Journal of Applied Physics, vol. 80, no. 4, May 1996, pp 2142-2155

[9] T. Allanche, "Effect of high radiation doses (MGy) on light Emitting Diodes and optical glasses", HAL id: tel-03215958, May 2021

[10] T. Shikama, T. Kakuta, M. Narui, T. Sagawa, H. Kayano, "Optical properties in fibers during irradiation in a fission reactor", J. Nucl. Mater 225 (1995) 324-327

[11] D. W. Cooke, B. L. Benett, E. H. Farnum, "Optical absorption of neutron irradiated silica fibers", J. Nucl. Mater, May 1996, pp 214-218

[12] T. Kakuta, T. Shikama, M. Narui, T. Sagawa "Behavior of optical fibers under heavy irriadiations", Fus. Eng. Des. 51-52 (2000) 179-183

[13] B. Brichard, P. Borgemans, A. F. Fernandez, K. Lammens, M. Decréton, "Radiation effect in silica optical fiber exposed to intense mixed neutron-gamma radiation field", IEEE Trans. Nucl. Sci, vol. 48, no. 6, Nov. 2001, pp2069-2073

[14] T. Kakuta, T. Shikama, T. Nishitani, B.Brichard, A. Krassilinikov, A. Tomashuk, S. Yamamoto, S.Kasai "Round-robin irradiation test of radiation resistant optical fibers for ITER diagnostic application", Journal of nuclear materials 307-311 (2002) 1277-1281

[15] G. Cheymol, H. Long, J. F. Villard, B. Brichard, "High level gamma and neutron irradiation of silica optical fibers in CEA OSIRIS nuclear reactor", IEEE. Trans. Nucl. Sci, vol. 55, no. 4, Aug. 2008, pp 2252-2258

[16] P. F. Kashaykin, A. L. Tomashuk, S.A. Vasiliev, A.D. Ignatyev, A.A. Shaimerdenov, Yu. V. Ponkratov, T. V. Kulsartov, Y. A. Kenzhin, Sh. Kh. Gizatulin, T. K. Zholdybayev, Y. V. Chikhray, S. L. Semjonov, "Radiation resistance of single-mode optical fibres with view to in-reactor applications", Nuclear Materials and Energy 27 (2021) 10098.

[17] C. Z. Tan, "Temperature dependence of refractive index of glassy SiO2 in the infrared wavelength range", Journal of Physics and Chemistry of Solids 61 (2000) 1315-1320.

First in-core gamma spectroscopy experiments in a zero power reactor

Oskari Pakari[1], Vincent Lamirand[1,2], Tom Mager[1], Axel Laureau[1], and Pavel Frajtag[1], and Andreas Pautz[1,2]

[1]Laboratory for Reactor Physics and Systems Behaviour
Ecole polytechnique fédérale de Lausanne

[2]Laboratory for Reactor Physics and Thermal-Hydraulics
Nuclear Energy and Safety Division
Paul Scherrer Institut, Villigen, Switzerland

Corresponding author: oskari.pakari@gmail.com

***Abstract*—**

Gamma rays in nuclear reactors, arising either from nuclear reactions or decay processes, significantly contribute to the heating and dose of the reactor components. Zero power research reactors offer the possibility to measure gamma rays in a purely neutronic environment, allowing for validation experiments of dose estimates, computed spectra, and prompt to delayed gamma ratios. The resulting data can contribute to models, code validation and photo atomic/nuclear data evaluation. To date, most experiments have relied on flux measurements using TLDs, ionization chambers, or spectrometers set in low flux areas. The CROCUS reactor allows for flexible detector placement in and around the core, and has recently been outfitted with gamma detection capabilities to fulfill the need for in-core gamma spectroscopy, as opposed to flux. In this paper we report on the experiments and accompanying simulations of gamma spectrum measurements inside a zero power reactor core, CROCUS. It is a two-zone, uranium-fueled light water moderated facility operated by the Laboratory for Reactor Physics and Systems Behaviour (LRS) at the Swiss Federal Institute of Technology Lausanne (EPFL). Herein we also introduce, in detail, the new LEAF system: A Large Energy-resolving detection Array for Fission gammas. It consists of an array of four detectors – two large ⌀127×254 mm Bismuth Germanate (BGO) and two smaller ⌀12×50 mm Cerium Bromide ($CeBr_3$) scintillators. We describe the calibration and characterization of LEAF followed by first in-core measurements of gamma ray spectra in a zero power reactor at different sub-critical and critical states, and different locations. The spectra are then compared to code results, namely MCNP6.2 pulse height tallies. We were able to distinguish prompt processes and delayed peaks from decay databases. We present thus experimental data from hitherto inaccessible core regions. We provide the data as validation means for codes that attempt to model these processes for energies up to 10 MeV. We finally draw conclusions and discuss the future uses of LEAF. The results indicate the possibility of isotope tracking and burn-up validation.

***Index Terms*—Gamma spectroscopy, Zero power research reactor.**

I. Introduction

GAMMA radiation as a by-product of fission and radioactive decay is a major component of a nuclear reactor's radiation field. It contributes to the heating of structures, the degradation of materials, and the overall dose rate around a reactor [1]. The characteristics of a reactor's gamma field is thus important for operation and safety. Accurate predictions are desired in operational planning, in the design of experiments in research facilities, and in the inception of new reactors [2]. The sources of gamma radiation in nuclear reactors include the prompt fission process, delayed radioactive decay, and nuclear capture reactions. Experiments often target the individual processes to single out their effect; examples range from spent fuel spectroscopy [3], [4] to fission product decay spectra and direct fission gamma spectra [5], [6].

Quantitative experiments on the spectra and fluxes of gamma radiation have become of interest with the introduction of coupled photon and neutron transport options in common Monte Carlo codes, such as Serpent [7] and MCNP [8] – recently compared for the specific case of CROCUS in [9]. Measurements of nuclear reactor gamma ray fields have included spectroscopy outside of the reactor shielding [10], and in-core measurements of fluxes using ionization chambers [11]–[13]. A recent focus has been the determination of delayed gamma contributions measured with mixed neutron-gamma flux measurements in research reactors [14], [15]. To the knowledge of the authors, no direct measurements of in-core gamma spectra have hitherto been performed. The CROCUS research reactor offers a flexible zero power environment for these types of experiments. To date, TLD measurements of the gamma field of CROCUS have been conducted in the context of activation and far field flux analysis. Furthermore, combined neutron/gamma measurements using a sCVD detector [16] have added to the radiation field flux characterization. With a variety of other experimental programs [17] that so far only exploited neutron detection [18], [19], we extended the capabilities of CROCUS to state of the art gamma detection capabilities.

This is the intent of the LEAF system: A Large Energy-resolving detection Array for Fission gammas to directly measure gamma ray spectra in CROCUS. We hereby aim at providing measurement data in previously unexplored param-

 https://doi.org/10.1051/epjconf/202125304022

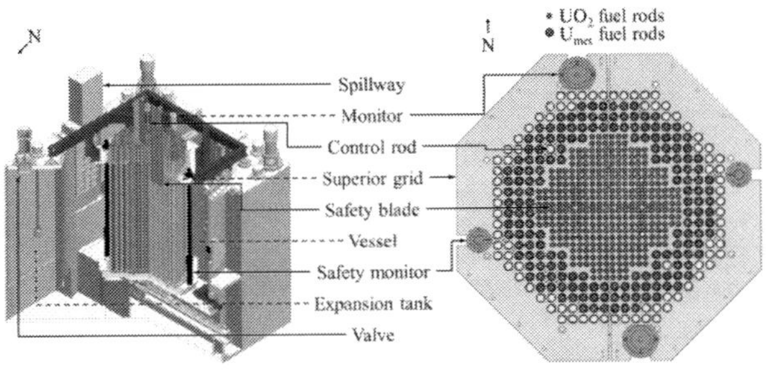

Fig. 1. Schematic isometric view of the CROCUS reactor (left), and top view of the core configuration (right).

eter regions, namely in-core and high energy (> 2 MeV) spectroscopy. In this paper we present the characterization and first measurements of said system - notably the first measurement of gamma ray spectra from an in-core location. We facilitate a possible code validation by providing a detailed overview of the geometry. In Section II we introduce the experimental details: the CROCUS reactor, a detailed description of LEAF and the energy calibration of all detectors, as well as the experimental setup used for spectroscopy. In Section III we show the results of measurements using LEAF in CROCUS – namely in-core and ex-core spectra – and compare the results to MCNP6.2 simulations and thereafter discuss their implications.

II. Methods and Materials

A. The Experimental Reactor CROCUS

A full description of the core can be found in the International Reactor Physics Experiments Handbook (IRPhE) [20], [21]. The CROCUS reactor is a two-zone, uranium-fueled, H_2O-moderated critical assembly operated by the Laboratory for Reactor Physics and Systems Behaviour (LRS) at the Swiss Federal Institute of Technology Lausanne (EPFL). It is a zero power reactor, with a maximum power of 100 W. The core is approximately cylindrical in shape with a diameter of about 58 cm and a height of 100 cm. Two different kinds of fuel rods make up the reactor core of CROCUS (see Fig. 1). The central zone is loaded with 336 UO_2 fuel rods (1.806 wt.%-enriched), set in a square lattice with a pitch of 1.837 cm. The peripheral zone is loaded with up to 176 thicker, U_{met} fuel rods (0.947 wt.%-enriched) with a pitch of 2.917 cm, also in a square lattice. All fuel rods are cladded with aluminum and are maintained in a vertical position by the upper grid and lower grid plates, spaced 100 cm apart. In the current COLIBRI configuration [22]–[25], both grid plates incorporate a cadmium layer with a thickness of 1 mm to limit axial thermal leakage to surrounding structures. This is also the configuration for the herein presented results. The active fuel length starts at the top surface of the lower cadmium layer and extends to 100 cm. The core is located in an aluminum water tank, its diameter is 130 cm and thickness is 1.2 cm. Demineralized light water is used as moderator and reflector. Reactivity is nominally controlled by a variation of the water level using a spillway with an accuracy of 0.1 mm (equivalent to 0.4 pcm) and optionally by means of two control rods containing naturally enriched boron carbide (B_4C) sintered pellets located in diagonal symmetry within the outer fuel zone.

B. The LEAF system: Overview and Calibration

We specifically intend to expand the capabilities of CROCUS by a dedicated gamma detection system. Scintillators were hereby the final choice when optimizing for radiation hardness, flexible use, price, and nonetheless high efficiency when compared to semiconductor based detectors. In total, an array of four detectors was acquired from Scionix Holland [26] to allow for symmetric placement of the system in and around the core (see Figure 7). The high voltage supply and photomultiplier tube (PMT) signal amplification was handled by the Mirion Technologies DSA-LX [27] (here designated 419 and 420 as per serial number) that theoretically allow for ~MHz count rates to be treated. In the following we introduce the respective detectors in detail and present their characteristics and calibration.

1) Cerium Bromide ($CeBr_3$) detectors

When the two control rods are removed, their guide tubes are a prominent location for in-core measurements, e.g. used for delayed gamma fraction estimation [28], intrinsic gamma noise [29], and induced neutron/gamma noise measurements [22], [23]. Cerium(III) Bromide ($CeBr_3$), as a comparatively new scintillation material [30], [31], was chosen as the most suitable material for a high photon flux zone with a comparatively fast decay time of 20 ns at a density of 5.2 g/cm^3 [32]. The radiation hardness was estimated to be well sufficient for use in CROCUS [9], [33]. In order to fit the guide tubes, the cylindrical crystals measure 13 mm in diameter and 15 mm in length, see Figure 2. Both detectors use a Hamamatsu Type R12421 PMT.

2) Bismuth Germanate (BGO) detectors

To measure gamma rays in ex-core locations, specifically photons with energies above 2 MeV, a material with high photon stopping efficiency was desired. Bismuth Germanate (BGO) is a well established scintillator in various detection applications [34], [35] with slower decay and lower light yield than the $CeBr_3$ [36], but with a higher density of 7.13 g/cm^3 and lower price per volume. To meet an absorption efficiency of above 95% for 10 MeV photons the two acquired cylindrical crystals are therefore 127 mm in diameter. The height was constrained by weight (25 kg) and price to 250 mm. Both are equipped with Photonis 5" Type XP4578 PMTs, see Figure 3.

3) Amplification settings and energy calibration of LEAF

For both detectors and all experiments presented herein the DSA-LX's settings were as follows:

- Rise time of 0.2 μs, 0.0 μs flat top.
- -610 V and -1260 V of HV for the $CeBr_3$ and BGO PMTs, respectively.
- Lower level discrimination at 0.5% of the maximum channel (2^{14}).
- Coarse gain of 6.4.

All four detectors were calibrated using standard Eu-152 sources – the resulting spectra are displayed in Figures 4

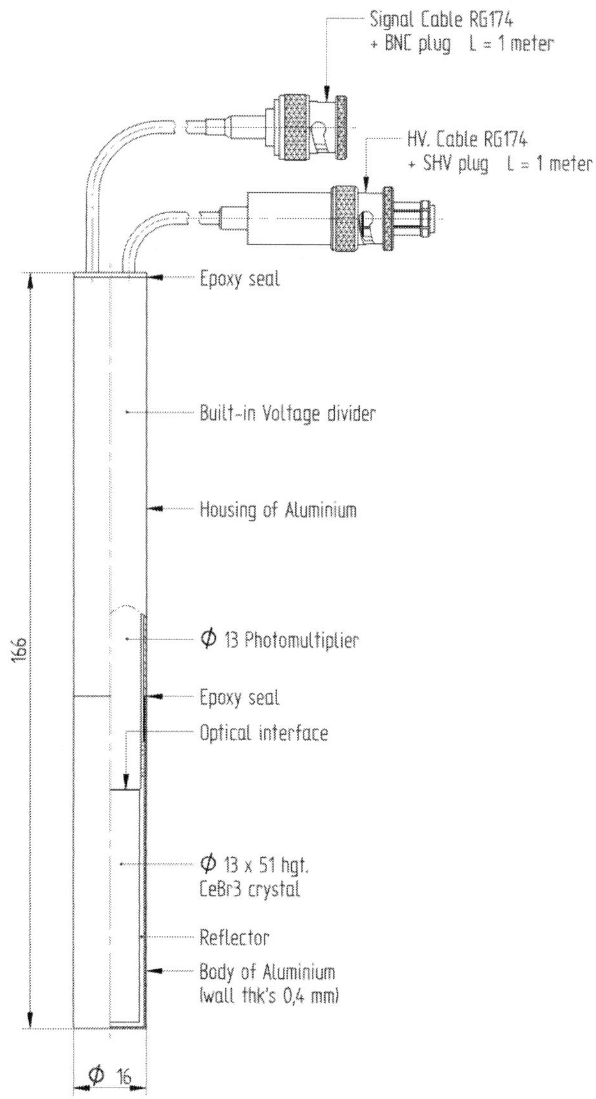

Fig. 2. Technical drawing of the $CeBr_3$ detector as provided by Scionix, used to create models for Monte Carlo transport codes. The PMT is a Hamamatsu Type R12421.

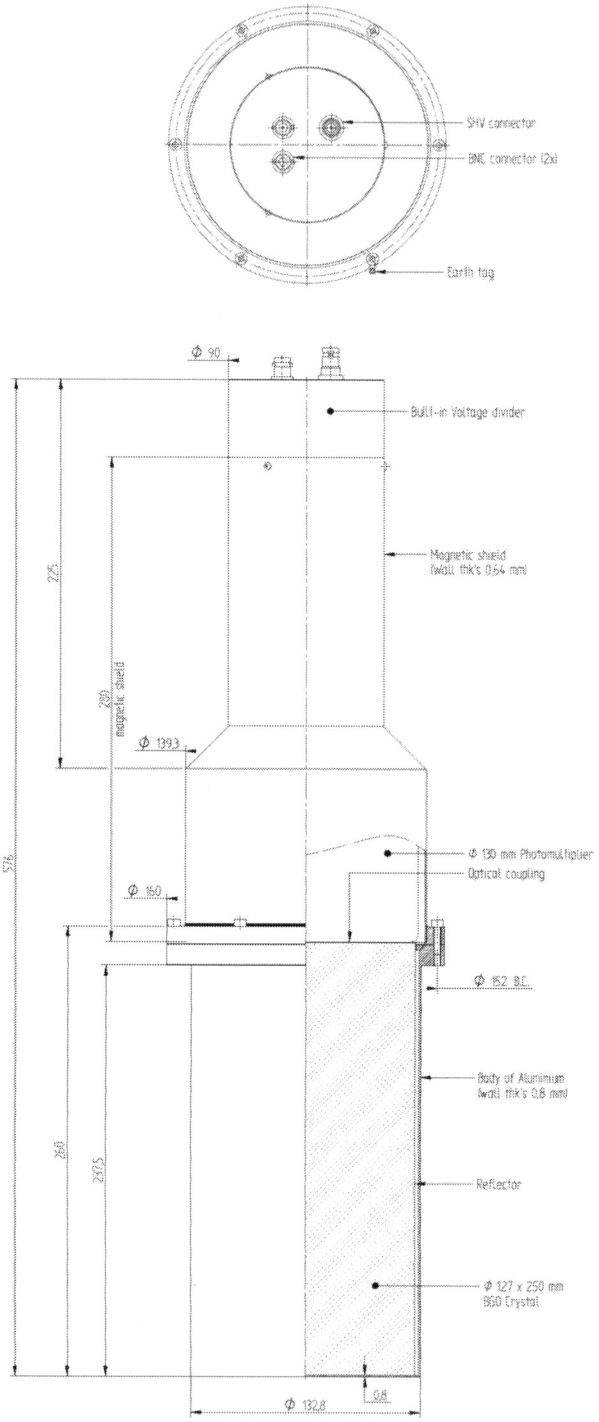

Fig. 3. Technical drawing of the BGO detector as provided by Scionix, used to create models for Monte Carlo transport codes. The PMT is a Photonis 5" Type XP4578 PMT.

and 5. The BGO, due to its comparatively higher full width at half maximum (FWHM), required additionally a Co-60 measurement to aid in the peak distinction. Each identified peak was fitted with a Gaussian to find the mean value used as input for the calibration. The final calibrations for all detectors is linear with an $R^2 > 0.99$. The calibration was purposefully undertaken using only the first quarter of channels as higher energy events were expected. We also examined the FWHM (see Figure 6) of all detectors with energy and found them to be comparable to other crystals in literature, e.g. [37].

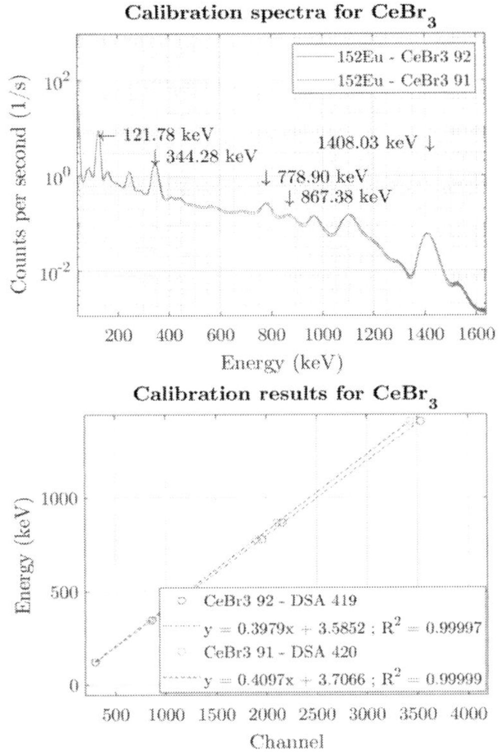

Fig. 4. **Top:** Calibration gamma spectra acquired with the $CeBr_3$ detectors. **Bottom:** Resulting linear calibration fits after extracting the peak center values via Gaussian fitting of each peak.

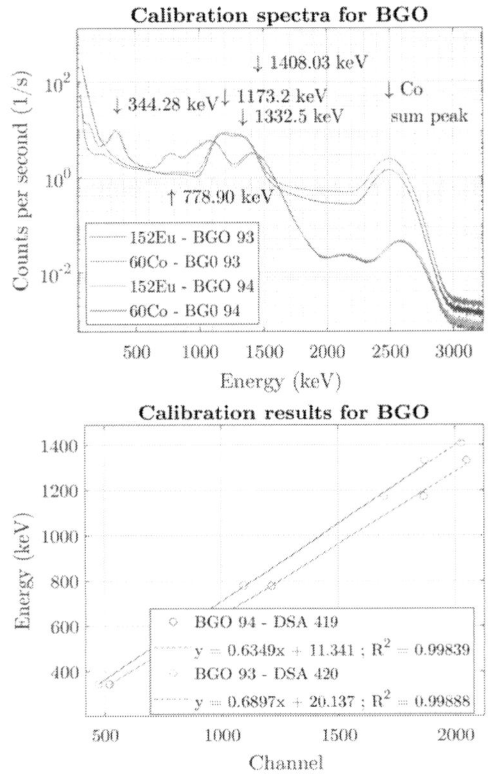

Fig. 5. **Top:** Calibration gamma spectra acquired with the BGO detectors. **Bottom:** Resulting linear calibration fits after extracting the peak center values via Gaussian fitting of each peak.

C. Experimental setup: LEAF in CROCUS

The $CeBr_3$ detectors were placed at mid core height in the control rod tubes, while the BGO detectors were set just outside of the main vessel, as displayed in Figure 7, also at mid core height. The acquisition layout with the electronics is sketched in Figure 8. CROCUS in its shutdown state contains no water in the vessel. In the start-up procedure water is filled step-wise until 800 mm, after which the operator has control over it. For the experiments in this work we chose a set of sub-critical and critical states to use LEAF in a variety of reactor states. The configurations, water levels and estimated fission powers are listed in Table I. The system shows – consistently across the detectors – an excellent counting response up to 0.2 MHz which in this setup corresponded to about 100 mW of power [38]. This is about two orders of magnitude better than the in-house developed neutron noise detection system based on regular gamma spectrometry electronics [39].

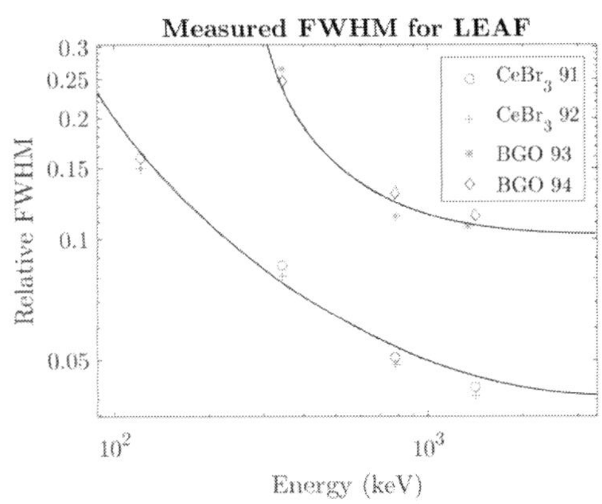

Fig. 6. FWHM for the LEAF detectors in dependence of energy, determined with a standard Eu-152 source. The fitted lines are for illustrative purposes.

D. Simulation of gamma spectra in CROCUS using MCNP6.2

MCNP6.2 offers the possibility to simulate coupled neutron photon transport with pulse height estimation tallies for detectors. These simulations herein include only prompt processes, i.e. fission and, if available in the nuclear data used (ENDF/B-7.1), (n,γ) photons. In this work we compare the results qualitatively to the energy deposition tallies without taking smearing into account, e.g. due to the scintillation process and the electronics .

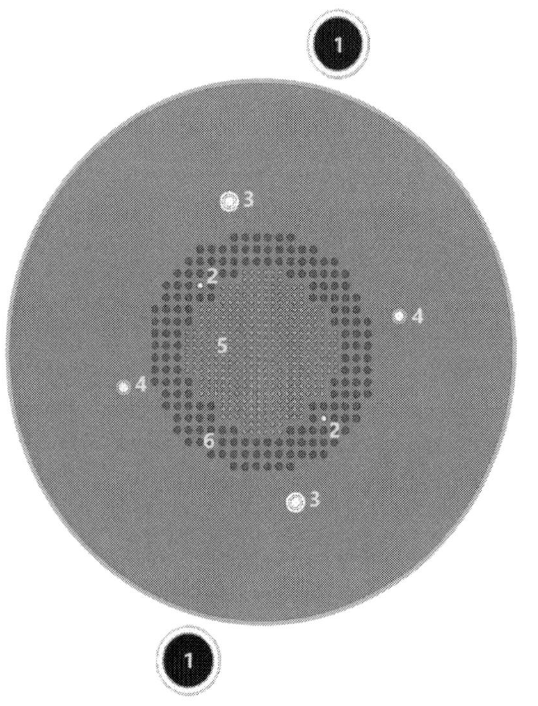

Fig. 7. Top view of the Serpent 2 model of CROCUS with LEAF at mid-core height. **1)** BGO detector, **2)** $CeBr_3$ detector, **3)** Compensated ionization chamber, **4)** Power monitors, **5)** UO_2 fuel lattice, **6)** U_{met} fuel lattice.

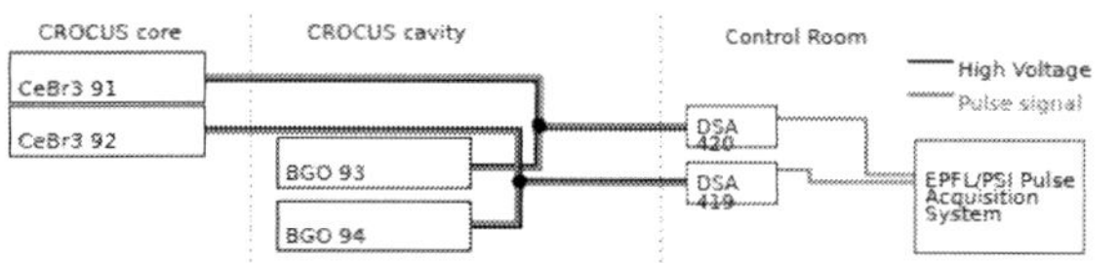

Fig. 8. Schematic of the acquisition layout for the gamma spectroscopy experiment in CROCUS using LEAF.

TABLE I
OVERVIEW OF THE CROCUS CONFIGURATIONS USED FOR THE GAMMA SPECTROSCOPY EXPERIMENTS.

Configuration	Water level mm	Fission power estimate [38] mW
Shutdown	Empty	0
H3	500.0	1.2
Sub-critical:	800.0	3.3
	850.0	3.9
	900.0	6.1
	925.0	9.4
Critical:	963.3	up to 100k

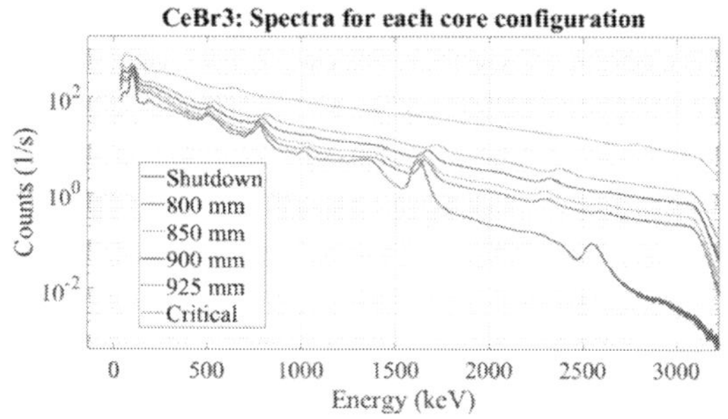

Fig. 9. Measured in-core gamma spectra using a $CeBr_3$ detector in the control rod position for sub-critical and critical configurations.

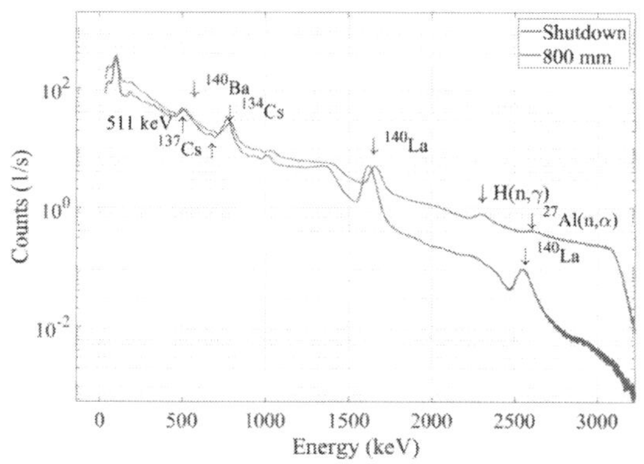

Fig. 10. Measured in-core gamma spectra using a $CeBr_3$ detector in the control rod position for 800 mm and the shutdown state. Preliminary identified peaks are indicated in the plot using the data from Table II.

III. RESULTS

A. $CeBr_3$: First recorded in-core gamma spectra

In Figure 9 we display $CeBr_3$ spectra acquired at different reactor configurations, ranging from shutdown to criticality. We most strikingly observe the expected exponential shape of the spectrum from fission with several resolved peaks from specific decays on top. We note a gain shift in the spectrum to higher energies with higher count rate, an effect often found in PMTs at high event rates [40]. The spectra notably all exhibit a cut-off at around 3.3 MeV. Using NIST standard data on photon mass attenuation lengths [41] and the density mentioned before, we find that photons of this energy have a mean free path of 3.8 cm – a value consistent with the detector geometry.

B. BGO: Coarse ex-vessel spectra up to 10 MeV

In Figure 11 we display BGO spectra acquired at 500 mm water level with and without the PuBe start-up neutron source of CROCUS below the core [42]. At higher water levels the spectra showed only an exponential shape induced by fission with no visually identifiable peaks, similar to the spectrum of the $CeBr_3$ at criticality. By comparing spectra with and

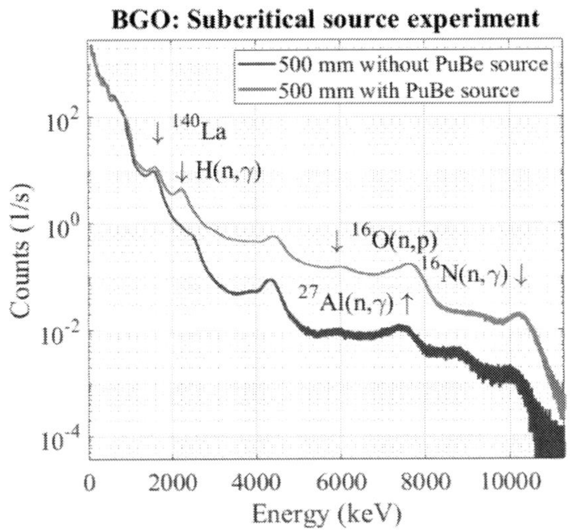

Fig. 11. Measured gamma spectra using a BGO detector in the core vessel periphery for the H3 sub-critical state, with and without the start-up neutron source below the core. Identified peaks are indicated in the plot.

TABLE II
OVERVIEW OF EMISSIONS USED FOR PEAK IDENTIFICATION. ENERGIES ROUNDED TO THE CLOSEST KEV. THE HALF LIFE IS INDICATED FOR BETA DELAYED GAMMA EMISSIONS.

Reaction	Energy (keV)	Half life	Reference
Annihilation	511	prompt	[43]
^{140}Ba	537	12.7 d	[44]
^{137}Cs	662	30.5 a	[44]
^{134}Cs	795	2.1 a	[44]
^{140}La	816	1.7 d	[44]
^{140}La	1596	1.7 d	[44]
^{27}Al(n,γ)^{28}Si	1779	2.3 min	[45]
^{1}H(n,γ)	2223	prompt	[43]
^{140}La	2521	1.68 d	[44]
^{27}Al(n,α)^{24}Na	2754	15.0 h	[44]
^{16}O(n,p)^{16}N	6140	7.1 s	[43]
^{27}Al(n,γ)	7724	prompt	[46]
^{14}N(n,γ)	10829	prompt	[43]

without source we attempted to distinguish neutron induced prompt lines from longer lived activation and fission products. Indeed, we find a general increase in spectral baseline due to fission gammas – and also a new line e.g. at around 2.2 MeV and 6.1 MeV. In the next section we discuss the gamma line identification and implications.

IV. DISCUSSION

Peak identification given the here presented data is challenging. To approach this problem we herein use two methods: Firstly, by accessing various decay and neutron capture databases we chose a list of likely causes based on expert judgement, listed in Table II. Secondly, we simulated the prompt gamma transport in MCNP6.2 in the respective detector volumes and overlaid the F8 pulse height tally over the experimental data - as shown in Figures 12 and 13 for the $CeBr_3$ and BGO, respectively. Note that the spectra are arbitrarily normalized to allow for qualitative comparison of the shapes.

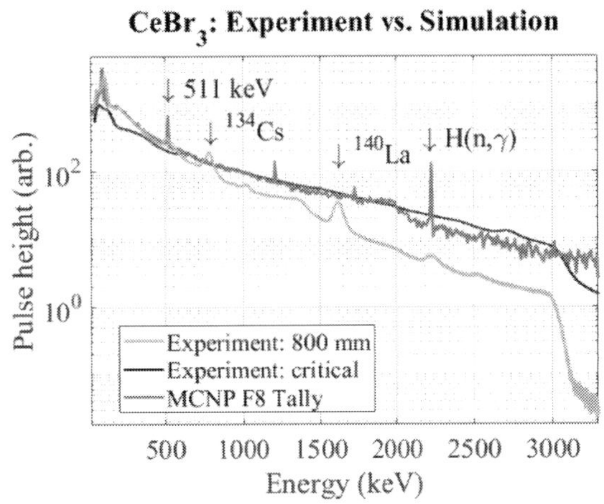

Fig. 12. Comparison of experimental in-core gamma spectra using a $CeBr_3$ detector in the control rod position to the MCNP pulse height tally (F8). Preliminary identified peaks are indicated in the plot.

We most notably identify across both in-core and ex-vessel spectra:

- Several unidentified lines below 500 keV that could correspond to a range of fission products but we refrain for attempting an identification in this work.
- Long lived fission products relevant for waste and fallout quantification [47] such as ^{137}Cs, ^{134}Cs, ^{140}La, and ^{140}Ba.
- Radiative neutron capture and activation, mostly from H(n,γ) and potentially the surrounding aluminum structures. In the high energy spectra of the BGO we furthermore observe the 6.1 MeV and 10.8 MeV ^{16}N decay lines.

We emphasize here that the identified peaks using the outlined methods (MCNP correlation and databases) is preliminary. Future work will focus on quantifying the decay behavior of the peaks to establish a more solid causal link. An unresolved inconsistency is for example the line at ~4.4 MeV as seen in the BGO spectra in Figures 11 and 13. According to MCNP, it is exactly twice the energy of the hydrogen capture gamma, yet it is found in both spectra with and without the PuBe neutron source. We nonetheless can state that the MCNP simulations qualitatively agree with the measurements for other expected prompt processes, such as annihilation and capture in hydrogen, and disagree when it comes to fission product decay such as ^{140}La. This highlights the need and opportunity for validation taking delayed gammas and signal smearing into account.

The results are overall promising: fission products could be directly measured in-core, opening the possibility to track their production on-line as well as validate burnup calculations of CROCUS [48].

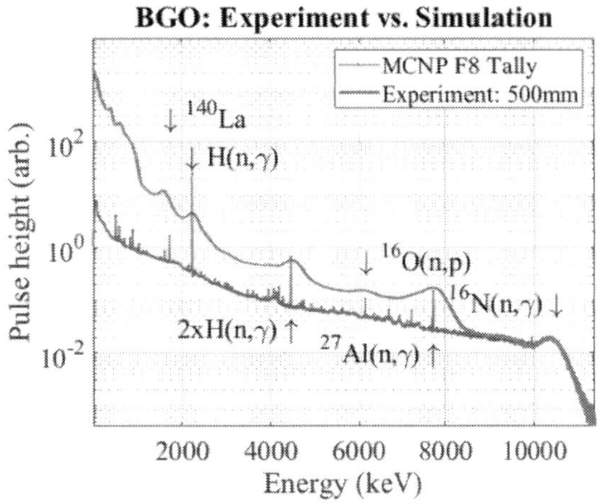

Fig. 13. Comparison of experimental BGO gamma spectrum to the MCNP pulse height tally (F8). Preliminary identified peaks are indicated in the plot.

V. Conclusion

In this work we present, to the best of the authors' knowledge, for the first time gamma spectra from a central region of a nuclear reactor. By using a modern scintillator material ($CeBr_3$) with comparatively low FWHM, we were able to distinguish peaks above the fission spectrum. We also present the measurement of ex-core gamma ray spectra up to 10 MeV using BGO detectors that also show several peaks.

We introduced the new gamma spectroscopy system for the CROCUS zero power reactor called LEAF. It consists of four scintillation based detectors, two $CeBr_3$ and two large BGO crystals, each with different specifications allowing for a wide range of experimental reactor applications. We presented detailed geometry, characteristics and calibration to allow for explicit modeling and ultimately aspire to provide benchmark quality measurements for validation purposes.

The measurements of in-core gamma spectra were presented notably at different sub-critical and critical states of CROCUS. We also presented sub-critical measurements with and without start-up neutron source to highlight the difference with and without fission and other neutron induced interactions. Additionally, the experiments are compared to pulse height tallies produced with MCNP6.2. The experiments qualitatively agree with the simulations with regards to prompt spectrum shape and preliminary (n,γ) line identification. Resolved peaks of delayed processes not modeled in aforementioned codes were tentatively identified using databases, most notably the fission products ^{134}Cs and ^{140}La, verifying the need for more sophisticated simulations taking delayed processes into account [49].

In the future, in addition to reactor gamma spectroscopy, LEAF will be used for reactor perturbation studies [50], reactor noise [51] in the gamma regime as well as further investigations into gamma ray spectra in- and ex-core with emphasis on delayed process identification and quantification.

References

[1] A.-C. Colombier, H. Amharrak, D. Fourmentel, S. Ravaux, D. Régnier, O. Gueton, J.-P. Hudelot, and M. Lemaire, "Nuclear data production, calculation and measurement: a global overview of the gamma heating issue," *EPJ Web of Conferences*, vol. 42, p. 04001, 2013.

[2] G. Rimpault, D. Bernard, D. Blanchet, C. Vaglio-Gaudard, S. Ravaux, and A. Santamarina, "Needs of accurate prompt and delayed γ-spectrum and multiplicity for nuclear reactor designs," *Physics Procedia*, vol. 31, pp. 3–12, 2012.

[3] D. Beddingfield and F. Cecil, "Identification of fissile materials from fission product gamma-ray spectra," *Nuclear Instruments and Methods in Physics Research Section A: Accelerators, Spectrometers, Detectors and Associated Equipment*, vol. 417, no. 2-3, pp. 405–412, nov 1998.

[4] C. Willman, A. Håkansson, O. Osifo, A. Bäcklin, and S. J. Svärd, "Non-destructive assay of spent nuclear fuel with gamma-ray spectroscopy," *Annals of Nuclear Energy*, vol. 33, no. 5, pp. 427–438, mar 2006.

[5] F. Maienschein, R. Peelle, W. Zobel, and T. Love, "Gamma rays associated with fission," Oak Ridge National Laboratory, Tech. Rep., oct 1958.

[6] S. Oberstedt, R. Billnert, A. Gatera, W. Geerts, P. Halipré, F.-J. Hambsch, M. Lebois, A. Oberstedt, P. Marini, M. Vidali, and J. Wilson, "Prompt fission γ-ray spectra characteristics - a first summary," *Physics Procedia*, vol. 64, pp. 83–90, 2015.

[7] T. Kaltiaisenaho, "Implementing a photon physics model in serpent 2; fotonifysiikkamallin kehittäminen serpent 2-koodiin," Master's thesis, Aalto University, 2016.

[8] M. C. White, "Development and implementation of photonuclear cross-section data for mutually coupled neutron-photon transport calculations in the monte carlo n-particle (MCNP) radiation transport code," Los Alamos National Lab. (LANL), Tech. Rep., jul 2000.

[9] O. Pakari, V. Lamirand, B. Vandereydt, F. Vitullo, M. Hursin, C. Kong, and A. Pautz, "Design and simulation of gamma spectrometry experiments in the CROCUS reactor," in *ANIMMA, Portoroz, Slovenia*, 2019.

[10] Y. Nakashima, S. Minato, M. Kawano, T. Tsujimoto, and K. Katsurayama, "Gamma-ray energy spectra observed around a nuclear reactor," *Journal of Radiation Research*, vol. 12, no. 3-4, pp. 138–147, 1971.

[11] D. Fourmentel, P. Filliatre, J. Villard, A. Lyoussi, C. Reynard-Carette, and H. Carcreff, "Measurement of photon flux with a miniature gas ionization chamber in a material testing reactor," *Nuclear Instruments and Methods in Physics Research Section A: Accelerators, Spectrometers, Detectors and Associated Equipment*, vol. 724, pp. 76–82, oct 2013.

[12] G. Žerovnik, T. Kaiba, V. Radulović, A. Jazbec, S. Rupnik, L. Barbot, D. Fourmentel, and L. Snoj, "Validation of the neutron and gamma fields in the JSI TRIGA reactor using in-core fission and ionization chambers," *Applied Radiation and Isotopes*, vol. 96, pp. 27–35, feb 2015.

[13] K. Ambrožič, A. Gruel, V. Radulović, M. L. Guillou, P. Blaise, C. Destouches, and L. Snoj, "Delayed gamma determination at the JSI TRIGA reactor by synchronous measurements with fission and ionization chambers," *Nuclear Instruments and Methods in Physics Research Section A: Accelerators, Spectrometers, Detectors and Associated Equipment*, vol. 911, pp. 94–103, dec 2018.

[14] D. Fourmentel, V. Radulovic, L. Barbot, J.-F. Villard, G. Zerovnik, L. Snoj, M. Tarchalski, K. Pytel, and F. Malouch, "Delayed gamma measurements in different nuclear research reactors bringing out the importance of the delayed contribution in gamma flux calculations," in *2015 4th International Conference on Advancements in Nuclear Instrumentation Measurement Methods and their Applications (ANIMMA)*. IEEE, apr 2015.

[15] K. Ambrožič, V. Radulović, L. Snoj, A. Gruel, M. L. Guillou, P. Blaise, C. Destouches, and L. Barbot, "Characterization of gamma field in the JSI TRIGA reactor," *EPJ Web of Conferences*, vol. 170, p. 04001, 2018.

[16] M. Hursin, C. Weiss, P. Frajtag, V. Lamirand, G. Perret, P. Kavrigin, A. Pautz, and E. Griesmayer, "Testing of a sCVD diamond detection system in the CROCUS reactor," *The European Physical Journal A*, vol. 54, no. 5, may 2018.

[17] V. Lamirand, M. Hursin, G. Perret, P. Frajtag, O. Pakari, and A. Pautz, "Future experimental programmes in the CROCUS reactor," *Conference proceedings of RRFM/IGORR 2016*, 2016.

[18] O. Pakari, V. Lamirand, G. Perret, P. Frajtag, and A. Pautz, "Kinetic parameter measurements in the crocus reactor using current mode instrumentation," *IEEE Transactions on Nuclear Science*, 2018.

[19] O. Pakari, V. Lamirand, G. Perret, L. Braun, P. Frajtag, and A. Pautz, "Current mode neutron noise measurements in the zero power reactor CROCUS," in *EPJ Web of Conferences*, vol. 170. EDP Sciences, 2018, p. 04017.

[20] U. Kasemeyer, R. Früh, J. M. Paratte, and R. Chawla, "Benchmark on kinetic parameters in the CROCUS reactor," *International Reactor Physics Experiments Handbook (IRPhE), no. 4440, OECD*, 2007.

[21] J. Paratte, R. Früh, U. Kasemeyer, M. Kalugin, W. Timm, and R. Chawla, "A benchmark on the calculation of kinetic parameters based on reactivity effect experiments in the CROCUS reactor," *Annals of Nuclear Energy*, vol. 33, no. 8, pp. 739–748, 2006.

[22] V. Lamirand, P. Frajtag, D. Godat, O. Pakari, A. Laureau, A. Rais, M. Hursin, G. Hursin, C. Fiorina, and A. Pautz, "The colibri experimental program in the crocus reactor: characterization of the fuel rods oscillator," in *EPJ Web of Conferences*, vol. 225. EDP Sciences, 2020, p. 04020.

[23] V. Lamirand, A. Rais, O. Pakari, S. Hübner, C. Lange, J. Pohlus, U. Paquee, C. Pohl, M. Hursin, A. Laureau, P. Frajtag, D. Godat, G. Perret, C. Fiorina, and A. Pautz, "Experimental results of the first COLIBRI neutron noise campaign in the CROCUS reactor for the european project CORTEX," in *Submitted to PHYSOR 2020*, 2020.

[24] V. Lamirand, A. Rais, S. Hübner, C. Lange, J. Pohlus, U. Paquee, C. Pohl, O. Pakari, P. Frajtag, D. Godat, M. Hursin, A. Laureau, G. Perret, C. Fiorina, and A. Pautz, "Neutron noise experiments in the AKR-2 and CROCUS reactors for the european project CORTEX," *EPJ Web of Conferences*, vol. 225, p. 04023, 2020.

[25] V. Lamirand, A. Rais, O. Pakari, M. Hursin, A. Laureau, J. Pohlus, U. Paquee, C. Pohl, S. Hübner, C. Lange, P. Frajtag, D. Godat, G. Perret, C. Fiorina, and A. Pautz, "ANALYSIS OF THE FIRST COLIBRI FUEL RODS OSCILLATION CAMPAIGN IN THE CROCUS REACTOR FOR THE EUROPEAN PROJECT CORTEX," *EPJ Web of Conferences*, vol. 247, p. 21010, 2021.

[26] "Scionix holland, mechanical, optical and scintillation properties, https://scionix.nl/scintillation-crystals/#tab-id-4," 2019.

[27] *DSA-LX, Digital signal analyzer, https://www.mirion.com/products/dsa-lx-digital-signal-analyzer.*

[28] O. Pakari, V. Lamirand, T. Mager, P. Frajtag, and A. Pautz, "Delayed gamma fraction determination in the zero power reactor CROCUS," *EPJ N*, 2021.

[29] O. Pakari *et al.*, "High accuracy measurement of the prompt decay constant in CROCUS using gamma noise and bootstrapped uncertainties," *Submitted to Annals of Nuclear Energy*, 2021.

[30] R. Billnert, E. Andreotti, F.-J. Hambsch, M. Hult, J. Karlsson, G. Marissens, A. Oberstedt, and S. Oberstedt, "Novel scintillation detectors for prompt fission γ-ray measurements," *Physics Procedia*, vol. 31, pp. 29–34, 2012.

[31] K. Shah, J. Glodo, W. Higgins, E. van Loef, W. Moses, S. Derenzo, and M. Weber, "CeBr/sub 3/ scintillators for gamma-ray spectroscopy," in *IEEE Symposium Conference Record Nuclear Science*. IEEE, 2004.

[32] W. Higgins, A. Churilov, E. van Loef, J. Glodo, M. Squillante, and K. Shah, "Crystal growth of large diameter LaBr3:ce and CeBr3," *Journal of Crystal Growth*, vol. 310, no. 7-9, pp. 2085–2089, apr 2008.

[33] O. Pakari, "Experimental and numerical study of stochastic branching noise in nuclear reactors," Ph.D. dissertation, École polytechnique fédérale de Lausanne (EPFL), 2020.

[34] M. J. Weber and R. R. Monchamp, "Luminescence of Bi4 Ge3 O12 : Spectral and decay properties," *Journal of Applied Physics*, vol. 44, no. 12, pp. 5495–5499, dec 1973.

[35] O. H. Nestor and C. Y. Huang, "Bismuth germanate: A high-z gamma-ray and charged particle detector," *IEEE Transactions on Nuclear Science*, vol. 22, no. 1, pp. 68–71, 1975.

[36] S. E. Brunner and D. R. Schaart, "BGO as a hybrid scintillator / cherenkov radiator for cost-effective time-of-flight PET," *Physics in Medicine and Biology*, vol. 62, no. 11, pp. 4421–4439, may 2017.

[37] M. Gierlik, T. Batsch, M. Moszynski, T. Szczesniak, D. Wolski, W. Klamra, B. Perot, and G. Perret, "Comparative study of large NaI(tl) and BGO scintillators for the EURopean illicit TRAfficking countermeasures kit project," *IEEE Transactions on Nuclear Science*, vol. 53, no. 3, pp. 1737–1743, jun 2006.

[38] V. Lamirand, A. Laureau, O. Pakari, P. Frajtag, and A. Pautz, "Power calibration methodology at the CROCUS reactor," *EPJ Web of Conferences*, vol. 225, p. 04022, 2020.

[39] G. Perret, "Tm-41-14-02 rev. 1: Decay constant and delayed neutron fraction measurements in crocus," ERP, LRS, Paul Scherrer Institut, Tech. Rep., 10 2014.

[40] G. Knoll, *Radiation detection and measurement*. John Wiley & Sons, 2010.

[41] J. Hubbell, "Photon mass attenuation and energy-absorption coefficients," *The International Journal of Applied Radiation and Isotopes*, vol. 33, no. 11, pp. 1269–1290, 1982.

[42] O. V. Pakari, "Development of current and fast neutron noise measurements in CROCUS," Master's thesis, École polytechnique fédérale de Lausanne (EPFL), 2016. [Online]. Available: http://infoscience.epfl.ch/record/253177

[43] *Database of prompt gamma rays from slow neutron capture for elemental analysis*. Vienna: International Atomic Energy Agency, 2007.

[44] M.-M. Bé, V. Chisté, C. Dulieu, M. Kellett, X. Mougeot, A. Arinc, V. Chechev, N. Kuzmenko, T. Kibédi, A. Luca, and A. Nichols, *Table of Radionuclides*, ser. Monographie BIPM-5. Pavillon de Breteuil, F-92310 Sèvres, France: Bureau International des Poids et Mesures, 2016, vol. 8.

[45] D. E. Alburger and G. A. P. Engelbertink, "Half-lives ofBe11,c15,n16,o19, andAl28," *Physical Review C*, vol. 2, no. 5, pp. 1594–1596, nov 1970.

[46] H. H. Schmidt, P. Hungerford, H. Daniel, T. von Egidy, S. A. Kerr, R. Brissot, G. Barreau, H. G. Börner, C. Hofmeyr, and K. P. Lieb, "Levels and gamma energies of Al28 studied by thermal neutron capture," *Physical Review C*, vol. 25, no. 6, pp. 2888–2901, jun 1982.

[47] K. Shozugawa, N. Nogawa, and M. Matsuo, "Deposition of fission and activation products after the fukushima dai-ichi nuclear power plant accident," *Environmental Pollution*, vol. 163, pp. 243–247, apr 2012.

[48] W. Kim, M. Hursin, A. Pautz, L. Vincent, F. Pavel, and D. Lee, "Determination of the activity inventory and associated uncertainty quantification for the CROCUS zero power research reactor," *Annals of Nuclear Energy*, vol. 136, p. 107034, feb 2020.

[49] J. W. Durkee, M. R. James, G. W. McKinney, L. S. Waters, and T. Goorley, "The MCNP6 delayed-particle feature," *Nuclear Technology*, vol. 180, no. 3, pp. 336–354, dec 2012.

[50] V. Lamirand, O. Pakari, F. Vitullo, K. Ambrožič, D. Godat, P. Frajtag, and A. Pautz, "Local and high distance neutron and gamma measurements of fuel rods oscillation experiments," *ANIMMA 2021*, 2021.

[51] O. Pakari *et al.*, "High distance measurements of the prompt decay constant in the CROCUS reactor using gamma noise," *Submitted to Annals of Nuclear Energy*, 2021.

Design of a 150-miniature detectors 3D core-mapping system for the CROCUS reactor

Fanny Vitullo[1], Vincent Lamirand[1,2], Klemen Ambrožič[1], Laurent Braun[1], Daniel Godat[1], Pavel Frajtag[1], and Andreas Pautz[1,2]
[1]École polytechnique fédérale de Lausanne (EPFL), Switzerland
[2]Paul Scherrer Institut (PSI), Switzerland
fanny.vitullo@epfl.ch, vincent.lamirand@epfl.ch

***Abstract*—The present article provides an overview of the design of a three-dimensional (3D) full-core mapping system for the CROCUS reactor, operated at the École polytechnique fédérale de Lausanne (EPFL), Switzerland. The system is composed of 149 miniature neutron detectors distributed within the core double lattice at three main axial levels. The miniature detector technology is based on the optimization of the well-proven coupling of a miniature ZnS:^{6}LiF(Ag) scintillator to a state-of-the-art silicon photomultiplier (SiPM) via jacketed optical fibers. The challenges in the mechanical design, the detector optimization, the core criticality, and the development of the acquisition electronics are strongly interconnected and their combination is addressed in this article. The 3D full-core mapping system is foreseen to be installed in CROCUS in autumn 2021 and it will pave the way for the investigation of 3D dynamic phenomena in nuclear reactor cores.**

***Keywords* — Reactor instrumentation, 3D core-mapping, miniature neutron detector, in-core neutron detection, CROCUS.**

I. Introduction

THE interest in studying space-dependent neutronics phenomena in nuclear reactor cores has grown in recent years among the scientific community. The knowledge of detailed three-dimensional (3D) power distributions, local flux perturbations due to dynamic events, and the spatialization of neutron noise induced by mechanical perturbations would allow to determine and optimize reactor operational safety margins in both normal and accidental conditions. As of today, the spatial and/or temporal behavior of a reactor core can be predicted with high-fidelity by advanced neutronics codes. However, the validation of their 3D high-fidelity capabilities is non-trivial due to the limited availability of extensive and localized in-core neutron flux measurements in reference reactor cores [1]. Indeed, the challenges in the installation of 3D detection systems to gather such data are numerous. First of all, positioning a great number of in-core neutron detectors requires good reactor accessibility and inter-pin spaces fitting neutron detectors of a suitable size. Secondly, both the selected detectors and their acquisition chain must be easily produced in series at an acceptable cost.

In this framework, the zero-power research reactor CROCUS, operated at the École polytechnique fédérale de Lausanne (EPFL) was selected as a perfect candidate for the installation of a 3D full-core mapping system. A 3D measuring system in CROCUS will provide valuable insight on the propagation of neutron modulation in the core, thanks to the presence of several mechanical perturbation experiments, including the COLIBRI fuel rod oscillator [2], [3], the PISTIL rotating absorber [4], and a linear oscillating absorber currently under development. In addition, CROCUS is a readily available reactor with low technological uncertainties [5], where it would be possible to accurately position ~150 neutron detectors within the whole core volume upon the design of a proper mechanical supporting structure. At the same time, the Laboratory for Reactor Physics and Systems Behaviour (LRS) at EPFL in collaboration with the Paul Scherrer Institut (PSI) had developed a prototype of a miniature detector with an active volume below 1 mm^3 [6], [7]. The detection technology is based on the well-proven coupling of a miniature ZnS:^{6}LiF(Ag) scintillator to a state-of-the-art silicon photomultiplier (SiPM) via jacketed optical fibers. After optimization in dimensions, processing electronics, and acquisition scheme, such detectors suit the needs in terms of size and scalability for the setup of a 3D mapping system in CROCUS.

In this article, the full design of the 3D core-mapping system for the CROCUS reactor is presented. The synergies between neutron detectors physics and size, mechanical design, core criticality, and electronics acquisition are highlighted. In the first section, the custom mechanical design of the system is described. Secondly, the optimization of the miniature neutron detector prototype is discussed, followed by a study on the reactivity impact of the whole system. Finally, the digital acquisition electronics, developed in-house at LRS and tailored for the 3D core-mapping system, is presented.

II. Mechanical design

A. The CROCUS reactor

CROCUS is a zero-power research reactor operated at EPFL and dedicated to education and research [8]. The cylindrical reactor core is composed of low-enriched and aluminum-cladding fuel rods inserted into an open pool containing demineralized water acting as moderator and reflector. In the current configuration, the core active zone is divided into two different fuel zones with distinct lattice dimensions. The inner zone is composed of 336 uranium dioxide (UO_2) rods arranged in a squared lattice with a pitch of 18.37 mm. In the surrounding outer zone, 176 metallic uranium rods (U_{met}) are organized in a square lattice as well, but with a 29.17-mm pitch. In the west part of the outer fuel zone, a fuel rod oscillator device was installed as part of the COLIBRI experimental program [2]. The device allows to oscillate up to 18 U_{met} fuel rods (called "COLIBRI rods" in this article) of up to ±2.5 mm around nominal with a frequency in the Hz range. All the rods in the core are kept in a vertical position by two octagonal aluminum grids positioned at 1 m from each other. A 1-mm cadmium layer is included in both grids to limit the axial neutron leakages.

The maximum allowed fission power is limited to 100 W, corresponding to a neutron flux of about 2.5×10^9 $cm^{-2}\cdot s^{-1}$ at the core center. The core criticality can be controlled in two different ways: by adjusting the water level using a spillway, or by the insertion and withdrawal of two B_4C absorber control rods in two aluminum tubes, which can also be used as experimental channels. Six independent safety systems are installed in the core: two cruciform cadmium safety blades falling into the UO_2 lattice and four expansion valves for quickly emptying the water tank. Each of the six systems is enough to shut down the reactor. A 3D model of CROCUS is shown in Fig.1.

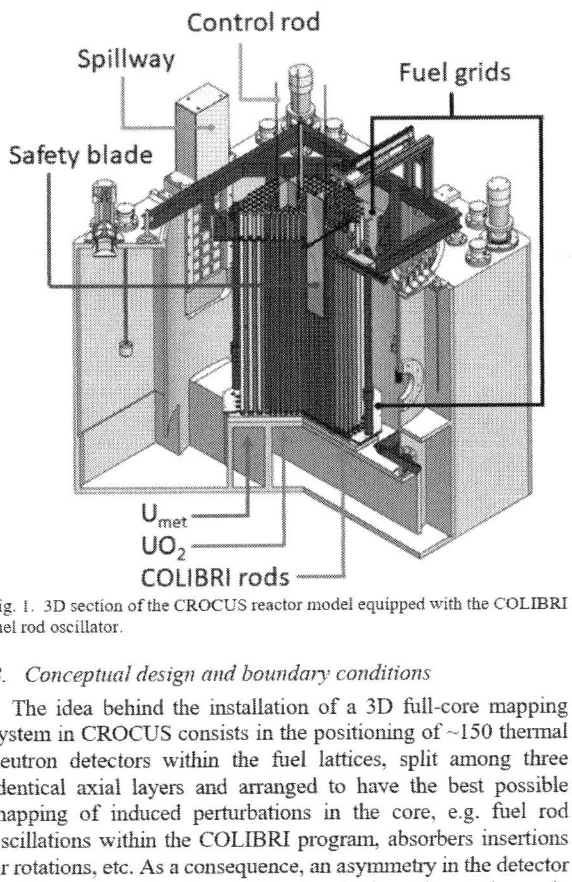

Fig. 1. 3D section of the CROCUS reactor model equipped with the COLIBRI fuel rod oscillator.

B. Conceptual design and boundary conditions

The idea behind the installation of a 3D full-core mapping system in CROCUS consists in the positioning of ~150 thermal neutron detectors within the fuel lattices, split among three identical axial layers and arranged to have the best possible mapping of induced perturbations in the core, e.g. fuel rod oscillations within the COLIBRI program, absorbers insertions or rotations, etc. As a consequence, an asymmetry in the detector distribution can be observed in the horizontal cut shown in Fig. 2, where a greater number of detectors is envisaged to be positioned within the 18 U_{met} oscillating rods. In the same optic, the addition of two more layers of detectors in the proximity of the control rods guide tubes (and on the symmetrical rods) would allow for a study of the localized perturbation induced by a control rod extraction. In conclusion, each of the three axial main layers will hold 47 detectors, plus two extra layers of 4 detectors each, for a total of 149 detectors in the CROCUS core.

However, installing 149 neutron detectors within the CROCUS core is a non-trivial goal. These detectors require a mechanical structure that holds and keeps them in the selected positions without having to modify the basic design of the reactor described in the previous section. The mechanical system must be designed to be as practical as possible for its out-of-vessel loading, of easy manufacturing, and respectful of the CROCUS safety standards. Indeed, there must be no contact or friction between the detectors or the structure and the aluminum cladding of the fuel rods. All the additional components must be chemically inert in demineralized water and they have to minimize the reactivity impact on the core criticality (see Section IV for a detailed reactivity study).

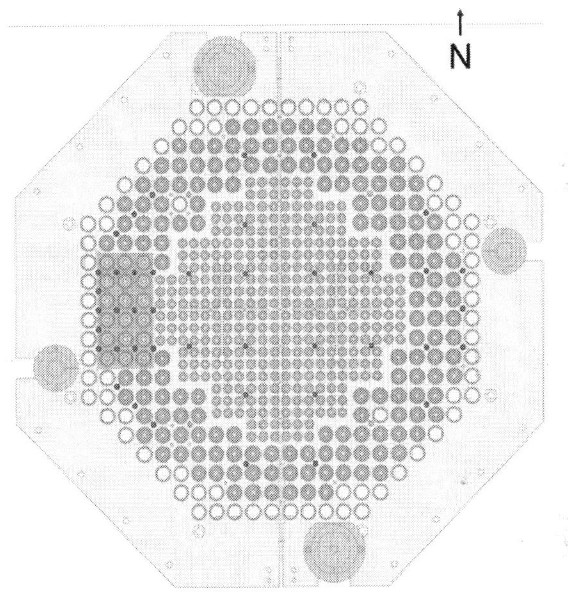

Fig. 2. Horizontal cut of CROCUS representing one of the three main layers of the 3D mapping system. Red dots represent the position of interest to install neutron detectors, while in yellow are indicated the positions where two supplementary intermediate layers of detectors are added.

C. Mechanical constraints

In the CROCUS reactor, a vertical insertion of the miniature detectors in the inter-pin space is made impractical by the presence of the upper fuel grid. Opening a passage for the optical fibers would require delicate machining of the grid to fit the purpose. Therefore, a horizontal insertion was preferred for the 3D mapping system, even though the peculiar double-lattice arrangement of CROCUS introduces limitations. Figure 3 shows the positions at which it is possible to cross both lattices and indicates the maximum available space in mm to perform full or partial straight crossings. Partial crossings arise from the presence of the cruciform safety blades in the core center, whose run must be kept clear at all times for safety reasons. However, these partial crossings can be performed from both north and south directions. Looking at Fig. 3, and excluding the partial crossing of 3.48 mm, the minimum space available between fuel rods is equal to 4.59 mm. Hence, the maximum allowed thickness of material cutting across the core was set to 4 mm to leave a margin for possible mechanical tolerances. The 4-mm limit remains valid also when the oscillating fuel rod device is displacing the 18 fuel rods indicated in pink in Fig. 3 by ±2.5 mm and reducing the 9.87 mm margin to 7.37 mm.

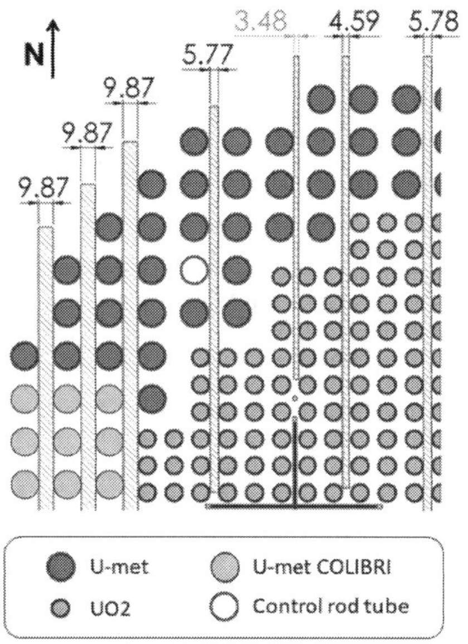

Fig. 3. North-west portion of the horizontal section of the CROCUS reactor with indication of the locations available to cross in a straight line the double fuel lattice. The maximum space available for the crossing is indicated in mm.

D. Final design of supporting structure

Based on the lateral insertion strategy of the 149 detectors and the 4-mm space limit, a tailored mechanical structure to load and hold the 3D full-core mapping system has been designed in-house at LRS. The SolidWorks model of the system installed in the CROCUS core is shown in Fig.4. The main elements composing the 3D full-core mapping system are:

- 50 rectangular "rulers" of polyoxymethylene (POM-C) with a thickness of 4 mm guiding and positioning the neutron detectors at the intended position. Black POM-C was selected because of its good resistance in water, its good light opacity and its density comparable to that of water (1.41 g/cm^3). Depending on the lattice arrangement, rulers can be crossing the whole core or being split in two to accommodate the presence of the cruciform safety blades;
- 52 horizontal supporting rods made of anodized aluminum for holding and positioning the rulers among fuel rods at 5 different axial layers. The three main levels (L1, L2 and L3 in Fig. 4) with 47 detectors each are positioned in a way that the detector's active part is located at 15, 50, 85 cm of core height respectively. The two additional layers close to control rods guide tubes (LM1 and LM2 in Fig.4) are at 30 and 70 cm;
- 8 vertical supporting rods of anodized aluminum in ex-core positions to support the horizontal rods at the different axial levels;
- 8 vertical supporting rods in anodized aluminum in in-core position to hold the not-crossing rulers. Dedicated adaptation pieces in aluminum and POM-C allow to attach these rods to the upper and lower fuel grid of CROCUS.

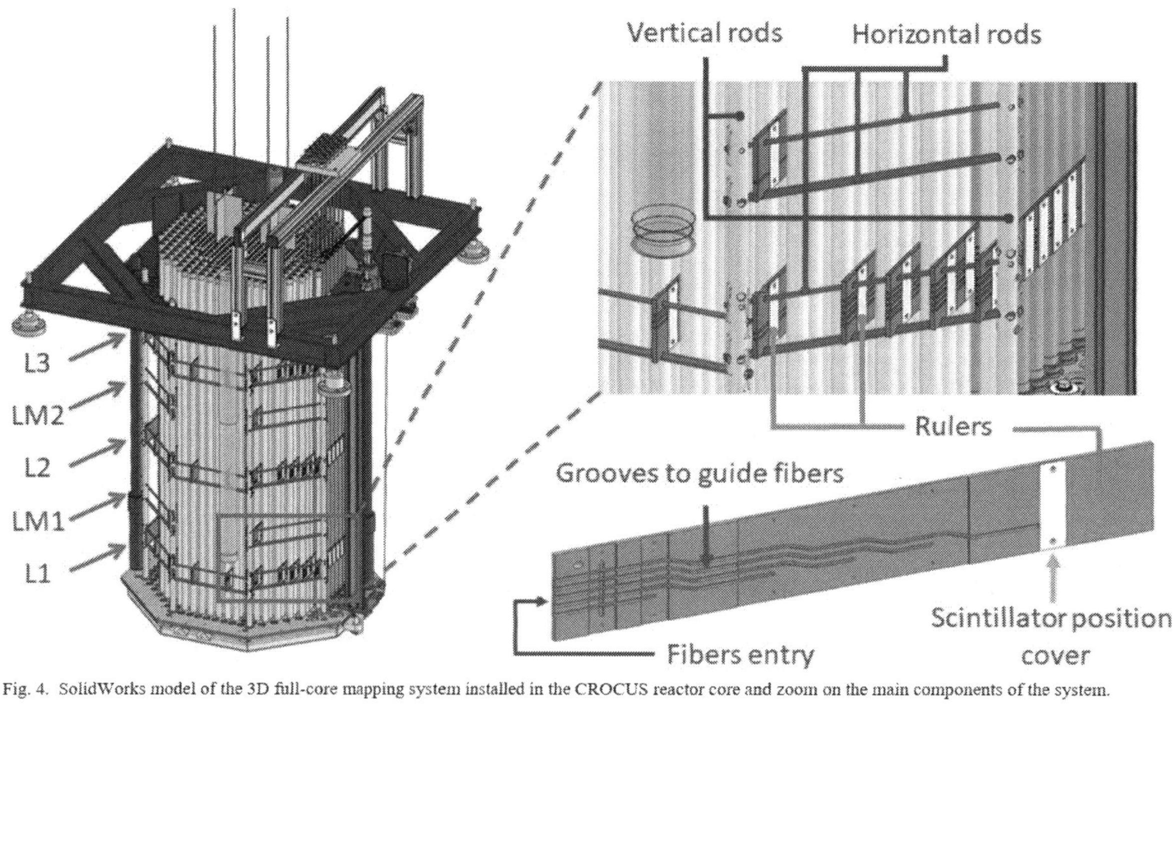

Fig. 4. SolidWorks model of the 3D full-core mapping system installed in the CROCUS reactor core and zoom on the main components of the system.

Aside from the standard components mentioned above, small variations are necessary to accommodate special design needs arising from the presence of numerous experiments and customizations in CROCUS. An example is represented by the vertical rod placed at the north side of the COLIBI rods, which cannot be inserted as the other 7 due to the presence of the fuel rod oscillator, and thus has two special grid attachments.

Mock-up pieces of both rulers and supporting structure have been produced to assess the mechanical stability of the assembly. Pictures of a test ruler and the mock-up mechanical structure are reported in Fig. 5.

Fig. 5. Mock-up mechanical structure to support a test ruler.

III. Optimization of detector design

A. *First prototype design*

The first prototype of a miniature neutron detector was developed and successfully tested at EPFL in collaboration with PSI in 2018 [6], [7]. This first development version was composed by a 1 mm^2 × 0.2 mm ZnS:^{6}LiF(Ag) screen [9] positioned at the tip of a 2-mm core optical fiber [10]. The blue light produced (with peak emission at 450 nm [9]) after a thermal neutron interaction with ^{6}Li in the screen is transported through the optical fiber to a silicon photomultiplier (SiPM) [11], where each photon event is translated into an electrical signal after proper processing. Neutron events are then discriminated from background on the basis of the photon-density variation in the stream of photon pulses. Indeed, assuming good dark conditions of the system, only few sparse pulses correspond to the background, while a step-wise and long-lasting (few tens of μs [9]) increase of the photon density is visible after a neutron interaction. In this first development version, the discrimination and the thermal neutron counting were performed using an analog processing electronics (see Section V for more details on processing electronics). Figure 6 summarizes the detection principle.

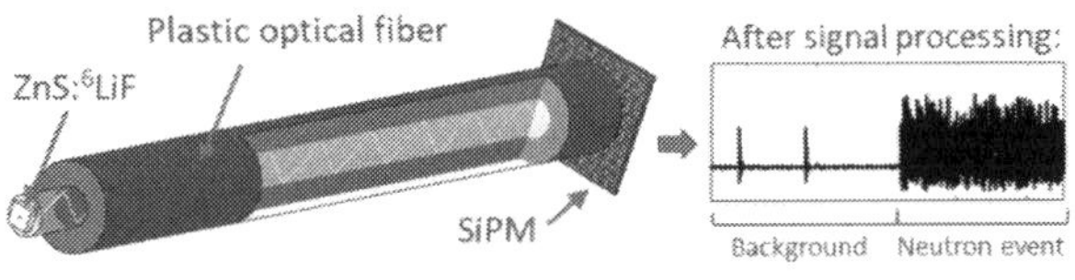

Fig. 6. Schematic of a miniature neutron detector working principle.

The first testing and characterization of the prototype showed that a detector of such reduced dimensions (< 1 mm^3) maintains a good sensitivity in terms of neutron counting in CROCUS, up to a total neutron flux of ~10^7 n/cm^2/s [6], [7]. It was proven that the detector is insensitive to gammas due to its small active volume and discrimination scheme [7], and the perturbation induced in the flux map is minimal [12], [13]. All these features, combined with the detector low-cost and easy-manufacturing, make the technology the perfect choice for a distributed core-mapping system, as long as it is possible to acquire data simultaneously from a large number of detectors in a cost-effective way (see Section V).

B. *Optical fiber optimization*

Although the excellent performance of the first prototype version proven in [7], [12], and [13], the outer fiber diameter of 3 mm represents a significant constraint when the miniature detectors are envisaged to be inserted in the CROCUS reactor core via the 4-mm thick POM-C rulers. A 3-mm fiber would leave a minimal plastic thickness that might compromise the mechanical stability of the rulers. Therefore, four different plastic optical fibers were tested and compared to the prototype version (called fiberA) to find a tradeoff between diameter and performances. The four fibers, whose characteristics are listed in Tab. 1, are all produced by the ESKA Mitsubishi Chemical Company [14].

TABLE I
Optical fibers tested for optimization.

Name	Model	Outer diameter (mm)	Core diameter (mm)	Jacket thickness (mm)
fiberA	SH8001	3	2	0.5
fiberB	SH4001	2.2	1	0.6
fiberC	SH4001-1.3	1.3	1	0.15
fiberD	SH2001	1	0.5	0.25

The four fibers were tested in blank conditions, meaning without counting neutrons, in order to assess their background level independently from the measurement conditions. Signals corresponding to background photons arise from the light-tightening quality of the fiber tip coverage, the fiber jacketing, and the coupling with the SiPM. Considering that all four optical fibers had the same front-end tip cover and an identical pairing with the SiPM, the number of background photons filtering from the polyethylene jacket in 1 ms was measured. The counting was performed using a CAEN V2495 FPGA [15] loaded with the PLUscaler firmware [16]. The test was repeated in two different light conditions to assess whether laboratory environmental conditions could bias the measurements.

The results are presented in Fig. 7 that clearly shows how the jacket thickness is a key parameter in the selection of the optical fiber. In dark conditions, the number of background photons is of the same order of magnitude for the four optical fibers. However, in daylight conditions the number of background photons increases significantly for optical fibers with a thin jacket with respect to their counting in the dark. The number

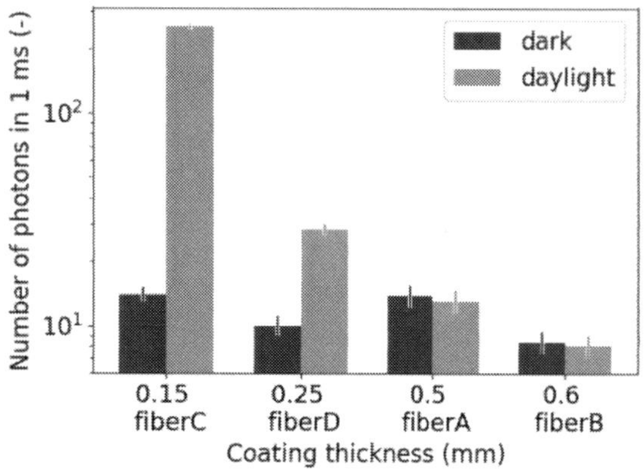

Fig. 7. Number of background light photons filtering from the optical fiber jacket in 1 ms as a function of the jacket thickness, in two different light conditions.

of photons measured in daylight decreases with the increase of the jacket thickness, up to a point where there is no difference between daylight or darkness conditions.

Therefore, above a jacket thickness of 0.5 mm, the optical fiber is properly light-shielded and unaffected by the external light conditions. Looking at Fig. 7, the only two optical fibers having a jacket thickness above 0.5 mm are fiberA (SH8001), corresponding to the prototype fiber in [7], and fiberB (SH4001), having a polymethyl methacrylate (PMMA) core of 1 mm in diameter and an outer diameter of 2.2 mm. According to the requirements, fiberB represents the perfect combinations between the optical fiber performances and the dimensional constraints set by the design of the 3D full-core mapping system.

C. Final detector design for 3D core-mapping system

The choice of fiberB, having a core diameter of 1 mm, lead to the optimization of the ZnS:^{6}LiF(Ag) screen so that the level of thermal neutron sensitivity is maximized. A new Scintacor ND2:1 foil with 0.225 mm thickness [9] was precisely cut at EPFL into square pieces of 0.66 mm × 0.66 mm using a wire-cutting technique. This screen size allows for a good coverage of the fiber tip and a possible reduction of the reactivity impact if compared to the prototype version (see next Section).

The 0.66 × 0.66 mm screen is glued at the sanded and uncovered tip of the optical fiber PMMA core. An epoxidic adhesive called Araldite [17] is used due to its excellent resistance in water, fast setting and good transparence. With this method, the ZnS:^{6}LiF(Ag) screen is embedded in the glue itself and both unwanted movements or loss of the screen in the CROCUS core are prevented.

Figure 8 shows the final optimized design chosen to build the 149 miniature neutron detectors for the 3D full-core mapping system in CROCUS. Although the good accuracy in the ZnS:^{6}LiF(Ag) screens cutting and gluing, a detector inter-calibration with a Pu-Be source available at LRS is envisaged to be carried out before the system installation in the CROCUS core.

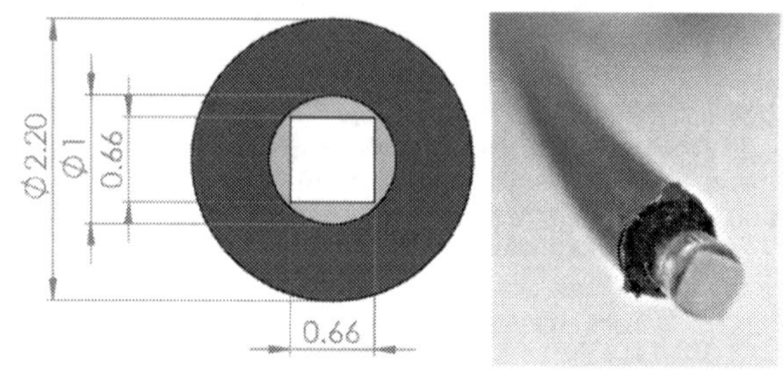

Fig. 8. Horizontal section in scale (a) and picture (b) of the SH4001 optical fiber with a squared 0.66 × 0.66 × 0.225 mm ND2:1 screen glued at the tip.

IV. REACTIVITY IMPACT

The 3D core-mapping system designed for the CROCUS reactor, composed by 149 optimized miniature neutron detectors and the mechanical structure, will have an impact on the reactor operations in terms of criticality. It is foreseen that the 149 ZnS:^{6}LiF(Ag) scintillator screens will introduce a strong negative reactivity insertion, which could be partially counteracted by the positive reactivity effect induced by the installation of the POM-C rulers having a slightly higher density than water. By understanding the balance between these concurrent effects it is possible to assess if the reactor can be operated within the authorized core configurations and in the respect of the safety limits. Hence, the core criticality is studied computationally via Monte Carlo simulations performed with the MCNPX-2.4 code [18], as required by the Swiss safety authority. A detailed MCNPX model of a typical configuration of CROCUS was used as a baseline and reference.

A key parameter to obtain reliable simulation results is represented by the material composition of the ZnS:^{6}LiF(Ag) screens, and in particular by the level of enrichment of ^{6}Li. Based on the supplier data and in order to obtain conservative results, a 95% ^{6}Li enrichment was used, and the resulting reactivity worth of the screen was compared with experimental results. Two miniature neutron detectors with different sensitive volume were inserted alternatively in the north-west control rod guide tube of CROCUS and the critical level variation with respect to the standard configuration was recorded. The results of both experiments and simulations are reported in Table II, where a good agreement in the reactivity worth range of few negative pcm is shown.

TABLE II
MEASURED AND SIMULATED RESULTS FOR THE REACTIVITY WORTH IN CROCUS OF TWO MINIATURE NEUTRON DETECTORS WITH DIFFERENT SENSITIVE VOLUME [21]. SIMULATIONS WERE PERFORMED WITH JEFF 3.1 [19].

Screen volume (mm^3)	Measured reactivity worth (pcm)	Simulated reactivity worth (pcm)
0.2	- 1.2 ± 0.2	- 1 ± 2
0.09	- 0.7 ± 0.2	- 1 ± 1

Having confirmed the material composition, the entire 3D core-mapping system was implemented in the MCNPX model. For the sake of simplicity, rulers are modeled as single parallelepipeds, with no distinction between the optical fiber and the ruler material because both are made of similar plastics. The 0.66 × 0.66 × 0.225 mm ZnS:^{6}LiF(Ag) screens are modeled in the predetermined positions (see Fig. 2). The rest of the supporting structure was included in the model following as much as possible the previously described mechanical design. Two different nuclear data libraries were used for the simulations: JEFF 3.1 [19] and ENDF/B-VII.1 [20]. The simulation results are presented in Table III.

TABLE III
SIMULATED REACTIVITY WORTH FOR THE 3D CORE-MAPPING SYSTEM [21].

Configuration	JEFF-3.1 K_{eff}	JEFF-3.1 worth (pcm)	ENDF/B-VII.1 K_{eff}	ENDF/ B-VII.1 worth (pcm)
Standard	1.00240 ± 1 pcm	0	1.00289 ± 2 pcm	0
With 3D core-mapping	1.00189 ± 2 pcm	- 51 ± 3	1.00245 ± 2 pcm	-44 ± 4

The results in Table III show that the general effect of mounting the 3D full-core mapping system in CROCUS is a negative reactivity insertion. The strongest negative reactivity insertion corresponds to -51 ± 3 pcm, which is well below the maximum available reactivity reserve of CROCUS of +200 pcm. Therefore, the CROCUS reactor with the 149-detectors system installed in its core can be safely operated within its standard limits. At the same time, a request for the addition of two U_{met} rods in the core periphery has been submitted to the Swiss safety authority to avoid that the future installation of additional absorbing devices would deplete the core's reactivity reserve.

V. DIGITAL ACQUISITION CHAIN

The first prototype model of miniature neutron detector described in Section III.A was using a processing electronics developed at PSI to transform SiPM currents into a stream of photon events, and an analog read-out electronics assembled at LRS to count and discriminate thermal neutron events from the background. Despite the good performance and excellent reliability of this electronic system (more details can be found in [7]), the latter presents some limitations whether to be employed for an extensive core-mapping system. In particular, the acquisition part was composed by standard and expensive electronic modules that are limiting the acquisition to only a few channels simultaneously. This restriction on the system parallelization at a reasonable price prompted the development of a new electronics.

In a first attempt of upgrading the system, a novel processing electronic board was developed internally at LRS to improve photon collection efficiency, pulse shortening, as well as light noise reduction. Three boards, composed by a SiPM and a fast pre-amplifier stage, were built and tested in [13]. Their good performances paved the way for the development of an extensive electronics tailored for the 3D core-mapping.

A. Novel 32-channels processing electronics

Six novel stand-alone electronic modules based on the three prototype boards were developed and built in-house at LRS. Each module is able to process simultaneously the incoming signal from 32 different miniature neutron detectors. The use of five modules together covers 160 channels and thus the needs of the 3D core-mapping system. The sixth module is a backup.

The module front-end, visible in Fig. 9(a), presents 32 openings where each optical fiber can be connected and coupled with its respective SiPM via a black plastic cylinder. The cylinders are separating and keeping in the dark the 32 SiPMs to minimize background noise and cross-talk events. Although the cylinders and the external structure are designed to minimize the light filtering, additional layers (a wardrobe and a black cloth) are foreseen to be added to maximize the coverage. Each SiPM output is sent to one of 32 separated processing boards produced following the prototype design and shown in Fig. 9(c). The boards are all mounted on a motherboard visible in Fig. 9(b) from above. Each board is dedicated to the conversion of the corresponding SiPM current signal into a photon counting stream through a user-selected threshold level (called "photon threshold") for electronic noise removal. Both the 32 SiPMs and the 32 processing boards are powered by an internal alimentation module. The bias voltage applied to the SiPMs and the photon threshold can be selected with knobs on the module front-end and their level can be visualized on a display.

The output of a single 32-channels module consists in the simultaneous photon counting stream in every of the 32 channels. These signals, using the LVDS standard, can be then sent to a dedicated FPGA module to discriminate neutron events and perform the neutron counting.

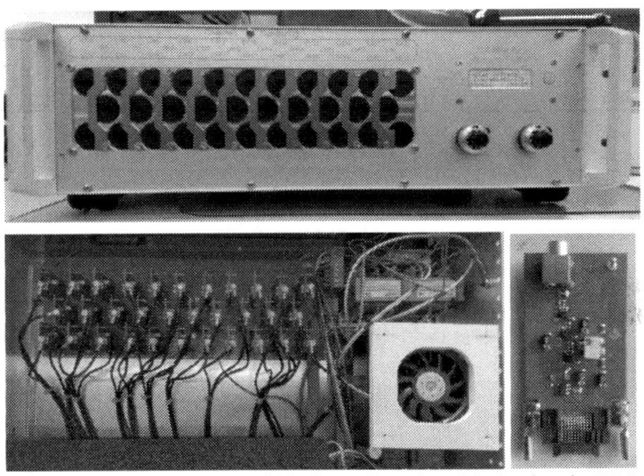

Fig. 9. Front-end (a) and internal (a) of a 32-channel module developed at LRS. One of the 32 fast-preamplifier board installed in the module (c).

B. FPGA acquisition

The 160 photon counting signal produced by the five 32-channels boxes can be treated at the same time by using an FPGA board. In this work, the general-purpose programmable CAEN V2495 [15] was used. The board has 64 build-in inputs that can be expanded to 160 adding three mezzanines boards in the expansion slots, accommodating the 160 photon counting signals. The board can be controlled and programmed through the USB interface, or as in this work via the V1718 VME bridge module [22].

A custom-made proprietary firmware to process photon counting into neutron counting was developed at PSI and in the present work is loaded into the V2495 User FPGA. The software counts the number of photons in fixed intervals of time, whose length can be user-selected and ranges from 5 ns to 1.125 μs. A moving-average based algorithm is then applied to the number of photons counted in each interval: the sum of photons in five consecutive intervals is compared with the sum of photons counted in the previous five. This moving-average technique allows to identify density variations in the number of photons and thus to recognize the dense and long stream of photons originating from a neutron event in the ZnS:^{6}LiF(Ag) screen. To perform this discrimination, a limit called "digital threshold" is set by the user to distinguish between low photon-density occurrences, i.e. background, and the high light-density events corresponding to neutrons. An additional parameter called "blocking window" can be tweaked by the user to avoid erroneous discriminations.

In conclusion, thermal neutron events can be counted and recorded according to the selected digital threshold and blocking window. The entire acquisition scheme is performed simultaneously by the FPGA in all the 160 channels available. Hence, the signals coming from the 149 miniature neutron detectors composing the 3D full-core mapping system can be acquired with a single affordable module.

VI. Conclusions

A 3D full-core mapping system was designed for the CROCUS reactor. The system features the installation of 149 optimized miniature neutron detectors distributed within the core double lattice at three main axial levels with the support of a mechanical structure. The system was mechanically designed taking into account the safety limits, the dimensional constraints of the core, and the need for mapping the propagation of mechanical perturbations in the core. The final design is composed of 50 4-mm thick plastic rulers guiding and positioning the detectors at the desired positions. The rulers are kept in position by an aluminum structure surrounding the core.

The miniature neutron detector design was optimized to accommodate them in the 4-mm rulers. An optical fiber with an external diameter of 2.2 mm and a 1-mm diameter core was chosen due to the excellent trade-off between dimension reduction and enhanced performances offered. The size of the detector sensitive part, the ZnS:^{6}LiF(Ag) screen, was adapted to the updated optical fiber by cutting it into squares of $0.66 \times 0.66 \times 0.225$ mm.

The reactivity impact of the entire 3D core-mapping system was assessed with MCNPX simulations. It was calculated that the entire assembly introduces a negative reactivity insertion of -51 ± 3 pcm, which allows to safely operate the reactor without exceeding its reactivity reserve.

Finally, the electronics detection chain was upgraded by developing six 32-channels processing modules and a dedicated FPGA acquisition scheme. The combination of the two allows for the simultaneous read-out, amplification, processing, and acquisition of up to 160 channels, corresponding to 160 miniature neutron detectors.

The installation of the full-core mapping system for the CROCUS reactor is planned for autumn 2021. After a proper detector inter-calibration, the system will mark a milestone in proving the feasibility of such a first-of-a-kind system and in providing valuable localized experimental data for the validation of high-fidelity neutronics codes.

Acknowledgment

The research leading to these results has received funding from the Euratom research and training programme 2014-2018 under grant agreement No 754316.

The authors would like to thank the IVICT Europe GmbH company for their availability and kindness in providing optical fiber samples.

References

[1] Z. I. Böröczki, G. Klujber, G. Tolnai, B. Molnár, D. Légrády, F. Gabrielli, A. Rineiski, and M. Szieberth, "Simulation of a research reactor reactivity transient with deterministic and GPU-assisted Monte Carlo reactor kinetics codes," *Eur. Phys. J. Plus 135, 281* (2020). Available: https://doi.org/10.1140/epjp/s13360-020-00280-4.

[2] V. Lamirand, P. Frajtag, D. Godat, O. Pakari, A. Laureau, A. Rais, M. Hursin, G. Perret, C. Fiorina and A. Pautz, "The COLIBRI experimental program in the CROCUS reactor: characterization of the fuel rods oscillator", *EPJ Web Conf., 225 (2020) 04020.* Available: https://doi.org/10.1051/epjconf/202022504020.

[3] V. Lamirand, A. Rais, S. Hübner, C. Lange, J. Pohlus, U. Paquee, C. Pohl, O. Pakari, P. Frajtag, D. Godat, M. Hursin, A. Laureau, G. Perret, C. Fiorina and A. Pautz, "Neutron noise experiments in the AKR-2 and CROCUS reactors for the European project CORTEX," *EPJ Web Conf., 225 (2020) 04023.* Available: https://doi.org/10.1051/epjconf/202022504023.

[4] Y. Jiang, B. Geslot, V. Lamirand, P. Leconte, "PISTIL, a reactivity modulation device to probe the transfer function of the research nuclear reactor CROCUS", presented at the ANIMMA2021 conference, Prague, Czech Republic, Jun. 21-25, 2021.

[5] U. Kasemeyer, R. Früh, J. M. Paratte, and R. Chawla, "International Reactor Physics Experiments Handbook (IRPhE)," no. 4440, OECD, Ed. 2007, p. 94.

[6] F. Vitullo, V. Lamirand, J.-B. Mosset, P. Frajtag, O. Pakari, G. Perret and A. Pautz, "Developing and testing a miniature fiber-coupled scintillator for in-core neutron counting in CROCUS," *EPJ Web Conf., 225 (2020) 04018.* Available: https://doi.org/10.1051/epjconf/202022504018.

[7] F. Vitullo, V. Lamirand, J.-B. Mosset, P. Frajtag, O. Pakari, G. Perret, and A. Pautz, "A mm 3 Fiber-Coupled Scintillator for In-Core Thermal Neutron Detection in CROCUS," IEEE Trans. Nucl. Sci., vol. 67, no. 4, pp. 625–635, Apr. 2020.

[8] V. Lamirand, "Installation nucléaire CROCUS - Rapport de sécurité," version 1.2, 2017.

[9] Scintacor, "Neutron screens," 2020. Available: https://scintacor.com/wp-content/uploads/2015/09/Datasheet-Neutron-Screens-High-Res.pdf. Accessed on: August 2021.

[10] ESKA Mitsubishi Chemical Corporation, "Data sheet SH-8001," 2019. Available: https://i-fiberoptics.com/pdf/sh-8001.pdf. Accessed on: August 2021.

[11] Hamamatsu Photonics K.K., Solid State Division, "MPPC (Multi Pixel Photon Counter) – S13360 Series," in particular SH13360-3050PE, 2016. Available: https://www.hamamatsu.com/resources/pdf/ssd/s13360_series_kapd1052e.pdf. Accessed on: August 2021.

[12] F. Vitullo, V. Lamirand, P. Frajtag, G. Perret and A. Pautz, "HIGHLY LOCALIZED AZIMUTHAL MEASUREMENTS IN THE CROCUS REACTOR TOWARDS THE VALIDATION OF HIGH-FIDELITY NEUTRONICS CODES," *EPJ Web Conf., 247 (2021) 08014.* Available: https://doi.org/10.1051/epjconf/202124708014.

[13] E. L. Brunetto, F. Vitullo, V. Lamirand, K. Ambrožič, D. Godat, M. Buck, G. Pohlner, J. Starflinger, A. Pautz, "High resolution measurements with miniature neutron scintillators in the SUR-100 zero power reactor," presented at the ANIMMA2021 conference, Prague, Czech Republic, Jun. 21-25, 2021.

[14] ESKA Mitsubishi Chemical Corporation. Cable products line-up. Available: https://www.pofeska.com/pofeskae/product/02/index.html. Accessed on: August 2021.

[15] CAEN S.p.A., V2495 - Programmable Logic Unit PLUS. Available: https://www.caen.it/products/v2495/. Accessed on: August 2021.
[16] CAEN S.p.A., FW2495SC - 160 Channels Latching Scaler for V2495 and DT5495. Available: https://www.caen.it/products/fw2495sc/. Accessed on: August 2021.
[17] Araldite. ARALDITE® RAPID 24ML. Available: https://www.go-araldite.com/products/epoxy-adhesives/araldite-rapid-24ml-syringe. Accessed on: August 2021.
[18] L. S. Waters, ed., MCNPX User's Manual, Version 2.4.0, Los Alamos National Laboratory report LA-CP-02-408 (Sept. 2002).
[19] OECD/NEA Data Bank, "The JEFF-3.1 Nuclear Data Library", JEFF Report 21, OECD/NEA Data Bank (2006).
[20] M.B. Chadwick, M. Herman, P. Obložinský, M.E. Dunn, Y. Danon, A.C. Kahler, D.L. Smith, B. Pritychenko, G. Arbanas, R. Arcilla, R. Brewer, D.A. Brown, R. Capote, A.D. Carlson, Y.S. Cho, H. Derrien, K. Guber, G.M. Hale, S. Hoblit, S. Holloway, T.D. Johnson, T. Kawano, B.C. Kiedrowski, H. Kim, S. Kunieda, N.M. Larson, L. Leal, J.P. Lestone, R.C. Little, E.A. McCutchan, R.E. MacFarlane, M. MacInnes, C.M. Mattoon, R.D. McKnight, S.F. Mughabghab, G.P.A. Nobre, G. Palmiotti, A. Palumbo, M.T. Pigni, V.G. Pronyaev, R.O. Sayer, A.A. Sonzogni, N.C. Summers, P. Talou, I.J. Thompson, A. Trkov, R.L. Vogt, S.C. van der Marck, A. Wallner, M.C. White, D. Wiarda, P.G. Young, "ENDF/B-VII.1: Nuclear Data for Science and Technology: Cross Sections, Covariances, Fission Product Yields and Decay Data", *Nucl. Data Sheets 112* (2011) 2887.
[21] V. Lamirand, F. Vitulo, K. Ambrožič, O. Pakari, L. Braun, D. GODAT, "Report on the development of fibre-based scintillator," CORTEX D2.3 report, 2021.
[22] CAEN S.p.A., V1718 - VME-USB2.0 Bridge. Available: http://primisdb.caen.it/products/v1718/. Accessed on: August 2021.

Local and high distance neutron and gamma measurements of fuel rods oscillation experiments

Vincent LAMIRAND[1,2], Oskari PAKARI[1], Fanny VITULLO[1], Klemen AMBROŽIČ[1], Daniel GODAT[1], Pavel FRAJTAG[1], Andreas PAUTZ[1,2]

[1]École polytechnique fédérale de Lausanne (EPFL), Switzerland
[2]Paul Scherrer Institut (PSI), Switzerland

vincent.lamirand@epfl.ch

***Abstract*— We report in the present article on the successful observation using noise analysis of the lateral oscillation of one fuel rod by ±2.5 mm around nominal at 0.1 Hz frequency, using an mm^3 miniature neutron scintillator *at the rod level*, and a BGO gamma detector *seven meters away from the reactor core center*. The experiment was conducted as part of the COLIBRI program in the CROCUS reactor, which is dedicated to the investigation of reactor noise induced by fuel vibrations. It consists in experiments on rod lateral displacement (static) and oscillation (dynamic) with different rods' numbers at various relevant amplitudes and frequencies. Its main motivation is the increased amplitudes in the neutron noise distributions recorded in ex- and in-core detectors that have been observed in recent years in Siemens pre-Konvoi type of PWR reactors. The obtained experimental data are used for the purpose of code validation, especially within the framework of the European project CORTEX on reactor noise applications. During the first phase of COLIBRI, the observation of a spatial dependence of the perturbation noise, also called neutron modulation, was demonstrated. In the second phase of COLIBRI starting 2021, it is planned to use a core mapping array of neutron detectors to record its propagation. It consists in about 150 miniature scintillators coupled to optical fibers and SiPM readouts, to be distributed in the reactor core. As a feasibility test, experiments were performed using a miniature scintillator prototype placed on a fuel rod, and oscillating the instrumented rod or the one directly adjacent to the detector. In addition, it is theoretically possible to measure branching or perturbation reactor noise using gamma radiation. Following recent developments on gamma measurements in CROCUS, the fuel oscillation was simultaneously recorded with a gamma detection array, LEAF. Its large BGO detectors were used by placing them at the maximum distance to the core, i.e. seven meters away with a clear line of sight using an experimental channel through the shielding of the reactor cavity.**

***Keywords* — Reactor instrumentation, miniature neutron detector, neutron scintillator, gamma scintillator, reactor noise, noise analysis, zero power transfer function.**

I. Introduction

An expertise on perturbation and branching neutron noise was progressively acquired at the Laboratory for Reactor Physics and System behaviour (LRS) at École polytechnique fédérale de Lausanne (EPFL), thanks to a variety of instrumentation developments and experiments since the restart of experimental activities a bit less than a decade ago [1]–[10]. In particular, the COLIBRI program is dedicated to the experimental investigation of reactor noise related to fuel vibrations in the CROCUS zero power reactor. It consists in experiments on fuel rod lateral displacement and oscillation with different rods' numbers at various relevant amplitudes and frequencies. Its original and main motivation is the increased amplitudes in the neutron noise distributions recorded in ex- and in-core detectors that have been observed in Siemens pre-Konvoi type of pressurized water reactors, as recently in the Swiss Gösgen nuclear power plant. Thanks to this experimental program, EPFL contributes to the Horizon 2020 European project CORTEX, which is dedicated to the understanding and simulation of reactor perturbations for the development of novel core monitoring techniques [11], [12].

During the first phase of COLIBRI, the observation of a spatial dependence of the perturbation noise, also called neutron modulation, was demonstrated [13]. In the second phase of COLIBRI starting 2021, it is planned to use a core mapping array of neutron detectors to record its propagation [14]. It consists in about 150 miniature scintillators coupled to optical fibers and silicon photomultiplier (SiPM) readouts, to be distributed in the reactor core. As a test of local neutron noise measurements, experiments were performed using a miniature neutron scintillator prototype developed at LRS in collaboration with the Paul Scherrer Institut (PSI). With the detector directly placed on a fuel rod, the instrumented rod or the one directly adjacent to the detector were successively oscillated within the lattice at an amplitude of ±2.5 mm around nominal and a frequency of 0.1 Hz.

In addition, it is theoretically possible to measure branching or perturbation reactor noise using gamma radiation. In-core and at distance branching gamma noise measurements were recently conducted in CROCUS, thanks to recent developments on gamma measurements at LRS [3], [15]–[18]. As a test to measure perturbation noise at distance, the fuel oscillation was simultaneously recorded with a gamma detection array, LEAF [15]. Its two large and high efficiency BGO detectors were used by placing them at the maximum distance to the core, i.e. seven meters away with a clear line of sight using an experimental channel through the reactor cavity, in the same configuration than the most distant branching noise measurements.

The article is structured as follows: in Section II, we introduce neutron modulation and the used methodology; in Section III, we present the experimental campaign, including the CROCUS reactor, the COLIBRI fuel rod oscillation device, the neutron and gamma detection systems, as well as the conducted experiments. We then present and discuss the results in Section IV, for both detection systems separately, and in comparison. Finally, we conclude in Section V.

II. Neutron modulation and methodology

When a small and periodic reactivity perturbation $\rho(t) = \rho_0 + \delta\rho(t)$ is applied to a critical assembly, its kinetic behavior with respect to the neutron population $N(t) = N_0 + \delta N(t)$ can be modelled by a linear time invariant system represented in the frequency domain by the zero power transfer function, or ZPTF [19]:

$$\frac{N(j\omega_0)}{\rho(j\omega_0)N_0} = H(j\omega_0) = \left(j\Lambda\omega_0 + \sum_i^n \frac{j\omega_0\beta_i}{j\omega_0+\lambda_i} - \rho_0\right)^{-1} \qquad (1)$$

Where j is the unit imaginary number, ω_0 is the angular frequency of the perturbation, $(\beta_i)_{i=1,2,\dots,n}$ the effective fractions and $(\lambda_i)_{i=1,2,\dots,n}$ the decay constants for the precursor group i. The function, as in the complex domain, is conventionally represented by its modulus (amplitude) and argument (phase).

The ZPTF is commonly determined experimentally by modulating periodically the reactivity at several frequencies and measuring its effect on the neutron population. Experimental data consist of time series obtained from radiation detectors sensitive to the core neutron flux, such as the miniature neutron scintillators used in this study and presented hereafter. In addition, it was already demonstrated that fission gamma rays can be used for reactor noise analysis, in the specific case of branching noise, as they are time stamping fission and decay events as neutrons do [3]. We hereby test the hypothesis using gamma scintillators on the specific case of modulation, or perturbation noise, which can be considered an extension of branching noise.

The time series are analyzed in the frequency domain by typically calculating Auto- and Cross Power Spectral Densities (APSD/CPSD) of respectively one or two detectors' signals. As a consequence, a fuel rod oscillation in COLIBRI at a given base frequency would induce a peak at this given frequency. The amplitude in the power spectral density would depend on the corresponding reactivity change, and frequency.

III. Experimental setup

A. The CROCUS reactor

CROCUS is an experimental zero-power reactor located at EPFL, uranium-fueled and water-moderated, dedicated to teaching radiation and reactor physics, and to research [20]. A complete description of the reference core can be found in the International Reactor Physics Experiments Handbook (IRPhE) [21], [22]. It has been licensed for operating at a maximum power of 100 W, i.e. a total neutron flux of ~$2.5\cdot10^9$ cm$^{-2}\cdot$s^{-1} at the core center. Criticality is controlled either by water level using a spillway, or by two B_4C absorber control rods, with an accuracy of ±0.1 mm (equivalent to approximately ±0.4 pcm) and ±0.5 mm (up to ±0.2 pcm), respectively. CROCUS operates at room temperature using a controlled water loop with secondary and tertiary circuits, two heat exchangers and an electrical heater.

The core is located in an Al-6060 grade vessel of 130 cm in diameter, 160 cm in height, and 1.2 cm in thickness. The vessel is filled with demineralized light water used as both moderator and reflector. The core active part has the approximate shape of a cylinder of 100 cm in height and about 60 cm in diameter. It consists of two interlocked fuel zones with square lattices of different pitches:

- an inner zone of 336 UO_2 rods with an enrichment of 1.806 wt.% and a pitch of 1.837 cm;
- an outer zone of 176 U_{metal} rods for these experiments, 0.947 wt.% and 2.917 cm;
- a varying water gap between the two zones because of the two different pitches.

A picture of the facility and core configuration is shown on Figure 1. Both uranium fuels consist of a 1-m pile of cylindrical pellets cladded in aluminum. The rods are maintained vertically by two octagonal aluminum grid plates spaced 1 m apart. In the COLIBRI program, the grids have a 1 mm cadmium layer to limit axial neutron leakage to the environment, i.e. structures activation, with the active zone of the fuel starting in the middle of the lower cadmium layer.

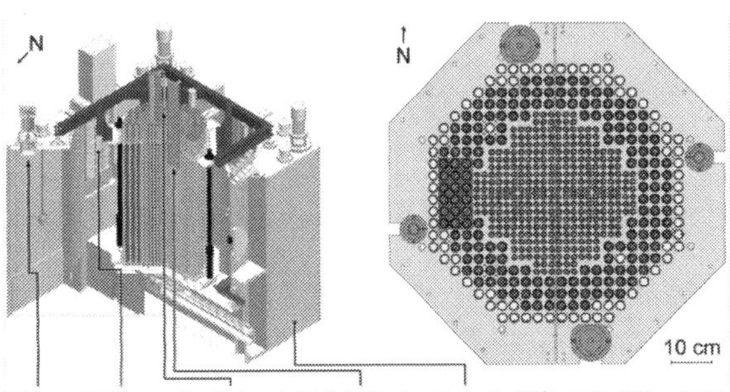

Fig. 1. View of the CROCUS reactor (left), and the core superior grid and configuration with the oscillation location (highlighted with green U_{met} rods).

B. The COLIBRI fuel rods oscillator

The COLIBRI fuel rods oscillator is designed to simultaneously oscillate laterally any of 18 metallic uranium fuel rods in the west region of the core periphery zone. It consists of two moving plates set above and below the core grids, and rigidly connected by an aluminum beam (see Fig. 2). Each plate carries an extremity of the fuel rods, top and bottom respectively. The top moving plate is fixed on the superior grid *via* gliders. Its oscillation is produced by a motor: the motor rotation is converted to a linear translation using an eccentric sheave and a connecting rod. The oscillation is transferred to the bottom moving plate *via* the aluminum beam. The bottom moving plate is not constrained by gliders, and is displaced only due to its connection to the transmission beam.

The selection of the moving fuel rods is performed by letting the rods lay on the reactor base plate (non-moving), or suspend them up 10 mm above the base plate to insert them in the moving plates. Top and bottom end caps are fixed to each rod to allow the insertion in the enlarged holes of either the static grids or the moving plates. The weight of the oscillating rods is supported by a platform. The amplitude of the oscillation is precisely tuned by changing the eccentricity of the sheave with calibration plates, 0.5 mm by 0.5 mm from 0 to ±2.5 mm. Its frequency depends on the speed of the motor.

The oscillation is controlled and monitored *via* a LabVIEW-developed software with 10 ms time-steps. An inductive captor is set at the rotation axis (i.e. at the top), which detects the actual movement of the motor by detecting the passage of four metallic pins per rotation. A cable coder is used to measure the displacement of the moving plate, i.e. at the bottom, with a 0.1 mm precision. The software produces a csv file output with the recordings of the motor position and speed, the signal of the inductive captor, and the position measurement of the cable. The inductive captor signal is also extracted for live and synchronized recording with the detection instrumentation.

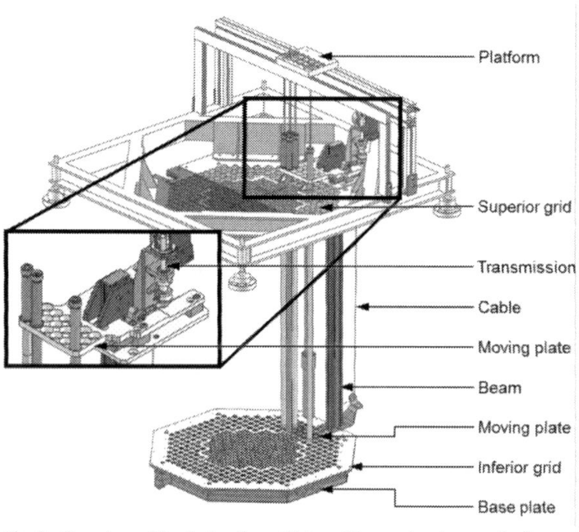

Fig. 2. Overview of the fuel rods oscillator with core structures and a few rods inserted in the device, and details of the top part with moving plate and motor.

C. Miniature neutron scintillator

The miniature fiber-coupled neutron scintillator used in this study has been developed at LRS in collaboration with the Paul Scherrer Institut [7]. Recent experiments demonstrated its applicability for the study of highly localized measurements, thanks to its miniature dimensions, as in this study [23]. In the used prototype version, the neutron sensitive area of the detector is a ~1 mm^2 Scintacor scintillator screen of zinc sulphide (ZnS) doped with ^{6}Li. The conversion of the incoming neutrons in charged particles is performed by the (n,α) reaction of ^{6}Li. The alpha and triton particles arising from the reaction induce the scintillation of the ZnS(Ag) molecules of the crystal at a peak wavelength of 450 nm. It has a typical light yield of 1.6×10^5 photons per neutron, and a decay time to 10% of about 80 μs. Through this two-step process, the incoming neutron radiation is converted to a light signal. An ESKA plastic optical fiber is coupled to the scintillator screen, guiding the scintillation photons to the read-out electronics. These consists of a silicon photomultiplier (SiPM) for photon detection, and in the present study of a standard analog detection chain setup, depicted in Fig. 3. Individual pulses from the SiPM signal are first pre-amplified, then converted into logic pulses (TTL). A second amplifier and a single channel analyzer (SCA) allow the conversion and discrimination of photon trains into individual neutron events, for neutron counting purposes.

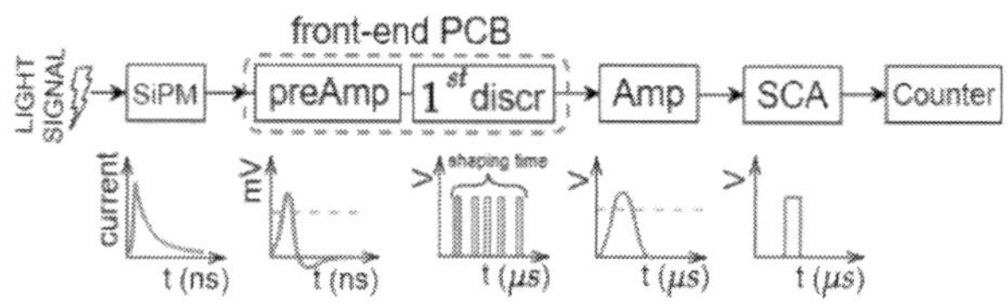

Fig. 3. Electronics and signal treatment of the miniature neutron scintillator, from light collection to neutron counting.

The detector consists of the scintillator, its fiber, and an aluminum cap. As represented in Fig. 4, it was set in a PTFE holder on one of the outmost rods on the second line (namely, A2) of COLIBRI, at an axial position of (525 ± 1) mm.

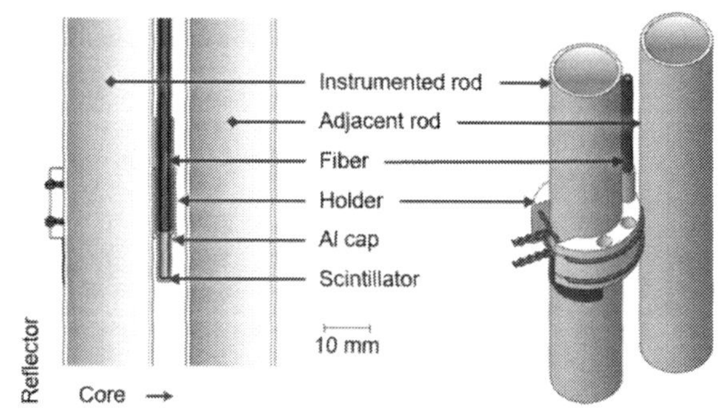

Fig. 4. East-West cross section and 3D view of the positioning of the miniature scintillator on one of the outmost fuel rods of COLIBRI (A2), with respect to its adjacent one (B2).

D. The LEAF gamma detection array

The LEAF gamma detection array consists of two pairs of gamma scintillators, namely two small cerium bromide ($CeBr_3$) and two large bismuth germanate (BGO) detectors [15]. Each detector is directly coupled to a photomultiplier tube (PMT) of matching dimensions. The two small $CeBr_3$ detectors are designed for in-core use in CROCUS, whereas the BGO detectors are those of interest in the present study. The crystal type and their important size of 127 mm diameter and 250 mm height are selected for efficiency purposes. These detectors house a Photonis 5" Type XP4578 PMT. The powering, signal treatment and acquisition are performed using the integrated system Canberra DSA-LX. For both detectors, the high voltage is -1260 V, the coarse gain is 6.4, and the lower level threshold is 0.5 % of the maximum channel (2^{14}). The rise time and flat top are set at 0.2 μs and 0.0 μs, respectively. For the noise application at hand, the analog output of the amplifier is used for synchronized acquisition with other signals.

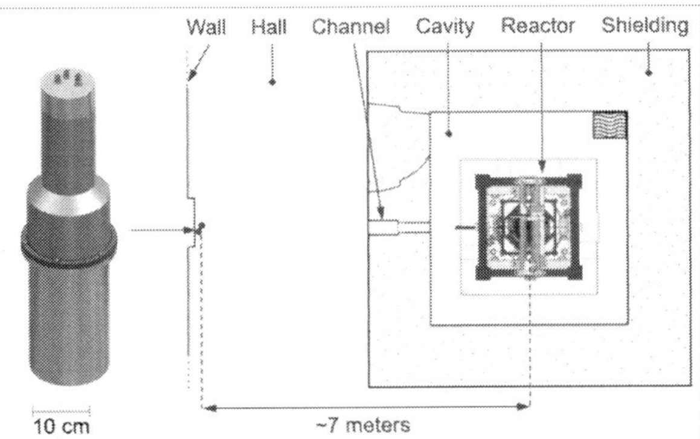

Fig. 5. CAD model of a BGO gamma scintillator (left), and positioning at about seven meters with respect to the reactor core center with a direct line of sight.

E. Conducted experiments

The experimental campaign was carried out on 19th and 20th December 2019. It consists of static fuel rod displacement and dynamic fuel rod oscillation experiments. Whereas COLIBRI is designed for displacing up to 18 rods, only individual rods are moved in these experiments, in order to single out the local effect of one rod only with the miniature neutron scintillator, and to test the smallest perturbation case with the large gamma scintillators set at distance. Thus only the rods on line 2 (North is line 1), and outmost (column A) and central (column B) positions are displaced individually: the instrumented rod is A2, the one adjacent to the detector is B2 (see Fig. 6). We hereby present the results for the fuel rod oscillation, whereas a corresponding static displacement study will be presented in a further article currently under preparation. The experiments consist of two 30-minutes oscillations at ± 2.5 mm amplitude and 0.1 Hz frequency carried out on 20th December: one oscillation of the B2 adjacent rod from 11:56, and one of the A2 instrumented rod from 15:17. The detectors signals and the inductive captor signal from COLIBRI are acquired together with the EPFL/PSI pulse acquisition system with a 0.5 ms dwell time [25].

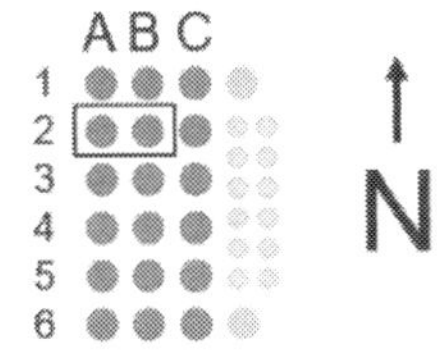

Fig. 6. Identification of fuel rods in COLIBRI. Fuel rods oscillated during this campaign are circled in black, with A2 the instrumented one.

IV. RESULTS AND DISCUSSION

The estimation of the Auto Power Spectral Densities (APSD) was performed with a MATLAB script using the Welch method. A rectangular window with 50% overlap was chosen. The 0.5 ms dwell time, i.e. a sampling frequency of 2 kHz, is sufficient for capturing the frequency of interest, 0.1 Hz. With a window length of 2^{18} samples, the spectral resolution is consequently 8 mHz.

In Fig. 7 are represented the two APSD obtained with the miniature neutron scintillator. Both oscillations of the instrumented rod and its adjacent one present a sharp peak at the expected frequency of 0.1 Hz, or 0.97 Hz more precisely. A slight but significant difference can be observed in amplitude between instrumented and adjacent rods' oscillation, 0.21 vs. 0.27 Hz^{-1}, respectively, with a maximum for the adjacent rod oscillation. The existence of a difference is consistent with the observation of local variations thanks to the miniature size of the detector. Harmonics seem to be present as well, at 0.2 Hz for instance, but are relatively lost in the baseline noise.

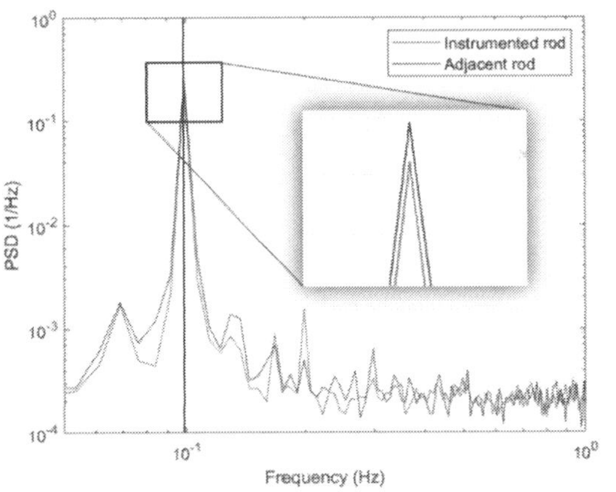

Fig. 7. Auto power spectral densities obtained with the miniature neutron scintillator for the oscillation of the instrumented fuel rod (red), and its adjacent one (blue).

In Fig. 8 are represented the APSD obtained with one of the BGO gamma scintillators. A sharp peak at 0.1 Hz is successfully observed, as well as significantly high harmonics, i.e. at 2 Hz and 4 Hz, and additional frequencies, e.g. 0.17 Hz. It is to be noted that only minimal differences are observed between the two oscillations cases (5.4×10^{-2} Hz^{-1}), which is expected due to the global nature of the measurement. Especially, numerous peaks are consistently observed for both oscillation cases at higher frequencies.

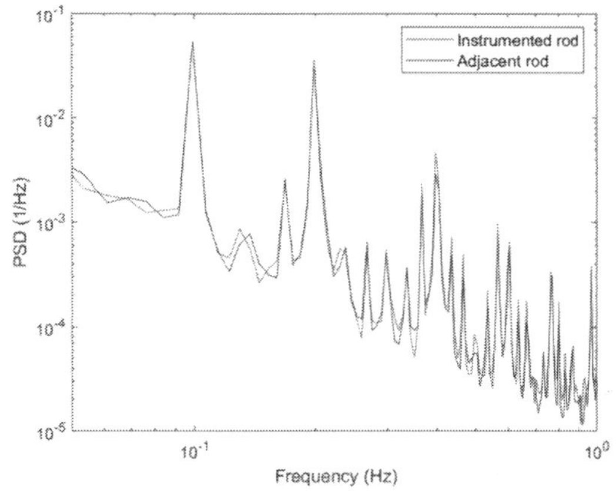

Fig. 8. Auto power spectral densities obtained with the large BGO gamma scintillator for the oscillation of the instrumented fuel rod (red), and its adjacent one (blue).

The comparison of local neutron and distance gamma APSD (see Fig. 9) for the instrumented rod case allows to complement the observation. First, as observed in previous noise experiments, the branching noise baseline is visible in the BGO results, but not in the miniature neutron scintillator ones: its sensitivity – per fission event – is too low to capture the reactor noise itself [26]. In addition, it would seem that the relatively small peak (9×10^{-4} Hz^{-1}) at 0.17 Hz in the neutron APSD remains significant, as both detectors detect it consistently.

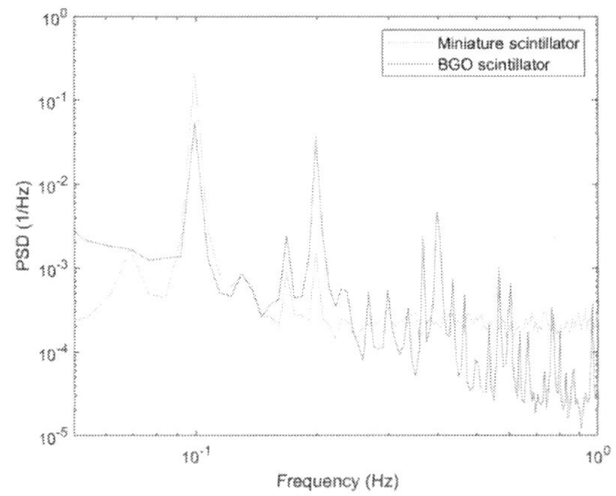

Fig. 9. Auto power spectral densities obtained with the miniature neutron scintillator (cyan) and the large BGO gamma scintillator (orange) for the oscillation of the instrumented fuel rod.

V. Conclusions

The COLIBRI program in CROCUS consists in experiments on the lateral displacement (static) and oscillation (dynamic) of fuel rods in their lattice. As a feasibility study for further experiments, we report here on the successful observation of individual fuel rods' oscillations, using neutron and gamma noise, locally and at meters-distance, respectively.

One of the outmost fuel rods of CROCUS and its adjacent one were oscillated successively, ±2.5 mm from core center to reflector around their nominal position, and at about 0.1 Hz frequency. Two different detection systems were employed: a miniature neutron scintillator set directly on the outmost fuel rod, and a BGO gamma scintillator set at about seven meters from core center, in front of an opening in the cavity shielding.

Both systems were able to detect the fuel rod oscillations. The miniature detector seemed to observe a difference between the two oscillations, with amplitudes of 0.21 vs. 0.27 Hz^{-1} for instrumented and adjacent rod oscillations, respectively. As a difference on the local neutron flux is indeed expected, this confirms its capability to differentiate highly localized effects. Limited harmonics and branching noise contents were observed, aside from a harmonic at 0.17 Hz which was confirmed thanks to the gamma noise results. On these latter, we successfully observed gamma oscillations that correspond to an in-core single rod perturbation at seven meters to the reactor core center. It confirms the possibility to monitor local perturbations and the corresponding very small changes in reactivity, at an unprecedented and unforeseen high distance. No visible differences were observed between both oscillations, as can be expected for the observation of a global effect of the perturbation. Numerous higher harmonics are visible too, as well as branching noise as observed previously in dedicated experiments.

These positive results on both local neutron noise and distance gamma noise open up for a variety of prospects. First, the campaign will be further analyzed to complement the neutron noise results with the static measurements of a single fuel rod's displacement, as well as providing quantitative results with propagated uncertainties [27]. The next planned outlook consists in the implementation of 149 detectors to map CROCUS in 3D. In addition to noise studies, the array will be used for static and dynamic experiments as well, e.g. control rod withdrawal, representing a unique experimental data set for validation purposes. On the gamma noise side, further publications are on-going on branching noise, and data analysis of a larger data set is ongoing for detailed and quantitative results of perturbation noise.

Acknowledgment

The authors thank the CROCUS operation team for its help and dedication to this series of demanding experiments. The research leading to these results has received funding from the Euratom research and training programme 2014-2018 under grant agreement No 754316.

References

[1] V. Lamirand, M. Hursin, G. Perret, P. Frajtag, O. Pakari, and A. Pautz, "Future experimental programmes in the CROCUS reactor," in *RRFM/IGORR 2016, Berlin (Germany) 13-17 March*, 2016, no. 02–2016, pp. 284–292.

[2] O. Pakari, V. Lamirand, G. Perret, P. Frajtag, and A. Pautz, "Kinetic Parameter Measurements in the CROCUS Reactor Using Current Mode Instrumentation," *IEEE Trans. Nucl. Sci.*, vol. 65, no. 9, pp. 2456–2460, Sep. 2018.

[3] O. V. Pakari, A. Pautz, and V. Lamirand, "Experimental and numerical study of stochastic branching noise in nuclear reactors," EPFL, Lausanne, 2020.

[4] O. Pakari, D. Mancusi, O. Petit, A. Zoia, V. Lamirand, and A. Pautz, "Towards the Validation of Noise Experiments in the CROCUS Reactor Using the TRIPOLI-4 Monte Carlo Code in Analog Mode," *EPJ Web Conf.*, vol. 247, p. 04007, Feb. 2021.

[5] M. Hursin, O. Pakari, G. Perret, P. Frajtag, V. Lamirand, I. Pázsit, V. Dykin, G. Por, H. Nylén, and A. Pautz, "Measurement of the Gas Velocity in a Water-Air Mixture in CROCUS Using Neutron Noise Techniques," *Nucl. Technol.*, vol. 206, no. 10, pp. 1566–1583, Feb. 2020.

[6] V. Lamirand, P. Frajtag, D. Godat, O. Pakari, A. Laureau, A. Rais, M. Hursin, G. Hursin, C. Fiorina, and A. Pautz, "The COLIBRI experimental program in the CROCUS reactor: characterization of the fuel rods oscillator," in *ANIMMA 2019*, 2019.

[7] F. Vitullo, V. Lamirand, J.-B. Mosset, P. Frajtag, O. Pakari, G. Perret, and A. Pautz, "A mm 3 Fiber-Coupled Scintillator for In-Core Thermal Neutron Detection in CROCUS," *IEEE Trans. Nucl. Sci.*, vol. 67, no. 4, pp. 625–635, Apr. 2020.

[8] P. Andričević, P. Frajtag, V. P. Lamirand, A. Pautz, M. Kollár, B. Náfrádi, A. Sienkiewicz, T. Garma, L. Forró, and E. Horváth, "Kilogram-Scale Crystallogenesis of Halide Perovskites for Gamma-Rays Dose Rate Measurements," *Adv. Sci.*, vol. 8, no. 2, pp. 1–9, 2021.

[9] Y. Jiang, B. Geslot, V. Lamirand, and P. Leconte, "PISTIL , a reactivity modulation device to probe the transfer function of the research nuclear reactor," in *ANIMMA 2021*, 2021, no. November 2020, p. 13108.

[10] K. Ambrožič, V. P. Lamirand, S. Hübner, M. Hursin, A. Rais, O. Pakari, A. Laureau, P. Frajtag, C. Fiorina, and A. Pautz, "Noise analysis techniques of in-core modulation experiments of the European CORTEX project," *29th Int. Conf. Nucl. Energy New Eur.*, vol. 2, no. September, pp. 1–8, 2020.

[11] C. Demazière, P. Vinai, M. Hursin, S. Kollias, and J. Herb, "Overview of the CORTEX project," in *Proceedings of PHYSOR 2018*, 2018, pp. 2971–2980.

[12] V. Lamirand, A. Rais, S. Hübner, C. Lange, J. Pohlus, U. Paquee, C. Pohl, O. Pakari, P. Frajtag, D. Godat, M. Hursin, A. Laureau, G. Perret, C. Fiorina, and A. Pautz, "Neutron noise experiments in the AKR-2 and CROCUS reactors for the European project CORTEX," *EPJ Web Conf.*, vol. 225, p. 04023, Jan. 2020.

[13] V. Lamirand, A. Rais, O. Pakari, M. Hursin, A. Laureau, J. Pohlus, U. Paquee, C. Pohl, S. Hübner, C. Lange, P. Frajtag, D. Godat, G. Perret, C. Fiorina, and A. Pautz, "Analysis of the first COLIBRI fuel rods oscillation campaign in the CROCUS reactor for the European project CORTEX," *EPJ Web Conf.*, vol. 247, p. 21010, Feb. 2021.

[14] F. Vitullo, K. Ambrožič, V. Lamirand, L. Braun, D. Godat, P. Frajtag, and A. Pautz, "Design of a 150-miniature detectors 3D core-mapping system for the CROCUS reactor," in *ANIMMA 2021*, 2021, pp. 3–4.

[15] O. Pakari, V. Lamirand, T. Mager, A. Laureau, P. Frajtag, and A. Pautz, "First in-core gamma spectroscopy experiments in the zero power reactor CROCUS," in *ANIMMA 2021*, 2021.

[16] O. Pakari, T. Mager, V. Lamirand, P. Frajtag, and A. Pautz, "Delayed gamma fraction determination in the zero power reactor CROCUS," *Submitt. to EPJ Nucl. Sci. Technol.*, 2020.

[17] O. Pakari, V. Lamirand, T. Mager, P. Frajtag, and A. Pautz, "High accuracy measurement of the prompt decay constant in CROCUS using Gamma Noise and bootstrapped uncertainties," *Submitt. to Ann. Nucl. Energy*, 2021.

[18] O. Pakari, T. Mager, V. Lamirand, P. Frajtag, and A. Pautz, "High distance measurement of the prompt decay constant using Gamma Noise," *Draft Ann. Nucl. Energy*, 2020.

[19] Y. Jiang, B. Geslot, V. Lamirand, and P. Leconte, "Review of kinetic modulation experiments in low power nuclear reactors," *EPJ Nucl. Sci. Technol.*, vol. 6, p. 55, 2020.

[20] V. Lamirand, "Experiments in a ZPR in the 2010s : Harvesting data in CROCUS," in *ANIMMA 2019 - Workshop N°1 "The coupling of experiments and modelling to enhance experiments in research reactors,"* 2019.

[21] J. M. Paratte, R. Chawla, and R. Früh, "Validation efforts for the neutronics of a plutonium–erbium–zirconium oxide inert matrix light water reactor fuel," *J. Nucl. Mater.*, vol. 274, pp. 120–126, 1999.

[22] U. Kasemeyer, R. Früh, J. M. Paratte, and R. Chawla, "Benchmark on Kinetic Parameters in the CROCUS Reactor," in *International Reactor Physics Experiments Handbook (IRPhE)*, no. 4440, OECD, Ed. 2007, p. 94.

[23] F. Vitullo, V. Lamirand, P. Frajtag, G. Perret, and A. Pautz, "Highly localized azimuthal measurements in the CROCUS reactor towards the validation of high-fidelity neutronics codes," *EPJ Web Conf.*, vol. 247, p. 08014, Feb. 2021.

[24] S. Yamamoto and H. Tomita, "Comparison of light outputs, decay times, and imaging performance of a ZnS(Ag) scintillator for alpha particles, beta particles, and gamma photons," *Appl. Radiat. Isot.*, vol. 168, no. November 2020, p. 109527, 2021.

[25] V. Roland, G. Perret, G. Girardin, P. Frajtag, and A. Pautz, "Neutron noise measurement in the CROCUS reactor," in *Proc. IGORR 2013*, 2013, pp. 2–7.

[26] O. Pakari, V. Lamirand, G. Perret, D. Godat, M. Hursin, P. Frajtag, and A. Pautz, "Investigation of spatial effects on neutron noise measurements in the zero power reactor CROCUS," in *Proceedings of PHYSOR 2018*, 2018, p. 12.

[27] K. Ambrožič, V. Lamirand, and A. Pautz, "Characterization of high harmonics frequencies in reactor noise experiments within the CORTEX," in *ANIMMA 2021*, 2021.

Calibration of CFUL01 fission chambers in the standard neutron fields of the BR1 reactor at SCK CEN

Anatoly Kochetkov, Antonín Krása, Luc Borms, Edouard Malambu, Guido Vittiglio, Jan Wagemans and Jeroen Willems
SCK CEN, Belgian Nuclear Research Centre, Mol, Belgium
anatoly.kochetkov@sckcen.be

***Abstract*— Recent subcritical VENUS-F experiments showed that fission chambers with a threshold deposit like U-238 can essentially improve the on-line sub-criticality measurments with the beam interruption method, which is currently supposed to be the main method for the ADS MYRRHA. To suppress the uncertainty caused by fissions in the U-235 impurities, the fraction of U-235 in the U deposit should be accurately known. Three PHOTONIS CFUL01 type fission chambers with U-238 deposit were purchased for sub-critical experiments in the VENUS-F reactor. To verify the purity of their deposits, the effective U-235 masses were measured in the empty cavity of the BR1 reactor with a well-known thermal neutron spectrum. It turned out that the measured effective U-235 mass in two fission chambers is lower than the declared mass (as it should be), but this is not the case for the third fission chamber. Then, the effective U-238 mass in these FCs was measured in the well-known fast spectrum of the MARK-III convertor in the BR1 reactor. Finally, the isotopic composition was obtained and it was found that the purity of two CFUL01 FCs is in agreement with the values declared in the certificates but it is not the case for the third fission chamber. As the length of the deposit is bigger than the length of the MARK-III convertor, necessary corrections were calculated with MCNP. The developed procedure using the BR1 standard irradiation fields can be applied for calibration and impurity determination of large fission chambers.**

***Keywords* — sub-critical monitoring, U-238 fission chambers, isotopic composition determination in BR1 neutron fields**

I. Introduction

THE FREYA FP7 project results show that the choice of the detector deposit and position in the core or reflector chosen for the reactor instrumentation can make reactivity monitoring easier or more difficult. Preliminary results show that fission chambers containing a U-238 deposit with a low U-235 content can be advantageously used for ADS sub-criticality measurements by applying the beam interruption method since they bring more flexibility as far as the duration of the necessary beam interruption is concerned [1].

Three PHOTONIS fission chambers of CFUL01 type with U-238 deposit were used for the sub-criticality measurements in a VENUS-F sub-critical core coupled with the GENEPI-III accelerator to simulate the ADS principle. The U-238 purity of the FCs according to the certificates is summarized in Table I. The total masses of the deposits as well as the mass fractions in the certificates are presented. No information on the uncertainties is available.

TABLE I
THE PHOTONIS CERTIFICATE DATA OF THE CFUL01 FISSION CHAMBERS WITH U-238 DEPOSIT.

Fission chamber ID #	673	676	677
Total deposit mass, g	1.00	1.15	1.15
U-238 wt %	99.964	99.9983	99.797
U-235 wt %	0.035	0.0012	0.2016

To verify the purity (U-235 content) in the deposits of these FCs, the effective U-235 and U-238 masses were measured in the standard thermal and fast neutron fields of the BR1 reactor [2, 3]. These fields are normally used to calibrate miniature fission chambers (typically 2-4 mm in diameter and 10-25 mm in length). Calibration of large fission chambers with 10× bigger dimensions represents a challenge.

II. Effective Masses Determination

The effective masses of U-235 (M_{eff}^{235}) of CFUL01 FCs were measured in the thermal flux in the empty cavity while the effective masses of U-238 (M_{eff}^{238}) were measured in the fast flux of the MARK-III convertor in the BR1 reactor.

The effective masses were obtained using standard procedures [4]:

$$M_{eff} = \frac{A}{6.02 \times 10^{23}} \times \frac{C_{FC}}{\Phi \times \bar{\sigma}_{eq}} \quad (1)$$

where A is the molar mass of a deposit, C_{FC} is the count rate of a FC with M_{eff} to be determined, Φ is the neutron flux in MARK-III or empty cavity at the position of the deposit, and $\bar{\sigma}_{eq}$ is the spectrum averaged fission cross-section (JEFF-3.1.1 [5]) of the deposit material (U-235 or U-238 with impurities in the center of the cavity or in the MARK-III convertor. Correction factors that take into account differences between the thermal and fast neutron spectrum as well as the spatial effects are described in the following section.

III. Results

A. U-235 Effective Mass

The M_{eff}^{235} (impurities) of CFUL01 FCs, with U-238 as the main isotope in the deposit, were measured in the thermal flux of the BR1 empty cavity and were reported in [6]. In Table II the results of the M_{eff}^{235} that were reported in [6] are multiplied by $K_{Meff}^{235} = 1.090 \pm 0.033$ to fit to the MARK-III conditions because it was observed in earlier measurements with FCs of different designs that calibration in MARK-III leads to 9 % bigger M_{eff}^{235} than when calibrated in the empty cavity. This was confirmed in the current experiments (May 2021) with CFUL01 and CFUM21 FCs. The uncertainties of the results presented in Table II are propagated as discussed for this method in [7] and include the uncertainty on K_{Meff}^{235}.

TABLE II
THE M_{EFF}^{235} IN CFUL01 FISSION CHAMBERS MEASURED IN BR1 EMPTY CAVITY AND MULTIPLIED ON COEFFICIENT K_{MEFF}^{235} = 1.090 TO FIT TO THE MARK-III CONDITIONS. UNCERTAINTIES ARE 3.6% .

FC #	673	676	677
Dis. level	Meff (μg)	Meff (μg)	Meff (μg)
0.15R	272	61	2020
0.2R	258	60	1938
0.25R	251	58	1885
0.3R	243	57	1829
0.35R	234	56	1770
0.4R	224	54	1704
0.45R	215	53	1633
0.5R	204	51	1558
0.55R	192	49	1478
0.6R	179	46	1393

The M_{eff}^{235} presented in Table II depends on the discrimination level R, which also can be seen in Fig.1.

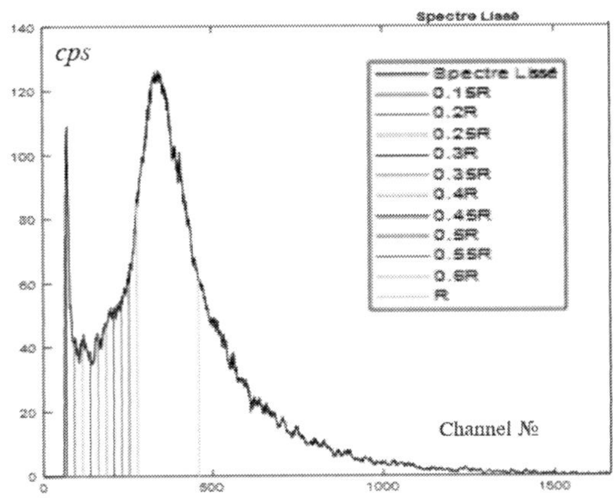

Fig. 1. Spectrum of U-235 fragments obtained with FC #676 in the BR1 empty cavity.

B. *U-238 Effective Mass*

It is not straightforward to measure the M_{eff}^{238} of CFUL01 in the fast flux of the MARK-III convertor because of its size. The length of the sensitive deposit (211 mm) is longer than the U-235 convertor (80 mm, see Fig.2). Moreover, there is a gradient of the fission rates in the radial direction. [3]. Therefore the M_{eff}^{238} of CFUL01 measured in MARK-III should be corrected for these spatial effects. The coefficient $K_{Meff\text{-}238}^{H.R}$ =1.787 ± 0.036 (the uncertainty include statistics and the systematic contributions), which is the ratio of the U-238 fission rate (F28) in the centre of the MARK-III convertor without presence of the fission chamber to F28 in the real fission chamber deposits, was obtained with MCNP6 [8].

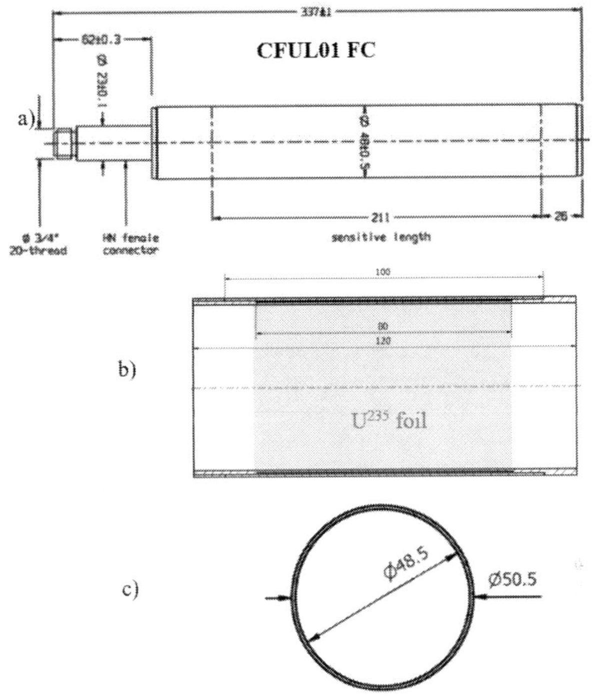

Fig. 2. The dimensions of the CFUL01 PHOTONIS FC (a) and the MARK-III U-235 90% convertor in vertical (b) and horizontal (c) view.

The calculated coefficient $K_{Meff\text{-}238}^{H.R}$ was verified with additional measurements and calculations of fission rates in axial F28(H) and radial F28(R) directions of the MARK-III convertor. The results are in a good agreement (see Fig.3 and Fig.4). These results confirm the calculated coefficient $K_{Meff\text{-}238}^{H.R}$.

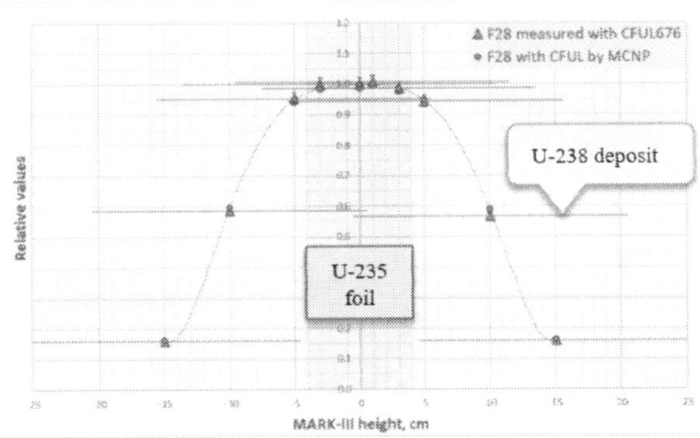

Fig. 3. Comparison of F28(H) measured with CFUL676 with F28(H) in CFUL01 calculated with MCNP in MARK-III.

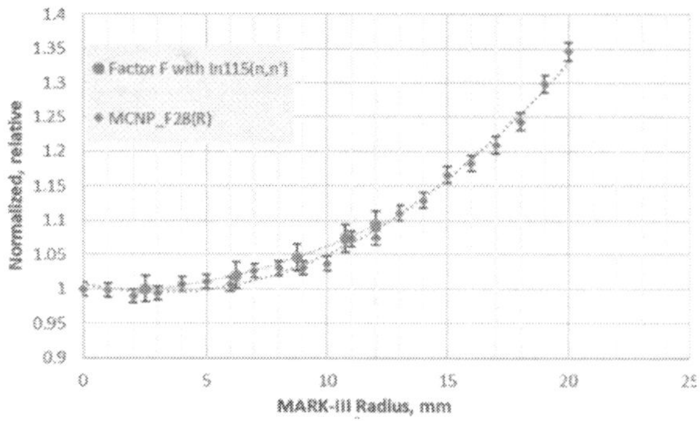

Fig. 4. Radial gradient in the MARK-III convertor (at axial position H = 0 cm) obtained with irradiation of In^{115} [3] and F28(R) calculated with MCNP.

The results of $Meff^{238}$ of CFUL01 measured in the MARK-III fast neutron flux using (1) and multiplied by $K_{Meff\text{-}238}{}^{H.R}$ are shown in Table III. The uncertainties of the results are ±2.9 % and they are propagated as discussed for this method in [6] and include the uncertainty on $K_{Meff\text{-}238}{}^{H.R}$.

TABLE III
THE $M_{EFF}{}^{238}$ IN CFUL01 FISSION CHAMBERS MEASURED IN MARK-III CONVERTOR AND MULTIPLIED ON COEFFICIENT $K_{MEFF\text{-}238}{}^{H.R}$ =1.787. UNCERTAINTIES ARE 2.9%.

FC #	673	676	677
Dis.level	Meff (mg)	Meff (mg)	Meff (mg)
0.15R	712	854	912
0.2R	706	814	893
0.25R	685	791	864
0.3R	662	766	838
0.35R	638	740	809
0.4R	612	712	779
0.45R	584	681	745
0.5R	552	648	710
0.55R	518	613	672
0.6R	479	574	631

C. Isotopic Composition

To obtain the enrichments or the impurities in the deposit of the fission chambers, the $M_{eff}{}^{238}$ shown in Table III or the $M_{eff}{}^{235}$ presented in Table II were divided by the sum $M_{eff}{}^{238}$ + $M_{eff}{}^{235}$. Since the M_{eff} depends on the discrimination level R, for each ratio (to obtain the enrichment or impurity) the $M_{eff}{}^{238}$ and $M_{eff}{}^{235}$ of the same R were taken. The results of the enrichments and the impurities obtained with the different discrimination levels are very close to each other (standard deviation when averaging <0.1%). The mean values are presented in Table IV and Table V. The uncertainties in these tables are experimental ones (including of the uncertainty on the corrections obtained with MCNP) without the uncertainty on the certificate values because these are not available.

TABLE IV
MEASURED AND CERTIFICATE U-238 ENRICHMENT (IN %).

FC#	Certificate (C), %	Measured (E), %	± Δ, abs.	C-E, %
673	99.9650	99.9633	0.0028	0.0017
676	99.9988	99.9924	0.0006	0.0064
677	99.7984	99.7818	0.0168	0.0166

TABLE V
MEASURED AND CERTIFICATE U-235 IMPURITIES (IN %).

FC#	Certificate (C), %	Measured (E), %	± Δ, rel %	C/E
673	0.0350	0.0367	3.8	0.95
676	0.0012	0.0076	3.8	0.16
677	0.2016	0.2182	3.8	0.92

IV. CONCLUSIONS

Three PHOTONIS FCs of CFUL01 type with U-238 deposit, which were used in VENUS-F sub-critical experiments for beam interruption method validation, were calibrated in the BR1 reactor because the method was found to be sensitive to impurities in the deposits.

The enrichments and impurities of the deposits were obtained by measuring the effective masses of U-235 and U-238 in the accurately characterized thermal and fast neutron fields of the BR1 reactor.

In particular it was found that:

- All the measured results of U-238 enrichment in FCs are smaller than declared in their certificates.
- The declared U-235 impurities of two FCs are in agreement with the measurement results within 2σ. The U-235 content in the third FC is 6 times bigger, which indicates a possible contamination of the base material in the deposit.
- It is possible to measure Meff, impurities and enrichments of big PHOTONIS detectors like CFUL01 in the BR1 reactor.

ACKNOWLEDGMENT

This work was initiated as a prolongation of the verification of the sub-criticality monitoring for ADS MYRRHA which was carried out in European Projects FREYA FP7 (2011-2016), MYRTE H2020 (2016-2017) and CNRS\SCK•CEN Project MYRACL (2017-2019).

REFERENCES

[1] J.-L. Lecouey et al., "Reactivity measurements with threshold fission chambers in the MYRRHA mock-up fast reactor VENUS-F", *in Proc. PHYSOR 2018*, Cancun, Mexico, April 22-26, 2018.
[2] J. Wagemans, E. Malambu, and L. Borms, "The Neutron Standard Fields at the BR1 Reactor at SCK•CEN," Journal of ASTM International 9, no. 3 (2012): 1-7
[3] L. Borms, "Characterisation of the fast neutron field in the BR1 MARK3 converter", Mol, Belgium, Internal report SCK•CEN, I-287, May 2015
[4] J. P. Hudelot et al., "Calibration des chambres à fissions EOLE/MINERVE et MASURCA dans les spectres de reference du reacteur BR1 du CEN/MOL", Cadarache, France, NT/DER/SPEx/LPE/02-037, 02.12.2002
[5] *OECD/NEA Data Bank,* "The JEFF-3.1.1 Nuclear Data Library", JEFF Report 22, Paris, 2009.
[6] A. Kochetkov et al., "Result of the tests of the new PHOTONIS CFUL01 fission chambers with high-purity U-238 deposit and Al containers with boron nitride cover", Mol, Belgium, SCK•CEN Report, R-6459, September 2017
[7] Adimir Dos Santos, "Guide to the expression of uncertainty", International Reactor Physics Experiments Evaluation Project (IRPhEP), Paris, OECD Report, NEA/NSC/DOC(2017), 2017
[8] C. J. Werner, MCNP User's manual Code version 6.2, LA-UR-17-29981, 2017.

Measuring gamma doses over the mGy-kGy range with a single type of TLD detector

Adrien Gruel[1], Alix Sardet[1], Vincent Chaussonnet[1], Maxime Houdouin-Quenault[1] and Daniel Garnier[1]
[1]CEA, DES, IRESNE, DER, Cadarache F-13108 Saint-Paul-Lez-Durance, France
adrien.gruel@cea.fr

***Abstract*— Thermo-luminescent detectors are currently used to measure gamma doses in the medical and environmental surveillance fields. During the past few years, the CEA Reactor Studies Division tested and validated the use of these detectors for gamma flux characterization and nuclear heating measurements in mixed neutron/gamma fields of low power reactors. Doses were comprised between a few mGy and a few Gy for dose rates up to a few Gy.h^{-1}. However, in MTR or TRIGA reactors, the gamma flux level is much higher (> 10^{12} n/cm²/s) and the TLD currently in use (CaF_2:Mn and ^{7}LiF:Mg,Ti) and their readout protocols were no longer suitable for the resulting doses. In order to extend the applicable dose range up to ~1 MGy (dose rate of a few kGy.h^{-1}), several options were explored. On one side, some adjustments were made to the readout protocols of CaF_2:Mn and ^{7}LiF:Mg,Ti, notably by testing the use of filters to reduce the amount of light received by the reader PMT to avoid saturation. On the other side, a new type of TLD (LiF:Mg,Cu,P) with different Li enrichments (natural or enriched in ^{7}Li) was tested.**

This paper presents the calibration measurements results performed in pure gamma fields, at the irradiation platform of the CEA Cadarache Radioprotection Division, between 250 mGy and 3 Gy for all detector types. In addition to the calibration, these measurements also studied the Mg,Cu,P-doped detectors response: reproducibility, dose rate dependence, incoming photon energy dependence, high temperature effect when reading TLD, etc.

Results show that at low doses Mg,Cu,P-doped TLDs are slightly less stable than CaF_2:Mn and ^{7}LiF:Mg,Ti. The sensitivity modification after a high dose exposure seems to indicate that a new protocol readout should be defined for Mg,Cu,P-doped sensors (high temperature peak).

***Keywords* —TLD, calibration, gamma dose.**

I. Introduction

Thermo-luminescent detectors are mainly used to measure gamma doses in the medical and environmental surveillance fields. These detectors can also be used for nuclear heating measurements in the mixed neutron/gamma field of low power reactors. To characterize doses and nuclear heating in Material Testing Reactors, their range of use should be extended to be able to reach kGy or even MGy. This would enable the use of a single type of detector to carry out measurements in various locations, from the core periphery to the outermost components.

In this work we compare the performances of four types of thermoluminescent detectors:

- CaF_2:Mn, (hereafter named TLD400)
- ^{7}LiF:Mg,Ti (hereafter named TLD700)
- ^{7}LiF:Mg,Cu,P (hereafter named MCP7)
- ^{nat}LiF:Mg,Cu,P (hereafter named MCPN)

The two first types of sensors are well characterized in the field of low dose measurements (≲ 1 Gy) in zero power experimental reactors [1]-[5]. They were also tested in the JSI's TRIGA reactor at doses up to 1 kGy [6], where TLD400 showed a good agreement against calculations. The two last types are LiF-based detectors but with different doping elements, that should increase their sensitivities, as well are make them suitable for high dose measurements [7][8]. They are labelled MCP in the following.

During irradiations, the mm-sized detectors are set in cylindrical aluminium pillboxes with 2-mm wall thickness to ensure the charged particle equilibrium (Fig. 1).

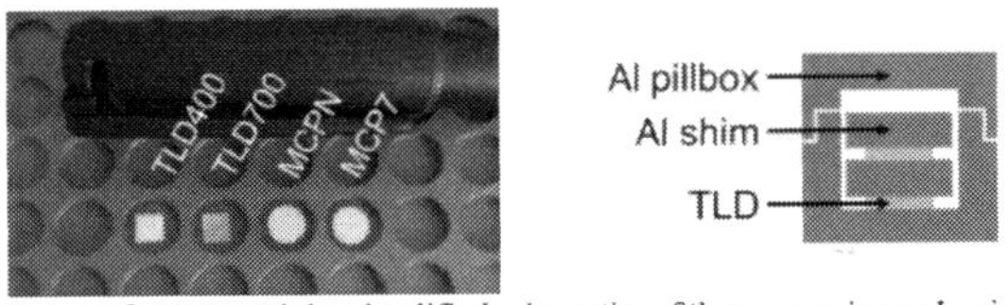

Fig. 1. Left: TLDs. Right: simplified schematics of the sensors in an aluminium pillbox.

The reader system is a Harshaw 3500. For each sensor, two temperature profiles for the readout phase were used. The first is the one recommended by the constructor. The second one is a test law, in order to emphasize possible modifications in the luminescence curves after the irradiations. Readout parameters are presented in Table I and illustrated in Fig. 2.

30 of each MCP-type TLDs were irradiated, 5 of which were read with the test law. 10 TLD400 and 10 TLD700 were irradiated, half of them being read with a test law.

For TLD400 and TLD700, annealing protocol consists in heating detectors at 400 °C for 1 hour and 100 °C for 2 hours. For MCP7 and MCPN, it consists of heating the detectors at 240 °C for 10 min.

Tests of those sensors at doses up to 3 Gy are presented in the second section of this work, while preliminary tests after a high dose exposure at 50 kGy are presented in the third part.

TABLE I
READOUT TEMPERATURE PROFILE PARAMETERS. BOLD VALUES EMPHASIZE DIFFERENCES AGAINST REFERENCE PARAMETERS

	Starting temperature (°C)	Temperature Increase Rate (°C.s-1)	Maximum tempereature (°C)	Total time (s)
TLD400	150	10	350	30
TLD400 test	150	**25**	**370**	30
TLD700	140	15	280	23.3
TLD700 test	140	**10**	280	23.3
MCP	140	15	240	30
MCP test	140	15	**500**	30

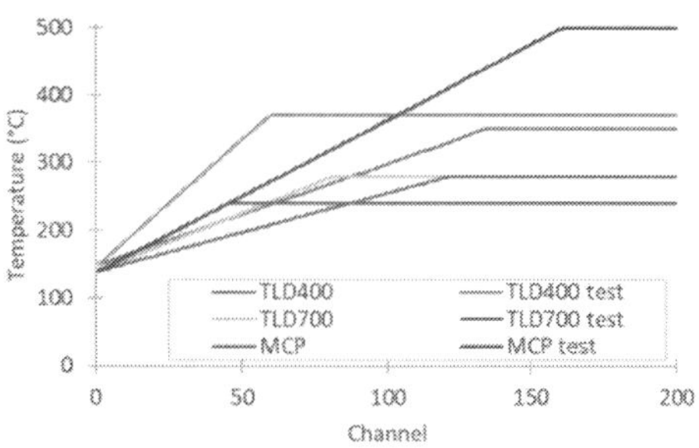

Fig. 2. Temperature profiles.

II. DETECTORS PERFORMANCE COMPARISON FOR LOW DOSES MEASUREMENTS

In this section, performance of different types of TLDs are studied against various parameters, at doses up to 3 Gy. Measurements were carried out at CEA Cadarache Radioprotection Division's irradiation platform. Several ^{60}Co and ^{137}Cs gamma sources are available, each one calibrated against the Henri Becquerel National Laboratory, a reference institute in the field of radiation measurements. On this platform, the detectors position can be set at any distance from 80 to 500 cm; and irradiation time is set to the second. The highest dose rate reachable is around 300 mGy.h^{-1} (^{60}Co source), with an uncertainty below 1 % at one standard deviation.

First, reproducibility studies are presented, then dose rate and gamma energy dependences, and finally linearity against integrated dose.

A. Reproducibility of the measurements

The aim of this study is to characterize the stability of the TLDs' response for a given experimental setup. Namely, the type of source, the source-detector distance (meaning the dose rate), the total integrated dose and the readout and annealing protocols are kept constant. Four irradiations at 500 mGy at a distance of 80 cm of a ^{60}Co are used to assess the sensors' reproducibility. The reproducibility over n measurements is evaluated using:

$$U_x = \frac{1}{\sqrt{n}}\sqrt{\frac{1}{n-1} * \sum_{i=1}^{n}(x_i - \bar{x})^2}, \quad (1)$$

x_i being the integrated charge of the ith measurement during the readout process, and $\bar{x}$ the average of the n measurements.

With the standard readout protocol, TLD400 and TLD700 show a very good reproducibility: 3.4 % and 4.8 %, respectively, consistent with past studies [3]. The test law worsen this reproducibility, especially for TLD700: 4.5 % and 7.0 %, respectively.

In this dose range, the reproducibility of the Mg,Cu,P-doped detectors is not as good as TLD400 and TLD700. With the recommended readout protocol, it is 9.8 % for MCPN and 5.8 % for MCP7. With the test law (heating up to 500 °C), the results deteriorate very significantly, the reproducibility being higher than 20 % for both types. Indeed, the luminescence curves could be very different from an irradiation to another. This effect cannot be attributed to irradiation conditions, since all detectors are irradiated simultaneously.

B. Response dependence against dose rate and photon energy

To test dose rate dependence of the TLDs' responses, irradiations targeting at the same total integrated dose were carried out, the distance between the source and the detectors being modified between 80 cm to 140 cm. Irradiations characteristics are presented in Table II, two doses were investigated: 250 mGy and 750 mGy.

TABLE II
CHARACTERISTICS OF IRRADIATIONS FOR DOSE-RATE INFLUENCE

Total dose	Distance (cm)	Dose rate (mGy.h^{-1})	Irradiation time (s)
250 mGy	80	330 ± 0.7 %	2727
	100	211 ± 0.65 %	4265
	140	107 ± 0.6 %	8411
750 mGy	80	327 ± 0.7 %	8257
	100	211 ± 0.65 %	12796
	140	107 ± 0.6 %	25234

For TLD400 and TLD700 the discrepancies between the integrated charges at different dose rates are below or of the order of magnitude of the reproducibility. From those irradiations, we can conclude that these sensors are not sensitive to the dose rate, at least in the tested range.

MCPN and MCP7 showed a good behavior for irradiations carried out at 750 mGy, the discrepancies between the integrated charges at different dose rates being below or of the order of magnitude of the reproducibility. Results after irradiation at 250 mGy are less consistent. Luminescence curves could be very different from an irradiation to another. It is therefore more difficult to conclude. Nevertheless, it seems that Mg,Cu,P-doped TLDs are less stable for low doses.

The gamma energy dependence of the detectors' response has been tested by comparing ^{137}Cs (662 keV) and ^{60}Co (1173 keV and 1332 keV) irradiations at the same dose level. In this study, all tested detectors and all readout protocols proved to be stable. Discrepancies between irradiations are of the same order of magnitude as the reproducibility (see previous section). It should however be noted that only two irradiations were carried out (one with ^{137}Cs and one with ^{60}Co), and that a more comprehensive study could bring interesting additional new results by studying lower and higher gamma energies.

C. Response linearity

In this section, individual calibration curves for each detector are obtained with irradiations between 200 mGy and 3000 mGy for TLD400 and TLD700, and between 200 mGy and 1000 mGy for MCPN and MCP7. Indeed, measurements of Mg,Cu,P-doped sensors above 1000 mGy lead to saturation of the TLD reader.

It is assumed that the detectors' response, meaning the integrated charge obtained as the integral of the luminescence curve, is linear versus dose.

For TLD400 and TLD700 the final uncertainty on the calibration factor is below 2 % on average. The dispersion of the calibration factor values is around 5 % for TLD400 and 7 % for TLD700. As this is slightly higher than the reproducibility observed, it is mandatory to carry out individual sensor calibration, unless having more homogeneous sensor batches. It should be noted that the dispersion is smaller using the recommended readout protocol.

For Mg,Cu,P-doped detectors, uncertainties on the calibration factor are slightly higher, 5.6 % and 3.6 % for MCPN and MCP7, respectively. The dispersion of the calibration factor values is consistent with that of TLD400 and TLD700: around 5 % for MCPN and 6 % for MCP7. Results are far less satisfactory with the test law (max. temperature of 500 °C): uncertainties are between 16 % to 20 % and dispersion around 15 %.

On this dose range, TLD700 are less sensitive (16 nC.mGy^{-1} on average) than TLD400, which are three times more sensitive (43 nC.mGy^{-1} on average). Mg,Cu-P-doped TLDs are about five times more sensitive than TLD400, with an average sensitivity of 200 nC.mGy^{-1} for MCPN and 230 nC.mGy^{-1} for MCP7.

Examples of luminescence curves obtained for calibration purposes for a TLD400 and for a MCP7 are presented Fig. 3 and Fig. 4.

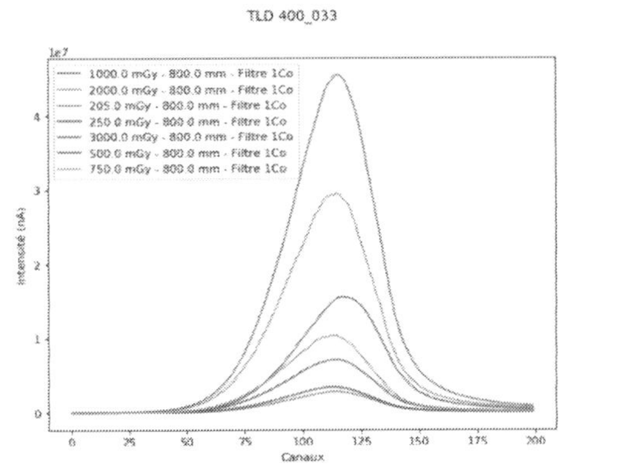

Fig. 3. Luminescence curves at different doses for a TLD400 after irradiations carried out for the calibration phase ("*Filtre 1Co*": measurement with 100%-transmittance filter, irradiation in front of a cobalt source).

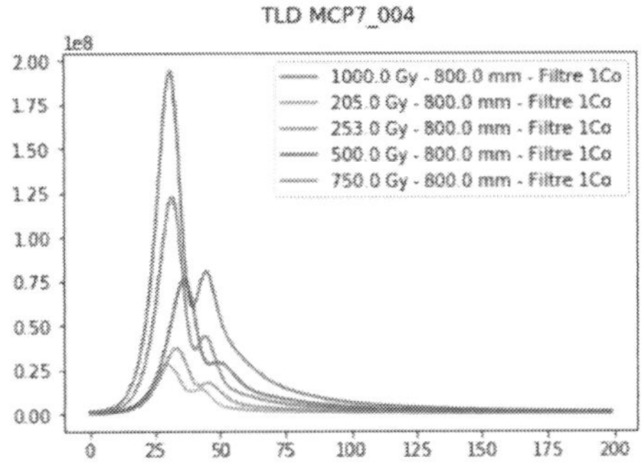

Fig. 4. Luminescence curves at different doses for a MCP7 after irradiations carried out for the calibration phase. ("*Filtre 1Co*": measurement with 100%-transmittance filter, irradiation in front of a cobalt source)

III. Behavior after high dose exposure

Detectors were exposed to a 50 kGy dose at the EPICUR irradiation facility, operated by the IRSN (French Institute for Radiation Protection and Nuclear Safety). This facility is devoted to physicochemical studies on confined iodine under irradiation. It consists of a small irradiation chamber (about 30 cm in width, 30 cm in length and 60 cm in height) with intense ^{60}Co sources, where the dose rate ranges from 0.1 kGy.h^{-1} to 20 kGy.h^{-1} given the irradiation position in the cavity.

For this first test, detectors were subjected to a 50 kGy irradiation, and they were split in two batches with different dose rates. In each batch, two detectors of each type were irradiated.

Reference dose rates were measured with alanine dosimetry, with a typical uncertainty of 10 %. Several alanine dosimeters were set around void aluminium pillboxes in each position, to catch the average dose rate inside pillboxes. The highest dose rate was 8.2 ± 0.2 kGy.h^{-1}, the lowest was 1.81 ± 0.04 kGy.h^{-1}. The uncertainty corresponds to the discrepancy of the reference dosimetry for each irradiation location.

The sensitivity modification of the sensors after this high dose irradiation was verified with a 750 mGy irradiation on the CEA Cadarache Radioprotection Division's irradiation platform. The evolution of luminescence curves obtained for a 750 mGy irradiation before and after the 50 kGy exposure are shown in Fig. 5 to Fig. 7.

There is a large sensitivity increase for TLD400: the total integrated charge was multiplied by a factor between 4 and 6.6. For TLD700 the sensitivity slightly decreased, by a factor about 1.5. For those two types of TLDs, the luminescence curve shape was not impacted by the high dose exposure. The temperature profile used (constructor or test) does not seem to have an impact on the sensitivity change before and after the high dose exposure

For MCPN and MCP7 sensors, two cases appear regarding the readout process:

- With the constructor law (maximum temperature of 240 °C, top graphs of Fig. 6 and Fig. 7), the glow curve shape cannot really be analyzed anymore, since no peak can be distinguished. It seems that the temperature is not high enough to release the trapped electrons, and therefore to be able to read the integrated dose.

- With the test law (maximum temperature of 500 °C, bottom graphs of Fig. 6 and Fig. 7), a peak at higher temperature appears, while the intensities of the peaks at 200 °C and 300 °C decreased by an order of magnitude. Their overall sensitivity (total integrated charge) decreased by a factor between 2.5 and 3 for MCPN, and by a factor between 4.5 and 5 for MCP7. With this test law, it has been verified that the high temperature peak was not due to the signal caused by the heating of the plate nor of the TLD itself. For comparison, this signal is plotted in yellow dashed lines on the bottom graphs in Fig. 6 and 7, and represent only a small fraction of the high temperature signal.

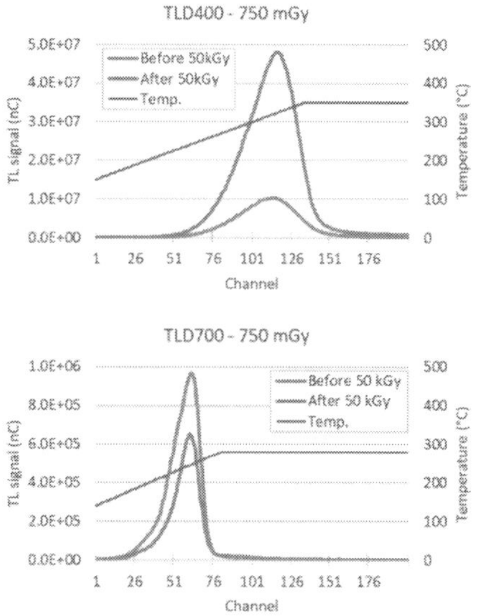

Fig. 5. Luminescence curves for TLD400 and TLD700 before and after 50 kGy exposure. Measurement after a 750 mGy irradiation in front of a ^{60}Co source with the standard readout protocol.

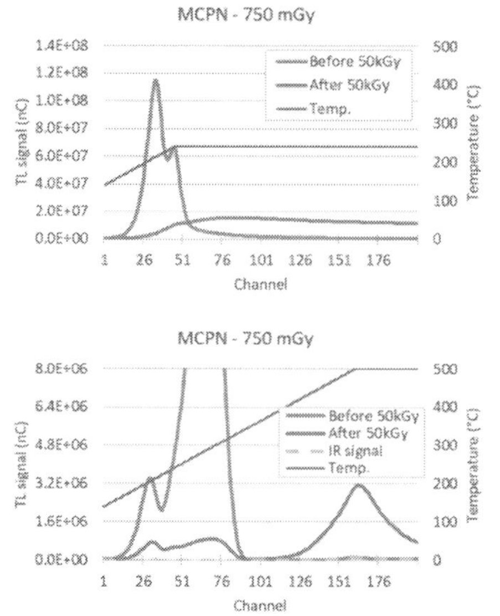

Fig. 6. Luminescence curves MCPN TLDs before and after 50 kGy exposure. Measurement after a 750 mGy irradiation in front of a ^{60}Co source. Top: constructor readout protocol, bottom: test readout protocol.

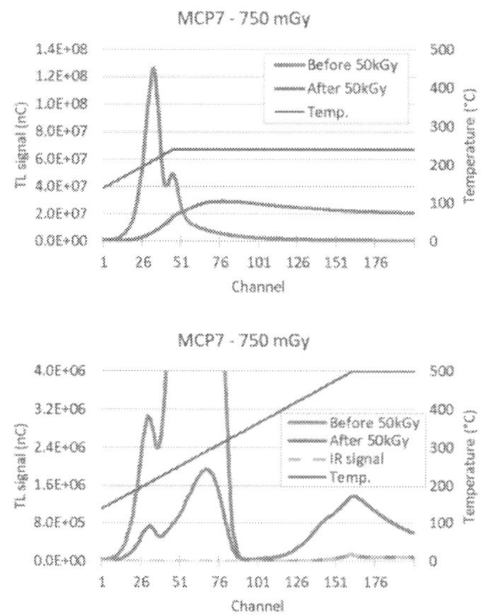

Fig. 7. Luminescence curves MCP7 TLDs before and after 50 kGy exposure. Measurement after a 750 mGy irradiation in front of a ^{60}Co source Top: constructor readout protocol, bottom: test readout protocol.

IV. Conclusions

This work presents the study of different TLDs behavior regarding reproducibility, dose rate influence and gamma energy dependence.

In the low dose range (< 3 Gy) TLD400 and TLD700 seem to present better characteristics. Indeed, reproducibility is higher for Mg,Cu,P-doped sensors. As the luminescence curves are less stable regarding irradiations conditions, individual calibration factors present higher final uncertainties for those detectors.

Sensitivity change was preliminary characterized, and different behaviors were described. This sensitivity change seemingly being definitive with the annealing protocol used, a new calibration for highly irradiated sensors should be carried out. High-dose irradiation for calibration up to several hundred of kGy should also be performed.

After high dose calibration, TLD detectors will be used to characterize the gamma field and nuclear heating in the TRIGA Mark II reactor at the Jožef Stefan Institute (JSI) in Slovenia. Measurements are expected to take place in the first half of 2022, in the frame of a CEA-JSI collaboration dedicated to the benchmark of the TRIGA reactor neutron and photon 3D transport model.

The large amount of data available will also be used to determine material physical properties to model TLDs luminescence curves.

ACKNOWLEDGMENT

The authors wish to thank the SPR/LRIR EDEPE group for the calibration phase at their irradiation platform.

REFERENCES

[1] M. Le Guillou, *et al.*, State of the art on nuclear heating measurement methods and expected improvements in zero power research reactors, *EPJ Nuclear Sci. Technol.*, vol. 3, p. 11, 2017.
[2] H. Amharrak, *et al.*, Analysis and recent advances in gamma heating measurements in MINERVE facility by using TLD and OSLD techniques, *IEEE Transactions on Nuclear Science*, vol. 59, no. 4, pp. 1360–1368, Aug. 2012.
[3] H. Amharrak, *Développement et optimisation de méthodes de mesures d'échauffements nucléaires et de flux gamma dans les réacteurs* expérimentaux *: identification, maîtrise, traitement et réduction des incertitudes associées*, thesis, Aix-Marseille University, 2012.
[4] H. Amharrak *et al.*, Development and Optimization of Nuclear Heating Measurement Techniques in Zero Power Experimental Reactors, *IEEE Transactions on Nuclear Science*, vol. 61, no. 5, pp. 2515–2526, Oct. 2014.
[5] M. Lemaire, *Validation des calculs d'échauffements photoniques en réacteur d'irradiation au moyen du programme experimental AMMON et du dispositive CARMEN*, thesis, Aix-Marseille University, 2015
[6] A. Gruel, *et al.*, Gamma-heating and gamma flux measurements in the JSI TRIGA reactor, results and prospects, *IEEE Transactions on Nuclear Science*, 67 (4), pp. 559-567.
[7] B. Obryk, *et al.*, Method of thermoluminescent measurement of radiation doses from micrograys up to a megagray with a single LiF:Mg,Cu,P detector, *Radiat Prot Dosimetry*, vol. 144, no. 1–4, pp. 543–547, Mar. 2011.
[8] B. Obryk, *et al.*, High-dose TL dosimetry of reactor neutrons, *Radiat Prot Dosimetry*, vol. 180, no. 1–4, pp. 235–239, Nov. 2017.

Characterization of the proton pulsed beam at CMAM

S. Viñals[1], D. Sánchez-Parcerisa[2], L.M. Fraile[2], S. España[2], G. García[1], M. García-Díaz[2], V. Sánchez-Tembleque[2], J.M. Udías[2]
[1]Centro de Micro-Análisis de Materiales, Madrid, Spain
[2]Grupo de Física Nuclear, Universidad Complutense, Madrid, Spain
silvia.vinnals@uam.es

***Abstract*— In this paper, the technicalities performed to obtain a pulsed beam at the CMAM facility will be explained. The pulsed beam has been characterized with an 8 MeV proton beam, using an existing equipment at CMAM: two pairs of electrostatic plates (RASTER) that deflect the beam, commonly used for homogeneous irradiation of large areas. A pulsed beam is used in many areas such as nuclear physics, material science and, in particular, for proton-therapy medical studies. Rectangular and pyramidal functions have been used to generate different pulses and characterize the response of the RASTER. The results point out that the pulses obtained are suitable for preclinical proton-therapy studies in the FLASH regime, which consists on fractionating the dose in time with short and intense pulses. The set-up for the characterization has been a function generator and a Si-PM outside the chamber.**

***Keywords* —Pulsed Beam in air, Electrostatic Deflection.**

I. Introduction

A. CMAM: The facility

The Centro de Micro-Análisis de Materiales (CMAM) is one of the two ion accelerators research centers in Spain. It belongs to the Universidad Autónoma de Madrid (UAM) and the building that hosts the laboratory is at the university campus.

The equipment of the facility consists of an electrostatic ion accelerator with a maximum terminal voltage of 5 MV and six beam-lines dedicated to various application areas such as the analysis and modification of materials, the study of the nuclear reactions or archaeometry studies. The accelerator is unique in Spain and singular in the European context [1]. It was installed in the early 2000s and has been in regular operation ever since.

The accelerator, built by High Voltage Engineering Europe (HVEE), is of the tandem type with a Cockroft-Walton acceleration system. It is provided with two sources: a plasma source for gaseous substances and a sputtering source for obtaining any element from H to U from a solid target.

The accelerator feeds up six beam-lines after the bending magnet at different angles: the standard multi-purpose line which ends in the internal micro-beam line, the time of flight line, the implantation line, the external micro-beam line and the nuclear physics line. For a more detailed description of each line, see Ref. [2]. Figure 1 shows a scheme of the accelerator with its spatial distribution (a) and a picture of the accelerator hall from the point of view of the end of the experimental lines (b). There are two lines where the beam can be extracted from vacuum to air: the implantation line (IMP) and the external micro-beam line (EuB). Until now, the former has been devoted to the implantation or irradiation of large area samples for material modification and the latter has been focused onto archaeometry studies. Now, a new project has started to use these two lines for proton-therapy preclinical studies.

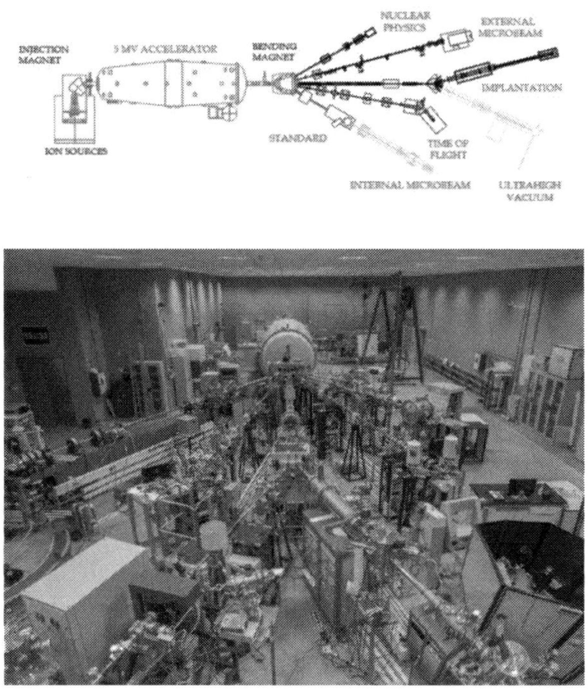

Fig. 1. (a) Scheme of the accelerator hall with the experimental lines identified. (b) Picture of the CMAM accelerator hall from point of view of the end of the experimental lines.

Nowadays, the facility has requested funds to upgrade the accelerator hall installing a pulsed light-ion source. It will be installed in the low energy part of the accelerator and will feed up the six experimental lines. However, this improvement will not be useful until a couple of years. In the following sections, the paper will describe how the facility has taken benefit of one of the existing elements in the center for pulsing the beam in air in one of the lines, the IMP, and thus opening the opportunity to some preliminary experiments, to be further developed once the upgrade of the sources will be fully implemented.

B. The Implantation line

The line is placed at the -20° exit port after the second high energy switching magnet, allowing experiments in a wide range of ions and ionic species with a reduced beam-current loss. The line has a total length of 6 meters and ends in a large irradiation vacuum chamber that offers the possibility of extracting the beam in air through various slits equipped with suitable output windows.

The main aim of this beam-line is to perform homogeneous implantations or irradiations in large areas (up to several cm^2). For this purpose, the line has a HVEE electrostatic scanner consisting of four independently polarized plates, designed specifically for the CMAM facility. This system is capable of scanning Hydrogen of 10 MeV at a size of 100x100 mm^2 when the two plate sets are used at the same time scanning at incommensurate frequencies, so that the scanning trajectory of the beam spot on the sample plane becomes ergodic. This is the most restrictive case and the one that will be considered for the purpose of this paper. Until now, this system called RASTER has been used merely for scanning the beam for irradiating in the mentioned large areas; nevertheless, it can open the door to obtain external pulsed beams.

C. The RASTER

The system used to pulse the beam is composed by two pairs of fast beam deflection electrostatic plates which accept an input signal of a maximum of 10 V in each of the plates. The voltage of the input signal is converted to kV to deflect the beam.

In the deflection process, the high voltage is applied to the electrode of the line to cast the beam. Then, the electric field is produced to deflect the beam. By modifying the conditions of the voltage induced in each of the plates, the deflection can follow an ergodic pattern, used for homogeneous irradiations of large areas, or deflect the beam at desired.

The deflection from the origin point (D) produced by one electrostatic plate is calculated following equation (1)

$$D = \frac{L \cdot \delta \cdot E_d}{2d \cdot E_a} \qquad (1)$$

where L is the length of the line from the beginning of the electrostatic plates to the irradiation point, δ the length of the deflection plates, E_d the voltage difference applied to the parallel plates, d the distance between the plates and E_a the energy divided by the charge state of the beam when passing through the plates.

The characteristics of the CMAM's RASTER are: δ=*447.7* mm, d=*132.5* mm and L=*6 m*.

D. Pulsed ion-beam applications

The capability of pulsing an ion beam in a facility opens the door for many applications such as material science, nuclear physics and medical physics.

For material science, which is one of the main research activities developed at CMAM, a pulsed ion-beam has been proved to be effective in annealing of ion implantations damage, in formation of p-n junctions in a single step doping process or in silicide formation as examples [3]. Moreover, the pulsed beam can be synchronized with the femtosecond (fs) pulsed laser installed at CMAM allowing to explore what is called soft materials [4].

In the nuclear physics case, being available to adjust the time structure of the pulsed beam will enhance the statistics by synchronizing the electronic acquisition with the beam as well as allow for performing time of flight spectroscopy. This will be a useful approach to reduce the dead-time of the acquisition systems and therefore obtain better statistics. Also, it will allow to study the nuclear structure through time of flight measurements [5].

Finally, in the medical physics field, a pulsed proton beam is of strong interest in the radiobiology and radiotherapy studies. In the last years, a new approach to irradiate tumors has been topic of research around the world: the FLASH treatment, where the given dose is fractionated in time with short and intense pulses. This will be explained in detail in the following subsection as the characterization presented in this document has the final objective of being used for this type of experiments.

E. Proton-therapy in FLASH regime

Proton-therapy is known as the radiotherapy technique for cancer treatment that uses proton beams for producing the damage. The main difference between the conventional radiotherapy treatments with X-Ray or electrons and the treatment with protons relays on how the deposition of energy is produced by each particle. In the case of the protons, most of the energy is delivered at the Bragg's peak region, which is a pronounced peak at a certain depth where the protons are completely stopped. One of the main advantages of the proton-therapy is that after the Bragg's peak, no dose is delivered assuring that healthy tissues behind the tumors are not radiated. The delivered dose curve for photons (or X-Ray) and electrons is wider at the superficial region and attenuates exponentially. Figure 2 presents the different curves of energy deposition for photons (blue), electrons (red) and protons (green).

Recent works in radiotherapy have raised interest in studying the outcome of very high dose rates, the so-called FLASH irradiations (>40 Gy/s), which are able to considerably reduce side effects compared to standard dose rates (~5 cGy/s). It is based on delivering ultra-high dose-rates inducing differential biological effects between normal and tumoral tissues and

 https://doi.org/10.1051/epjconf/202125304027

reducing the radiation-induced side effects [6-7]. The effectiveness of the FLASH-RT compared with the conventional radiotherapy plan has been demonstrated with electrons [8] and X-Ray [9]. Recently, its feasibility has been demonstrated with flash proton irradiation [9]. For the FLASH-RT, high intensities in short periods of time are required.

Relative dose

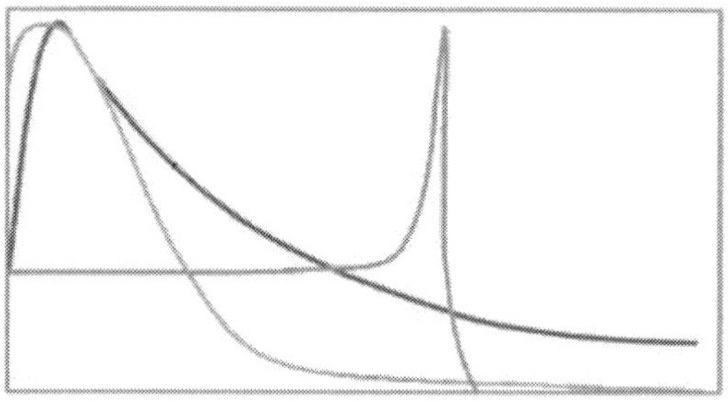

Depth

Fig. 2. Relative dose deposition as a function of the interaction depth for photons (red), electrons (blue) and protons (green).

II. Characterization of the Pulses

A. Experimental conditions

The experiment aimed at the characterization of the pulsed beam for proton-therapy pre-clinical studies. The beam used was a proton-beam of 8 MeV and 5 mm of diameter. These conditions were kept fixed during the experiment. The chamber was prepared to reduce the possible activation of the elements that interact with the beam by covering them with Ta foils of 250 µm. The beam was visible thanks to 2 arrays of scintillator pixels of SiO_2 that emit blue light as a response to the interaction with the beam. With the RASTER, the beam is moved along the silicon scintillator and only when it crosses the aperture, the beam exits the vacuum chamber and irradiates the sample placed in air. In Fig. 3 a picture of a SiO_2 placed inside of the chamber is shown, where the two blue spots are the positions where the beam reaches its maximum deflection and stays most of the time.

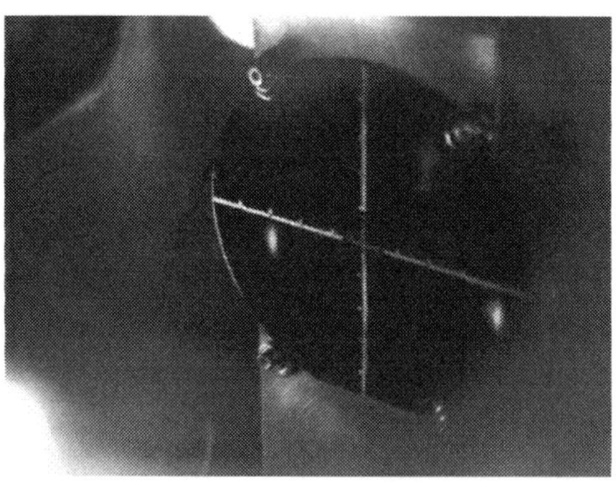

Fig. 3. Picture of a SiO_2 scintillator placed inside the IMP chamber. The two blue spots are the positions where the beam is most of the time.

Once the beam is extracted outside the vacuum chamber through the aperture and output window, it is characterized with a setup that consists of a CeGAG scintillator (1x1x3 cm) coupled to a S13360-6075CS SiPM from Hamamatsu. The signal is read directly with a digital oscilloscope, using the Picospe 6403D, with 8 bit resolution and a bandwidth of 350 MHz.

The function generator was connected to one of the X-axis deflection plates, moving the beam to the right. From the RASTER, the voltage signal produced was sent to the *Picoscope* to be processed and used as a trigger for the time signal. Due to the capacitance of the plate, a minimum time is needed to charge the plate. The minimum time for charge the electrostatic plate with 7 kV is 20 µs and, for rectangular functions, the resultant signal at RASTER will have a raise time. Different functions were tested in order to obtain the best conditions, optimizing the time of the pulses and avoiding the overshooting effect produced by forcing the plate to be charged/discharged faster than the time required by the capacitor.

B. Results

During the experiment, other functions were tested such as exponential rise, however, the function that best matches with the objectives of pulsing a beam for a FLASH proton-therapy treatment was the pyramidal one. In Table 1 it is detailed the parameters of the two functions analyzed: the rectangular and the pyramidal.

Function	Amplitude	Width
Rectangular	7 V	130 µs
Pyramidal	8 V	120 µs (ramp up)

Table 1. Parameters for the definition of the functions

Figure 4 summarizes the results of the experiment, comparing the pulse shape of the two functions studied. The first column is the input voltage of the RASTER generated with a function generator. The second, the output voltage from RASTER, which indicates how the plate has been charged. The third column is the signal detected on the Si-PM, which would be the pulse used to irradiate. Fig. 4a (first row) are the voltages for a rectangular pulse and Fig. 4b (second row), the results for the pyramidal function. It can be seen that for the latter, the pulse is faster and the beam only exits the chamber once. When using the rectangular function, the beam exits twice from the chamber due to the overshoot. The discharge of the pyramidal function also produces an overshoot at the end of the pulse; however it does not affect the pulse signal detected on the Si-PM as the beam only exits the chamber at the maximum voltage.

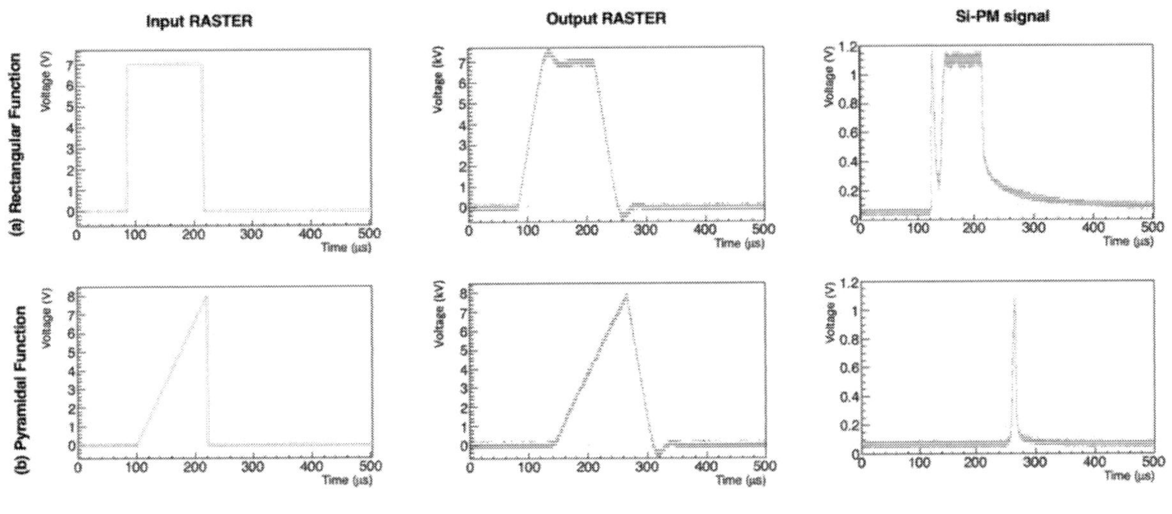

Fig. 4. Input, Output and Si-PM signal for two functions generated for pulsing the beam. First row (a) are the voltages obtained in each step for a rectangular function of 7V. The second row (b), for a pyramidal function of 8V. In green, the function generated used as input for the RASTER. In blue, the voltage induced on the RASTER's plate. In red, the signal detected in the Si-PM outside the chamber.

With the results obtained, the duration of the shortest pulse that can be obtained with this system has been determined: 10 µs, using the pyramidal function. With the rectangular function, the length of the pulse can be modified but due to the overshoot effect, the pulse produced by a rectangular function will be difficult to characterize due to the fact that it has a double peak.

C. Application of the pulses to a FLASH proton-therapy irradiation

In a conventional proton-therapy treatment, the dose given in each session ranges from 3 to 9 Gy. For a FLASH proton-therapy planning, the minimum beam intensity required with an 8 MeV proto-beam is 40 pA. The beam operation conditions allow to increase the intensity up to 10 nA, which will produce a dose rate ($\dot{D}$) of 1000 Gy/s. These conditions are the most suitable operation conditions due to beam stability and activation thresholds.

Considering the beam conditions described above, to deliver a treatment dose (3 Gy) in the FLASH regime with pulses of 10 µs, the number of pulses required is 296. Even though it seems a large number of pulses, the separation between the pulses can be less than 5 ms, irradiating the sample in less than 2 seconds. Due to the shape of the signal detected on the Si-PM, it is not recommended to use the rectangular function, that will allow to produce longer pulses but with an overshoot difficult to quantify.

III. Conclusions

In this paper we present the characterization of the first pulsed beam in air obtained at the CMAM facility. The pulsing of the beam is obtained deflecting it using an electrostatic plate commonly used for homogeneous irradiations of large areas.

The characterization of the pulses has been done with a generator function, a Si-PM installed outside the irradiation chamber of the IMP line and a *picoScope* to register and trigger the voltage signals in each of the three steps: the input function, the output voltage at the electrostatic plates and the signal detected in the Si-PM.

Rectangular and pyramidal functions have been used to characterize the system. The former has the advantage of being able to modify at will the duration of the pulses by increasing the width of the signal. However, the pulse is not optimal as the beam exits the chamber twice per cycle. This effect is due to the overshoot produced by using a function that increases the voltage faster than the time that the capacitor requires to charge. With the pyramidal function, pulses of 10 µs are obtained, being suitable for many applications and, in particular, for proton-therapy treatments in FLASH regime.

Acknowledgment

The authors want to thank the effort of the technician team of CMAM and the help provided by J. Olivares, V. Joco and F. Arias-Valcayo.

References

[1] D.J.W. Mous et al., "Performance and Applications of the first 5MV Tandetron at the University of Madrid." Proceedings CAARI 2002, AIP Conference Proceedings, August 26, 2003, Volume 680, Issue 1, pp. 999-1002.

[2] A. Redondo-Cubero et al., "Current status and future developments of the ion beam facility at the centre of micro-analysis of materials in Madrid" Eur. Phys. J. Plus, 2021, pp. 136-175

[3] J. Piekoszewski and J. Langner, "High intensity pulsed ion beams in material processing: Equipment and applications" in NIM B: Beam Interactions with Materials and Atoms, vol. 53, Issue 2, 1991, pp.148-160.

[4] Pavel Y. Apel, "Fabrication of functional micro- and nanoporous materials from polymers modified by swift heavy ions", Radiation Physics and Chemistry 159 (2019) 25–34.

[5] J. RD Copley and Terrence J. Udovic. "Neutron time-of-flight spectroscopy." Journal of research of the National Institute of Standards and Technology 98.1 (1993): 71.

[6] P. Montay-Gruel et al. "Irradiation in a flash: Unique sparing of memory in mice after whole brain irradiation with dose rates above 100 Gy/s." Radiotherapy and Oncology 124.3 (2017) 365–369.

[7] J. Bourhis et al. "Treatment of a first patient with FLASH-radiotherapy." Radiotherapy and Oncology 139 (2019): 18-22.

[8] P. Montay-Gruel et al. "X-rays can trigger the FLASH effect: Ultra-high dose-rate synchrotron light source prevents normal brain injury after whole brain irradiation in mice." Radiotherapy and Oncology 129.3 (2018): 582-588.

[9] E. Beyreuther et al. "Feasibility of proton FLASH effect tested by zebrafish embryo irradiation." Radiotherapy and Oncology 139 (2019): 46- 50.

Design of an acoustic sensor for fission gas release characterization devoted to JHR environment measurements

F. Baudry[1,2], E. Rosenkrantz[1,2], P. Combette[1,2], D. Fourmentel[3], C. Destouches[3], J.Y. Ferrandis[1,2]

[1] University Montpellier, IES, UMR 5214, F-34000, Montpellier, France
[2] CNRS, IES, UMR 5214, F-34000, Montpellier, France
[3] CEA, DES, IRESNE, DER, SPESI, LDCI Cadarache F-13108 Saint-Paul-Lez-Durance, France
Corresponding author: Jean-yves.Ferrandis@umontpellier.fr

***Abstract* - Among numerous research projects devoted to the improvement of the nuclear fuel behaviour knowledge, the development of advanced instrumentation for in-pile experiments in Material Testing Reactor is of great interest. In the frame of JHR reactor, new requirements have arisen creating new constraints.**
An acoustic method was tested with success during a first experiment called REMORA 3 in 2010 and 2011, and the results were used to differentiate helium and fission gas release kinetics under transient operating conditions. This experiment was leading at OSIRIS reactor (CEA Saclay, France). The maximal temperature during the irradiation test was about 150 °C. [1], [2]. We have developed thick film transducers produced by screen-printing process. They offered a wide range of possible application for the development of acoustic sensors and piezoelectric structure for harsh temperature environment measurements [3]. We proposed a screen-printed modified Bismuth Titanate piezoelectric element on alumina substrate allowing acoustic measurements [4] for JHR environment. In this paper we will focus on the mechanical design of the new sensor. This acoustic sensor is composed of an acoustic element for generation and detection of acoustic waves propagating into a cavity filled with gaz. We will detail the choice of piezoelectric materials, the thickness of the different layers, the cavity shapes, the electrical connections, the means of assembly of the different parts. Theoretical and experimental results will be given. All that point will be discussed in terms of acoustic sensor sensitivity versus dimensional constraints, in the case of a high temperature range working.

***Keywords* — Acoustic sensor, Gaz composition, Screen-printing, JHR**

I. Introduction

The IES laboratory (Montpellier, France), and more precisely the acoustics department, develop acoustic non-destructive testing methods for the nuclear field. And more specifically, a method to control the gas mixture in a fuel rod. Indeed, during a fission reaction, gaseous species such as Krypton (Kr) and Xeon (Xe) and helium (He) can be released and the knowledge of the quantity released can reveal information on the state of the fuel, but also influence the thermodynamic characteristics of the fuel elements.

In 2010 [2], a device called CACP-1 (in the frame of the Remora 3 project) was tested for the first time in operational condition in the OSIRIS reactor formerly located in Paris Saclay (CEA Saclay, France), at a temperature of 150°C and a thermal neutron flux up to 3.5×10^{19} n.cm². This first prototype used, as piezoelectric elements, a lead zirconate titanate (PZT) bulk piezoceramic commercialised by Meggit® (PZ27 Ferroperm) which had the advantage of having excellent piezoelectric properties ($d_{33\text{-}PZT}$ = 370-420 Pc/N [5]) compared to others piezoelectric ceramics. It made it interesting for characterisation in a gaseous medium, but had the disadvantage of having a limited Curie temperature of 350°C. However, this experiment allowed to follow the evolution of the gaseous composition of the fuel element with a precision of about 0.5 % of the molar fraction between helium and the gases coming from fission. The REMORA 3 experiment succeeded in its objective, namely to follow the evolution of the gas mixture in a fuel element from an acoustic measurement. But Improvement and optimization works have been undertaken with the following objectives:

- Increasing the maximum operating temperature of the sensor by using piezoelectric elements with a higher Curie temperature ($T_{curie\text{-}NBT}$ = 680 °C) such as Sodium Bismuth Titanate (NBT) in order to obtain a sensor with an operating temperature around 350°C.

- Direct integration of the piezoelectric elements on a ceramic substrate by screen printing process has been validated. This method avoids the use of metal soldering, which causes many technological issues such as the non-parallelism, the metal migration in the piezoelectric elements as well as the limitation of the maximum operating temperature.

Between 2011 and 2018 research has been conducted to produce NBT elements by screen printing on alumina substrates [4], [5]. Indeed, the NBT marketed by the company Meggitt® (trade name PZ46) is no longer available, it was necessary to formulate our own NBT. The NBT used until now is manufactured using the hydrothermal method followed by a crystallisation process. The manufacturing process is carried out by the company é-Ma. After various tests, the NBT formulated and screen printed in our laboratory has reached performances close to bulk PZ46 of Meggitt® ($d_{33\text{-NBT-ies}}$ = 14.1 Pc/N VS $d_{33\text{-PZ46 Bulk}}$ = 20 Pc/N [5]). Piezoelectrical elements has been screen-printed elements with a thickness of 40 micrometers of NBT ceramics. These devices have shown encouraging results with the generation of acoustic wave in a gas mixture of helium at 48 to 61 Bar [6], which can eventually be exploited for the estimation of the gas composition. However, this work has only focused on the design of the "active" part of the device, more precisely the piezoelectric element.

The work that I will present in this conference publication will serve to present the technological choices that have been retained in order to realise the acoustic device as a whole. Namely, the realisation of the piezoelectric element on a ceramic substrate, but also the realisation of the acoustic cavity. In a first long part named materials and methods, I will briefly present the principle of working of the device, and the choice of the materials composing the device by detailing its manufacturing process.
Then I will present the results of measurements of the impedance on the piezoelectric elements and the tests in a liquid of the device.

II. Material and method

In this part, material and method will be presented in a first time the techniques of manufacture employed for the realization of the piezoelectric elements. In a second and third part, I will detail the design of the sensor as well as the choices made for the materials which constitute the sensor as a whole. Finally, a instrumentation part will describe the material used for the measurements and tests.

A. Fabrication of piezoelectric elements by screen printing

The fabrication of piezoelectric element is done by screen printing deposition, of a piezoelectric ink. The first step is to formulate this ink that will be deposited by screen printing. This mixture is composed of crystallized NBT powder, mixed with a binder agent in order to obtain a liquid composition. The mixture is passed through a roll mill machine to obtain on the first hand a best mix as possible and on the second-hand particles size reduction of the crystallized NBT.

Then a succession of steps of deposition by screen printing allows to deposit successively a layer of electrodes then the multiple layers of NBT ink on an alumina substrate. Between each deposition step, a drying step is carried out in an oven in order to eliminate a part of the organic binder contained in the ink. On Figure 1.1 is represented a picture of a screen-printed element before the baking step and below a picture allowing to better realize the composition of an electrode-NBT-electrode stack (Figure 1.3). These stacks are realized in a grouped way on 2 inches square substrates allowing to realize up to 8 elements simultaneously.

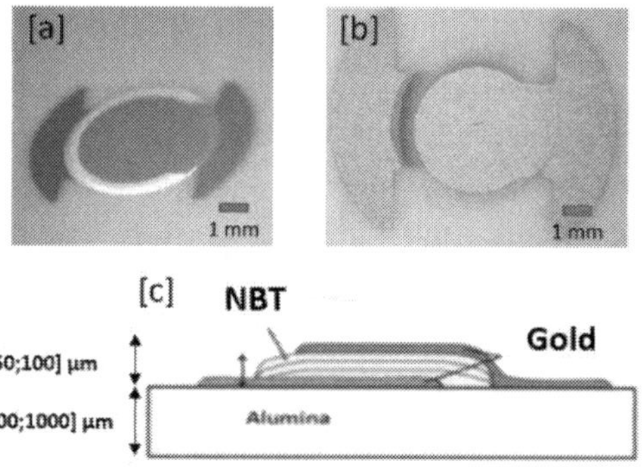

Figure 1 : (a) Piezoelectric element before sintering (b) Piezoelectric element after sintering (c) Cut view of the sensor showing the stacking of gold-NBT-gold

Once the different layers are stacked, an isostatic pressing step in water is performed. The samples are hermetically sealed in a plastic package under vacuum, then pressed at pressures higher than 1500 Bars. This step densifies the material by reducing the porosity of the NBT, which allows to improve the values of d_{33} and dielectric permittivity compared to the same element not pressed [7]. After pressing the stack, it is fired in a natural convection oven at 850°C, in order to sinter the powder and give the final microstructure. This sintering step can be preceded by a long firing step at a lower temperature (500°C) in order to eliminate the traces of organic solvents present in the binder. On the -Figure 1.2, we have the final form of the piezoceramic element directly printed on the alumina.

The thickness of the piezoelectric element and the substrate has an influence on the resonance frequency of the piezoelectric element which emits the acoustic wave, a mathematical model called KLM [2] makes it possible to estimate this value according to the thicknesses and the acoustic impedances of the substrate's materials and the piezoelectric ceramics. The thickness of the NBT layers produced by screen printing after firing varies from 50 to 100 micrometers and the thickness of the substrate varies between 0.7 and 1 mm. The choice of the thickness depends partly on the desired resonance frequency for the device, which should be between 3 and 7 MHz for acoustic characterization in a gaseous medium [1]. This aspect will be developed in the part III.A of the paper.

B. Acoustic measurements and design of the device

Before addressing the design and manufacture of the sensor itself, it is necessary to explain in detail the operation of the device. We can distinguish two parts to this device, the first is its active part, namely the piezoelectric element screen printed on alumina and the second is the acoustic cavity which is a cylindrical volume filled with gaz. The acoustic waves propagate in this cavity and realize multiple reflections on the reflector. This phenomenon is illustrated on Figure 2 by the blue arrows on the two different cavity designs.

By knowing the length of the cavity (e), and the time of flight of the acoustic wave (Δt), it is possible to determine the celerity of the acoustic wave (c) by the following equation:

$$c = \frac{2e}{\Delta t} \quad (1)$$

The celerity of an acoustic wave in a gas is related to the thermodynamic parameters of this one, namely its pressure and its temperature, but also to its molar mass. Thus, previous work [2] allowed to link by an analytical model the velocity of the acoustic wave to the ratio of quantity between the gas produced during the fission reaction (Xeon and Krypton) and the helium initially present in the instrumented fuel rod. The devices will be integrated in an instrumented fuel rod, more specifically in a hermetic cavity directly connected to the stack of fuel pellets. The devices have to fit in a volume equivalent to cylinder of 15 mm of diameter and 15 mm of high.

On Figure 2, are represented in a cross-sectional view of the two designs of sensor references. They are both cylindrical in shape with a diameter not exceeding 15 mm and 15 mm in height.

What distinguish the two designs are the fixing methods between the sensor body, the reflector and the alumina pellet on which the piezoelectric element is screen printed.

The first design consists of a hollowed-out cylinder on which the reflector and the disk and the piezoelectric element are glued with alumina cement (not showed on the figure). The second design uses a screw and nut system between which the reflector and the piezoelectric element are tightened.

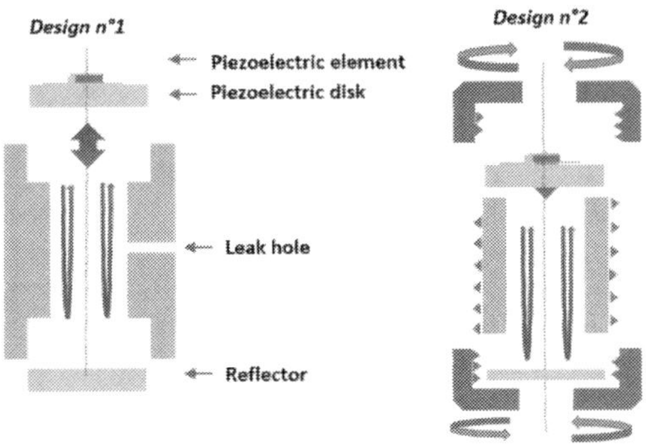

Figure 2 : (left) first design of the acoustic device (right) second design of the acoustic device

In the following part, I will detail the environmental conditions of the device and then the materials that have been selected for the manufacturing of the device.

C. *Technological and material choices for the sensors*

In this part I will detail the different materials and shaping techniques used for the realisation of the device, justifying the technological choices in relation to the technical constraints which are:

- The device has to resist to temperatures of at least 350°C [5]
- The elements must resist to radiative neutrons fluxes, the values of neutron and gamma fluxes are not well defined. But in 2019, NBT screen printed on alumina have been tested with a total dose of 1.75×10^{17} n/cm^2 (sum of thermal, epithermal and fast) with no apparent damage and critical evolution of its properties [8].

For the transducer, several piezoelectric materials have been investigated among the ceramics of Aurivillius structure group [5]. These ceramics belong to the family of "hard ceramics", with an average curie temperature above 600°C. Thus, two types of ceramics were tested, namely Bismuth titanate (BIT) $Bi_4Ti_3O_{12}$ and sodium Bismuth titanate (NBT) $Na_{0.5}Bi_{4.5}Ti_4O_{15}$. Moreover, it is important to specify that the performances of these piezoelectric materials depend largely on the choice of doping elements. And the method of shaping the piezoelectric elements which is for us screen printing on alumina substrate. The choice of NBT was motivated by the fact that it was the best compromise between a Curie temperature higher than 350°C and a weak influence to neutron fluxes (Thermal and fast). The d_{33} performances sufficient for our applications ($d_{33\text{-BIT-ies}}$ = 4 Pc/N VS $d_{33\text{-NBT ies}}$ = 14.1 Pc/N) for ceramics obtained by screen printing [4].

Regarding the electrodes, the choice was made for materials such as gold or platinum. Indeed, the use of silver as an electrode can create problems of electro migration. The passage of metallic silver in the dielectric layer could lead to breakdown phenomena [9]. In order to limit not only this problem and to resist to the different environmental conditions the choice was made to use gold electrodes deposited by screen printing. Screen printing allows to realize layers of 10 micrometers thickness more easily than sputtering method for instance. The reference used comes from the Ferro corporation (Reference: 8844-GI), and has a firing temperature of 850°C and a maximum operating temperature of 600°C.

For the realization of the cavity of cylindrical form of the two designs visible on the Figure 2, the choice fell on alumina and glass-ceramic. The glass-ceramic used is marketed by Corning inc ®, it has as main interest a maximum operating temperature of 1000°C, with no mechanical loading and the possibility of being machined with conventional means. Concerning the substrates of deposits, alumina or MACOR® can be retained, indeed they have both temperatures of use well higher than 350°C [10]. The cutting of the transducers in the form of pellets being ensured by a UV laser (Figure 3), it is completely possible to use the two types of materials. Moreover, the literature studies [10], have not indicated major problems as long as the use of MACOR® as a substrate of deposit.

https://doi.org/10.1051/epjconf/202125304028

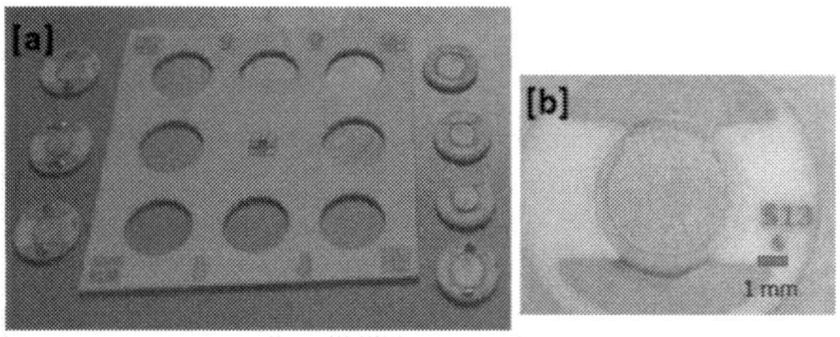

Figure 3 :[a] Alumina substrate with 8 elements cutting by UV LASER [B] Individual piezoelectric element ready to be integrated on a cavity

For the bonding of the elements, the choice was made to use alumina cements marketed by the brand Final material (reference: Final 940HT). This cement can be used at temperatures of more than 1500°C. A summary of the respective properties of alumina MACOR® and alumina cement are presented in Table 1. An important parameter to take into account is the difference of thermal coefficient expansion (TCE) between the different materials that form the sensor, the MACOR®, the alumina, the alumina cement and the NBT. In order to prevent the generation of crack and delamination between NBT ceramic and alumina, the mismatch of TCE between all elements have to be as low as possible.

Table 1 : TCE and maximum operating temperature of the different sensor elements

	Alumina (96%)	**Alumina cement**	**MACOR®**	**NBT screen printed**
TCE (K^{-1})	8.6E-6	7.2E-6	9E-6	7E-6
T_{max} (K)	1700	1540	1000	600

T_{max} indicates the maximum operating temperature of the elements. The TCE of elements is taken for a range of temperature between [25;300] °C. Data coming from their data sheet

As we can see on Table 1, the materials were chosen in order to limit these differences in TCE between the different parts of the sensor. On Figure 4, a photo represents respectively the design n°1 and n°2 assembled and ready to be tested.

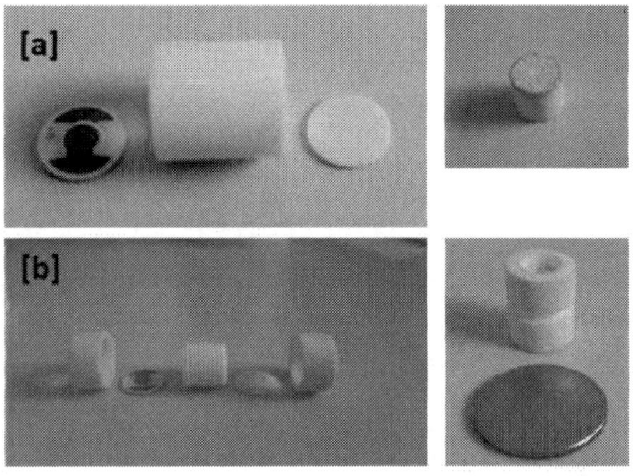

Figure 4 : [a] Design n°1 before and after the sealing step with alumina cement [b] Design n°2 of the screwed version before and after its assembly

D. *Instrumental bench*

Before assembling the piezoelectric elements in the cavity, the devices are polarized in order to give the NBT piezoceramic their ferroelectric properties. The elements are polarized at electric fields up to 150 kV/cm in a silicone oil bath.
Once the devices are polarized, an impedance measurement as a function of frequency is performed on each device, this measurement is performed on a Keysight E4990 impedance analyzer. Figure 5 shows the impedance analyzed with the sample holder under electrical contact tips to perform the measurements.

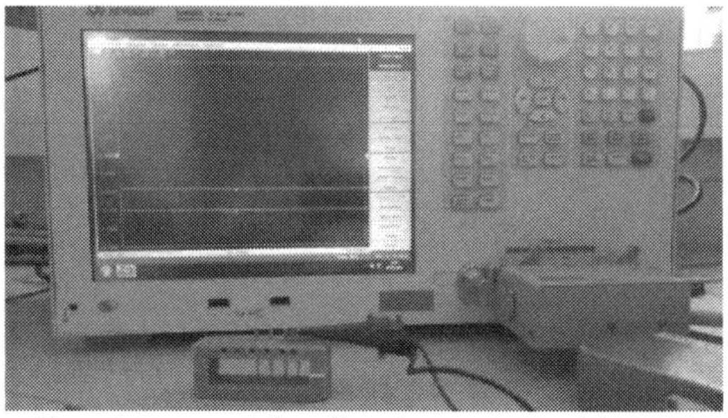

Figure 5 : Instrumentation of impedance meter Keysight E4990

Then tests are carried out in order to check the presence of reflected echoes in the devices n°1 and n°2 by filling them with ethanol. The objective is to ensure the proper functioning of the piezoelectric elements and the good parallelism of the cavities. For this, the transducers will be powered with an Olympus pulser/receiver model 5072R coupled to an oscilloscope allowing to realize an acquisition.

III. Model and Results

A. *Estimation of the optimal dimension by KLM modelling*

As previously stated, the KLM model allows to calculate an equivalent electrical impedance from the piezoelectric parameters of the materials (the mechanical quality factor, the coupling factor, the dielectric permittivity of the material and the dielectric losses) and from the properties of the layers (mainly the thickness, the density and the acoustic velocity). In the framework of our research, this model allows us to estimate the resonant frequencies of screen-printed piezoelectric transducer on alumina as a function of the thickness of the screen printed NBT and the thickness of the alumina substrate.

The screen-printing manufacturing process limits the thickness of the elements to 100 microns of post-fired ceramic. Thus, the KLM model, allowed to test the following conditions, namely three different thicknesses of alumina substrates (508 µm and 762 µm and 1000 µm) and two different thicknesses of NBT ceramics (50 µm and 100 µm). The objective is to understand the influence of these two parameters on the resonance frequency of the screen-printed element.

For this purpose, 4 combinations of substrates thickness have been tested, namely:

- Alumina: 508 μm and NBT: 50 μm
- Alumina: 762 μm and NBT: 50 μm
- Alumina: 762 μm and NBT: 100 μm
- Alumina: 1000 μm and NBT: 100 μm

The results are presented just below (Figure 6).

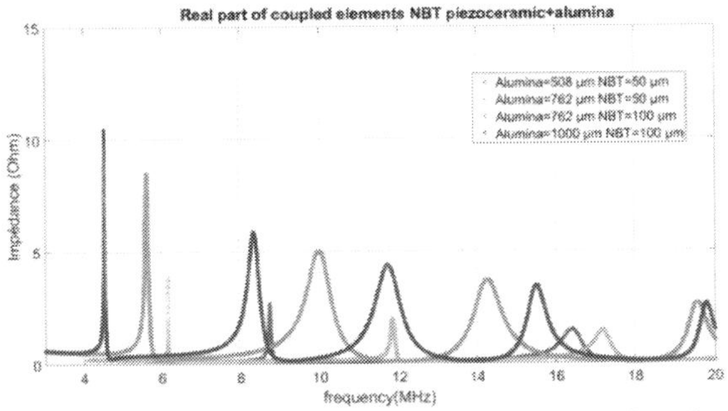

Figure 6 : Estimated real part of electrical impedance Vs frequency for various Alumina-NBT thickness

Two trends emerge from these simulations, the first is that for the same frequency thickness of NBT the more the thickness of the alumina substrate is important, the more the fundamental shifts towards low frequencies. As an example, for a thickness of 1000 μm of alumina and 100 μm of NBT the fundamental is 4.5 MHz Similarly, the greater the thickness of NBT, the more the fundamental frequency shifts to the low frequency.
It is then possible to estimate a couple of thickness of alumina substrate and NBT "optimal dimension" for our applications in order to obtain a fundamental frequency between 4 and 7 MHz So substrates of thickness 762 μm or 1000 μm seem quite indicated, with NBT elements whose thickness will be between 50 μm and 100 μm.

B. *Impedance of the NBT Layers*

As previously stated, the transducers must resonate at a frequency between 4 and 7 MHz Experimentally, several samples have been manufactured, their characteristics are given in the Table 2.

Table 2 : Characteristics of various sample tested

ID	Thickness alumina (μm)	Average Thickness of NBT (μm)
S4	508	50
S12	762	100

The objective of these measurements is to validate, the trends of the model. More specifically, the increase and decrease of the frequencies of the fundamental by changing the thickness of the substrate and NBT. The impedance spectra between 1 MHz and 20 MHz of two piezoelectric elements, namely S4 and S12, are shown in Figure 7.

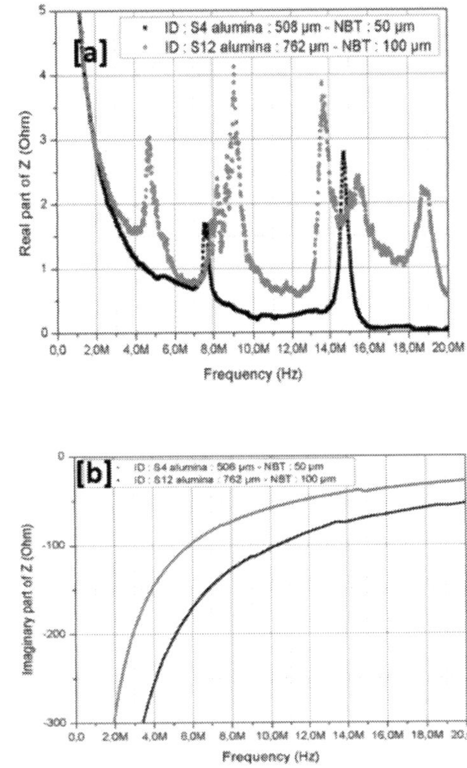

Figure 7 : Electrical impedance real part [a] and imaginary part [b] of S4 and S12 sample

The impedances confirm the trend of an increase in the resonant frequency of the fundamental by jointly increasing the thickness of the alumina substrate and the NBT ceramic. Thus, the couple 762 micrometers of alumina and 100 micrometers of NBT is validated for the fabrication of devices because its fundamental is about 4 MHz. However, it remains differences in amplitudes and quality factors concerning the resonances. This can be due to several parameters, namely:

- The piezoelectric parameters of the NBT used for the simulation do not exactly match the NBT used for the S4 and S12. Indeed, parameters are taken from the publication of O.Gatsa and other [4]. The reasons of differences is the addition of chemical elements in our NBT slightly different than the NBT tested by O.Gatsa and other.

- Moreover, the values of the impedances depend on other parameters which linked on the process of manufacture of the elements. For instance, the time and the duration of the sintering which will define the state of the ceramic microstructure. But also, the presence or not of a densification step during the manufacturing process.

https://doi.org/10.1051/epjconf/202125304028

C. *Acoustic test with ethanol*

Impedance testing showed that a thickness of 762 µm alumina and 100 µm NBT is a potentially satisfactory combination for gas composition study. The devices will be tested in a non-conductive liquid such as ethanol with two cavity designs, the cavity sealed by alumina cement bounding (design n°1) and the screwed cavity (design n°2). The objectives of these measurements are the following:

- To verify the proper functioning of the device by observing the emission of echoes. Indeed, if in a liquid medium the device does not manage to emit echoes, then it will surely not be able to emit and measure any echoes in a gas. Because gases are less dense than liquids, the acoustic waves for frequencies between 1 and 20 MHz are more attenuated than liquid middle.

- To validate the good functioning of the acoustic cavities. More specifically, the flatness between the piezoelectric element and the reflector must be the best as possible. It is possible to observe these problems on an echogram, with in particular the appearance of non-periodic echoes and a sudden variation of reflected echoes.

In Figure 8 below, the devices are mounted with a functional piezoelectric element. For the first design the reflector at the bottom of the device is glued to the bottom with alumina cement (Figure 4 [a]) and the active element is only maintained against the cavity. The absence of glue allows to test several elements per cavity, but also simplifies the filling of the cavity with ethanol. The second device is screwed on. The devices are electrically connected using single stranded wires connected with Sn/Pb solder. Nevertheless, connection method is not representative of the final device.

Figure 8 : Acoustics elements ready to be tested

In Figure 9, are presented the results of several measurements made successively with the design n°1 and n°2 using the same piezoelectric transducer. 7 measurements were made, 4 on design n°1 and 3 on design n°2. Between each measurement the device is dismounted and purged of ethanol and then fill a new time in order to validate the reproducibility of the measurements.

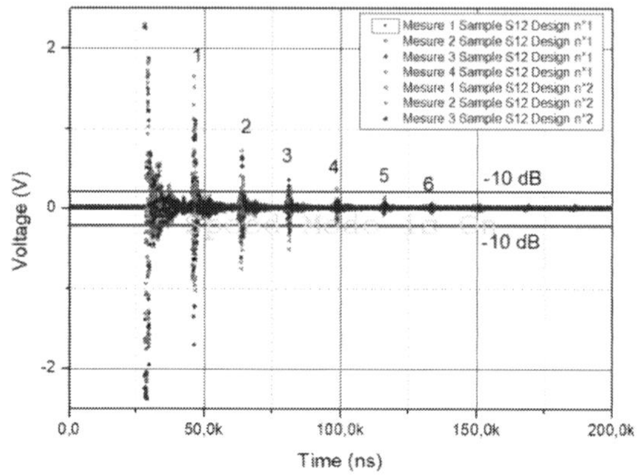

Figure 9 : Acoustics echoes measuring in Pulse/echo mode for designs n°1 and n°2

The measurements show several things: the first thing is a series of several reflected echoes, whose amplitude decreases exponentially in agreement with the acoustic models. The attenuation mechanism is linked to the phenomena of propagation and reflection. When the acoustic wave propagates in a medium, its energy decreases because of the transmission of acoustic energy in the material medium to set in motion the particles constituting this medium.
Thus, a medium such as a gas even at pressures of several hundred bars will have a density less important than a liquid at an equal pressure and temperature. And thus, the liquid will be far less attenuating because requiring less energy for its setting in movement. Furthermore, gas and liquid being dispersive media, the losses also depend on the frequency of the acoustic signal. Various model existing so as to estimate the acoustic attenuation in a liquid or gas mixture, but not presented in this paper.
The second type of notable losses are those by reflection, the acoustic energy is lost during each interaction with a solid medium. They can be modelled as a function of the acoustic impedances (Table 3 (2)) specific to the two media, with the formula Table 3 (3) representing the share of energy reflected during each reflection. For example, for the acoustic impedance of alumina ($Z_{Al2O3} = 3.89\times10^7$ Pa.m. s-1) and that of ethanol ($Z_{C2H5OH} = 9.16\times10^5$ Pa.m. s-1) the reflected acoustic energy corresponds to nearly 95% of the initial energy. While for a gas such as helium at 25 ° C and 200 Bar, the acoustic impedance is much lower than alumina ($Z_{He} = 600$ Pa.m. s-1), so that the reflected energy is reflected to more than 99.99%.

Table 3 : Definition of acoustic impedance and Rate of reflected energy

Acoustic impedance	$Z = \rho c$	Pa.m.s^{-1}	(2)
Rate of acoustic energy reflected	$R = \dfrac{Z_2 - Z_1}{Z_2 + Z_1}$		(3)

ρ and c represent respectively the volumic mass and the celerity of sound of the material, and Z_2 and Z_1 represent respectively the acoustic impedance of alumina and gaz (or liquid)

Thus, in ethanol and for the fourth echo, the acoustic wave has made nearly 7 reflections on the alumina which corresponds to $0.95^7= 0.69$ of the initial energy at the end of the fourth echo just by the reflection of the waves. And on the fourth echo almost 70% (-10 dB) of the energy is lost compared to the initial signal.
Thus, it is for the moment impossible to validate at least analytically the functioning of the device in gas medium. Because the types of acoustic losses are different and the acoustic attenuation complex to model in a gas. Nevertheless, this experiment has permitted to validate the emission of acoustic wave in a liquid medium. The first experiments in gaz medium are planned for the third quarter of 2021.

Besides, the time difference between each echo of the measurements on S12 is an average of 17.47 µs +/-0.37. The speed of acoustic wave estimated for this time of flight is 1144 m/s for a cavity of 10 mm height (equation 1). This is within the measurement interval of 1142.9 m/s +/- 1 measured by Erich C Meister [11] for pure ethanol at 25°C by another acoustic method. Finally, the echogram does not show any parasite echoes between the different reflections, which indicates a satisfactory parallelism for our applications.

IV. Conclusions and perspectives

This work showed the study, the realization and the test of an innovative acoustic sensor usable in an experimental fuel rod of the future JHR. The choices concerning the materials of the sensor are respectively, NBT for the piezoelectric ceramic which is deposited by serigraphy on alumina. Glass-ceramic such as MACOR® is used for the fabrication of the acoustic cavity and the reflector. These choices were motivated by the low thermal expansion of these materials taken individually, their resistance to the radiative fluxes and their maximum operating temperature well higher than 300°C

In parallel to the technological aspects, with the help of an analytical model, the choice of the thickness of the substrate and the piezoelectric element were fixed at respectively a thickness of 762 micrometers of alumina and 100 micrometers of NBT. These choices have been confirmed by the experimental results which showed the presence of an acoustic wave reflected in the cavity filled by ethanol. The shape of the signal and the time between each echo seem coherent in comparison to the state of the art.

Next results will concern, the tests in a gas to validate the sensitivity of the device immersed in a gas mixture of helium. The device must also be tested at a temperature of at least 300°C to verify that its piezoelectric performance remains stable despite the temperature.

References

[1] D. Fourmentel, Jf. Villard, Jy. Ferrandis, F. Augereau, E. Rosenkrantz, et M. Dierckx, « Acoustic sensor for in-pile fuel rod fission gas release measurement », in *2009 1st International Conference on Advancements in Nuclear Instrumentation, Measurement Methods and their Applications*, Marseille, France, juin 2009, p. 1-5. doi: 10.1109/ANIMMA.2009.5503679.

[2] E. Rosenkrantz, J. Y. Ferrandis, F. Augereau, T. Lambert, D. Fourmentel, et X. Tiratay, « An Innovative Acoustic Sensor for In-Pile Fission Gas Composition Measurements », *IEEE Trans. Nucl. Sci.*, vol. 60, n° 2, p. 1346-1353, avr. 2013, doi: 10.1109/TNS.2013.2252624.

[3] F. Very *et al.*, « Acoustic Sensors for Fission Gas Characterization in MTR Harsh Environment », *Phys. Procedia*, vol. 70, p. 292-295, 2015, doi: 10.1016/j.phpro.2015.08.157.

[4] O. Gatsa, P. Combette, D. Fourmentel, C. Destouches, et J. Y. Ferrandis, « High temperature ultrasonic sensor for fission gas characterization in MTR harsh environment », p. 9.

[5] O. Gatsa, « Development of acoustic sensors for the extension of measurements to high temperature in the experimental reactors », p. 177.

[6] O. Gatsa, E. Rosenkrantz, D. Fourmentel, C. Destouches, P. Combette, et J.-Y. Ferrandis, « Validation of the first prototype high temperature ultrasonic sensor for gas composition measurement », Honolulu, Hawaii, USA, 2017, p. 030001. doi: 10.1121/2.0000679.

[7] H. Debeda-Hickel, C. Lucat, et F. Menil, « Influence of the densification parameters on screen-printed component properties », *J. Eur. Ceram. Soc.*, vol. 25, n° 12, p. 2115-2119, janv. 2005, doi: 10.1016/j.jeurceramsoc.2005.03.206.

[8] Jy. Ferrandis *et al.*, « Acoustic instrumentation of the new generation of MTR: effect of nuclear radiation on modified Bismuth Titanate piezoelectric elements », *EPJ Web Conf.*, vol. 225, p. 04012, 2020, doi: 10.1051/epjconf/202022504012.

[9] G. S. Kulikov *et al.*, « Diffusion and electromigration of silver in PZT and HTSC ceramics », *Ferroelectrics*, vol. 144, n° 1, p. 61-70, juill. 1993, doi: 10.1080/00150199308008625.

[10] L. Seveyrat, « Elaboration et caractérisation de films épais piézoélectriques sérigraphiés sur alumine, silicium, aciers inoxydables et vitrocéramiques », p. 193.

[11] E. C. Meister, « Measurement of the Temperature and Concentration Dependent Sound Velocity in Ethanol-Water Liquid Mixtures », p. 7.

High resolution measurements with miniature neutron scintillators in the SUR-100 zero power reactor

Edoardo L. BRUNETTO[1,2], Fanny VITULLO[1], Vincent LAMIRAND[1,3], Klemen AMBROŽIČ[1], Daniel GODAT[1], Michael BUCK[4], Georg POHLNER[4], Jörg STARFLINGER[4], Andreas PAUTZ[1,3]

[1]École polytechnique fédérale de Lausanne (EPFL), Switzerland
[2]Politecnico di Torino, Italy
[3]Paul Scherrer Institut (PSI), Switzerland
[4] University of Stuttgart, Germany

Corresponding author: vincent.lamirand@epfl.ch

***Abstract*—Three 1-mm^3 miniature fiber-coupled scintillators have been used to perform cm-wise resolution measurements of the thermal neutron flux within experimental channels of the SUR-100 facility, a zero power thermal reactor operated by the Institute of Nuclear Technology and Energy Systems at the University of Stuttgart. The detection system is developed at the École Polytechnique Fédérale de Lausanne in collaboration with the Paul Scherrer Institut. Thermal neutrons count rates were measured along the experimental channels I and II, which cross the reactor at the center and tangentially to the core, respectively. The reactor was modelled with the Monte Carlo neutron transport code Serpent-2.1.31. The comparison of experimental and computed reaction rate distributions showed a good agreement within the core region, with discrepancies within 2σ. An unexpected discrepancy, probably caused by a geometric inconsistency in the computational model of the reactor, was observed in the reflector region of the experimental channel I, where a 20% difference (i.e. 8σ) was found between experimental and simulated results. Significant discrepancies, respectively worth 10σ and 15σ, were noticed at distance, in the lead shielding region, for both experimental channels I and II. In addition, reaction rate gradients across the 2.6 cm and 5.4 cm diameters of both channels were measured. A horizontal reaction rate gradient of (9.09 ± 0.20) % was measured within 2.4 cm across the diameter of the experimental channel II, with a difference from computed results of 2%. The absence of a vertical reaction rate gradient inside the experimental channel I was confirmed by measurements.**

***Keywords* — Scintillation detectors, Monte Carlo simulations, Reaction rate profiles.**

I. Introduction

The objective of the campaign presented hereafter was to perform high spatial resolution measurements to characterize the neutron thermal flux distribution inside experimental channels of the SUR-100 reactor, and to compare the measurement results with Monte Carlo simulations. The study benefited of the innovative miniature detector developed by the Laboratory for Reactor Physics and Systems Behaviour (LRS) at EPFL in collaboration with the Paul Scherrer Institut (PSI) [1]. Three units of the fiber-coupled scintillator detector were used in different arrangements to perform the flux characterization of the experimental channels of the reactor. Their small size results in a neutron flux perturbation lower than any other conventional detector and allows for measurements with high spatial discretization. All the profile measurements presented in this paper are given in relative reaction rates units. In addition, a preexisting MCNP [2] model of the reactor has been converted into Serpent2 [3] and used to run particle transport simulations. Monte Carlo simulations have been performed both to design the experiments, and to compare their results to the acquired measurements. The experimental campaign is described in the first section, followed by the modelling of the reactor, and finally the comparison and discussion of experimental and simulation results are presented.

II. Experimental campaign

A. The SUR-100 reactor

The Siemens Training Reactor SUR-100 (in Fig. 1) is a thermal zero-power reactor built in 1964 and located in the University of Stuttgart. It is used for mainly for educational purposes by the Institut für Kernenergetik und Energiesysteme (IKE). The reactor has a maximum thermal power output of 1 W, with a nominal power of 100 mW. It is equipped with a homogeneous core, composed of solid cylindrical slabs of moderator with fuel particles dispersed inside. The core is provided with a lifting mechanism which can move axially the lower core section away from the upper one. The core halves are separated for the shutdown and brought together for the startup of the reactor. The reactivity regulation is performed with two cadmium control plates, which are steered into the core from below. The reactor is provided with various experimental channels, which allow the insertion of nuclear instrumentation into the core or in its proximity.

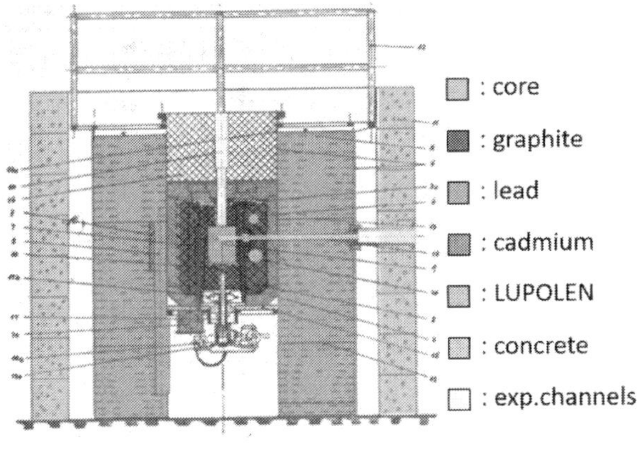

Fig. 1 - Scheme of the SUR-100 reactor

B. *Miniature neutron scintillator*

The miniature fiber-coupled neutron scintillator used in this study has been developed at the LRS-EPFL in collaboration with the Paul Scherrer Institut (PSI) [1]. Its miniature dimensions allow for highly localized measurements, while showing a good neutron sensitivity. Indeed, past experiments demonstrated improved performances with respect to others in-core detecting devices, e.g. by showing a sensitivity one order of magnitude higher than a miniature fission chamber or a sCVD diamond detector [4]. The neutron sensitive area of the detector is a ~ 1 mm^2 scintillator screen of zinc sulphide doped with ^{6}Li having a thickness of 0.28 mm [5]. The conversion of detected neutrons in charged particles is performed by the (n,α) reaction of ^{6}Li. The alpha and triton particles arising from the reaction induce scintillation of the ZnS molecules of the crystal. Through this two-step process, the incoming neutron radiation is converted to a light signal. A plastic optical fiber [6] is coupled to the scintillator screen, guiding the scintillation photons to the read-out electronics. These consists of a SiPM for photon detection, and in the present study of a standard analog detection chain setup, depicted in Fig. 2. As a preliminary step to the experimental campaign in the SUR-100 reactor, the instrumentation was previously tested in the CROCUS zero power reactor of EPFL.

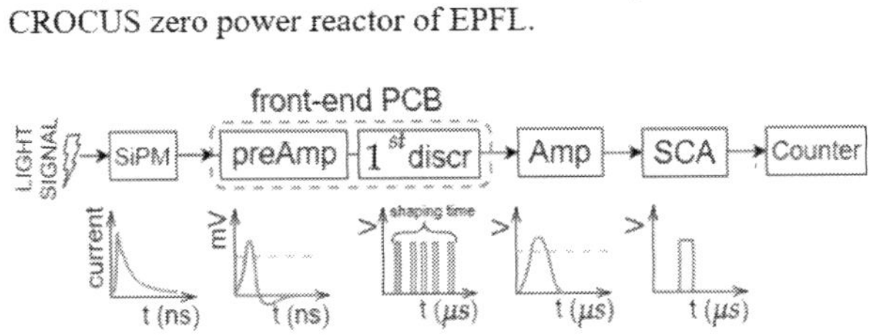

Fig. 2 – Standard analog detection chain.

C. *Detector arrangement*

The experimental instrumentation consisted in a multiple acquisition neutron detection system: three fiber-coupled scintillators have been used as detectors, with minor modifications with respect to the initial design of past studies [4]. The main improvement regards the aluminum cap, which has been reshaped in a new two-diameter configuration to improve the coupling with the fiber. An aluminum holder housing the three detectors tips was manufactured at a diameter of 26 mm corresponding to the internal diameter of the experimental channel I. Given the difference in diameter of the experimental channels, the diameter of the holding tube had the possibility to be adapted to a larger diameter (54 mm) by means of aluminum spacer rings. To move the detectors along the experimental channels, the front-end holder was mechanically coupled to an aluminum tube manufactured by the IKE laboratory, which was graduated to select the detectors' position inside the channels. The tube's handle allowed to adjust the angular orientation of the detectors inside the experimental channels. The choice of aluminum for the holding equipment was straightforward, given its small influence on thermal neutron flux, its availability, its low neutron absorption cross section and rapid decay of the induced activity.

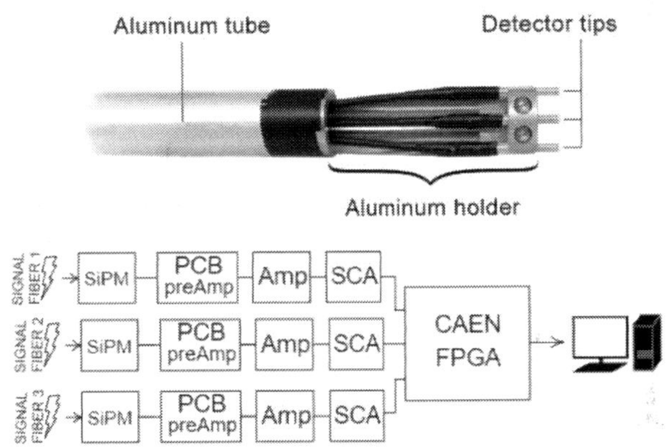

Fig. 3 - Fibers holder for measurements in SUR-100 (top) and read-out detection chain (bottom).

D. *Conducted experiments*

Steady-state and dynamic measurements were carried over along the campaign. We focus in the present article on the thermal flux characterization through the use of the miniature detectors, namely the measurements of thermal neutron flux profiles:

- along the central experimental channel I: [-40,+40] cm around core center,
- along the tangential channel II: [-8,+46] cm around core center,
- across the diameter of both channels I (vertically) and II (horizontally).

An explanatory scheme of the performed measurements is presented in Fig. 4. Depending on the experiment, the tube was slid into the channels with one of the possible configurations shown in Fig. 5.

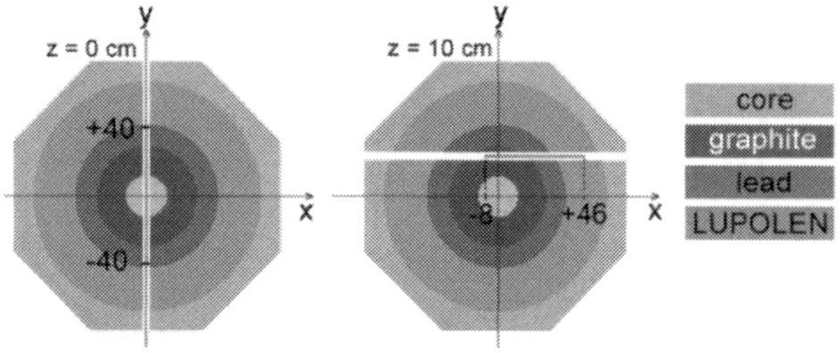

Fig. 4 - Simplified schemes of the profile measurements inside channel I (left) and channel II (right). The red lines indicate the spatial range in which measurements are performed.

The number of the detectors used was varied in dependence of the needs of each experiment. In most of the experiments, one of the miniature detectors (fiber 2) was removed from the holder and used for power monitoring. Count rates were recorded at each position for the time necessary to have statistical errors in the order of 0.1%.

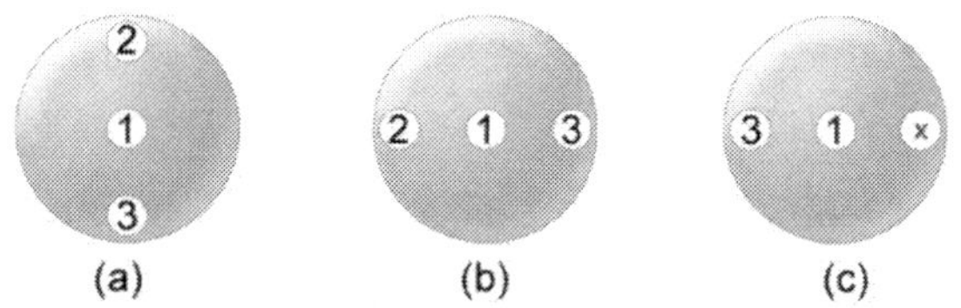

Fig. 5 - Different positioning of the detectors inside the experimental channels of SUR-100.

An additional measurement of the thermal flux profile inside the central channel was obtained with a neutron activation experiment which is conducted on a regular basis for teaching purposes at the SUR-100 reactor. The flux profile was retrieved by measuring with a High-Purity Germanium (HPGe) semiconductor detector the gamma-radiation emitted by some manganese samples irradiated in the reactor central channel for about 45 minutes. A cylindrical holder with the same diameter than the experimental channel was used to position 16 samples of ^{55}Mn inside the reactor core. The samples, consisting in tiny cylindrical rods of equal diameter with a length varying from 16.6 to 18.0 mm, were positioned in the 2 cm spaced housings of the holder. The stable ^{55}Mn isotopes interact with thermal neutrons through the ^{55}Mn(n,γ) neutron capture reaction, leading to the measurable decay of ^{56}Mn nuclei.

III. Serpent 2 Model of the SUR-100 reactor

The Serpent2 model hereafter presented results from the conversion of an already existent MCNP input: all the geometrical and materials details of the reactor have been taken from this reference, thus the same assumptions are considered. The main simplifications of the computational model concern:

- the absence of all the drive mechanisms, as they are not impacting the neutron economy relevantly;
- the absence of the experimental channel 10/V, being it filled with graphite and therefore considered part of the reflector;
- the absence of the detector housings and the detectors themselves, since deemed non-impacting for the studies performed in this work because of their distance from the experimental channels;
- the absence of the aluminum coating of control plates, since they are assumed to not perturb the neutron flux in the thermal range of interest.

To improve its usability and its level of accuracy, the computational model has been adapted with minor modifications beyond its bare conversion, concerning shielding and control systems.

Despite its absence in the model provided by IKE, the concrete shielding has been modeled in Serpent2. Although the presence of the outermost layer of the shielding might not affect the neutronics in the center of the core, it is key for possible radiation protection studies e.g estimation of dose rate delivered to the operators. While the geometric dimensions of the concrete shielding have been taken from the official documentation from Siemens [7], its composition has been assumed "regular concrete" from the MCNP material compendium [8]. The concrete shielding has been shaped with a cylindrical shape, being this a simplification motivated by the absence of geometrical data on the real octagonal shape of the component at the time of the modeling procedure.

In addition, and to facilitate the modeling task of moving the control plates and the lower core section in different positions, the model is provided with a dedicated cell for each of them. This, together with the use of the translation command of Serpent2, enables to easily displace the elements without the need of a significant revision of the surfaces definition and opens to the use of apposite scripts to automatize the calculations. Thanks to these enhancements, the two control plates can be steered independently in the range [0,-250] mm and the lower core section (together with the lower section of the inner reflector) can be moved within the interval [0,-45] mm.

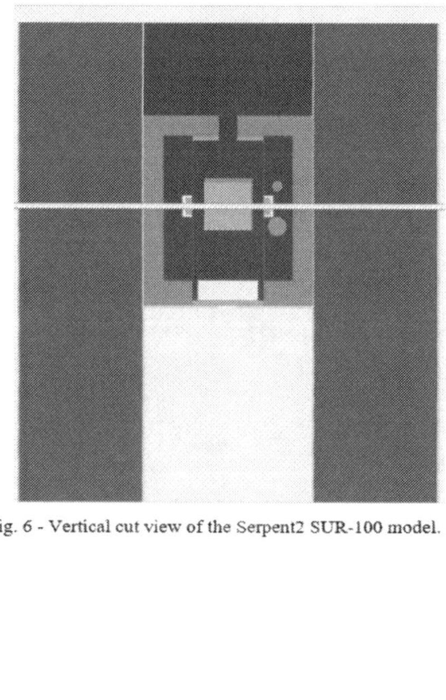

Fig. 6 - Vertical cut view of the Serpent2 SUR-100 model.

IV. Results

A. Neutron thermal flux profile along central channel I

For this experiment, the miniature detectors 1 and 3 were moved along the length of channel I through the aluminum tube in the configuration (c) of Fig. 5, while the detector 2 was used as a monitor at a fixed position along the whole duration of the measurements in the tangential channel II. The results of the measurements recorded using detector 1 are depicted in Fig. 7. The count rates are normalized twice: first by detector 2, then by the maximum measured value. The profile is symmetric with respect to the core center, given the overall agreement of the data measured on each side of the core. A slight difference is found to exist between the points lying in the interval from 20 to 28 cm far from the core center, with a disagreement in the order of 10 %.

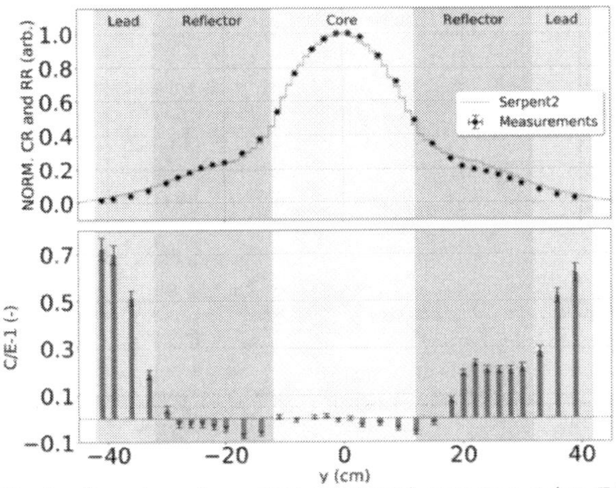

Fig. 7 - Comparison of normalized experimental count rates against 6Li reaction rates predicted by simulations along experimental channel I (top) and observed discrepancies (bottom).

The measured data are then compared with computational results obtained through the Serpent2 model. The real configuration of the reactor (i.e. power level, control plates position and filling of experimental channels) of the day of the measurement is implemented in the computational model, in order to reproduce the experiments as faithfully as possible. Cylindrical virtual detectors with the same radius of the channel (26 mm) and a thickness of 1 cm are used to score the neutrons reaction rate with the Lithium-6 of the scintillator in every cm of the experimental channel in the range [-45;45] cm. The modeling of real-size detectors, given the miniature dimensions of the scintillators, was deemed to be unnecessary for the scope of the work at hand. Indeed, simulating real-size scintillator would cause a significant increase in the computational cost of the simulations, given the lower occurrence of neutrons interaction with a much smaller volume. Despite this, the smearing of the detectors is expected to introduce a discrepancy with the measured data, since the Monte Carlo results are in this way averaged on larger volumes. A depiction of the comparison of experimental and simulated data is also given in Fig. 7, as well as residuals.

The experimental results are seen to be particularly in agreement with the simulated data in the spatial range [-30,18] cm, with a C/E-1 value lower than 10 %. It is worth noting that this overall good agreement proves the negligible neutron flux perturbation induced by the miniature detectors, not modeled for the computational simulation.

An asymmetry in the radial profile is found in the zone of the graphite reflector between +20 and +30 cm away from the core, and it is not foreseen by the simulations. This discrepancy, which causes a disagreement of about 20 %, could be caused by localized change in material composition or some geometric asymmetry not considered in the computational model. Further away from the core center, a higher discrepancy is found to exist symmetrically from 30 to 40 cm, with a C/E-1 value that reaches about 70 %. Given the presence of the lead shielding in the zone of highest disagreement, such a discrepancy could be motivated with the uncertainties in material composition or in its cross section. Since the data series are normalized for their maximum values (which is located in the core center), the disagreement observed could not necessarily be attributed to the peripheral zones and therefore at the lead shielding region. Nevertheless, the main suspicions fall on the lead shielding area, as it is mechanically and chemically less known than the core center region.

The results are compared as well with the activation experiment in Fig. 8 (top), where a substantial disagreement is found between the fiber measurements and the count rates retrieved with the activation technique.

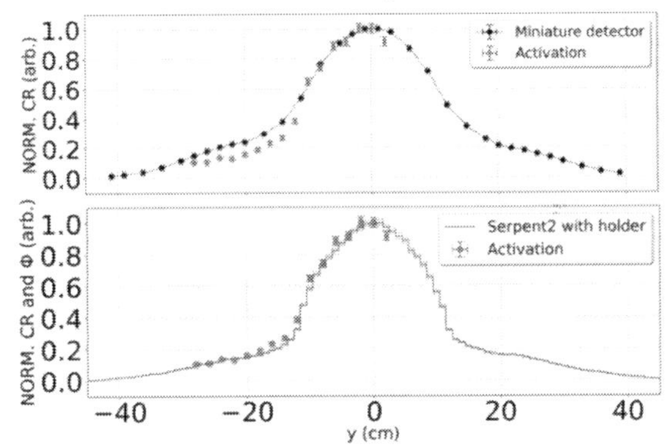

Fig. 8 - Direct comparison of the activation profile with measurements of miniature detector (top); Comparison of the activation profile with Serpent2 computed results (bottom).

The two measured neutron flux distributions differ particularly in the region outside the core, where a discrepancy of about 30 % is observed. The reason behind this relevant discrepancy is found to be due to the presence of the plastic-graphite holder. The use of the activation holder causes a perturbation in the shape of the neutron flux distribution due to its constituent materials, which increase the thermal neutron flux in the core center. Indeed, the presence of PMMA enhance the neutron moderation while the graphite promotes the neutrons reflection towards the inner core region. Therefore, if the two flux distributions are plotted by normalizing the measurements for their respective maximum value (as in Fig. 8

top), a meaningless comparison is obtained, with differences up to 50 %. A fair comparison between the activation measurements and the Serpent2 results are performed by including in the computational model the activation holder. The comparison is shown in in Fig. 8 (bottom), where a good agreement is found between experimental and computational data, with C/E-1 values lower than 20 %.

B. Neutron thermal flux profile along channel II

Similarly to the central channel I case above, the upper tangential channel was deployed to characterize the thermal neutron flux distribution along its length. No experimental data on the flux distribution was available beforehand, due to the difficulties to perform activation measurements in a region too far from the core center.

The detector 2 was used as a monitor as well, and it was positioned in the experimental channel I, approximately in correspondence of the core center. The remaining two detectors were configured as shown in Fig. 5(c). In addition, to fit the holder tube in a channel with larger diameter, apposite ring spacers made out of aluminum were used. The experimental results are shown in Fig. 9, together with a comparison against simulated data. For the Serpent2 simulation, virtual track-length detector of cylindrical dimension as big as the channel (54 mm diameter) are used to characterize the thermal flux profile.

The experimental results show excellent agreement with most of the simulated points, exception made for the data lying in the interval [20;35] cm, where the disagreement reaches values of about 50 %. Again, the discrepancy is localized in the zone occupied by the lead shielding, which might have substantial differences in composition with respect to the modeled one.

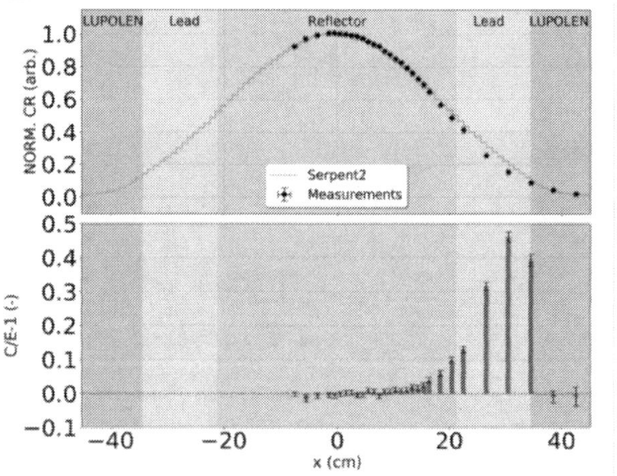

Fig. 9 - Comparison between experimental and simulated results for the thermal neutron flux distribution in tangential channel II.

C. Thermal neutron flux profiles across channels diameters

For the experiments on flux profiles across both central and tangential channels, the detectors were positioned in the center of the channel according to the configuration (a) and (c) of Fig. 5 respectively, and then rotated 180° to repeat the measurement. Data from detector 2 are used for the central channel, with a power normalization using detector 1, since its position remains unchanged even after the 180° rotation. Data from detector 3 are used for the tangential channel, with a power normalization using detector 2 set in the core center.

For the central channel, both experimental and simulation results suggest the absence of a vertical gradient, being the reaction rate distribution almost flat in the core center. However, for the tangential channel II, the experiment shows the existence of a horizontal gradient of (9.09±0.20) % from the innermost to the outermost radial position measured with the miniature detectors.

V. CONCLUSIONS

High spatial resolution measurements of the thermal neutron flux have been performed with three 1-mm^3 miniature fiber-coupled scintillators within the experimental channels of the SUR-100 facility, a zero-power thermal reactor operated by the IKE of the University of Stuttgart. The miniature detection system was developed at EPFL in collaboration with PSI.

Relative flux distributions were measured along the experimental central (I) and tangential (II) channels of SUR-100 and compared against Monte Carlo simulations performed with Serpent2. The comparison of experimental and computed neutron distributions showed a good agreement within the core region, with discrepancies within 2σ. An unexpected discrepancy, probably caused by a geometric inconsistency in the computational model of the reactor, was observed in the reflector region of the experimental channel I, where a 20% difference (i.e. 8σ) was found between experimental and simulated results. Significant discrepancies, respectively worth 10σ and 15σ, were noticed at distance, in the lead shielding region, for both experimental channels I and II.

The results for the experimental central channel I are compared to neutron activation measurements, showing a strong perturbation of the neutron flux shape (35% increase of thermal neutron flux in the core region of experimental channel I) due to the plastic-graphite holder used for the samples irradiation. The negligible flux perturbations induced by the miniature fiber-coupled detectors has been found to be a great advantage compared to the previously used manganese activation technique. In addition, reaction rate gradients across the 2.6 cm and 5.4 cm diameters of the channels are successfully measured.

An horizontal reaction rate gradient of (9.09 ± 0.20)% was measured within 2.4 cm across the diameter of the experimental channel II, with a difference from computed results of 2%. The absence of a vertical reaction rate gradient inside the experimental channel I was confirmed by measurements.

In conclusion, the experimental campaign carried out in the SUR-100 zero-power reactor highlighted the suitability of the miniature fiber-coupled scintillators for high resolution thermal neutron flux distribution measurements and for the characterization of highly localized gradients in the range of few cm.

REFERENCES

[1] F. Vitullo *et al.*, "A mm3 Fiber-Coupled Scintillator for In-Core Thermal Neutron Detection in CROCUS," *IEEE Trans. Nucl. Sci.*, vol. 67, no. 4, pp. 625–635, Apr. 2020, doi: 10.1109/TNS.2020.2977530.

[2] C. J. Werner *et al.*, "MCNP Version 6.2 Release Notes."

[3] J. Leppänen, M. Pusa, T. Viitanen, V. Valtavirta, and T. Kaltiaisenaho, "The Serpent Monte Carlo code: Status, development and applications in 2013," *Ann. Nucl. Energy*, vol. 82, pp. 142–150, Aug. 2015, doi: 10.1016/J.ANUCENE.2014.08.024.

[4] F. Vitullo *et al.*, "Developing and testing a miniature fiber-coupled scintillator for in-core neutron counting in CROCUS," *Adv. Nucl. Instrum. Meas. Methods Their Appl. (Animma 2019)*, vol. 225, p. 04018, Jan. 2020, doi: 10.1051/EPJCONF/202022504018.

[5] SCINTACOR, "SCINTACOR - Neutron Screens." [Online]. Available: www.scintacor.com. [Accessed: 29-Jul-2021].

[6] ESKA, "High-Performance Plastic Optical Fiber Specification Sheet SH-8001."

[7] H. Von Gunter and E. Schwartz, "Siemens-Unterrichts-Reaktor SUR Part 2: Reactor description," 1963.

[8] R. J. McConn, C. J. Gesh, R. T. Pagh, R. A. Rucker, and R. Williams III, "Compendium of Material Composition Data for Radiation Transport Modeling," Mar. 2011, doi: 10.2172/1023125.

Numerical simulations in support of the design of an ultrasonic device for sub-assembly identification

Kevin Paumel, Tom Maurel and Christian Lhuillier

***Abstract*—In this paper, it is shown how numerical simulations can help designing an ultrasonic instrument operating in harsh conditions. To prevent fuel handling errors in sodium cooled fast reactors, the identification of fuel sub-assemblies using ultrasound is being investigated. It is based on the interpretation of a code (aligned notches) engraved on the sub-assembly head using an emitting/receiving ultrasonic sensor. This reading is performed in liquid sodium with high temperature (up to 600°C) transducers. A first experiment in liquid sodium demonstrated the feasibility of this method. The reading quality and robustness depend on various parameters related to the ultrasonic beam (spectral response, focal distance, focal spot size), the code geometry (especially the notches' dimensions) and geometrical alignments. In order to avoid numerous experiments, two numerical models are developed. The first one is a finite element simulation of the sensor providing its radiated field. This model is validated with the well-known analytic solution of the Rayleigh integral; then it is applied to the sensor used in the sodium experiment. The focal distance and focal spot diameter are close to the expected values. The second simulation, using CIVA software, provides the ultrasonic scan of the code. The result is in good agreement with the sodium experiment and a first comparison with a water experiment shows that this numerical tool is relevant for easily taking into account misalignment and misorientation of the scan.**

***Index Terms*—Focusing transducer, harsh conditions, Finite Element Method, ray-tracing code.**

I. Introduction

In order to prevent handling errors in French sodium cooled fast reactors [1], methods identifying fuel sub-assemblies are currently considered. Liquid sodium being opaque, using ultrasound appears as a suitable solution. For decades, ultrasonic techniques have been mentionned for sub-assembly identification in liquid metal [2]–[5]. Nonetheless, as far as we know, the identication system developed for the MYRRHA reactor [6] is the only one which was described in details in the literature. This robust method uses a circular setup of twelve transducers installed on the robotic fuel manipulator, each aiming vertically at its associated notch (with varying depth) machined on the nozzle of the fuel assembly. In French sodium fast reactors, the limited available volume and dedicated sites for the transducers prompts to study in priority configurations with a single transducer aiming horizontally.

This method is based on reading, by mechanical scan, an engraved code (aligned notches) on the sub-assembly using an emitting/receiving ultrasonic sensor working in pulse-echo mode. This code can be compared to a bar binary code expressing the sub-assembly reference number. This reading is carried out by a sensor directly immersed in liquid sodium. It means that the sensor must withstand fuel handling conditions—about 200°C. However, as the sensor stays in the reactor during operation, it must withstand temperature up to 580°C. That is why a high temperature ultrasonic transducer (TUSHT® [7], [8]) is used.

In a first study [9], the feasibility of the method was demonstrated with a water experiment. This experiment also determined the resolution of the reading, i.e. the legible minimal dimensions of the notches of the code. Lately, an experiment in liquid sodium at 200°C [10] was carried out to demonstrate the feasibility of the method in conditions which are representative of fuel handling operations in the reactor.

In this paper, two numerical tools are studied in order to, in the future, replace numerous experiments by simulations. Indeed, the reading quality and robustness of this method depends on various parameters. Some are related to the sensor and its ultrasonic beam: spectral response, focal distance and focal spot size. Some rely on the code geometry, especially the notches' dimensions. Furthermore, the reading is sensitive to misalignment and misorientation between the ultrasonic beam and the sub-assembly head.

The first numerical tool is a model of the sensor and its beam using the finite element Method (FEM). Modeling ultrasonic transducers using the FEM has been carried out for more than two decades [11]. Unlike well-known analytic one-dimensional models [12], [13], the FEM is able to take into account the curved geometry of a focusing sensor. Even if some researchers have developed an extended KLM model [14] to efficiently compute the pressure field generated by a lens-focused axisymmetric transducer, this model neglects radial modes in the sensor, reflection angle of longitudinal waves and radial displacement at the curved interface [15]. Moreover, the FEM allows to model complex geometries of transducer. Finally, with the continuous growth of computing power, the FEM allows the calculation of the radiated field with reasonable computational cost. Besides the computation time, the FEM is criticized for its inaccuracy when simulating the propagation of acoustic waves over large distances [16]. A coupling with the Rayleigh integral [17] or using pseudo-spectral methods are often preferred. Nevertheless, for the short distance of interest of a focusing transducer and with a adequatly refined mesh, the accuracy may be sufficient [18],

Kevin Paumel, Tom Maurel and Christian Lhuillier are with the CEA, DEN, Nuclear Technology Department, F-13108 Saint-Paul-lez-Durance, France, e-mail: kevin.paumel@cea.fr

as it will be demonstrated in this paper.

The second model aims to assess the sensitivity of the reading method to the code geometry and its robustness to imperfect settings, such as misalignment and misorientation of the sub-assembly head. The ultrasonic scan is computed by an ultrasonic ray tracing software (CIVA [19]), using the pencil method to derive echoes with their amplitude [20], which allows computational cost much lower than the FEM.

Firstly, the reading principle of the sub-assembly identification in reactor is introduced. Then the experiment in sodium is reminded. After the parameters of the reading method are identified, and considering the heaviness of experiments, numerical simulations appear as useful tools. Thus, a FEM simulation is proposed to predict the spectral response and the radiated field characteristics of the focusing transducer. This simulation is firstly validated with the well-known analytic solution of the Rayleigh integral; then it is applied to the sensor used in the sodium experiment. Finally, a numerical simulation of the ultrasonic scan is presented. It is validated with a water experiment and then applied to the scan carried out in the sodium experiment.

II. Reading principle

Several approaches for reading a code using ultrasound can be considered. This paper focuses on a robust one, i.e. that with a "favorable" reading configuration (Fig. 1). The reading consists in an upward vertical translation of a code engraved on the sub-assembly head in front of a emitter/receiver focusing ultrasonic transducer. It is performed while the sub-assembly is lifted high enough by the handling gripper so that it is completely separate from the lattice and hangs freely (no more contact). This configuration minimizes off-centering and tilt with respect to the axis of the handling machine sheath. This full separation prevents friction and jamming when moving the sub-assembly's code in front of the ultrasonic sensor. It is assumed that pendular oscillations are insignificant. Lastly, the reading is performed at the top of the sub-assembly to take advantage of less "agitated" sodium than near the core outlet.

Similarly to a bar code, the code is made of an arrangement of notches. The presence or absence of notch means respectively the 0 or 1 binary value. The cross section of the notches is triangular. In the absence of notch, when the beam hits the generatrix of the cylindrical head, the reflected signal is fully directed towards the sensor which records a significant echo. When the beam hits a notch, the oblique face of the notch deflects the echo away from the sensor which therefore records no echo.

By scanning along the vertical axis, it is possible to obtain a one-dimensional C-scan of the amplitude along the generatrix. The interpretation of the code requires post-processing this C-scan (not discussed in this study).

A focusing sensor is used to obtain better lateral resolution than a flat sensor. The distance between the sensor's face and code must equal the sensor's focal distance so that the focal spot is located right on the code.

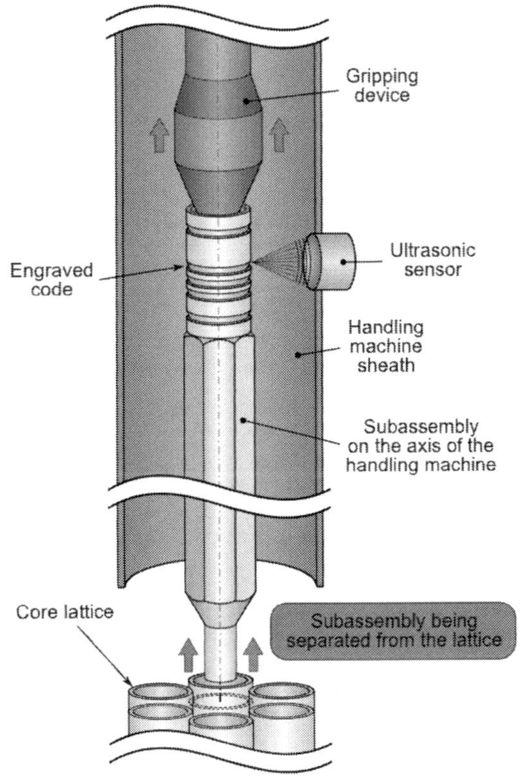

Fig. 1. Schematic diagram of a favorable reading configuration for which the sub-assembly is high enough to be completely free from the lattice.

III. Remind of the in-sodium experiment

The experiment objective was to demonstrate the feasibility of reading a code in conditions which are representative of handling operations in the reactor, i.e. in liquid sodium at 200°C.

A. *Experimental set-up*

This experiment involves scanning along the engraved code, i.e. down the generatrix of a mock-up cylindrical tube (or a cylinder arc for the mock-up in the sodium experiment) representing the sub-assembly head (Fig. 2). It is carried out under these favorable conditions:

- weak off-centering and tilt;
- straight translation of the code in relation to the ultrasonic beam;
- well-controlled displacement increment between each ultrasonic pulse emission.

The vertical movement is done by the sensor and not by the mock-up in order to simplify the process. A displacement increment of 0.5 mm is chosen.

This experiment was performed in a specific facility [10] in *Cadarache* (France) consisting of a 1,400 liters vessel of liquid sodium. Thanks to heating cables surrounding the vessel, the sodium temperature is regulated at 200°C.

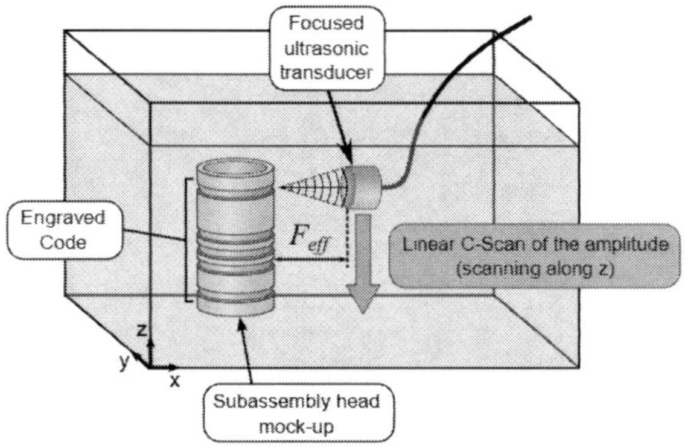

Fig. 2. Schematic diagram of the in-sodium experiment.

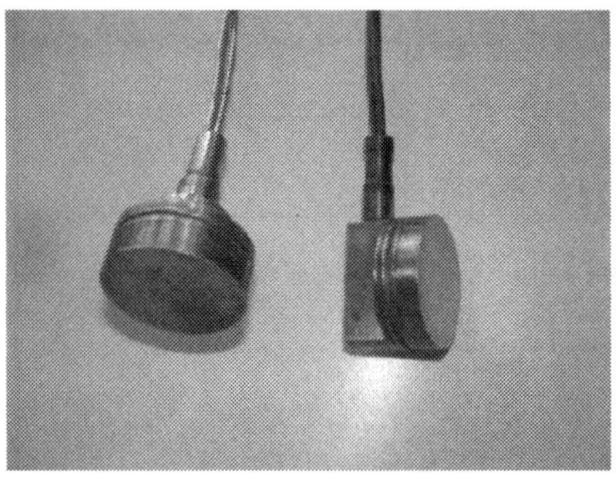

Fig. 3. High temperature ultrasonic transducer: TUSHT® ⌀40 mm with straight (left) and 90° (right) electrical wire outputs.

B. Ultrasonic transducer

1) TUSHT® ⌀40 mm: A high-temperature ultrasonic transducer, called in French: TUSHT® (Fig. 3), is used. This sensor is able to withstand the high temperature (up to 580°C) and high radiations of the reactor core outlet under nominal power operating conditions. Other transducers withstanding only 200°C and offering a better damping (so a better axial resolution) could have been used for the experiment. However, as opposed to TUSHT®, those could not be permanently installed in the core outlet region.

The TUSHT® uses high Curie temperature (about 1150°C) lithium niobate ($LiNbO_3$) crystal as piezoelectric material. Its resistance against neutronic damage is improved thanks to a special ^{7}Li enrichment (or ^{6}Li depletion). It can withstand a 10^{21} fast neutron fluence and a 10^6 Gy/h dose rate. The crystal is bound mechanically, acoustically and electrically to the stainless steel casing and electrode by using a hard soldering technique [7], [8], [21].

The TUSHT® is currently available in two piezoelectric crystal diameters: 15 and 40 mm. The 40 mm diameter transducer was selected because it has a higher sensitivity. During the experiment, the sensor is excited with an impulse of 250 V.

2) Focal distance: The lens of a focusing TUSHT® is at the same time an electrode, the interface layer between the crystal and sodium and an integral part of the housing. Among such focusing TUSHT®, only a 40 mm-diameter one with a curvature radius of $R_c = 148$ mm was available. Its focal distance is not compatible with the confined space of the device in the vessel. Consequently, a flat transducer with a separate lens, of lower curvature radius: $R_c = 90$ mm, is used. The interface between the planar face of the TUSHT® and the lens is a liquid sodium layer. A specific device was designed to press the lens against the sensor face to guarantee a uniform and as thin as possible sodium layer.

From the Rayleigh integral, O'Neil [17] proposed a simple theoretical expression for the acoustic pressure amplitude p_A along the propagation axis, for sinusoidally vibrating focusing transducers with uniform normal velocity distribution on their surfaces:

$$p_A(z) = \frac{\rho_m c_m u_n e^{-i\omega t}}{1 - z/F}(e^{ikz} - e^{ikR}). \tag{1}$$

ρ_m and c_m are the density and sound velocity of the propagation medium, u_n is the normal velocity amplitude on the radiator surface, ω is the angular frequency, $k = \omega/c$ is the wave number, z is the distance along the propagation axis, $R = F\sqrt{1 + (1 - z/F)^2 - 2(1 - z/F)cos(\alpha_{max})}$ is the distance between the observation point and the radiator edge, where α_{max} is the half-aperture angle of the radiator. In this purely theoretical configuration, F corresponds to the radius of the radiator's curvature R_c. For a plane-concave lens excited on its plane side, F corresponds to the geometrical focal distance and is given by the well-known formula:

$$F = \frac{R_c}{1 - c_m/c_l} \tag{2}$$

with c_l the longitudinal wave velocity in the lens.

More precisely, the effective focal distance F_{eff} depends on the near-field limit L_{nf}:

$$L_{nf} = \frac{ka^2}{2\pi} \tag{3}$$

where a is the radius of the excitating flat surface. Maréchal et al. [15] give an approximate relation for $F/L_{nf} < 2$:

$$F_{eff} = \frac{F}{1 + \frac{2}{3}(\frac{F}{L_{nf}})^{4/3}}. \tag{4}$$

To choose the best frequency range for the reading, a compromise between a high resolution (corresponding to a high frequency) and high sensitivity is necessary. As shown by the Fourier transform of the TUSHT pulse-echo response (Fig. 4), the 4.5 MHz frequency appears to be a good choice. This also applies to the sensor used for the experiment in water (section V). Hence the distance between the sensor and mock-up generatrix is set to $F_{eff} = 147$ mm.

3) Reading resolution: The reading resolution is given by the focal spot diameter of the ultrasonic beam. Assuming a non-attenuating fluid, the lateral profile in the geometric focal plane can be approximated by a Jinc function [17], [22], [23]:

$$\text{Jinc}\left(\frac{kax}{\sqrt{F^2 + x^2}}\right) = \frac{2J_1\left(\frac{kax}{\sqrt{F^2+x^2}}\right)}{\frac{kax}{\sqrt{F^2+x^2}}} \tag{5}$$

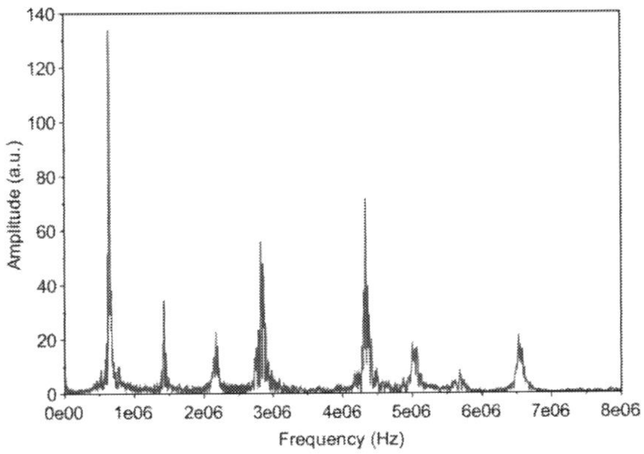

Fig. 4. Fourier transform of the echo reflected by the mock-up (in absence of notch).

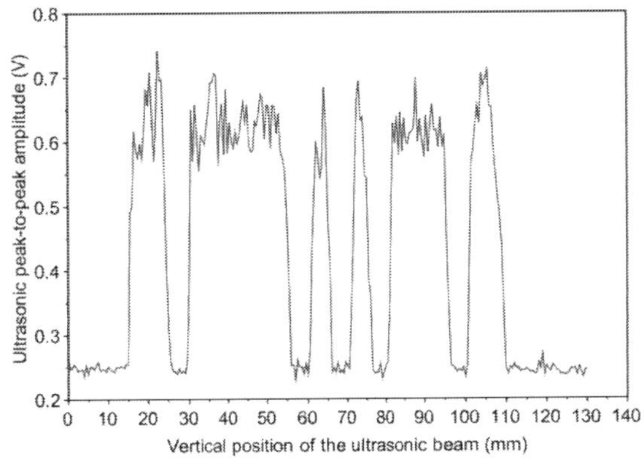

Fig. 6. One-dimensional C-scan of the sodium experiment.

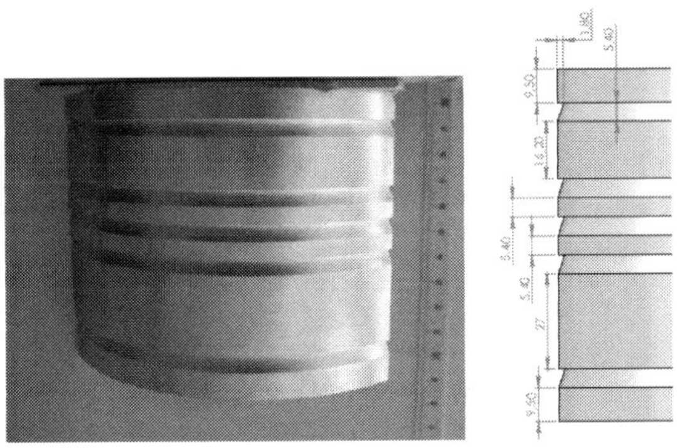

(a) Photograph of the mock-up on its supporting arm

(b) Mock-up definition drawing

Fig. 5. Sodium experiment mock-up.

with x the lateral coordinate (distance from the axis) of the beam. The focal spot diameter can be assimilated to the beam first lobe width. The coordinate $x = \sigma$ which represents the half-width of the first lobe, is given by the first zero of Bessel function J_1, which is: 3.8317. The diameter d of the focal spot is defined as follows:

$$d = 2\sigma = 2F\sqrt{\frac{1}{(\frac{ka}{3.8317})^2 - 1}} \tag{6}$$

Furthermore, using the approximation $\sin\theta \approx \tan\theta$, the widely used formula for the first lobe radius [17] is found:

$$\sigma = \frac{3.83F}{ka}. \tag{7}$$

Using (2) in (6) gives a focal spot diameter of 5.4 mm.

C. Mock-up sub-assembly head

The mock-up of the sub-assembly head (Fig. 5) only represents the cylindrical part at the top of the sub-assembly (only a cylinder arc in this sodium experiment due to limited volume). The notch's cross-section is a right triangle presenting an oblique surface to the ultrasonic beam. The notch is characterized by its width l and its depth p.

This code only has eight notch spots, not all occupied by a notch. All notches have the same width and the notch spots are equally spaced by a distance equal to the notch width. The code is 01100010. To ensure a good separation between the two binary states with respect to specular reflection of the signal received by the ultrasonic sensor, while limiting the code height, the notch's width l is set to the focal spot diameter $d = 2\sigma = 5.4$ mm.

The height of the code is the sum of all notch spots and all spacings between them. The number of bits required for the identification is expected to be greater, especially if additional criteria of redundancy or error detection are considered.

D. Result

The output data is a one-dimensional amplitude C-scan for a full vertical sweep along the mock-up (Fig. 6). More precisely, each point of this C-Scan corresponds to the peak-to-peak amplitude of the filtered A-Scan at a given position of the transducer. The filtering consists in a high-pass Butterworth's filtering (cut-off frequency at 4.5 MHz) in order to remove low frequencies which lead to a loss of resolution.

The code appears clearly on the graph. However, while the 0-bit (presence of a notch) is well delimited in amplitude, the 1-bit is subjected to noise. This is due to the interface between the sensor and the lens which induces an acoustic impedance discontinuity and so a weak signal-to-noise ratio. A much cleaner result is expected with a real focusing TUSHT® (i.e. with a lens directly incorporated to the sensor).

With a relevant choice of amplitude thresholds, a relatively simple algorithm could convert this C-Scan into a binary code expressing the sub-assembly type and number. Hence, the reading quality depends on the contrast between amplitude of "notch" and amplitude "of no-notch". This contrast is strongly affected by the misalignment and the misorientation between the ultrasonic beam and the sub-assembly head.

IV. Finite element modelling of the sensor and its radiated field

In order to verify the analytic assessments derived in sections III-B2 & III-B3 and to have a powerful tool available for designing sensors, a finite element model of the sensor and the propagation medium is implemented.

A. Model set-up

The model is based on the mechanical and electrical behaviors of the sensor's components. It computes the ultrasonic field given a harmonic electrical excitation applied at the ports of the piezoelectric cristal. It's a frequency domain study—simplifying the mathematical expression of the problem.

The model, implemented on COMSOL®, uses three constitutive relations :

- The stainless steel (AISI 304L) of the two electrodes and the lens is given a visco-elastic isotropic behavior.
- The Lithium Niobate of the piezoelectric cristal is given a coupled piezoelectric and elastic behavior (which is anisotropic and depends on the crystal's orientation).
- The liquid sodium of the propagation medium is given a non-viscous fluid-acoustic behavior.

Given the symmetry of the system, a 2D axisymetric geometry is chosen to reduce the computing cost.

The mesh is made of triangular elements with quadratic interpolation functions. The elements' size is chosen so that a wavelength is discretized by at least 8 elements in any material. Some convergence tests have shown that this criterion offers a satisfying compromise between computing cost and accuracy. As a consequence, the element's size is determined by the medium in which the celerity of transverse or longitudinal waves is minimal. Here, the celerity of longitudinal waves in sodium is the lowest (2472 $m.s^{-1}$ at 200°C), even lower than the transverse waves in stainless steel (greater than 3000 $m.s^{-1}$). As an example, the mesh consists of five million elements for a sodium domain 55 mm in diameter and about 200 mm long at a 4.5 MHz frequency. The associated maximum element size is around 70 μm.

The interfaces between materials are given a continuous displacement boundary. Perfectly Matched Layers (PML) are used to reproduce an infinite volume by preventing reflections at the edge of the propagation domain. Every other external boundary is considered free. A voltage amplitude of 200 V is applied at the back of the piezoelectric crystal while the other port is grounded.

For any given frequency and any geometry of the sensor-lens system, the simulation computes the associated radiated field. The focal distance and shape of the focal spot can then be derived.

B. The Rayleigh's configuration

To validate the model for wave propagation in sodium, a comparison to the analytic solutions—equations (1) and (5)—has been carried out. Its principle is to compute the acoustic pressure field radiated by the curved interface of the lens for a uniform normal velocity. The FEM simulation gives the same pressure field as the analytical model (Fig. 7).

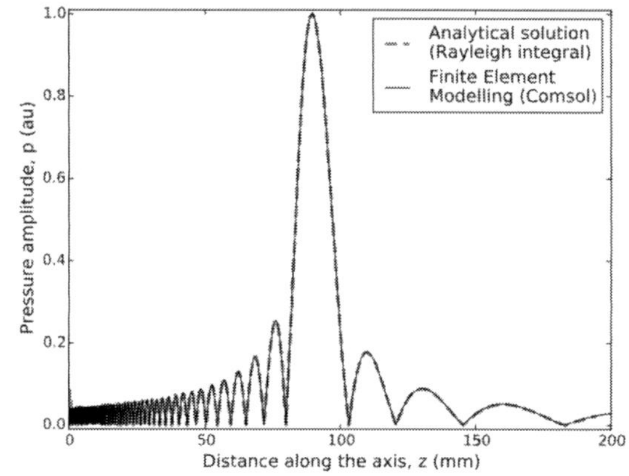

(a) Along the propagation axis Z

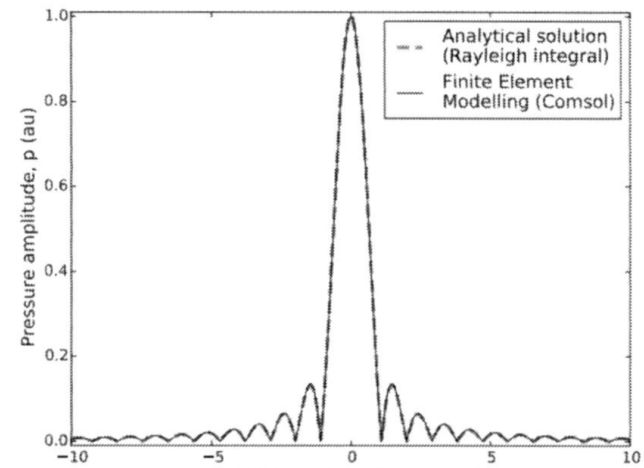

(b) Along the radial axis on the focal spot

Fig. 7. Evaluation of the pressure field's amplitude for the Rayleigh's configuration (2D field and evaluation axes on figure 9).

This validates the hypotheses of the FEM model, mesh and boundary conditions. It also confirms that the error accumulated—inherent to finite elements methods—along distances relevant for this study (typically focal length) remains negligible.

C. The whole TUSHT® configuration

The model is then applied to the whole TUSHT®. The resulting pressure field along the propagation axis shows that the maximum of pressure amplitude beyond the near field is about 146 mm from the sensor (Fig. 8). This focal distance is close to the approximate analytical value 147 mm, given by the equation (4) in section III-B2. Besides, the width of the first lobe observed on the radial axis at this focal distance is in good agreement with the focal spot diameter of 5.4 mm calculated by the equation (6) in section III-B3.

Figure 9 shows a 2D view of the pressure field in both cases: the Rayleigh's configuration and the whole TUSHT®. This representation offers some information on the spatial distribution of acoustic energy.

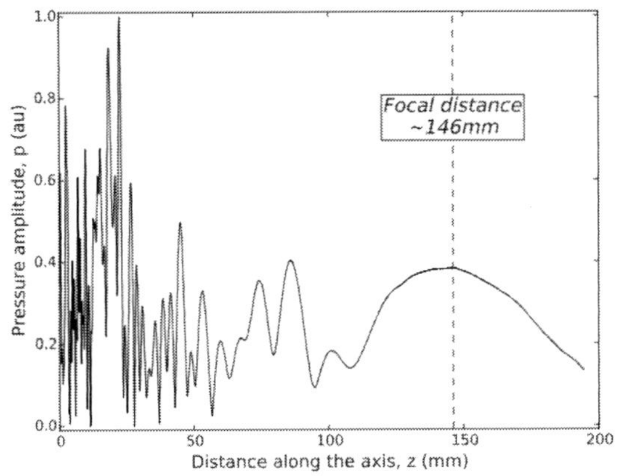

(a) Along the propagation axis Z

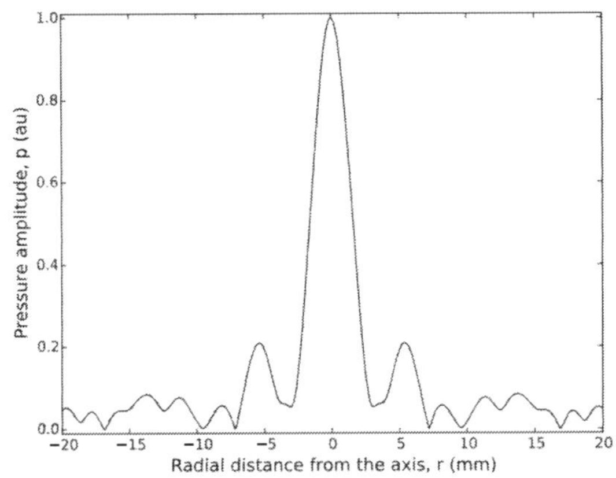

(b) Along the radial axis on the focal spot

Fig. 8. Evaluation of the pressure field's amplitude for the whole TUSHT® configuration (2D field and evaluation axes on figure 9).

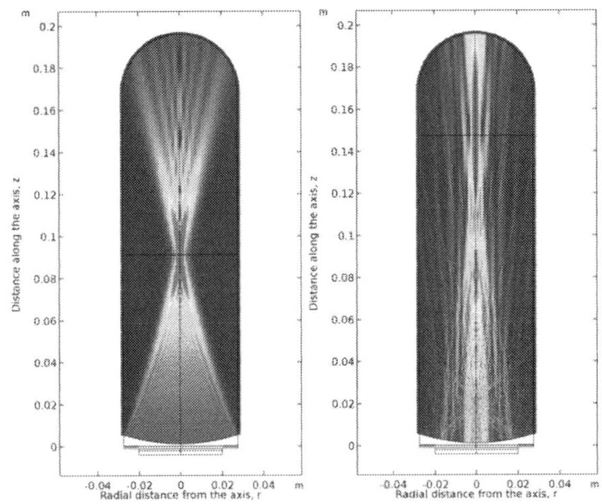

(a) Rayleigh's configuration (b) Whole TUSHT® configuration

Fig. 9. 2D view of the radiated field of the sensor at 4.5 MHz with depiction of the evaluation axis (z-axis and focal spot r-axis).

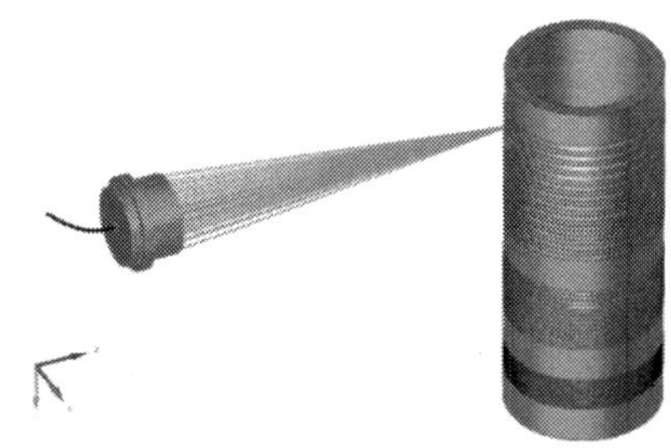

Fig. 10. 3D view of the CIVA simulation.

D. Conclusion on FEM model

These results show the relevance and accuracy of the FEM simulation to evaluate the radiated field of a non-standard focusing transducer. In this first harmonical approach, the agreement between analytic formulae and FEM simulation confirms the consistency of both tools for designing the sensor. The FEM simulation will bring its benefit in particular for complex geometries of the sensor's components, for example non-axisymmetric lenses. The direct 2D representation already provides useful information about the energetical topology of the beam.

The next step is to upgrade the model to take into account the multifrequency feature of the excitation and to obtain a pulse response of the transducer.

V. Numerical simulation of the scan with CIVA

Numerical simulation allows to modify reading parameters of the code (focal distance of the sensor, geometry and size of the code, size of the assembly head, etc.) very easily while obtaining results very quickly and at low cost. Moreover, a numerical model simplifies the implementation misalignment and misorientation to determine the maximal admissible errors of the reading method. The reading configuration is modeled with CIVA software (Fig. 10).

First, the model needs to be validated. Simulation results are compared to a previous water experiment [9] in which a real focusing TUSHT (with $R_c = 148$ mm) was used. In this experiment, the mock-up consists of three codes of twelve notches with various notch geometries. All the spots are taken by a notch. The notches have the following dimensions:

- Width l (same as spacing between notches) with the following values: $0.2d = 0.8$ mm, $0.4d = 1.6$ mm and $0.8d = 3.2$ mm, with d given by (6) using (2). l varies from one code to the next.
- Depth p so that the ratio l/p corresponds to the following values: 2, 3 and 4 (but its influence was found to be insignificant). p varies in a same code.

An experimental signal is implemented as input data in the CIVA model. It should be noted that, for future studies, the objective is to implement a computed signal derived from the FEM model to avoid the need of experimental recording.

Figure 11 shows the one-dimensional C-Scan. A good agreement is observed between experimental data and numeri-

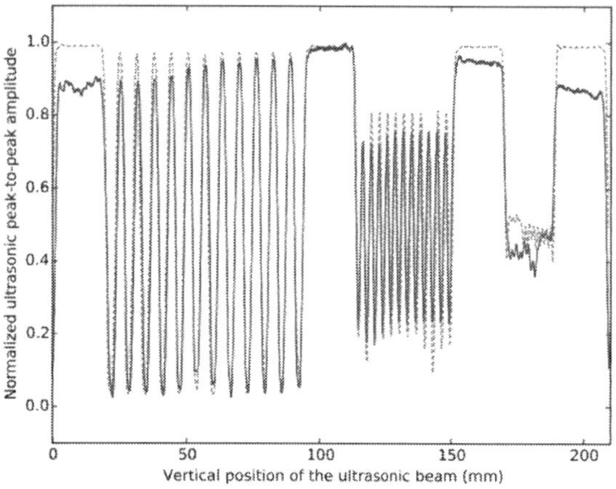

Fig. 11. One-dimensional C-scan of the amplitude of the water experiment. Red dotted: CIVA simulation, blue: experiment.

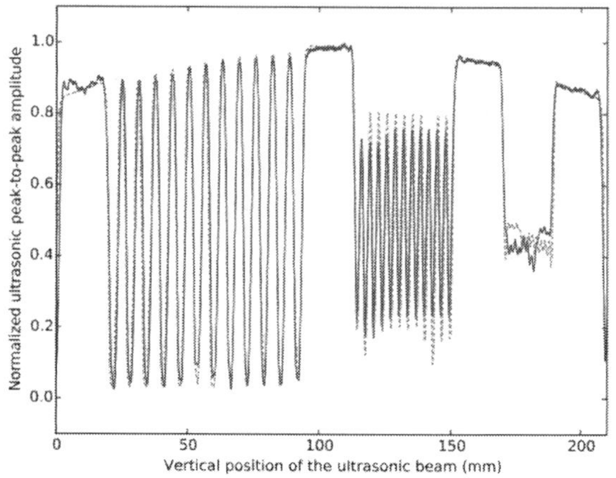

Fig. 12. One-dimensional C-scan of the amplitude of the water experiment. Red dotted: CIVA simulation with 0.7 degree rotation of the mock-up around the beam axis at middle position of the sweep, blue: experiment.

cal data from CIVA simulation with full incident beam setting. This result is better than the former one [10] obtained with the plane wave approximation setting. The noise of the signal observed in the sodium experiment has been greatly reduced by using a real focusing TUSHT® in the water experiment. Moreover, the reading exhibits a very good quality, for the largest notch width at least, although the width is smaller than the focal spot.

A deviation for the global envelope is observed though in the form of a slight curvature of the code. This deviation between simulation and experiment could be caused by slight misalignment and misorientation in the experiment. It is indeed very difficult to accurately set the orientation of the sensor with respect to the mock-up generatrix for every position of the sweep. In an updated computation with CIVA, a 0.7 degree rotation of the mock-up around the beam axis at middle position of the sweep is introduced. With that correction, the numerical result is even closer to the experiment (Fig. 12).

Hence, the CIVA model can reasonably be considered as validated. It is then used to simulate the sodium experiment (Fig. 13). Again, for this experiment, a satisfying agreement is obtained between the model and the experiment. The low signal to noise ratio in the experiment may explain the over-estimation of the contrast—between the presence and absence of notch—by the simulation.

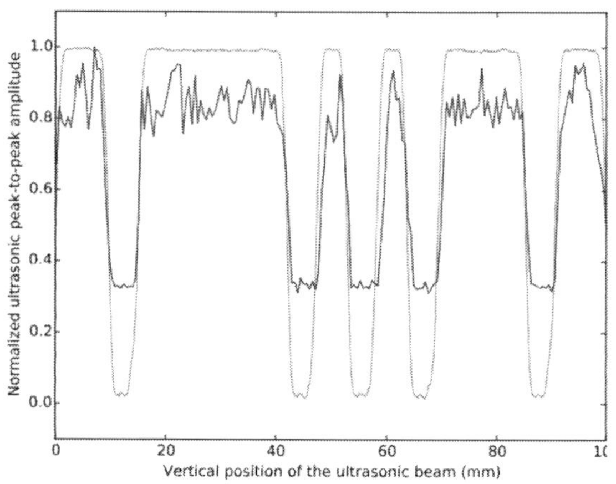

Fig. 13. One-dimensional C-scan of the amplitude of the sodium experiment. Orange: CIVA simulation, blue: experiment.

VI. Conclusion

This study showed the relevance of numerical simulation in support of the design process of an ultrasonic instrument. It allows to save experiments in harsh conditions.

A method for sub-assembly identification in liquid sodium using ultrasound has been proposed. It is based on the reading of a code engraved on the sub-assembly head by an emitting/receiving ultrasonic sensor. This code, whose height doesn't exceed two hundreds millimeters, comprises aligned notches (of a few millimeters of width) deflecting the ultrasonic beam. This reading is performed in sodium with a high temperature transducer, the TUSHT®. The resulting scan can be compared to a binary code expressing the sub-assembly reference number.

A representative experiment in liquid sodium at 200°C demonstrated the feasibility of this method for a notch width equaling the focal spot diameter of the focusing sensor.

In order to avoid numerous experiments, two numerical tools have been studied. The first is a finite element model of the sensor and its radiated field. Concerning wave propagation in sodium, the model has been validated by comparison with well known analytic expressions. The model has then been applied to the TUSHT®. The focal distance and spot diameter obtained are in agreement with the analytic formulae. The second numerical simulation uses CIVA software. It provides the ultrasonic scan of the code. An in-water experiment validated the model and proved its efficiency to take into account misorientation of the ultrasonic beam with respect to the code (misalignment may be treated as well). Concerning the in-sodium experiment, the simulation also gives results with a satisfying agreement despite the low signal to noise ratio.

In a near future, the FEM model might be upgraded to take into account the multi-frequency feature of the excitation. The resulting computed signal could then be used as an input data in the CIVA simulation.

REFERENCES

[1] F. Dechelette, F. Morin, G. Laffont, G. Rodriguez, E. Sanseigne, S. Christin, X. Mognot, and A. Morcillo, "Study and Evaluation of Innovative Fuel Handling Systems for Sodium-Cooled Fast Reactors: Fuel Handling Route Optimization," *Science and Technology of Nuclear Installations*, 2014.

[2] J. Spanner, "Preliminary development of in-service inspection methods for LMFBRs," in *Specialists Meeting on "Inservice Inspection and Monitoring of LMFBR's"*, Bensberg, Germany, Mar. 1976.

[3] C. C. Scott, P. R. Huebotter, and R. C. Callen, "Potential applications for an under-sodium ultrasonic scanning device," Tech. Rep. Atomic Power Development Associates Report APDA-205, Jul. 1967.

[4] R. N. Ord and R. W. Smith, "Development of Under- Sodium Ultrasonic Scanner for in- Reactor Surveillance," Hanford Engineering Development Laboratory, Richland, Washington, Tech. Rep. Technical Report: HEDL-TME 72-91, 1972.

[5] C. K. Day and R. W. Smith, "Under-Sodium Viewing," in *IEEE Ultrasonics Symposium*, Monterey, California, USA, Nov. 1973, pp. 191–194.

[6] D. V. Dyck and M. Dierckx, "An Ultrasonic Fuel Identification System for Liquid Metal Cooled Reactors Resilient Against Multiple Transducer Failures," *IEEE Transactions on Nuclear Science*, vol. 61, no. 4, pp. 2291–2299, Aug. 2014.

[7] C. Lhuillier, B. Marchand, J. M. Augem, J. Sibilo, and J. F. Saillant, "Generation IV nuclear reactors: Under sodium ultrasonic transducers for Inspection and Surveillance," in *2013 3rd International Conference on Advancements in Nuclear Instrumentation, Measurement Methods and their Applications (ANIMMA)*, Jun. 2013, pp. 1–6.

[8] C. Lhuillier, "Method for Manufacturing a High-Temperature Ultrasonic Transducer Using a Lithium Niobate Crystal Brazed with Gold and Indium," France Patent US20 140 215 784 A1, Aug., 2014.

[9] K. Paumel and C. Lhuillier, "Identifying subassemblies by ultrasound to prevent fuel handling error in sodium fast reactors: First test performed in water," in *2015 4th International Conference on Advancements in Nuclear Instrumentation Measurement Methods and their Applications (ANIMMA)*, Apr. 2015, pp. 1–7.

[10] K. Paumel, "Subassembly identification by ultrasound in sodium cooled fast reactors," in *2018 IEEE International Instrumentation and Measurement Technology Conference (I2MTC)*, May 2018, pp. 1–6.

[11] N. N. Abboud, G. L. Wojcik, D. K. Vaughan, J. Mould, D. J. Powell, and L. Nikodym, "Finite element modeling for ultrasonic transducers," in *Ultrasonic Transducer Engineering: Medical Imaging 1998*, K. K. Shung, Ed. Bellingham: Spie-Int Soc Optical Engineering, 1998, vol. 3341, pp. 19–42, wOS:000074538600002.

[12] M. Redwood, "Transient Performance of a Piezoelectric Transducer," *The Journal of the Acoustical Society of America*, vol. 33, no. 4, pp. 527–536, Apr. 1961. [Online]. Available: https://asa.scitation.org/doi/10.1121/1.1908709

[13] R. Krimholtz, D. A. Leedom, and G. L. Matthaei, "New equivalent circuits for elementary piezoelectric transducers," *Electronics Letters*, vol. 6, no. 13, pp. 398–399, Jun. 1970.

[14] P. Maréchal, F. Levassort, L.-P. Tran-Huu-Hue, and M. Lethiecq, "Lens-focused transducer modeling using an extended KLM model," *Ultrasonics*, vol. 46, no. 2, pp. 155–167, May 2007. [Online]. Available: http://www.sciencedirect.com/science/article/pii/S0041624X07000194

[15] ——, "Effect of Radial Displacement of Lens on Response of Focused Ultrasonic Transducer," *Japanese Journal of Applied Physics*, vol. 46, no. 5R, p. 3077, May 2007.

[16] E. Filoux, S. Calle, R. Lou-Moeller, M. Lethiecq, and F. Levassort, "3-D numerical modeling for axisymmetrical piezoelectric structures: application to high-frequency ultrasonic transducers," *IEEE Transactions on Ultrasonics, Ferroelectrics, and Frequency Control*, vol. 57, no. 5, pp. 1188–1199, May 2010.

[17] H. T. O'Neil, "Theory of Focusing Radiators," *The Journal of the Acoustical Society of America*, vol. 21, no. 5, pp. 516–526, Sep. 1949.

[18] P. N. Bilgunde and L. J. Bond, "A 2d finite element simulation of liquid coupled ultrasonic NDT system," *AIP Conference Proceedings*, vol. 1650, no. 1, pp. 1543–1552, Mar. 2015. [Online]. Available: https://aip.scitation.org/doi/abs/10.1063/1.4914773

[19] S. Mahaut, S. Chatillon, M. Darmon, N. Leymarie, R. Raillon, and P. Calmon, "An Overview of Ultrasonic Beam Propagation and Flaw Scattering Models in the Civa Software," in *Review of Progress in Quantitative Nondestructive Evaluation, Vols 29a and 29b*, D. O. Thompson and D. E. Chimenti, Eds., vol. 1211. Melville: Amer Inst Physics, 2010, pp. 2133–2140.

[20] N. Gengembre and A. Lhémery, "Pencil method in elastodynamics: application to ultrasonic field computation," *Ultrasonics*, vol. 38, no. 1, pp. 495–499, Mar. 2000.

[21] M. Giot, L. Vermeeren, A. Lyoussi, C. Reynard-Carette, C. Lhuillier, P. Mégret, F. Deconinck, and B. S. Gonçalves, "Nuclear instrumentation and measurement: a review based on the ANIMMA conferences," *EPJ Nuclear Sciences & Technologies*, vol. 3, p. 33, 2017.

[22] B. G. Lucas and T. G. Muir, "The field of a focusing source," *The Journal of the Acoustical Society of America*, vol. 72, no. 4, pp. 1289–1296, Oct. 1982.

[23] A. Goldstein, "Steady state spherically focused, circular aperture beam patterns," *Ultrasound in Medicine & Biology*, vol. 32, no. 10, pp. 1441–1458, Oct. 2006.

Gas detection in sodium cooled fast reactors: determination of a transfer function

C. Ding[1], P. Filliatre[1] and L. Desgranges[2]
[1]CEA, DES, IRESNE, DER, SPESI, LDCI, Cadarache, F-13108 St-Paul-Lez-Durance, France
[2]CEA, DES, IRESNE, DEC, Cadarache, F-13108 St-Paul-Lez-Durance, France
congjin.ding@cea.fr

***Abstract*—To keep the dependability of Sodium Cooled Fast Reactor, the "clean sodium concept" is demanded, which means that the sodium is free from contamination. The release of fission products is searched for by a contamination measuring system. We need to have a comprehensive description of cladding failures and the detection of contamination, including the failure occurrence on the fuel pin, the transfer process through the sodium and cover gas, the measurement efficiency, etc. We aim to identify the important parameters of physical phenomena, with modelling and simulations based on the return of experiments from past reactors such as PHENIX. There have been a total of 15 open pin failures in PHENIX reactor. Through studying these detected signals, we can get a better physical explanation and description of the evolution of failures. The detection system is related to different stages of the evolution of fuel pin, with different types of fission products, various release mechanism and physical properties. During the evolution of the failed fuel pin, gaseous fission products is released on the first stage of failure and the gas detection system is aimed for the gaseous fission products monitoring. We proposed a quantitative modeling of transfer function to describe the time broadening of the gas release from the fuel pin to the detector. The result matches well with PHENIX experiment data, with the same order of magnitude of the time broadening and the same shape of exponential decreasing. A determination of the transfer function of this gas detection system is validated in this paper.**

***Keywords*—** Sodium fast reactors, PHENIX reactor, cladding failure detection, transfer function

I. Introduction

Fast reactors have the unique capability of improving the usage of natural resources and reducing of volume and heat load of high-level waste thanks to the closed fuel cycle [1]. Among the six types of fast reactor systems, the Sodium cooled Fast Reactor (SFR) has many experience from operating experimental, prototype and commercial size reactors [2].

The "clean sodium concept" is demanded in SFR to keep the dependability of the reactor. A contamination measuring system was designed in order to continuously search for the release of fission products by a fuel pin failure [3]. A fuel pin failure [4] represents a crack or a rupture on the cladding which is the first confinement barrier of the fuel inside the reactor core [5]. Thanks to the 50 years experience of designing, building and operating SFR in France, the fuel pin failure was studied in many fast experimental reactors. For example, during the operational period of the PHENIX reactor (1973-2009), there were a total of 42 gas pin failures, 15 open-pin failures [6]. In the detection system of the reactor, gas detection system (DRG system) system is dedicated to detect the failures with gaseous fission products. In this paper, we aim at a more physical interpretation of these data that could be helpful for the design of future designs of SFR.

The transfer process of the detection system is described as a transfer function which convolutes the entry signal to form the observed signal. The main effect of the transfer function is to introduce a time broadening between the fission products released at the pin failure and the detected signal in detectors, which needs to be assessed.

We observe some gas signal in the experimant cases is expected to be nearly instantaneous release according to the gas signal time evolution. It is foreseen that these observed signals are mostly given by the transfer function. We can justfy the assumption of a Dirac peak for this kind of burst and then the the experiment data can be regarded as an experimental evaluation of the transfer function.

The goal of the paper is to set up, using the specification of the PHENIX reactor, a model for the transfer function in order to predict the time broadening of the observed signal. Then we compare the experiment detection to assess how quantitatively predictive the model is. Although the paper will make use of PHENIX data, the model is a general formulation which can be applied to other reactors and to guide the conception choice for the future design.

The paper is organized as follows: the fuel pin failure evolution and the detection system are introduced in Sec. II; the determination of the transfer function is described in Sec. III; the conclusion is summarised in Sec. IV.

II. Fuel pin failure and the detection system

In this section, the background of the research is introduced including the evolution of the fuel pin failure and the detection system for gaseous fission products in PHENIX reactor.

A. Evolution of the fuel pin failure

There are four stages of the fuel pin failure evolution [7], as shown below:

- Normal: This stage is defect-free. The sodium coolant is clean.
- Crack: This stage is only with gaseous fission products. The detection system can only detect gas emissions. During the burn up, most of the gaseous fission products,

about 80% to 90 % [8], will be released from the fuel pin and be accumulated in fission gas plenum, which is the gas expansion volume inside the fuel pin. When there is a crack on the cladding, the gaseous fission products in the gas plenum will be released outside the cladding through the crack instantaneously because of the pressure difference between the fuel and the coolant.

- Narrow rupture: A narrow rupture may be subsequent to a crack in case it undergoes several thermal expansions due to various power variations. The detection system can detect neutron signals emitted by delayed neutron precursors. The oxide-sodium reaction (OSR) happens [9].
- Wide rupture: The oxygen atoms feeding the oxide-sodium reaction mainly come from the sodium coolant, which continuously flows along the fuel oxide at the rupture location. There are particles of fissile material released into primary circuit. An emergency shut down is immediately ordered.

In summary, gas signal can be detected on the crack stage. If the failure turns to more severe situation, the neutron signal can be detected.

B. Detection system in PHENIX

The detection system in PHENIX reactor is related to the fuel pin failure evolution, which consists of mainly two parts:

- DRG system (Détection des Ruptures de Gaine) (Gas detection system): DRG system is located in the cover gas. Because gas signal is the first signal can be detected when there is a cladding failure, the aim of DRG system is to know the failure as soon as possible.
- DND system (Détection des Neutrons Différés) (Neutron detection system): DND system is located in the sodium coolant. Because DND signal is concerning the breached fuel area exposed to the coolant, the DND signal is monitored under a threshold.

In addition, there is a LRG (Localisation des Ruptures de Gaine) system for locating the failed pin for both DRG and DND system. This paper will focus on DRG system study. The scheme of DRG-LRG system of PHENIX reactor is shown in Fig. 1. The main detectors for gaseous fission products are shown below:

- MX 32 high activity ionization chambers: they are assigned to the DRG cover gas to supply continuous overall value of the activity in the accident situation.
- Gamma chromatography: it is identification detection and the measurement is periodical (1 time a day) and on demand.

Because MX 32 is the continuous measurement, only the gas signal detected in MX 32 can be applied to transfer function modeling.

III. Determination of the Transfer Function

The transfer process of the detection system is described as a transfer function which convolutes the entry signal to form the observed signal. The main effect of the transfer function is to introduce a time broadening between the fission products released at the pin failure and the detected signal in detectors, which needs to be assessed. To describe the transfer process from the cladding failure to the detector, a scheme of fission gas release is built and a transfer function model is proposed in this section. An experiment case RG 14 in PHENIX reactor with instantaneous release is chosen as an experimental evaluation of the transfer function.

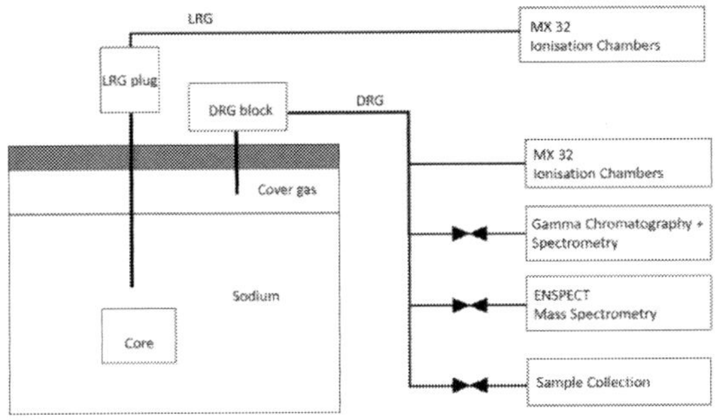

Fig. 1. General scheme of DRG-LRG system in PHENIX reactor.

A. Scheme of fission gas transfer and the model of transfer function

To describe the fission gas release (FGR) transport process from the fuel pin to the detection system, a simplified scheme is built by taking into account fission gas dilution and time broadening.

When there is a crack on the cladding, the fission gas will be released into the sodium coolant circuit rapidly because of the pressure difference. There will be two forms of gas, either in solution in the sodium or in bubbles which eventually go into the cover gas. The solubility of Krypton and Xenon in liquid sodium at 700 K and 0.1 MPa is calculated by Eq. 1. According to the Henry's law, the amount of dissolved gas in a liquid is proportional to its partial pressure above the liquid. The amount of dissolved Xenon and Krypton in sodium depends on the partial pressure of Xenon and Krypton above sodium.

$$x(g)P = x(l)K \quad (1)$$

In Eq. 1, where $x(g)$ is the number of moles of Xenon and Krypton per number of moles of Argon in the cover gas, P is the absolute pressure above the sodium, $x(l)$ is the number of moles of Xenon and Krypton per number of moles of Sodium, K is Henry constant for Xenon and Krypton, which is derived from the data in [10], which is 4.8×10^{11} and 1.4×10^{10} respectively.

The calculated fraction of Xenon and Krypton dissolved in sodium is 1.0×10^{-8} and 4.0×10^{-7} respectively, which are both very low. We can conclude that most of the fission gas will be released into the cover gas. To simplify the transport process, for the predictive calculation we can make the hypothesis that all of the FGR goes into the cover gas. Finally the fission gas in cover gas will be sampled into the DRG gas block and be transported to the detection chamber. The activity will be measured by the detectors.

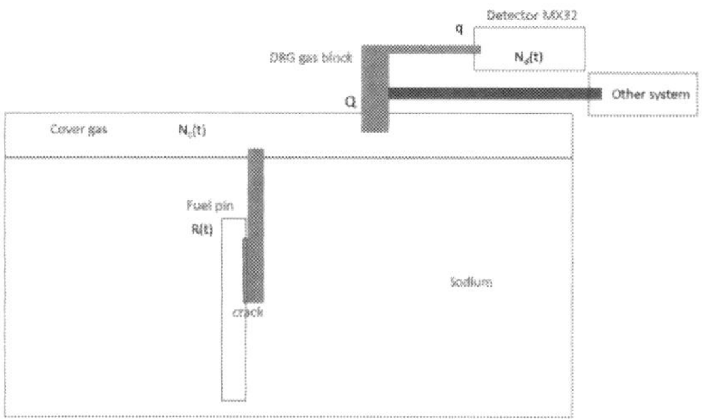

Fig. 2. The scheme of fission gas release from the fuel pin to DRG detection system.

The scheme of FGR from the fuel pin to DRG detection system is shown in Fig. 2. $R(t)$ is the FGR from the fuel pin, in the unit of number of atoms per second. The FGR will go through the crack, which is present in orange in Fig. 2, to the sodium coolant, then will be transported into the cover gas. With the simplification mentioned above, this process is represented by the blue tube leading to the cover gas directly. Then the FGR will dilute in the cover gas. N_c is the number of atoms released in cover gas. The total flow rate in the cover gas is Q. The sipping system of DRG block samples the FGR from the cover gas, thus the FGR will then flow into the detection chamber, represented by the blue tube leading to the detector MX 32, at the flow rate q.

Based on the scheme of FGR from the fuel pin to DRG detection system, we derive the conservation equation of the number of atom of FGR in the cover gas at moment t using the framework of MERARG facility [11], which can be expressed in:

$$\frac{\mathrm{d}N_c}{\mathrm{d}t}(t) = R(t) - \frac{Q}{V_c}N_c(t) \tag{2}$$

The same balance in the detector can be expressed in:

$$\frac{\mathrm{d}N_d}{\mathrm{d}t}(t) = \frac{q}{V_c}N_c(t) - \frac{q}{V_d}N_d(t) \tag{3}$$

In Eq. 2 and Eq. 3, R represents the FGR from the fuel pin; Q is the total gas flow rate in the cover gas; q is the gas flow rate to the detector; N_c is the number of atoms released in the cover gas; N_d is the number of atom in the detector; V_c is the volume of cover gas; V_d is the volume of detector chamber.

We can define $q = \alpha\, Q$ $(0 < \alpha < 1)$ and apply the Laplace transform to Eq. 2 and Eq. 3, we can express R as a function of N_d:

$$R(s) = (\frac{V_c}{\alpha Q}s^2 + [\frac{V_c}{V_d} + \frac{1}{\alpha}]s + \frac{Q}{V_d})N_d(s) \tag{4}$$

Applying the inverse Laplace transform:

$$R(t) = \frac{V_c}{\alpha Q}\frac{\mathrm{d}^2 N_d}{\mathrm{d}t^2}(t) + (\frac{V_c}{V_d} + \frac{1}{\alpha})\frac{\mathrm{d}N_d}{\mathrm{d}t}(t) + \frac{Q}{V_d}N_d(t) \tag{5}$$

We define the transfer function $F(s)$ to express the relation between R and N_d:

$$N_d(s) = F(s)R(s) \tag{6}$$

$$F(s) = \frac{1}{As^2 + Bs + C} = \frac{1}{A[(s+a)^2 - b^2]} \tag{7}$$

With A = $\frac{V_c}{\alpha Q}$, B = $\frac{V_c}{V_d} + \frac{1}{\alpha}$, C = $\frac{Q}{V_d}$, a = $\frac{B}{2A}$, b = $\sqrt{(\frac{B}{2A})^2 - \frac{C}{A}}$.

Then we can induce the transfer funtion formula:

$$f(t) = \frac{1}{Ab} \times e^{-at} \times \sinh(bt) \tag{8}$$

We can also simplify Eq. 8 and obtain:

$$f(t) = \frac{\alpha V_d}{\alpha V_c - V_d} \times (e^{-\frac{Q}{V_c}t} - e^{-\frac{\alpha Q}{V_d}t}) \tag{9}$$

The transfer function Eq. 9 is a general formulation of the problem, which is derived from the physical descriptions, can be applied for different detection setups.

For the cases with instantaneous release, we can justify The FGR from the fuel pin is a Dirac peak, then we can obtain:

$$N_d(t) = f(t) \tag{10}$$

Then we can apply the transfer function to a specific case to compare with the experiment detection to assess how quantitatively predicted the model is.

B. The experiment case RG 14

The RG 14 fuel pin failure occurred on the experimental assembly DCI 3124, SPHINX 5 in 1988. The first gas emission was detected on February 26 at 13h00. The first major release of fission gas corresponds to the instantaneous leakage of a cladding. Then the internal pressure suddenly drops and the cover gas activity rapidly rises to a maximum of 1.5 Ci/m^3 due to the arrival of fission gases.

The signal of the first period fission gas release detected by MX 32 is shown in Fig. 3. We derived the data according to the original curve and the statement in experiment report. We can observe that the gas signal is expected to be an instantaneous release with time broadening during the first release and then has some time evolution. In order to assess the transfer function, only the first period gas release is studied in this paper.

The post irradiation examination photo for the crack in RG 14 is shown in Fig. 4, taken from the PHENIX internal

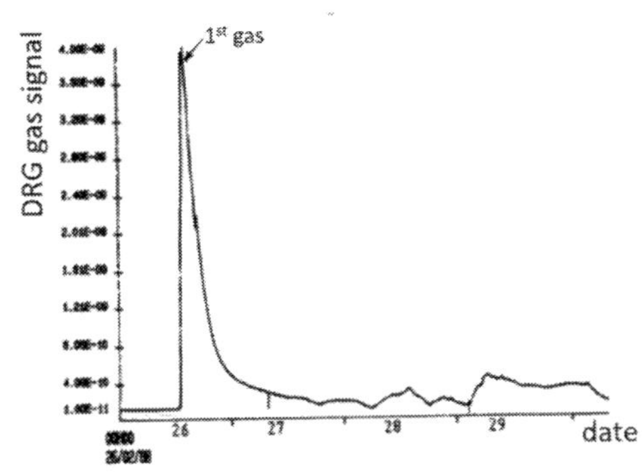

Fig. 3. The signal of the first period fission gas release detected by MX 32.

report, which is a long and straight breach, with length around 480 mm and with relatively large width around 0.5 mm. The pin failure happened in the end of life with high burn up of 11.56 %, probably due to embrittlement of cladding by the radiation damage. The crack was brittle and large which means the crack occurred rapidly.

For fast reactor, because of the high temperature, steep temperature gradient and the high burn up, the fraction of fission gas release (FGR) in the fission gas plenum is about 80 % to 90 % [8]. In other words, during the burn up without any failure, most of the gaseous fission products will be released from the fuel pin and be accumulated in fission gas plenum, which is the gas expansion volume inside the cladding. When there is a failure on the cladding, the gaseous fission products in the gas plenum will be released outside the cladding through the crack instantaneously because of the pressure difference between the fuel and the coolant. Thus the crack observation is consistent with the instantaneous gas signal detection.

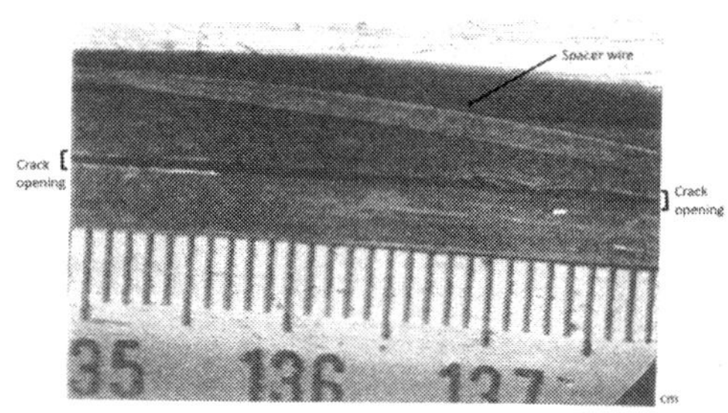

Fig. 4. Post irradiation examination photo (from PHENIX internal report, 24/11/1995).

We can thus justify the assumption of a Dirac peak for the burst release in the case of RG 14. As a consequence, the observed signal of RG 14 can be regarded as an experimental evaluation of the transfer function. Then we can apply the transfer function derived in Sec. III-A to RG 14 case and compare the modeling results with RG 14 experiment data.

C. *Experiment evaluation of transfer function*

In order to do the assessment of the transfer function, the transfer function modeling results and the comparison with the experiment data in RG 14 are shown in this section.

We applied the transfer function to our specific case RG 14, with the total flow rate in the cover gas Q (5.0 l/s), the flow rate to MX 32 detector q (0.2 l/s),as well as volume data V_c (40 m^3) of the cover gas and V_d (0.09 l) in the detector chamber.

As shown in Fig. 5, the red and pink curves are the transfer function modelings taking from Eq. 9, with Q equal to 5.0 $\mathrm{l/s}$ and 1.94 $\mathrm{l/s}$ respectively. Because in accident condition, the flow rate in the cover gas is risen from 1.94 $\mathrm{l/s}$ to 5.0 $\mathrm{l/s}$.

The blue curve is the experiment data for the gas release detected by MX 32. The x-axis represents the time in hours. To make the comparison, we multiplied a coefficient to make these curves have the same maximum activity value, which is 1.5 $\mathrm{Ci/m^3}$. Then we can observe that within a certain flow rate range in the cover gas, the shape of the transfer modeling curve is consistent with the experiment data, which has the same order of magnitude of the time broadening and the same shape of exponential decreasing. We can conclude that the model proposed is able to model the time broadening quantitatively and the transfer function is validated.

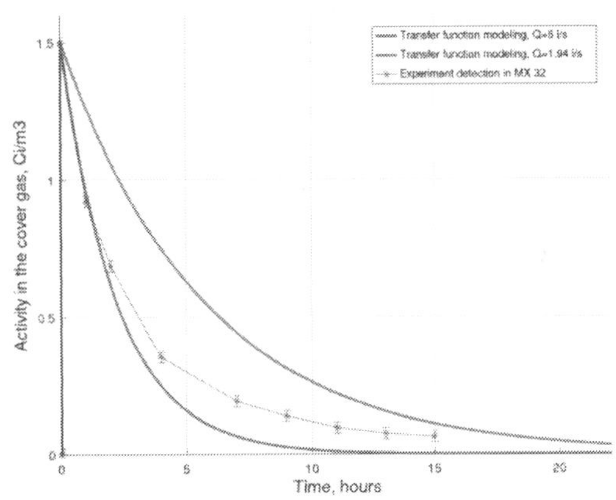

Fig. 5. The fission gas activity in MX 32 and the transfer function modeling results comparison.

IV. Conclusions

In this paper, a model for the transfer function of the detection process, making it possible to predict the time broadening of the observed signal, has been proposed. Taking the actual specifications of the reactor, it matches well with the PHENIX reactor experimental data, with the same order of magnitude of the time broadening and the same shape of exponential decreasing. As a result, the transfer function is validated by a comparison between the modeling result and the experimental data. Although the study was based on an experimental case in PHENIX reactor, the application for this model can contribute to an improved understanding of fission gas transfer in sodium cooled fast reactor and a guidance for the future design.

REFERENCES

[1] S.M. Goldberg, R. Rosner, Nuclear Reactors: Generation to Generation, Nuclear Reactors. (n.d.) 40.
[2] G.L. Fiorini, A. Vasile, European Commission – 7th Framework Programme: The Collaborative Project on European Sodium Fast Reactor (CP ESFR), Nuclear Engineering and Design. 241 (2011) 3461–3469. https://doi.org/10.1016/j.nucengdes.2011.01.052.
[3] E. Rohée, R. Coulon, C. Jammes, P. Filliatre, S. Normand, F. Carrel, F. Lainé, H. Hamrita, Delayed Neutron Detection with graphite moderator for clad failure detection in Sodium-Cooled Fast Reactors, Annals of Nuclear Energy. 92 (2016) 440–446. https://doi.org/10.1016/j.anucene.2016.02.003.
[4] J.R. Matthews, T. Preusser, On the failure of fast reactor fuel pins, Nuclear Engineering and Design. 101 (1987) 281–303. https://doi.org/10.1016/0029-5493(87)90056-2.
[5] C. Jammes, N. Chapoutier, P. Filliatre, J.-P. Jeannot, F. Jadot, D. Verrier, A.-C. Scholer, B. Bernardin, Neutron flux monitoring system of the French GEN-IV SFR: Assessment of diverse solutions for in-vessel detector installation, Nuclear Engineering and Design. 270 (2014) 273–282. https://doi.org/10.1016/j.nucengdes.2013.12.057.
[6] D. Tenchine, Some thermal hydraulic challenges in sodium cooled fast reactors, Nuclear Engineering and Design. 240 (2010) 1195–1217. https://doi.org/10.1016/j.nucengdes.2010.01.006.
[7] Herbreteau, K., Marie, N., Bertrand, F., Seiler, J.-M., Rubiolo, P., 2018. Sodium-cooled fast reactor pin model for predicting pin failure during a power excursion. Nuclear Engineering and Design 335, 279–290. https://doi.org/10.1016/j.nucengdes.2018.05.023
[8] C.A. Friskney, J.A. Turnbull, The characteristics of fission gas release from uranium dioxide during irradiation, Journal of Nuclear Materials. 79 (1979) 184–198. https://doi.org/10.1016/0022-3115(79)90446-X.
[9] Barkia, B., Courouau, J.L., Perrin, E., Lorentz, V., Rivollier, M., Robin, R., Nicolas, L., Cabet, C., Auger, T., 2018. Investigation of crack propagation resistance of 304L, 316L and 316L(N) austenitic steels in liquid sodium. Journal of Nuclear Materials 507, 15–23. https://doi.org/10.1016/j.jnucmat.2018.04.036
[10] E.E. Shpil'rain, S.N. Skovorod'ko, A.G. Mozgovoi, New Data on the Solubility of Inert Gases in Liquid Alkali Metals at High Temperature, 40 (n.d.) 7.
[11] P. Menegon, L. Desgranges, Y. Pontillon, A. Poulesquen, Evidence of two gas release kinetics during the oxidation of an irradiated PWR UO2 fuel, Journal of Nuclear Materials. 378 (2008) 1–8. https://doi.org/10.1016/j.jnucmat.2008.04.008.

Novel Model-Based approach for instrumentation and control of nuclear reactors

Bassem Ouni[1*], Christophe Aussagues[1], Saadia Dhouib[1] and Chokri Mraidha[1]
[1]CEA LIST, University of Paris-Saclay, 91120 Palaiseau, France
*corresponding author: bassem.ouni@cea.fr

***Abstract*— Technological platforms dedicated for digital instrumentation and control of nuclear reactors are quite complex in terms of functionalities and devices. Hence, the design of these platforms requires high-level abstraction layers able to reduce the complexity, to rise the automation and to check the consistency between different development stages. The development of such systems is a challenging task that requires modeling of various components at different levels of abstraction and viewpoints, notably functional, hardware and software levels. In this paper, a new system engineering methodology is proposed to provide high-level models of different components and inter/intra-communication between them. These models are used for system specification, architecture design, performance evaluation or verification and validation. This approach focuses on the internal behavior of different components at different levels of abstraction in order to enable the interoperability of these components and to enhance cooperation between different stakeholders of the development process. An experimental setup has been carried out to validate this approach by customizing an open source model based engineering tool, Eclipse Papyrus, towards a significant reduction of system development cost in terms of engineering resources and equipment devices.**

***Keywords* — System Engineering; Modeling; Instrumentation and Control; Monitoring nuclear reactors; model driven engineering**

I. Introduction

With the increasing complexity of Instrumentation and Control (I&C) systems, developers faced challenges in terms of functionalities, architecture definition, and implementation. Consequently, they deployed a Document-Based System Engineering (DBSE) approach carrying out text-based specifications. However, the DBSE approach raised several issues related to product definition and management, system development process, specifications, trace-ability, collaborative work and standards compliance. To overcome these weaknesses, system designers adapt the Model-Driven Engineering approach, known as MDE [1]. MDE attracted a lot of research interest in recent decades due to its ability to overcome the complexity challenges of systems. Moreover, MDE-based approaches allow analyzing the behavior, implementing and checking the safety of these systems where models act as the main development feature. The MDE area covers various standardized modeling languages, such as the Unified Modeling Language (UML) [2], and domain specific languages (DSLs) [3]. DSLs can be based on the UML extension mechanism, called UML profiles, or on meta-models. Considering this, modeling stakeholders harness the capabilities of MDE to reduce a system's complexity in many research fields [4-6].

The architecture of I&C devices dedicated to the control and protection of nuclear reactors is quite complex. To overcome this challenge, we aim to significantly reduce the development costs of these systems in terms of engineering time and equipment aspects. In order to address the cost improvement objectives of the engineering part, a new MDE-based methodology is proposed. To validate this approach, we extend the specific use of the open-source system engineering tool base, Papyrus [7], for the control and protection of nuclear reactors. The integration of this methodology within this tool encapsulates the main stages of use-case system engineering and carries out many automation steps and consistency checks between different design stages. In addition, it also improves the links between different tasks during the specification, design and development of the system.

This paper is organized as follows: next section introduces Model-Driven proposed approach. Section 3 refines the experimental setup. Finally, the conclusion and future works are drawn in the last section.

II. Proposed approach

The proposed methodology aims to reduce the development cost of I&C systems, automate, and facilitate the consistency checks at different levels of abstraction. The Papyrus-based system engineering approach will replace the existing process where optimization in terms of resources has been identified. The first step of the approach consists of the description of the system functional architecture and the refinement of its behavior as a sequence of functions, their sub-functions and their interactions. The physical architecture is described subsequently as hardware equipment blocks and the network architecture is specified. After that, the system's physical behavior and the allocation of its functions on the hardware platform are detailed. Throughout different phases of this approach, a database of I/O, including data at different levels, is updated. The developed models are exploited for functional simulation, document generation and exports to other tools. The modeling language tool used for developing the architectural model-based framework is Papyrus. Papyrus is an open-source model-driven engineering tool and available in the Eclipse platform [8].

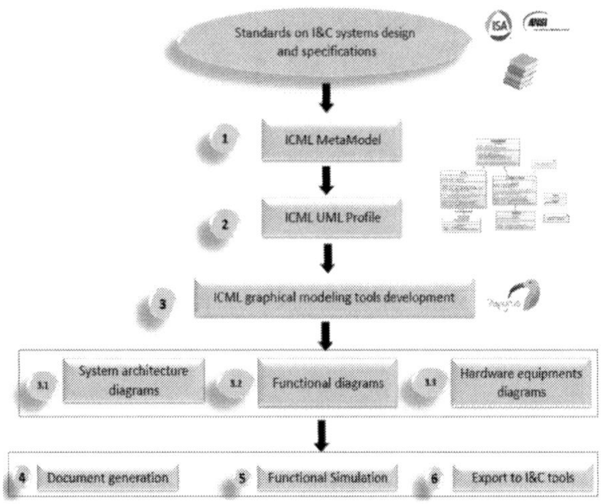

Fig. 1- Model-based tools development process.

The development of proposed tool and the associated graphical editors follows the flow depicted in Figure 1. Considering the I&C specification and the study of existing standards for I&C systems graphical notation [9], the tool is designed based on these two syntaxes:

- An abstract syntax or Metamodel: the metamodel defines different concepts manipulated in the domain and structures the relationships between these concepts and their semantics into a coherent unit. UML is used to formalize these concepts and define the relationships between them. The concept's semantics can be modeled textually or formally via a dedicated language, such as OCL (Object Constraint Language) [10].
- A concrete syntax: This defines the textual or graphical notations for the modeling language. The links between the abstract syntax concepts and the concrete syntax notations are also defined. Several concrete syntaxes can be defined for the same abstract syntax, which allows several representations for the same metamodel. To develop the concrete syntax, the UML [11] profile is used, which is to say, we define stereotypes extending the UML meta-classes. The major advantage of using the UML profiling approach is to benefit from a set of existing tools, both commercial and open-source.

Figure 2 details different levels of abstraction of the proposed framework. The first level of design covers the expected functions and their classification according to their safety category. Then, the behavior of these functions is refined: each system is an arrangement of functions, their sub-functions and their interactions with whatever their technological implementation. Furthermore, the system I/O data are modeled, considering the naming conventions, during this stage. For instance, in the example of Figure 2, we have three systems: system 1, system 2 and system 3. System 3 is not represented at the functional behavior description level for space reasons.

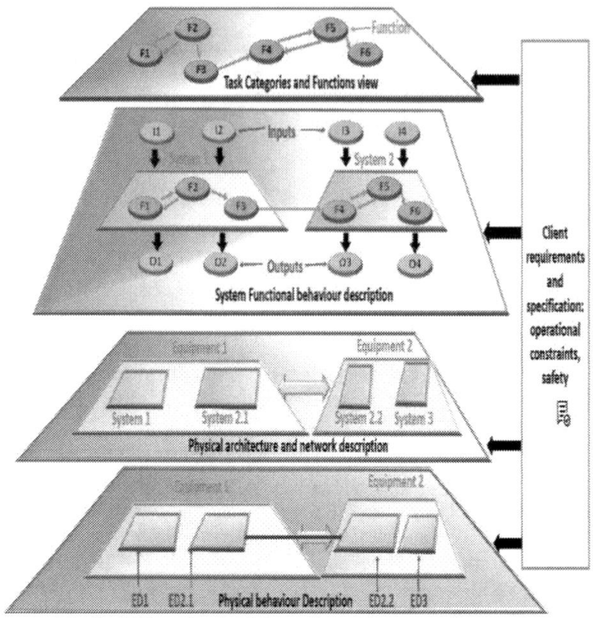

Fig. 2 - Different levels of abstraction of the proposed approach

The third level of abstraction defines the allocation of system functions on the hardware platform. The latter is modeled as a set of equipment. A system can be mapped on two different equipment: for example, System 2 is implemented on two equipment: the sub-systems system2.1 and system2.2 are mapped, respectively, in Equipment1 and Equipment2.

Finally, the detailed specification and equipment behavior (ED) are modelled, taking into account the technological implementation (failure processing, data definition for the monitoring system). A system's ED can describe the behavior of two different equipment. In Figure 2, ED2.1 and ED2.2, respectively, represent the behavior of Equipment1 and Equipment2, implementing System 2.

III. Experimental setup

The use case system is a Reactor Protection System (RPS) gathering various safety and security sensor-based devices embedded in a civil nuclear power plant aiming to safely shut down the civil nuclear reactor if radioactive materials leak. The (RPS) is composed of various safety functions placed on different hardware devices. The use case digital (RPS) is designed by experts from the Tsinghua University of China and described thoroughly in [12]. It is based on electronic boards, sensors, actuators, computers and software tools that are deployed within various sub-systems to ensure the safety and security of a 10MW high-temperature gas-cooled experimental reactor (HTR-10). The first part of this RPS consists of redundant protection logic units in three channels (trains); the second part gathers the surveillance stations and Post Accident Monitoring System located in the main control room; the third part is the monitoring system located in the auxiliary shutdown point.

The system outputs are the protection signals resulting from a

hard-wired 2/3 voting logic operation of the triggering signals in three redundant channels.
The RPS outputs a set of warnings and protection actions for each protection variable. For better safety management, developers divided the protection variables into two groups A and B. In each group, at least one protection variable is set up to protect each initial event, and each group is independently implemented in different processors to act as diversity within each channel. These protection actions are classified into four classes: Protection A (PA) includes the generation of an emergency trip signal to drop the control rods, trip the helium blower and isolate the secondary loop. The protection action B (PB) is referred to the relief of the steam generator. The protection action C (PC) consists of the isolation of the fuel loading/unloading system and the helium purification system from the primary loop. Finally, the protection action D (PD) targets isolating the thermal measurement system from the primary loop.

The proposed tool allows the modeling of a package called "SafetyFunctionLibrary" to define various safety functions; the modeled library has three graphical representations: tree, tabular or class diagram representations. A safety function can include many sub-functions. After that, we define the set of systems. Each system is placed on a rack and runs one or more safety functions. At a higher level of abstraction, the RPS project is modeled as a class with a structural diagram describing its different systems, sensors and actuators, as shown in Figure 3.

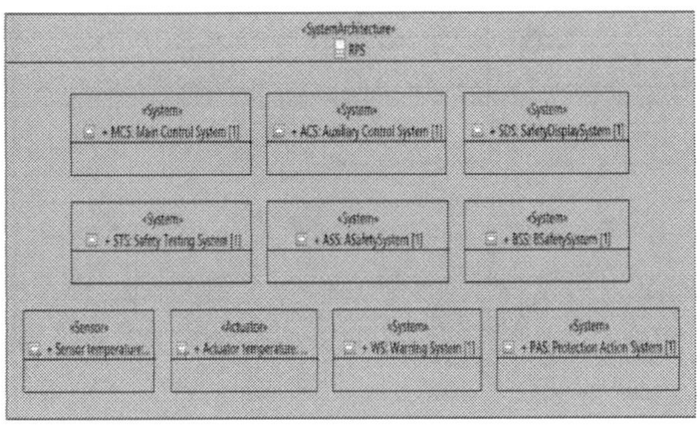

Fig.3- . RPS sub-systems, sensors and actuators.

The input/outputs of all systems are modeled as stereotypical "Sensor" or "Actuator" classes. Their assignment to systems can be edited using an attribute. Communications between systems are represented by links specifying the nature of communication, as depicted by Figure 4. Furthermore, each functional signal is characterized by a set of attributes to specify its type, binary or analog, whether it is an input or output, its I/O board, etc.

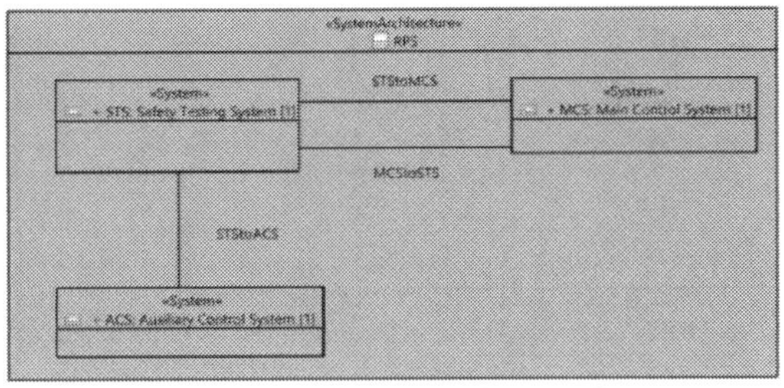

Fig.4- Communications between the RPS' sub-systems.

Moreover, communications between the equipment realizing the same functions or different functions are modeled at this stage. Figure 5 highlights the communication links between the equipment implementing different functions, which are the cores of the APIC and a temperature adapter device, through oriented communication links holding physical signals.

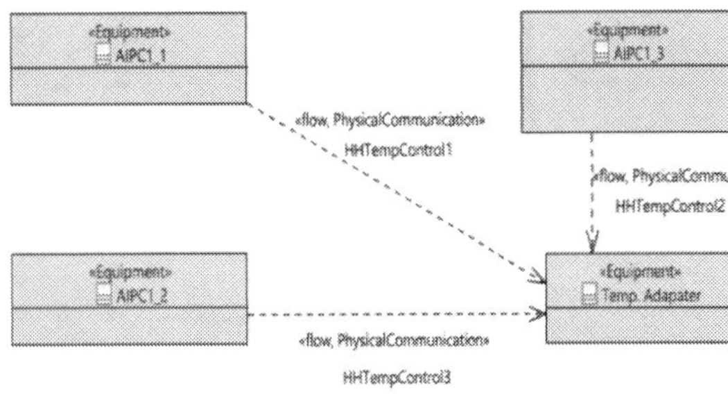

Fig. 5- Communication links between the equipment implementing different functions

IV. Conclusions

In this paper, we proposed a new model driven methodology to model at different levels of abstraction different I&C components and communication between them. These models are used for I&C system design, performance evaluation, verification and validation. Future works will focus on the enhancement of the synchronization between components at different levels of abstraction, the integration of simulation within the proposed framework to execute UML models, provide control, observation and animation facilities over these executions.

Acknowledgment

This work was supported by Bpifrance institution under the project ORION.

References

[1] Adedjouma, M.; Thomas, T.; Mraidha, C.; Gerard, S.; Zeller, G. From Document-Based to Model-Based System and Software Engineering. In Proceedings of the OSS4MDE 2016—Open Source Software for Model-Driven Engineering, Saint Malo, France, 2–7 October 2016; pp. 27–36.
[2] Object Management Group. OMG Unified Modeling Language TM (OMG UML). In Technical Report; Object Management Group: Needham, MA, USA, 2015.
[3] Fowler, M. Domain-Specific Languages; Pearson Education: London, UK, 2010.
[4] Reyes C.R.P.; Vaca, H.P.; Calderón, M.P.; Montoya, L.; Aguilar, W.G. MilNova: An approach to the IoT solution based on model-driven

engineering for the military health monitoring. In Proceedings of the 2017 CHILEAN Conference on Electrical, Electronics Engineering, Information and Communication Technologies (CHILECON), Pucon, Chile, 18–20 October 2017; pp. 1–5.

[5] Prasinos, M.; Spanoudakis, G.; Koutsouris, D. Towards a model-driven platform for evidence based public health policy making. In Proceedings of the SEKE 2017—29th International Conference on Software Engineering & Knowledge Engineering, Pittsburgh, PA, USA, 5–7 July 2017; KSI Research Inc. and Knowledge Systems Institute: Pittsburgh, PA, USA, 2017; pp. 566–571.

[6] Van Lingen, F.; Yannuzzi, M.; Jain, A.; Irons-Mclean, R.; Lluch, O.; Carrera, D.; Perez, J.L.; Gutierrez, A.; Montero, D.; Marti, J.; et al. The Unavoidable Convergence of NFV, 5G, and Fog: A Model-Driven Approach to Bridge Cloud and Edge. IEEE Commun. Mag. 2017, 55, 28–35.

[7] Eclipse Papyrus ™ Modeling Environment. Available online: https://www.eclipse.org/papyrus/

[8] Eclipse Environment. Available online: https://www.eclipse.org/

[9] American National Standard. ANSI/ISA-5.1-2009 Instrumentation Symbols and Identification; American National Standard: Washington, DC, USA, 2009.

[10] Object Management Group. Object Constraint Language; Technical Report; Object Management Group: Milford, CT, USA, 2014.

[11] Object Management Gruop. OMG Unified Modeling Language TM (OMG UML). In Technical Report; Object Management Group: Needham, MA, USA, 2015.

[12] Li, F.; Yang, Z.; An, Z.; Zhang, L. The first digital reactor protection system in China. Nucl. Eng. Des. 2002, 218, 215–225.

Radiation Hardness Test of a Silicon Detector under Radiation Dose Rate of Nuclear Power Plant for In-Containment Coolant Leakage Detection System

Jongheon Kim[1], Kihong Pak[1], Junesic Park[1] and Yong Kyun Kim[1*]
[1]Department of Nuclear Engineering, Hanyang University, Seoul, 04763, and the Republic of Korea
*Corresponding author`s e-mail : ykkim4@hanyang.ac.kr

***Abstract*—An influence evaluation by background radiation on a silicon detector, which will be used to detect a coolant leakage, and installed in a containment building of a nuclear power plant, was performed and the result was discussed. The detector that consists of a silicon sensor and preamplifier mounted in a shielding structure which composed of a 5 cm lead cylinder will be installed in an annulus zone that is influenced by background radiation (neutron and gamma ray) from an operation of a nuclear reactor. Absorbed dose rates on a silicon sensor and preamplifier were calculated as 2.15 mGy/hr and 1.05 mGy/hr, respectively, by Monte Carlo N-Particle (MCNP) simulation. Data of background radiation had referred to a Final Safety Analysis Report (FSAR) of a nuclear power plant in the Republic of Korea.**

A silicon sensor and preamplifier were irradiated by a Co-60 gamma radiation source equipped in a facility of Korea Atomic Energy Research Institute Advanced Radiation Technology Institute (KAERI ARTI) of the Republic of Korea. A Po-210 alpha source was used as a check source to evaluate a state of a function of the detector during gamma irradiation. Absorbed dose rates were about 22.92 mGy/hr and 6.6 mGy/hr on silicon sensor and preamplifier, respectively. Before and during gamma irradiation, a count rates from the check source wasn`t changed (from 18.4 cps to 18.4±0.2 cps after irradiation), and any degradations of function also weren`t observed. Even more harsh condition than calculated dose rates referred by the condition of background radiation of in-containment, the silicon detector maintained the ability of function of charged particles detection. Based on the result, it has been demonstrated that a silicon detector is a suitable detector for detecting charged particles from a leaked coolant even during interfered by the background radiation of a primary system of a nuclear power plant.

***Keywords* — Silicon detector, Irradiation test, Coolant leakage detection system**

I. Introduction

A silicon detector, which has a thin depletion layer, has a characteristic that has small reactivity with gamma ray, and has high reactivity with charged particles. Therefore, the detector could be considered an option to use in a nuclear power plant that has a high dose of background gamma radiation during an operation. Therefore, we had suggested the silicon detector as a radiation detector of a project for developing a detection system of coolant leakage less than 0.5 gallons per minute in one hour (0.5 gpm/hr) [1].

The annulus zone of a nuclear power plant, in which the detector would be installed, has high dose background radiation during a plant`s operation. In the case of Advanced Power Reactor-1400 (APR-1400), the most advanced reactor model of the Republic of Korea, a design criterion of a dose rate at the location is 10 mSv/hr. The detection system would be needed to have the ability to detect coolant leakage under high dose background radiation conditions. To protect the detector from background radiation, a 5 cm thickness lead cylinder structure was composed [1]. Even though the absorbed doses on the silicon sensor and preamplifier were dramatically reduced because of the structure, background radiation could impact the detector such as interfering signal, etc.

There are many studies to investigate a radiation hardness of a silicon detector especially specific conditions such as using as a detector for accelerators, or very high dose conditions to verifying damage of the detector [2,3]. However, there is no study to investigate the influence of a silicon detector under dose rate conditions of a nuclear power plant.

In this study, to investigate the effects of background radiation on the detector during an operation of the reactor, absorbed dose rates to the silicon sensor and preamplifier were calculated through MCNP simulation. Based on the results of the simulation, a gamma irradiation test was conducted at the KAERI ARTI facility using a Co-60 gamma source.

II. Material and Method

A. Simulation

To determine the dose rate for exposure on a silicon sensor and preamplifier, MCNP simulation was used. The calculated absorbed doses in the annulus zone on the silicon sensor and preamplifier were about 2.15 mGy/hr and 1.05 mGy/hr, respectively. Data of background radiation had referred Final Safety Analysis Report (FSAR) of a nuclear power plant in the Republic of Korea [4]. Using the data and MCNP simulation, fluxes of background radiation in an annulus zone were derived. Fig.1 shows the fluxes of gamma and neutron background radiation in the annulus zone of the nuclear power plant. Fig.2 shows the geometry of MCNP simulation to calculate absorbed doses. To calculate the intensities of background radiations, the source section was set up as F4 Tally, and outputs of F4 Tally was matched with the data of Fig.1.

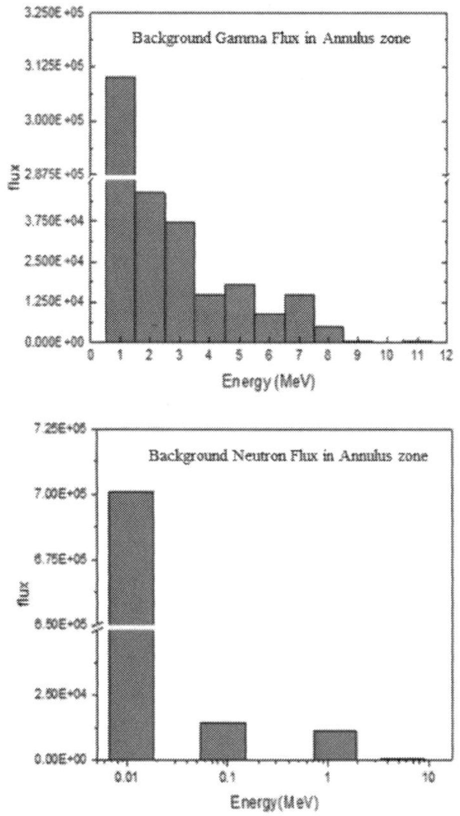

Fig. 1. The background radiation in an annulus zone of a nuclear power plant (top) gamma (bottom) neutron

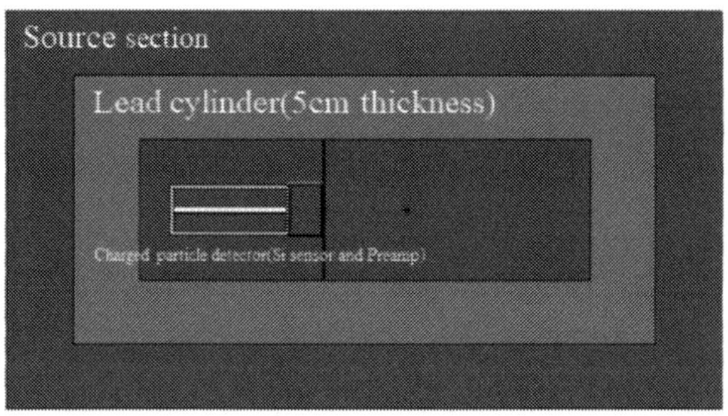

Fig. 2. MCNP geometry to calculate the absorbed dose rates on silicon sensor and preamplifier

B. *Experiment setting*

To investigate the influence of radiation exposure similar absorbed dose with an annulus zone of a nuclear power plant on a silicon sensor and preamplifier, an ULTRA ion implanted silicon sensor (ORTEC) was used. The silicon sensor had a 300 μm thickness of depletion depth and 1200 mm^2 of an active area. As electronics devices, 142A preamplifier, 460 amplifier, 428

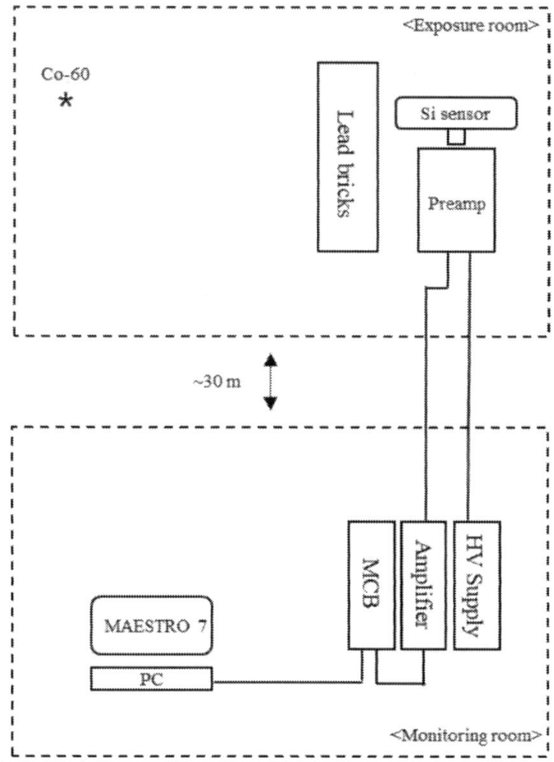

Fig. 3. Block diagram of the setting of the detector and electronics system

bias supply, 928 MCB, and MAESTRO® 7 application software made by ORTEC were used for identifying spectra during radiation exposure.

The set value of electronics were 0.1 μsec of shaping time, 50 of amplifier gain, 90 V of bias voltage, and 1024 channel of conversion gain.

The activity of the Co-60 gamma source for gamma irradiation was about 2.4 kCi. Therefore, to satisfy the situation of an absorbed dose rate of the detector in the annulus zone, the experiment was conducted using 10 cm lead bricks and sufficient distance from the source. Furthermore, to identify absorbed doses accumulated in the silicon sensor and preamplifier, alanine checkers were used. In addition, a Po-210 alpha source was used as a check source for identifying functional states of the silicon detector during gamma irradiation. Fig. 3 and Fig.4 present a block diagram and the setting of the experiment.

Because the reliability of absorbed dose rate data by alanine checker could be reduced under several Gy/hr, an absorbed dose at the point before shielding by lead bricks was measured. After that, using MCNP simulation, absorbed dose rates were calculated. Fig. 5 shows the geometry to calculate the experimental absorbed dose rates for each component.

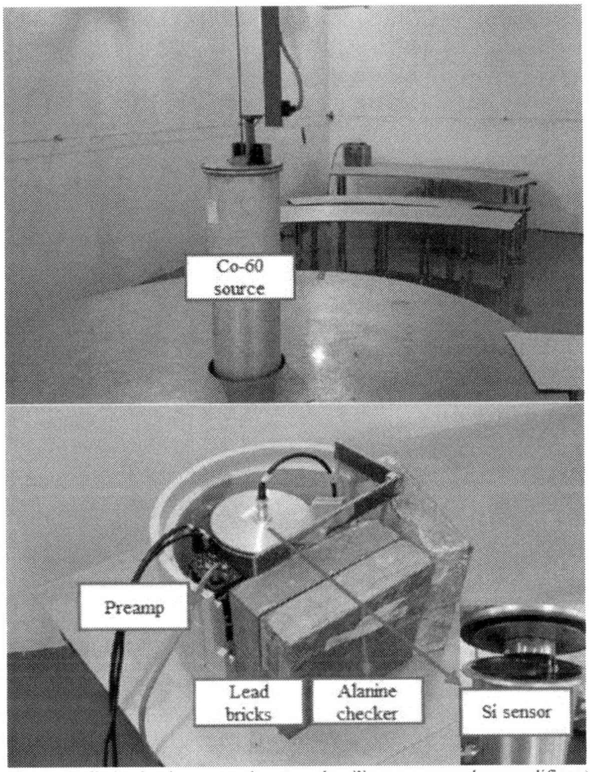

Fig. 4. Radiation hardness experiment on the silicon sensor and preamplifier at KAERI ARIT facility

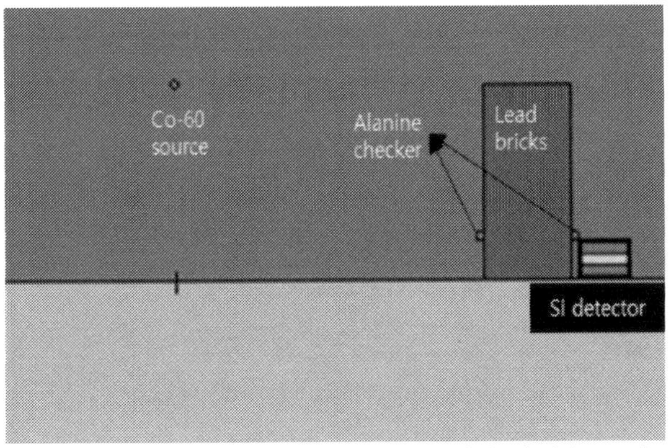

Fig. 5. MCNP geometry to calculate the experimental absorbed dose on the silicon sensor and premplifier

TABLE I
THE EXPERIMENTAL ABSORBED DOSE RATES ON EACH COMPONENT

Components	F6 Tally	Absorbed dose rate
Silicon	3.5207E-08	22.92 mGy/hr
Preamp	1.0107E-08	6.6 mGy/hr
Alanine	4.1311E-06	2.690 mGy/hr

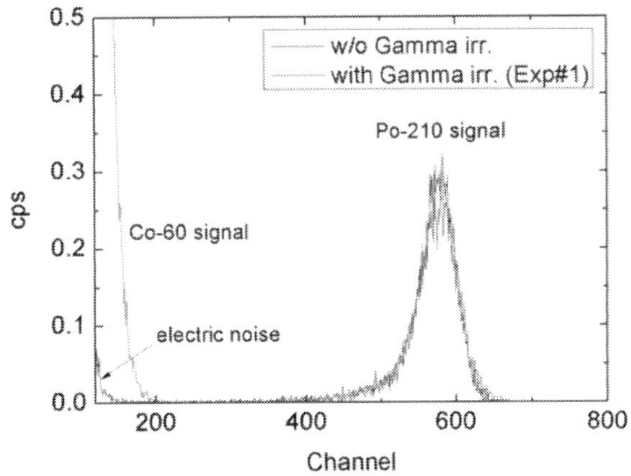

Fig. 6. The alpha and gamma signals before and during gamma irradiation

III. RESULT AND DISCUSSION

The experimental absorbed dose rates on the silicon sensor and preamplifier were 22.92 mGy/hr, and 6.6 mGy/hr, respectively. Table I shows the results of F6 Tally and absorbed dose rates at the silicon sensor, preamplifier, and alanine checker. The absorbed dose rate of the alanine checker is experimental data, and the absorbed dose rates of the silicon sensor and preamplifier are calculated value using the experimental data.

During gamma irradiation on the silicon sensor and preamplifier, a total of 12 repeated measurements were performed at 10-minute intervals to obtain spectra and to investigate the influences of gamma irradiation on charged particle detection. Fig. 6 shows the spectra before and during gamma irradiation. The first experiment shows that the counts of alpha peak slightly decrease compared with before gamma irradiation. In case of high dose background radiation conditions, the gross counts could be increased because of signals by the background radiation, and it also could make increase dead time. Therefore, the peak signal of alpha could be decreased by the increased the dead time. In the following experiments, however, there was no degradation by the dead time. In other words, the situation of the experiment in which gamma absorbing about 6 to 23 mGy/hr on the silicon detector, is not a condition to decrease the functional states of the detector.

Fig. 7 and Fig. 8 show the alpha and gamma spectra during the 12 times experiments, and the alpha counts fluctuation during gamma irradiation. There are no remarkable fluctuations and degradation of alpha counts in the spectra, such as shifting of alpha peaks, and a remarkable decrease of alpha counts, during gamma irradiation. The average count rates of alpha before and during irradiation are 18.4, and 18.4±0.2, respectively.

Even though the experiment's conditions were harsher than the conditions of the annulus zone of a nuclear power plant, the

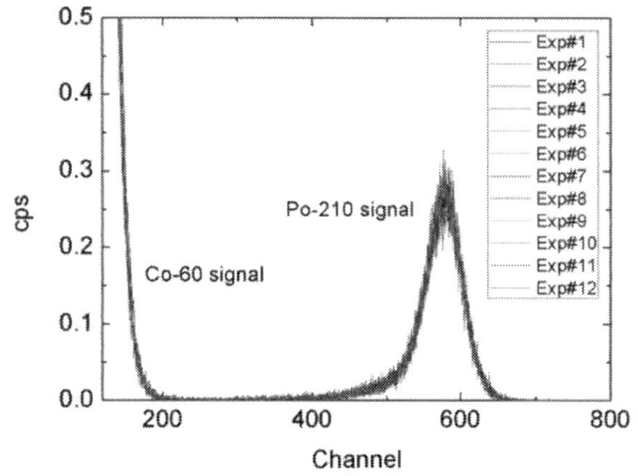

Fig. 7. The alpha spectra of the total of 12 repeated experiment that performed at 10-minute intervals

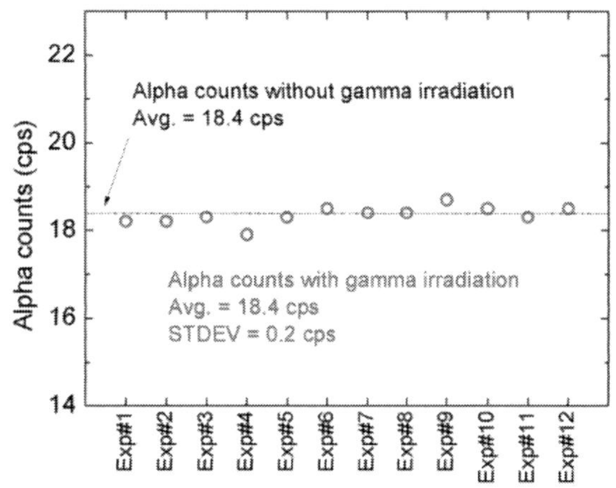

Fig. 8. Alpha counts before and during gamma irradiation

functional states of the silicon detector for detecting the ability of charged particles were maintained. Therefore, it is verified that the silicon detector could be used as a detector at the annulus zone for detecting charged particles from leaked coolant.

IV. Conclusion

To develop a coolant leakage radiation detection system, the radiation hardness test under the condition of dose rates of the annulus zone of a nuclear power plant was conducted and the result was discussed. The absorbed dose rates on the silicon sensor and preamplifier were 22.92 mGy/hr, and 6.6 mGy/hr, respectively, which are higher compared with under the simulated condition of the annulus zone of the nuclear power plant. In other words, in more harsh conditions than the annulus zone, the detector remains functional integrity. Especially, it is verified that the ability to detect charged particles, such as alpha, could be maintained even during gamma irradiation. Based on the result, it has been demonstrated that a silicon detector is a suitable detector for detecting charged particles from a leaked coolant even during interfered by the background radiation in the annulus zone of the nuclear power plant.

Acknowledgment

This work was supported by Korea Institute of Energy Technology Evaluation and Planning (KETEP) grant funded by the Korea government (MOTIE) (Development of a real time detection system for unidentified RCS leakage less than 0.5 gpm)

References

[1] J. Y. Jang, T. schaarschmidt, Y. K. Kim, "Feasibility Study of Beta Detector for Small Leak Detection inside the Reactor Containment", Journal of Radiation Protection and Research, Vol. 43, No. 4, pp.154-159(2018)

[2] P. J. Sellin, J. Vaitkus, "New materials for radiation hard semiconductor detectors", Nuclear Instruments & Methods in Physics Research Section A, Vol.557, Issue 2, pp. 479-489(2006)

[3] S. Metzger, H. Henschel, O. Köhn, and W. Lennartz, "Silicon Carbide Radiation Detector for Harsh Environment", IEEE Transactions on Nuclear Science, Vol. 49, Issue 3, pp.1351-1355(2002)

[4] Korea Hydro & Nuclear Power Co., Ltd. Final Safety Analysis Report of Shin-Kori Nuclear Unit 3,4, Republic of Korea(2015)

Position evaluation of ex-core neutron flux measurement in new type graphite reactors

Eva Vilimova[1*], Tomas Peltan[1,2], and Radek Skoda[3]
[1]University of West Bohemia, Faculty of Electrical Engineering, Univerzitni 8, 30100 Pilsen, Czechia
[2]Research Center Rez, Hlavni 130, Rez, 250 68 Husinec, Czechia
[3]Czech Institute of Informatics, Robotics and Cybernetics, Jugoslavskych partyzanu 1580/3, 160 00 Prague, Czechia
*vilimova@fel.zcu.cz

***Abstract*—Several concepts of new reactors use graphite. Some of them use graphite as a moderator, some of them as a reflector. There are at least two concepts of these graphite-type reactors under development in the Czech Republic. Both reactors use graphite as the reflector. An in-core measurement might be impossible to use because of various reasons, for instance, high temperature or aggressive environment. Therefore, this article focuses on ex-core neutron flux measurement system placed in the graphite reflector and the optimization of ex-core detector position. A set of experiments were performed at LR-0 reactor. The LR-0 is a light water reactor with a well-defined neutron field, which can be used for different material insertion tests and testing of its influence on criticality. Several modifications of LR-0 cores were modelled in Monte Carlo codes Serpent and KENO. A set of calculations were performed for verification of the criticality and neutron flux course in the reactor core and graphite reflector. Further investigation was focused on the influence of the presence of a graphite reflector on the neutron distribution in the reactor core. The LR-0 graphite experiments were also used to verify the calculations. Based on the results of this article, the optimal position of ex-core detectors in the graphite reflector is proposed.**

***Keywords*—ex-core measurement, neutron flux, graphite reflector, Serpent, KENO**

I. Introduction

In the context of global interest in small modular reactors and new 4th generation reactors, the Czech Republic thinks about their utilization in its own energy sector. Czech researchers propose a few designs of small reactor concepts. One of the main nuclear core components of these concepts is graphite either as a moderator or as a reflector.

In addition to the power reactors, all small reactor concepts need to be equipped with neutron flux instrumentation to control reactor power and ensure reactor safety functions. Hence, the appropriate neutron flux instrumentation should be designed and implemented in the concept. The neutron flux can be measured by in-core detectors or by ex-core detectors. In practice, the combination of both methods is usually used. This paper focuses mainly on ex-core neutron flux instrumentation in reactors with the graphite reflector. The research will be applied to the molten salt modular reactor.

Reference [1] discusses parameters and operating conditions for 4th generation reactors and also the challenges of positioning of the neutron flux measurement. This work presents the following research of the optimal detector position in the graphite reflector. The positioning of the detectors has to be determined with respect to the axial and radial neutron flux distribution and the whole reactor core design.

Findings from the research in reference [1] were applied to the research reactor LR-0 in the Research Centre Řež. The research reactor LR-0 is a pool-type light-water reactor with zero power, which serves mainly for measuring neutron-physical characteristics of VVER type reactors. To design the appropriate neutron flux apparatus, the graphite reflector and its behavior in neutron flux were investigated. A graphite prism was prepared and a set of experiments were carried out in the LR-0 reactor. These experiments serve as a source of experimental data for the validation of neutron flux models, which were created in the Monte Carlo codes Serpent and SCALE-KENO [2-3]. At the same time, these experiments will serve for the development of the new ex-core measurement system and validation of its properties. Thus, the performed calculations investigate the optimal position of the ex-ore detectors in the graphite prism and a neutron flux distribution.

II. Experiments and Models

A. Critical experiments in reactor LR-0

Recently, four critical experiments have been performed by the expert team at the LR-0 reactor. The reactor LR-0 is an experimental light-water-moderated zero-power reactor, originally designed for research of the VVER-1000 and VVER-440 type cores, spent-fuel storage lattices and benchmark experiments. The reactor reaches criticality by changing the moderator level. These experiments have been evaluated and the Research center Řež has provided the data for validation of created models.

The first experiment includes two variants with different insertions: CASE 1A and CASE 1B, as shown in Fig. 1. The core consists of six shortened VVER-1000 fuel assemblies with enrichment 3.28%, 3.29%, and 3.30% of ^{235}U. An experimental module, where the insertions are installed, stands in the center of the six fuel assemblies. In CASE 1A, the central module is empty and filled with dry air. In CASE 1B, the central experimental module contains six graphite blocks. These experiments are proposed as benchmark experiments. For CASE 1A, the critical height of the moderator H_{cr} is 44.381 cm. In CASE 1B, the configuration became critical at the moderator level H_{cr} = 55.600 cm. [4], [5]

The second core configuration, CASE 2 in Fig. 1, examines graphite as the reflector. Twelve graphite blocks surround the reactor core, which is the same as the core in CASES 1A and 1B. The center of the reactor core is filled with dry air. This case represents the reference core configuration with the graphite reflector, where the ex-core detectors will be situated. In this critical experiment, the critical moderator level H_{cr} is 41.661 cm. [6]

The last critical experiment composes of seven graphite insertions as a central reflector and twelve radial fuel assemblies enriched to 3.6% and 3.59% of ^{235}U. The critical height of the moderator is 39.940 cm in this case. [7]

B. Models and calculations

All criticality calculations and neutronic analyses were done in two Monte Carlo calculation codes. One of the codes is Serpent [2] developed by VTT Technical Research Centre of Finland. The second calculation code is SCALE-KENO developed and maintained by Oak Ridge National Laboratory [3]. All cases were modeled in accordance with the corresponding benchmark specifications, for instance void replaces the dry air in the reactor.

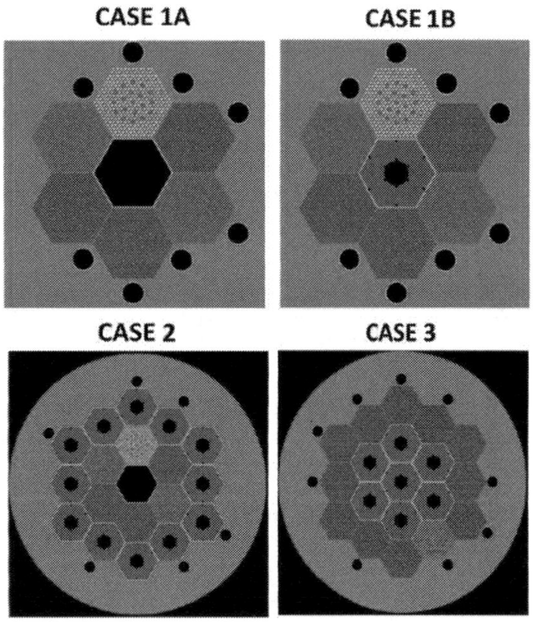

Fig. 1. The core configurations performed at the LR-0 reactor. CASE 1A and CASE 1B, CASE 2 and CASE 3.

The Serpent 2.1.31 code was used as a reference for neutronic calculations as criticality and neutron spectra. The neutron transport in criticality calculations was simulated with 50 000 neutrons per cycle and 10 000 active cycles (100 inactive cycles). The neutron spectrum analyses in the core center ran with 500 000 neutrons per cycle in 50 000 active cycles with 100 inactive cycles because of a very fine energy group structure. The energy group structure includes 640 groups with a total energy span of 10^{-10} MeV to 20 MeV. This group structure has 45 energy groups per energy decade below 1 MeV and a group width of 100 keV above 1 MeV [8]. The dimensions of the modelled detector in the center are 6 x 6 cm, which corresponds with the stilbene detector used in real experiments.

For the analyses of the radial neutron spectrum, the predefined 7-group CASMO energy structure was used. The energy structure was chosen taking into account the microscopic cross-section of the graphite, which is smooth in the desired thermal energy. For this reason, the calculations were carried out with 200 000 neutrons in 50 000 histories with 100 inactive cycles.

In addition, all four critical configurations were created in the code SCALE-KENO. The criticality calculation results from KENO serve as a code-to-code comparison. For the neutron transport simulation, KENO used 50 000 neutrons per generation in 5 000 generations and 100 inactive generations.

TABLE 1: COMPARISON OF CRITICALITY CALCULATIONS IN SERPENT (σ = 0.00004) AND SCALE-KENO (σ = 0.00005)

ENDF/B-VII.1	Serpent k_{eff}	KENO k_{eff}	Serpent-KENO [pcm]	H_{cr} [cm]
CASE 1A	1.00056	0.99991	50	55.600
CASE 1B	1.00077	1.00057	20	44.381
CASE 2	1.00295	1.00326	-31	41.661
CASE 3	1.00311	1.00365	-54	39.940

All calculations were performed with the ENDF/B-VII.1 nuclear data library [9]. The uncertainty of the multiplication factor k_{eff} is σ = 0.00004 for Serpent and σ = 0.00005 for KENO.

Table 1 shows that the results obtained from Serpent and KENO are in good agreement with each other. Note that CASE 1B and CASE 2 indicate the effect of the position of the graphite reflector in the reactor core. Despite the different fuel configurations in CASE 3, there is an obvious indirect dependency of the critical height of the moderator level and the amount of graphite in all cases.

III. NEUTRON SPECTRA AND FLUX DISTRIBUTION

A. Neutron spectra in the core center

Gradual insertion of the graphite blocks inside or outside the fuel geometry enables an investigation of the influence of graphite on the neutron spectrum in the core center by comparison of four neutron spectra in the core center in all cases. This comparison shows Fig. 2. In this figure, the effect of gradual enlargement of the central reflector is clearly visible in CASE 1B and CASE 3.

Note that the number of thermal neutrons in CASE 3 is much higher than in the other cases and the neutron spectrum is flat at higher energies, the neutrons are very well thermalized. On the other hand, the difference between CASE 1A and CASE 2 is negligible, the contribution of the reflector to the detector response is very small.

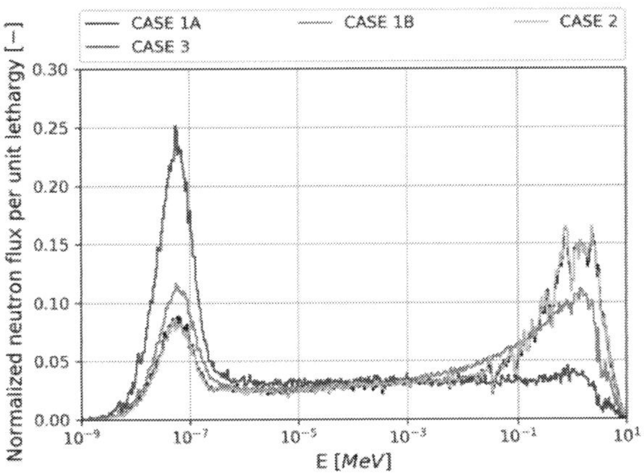

Fig. 2. Neutron spectra in the core center in all cases.

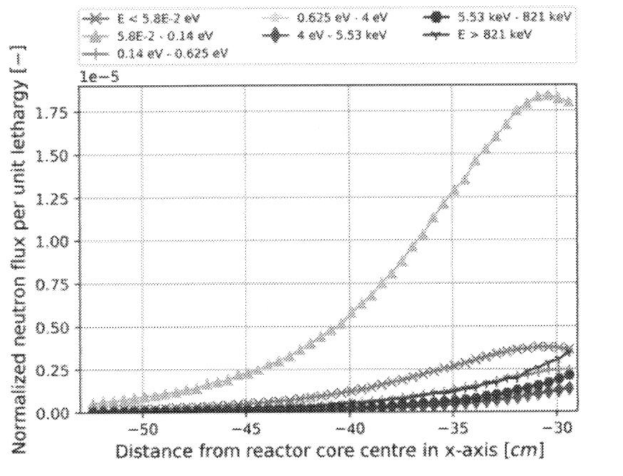

Fig. 4. Neutron flux profile in light water - CASE 1A. The same position of the detectors as in the CASE 2. 7-group structure, ENDF/B-VII.1 [6]

B. *Neutron spectra in the position of the reflector*

To evaluate the optimal position of ex-core detectors in the graphite reflector, the neutron flux distribution profile was studied in the core configuration called CASE 2. This core configuration is most similar to the reactor core, where the neutron detection system under development will be tested. The detectors score in the width of the graphite block, which is 25 cm on the left-hand side of the reflector in the negative x-axis direction. The calculated neutron flux profile in the radial direction shows Fig. 3. The flux of thermal neutrons appears to be constant, or rather it decreases very slowly through the graphite reflector. For comparison, the radial neutron flux profile was calculated also in CASE 1A without graphite i.e., in light water (Fig. 4). The detectors score in the same position as in CASE 2.

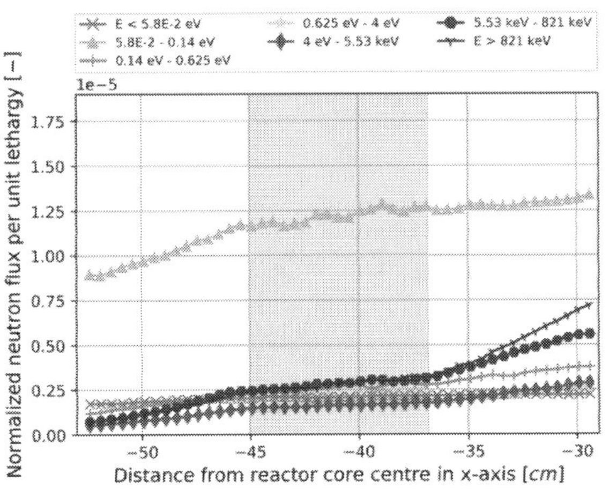

Fig. 3. Neutron flux profile in the graphite reflector – CASE 2. The 7-group structure, ENDF/B-VII.1 [6]

C. *Ex-core detector position evaluation*

Due to very high slowing-down power of light water, the thermal reflector peak is sharp, clearly visible and obvious and it locates very close to the fuel. The thermal neutron flux in the graphite reflector is comparatively flatter, when compared to light water. Moreover, there is an obvious effect of the center holes in the graphite blocks (gray background in Fig. 3), which slightly disturbs the neutron distribution in the graphite reflector, and the reflector peak is difficult to recognize.

The results show that the optimal position of the ex-core detector is not as strictly defined as for a light-water reflector. The position in the graphite reflector could be adapted to detector requirements such as a temperature limit or maximal neutron flux. However, the optimal position of the ex-core detectors, in terms of good neutron flux response of the detectors, is in the reflector peak, which is not clearly visible in the results.

Regarding the obtained results, another experiment will be carried out. This experiment will have the same reactor core arrangement as in CASES 1A, 1B, and 2, but the width of the graphite reflector will be doubled and the holes in the graphite block will be plugged with graphite cylinders. This new experiment is expected to show better results in terms of a clearer neutron flux profile in the reflector and, hopefully, hint of the reflector peak. This will help to determine the ex-core detectors' position more precisely.

IV. Conclusions

This work focuses on the design of an ex-core measuring system for the new small reactor concepts with a graphite reflector. These reactor concepts need to be equipped with a neutron flux instrumentation to control the reactor power and to ensure the safety functions. Hence, a neutron flux detection system should be designed, so this work presents the research of an optimal detector position in the graphite reflector.

A set of experiments were performed at research reactor LR-0. The critical levels of the moderator, which were obtained from experimental data, were used to model these core arrangements in Serpent and SCALE-KENO. Then, the neutron properties of the graphite were studied to determine the optimal position of the ex-core detectors in the graphite reflector. The study of graphite insertions in different reactor core geometry showed a noticeable effect on neutron spectra in the centers of the core.

Optimal position of the ex-core detectors should be specified from the radial neutron distribution in the graphite reflector. However, the calculated results of the radial neutron distribution do not show the exact position of the detectors, so another experiment with a double-width reflector will be carried out in future research to determine the position more accurately.

ACKNOWLEDGMENTS

The research has been supported by the TK01030103 Ex-core Neutron Flux Measurement for the 4th Generation Nuclear Reactors project and by project no. SGS 2021-018 granted by Technology Agency of the Czech Republic (TA CR) and by Ministry of Education Youth and Sports of Czech Republic.

REFERENCES

[1] A. Fořtová, R. Škoda, and J. Závorka, "Ex-core neutron flux monitoring system in graphite prism for gen. IV Reactors," in Proc. ICONE19, Track 16, Tsukuba, Japan, 2019, doi: 10.1299/jsmeicone.2019.27.1627.
[2] J. Leppänen, M. Pusa, T. Viitanen, V. Valtavirta, and T. Kaltiaisenaho, "The Serpent Monte Carlo code: Status, development and applications in 2013," Ann. Nucl. Energy, vol. 82, pp. 142–150, 2015, doi: 10.1016/j.anucene.2014.08.024.
[3] W. A. Wieselquist, L. A. Lefebvre, and M. A. Jessee, "SCALE Code System." Oak Ridge National Laboratory, Oak Ridge, Tennessee, 2020.
[4] M. Košťál, V. Rypar, E. Losa, M. Schulc, E. Novák "VVER-1000 Physics Experiments Hexagonal Lattices (1.275 cm Pitch) of Low Enriched U(3.3 wt.% U235)O2 Fuel Assemblies in Light Water with Graphite and Fluoride Salt Insertions in Central Assembly." IRPhE benchmarks database (2016)
[5] M. Košťál, M. Schulc, E. Losa, J. Šimon, N. Burianová, et al, "A reference neutron field for measurement of spectrum averaged cross sections," Ann. Nucl. Energy, vol. 140, 2020 , ISSN 0306-4549,doi: 10.1016/j.anucene.2019.107119.
[6] T. Peltan and M. Košťál, "Study of reflection properties of silica sand," in 2020 ANS Virtual Winter Meeting, 2020, vol. 123, pp. 1308–1311, doi: 10.13182/T123-33334.
[7] T. Czakoj, E. Losa, and M. Košťál, "Design of a special core for large graphite insertion studies in the LR-0 reactor," in PHYTRA4 – Fourth Int. Conf. Phys. Technol. React. Appl., pp. 238–249, Marrakech, Morocco, September 17-19, 2018.
[8] W.L. Zijp, H.J. Nolthenius, G.C.H.M. Verhaag "CROSS-SECTION LIBRARY DOSCROS84 (in a 640 group structure off the SAND-II type)," Netherlands Energy Research Foundation ECN, vol. 84, October, 1984.
[9] M. B. et al. Chadwick, "ENDF/B-VII.1: Nuclear Data for Science and Technology: Cross Sections, Covariances, Fission Product Yields and Decay Data," Nucl. Data Sheets, vol. 112, 2011.

Development of a Fast-Spectrum Self-Powered Neutron Detector for Molten Salt Experiments in the Versatile Test Reactor

K. C. Goetz[1*], S. M. Cetiner[2], C. Celik[1] and J. Hu[1]
[1]*Oak Ridge National Lab, USA*
[2]*Idaho National Lab, Massachussets Institute of Technology, USA*
Corresponding author: goetzkc@ornl.gov

***Abstract*—The self-powered neutron detector (SPND) is a widely used flux monitor in thermal nuclear reactors. Although this is a mature technology, the current state of the art is tuned for a thermal neutron spectrum, so many of the devices currently in use lack sensitivity to fast neutrons. Because current in SPNDs is produced through nuclear reactions with the neutron flux inside a reactor, sensitivity in SPNDs is determined by the neutron cross section of the neutron-sensitive portion of the detector, termed the emitter. This neutron cross section drops by orders of magnitude between thermal and fast neutron energies for many emitters in currently used SPNDs, with a corresponding drop in current from the detector. This paper discusses efforts to develop a fast-spectrum self-powered neutron detector (FS-SPND) that is sensitive to neutrons with energies ranging from 0.025 eV up to 1 MeV. An in-depth analysis of Evaluated Nuclear Data File (ENDF)/B-VII.1 neutron-capture cross sections was performed, and four new materials were identified that are suitable emitter candidates for use in measuring fast neutrons. All four materials are stable mid-shell nuclei in the region between doubly magic ^{132}Sn and ^{208}Pb. Each candidate was simulated with the Geant4 Monte Carlo simulation toolkit to optimize overall detector efficiency.**

***Index Terms*—fast-spectrum nuclear reactor, instrumentation and controls, nuclear reactor, self-powered neutron detector, sodium-cooled fast reactor, Versatile Test Reactor**

I. Introduction

Self-powered neutron detectors (SPNDs) have been a common diagnostic tool for intra-core neutron flux mapping in thermal nuclear reactors for more than 60 years [1], [2]. They are useful flux monitors because they are durable, compact, and simple, and they produce a signal proportional to local neutron flux without requiring an external source of power. Signal in these detectors is a current source driven by the electrons generated from nuclear reactions within the emitter that make it out to the collector. A diagram of a generic SPND is shown in Figure 1.

Some common SPND emitters are vanadium, rhodium, silver, and cobalt. SPND insulators are generally made of magnesium oxide, aluminum oxide, or are a vacuum. SPND collectors and signal wires are most often made of Inconel steel.

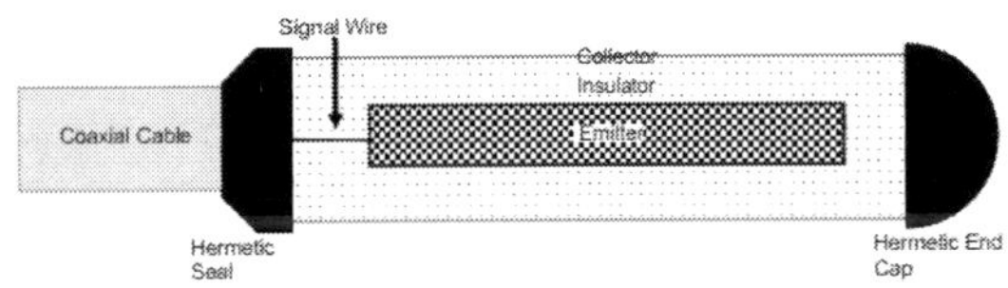

Fig. 1. Diagram of an SPND. It is expected that neutrons will be captured or scattered within the emitter (checked region), resulting in nuclear or ionization reactions that produce electrons with energies high enough that they can escape the emitter and travel through the insulator (white dotted region) to the collector (gray region).

This manuscript has been authored by UT-Battelle, LLC, under contract DE-AC05-00OR22725 with the US Department of Energy (DOE). The US government retains and the publisher, by accepting the article for publication, acknowledges that the US government retains a nonexclusive, paid-up, irrevocable, worldwide license to publish or reproduce the published form of this manuscript, or allow others to do so, for US government purposes. DOE will provide public access to these results of federally sponsored research in accordance with the DOE Public Access Plan (http://energy.gov/downloads/doe-public-access-plan).

II. Self-Powered Neutron Detector Types

The two types of SPNDs—delayed and prompt—are distinguished by the reaction mechanism producing the majority of the electrical signal in the detector. A single SPND will have a delayed component to its signal, as well as a prompt component. Delayed-type SPNDs produce a signal through a (n, β) reaction. In this reaction, the incoming neutron is captured, and the subsequent reaction product β decays, emitting a β particle (electron). The electron produces signal in the detector when it exits the emitter and travels to the collector. SPNDs are generally read out by a current meter at the end of a long coaxial cable well outside reactor containment. Delayed-type SPNDs are so named because the response time of the detector to changed in neutron flux depends on the β-decay half life of the neutron capture reaction product.

Alternately, prompt-type SPNDs generate an electrical current through a $(n, \gamma)(\gamma, e^-)$ reaction series. In this reaction series, a neutron is captured or scattered, and the resulting nucleus de-excites through γ-ray emission. Current is produced when the resultant γ ray knocks an electron out of the neighboring atom, and that electron is absorbed in the collector. Prompt-type SPNDs respond instantaneously to changes in neutron flux.

III. Fast-Spectrum Self-Powered Neutron Detectors

Next-generation reactors are on the horizon. Because some next-generations reactors are designed to operate with harder neutron spectra peaking around 0.5 MeV, diagnostic tools that are tuned to their neutron spectra must be developed. Faster neutron spectra are of salient concern for detectors such as SPNDs, in which the signal is driven by nuclear reactions, and for which the state-of-the-art is tuned for sensitivity to a thermal neutron spectrum. This is because the neutron cross section for many materials can drop by orders of magnitude between thermal and fast neutron energies.

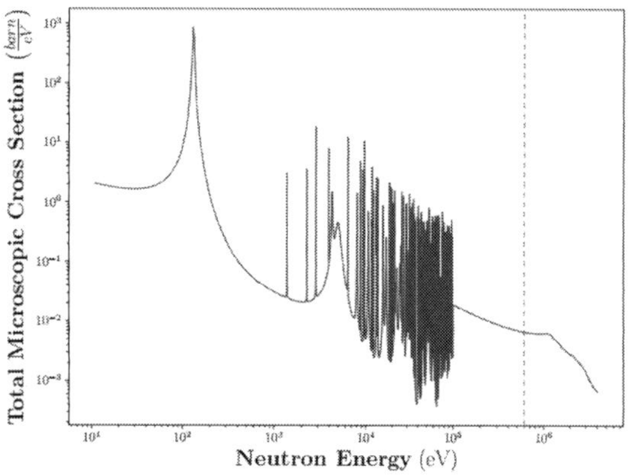

Fig. 2. The total microscopic (n,γ) cross section for ^{59}Co from the ENDF/B-VII.1 library, courtesy of the NNDC [3], [4]. The gray dotted line represents the peak neutron energy in an SFR spectrum at 0.5 MeV. Peaks in the neutron cross section, also termed resonances, correspond to nuclear levels. ^{59}Co displays the typical structure wherein at low energies the distance between levels is large (resonance at ~100 eV) followed by an increase in level density (1 - 100 keV) resulting in an eventual continuum (¿ 100 keV) where individual levels can no longer be resolved.

One example of this is shown in Figure 2 for ^{59}Co, a commonly used emitter material in SPNDs for thermal nuclear reactors. The neutron cross section for ^{59}Co drops by multiple orders of magnitude between thermal energies and 0.5 MeV with a corresponding decrease in detector sensitivity.

IV. Analysis and Material Selection

A. Analysis of ENDF Cross sections

The object of this work is to identify new and currently used emitter materials that will have the requisite sensitivity for operation within a fast-spectrum nuclear reactor. Requisite sensitivity is defined here as having an expected output current while operating in a fast reactor flux that is comparable to that of a state-of-the-art thermal-spectrum SPND operating in a thermal reactor flux. To that end, as part of the emitter selection process a generic neutron spectrum from a sodium-cooled fast reactor was used [5]. This flux spectrum was used in conjunction with (n,γ) cross section data from the ENDF library for the full range of stable nuclei to calculate weighted total microscopic (n,γ) cross sections as shown in Eq. (1) [6], [7]:

$$\sigma_\xi = \frac{\int\limits_{E_{SFR}} dE\, \phi(E)\sigma_\xi(E)}{\int\limits_{E_{SFR}} dE\, \phi(E)}, \tag{1}$$

where σ_ξ is the microscopic cross section for reaction type ξ integrated over the full energy range of the sodium-cooled fast reactor (SFR) neutron flux, E_{SFR}. $\sigma_\xi(E)$ is the energy-dependent microscopic cross section for reaction type ξ, in this case total (n,γ), from the ENDF/B-VII.1, ENDF/B-VIII.0, or JEFF-3.3 libraries [3]. $\phi(E)$ is the scalar neutron flux [5]. An example of the result of the process is shown in Figure 3 for ^{51}V, another common SPND emitter.

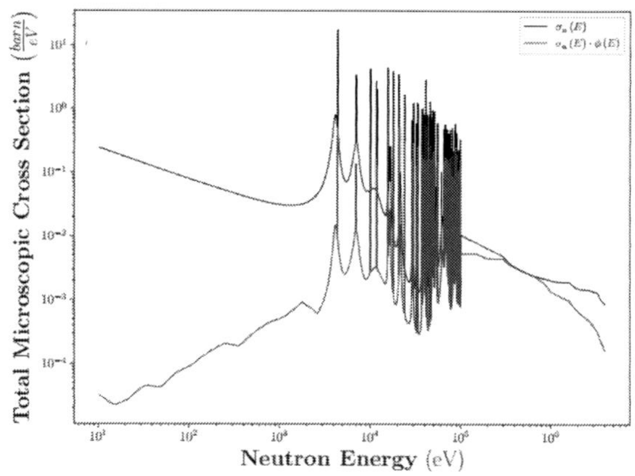

Fig. 3. Microscopic cross section data for ^{51}Va from ENDF/B-VII.1 (black) and the resultant calculated cross section data (blue) taking into account the SFR neutron flux spectrum [3], [5].

The macroscopic cross section Σ_ξ representing all (n,γ) interactions per unit path length traveled in a given medium and integrated over the entire SFR spectrum was then calculated via the following expression:

$$\Sigma_\xi = N\sigma_\xi, \tag{2}$$

where N is atomic number density of the element. For elements with more than one stable isotope, Σ_ξ was calculated for each individually.

Reaction and burnup rates are gauges for the sensitivity of an eventual FS-SPND. To that end, the time-dependent reaction rate $R_\xi(t)$ for a given reaction-type ξ was calculated with

$$R_\xi(t) = \int\limits_{E_{SFR}} dE\, N(t)\phi(E,t)\sigma_\xi(E), \tag{3}$$

where $N(t)$ is the time-dependent atomic number density for the material, and $\phi(E,t)$ is the time-dependent scalar flux.

The time-dependent burnup rate, $b(t)$, of the emitter material at a given time t is defined as

$$b(t) = \int\limits_{E_{SFR}} dE\, N(t)\phi(E,t)\sigma_a(E), \tag{4}$$

where $\sigma_a(E)$ is the neutron absorption cross section for a given stable isotope. Similarly, the time-integrated burnup,

$B(t)$, at a given time t is given by

$$B(t) = \int_0^t d\tau\, b(\tau) = \int_0^t d\tau\, N(\tau) \int_{E_{SFR}} dE\, \phi(E,\tau)\sigma_a(E). \quad (5)$$

The analysis was verified through a comparison of the calculated burnup rate for ^{103}Rh in a thermal neutron flux to that in literature. A burnup rate of 0.39% was calculated, which matches the published value [8].

This stage of the analysis yielded a list of elements with a sensitivity to the neutron flux in an SFR on par with rhodium, a highly sensitive emitter, in a thermal nuclear reactor. Criterion included total macroscopic cross section (Equation 2) and burn up rate (Equation 4). The list is as follows: rhenium, iridium, osmium, europium, lutetium, tantalum, gadolinium, terbium, tungsten, platinum, palladium, silver, rhodium, technetium, thallium, thulium, ytterbium, erbium, dysprosium, and hafnium.

B. General Selection Criteria

However, sensitivity to fast neutrons is not the only concern when choosing an appropriate emitter material for use in next-generation nuclear reactors. The materials must have a high enough melting point to survive within an operating nuclear reactor core. This requirement led to selection of materials with melting points above 700°C [9]. Because a simpler detector response to changes in neutron flux is preferred, materials with one or two single isotopes were selected. The final general criterion was that all materials must be both readily available and machinable to facilitate fabrication of the end detector. These constraints pared down the list to rhenium, iridium, lutetium, tantalum, terbium, silver, rhodium, and thulium.

C. Physics-based Selection Criteria

Following down-selection according to the general criteria listed above, further constraints must be applied to minimize background noise and maximize efficiency in the detector.

For prompt-type SPNDs, the β-decay half life of the neutron-capture reaction product should be maximized, on the order of months. With an ideal prompt-type emitter having a delayed response time that is on the same order of magnitude as the expected reactor cycle. This timeframe was chosen to minimize the introduction of an ever-growing delayed component to the detector response as reaction products continue to build up and decay within the emitter. In an improperly chosen emitter material, this delayed component would act as undesirable background noise that could eventually overwhelm the preferred (γ, e^-) reaction channel. Additionally, to maximize detector efficiency, prompt SPNDs should have a large atomic number (Z). This serves to have an appropriate number of electrons available for the signal-producing reaction in the detector.

In delayed-type SPNDs it is desirable to minimize the response time of the detector to changes in neutron flux. Because signal in the detector is produced through an (n, β) reaction, this is achieved through selecting materials for which the neutron-capture reaction product has as a β-decay half life that is as short as possible, preferably on the order of seconds or less. For this SPND type, detector efficiency is maximized by selecting emitter materials that will produce higher energy β particles, thus increasing the likelihood that the electron will have enough energy to escape the emitter and travel out to the collector. This is achieved by selecting materials for which the neutron-capture reaction product has a high Q_β value and strong feeding to the ground or low excited states in the β-decay daughter.

Rhenium was rejected at this stage because the β-decay half lives of the neutron-capture reaction products were unsuitable for either a prompt or delayed SPND. The two stable rhenium isotopes, ^{185}Re and ^{187}Re, neutron capture to ^{186}Re and ^{188}Re, respectively. ^{186}Re has a β-decay half life of 3.7 days, whereas the β-decay half life of ^{188}Re is 17 hours. This disqualifies rhenium as a candidate, because these response times are far too short to be a reasonable prompt-type SPND and far too long for a delayed-type SPND.

It should be noted however, that many of the rejected candidates for this work are currently used in SPNDs, i.e. hafnium and platinum. Therefore the disqualifications of materials discussed here should be viewed more as guidelines with materials rejected based on the understanding that they are less ideal SPND emitters than the elected candidates.

D. Geant4 Model of SPNDs

A detector inside the core of an operating nuclear reactor will experience more than just the neutron flux accounted for in the analysis up to this point. The γ flux inside the core will induce a prompt signal in both prompt- and delayed-type SPNDs through a (γ, e^-) reaction. Depending on the application and detector type, this signal may or may not be a desirable component of the overall detector response. Additionally, the high-temperature environment inside the core will cause changes to the neutron cross section through Doppler broadening [10]. To maximize efficiency for each SPND emitter, the geometry of each SPND must be tuned to account for the difference in its respective electron spectrum. Therefore, to appropriately optimize any future FS-SPNDs ahead of a costly, time-consuming fabrication process, it is beneficial to first simulate each in a Monte Carlo code such as Geant4.

To that end, a model of an generic SPND was developed using Geant4 version 10.6.1 [8]. A visualization of the model is shown in Figure 4.

This model takes prompt γ and neutron spectra as an input to the simulation, as well as irradiation "time." This "time" allows for the grow-in and decay-of-neutron-capture reaction products to reproduce the effects of burn-up within the detector. An additional feature in the simulation is the use of the neutron cross section database generated for Geant4 using ENDF/B-V11.1 by Mendoza et al. [3], [11]. The data are further broadened using AMPX-6, allowing an irradiation temperature to be established within the simulation space to more accurately reproduce material changes inside an operating SFR [10], [12]. The model was verified by simulating a

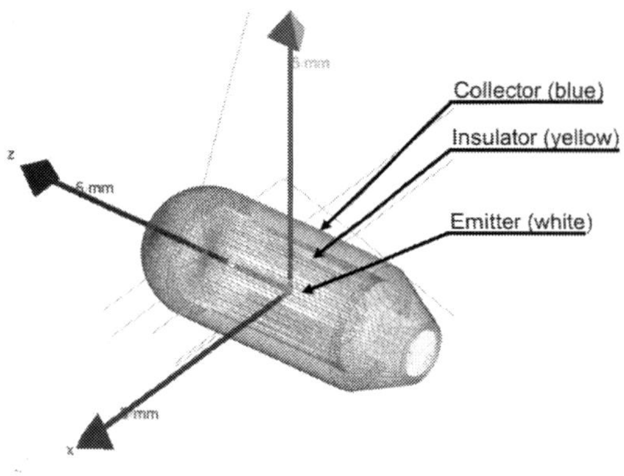

Fig. 4. Visualization of an SPND from within the simulation shown as a wire frame to allow a view inside the detector. The white at the center of the detector is the emitter, the yellow is insulator, and the blue is the collector. The green lines are reactor γ rays impinging the detector.

rhodium SPND in a field of thermal neutrons. The calculated response time of the detector was 4.8 minutes, falling within the expected 3–5 minute window [2].

E. Geant4 Analysis

After the candidate list was narrowed down to iridium, lutetium, tantalum, terbium, silver, rhodium, and thulium, each material was simulated using the Geant4 SPND model, assuming natural elemental abundances. The insulator and collectors were the same materials for all runs, MgO and Inconel 600, respectively. The environment immediately around the detector was simulated as a pipe filled with magnesium salt. An SFR environment was assumed, with prompt neutrons and γ fluxes from Verma et al., and an irradiation temperature of 400 °C [5]. All incoming particles were shaped into a rectangular beam and aimed at the emitter to reduce the number of events needed in the simulation. The irradiation time was set to 0 seconds to act as a reactor impulse at time $t = 0$. Reaction products were allowed to grow in and decay away until $t=\infty$. All reaction channels were allowed in the simulation, including any that might compete with the desired signal. At this stage, no attempt was made to monitor electron movement between the emitter to the collector, as that optimization is beyond the scope of this work. The results from these efforts are a tally of the number of electrons generated in the emitter for each SPND to provide a direct comparison of emitter materials.

V. Results

Five emitter candidates were identified for prompt-type SPNDs, four of which are new, and two of which are suitable for delayed-type SPNDs, and both of which are well known SPND emitters. The prompt-type emitter materials are shown in Table I, and the delayed-type materials are shown in Table II. The results are listed in order of preference. Suitability is judged based on the total macroscopic cross section, the number and relative abundance of stable isotopes, the β-decay half lives of the reaction products, as well as calculated reaction and burnup rates.

TABLE I
Top emitter candidates for prompt-type SPNDs listed in order of suitability. All candidates are newly identified materials. $T_{\frac{1}{2}}$ represents the β-decay half life of the reaction product, burnup (Δ) is calculated assuming $\phi = 5 \times 10^{15}$ N/s/cm^2.

Nucleus	$\Sigma\left(\frac{1}{cm^2}\right)$	Δ/month at 0.5 MeV (%)	$t_{\frac{1}{2}}$
^{181}Ta	13372	0.21	114 days
^{159}Tb	13423	0.40	72 days
^{169}Tm	8043	0.25	128 days
^{176}Lu	16500	0.75	6.6 days
^{191}Ir	25281	0.34	73 days

TABLE II
Top emitter candidates for delayed-type SPNDs listed in order of suitability. $T_{\frac{1}{2}}$ represents the β-decay half life of the reaction product, burnup (Δ) is calculated assuming $\phi = 5 \times 10^{15}$ N/s/cm^2.

Nucleus	$\Sigma\left(\frac{1}{cm^2}\right)$	Δ/month at 0.5 MeV (%)	$t_{\frac{1}{2}}$
^{103}Rh	12694	0.17	42 seconds
^{109}Ag	11559	0.20	24 seconds

Results of the Geant4 simulation, showing the ratio of electrons produced in the emitter for each material compared to rhodium, are in Table III. Results are listed in order of suitability with tantalum being the top candidate, followed by terbium, thulium, lutetium, rhodium, silver, and finally iridium.

TABLE III
Geant4 simulation results showing ratio of electrons produced in the emitter to that of rhodium for each emitter candidate. Candidates are listed in order of judged suitability.

Element	Ratio of e^- in emitter compared to Rh
Ta	2.3
Tb	1.2
Tm	1.0
Lu	1.5
Rh	1.0
Ag	0.9
Ir	2.7

VI. Discussion

Overall, tantalum is the most promising emitter candidate for FS-SPND fabrication. It produces a relatively large number of electrons per unit volume and is readily available. It has one naturally occurring isotope, ^{181}Ta, and the β-decay half life of the (n, γ) reaction product, ^{182}Ta, is 114 days. 114 days is a long enough β decay half life as to not add appreciably to the detector response during one fuel cycle of a nuclear reactor. Terbium and thulium will also produce excellent SPND emitters for these same reasons. They are rated lower because they will produce less efficient detectors than tantalum in terms of current produced per unit flux.

While iridium does produce more electrons per unit volume in an SFR flux, it will likely produce a noisier detector response than tantalum. It is for this reason that iridium was not rated as the first choice material. Iridium has two stable isotopes, ^{191}Ir and ^{193}Ir, with 37.3 and 62.7% natural abundances, respectively. ^{194}Ir, which is the neutron-capture reaction product of the more abundant ^{193}Ir, has a β-decay half life of 19 hours. This is far too short to make iridium an ideal prompt-type emitter candidate and much too long to qualify as an acceptable delayed-type candidate. Additionally, the signal from an iridium detector would likely be significantly more complex than that of tantalum by having a prompt component as well as a 19-hour component and a 7-day component.

At first, lutetium appears to be similar to iridium. It has two stable isotopes: ^{175}Lu, which is 97.4% abundant, and ^{176}Lu, which is 2.6% abundant. Because ^{175}Lu neutron captures to stable ^{176}Lu, the signal would for the most part be entirely prompt. This is because it would take a significant number of neutron captures for the bulk of the material to transmute to ^{177}Lu, which has a β-decay half life of 6.6 days. It is for this reason that lutetium would still make a suitable SPND emitter.

The two delayed-type candidates, rhodium and silver, produce a lower number of electrons per unit volume than most of the other prompt-type candidates, a fact that runs contrary to the rule of thumb for existing thermal-spectrum SPNDs. Rhodium is the best candidate of the two because it has only one stable isotope, ^{103}Rh, the neutron-capture reaction product of which, ^{104}Rh, has a sufficiently short half life of 42 seconds for a delayed-type SPND. The β-decay daughter of ^{104}Rh has a Q_β value of 2.44 MeV and 98% feeding to the ground state. Any β particles produced by this branch of the decay will likely have enough energy to escape to the collector to produce a signal in the detector.

Alternatively, silver will produce less electrons per unit volume, and it also has two stable isotopes: ^{107}Ag (51.8% abundant) and ^{109}Ag (48.2% abundant).

Both of the neutron-capture reaction products have short enough β-decay half lives for suitable delayed-type SPND at 2.4 minutes and 24.6 seconds, respectively. However, the detector response function from this emitter would be more complicated than rhodium, so it is less ideal as a delayed type emitter.

VII. Conclusions

The calculated burnup rates for these chosen emitter candidates in a fast-neutron flux are comparable to those of existing high-sensitivity emitters operating in thermal-neutron fluxes [8]. Therefore, it is expected that each of the identified candidates will produce a suitably sensitive detector for this application. However, these results are preliminary, as further work remains to be done to simulate these detectors with optimized geometries and to examine the radiation hardness of these materials.

Acknowledgments

This work was conducted in conjunction with the VTR project and is based on work supported by the US Department of Energy under Prime Contract No. DE-AC05-00OR2272 to Oak Ridge National Laboratory. Any opinions, findings, and conclusions or recommendations expressed in this publication are those of the authors and do not necessarily reflect the views of the US Department of Energy or Oak Ridge National Laboratory.

References

[1] G. F. KNOLL, *Radiation Detection and Measurement*, John Wiley & Sons (2010).

[2] N. GOLDSTEIN and W. TODT, "A Survey of Self-Powered Detectors—Present and Future," *IEEE Transactions on Nuclear Science*, **26**, *1*, 916–923 (1979).

[3] M. CHADWICK, M. HERMAN, P. OBLOŽINSKÝ, ET AL., "ENDF/B-VII.1 Nuclear Data for Science and Technology: Cross Sections, Covariances, Fission Product Yields and Decay Data," *Nuclear Data Sheets*, **112**, *12*, 2887 – 2996 (2011), Special Issue on ENDF/B-VII.1 Library.

[4] BNL, *NNDC*, https://www.nndc.bnl.gov/nudat2/.

[5] V. VERMA, L. BARBOT, P. FILLIATRE, C. HELLESEN, C. JAMMES, and S. J. SVÄRD, "Self powered neutron detectors as in-core detectors for sodium-cooled fast reactors," *Nuclear Instruments and Methods in Physics Research Section A: Accelerators, Spectrometers, Detectors and Associated Equipment*, **860**, 6–12 (2017).

[6] C. ALLAN, "Experimental measurements of neutron and gamma-ray sensitivities of flux detectors having a mixed response," *IEEE Transactions on Nuclear Science*, **29**, *1*, 660–664 (1982).

[7] J. J. DUDERSTADT, *Nuclear Reactor Analysis*, Wiley (1976).

[8] W. TODT, "Characteristics of self-powered neutron detectors used in power reactors," *Core Instrumentation and Core Assessment, Nuclear Energy Agency, Boulogne-Billancourt, France* (1996).

[9] S. J. ZINKLE and G. WAS, "Materials challenges in nuclear energy," *Acta Materialia*, **61**, *3*, 735–758 (2013).

[10] S. GLADSTONE and A. SESONSKE, "Nuclear Reactor Engineering: Reactor Design Basics," (1994).

[11] E. MENDOZA and D. CANO-OTT, "Update of the Evaluated Neutron Cross Section Libraries for the Geant4 Code (see also INDC (NDS)-0612)," Tech. rep., International Atomic Energy Agency (2018).

[12] D. WIARDA, M. DUNN, N. GREENE, M. WILLIAMS, C. CELIK, and L. PETRIE, "AMPX-6: a Modular Code System for Processing ENDF/B," (2016).

 https://doi.org/10.1051/epjconf/202125306001

Analysis of the Signal over Noise Ratio of the hodoscope determined by Monte Carlo calculation

J. Di Salvo*,****, S. Mirotta**, V. Chevalier***
****Corresponding author: jacques.disalvo@irsn.fr

***Abstract*— The CABRI experimental pulse reactor, located at the Cadarache nuclear research center, southern France, is devoted to the study of Reactivity Initiated Accidents (RIA) for the purpose of the CABRI International Program (CIP), managed by IRSN in the framework of an OECD/NEA agreement. The hodoscope equipment installed in the CABRI reactor is an almost unique online fuel motion monitoring system, thanks to the measurement of the fast neutrons emitted during a power pulse by a tested rod positioned inside a dedicated test loop reproducing PWR conditions. This system is dedicated to the analysis of fuel displacement. Hence, one of the most important parameter measured by the hodoscope detectors is the Signal over Noise Ratio (SNR), characterizing the fraction of neutrons directly coming from the test rod ("signal") over neutrons coming from the core ("noise").**

It is interesting to calculate the SNR in order to define some quantitative criterions to improve hodoscope measurements and to understand if any modification linked to the test loop may significantly change this essential parameter.

Another parameter of interest is the so-called "scattering coefficient", which corresponds to the fraction of neutrons coming from the test rod and being scattered between their birth and their detection. This parameter is used to enhance the analysis of the fuel displacement which may happen during the power transient.

In this article, the method used to calculate the SNR using MCNP6.2 Monte Carlo code will be detailed. Because the hodoscope detectors are located far away from the test rod (up to 4 meters), a 2D model of CABRI core and instrumentation has been implemented. No variance reduction techniques have been used to solve this problem in order to record the place of birth of neutron which contributes to the different scores with the goal to perform a detailed analysis of the SNR. This strategy allows to access numerically to the "scattering coefficient".

Finally, the comparison between calculated and measured SNR for a case study will be carried out. A quite good agreement between the 2D simulations and experiments recently performed in the CABRI reactor has been obtained.

***Index Terms*— CABRI, Hodoscope, MCNP, RIA, Neutron transport.**

*J. Di Salvo Author, was with IRSN/PSN-RES/SEREX/L2EP, Cadarache, Saint Paul Lez Durance 13115 FRANCE (email: jacques.disalvo@irsn.fr),
**S. Mirotta Author, was with IRSN/PSN-RES/SEREX/L2EP, Cadarache, Saint Paul Lez Durance 13115 FRANCE (email: salvatore.mirotta@irsn.fr),
***V. Chevalier Author, was with IRSN/PSN-RES/SEREX/L2EP, Cadarache, Saint Paul Lez Durance 13115 FRANCE (email: vincent.chevalier@irsn.fr)

I. Introduction

For enhancing safety of nuclear power plants (NPP), the French Nuclear Safety and Radiation Protection Institute (IRSN) carries out some experimental programs in order to improve the understanding of the fuel behavior under accident conditions. One of them is the CABRI International Program, managed and funded by IRSN under OECD/NEA umbrella, devoted to study of Reactivity Initiated Accidents (RIA) in representative PWR conditions [1].

For this purpose, the renovation of the CABRI facility has been conducted by CEA, its operator. This pool type reactor is designed to submit to a RIA a test rod placed into the center of the core.

The power transients are generated by the unique reactivity injection system of CABRI [2]. 96 tubes (so-called "transient rods", visible in Figure 1) located in 4 banks among the CABRI fuel rods are pressurized with ^{3}He gas. The very fast depressurization of this strong neutron absorber into a discharge tank, through two flow channels (low and high flow rates), is then finely adjusted to trigger the desired power pulse.

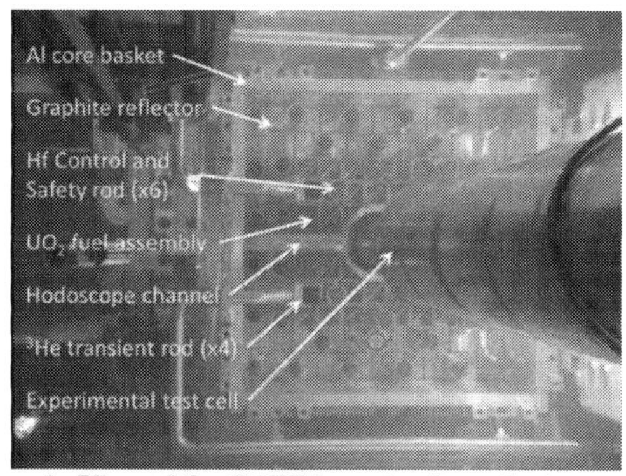

Figure 1 –View of the CABRI reactor from the top

The CABRI facility is equipped with two nondestructive measurement systems operated by IRSN:

- The IRIS facility, for performing X-ray radiography and tomography imaging before and after a power transient thanks to a linear electron accelerator, as well as quantitative gamma scanning analyses;

- The Hodoscope, an online fuel motion measurement system, which aims at analyzing the fuel motion deduced from the detection of fast neutrons emitted by the tested rod, in real time (with a time step of 1ms) during the transient [3].

This article present the numerical model for the evaluation of the Signal over noise ratio of the hodoscope measurement system which is a fundamental parameter to observe the fuel displacement (if any) during the power excursion. More in details, in section II of this article, a brief description of the meaning of the signal over noise ratio will be given. In section III the 2D model developed with MCNP6 [4] will be discussed. Furthermore, in section IV results and a comparison to experimental values will be analyzed. Finally, current studies will be presented, and conclusion is given in section V. Some definitions are reminded in section VI.

II. Signal Over Noise Ratio definition

The hodoscope measurement system, installed on the CABRI reactor, is used to detect and measure online, during power transients, fuel displacement which may occur on the test rod. The hodoscope system is made by a huge collimator about 3 m long, which is placed in front of the test rod 1 m far and allows to filter in space and energy both, neutrons coming from the core and those coming from the test rod. The collimator has 153 lines of sight arranged in a matrix of 51 rows and 3 columns in order to assure both axial and radial discretization of the test rod as well as the possibility to use the signal coming from one of these columns to normalize the signal during power excursion. Beyond each line of sight of the collimator two different detectors are installed. The first is a ^{237}Np fission chamber (energy threshold cross section at 600 keV) and the second one is proton recoils detectors (energy threshold set by electronic system at 400 keV) [5].

In this context the observable used to detect fuel displacement during power transient and therefore the signal over noise ratio (SNR) are fast neutrons coming from the test rod [6].

Only these neutrons are useful for our purpose, all other neutrons (thermal neutrons, fast neutrons produced in CABRI rods, etc…) do not contain any useful information for fuel displacement evaluation. Hence, for this reason, they contribute to the "noise" on the measurement. With both, detector's technology and collimator geometry it is possible to measure only fast neutrons coming from the test rod (which contribute to the useful "signal").

The main difficulty in the measurement of the SNR is due to the small amount of these neutrons coming from the test rod compared to all other neutrons present in the reactor.

A specific methodology is used to evaluate this parameter by the means of a Monte-Carlo simulation.

In Figure 2, a schematic cross section of the CABRI reactor and the line of sight of the hodoscope collimator are represented. In this figure it is possible to see that if the center column of the hodoscope collimator is placed in front of the test rod, detectors placed beyond the center column aim the test rod and therefore capture neutrons coming from it.

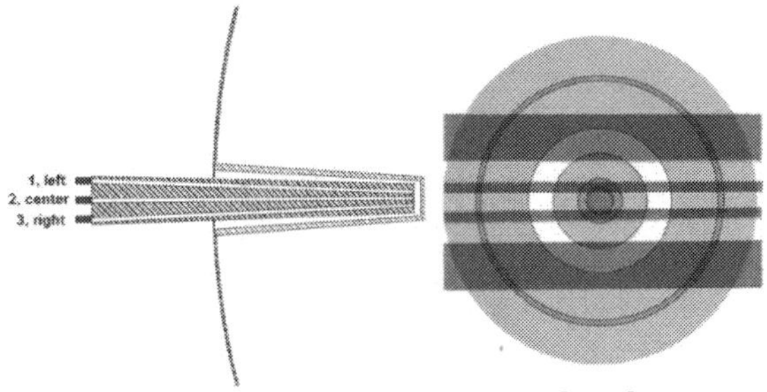

Figure 2 –Lines of sight of the collimator when the center column aims entirely the test rod

The system of Equation 1 for the detectors counting rate can be written when the collimator is placed in its centered position. The left column detector as well as the right column detector can measure fast neutrons coming from the reactor (n) and those coming from the test rod scattered into the structural materials of the reactor (αs). Only the center column can measure the fast neutrons coming from the test rod (s), in complement with the noise and the scattered neutrons.

$\begin{cases} c_l = n + \alpha s \\ c_c = n + \alpha s + s \\ c_r = n + \alpha s \end{cases}$	Equation 1

Using Equation 1, Equation 2 can be written as follow.

$R = \dfrac{c_c}{\dfrac{c_l + c_r}{2}} = \dfrac{s}{n + \alpha s} + 1 = \dfrac{s}{N} + 1$	Equation 2

The signal over noise ratio can be measured by the ratio between the counting rates of the hodoscope detectors. It is important to note that it is not possible to measure directly the signal over pure noise ratio s/n, because the scattering coefficient has to be determined firstly.

III. Numerical Model

A. Geometry description

The CABRI reactor model was built using MCNP6.2. In Figure 3, a radial cross section of the reactor at z=0 plane is shown. It is a 2D model which does not take into account the axial control rod position nor the axial power flux. This simplification was needed because of the poor statistic obtained with the 3D model in the vicinity of the hodoscope detectors. In fact, detectors are located very far from the reactor and a non-analog simulation did not give useful results since the place of birth of the neutron is lost. More precisely, following simplifications were implemented in the 2D model:

- Axial grids were not modeled;
- Control and security rods were completely inserted into the reactor;
- The Helium 3 pressure into the transient rod was adjusted in order to keep the system critic;
- The overall water pool was not modeled.

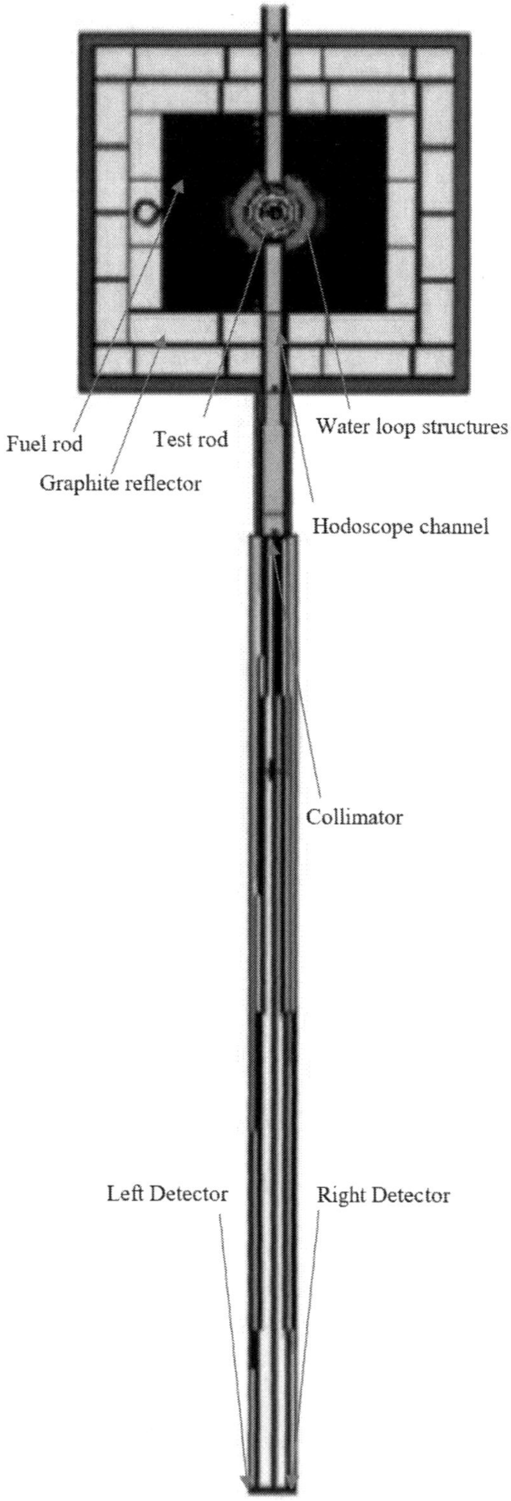

Figure 3 - Radial cross section of CABRI reactor

Once the CABRI model implemented, two different test rods were put in place in the center of the reactor: a fresh UO_2 enriched up to 4.25% in ^{235}U, and an aluminum one.
Moreover, several fictive control volumes were placed in the line of sight of the center hodoscope channel in order to observe the attenuation of the neutrons current through the collimator

B. Tally description

Tallies are used in MCNP6 to obtain an estimation of a physical observables into the model. For this study F4 tally was considered. F4 tallies were split in two energy bins:

- The first one collects all neutrons with energy below 600 keV;
- The second group takes into account neutrons with energy above 600 keV.

In this article fast fluxes which will be enounced, correspond to all neutrons with an energy above 600 keV which is the threshold value for ^{237}Np fission cross section. In the simulation, we collect the neutrons above 600 keV coming from both the core and the test rod into the three detectors region. Scoring the fast fluxes all over three detectors region is enough to estimate the SNR. It important to observe that the scattering coefficient is not available using this method.
Hence, another option of the MCNP6 code was used to evaluate the proportion of the neutrons produced into the test rod and those scattered into the water loop structures.
For this evaluation, the knowledge of the birthplace of each neutron contributing to the tally results is required. To do that, FU tally tagging option was activated and the tally results were split into several contributions.
Two different ways to treat elastic scattering contribution can be used within this option:

- scattered particles lose their birthplace tag and they are putted into elastic scattered bin (option TAG 1);
- scattered particles still keep the birthplace tag, elastic scatter bin is empty in this case (option TAG 3).

Hence, for TAG 1 calculation, four different bins were defined for this study:

1. Bin 1, it takes into account source neutrons born by fission into the test rod (the signal s, red line in Figure 4);
2. Bin 2, which takes into account (n,xn) or (n,n') neutrons produced into the test rod by other neutrons (they are also part of the signal s, black dash-dotted line in Figure 4);
3. Bin 3, takes into account all scattered neutrons independently from where they were born (which contains the scattered part of the signals αs - blue dashed line in Figure 4 -, a part of the signal s itself - black dashed line in Figure 4 -, and all the neutrons coming from everywhere in the core and being scattered once time in their history);
4. Bin 4 takes into account all other neutrons which come from the reactor (not represented in the Figure 4), and (n,n') reactions of neutrons coming from the test rod (grey dash-dotted line in Figure 4);

In the case of TAG3, the filling logic for those bins is quite different: the 3rd bin is always empty because all scattered neutron keep the birthplace tag and they are placed into bin 1, bin 2 or bin 4.

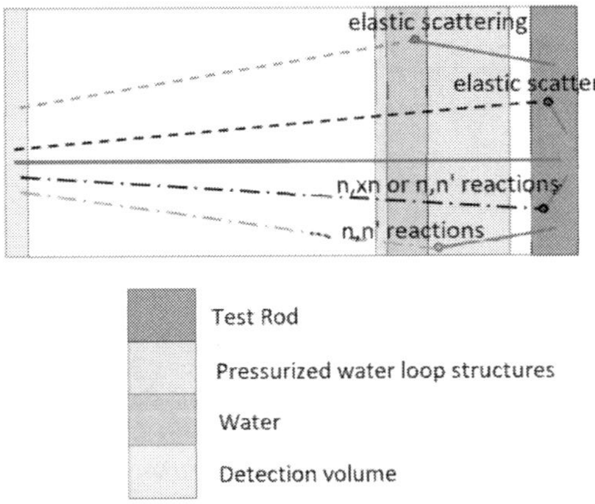

Figure 4 – All different neutron reactions produced by neutrons coming from the test rod (axial cut)

Therefore, for TAG3 calculation, Equation 3 can be written. The sum of the contributions of bin 1 and bin 2 takes into account all neutrons produced into the test rod.

$s + \alpha s = b_{1_{c,TAG3}} + b_{2_{c,TAG3}}$	Equation 3

For TAG1 calculation, Equation 4 can be written.

$s = b_{1_{c,TAG1}} + b_{2_{c,TAG1}} + \varphi_{el,c}$	Equation 4

This equation takes into account the term $\varphi_{el,c}$, which corresponds to the fraction of neutrons produced into the test rod and elastically scattered in itself, participating to the signal in addition with bin 1 and bin 2.

The evaluation of this fraction was possible using the neighboring channels using Equation 5, because they don't aim the test rod so that the signal is not contained in the counting rate of these channels.

$\varphi_{el,c} = b_{3_{c,TAG1}} - \frac{b_{3_{r,TAG1}} + b_{3_{l,TAG1}}}{2}$	Equation 5

Finally it is possible to write Equation 6 to obtain the scatter coefficient for the CABRI reactor.

$\alpha = \frac{b_{1_{c,TAG3}} + b_{2_{c,TAG3}}}{b_{1_{c,TAG1}} + b_{2_{c,TAG1}} + \varphi_{el,c}} - 1$	Equation 6

In conclusion, with TAG 3 option it is possible to split the contribution to the tally which comes from the core (noise n) and the contribution to the tally due to the test rod (signal $s + \alpha s$) within a slight approximation due to neutrons coming from the test rod and scattered by inelastic collisions from the water loop structures. Hence, with TAG 1 option it is possible to split the contribution of s from that due to neutrons coming from the test rod and elastically scattered into the water loop structures (αs).

IV. Results

Results are presented in this section with uncertainties evaluated at 2σ, which means a confidence interval of 95%. Three different results will be presented in this section:

- The collimator importance for the increasing of SNR;
- The evaluation of the SNR for two different test rods;
- The preliminary evaluation of the scattering coefficient.

A. Collimator role

The importance of the hodoscope collimator is shown in this section. In Figure 5 the radial distribution of the fast flux (>600keV) is presented. Two different simulations were performed, the first one without the collimator and the second one with the collimator. In the first case, without the collimator the flux into the detectors zone is reduced by a factor 100. With the collimator, the reduction factor is more important: the neutrons flux is reduced by a factor of 500. Comparing the two calculations, the flux loss is quite important, but the interesting result is that the Signal over Noise ratio is equal to 11,4 % ± 0,4 % without the collimator and reaches 40,2 % ± 1,1 % with the collimator. This means that the collimator is crucial for a good SNR evaluation, complementary to the axial discretization.

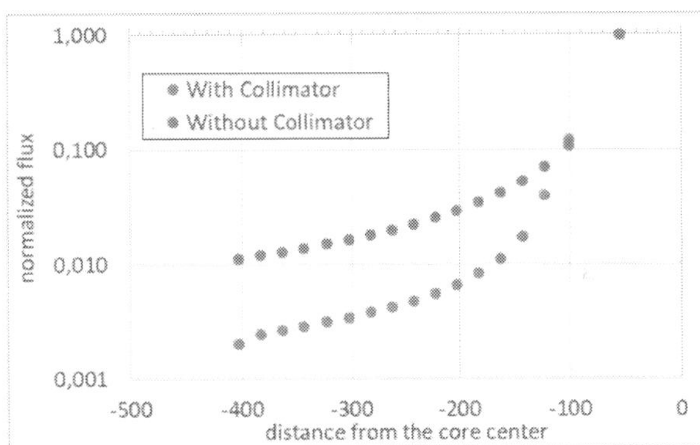

Figure 5 – Fast neutron flux with and without the hodoscope collimator

B. SNR evaluation

Thanks to the methodology presented in II, it is possible to evaluate the SNR ratio.

Table 1 shows results obtained for fresh UO_2 test rod for the three detectors beyond the collimator.

Table 1 – Tally score for the three channels for fresh UO_2 test rod loaded into the experimental cell

	Flux (a.u.)
Neutron Flux for right channel c_r	$4.14\ 10^{-08} \pm 1.4\%$
Neutron Flux for center channel c_c	$5.33\ 10^{-08} \pm 1.2\%$
Neutron Flux for left channel c_l	$4.17\ 10^{-08} \pm 1.4\%$

$s/N_{UO2} = 28.3\% \pm 1.7\%$

The SNR is contained in the interval [26.5% ; 30.2%] for fresh UO_2 test rod. The measured value obtained during a reactor power plateau, by using the same technique is 27.0% ± 2.0%. Comparison of calculated result versus experimental one in the case of the fresh UO_2 fuel are quite in good agreement.
The second case study concerns the simulated signal over noise ratio for a non-fissile material. In this theoretical case the value should be 0, because no neutrons (or at least very few) come from the test rod.

Table 2 – Tally score for the three channels for Al test rod loaded into the experimental cell

	Flux (a.u.)
Neutron Flux for right channel c_r	$3.73\ 10^{-08} \pm 1.4\%$
Neutron Flux for center channel c_c	$3.71\ 10^{-08} \pm 1.4\%$
Neutron Flux for left channel c_l	$3.72\ 10^{-08} \pm 1.4\%$

$$s/N_{ALU} = -0.3\% \pm 1.7\%$$

The SNR is contained in the interval [-2.0% ; 1.5%]. As shown from this results, in the case of the aluminum test rod, the computed SNR is close to 0, as expected.
This was considered as promising results for this preliminary modelling, and further analysis has to be realized with irradiated fuel pins, in order to increase the validation database of the model.

C. Scattering coefficient

It is also possible to evaluate the scattering coefficient α, representing the effect of water loop structures on neutrons coming from the test rod, with the methodology described in section III.B. In Table 3 results obtained with the two options for the fresh UO_2 fuel are presented.

Table 3 – Tally score for the central channel using different elastic scattering contribution evaluation options

s (TAG1)	Direct contribution	$9.63\ 10^{-09} \pm 2.8\%$
	(n,xn) and (n,n') reactions on the test rod contribution	$2.74\ 10^{-10} \pm 15.0\%$
	Elastic scattering contribution into the test rod	$2.02\ 10^{-09} \pm 17.0\%$
$s + \alpha s$ (TAG3)	Direct contribution	$1.52\ 10^{-08} \pm 2.2\%$
	(n,xn) and (n,n') reactions on the test rod contribution	$4.58\ 10^{-10} \pm 11.6\%$

$$\alpha = 31.6\% \pm 3.3\ \%$$

Therefore the scattering coefficient is contained in the interval [29.5% ; 33.6%] with this method. This value is consistent to the one find experimentally for the previously sodium loop implemented in CABRI, which was equal to 24%. More scattering effect can be expected in the water loop than in sodium loop. As it is not foreseen to measure the scattering coefficient in the CABRI water loop, this new methodology will be applied in a next step to the previous sodium configuration in order to analyse the discrepancy to experimental data.

V. Conclusion

In conclusion, this work presents results obtained with a newly implemented, simplified, 2D model of the CABRI reactor. The purpose was to determine the "Signal to Noise Ratio" as well as the scattering coefficient, both involved in the treatment of the hodoscope measurements.
Thanks to this work, it is possible to predict with a good accuracy the SNR measured for the different pins soon tested in the framework on the CIP program. This new model can also play a tremendous role to control any modification linked to the test loop towards the hodoscope measurements, particularly in the phase of conception of new experimental programs.
A particular skill of the hodoscope is also to perform a quantitative analysis thanks to a fuel signal-to-mass conversion model for the fuel ejection into the coolant channel at pin failure. Histograms of masses that are built require the knowledge of the scattering coefficient [6]. For the previous experiments (sodium loop configuration), this parameter has been determined thanks to a specific experimental campaign. For the Pressurized Water Loop configuration, the parameter introduced in the analysis will be the calculated one. Hence this work shows the possibility to access to such kind of information with acceptable uncertainties and reasonable calculation time.
Results presented in Table 4 show the quite good consistency between these preliminary simulated values and the measured ones. A more complete analysis of the discrepancy between measures and calculation has now to be done, by performing sensitivity studies and increasing the experimental validation database.

Table 4 – Synthesis of the results

		Simulation	Measurement
s/N	Fresh UO_2	28.3 % ± 1.7 %	27.0 % ± 2.0 %
	Aluminum	-0.3 % ± 1.7 %	NA
α		31.6 % ± 3.3 %	23 % (sodium loop conditions)

VI. Nomenclature

s	Signal of neutrons coming from the test rod
s_c	Scattered neutrons coming from the test rod but useless for fuel displacement characterization
α	The scatter coefficient, $s_c := \alpha s$
n	Noise, neutrons coming for the CABRI reactor
N	$n + s_c$
c_i	Counting rate of the ith column (r=right, c=center, and l=left)
$b_{ji,TAG\ option}$	The value into bin j[th]: - 1=test rod signal, - 2= test rod (n,xn) or (n,n'), - 3=elastic scattered neutrons, - 4= all other contributions; of the ith column (r=right, c=center, and l=left) evaluated with TAG option (TAG1 or TAG3)

REFERENCES

[1] "Reactivity Initiated Accident transient testing on irradiated fuel rods in PWR conditions: The CABRI International Program". Biard B., et al. Annals of Nuclear Energy, Vol **141**, 15 June 2020, 107253. https://doi.org/10.1016/j.anucene.2019.107253

[2] "Renovation, improvement and experimental validation of the Helium-3 transient rods system for the reactivity injection in the CABRI reactor". Duc B. et al. Proc of IGORR 2014 conference, Bariloche, Argentina, November 2014

[3] "The CABRI fast neutron Hodoscope: Renovation, qualification program and first results following the experimental reactor restart". Chevalier V., Mirotta S., Guillot J., Biard B. EPJ Web of Conferences, Volume **170**, 10 January 2018

[4] "MCNP® user's manual. Code version 6.2. Manual Rev. 0" Werner C. J. editor. LA-UR-17-29981. October 27, 2017

[5] "The CABRI fast neutron Hodoscope: calibration campaign results". Chevalier V., Mirotta S., Monchalin N., Guillot J. The European Research Reactor Conference, the annual gathering of the research reactor community in Europe, RRFM, Oct 2020, HELSINKI (on line), Finland. (https://hal.archives-ouvertes.fr/hal-03080772).

[6] "Quantitative fuel motion determination with the CABRI fast neutron hodoscope: Evaluation methods and results." Baumung K., Augier G., Nuclear Technology, vol **96**, p. 302, June 1991

The VINON-LOCA test facility: exploring the LOCA phenomenology through an out-of-pile thermal sequence on irradiated pressurized fuel rod

B. Biard [1], C. Colin [1], S. Bernard [1], V. Marty [1], G. Volle [1], F. Martin [1], P. Charmasson [1], K. Ronné [1], I. Moysan-Lavoine [1], J.-Y. Ferrandis [2], V. Schoepff [3], G. Amoyal [3], E. Fédérici [1]
[1] CEA, DES, IRESNE, DEC, Cadarache F-13108 Saint-Paul-Lez-Durance, France
[2] CNRS, IES-UMR 5214, 34000, Montpellier, France
[3] CEA, LIST, Sensors and Electronic Architectures Laboratory, CEA-Saclay, 91191 Gif-Sur-Yvette cedex, France
bruno.biard@cea.fr

***Abstract*— Since the out-of-pile semi-integral tests performed at Studsvik in 2011 for the NRC [1] and the Halden Loss-Of-Coolant Accident (LOCA) test series IFA-650 [2], a major safety interest has raised for Fuel Fragmentation, Relocation and Dispersal (FFRD) during a LOCA sequence. In addition to the characteristics of the fuel ejected from the rod after the clad failure, the fuel behaviour before the clad failure is still to be investigated, especially its fragmentation and its possible relocation within the rod during the clad ballooning phase. Furthermore, the chronology and the sequencing of these phenomena is of particular interest.**

For this purpose, the VINON-LOCA program, lying in the framework of a trilateral agreement between EDF, Framatome and CEA, is aimed at performing Out-Of-Pile heating tests on irradiated repressurized fuel rods, reproducing a typical Loss Of Coolant Accident thermal sequence. The VINON-LOCA experimental set-up is located in the so-called VERDON lab of the LECA-STAR hot cell complex. This lab was dedicated to the VERDON-ISTP program [3]. The VINON-LOCA set-up is thus largely instrumented for addressing not only these FFRD topics, but also Fission Gas Release (FGR), combining both online measurement (gamma stations, gamma camera, acoustic sensor, pressure, temperatures, flow meters, microGC…), and pre- and post-test characterization (gamma scanning, tomography, metrology, fuel fragments weighing and sieving, gas analyses…).

An extensive and substantial qualification campaign has been performed to validate the furnace design regarding the desired test conditions, and to qualify the instrumentation. Following some preliminary modelling and calculations, it has included tests on an out-of-cell twin mockup and tests on dummy inactive rods in the hot cell. This allowed achieving successfully the first experimental qualification test of the program end of 2019 on an irradiated UO_2 fuel rodlet. A second irradiated experiment is planned with increased instrumentation capabilities, notably a 2D gamma camera for online fuel motion detection.

***Keywords* — Nuclear fuel / experimentation / gamma imaging / FFRD / LOCA / Hot lab.**

I. Introduction

A loss-of-coolant accident will impact both the cladding and the fuel of a rod. What is mainly investigated in VINON-LOCA experiments is the chronology and sequencing of the phenomena occurring during the first phase of a LOCA sequence: clad ballooning, fuel relocation, fission gas release, clad failure, fuel ejection. When do these phenomena start, and is there a specific occurrence order in between them? The topics that are addressed include thus clad behaviour, Fuel Fragmentation, Relocation and Dispersal (FFRD) and Fission Gas Release (FGR). Results are expected to provide some relevant experimental data for a better understanding of physical phenomena and for code validation, especially the ALCYONE code of the PLEIADES platform [4]. These developments are therefore conducted within a trilateral collaboration between EDF, Framatome and CEA.

The LOCA-type sequence that has been chosen (Fig. 1) starts from a conditioning plateau at 300°C, under inert atmosphere (argon circulation), and follows a desired thermal ramp at 5°C/s, up to a maximal possible temperature of 1200°C. The absence of external cooling leads to an increase of the rod inner temperature and pressure (in case of a closed volume), up to, eventually, the clad ballooning and failure, which should occur between 800 and 1000°C. The experimental sequence will then be ended by a natural cooling.

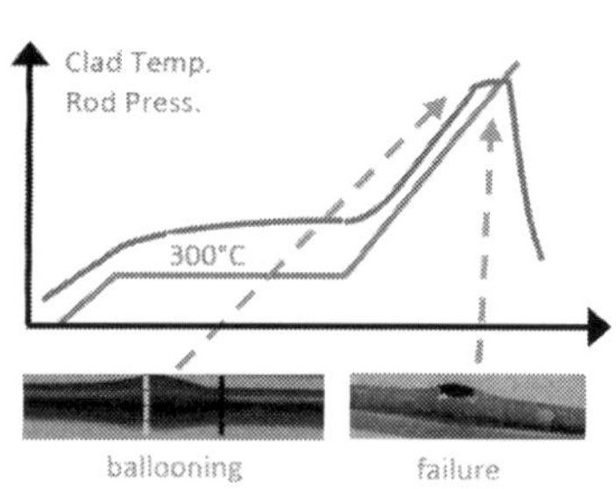

Fig. 1. LOCA sequence and clad expected behaviour.

https://doi.org/10.1051/epjconf/202125306002

The rods to be tested are refabricated from rods irradiated in power plant, shortened to approximately 20cm of fissile column height, and with a maximum burn-up of 90GWd/t (limitations due to furnace dimension and safety authorization for the facility). Two Zircaloy-4 half-pellets are placed at each side of the fissile stack, and the rod is equipped with two welded plugs. The upper plug is axially drilled so as to connect the inner free volume of the rod to an extended plenum instrumented with a pressure sensor. Before the test, the test rod is pressurized at a pressure representative of End-Of-Life conditions (typically 50bar at ambient conditions), either with pure He or with a He/Ar calibrated mix.

II. Facility and instrumentation

The LECA-STAR fuel hot cell complex, located in the CEA Cadarache Research Center, welcomes the VERDON lab comprising two adjacent hot cells, a glove box and their associated utilities. This lab initially hosted the VERDON-ISTP program [3], and VINON-LOCA takes benefits from the equipment dedicated to this previous program.

A High Frequency inductive furnace generates heat in a tungsten susceptor tube, providing indirect heating to the test rod (Fig. 2). The HF generator was already available, but a specific inductive coil, helicoidal plate shape, has been designed and qualified. The rod is held in its upper part and centered in the furnace, in the susceptor tube, by its instrumented line through a metal piece ensuring the furnace air tightness and the crossing for the thermocouples attached to the cladding (rather than welded) for monitoring the clad surface temperature.

The instrumentation and the furnace configuration are illustrated on Fig. 2.

Several gamma stations are available with High Purity Germanium (HPGe) detectors:

- The Furnace Sighting looks online at the rod center for detecting fuel motion in the expected ballooning and failure region through ^{137}Cs activity measurement.
- The Plenum Sighting looks at the plenum cavity (empty Zircaloy cladding tube), following the fission gas release before clad failure through ^{85}Kr activity measurement.
- The Gas Sighting focuses at the experimental circuit downwards the furnace, following the fission gas release and transport after clad failure, once again through ^{85}Kr activity measurement.

These gamma examination systems will be complemented by a new 2D gamma camera (Fig. 3), which is currently under implementation in the hot cell. This camera will record the fuel motion in the same ballooning region as the Furnace gamma sighting, but at a higher speed and in two dimensions (creating online gamma mapping instead of a mean local measurement). This fast gamma imaging system has been specifically designed for the VINON-LOCA needs (detector choice, collimator design, mechanical integration…) in collaboration with CEA/LIST at Saclay [5]. The 2D active panel is an ADVACAM WidePix sensor with a detection window size of 28.2x71.5mm composed of 512x1280 pixels of 55x55µm. This sensor, cooled by water circulation, is placed behind a 6cm-thick tungsten collimator comprising approximatively 500 cylindrical holes of 1mm diameter. A circular rail support allows the azimuthal positioning around the rod in the furnace. The acquisition rate for VINON-LOCA will be around 10 to 50 frames/s.

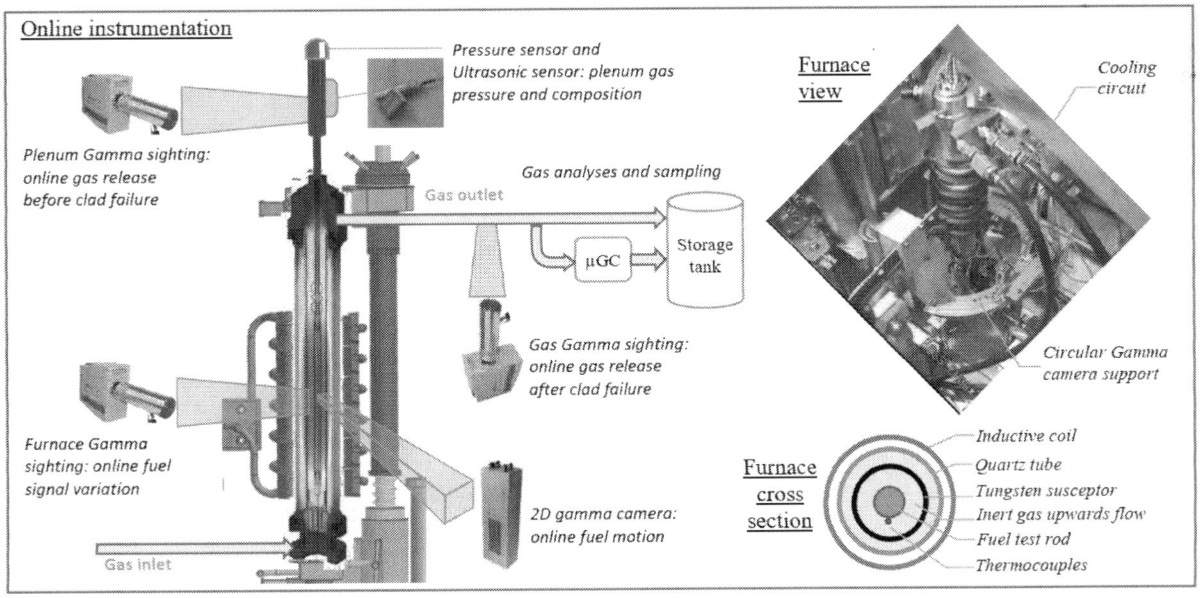

Fig. 2. VINON-LOCA online instrumentation, furnace view and furnace cross section

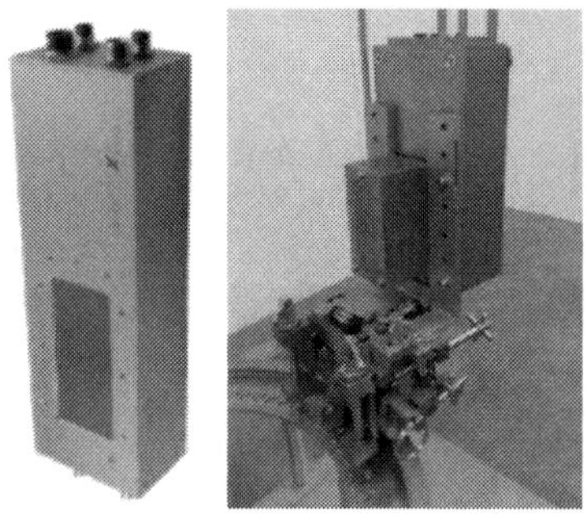

Fig. 3. 2D gamma imaging camera. Left: Advacam WidePix sensor. Right: sensor housed in the collimation box and mounted on its alignment support.

In addition to the clad thermocouples and to the gamma systems, the rod inner pressure is recorded in the plenum cavity by a piezoelectric pressure sensor couples to a Pt100 temperature sensor. It monitors the rod pressure evolution during the thermal ramp and detects the onset of ballooning and clad failure.

The plenum cavity is also equipped with an acoustic ultrasonic sensor, developed and patented by IES/CNRS and EDF [6] and fabricated by Sonaxis. This sensor estimates the gas composition through sound speed measurement, until the clad fails and the pressure becomes too low (Fig. 4). The sensor is stuck onto the cavity with an acoustic coupling material for ensuring proper sound wave propagation. The HORTUS software, specifically developed by IES for this experimentation, ensures on-line measurement and analysis. Post-acquisition analyses based on other set of parameters can be performed, depending on the measurement aim: estimating the Xe fraction within He, or the fraction of a mix Xe/Kr within He for example.

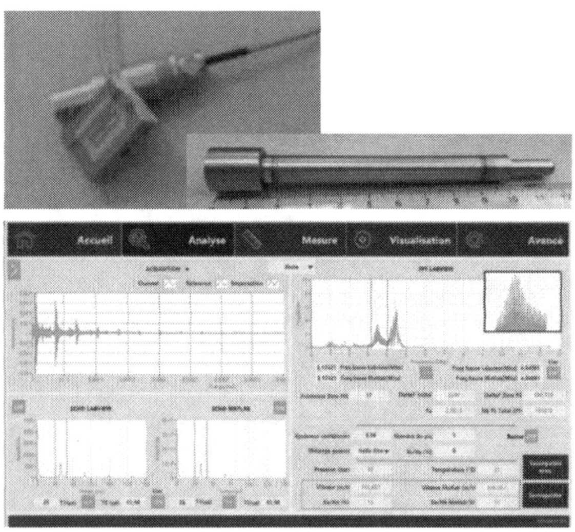

Fig. 4. Ultrasonic sensor. Top left: the sensor and its connector. Top right: the plenum cavity on which the sensor is stuck. Bottom: the associated HORTUS software (example of measurement on a cavity filled with a 10%Xe in He standard gas mixture at 55bar and room temperature).

During the test, a laminar flow of inert gas (classically argon at a 60sccm flowrate) circulates upward in the furnace around the fuel rod, and is guided through a high efficiency Poral filter to a glove box outside the hot cell. The latter comprises a micro gas chromatograph and a storage tank. Once the hydraulic conditions are established, the thermal sequence starts with a relatively slow heating up to 300°C at 1°C/s. A conditioning plateau lasts between 20 to 30 minutes in order to reach a correct thermal equilibrium for the furnace and its components, and for the test sample and its instrumented line. This plateau is also intended for a final checking of the instrumentation, and for acquiring the initial state of the test rod (2D gamma image, gamma emission level, thermal profile). The final ramp is then performed from 300°C either to a desired pre-defined value (e.g. 1000°C, up to 1200°C) or to the rod failure (test stopped manually based on the rod inner pressure measurement). The heating is then cut and the thermal sequence is generally terminated by a free cooling, although it could be followed by a high temperature plateau or a controlled cooling rate.

After the rod failure, the gas circulation continues until no more radioactive krypton (^{85}Kr isotope) is detected by the Gas gamma sighting. A fraction of the circulating gas is driven to a micro gas chromatograph, for estimating the release kinetics of He, Kr, Ar or Xe. All gases are collected in a storage tank during the experiment. Post-test samplings allows quantitative gamma measurement of the ^{85}Kr total released fraction, and possible quantitative analyses.

Pre-test and post-test examinations are performed for characterizing the rod final state. First of all, before moving the failed rod out of the furnace, a gamma scanning is carried out with the Furnace gamma sighting which is mobile along the vertical axis. The 2D gamma camera may also acquire static images from different orientations around the rod. The failed rod is then extracted from the furnace, separated from its instrumented line and thermocouples, visually examined, and weighted for mass loss estimation. It is then transferred to the adjacent hot cell equipped with a 3-dimension examination bench (Fig. 5).

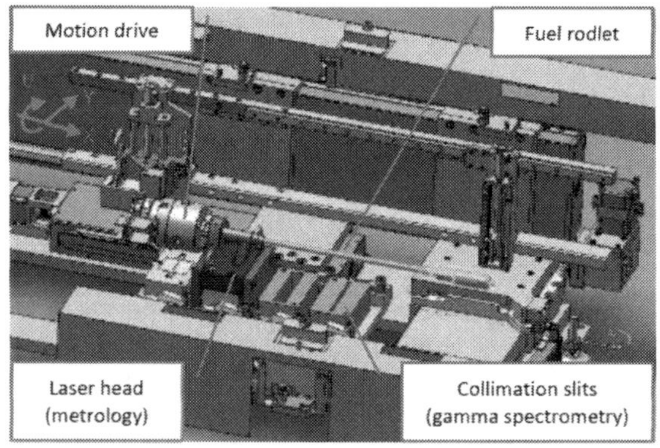

Fig. 5. 3D examination bench for gamma scanning, gamma emission tomography and laser metrology.

The different examinations capabilities of this bench are:

- Axial gamma scanning, with adjustable collimation slits (the thinnest one being 0.6mm large);
- Gamma emission tomography, with a crossed collimation slit of 0.2x10mm length (meaning that the gamma signal is an average on 10mm along the fuel stack axis);
- Laser measurement of the clad, for estimating its diameter and deformation, notably in the balloon region.

The fuel mass balance is closed thanks to the collection and weighing of the following elements:

- the test rod itself;
- the fuel fragments ejected during the thermal sequence in the furnace, collected in the susceptor tube and in the ashtray at the furnace lower end;
- the fuel fragments recovered during the rod handling after the test (drip pans under the different equipment).

The granulometry of these fuel fragments is finally estimated by sieving.

III. Simulation, qualification and first results

The assessment and qualification of the furnace have included modelling studies, tests in a twin mockup named EXPAT, available out of the hot cell, and finally in-cell qualification tests carried out on dummy rods. The coil design required numerous tests before retaining the final helicoidal plate configuration. The objective was to reach a rather homogeneous axial thermal gradient, optimized as a compromise between two opposite goals: on one hand the flatter the profile, the more realistic it is, but on the other hand, a peak profile is required for controlling the ballooning and failure region and positioning the online instruments. Moreover, the performances are obviously limited by the physics, and of course by the heat leaks of the system. Fig. 6 presents the final recordings of this profile, monitored with 6 thermocouples along a dummy rod cladding. Simulant pellets (Zirconia or Alumina) have been used, sometimes drilled for inserting a thermocouple. PID settings for the furnace HF generator were finely adjusted to reach the desired thermal gradient during the ramp. Finally, COMSOL calculations validated the thermal aspects (cooling impact at both quartz tube ends, inner rod thermal gradient for examples).

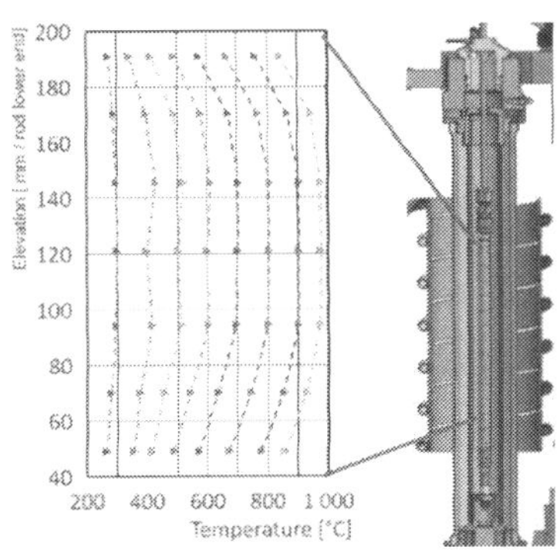

Fig. 6. Axial thermal profile in the VINON furnace (qualification test)

The qualification tests proved the capability of the VINON-LOCA experimental setup and instrumentation to control and monitor the rod failure region. Many dummy rods were tested, some of them undergoing several thermal ramps, which gave to the experimenters confidence in the clad balloon and failure position. This allowed a correct alignment for the furnace gamma sighting so as to record the gamma activity signal of the fuel in this specific region. Some of these dummy rods were also tested first in the EXPAT furnace then in the VINON-LOCA one to check the reproducibility of the thermal performances of both devices.

The first experimental test on irradiated UO_2 was then successfully achieved by the end of 2019. Results were in good accordance with preliminary calculations performed with the CEA ALCYONE code to estimate the experimental rod behavior in such conditions. The targeted ramp was reached, the balloon and clad failure occurred at the expected location and time, and the fuel signal variation was correctly detected by the furnace gamma sighting. The ^{85}Kr gas release was properly detected both by the gas gamma sighting and by the post-test samplings, showing very consistent results. The rod was fully characterized after the test. Examples of results are shown on Fig. 7 and Fig. 8.

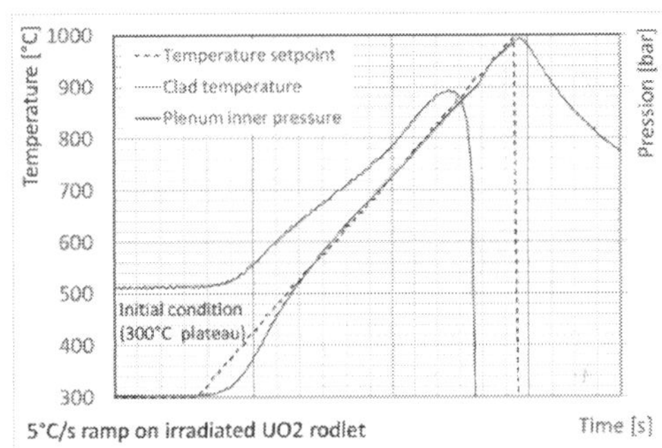

Fig. 7. Clad temperature and rod inner pressure recordings during the first test on irradiated UO_2.

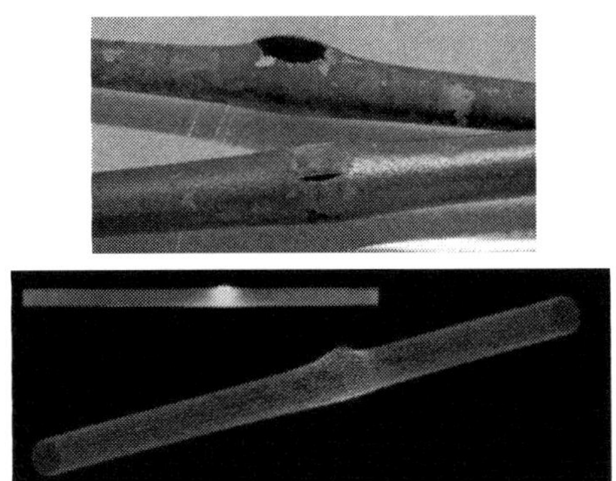

Fig. 8. Post-test examination results. Top: final aspect of burst dummy rods (pre-oxidised Zircaloy cladding, Al_2O_3 pellets, after several ramps). Bottom: 3D post-test laser metrology.

IV. Perspectives and conclusion

The VINON-LOCA facility has been successfully used to perform its first experimental qualification test on an irradiated UO_2 fuel rodlet at the end of 2019. It will soon be equipped with enhanced instrumentation capabilities, notably the fast 2D gamma camera and upgraded electronic modules for the furnace gamma sighting. Reflections are underway for installing, if possible, other instrumentation such as a confocal imaging system for rod/clad characterization, a microphone for recording the clad failure sound (and more), or contactless pyrometric temperature measurement. Future tests to be performed in this facility might study the impact of burnup, initial pressure or free volume, various parameters of the thermal sequence (ramp stopped at clad failure or extended with a final plateau), or the behavior of innovative fuel rods (e.g. ATF) under the first phase of a LOCA transient.

Acknowledgment

Technical support and financial contribution of EDF and Framatome are gratefully acknowledged.

References

[1] P. Raynaud, Fuel Fragmentation, Relocation, and Dispersal during Loss-of-Coolant Accidents, US NRC, NUREG-2121, March 2012
[2] Report on Fuel Fragmentation Relocation, Dispersal, Organisation for Economic Co-operation and Development, Nuclear Energy Agency, Committee on the Safety of Nuclear Installations, NEA/CSNI/R (2016)/16
[3] A. Gallais-During, et al., Overview of the VERDON-ISTP Program and main insights from the VERDON-2 air ingress test, Annals of Nuclear Energy, Vol. 101, pp. 109-117, ISSN 0306-4549, 2017, https://doi.org/10.1016/j.anucene.2016.09.045
[4] B. Michel et al., Simulation of Pellet-Cladding Interaction with the PLEIADES Fuel Performance Software Environment, Nuclear Technology, 182:2, 124-137, 2013, DOI: 10.13182/NT13-A16424
[5] G. Amoyal et al., Design and characterization of a gamma imaging system for fuel rod deformations, ANIMMA 2021 Proceedings
[6] E. Rosenkrantz, Jean-Yves Ferrandis et al., Ultrasonic measurement of gas pressure and composition for nuclear fuel rods, Nuclear Instruments and Methods in Physics Research Section A: Accelerators, Spectrometers, Detectors and Associated Equipment, Vol. 603, Issue 3, pp. 504-509, 2009, https://doi.org/10.1016/j.nima.2009.02.028

Long-term Transmission Characteristics of CYTOP Fiber Exposed to Gamma Radiation

Ivan Chapalo[1], Andrei Gusarov[2], Damien Kinet[1], Karima Chah[1], Ying-Gang Nan[1] and Patrice Mégret[1]
[1]Electromagnetism and Telecom Department, University of Mons, Belgium
[2]SCK-CEN, Belgium
Ivan.Chapalo@umons.ac.be

***Abstract*—We investigated post-irradiation variations of the radiation induced attenuation (RIA) of polymer perfluorinated (CYTOP) optical fiber exposed to gamma radiation. The RIA measured five months after the irradiation was qualitatively similar to that measured several hours after irradiation. However, quantitatively the RIA post-irradiation evolution significantly depends on the spectral range: in the visible range, the transmission partly recovers, while in the infrared, at wavelengths above 1300 nm the RIA grows and then saturates becoming permanent. The latter feature is prospective for gamma radiation dosimetry as a permanent long-term indicator of received radiation dose.**

***Keywords* —CYTOP, perfluorinated fiber, polymer optical fiber, gamma radiation, radiation induced attenuation, fiber dosimetry.**

I. Introduction

POLYMER optical fibers (POFs) have been attracting substantial attention from research and industry thanks to particular mechanical properties, simple handling, biomedical compatibility and safety [1,2]. Among POFs, cyclic transparent amorphous fluoropolymer (CYTOP) fiber demonstrates radically low attenuation in the telecom transparency windows of 850 and 1310 nm, thus becoming an efficient solution for short distance communication links [3]. Besides fiber communications, CYTOP fiber has also been intensively studied for various sensing applications: fiber Bragg gratings [4], Brillouin scattering [5] and intermodal interference [6] are good examples of investigated sensing principles.

A particular topic of research is the use of POFs in radiation environments, including possible solutions for gamma radiation dosimetry [7,8]. A basic step in this direction is the study of gamma radiation influence on fibers' properties aiming for evaluation of their transmission characteristics degradation. Recently, the gamma-radiation induced attenuation (RIA) in CYTOP fibers has been extensively investigated. The RIA was measured in the visible and near-infrared spectral bands [9,10]; the influence of temperature and relative humidity on irradiated fibers was examined [11], and the distributed gamma radiation detector was proposed [7]. However, evolution of the RIA properties versus time was not considered in detail.

In this work, we investigate the long-term (residual) RIA of CYTOP fiber. We irradiated several POF samples by gamma radiation to 1, 5, 20, 50 kGy doses and measured the RIA five months after irradiation in the range of 500-1700 nm. Additionally, we investigate and analyze the RIA time-evolution measured starting shortly after irradiation and during up to 80 hours.

II. Experimental Setup

In experiments, we used multimode graded-index fiber (MMF) GigaPOF-50SR with 50-µm CYTOP core and 490-µm polycarbonate overclad (Chromis Technologies). Several 2-m POF samples were irradiated up to 1 kGy, 5 kGy, 20 kGy and 50 kGy gamma radiation doses at a 7.5 kGy/h dose-rate. Irradiation was conducted using a calibrated setup (Brigitte facility, SCK-CEN, Belgium) based on ^{60}Co irradiation sources located at a depth of seven meters in a water pool (Fig. 1). The sources formed a cylindrical volume where the stainless steel hermetic container with fiber samples was placed for a specified time according to the required dose. After irradiation, the samples were stored at uncontrolled laboratory conditions during five months.

To measure POF transmission characteristics in a wide spectral range, we developed the scheme based on a supercontinuum broadband source (BS) NKT SuperK Compact 450-2400 nm and an optical spectrum analyzer (OSA) Yokogawa AQ6374 350-1750 nm (Fig. 2). The BS was

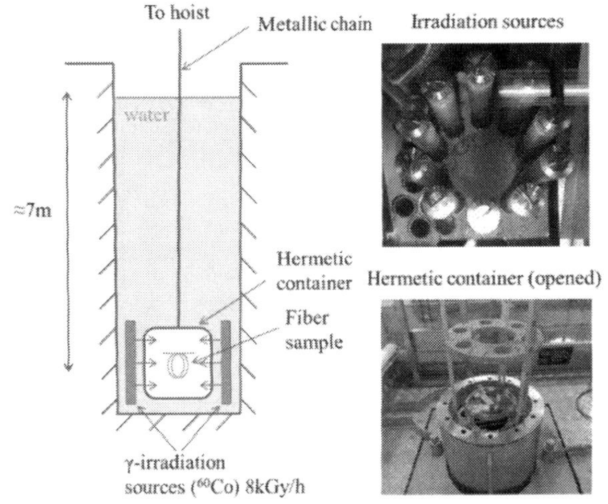

Fig. 1. Schematic and photographs of the irradiation setup.

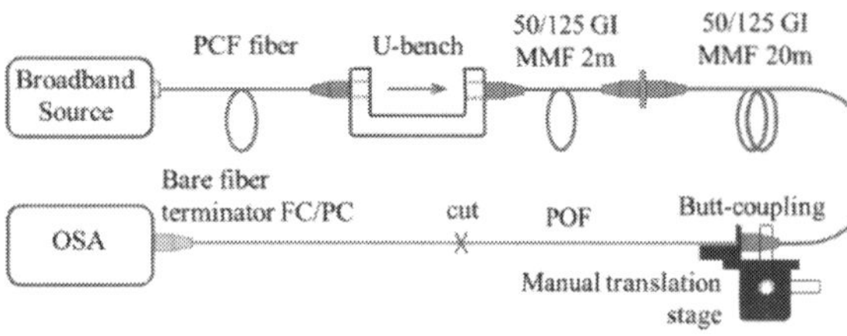

Fig. 2. Schematic of the experimental setup.

equipped with a "U-bench" with collimating optics and FC/APC connectors to avoid mechanical contact and possible damage of delicate PCF-fiber patchcord serving as a BS output. Light was introduced to the POF sample from the FC/PC connector of silica graded-index 50/125 MMF patchcord (GIF50C, Thorlabs) by butt-coupling method at the manual translation stage (NanoMAX 313D, Thorlabs).

The co-axial positioning of the fibers was obtained by observing the speckle pattern of the near-field image at the output POF facet and minimizing the number of interference spots at the speckle pattern. For this, we used visible light source (635 nm) and silica SMF patchcord (SM600 fiber, Thorlabs) connected to the silica MMF to provide near single-mode launch, when the POF and the MMF were well-aligned. Additionally, fiber positioning was adjusted by maximizing output power from the POF using the OSA in the single wavelength mode. Index-matching gel was utilized at the POF and silica fiber joint. The output end of POF was connected to the OSA using FC/PC bare fiber terminator with a ferrule's internal diameter of 500 µm.

The losses measurement for each POF sample was conducted by cut-back method: we measured the transmission spectrum at the output of 2-m POF sample, then we cut one meter of the sample and measured the transmission spectrum again. When it was required (for example, in the case of very low signal), the procedure was repeated for 1 m and so on. Since the output end of the sample was connectorized by the bare fiber terminator, it can bring some measurement inaccuracies. To avoid significant measurement errors, each time we connectorized the POF end-face 3-5 times and measured the transmission spectrum for each connectorization. To obtain the final transmission spectrum, we excluded spectra that demonstrated noticeably less power than others and we averaged the remaining spectra.

We cut POFs by a FiberFin razor cutter. After cutting, fiber facets were cleaned by alcohol and inspected with a microscope. The cutting was repeated if visible disturbances of fiber facets were observed.

Multimode nature of CYTOP fiber caused some difficulties with the OSA operation: the spectrum contained parasitic intensity modulation as a result of intermodal interference. The parasitic effect was minimized by two parameters: the fiber length and the OSA resolution. We used additional 20-m 50/125 silica MMF patchcord prior to the POF to increase the fiber length. This made the intermodal interference signal more sensitive to the optical frequency and a modulation in the OSA spectrum became more quickly changing with a wavelength. Then, we choose a resolution of the OSA of 2 nm so that the parasitic modulation was effectively averaged on the one hand, and spectral peaks inherent in irradiated fibers were not hidden on the other hand. We also supposed that light entering the OSA from the POF end facet was less subjected to spatial filtering by the OSA internal slits when using lower resolution (2 nm) and therefore the intermodal interference signal was less pronounced.

The RIA was calculated as follows. First, the loss spectrum of irradiated fiber $\alpha_{irr.}$ was calculated from cut-back measurement by subtracting 2-m sample spectrum S_1 from 1-m sample spectrum S_2 and dividing by cut length l (1 m in this case):

$$\alpha_{irr.}\ [\mathrm{dB/m}] = (S_2\ [\mathrm{dBm}] - S_1\ [\mathrm{dBm}])/l\ [\mathrm{m}]. \quad (1)$$

Then, the RIA spectrum was calculated by subtraction the POF loss spectrum α_{CYTOP} from the irradiated POF loss spectrum $\alpha_{irr.}$:

$$\mathrm{RIA}\ [\mathrm{dB/m}] = \alpha_{irr}\ [\mathrm{dB/m}] - \alpha_{CYTOP}\ [\mathrm{dB/m}]. \quad (2)$$

The POF loss spectrum α_{CYTOP} was measured in advance by cut-back method using 35-m fiber cut to a length of 2 m.

III. Experimental Results and Discussion

Figure 3 presents typical measured raw spectra: the spectrum of the light source (dashed black line), transmission spectrum of 2-m pristine POF sample, and transmission spectra of the POF sample that received a dose of 20 kGy (red and green lines). The irradiated 1-m POF sample was cut to 0.5 m for the cut-back measurement in this example. It is seen that the light source power spectral density rapidly decreases below 550 nm; irradiated POF demonstrates strong losses in this range as well. Moreover, in the range of 350-550 nm some oscillating structure independent on fiber length was observed. We interpreted it as a "ghost" spectra according the user's manual of the OSA. Taking the above into account, the lower wavelength of the RIA curves was varied depending on loss level in this spectral region. It also should be mentioned that the light source' power was unstable in a range of 1000-1130 nm, therefore, at the RIA graphs, we did not plot data in this range.

Figure 4 demonstrates RIA graphs measured 5 months after irradiation for doses 1, 5, 20, 50 kGy. These measurements were performed at UMons. For convenience of total loss estimation, the pristine CYTOP fiber loss curve is presented as well using additional vertical axis. Three typical zones of RIA spectra can be distinguished: visible and the initial part of infrared with strong RIA increasing towards shorter

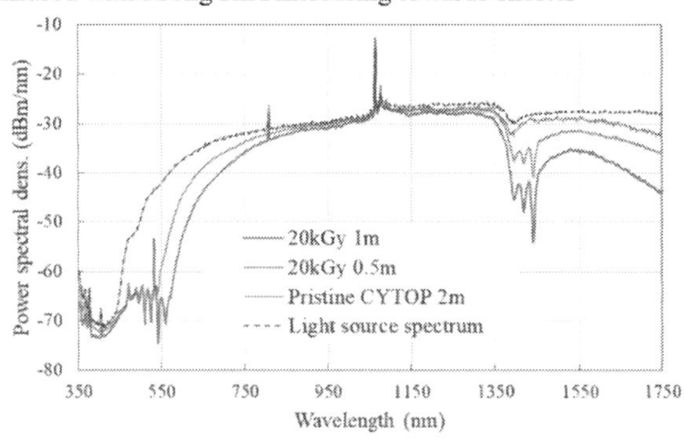

Fig. 3. Typical spectra of the light source, pristine 2-m POF, and irradiated POFs during the cut-back loss measurement.

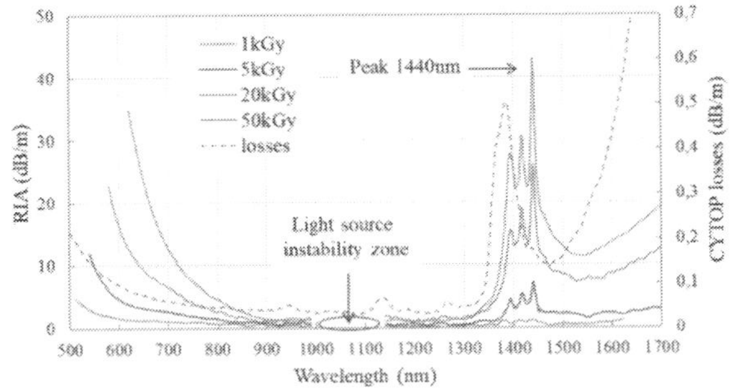

Fig. 4. CYTOP fiber RIA measured 5 months after irradiation.

wavelengths, the zone 900-1300 nm with a relatively weak RIA, and the zone above 1300 nm having complex behavior with a few sharp attenuation peaks in the vicinity of 1400 nm. The RIA dependence on the irradiation dose (measured 5 months after irradiation) is presented in Fig. 5 for two representative wavelengths: 650 nm chosen in the zone of high RIA in the visible part of the RIA spectrum, and the RIA maximum peak of 1440 nm. The dependence at 650 nm is well approximated by a linear function with a slope 0.46 dB/(m•kGy). For the peak wavelength of 1440 nm, the results demonstrate more complex behavior: at the lower doses (1, 5, 20 kGy) the dependency is close to linear (estimated as 1.27 dB/(m•kGy), while at 50kGy a lower sensitivity was observed. This behavior is not completely clear and should be investigated in more detail with a higher number of irradiation doses. It should be mentioned that at other wavelengths in the visible range, the RIA also demonstrated linear dependency on the irradiation dose but with different sensitivity: the sensitivity increases with decreasing wavelength.

We investigated the influence of launching conditions on the measured RIA. For this, we utilized the superluminescent diode (S5FC1005S, Thorlabs) operating at 1550 nm with a single-mode output and the mode scrambler (ModCon, Arden Photonics) providing stable multimode regime of light propagation. The RIA measurements using underfilled launch (by the single-mode patchcord) and multimode launch (using the mode scrambler) did not demonstrate noticeable differences in the RIA.

We also investigated the RIA evolution during first hours after irradiation. The irradiation was performed at the same irradiation facility described earlier. To allow for comparison with the previously analyzed long-term RIA, we re-assembled

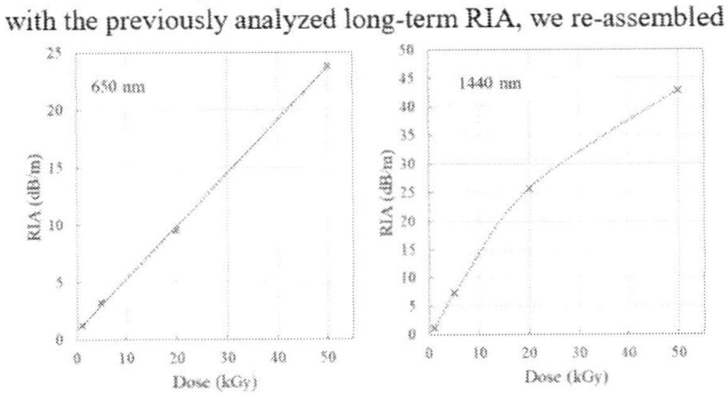

Fig. 5. RIA dependency on the irradiation dose at two wavelengths.

the same measurement setup (Fig. 2) at the irradiation site of SCK-CEN. This allowed us to start experiment within 1-2 hours after irradiation. The delay was defined by the safety procedures. We irradiated two 2-m fiber samples up to 5 and 17.5 kGy doses and measured the RIA during more than 20 hours. During the experiment, the temperature was in the range of 27 to 29 ^{0}C. The RIA was calculated by subtraction the transmission spectrum of irradiated fiber sample $S_{irr.}$ from the spectrum of the same sample measured before irradiation $S_{ref.}$ and normalizing by the sample length l:

$$RIA(t)\ [dB/m] = (S_{ref.}\ [dBm] - S_{irr.}(t)\ [dBm])/l\ [m]. \quad (3)$$

After irradiation and the RIA measurements, the samples were again stored at room conditions during several months in order to measure the long-term RIA.

Measured RIA is presented in Fig. 6. It is seen that after irradiation, significantly stronger RIA (comparing to results obtained 5 months after irradiation) is observed at the short-wavelength part of the spectrum. It, however, drastically decreases with time as a result of annealing process. For example, for 5 kGy dose, the RIA at 850nm decreased from 10 dB/m measured 1.5 hours after irradiation to 2.5 dB/m measured 80 hours after irradiation. Moreover, 5 months after irradiation, the RIA was stabilized at the value less than 1 dB/m.

In contrast, opposite behavior was observed at the long-wavelength part: the RIA continued to rise after irradiation and saturated at levels significantly higher than initial ones in both 1400 nm sharp peaks region and wavelengths longer than these peaks. This is especially pronounced for the higher dose of 17.5 kGy (Fig. 6b). Unlike the effect of the transmission recovery with time in the visible spectral range, which is typical for various types of fibers and for CYTOP in particular, the

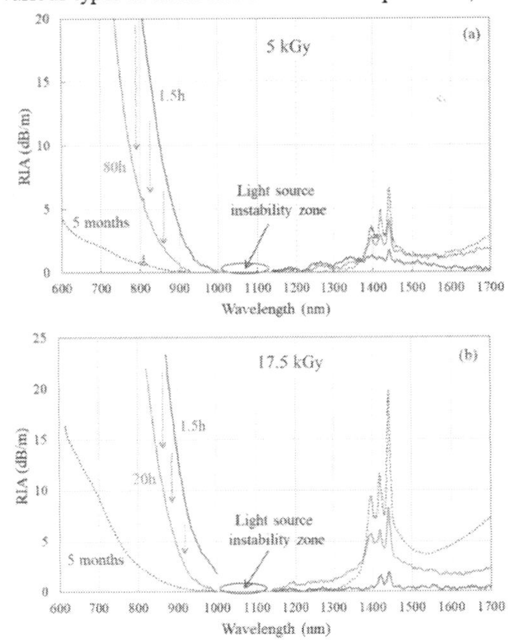

Fig. 6. RIA measured 1.5 h, 80 h and 5 months after irradiation for 5 kGy dose (a) and 1.5 h, 20 h and 5 months after irradiation for 17.5 kGy dose (b).

continued rise of attenuation after irradiation is quite unusual. As discussed in [11], RIA in 1440 nm region can be related with a formation of conjugated systems that are stable and expected to cause permanent RIA. Nevertheless, this phenomena's origin should be additionally investigated.

The dynamics of the above mentioned effects is more clearly seen on the graphs of the RIA evolution versus time passed after the end of the irradiation (Fig. 7). The graphs are plotted for two wavelengths: 880 nm representing the left side of the spectrum, and the RIA peak wavelength of 1440 nm. The transmission recovery with time inherent for the left side of the spectrum is clearly seen in Fig. 7a. Moreover, a strong decrease of the RIA is observed during approximately five hours after irradiation. At 1440 nm (Fig. 7b), strong rise of the RIA also during approximately five hours after irradiation is clearly seen for the dose of 17.5 kGy; then the RIA rises more slowly tending to be saturated. The lower dose of 5 kGy does not demonstrate rapid RIA growth in the beginning part of the graph and rises more smoothly, also tending to be saturated. It should be mentioned that during 2-2.5 hours after irradiation, the RIA at 1440 nm (Fig. 7b) demonstrates lower values in the case of 17.5 kGy dose than the 5 kGy case. We also can notice that the strong growth of the RIA occurs after but not during irradiation. These effects are not clear and future RIA investigation not only after, but also during irradiation process would clarify this behavior.

IV. Conclusions

In this work, we experimentally investigated long-term post-irradiation RIA variations of the CYTOP fiber. RIA measurements performed five months after irradiation demonstrated spectra qualitatively similar to those measured during and after irradiation and reported in previous works [7,11]. The main features of the long-term RIA are (1) significant decrease in time in visible range (however residual RIA is still present in the fiber even few months after irradiation), and (2) post-irradiation rise and saturation at wavelengths >1300 nm (this RIA can be considered as permanent). The first feature should be taken into account when considering CYTOP fiber dosimetry applications based on the RIA: the RIA values strongly depend on time passed after irradiation. The second feature suggests to consider alternative wavelength range for dosimetry application where the RIA stabilization takes time, but the RIA values become permanent.

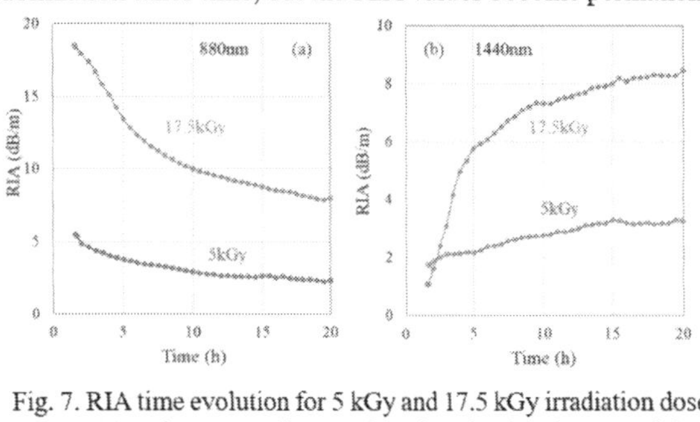

Fig. 7. RIA time evolution for 5 kGy and 17.5 kGy irradiation doses at 880 nm (a) and 1440 nm (b) wavelengths. The time is counted from the end of the irradiation.

Considering data transmission, the wavelength range 1000-1300 nm has minimum attenuation in long-term taking into account both RIA and pristine fiber losses. Partial transmission recovery with time in 850 nm window gives the opportunity to use this transparency window as well, if the method of further RIA reduction could be found.

References

[1] K. Peters, "Polymer optical fiber sensors—a review," Smart Mater. Struct., vol. 20, no. 1, 2011, Art. no. 013002.

[2] L. Bilro, N. Alberto, J. L. Pinto, and R. Nogueira, "Optical sensors based on plastic fibers," Sensors (Switzerland), vol. 12, no. 9, pp. 12184-12207, 2012.

[3] Y. Koike and M. Asai, "The future of plastic optical fiber," NPG Asia Mater., vol. 1, no. 1, pp. 22-28, 2009.

[4] A. Theodosiou and K. Kalli, "Recent trends and advances of fibre Bragg grating sensors in CYTOP polymer optical fibres," Opt. Fiber Technol., vol. 54, 2020, Art. no. 102079.

[5] Y. Mizuno and K. Nakamura, "Potential of Brillouin scattering in polymer optical fiber for strain-insensitive high-accuracy temperature sensing," Opt. Lett., vol. 35, no. 23, pp. 3985-3987, 2010.

[6] I. Chapalo, A. Theodosiou, K. Kalli, and O. Kotov, "Multimode Fiber Interferometer Based on Graded-Index Polymer CYTOP Fiber," *J. Lightw. Technol.*, vol. 38, no. 6, pp. 1439-1445, 2020.

[7] P. Stajanca and K. Krebber, "Radiation-induced attenuation of perfluorinated polymer optical fibers for radiation monitoring," Sensors (Switzerland), vol. 17, no. 9, 2017, Art. no. 1959.

[8] C. Broadway, D. Kinet, A. Theodosiou, K. Kalli, A. Gusarov, C. Caucheteur, and P. Mégret, "CYTOP fibre Bragg grating sensors for harsh radiation environments," Sensors (Switzerland), vol. 19, no. 13, 2019, Art. no. 2853.

[9] P. Stajanca and K. Krebber, "Towards on-line radiation monitoring with perfluorinated polymer optical fibers," *Proc. SPIE*, vol. 10323, 2017, Art. no. 103231O.

[10] P. Stajanca, L. Mihai, D. Sporea, D. Negut, H. Sturm, M. Schukar, and K. Krebber, "Impacts of gamma irradiation on Cytop plastic optical fibres," in *The 25th International Conference on Plastic Optical Fibers (POF2016)*, Birmingham, UK, 2016, pp. 114-117.

[11] P. Stajanca, L. Mihai, D. Sporea, D. Negut, H. Sturm, M. Schukar, and K. Krebber, "Effects of gamma radiation on perfluorinated polymer optical fibers," Opt. Mater., vol. 58, pp. 226-233, 2016.

Optimized High-Temperature Irradiation-Resistant Thermocouple for Fast-Response Measurements

Richard Skifton[1], Joe Palmer[1], and Alex Hashemian[2]
[1]Idaho National Laboratory, USA
[2]Analysis and Measurement Services, USA
Richard.Skifton@inl.gov

***Abstract*—The high-temperature irradiation-resistant thermocouple is the only temperature probe proven to withstand the high-temperature (>1290°C), high-radiation (a fluence of up to ~1 × 10^{21} n/cm^{2}) environments of nuclear reactor fuel design testing and/or over-temperature accident conditions. This report describes the improved performance of a molybdenum and niobium thermocouple by utilizing a coaxial design (i.e., a single wire grounded to the outer sheath). This optimized high-temperature irradiation-resistant thermocouple features a simplified design yet allows for more robust individual components. The niobium and molybdenum thermoelements can be used interchangeably in either the sheath or wire, depending on the intended application. Via a plunge test in flowing water, the response time of the coaxial build of the high-temperature irradiation-resistant thermocouple was determined to be 30x faster than that of the comparable ungrounded type-K thermocouples, and 10x faster than the grounded type-K thermocouples and traditional ungrounded high-temperature irradiation-resistant thermocouples (i.e., two-wire configurations). Furthermore, by capitalizing on the coaxial design, a multi-core high-temperature irradiation-resistant probe with multiple "single-pole" wires along the length of the sheath was proven feasible. This multi-core, thermocouple design was dubbed a "demicouple." The high-temperature irradiation-resistant demicouple is primarily applied during fuel experiments to record multiple fuel-pin centerline temperature measurements using a single compact sensor. Furthermore, the shared "common" leg between demicouple junctions reduces error propagation in secondary measurements such as temperature differentials.**

***Keywords* —Thermocouple, Irradiation-Resistant, In-pile, Sensor**

I. Introduction

THE high-temperature irradiation-resistant thermocouple (HTIR-TC) is proven to survive periods of temperature and irradiation exposure never approached by any other thermocouple (TC). Recent irradiation tests during the Advanced Gas Reactor 5/6/7 experiment demonstrated the HTIR-TC's lifetime to be on the order of 10^{21} n/cm^{2} at temperatures of up to 1500°C [1]. Such a lifetime is sufficient for commercial application during a nuclear power plant's typical 18-month refueling cycle, as well as in consideration of the high temperatures expected to be generated by advanced nuclear reactor concepts.

There is a need to continually improve the HTIR-TC so as to achieve longer lifetimes, improved form factors, faster time responses, etc. All deviations from the traditional HTIR-TC concept must be built, tested, and qualified for nuclear application—adding to the repertoire of high-temperature sensors capable of withstanding high doses of radiation.

II. Background

The basic working principle behind any TC is the generation of electromotive force (EMF) by heating one end of a circuit of two dissimilar metals [2, 3]. The EMF is a measurement of electrical potential (usually in mV), as determined via the material specific Seebeck effect. For a basic overview of how a TC operates and behaves during an irradiation test, see the appendix at the end of this report.

Conventional base metal TCs (e.g., type-K TCs) are commonly used in pressurized-water reactors (PWRs) to measure the reactor core exit temperature. These core exit TCs (CETs) generally do not provide input to the reactor trip system; thus, they are not subject to the same stringent response time and calibration requirements as, for example, safety-related narrow-range reactor coolant system resistance temperature detectors. At minimum, PWRs require two operable CET channels per reactor quadrant to enable radial temperature gradient indications. With two CETs per channel, this means that a minimum of 16 operable CETs are required from an inventory of more than 60 installed CETs [4]. In general, CETs are declared inoperable or out of service if they read an unrealistic temperature (e.g., one that is off the scale) or suffer from erratic spiking on the output signal. The cause of the CET failure may be due to the TC itself, or to a cable/connector issue—as is often the case [5]. Aside from these failure modes, CETs have not been shown to suffer from systematic calibration drift under normal PWR operation conditions—mainly due to the low temperature and low neutron flux regions they reside. However, it is generally understood within the advanced nuclear reactor community that both base metal and noble metal TCs may experience calibration drift (i.e., decalibration); thus, they may be unsuitable for long-term operation in the high-temperature, high-radiation environments expected in next-generation nuclear power plants.

Further, direct temperature measurements of core fuel—either at a centerline or cladding location—may reduce temperature uncertainties by an order of magnitude. This could provide reactor operators a better envelope in which to operate. Of course, temperature is not the only metric to be considered, but it plays a significant role in the thermal, mechanical, and neutronics of the core.

III. Optimized HTIR-TC Design

The HTIR-TC was recently redesigned as an optimized coaxial cable—a TC built using an outer sheath wrapped around an individual wire. By using the sheathing material as a "live" thermoelement, the sheathing is integral to the generated EMF. In contrast, the traditional HTIR-TC build utilizes two dissimilar wire thermoelements housed in an outer "dead" sheath housing.

The coaxial build was then further iterated by forming a multipoint TC, referred to here as a "demicouple" (DC). Up to five "coaxial" wires can be integrated into the sensor while still maintaining the same overall diameter as the traditional single-wire HTIR-TC build.

A. Coaxial HTIR-TC

A cartoon schematic/cutaway of the coaxial HTIR-TC is shown in Fig. 1, with the outer sheath (shaded red) composed of the niobium (Nb) thermoelement, and the inner wire (shaded green) composed of the molybdenum (Mo) thermoelement. (Though not shown, the sensor can also be built in an inverted manner, with Mo for the outer sheath and Nb for the inner wire). With the Nb sheathing thermoelement now a "live" part of the sensor, the EMF generated within the sheath is part of the overall sensor EMF output. In Fig. 1, the metal-oxide insulation between the thermoelements is shown as yellow, and its purpose is to electrically insulate the two thermoelements from each other along the length of the cable. At one end of the TC, a single junction would be formed by swaging the outer sheath onto the inner wire. This junction—to the right of the schematic shown in Fig. 1—would be the location of the temperature measurement. The cabling and lead wires going to the data acquisition (DAQ) module is shown as going to the left of the schematic in Fig. 1.

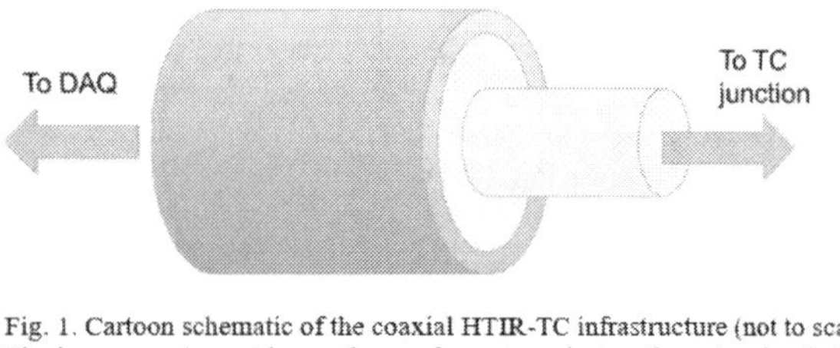

Fig. 1. Cartoon schematic of the coaxial HTIR-TC infrastructure (not to scale). The inner core (green) is one thermoelement conductor; the outer sheath (red) is another. The insulation (yellow) between the two conductors is typically comprised of a metal oxide (i.e., alumina). The sensor would continue on to the right to the measurement location and to the left to the Data Acquisition System (DAQ).

There are several advantages to building the HTIR-TC according to a coaxial design, rather than the traditional two-wire design: (1) the smaller outer diameter of the sensor can lead to less disruption of the sensing environment; (2) the additional material (i.e., mass) comprising both thermoelements, individually, within the sensor cross section enhances their robustness; (3) less overall lengths of material are needed to manufacture the TC, thus reducing costs and lead times; (4) manufactured spools of coaxial HTIR-TC cabling are mass-producible, (5) the sensor can be cut to length from the mass-produced spool, as opposed to the one-off builds for the traditional two-wire HTIR-TC. One major disadvantage—contrary to these five advantages listed—is the inability to be electrically isolated from the surroundings due to being inherently grounded. As with all grounded thermocouples, this must be considered to avoid erratic noise in the measurement.

An abbreviated cross-section view of the coaxial design is shown in Fig. 2, taken from an x-ray radiograph. In the figure, the outer sheath—seen at the top and bottom of the image—is comprised of Nb, while the inner wire is Mo. The lighter gray area between the outer sheath and inner wire is alumina—an electrical insulator.

Fig. 2. Abbreviated cross-section view—using x-ray radiography—of the coaxial HTIR-TC. The inner-wire core is composed of Mo, while the outer sheath is made of Nb (top and bottom of image). The insulation between the two conductors is comprised of alumina.

B. Multipoint HTIR-DC

The coaxial build can be integrated into a DC by using only "half" (corresponding to the prefix "demi-") the number of wires required for a traditional TC. Multiple identical thermoelement legs are inserted into the insultation and are terminated at the desired locations for temperature measurement. A cutaway example of a 5-point DC is seen in Fig. 3, with one of the five wires shown as a TC junction being swaged onto the inner wall of the TC sheath. The other four wires continue along down the cabling to form junctions elsewhere.

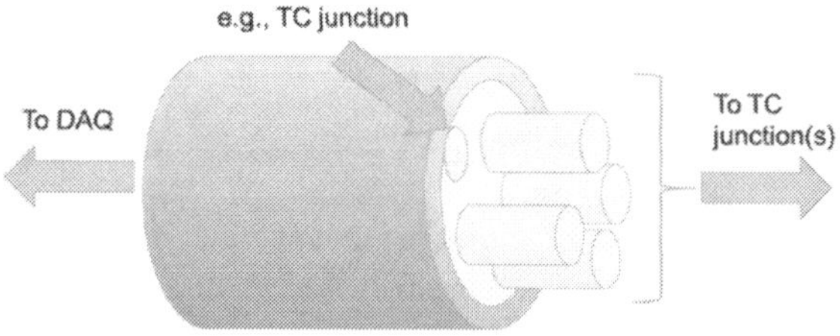

Fig. 3. Cartoon schematic/cutaway of the HTIR-DC (not to scale). An example TC junction is seen at the figure cut plane, while four other unique thermoelements continue along down the length of the cable to form TC junctions elsewhere.

IV. Experimental Setup, Testing, and Results

Two major tests (i.e., a long-term drift test and a time response plunge test) were performed on the coaxial HTIR-TCs, along with the expected proof-of-concept tests.

A. Long-term drift test

The coaxial HTIR-TC was implemented in Idaho National Laboratory's high-temperature longevity furnace, remaining there for thousands of hours at 1250°C. The furnace control TC was periodically checked by inserting a type-B TC with National Institute of Standards and Technology (NIST) traceable calibration into the furnace thermal well. This method alleviated drift within the furnace Type-B TC and ensured that the furnace itself did not drift over time.

The results of the longevity furnace test for the coaxial HTIR-TC are seen in Fig. 4. HTIR-TC A and B were both identical in build using Mo as the center wire, alumina (Al_2O_3) insulators, and a Nb alloy with 1% zirconium (Zr) balance for the outer sheath. HTIR-TC C varied only in the insulator material where hafnia (HfO_2) was utilized instead. Both HTIR-TC A and B drifted upwards no more than 5% over the course of the test—with HTIR-TC B performing slightly better. For HTIR-TC C, the drift was essentially 0%. This appears to show a slight edge using hafnia over alumina for the coaxial build.

For all three TCs, a downward trend was expected, such normally being the case when these TC types drift due to exposure to excessive heat. These drift results show that the designated channels that the TCs were installed in were perhaps slowly increasing in temperature (e.g., gas flow, TC movement, or conduction paths), or perhaps the thermocouple cabling was becoming contaminated with time due to its surroundings through solid state diffusion at these elevated temperatures. As a final note to the observed temperatures, a great deal of confidence is found in the overall furnace set up with the type-B TC temperature spot check not showing any sign of temperature variation.

B. Plunge test

A TC's response time can be directly measured by suddenly plunging it from air into flowing water—both being at different temperatures. This is known as the "plunge test." Using the definition of a first-order system, the TC's response time can be obtained as a time constant, τ, by simply calculating the time it takes the TC output signal to reach 63.2% of its final value, following this step change in temperature. The plunge test principal concept is illustrated in Fig. 5, with the TC being plunged from air into stirred or flowing water (left). The time constant, τ, of the new plunge temperature is accurately measured [6].

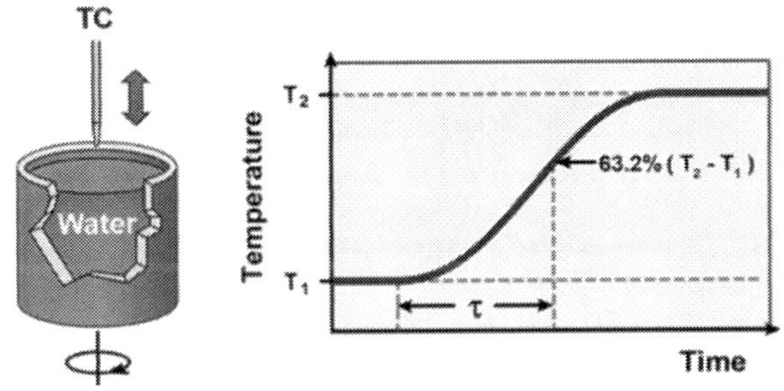

Fig. 5. A typical plunge test for determining the time constant, τ, of a new TC sensor to obtain a baseline comparison with other TCs available on the market.

Note that the TC's response time under these test conditions will not equal that when subjected to other conditions (e.g., different fluids, temperatures, or flow rates). However, this test is typically performed by sensor vendors wanting to establish a baseline comparative response time for a given sensor design, in hopes of potentially guiding potential buyers.

Three different types of HTIR-TCs were tested in the plunge test apparatus, along with two standard type-K TCs. The sensors varied in terms of the ground state, diameter, number of junctions, and base metals used. One HTIR-TC followed the optimized design, which was inherently grounded. Two HTIRs were DCs with two junctions each: one at the tip of the TC and another ~25 cm (10 in.) away from the tip. And one was a standard HTIR-TC ungrounded from the sheath. For comparative purposes, one of the type-K TCs was used in a

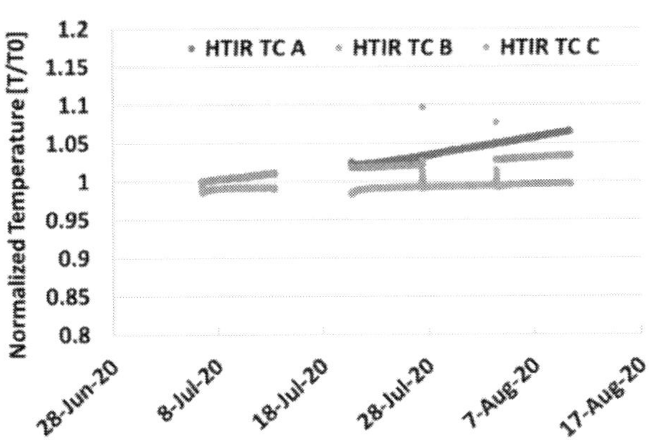

Fig. 4. Longevity test for the coaxial HTIR-TCs, showing <5% drift over a period of 1000+ hours. One set of TCs included Nb1%Zr.

TABLE I
RESULTS OF THE HTIR-TC PLUNGE TEST

TC #	Number of Junctions	Junction Type	Probe Outer Diameter	Response Time [s]
HTIR				
#1	2*	Grounded	1.6 mm (0.0625 in.)	0.013
#2	1	Grounded	1 mm (0.040 in.)	0.038
#3	1	Ungrounded	1.6 mm (0.0625 in.)	0.144
#4	2*	Grounded	1.6 mm (0.0625 in.)	0.013
Type K				
#1	1	Grounded	1.6 mm (0.0625 in.)	0.160
#2	1	Ungrounded	1.6 mm (0.0625 in.)	0.440

*Tip and midpoint junctions are assumed to be similar.

ground state, whereas the other was ungrounded. Baseline response time test measurements were conducted in 20°C water flowing at 1 m/s.

The results of the plunge test show the response times of grounded coaxial HTIR-TCs to be 10–30x faster than those of comparative technologies. Though the coaxial design is inherently grounded, when compared with type-K grounded TCs, it is an order of magnitude faster. Even when compared to its ungrounded counterparts, the TCs are still much slower.

V. Conclusions

The coaxial HTIR-TC was demonstrated to be long-lasting and fast-responding. The drift test revealed a drift of ~0% throughout the entire course of the test by utilizing hafnia as the electrical insulator of the thermoelements (see HTIR-TC C above). The results of the plunge test show the coaxial HTIR-TCs/DCs to be 10–30x faster, respectively, than traditional type-K and HTIR TCs.

Appendix

The TC operates under the basic principle of EMF being generated as per the material specific Seebeck effect. By completing a circuit of two dissimilar metals, then heating one end of the circuit, electrical potential is generated, usually on the order of mV. The general equation is as follows:

$$EMF(T) = \int_0^L [S_1(T,x) - S_2(T,x)] \frac{dT}{dx} \partial x, \qquad \text{(A1)}$$

where $S_{1\&2}$ are the material dependent local Seebeck coefficients for the respective thermoelements; T is the local temperature along the length, x, of the thermocouple (TC) cable; and L is the overall length of the TC. The main takeaway from Eq. A1 is that it is necessary to calculate the local EMF generated in the TC thermoelement if material properties and/or temperature gradients are not constant.

Under irradiation, the Seebeck coefficient can be affected (i.e., "damaged") by thermal and fast neutrons [7], leading to sensor decalibration. The Seebeck coefficient can be substituted in Eq. A1 by:

$$S^*(T,x) = S(T,x) \times Reduction\ Factor, \qquad \text{(A2)}$$

where S^* is the local irradiated Seebeck coefficient and the reduction factor is defined by:

$$Reduction\ Factor = e^{-(C_1\varphi_{Thermal}+C_2\varphi_{Fast})t}, \qquad \text{(A3)}$$

where $C_{1\&2}$ are constants that heavily depend on the thermal and fast neutron cross sections, respectively; φ is the local neutron flux, as labeled; and t is the total time under irradiation.

It is important to note that equations A1–A3 imply that the neutron irradiation present in any neutron field affects the entire length of TC cabling. The TC junction is simply the location of the temperature measurement, but the EMF is generated along the entire length of the TC cabling, which is bombarded by thermal and/or fast neutrons during reactor operation.

References

[1] Palmer, J., Skifton, R. S., Pham, B., Hawkes, G., "Summary of Thermocouple Performance in the Advanced Gas Reactor Experiment AGR-5/6/7 during Irradiation in the Advanced Test Reactor," ANIMMA, Prague, Czech Republic, 2021.

[2] Moffat, R. J., "Experimental Thermal and Fluid Science," 3, ISBN 0894–1777, 14–32, 1990.

[3] Schley, R. and Metauer, G., (1982) "Thermocouples for Measurements under Conditions of High Temperature and Nuclear Radiation," in Temperature: Its Measurement and Control in Science and Industry. Vol. 5, Part 2, J. F. Schooley, ed., American Institute of Physics, New York, NY, pp. 1109-1113.

[4] Standard Technical Specifications – Westinghouse Plants: Bases, NUREG-1431, Revision 4, Volume 2, 2012.

[5] Hashemian, H. M., "Maintenance of Process Instrumentation in Nuclear Power Plants," Springer, Berlin, 10.1007/978-3-540-33704-1, 2006.

[6] Hashemian, H. M., "Determination of Installed Thermocouple Response." U.S. Air Force, Arnold Engineering Development Center, Report Number AEDC-TR-86-46, 1986.

[7] Scervini, M., Rae, C., Lindley, B., "Transmutation of thermocouples in thermal and fast nuclear reactors," 3rd International Conference on Advancements in Nuclear Instrumentation, Measurement Methods and their Applications (ANIMMA), Marseille, (June 2013): 1-8, https://doi.org/10.1109/ANIMMA.2013.6727900.

X-Ray Imaging Calibration for Fuel-Coolant Interaction Experimental Facilities

Christophe Journeau[1], Michael Johnson[1,2], Shifali Singh[1,3], Fréderic Payot[1], Ken-ichi Matsuba[4], Yuki Emura[4] and Kenji Kamiyama[4]
[1]CEA, DES, IRESNE, Cadarache, 13108 St-Paul-lez-Durance, France
[2]SCALIAN DS, 399 Avenue du Château de Jouques, 13420 Gémenos, France
[3] STROBE, INSERM (UA 7), Université Grenoble Alpes, Grenoble, France
[4] JAEA, Oarai Research and Development Center, 4002 Narita, Oarai, Ibaraki 311-1393, Japan

christophe.journeau@cea.fr

***Abstract*— During a severe accident in either sodium-cooled or water-cooled nuclear reactors, jets of molten nuclear fuel may impinge on the coolant resulting in fuel-coolant interactions (FCI). Experimental programs are being conducted to study this phenomenology and to support the development of severe accident models. Due to the optical opacity of the test section walls, sodium coolant, and the apparent optical opacity of water in the presence of intense ebullition, high-speed X-ray imaging is the preferred technique for FCI visualization. The configuration of these X-ray imaging systems, whereby the test section is installed between a fan-beam X-ray source and a scintillator-image intensifier projecting an image in the visual spectrum onto a high-speed camera, entails certain imaging artefacts and uncertainties. The X-ray imaging configuration requires precise calibration to enable detailed quantitative characterization of the FCI. To this end, 'phantom' models have been fabricated using polyethylene, either steel or hafnia powder, and empty cavities to represent sodium, molten fuel and sodium vapor phases respectively. A checkerboard configuration of the phantom enables calibration and correction for lens distortion artefacts which magnify features towards the edge of the field of view. Polydisperse steel ball configurations enable precise determination of the lower limit of detection and the estimation of parallax errors which introduce uncertainty in an object's silhouette dimensions. Calibration experiments at the MELT facility determined lower limits of detection in the order of 4 mm for steel spheres, and 1.7-3.75 mm for vapor films around a molten jet.**

***Keywords* —X-Ray imaging, calibration, phantom, jet fragmentation, fuel-coolant interaction.**

I. Introduction

During a postulated severe accident in either a Sodium-cooled Fast Reactor or a Light Water Reactor, molten core material (corium) may be discharged to the lower plenum and interact with the coolant. The progression of the subsequent fuel-coolant interaction (FCI) is likely to differ between corium-water [1-3] and corium-sodium interaction [4-6], however in each case the jet fragmentation will dictate the conditions for fuel-coolant contact and the mode of heat transfer, and may lead in some conditions to vapor explosions.

Due to the optical opacity of sodium, and of water in the presence of a large flux of bubbles, high-speed X-ray imaging has been the preferred approach for the visualization of FCIs [7-9]. CEA has developed SPECTRA (Software for Phase Extraction and Corium TRacking Analysis) [6] for the analysis of X-ray images acquired during FCI experiments, including the suppression of imaging artefacts and the quantitative characterization of the melt fragment population, in support of the development of the corium-sodium interaction code SCONE [10] .

The use of fan-beam X-ray imaging entails certain imaging artefacts, not limited to vignetting, due to the radial decay in photon intensity away from the axis of the fan-beam [9] [11] , lens distortion and additive white noise. In order to quantitatively interpret the progression of an FCI from X-ray image processing, calibration of the X-ray imaging system is recommended. The calibration of high-speed X-ray imaging systems using various 'phantom' models is presented herein. These phantoms provide images containing features of known geometries and with attenuation properties representative of the phases observed during the FCI. This calibration procedure enables a geometric calibration of the imaging system, achieving a homogeneous pixel resolution across the image, as well as the estimation of the observed X-ray energy within the test section and the lower limit of detection for small particles presenting with intensities close to the threshold contrast to noise ratio. Calibration results are presented for the Japanese MELT facility and the French KROTOS facility.

II. Geometric Calibration

A. Phantom design

The choice of analogue materials for the phantoms is largely dictated their linear X-Ray attenuation coefficients and their suitability for the fabrication of complex geometries. At high X-ray energies, the energy dependence of mass attenuation coefficients diminishes and the disparity in mass attenuation coefficient between potential analogues is reduced, and so finding analogue materials with appropriate linear attenuation coefficient becomes increasingly dependent on the material density [12] . Inert analogue materials were chosen with linear attenuation coefficients as close as possible to those of the relevant phases observed during FCI experiments:

- Thermoplastics, such as polyoxymethylene (POM) and polyethylene (PE), have attenuation behavior comparable to liquid sodium at photon energies >100 keV, while offering sufficient versatility for the construction of complex phantom geometries. PE also represents a good simulant of liquid water in this energy spectrum.
- The melt phase can be represented by stainless steel 316 for metallic fuels or hafnia for oxidic fuels, due to its comparable density to UO_2.
- Coolant vapor is represented by air.

At the 9 MeV X-ray source energy employed at the KROTOS facility, the mass attenuation coefficient of a 50:50 uranium dioxide-stainless steel melt mixture is 0.038 $cm^2\ g^{-1}$ while that of stainless steel is 0.030 $cm^2\ g^{-1}$. The mass attenuation coefficient of sodium will be 0.0232 $cm^2\ g^{-1}$ compared to 0.0238 $cm^2\ g^{-1}$ for polyethylene [12] .

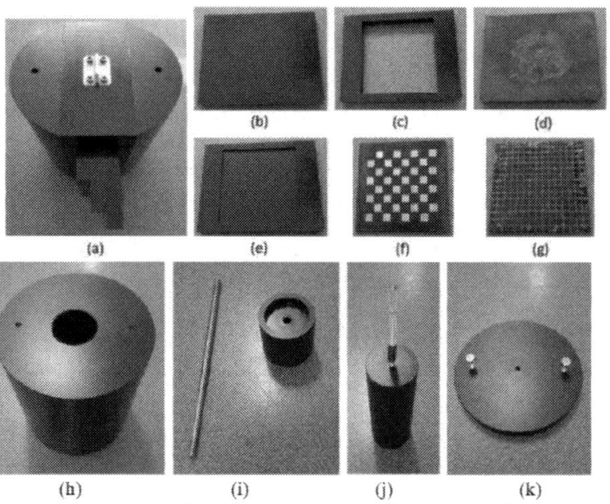

Fig. 1. Components of phantom 1 (a-g) and phantom 2 (h-k) used for calibration of the X-Ray imaging system at the MELT facility.

Fig. 1 presents the components of two phantoms which have been used for the calibration of the X-ray imaging system employed at the MELT facility at the JAEA's Oarai Center [6] . Both phantoms have 250 mm diameter, 280 mm length cylindrical geometries. Phantom 1 (Fig. 1a-g) has been designed such that interchangeable inserts can be installed within a void at the center of the cylinder, each designed with the purpose of calibrating a unique aspect of the imaging system. For example, the checkerboard inserts in Fig. 1j-g are designed to calibrate lens distortion artefacts, while an insert embedded with polydisperse steel balls (Fig. 1d) is used to determine the lower limit of detection, or the smallest detectable spherical melt objects. The interchangeability of the inserts allows objects of interest to be moved towards, or away from the X-ray source for the calibration of parallax uncertainties. Phantom 2 (Fig. 1h-k), has a smaller cylindrical void at its center for the insertion of a smaller phantom consisting of four concentric annular rings. Each of these annular rings can be selectively filled with fuel analogue to mimic different stages of jet pre-mixing.

B. Noise and vignetting artefacts

Dark current images are acquired without X-ray illumination and provide information on pattern and shot noise associated with the camera electronics. The time-averaged and denoised dark current images reveal the pattern noise which is used to offset the X-ray images prior to further image processing. Local denoising filters (mean, median or Gaussian) are generally avoided as smaller objects are likely to be obscured, phase boundaries blurred and the fine textures of complex melt fragment geometries are likely to be lost. Video-block matching and 3D filtering (V-BM3D) [13] and non-local means, NLM [14] [15] filters are preferred due to their selectivity in distinguishing between noise and real textures, rather than averaging local variations indiscriminately.

Flat field images represent X-ray images of the test section filled with coolant at rest, or a solid cylinder of thermoplastic coolant analogue for the purpose of the calibration experiments. Vignetting artefact represents the observed decay of pixel intensity away from the image center towards the border in the radiography of a homogeneous object, and is caused by the decay in photon flux away from the axis of the fan-beam X-ray source. Homogeneous illumination of the image is essential to achieve accurate segmentation of melt fragments and vapor clouds, independent of their location. Normalization of the dark current offset X-ray images by the flat field corrects for the heterogeneity in the illumination of the field of view [16] :

$$I_{FFN} = k_{eq} \ln \left| \frac{I - I_{DC}}{I_{FF} - I_{DC}} \right|$$

where I_{FFN} is the log-flat field normalized image, I is the original projected X-ray image, I_{DC} is the pattern noise of the dark current, I_{FF} is the flat field image and k_{eq} is an equalization factor employed to broaden the intensity histogram so that the image can be cast as an integer class (usually 8-bit or 16-bit) in order to reduce the computational expense of subsequent image processing.

Fig. 2a presents a schematic of phantom 2 installed within the test section and Fig. 2b is a corresponding X-ray image (denoised and temporally-averaged for ease of visualization) of the phantom with its two inner annuli filled with stainless steel balls, while the two outer rings are void, thus representing a cloud of fragmented melt particles surrounded by vapor. Fig. 2b demonstrates the decay in intensity towards the periphery of the circular field of view illuminated by the X-ray source. This is better demonstrated by Fig. 3a-b, which present the intensity signal over the horizontal analysis line shown in green Fig. 2b, before and after denoising respectively. The intensity signal of the coolant decays from around 60 near the center of the image to zero at the right-hand border and 20 at the left-hand border, demonstrating the asymmetry of the vignetting artefact.

https://doi.org/10.1051/epjconf/202125306005

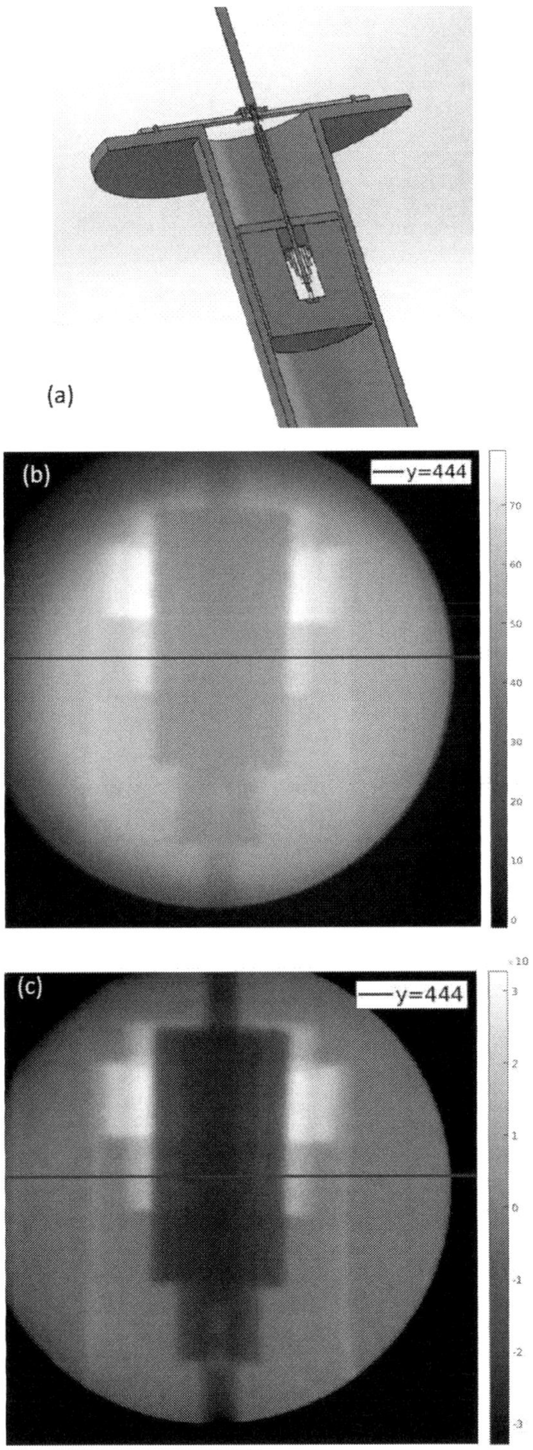

Fig. 2. (a) A schematic of phantom 2 installed in the test section, (b) a denoised X-ray image of phantom 2 with its 2 inner rings filled with steel balls, and (c) a log-flat field normalized version of (b) equalized by a factor of 10^5.

The resulting image, following log-flat field normalization, is presented in Fig. 2b, with the corresponding intensity signal shown in Fig. 3c. The plastic *coolant* regions of the log-flat field normalized image present with an intensity close to zero. The mild residual vignetting in Fig. 3c is due to subtle misalignment of the phantom between the calibration image and the flat field image, which would not be observed in the FCI experimental images. The void regions either side of the melt analogue exhibit elevated intensities and the steel-filled regions present with negative intensities. The three phases are much more easily distinguished in Fig. 2c and Fig. 3c following denoising and log-flat field normalization.

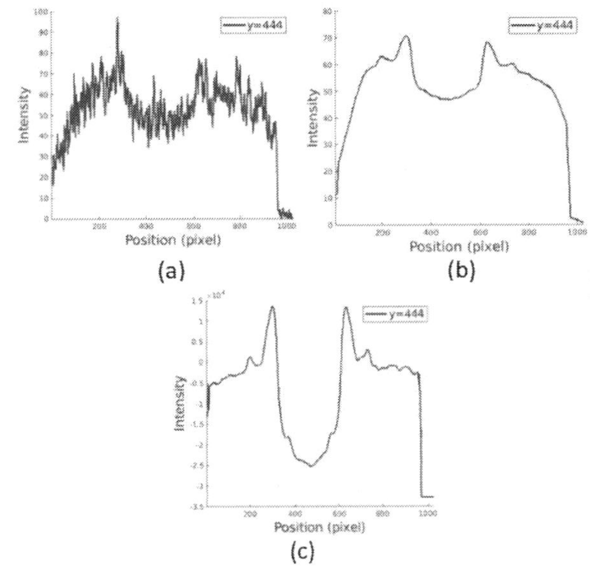

Fig. 3. Intensity profiles across the width of the phantom 2 images presented in Fig. 2b-c. The intensity signals correspond to images of the phantom (a) prior to denoising, (b) after denoising, and (c) after denoising and log-flat field normalization.

C. Distortion artefacts

In order to correct for distortion artefacts, X-ray images are acquired of checkerboard inserts installed in phantom 1. The coarse checkerboard (Fig. 1f) is constructed of 14.5×14.5×4 mm plastic and steel cuboids. The fine-wire checkerboard (Fig. 1g) is made of 14 horizontal and 14 vertical 1 mm diameter steel wires embedded in POM at regular 10 mm intervals. X-ray images of the two checkerboards are presented in Fig. 4. The fine-wire checkerboard in Fig. 4b, while fainter in contrast between the steel and plastic, was more effective in detailing the geometric distortion of the X-ray images. From the observed curvature of the vertical and horizontal wires in Fig. 4b, the origin of radial distortion was identified. The radial component of the lens distortion can be corrected using the approach detailed by Vass and Perlaki [17] . If r and $\hat{r}$ are the distorted and undistorted radii from the center of distortion, the translation function to correct for radial distortion is found from a truncated Taylor-series expansion:

$$\hat{r} = r(1 + k\, r^2).$$

A distortion coefficient of k=0.2 and the origin of radial

distortion at pixel (355,355), indicated in Fig. 4b, resolved the apparent curvature of the steel wires; however, the images demonstrated broadening, or magnification, of the checkerboard squares towards the top of the image. This is consistent with projection onto a screen, or camera, which is mounted lightly above the projector, commonly referred to as a 'keystone' artefact. The estimation of transformation functions for the correction of projection artefacts is comprehensively detailed in Goshtasby 1988 [18] .

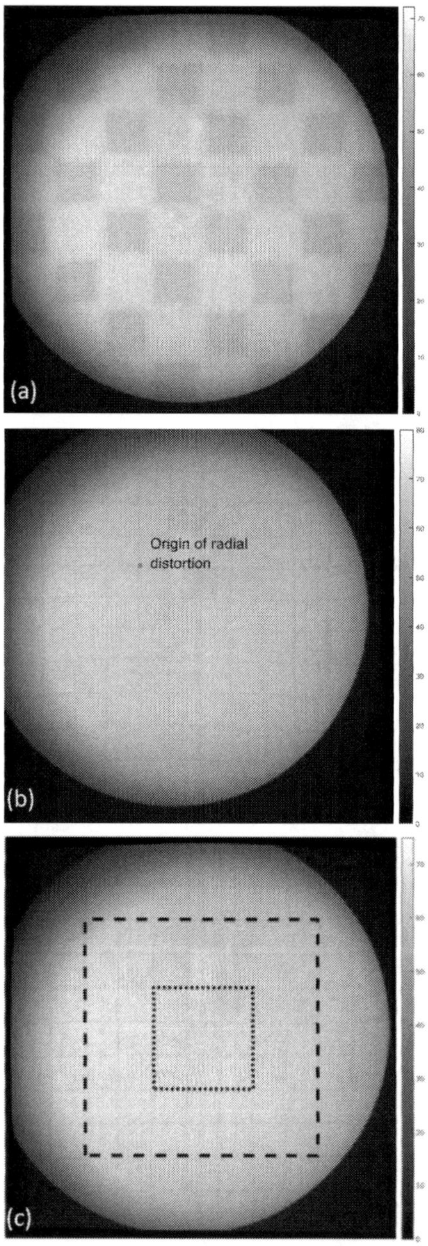

Fig. 4. Images of (a) the coarse checkerboard, (b) the fine-wire checkerboard indicating the origin of radial distortion, and (c) the fine-wire checkerboard following correction for radial and projective distortion, showing the corners of 5×5 and 7×7 square grids to demonstrate the uniformity of the corrected image.

The fine-wire checkerboard is presented following correction for radial and keystone distortion in Fig. 4c. Dashed and dotted outlines of 7×7 and 5×5 square regions respectively are overlaid to demonstrate the uniformity of the checkerboard following the two-stage transformation. The geometric transformation functions will remain valid for the experimental facility as long as the relative positioning of the X-ray source, scintillator/image intensifier and camera remain unchanged.

D. Geometric calibration and parallax errors

Following correction for distortion artefacts, the 7×7 square region of the checkerboard outlined in Fig. 4c (69×69 mm in dimension) measured 692×693 pixels, corresponding to a pixel resolution of 100.7 μm. The observed cross-sectional areas of multiple phantom geometries of different sizes, including the steel-filled regions of phantom 2 presented in Fig. 2, are compared with their known dimensions in Fig. 5a. Measurements from 7 different calibration experiments indicated an average pixel resolution of 101.8 μm. The three largest objects measured (including the 7x7 checkerboard shown in Fig. 4b and the filled rings of phantom 2, see Fig. 2b) lie within 6.5% uncertainty of the calibration line. This uncertainty escalates as the size of the object diminishes, increasing to ±24% for 8 mm steel balls.

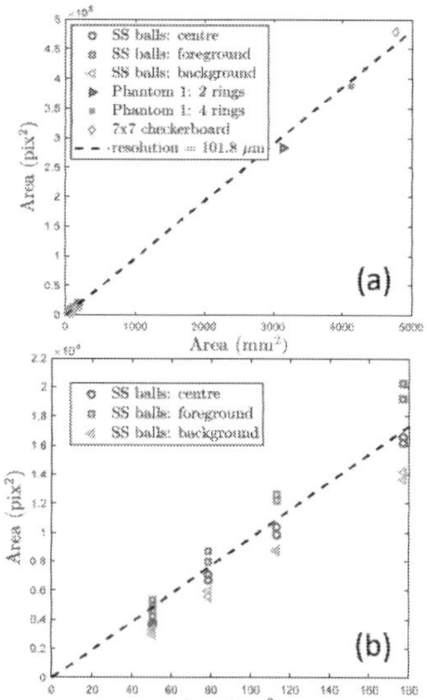

Fig. 5. (a) Calibration of pixel resolution and (b) the effects of parallax errors on the observed particle silhouette area.

The error in an object's silhouette area is liable to increase further as a result of the uncertainty in its depth location within the test section. Fig. 5b demonstrates the apparent magnification of an object's observed silhouette as it is displaced within the test section towards the X-ray source. For a 15 mm diameter steel ball at the center of the field of view, a 40 mm displacement towards the foreground manifests in a 10.5

% increase in observed diameter, while displacement 40 mm towards the background results in an observed 7.5 % reduction in measured diameter. These measurements indicate the X-ray source to be installed around 0.46 m from the central axis of the test section and suggest that the observed dimension of millimetre-scale objects could be subject to a parallax uncertainty in excess of ±10 %.

III. Assessment of Detection Thresholds

A. Numerical Simulation

The anticipated signal to noise ratio, contrast between the various phases and the limits of detection can be estimated by numerical simulation of the experimental facility in question and the associated imaging system. The MODHERATO code [9] enables the definition of a virtual representation of the X-ray source, test section, test section contents, scintillator and camera configuration, generating simulated X-ray images, inclusive of statistical noise, swank noise, shot noise, dark noise, readout noise and quantization noise.

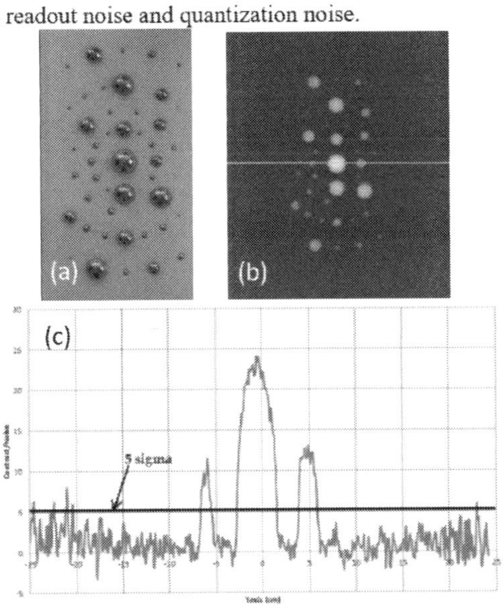

Fig. 6. Detection of metallic balls (reproduced with permission from Singh [12]) including (a) the phantom embedded with steel balls, (b) a simulated X-ray image generated using MODHERATO and (c) the contrast to noise ratio along the analysis line in green in the top right image.

Fig. 6a presents a phantom model consisting of a plate embedded with polydisperse steel balls, which has previously been used for calibration of the KROTOS facility. This geometry, installed within the KROTOS test section, has been simulated using MODERHATO, producing the theoretical radiographic image presented in Fig. 6b. The intensity profile along the green analysis line is shown in blue in Fig. 6c. A relatively conservative threshold at 5 times the standard deviation of the imaging noise corresponded to a detection limit equivalent to 5 mm diameter steel balls.

B. Determination of the Lower Detection Limits using Phantoms

While numerical simulation is valuable during the design a new radioscopic system, physical phantoms are preferred for the assessment of existing experimental facilities and X-ray imaging systems. A POM insert embedded with 16 polydisperse stainless steel balls, of 1 to 15 mm in diameter (see Fig. 1d) was installed in phantom 1 and imaged at the MELT facility. The resulting X-ray image, presented in Fig. 7a, is shown following log-flat field normalization and following segmentation in Fig. 7b and c respectively. A schematic of the steel ball locations and sizes is presented in Fig. 7d. Based on the 10 steel balls identifiable in the processed images, the lower limit of detection was found to correspond to a steel ball diameter of 4 mm.

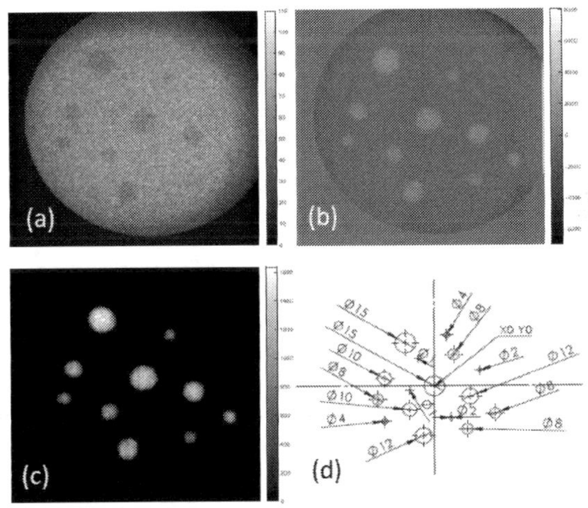

Fig. 7. The detection of a lower detectability limit using polydisperse steel ball phantoms, including (a) an unprocessed X-ray image of the phantom, (b) a denoised, log-flat field normalized and inverted image, (c) a segmented image showing the detected steel balls, and (d) a schematic of the steel ball sizes and locations.

During FCI experiments, a significant proportion of the melt fragment population is likely to be smaller than this lower limit of detection. However, given a sufficient concentration of fine particles it may be possible to detect clouds of fine particles. Fig. 8 presents a phantom employed at the KROTOS facility[8] [12] used to simulate clouds of small particles; this is the same phantom of concentric rings installed in the central annulus of phantom 2 to obtain the images acquired at the MELT facility shown in Fig. 2. The width of the concentric annuli is 20 mm. The phantom has a steel rod at its center, representing a molten steel jet. Melt fragments are represented either by 4 mm diameter steel balls (Fig. 8b) with a packing fraction of 52 %, corresponding to a cloud density of around 3780 kg m^{-3}, or hafnia powder (Fig. 8c), with 95% of grains below 510 nm and a packing fraction of 23 %, presenting with a cloud density of 2200 kg m^{-3}.

Several configurations of the phantom with 1, 3 or 4 of the concentric annuli filled with steel balls or hafnia powder are presented in Fig. 8e-g and Fig. 8h-j respectively. Despite the reduced contrast with the thermoplastic coolant analogue in the case of the less dense hafnia powder, the fragment clouds occupying a single 20 mm wide annulus were clearly detectable, even within a vapor cloud of three-times the thickness of the particle cloud. Furthermore, the boundaries between particle clouds with 40 mm difference in thickness were clearly distinguishable at the KROTOS imaging facility.

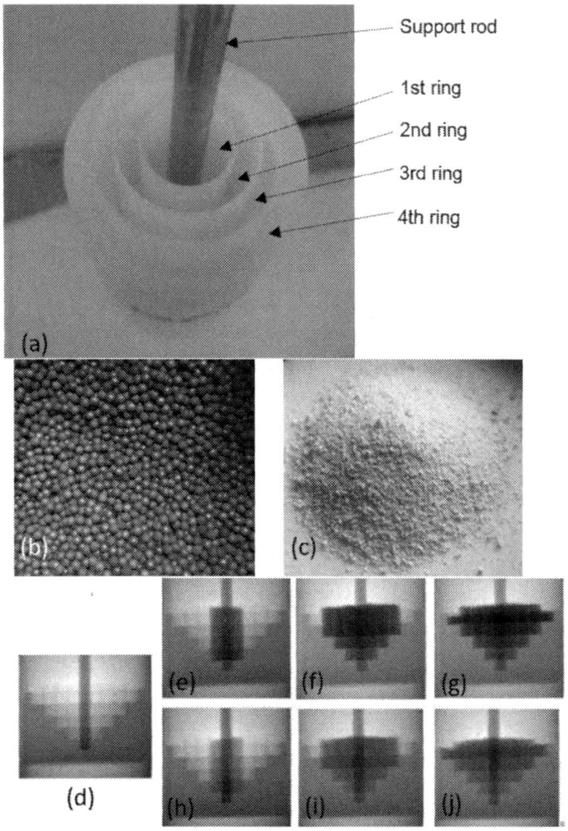

Fig. 8. Investigation of clouds of particles using an annular phantom (reproduced from [12] with permission) including (a) the phantom model, (b) 4 mm diameter steel balls, (c) fine HfO_2 powder, and X-ray images of (d) the empty phantom, the phantom with (e-g) 1, 3 and 4 rings filled with steel balls, and (h-j) 1, 3 and 4 rings filled with HfO_2 powder.

An additional phantom has been designed to assess the detectability of gas films around the melt jet, for investigation using the X-ray imaging system employed at the MELT facility. The phantom mimics a gaseous film around a steel jet. The tapered, or conical, steel rod at the center of the phantom is surrounded by a void film which increases from a thickness of 0 mm at the base, to 6.75 mm at the top. The phantom design and a corresponding X-ray image of the phantom are presented in Fig. 9. It was determined that vapor films of as little as 1.7 mm thickness were detectable, however the contrast to noise ratio fell significantly for film thicknesses below 3.75 mm.

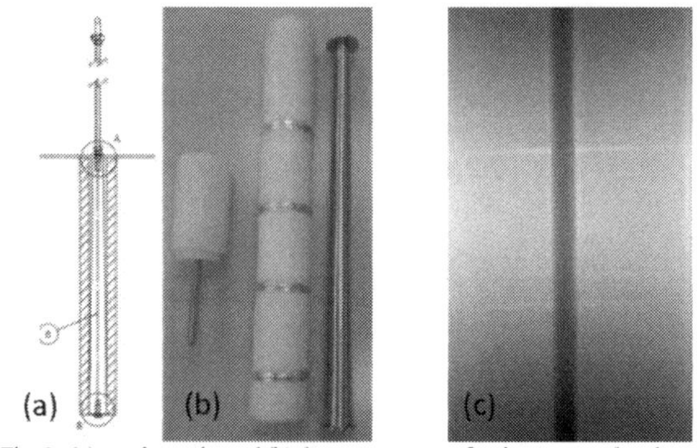

Fig. 9. (a) A schematic, and (b) the components of a phantom used to determine the detection limit for vapor films, with (c) the corresponding X-ray image acquired at the MELT facility.

IV. Conclusions

High-speed fan beam X-ray imaging systems employed for the visualization of transient thermal hydraulic experiments such as fuel-coolant interactions are subject to significant imaging artefacts, such as noise, loss of intensity towards the border of the image, and lens distortion. Image processing is necessary to suppress these artefacts and achieve homogeneous illumination of the imaging window to detect melt fragments and to achieve a homogenous pixel resolution for quantitative assessment of the images. Image calibration is highly recommended to determine the extent of these imaging artefacts, the uncertainties associated with quantitative measurements and the lower limit of detectability for small particles and vapor films. A series of versatile phantom model designs have been presented to assess the various capabilities of such X-ray imaging systems, using steel balls and hafnia powder as analogues for different melt phases, thermoplastic materials such as polyoxymethylene or polyethylene to represent liquid sodium coolant, and voids for the coolant vapor.

Calibration experiments using phantoms enabled the determination of transform functions to correct for the lens distortion at the MELT facility and achieve images with a homogeneous resolution of 102 µm. The lower limit of detectability was found to correspond to individual melt fragments in the order of 4 mm diameter and vapor films of around 1.7-3.75 mm thickness. Characterization of the size of the melt fragment population was subject to uncertainties which increased with diminishing particle size, rising to ±24 % for 8-mm diameter steel spheres. Assuming the jet to disperse to a radius of 40 mm during pre-mixing, measurements may be subject to a further parallax uncertainty in the order of ±10 % due to the apparent magnification of foreground objects.

Acknowledgment

This work was carried out within the arrangement between CEA, FRAMATOME, JAEA, MHI and MFBR entered into force January 1, 2020.

The calibration experiments at the MELT facility were supported by the Japanese Ministry of Economy, Trade and Industry through the "Technical Development Program on Fast

Reactor International Cooperation" and the "Technical Development Program on a Common Base for Fast Reactors".

The authors wish to thank the JAEA's MELT team for performing the X-ray image acquisition and Frédéric Compagnon for his work on the design of the phantom models used in this study.

References

[1] M.L. Corradini, B.J. Kim and M.D. Oh, "Vapor explosions in light water reactors: A review of theory and modelling", Prog. Nucl. Ener. vol. 22, no. 1, pp. 1-117, 1988, 10.1016/0149-1970(88)90004-2.

[2] R. Meignien, B. Raverdy, M. Buck, G. Pohlner, P. Kudinov, W. Ma, C. Brayer, P. Piluso, S.-W. Hong, M. Leskovar, M. Uršič, G. Albrecht, I. Lindholm and I. Ivanov, "Status of steam explosion understanding and modelling", Ann. Nucl. Ener., vol. 74, pp. 125-133, 2014, 10.1016/j.anucene.2014.07.008.

[3] P. Shen, W. Zhou, N. Cassiault-Louis, C. Journeau, P. Piluso and Y. Liao, "Corium behavior and steam explosion risks: a review of experiments", Ann. Nucl. Ener. vol. 121, pp. 162-176, Nov. 2018, 10.1016/j.anucene.2018.07.029.

[4] G.Berthoud, H. Jacobs and B. Knowles, "Analysis of large scale UO_2 Na interactions performed in Europe", *IAEA-IWGFR technical committee meeting on material-coolant interactions and material movement and relocation in liquid metal fast reactors*, O-arai, Ibaraki (Japan), June 1994, pp. 127-165.

[5] K. Matsuba, K. Kamiyama, J. Toyooka, Y. Tobita, V.A. Zuyev, A. A. Kolodeshnikov and Y.S. Vassiliev, "Experimental discussion on fragmentation mechanisms of molten oxide discharged into a sodium pool", Mech. Eng. J. vol. 3, no. 3, pp. 15-00595, 2016, 10.1299/mej.15-00595.

[6] M. Johnson, C. Journeau, K. Matsuba, Y. Emura and K. Kamiyama, Characterization of high-temperature nuclear fuel-coolant interactions through X-ray visualization and image processing", Ann. Nucl. Ener., vol. 151, pp. 107881, Feb. 2021, 10.1016/j.anucene.2020.107881.

[7] H.S. Park, R.C. Hansson, B.R. Sehgal, Fine fragmentation of molten droplet in highly subcooled water due to vapor explosion observed by X-ray radioscopy, Exp. Therm. Fluid Sci., vol. 29, pp. 351-361, 2005.

[8] C. Brayer, D. Grishchenko, N. Cassiaut-Louis, P. Piluso, "Application of X-Ray radioscopy for investigations of the 3-phase-mixture resulting from the fragmentation of a high temperature molten material jet in water", in: *8th Int. Conf. Multiphase Flow, ICMF-2013*, Jeju, Korea, 2013, 10.1016/j.expthermflusci.2004.05.013.

[9] L. Berge, N. Estre, D. Tisseur, E. Payan, D. Eck, V. Bouyer, N. Cassiaut-Louis, C. Journeau, R. Le Tellier, S. Singh, E. Pluyette, "Fast Energy X-Ray Imaging for Severe Accidents on the Future PLINIUS-2 Platform", IEEE Trans. Nucl. Sci. vol. 65, pp. 2573-2581, 2018.

[10] M. Zabiego, C. Fochesato, "Corium-sodium interaction the development of the SCONE software", in: NURETH-17 - 17th International Topical Meeting on Nuclear Reactor Thermal Hydraulics, Xi-An, China, 2017.

[11] A.R. Cowen, A.G. Davies, M.U. Sivananthan, "The design and imaging characteristics of dynamic, solid-state, flat-panel X-ray image detectors for digital fluoroscopy and fluorography", Clinical Radiology, voL 63, pp. 1073-1085,

[12] S. Singh, "X-Ray diagnostics on corium-sodium interactions during a severe accident scenario", PhD dissertation, University Paris-Saclay, 2019, https://hal.archives-ouvertes.fr/tel-02897391

[13] A. Danielyan, V. Katkovnik, K. Egiazarian, K.. BM3D frames and variational image deblurring. IEEE Trans. Image Proc. vol. 21(4), pp. 1715–1728, 2012, https://doi.org/10.1109/TIP.2011.2176954

[14] A. Buades, B. Coll, J. M. Morel, "A Review of Image Denoising Algorithms, with a New One", Multiscale Model. Simul., vol. 4, no. 2, pp. 490–530, 2005.

[15] I. Frosio J. Kautz, "Statistical Nearest Neighbors for Image Denoising," IEEE Trans. Image Process., vol. 28, no. 2, pp. 723–738, 2019.

[16] V. Van Nieuwenhove, J. De Beenhouwer, F. De Carlo, L. Mancini, F. Marone, J. Sijbers, "Dynamic intensity normalization using eigen flat fields in X-ray imaging," Opt. Express, vol. 23, no. 21, Art no. 27975, 2015.

[17] G. Vass and T. Perlaki, "Applying and removing lens distortion in post production," in 2nd Hungarian Conference on Computer Graphics and Geometry, 2003, pp. 9–16.

[18] A. Goshtasby, "Image registration by local approximation methods," Image Vis. Comput., vol. 6, pp. 255–261, 1988.

Neutron Coincidence Measurements and Monte Carlo Modelling of Waste Drums Containing Reference Nuclear Material

A. Borella, R. Rossa, S. Boden, C. Bruggeman, B. Rogiers, S. Smets, E. Valcke
Belgian Nuclear Research Centre, Belgium
corresponding author: alessandro.borella@sckcen.be

***Abstract*— Within the EC-funded CHANCE project several non-destructive techniques are being considered for the assay of waste bearing drums. Such techniques include calorimetry, gamma-ray spectrometry and neutron coincidence counting. The aim is to quantify uncertainties on the inventory of radionuclides, and how these are potentially reduced by combining the signatures from different techniques in the data analysis.**

In this framework, neutron coincidence measurements were carried out with two slab counters based on ^{3}He detectors coupled to shift register electronics. Such a system consists of two identical slabs with 6 detectors each, and is transportable, rather compact and flexible in terms of sizes and geometries that can be measured. With this system three 200 L drums containing certified reference nuclear material and different filling materials were measured. The certified nuclear material was in the form of 21 pellets of mixed oxide of U and Pu with a total mass of about 10.5 g; in addition, a single pellet of about 10.05 g was also available. The pellets could be placed in predefined positions within the drum in a reproducible way. The geometry and composition of the three drums was well characterized and consisted of Ethafoam, a mixture of Ethafoam, stainless steel and PVC, and mortar with an inner core of extruded polystyrene. The measurement setup was arranged such that the drum was placed between the two slab counters. The positions of the slab counters relative to the drum were accurately measured before each measurement, and a dedicated system was used to minimize the uncertainty on the detector positioning.

The measurement data were first analysed by applying the point model of Hage and the mass of nuclear material in the drum was determined from the rate of totals and reals and the radionuclide composition. Due to the fact that not all the point model conditions were met, we found that the point model overestimates the mass up to about 50%. In addition, a Monte Carlo model of the measurement geometry was developed using the MCNP code. The model was used to determine a calibration factor between the reals rate and the mass of the sample. Measurements with a calibrated ^{252}Cf source were used to verify the model. With a Monte Carlo based approach the mass of the mixed oxide pellets is within a few percent from the nominal values, except for strongly asymmetrical configurations where the deviation is up to about 20%. The results reveal the importance of an accurate background correction and of accounting for surrounding materials of the building such as walls, floor and ceiling in the Monte Carlo model.

***Keywords* —Neutron coincidence counting; reference nuclear material; waste characterization.**

I. Introduction

The EC-funded CHANCE project (Characterization of Conditioned Nuclear Waste for its Safe Disposal in Europe) aims to establish an understanding of current characterization methods and quality control schemes for conditioned radioactive waste in Europe. Furthermore, the CHANCE project develops, tests and validates already identified and novel nondestructive assay (NDA) techniques, to improve the radiological characterization of conditioned radioactive waste by reducing the uncertainty on the inventory of radionuclides. One of the innovative technologies for conditioned radioactive waste packages containing high alpha activities and relatively low beta/gamma activities is calorimetry. Currently, an experimental programme is being run to demonstrate in which cases calorimetry combined with various NDA techniques such as gamma-ray spectrometry and passive neutron coincidence counting could reduce the global uncertainty on the inventory of radionuclides. Within this program, three different 200 L mock-up drums containing certified reference nuclear material and one 200 L drum filled with real unconditioned decommissioning waste, are measured. Once all results are available, the different uncertainty sources for each individual technique will be quantified, followed by a combination of the three techniques using a Bayesian approach. This paper exclusively reports on the neutron coincidence counting measurement results of the three mock-up drums containing reference material. The use of neutron coincidence counting for waste characterization is also investigated in the MICADO project [1].

In this paper we will first outline how we apply neutron coincidence counting to determine the mass of material in the assayed drums. We will then present the measurement campaign by describing the used detectors and electronics, the measured drums and sources; hereafter we describe the Monte Carlo model; finally, we will discuss the obtained results in terms of masses and compare them with the nominal ones.

The uncertainties in this paper are quoted at a 68%

confidence limit. We quote the uncertainties between parentheses as recommended in paragraph 7.2.2 of Ref. [2].

II. Neutron Coincidence Counting

In general, radioactive materials mainly emit neutrons according to three processes: (α,n) reactions, spontaneous fissions, and neutron-induced fissions. No difference is shown in a neutron detector when neutrons emitted from the different processes are detected. However, (α,n) reactions always emit only one neutron per event, whereas more than one neutron may be emitted in case of fission events. Neutron coincidence counting uses this difference to estimate the quantity of material undergoing spontaneous fission in the sample being measured [3].

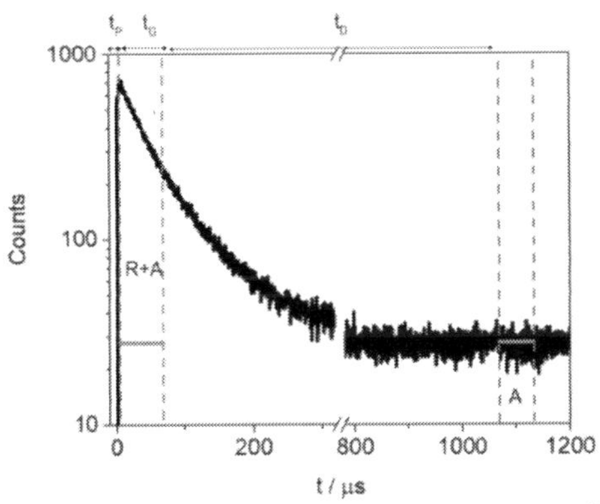

Figure 1: Rossi-alpha distribution from the measurement of a ^{252}Cf source.

Figure 1 shows the Rossi-Alpha distribution (RAD) for a ^{252}Cf source, which represents the time distribution of neutron detections after an arbitrary neutron detection.

If the conditions of the point model are satisfied [4], the Rossi-alpha distribution can be described by the sum of an exponential term and a constant. The constant term is related to so-called accidental time-uncorrelated events and the exponential term is related to the so-called real coincident events. The exponential term is characterized by a die-away time (τ) which is the mean neutron lifetime in the detector and is determined mainly by the detector geometry. The area of the exponential term represent the total number of coincident events.

In a neutron coincidence measurements, one typically does not measure the RAD over the whole time range but rather in two time intervals (gates) that are open after the detection of each event. Right after the detection of the arbitrary neutron event (at time t=0 s) a short time interval, also called pre-delay t_P, is taken to account for the effects of electronic dead-times. After the pre-delay, a gate t_G is taken to count the $R+A$ events, followed by a long delay t_D, and by a second gate to count the A events. The gates for the calculation of the $R+A$ and A events have the same duration. It is common practice to choose a pre-delay in the order of a few µs (e.g. 4.5 µs), the gate width between 32 and 64 µs, and the long delay around 1000 µs [5]. The determination of the quantity of material undergoing spontaneous fission from the observables $R+A$ and A is explained in Sections IV and V.

III. Experiments

A. Slab counters and electronics

Two Canberra WM3400 slab counters [6] were used for the measurements within the CHANCE project. A picture of the detectors during the measurement of a drum is shown in Figure 2.

Figure 2: Picture of a measurement with the WM3400 slab counters and the ESARDA drum 206.

Each WM3400 counter measures 104 cm (height) × 41 cm (width) × 16 cm (depth) and has an approximate weight of 50 kg. To facilitate the movement and positioning of the WM3400 each slab counter has been fixed to a moving cart as shown in Figure 2. The WM3400 counter contains six ^3He tubes (active length of 91.44 cm and diameter of 2.54 cm) which are positioned in-line and embedded in a polyethylene slab. The polyethylene slab is then surrounded by 1-mm Cd sheet that covers five sides of the slab. During a measurement the WM3400 should be placed such that the side without the Cd cover is facing the sample being measured. An additional aluminum sheet covers the slab counter and provides further protection to the inner components.

The default electronic system of the WM3400 has been replaced to reduce the electronic noise and gamma-ray sensitivity [7]. The updated electronic system includes a dedicated preamp for each ^3He tube, the possibility to adjust for gain differences between the ^3He tubes, and an optimised shaping time for ^3He detectors. During the CHANCE measurements two types of data acquisition systems were used in parallel. First a system based on JSR-12 shift registers [8] was used. One JSR-12 shift register was connected to each slab counter, and one JSR-12 was used to process the logical OR of the signals coming from both slab counters. In addition, the time stamps of the neutron detections were recorded with the MCA-527 multi-channel analyzer equipped with time stamp processing firmware [9]. The shift registers recorded the measurement time t, the Total events T, Real+Accidental events $R+A$, and Accidental events A. The time stamps recorded by the

MCA-527 were further processed off-line with a dedicated software to determine *T*, *R+A*, and *A* events, the multiplicity distribution, as well as to remove system instabilities due to high voltage and fluctuations or interferences.

B. ESARDA drums

The European Safeguards Research and Development Association (ESARDA) constructed a set of 16 so-called reference waste drums (also called ESARDA drums) [10]. Due to the high-level of design information and certification the ESARDA drums were conceived for intercomparison exercises and calibration of NDA equipment devoted to the characterization of plutonium-bearing waste drums.

Three ESARDA drums with the following filling materials were selected for the measurements within the CHANCE project:

- ESARDA drum 201: homogeneous filling with ethafoam
- ESARDA drum 205: heterogeneous filling with ethafoam, PVC, and steel
- ESARDA drum 206: central section with extruded polystyrene and outer layer of mortar

ESARDA drums 201 and 205 were used in their original state, while drum 206 was modified according to the characteristics mentioned above. A central section with extruded polystyrene with a density of 37 kg/m^3 was added to ESARDA drum 206 to limit its total weight to about 300 kg.

The ESARDA drums were designed with 7 vertical channels where sources can be inserted. A schematic representation of these channels is shown in Fig. 3 and Fig. 4.

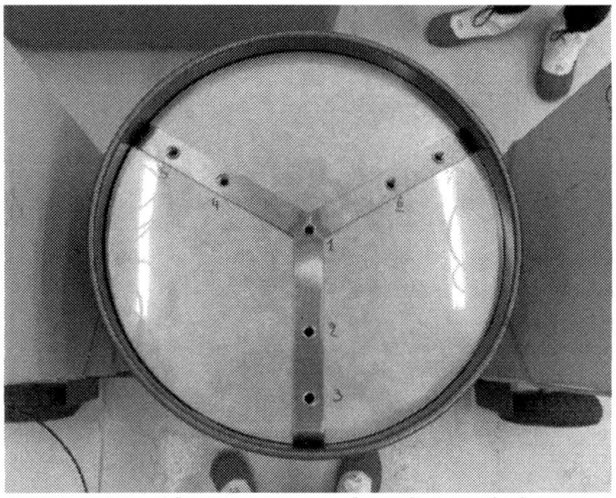

Figure 3: Top view from an ESARDA drum showing the locations of the vertical channels for the positioning of radioactive sources.

To ensure an accurate positioning of the drum and detectors a square metal frame, was placed around the drum. The two slab counters were positioned in contact with the metal frame on opposite sides. By using the metal frame the distance between the slab counters was either 74 or 76 cm in the two different measurement campaigns. The detectors had to be repositioned manually and the positioning of the slab counters could be reproduced with an accuracy of less than 1 cm. In addition, we verified that the distance between the detectors at the top and at the bottom varied, depending on the measurement point, within 1 cm. The metal frame can be seen in Figure 2 at the bottom of the ESARDA drum.

C. ESARDA sources

Two sets of PuO_2 sources [11] were used during the CHANCE measurements:

- C80180: 1 pin with approximately 10.05 g PuO_2 in a single point
- C10100: 7 pins, each with approximately 1.5 g PuO_2 distributed over three axial points

TABLE I: ISOTOPIC COMPOSITION OF THE PU SOURCES (REFERENCE DATE: END OF 1996).

Pu isotope	Mass fractionition
^{238}Pu	1.32%
^{239}Pu	64.45%
^{240}Pu	23.83%
^{241}Pu	5.75%
^{242}Pu	4.60%
^{241}Am	4.10%

The mass fractions for different radionuclides of the Pu sources is included in Table I [12].

Given the structure of the ESARDA drums the pin with ~10.05 g of PuO_2 was placed in several positions to assess the sensitivity of the measurement technique to the source location. The second set of Pu sources was used to simulate the condition of a distributed source throughout the drum volume as shown in Fig. 4.

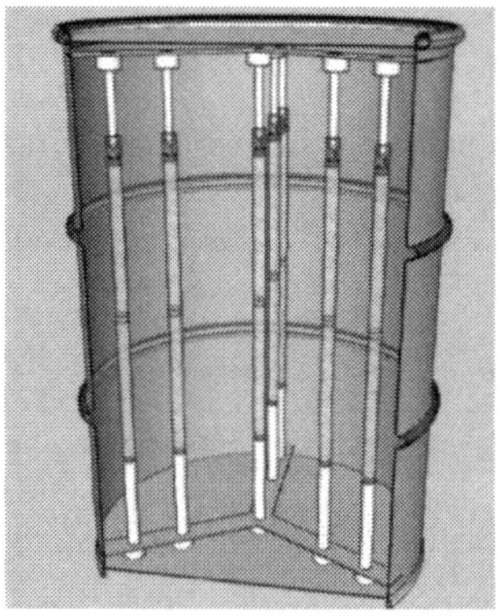

Figure 4: Positioning of the radioactive sources for the 7 pins configuration in a 3D model of a 200 L mock up drum.

In addition to the Pu sources, a ^{252}Cf source with a nominal activity of 37 kBq on 15 August 2015 was used also as calibration source. The ^{252}Cf source was used in the standard measurement setup without the presence of a drum or inserted in one of the ESARDA drums.

D. Considered measurements

The list of measurements carried out with the slab counters within the CHANCE project is summarized in Table II. The measurement campaign took place in two periods, March 2020 and March 2021, due to an interruption related to the Covid-19 outbreak. For the first two measurements in Table II, no drum was present in the room and the sources (when present) were placed between the detectors. Equal distancing between source and detector was ensured with the metal frame shown in Fig. 2.

The measurements with the ESARDA drum 201 consisted in a background measurement with only the drum placed between the detectors, measurements with the ^{252}Cf source or C80180 source placed in different positions within the drum, and measurement with the C10100 sources placed in all available positions. A similar series of measurements were conducted with the ESARDA drum 206, but in this case a total of 7 measurements were carried out with the C80180 source considering all measurement positions except for position 4. The measurement in position 3 was carried out twice to assess the repeatability of the measurement approach. The measurements with the ESARDA drum 205 consisted in the measurement of background, ^{252}Cf source, and the C10100 sources.

TABLE II
Background subtracted totals and reals rate values for the measured combinations of drums and sources at different positions. The uncertainty on the totals rate was always at most 0.1 s^{-1}.

Drum	Source	Position	T_{exp} / s^{-1}	$R_{exp,m}$ / s^{-1}
None	^{252}Cf	1	27.1	0.77(<0.01)
	C80180	1	242.3	1.75(0.01)
201	^{252}Cf	1	29.8	0.76(<0.01)
		3	25.9	0.43(<0.01)
		7	47.7	1.58(0.01)
	C80180	1	262.3	1.54(0.01)
		2	226.7	1.12(0.01)
		3	176.4	0.67(0.01)
		5	333.0	2.72(0.02)
		7	320.8	2.54(0.02)
	C10100	All	260.5	1.53(0.01)
205	^{252}Cf	1	26.0	0.64(<0.01)
	C10100	All	239.7	1.39(0.01)
206	^{252}Cf	1	24.5	0.50(<0.01)
		7	35.1	1.07(0.01)
	C80180	1	218.9	0.99(0.01)
		2	182.0	0.66(0.01)
		3	91.1	0.14(>0.01)
		3	90.9	0.14(<0.01)
		5	240.8	1.19(0.01)
		6	341.6	2.66(0.02)
		7	358.8	2.93(0.02)
	C10100	All	222.6	1.08(0.01)

The measurement data from the time stamps recorded with the MCA527 were analyzed by means of Python [13] scripts, in the Spyder environment version 4 [14] . The scripts first carried out stability checks of the data by monitoring the total count rate over 10 seconds interval. Then the data were analyzed to determine the totals and the reals rate for each measurement. The background subtraction was then carried out and the net total and net real count rates, indicated with T_{exp} and $R_{exp,m}$ respectively, were determined.

The obtained data were used to determine the mass of the assayed sample by applying two methodologies, one based on the point model approximation and one based on Monte Carlo modeling.

IV. POINT MODEL APPROACH

A. Theory

Starting from basic principles, theories were developed [15,16, 4] that express the observables of neutron coincidence or multiplicity measurement as function of the mass of neutron emitting radionuclides and other parameters. This model, called the point model, can be applied in case the assayed sample has point geometry, no neutrons return from the detector to the sample, the time response function of the moderator detector assembly is a pure exponential function with decay constant, and there are no dead time losses of signals. Other conditions also apply but we refer to the reference for the full description.

In absence of multiplication, the theoretical estimates from the point model of Hage [4] are the following for the totals and reals rate:

$$T_{pm} = \varepsilon F_s M \nu_{s(1)}[1+\alpha] \tag{1}$$

$$R_{pm} = \varepsilon^2 F_s M^2 \nu_{s(2)} \tag{2}$$

Where

- ε is the (total) detection efficiency
- F_s the spontaneous fission rate of the test item
- $\alpha = \frac{F_s}{S_\alpha}$
- S_α the (α, n) neutron emission rate of the test item
- $\upsilon_{j(\mu)} = \sum_{\mu=\upsilon}^{\infty} \binom{\mu}{\nu} P_{j\mu}$ is the μ_{th} factorial moment of the neutron emission distribution P by spontaneous fission

The experimental observables from the recorded pulse train can be obtained by looking at the signal triggered multiplets. In our case we limit to multiplets of order 0 and 1 and the experimental observables are the total count rate T_{exp} and the measured reals rate $R_{exp,m}$:

$$R_{exp} = \frac{(R+A)-A}{tf} = \frac{R_{exp,m}}{f} \tag{3}$$

Where

$$f = \int_{t_p}^{t_p+t_g} \frac{e^{-t/\tau}}{\tau} dt = e^{-t_p/\tau}\left(1 - e^{-t_g/\tau}\right) \quad (4)$$

and t is the measurement time.

The f factor accounts for the finite duration of the trigger gate and the presence of a predelay. The given expression for f holds only for an exponential RAD.

B. Data Analysis

In this work we applied the model equation (1) and (2) using the measured observables T_{exp} and $R_{exp,m}$ to derive the detection efficiency ε and the spontaneous fission rate F_s. The term α as well as the factorial moments of the neutron emission distribution associated to the assayed item were derived from the radionuclide composition of the sample. The f factor was obtained from the average of three values determined from the RAD: one from fitting the RAD with 2 exponentials, one from the observed RAD and correction for predelay assuming a 1 exponential function, one from the observed RAD and correction for predelay assuming a 2 exponentials function.

From the F_s one can determine the mass of the sample by dividing by F'_s , the number of spontaneous fission per second and per gram.

$$m_c = \frac{F_s}{F'_s} \quad (5)$$

V. Monte Carlo approach

A. Description of the MCNP model

The Monte Carlo code MCNP6 [17] was used to carry out a full scale modelling of the measurement setup. The model included the detectors and the used source(s) and drum, as well as the room containing the measurement setup (walls, ceiling and floor).

Since the specifications of detectors, drums and sources were very well known, the geometry and material composition could be modelled with a high degree of fidelity. A cross section of an example of a modeled geometry is shown in Fig. 5.

MCNP6 was run in the so-called analog mode i.e. tracking event-by-event the interactions of neutrons rather than their average behaviors. A tally of the type F8 with the CAP special treatment [18] was used. Such a treatment allows to record neutron absorption in given radionuclides and cells in given time intervals after a neutron absorption has taken place at time zero. The neutron multiplicity is scored in user defined intervals. It was found that neglecting walls, ceiling and floor resulted in an underestimation between 7% and 28% on the simulated reals rate. The simulations only account for neutrons emitted by the spontaneous fission and do not account for reals induced by (α,n) reaction which are present only in the presence of a multiplying sample. For the Cf source the spontaneous fission of ^{252}Cf was considered while for the PuO_2 ESARDA sources we considered ^{240}Pu.

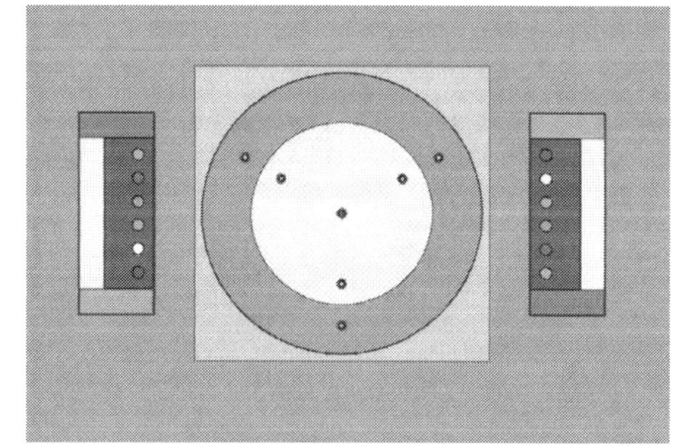

Figure 5: Horizontal cross section of the Monte Carlo model of the measurement setup for the 206 drum and the slab counters.

B. Post processing

The neutron multiplicity due to spontaneous fission were simulated in two time intervals both of duration of 64 μs: the first one called *R+A gate* and starting after 4.5 μs from the spontaneous fission event; the second one called *A gate* and starting 1,000 μs after the end of *R+A gate*. From the multiplicity we determined the reals $R_{sim,m}$ per spontaneous fission event as:

$$R_{sim,m} = \frac{S \times (m_{R+A(1)} - m_{A(1)})}{N_F} \quad (6)$$

Where:

- S is the total count rate
- N_F is the number of fission events in the MCNP simulation
- $m_{R+A(k)} = \sum_{x=k}^{\infty} \binom{x}{k} n_x$
- $m_{A(k)} = \sum_{x=k}^{\infty} \binom{x}{k} b_x$
- n_x is the normalised multiplicity distribution in the R+A gate.
- b_x is the normalised multiplicity distribution in the A gate.
- $\binom{x}{k} = \frac{x!}{(x-k)!k!}$ is the binomial coefficient

Note that the reals rate $R_{sim,m}$ was calculated with the predelay and trigger gate used in the measurement to determine $R_{exp,m}$ and therefore a correction for the f factor defined in equation 3 is not needed.

C. Data Analysis

From the reals per fission event $R_{sim,m}$ obtained from the MCNP6 simulations, the number of ^{240}Pu fission per second was determined. By using the spontaneous fission rate of ^{240}Pu $F'_{s,240Pu}$ we then determined the mass of ^{240}Pu-equivalent.

$$m_{^{240}Pu_{eq}} = \frac{\frac{R_{exp,m}}{R_{sim,m}}}{F'_{s,240Pu}} \tag{7}$$

The mass of ^{240}Pu-equivalent, $m_{^{240}Pu_{eq}}$, is the mass of ^{240}Pu that results in the same number of reals rate, $R_{exp,m}$.

The estimate mass m_c was then determined by [19]:

$$m_c = \frac{m_{^{240}Pu_{eq}}}{w} \tag{8}$$

where

$$w = \sum_i \frac{Q_i}{Q_{^{240}Pu}} N_i \tag{9}$$

with

$$Q = F'_s \nu_{s(2)} = \lambda \beta_{sf,i} \frac{N_{av}}{A} \nu_{s(2)} \tag{10}$$

- F'_s the spontaneous fission rate per gram
- λ the decay constant
- β_{sf} the branching ratio for spontaneous fission
- N_{av} is the Avogadro constant
- A is the atomic mass
- $\nu_{s(2)}$ is the second factorial moment of the neutron emission distribution by spontaneous fission
- N_i mass fraction of the considered radionuclide

The factorial moment of the neutron emission distribution by spontaneous fission were taken from a compilation of different sources e.g. [20,21,22].

In this work, we did not simulate the neutron emitted by (α,n) reactions. In absence of multiplication, (α,n) neutrons are time uncorrelated and therefore their impact is limited to the term of order zero in the multiplicity distributions. Such a term cancels out when calculating the reals rate.

The impact of accounting for other radionuclides present in the sources such as ^{250}Cf in the Cf source, or other even Pu isotopes in the PuO_2 sources was studied. Due to the different emission spectrum and multiplicity distributions the calculated $R_{sim,m}$ may slightly change and hence the calculated mass m_c. The maximum observed deviation in the calculated mass m_c was 3.3% for the bare ^{252}Cf source.

VI. Results and Conclusions

The obtained masses m_c with the previously outlined procedures were compared with the nominal masses m_n of the measured samples. The obtained ratios m_c/m_n for the considered cases are given in Table III for the point model and Monte Carlo approaches. The quoted uncertainties result from count statistics on the net totals rate T_{exp}, the net reals rate $R_{exp,m}$ and on the reals per spontaneous fission event $R_{sim,m}$ from Monte Carlo simulations, and have been obtained by first-order Taylor expansion-based error propagation.

The masses obtained by applying the point model are in good agreement with the nominal ones for the ^{252}Cf measurements both in the bare configuration and for the measurement with a drum. For the measurements with the ESARDA sources we observe a systematic overestimation of the mass. The reason of such a difference could be related to the fact for the ^{252}Cf the (α,n) contribution is negligible, when compared to the spontaneous fission term, while the contribution from (α,n) is significant for the PuO_2 sources and directly impacts the total count rate as indicated in Eq. 1. Given the open measurement geometry, the measured total count rate is also affected by neutron scattering in the surroundings and should be properly accounted for to correctly meet the point model conditions.

TABLE III
m_c/m_n values for the measured combinations of drums and sources at different positions

Drum	Source	Position	m_c/m_n Point Model	m_c/m_n Monte Carlo
None	^{252}Cf	1	0.95(0.01)	1.00(0.01)
	C80180	1	1.21(0.03)	1.05(0.01)
201	^{252}Cf	1	1.03(0.01)	0.98(0.01)
		3	1.08(0.02)	0.98(0.01)
		7	1.04(0.01)	0.86(<0.01)
	C80180	1	1.38(0.03)	0.98(0.01)
		2	1.46(0.05)	0.99(0.01)
		3	1.50(0.07)	1.01(0.01)
		5	1.36(0.03)	0.96(0.01)
		7	1.37(0.03)	0.90(0.01)
	C10100	All	1.41(0.04)	1.00(0.01)
205	^{252}Cf	1	0.99(0.01)	0.97(0.01)
	C10100	All	1.31(0.03)	0.98(0.01)
206	^{252}Cf	1	1.01(0.01)	0.99(0.01)
		7	1.01(0.01)	0.88(<0.01)
	C80180	1	1.42(0.04)	1.02(0.01)
		2	1.49(0.05)	1.03(0.01)
		3	1.60(0.09)	1.07(0.02)
		3	1.56(0.08)	1.04(0.02)
		5	1.45(0.04)	1.01(0.01)
		6	1.41(0.03)	1.14(0.01)
		7	1.38(0.03)	1.26(0.01)
	C10100	All	1.37(0.04)	1.00(0.01)

For the masses obtained by applying the Monte Carlo approach, the overall agreement between the obtained mass and nominal mass is very good. However, we observe that for asymmetrical measurement positions, such as position 7, deviations up to 26% are observed. Such a deviation is present both for the ^{252}Cf and the C80180 source although the sign is not always consistent: the C80180 mass is overestimated in position 7 in drum 206 and underestimated in drum 201, while the ^{252}Cf mass is always underestimated. The reason of these deviations is not yet understood. While in symmetrical measurements the contributions to the total and reals count rate from the two detectors are essentially the same, in an asymmetrical configuration one detector can contribute much more to the reals rate as, in the point model approximation, the reals rate is proportional to the square of the neutron detection efficiency [4]. Therefore the accurate position of the detector and source is even more important in an asymmetrical configuration.

Future work will focus on integrating these results with similar studies, based on the same sources and drums, for gamma spectrometry and calorimetry, where uncertainty quantification will be performed using state-of-the-art Bayesian probabilistic modelling. Additionally, application to a real unconditioned waste drum, with unknown radionuclide inventory, is also envisioned, to test the methodology on a case more representative of daily practice.

ACKNOWLEDEMENTS

This project has received funding from the Euratom research and training programme 2014-2018 under grant agreement No 755371.

REFERENCES

[1] Available online: https://www.micado-project.eu/. Last accessed 28/04/2021.
[2] Evaluation of measurement data — Guide to the expression of uncertainty in measurement, JCGM 100:2008, First edition September 2008, available at https://www.iso.org/sites/JCGM/GUM/JCGM100/C045315e-html/C045315e.html?csnumber=50461
[3] Ensslin E., 1991. "Principles of Neutron Coincidence Counting" in Passive Nondestructive Assay of Nuclear Materials, NuREG/CR-5550, LA-UR-90-732.
[4] D.M. Cifarelli & W. Hage Nucl. Instr. & Meth. A251 (1986) 550-563
[5] Menlove H. O., 1991. "Neutron coincidence instruments and applications". in Passive Nondestructive Assay of Nuclear Materials, NuREG/CR-5550, LA-UR-90-732.
[6] https://www.mirion.com/products/wm3400-passive-neutron-slab-counter. Last accessed 28/04/2021
[7] Boogers E., et al., 2019. "Improved electronics for 3He based neutron counters". Proceedings of the 2019 ANIMMA conference.
[8] Canberra, 2006. "Model JSR-12 Neutron Coincidence Analyzer User's Manual", Canberra Industries Inc.
[9] GBS, 2021. https://www.gbs-elektronik.de/media/download_gallery/mca527_neutron.pdf. Last accessed 28/04/2021
[10] Bignan G. et al., 1995. "Proposal of an intercomparison exercise for the characterization of waste containing plutonium by NDA methods." Working document of the 17th annual ESARDA meeting.
[11] Bignan G., et al., 1996. "Preparation of standard drums for the calibration of nuclear waste measuring facilities. Final report". CEA note technique NT-SSAE-LSMR 98/0016.
[12] A. Verbruggen, R. Garcia-Galan, Certification of BC1413, GE/R/SIM/7/98 June 1998, IRMM Stable Isotope Measurement Unit
[13] https://www.python.org/]
[14] https://docs.spyder-ide.org]
[15] R. Dierckx & W. Hage (1983) Neutron Signal Multiplet Analysis for the Mass Determination of Spontaneous Fission Isotopes, Nuclear Science and Engineering, 85:4, 325-338,
[16] W. Hage & D.M. Cifarelli Nucl. Sci & Eng. 89 (1985) 159-176
[17] MCNP6 Users Manual - Code Version 6.1.1 beta, LA-CP-14-00745, June 2014.
[18] M. T. Swinhoe, J. S. Hendricks, and D. R. Mayo, "MCNPX for Neutron Multiplicity Detector Simulation," Los Alamos National Laboratory report LA-UR-04-8025 (2004).
[19] Report D8.2 "Identification of the sources of uncertainty related to individual methods", 2020, MICADO, H2020 NFPR 10 -2018, #847641
[20] Safeguards and Fissile Materials Management, a Preliminary procedure for the treatment of time correlation analyser measurement data for non-destructive assay of P, W. Hage K. Caruso, Technical Note I.07.C3.86.115, 1987
[21] Norman E. Holden & Martin S. Zucker (1986) Prompt neutron multiplicities for the transplutonium nuclides, Radiation Effects: Incorporating Plasma Science and Plasma Technology, 96:1-4, 289-292, DOI: 10.1080/00337578608211755
[22] Langner, D G, Stewart, J E, Pickrell, M M, Krick, M S, Ensslin, N, and Harker, W C. Sun . "Application Guide to Neutron Multiplicity Counting". United States. https://doi.org/10.2172/1679. https://www.osti.gov/servlets/purl/1679.

Peak Area Consistency Evaluation in Gamma Spectrometry

Henrik Persson[1] and Kara Phillips[1]
[1]Mirion Technologies (Canberra) Inc., USA
hpersson@mirion.com

Abstract— **Quantification of radionuclide activities in gamma spectrometry can be a challenging task. It depends on efficiency calibration, peak area calculation, nuclide decay data and correction factors, such as attenuation correction or true coincidence summing corrections. These quantities can present significant challenges to an accurate analysis. It is therefore desirable to have a way of assessing the quality of the radionuclide quantification that can be applied to samples with unknown activities and radionuclide compositions. A verification of the self-consistency of the analysis is one possible way of accomplishing this. In gamma spectrometry it is possible to calculate radionuclide activities using information from multiple gamma emission energies. This leads to an overdetermined system for which the solution can be used to look for inconsistencies. By calculating the recovered peak areas from the radionuclide activities and comparing these to the measured peak areas, outliers can be identified and by resolving these inconsistencies the analysis of the spectrum can be improved. This peak area consistency evaluation can be used to find incorrect shape of the efficiency calibration, missing interferences in the nuclide decay data, and point to peaks where the peak area calculation needs to be optimized. The performance of the method has been shown on a simple spectrum consisting of three radionuclides that are interfering with each other as well as a complex spectrum with unknown radionuclide composition and activities. The method will be integrated into a future version the Genie 2000 Gamma Spectroscopy Software.**

***Keywords* —Gamma Spectrometry, Self-consistency check, Genie 2000.**

I. Introduction

GAMMA spectrometry is a non-destructive measurement method for identifying and quantifying the activities of gamma-ray emitting radionuclides in a wide variety of samples including radioactive waste, environmental, and radio pharmaceutical.

A typical gamma spectrometry analysis contains several components: peak location, peak area calculation, background subtraction, nuclide identification and activity evaluation, as well as detection limits determination [1]. The main inputs to the nuclide identification and activity calculation are the peak energies and background corrected peak areas, nuclide decay data, and efficiency calibration. Each of these present a significant challenge to the gamma spectrometrist and can be the cause of inconsistencies in the analysis. Some of the considerations that need treatment are, peak area calculations need to accurately account for the shape of continuum under the peak. The detection efficiency of the sample can be different than the calibration standard or mathematical model used for the efficiency calibration. Finally, care needs to be taken when choosing the radionuclides and their emission energies to include in the nuclide library used in the analysis.

Many radionuclides emit gamma rays with different energies which creates peaks at different locations in the spectrum. The activity can be calculated from each of these peaks, and it should be the same within the uncertainties of the method. The line activities calculated from the peaks under the assumption that all counts in the peak originates from the radionuclide in question has previously been used for consistency test. However, this assumption is not fulfilled when two or more radionuclide emits energies that cannot be resolved by the detector. The radionuclide activity is in this case calculated from a least-squares fit of the activities that reproduces the measured peak areas. The peak area consistency evaluation (PACE) method relies on how well the measured peak areas are reproduced from the contribution from all radionuclides contributing to the peak. The recovered peak area is then compared to the measured peak area for each peak, and they need to be within the uncertainties for the analysis to be self-consistent. Self-consistency is a necessary but not sufficient condition for a correct analysis of the gamma spectrum. This method is applicable for complex spectra where more than one peak contributes to the peaks in the gamma spectrum.

The PACE method only uses information from input calibrations and the measured spectrum, and therefore it can be used on measured spectra with unknown activities and radionuclide compositions. The method will be included in the Genie 2000 Gamma Spectroscopy Software.

II. Method

The weighted mean activity, A, of radionuclide emitting n gamma-rays with different energies and that do not have any interferences are calculated from

$$A = \frac{\sum_{i=1}^{n} \frac{L_i}{\sigma_i^2}}{\sum_{i=1}^{n} \frac{1}{\sigma_i^2}}, \quad (1)$$

in Genie 2000 [2] where L_i is the line activity of line at peak i with an uncertainty of σ_i. The recovered peak area for the ith peak, P_i, is then calculated from

$$P_i = A\varepsilon_i I_i T, \quad (2)$$

where ε_i is the efficiency at the energy of the peak, I_i is the intensity of the line contributing to the peak and T is the live time of the acquisition. The uncertainty in the recovered peak area is

$$\sigma_{P_i} = \text{P}\sqrt{\left(\frac{\sigma_{A_i}}{A}\right)^2 + \left(\frac{\sigma_{\varepsilon_i}}{\varepsilon_i}\right)^2 + \left(\frac{\sigma_{I_i}}{I_i}\right)^2}. \quad (3)$$

For radionuclides that have at least one of its line interfering with one or more other radionuclides the activity is obtained from minimizing the χ^2 function

$$\chi^2 = \sum_{i=1}^{n} \frac{\left(y_i - \sum_{j=1}^{m} I_{ij} x_j\right)^2}{\sigma_i^2}, \quad (4)$$

where n is the number of peaks in the interference, m is the number of radionuclides in the interference, $y_i = \frac{N_i}{\varepsilon_i T}$, N_i is the measured peak area for peak i, I_{ij} is the intensity of radionuclide j for peak i. The recovered peak areas are calculated from the minimum of (4).

$$P_i = \varepsilon_i T \sum_{j=1}^{m} I_{ij} x_j, \quad (5)$$

With an uncertainty of

$$\sigma_{P_i} = \sqrt{\left(T \sum_{j=1}^{m} I_{ij} x_j\right)^2 \sigma_{\varepsilon_i}^2 + \varepsilon_i^2 T^2 \sum_{j=1}^{m} (I_{ij}^2 \sigma_{x_j}^2 + x_j^2 \sigma_{I_{ij}}^2)} \quad (6)$$

Finally, a z-score can be defined as the difference between the recovered and measured peak area divided by their combined uncertainty

$$z_i = \frac{P_i - N_i}{\sqrt{\sigma_{P_i}^2 + \sigma_{N_i}^2}} \quad (7)$$

which makes it possible to define an outlier peak if the absolute value of the z-score is larger than 3.

This makes it possible to find the peaks for which the analysis is not accounting for the counts for a peak while minimizing the number of falsely labeled peaks that are not outliers.

The outliers that are found using this method should be investigated for possible causes of the inconsistency and the reason for the inconsistency should be resolved. Failing that a conscious decision can be made to exclude the peak from the analysis if the reason cannot be resolved so that the activities of the identified radionuclides are not affected by the outlier.

III. Analysis

The PACE method has been applied to two example gamma spectra, the first is a known spectrum containing three radionuclides that all forms an interference and no other radionuclides. The purpose of this spectrum is to give examples of signatures that show up in the PACE analysis. The second spectrum is a complex spectrum with unknown activities and radionuclide composition and shows how PACE can be used to improve the analysis of unknown spectra.

A. Known spectrum

The first example that the PACE method was applied to is a spectrum containing Co-57, Se-75, and Eu-152. The spectrum was generated by MCNP-CP [3]. The advantage of using a simulated spectrum is that the spectrum is fully known and there are no unknown radionuclides contributing to the spectrum. All three radionuclides contribute to the 121 keV peak and therefore consist of an interference set. Table 1 shows the energy of the peaks included in the analysis and which radionuclide contribute to each peak. Both peaks of Co-57 have contribution from at least one of the other two radionuclides, and the activity is inferred by excess counts in the 121 and 136 keV peaks.

TABLE I
THE PEAKS USED FOR ACTIVITY DETERMINATION AND THE ASSOCIATED RADIONUCLIDES FOR THE CO-57, SE-75, AND EU-152 SPECTRUM

Peak Energy (keV)	Co-57	Se-75	Eu-152
40			X
66		X	
99		X	
121	X	X	X
136	X	X	
244			X
264		X	
279		X	
344			X
400		X	
778			X
964			X
1085			X
1408			X

The peak area recovery z-score was calculated using (7) for the peaks in Table 1 and is shown as a function of energy in Figure 1a. The absolute peak area recovery z-score is less than 2 for all peaks and there is no trend in the ratio as a function of energy, indicating that the analysis is self-consistent, as expected for a known spectrum. Figure 1b shows the line activity ratio, defined as the line activity divided by the weighted mean activity for the three radionuclides. The line activity ratio is not within the uncertainties for the 121 and 136 keV peaks. These are the peaks for which multiple radionuclides contribute to. This demonstrates that the PACE method is a better measure of self-consistency for radionuclides that interfere with each other.

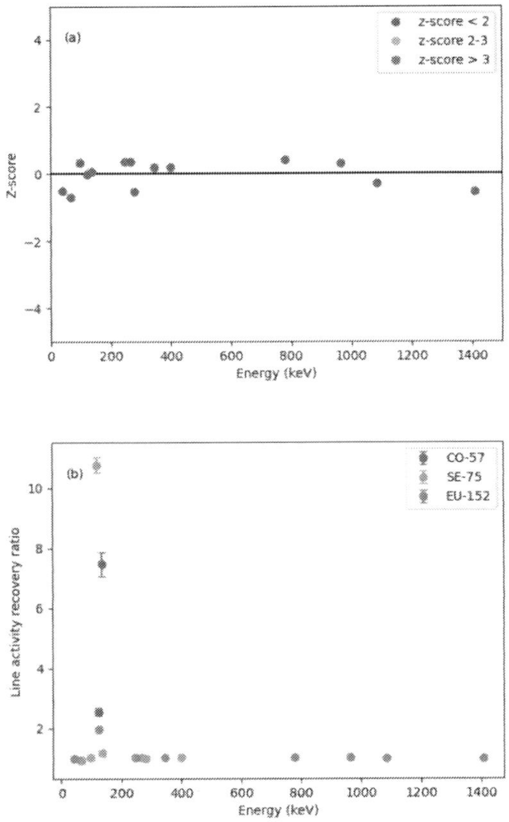

Figure 1. The Peak Area recovery z-score (a) and Line activity ratio (b) for the interference consisting of Co-57, Se-75, and Eu-152.

Different causes for inconsistencies can create different signatures in the peak area recovery ratio. One possible cause of inconsistency is that the shape of the efficiency calibration is different than the shape of the sample efficiency. To demonstrate this signature, the efficiency calibration used for the analysis of the Co-57, Se-75, and Eu-152 was increased for lower energies emulating analyzing a sample with increased attenuation compared to the efficiency calibration. Figure 2 shows the z-score for the analysis with the altered efficiency calibration. The z-score for low energies decreases for when the efficiency calibration is higher than the sample efficiency, which is seen in the decrease in the ratio for the 40, 66, and 99 keV peaks compared to the ratios in Figure 1a. The uncertainties for the 66 and 99 keV are large, which brings up the z-score to a value between negative 2 and 3 while the z-score for the 39 keV is clearly below negative 3. This shows that the method works better when the uncertainties are small for most of the peaks. The least squares solution decreases the activity of Co-57, hiding the effect for these two peaks, demonstrating that radionuclides that do not have any peaks that are free from interferences from other radionuclides can hide the signature in the z-score. To a smaller degree the z-score increases at higher energies because the peaks at low energy decreases the radionuclide activity which increases the recovery ratio. The solution to this signature is to adjust the modeled geometry if the efficiency calibration is done mathematically. For measured efficiency calibrations, an attenuation correction is required.

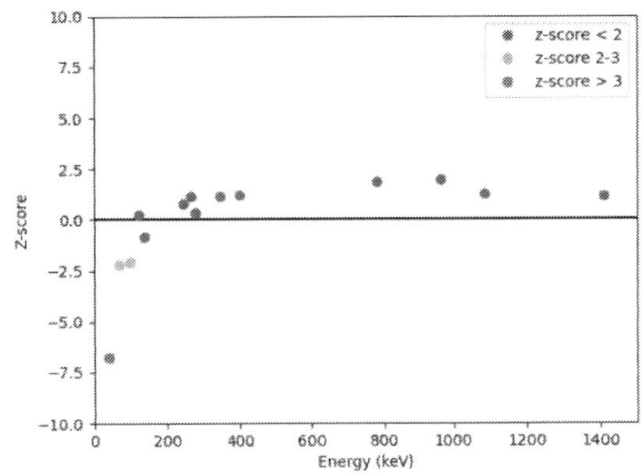

Figure 2. The peak area z-score for an analysis where the shape of the efficiency calibration is different from the shape of the sample efficiency

Another common cause for inconsistency is missing interferences in the nuclide library. Missing interferences can be missing lines from identified radionuclides, such as low intensity lines omitted from high activity radionuclides Alternatively, they could be radionuclides missing from the nuclide library. This is demonstrated in Figure 3 where the same spectrum has been analyzed with Co-57 removed from the nuclide library. The signature of missing interferences is a z-score larger than 3 for the peaks that are missing an interference and there is no correlation to peaks close in energy, as was the case for the shape of the efficiency calibration. The size of the increase in the z-score depends on the activity and intensity of the missing radionuclide, the contribution from the identified radionuclides, and the uncertainty of the peak area. The increase is largest for high intensity lines where the missing radionuclides have a large contribution. For Co-57 the increase is highest for the 121 keV peak because of the much higher intensity compared to the 136 keV line. Another consequence of missing interferences is that the other radionuclides contributing to the peak will have an overestimated activity and therefore the z-score will be lower than 1 as can be seen in Figure 3 for the peaks without contribution from Co-57 causing the 279 keV and 1408 peaks to have a z-score below negative 2. There are several other possible causes of inconsistency in gamma spectrometry analysis and some of them are listed with a suggested solution in Table 1.

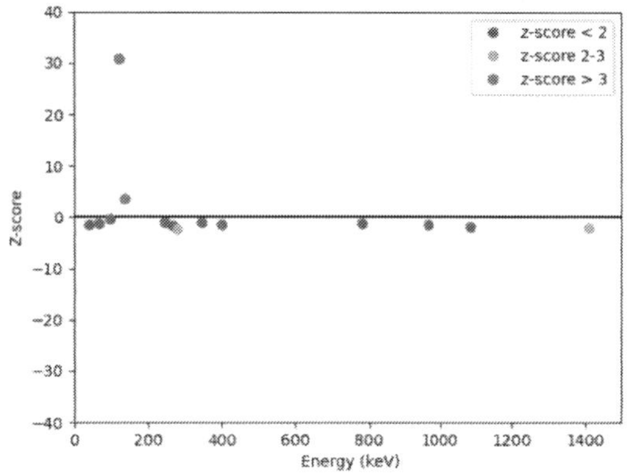

Figure 3. The peak area recovery z-score for an analysis of an interference where a contributing interference is missing.

TABLE 2 COMMON CAUSES FOR INCONSISTENCIES AND THEIR SIGNATURES IN THE PEAK AREA RECOVERY RATIO.

Cause	Signature	Suggested solution
Efficiency shape high/low at low energies	Decrease /increase at low energy	Refine model or perform attenuation correction
Missing interference	Increase at the peak missing the interference	Add line or nuclide to library
Missing true coincidence summing correction	Decrease /increase in ratio when summing in/out is dominating	Add correction factor or don't include in analysis
Peak fitting	Increase or decrease of recovery ratio	Improve peak fitting
Single/double escape interference	Increase at the escape interference	Calculate escape efficiency or exclude from analysis
Magnitude of efficiency calibration	No signature	N/A

B. Unknown spectrum

To demonstrate how the PACE method can be used to improve the analysis of an unknown spectrum, it was applied to the analysis of a complex spectrum of primary reactor coolant measured by a HPGe detector shortly after leaving the reactor. The radionuclide composition and their activities were unknown at the time of analysis. The measured spectrum is shown in Figure 4.

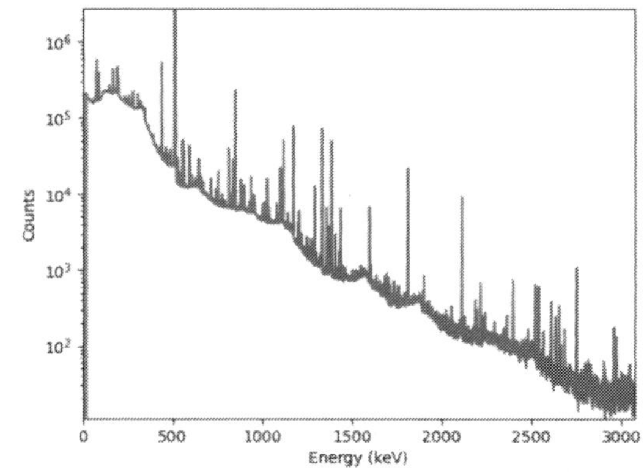

Figure 4, the reactor coolant spectrum.

There were 278 peaks found in the spectrum, with a large interference containing 139 peaks and 30 radionuclides. The peak area recovery ratios for the interference are shown in Figure 5. There are 22 outlier peaks where the absolute value of the z-score is larger than 3, 19 of these have a z-score larger than 3 and 3 had a ratio below negative 3. All the peaks that were identified as outliers were investigated for common reasons according to Table 2.

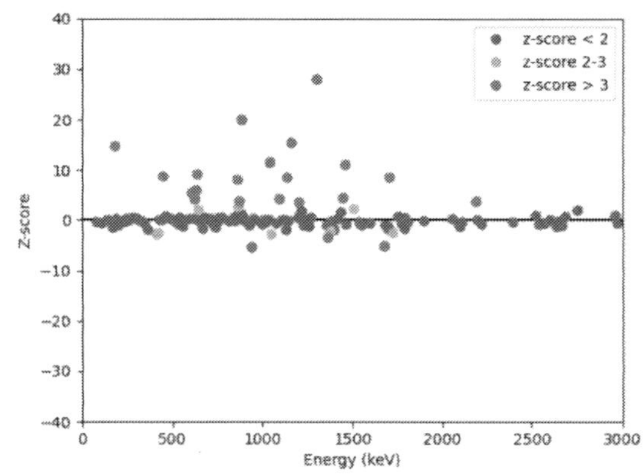

Figure 5. The peak area recovery z-score for the complex spectrum.

There is no noticeable energy dependence in the peak area recovery ratio indicating that the shape of the efficiency calibration, performed using ISOCS efficiency modeling software [4], is self-consistent. The efficiency is low enough that true coincidence summing can be neglected. The peaks were checked for peak fitting, missing interferences, and single and double escape interferences. The cause for the outliers is listed in Table 3. For one of the peaks, peak fitting was determined to be the cause of the inconsistency. It was determined that 12 peaks had missing interferences from radionuclides that were already identified in the nuclide library. Six of these were from low intensity lines from Sr-93. More importantly, the cause of 3 of the outliers were determined to be interference from radionuclides that were missing from the nuclide library and these radionuclides would not have been identified if the self-consistency analysis had not been performed. Additionally, the cause for one of the peaks were interference from single escape from a higher energy peak. For the peaks with a z-score lower than negative 3 the cause was determined to be overestimated activity of Ag-110m because of missing interferences for the 446, 620, and 884 keV peaks and overestimation of La-142 because of missing interference for the 1160 keV peak, and overestimation of I-135 caused by missing interference for the 1040, 1458, and 1706 keV peaks.

TABLE 3 THE OUTLIER PEAKS AND THE CAUSE DETERMINED FOR THE OUTLIER

Peak energy (keV)	Cause
184	Peak fitting
446	Missing line from Sr-93, I-132
602	Cs-140 missing from library
620	Missing line from I-132
630	Missing line from Sr-91, Sr-93
635	Missing line from Ba-141
857	Missing line from Sr-91
871	Rb-90m missing from library
884	I-134 missing from library
937	Overestimation of Ag-110m activity
1040	Missing line from Sr-93
1093	Missing line from Sr-93 and Ba-142
1136	Missing line from Sr-93 and I-134 missing from library
1160	Missing line from I-135 and Ba-141
1203	Missing line from Y-91 and Cs-138
1299	Interference from single escape peak
1362	Overestimation of La-142 activity
1444	Missing lines from Sb-124, I-132, Cs-138 and La-142
1458	Missing line from I-135
1677	Overestimation of I-135
1706	Missing line from Sr-93
2187	Unknown

IV. CONCLUSIONS

PACE is a method for analyzing self-consistency in gamma spectrometry that has been developed for inclusion in the Genie-2000 Gamma Spectroscopy Software. The method relies on the fact that it is possible to calculate radionuclide activities from multiple inputs from a single spectrum. The radionuclide activities are then used to calculate the peak areas recovered from the radionuclide activities and compared to the measured peak areas. The method has been demonstrated on a known sample and signatures as well as suggested solutions for several common causes for inconsistencies were discussed. The method uses no other information than what is available from the calibrations and the spectrum and can therefore be applied to spectra with unknown radionuclide activity and composition. This was demonstrated for an analysis of an unknown sample of primary coolant measured shortly after being removed from the reactor. The analysis revealed that 3 radionuclides were missing from nuclide library and that several radionuclides were missing lines contributing to the peaks in the spectrum. It also revealed that the activity was overestimated for three radionuclides.

REFERENCES

[1] J Verplancke, *et al*, "Semiconductor Detectors," in Handbook of Radioactivity Analysis, M. F. L'Annunziata, 3rd ed, Amsterdam, The Netherlands: Elsevier, 2012, pp. 342-360.

[2] *Genie 2000 Spectroscopy Software Customization Tools*, Mirion Technologies (Canberra) Inc., Meriden, CT, USA, 2013.

[3] A.N. Berlizov, "MCNP-CP: a correlated particle radiation source extension of a general purpose Monte Carlo N-Particle transport code" ACS Symposium series., vol. 945, pp. 183-194, Nov. 2006. https://pubs.acs.org/doi/abs/10.1021/bk-2007-0945.ch013

[4] R Venkataraman, F Bronson, V Abashkevich, B.M Young, M Field, "Validation of in situ object counting system (ISOCS) mathematical efficiency calibration software", NIMA, vol. 422, pp 450-454, Feb 1999, https://doi.org/10.1016/S0168-9002(98)01115-2

Design of 4π Directional Radiation Detector based on Compton Scattering Effect

Ghelman Max[1,2], Kopeika Natan[2], Rotman Stenley[2], Edvabsky Tal[1], Vax Eran[1], Osovizky Alon[1,3]

[1]Electronics & Control Laboratories, Nuclear Research Center, Negev, Israel
[2]Department of Electrical and Computer Engineering, Ben-Gurion University of the Negev, Israel
[3]Health Physics Instrumentation Department, Rotem Industries Ltd., Israel
maxghelman@gmail.com

***Abstract*—Obtaining directional information is required in many applications such as nuclear homeland security, contamination mapping after a nuclear incident and radiological events, or during the decontamination work. However, many directional radiation detectors are based on directional shielding, made of lead or tungsten collimators, introducing two main drawbacks. The first is the size and weight, making those detectors too heavy and irrelevant for utilization in handheld devices, drone mapping, or space applications. The second drawback is the limited field of view, which requires multiple detectors to cover the whole required field of view or machinery to rotate the narrow field of view detector. We propose a novel 4π directional detector based on a segmented hollow cubic detector, which uses the Compton effect interactions with no heavy collimators. The symmetrical cubical design provides both higher efficiency and 4π detection ability. Instead of traditional two types of detectors (scatterer and absorber) structure, we use the same type of detector, based on GAGG(Ce) scintillator coupled to silicon photomultiplier. Additional advantage of the proposed detector obtained by locating the photon sensors inside the detector, behind the scintillators, which improves the radiation hardness required for space applications. Furthermore, such arrangement flattens the temperature variation across the detector, providing better gain stability. The main advantage of the proposed detector is the ability of 4pi radiation detection for high energy gamma-rays without the use of heavy collimators.**

***Keywords* — Silicon photomultiplier (SiPM), Compton scattering, directional detector, scintillator.**

I. Introduction

DETECTION of gamma ray sources is a challenging task in many applications, where monitoring and mapping of radiological and nuclear materials is required. Obtaining directional information is required for nuclear homeland security (HLS) and safety [1], where directional indication decreases source localization time by a factor of 2 to 10 [2]. Moreover, directional information is required for mapping of post-accident decontamination of nuclear incident such as Fukushima Daiichi and other radiological events [3][4].

Many directional radiation detectors are based on directional shielding [5][6] or coded mask [7] made of lead or tungsten collimators. Such concepts introduce three main drawbacks. The first is the size and weight that make those detectors too heavy and irrelevant for utilization in HLS handheld devices, drone mapping, or space applications. The second drawback is the small Field Of View (FOV). Proposed solutions that require multiple detectors to cover all the required FOV or machinery to rotate the limited FOV detector [8][9] increase size and weight even further. The third is the loss of events interacted in the collimator, which leads to lower of efficiency.

An efficient approach might be using mutual collimation detectors. As demonstrated by [10], 7 cylindrical 2" BGO scintillators provide sufficient shielding for high energies required for post-nuclear accident, such as 662 keV emitted by ^{137}Cs. Furthermore, [11] used the minimal possible three mutual collimated scintillates to provide directional information in two axes (horizontal). Opposite approach [12] used 90 to 340 CsI(Tl) scintillators in random and mutually shielding 3D structure to obtain 4π directional detection. However, the efficiency drops significantly with increasing energy, practically limiting the directional efficiency up to about 100 keV, which is insufficient for most post-nuclear accident isotopes. A different approach employs a 3D position-sensitive CZT Compton-based detector [13]. Such a CZT detector is successfully used for imaging, but the total efficiency is very low due to the small volume of about 2 cm^3.

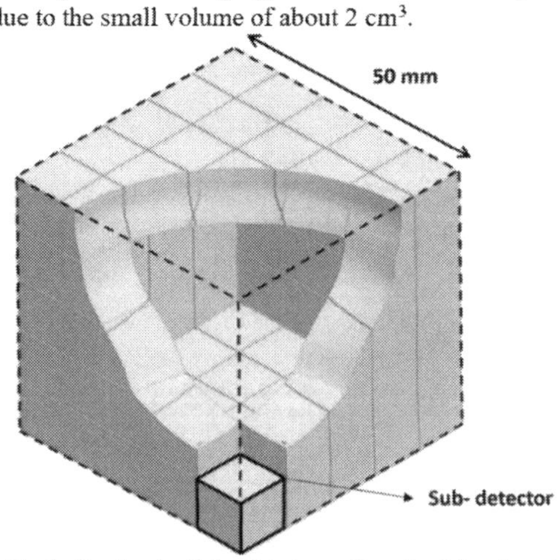

Fig. 1. The 4π directional radiation detector configuration is based on a hollow cubical structure with a segmented outer shell. Each segment made of GAGG scintillator. The whole detector size is 50 mm X 50 mm X 50 mm.

II. Measurement Principle

We propose a novel 4π directional radiation detector based on Compton effect interactions. The basic segmented structure constructed of sub-detectors on a surface of a hollow cube is shown in Fig. 1. For lower energies, the photoelectric effect is dominant, and the approach of mutual shielding is preferred. At higher energies, instead of using various shielding methods for directional information, we make use of the Compton scattering events, see Fig. 2. Such events provide directional information in the form of scattered angle ϑ as defined in (1).

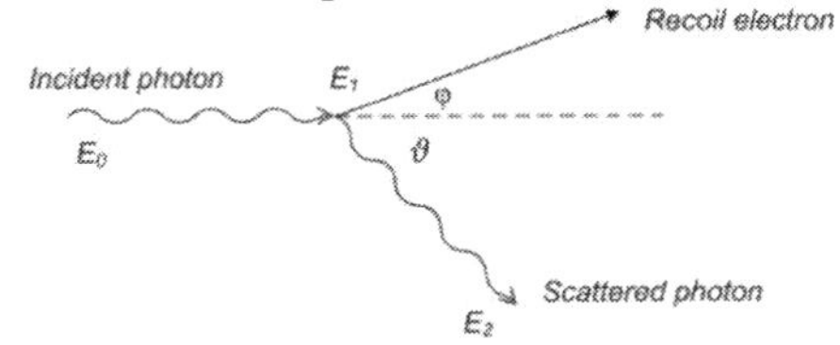

Fig. 2. Compton scattering of a gamma-ray photon and a recoil electron. The incident photon energy E_0, the electron energy E_1, and the angle ϑ dependence are defined by (1), scattered photon energy E_2 is not relevant.

$$E_1 = \frac{E_0}{1+\frac{h\nu}{m_0c^2}(1-cos\vartheta)} \quad (1)$$

Where, m_0c^2 is the electron rest energy, E_1 is the recoiled electron energy, and E_0 is the initial gamma photon energy, which is known for predefined isotopes. Measuring the scattered photon interaction location is required to establish the direction vector, see the dashed line in Fig. 3, which, in turn, is necessary to measure the angle ϑ. Therefore, events with two or more interactions within the detector are required, see Fig. 3.

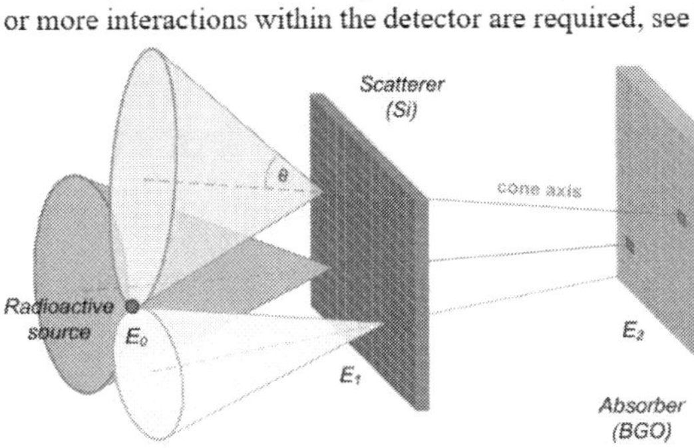

Fig. 3. Compton events with at least two simultaneous interactions within the detector. The conic probabilities define ellipses on a plane. The red dot in the intersection of those ellipses represents the radioactive source location.

For such interactions, when both the first Compton interaction, creating recoiled electron energy of energy E_1, and the second scattered gamma interaction locations were measured, the directional information embedded in the scattered angle ϑ is given by (2):

$$cos(\vartheta) = 1 - m_0c^2\left(\frac{1}{E_0-E_1} - \frac{1}{E_0}\right). \quad (2)$$

When the isotope type is unknown, the scattered gamma energy must be measured in the absorber as well [14]. Equation (3) rearranges (2), so the missing isotope energy E_0 is replaced with the scattered photon energy E_2.

$$cos(\vartheta) = 1 - m_0c^2\left(\frac{1}{E_2} - \frac{1}{E_1+E_2}\right). \quad (3)$$

Based on the interaction's locations of Compton-pairs events, we can back-project a probability cone of source locations. A series of events, generated by GAMOS software [15], was used to calculate a series of cones. These cones have one intersection point, which is the radioactive source location. Several algorithms were developed to overcome the uncertainties, noisy measurements, and limited resolution to converge the measured events into a single point source. The Maximum Likelihood Expectation Minimization ML-EM [16], provided the best results at this work.

A. *Material Selection*

A standard Compton camera is designed to have two types of materials. The scatterer material, which is selected for the first interaction, is a high energy-resolution detector with a low atomic number material, such as ^{14}Si. Low atomic number decreases the photoelectric effect and maximizes the Compton effect probability. On the contrary, the absorber is selected to have maximal detection efficiency, without energy resolution requirements when using (2), such as BGO or LSO [17], or with energy resolution abilities when using (3) such as CdTe or CZT [18]. Maximal detection efficiency is achieved at high atomic number materials, which is opposite to the previously described qualities for the absorber. Nevertheless, we designed a symmetrical structure to provide a 4π FOV; thus, we use the same material for both the absorber and the scatterer. Therefore, there is an inherent disagreement between some of the scintillation-materials properties requirements. Semiconductor materials such as Si or CZT, similarly, are not suitable for both types of detectors, due to the high noise at high temperatures of 50°C [19] required HLS applications.

Scintillators with low intrinsic radiation are preferable, since when such intrinsic gamma ray event occurs at the same measurement time-window with an external gamma event, they are erroneously considered as a Compton-pair.

The decay time property has a somewhat similar influence as intrinsic radiation. Longer decay times require a longer coincidence window, causing a higher probability to measure a false pair event. Scintillator energetic resolution has a direct influence on the measurement uncertainty of ϑ, as calculated in [20], see (4). Hence, higher energy-resolution is advantageous for higher angular resolutions.

$$\sigma(\vartheta) = \sigma(E_1) \cdot \frac{m_0c^2}{sin(\vartheta)\cdot(E_0-E_1)^2}. \quad (4)$$

Scintillators in Table I with less than 4% FWHM resolution ($LaBr_3$, SrI_2, and $CeBr_3$) are both hygroscopic and have intrinsic radiation, which makes them unsuitable. Scintillators with higher density, above 7 g/cm³ (BGO and $CdWO_4$), provide lower resolution, about 10%. Considering all the scintillation properties and their influence on the performance with the considered scintillators in Table I, only the GAGG(Ce) scintillator provides fast decay time, high density, good FWHM resolution, and has no hygroscopic or intrinsic radiation deficiencies. The main drawback of the GAGG scintillator is that its maximal wavelength light yield, peaking at 530 nm, is not compatible with the selected photon sensor, the Silicon Photo Multiplier (SiPM) discussed in the next chapter. The SiPM maximal quantum efficiency occurs at 420 nm.

TABLE I
SCINTILLATION CRYSTALS PROPERTIES

Scintillator	CsI(Tl)	Na(Tl)	GAGG(Ce)	BGO	$CeBr_3$	$LaBr_3$	SrI_2(Eu)	$CdWO_4$
Light yield (Photons / MeV)	54000	38000	57000	8000	60000	63000	11500	15000^b
Resolution FWHM (%)	5.7^a - 7.1^c	6.8^a - 8.7^c	6.6^a – 10.5^d	9.5^c – 12^h	4.1^e - 4.4^a	2.9^a - 5.3^c	3.1^g – 4.4^f	8.5^a – 15.6^c
Hygroscopic	Slightly	Yes	No	No	Yes	Yes	Yes	No
Density (g/cm^3)	4.51	3.67	6.63	7.13	5.1	5.2	4.6	7.9
Effective atomic number	54	50	54/51 fur	72	46	45	50	64
Average time constant (ns)	1000	250	90	300	20	25	1500	13000
Peak emission (nm)	550	415	530	480	380	380	435	480
Intrinsic radiation	No	No	No	No	Slightly	Yes	Slightly	No

The relevant material properties of selected scintillators. ^{a} – Measured in [21] with a Hamamatsu R6231-100 PMT; ^{b} – by [22] with PMT; ^{c} – by [23] with a Hamamatsu S11827-3344MG MPPC or XP 2020Q PMT; ^{d} – by [24] with a SensL MicoFC SiPM; ^{e} – by [25] with Hamamatsu R6231 PMT; ^{f} – by [26] with a Hamamatsu TSV MPPC S12642-0404A-50; ^{g} – by [27] with a Hamamatsu R6231-100HA PMT; ^{h} – by [28] with a SensL MicroFC-30035-SMT SiPM.

Such emission to absorption incompatibility lowers the number of measured photons and consequently decreases the measured resolution.

B. Photon Sensor Selection

Scintillation photons can be measured and amplified by a few types of photon sensors. The prime mandatory requirement for the segmented detector is compactness. Photomultiplier Tubes (PMT) are well based technology, with good noise performance at the required high temperatures but large and cumbersome, making the PMTs not suitable to be coupled to each of the small sub-detectors. The PIN diode and the Avalanche Photodiode (APD) are compact but undergo noise and gain degradation reaching 50°C [29]. The Silicon Photomultiplier (SiPM) is a compact sensor with good tiling ability that enables utilization of multiple sensors in proximity, without significant dead space between them. Furthermore, while the noise degradation with temperature is exponential, at 50°C, noise levels are acceptable, degrading resolution by a factor of two, as measured with CsI(Tl) and ArrayJ-600035-64P SiPM [30].

The desired requirements for the light sensor include microphonic insensitivity. Sensors such as PIN diode that have no intrinsic gain are classically amplified by charge-sensitive amplifiers. These amplifiers are sensitive to microphonic and piezoelectric noise. The SiPM provides a gain of millions of electrons for each measured photon. Such a high gain at the first stage is beneficial from electronic measurement SNR considerations and enables simpler and lower electronic amplification in the next stages. The bias voltage is only a few tens of volts instead of a few hundred volts for APDs and up to a few thousands of volts for PMTs. SiPMs are insensitive to electromagnetic interference [31, [32], which is an additional advantage over the classical PMT.

Nevertheless, there are three significant drawbacks of the SiPM technology that needs to be addressed. First, the radiation hardness of SiPMs is lower than PMTs, and they undergo a severe degradation in space, after a few years of operation at the international space station [33]. The other two drawbacks are gain variation with bias voltage and with temperature.

Bias variation requires stable and accurate voltage supply, with less than a few tens of mV variation over a few tens of volts bias voltage. Temperature-dependent gain stability can be achieved by stabilizing the temperature, which is not applicable for low power or handheld devices. A more practical approach is measuring the temperature T and correcting the bias voltage, V_B, according to (5,6).

$$V_B = V_{BR}(T) + V_{OV} \, , \quad (5)$$

$$V_{BR}(T) = V_{BR}(T_0) + \beta \cdot (T - T_0) \, , \quad (6)$$

Where V_{BR} is the temperature-dependent breakdown voltage, T_0 is a reference temperature, β is the temperature coefficient, and V_{OV} is the overvoltage. Since the gain of SiPM is V_{OV} dependent, measuring T and keeping $V_{OV} = V_B - V_{BR}(T)$ constant will stabilize the gain.

C. Geometrical Considerations

To achieve a high geometrical resolution of radioactive source location, accuracy of scattered angle ϑ and the established directional vector is required. The scattered angle is primarily depended on the energy-resolution of the detector. The accuracy of the established directional vector, as defined by the dashed line in Fig. 3, requires high definition of the difference between the integration locations. Consequently, small detection segments and large distance between the first and the second interactions is desirable. Classically, this can be achieved by setting a few thin, low atomic number, pixelated scattering detectors in a distance from a high atomic number absorber, obtaining a high resolution in a narrow FOV [18] in one direction of the vector.

We propose using the same material for both the absorber and the scatterer to enable detection at both directions of the directional vector. A symmetrical cubic structure is formed by using three opposing pairs, providing a 4π FOV directional detector. Furthermore, there is no requirement for the first and second events to occur in opposite pairs. Including the other four cube plains substantially increases the total detector efficiency.

The sub-detectors are located at the outer shell to obtain the maximal distance between interactions, producing a hollow cubical structure with a segmented scintillator surface (see Fig. 1). The whole detector size is 50 mm X 50 mm X 50 mm. Each sub-detector is made of GAGG(Ce) scintillator coupled to an SiPM photon sensor. Each of the SiPMs needs to be measured thus requires multi-channel electronics [34,[35]. The results in this work are for a cube of 10^3 size requiring 488 of possible 512 channels.

The detector can be encapsulated with a rigid aluminum structure for the Compton scattering energy range without significant influence on the interaction efficiency. On the other hand, for the lower energies, where the photoelectric effect is predominant, adding structural material lowers the efficiency significantly. Using hygroscopic scintillators, especially for low-pressure space applications, would create additional encapsulation demands. These will reduce the efficiency even further. Hence, as mentioned before, they are not considered feasible for the proposed detector.

Furthermore, for the lower energy range used in directional radiation detectors, even thin PCBs are of much consideration for loss of efficiency [12]. The proposed hollow cube configuration's outer layer is rigid by itself, made of GAGG scintillators, so there is no need for additional significant structure. The PCBs and the SiPM sensors are located inside the cube volume, so there is no shielding at the front of the scintillators.

Furthermore, positioning the SiPMs inside the detector provides a significant gain stability advantage. Since SiPM gain is temperature-dependent, the temperature is measured, and the bias voltage is adjusted according to (5) for gain stability.

An additional advantage achieved by locating the SiPMs inside is enhanced radiation hardness. As stated in the properties of the photon sensors in Table I and demonstrated in [33[36] when irradiated, SiPMs experience a large increase in dark current, increasing power consumption and noise, limiting the detection of low energy events. Hence, shielding the SiPMs from external radiation with a dense scintillator could provide prolonged lifetime of the detector.

III. Results

The presented results are for the 10^3, cube size, using a GAGG(Ce) based 5mm X 5mm X 5mm sub-detectors with 7% FWHM energy resolution. Results of one side projection are shown on the right side of Fig. 4 (colored estimated distribution on the black background). The detector is located at the origin (blue), with the directional vectors (green) shown inside the detector. The red arrow represents the location and direction of the estimated radioactive source.

The performance of the proposed detector was evaluated for the 662 keV of the ^{137}Cs isotope. The final estimation is based on five iterations of ML-EM, reaching an estimation accuracy of $\sigma(\vartheta) = 2°$.

IV. Conclusions

This work presents a novel 4π directional radiation detector. We designed a detector based on a Compton scattering effect as a hollow cube-shaped segmented scintillator. We have demonstrated that without using collimators, a 4π directional detection of high-energy isotopes is feasible. Typically, directional detection for nuclear incidents requires not only high gamma energy measurement capability, but also composed of a few radioactive sources.

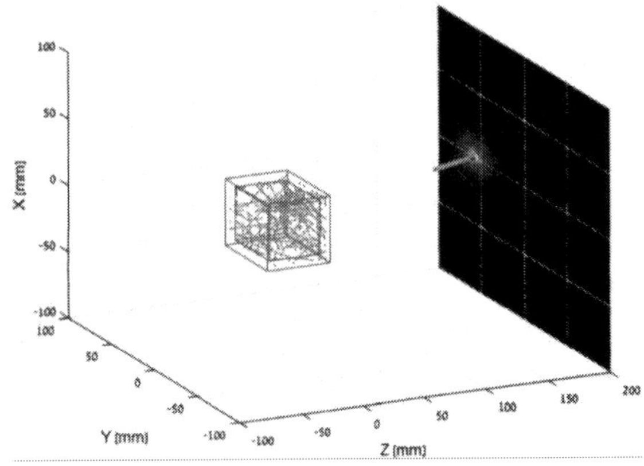

Fig. 4. Estimation of 662 keV ^{137}Cs isotope. The detector located at the origin (blue), with the directional vectors (green) shown inside the detector. The red arrow represents the location and direction of the estimated radioactive source. Colored estimation distribution is shown on the black background.

While using the same type of scintillators for both the absorber and the scatterer detectors is not optimal, such an approach enables us to double the measurable field of view. Moreover, the selected GAGG scintillator is sufficient to obtain a high directional accuracy of about 2°.

Utilizing any one of the six cube planes as scatterer detector, while using the other five as absorbers, provides 4π FOV and increases the efficiency, since more Compton pairs are available in comparison to the conventional two planes design (one scatterer and one absorber), which limits the direction of arrival of photons.

The GAGG(Ce) scintillator coupled to an SiPM photon sensor as a sub-detector segment provides the most suitable overall properties for the proposed hollow cube configuration detector.

Hollow cube configuration with the SiPMs located inside provides the ability to stabilize the temperature and thus stabilize the gain while protecting the delicate photon sensors from radiation, providing an extended lifetime.

References

[1] "AccuRad™ PRD, Personal Radiation Detector", [Online]. Available: https://www.mirion.com/products/accurad-prd-personal-radiation-detector. Accessed on: June 05, 2021.

[2] C. L. Larsson, T. Jones, "Testing a new directional gamma survey meter", Defense R&D Canada – Ottawa, 2012, [Online]. Available: https://cradpdf.drdc-rddc.gc.ca/PDFS/unc240/p803606_A1b.pdf. Accessed on: June 05, 2021.

[3] Y. Sato, Y. Tanifuji, Y. Terasaka, H. Usami, M. Kaburagi, K. Kawabata, et al., "Radiation imaging using a compact Compton camera inside the Fukushima Daiichi Nuclear Power Station building", Journal of Nuclear Science and Technology, 2018, Vol. 55 No. 9, 965-970, DOI:10.1080/00223131.2018.1473171.

[4] J. Jiang, K. Shimazoe, Y. Nakamura, H. Takahashi, Y. Shikaze, Y. Nishizawa, et al., "A prototype of aerial radiation monitoring system using an unmanned helicopter mounting a GAGG scintillator Compton camera", Journal of Nuclear Science and Technology, 53:7, 1067-1075, DOI:10.1080/00223131.2015.1089796.

[5] V. Batiy, O. Stoyanov D. Fedorchenko, S. Prohoretz, M. Khazhmuradov, "Mathematical modeling to support gamma

radiation angular distribution measurements", Waste Management Symposium, Tucson, AZ (United States), January 2007; https://www.researchgate.net/publication/237505414.
[6] K. Kojima, T. Nakamori, D. Nemoto, S. Gunji, H. Sato, S. Ito, S. Kato, M. Yoshino, Y. Usuki, J. Kataoka, "Development of Gamma-Ray Detector Sensitive to Source Directions using GAGG(Ce) Scintillators and MPPCs" , IEEE- NSS October 2016, DOI: 10.1109/NSSMIC.2016.8069923.
[7] O. Gala, M Gmara , O. P. Ivanovb , F. Laine, F. Lamadiec , C. Le Goallerc, et al., "Development of a portable gamma camera with coded aperture", 563(1):233-237, NIM A 563 (2006) 233–237 DOI:10.1016/j.nima.2006.01.119.
[8] O. Gala Bryan V. Egner, Darren E. Holland, Larry W. Burggraf, James E. Bevins, "Development of a modular mixed-radiation directional rotating scatter mask detection system". 987, NIM A, 2021, 164820, DOI:10.1016/j.nima.2020.164820.
[9] G. Cao, B. Liu, H. Gong, H. Yu and G. Wang, "A Stationary-Sources and Rotating-Detectors Computed Tomography Architecture for Higher Temporal Resolution and Lower Radiation Dose", IEEE Access, vol. 2, pp. 1263-1271, 2014, DOI: 10.1109/ACCESS.2014.2363367.
[10] A. Pernick, I. Orion, D. Ilzycer, H Zafrir, "Applications of a Self-Collimating BGO Detector System to Radiological Emergency Response", Health Phys. 1997 Jan; 72(1):136-40. DOI:10.1097/00004032-199701000-00019.
[11] Y. Shirakawa, T. Yamano and Y. Kobayashi, "Remote sensing of nuclear accidents using a direction-finding detector," 2009 35th Annual Conference of IEEE Industrial Electronics, 2009, pp. 1917-1922, DOI:10.1109/IECON.2009.5414850.
[12] R. Rahin, L. Moleri, A. Vdovin, A. Feigenboim, S. Margolin, S. Tarem, et al., "GALI: a gamma ray burst localizing instrument", Proc. SPIE 11444, Space Telescopes and Instrumentation 2020: Ultraviolet to Gamma Ray, 114446E, 2020, DOI:10.1117/12.2576126.
[13] Dan Xu, Z. He, C. E. Lehner, F. Zhang, "4π Compton imaging with single 3D position sensitive CdZnTe detector", 2004, Proceedings Volume 5540, Hard X-Ray and Gamma-Ray Detector Physics VI, DOI: https://doi.org/10.1117/12.563905.
[14] Daniel W. Mundy and Michael G. Herman, "Uncertainty analysis of a Compton camera imaging system for radiation therapy dose reconstruction", 2010, American Association of Physicists in Medicine, DOI: 10.1118/1.3399777.
[15] P. Arce, P. Rato, M. Canadas, and J. I. Lagares, "GAMOS: A Geant4-based easy and flexible framework for nuclear medicine applications," IEEE Nuclear Science Symposium Conference Record, pp. 3162-3168, 2008.
[16] S. J. Wilderman, N.H. Clinthorne, J.A. Fessler, W. L. Rogers, "List-Mode Maximum Likelihood Reconstruction of Compton Scatter Camera Images in Nuclear Medicine", 1998, IEEE Nuclear Science Symposium and Medical Imaging Conference, DOI: 10.1109/NSSMIC.1998.773871.
[17] F. Hueso-Gonzalez, A. K. Biegun, P. Dendooven, W. Enghardt, F. Fiedler, C. Golnik, at. al., "List-Mode Maximum Likelihood Reconstruction of Compton Scatter Camera Images in Nuclear Medicine", 2015 *JINST* 10 P09015, DOI: http://dx.doi.org/10.1088/1748-0221/10/09/P09015.
[18] T. Takahashi, M. Kokubun, K. Mitsuda, R. L. Kelley, T. Ohashi, F. Aharonian, et. al., "The ASTRO-H X-ray Astronomy Satellite", 2018, J. of Astronomical Telescopes, Instruments, and Systems, 4(2), 021402, DOI:https://doi.org/10.1117/1.JATIS.4.2.021402.
[19] S. H. Park, J. H. Ha, J. H. Lee, H. S. Kim, Y. H. Cho, et. al. "Effect of Temperature on the Performance of a CZT Radiation Detector", Journal of the Korean Physical Society, Vol. 56, No. 4, April 2010, pp. 1079~1082.
[20] E. Munoz1, J. Barrio1, A. Etxebeste, P. G. Ortega, C. Lacasta, J. F. Oliver, et. al.," Performance evaluation of MACACO: a multilayer Compton camera", Phys. Med. Biol. 62 (2017) 7321–7341, DOI:https://doi.org/10.1088/1361-6560/aa8070.
[21] P. Sibczynski, A. Broslawski, A. Gojska, V. Kiptily, S. Korolczuk, R. Kwiatkowski, et.al. "Characterization of some modern scintillators recommended for use on large fusion facilities in g-ray spectroscopy and tomographic measurements of g-emission profiles", Jun. 2017 Nukleonika 62(3):223 DOI:10.1515/nuka-2017-0032.
[22] L. Nagornaya, S. Burachas, Y. Vostretsov, V. Martynov, V. Ryzhikov, "Studies of ways to reduce defects in CdWO4single crystals", Journal of Crystal Growth 198/199 (1999) 877–880, DOI: 10.1016/S0022-0248(98)01236-6.
[23] M. Grodzicka, M. Moszynski, T. Szczesniak, M. Szawlowskia, J. Baszak, "Characterization of 4×4 ch MPPC array in scintillation spectrometry", 2012 IEEE Nuclear Science Symposium and Medical Imaging Conference Record (NSS/MIC), DOI:10.1109/NSSMIC.2012.6551118.
[24] B. Seitz, N. Campos Rivera, and A. G. Stewart, "Energy Resolution and Temperature Dependence of Ce:GAGG Coupled to 3 mm × 3 mm Silicon Photomultipliers", IEEE TRANSACTIONS ON NUCLEAR SCIENCE, VOL. 63, NO. 2, APRIL 2016 DOI: 10.1109/TNS.2016.2535235.
[25] F.G.A.Quaratiab, P.Dorenbosa, J. van der Biezenc, A Owensc, M. SelledL, P. Schotanusf, "Scintillation and detection characteristics of high-sensitivity CeBr3 gamma-ray spectrometers", 2013 NIM A, 729 596-604, DOI: 10.1016/j.nima.2013.08.005.
[26] K. Shimazoe, , A. Koyama, H. Takahashi, S. Sakuragi, Y. Yamasaki, "Fabrication and characterization of rectangular strontium iodide scintillator coupled to TSV-MPPC array", NIM A 845 503-506, 2017, DOI: 10.1016/j.nima.2016.06.077
[27] M. Takabe a,n , A. Kishimoto a , J. Kataoka a , S. Sakuragi b , Y. Yamasaki, "Performance evaluation of newly developed SrI2(Eu) scintillator", NIM A 831 260-264, 2016, DOI:10.1016/j.nima.2016.04.043.
[28] A. J. González, F. Sanchez, S. Majewski, A. Aguilar, A. González-Montoro, et. al., "Performance of large BGO arrays coupled to SiPM photosensors — Continued study", 2015 IEEE Nuclear Science Symposium and Medical Imaging Conference (NSS/MIC), 2015, pp. 1-4, DOI: 10.1109/NSSMIC.2015.7582017.
[29] Si APD, MPPC, Handbook chapter 3, E03, [Online]. Available: https://www.hamamatsu.com/resources/pdf/ssd/e03_handbook_si_apd_mppc.pdf. Accessed 14.07.2021.
[30] J. Valencia, B. Godfrey, M. Hamel, B. Maestas, E. Padilla, "Evaluation of a Silicon Photomultiplier Array as a Photomultiplier Tube Replacement", 2019, SAND2019-6405C. Sandia National Laboratories2, Albuquerque, NM, USA. [Online]. Available: https://www.osti.gov/servlets/purl/1640659. Accessed 14.07.2021.
[31] M. Ren, J. Zhou, B. Song, C. Zhang, M. Dong, R. Albarracin, "Towards Optical Partial Discharge Detection with Micro Silicon Photomultipliers", Sensors 2017, 17(11), 2595, DOI: https://doi.org/10.3390/s17112595.
[32] R. Hawkes, A. Lucas, J. Stevick, G. Llosa, S. Marcatili, C. Piemonte, "Silicon photomultiplier performance tests in magnetic resonance pulsed fields", 2007, IEEE Nuclear Science Symposium Conference Record, DOI:10.1109/NSSMIC.2007.4436860.
[33] A. Ulyanov, D. Murphy, J. Mangan, V. Gupta, W. Hajdas, at. al., "Radiation damage study of SensL J-series silicon photomultipliers using 101.4 MeV protons", NIM A, 976 (2020) 164203, DOI:https://doi.org/10.1016/j.nima.2020.164203.
[34] D. Impiombato, S. Giarrusso, T. Mineo, , O. Catalano, C. Gargano, G. La Rosa, et. al., "Characterization and performance of the ASIC (CITIROC) front-end of the ASTRI camera", NIM A, 794 (2015) 185-192, DOI:10.1016/j.nima.2015.05.028.
[35] V. Nadig, D. Schug, B. Weissler, V. Schulz, "Evaluation of the PETsys TOFPET2 ASIC in multi-channel coincidence experiments", EJNMMI Physics volume 8, Article number: 30 (2021), DOI:https://doi.org/10.1186/s40658-021-00370-x.
[36] K. Lacombe, J. Knodlseder, B. Houret, T. Gimenez, J. F. Olive, P. Ramon, "Characterization of silicon photomultipliers for new high-energy spacetelescopes", NIM A, 912 (2018) 144-148, DOI:https://doi.org/10.1016/j.nima.2017.11.005.

Toward UO_2 micro/macro machining: a laser processing approach

Thomas Doualle[1], Matthieu Reymond[1,2], Yves Pontillon[1] and Laurent Gallais[2]
[1]CEA, DES, IRESNE, DEC, Cadarache F-13108 Saint-Paul-Lez-Durance, France
[2] Aix Marseille Univ, CNRS, Centrale Marseille, Institut Fresnel, Marseille, France
thomas.doualle@cea.fr

***Abstract*—Linked to experimental data acquisition and to development of improved models, a better detailed description of the behaviour of the nuclear ceramics as regard to the fission gases release during thermal transient representative of nuclear accidents such as RIA (Reactivity Initiated Accident) and or LOCA (LOss of Coolant Accident) requires access to local information within the fuel pellet, and no longer averaged over the whole of the pellet. One of the major challenge in this context is the sample size, which depends on the main objective of the study, typically from the order of a few hundred microns to millimeters. Few techniques allow this dynamic while being compatible with irradiated fuel constraints. Laser micromachining is a high precision non-contact material removal process that would be adapted to this dynamic. We present experimental and numerical studies, carried out in order to evaluate the possibility to apply this process for the preparation of irradiated UO_2 samples of various dimensions. First, preliminary experimental and numerical works conduced on graphite, as model material, which have comparable properties (in particular their behaviours under laser irradiation and their melting point) in order to validate the feasibility, will be detailed. Afterwards, based on these results, we present our first results on UO_2. The objective is to transfer the technique to non-irradiated UO_2 and then to the irradiated material.**

***Keywords* —UO_2, graphite, laser ablation, laser cutting finite element model**

I. Introduction

Lasers are essential tools for macro and micro-processing for a whole range of materials. Processes involving lasers provide a unique solution with minimum mechanical and thermal influence on the processed part due to their selective energy control and deposition and allow generally high processing speeds.

For instance, Pulsed Laser Ablation (PLA) is a technique of vast scientific and significant industrial interest because of its potential applications in micromachining (drilling and cutting), deposition of coatings and thin films, synthesis of a wide range of nanostructures and nanoparticles or surface patterning. With PLA approach, the quality of the processing result is reduced by thermal damage of the workpiece and therefore the quality is increased with shorter pulse duration: the Heat Affected Zone (HAZ) is reduced, leading to higher accuracy. However, ablation efficiency decreases as well. Thus, if high precision is needed, short and ultrashort (femto-, pico- and nanosecond laser) in combination with adequate drilling strategies are applied (such as trepanning). However, single pulse drilling using milli- and microsecond pulses are much more efficient and allow the production of many craters within a short time. The drawback however is the extension of the HAZ, leading to potentially low quality and reproducibility of the craters. In this work, we use a microsecond pulsed laser and the objective was to find the best compromise between the efficiency of the process (ablation rate) and the quality (reduced heat affected zone).

Our approach is based on the one hand on experiments involving a monomode kW Ytterbium fiber laser, and on the other hand on a numerical model to improve the understanding of the underlying physical phenomena and processes involved. These results, obtained on graphite, will be used as a first step towards machining nuclear fuels, such as UO_2, allowing sample preparation for various applications (annealing tests, mechanical behaviour or thermophysical properties).

In this paper, we present in Sec. II the experimental details which enable us to heat up graphite material above sublimation temperature and hence locally ablate material from graphite surface. Section III is dedicated to the description of the model developed to understand the typical crater profiles obtained with our microsecond laser ablation/cutting approach. In Sec. IV, we discuss about the results. First, we describe single-pulse laser ablation and the evolution of the crater dimensions as a function of the laser power. To improve the understanding of the experimental results (craters dimensions and morphology), we compare the simulations results to the experimental crater dimensions. Secondly, we describe a laser cutting technique and its performances which could be applied for graphite manufacturing. Finally, we present the first single pulse ablation experiments on UO_2.

II. Experimental setup

A. Samples

The experiments were conducted on the EDM-3 graphite from POCO Graphite Inc., which is a polycrystalline graphite produced by sintering [1]. The UO_2 samples were fabricated at CEA. Samples have a disk morphology with a diameter of 8 mm and their thicknesses vary from 0.5 to 1.5 mm.

https://doi.org/10.1051/epjconf/202125307004

B. Experimental setup

For the experiments, an Ytterbium fibre laser (redPOWER Qube 1.5 kW, SPI Lasers), which has a central wavelength of 1080 nm with a bandwidth < 4 nm, with a maximum output power of 1500 W is used. A high power laser collimator generates a 13.5 mm diameter Gaussian beam that is injected in a 2D scanning head composed by two galvanometer mirrors (intelliSCAN 20, ScanLab). The experimental configuration is represented in Figure 1 (a). The laser beam is focused using a F-theta lens with a focal length of f = 255 mm and the working distance, i.e. the distance between the exit of the focusing lens and the nominal working plane, is 303 mm. In this plane, the laser beam was measured with a beam profiler (Beamage 4M, Gentec) and the diameter at $1/e^2$ is 40 µm. The typical image field size is 110 x 110 mm^2. The maximum marking speed of our scanning head is 1 m/s with a repeatability < 2 µrad. A commercial laser machining software (LaserDesk, Scanlab) is used to control and generate the movements of the mirrors and ensure the synchronization with the laser irradiation. The laser power was calibrated with a power calorimeter (Gentec, UP55C-2.5KW-HD-D0). For the experiments conducted under vacuum atmosphere (on UO_2 samples), the laser beam is focused in a vacuum chamber (X40K, Neyco), equipped with a UV fused silica window (WG41050-C, Thorlabs) which has an antireflection coating for the laser wavelength. The operating pressure is close to 10 mbar. The experiments on graphite were carry on under air.

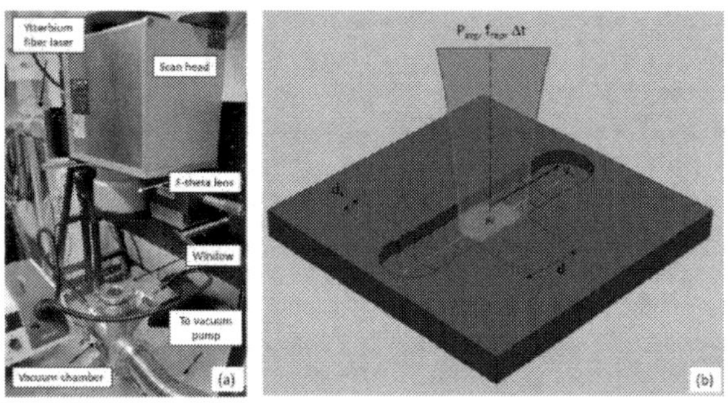

Fig. 1. (a) Experimental configuration (b) Illustration of the process parameters involved during laser processing

The main process parameters for ablating and cutting graphite sample with pulse laser irradiation are shown schematically in Figure 1 (b). The laser beam has a power of P, a pulse duration Δt, a repetition rate f_{rep}, and is focused on a diameter d (gaussian beam, defined at $1/e^2$). It is moved across the graphite surface at the scan speed v_s in a unidirectional scan pattern resulting in a distance d_x between each pulse:

$$d_x = \frac{v_s}{f_{rep}}, \quad (1)$$

The ablation depth z_{ab} is determined as the height difference between the initial surface and the bottom of the ablated area and the ablated width w_{ab} corresponds to the diameter of the individual hole generated by each pulse.

C. Characterizations

Regarding sample characterizations, surface observations are carried out employing an optical microscope (Axiotech, Zeiss) and a scanning electron microscope (TM 1000, Hitachi) to investigate the morphology of craters and structures formed and measure their dimensions.

III. Modeling

Simulations are conducted with the commercial software COMSOL, which is based on the Finite Element Method, in axisymmetric configuration, with a two dimensional meshing, to go further in the investigation and understand the laser ablations mechanisms. Heat transfer by conduction associated with deformed geometry and ray-tracing modules are employed for the analysis of the thermal effects, material evaporation and laser propagation on the fabricated structures.

The heat transfer model relies on several assumptions. First, the graphite is considered homogeneous and isotropic. Second we take into account the heat transfers by conduction combined with thermal radiation to take into account losses at the surface (convection is neglected). Third, as the energy is absorbed on the surface, there is no heat source in the material but an incoming heat flux on the boundary, deduced from a ray tracing approach. The corresponding non-linear heat transfer equation can be expressed as:

$$\rho(T)C_p(T)\frac{\partial T}{\partial t} + \nabla(-k(T)\nabla T) = 0, \quad (2)$$

where ρ is the density, C_p is the specific heat capacity under constant pressure, T is the temperature, k is the thermal conductivity. The thermal model and its associated input parameters presented in this paper are based on the model described and validated up to 3800 K in previous study [2].

The laser beam is discretized in rays transporting an amount of energy and computed, one by one, through ray propagation and interactions between the material and the geometry.

To simulate the crater formation, we have taken into account in the numerical model the velocity of the evaporation front at the sample surface, given by [3]:

$$v(T) = \beta\sqrt{\frac{M}{2\pi RT}}\frac{p_{sat}(T)}{\rho(T)}, \quad (3)$$

To investigate the crater formation dynamics, we have used a moving boundary condition at the sample surface, controlled by the local evaporative velocity, associated with a moving mesh functionality in the 2D numerical model.

Evaporative cooling is included as a surface boundary condition since it can be a significant heat transport mechanism as the temperature comes closer to the sublimation point. The outgoing heat flux at the boundary can be obtained as [2]:

$$\varphi(T) = \beta\sqrt{\frac{M}{2\pi RT}\frac{\Delta H_v}{R}}p_0 exp\left[\frac{\Delta H_v}{R}\left(\frac{1}{T_v} - \frac{1}{T}\right)\right], \quad (4)$$

with M being the molar mass of carbon, R the perfect gas constant, ΔH_v the enthalpy of the phase change, T_v the temperature of the phase change at ambient pressure p_0. The input parameters used in this last equation are given in [4].

IV. Results and discussions

A. Single pulse ablation on graphite

Pulse laser ablation results in a crater formation on the solid matter surface. The observation and the description of crater is an interesting method to understand the physics of the ablation on the material of interest. We have coupled this experimental

 https://doi.org/10.1051/epjconf/202125307004

study with numerical simulations to improve the understanding of the results.

Figure 2 shows craters that are produced on graphite by using single-pulse ablation process with increasing laser power. The pulse duration was 50 µs. This is the shortest pulse possible duration that the laser can produce and we have chosen it to minimize the heat effects.

The top-views show that the entrance craters are extremely clean and present an excellent definition with very little evidence of collateral thermal effects. It is interesting because for most of materials, microsecond pulses involve many collateral effects that can affect the quality of the structure like a large HAZ, a recast layer, micro-cracks, shock waves surface damages and debris from ejected material.

The cross-sections observations highlight the "channel" shape of the hole. The Figure 3 presents the results of the simulations in the case of laser power during 50 µs, which corresponds to the experimental crater represented in the Figure 2 (a) and (f).

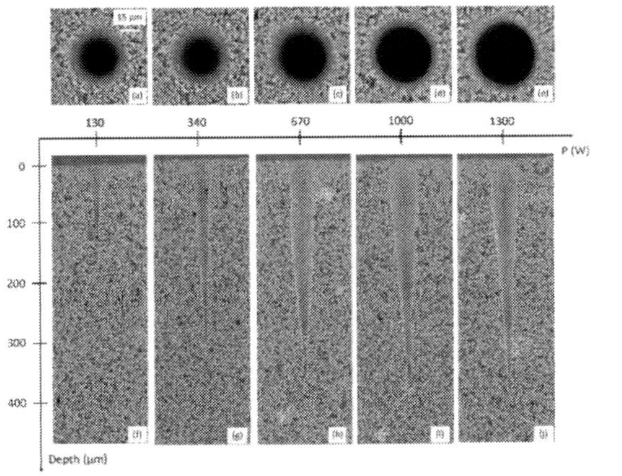

Fig. 2. SEM measurements of craters obtained at different laser powers. (a) to € and (f) to (j) are respectively top and cross-section views. The duration of the laser pulses is 50 µs.

Although, the simulated crater at the end of the irradiation does not have the tapered morphology, we note that dimensions (width and depth) obtained are comparable. Furthermore, as represented on the right part of the Fig. 3, the rays, after one reflection (black rays), which still contain a non-negligible energy (reflection is 8.5 + 0.5%), propagate in the direction of the centre of the hole. This can explain the high-aspect ratio of the crater and its tapered shape observed experimentally.

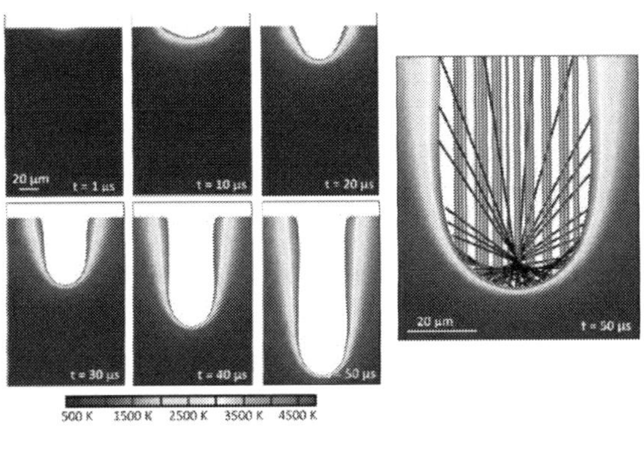

Fig. 3. Simulations of the distribution of temperature, the geometry deformation and the ray propagation in the graphite material at different time (from 1 µs to 50 µs) for a laser power of 130 W.

To go further, we have extended this numerical study by increasing the laser power. The Figure 4 represents the simulated (blue and black dashed lines) and experimental (blue and black dots) results of the ablated depth z_{ab} and width w_{az} of the craters as a function of laser power applied.

We note that the crater width increases with the laser power to reach a value close to the beam diameter. Considering experimental uncertainties, the calculated crater width (represented by the black dashed line) are close and follows the same trend as the experimental ones.

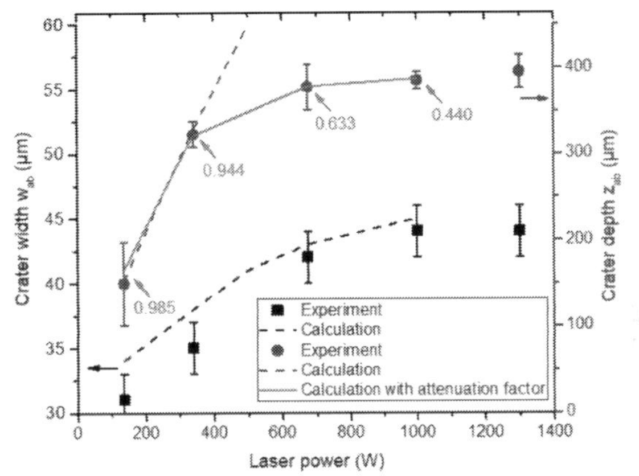

Fig. 4. Experimental crater depth z_{ab} (blue dots) and diameter w_{ab} (black dots) as a function of laser power. The black and blue dashed lines are the corresponding numerical results. The red line indicates the depths obtained with the model considering the attenuation factors (the numerical values are indicated for each power).

Regarding the experimental crater depth evolution, it increases with laser power then shows a trend of saturation. The good agreement with experiments is obtained in the lowest power range, leading to crater depth up to 300 µm. For higher powers, discrepancy is observed. This is probably correlated to a strong material ejection and the formation of a vapour plume which leads a partial screening of the laser power at the sample surface. This phenomenon is not considered in our model and it is certainly a limitation of our approach. However, to take into account the attenuation of the laser resulting from the vapour plume, we could incorporate in the incident laser beam equation an attenuation factor which would depend, for example, on the ablation depth or the laser power. Sinha [5] presented a theoretical model which takes into account a residual attenuation of the laser (resulting of the ejected vapour plume). In our case, we can deduce an attenuation factor, for each laser power. It corresponds to the coefficient which should have been multiplied to the laser power in the entry of the model to obtain the experimental depth. The numerical values of these factors are indicated in Figure 4 and the corrected depths, considering these attenuation factors, are represented by the red line.

B. *Cutting on graphite*

From the single pulse ablations presented in Sec. IV. A, we propose a laser cutting technique which could be applied for graphite manufacturing. The cutting process technique

developed in our laser facility involves overlapping series of individual craters generated by each pulse.
The spot overlap is crucial in laser cutting because it has a direct impact on the cut quality or on the width of laser cut. We conducted a parametric study which optimized the spot overlap to cut 1.5 mm thick graphite samples. The laser processing parameters are: laser of 1300 W, laser frequency of 0.25 kHz, scan speed of 0.25 mm/s and pulse duration of 50 µs.
We fabricated high aspect ratio shapes in graphite samples of 1.5 mm thickness, as represented in Figure 5.

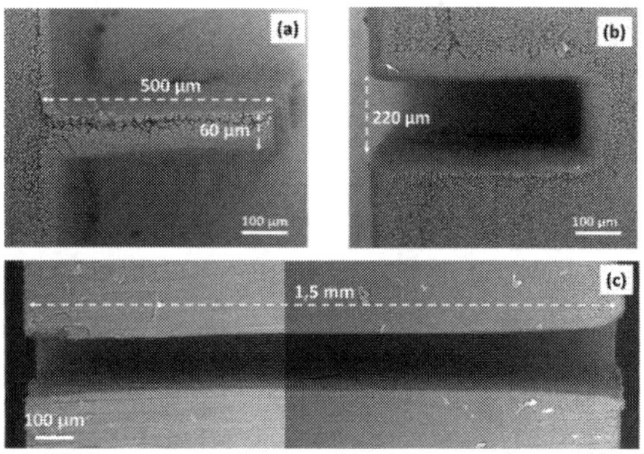

Fig. 5. SEM measurements of a pattern. (a) and (b) are top-views and (c) is the corresponding cross-section view of the part observed in (b).

The desired pattern was a narrow trench with a width of 150 µm and a depth of 500 µm in the 1.5 mm thickness graphite sample. The top-views SEM observation of the two cut parts show the proper contour of the cutting in Figure 5 (a) and (b). We note the difference between the width of the two parts: 220 µm – 60 µm = 160 µm which corresponds to the kerf width during the laser cutting.
The cross-section view, on the Figure 5 (c), shows the homogeneity of the trench width throughout the sample thickness. These parts, and particularly the one on Figure 5 (a) demonstrate the capacity of the laser cutting technique to fabricate high aspect ratio parts (here more than 8) with a good balance between the relative simplicity of technique implementation and its high cutting speed. For this thickness sample graphite of 1.5 mm, the cutting pattern is realized in around 4.5 s.

C. Pulse laser ablation on UO_2

From these results on graphite, we transferred the laser ablation technique to non-irradiated UO_2. Our approach is similar: study the behavior of the material with single-pulsed ablation and then overlap the craters to form a trench. We present here the first step, *i.e.* single pulse laser ablation on UO_2.
The experimental configuration remains identical except that we use a chamber to carry out the ablations under vacuum atmosphere to avoid oxidation and to control the evaporating gases during the processing.
The Figure 6 are top-views SEM observations of single pulse ablation on UO_2 with an increasing laser power. The laser pulse duration was 50 µs, still with the idea to minimize the thermal effects. As observed on the graphite, the higher is the laser power, the larger is the crater until reaching a trend of saturation which corresponds to the diameter of the laser beam.

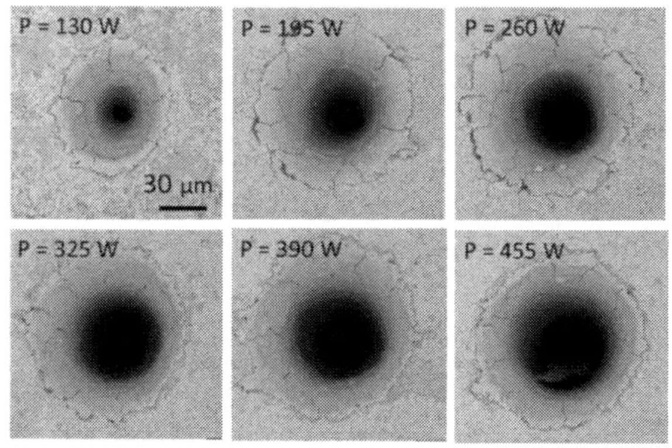

Fig. 6. Top-view SEM measurements of craters obtained at different laser powers.

Although the crater quality is not directly comparable with the craters obtained on the graphite, these results are encouraging because the collateral effects are limited: there is observable no micro-cracks or shock waves surface damages or debris from ejected material. However, we can observe than the HAZ is larger and an observable recast layer around the crater that cannot be neglected.

V. Conclusion and perspectives

As a conclusion, laser micro-processing of graphite employing an Ytterbium fibre laser delivering microsecond pulses at 1080 nm has been experimentally investigated, with the support of a numerical model to investigate and understand the laser ablation mechanisms processes. Firstly, studies were conducted on single-pulse laser ablations with a minimum pulse duration of 50 µs and various power levels up to 1300 W with a beam diameter of 40 µm. The crater depth was found to increase firstly linearly with laser power, in good agreement with the simulation results, then followed by a saturation trend that we attribute to the effect of screening by the vapour plume. Based on these results, we have shown that such laser cutting technique can produce parts with excellent quality (reduced affected area, no recast layer, no micro-cracks and no debris from ejected material) with high efficiency. Production of parts with high aspect ratio were demonstrated, with a cutting depth of 1.5 mm.
From this works on graphite, we consider to transfer this laser ablation technique to nuclear fuels. In this context, we presented first micro-second pulse laser ablation experiments on UO_2. We observe that the crater quality is correct which encourages us to continue the experiments especially using the same approach than of graphite: perform a parametric study of the laser processing parameters to optimize the overlap between two craters and find a UO_2 laser cutting procedure which gives the smallest thermal affected area and the highest precision machining kerf. It can be also envisaged to perform experiments with shorter pulsed lasers (pico and femtosecond) in order to reduce the improve the precision and minimize the HAZ.

References

[1] Poco Graphite, Inc. "Properties and Characteristics of Graphite for the EDM Industry", Edited by R. G. Sheppard, Dwayne Morgan, D. M. Mathes, D. J. Bray, 5th Printing (2002).

[2] L. Gallais, T. Vidal, E. Lescoute, Y. Pontillon and J.L. Rullier, "High power continuous wave laser heating of graphite in a high temperature range up to 3800 K", Journal of Applied Physics, vol 129, pp. 043102, 2021, doi.org/10.1063/5.0033530

[3] L. Robin, P. Combis, P. Cormont, L. Gallais, D. Hébert, C. Mainfray and J.L. Rullier, "Infrared thermometry and interferential microscopy for analysis of crater formation at the surface of fused silica under CO2 laser irradiation", Journal of Applied Physics, vol 111, pp. 063106, 2012, doi.org/10.1063/1.3695375

[4] D. Lee, R. Patwa, H. Herfurth, J. Mazumder, "High speed remote laser cutting of electrodes for lithium-ion batteries: Anode", Journal of Power Sources 240, pp. 368 – 380, 2013, doi.org/10.1016/j.jpowsour.2012.10.096

[5] S. Sinha, "Nanosecond laser ablation of graphite: A thermal model based simulation", Journal of Laser Applications 30, pp. 012008, 2018, doi.org/10.2351/1.5021520

Development of UO_2 thermal diffusivity measurement with laser techniques

Thomas Doualle[1], Vincent Le Guillous[1], Vincent Klosek[1], Claire Onofri-Marroncle[1], Matthieu Reymond[1,2], Laurent Gallais[2], and Yves Pontillon[1]
[1]CEA, DES, IRESNE, DEC, Cadarache F-13108 Saint-Paul-Lez-Durance, France
[2] Aix Marseille Univ, CNRS, Centrale Marseille, Institut Fresnel, Marseille, France
thomas.doualle@cea.fr

***Abstract*—The knowledge of the thermal conductivity of nuclear fuel and its evolution as a function of temperature and burn up is a major challenge in the context of the evaluation and understanding of irradiated fuel performances in current reactors. It is also the case for the development and qualification of fuel for future reactors. Indeed, numerical simulations of the fuel behaviour under various conditions require the accurate knowledge of thermal conductivity over a wide range of temperature (from ambient to melting point temperature) but also at the scale of few tens of micrometres to take into account the microstructural effects on the thermomechanical evolution of the fuel in normal or incidental irradiation conditions. Different methods, using laser matter interactions, can deduce the thermal conductivity from a thermal diffusivity measurement. In this paper, the potential of two techniques, which present spatial resolution from millimetre to few tens microns, are discussed in the context of the determination of the fuel thermal conductivity: laser flash method and infrared microscopy. Experiments on graphite, as material model, have been conducted and validate these two thermal diffusivity measurement techniques. We present a measurement example for both methods on graphite and then a first experiment carried out with the infrared microscopy technique on UO_2.**

***Keywords* —UO_2, thermal diffusivity measurements, laser**

I. Introduction

In power reactors, thermal conductivity is a key parameter for understanding nuclear fuel performance and its evolution as a function of temperature and burn up. Up to now, only few data are available in the very high temperature range (i.e. close to melting point). The laboratory reference method can deduce the thermal conductivity from a thermal diffusivity measurement.
One of the most common and non-intrusive method is the laser flash method. A device employing this method has been in use at JRC Karlsruhe since 1999 [1]. This technique is based on a laser-flash excitation and infrared thermography non-contact measurement of the thermal response of a sample. The sample can be heated to different temperatures (for instance with lasers) to determine the dependence of thermal conductivity with the temperature. The thermal diffusivity measurements obtained with this technique typically have a millimeter spatial resolution. Modelling fuel behavior requires knowledge of thermal conductivity on a scale of a few tens of micrometers to take into account microstructural effects on thermomechanical behavior of the fuel under normal or off-normal conditions. To date, the flash method usually used has not provided this resolution [1,2].
A laser method with few tens of microns resolution is the Infrared Microscopy technique (IRM) [3], based on the detection, by means of infrared thermography, of the sample surface temperature distribution induced by the absorption of an intensity-modulated focused laser beam.
As an example of a method which has a micrometric spatial resolution, which can be considered for our application, is the photoreflectance microscopy (PM) [4], based on the measurement and analysis of the periodic temperature increase induced by the absorption of an intensity modulated laser beam. All of these approaches, using laser-induced heat source, enable thermal diffusivity measurements for temperatures up to at least melting temperature on a scale which has never yet been reached for active materials. The potential of laser flash and the IRM methods will be discussed, in this paper. All of the presented first experiments (and associated simulations) were conducted on the ChauCoLase platform (Chauffage Controlé par Laser / laser controlled heating), available at the Institut Fresnel [5] [6]. In a first time, experiments with these two techniques were carried out on graphite, that we consider as model sample, and measurement examples will be presented. They allowed us to verify that a signal can be measured with the experimental devices envisaged, to validate the methodology and the potentiality for each techniques. The flash method was realized at different temperatures (under 1000 °C) and we present one at 900 °C (with laser heating). For IRM, only experiments at ambient temperature have been carried out up to now. In a second time, we present in this paper a first experiment with the infrared microscopy technique on UO_2. Thereafter, the idea is to extend the temperature range for both techniques and determine the dependence of thermal conductivity with the temperature. In future, the next step will be to apply the methods on non-irradiated UO_2 in a glovebox in the experimental bench being developed at CEA Cadarache before its transposition in a hot cell for measurements on irradiated fuels.

II. THERMAL DIFFUSIVITY DETERMINATIONS: METHOD AND EXPERIMENTAL SETUPS

This part describes the experimental benches, under development at the Institut Fresnel, implementing the two techniques previously described to obtain thermal conductivity, from thermal diffusivity measurements, with two different scales of interest: laser flash method and infrared microscopy.
The well-known technique, laser flash method, gives a global thermal diffusivity value of a sample, as a function of temperature. This method was already applied on irradiated sample in hot cell [7]. In order to go further and take into account the heterogeneity of the fuel material (like in mixed oxides fuel) and to consider the microstructural effects on thermomechanical behavior of the fuel under normal or off-normal conditions, IRM method seems suitable, and make it possible to calculate an equivalent thermal conductivity which can be used in the models.
Experiments described below have been carried out on EDM-3 graphite from POCO Graphite Inc. and UO_2 samples fabricated at CEA. Graphite is a polycrystalline graphite produced by sintering and we use it as a model material to establish the methodology of the measurement techniques. The graphite and UO_2 samples have a disk morphology with a diameter of 8 mm and thicknesses of 1.66 mm and 1.44 mm respectively. They are placed in a sample holder in a dedicated experimental chamber operated in vacuum (10^{-2} mbar) which allows monitoring and optical access through multiple viewports.

A. Laser Flash Technique (LF)

During a LF experiment, the sample is subjected to a high-intensity, short pulse duration. The energy is absorbed on the front surface and the thermal diffusivity value is computed from the resulting temperature response on the back surface sample, typically measured by an infrared detector. The variation in back surface temperature is referred to as a ''thermogram''. To determine the thermal diffusivity of graphite as a function of temperature, our approach was to set up an experimental bench to carry out experiments and measure thermograms and develop a numerical model to calculate the thermal diffusivity value, through a parameter estimation method.

The schematic diagram of the experimental setup for the LF method bench developed is given in Figure 1. A laser diode (LIMO32, Limo), operating at 880 nm, is used to heat the sample at different temperatures. The thermal perturbation, *i.e.* the flash, is created by an another laser: an Ytterbium fibre laser (redPOWER Qube 2 kW, SPI Lasers), which has a central wavelength of 1080 nm with a bandwidth < 4 nm, with a maximum output power of 2000 W. The temporal evolution of the temperature of the back face is obtained by a two-colors (channel 1 : 1,65 – 1,8 μm and channel 2 : 1,45 – 1,65 μm) pyrometer (Metis H322, Sensortherm GmbH), which has a response time of 80 μs.

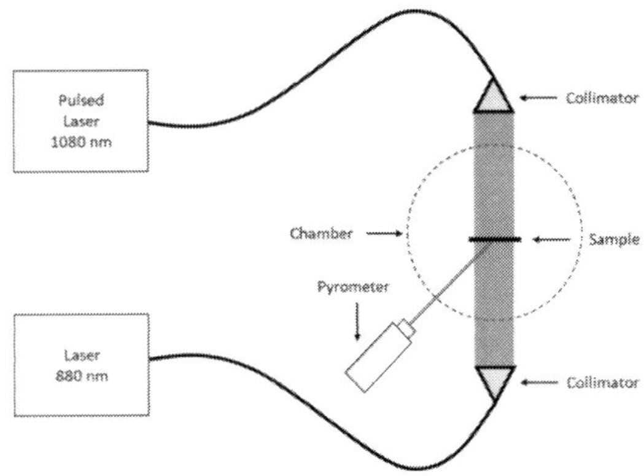

Fig. 1. Schematic of laser flash method setup

The simulations are conducted with the commercial software COMSOL which is based on the Finite Element Method in axisymmetric configuration, with a 2 dimensional meshing. The heat transfer model relies on several assumptions. First, the graphite is considered homogeneous and isotropic. Second we take into account the heat transfers by conduction combined with thermal radiation to take into account losses at the surface (convection is neglected). Third, as the energy is absorbed on the surface, there is no heat source in the material but an incoming heat flux on the boundary. The corresponding non-linear heat transfer equation can be expressed as:

$$\rho(T)C_p(T)\frac{\partial T}{\partial t} + \nabla(-k(T)\nabla T) = 0, \quad (1)$$

where ρ is the density, C_p is the specific heat capacity under constant pressure, T is the temperature, k is the thermal conductivity. The absorbed laser energy induces a heat flux F expressed as:

$$F(r,t) = A(T).I(r,t), \quad (2)$$

where A is the temperature dependence absorption and I the laser intensity (the measured spatial laser profile is used).
This thermal model is associated to a parameter estimation study which enables to estimate the value of one or more parameters. In our case, we calculate the value of the specific heat capacity and the thermal conductivity, so that the computational results match the experimental reference data, i.e. the thermogram. A least-squares optimization, with a Levenberg-Marquardt method, is used. It defines an objective function representing the sum of squared differences between back surface sample temperature measurements stored in an experimental data file (and acquired through the pyrometer) and corresponding expressions evaluated in the thermal model: the specific heat capacity and the thermal conductivity.

B. Infrared Microscopy Method (IRM)

Infrared Microscopy (IRM) is based on the detection of the surface sample temperature rise distribution induced by the absorption of an intensity-modulated laser beam. We assume a homogeneous, isotropic and semi-infinite material is heated at its surface by a point source with a modulated intensity at a frequency f. The expression of the spatial and temporal evolution of temperature δT is [8]:

$$\delta T(r,t) = \frac{Q}{4\pi k t}.\exp\left(-\frac{r}{\mu}\right).\cos\left(2\pi f t - \frac{r}{\mu}\right), \quad (3)$$

with Q the power of the heat source, r the radial distance from the heating point, k the thermal conductivity of the material and μ the thermal diffusion length. The thermal diffusion length of a material is expressed as:

$$\mu = \sqrt{\frac{a}{\pi f}} \quad (4)$$

with a the thermal diffusivity and f the modulation frequency of the pump beam.

The equation (3) describes a wave that dampens with the increasing of the frequency f, as represented in Fig. 2.

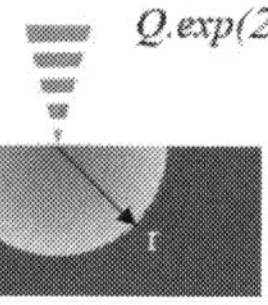

Fig. 2. Schematic representation of the spatio-temporal evolution of the temperature in a material for a source point on the surface with an intensity modulated source.

The temperature variation can be considered completely attenuated after a radial distance of approximately 3 times the thermal diffusion length. Thus, the modulated temperature of a material is confined in a volume approximately equal to 3 thermal diffusion lengths. This distance corresponds to the spatial resolution of the estimated thermal diffusivity. Therefore, by choosing an appropriate excitation frequency, it is possible to adapt the characterization scale compatible with the dimensions of the different constituents of a multi-scale (or multi-material) material. It can be seen that the phase shift of this thermal wave decreases linearly as a function of the distance r from the heat source with the inverse of the diffusion length:

$$\frac{\partial \varphi}{\partial r} = -\frac{1}{\mu}, \quad (5)$$

The signal phase measurement makes it possible to identify the thermal diffusivity through a linear adjustment using the Eq. (4). Thus, the slope calculation of the phase curve is a very easy way of evaluation of the thermal diffusivity by linear fit. Thus, we use an infrared camera to measure the periodic evolution of the temperature on the surface sample and deduce amplitude and phase-shift images. However, convolution effects, linked to a non-ideal point heat source, can limit the spatial resolutions of the measurement and must be taken into account to estimate accurately the thermal diffusivity.

A schematic diagram of our experimental bench is represented in Figure 3. The pump beam is a green continuous laser (Samba, Cobolt), which has a central wavelength of 532 nm with a bandwidth < 0.3 nm, with a maximum output power of 1 W. This laser wavelength was chosen to ensure efficient coupling and low penetration depth (photon energies are higher than the UO_2 bandgap). The laser beam is transported through an optical fiber (PMC-400, Schäfter Kirchhoff) and focused onto the sample surface with an adjustable aspheric collimator (CFC11A-A, Thorlabs). A function generator is used to send a modulated signal to a frequency generator-driven acousto-optic modulator (MT110-A1-VIS, AA Opto-Electronic) which modulate the laser power intensity deposited on the surface of the sample. The infrared (IR) signal on the sample surface is measured to interpret the temperature distribution and is collected by a high speed IR camera (X6901SC, FLIR).

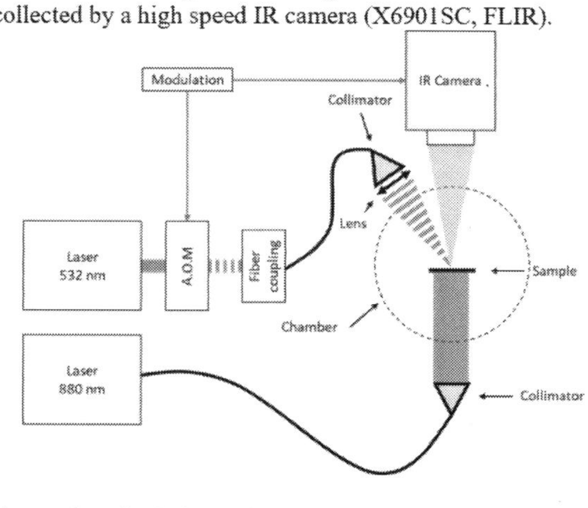

Fig. 3. Schematic of infrared microscope setup

III. Results and discussions

In this section we present the first experimental results obtained with our two thermal diffusivity measurement techniques under development. The studies were conducted on graphite sample, considered as model material. Thereafter, we show a first result with IRM technique at ambient temperature on UO_2.

A. Laser Flash Method

The laser parameters are: a power of 600 W, a pulse duration of 1 ms and a laser beam diameter on the sample surface of 3 mm (defined at $1/e^2$) with a super-Gaussian intensity profile (super-Gaussian function of power 4). Different initials temperatures were tested and we present below an example at around 900 °C. The temperature elevation has been recorded through the pyrometer and is represented by the red dashed line in the Fig. 4. Just before the laser flash, the initial temperature on the sample was 900 °C ± 10 °C. We note that the temperature rise is due to the thermal diffusivity and conductivity, the amplitude is dependent on the specific heat capacity and density of the material and the temperature drop is dominated by the heat losses (convective and radiative heat transfer).

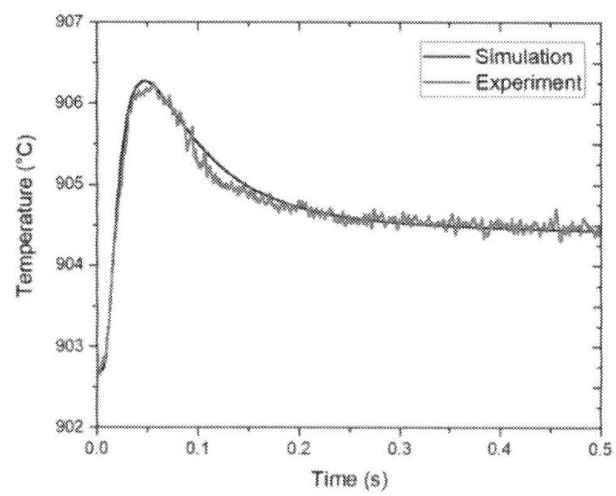

Fig. 4. On-axis surface temperature measured (red line) as a function of time and model thermogram (black line) obtained via least square optimization

From this experimental thermogram, we have conducted the optimization algorithm to find the least squares minimum between the experiment and model thermograms. The simulated temperature elevation found is represented by the black line in Fig. 4. The maximum temperature rise of 3.5 °C, on the back surface graphite sample, is at 47 ms after the laser irradiation on the front surface. The Fig. 5 represents the simulated temperature distribution at this time.

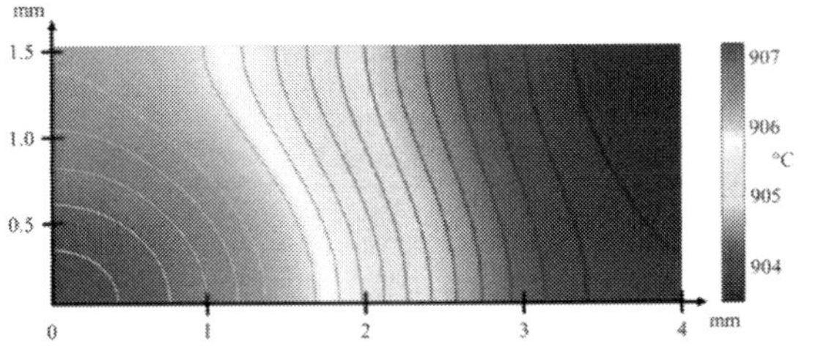

Fig. 5. Calculated temperature distribution and isotherms in graphite at 47.21 ms

The fitted parameters thermal conductivity *k*, the specific heat capacity C_p and the deduced thermal diffusivity, at the temperature of 900 °C, are resumed and compared with the theoretical values in the Table 1.

TABLE I
COMPARISON BETWEEN THE THEORETICAL AND THE CALCULATED THERMAL PROPERTIES OF THE GRAPHITE AT 900 °C

Property	Theoretical [9]	Calculated	Relative measurement errors
Thermal conductivity ($W.m^{-1}.K^{-1}$)	59	58.2	1.4%
Specific heat capacity ($J.kg^{-1}.K^{-1}$)	1879	1955	4.0%
Thermal diffusivity ($m^2.s^{-1}$)	1.84x10^{-5}	1.75x10^{-5}	4.9%

As we can see in the Fig. 4, the model is in very good agreement with the presented experimental thermogram example. Others initial temperatures were studied (under 1000°C) and also have a concordance with theoretical values. To go further, we plan to continue experiments on graphite at higher initial temperatures to characterize the parameters as a function of the temperature in a large range and to apply the method on UO_2.

Meanwhile, we will conduct a sensitivity analysis, as presented by Pavlov *et al.* [10], to study the relative importance of the different model parameters on the numerical parameter estimation method (sample dimensions and particularly the thickness, deposited and absorbed laser power, laser spot size, emissivity).

B. Infrared Microscopy Method (IRM)

Experiments using IRM measurement technique were conducted on the same graphite sample than in the previous section, with LF. We present below a measure that was carried out at ambient temperature. The pump laser (green laser in Fig. 3) power was 50 mW and the beam was focused on the surface of the sample with a beam diameter of 120 µm (Gaussian distribution with diameter defined at $1/e^2$). The modulation frequency was chosen at 800 Hz. In these conditions, the spatial resolution of the estimated thermal diffusivity is around 500 µm. From two-dimensional temperature measurements obtained with the infrared camera, we deduce the phase-shift image, given in Fig. 6 (a).

The phase-shift images are used to access to the local thermal diffusivity, knowing the frequency of the modulated laser intensity. Indeed, a radial profile, as represented in Fig 6 (a) (in black line) and plotted in Fig. 6 (b), gives the phase shift of the thermal wave as a function of the radius from the center of the laser beam. The calculation of the profile slope, represented by the red dashed line in Fig. 6 (b), represents the inverse of the thermal diffusion length (Eq. 5)). To use this approach, the points to calculate the slope have to be judiciously chosen. Indeed, they have to be sufficiently far from the center of the laser beam to consider it as a point source and the number of points need to be enough to plot a representative slope and describe it accurately.

In the case presented in Fig. 6 (b), the deducted thermal diffusion length is 165 µm. From the Eq. (4), we are able to calculate the thermal diffusivity:

$$a = \mu^2.\pi.f = 6.84 \times 10^{-5}\ m^2/s, \qquad (6)$$

At ambient temperature, the theoretical value is:

$$a(T) = \frac{k(T)}{\rho(T).C_p(T)} = \frac{112}{880.1737} = 7.32 \times 10^{-5}\ m^2/s, \qquad (7)$$

which give a relative measurement error *E* of 6%.

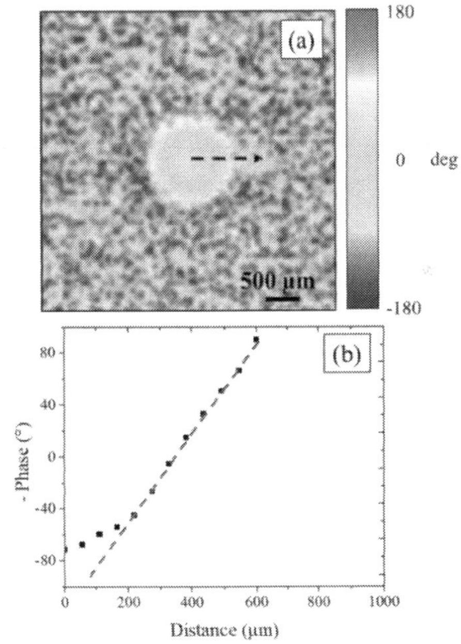

Fig. 6. (a) Phase-shift image and (b) the corresponding plot of the black dashed arrow of (a), scan at 800 Hz. The red dashed line in (b) is the linear fit used to calculate the thermal diffusivity of the graphite.

This local thermal diffusivity measurement on graphite confirms the potential of this method. Therefore, we have carried out first IRM experiment on UO_2 sample. The laser power and beam diameter are the same. The modulation frequency was chosen at 30 Hz which gives a spatial resolution of around 570 µm. The phase-shift image is calculated from the two-dimensional temperature measurement and the deduced

thermal diffusion length is 203 µm, obtained from the slope. Thus, the calculated thermal diffusivity a is 3.89×10^{-6} m²/s.
At ambient temperature, the theoretical value of the UO_2 thermal diffusivity is 3.42×10^{-6} m²/s [11] which give an acceptable correct relative measurement error E of 14%.
The accuracy of our evaluation with IRM technique can be improved through different ways that will be tested in the next experiments regarding: the laser parameters (the decreasing of the laser spot size, the optimization of the laser power to enhance the signal), the IR camera performances (spatial resolution, acquisition time and frame rate) and conduct tests on other materials. Thereafter, next experiments will be carried out at different temperatures on graphite and we will continue to transfer the technique on UO_2 samples. We intend to increase the spatial resolution up to few tens of microns to take into account the heterogeneity of the fuel material (like in mixed oxides fuel).

IV. CONCLUSION AND PERSPECTIVES

This paper presents thermal diffusivity measurement techniques that we are testing in Institut Fresnel and that we consider to deploy on non-irradiated UO_2 in glovebox at CEA Cadarache before its transposition in a hot cell for measurements on irradiated fuels. The idea is to reach measurements on a scale which has never yet been reached for active materials. For that, we are carrying out two different methods, using photothermal methods, enabling thermal diffusivity measurements at high temperature: the Laser Flash and the Infrared Microscopy methods for respectively a millimeter and a few tens microns resolutions. We have presented in this paper experiments on graphite samples at 900°C (for LF) and at ambient temperature (for IRM). A first encouraging measurement has been carried out and is presented with the infrared microscopy technique on UO_2 at ambient temperature. These techniques are validated and their developments will be pursued, to have thermal properties as a function of temperature in a first time on model materials and in a second time on UO_2 samples.
As a perspective of this work, we can consider an enhancement of the spatial resolution of the thermal diffusivity measurement, up to microns. A technique that would allow us to reach this resolution could be the use the Photoreflectance Microscopy (PM) [7]. The principle is similar to the IRM: a measurement and an analysis of the periodic temperature increased induced by the absorption of an intensity-modulated laser beam (pump beam). The main difference is the measurement: the local temperature increase at the sample surface is measured with the help of a secondary continuous laser beam (probe beam) through the variations of the thermally induced reflection coefficient.

ACKNOWLEDGMENT

The authors wish to thank EDF and Framatome for supporting this work in the frame of Tripartite Institute (CEA/EDF/FRA).

REFERENCES

[1] C. Ronchi, W. Heinz, M. Musella, R. Selfslag and M. Scheindlin, "A universal laser-pulse apparatus for thermophysical measurments in refractory materials at very high temperatures," International Journal of Thermophysics, vol 20, pp. 987-996, 1999, 10.1023/A:1022607924536.
[2] W. Parker, R. Jenkins, C. Butler and G. Abbot, "Flash method of determining thermal diffusivity, heat capacity and thermal conductivity," Journal of Applied Physics, vol 32, pp. 1679-1684, 1961, 10.1063/1.1728417.
[3] J. F. Bisson, "Realisation d'un diffusivimetre a haute resolution spaciale utilisant la thermographie infrarouge stimulee," M.S. thesis, Université Paris VI, 1999.
[4] J. Jumel, "Microscopie photothermique : application à la caractérisation des propriétés thermoélastiques microscopiques de composites Carbone/Carbone et des barrières thermiques," M.S. thesis, Ecole Normale Supérieure de Cachan, 2003.
[5] T. Vidal, L. Gallais, J. Faucheux, H. Capdevila, J. Sercombe and Y. Pontillon, "Simulation of Reactivity Initiated Accident thermal transients on nuclear fuels with laser remote heating, " Journal of Nuclear Materials vol 530, 151944 (2020).
[6] T. Vidal, L. Gallais, J. Faucheux, H. Capdevila and Y. Pontillon, "Using laser remote heating to simulate extreme thermal heat loads on nuclear fuels," EPJ Web of Conferences, vol 225, 08002, 2020, doi.org/10.1051/epjconf/202022508002
[7] D. Staicu, C. Cozo, G. Papaioannou, S. Bremier, V.V. Rondinella, C.T. Walker and A. Sasahara, "Thermal conductivity of homogeneous and heterogeneous MOX fuel with up to 44 MWd/kgHM burn-up", Journal of Nuclear Materials 412, pp. 129-137, 2011, doi.org/10.1016/j.jnucmat.2011.02.042
[8] D. Rochais, H. Le Houëdec, F. Enguehard, J. Jumel and F. Lepoutre, "Microscale thermal characterization at temperatures up to 1000°C by photoreflectance microscopy. Application to the characterization of carbon fibres," Journal of Physics D : Applied Physics 38, pp. 1498-1503, 2005, doi:10.1088/0022-3727/38/10/002
[9] Poco Graphite, Inc. "Properties and Characteristics of Graphite for the EDM Industry", Edited by R. G. Sheppard, Dwayne Morgan, D. M. Mathes, D. J. Bray, 5th Printing (2002).
[10] T. Pavlov, L. Vlahovic, D. Staicu, R.J.M. Konings, M.R. Wenman, P. Van Uffelen and R.W. Grimes, "A new C. Ronchi, W. Heinz, M. Musella, R. Selfslag and M. Scheindlin, "A universal laser-pulse apparatus for thermophysical measurments in refractory materials at very high temperatures," Thermochimica Acta, vol 652, pp. 39-52, 2017, 10.1016/j.tca.2017.03.004..
[11] J.J. Carbajo, G.L. Yoder, S.G. Popov, V.K. Ivanov, "A new C. Ronchi, W. Heinz, M. Musella, R. Selfslag and M. Scheindlin, "A review of the thermophysical properties of MOX and UO_2 fuels," Journal of Nuclear Materials, vol 299, pp. 181-198, 2001, 10.1016/S0022-3115(01)00692-4

Qualification Test System for Radiation Detection Devices - QuTeSt

Monika Risse, Peter Clemens, Jeannette Glabian, Olaf Schumann, Theo Koeble, Hermann Friedrich, Wolfram Berky, Marie Charlotte Bornhoeft, Sebastian Chmel
Fraunhofer-Institut für Naturwissenschaftlich-Technische Trendanalysen, Germany
monika.risse@int.fraunhofer.de

***Abstract*— Measurement equipment for the detection and identification of radioactive and nuclear (RN) material has a wide application area. The main application aspects are monitoring, search, and identification. A common goal is to gain reliable measurement results. In the past, the only way to assess the performance of a measuring device was to rely on the data given by the manufacturer of the device itself. Reliable test results from an independent third party are more than welcome. These tests can be performed against consensus standards in order to have reproducible test results, independent of the testing location and the performing laboratory. Fraunhofer INT has conceived and built a test environment to perform dynamic and static test measurements using neutron and gamma sources. Tests can be performed in accordance with the IEC and ANSI standards as well as the ITRAP+10 test procedures. This includes qualification tests of truck portal monitors with the dynamic test system. Generally, the effects of one test parameter on other test parameters are not considered in the test procedures. For example, the accuracy of the dose rate may depend on the energy range of the radioactive source used. Besides the overview of the test systems the paper will address restrictions, problems and limitations of the possible qualification measurements as well as potential limitations arising from the given test procedures themselves.**

***Keywords* —Qualification tests, RIID, PRD, SPRD, D3S, QuTeSt, Standards, Accuracy tests, Time to alarm tests.**

I. Introduction

In the field of the detection of nuclear and radioactive material a large variety of measurement systems exists. These are handheld systems, like Radiation Isotope Identifiers (RIID) or Personal Radiation Detectors (PRD), also in the spectroscopically version as SPRD, or stationary systems like portal monitors (RPM). They all have the common requirement to provide reliable results within a short time.

For decision making processes and the assessment of measurement results one needs a reliable statement about the quality of the devices. The quality of the systems shall be determined by performing qualification tests in accordance with existing test methods.

Different international standards for testing exist. The IEC standards from the International Electrotechnical Commission, the ANSI standards from the American National Standards Institute and ITRAP+ 10 standards.

In the "Illicit Trafficking Radiation Assessment Program (ITRAP+10)", initiated by the European Commission, several laboratories were upgraded to be able to carry out these tests. This is where the Fraunhofer INT started to build the QuTeSt facility. Here tests were performed in accordance with the ITRAP+ 10 test procedures see [9] and [10] .

This paper will introduce the QuTeSt facility and give some exemplary results gained with handheld systems. It will also address aspects arising from the different existing standards.

II. QuTeSt

The QuTeSt consists of two parts: the static test system and the dynamic test system.

A. Static Test System

The static test system consists of a central cube for placing the sources. The source can be lifted from a shielded position by a compressed air system in front of the device to be tested.

Fig. 1 shows the system with one guide rail system attached and held by lateral guides on the central cube. We have four guide rails which can each be placed on the sides of the cube or can be placed in a row in order to enlarge the distance from the source. The length of each guide rail is 1.5 m.

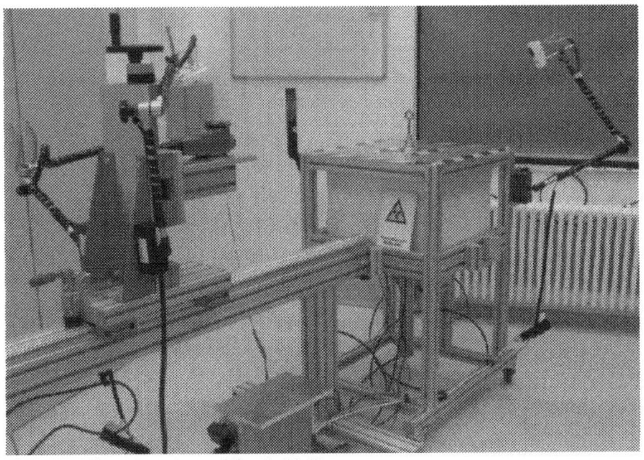

Fig. 1. Static test system of the QuTeSt.

On a roller carriage the device to be tested is placed. Its height can be adjusted which enables us to bring the center of the detectors to the same height as the center of the source. The roller carriages are fully adjustable via setting wheels. The roller carriages and the guide rails are equipped with scales for measuring distances.

Together with the device to be tested, a timer is placed and observed by a video camera. The picture, as well as a video obtained from the source, is transferred to and stored at a computer with the software Blue Iris for observation and data storage. Fig. 2 shows the screenshot of the timer camera: on the left the timer and on the right the display part of the measurement device D3S from Kromek, which is introduced in section IV. On the timer the line marked with "U" indicates the duration in the upper position after stepping out. The line marked with "D" indicates the time in the down position inside the shielding.

The stepping times are also given, they are below 0.35 s and a fast step change is possible.

Fig. 2. Exemplarily screenshot of a test. On the left the timer and on the right the display part of the device to be tested, the D3S (see section IV).

B. *Dynamic Test System*

Fig. 3 shows the dynamic test system, which consists of a battery powered automated trolley which runs on a rail system mounted on wooden boards. Altogether a track length of 45 m exists. The velocity can be chosen like needed according to the given standards between 0.02 m/s to 2.2 m/s. Fig. 3 shows the trolley with the source holder in the highest position.

Fig. 3.. Dynamic test system with portal monitor from Arktis.

The same interchangeable source system as for the static system is used and the sources can be placed easily and fast.

The system is portable and can be flexibly set up and operated at any suitable location, as well as adaptions to special needs are possible.

The dynamic system is completed with the video system for observation and documentation of the measurement results and the operating system of the trolley (see Fig. 4).

Fig. 4. Left laptop: operation of the source trolley, middle laptop: video system and right laptop: software of the device to be tested.

The left laptop is for operating the source trolley, the middle one for the video system, and if needed we use another one for the test system. For example, for portal monitors the results are displayed on this third laptop and included via HDMI in the video system.

In Fig. 5 an example of a screenshot from a test with handheld devices is given. The source is attached to an extension to get closer to the devices. It is moving as indicated by the direction of the yellow arrow. The display gives the time prior and after passing the zero position which is marked with the red line. The velocity is given as well. Three handheld devices are placed for tests.

Fig. 5. The source is marked with the yellow ellipse, it is moving in the direction of the yellow arrow. The D3S from Kromek as well as two RadEYE devices from Thermo are tested. Below the source the control display is placed. The red line indicates the zero position.

III. Standard specifications for handheld devices

https://doi.org/10.1051/epjconf/202125307006

A. Time to Alarm Tests for Neutrons

The specifications for time to alarm tests for neutrons are listed in TableTABLE *I*. All standards state that a ^{252}Cf source with N_R = 20 000 n/s surrounded by 4 cm high density polyethylen HDPE shall be used in d_0 = 25 cm distance. The devices shall be placed on a polymethyl methacrylate PMMA phantom.

If the source used does not fulfill the foreseen criterion, the conditions can be changed. With a lower source count rate N_T, the distance has to be reduced to d using (1). A change in the distance has to be compensated in the dynamic tests by changing the velocity v_0 stated in the standard (see Table TABLE *I*) to v using (2).

$$d = d_0 \times \sqrt{N_T/N_R} \quad (1)$$

$$v = v_0 \times d/d_0 \quad (2)$$

In Table I the allowed times to alarm are listed as well as the conditions and the number of trials and the criterion for passing. Three cases foresee a step change test, this is done with the static system. All other cases foresee dynamic tests but partly with the source stopping in front of the device and partly with the source going back and forth. The latter case is for RIID devices with the same conditions for ANSI and IEC. Results we obtained in such a test are given in section V.B

A variety of different test conditions for the time to alarm tests are stated in the common standards. Timing systems are required for the dynamic and the static test systems.

TABLE I
TIME TO ALARM SPECIFICATIONS FOR NEUTRONS

Device class	ANSI			IEC			ITRAP +10		
	[s]	conditions	trials	[s]	conditions	trials	[s]	conditions	trials
RIID	1	0.5 m/s passing	9/10	2	0.5 m/s passing	9/10	5	step change	29/30
PRD	5	1.2 m/s source stops	19/20	5	1.2 m/s source stops	8/10	5	step change	29/30
SPRD	5	1.2 m/s source stops	19/20	20	step change	8/10	-	-	

All test to be performed with ^{252}Cf, 20 000 n/s, surrounded by 4 cm HDPE, 25 cm distance. The device shall be on a PMMA phantom.
ANSI values taken from [4], [5] and [6].
IEC values taken from [1], [2] and [3].
ITRAP+10 values taken from [7] and [8].

B. Time to Alarm Tests for Photons with Step Change

Similar to the time to alarm test for neutrons, only the ITRAP+10 and the IEC method of test for photons for SPRDs foresee a step change. In here a ^{137}Cs source with 0.5 µSv/h above background shall be brought in front of the device to be tested. The alarm shall occur within 3 s for the IEC SPRD and the ITRAP+10 RIID test. For the ITRAP+10 PRD test it shall be within 2 s.

C. Accuracy Tests for Photons

In TableTABLE *II* the accuracy and test dose rate values for the three different standards and the different device classes are shown. The allowed accuracy, which is the percentage deviation from the measured gamma dose rate to the dose rate measured with a reference device, is given as well as the dose rate values which shall be used for the tests. The ANSI tests for RIIDs and PRDs only foresee tests with ^{137}Cs, in all other cases the whole energy region is covered and in addition tests with ^{241}Am and ^{60}Co have to be done.

It is important, that prior to the tests it has to be decided which standard shall be used as basis for the tests. As the QuTeSt system was started in the ITRAP+10 program, up to now most of the tests have been carried out under that conditions, whereas the system is not limited to that.

The accuracy tests are performed with the static system.

TABLE II
ACCURACY SPECIFICATIONS FOR PHOTONS FROM STANDARDS

Device class	ANSI		IEC		ITRAP +10	
	[%]	test DR values[a]	[%]	test DR values[a]	[%]	test DR values[a]
RIID	± 10	1 µSv/h – *M*	± 30	1 µSv/h 10 µSv/h 70 % *M*	± 30	0.5 µSv/h 5 µSv/h 50 µSv/h 100 µSv/h
PRD	± 30	4 µSv/h 10 µSv/h 16 µSv/h 70 % *M*	± 50	1 µSv/h 10 µSv/h 50 % *M*	± 30	0.5 µSv/h 5 µSv/h 50 µSv/h
SPRD	± 30	1 µSv/h 10 µSv/h 70 µSv/h 80 % *M*	± 50	1 µSv/h 10 µSv/h 70 µSv/h 70 % *M*	-	-

M indicates the maximum dose rate value stated by the manufacturer.
ANSI RIID and PRD tests with ^{137}Cs, all other with ^{241}Am, ^{60}Co and ^{137}Cs.
ANSI values taken from [4], [5] and [6].
IEC values taken from [1], [2] and [3].
ITRAP+10 values taken from [7] and [8].
[a]Dose rate values at which tests are performed.

IV. TEST SYSTEM D3S

The device which is used to show the performance of the QuTeSt system is the D3S from Kromek [11]. It is a system which consists of two parts, which are shown in Fig. 5: the detection unit, which can be worn on a belt, and a smartphone for data collection and display. Both are connected via Bluetooth. The detection unit has as detectors a CsI crystal for gamma detection and a Lithium based one for neutron detection. As the unit is rather small it uses a SiPM. The D3S can either function as PRD or as RIID. Therefore, the different criteria for the different device classes have to be considered when testing.

V. EXAMPLE TESTS

A. Time to Alarm – Static

In Fig. 6 results for 30 separate trials for a time to alarm test with the static test system are given. A ^{137}Cs source is brought in front of the D3S detector via a fast step change with < 0.35 s duration. The IEC standard for SPRDs as well as the ITRAP+10 method of test for RIID devices state that the alarm shall be within 3 s, which is fulfilled in all cases and the test is passed. The criterion for the ITRAP+10 PRD states, that the device shall alarm within 2 s, which is not fulfilled.

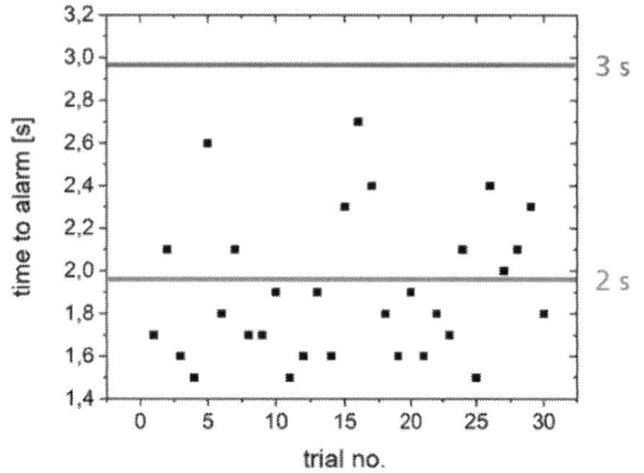

Fig. 6. ^{137}Cs with 0.5 μSv/h above background brought in front of the D3S via a step change. 3 s is the criterion for IEC SPRD and ITRAP+10 RIID testing, 2 s is the criterion for ITRAP+10 PRD testing.

B. Time to Alarm – Dynamic

Fig. 7 shows the setup of the time to alarm test for neutrons for the D3S. The detector part is fixed to the phantom and the display part is placed next to the display in order to observe the temporal correlation. In this configuration we could not place the HDPE around the source. This is why we placed it in front of the D3S.

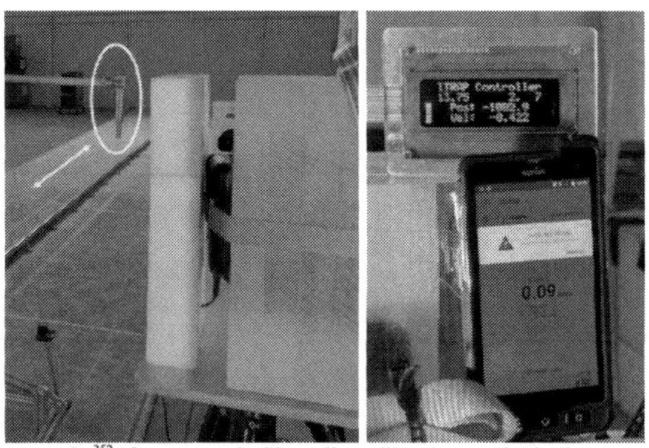

Fig. 7. Left: ^{252}Cf source marked by yellow ellipse is moved. The detector part of the D3S is fixed to the phantom. HDPE is placed in front. Right: The display part of the D3S next to the control display.

Fig. 8 shows the measured time to alarm for 20 trials. The ANSI standard for RIIDs specifies that the alarm shall be no later than 1 s. This happened in only 4 out of 20 cases, which means the device would fail the test. In the case of the IEC standard the alarm shall be within 2 s, which is fulfilled in more cases but still not enough and the device would fail this test, too.

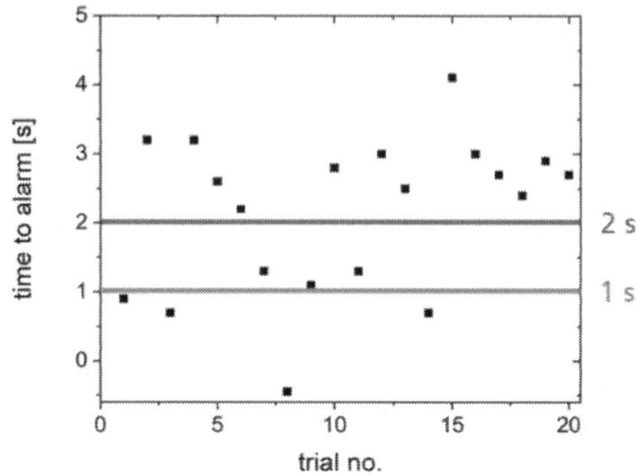

Fig. 8. ^{252}Cf with 14 200 n/s in 20.8 cm distance and 0.42 m/s. 2 s is the criterion for the IEC test, 1 s is the criterion for the ANSI RIID test.

C. Accuracy

For the accuracy tests, the results of one test are shown exemplarily. As stated in section III.C a large variety of criteria exists. The results for the lowest test point of the ITRAP+10 tests with ^{137}Cs, which is 0.5 μSv/h above the background level, are given in Fig. 9. The values are obtained after stabilization after a step change. The results obtained during 30 individual trials are shown. The value including background is 0.6 μSv/h. The ITRAP+10 gives an accuracy of ± 30 %, which is clearly fulfilled. Even the ± 10 %, which is stated for RIIDs in the ANSI Standard, would be fulfilled. This would be the green area in the diagram. But this is only one single criterion from the accuracy tests which does not show the overall performance of the D3S for the accuracy test.

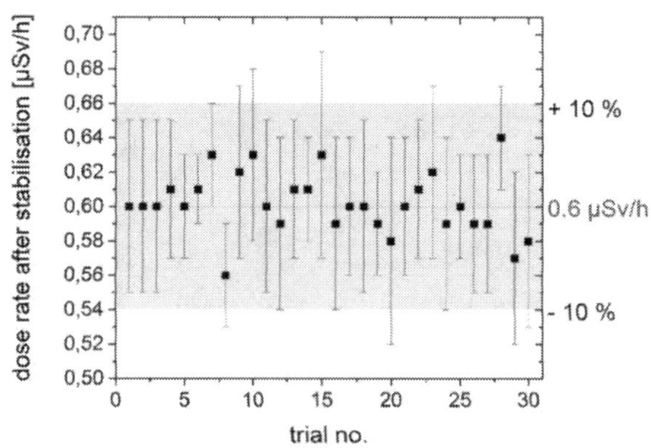

Fig. 9. ^{137}Cs with 0.5 μSv/h above the background level which is 0.6 μSv/h brought in front of the D3S after step change. The results obtained during 30 individual trials, i.e. the dose rate values after stabilization, are given.

VI. Conclusions

Testing of radiation detection devices can be done according to different standards. We were able to show with the time to alarm tests that accurate time measurement is necessary. A static as well as a dynamic test system is necessary. A good

 https://doi.org/10.1051/epjconf/202125307006

observation and documentation system are required, too. Several different radioactive sources are also needed. This is a very important point for testing, which will be addressed in further works.

Our Fraunhofer INT QuTeSt enables us to perform such radiation detection device testing. It is self-made and we can easily adapt it to special needs.

References

Standards:

[1] Radiation Protection Instrumentation - Hand-held Instruments for the Detection and Identification of Radionuclides and for the Estimation of Ambient Dose Equivalent Rate from Photon Radiation, IEC Standard 62327, 2017

[2] Radiation Protection Instrumentation - Alarming Personal Radiation Devices (PRDs) for the Detection of Illicit Trafficking of Radioactive Material, IEC Standard 62401, 2017

[3] Radiation Protection Instrumentation - Spectroscopy-based Alarming Personal Radiation Detectors (SPRD) for the Detection of Illicit Trafficking of Radioactive Material, IEC Standard 62618, 2013

[4] American National Standard Performance Criteria for Handheld Instruments for the Detection and Identification of Radionuclides, ANSI Standard N42.34-2015

[5] American National Standard Performance Criteria for Alarming Personal Radiation Detectors for Homeland Security, ANSI Standard N42.32-2016

[6] American National Standard Performance Requirements for Spectroscopic Personal Radiation Detectors (SPRDs) for Homeland Security, ANSI Standard N42.48-2018

[7] C. Carrapiço, G. Eklund, M. Marin-Ferrer, P. Peerani, F. Rosas, A. Tomanin and H. Tagziria, "Illicit Trafficking Radiation Detection Assessment Program ITRAP+10 Phase II - Round Robin: Radiation Isotope Identifiers Device (RIID) - Test Method", JRC Technical Reports, 2017

[8] C. Carrapiço, G. Eklund, M. Marin-Ferrer, P. Peerani, F. Rosas, A. Tomanin and H. Tagziria, "Illicit Trafficking Radiation Detection Assessment Program ITRAP+10 Phase II - Round Robin: Personal Radiation Detector (PRD) - Test Method", JRC Technical Reports, 2017

[9] M. Risse, J. Glabian and T. Köble, "Qualification measurements of handheld radiation detectors for homeland security purposes", in INMM 61st Annual Meeting, Virtual Event, 2020, pp. 1172-1181

[10] M. Risse, C. Bornhöft, J. Glabian and T. Köble, "D3S - results of qualification measurements of a wearable RIID for homeland security", in INMM 60th Annual Meeting, Palm Desert, California, USA, 2019, pp. 432-441.

[11] Kromek Limited, "Kromek 3DS ID User Manual," Revision 4.0. [Online] Available: http://www.kromek.com/wp-content/uploads/2019/07/D3S-ID-User-Manual-A5.pdf. Accessed on: Jun. 30, 2021

Tests of Various Scintillator Detectors in Selected Mono-Energetic Neutron Beams

Aleš Jančář[1], Jiří Čulen[1], Filip Mravec[1], Michal Košťál[3], Daniel Dlhopolček[2], Elisa Pirovano[4], Ralf Nolte[4], František Cvachovec[2], Václav Přenosil[2], Zdeněk Matěj[2]

[1]VF, a.s., Svitavská 588, 67921 Černá Hora, Czech Republic
[2]Masaryk University, Botanická 68a, 612 00 Brno, Czech Republic
[3]Research Center Řež Ltd, 250 68 Husinec-Řež 130, Czech Republic
[4]Physikalisch-Technische Bundesanstalt, Bundesallee 100, 38116 Braunschweig, Germany
ales.jancar@vfnuclear.com

***Abstract*—In this paper we study quality of three types of organic scintillators - stilbene, p-terphenyl and EJ-299-33. We used monoenergetic neutron fields with a wide range of neutron energies in PTB Braunschweig. All the tests were carried out with NGA-01 spectrometer. The results of the measurements are evaluated spectra from the spectrometer.**

We discuss the quality of each scintillator, such as the FWHM of the peaks.

***Keywords* —Scintillation detectors, Mono-energetic neutron sources, Pulse-shape discrimination, FWHM, Digital-signal processing.**

I. Introduction

THIS paper presents preliminary results from measurements at the PTB facility in Germany. Several monoenergetic neutron beams were used together with three scintillators connected to fast digitizer NGA-01: stilbene, p-terphenyl and EJ-299-33 [5].

II. Experimental Setup

The irradiations were performed at the PTB Ion Accelerator Facility, where monoenergetic neutron fields are produced via selected reactions of proton and deuteron beams with light or medium-weight target nuclei. The measurements were carried out in open geometry in the low-scattering hall where the contribution of room-return neutrons is minimized by having grid floors [1, 2]. The three neutron energies considered in this campaign 1.5, 2.5 and 19 MeV, see Tab. 1.

TABLE I
MEAN NEUTRON ENERGY E_N OBTAINED AT THE NEUTRON EMISSION ANGLE OF ZERO DEGREES RELATIVE TO THE DIRECTION OF THE INCIDENT BEAM, FOR THE GIVEN TARGETS AND ION BEAM (PROJECTILE) ENERGIES E_{PROJ}.

E_{proj} [keV]	Target	Reaction	En [MeV]
2336	Ti(T), 961 μg/cm^2	$^3H(p,n)^3He$	1.5
3350	Ti(T), 1831 μg/cm^2	$^3H(p,n)^3He$	2.5
2676	Ti(T), 1831 μg/cm^2	$^3H(d,n)^4He$	19.0

Fig. 1. Position of the detector (on the right-front) and the beam target (on the left-rear) as used for the measurements. In between there is a shadow cone.

Neutron-Gamma spectrometric system NGA-01 has been used for the measurements [4]. The input analog signal from the detector was digitized by two fast A/D converters working with a sampling frequency of 500 MHz. Digital signal processing is implemented in FPGA Xilinx Virtex-6. Integration method based on the pulse charge comparison was implemented in FPGA. Separation parameters for various scintillators calculated with this method are shown below in PSD matrices. Hamamatsu R329-02 photomultiplier was used for these measurements.

^{137}Cs and ^{60}Co sources were used for calibration.

Fig. 2. Neutron gamma analyser NGA-01.

III. Evaluation

NGA-01 has been used for the measurements of the apparatus proton recoil spectra. These spectra were evaluated using Maximum Likelihood Estimation method of Expectation Maximization [3]. Response functions needed for this evaluation were calculated with Monte Carlo simulation code using ENDF/B-VII.1 nuclear data.

For the evaluation of the measurements we developed spectrometric software. MCNP simulations of response functions using ENDF/B-VII.1 nuclear data were used. Neutron spectral flux densities are displayed.

The 19 MeV field contained parasitic low-energy neutrons from D(d, n), Ti(d, n) and Ag(d, n) reactions which were not subtracted using a non-tritiated target. For this further analysis will be performed.

IV. RESULTS

Stilbene scintillators of the sizes of 10x10 mm and 45x45 mm have been used for measurements of neutron energies of 1.5, 2.5 and 19 MeV. In all measurements corresponding peaks are identified in evaluated spectra, Fig. 3 - 14. Measurements with neutron energies of 2.5 and 19 MeV were carried out with EJ-299-33 scintillator (Fig. 15 - 18) and with p-terphenyl scintillator (Fig. 19 - 22). We compared spectra from EJ-299-33 and p-terphenyl scintillators with 45x45 mm stilbene scintillator (shown in dotted gray line in Fig. 17, 18, 21, 22). Arbitrary units have been used for the y-axis in the graphs of evaluated spectra. Tab. 2 shows resolution values calculated from evaluated spectra.

TABLE II

THE ENERGY RESOLUTION FWHM/E FOR NEUTRON ENERGIES 2.5 AND 19 MEV.

E_n [MeV]	Stilbene 45 mm	EJ-299-33	p-terphenyl	Stilbene 10 mm
2.5 MeV	0.216	0.361	0.280	0.232
19.0 MeV	0.102	0.184	0.184	0.063

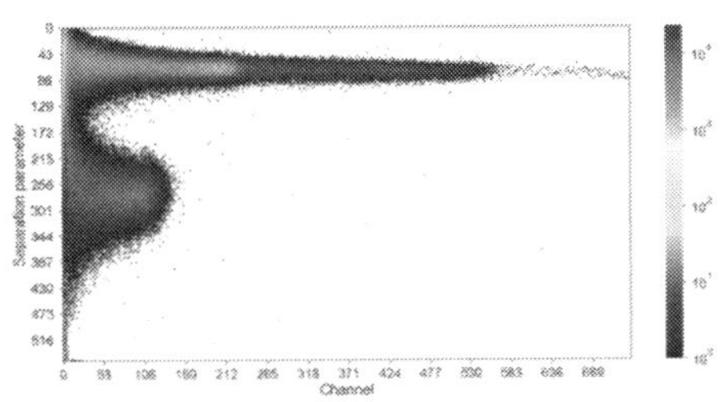

Fig. 3. PSD matrix for 1.5 MeV measurement with stilbene 10x10 mm.

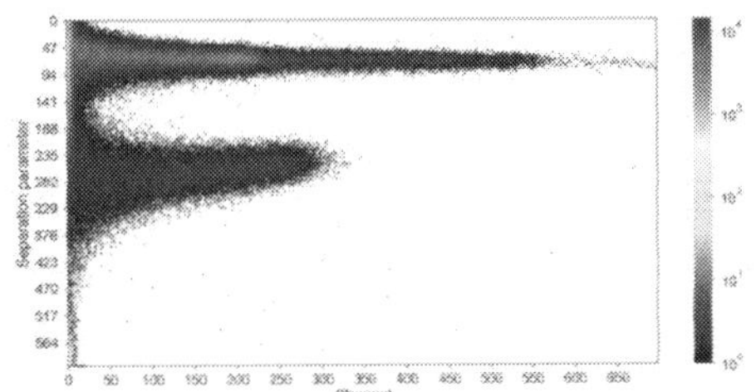

Fig. 4. PSD matrix for 2.5 MeV measurement with stilbene 10x10 mm.

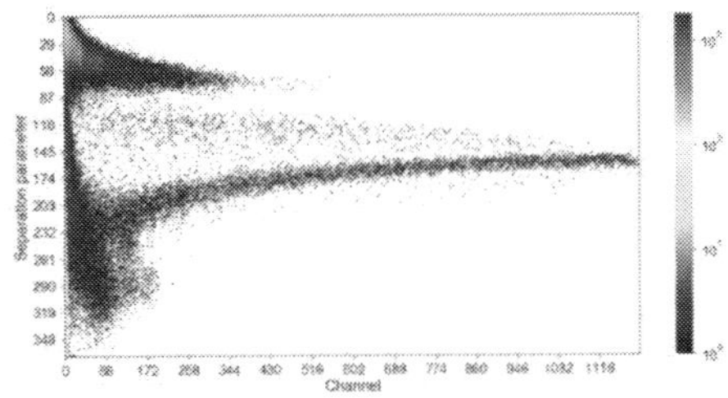

Fig. 5. PSD matrix for 19 MeV measurement with stilbene 10x10 mm.

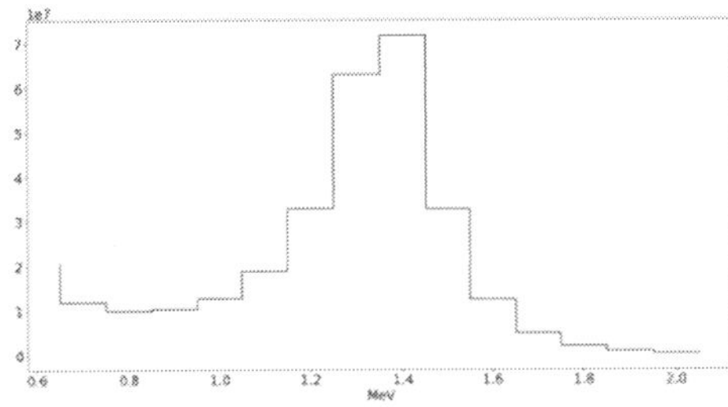

Fig. 6: Neutron spectral flux density for 10x10 mm stilbene for 1.5 MeV neutrons.

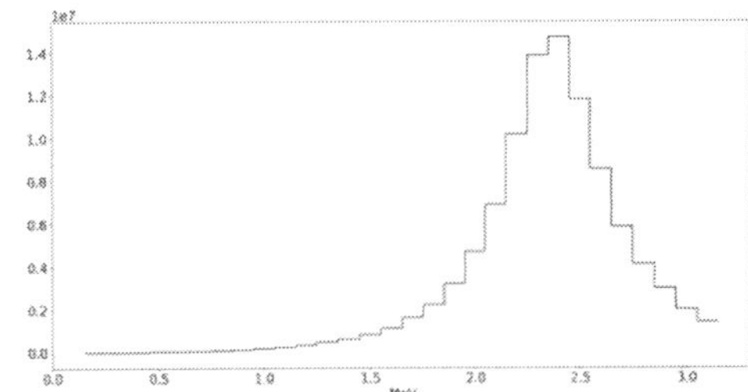

Fig. 7: Neutron spectral flux density for 10x10 mm stilbene for 2.5 MeV neutrons.

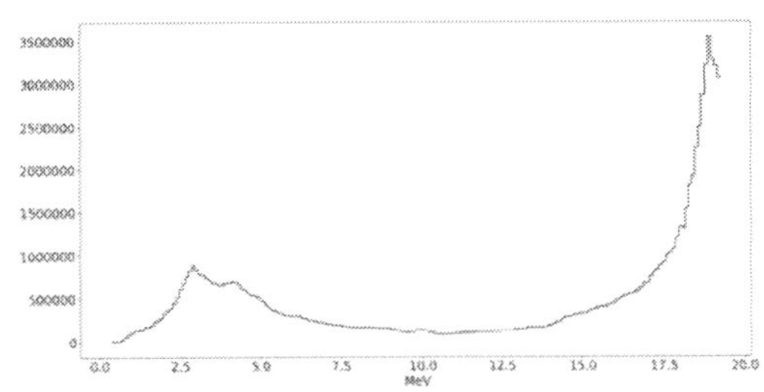

Fig. 8: Neutron spectral flux density for 10x10 mm stilbene for 19 MeV neutrons.

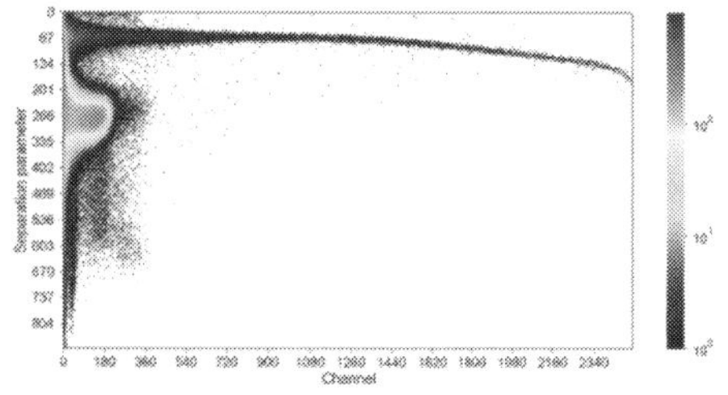

Fig. 9. PSD matrix for 1.5 MeV measurement with stilbene 45x45 mm.

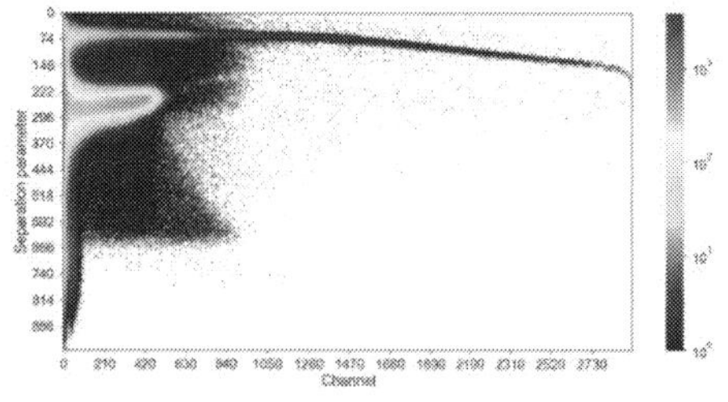

Fig. 10. PSD matrix for 2.5 MeV measurement with stilbene 45x45 mm.

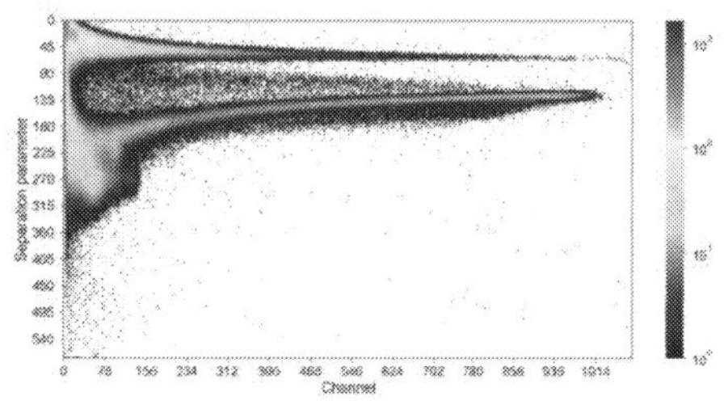

Fig. 11. PSD matrix for 19 MeV measurement with stilbene 45x45 mm.

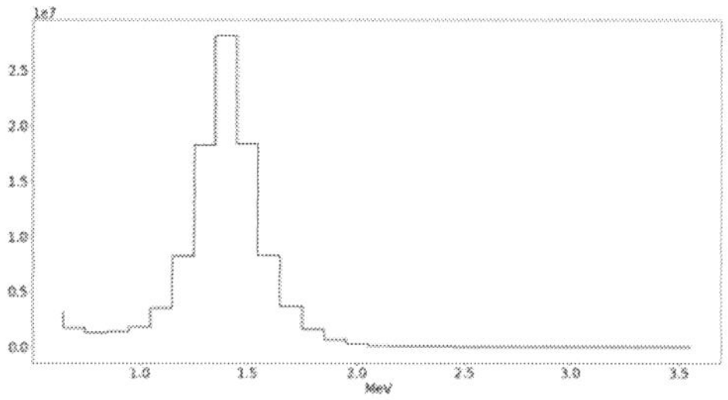

Fig. 12: Neutron spectral flux density for 45x45 mm stilbene for 1.5 MeV neutrons.

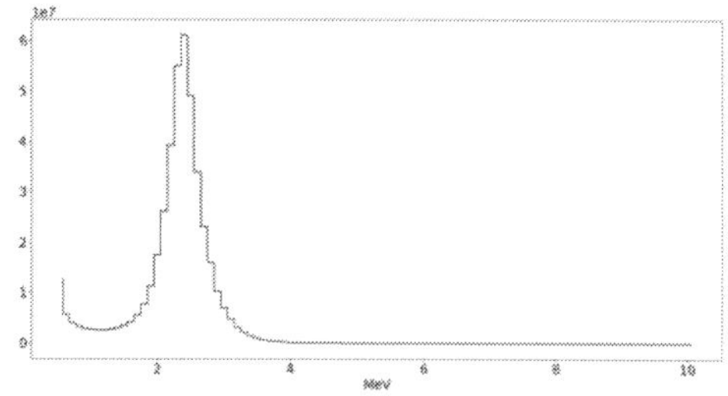

Fig. 13: Neutron spectral flux density for 45x45 mm stilbene for 2.5 MeV neutrons.

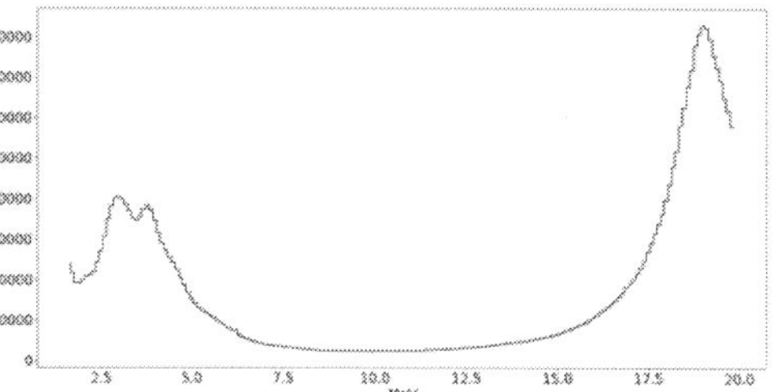

Fig. 14: Neutron spectral flux density for 45x45 mm stilbene for 19 MeV neutrons.

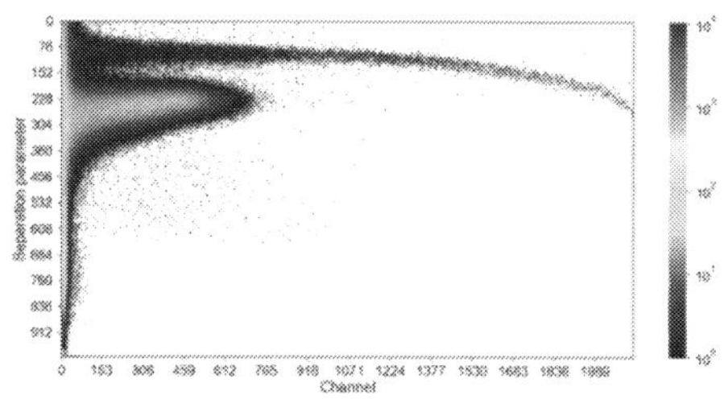

Fig. 15. PSD matrix for 2.5 MeV measurement with 1x1 inch EJ-299-33.

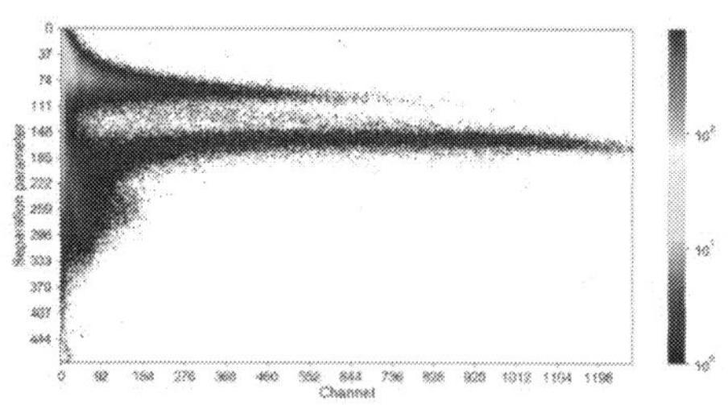

Fig. 16. PSD matrix for 19 MeV measurement with 1x1 inch EJ-299-33.

https://doi.org/10.1051/epjconf/202125307007

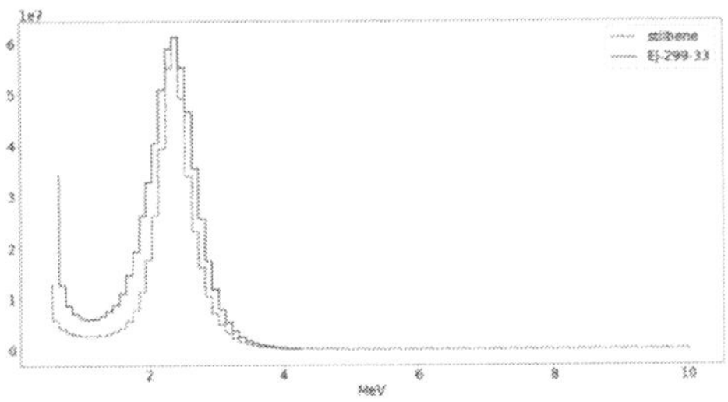

Fig. 17: Neutron spectral flux density for 1x1 inch EJ-299-33 for 2.5 MeV neutrons. Results for stilbene 45x45 were added for comparison.

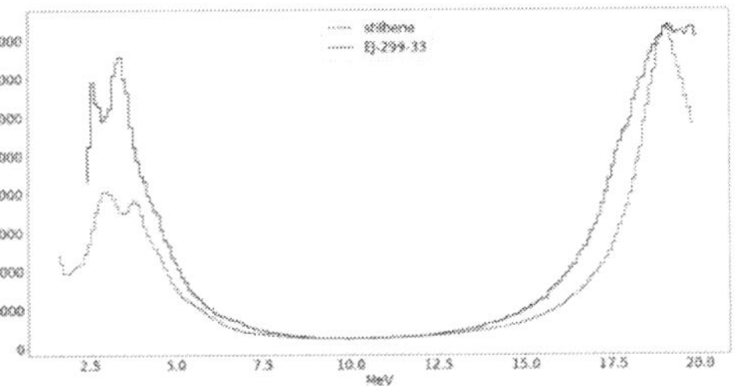

Fig. 18: Neutron spectral flux density for 1x1 inch EJ-299-33 for 19 MeV neutrons. Results for stilbene 45x45 were added for comparison.

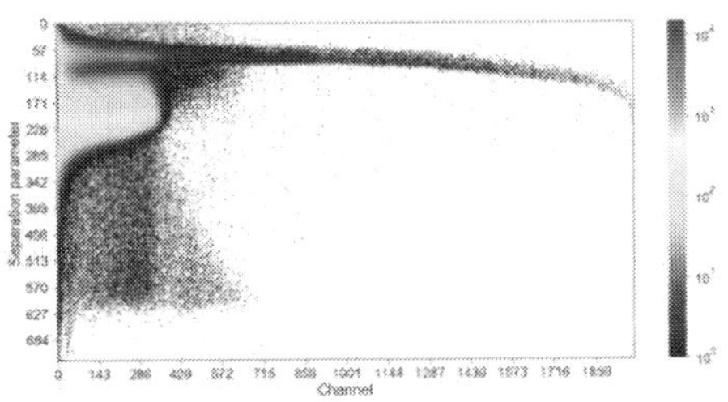

Fig. 19. PSD matrix for 2.5 MeV measurement with p-terphenyl 45x45 mm.

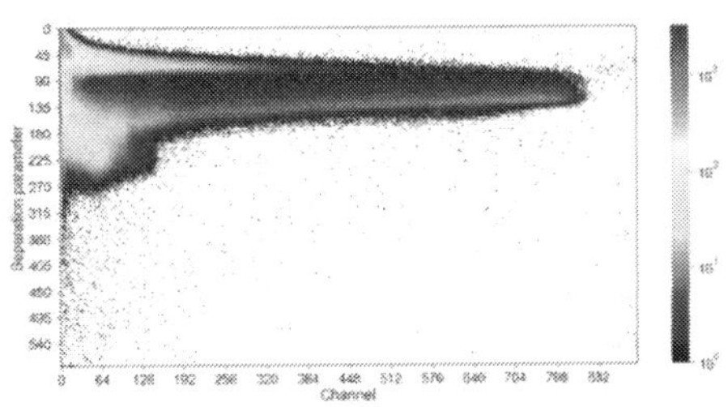

Fig. 20. PSD matrix for 19 MeV measurement with p-terphenyl 45x45 mm.

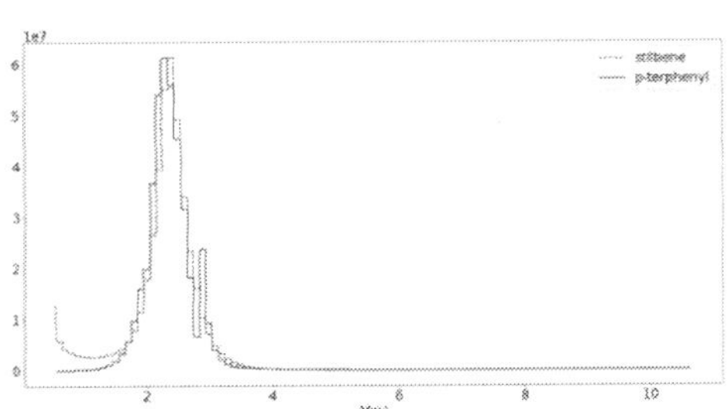

Fig. 21: Neutron spectral flux density for 45x45 mm p-terphenyl for 2.5 MeV neutrons. Results for stilbene 45x45 were added for comparison.

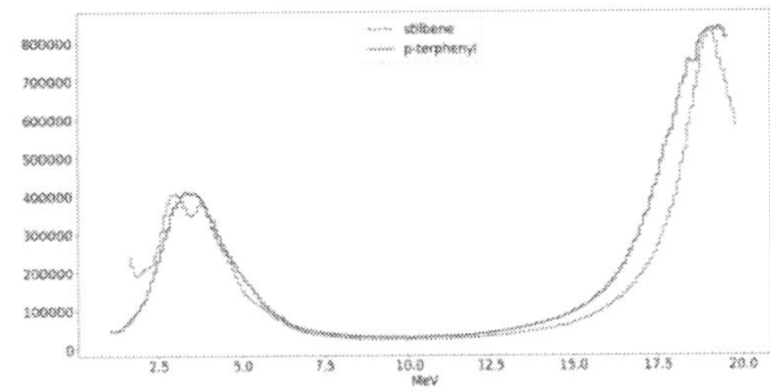

Fig. 22: Neutron spectral flux density for 45x45 mm p-terphenyl for 19 MeV neutrons. Results for stilbene 45x45 were added for comparison.

V. Conclusions

All presented scintillators show satisfactory spectrometric and pulse shape discrimination properties. Using unfolding it is possible to find characteristic peaks at expected energies.

The next step of the measurement processing would be the subtraction of the measurements with the shadow cone and with the blank target (for the 19 MeV experiment). The data can be also used to improve the light output functions for tested scintillation materials.

Acknowledgment

The work presented in this paper has been supported by the research project "Methods of transmission and detection of scintillation radiation with optical fibers and energy resolution of ionizing radiation sources", No. VI20192022116 of the Ministry of the Interior of the Czech Republic and research project "Integrated spectrometric release system radioactive waste into the environment", No. FW01010063 of Technology Agency of the Czech Republic. This project has received funding from the Euratom research and training programme 2014-2018 under grant agreement No 847594 (ARIEL).

References

[1] R. Nolte and D. J. Thomas, Monoenergetic fast neutron reference fields: I. neutron production, Metrologia, vol. 48, p. S263S273, 2011.
[2] R. Nolte and D. J. Thomas, Monoenergetic fast neutron reference fields: II. field characterization, Metrologia, vol. 48, p. S274S291, 2011.
[3] J. Cvachovec and F. Cvachovec. Maximum Likelihood Estimation of a Neutron Spectrum and Associated Uncertainties. Advances in Military Technology. Brno: Univerzita obrany, 2008, Vol. 3, No. 2, 14 s. ISSN 1802-2308.
[4] Z. Mat'ej at al., Progress in Nuclear Science and Technology, 4 (2014), p. 670.
[5] Eljen EJ-276 scintillator [Online] Available: https://eljentechnology.com/products/plastic-scintillators/ej-276

Development of Prototype Simplified Neutron Scatter Camera for Nuclear Safeguards Applications

Taylor Harvey[1] and Andreas Enqvist[1]
[1]University of Florida, Gainesville, Florida, United States of America
Corresponding author: taylor.harvey250@ufl.edu

***Abstract*—Neutron scatter cameras are a type of directional neutron detectors that rely on multiple neutron scatters to generate images that can reveal the direction and distribution of neutron sources. Fast neutron cameras which have recently been developed rely on multiple detector volumes and make use of neutron time-of-flight measurements. These designs, though effective in localizing the source direction, relies on a large amount of detection and electrical equipment, thus increasing size, cost, and complexity of the systems to unreasonable levels for some applications. This project seeks to develop a compact scatter camera that is less expensive than systems relying on multiple detector volumes. Crucially, two components and capabilities are needed to achieve this: fast scintillation detection materials and picosecond electrical pulse timing. Utilizing such electronics, distinguishing between scintillation light pulses generated by the same neutron within one detector volume is possible. An MCNPX-PoliMi model of such a system has been developed to guide prototype designs. A cube of EJ-230 fast plastic scintillator and six photomultiplier tubes (PMTs) were used to construct the prototype camera that localizes neutron sources based on the principle of cone back projection. Prototypes of the system in one, two, and three dimensions have shown promising initial results when coupled with a script that algorithmically identifies candidate neutron double scatter events and back projects probability cones in the direction of possible sources. Imaging resolution/quality, double scatter efficiency, and cost for the system are quantified. Paths forward for further improvement of a future system based on this camera' operating principles are discussed.**

***Keywords* — neutron imaging, neutron detection, international safeguards**

I. Introduction

The field of nuclear safeguards seeks to stop the spread of nuclear weapons by developing institutional, legal, and technical mechanisms intended to prevent the misuse of nuclear materials and technology [1,2]. Essential to the nuclear safeguards paradigm is the use of nuclear material measurement technologies to monitor and verify the quantity, identity, and movement of radioactive sources relevant to both peaceful and weapons programs. Of particular interest is Special Nuclear Material (SNM), such as uranium and plutonium, which can be monitored and identified using a variety of neutron detection techniques [3]. Much of the focus of nuclear material measurement technologies for safeguards is on destructive analysis (DA) and nondestructive assays (NDA) of nuclear material. Both DA and NDA techniques seek to determine properties of nuclear material through chemical, spectroscopic, or other means to make quantities known for material and accounting purposes. This suite of techniques can reveal, often in fine detail, the "what" and "how much" of a material of interest but cannot give information about the location or distribution of a source if these facts are unknown. A new class of radiation detectors, known as directional or imaging detectors, seek to further the international safeguards technical repertoire by providing spatial information on radioactive sources.

Neutron scatter cameras (NSCs) operate by detecting two consecutive elastic scatters of a single fast neutron emitted by a nearby source. The kinematic principle involved in determining the original particle trajectory from two consecutive scattering events is similar to the operating principle of Compton cameras, though NSCs use fast neutrons rather than gamma rays [4]. The detector active volumes present in all currently existing NSCs use either organic liquid or plastic scintillators. The initial trajectory of individual neutrons is determined by finding the approximate (x, y, z) position of both the first and second scatter within the scintillator volume. The position of each scatter is found by either using spatially separated volumes of scintillators, or by comparing the relative position and timing of the arrival of light to a series of photodetectors coupled to the same scintillator volume. Spatially separated arrays of scintillators may be arranged in either multiple planes or in a radially-symmetric arrangement. Mascarenhas *et al.* [5], Goldsmith *et al.* [6], and Weinfurther *et al.* [7] provide detailed technical discussions of the kinematics of scattering in plane-based, radial, and single volume designs respectively. After the determination of two consecutive scatter positions and the determination of the energy deposited in the first scatter, probability cones encompassing the possible trajectories of individual neurons can be back projected in 3-D space. The region of space where the surfaces of the all the back projected cones overlap is interpreted to be the most likely direction of the neutron source.

Advances in both fast light pulse plastic scintillator materials and sub-nanosecond digitizer have allowed for the construction of compact or single volume scatter cameras with high efficiencies and accurate angular imaging resolutions. Currently developed systems rely on photoelectronics and/or segmented designs that lead to systems with high unit costs and high amount of data throughput from many digitized channels. The "simplified" prototype neutron scatter camera discussed in this paper proposes a design that is mobile, simple to use, and affordable while still providing useful data regarding the location, distribution and identity of neutron-emitting sources.

II. Experimental design

A. Detector Construction and Calibration

An MCNPX-PoliMi model of a cube-shaped plastic scintillator volume was developed to guide the parameters of the simplified neutron scatter camera system. Simulation guided the selection of a 6'' side length cube of Eljen's EJ-230 fast plastic scintillator for the detection medium, six Electron Tube 9821KB photomultiplier tubes for the photodetectors, PSI's DRS4 digitizer for pulse digitization, and a custom MATLAB script for the data processing. These materials were selected to allow for the resolution of consecutive same-neutron scatters that appear only nanoseconds apart in time.

To create the back-projected probability cones needed to image neutron sources, the camera must be able to determine the scatter positions of neutron scatters one and two, the energy deposited in the first neutron scatter, and the time of flight between scatters one and two. In the simplified design, determination of the neutron scatter positions is done by comparing the amount of light arriving at photomultiplier tubes coupled to opposing scintillator cube faces for voltage pulses algorithmically tagged as neutron double scatters. The double scatter pulse heights are proportional to the amount of light arriving at each PMT, which is itself proportional to the sighting factor from the scatter position to PMT photocathode, as described by the equation:

$$P_n = L_n log(NP_{n1}/NP_{n2}) \quad (1)$$

Where P_n is the position along the n-dimension of the cube scintillator, L_n is the length of the scintillator in the n-dimension, and NP is the number of photons arriving at the two photodetectors oriented normal to the n-dimension. The scintillation position is localized in all three coordinates when this light intensity ratio comparison is done simultaneously across all six photodetectors. The energy deposited in the first neutron scatter is found by summing the light arriving at all six PMTs, determining the electron equivalent energy deposition corresponding to that amount of light, and calculating the proton recoil energy by using the electron equivalent value with the characteristic light curve for the scintillator material. The time of flight between the two neutron scatters is found by averaging the time between "peaks" corresponding to primary and secondary same neutron scatters in the pulse streams produced by each of the six PMTs. The relevant kinematic equation for cone back-projection are:

$$E_s = \frac{1}{2}m(\frac{d}{t})^2 \quad (2)$$

$$E_i = E_f + E_s \quad (3)$$

$$\theta_n = \sin^{-1}\sqrt{\frac{E_f}{E_i}} \quad (4)$$

Where E_f is the energy deposited by the first neutron scatter, E_s is the neutron energy after the first scatter, E_i is the kinetic energy of a neutron from the source, m is the mass of the neutron, t is the time between scatters 1 and 2 , d is the distance between scatter positions 1 and 2, and θ_n is the central angle of the back projected cone of possible source locations. Figure 1 summarizes the kinematics of double neutron scattering in the simplified design, and Figure 2 shows a photograph of the prototype simplified neutron scatter camera system.

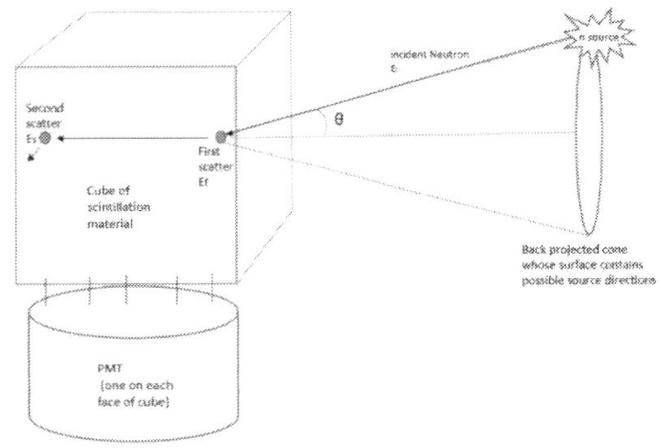

Fig. 1. Basic operating principle of a monolithic neutron scatter camera. Neutrons emitted by nearby sources must scatter at least twice in the scintillator volume. The scattering positions are determined by comparing the amount of light arriving at each of the six PMTs mounted to the six surfaces of the scintillator. By solving a series of kinematic equations, a probability cone whose surface contains the possible trajectories of the neutron can be back-projected into 3-D space. Many correctly back-projected cones will converge to the source location.

Fig. 2. Three-dimensional neutron scatter camera prototype. The PMT coupled to the bottom face is not visible as it is resting inside of the cardboard box support. Photo courtesy of the authors.

The camera was gain calibrated for neutron scatter position by finding maximum and minimum light ratios through simulation and assuming a log-linear relationship across the detector dimensions. Energy calibration was performed by plotting the 90 percent height Compton edge values for ^{60}Co, ^{137}Cs, and ^{22}Na gamma spectra measured with each of the PMTs used with the camera. Proton recoils for each neutron scatter event were found by performing neutron time-of-flight measurements using the EJ-230 material and comparing them with published results with the same material [8]. The system was time calibrated by comparing the pulse rise time response between two of the system PMTs, resulting in a system timing uncertainty of σ_t=0.82 ns. Figures 3, 4, and 5 display the relevant calibration curves used in finding the kinematic data from each neutron double scatter event.

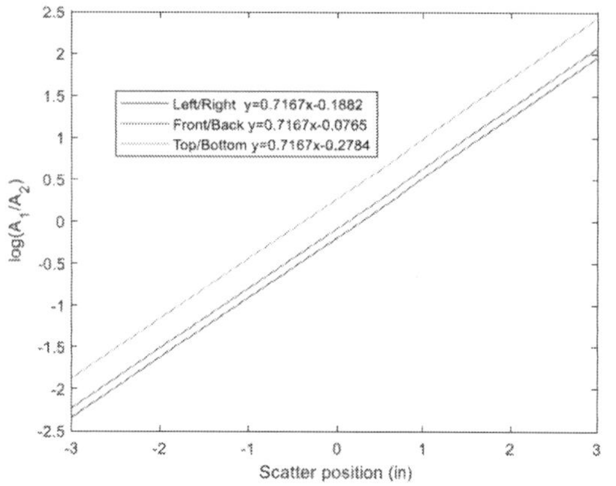

Fig. 3. Position calibration curves for opposite PMT neutron scatter events for the left/right (x), front/back (y), and top/bottom (z) PMT pairs. The linear curves were fit to the maximum/minimum achievable light ratios from the geometric end points at -3 and 3 inches determined from simulation.

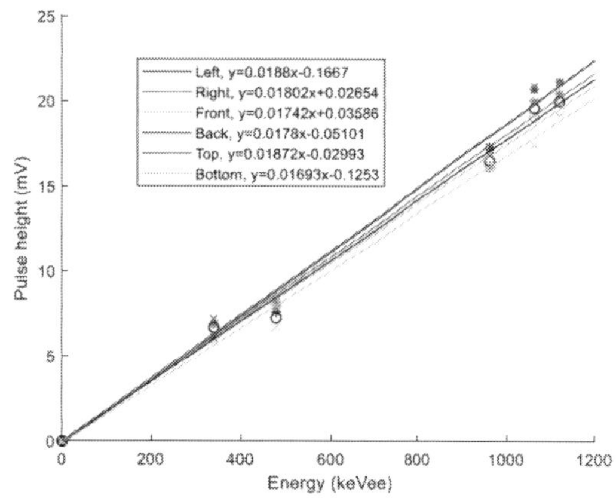

Fig. 4. Linear energy calibration curves for each of the six PMT used in the scatter camera. The five 90% height Compton edge values for each PMT are plotted along with lines of best fit forced to pass through zero.

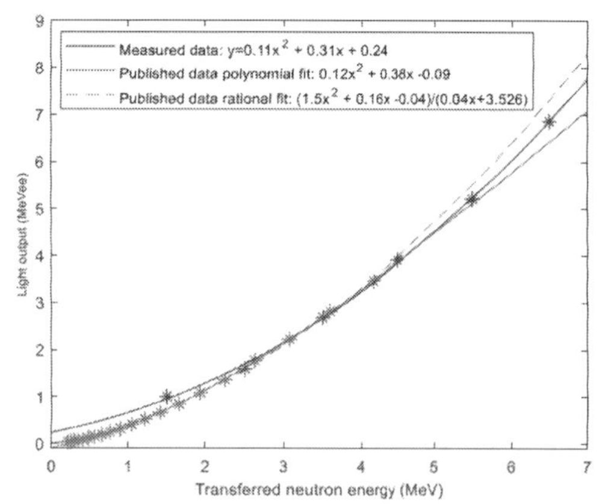

Fig. 5. Light functions for EJ-230 material from measured data and previously published data [8]. Second-degree polynomial and two-over-one rational fits were used to generate the light functions. Deviation between the two data sets can be attributed to different detector shapes and different methods of PHD edge finding.

B. Double Scatter Identification Algorithm

The overall goal of the algorithm used in tandem with the simplified NSC is to read in voltage pulse arrays produced by the system, determine parameters of the pulses based on their shape, height, timing, and relative prominence, use these parameters to estimate the positions and energy depositions associated with neutron double scatter events, and finally use these scattering parameters to generate neutron images based on the principle of back-projection. The data files needed to complete this sequence are necessarily large, with a six-channel pulses set with one million pulses at a sampling rate of 5 GSPS per channel resulting in a data file size of approximately 12 GB. To prevent memory allocation issues when running the script in MATLAB, the script was broken into three main sections which can be run sequentially and independently. The first section reads the data files output by the PSI DRS4 digitizer into pulse arrays that can be manipulated by MATLAB. The second section takes those arrays and extracts energy, timing, and shape data from them to identify pulses that may have resulted from double scatter events. The third section uses the double scatter candidate pulses to form a neutron image through cone back-projection and improves the image through a process of cone down-selection based on an initial source direction guess vector. The overall script runs most efficiently and experiences the least memory allocation related slowdown when the first script section is used to read in a small fraction of the total pulses from the data file and transfers the read pulses to the second section for selection of double scatter candidates. Once the second section finds the double scatter candidates for the sample of the total pulses, the non-candidate pulses can be deleted, and the first section can be run again to import the next sample of pulses. This loop between sections I and II can continue until the original data file has been completely read. Next, with an array of double pulse candidates found, the imaging section can be run once to generate a single neutron image or be run iteratively to generate a series of images with different down-selection parameters, such as energy disposition cutoff, choice of proton recoil light curve, second scatter criterion, second pulse timing window length, and size of guess vector down-selection region. A flowchart demonstration the overall data flow in the script is shown in Figure 6.

https://doi.org/10.1051/epjconf/202125307008

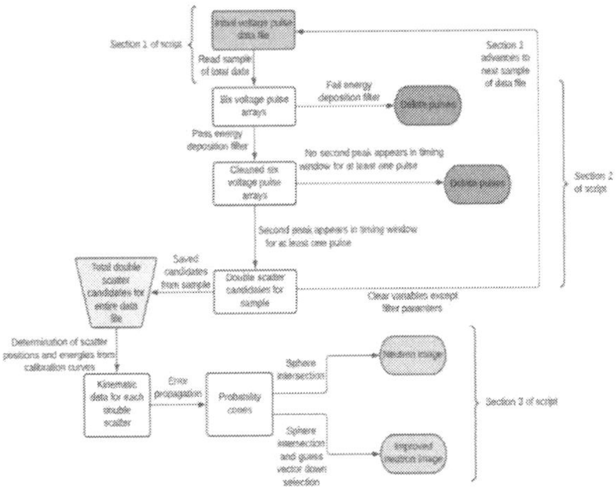

Fig. 6. Flowchart of data in the MATLAB script used in the algorithm that reads and processes neutron imaging data from the prototype simplified NSC.

The second section of the script is the most computationally taxing, mostly due to the 'findpeaks' function used in identification of pulse features that could correspond to second neutron scatters. This section of the script opens with the definition of a set of user-set parameters for the sensitivity of the second scatter feature search. These include the selection of which data set and fit to use for the light curve, the peak prominence ratio cutoff for the two most prominent peaks in a pulse, the first scatter energy deposition cutoff, a pure pulse height cutoff value to remove noise, and second peak window start and ending time values. After defining these parameters, the raw pulses are inverted, adjusted for gain pedestal to move each pulse to a baseline of zero, and cut down from 1024 data points to 351 data points each. These steps are taken to reduce the amount of data processing necessary for each pulse and to ensure the built-in MATLAB functions work properly. The maximum value for each pulse is then found and saved in six separate arrays. These maximums are then used to find log pulse height ratios for the left/right, front/back, and top/bottom PMT pairs. The pulse height values are then converted from digitizer units to millivolts and the maximum peak values are applied to the energy calibration curves presented in Figure 4 to yield first scatter deposition values in MeVee for each PMT. The MeVee value corresponding to each of the six PMTs for a single event is then averaged to yield the "true" electron-equivalent energy deposition value for the scattering events. Averaging across the PMTs removes any scatter position-related bias related to asymmetric light arrival at the various photocathodes. The electron-equivalent energy values are then applied to the inverse of one of the light curve fits presented in Figure 5. This yields an array of first scatter neutron energies. In actuality, some portion of these pulses are caused by gamma scatters or cosmic rays, though those events should be removed by the later double scatter and source direction guess vector filters. The script then removes all pulses correlated with neutron deposition energies of less than the cutoff value. Altering this cutoff value greatly changes the ultimate image quality, as requiring a high (>0.5 MeV) cutoff results in few cones with low uncertainty per cone, while a low cutoff results in many cones with high uncertainty per cone.

The pulses remaining after the application of the energy and timing filters are fed into a 'for' loop to identify neutron double scatter candidates. A voltage pulse resulting from a double neutron scatter in a small, ideal scintillator detector should resemble a large voltage pulse followed by a second pulse several nanoseconds later. The second pulse rise will begin before the first pulse has a full time to decay to baseline. In most scintillator detectors, the second pulse will not be distinguishable from the first, both because of the long light scintillation time and/or because the sampling rate is too low to digitize the second scatter features. Both of these scenarios result in double scatters that resemble single pulses. The simplified NSC system can distinguish pulses that occur several nanoseconds apart through the use of a fast timing scintillator material and a sub-nanosecond digitizer. The first step in searching for these double scatter events is establishing a timing window in which it is possible for secondary scatters to appear. The reported pulse width for the EJ-230 material is 1.3 ns, and it is known that any photodetectors will add some amount of timing blur to the that width. Additionally, even if pulses very close together in time can be resolved, high uncertainty due to their low times-of-flight will result in probability cones with high errors in kinematic properties. Due to these facts, it was assumed that the window for possible second peaks should begin at 2 ns after the first peak in each voltage pulse. The distribution in time difference between scatter positions 1 and 2 is known from Monte Carlo simulation. This distribution provides guidance on the end of the time window. Virtually no double scatters with a time separation of greater than 12 ns occur in a detector of this size. This means that a window for a second peak beyond that upper value would likely be incorporating secondary particle coincidences, which results in erroneous cones. Based on these material and empirical considerations, the timing window for the system was set at 2 to 12 ns after the first scatter peak. Figure 7 shows a candidate double pulse as seen by one PMT with the timing window demonstrated.

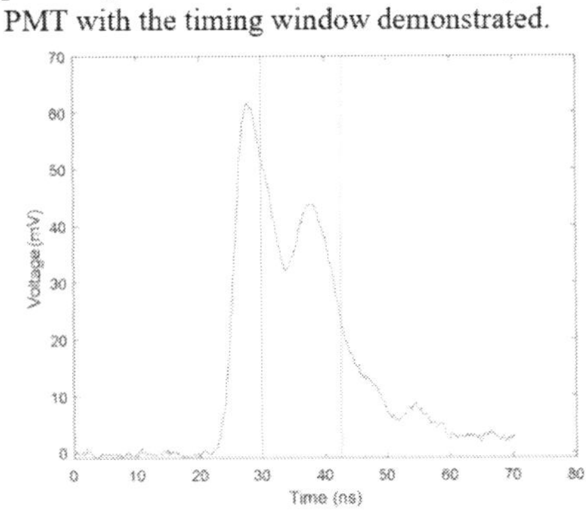

Fig. 7. Timing window for neutron double scatter candidates. The peak of the secondary scatter feature must appear between 2 and 12 ns after the primary peak.

With the timing window established, the next parameter to determine is the empirical value related to the secondary feature necessary for the event to be considered a true second neutron scatter. We will call this the second scatter criterion. It is relevant to determine whether the second scatter criterion should be applied to each pulse individually, or only to a select few from each sestet of coincident pulses. A number of definitions for the second scatter criterion were applied in a trial-and-error manner to data sets, as there is little guidance in this area in the literature.

For the testing of each second scatter criterion definition, the NSC system measured a ^{252}Cf source at a known location for six hours. Each second scatter criterion definition was applied to the data set to yield an array of double scatter candidates. Section III of the script was then run on each resulting double scatter candidate array to generate neutron images. The neutron images resulting from the differing definitions were analyzed for convergence to the neutron source position over a range of energy deposition cutoff values and second scatter criterion values according to their definitions. The results from these trials are summarized in Table I. It was found that the two peak prominence ratio method yielded images that converged to the correct source position most often. A visualization of this method is shown in Figure 8.

TABLE I
DESCRIPTION AND PERFROMANCE OF SECOND SCATTER CRITERIA

Second scatter criterion	Description	Performance
Second peak absolute prominence	Prominence of highest peak in the time window must pass an absolute cutoff value.	Fair
Positive gradient	Derivative of voltage values in the time window is found and must pass cutoff indicating that the voltage in at least part of the window has a positive slope	Poor
Two peak prominence ratio	Prominence of pulse maximum and peak maximum in the time window are found. Ratio of the peak prominence of the two must pass a cutoff value indicating the second peak has a prominence that is greater than or equal to a certain fraction of the prominence of the first.	Good
Forced Gaussian fit	Each pulse with a peak in the time window is fit with a Gaussian centered at that peak. The goodness-of-fit value for the fit must pass a cutoff to be considered a true second scatter.	Poor
Selection by visual inspection	Pulses with a peak in the time window are saved, and then each remaining is visually inspected. Pulses that visually look like noise or coincidence are eliminated.	Fair, but time consuming

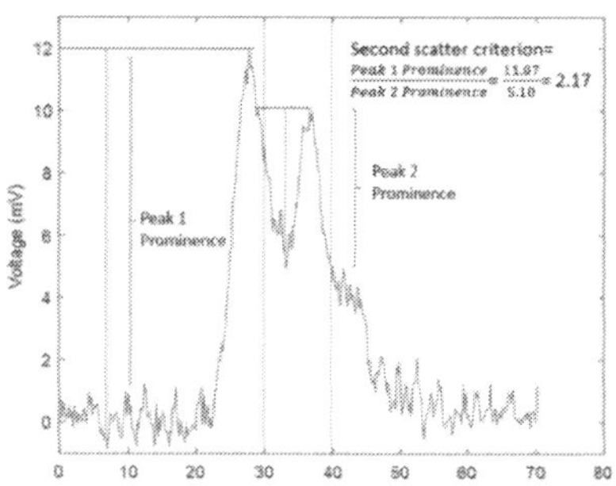

Fig. 8. Demonstration of second scatter criterion using the two peak prominence ratio method. In generating the neutron images to evaluate this method, the second scatter criterion for this method was ranged over values between 1.5 and 10.

This method for second scatter identification was implemented in the code so that only one pulse each sestet of coincident pulses needed to fulfill the second scatter criterion for the entire sestet to be accepted as a double scatter candidate. It was found that when generating test images using this method, requiring all six or some combination of pulses to meet the second scatter criterion resulted in double scatter candidate data sets with much fewer pulses than the simulated double scatter efficiency suggested, indicating a high rate of false negatives. Only requiring any one of the pulses in each sestet to meet the criterion greatly lowered this false negative rate.

Determining scatter position 1 begins by finding the voltage maximum in the timing window where scatter 1 is expected to be for each pulse. This window is defined as plus or minus 2 ns from the average pulse maximum timing bin location for that digitizer channel. The maximum voltage values are then compared with the values generated by the same event in the opposite PMT channel by taking the log ratio of the pulse heights and then plugging those value into the position calibration curves presented in Figure 3. Doing this for each pair of PMTs yields scattering position of first scatter events for each double scatter candidate, labeled x_1, y_1, and z_1 in the script.

Second scattering positions are found in a similar manner, though they require the extra step of subtracting the elevated baseline caused by the first scatter just nanoseconds before. For each double scatter candidate pulse, a Gaussian fit is performed on the second peak to estimate the pulse height that would be caused by that scattering event had the first scatter not occurred. The height from the signal baseline to the maximum of the Gaussian is then used as the second peak height, which is subsequently applied to the log ratio and then the calibration curve. A visual demonstration of this process for double scatter candidate pulses from opposite PMTs is shown in Figure 9.

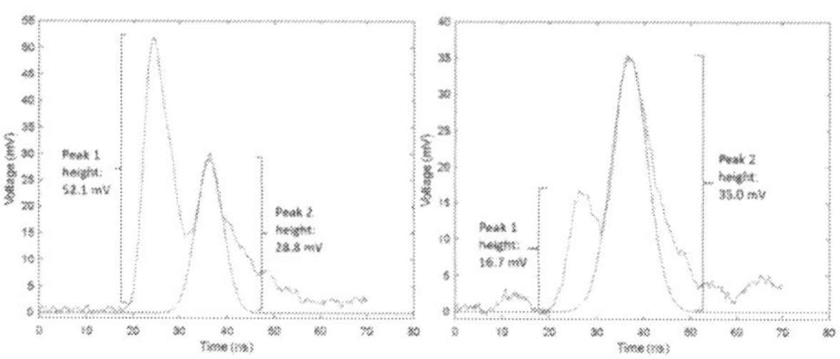

Fig. 9. Peak height determination for neutron double scatter candidate pulses. These pulses were produced by the pair of PMTs mounted to opposite ends of the camera along the x-axis of measurement. Taking the log ratios of the peak 1 and peak 2 heights yields scatter positions of +1.85 inches and -0.20 inches from the detector centerline in the x direction for scatters 1 and 2, respectively.

The energy deposited by the first neutron scatter, E_f, is determined running the peak 1 height values for a sestet through the energy calibration curves presented in Figure 4, averaging over each of the six-correlated pulses for the event, and then applying the electron-equivalent energy deposition values to the neutron light curve presented in Figure 5. The error for each E_f value is found by assuming that the production of light in the scintillator is a Poisson process, and thus the number of photoelectrons generated and the voltage pulse heights have an error proportional to the square root of their respective values. The errors of the pulse height are then propagated through the energy calibration curve and the light curve to yield an array of σ_{Ef} values.

The proton recoil energies and the first and second neutron scatter positions can be used with Equations 2, 3, and 4 and standard error propagation rules to find all the values necessary for probability cone back-projection.

III. RESULTS

A. Point Source Measurements

A small 45 μCi ^{252}Cf spontaneous fission source was measured at a variety of angles relative to the camera to test the system's localization ability. Measurements of 5 million total counts at two different spatial positions resulted in the raw neutron images shown in Figures 10 and 11. The high amount of noise resulting from erroneously oriented cones leads the raw images to converge to incorrect positions or feature multiple probability hotspots, reducing the image usefulness. Erroneous cones can be eliminated by down-selecting the cones used to produce the raw images according to an initial source direction guess vector. A guess vector can predict a likely direction of a neutron source by comparing the skew in the distribution of opposite PMT light ratios for all events detected in the scintillator volume during the measurement time. The skewness of log light ratios can be compared to values in a lookup table based on reference measurements of the camera with the same calibration settings. Once a guess vector is made, a region around it is formed on the spherical projection surrounding the camera system. All back-projected probability cones that do not intersect the guess vector region are then removed from the neutron image, resulting in cleaner and better resolved images free of incorrectly oriented cone. Figure 12 visually demonstrates the process of image improvement using guess vector down-selection. Figures 13 and 14 show "improved" neutron images that use the guess vector down-selection technique to create symmetric and well-resolved images.

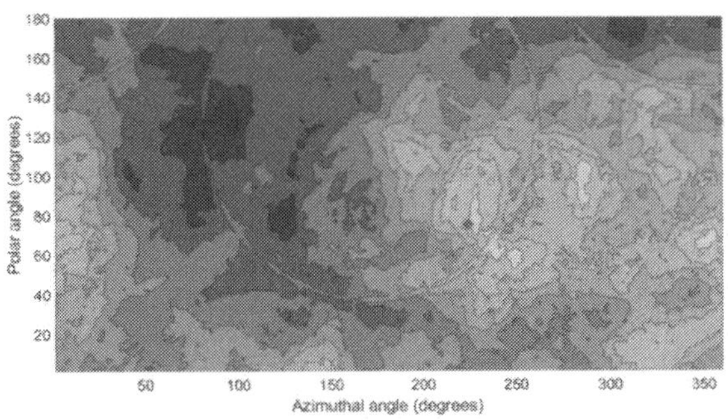

Fig. 10. Raw neutron image of point source measured at (224, 75). The image was generated with a first scatter energy deposition cutoff of 0.2 MeV and a peak prominence ratio cutoff of 4. 154 cones were used to generate the image.

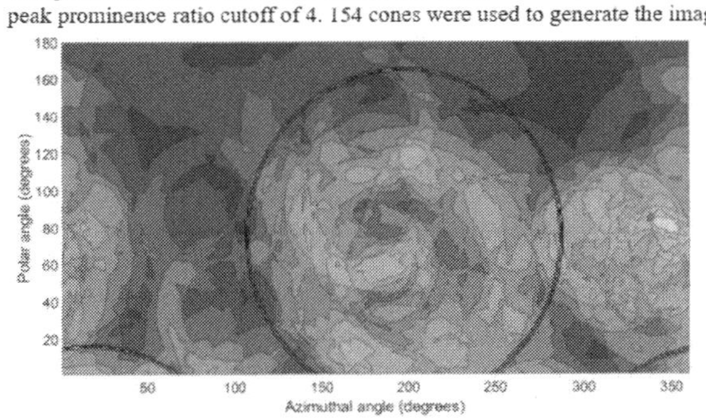

Fig. 11. Raw neutron image of point source 2 measured at source position 2. The image was generated with a first scatter energy deposition cutoff of 0 MeV and a peak prominence ratio cutoff of 2. 42 cones were used to generate the image.

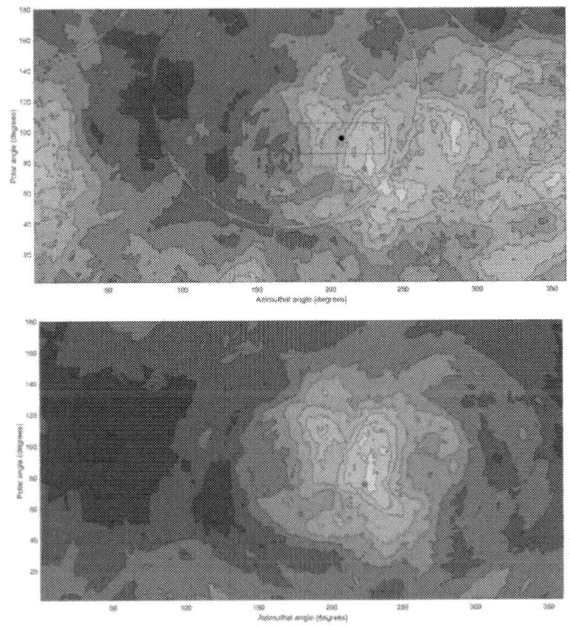

Fig. 12. Demonstration of cone down-selection using source direction guess vector region. All cones that do not overlap the region outlined in black in the first image are removed to form the improved image.

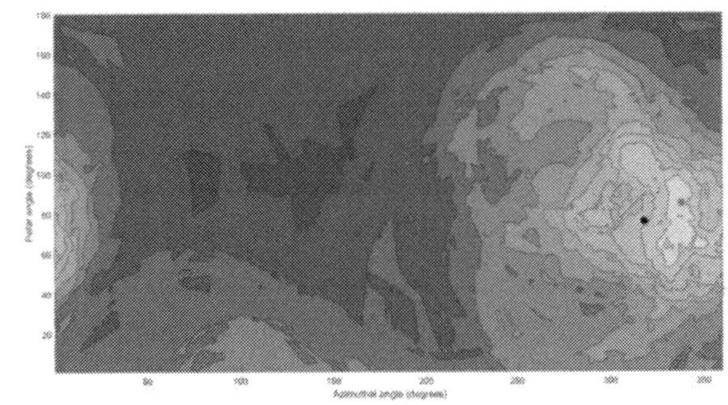

Fig. 13. Improved neutron image of a point source located at (338, 85). The initial guess vector points towards (318,76). The image was made with an initial energy deposition cutoff of 0.5 MeV, a second scatter criterion prominence ratio of 6.5, and 48 cones. The guess vector region was set to 60° azimuthal width and 20° polar height.

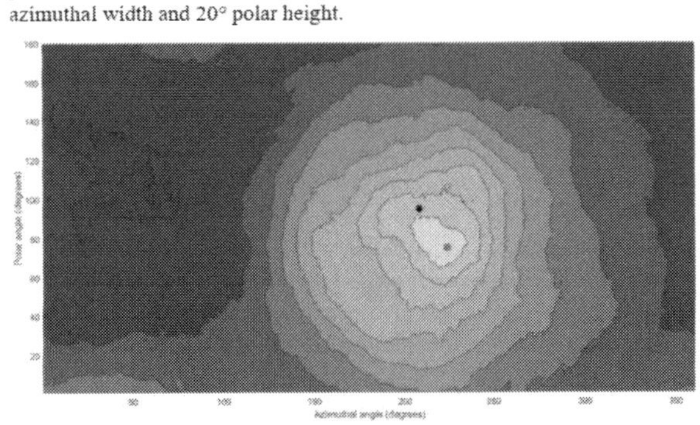

Fig. 14. Improved neutron image of a point source located at (224, 75). The

https://doi.org/10.1051/epjconf/202125307008

initial guess vector points towards (208, 95). The image was made with an initial energy deposition cutoff of 0 MeV, a second scatter criterion prominence ratio of 4, and 654 cones. The guess vector region was set to 60° azimuthal width and 20° polar height.

B. Camera Performance

The imaging resolution of a camera-type radiation detector can be assessed through source position cross-cutting. The camera system does not have a single imaging resolution but rather a range of possible resolutions that vary with energy deposition cutoffs, second scatter criterion, guess vector down-selection use, and guess vector down-selection region size. To assess this range of resolutions, the data from the point source measured at (18.3, 224, 75) was iteratively run through the neutron image creation portion of the script while varying both the energy deposition cutoff and the second scatter criterion. A series of images corresponding to each cutoff value was saved for both the raw images and improved images (those that used guess vector down-selection). The resolution of each image was then determined by crosscutting each of the images in the polar and azimuthal directions at the known source location. In many cases, the peak of these crosscuts did not correspond to the true source direction. In these cases, the azimuthal and polar error between the true source position and the images' maximum probability pixel are recorded. Table II lists the resolution values, localization errors, and number of probability cones for a range of improved neutron images. In cases where multiple probability hotspots appeared along a crosscut, the peak corresponding to the hotspot nearest to the true source position was used for resolution and error determination. Images were determined to have "no convergence" if a hotspot of at least 80% of the probability of the hottest projection bin was not within 60 degrees of the true source position.

TABLE II
CAMERA RESOLUTION AND CONVERGENCE ERRORS

Peak prominence ratio cutoff	First scatter energy deposition cutoff (MeV) 0	0.2	0.4	0.6
5	(44,1,38,9,413)	(49,16,41,4,183)	(54,33,41,17,55)	(30,30,20,13,16)
4.5	(42,13,40,5,289)	(39,20,39,11,116)	(56,18,29,14,33)	(38,9,26,13,12)
4	(31,20,43,15,176)	(43,16,29,3,87)	(44,23,29,17,25)	No convergence
3.5	(31,1,31,3,104)	(30,22,24,3,42)	No convergence	No convergence
3	(43,23,34,10,67)	(40,23,31,16,30)	(31,8,25,7,9)	No convergence
2.5	(54,22,37,6,35)	(37,19,23,17,17)	No convergence	No convergence
2	(30,3,45,2,23)	No convergence	No convergence	No convergence
1.5	(20,3,23,10,13)	No convergence	No convergence	No convergence

Resolutions, convergence position errors, and number of cones for raw neutron images generated at a range of energy deposition and peak prominence ratio cutoffs. The data in each cell is reported in the form: (azimuthal resolution, azimuthal error, polar resolution, polar error, number of cones). The resolutions and errors are reported in degrees.

Like the resolution determination, the camera system will not have a single double scatter efficiency but rather a range of achievable efficiencies determined by the cutoff values used in determination of which pulse sestets are used in cone generation.

To test the camera system's useful cone efficiency, a measurement of 12 million counts generated from the two ^{252}Cf point sources together was run iteratively over a range of cutoffs. The sources were position together and shielded with three lead bricks to simulate a "pure" neutron source free from non-background gamma rays. The datafile was iteratively run through section 2 of the double pulse identification algorithm, with either the peak prominence ratio cutoff or the first scatter energy deposition cutoff changed at each step. At each iteration, the useful cone efficiency for those settings was determined by divided the double scatter candidate identified by the total counts in the datafile. Figure 15 shows a series of useful cone efficiency curves at different cutoff values. As the first scatter energy depostion cutoff value is increased, the useful cone efficency decreases in a log linear fashion across the range of peak prominence ratio cutoffs. These curves can be used to guide the selection of cutoff values for a desired count rate or measurment time as needed.

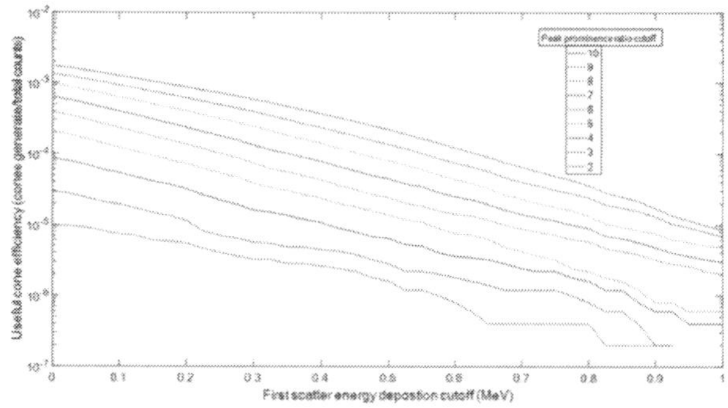

Fig. 15. Useful cone efficiency curves for the camera system prototype.

IV. CONCLUSIONS

The prototype simplified neutron scatter camera described in the proceeding sections has shown promising results for the localization of neutron sources given measurement times on the order of several hours. Based on the data shown in the proceeding sections and observations made during data analysis, the following is a non-exhaustive list of the successes of the simplified neutron scatter camera prototype:

- The combination of a short light pulse scintillator material with sub-nanosecond digitizer timing allows the system to resolve scattering events just nanoseconds apart across six PMTs simultaneously.
- Ratios of the amount of fluorescence light arriving at opposite PMTs coupled to a single volume can be used to reconstruct neutron scatter positions within the scintillator volume with errors on the order of millimeters. No use of pixelated photomultipliers or optical segmentation is necessary for position reconstruction.
- Simulation scaling studies show that scatter position reconstruction using the light ratios method functions at variety of scintillator cube volumes, making it possible to construct detectors with this design with a range of sizes.
- The algorithm developed to identify neutron double scatter candidates based on analysis of pulse characteristics was successful. Several different criteria for second scatter identification showed some successes, with a comparison of first and second peak prominence empirically the most successful.
- Double scatter candidate pulses successfully localize neutron point and volume sources over a range of spatial positions, and at a variety of energy and double scatter criterion cutoffs.

- Guess vector probability cone down-selection improves neutron images by removing erroneous cones, leading to images with better resolution and fewer incorrect localization positions.
- Guess vector method can give users of the simplified NSC a general idea of an unknown source direction before any cone back-projection is performed.
- The entire system is significantly less expensive per unit that other designs for neutron scatter cameras due to the use of fewer photomultipliers and a single unsegmented scintillator volume.

With further refinement of the double scatter identification algorithm, optimizing of the detector volume and photomultiplier size, explorations of different scintillator materials, reflectors, and photomultipliers, and the honing of the imaging down-selection method, a detector using the Simplified Neutron Scatter Camera design could become a valuable tool for international nuclear safeguards.

References

[1] International Atomic Energy Agency. "Basics of IAEA Safeguards," [Online] Available: https://www.iaea.org/topics/basics-of-iaea-safeguards. Accessed on" June 30, 2021.

[2] J.E. Doyle. "Introduction: Nuclear Security and Global Challenges," in *Nuclear Safeguards, Security, and Nonproliferation,* 1st ed. Burlington, MA, USA: Butterworth-Heinmann, 2008.

[3] R.C. Runkle, A. Bernstein, and P.E. Vanier, "Securing Special Nuclear Material: Recent Advances in Neutron Detection and Their Role in Nonproliferation," Journal of Applied Physics, vol. 108, no. 11, 2010, 10.1063/1.3503495.

[4] D. Xu, Z. He, C.E. Lehner, F. Zhang. (2004, August). 4-pi Compton Imaging with Single 3D Position Sensitive CdZnTe Detector. Presented at SPIE 5540, Hard X-Ray and Gamma-Ray Detector Physics VI. [Online]. Available: doi.org/10.1117/12.563905. Accessed on: June 30, 2021.

[5] N. Mascarenhas, J. S. Brennan, and K. Krenz. (2006, October). Development of a Neutron Scatter Camera for Fission Neutrons. Presented at: 2006 IEEE Nuclear Science Symposium. [Online]. Available: https://www.semanticscholar.org/paper/Development-of-a-Neutron-Scatter-Camera-for-Fission-Mascarenhas-Brennan/a3544f4fbce0f4c048d11c2577ad2ab3d0ddb0e7. Accessed on: June 30, 2021.

[6] J. E. M. Goldsmith, M.D. Gerling, J.S. Brennan, "A compact neutron scatter camera for field deployment," Review of Scientific Instruments, Vol. 87, August 2016, DOI. 10.1063/1.4961111.

[7] K. Weinfurther, J. Mattingly, E. Brubaker, J. Steele, "Model-Based Design Evaluation of a Compact, High-Efficiency Neutron Scatter Camera," Nuclear Instruments and Methods in Physics Research Section A, Vol. 883, pp. 115-135. March 2018. DOI. 10.1016/j.nima.2017.11.025.

[8] J.J. Manfredi, B.L. Goldblum, T.A. Laplace, G. Gabella, J. Gordon, A. O'Brien, S. Chowdhury, J.A. Brown, and E. Brubaker, "Proton Light Yield of Fast Plastic Scintillators for Neutron Imaging," IEEE Transactions in Nuclear Science, vol. 67, no. 2, pp. 434-442, December 2019. DOI. 10.1109/TNS.2019.2959979.

EPJ Web of Conferences **253**, 07009 (2021) https://doi.org/10.1051/epjconf/202125307009
ANIMMA 2021

μRANIA-V: an innovative solution for neutron detection in homeland security

R. Farinelli,[a,*] I. Balossino,[a,b] G. Bencivenni,[c] G. Cibinetto,[a] G. Felici,[c] S. Fiore,[d] I. Garzia,[a] M. Gatta,[c] M. Giovannetti,[c], R. Hall-Wilton,[e] C. C. Lai,[e] L. Lavezzi,[f] G. Mezzadri,[a,b] G. Morello,[c] E. Paoletti,[c] G. Papalino,[c] A. Pietropaolo,[d] M. Pillon,[d] M. Poli Lener,[c] L. Robinson,[e] M. Scodeggio,[a] P.O. Svensson.[e]

[a] INFN-Ferrara, via Saragat 1, 44122 Ferrara, Italy
[b] Institute of High Energy Physics, Chinese Academy of Sciences, 19B Yuquan Road, 100049 Beijing, China
[c] INFN-National Laboratory of Frascati, via Enrico Fermi 54, 00044 Frascati (Roma) Italy
[d] ENEA, via Enrico Fermi 45, 00044 Frascati (Roma), Italy
[e] Detector Group, European Spallation Source ERIC (ESS) Box 176, SE-221 00 Lund, Sweden
[f] INFN-Torino, via P.Giuria 1, 10125 Torino, Italy
[*] Corresponding author

rfarinelli@fe.infn.it

Abstract—**Detection of neutrons is becoming of the utmost importance, especially in the studies of radioactive waste and in homeland security applications. The crisis of ^{3}He availability has required the development of innovative techniques. One solution is to develop light gas detectors for neutron counting to be used as portals for ports and airports. The neutron is converted on the Boron-coated cathode, releasing a charged particle, whose passage can be identified by the gas detector. While several technologies have been deployed in the past, the project μRANIA-V (μRwell Advanced Neutron Identification Apparatus) aims to detect thermal neutrons by means of the μRwell technology, an innovative gas detector. The goal is to produce tiles to operate as portals in homeland security or for radioactive waste management. The technological transfer towards the industry has started, thus the production can be cost-effective also owing to a construction process relatively easier compared to similar apparatus. By reading directly the signals from the amplification stage, the neutrons can be counted with simplified electronics further reducing the total cost. In this paper, the project will be described, with details on the μRwell technology and on the neutron counting, on the test beam performed, and on the future plans.**

Keywords — **MPGD, neutron detection, radioactive waste, homeland security.**

I. INTRODUCTION

Neutrons are subatomic particles with neutral electric charge and a mass slightly greater than that of a proton. Neutrons have high penetration power and together with neutron detectors are used to probe the structure and the motion of the matter in a complementary way to X-ray imaging [1]. Neutrons play an important role with radioactive materials: there a neutron detector can be used as a monitor for those materials or even as a portal monitor for homeland security. Neutrons are involved in nuclear transmutations such as the nuclear fission and free neutrons cause ionizing radiation with biological hazard, depending on dose. On Earth, a natural neutron background flux exists caused by cosmic ray showers and natural radioactivity. Free neutrons can be produced by neutron sources like neutron generators, research reactors and spallation sources [2-3]. In literature, neutron detection is categorized according to the nuclear processes, mainly neutron capture and elastic scattering (not threatened in this paper). The neutron detection happens by means of the neutron caption reactions into electrical signals through particles and energy released. Nuclides such as ^{3}He, ^{6}Li, ^{10}B and other heaviers like ^{235}U have a high neutron capture cross section and a larger probability of absorbing a neutron, as reported in Table 1.

Reaction	Q-value [MeV]	Cross section for thermal neutrons [barns]
$^{10}B + n \rightarrow {}^{7}Li + \alpha$	2.31	3840
$^{6}Li + n \rightarrow {}^{3}H + \alpha$	4.78	940
$^{3}He + n + {}^{3}H + p$	0.754	5330
$^{235}U + n + X + Y$	~200	575

Table 1. Reactions of interest involving neutron capture, the released energy (Q-value) and their cross-section for thermal neutrons are shown, where *n* represents a neutron, *p* a proton, α an alpha particle and *X/Y* the fission fragments.

The cross section of the neutron capture strictly depends on the neutron energy[4]: lower cross sections are measured for fast neutrons and higher values are observed for the slow ones. Calorimeters are used in the high energy regime where the neutron is totally absorbed and the deposited energy is measured; scattering with protons is exploited in a moderate energy regime; while in the low energy range mediums with large capture cross-section are used. A well known converter used for neutron detection is the ^{3}He due to its large cross-section (5330 at 25 meV) but in the latest years a shortage of this element increased the ^{3}He cost with a large impact on the neutron detector production and their application in the fields of interest. ^{3}He is not extracted by natural resources but it is of industrial origin, precisely stemming from the decay of Tritium. In the Cold War a large amount of Tritium was accumulated but at present days the increasing interest in neutron detection and the reduced amount of ^{3}He in the storage addressed its shortage. A call for alternative solutions triggered the URANIA-V project[5].

II. Boron-coated gaseous detectors

Thanks to the neutron capture process, the neutron detection can be achieved with gaseous detectors through a thin surface of a converter: the reaction products are generated on the converter. Once they interact with the gas, the ionization path is amplified by a Micro Pattern Gaseous Detector (MPGD). The MPGD technology exploits the modern photolithography and thin layer polyimide depositions to develop a new design to amplify the ionization charge and to readout the signal. The granularity of the detector improves the intrinsic rate capability and the spatial resolution below to 50-100 μm. Triple-GEM, MicroMegas and μRWELL are examples of MPDG of success used for the particle detection; within the URANIA-V project the neutron detection is performed with a μRWELL-based detector.

The μRWELL is an innovative MPGD that combines the best features of the GEM amplification stage with a resistive layer such as in the MicroMegas detector. This led to a compact, spark-protected, resistive MPGD with a single amplification stage. The μRWELL detector is composed of two elements, cathode and the readout-PCB, to simplify its production process and to reduce the gap with the technological transfer.

The cathode is based on a Cu layer on a glass epoxy plate. Applications for neutron detection need a B_4C deposition of a few μm to convert the thermal neutron into α or ^{7}Li particles. The readout-PCB combines the anode, the amplification stage and the segmented readout. A WELL patterned Apical foil with a thin copper layer is used to amplify the ionization signal up to a gain of 10^4. A Diamond-Like-Carbon (DLC) layer of resistivity of about 80-100 MΩ/❐ is used to quench the sparks and it does not affect the detector gain up to 1-10 MHz/cm^2. The detector reaches an efficiency of 98%, a spatial resolution of about 50 μm and a time resolution of 5 ns. Those performances match with the purpose of the project to develop a nuclear waste monitor, a homeland security portal and a neutron imaging detector.

III. Neutron converter designs

The chosen converter in the URANIA project is the Boron. A B_4C deposition on the cathode with sputtering technique was studied together with the colleagues of the ESS Coating Workshop in Linkoping, Sweden. A *planar converter* layer was used for the first design of a neutron detector together with the μRWELL detector. The B_4C thickness is the first parameter to optimize the neutron detection efficiency: large amounts of Boron increase the conversion efficiency but as the thickness increases too much, then the reaction production (α and ^{7}Li) can not enter into the gas volume, reducing the detection efficiency. The detection efficiency depends also on the neutron impinging angle with respect to the B_4C surface: as long as the angle goes from a orthogonal to parallel then the neutron path inside the Boron layer is larger and it increases the conversion efficiency. The planar cathode design is the simpler one to be assembled. The performance of this converter was studied through simulation and experimental measurements as described in the next sections.

A second cathode design was studied to increase the detection efficiency of the detector. The neutron detection depends on their impinging angle then a *grooved cathode* was designed to develop a compact histrument with one cathode and a readout-PCB. Machining operation shapes the cathode with peaks of about 2 mm and a spacing of 0.5-1.5 mm then a B_4C sputtering procedure is performed.

In addition, a third cathode design is under study to improve the detection efficiency. By means of a *mesh* between the cathode and the readout-PCB it is possible to introduce another electrodes within the active volume and with a proper electrical configuration it is possible to collect the primary ionization electron through the mesh up to the readout while the mesh itself acts as another neutron conversion stage thanks to a B_4C layer sputtered on it. A mesh with a wire of 53 μm and a pitch of 127 μm was chosen for the measurements to prove the concept.

The second and the third designs are more complex because they have to take into account the geometry for the mesh or the grooved cathode to optimize the neutron conversion and the electron transparency of the detector to collect as much signal as possible. The three detector designs are summarized in Fig. 1.

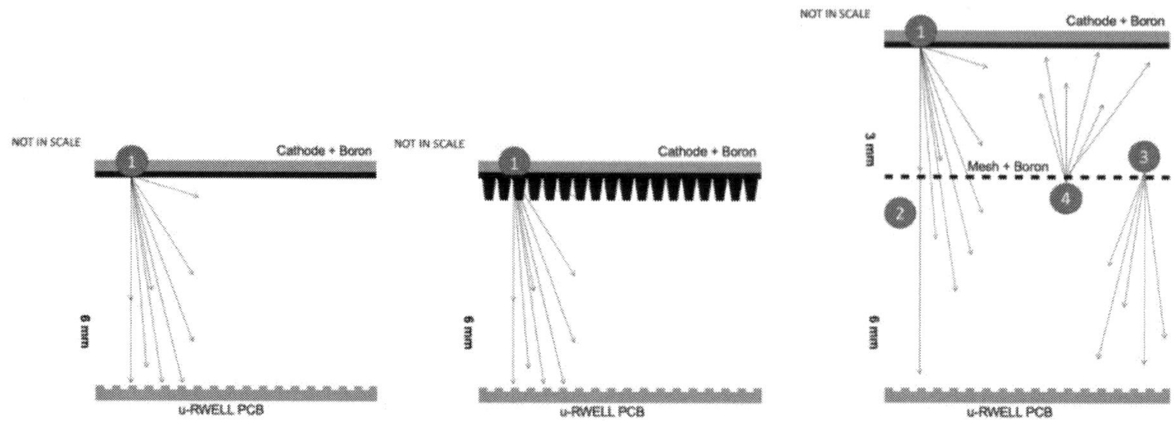

Fig. 1. Summary of the three detector designs under study in the URANIA-V project. From the left ot the right: planar cathode, grooved cathode and planar cathode with a mesh. Cathode and mesh are boron-coated. The numbers from 1 to 4 summarize the different processes.

IV. Simulations

The first tool to approach neutron detection is simulation. Nowadays, computer software can accurately simulate the particles interaction through gas and matter, i.e. neutrons, alpha, lithium and electrons. The main simulators used are Geant4 and Garfield++ [6,7]. The first one is a toolkit for the simulation of the passage of particles through matter. The second is a toolkit for the detailed simulation of particle detectors based on ionisation measurement in gases and semiconductors. These tools allow us to measure the interaction probability of the neutron with the Boron layer and the other detector material that may participate in the interactions with the neutrons. Once the neutron capture happens, then the reaction products are studied to determine their path in the gas mixture, the active volume of the μRWELL detector. Once the primary ionization takes place, the electrons have to be properly collected on the readout plane: simulations can provide detailed knowledge of the processes. The first study is focussed on the optimization of the B_4C thickness in the conversion layer sputtered on a *planar cathode*: by means of thermal neutron impinging orthogonally on the layer, several thickness is studied and the amount of α and ^7Li particles reaching the gas volume is measured, as shown in Fig. 2. Energy distributions are described by the two reactions involved in (1).

$$n + {}^{10}B \rightarrow {}^{11}B^* \rightarrow \alpha + {}^{7}Li\ (6\%)$$
$$n + {}^{10}B \rightarrow {}^{11}B^* \rightarrow \alpha + {}^{7}Li + 477\ keV\ \gamma(94\%) \qquad (1)$$

No processes observe α and ^7Li at the same time because their emission is almost back to back, therefore only one enters the gas volume. The production of γ ray particles through the detector is much higher with respect to the charged particles but their impact is negligible up to detector gain of 10^4.

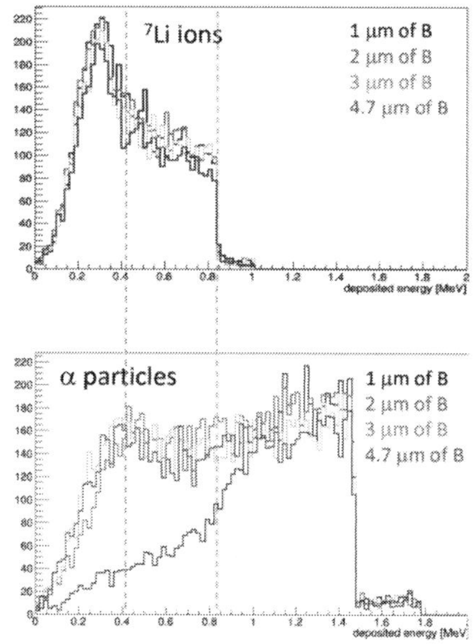

Fig. 2. Energy released for different Boron thickness for ^7Li (top) and α (bottom) in four different simulations of 1M thermal neutrons impinging on the Boron layer.

The planar converter shows its maximum neutron conversion between 2 and 3 μm B_4C thickness.

The second design under study is the one with the *grooved cathode*. This design has to maximize the neutron conversion with sharp spikes and a small spacing from one side and the electron extraction from the depth from the other side. The first prototype of the grooved cathode has depths of 2.5 mm, a slope of 10° and a spacing of 0.5 mm. The Boron sputtering thickness of this layer is 2.5 μm. The simulation of the deposited energy as a function of the production region (spacing, slope and peak) is shown in Fig. 3: the differences in

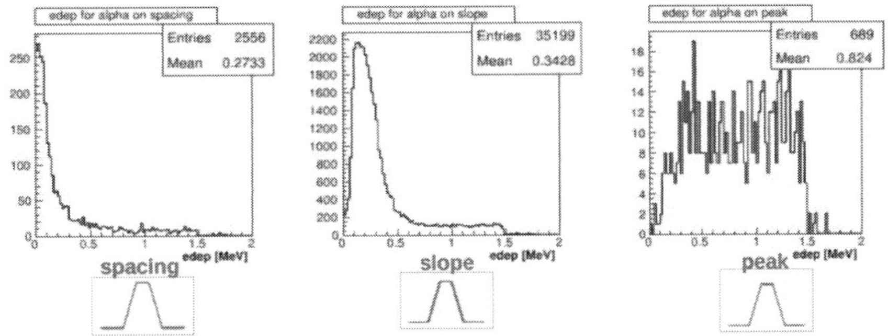

Fig. 3. Deposited energy distribution of α particles produced by 1M thermal neutrons on the grooved cathode for three different regions: spacing, slope and peak. Similar distribution can be found for ^{7}Li particles too.

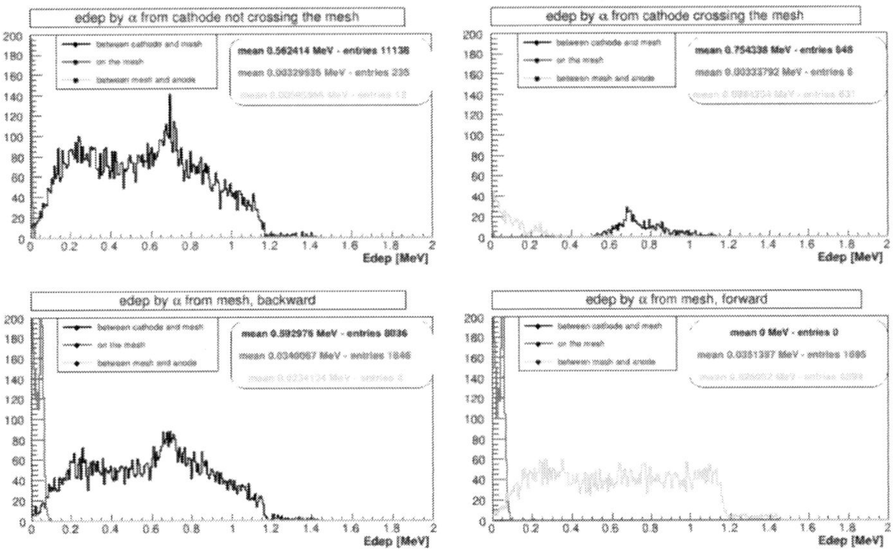

Fig. 4. Deposited energy distributions for α particles produced on a μRWELL having a boron-coated cathode and a boron-coated mesh. The distributions are separated in regions to define the amount of expected primary electrons as a function of the ionization regions. Four different conditions were simulated: α particles from the cathode not crossing the mesh, α particles from the cathode crossing the mesh, α from the mesh to the cathode and α from the mesh to the readout.

the distributions are related to the path length of the particles and to the cathode geometry. Outside the peak regions, the probability to impact the next peak is higher and this reduces the particle path, therefore the deposited energy.

The third design under study is the one with *boron-coated mesh*. The one used in the measurements has an optical transparency of 33% and with a proper electrical configuration it is possible to extract the largest number of electrons generated in the ionization. The contribution of the neutron conversion coming from the cathode and the mesh are measured separately, as a function of the α and ^{7}Li directions, as shown in Fig. 4.

The simulations described above are used to measure the neutron conversion probability as a function of the Boron thickness and the effects related to the involved geometry (planar, grooved or mesh). This information together with the detector settings will provide the expected performance of the μRWELL in the experimental measurement proposed in the next sections.

V. Experimental setup

A preliminary characterization of the prototypes was done at the ENEA HOTNES in Frascati [8]. The source is a calibrated ^{241}Am-B thermal neutron source placed in a cylindrical cavity delimited by polyethylene walls. HOTNES exploits a polyethylene shadow bar that prevents fast neutrons from directly reaching the samples. The effect of the shadow bar and of the cavity walls combined in such a way is that the thermal neutron fluency is nearly uniform. The resulting fluence rate at the HOTNES reference irradiation plane is

758±16 $cm^{-2}s^{-1}$. Experimental measurements and simulations evaluate the neutron fraction absorbed by the mechanical structure of the detector: a fraction of 17.7% is removed from the neutron fluence rate. A representation of the HOTNES setup is shown in Fig. 5.

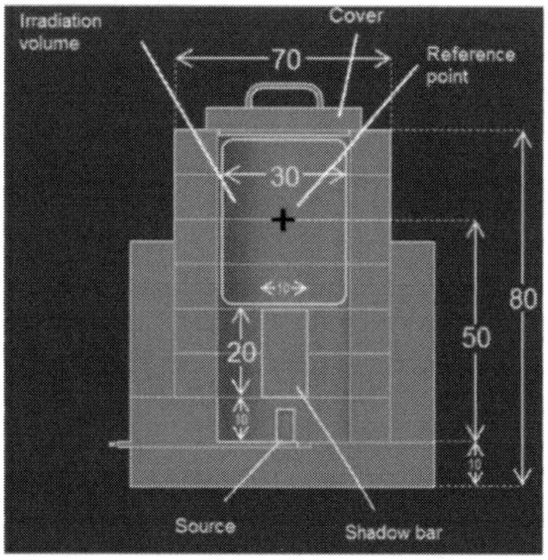

Fig. 5. Representation of the HOTNES setup with the radioactive source and the shadow bar in the middle The polyethylene blocks the neutrons to remove any hazard.

The energy distribution of the neutron generated by HOTNES is not centered around 25 meV but it is slightly higher. The energy peak is around 100 meV and the FWHM is around 290 meV. This has an impact on the experimental and the simulation results because the neutron cross section depends mainly on their energy: higher energy leads to smaller efficiency. A factor two of reduction is expected with the HOTNES setup with respect to the one evaluated with thermal neutrons.

VI. Measurement techniques

Neutrons release in the gas volume a large number of electrons and a proper amplification can be used to increase it up to several hundreds of fC. The two measures used in this paper exploit the same signal with different techniques: the *current mode* and the *counting mode*. Thanks to an A1561HDM CAEN power supply with a 50-100 pA sensitivity current monitor, it is possible to measure the neutron detection efficiency. The current flowing to the resistive layer of the μRWELL is proportional to the electrons released in the gas and the detector gain as shown in (2).

$$i = \Phi * \varepsilon * N * G * S \quad (2)$$

where i is the current, Φ is the neutron flux at the reference point, ε is the neutron conversion efficiency, N is the average number of electrons coming from the ionization (contribution evaluated from simulation), G is the detector gain and S is the detector surface.

The setup with the mesh has four different contributions as shown in Fig. 1 and 4. The proper electron transparency (evaluated from simulation) has to be taken into account as shown in (3).

$$\begin{aligned} i_1 &= \Phi * \varepsilon_1 * N_1 * G * S * T_{elec} \\ i_2 &= \Phi * \varepsilon_2 * N_2 * G * S \\ i_3 &= \Phi * \varepsilon_3 * N_3 * G * S \\ i_4 &= \Phi * \varepsilon_4 * N_4 * G * S * T_{elec} \end{aligned} \quad (3)$$

where T_{elec} is the electron transparency and the cases from 1 to 4 are the different production layers inside the detector as shown in Fig. 1 right. A detector gain calibration was made using an X-ray gun, covering the operating range of the measurements (gain from 300 to 700).

The other measurement technique is the counting mode developed with several custom electronic boards based on CAEN A1422 or CREMAT CR-110 to cope with the signal produced by a μRWELL and α or ^{7}Li particles, as shown in Table 2. This electronics readout has to read a signal of about hundreds pC from all the strips of the μRWELL. Several tests were performed to optimize the electronics with the detector before the source test by means of cosmic rays and Strontium radioactive source.

Pre-amplifier chip	CAEN A1422	CREMAT CR-110
gain	0.2 mV/fC	0.5 mV/fC
noise	20 mV	20 mV
signal duration	5 μs	5 μs

Table 2. Custom electronics' main features for the counting mode are listed.

Each signal is then acquired with a Picoscope digitizer. The neutron flux is measured directly and the ratio with the nominal values returns the neutron detection efficiency.

VII. Results

The neutron efficiency is evaluated in the current mode by means of simulation results. This technique is very simple and it does not need any electronics. The current expected from simulation is about a few nA, depending on the gain and the design chosen. The experimental neutron efficiency evaluated with the planar cathode design shows a variation as a function of the Boron thickness. The first measurement, reported

shown in Fig. 6, compares the experimental results using (2) with the simulations[5].

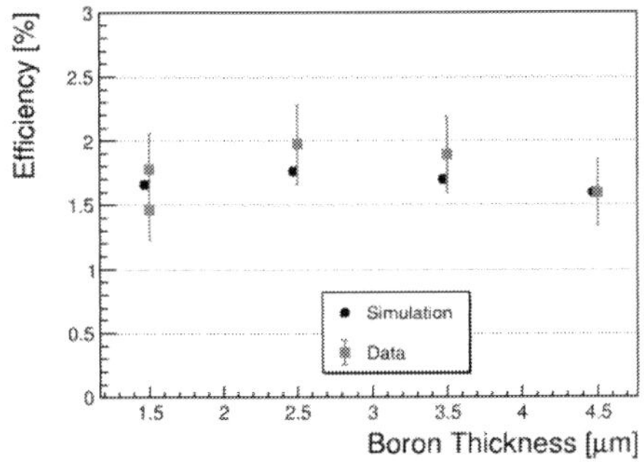

Fig. 6. Neutron detection efficiency of a µRWELL detector with planar boron-coated cathode as a function of the boron thickness is shown.

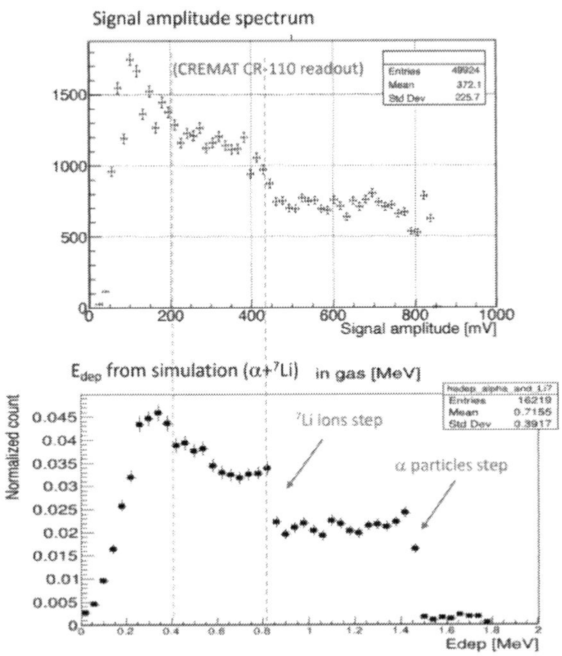

Fig. 7 Signal amplitude spectrum from experimental results (top) and deposited energy from α and ^{7}Li together from simulations (bottom) are shown. Dotted lines are drawn to connect similar shapes of the two connected quantities. Both plots have been evaluated with the planar cathode design.

The current mode measurement was applied also to the mesh design, where an efficiency of 4.6 ± 1.0 was measured using (3). Systematic measurements were performed to validate the current mode: a µRWELL without a coated cathode was used to measure the current measurement contribution from the HV. No current contributions above 0.1 nA have been found up to a gain of 1000 both inside and outside HOTNES. A contribution up to 1 nA has been found at a gain of 10^4 inside HOTNES probably due to the photon's contribution. All the measurements in current mode have been performed with a gain between 300 and 700.

The counting mode measurements were performed for the planar design and the grooved cathode. These measurements are independent from the simulation and any gain calibration. An efficiency of 2.4 ± 0.1 has been measured for the planar cathode and a value of 3.2 ± 0.1 for the grooved cathode. Together with the neutron rate, thanks to the Picoscope it is possible to acquire the induced signal then its amplitude. The signal amplitude distribution with a planar cathode is compared with the one from simulations and the two shapes show similar behaviors, as shown in Fig. 7.

The systematic errors in the counting mode depend on the threshold chosen and the electronic dead-time. A threshold scan has been performed to properly set the threshold above the noise level; while from the electronics side a 20 µs dead time has been chosen to reject any other contribution due to noise correlated to signal contributions.

VIII. Conclusions

A neutron detector based on µRWELL technology was designed, manufactured and tested. Coupling a Boron converted to an MPGD makes it possible to develop a low cost and a high precision detector. Radioactive waste monitors and national security portals can be developed based on this technology with the ongoing technological transfer for large area detector construction. In this paper several designs have been studied by means of a neutron source with a neutron conversion efficiency ranging from 2 to 4.6% with the HOTNES neutron energy distribution. A 10% neutron conversion efficiency is expected with a combination of the different designs and thermal neutrons at 25 meV. No contribution comes from the photons' background. The project's next step will be focused on the finalization of the detector design and the electronics for the counting mode.

References

[1] J. Banhart et al., "X-ray and neutron imaging – Complementary techniques for materials science and engineering". International Journal of Materials Research, vol. 101, no. 9, 2010, pp. 1069-1079.

[2] Reijonen, J. "Compact Neutron Generators for Medical, Homeland Security, and Planetary Exploration". Proceedings of 2005 Particle Accelerator Conference, Knoxville, Tennessee: 49-53.

[3] The Unique Capabilities of ESS: https://europeanspallationsource.se/unique-capabilities-ess

[4] R. Bauer et al., "Application of simple ramsauer model to neutron total cross sections" United States. https://www.osti.gov/servlets/purl/641282.

[5] I. Balossino et al., "u-RANIA: a neutron detector based on µ-RWELL technology" 2020 JINST 15 C09029

[6] J. Allison, et al., Recent developments in Geant4, Nucl.Instr. and Meth. A 835 (2016), 186-225

[7] R. Veenhof, Garfield, recent developments, Nucl. Instr. and Meth. A 419 (1998), 726-730

[8] A. Sperduti, et al., Results of the first user program on the Homogeneous Thermal Neutron Source HOTNES (ENEA/INFN), JINST_12_P12029 (2017)

Timepix3 detector network for nuclear waste monitoring

Bartolomej Biskup[1], Benedikt Bergmann[1], Pavel Broulim[2], Petr Burian[1,2], Milan Malich[1], Petr Manek[1,3], Lukas Meduna[1,2], Yesid Mora[1], Martin Pichotka[1,4], Lukas Pusman[2], Peter Rubovic[1], Tomas Slavicek[1], and Petr Smolyanskiy[1]
[1]Institute of Experimental and Applied Physics, Czech Technical University in Prague, Czech Republic
[2]Faculty of Electrical Engineering, University of West Bohemia, Czech Republic
[3]Department of Physics and Astronomy, University College London, United Kingdom
[4]Clinic for Nuclear Medicine, University hospital Freiburg, Germany
Bartolomej.Biskup@utef.cvut.cz

Abstract— **In the framework of the MICADO project we proposed an active long-term nuclear waste monitoring system based on Timepix3 technology. Particle detectors based on Timepix3 are capable of particle type discrimination and provide spectrometric information in combination with precise timing information. To maximize the versatility of the setup, the Timepix3 detectors employed are equipped with a range of different sensors. To enable measurement of neutrons, two quadrants of the detectors are covered by neutron converters, allowing detection of thermal neutron by capture in ^{6}Li, as well as detection of fast neutrons through recoil protons generated in a polyethylene converter. Within this project a new housing for the detector unit and the readout was developed, with the network currently comprising of 9 such detectors. A specifically developed in-house software handles simultaneous operation of all devices, cluster analysis and data visualization. Energy calibration was performed using an X-ray tube with several fluorescence foils as well as photons from an ^{241}Am source. The neutron and gamma/X-ray detection efficiency calibration was carried out at the Czech Metrology Institute and the results are presented here. Tests on simultaneous operation of all 9 devices were performed using uraninite and ^{241}Am sources.**

Keywords **—Timepix3 detector, nuclear waste monitoring, neutron detection.**

I. Introduction

THE Horizon 2020 funded MICADO [1] project (Measurement and Instrumentation for Cleaning and Decommissioning Operations) aims at proposing a cost-effective solution for non-destructive characterization of nuclear waste, implementing a digitization process that could become a referenced standard facilitating and harmonizing the methodology used for the in-field Waste Management and Dismantling & Decommissioning operations.

Distributed networks of hybrid semiconductor pixelated detectors (HPDs) have proven their efficiency and robustness in characterization of mixed radiation fields in various long-term investigations. In particular the ATLAS TPX network [2,3] has provided valuable data on particle flux within the ATLAS cavity at LHC since 2015. In 2018 this network, which was originally based on the Timepix1 ASIC, has been amended by 4 Timpix3 [4,5] devices, resulting in improved particle discrimination capabilities as well as improved energetic and temporal resolution.

For this reason, in the framework of the MICADO project, we propose a long-term nuclear waste monitoring system based on Timepix3 technology as an alternative/complementary solution to the conventionally used detectors.

II. Timepix3 network

Timepix3 HPDs are capable of efficient particle discrimination and provide spectrometric information, in parallel with nano-second resolution timing information. The active area of the sensor of 1.98 cm^2 is formed by 256 × 256 pixels with a pixel pitch of 55 μm.

The particle flux encountered by these detectors, depending on nuclear waste composition, volume, geometry and the distance between the waste packages and detectors, is expected to cover a challenging range and to be composed of a mixed radiation field of ionizing and non-ionizing radiation. Moreover, in some cases, part of the detectors will be attached directly to the waste drums, while others are fixed to the storage walls. To maximize the versatility of the setup, the Timepix3 detectors employed are equipped with a range of different sensors, namely Si sensors with thicknesses of 100, 300, and 500 μm, as well as CdTe sensors with a thickness of 1 mm. The network currently comprises of 9 detectors.

To enable measurement of neutrons, two quadrants of the silicon detectors are covered by neutron converters. Thermal and epithermal neutrons are detected below a ^{6}LiF foil through products of a ^{6}Li(n,α)^{3}H reaction and fast neutrons are detected through recoil protons under a polyethylene (PE) layer. In addition, an aluminum foil covering half of the sensor acts as an energy threshold. A schematic drawing of the converter structure together with the sensor, bump-bonds, and readout chip is shown in Fig. 1a. Since cadmium has a significant thermal

 https://doi.org/10.1051/epjconf/202125307010

neutron capture cross section, no convertors were used in CdTe equipped detectors. Katherine [6], an Ethernet Embedded Readout Interface for Timepix3, allows to operate our detectors via UDP protocol in distributed networks using standard networking hardware.

Within the MICADO project, a new housing for the readout and detector unit (Fig. 1b.), optimized towards this specific task, was developed.

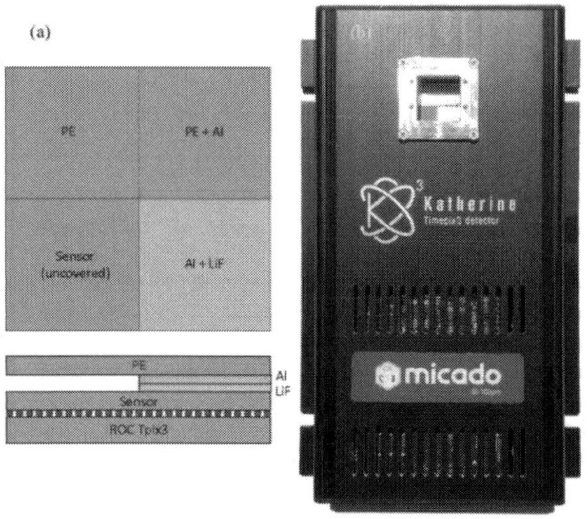

Fig. 1. (a) Structure of the converter (top and side view). (b) Housing of the Timepix3 detector and Katherine readout with the sensor window and the filter-holder in the upper part. Housing dimensions are 13 cm × 20.4 cm × 3 cm.

A new plugin-based software, MM Track Lab, was developed in-house for data acquisition and processing. A specific MICADO detector network plugin handles simultaneous operation of multiple devices. Clustering of pixels, event analysis and data visualization (as visible in Fig. 2) is performed in real-time. An automated warning system, giving detailed characteristics of the potential nuclear hazard in case certain thresholds are surpassed, is integrated here as well.

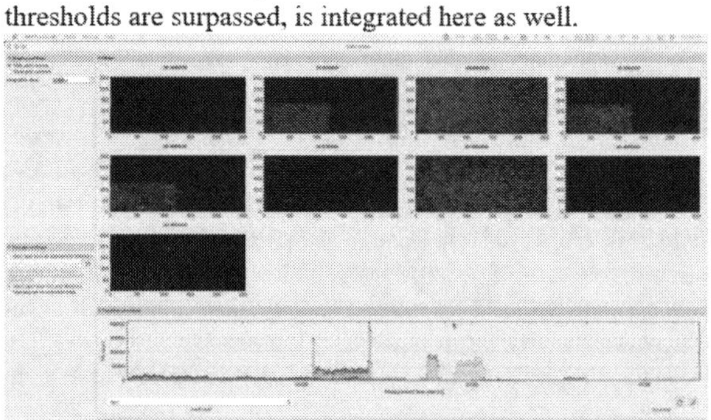

Fig. 2. Graphical user interface of the MM Track Lab MICADO detector network plugin. Individual matrices for all nine detectors measuring simultaneously are in the top part and the common timeline represents the number of events.

III. CALIBRATION AND MEASUREMENTS

A pixel-by-pixel energy calibration was performed using an X-ray tube with several fluorescence foils as well as photons from an ^{241}Am source. An ^{241}Am source was used also for the time-walk calibration [7].

The neutron and gamma/X-ray detection efficiency calibration was carried out at the Czech Metrology Institute [8] using well defined radiation sources and reference fields.

Thermal neutrons were measured for each detector separately using neutrons from a ^{238}PuBe source, moderated in a graphite pile. Fluence at the detector position was $(2.92{\cdot}10^4 \pm 1.80\%)$ cm$^{-2}{\cdot}$s^{-1}, corresponding to an ambient dose equivalent (*H*(10)*) of $(1.75 \pm 1.80\%)$ mSv/h. Sufficient statistics was reached for a measurement duration of 10 min/detector for Si sensors and 5 min for CdTe. For a fast neutron calibration, ^{252}Cf (ISO 8529-1) and ^{241}AmBe (ISO 8529-1) reference sources were used. A four-detector holder visible in Fig. 3 was used for this and most of the following measurements for efficiency. The ^{241}AmBe source was shielded by a lead capsule to minimize gamma background. Neutron fluences at the holder central positions were $(4.49{\cdot}10^3 \pm 1.30\%)$ cm$^{-2}{\cdot}$s^{-1} and $(4.32{\cdot}10^3 \pm 1.50\%)$ cm$^{-2}{\cdot}$s^{-1} for the ^{241}AmBe and ^{252}Cf sources, respectively, corresponding to ambient dose equivalents of $(6.32 \pm 2.40\%)$ mSv/h and $(5.99 \pm 1.80\%)$ mSv/h. While the ^{252}Cf source can be approximated as a point-like source, the actual shape of the ^{241}AmBe must be considered when estimating systematic errors. Standard fast neutron measurement times were between 18 min and 90 min. Several control measurements were executed overnight.

Fig. 3. Fast neutron calibration setup with an ^{241}AmBe source and five detectors operated simultaneously.

Photon calibration was conducted using gamma rays from ^{60}Co (with a mean energy E_{mean} of 1174 keV) and ^{137}Cs ($E_{mean} = 556$ keV) and standard X-ray beams N60

(E_{mean} = 48 keV), N150 (E_{mean} = 118 keV), and N300 (E_{mean} = 250 keV). For each silicon detector, we collected 1 mSv during a 6 min measurement for ^{60}Co and all X-rays and during 3 min for ^{137}Cs. While for higher energies of ^{60}Co we roughly reached a data transfer limit of the CdTe setup with a 15 m Ethernet cable, for lower energy sources (^{137}Cs at 20 mSv/h and X-rays at 10 mSv/h) the data rate was so high that the system stopped working. Therefore, in order to avoid bandwidth problems and to reduce the overall amount of data taken to a reasonable amount, for the CdTe detectors we had to set up a lower beam intensity and decrease measurement time. With the new settings resulting in a collection of only 0.031 mSv, each output file still occupies several gigabytes. An angle scan was performed for one 300µm Si detector from -90° to +90° by 10° rotation for the gamma sources and from 0° to +90° by 15° for the X-ray beams.

IV. Results

To determine neutron fluences in the real environment, we must assess the neutron conversion efficiencies for each detector. The thermal and fast neutrons are observed below the corresponding neutron converters through the products of the abovementioned neutron reactions as roundish pixel clusters, so called "heavy blobs". By applying cuts based on the shape and deposited energy of the respective clusters, we can filter out most of the background, as depicted in Fig. 4a vs. Fig. 4b. Since the fast neutrons interact also directly in the silicon material, the heavy blobs are visible in the whole sensor area. The stronger signal below the PE+Al quadrant in Fig. 4c suggests that the neutron spectrum is harder than the one in Fig. 4d. Such that we can distinguish between the signal from ^{241}AmBe with a mean energy of 4.16 MeV and from ^{252}Cf with a mean energy of 2.13 MeV.

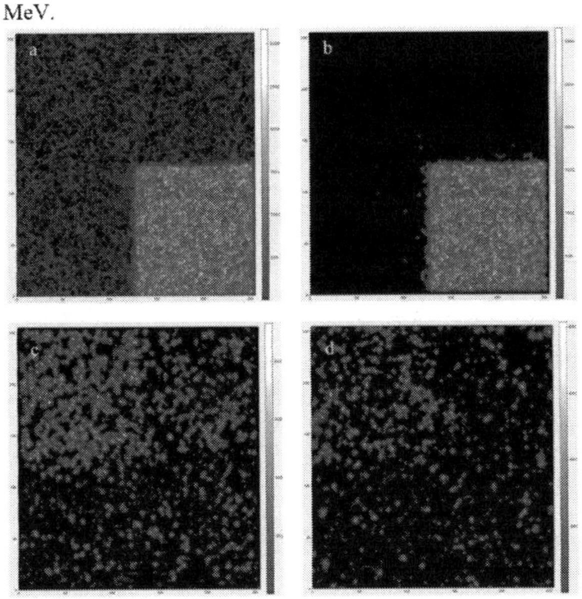

Fig. 4. (a) Integration window of thermal neutron measurement without any cluster cuts. (b) The same integration window as (a) but after cuts were applied. Almost no signal outside the LiF region. (c) Measurements with AmBe source after cluster cuts. More signal visible below PE and PE+Al regions. (d) Measurements with Cf source after cluster cuts. Dominant PE region.

The neutron detection efficiencies can be calculated using

$$\varepsilon_A = \frac{\frac{N_A}{S_A} - \frac{N_{Si}}{S_{Si}}}{\phi \cdot t}, \quad (1)$$

where N_A and N_{Si} are numbers of clusters detected during the measurement time t in a region A (i.e., below ^{6}LiF for thermal neutrons and below PE for fast neutron measurement) and in the uncovered region, respectively. S_A and S_{Si} are the areas corresponding to these regions, and Φ is the incident neutron fluence. The conversion efficiencies for silicon detectors are listed in Table I. The probability of registering a thermal neutron is slightly above 1% for all detectors, while for fast neutrons it is more than factor of ten lower for the ^{241}AmBe source and another factor of 4-13 lower for the ^{252}Cf source. The resulting errors below 5% for the thermal neutron detection efficiency and below 15% for fast neutron efficiency are dominated by systematics (mainly geometry and position uncertainties) and are not listed individually.

TABLE I
NEUTRON CONVERSION EFFICIENCY

Detector	$\varepsilon_{Li(thermal\ n)}$	$\varepsilon_{PE(AmBe)}$	$\varepsilon_{PE(Cf)}$
I6 100µm Si	1.30%	0.080%	0.008%
M7 100µm Si	1.28%	0.080%	0.006%
I5 300µm Si	1.41%	0.088%	0.020%
C5 300µm Si	1.29%	0.079%	0.019%
K3 500µm Si	1.04%	0.093%	0.020%
K5 500µm Si	1.16%	0.098%	0.020%

Efficiency of neutron detection in the Si sensors through the products of neutron reactions with the converter materials (i.e., ^{6}LiF and PE).

While these efficiencies can be directly used to calculate neutron fluences in real radiation environments (using the same cluster cuts as for the calibration), in case of CdTe sensors it is not a straightforward procedure. Table II summarizes only a ratio between number of impacted neutrons and number of clusters induced in CdTe by neutrons and accompanying photons. These data are nevertheless used to train our algorithms and still give a valuable information together with the cluster deposited energies, and other cluster features.

TABLE II
NEUTRON AND PHOTON INDUCED CLUSTERS IN CDTE PER IMPACTED NEUTRON

Detector	n_{cl}/n_{th}	n_{cl}/n_{AmBe}	n_{cl}/n_{Cf}
B7 1mm CdTe	0.15	0.93	0.95
I8 1mm CdTe	0.15	1.08	0.86
K8 1mm CdTe	0.16	0.96	1.00

Number of clusters caused by neutrons and accompanying photons inside the CdTe sensor material per number of impacted neutrons.

The neutron energy spectra for the three calibration sources inside the CdTe detector are plotted as Time over Threshold in Fig. 5a. Fig. 5b and Fig. 5c show a comparison of Time over Threshold ^{60}Co and ^{137}Cs spectra between the CdTe and Si detectors. These spectra can be comprehended as a justification for using CdTe alongside the Si sensors, as even the thickest Si sensor employed in the network is not capable of absorbing the high-energy photons of ^{60}Co (at 1.17 MeV and 1.33 MeV).

Number of clusters per ambient dose equivalent *H*(10)* was studied as a function of the photon source energy for all detectors. Fig. 6 shows such a dependency for each of the four sensor types used in our network. The behavior of all detectors is as predicted and follows the same tendency.

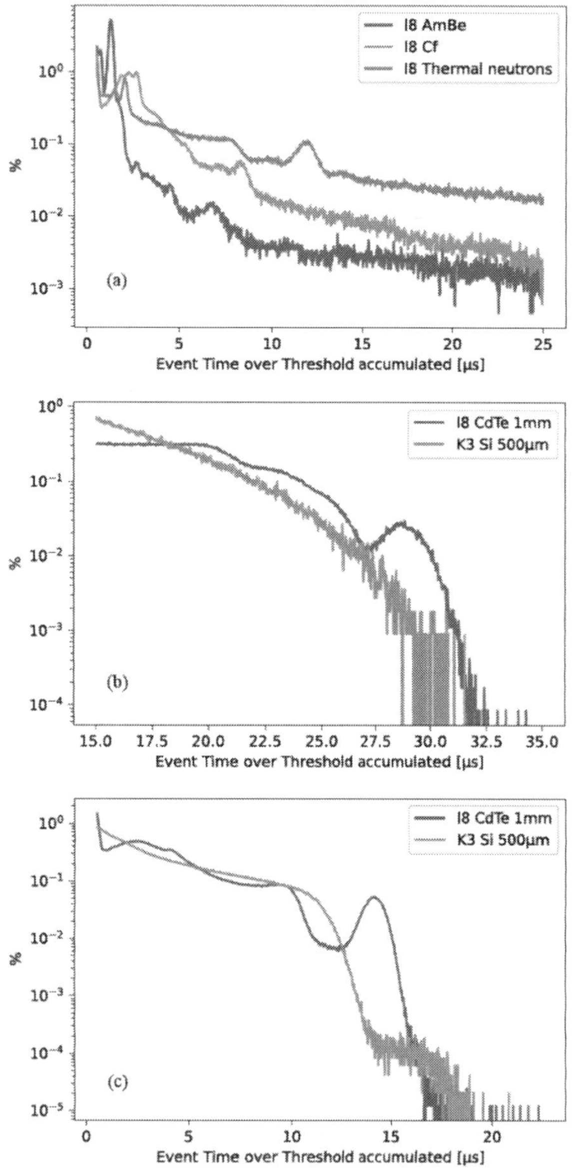

Fig. 5. Time over Threshold spectra of (a) three calibration neutron sources in the CdTe detector, (b) ^{60}Co in the CdTe and Si detectors, and (c) ^{137}Cs in the CdTe and Si detectors.

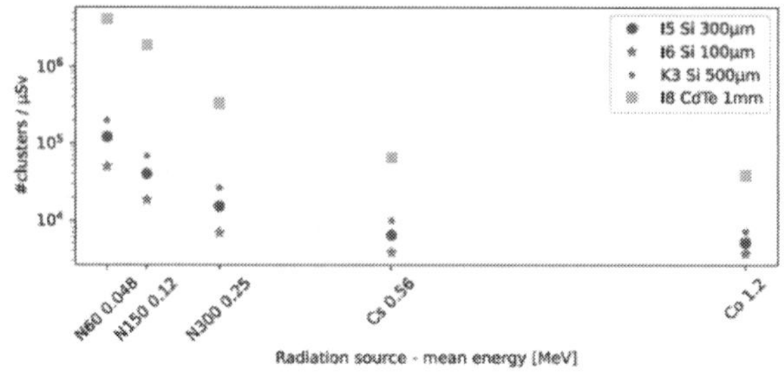

Fig. 6. Number of clusters per ambient dose equivalent as a function of the photon source energy.

Long-term simultaneous operation testing of all nine devices was performed using uraninite and ^{241}Am sources and exhibited good stability of the system.

V. Conclusions

In the framework of the MICADO project we devised an active long-term nuclear waste monitoring system, based on Timepix3 technology. To maximize the versatility of the network, 9 detectors, featuring four different sensor and neutron converter configurations, were integrated. A specifically developed in-house software handles simultaneous operation of all these devices, as well as cluster analysis and data visualization. Calibration, characterization and testing of the system exhibited high particle discrimination efficiency and sufficient stability for real world application. Measurements on Radwaste sites are currently being prepared.

Acknowledgment

This work was funded within the framework of European Union's Horizon 2020 research and innovation programme under grant agreement No 847641, project MICADO (Measurement and Instrumentation for Cleaning And Decommissioning Operations).

References

[1] MICADO project official website. [Online]. Available: https://www.micado-project.eu/. Accessed on: August 10, 2021.
[2] B. Bergmann, I. Caicedo, C. Leroy, S. Pospisil and Z. Vykydal, "ATLAS-TPX: a two-layer pixel detector setup for neutron detection and radiation field characterization," JINST, vol. 11, Oct. 2016, Art no. P10002.
[3] B. Bergmann, T. Billoud, C. Leroy and S. Pospisil, "Characterization of the Radiation Field in the ATLAS Experiment With Timepix Detectors," IEEE Trans. Nucl. Sci., vol. 66, no. 7, pp. 1861-1869, July 2019.
[4] T. Poikela et al., "Timepix3: a 65K channel hybrid pixel readout chip with simultaneous ToA/ToT and sparse readout," JINST, vol. 9, May 2014, Art no. C05013.
[5] P. Burian, P. Broulim, B. Bergmann, V. Georgiev, S. Pospisil, L. Pusman and J. Zich, "Timepix3 detector network at ATLAS experiment," JINST, vol. 13, Nov. 2018, Art no. C11024.
[6] P. Burian, P. Broulim, M. Jara, V. Georgiev and B. Bergmann, "Katherine: Ethernet Embedded Readout Interface for Timepix3," JINST, vol. 12, Nov. 2017, Art no. C11001.
[7] D. Turecek, J. Jakubek and P. Soukup, "USB 3.0 readout and time-walk correction method for Timepix3 detector," JINST, vol. 11, Dec. 2016, Art no. C12065.
[8] Czech Metrology Institute website. [Online]. Available: https://www.cmi.cz/?lang=2&language=en. Accessed on: August 10, 2021.

A comparison of bounding approach with isotopic correction factors and Monte Carlo sampling in burnup credit method

Pavlína Haroková[1], Martin Lovecký[1]
[1] Škoda JS a.s., Czech Republic
pavlina.harokova@skoda-js.cz

***Abstract*—One of the methodologies used in criticality safety analysis is burnup credit method, which allows considering fuel burnup in models with spent fuel. This removes excessive conservatism from the analysis, but it also brings new uncertainties originating from computational prediction of spent fuel composition. The burnup credit method offer several possibilities of how to deal with this problem, e.g. using bounding approach with correction factors on nuclide concentrations, which is simple, but still very conservative approach. Another option is Monte Carlo sampling, which aims at receiving the most realistic result as possible, but is very computationally demanding. In this work, we have analyzed correction factors for selected nuclides and compared the results of both methods on model of spent fuel storage pool. The results show how much conservative the bounding approach is – in this case, the multiplication factor was higher by almost 0.03 than in Monte Carlo sampling, exceeding the standard deviation by more than 5.4 times.**

***Keywords* —Bounding approach, Monte Carlo sampling, Burnup credit, SCALE**

I. Introduction

When evaluating the nuclear safety of a system, a criticality analysis is performed in order to determine whether the system meets the requirements of the local regulations. The regulation requirements could be summarized in simplified form as:

$$k_{eff} + \sum \Delta_i \leq k_{limit},$$

where k_{eff} is calculated value of effective multiplication factor of the evaluated model, which should be less than or equal to limit value k_{limit} (usually defined by legislation), if all biases and uncertainties marked as Δ_i are included.

Burnup credit is an approach that credits the reduction in reactivity due to fuel burnup. The calculation using burnup credit method consists of two main steps: a burnup calculation, which estimates nuclide concentrations in the spent fuel, and criticality calculation, which uses the nuclide concentrations determined in the first step. However, computational prediction of fuel composition introduces an additional source of bias or uncertainty. There are several methods that deal with these uncertainties and in general, they can be divided into two main categories: bounding methods and best-estimate methods.

The bounding approach aims to get the most conservative result by adjusting the concentrations of nuclides in a system. The concentration of selected nuclides in the model is multiplied by correction factors (referred to also as the isotopic correction factors) so that concentrations of fissile nuclides are increased and concentrations of absorbing nuclides are decreased. This leads to a more reactive system with the conservative estimate of multiplication factor k_{eff}.

The best-estimate methods aim to determine the uncertainty in spent fuel in a more accurate and realistic way. This can be done by using Monte Carlo sampling of the different parameters in a model based on their uncertainty and selected distribution. Other options, such as direct-difference method or sensitivity calculations, are also possible.

In this paper, we have tried to investigate how different the results of both methods are and what impact this could have on the possible compliance with regulatory limits.

II. Correction Factors

Correction factors (CF) are part of validation of each calculation code. The factors are calculated from difference between the experiment and the calculation (E/C) based on PIE experiments results. The mean E/C value of a given nuclide is increased or decreased (depending whether it is fissile or absorbing nuclide) by the value of its standard deviation multiplied by the tolerance-limit factor.

The value of tolerance-limit factor is determined by the number of measured samples in PIE experiments. If more than 10 measurement data are available for a nuclide, a two-sided tolerance-limit factor is used at statistics with 95 % certainty and 68.3 % population. For fewer than 10 measurement data, one-sided tolerance-limit factor is used with 95 % certainty and 95 % population. In our case, the number of samples varied from 8 to 46, which is equivalent to tolerance factor of 3.14 (for 8 samples) to 1.22 (for 46 samples).

In this work, the correction factors were determined for a set of 40 nuclides based on PIE experiments [5], [6] and [7]. The correction factors were analyzed for calculation codes from SCALE code system, the results are shown in Table I.

During this analysis, some modifications have been made. As the nuclide ^{103}Rh was not measured in PIE experiments, its correction factor was estimated to be the same as in the case of

^{109}Ag, which is the closest fission product in terms of weight while also having low value of its correction factor. Another problem was caused by isotopes ^{151}Eu and ^{153}Eu, which were measured with high deviation and thus leading into negative value of their correcting factor. Their correction factor was therefore set to zero.

Additional analysis was also made for isotopes of gadolinium contained in pins with Gd_2O_3 burnable absorber, as these data are not part of PIE experiments. The Gd cross-sections were conservatively modified by their 3σ deviation from uncertainty library, the impact of modified cross-sections was then studied during burnup calculation and correction factors values were determined as a function of burnup, as can be seen in Fig. 2.

III. Calculation

The comparison of both methods was made on TVSA-T Gd-2M+ 4.76 fuel assemblies. In burnup calculation, a 2D model of the fuel assembly was used. The fuel pins were divided into 15 groups with respect to 30°symmetry. The isotopic composition of the fuel was then calculated by TRITON deterministic code from SCALE code system and ENDF/B-VII.1 library in 252-group broadened energy structure.

In criticality calculations, only 47 nuclides were considered in spent fuel composition – this includes 40 nuclides listed in Table I, 6 more isotopes of gadolinium and nuclide ^{16}O, which is part of UO_2 and Gd_2O_3 compounds. The calculations were made on simplified 3D model of spent fuel pool. It consists of lattice filled with fuel assemblies with same symmetry as in burnup calculation. Each assembly is inserted into boron steel hexagonal tube, as is pictured in Fig. 2. The lattice is placed on stainless steel support plate, under which is a layer of moderator and concrete with steel lining. The pool is flooded with clean water with density conservatively assumed as 1.0 g/cm^3.

Fig. 1. Detail of fuel assembly in spent fuel pool calculation model.

TABLE I
CORRECTION FACTORS ANALYZED FROM PIE EXPERIMENTS

Nuclide	Mean E/C	Deviation E/C	CF	CF (used)
Mo-95	0.98	0.02	0.912	0.912
Tc-99	0.99	0.02	0.934	0.934
Ru-101	1.00	0.02	0.939	0.939
Rh-103	0.00	0.00	0.311	0.311
Ag-109	0.56	0.08	0.311	0.311
Cs-133	1.01	0.02	0.938	0.938
Cs-134	1.09	0.13	0.678	0.678
Cs-135	1.01	0.03	0.922	0.922
Cs-137	1.02	0.02	0.952	0.952
Ce-140	1.00	0.02	0.932	0.932
Ce-142	0.99	0.03	0.913	0.913
Nd-142	1.03	0.09	0.749	0.749
Nd-143	0.99	0.02	0.912	0.912
Nd-144	1.01	0.03	0.926	0.926
Nd-145	1.08	0.04	0.962	0.962
Nd-146	0.93	0.01	0.893	0.893
Nd-148	1.01	0.02	0.963	0.963
Nd-150	1.00	0.02	0.944	0.944
Sm-147	0.99	0.07	0.769	0.769
Sm-148	1.10	0.06	0.920	0.920
Sm-149	0.91	0.08	0.654	0.654
Sm-150	0.99	0.05	0.814	0.814
Sm-151	0.93	0.07	0.710	0.710
Sm-152	0.93	0.06	0.757	0.757
Sm-154	0.95	0.06	0.774	0.774
Eu-151	0.45	0.36	-0.666	0.000
Eu-153	0.18	0.06	-0.021	0.000
Gd-155	0.16	0.02	0.086	0.086
U-234	0.99	0.16	0.801	0.801
U-235	0.97	0.06	1.048	1.048
U-236	1.05	0.08	0.945	0.945
U-238	1.00	0.00	1.000	1.000
Np-237	1.47	0.59	0.627	0.627
Pu-238	1.20	0.26	0.879	0.879
Pu-239	0.97	0.10	1.089	1.089
Pu-240	1.00	0.07	0.906	0.906
Pu-241	1.02	0.11	1.158	1.158
Pu-242	1.06	0.22	0.787	0.787
Am-241	1.02	0.04	0.880	0.880
Am-243	0.87	0.18	0.654	0.654

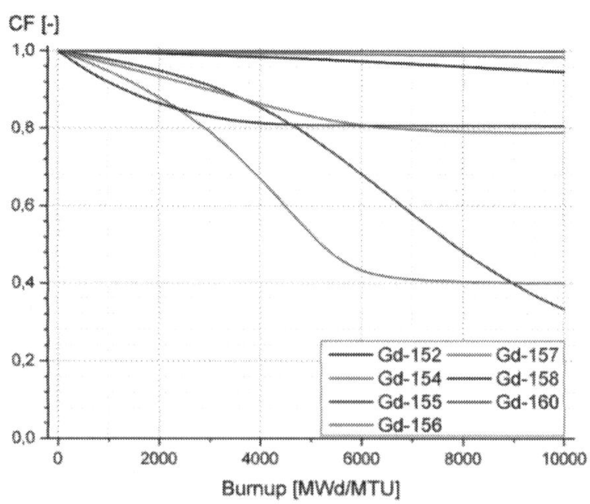

Fig. 2. Correction factors for isotopes of gadolinium in pins with Gd_2O_3 burnable absorber as a function of burnup.

A. Bounding Approach

In bounding approach, the concentrations of nuclides were adjusted using correction factors CF mentioned above. The criticality calculation was then made for 21 steps of fuel burnup, calculated by KENO-VI code from SCALE system. The results are shown in Fig. 3. The maximum of effective multiplication factor is in this case at 8000 MWd/MTU, with a value of:

$$k_{eff} = 0.94309 \pm 0.00016.$$

As this is the highest achievable k_{eff} value, it is also a crucial point from the nuclear safety perspective and was therefore investigated further in case of Monte Carlo sampling.

B. Monte Carlo Sampling

In Monte Carlo sampling, the correction factors were not used. Instead, the concentrations of nuclides were multiplied by a factor from a set of pseudorandom numbers, which were generated with normal distribution based on E/C value and its standard deviation as shown in Table I. Each nuclide in each pin has therefore its own unique concentration, while also satisfying its probability distribution (see Fig. 5).

As the sampling method is more computationally demanding than the standard bounding approach described above, we have analyzed only the case with maximum multiplication factor, which is according to the previous calculation at burnup of 8000 MWd/MTU. A total of 5000 calculation cases were created, from which the first case was sampled in unperturbed state, where nuclide concentrations were multiplied only by mean E/C value and the deviations were not considered. All other cases are perturbed, i.e. they were sampled according to the procedure described above.

The samples were calculated in criticality calculation by transportation code KENO-VI from SCALE code system. The results are individual values of multiplication factor k_i of sampled cases i. The mean value $\bar{k}$, as well as corrected sample standard deviation s, of all perturbed cases was determined:

$$\bar{k} = \frac{1}{n}\sum_{i}^{n} k_i,$$

$$s = \sqrt{\frac{1}{n-1}\sum_{i}^{n}\left(k_i - \bar{k}\right)^2}.$$

This leads to a final result – the effective multiplication factor of all perturbed cases:

$$k_{eff} = 0.91374 \pm 0.00542,$$

which is close to multiplication factor of unperturbed case k_0:

$$k_0 = 0.91872 \pm 0.00013.$$

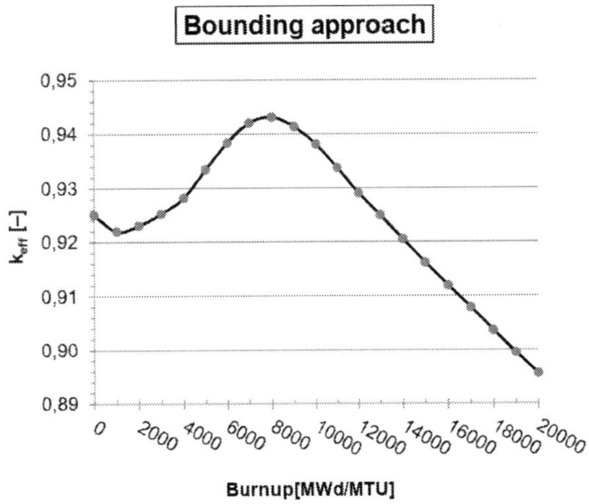

Fig. 3. Multiplication factor based on fuel burnup calculated by bounding approach with correction factors on spent fuel pool model.

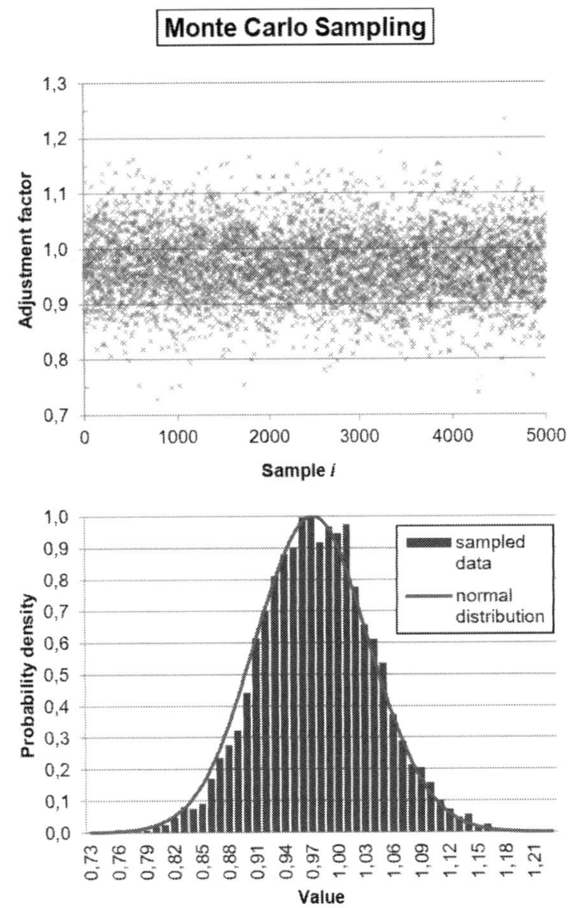

Fig. 4. Example of adjustment factor sampling for U-235 in one of the pins and its probability density function. In this case, the factor was sampled according to normal distribution with mean μ = 0.97 and deviation σ = 0.06.

IV. Results

The results of both methods can be summarized as follows:

- Bounding approach:

$$k_{eff}\ (\text{BA}) = 0.94309 \pm 0.00016.$$

- Monte Carlo sampling:

$$k_{eff}\ (\text{MC}) = 0.91374 \pm 0.00542.$$

In this case, the multiplication factor calculated by bounding approach is higher by 0.02935, which is about 5.4× more than the value of the standard deviation of the sampled Monte Carlo tasks. The comparison shows how much conservative the bounding approach is.

The using of Monte Carlo sampling method can reduce the maximum multiplication factor for models where the coefficient is slightly above the legislative limit based on the current conservative burnup credit methodology. This could lead to an alleviation of possible restrictions resulting from the limit exceeding. However, it should be kept in mind that Monte Carlo sampling approach requires a calculation of large number of tasks and is therefore more computationally demanding.

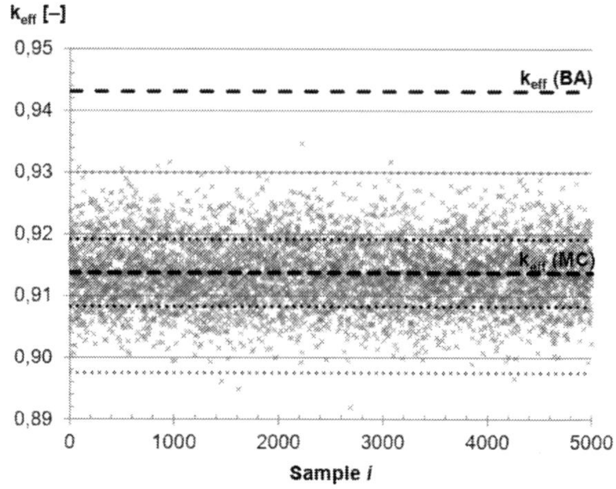

Fig. 5. Comparison of results calculated by bounding approach – k_{eff} (BA), and Monte Carlo sampling – k_{eff} (MC), with its standard deviation represented by dotted line. Results of individual samples k_i are also shown (represented by red crosses).

References

[1] P. Haroková, M. Lovecký, "*Comparison of bounding approach with correction factors and Monte Carlo sampling for determination of isotopic composition of spent fuel in burnup credit method.* ", Škoda JS report Ae19117, 2019 [in Czech].

[2] M. Lovecký, "*Comparison of results of spent fuel inventory calculations with ORIGEN-S code with experiments.* ", Škoda JS report Ae14561, 2012 [in Czech].

[3] G. Radulescu, I. C. Gauld, G. Ilas, J. C. Wagner, "*An Approach for Validating Actinide and Fission Product Burnup Credit Criticality Safety Analyses — Isotopic Composition Predictions*". US NRC Report NUREG/CR-7108, Oak Ridge, 2012.

[4] I. C. Gauld, "*Strategies for Application of Isotopic Uncertainties in Burnup Credit*". US NRC Report, NUREG/CR-6811, Oak Ridge, 2003.

[5] L. J. Jardine, "*Radiochemical Assays of Irradiated VVER-440 Fuel for Use in Spent Fuel Burnup Credit Activities*". LLNL Report, UCRL-TR-212202, Livermore, 2005.

[6] Yu. N. Kravchenko, "*Benchmark Calculation of Fuel Burnup and Isotope Composition of VVER-440 Spent Fuel*". Kurchatov Institue Report, Moscow, 1998.

[7] T. P. Makarova, B. A. Bibichev, V. D. Domkin, "*Destructive Analysis of the Nuclide Composition of Spent Fuel of WWER-440, WWER-1000, and RBMK-1000 Reactors*". Radiochemistry 50 (2008) 414-426.

Study of natural uranium fuel for a new reactor design TEPLATOR

Tomas Peltan[1], Eva Vilimova[1], and Radek Skoda[1]
[1]University of West Bohemia, Faculty of Electrical Engineering, Univerzitni 8, 30100 Pilsen, Czechia
peltan@fel.zcu.cz

***Abstract*—The TEPLATOR is a new type of nuclear reactor which the main purpose is producing heat for district heating. It is designed as a special thermal reactor with 55 fuel channels for fuel assemblies, which is moderated and cooled by heavy water and operated around atmospheric pressure. The TEPLATOR DEMO is designed for the use of irradiated fuel from PWR or BWR reactors. Using heavy water as the moderator and coolant in this reactor concept allows to use natural uranium as an alternative fuel in case that the irradiated fuel is not available for some reason. This solution is suitable because of the price of natural uranium and the absence of costly fuel enrichment. This article is focused on deeper analyses of alternative suitable fuel for TEPLATOR based on natural uranium and new fuel geometries. This work builds on previous research on alternative fuel material and geometry for the TEPLATOR. It is mainly concerned with the neutronic development of fuel assemblies, the possibility of manufacturing of developed fuel types, and optimization of fuel management and uranium consumption. This article contains predetermined candidates for suitable fuel geometries and new untested fuel geometry types with some new advantages. Finally, optimization of the whole reactor core and number of fuel channels was made in terms of increased safety and higher fuel burn-up. Presented calculations were performed by Monte Carlo code Seprent.**

***Keywords* — TEPLATOR, Natural uranium, Serpent code**

I. Introduction

THE TEPLATOR is a new reactor concept developed in the Czech Republic for district heating purposes. It is developed in cooperation between the University of West Bohemia in Pilsen and the Czech Institute of Informatics and Cybernetics of the Czech Technical University in Prague. The main purpose is producing only heat energy for district heating or for process heating for different types of factories, for chilling and other uses. The TEPLATOR is developed for using already irradiated VVER-440 fuel, which has optimal burnup [1] . The idea of TEPLATOR works with different spent PWR or BWR fuel types, TEPLATOR DEMO is designed namely with VVER-440 spent nuclear fuel. The reactor contains 55 hexagonal fuel channels where the fuel assemblies are placed. These 55 fuel channels are situated in a calandria filled with heavy water, and the heavy water is used also for cooling of the fuel assemblies. These two loops are separated by fuel channels, which have the same dimension as the VVER-440 fuel. The fuel channel consists of two concentric hexagonal tubes, between these tubes is CO_2 as a thermal shielding. Around the heavy water reactor core, a graphite reflector is placed in radial and axial position. TEPLATOR can run with a fresh fuel in VVER-440 geometry, this fuel should be slightly enriched up to 1.2% of ^{235}U [1] .

TEPLATOR DEMO is designed with 50 MWth output power with output temperature of the coolant around 98°C with a slightly higher atmospheric pressure. Later, after the DEMO testing, the concept of TEPLATOR aims up to 150MWth – TEPLATOR FULL. The operation time of one fuel loading is around 2 years for a 50MW unit with irradiated fuel. The visualization of TEPLATOR DEMO can be seen in Figure 1.

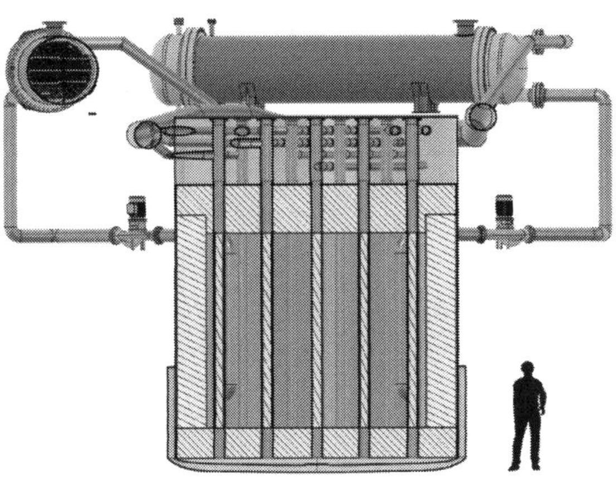

Figure 1. TEPLATOR DEMO layout in comparison with human size - side view.

II. Preliminary study of natural uranium

The TEPLATOR is designed with using of heavy water as the coolant and the moderator, which determine possibility of using a natural uranium as an alternative fuel. This solution can be suitable for countries, which do not have an irradiated fuel in storage and for which the enriching of uranium is expensive. Natural uranium is safer than enriched uranium due to possibility of criticality accidents. In previous work [2] , the sets of calculations with different fuel geometries with natural uranium were performed. Based on these results we assume that natural uranium can be used as an alternative fuel for

TEPLATOR due to sufficient reactivity excess. In previous research work, some suitable fuel geometries were found, and two fuel materials UO_2 and metal uranium were tested. During calculations, issues with positive void coefficient were found. This phenomenon was solved by developing a new coolant displacer with special construction features, which mitigates this problem [3] . Developed geometries with promising results are shown in Figure 2.

Two slightly different geometries, tubular and hexagonal tubes, were designed in previous work and this type of fuel was used for the next calculations.

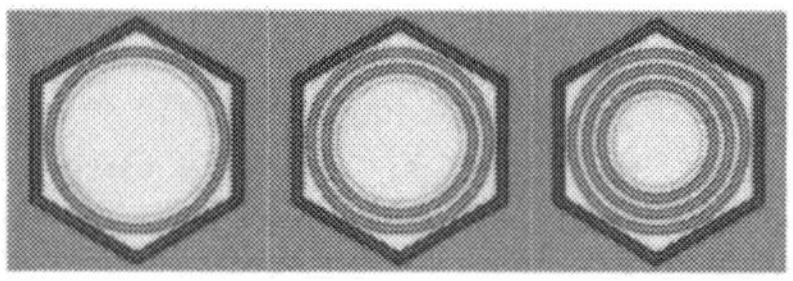

Cylindrical tubes with displacer – one, two and three tube modification

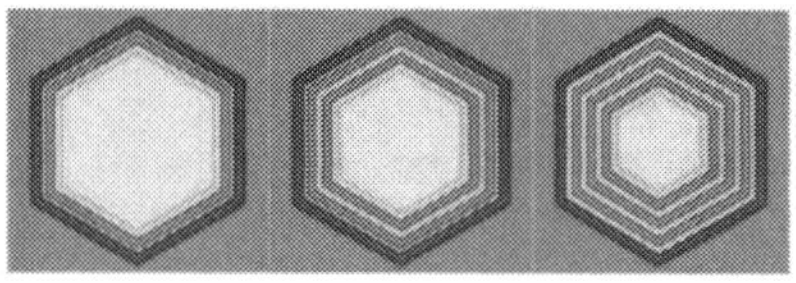

Hexagonal tubes with displacer – one, two and three tube modification

Figure 2. Two different geometries of fuel – cylindrical and hexagonal tubes with modifications of fuel tubes.

For safety reasons, this fuel types are crucial to operate it in optimal moderation condition. The optimal moderation condition means that there is no over-moderated arrangement of the reactor core from the H_2O/U ratio point of view. The all calculated cases with new developed fuel types in the designed TEPLATOR core layout with 55 fuel assemblies with fuel pitch for irradiated VVER-440 fuel were always slightly over-moderated, which is unacceptable from safety reasons point of view due to positive feedback of reactivity.

III. Optimal fuel pitch calculations

Based on the previous results, all six fuel geometries with the different fuel layer thickness described in [2] were recalculated for finding the optimal fuel channel pitch in the TEPLATOR core. The fuel layer thicknesses were obtained from the previous calculations, and for purposes of this paper, only the layers with the best *keff* were considered. A few UO_2 modifications were recalculated and mainly the modifications with the U metal fuel layer due to the higher *keff* and better heat conductivity of metal uranium in thick fuel layers.

The dimensions of the core and the fuel channels are the same as for the VVER-440 irradiated fuel, this study is modification of the designed reactor core of TEPLATOR, not development of a new reactor core. The seven different fuel channels pitches were calculated in step by 5 cm from 25 cm to 55 cm. The more precise fuel pitch of the new developed fuel is not part of this paper due to the early stages of the fuel development and the whole reactor arrangement. The construction materials of the cooling channels and the entire reactor, which are now under calculation testing, could affect the moderation properties in the reactor core, so the preliminary fuel pitch is assumed as sufficient. After finding the precise core arrangement, more accurate calculations with other modifications will be performed.

With the decreasing pitch of the fuel channels, the number of fuel assemblies in the reactor core increases, and directly affects the H_2O/U ratio. The number of fuel assemblies depending on the fuel pitch can be seen in Table 1.

TABLE 1. DEPENDENCE BETWEEN THE NUMBER OF FUEL ASSEMBLIES (FA) IN THE REACTOR CORE AND THE FUEL ASSEMBLIES PITCH

Fuel pitch [cm]	25	30	35	40	45	50	55
Number of FA [-]	185	121	87	73	55	43	37

An increasing number of the fuel elements in the reactor caused a higher weight of uranium that was used, which can be utilized for heat production. For each fuel geometry modification and for specific fuel layer thickness, the best fuel pitch was chosen. The radial section of the reactor core for imagination shows Figure 3.

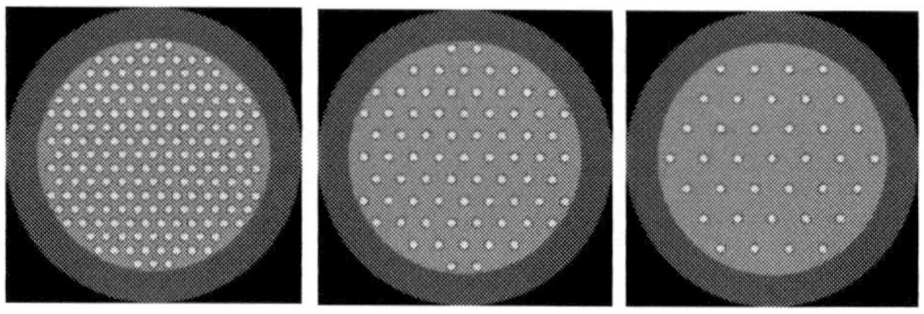

25 cm FA pitch 40 cm FA pitch 55 cm FA pitch

Figure 3. The layout of TEPLATOR core with the different fuel channels pitch

The course of *keff* depending on the fuel assembly pitch in the TEPLATOR core can be seen in Figure 4 and Figure 5. These courses were made for all six fuel geometry modifications, for each modification just some fuel layers thicknesses were considered.

IV. Burnup calculations and fuel utilization

After finding an optimal fuel pitch for each fuel modification, the burnup calculation was caried out. The burnup of fuel was calculated with the same steps for all cases, the predefined cumulative burnup sequence was used - *butot*. Burnup was calculated with 10 radial subdivisions of the fuel layer and with an average power density in the reactor core. The axial subdivision of the fuel was not considered. The output power of reactor was set to 50 MWth. The calculation proceeded at least until *keff* = 1, then the reactor became subcritical. During burnup the temperature of all materials and fuel was not changed. Obtained result from this type of calculation is average burnup of the fuel in MWd/kgU [4] . Higher burnup of the fuel means more utilized fuel and lower price of produced heat.

V. Calculations and models

The all mentioned modifications with different fuel pitch were calculated using Serpent 2.1.31 [4] code. For all materials with corresponding temperatures, the ENDF/B-VII.1 [5]

nuclear data library was used. The Thermal Scattering Library (TSL) for heavy water and for graphite was also obtained from the same nuclear data library. The temperature for all construction materials was conservatively set at 371 K, due to the output temperature of the coolant. The fuel material temperature was set to 600 K, which is relatively close to raw calculations. Temperature of the fuel layer is based on simplified thermohydraulic calculations with 10 mm thick layer of metal uranium.

Criticality calculations were performed with 20,000 neutrons per generation and 2,000 generations (2,000 active plus 100 inactive generations). The resulting standard deviation is below 8 pcm for all cases, which is sufficient in this preliminary study.

After criticality calculations, the cases with designated optimal fuel pitch with the highest *keff* were burned out. For the burnup calculations, 40,000 neutrons per one generation and 2,000 histories (2,000 active plus 100 inactive histories) were used. The standard deviation of all cases and calculated steps does not exceed 6 pcm.

VI. RESULTS

The course of *keff* depending on the fuel assembly pitch in the TEPLATOR core, the fuel layer thickness, and the fuel layer material can be seen in Figure 4 and Figure 5. These courses were made for all six geometry fuel modifications, for each modification only some fuel layer thicknesses were considered.

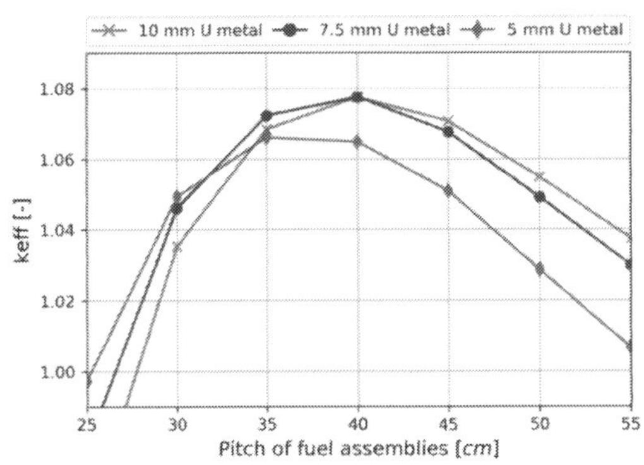

Figure 4. The results of *keff* for different fuel assembly pitch - one tubular tube geometry with metal uranium fuel layer

From the Figure 4 and Figure 5 can be seen, that the metal uranium reaches much higher *keff* compared to UO_2 in all calculated cases for all fuel layer thicknesses. This trend can be observed for all fuel modifications, suggesting that uranium metal should be a more suitable fuel material for these purposes.

The second advantage of U-metal is that the thickness of the fuel layer is not problematic and the temperature peak inside the fuel layer is not so high due to thermal conductivity of U-metal. The metal uranium heat conductivity is slightly dependent on temperature, the thermal conductivity of uranium increases with increasing temperature. The heat conductivity at 20 °C is around 26 W/m·K, for 300 °C the conductivity rises up to 33 W/m·K [6]

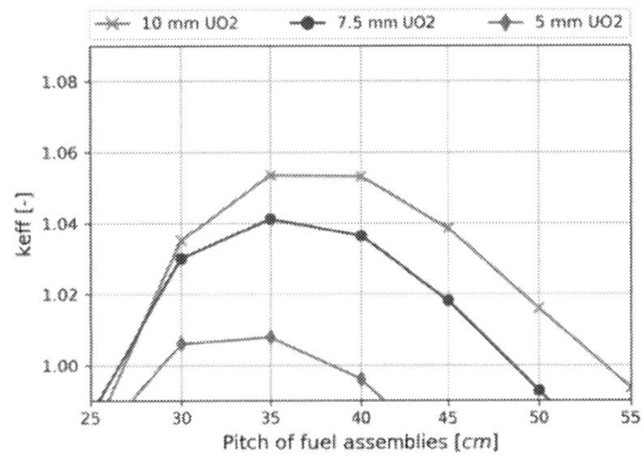

Figure 5. The results of *keff* for different fuel assembly pitch - one tubular tube geometry with UO2 fuel layer

The disadvantage of UO_2 is that it has more than 8 times lower compared to U metal, so operating thicker layers than 10 mm of UO_2 can be problematic. This is the reason why the 10 mm fuel layer was chosen as a maximal thickness for UO_2 material. For metal uranium, in some special cases, 15 mm thickness of fuel layer is still possible to use.

Figures Figure 4 and Figure 5 prove that the reactivity at the beginning of the fuel cycle with fresh natural uranium fuel is sufficient for both fuel materials, and both fuel modification were recalculated by burnup sequence. Figure 4 and Figure 5 show the under-moderated region – on the left side from the maximal *keff* value for each case, and the over-moderated region on the right side from the optimal condition. The over-moderated modifications of fuel pitch cannot be used due to positive feedback during abnormal operation. For our burnup calculations, the optimal moderation point was chosen for each geometry. The optimal pitch of the fuel assemblies was chosen to be 40 cm for all cases.

The calculated results from burnup calculations for more promising fuel modifications are given in the tables below. The results for hexagonal fuel geometry with various number of fuel tubes and materials can be seen in Table 2.

TABLE 2. CALCULATED BURNUP FOR HEXAGONAL FUEL GEOMETRY WITH THE DIFFERENT FUEL LAYER THICKNESS OF THE FUEL MATERIAL. THE PITCH OF THE FUEL ASSEMBLIES IS FOR ALL CASES SET TO 40 CM.

Number of tubes and material	Fuel layer thickness [mm]	EFPD [days]	Fuel Weight [kg]	Burnup [MWd/kgU]
1 - metal	10	2352.7	19177.7	6.15
1 - metal	7.5	1587.0	14622.8	5.42
2 - metal	6	2380.5	21302.6	5.58
2 - oxide	6	727.4	11815.8	3.07
3 - metal	3	1168.9	15430.7	3.77

From the Table 2 can be noticed that the highest burnup was reached for a single hexagonal tube with a 10 mm thick layer of U-metal fuel. Comparing burnup of single-tube modifications with various fuel thicknesses shows that the thicker layer is better than the thinner modification despite the previous criticality calculated results (from Figure 4 it is obvious that the maximum *keff* of both modifications reaches a very close value of *keff* around 40 cm pitch). This phenomenon is probably caused by different amount of uranium in the reactor core and the fact, that both modifications have a slightly different course of *keff* depending on the fuel assembly pitch. The layers with more than 10 mm fuel layer thickness were not considered as a suitable fuel geometry due to problematic fabrication, relatively high peak temperatures in the center of the fuel layer and due to huge amount of uranium per one loading.

The highest obtained burnup – 6.15 MWd/kgU was reached for one hexagonal tube modification with 10 mm fuel layer thickness. This modification can run for more than 6.44 years per one fuel loading at full output power – 50 MWth. The average fuel burnup 6.15 MWd/kgU is a slightly lower value of burnup than for commercially used CANDU fuel [7] . The burnup can be increased by using a Be reflector placed to special displacer in each fuel assembly, but this solution is very expensive so we did not consider it.

From the Table 2 can be noticed that using UO_2 instead of metal uranium in two-tube case caused rapid fall of burnup and operation time per one loading. This behavior is valid for all UO_2 cases. Using a UO_2 as the fuel material is not economically reliable due to the low fuel utilization in this fuel geometry and core arrangement. The comparison between cylindrical and hexagonal modifications for the singe-tube case was done, and the comparison can be seen in Table 3.

TABLE 3: COMPARISON OF HEXAGONAL (HEX) AND CYLINDRICAL (CYL) FUEL GEOMETRY WITH SAME FUEL LAYER THICKNESS.

Fuel modification	Weight of U [kg]	*keff* [-]	Burnup [MWd/kgU]
1 CYL 10 mm	17888.2	1.07748	5.68
1 HEX 10 mm	19177.7	1.07827	6.15

This table shows that the modification with hexagonal fuel shape is more promising than the cylindrical fuel shape. It can be caused by lower amount of fissile material in fuel assemblies and different geometry of fuel. To achieve the same amount of fuel, the fuel layer thickness should be higher. Advantage of hexagonal fuel shape compared to cylindrical fuel shape is fact that fabrication of hexagonal fuel is easier than cylindrical fuel. The fuel layer plates can be relatively simply cast as one uranium plate with corresponding dimensions. This plate can be inserted to the cladding box and tightly welded to prevent fission product release. After that, the hexagonal fuel assembly can be composed of six independent fuel plates with cladding, which can be welded to one fuel assembly. This method can be applicable for all hexagonal geometries with different number of fuel tubes.

The Figure 6 shows the *keff* course during reactor operation for different hexagonal fuel geometry for imagination.

Based on the Figure 6 the different operation time should be observed for changing the number of fuel tubes in one fuel assembly. It appears that the two-tube modification, the blue curve in Figure 6, has a longer operation time, which aims to achieve higher burnup and fuel utilization than for the one-tube modification, the green curve. In fact, more than 2.000 kg of uranium fuel is in two-tube modification compared to one-tube modification, which implicate a lower fuel burnup. Table 2 presents the reached burnup, operation time and other

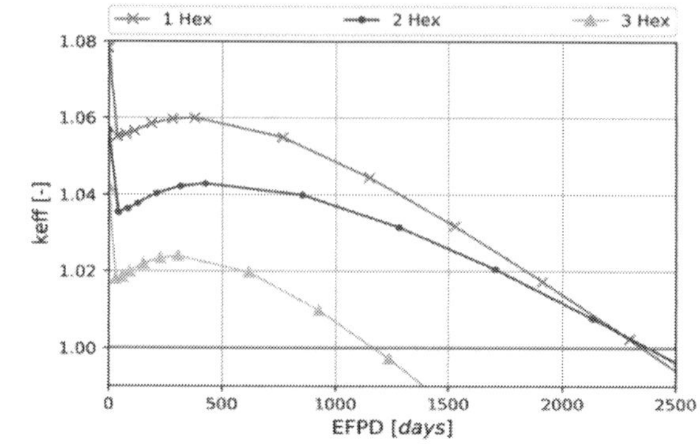

Figure 6: Comparison of *keff* courses and full power operation time for three cases of hexagonal fuel geometries with one hexagonal tube (1 HEX), two tubes (2 HEX) and three tubes (3 HEX)

information for the main calculated cases.

The motivation for calculations with more than one-tube fuel modifications is that when we would like to use this type of fuel for next TEPLATOR units with higher thermal output power, the one-tube modification cannot be used due to relatively high heat flux per one square meter. This problem is reduced using two or three-tube geometries. Using more than three concentric tubes with thinner fuel layers is not possible due to insufficient excess of reactivity at the beginning of the fuel cycle. This phenomenon is probably caused by self-shielding and higher amount of non-fissile material per fuel assembly. Higher fuel utilization is achieved with two-tube modification compared to three-tube modification, see Table 2.

After all these calculations we verified our results for new developed fuel geometries. We calculated two cases using a standard CANDU fuel geometry. The case was calculated with standard CANDU fuel bundles with natural UO_2 enrichment, which were placed in TEPLATOR fuel channels. The second option used a CANDU geometry, but the UO_2 was replaced by U-metal. We recalculate the results for these calculations, which can be seen in Table 4.

TABLE 4: CALCULATED BURNUP FOR THE CANDU FUEL GEOMETRY WITH DIFFERENT FUEL MATERIALS. THE PITCH OF THE FUEL ASSEMBLIES IS SET TO 35 CM.

Fuel material	EFPD [days]	Fuel weight [kg]	Burnup [MWd/kgU]
UO_2	690.1	12613.7	2.7
U-metal	1412.8	22741.0	3.1

Comparison of the results of achieved burnup for CANDU fuel and the results of the newly developed tube geometries shows that the new fuel design is more suitable for TEPLATOR and achieves higher burnup than the CANDU fuel. Reaching higher burnup than for the CANDU fuel type proves that using this new developed type of fuel is more meaningful in terms of economy and uranium consumption.

VII. Conclusions

This work focuses on the development of new natural uranium fuels for a new reactor design, the TEPLATOR. Sets of calculations in Serpent code were performed for verification of preliminary results. The new developed fuel geometries were tested with different fuel assembly pitches in TEPLATOR core for finding an optimal fuel channel pitch for safety and economic reasons. After that, the burnup calculations were done for more promising modifications with higher *keff*.

The burnup calculation shows that the use of natural uranium in metallic form is possible and economically sustainable. The best fuel arrangements in the form of single-hexagonal-tube with a 10 mm thick fuel layer of U-metal reach fuel burnup of more than 6 MWd/kgU. This burnup is quite high, considering the size of the entire reactor core, the amount of natural uranium and construction of the TEPLATOR core. This type of fuel can be marked as an alternative fuel for TEPLATOR.

The following research will be focused on deeper thermohydraulic analyses of the developed fuel and on following calculations of power distribution in the reactor core. The detailed behavior during fuel operation and the detailed burnup of the fuel will be determined.

ACKNOWLEDGMENTS

Research and Development has been funded by the Ministry of Education, Youth and Sports of the Czech Republic through the University specific research project no. SGS-2021-018.

REFERENCES

[1] J. Závorka, M. Lovecký, and R. Škoda, "Basic design of the TEPLATOR core - construction" in Proc. 29th International Conference Nuclear Energy for New Europe (NENE 2020). Ljubljana: Nuclear Society of Slovenia, 2020. s. 402.1-402.7. ISBN 978-961-6207-49-2.
[2] T. Peltan, E. Vilímová, and R. Škoda, "Natural uranium as alternative fuel for TEPLATOR". in: Proc. 29th International Conference Nuclear Energy for New Europe (NENE 2020). Ljubljana: Nuclear Society of Slovenia, 2020. s. 406.1-406.8. ISBN 978-961-6207-49-2.
[3] Special coolant displacer in nuclear fuel mitigating LOCA accident, by T. Peltan, E. Vilímová, R. Škoda (2020, March 20). Patent: PV 2020-498
[4] J. Leppänen, M. Pusa, T. Viitanen, V. Valtavirta, and T. Kaltiaisenaho, "The Serpent Monte Carlo code: Status, development and applications in 2013," Ann. Nucl. Energy, vol. 82, pp. 142–150, 2015, doi: 10.1016/j.anucene.2014.08.024.
[5] M. B. et al. Chadwick, "ENDF/B-VII.1: Nuclear Data for Science and Technology: Cross Sections, Covariances, Fission Product Yields and Decay Data," Nucl. Data Sheets, vol. 112, 2011.
[6] L. H. Ortega, B. Blamer, K. M. Stern, J. Vollmer, S. M. McDeavitt, "Thermal conductivity of uranium metal and uranium-zirconium alloys fabricated via powder metallurgy" Jour. of Nucl. Materials, Vol. 531, 2020, ISSN 0022-3115, doi: 10.1016/j.jnucmat.2019.151982.
[7] B. Rouben, "CANDU Fuel-Management Course" Atomic Energy of Canada Ltd.

The Mini Labyrinth – a Simple Benchmark for Radiation Protection and Shielding Analysis

Branislav Vrban[1], Štefan Čerba[1], Filip Osuský[1], Jakub Lüley[1], Vladimír Nečas[1], Karel Katovský[2], Ondrej Šťastný[2], Marko Gloginjić[3], Marko Erich[3], Željko Mravik[3], Srdjan Petrović[3]
[1]Slovak University of Technology in Bratislava, Institute of Nuclear and Physical Engineering, Slovakia
[2]Brno University of Technology, Department of Electrical Power Engineering, Czech Republic
[3]Vinča Institute of Nuclear Sciences Laboratory of Physics, Serbia
branislav.vrban@stuba.sk

***Abstract*— The Mini Labyrinth experiment is a simple neutron and gamma shielding experiment developed at STU, inspired by the ALARM-CF-AIR-LAB-001 ICSBEP benchmark experiment. The STU Mini Labyrinth is approximately ten times smaller and consists of NEUTRONSTOP shielding blocks. This paper describes the second version of the Mini Labyrinth experiment and presents the results of the neutron and gamma fields simulation and measurement. The PuBe neutron source with the emission rate of 1.0E7 n.s^{-1} was utilized in the experiment. The measurement of gamma ambient dose equivalent H*(10) and neutron count rates is performed by the Thermo Scientific RadEye portable survey meter. The simulation part was carried out using the state-of-the-art MCNP6 and SCALE6 MONACO stochastic calculation tools taking into account the detailed geometry of the labyrinth and a combined neutron and gamma source of particles. The comparisons were performed between codes and an experiment, based on the dose rate in the unique detection positions and using a 2D map of neutron and photon fluxes. The propagation of cross-section uncertainties was investigated through the shielding analysis. Partial agreement between codes and measurement was achieved, however serious discrepancies near the PuBe source were identified.**

***Keywords* — radiation shielding, experiment, calculation, cross-section uncertainties, labyrinth.**

I. Introduction

SINCE World War II there has been significant development of methods and approaches used in the calculation of radiation shielding. This development was directly supported by the needs of industry (military technology, nuclear power plants, food processing, medical applications, accelerators, etc.). Over time, modeling and simulation of relevant effects shifted from analytical modeling to methods based on the so-called primary principles and their stochastic nature. Even nowadays it is necessary to know the accuracy of available computation codes, used nuclear data and it is desirable to evaluate the influence of the user on the final calculated parameter. One of the most effective ways of gaining user experience and minimizing user effects on the results of calculation is international collaboration comprising the designing and construction of relevant benchmark experiments. The construction phase is usually followed by simulation with tools available at the workplace (engineering tools vs. high-fidelity methods), comparison of workgroup results and subsequent identification of the source of observed deviations from the experiment. Thus, this paper comprises a definition of the simple neutron and gamma shielding benchmark, inspired by the ALARM-CF-AIR-LAB-001 ICSBEP experiment [1]. The purpose of this experiment is to validate the computer codes of involved partners against real experimental data. In the following sections the measured quantities are compared to values calculated by MONACO (as a part of SCALE 6.2.4 system [2]) and MCNP 6 [3] stochastic codes. Moreover, the influence of different cross-section libraries and cross-section induced uncertainties are studied through the shielding analysis.

II. Experiment Setup

Several experimental setups were created during the development of the experiment [4]. The Mini Labyrinth V02 version of the experiment with PuBe neutron source is utilized in this paper. The experimental setup comprises the PuBe neutron source with the emission rate of 1.0E7 n.s^{-1}, several NEUTRONSTOP 6x12x25 cm C-shape shielding blocks (polyethylene with 5 % boron), H_2O filled PLA tank, plastic source holder, and active and passive detectors. The STU Mini Labyrinth, is a mini version of the original ICSBEP Labyrinth, currently with dimensions of 96 x 60 x 25 cm. The whole configuration is placed on a special desk, equipped with additional shielding material and gear. One of the main advantages of the experiment is the versatility in modifying the geometry configuration, the wall thickness, the height, and the width of the Mini Labyrinth. To allow a change of neutron spectra the graphite prism with a movable cavity for neutron source is placed on a desk near one of the entrances to the Mini Labyrinth. Due to the significant slowing down properties of graphite prism, it should be included in the simulation model even though it was not directly used in the specific experiment presented in the paper. The top view of the STU Mini Labyrinth is shown in Fig. 1.

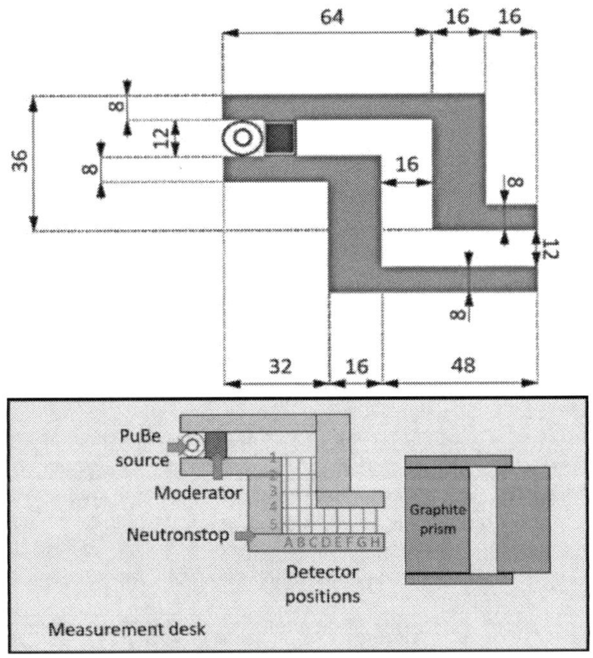

Fig. 1. STU Mini Labyrinth schematics (all dimensions are in cm).

III. Measurement Technique

The measurements were carried out at the Mini Labyrinth experimental facility shown in Fig. 1. The whole measurement process consisted of the loading of the PuBe neutron source in the plastic holder placed behind the plastic tank filled with 2.4 liters of demineralized light water. Next, the measurement of neutron and gamma dose rates in 17 positions was carried out. The measurement positions were chosen on the 2D grid. The grid consists of 5 positions in the vertical direction (1-5) and 8 positions in the horizontal direction (A-H), while the first measurement position is referred to as A1 and the last one as H5. Each grid position (window) corresponds to the size of the detector base. However, due to detector dimensions, the fifth line was omitted in measurement, thus only A1 – H4 positions were effectively used.

In each measurement position, the Thermo Scientific RadEyE SPRD-GN [5] digital dose meter was used. RadEye is a CLYC (Cesium Lithium Yttrium Chloride) scintillation detector calibrated to measure neutron and gamma count rate in units of CPS or CPM and ambient dose equivalent H*(10) of gamma radiation in units of μSv/h. The neutron measurement range of this detector is 0 – 1.0E4 CPS and the gamma measurement range is from 10 nSv/h to 250 μSv/h. It operates in two modes, "ratemeter" and "scaler".

At each position, the detector was placed in an upright position on the plastic holder ensuring that the crystal of the detector and the neutron source are in the same vertical position. 3 x 20 min measurement was performed, the ambient dose equivalent was obtained from the display and the neutron and gamma count rates were extracted from the RadEye Memory file. The final result per each position was evaluated as the average from the 3 measurements.

IV. Simulation

A. Radiation shielding

The simulation part of this analysis was performed using two stochastic Monte Carlo calculation codes, MCNP6 and MONACO, which is part of the SCALE 6.2.4 package. For both codes, the detailed 3D model of the Mini Labyrinth was created. The PuBe neutron source was modeled as a metallic cylinder. The emission spectra of the source are shown in Fig. 2 and Fig. 3.

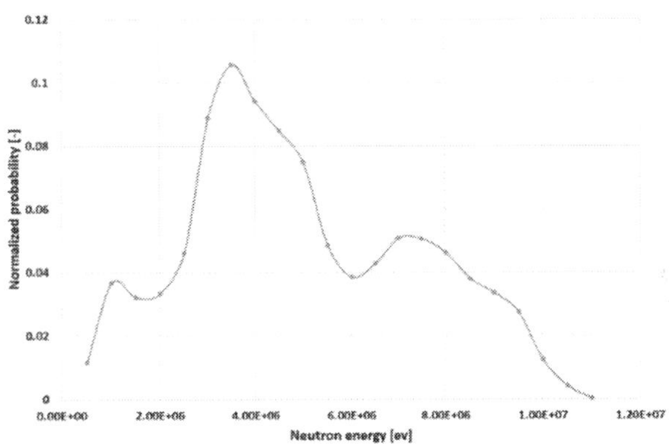

Fig. 2. PuBe neutron emission spectrum [6]

It should be noted, that the gamma emission spectra was modeled just as primary gammas from the (n,α) reaction. In this analysis, the angular distribution of the source was not considered.

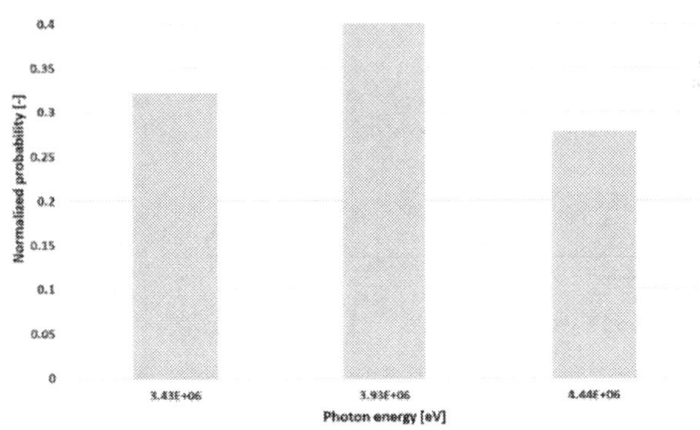

Fig. 3. PuBe photon emission spectrum

In both Monte Carlo codes, the calculations were performed as fixed source problems with the total number of 1.0E9 histories. In MCNP continuous energy cross-section libraries were used based on ENDF/B-VII.0 [7] and ENDF/B-VII.1 [8] evaluated data, processed by the STU staff. To estimate the gamma and neutron count rates, F5 type tallies were used both as mesh tallies and as point detectors in the four specified positions. The mesh tallies were modeled as a grid of 98 x 98 x 62 x 1cm parallelepiped elements. In MONACO the calculations were carried in forward mode, without using variance reduction techniques, and with standard 27n19g multi-group cross-section library, based on ENDF/B-VII.0 evaluated data. To allow comparison with MCNP, point detectors and mesh tallies were used in the calculations. The results were also multiplied by the source strength. To get comparable quantities with measurement the used detector flux-to-dose conversion factors were based on ICRU-57 [10] (ID 9036 for neutrons and ID 9510 for photons). The first response stands for the effective neutron dose and the second one represents the effective photon dose. The flux-to-dose conversion factors were chosen because of their ability to reasonably estimate the ambient dose equivalent H*(10) for photons under 5 MeV [11].

B. Cross-section induced uncertainty

The nuclear data used in a SCALE calculation are processed from evaluated nuclear data files (ENDF/B) produced by the National Nuclear Data Center through the AMPX code system. The ENDF/B data are based on a large set of nuclear data measurements, that have been evaluated by experts. Nuclear data are treated following the laws of probability and statistics. As such, they are represented by a central (recommended) value and are mostly burdened with uncertainty (in the form of covariance matrices) arising from measurements, modeling fits applied, or consistency adjustments. However, ENDF/B uncertainty data (including correlations) is not available for many nuclides or some type of data [11]. The present study aims to quantify nuclear data-induced uncertainties of the investigated quantities in the MONACO calculation explained above. To estimate the portion of simulation uncertainties induced by cross-section data, the stochastic procedure assuming that cross-section data used as an input to calculation may have a probability distribution, which is then propagated through a complex computational sequence. The SCALE system and super sequence SAMPLER with 200n47g cross-section library based on ENDF/B-VII.1 evaluated data is used in our approach. The SAMPLER module utilizes the previously generated perturbation factors for the 1D cross-sections on the multigroup library, assuming that the probability density functions are multivariate normal distributions with covariances given in the SCALE nuclear data covariance library. The SCALE 6.2 covariance library is based on ENDF/B-VII.1 data for 187 nuclides, combined with previous SCALE 6.1 covariances from other sources for another 215 nuclides [12]. Due to the stochastic approach applied, as the first attempt 40 independent calculations were performed. The evaluation of the final uncertainty in an individual requested output parameter p was calculated by an auxiliary tool according to (1)

$$\Delta p = \sqrt{\frac{1}{S-1}\sum_{s=1}^{S}\left(p_s^{MC} - \overline{p_s^{MC}}\right)^2}\,, \qquad (1)$$

where Δp is one standard deviation (1-sigma) of the investigated parameter due to uncertainties in the input parameters. The sign p_s^{MC} stands for the result of the sth independent simulation (from the S simulations in total) based on the Monte Carlo sampled input parameters which have been randomly varied within the specified distribution. In the calculation, the A1 detector calculation was assumed.

V. Results

A. Comparison of the simulated values

The neutron flux mesh tallies (the top view at the level of neutron source) calculated in the SCALE 6.2.4 MONACO and MCNP6 can be seen in Fig. 4 and Fig. 5. Their relative change, shown in Fig. 6, highlights deviations mainly in the close vicinity of the source and at the entrance corridor to the Labyrinth. Here the relative change reaches values between 50 to 91 %. Farther distances show smaller deviations about the level of 10 % We believe that these discrepancies are emerging from the different continuous energy and the multigroup approach utilized in MCNP and SCALE. The effect of the multigroup approach may be amplified by the use of just 27 neutron and 19 photon groups in the MONACO calculation.

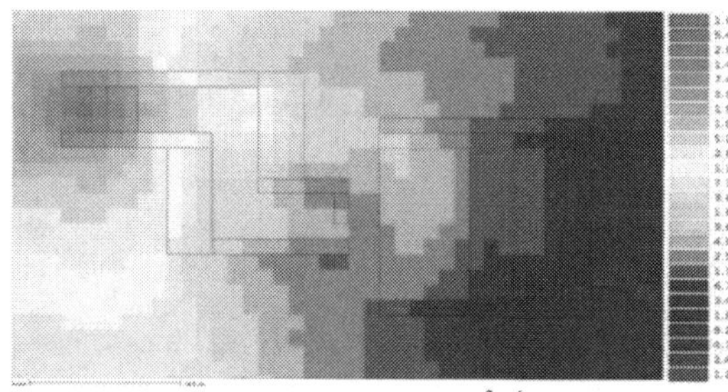

Fig. 4. Total neutron flux - SCALE [par.cm^{-2}.s^{-1}]

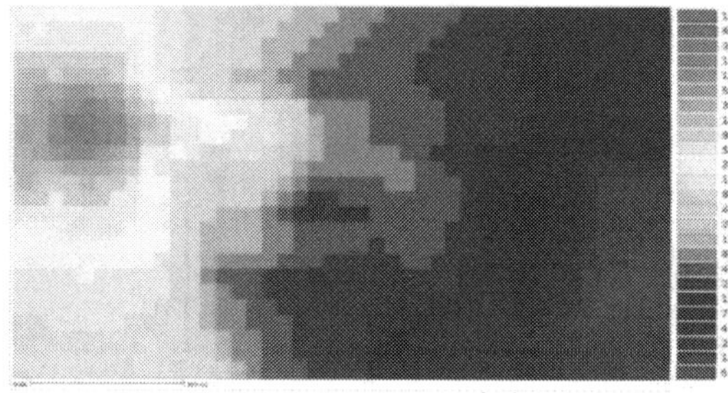

Fig. 5. Total neutron flux - MCNP [par.cm^{-2}.s^{-1}]

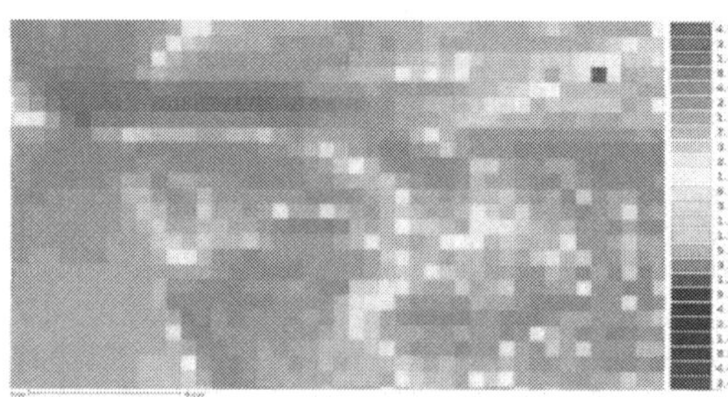

Fig. 6. Relative change of total neutron flux (MCNP as reference)

The gamma results in Fig. 7 to Fig. 9, show similar behavior. Nevertheless, the better gamma attenuation of the NEUTRONSTOP blocks in MCNP can be observed. The relative changes are lower than in the case of neutrons, however, the localization of the biggest discrepancies perfectly match. Again, we believe that the source of discrepancies relies in the different energy structures used within both codes.

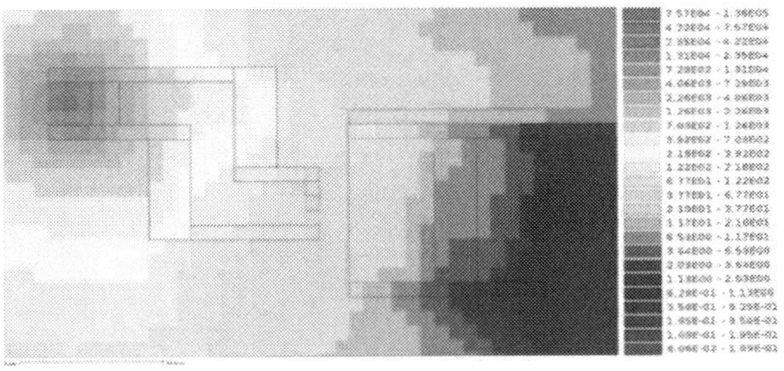

Fig. 7. Total photon flux - SCALE [par.cm^{-2}.s^{-1}]

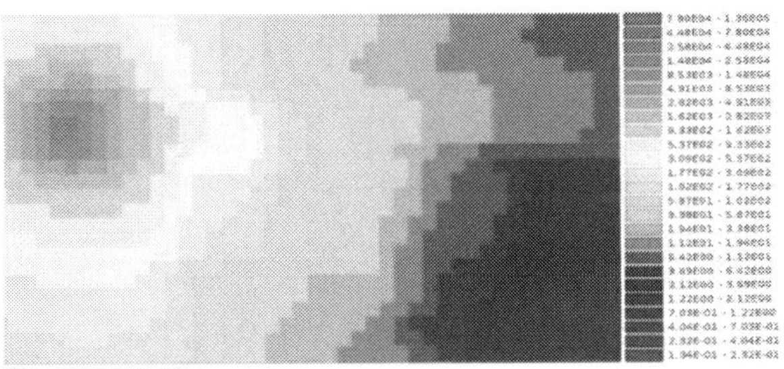

Fig. 8. Total photon flux - MCNP [par.cm^{-2}.s^{-1}]

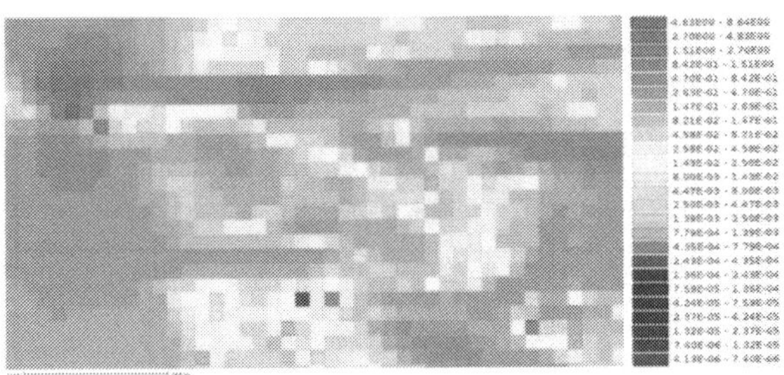

Fig. 9. Relative change of total photon flux (MCNP as reference)

The summary of calculated neutron and gamma fluxes in MCNP and SCALE for 17 detector positions can be found in Table I and Table II. The tables include both neutron and gamma fluxes, the statistical error resulting from the Monte Carlo calculation, the cross-section induced uncertainty determined in the next section, and the relative deviation from the MCNP results. Although the cross-section induced uncertainty was calculated in SCALE, as the first approximation, these values were applied also to MCNP.

TABLE I
CALCULATED NEUTRON FLUXES

Meas. Pos.	MCNP [par.cm^{-2}.s^{-1}]	SCALE [par.cm^{-2}.s^{-1}]	Δ_{MCNP} [%]
A1	9.42E+01 ± 1.81E+00	1.03E+02 ± 2.33E+00	8.94 ± 0.01
A2	4.00E+01 ± 7.93E-01	4.46E+01 ± 1.10E+00	11.34 ± 0.02
A3	2.73E+01 ± 5.52E-01	3.01E+01 ± 7.89E-01	10.04 ± 0.03
A4	2.22E+01 ± 4.55E-01	2.42E+01 ± 6.52E-01	8.76 ± 0.04
B1	1.12E+02 ± 2.14E+00	1.19E+02 ± 2.60E+00	5.74 ± 0.01
B2	4.39E+01 ± 8.70E-01	4.77E+01 ± 1.15E+00	8.58 ± 0.01
B3	2.73E+01 ± 5.52E-01	2.98E+01 ± 7.58E-01	8.96 ± 0.03
B4	2.04E+01 ± 4.21E-01	2.24E+01 ± 5.90E-01	9.70 ± 0.04
C1	1.30E+02 ± 2.47E+00	1.34E+02 ± 2.91E+00	2.85 ± 0.01
C2	4.68E+01 ± 9.22E-01	5.02E+01 ± 1.23E+00	7.37 ± 0.01
C3	2.73E+01 ± 5.52E-01	3.00E+01 ± 7.76E-01	9.61 ± 0.03
C4	1.86E+01 ± 3.86E-01	2.11E+01 ± 5.61E-01	13.05 ± 0.05
D4	1.34E+01 ± 2.83E-01	1.44E+01 ± 4.17E-01	8.06 ± 0.09
E4	8.36E+00 ± 1.84E-01	9.33E+00 ± 2.92E-01	11.65 ± 0.17
F4	5.90E+00 ± 1.35E-01	6.60E+00 ± 2.21E-01	11.83 ± 0.28
G4	4.65E+00 ± 1.09E-01	5.49E+00 ± 2.01E-01	17.92 ± 0.42
H4	4.48E+00 ± 1.05E-01	5.20E+00 ± 1.91E-01	16.22 ± 0.44

TABLE II
CALCULATED GAMMA FLUXES

Meas. Pos.	MCNP [par.cm^{-2}.s^{-1}]	SCALE [par.cm^{-2}.s^{-1}]	Δ_{MCNP} [%]
A1	4.25E+02 ± 2.65E+00	4.21E+02 ± 2.23E+00	-1.04 ± 0.01
A2	3.10E+02 ± 2.03E+00	3.12E+02 ± 1.73E+00	0.80 ± 0.01
A3	2.56E+02 ± 1.73E+00	2.60E+02 ± 1.47E+00	1.68 ± 0.01
A4	2.19E+02 ± 1.55E+00	2.24E+02 ± 1.28E+00	2.18 ± 0.01
B1	3.82E+02 ± 2.42E+00	3.78E+02 ± 1.96E+00	-1.23 ± 0.01
B2	2.70E+02 ± 1.83E+00	2.71E+02 ± 1.46E+00	0.49 ± 0.01
B3	2.17E+02 ± 1.52E+00	2.21E+02 ± 1.22E+00	1.54 ± 0.01
B4	1.85E+02 ± 1.34E+00	1.89E+02 ± 1.05E+00	2.15 ± 0.01
C1	3.70E+02 ± 2.37E+00	3.65E+02 ± 1.96E+00	-1.30 ± 0.01
C2	2.56E+02 ± 1.76E+00	2.57E+02 ± 1.41E+00	0.66 ± 0.01
C3	1.99E+02 ± 1.43E+00	2.03E+02 ± 1.15E+00	2.15 ± 0.01
C4	1.60E+02 ± 1.20E+00	1.69E+02 ± 9.45E-01	5.27 ± 0.01
D4	1.38E+02 ± 1.06E+00	1.42E+02 ± 8.23E-01	3.14 ± 0.01
E4	1.12E+02 ± 9.14E-01	1.17E+02 ± 7.06E-01	4.46 ± 0.01
F4	8.50E+01 ± 7.36E-01	9.06E+01 ± 5.73E-01	6.53 ± 0.01
G4	6.60E+01 ± 6.15E-01	7.14E+01 ± 4.62E-01	8.09 ± 0.01
H4	4.91E+01 ± 4.93E-01	5.86E+01 ± 4.08E-01	19.27 ± 0.02

The comparison of the energy integrated neutron and gamma fluxes showed just partial agreement between the calculation tools. In terms of the total neutron flux, MONACO significantly overestimated the MCNP results with the average relative change from MCNP of 10.04 %. The maximal relative change is 17.92 %. The largest deviations are achieved in the case of detector positions G4 and H4, where the neutron flux values are the lowest. In the case of gammas, the average relative change is only 3.23 % and the maximal error was 19.27 %, which is, again, achieved in measurement position H4. If we compare the gamma fluxes only for positions between A1 and C3, the relative error is only 1.11 %. Based on the partial results achieved, it can be concluded, that the results of both codes are showing the same trends, at detector positions may be considered partially acceptable, but cannot be judged as acceptable near the PuBe source and at the entrance corridor to the Labyrinth.

The comparisons of the neutron and gamma spectra per detector position calculated by MONACO are shown in Fig. 10 and Fig. 11. Each graph consists of data for 3 detector positions (A1, D4, and H4). In terms of the neutron count rate, and in accordance with our expectations, more slowed-down neutrons occurred at position H4 located farther from the source. In this position, the portion of thermal (E < 1 eV) neutrons is 8.98 %. This is caused by the reflection of thermal neutrons from the graphite prism. This increase of thermal neutrons is compensated by a significant, 9.34 % decrease of neutrons with energy above 100 keV. The hardest neutron spectrum was achieved in position D4, here the thermal neutron rate was only 6.11 %. This is caused by the absorption of thermal neutrons by the NEUTRONSTOP shielding blocks.

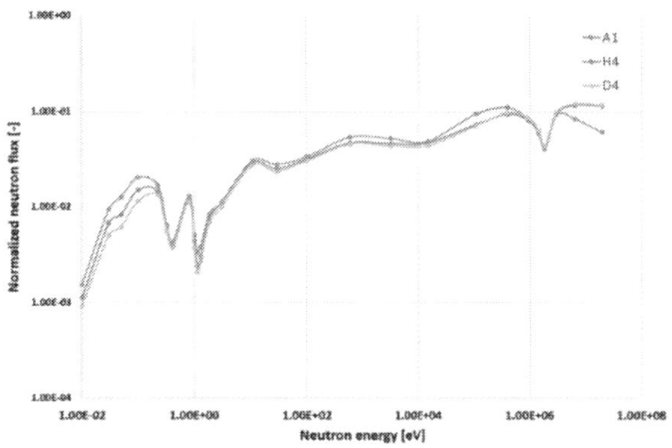

Fig. 10. Neutron spectra in A1, D4 and H4 positions

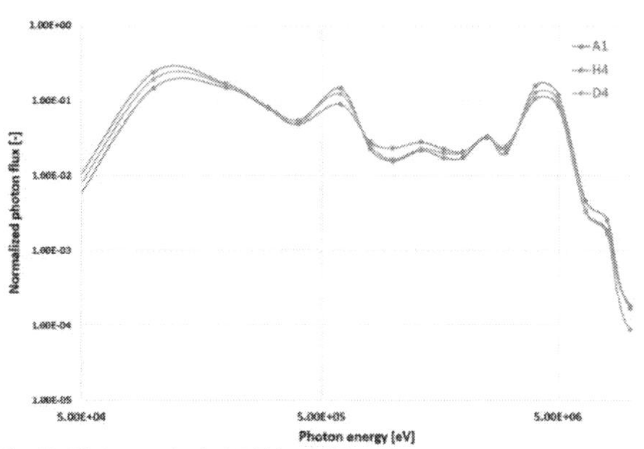

Fig. 11. Photon spectra in A1, D4 and H4 positions

The gamma spectra are showing considerably smaller deviations between detector positions. The largest discrepancies can be observed mainly in the energy intervals above 4 MeV and below 400 keV. The relatively small difference of spectra (and thus low attenuation) is caused by the presence of mainly light nuclei (H, B, and O) in the Mini Labyrinth materials.

B. Results of the cross-section induced uncertainty

The results presented in this section demonstrate the cross-section induced uncertainties to the effective doses from neutrons and photons. As illustrated in Fig. 12 and Fig. 13, the one sigma range of cross-section induced uncertainty to the effective dose from neutrons is 2.74 %, while the one sigma range of uncertainty to effective dose from photons of 0.62 % is considerably smaller.

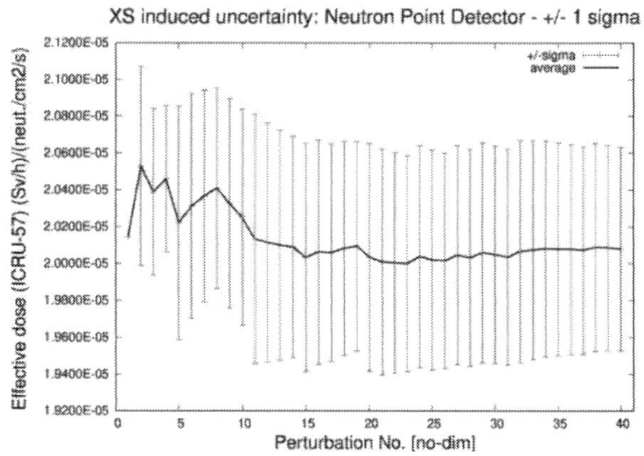

Fig. 12. Converge plot of the mean neutron effective dose and its uncertainty

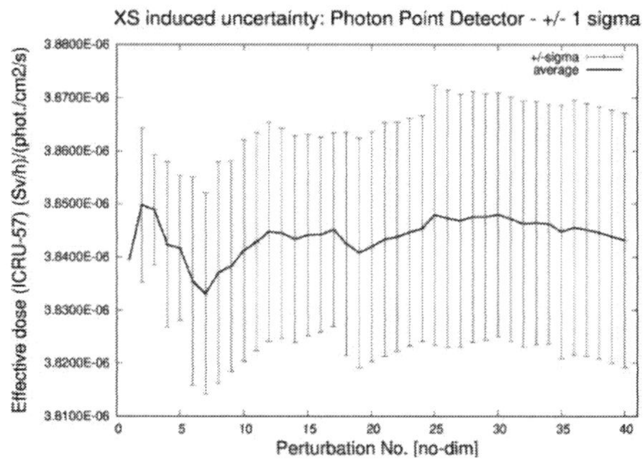

Fig. 13. Converge plot of the mean gamma effective dose and its uncertainty

The Monte Carlo uncertainty of the simulation was not taken into account due to its uniformity between cases and due to its size, which is by two orders of magnitude smaller than the cross-section induced uncertainty itself.

Interestingly, due to the energy dependence of flux-to-dose factors, the cross-section induced uncertainties to the integral neutron and photon fluxes are approx. 1.5 times smaller than in the case of effective doses. Based on the checked convergence plots, 40 independent calculations appear to be sufficient to obtain the first acceptable results. Anyhow, more perturbations are going to be calculated in near future.

C. Comparison of the simulation with measurements

The comparison of the measured and simulated neutron count rates is shown in Fig. 14. The figure includes the results from both calculation codes and the RadEye detector. It should be noted that the presented values are normalized in the sense that the measured count rate in the case of RadEye and the neutron fluxes in the case of the simulations are divided by the sum of count rates or fluxes per each detection position. It should be also noted, that while in the case of MCNP and SCALE, the uncertainty represents the statistical uncertainty of the Monte Carlo game plus the cross-section induced uncertainties, in the case of the RadEye detector it represents the mean squared error of the measured values.

As it can be seen from the Fig. 14, for some measurement positions, such as A4, B4, C3, or C4, the agreement between calculation and simulation is satisfactory, but for the rest of the positions, the results are not convincing. The most significant deviations were achieved for positions A1, B1, and C1, which are close to the wall of the Mini Labyrinth, and for positions D4 - H4, which are very far from the source. These deviations can be attributed to the spectral effect (and thus to the energy response function of the detector) of scattered neutrons on the wall of the mini labyrinth and also the possible evaporation of light water moderator during the measurement.

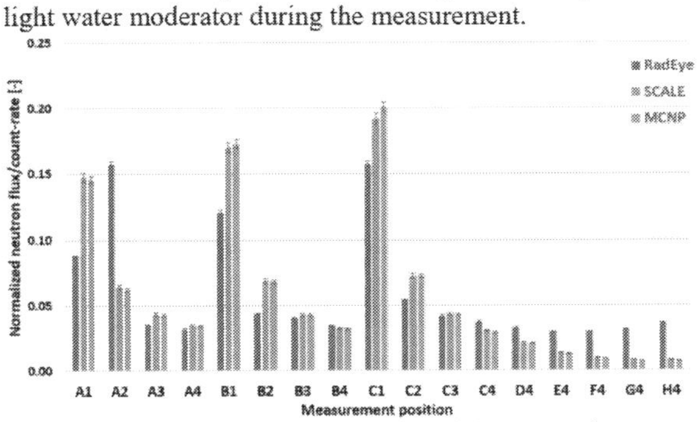

Fig. 14. Comparison of neutron count rates – simulation vs. experiment

The comparison of the measured ambient dose equivalent H*(10) rates and effective dose rates calculated by the SCALE system is presented in Fig. 15. The comparison showed an average relative change of 6.65% with the maximal relative change of ± 16 %. The largest discrepancies were achieved for positions A4, A3, B4, D4, and H4 measurement positions.

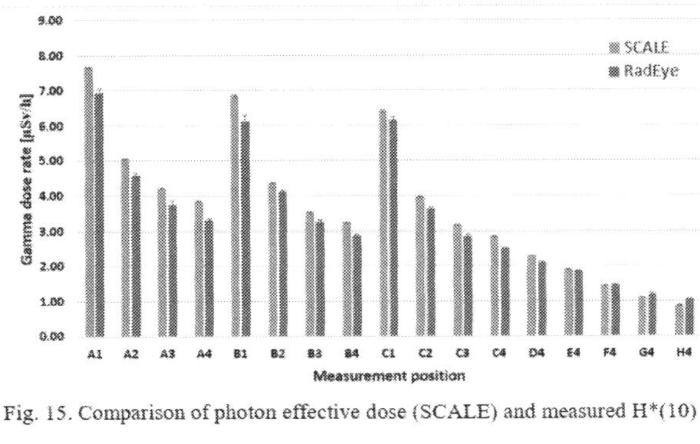

Fig. 15. Comparison of photon effective dose (SCALE) and measured H*(10)

Considering the slightly different units compared and the coarse energy group structure used in the calculation, we judge these results as acceptable. In the future H*(10) flux-to-dose factors will be used in simulation to be able to achieve more credible results. Moreover, to minimize the relative change between the calculations and measurements, the source definition should be thoroughly reviewed.

VI. Conclusions

The paper describes the Mini Labyrinth experiment developed at STU in Bratislava designed to assess the performance of computed codes in neutron and photon transport simulations. The first simulation results achieved by SCALE and MCNP show the discrepancies mainly at the entrance corridor to the Labyrinth and near the graphite prism. The results at detector positions are partially acceptable. We assume, that these discrepancies are emerging from the different continuous energy and the multigroup approach utilized in both codes. The effect of the multigroup approach may be amplified by the use of just 27 neutron and 19 photon groups in the MONACO calculation. In the future, also finer energy structures will be investigated. To partially understand the uncertainties included in the simulations, the cross-section induced uncertainties of effective doses from neutrons and photons were determined by the SAMPLER supersequence. The determined one sigma range of cross-section induced uncertainty to the effective dose from neutrons is 2.74 %, while the one sigma range of uncertainty to effective dose from photons is 0.62 %. These values may be considered in further designs of radiation shielding protection equipment and in case of comparison with measured values. The comparison of the measured and simulated neutron count rates is not convincing. Some measurement positions are in good agreement with simulations (A4, B4, C3, and C4) others are not satisfactory. These deviations have to be further investigated, but can be attributed to the spectral effect and the unknown energy response function of the detector used in the measurement. The comparison of the measured ambient dose equivalent H*(10) and effective dose calculated by the SCALE system showed acceptable agreement. In the future H*(10) flux-to-dose factors will be used in simulation to be able to achieve more credible results. The presented results and approaches serve us as a stepping stone for our further investigations and improvement of the Mini Labyrinth experiment at STU.

Acknowledgment

This study has partially been financially supported by the Slovak Research Development Agency No. APVV DS-FR-19-0014, No. APVV 16-0288 projects and by the Scientific Grant Agency of the Ministry of Education of Slovak Republic No. VEGA 1/0615/21.

References

[1] OECD NEA, "International Handbook of Evaluated Critical Experiments," Paris, 2020.

[2] W. A. Wieselquist, *et al*., "SCALE Code System," ORNL/TM-2005/39, Version 6.2.4, Oak Ridge National Laboratory, Oak Ridge, TN, 2020.

[3] C. H. Warner, *et al*., "MCNP 6.2 Release Notes," LA-UR-18-20808, Los Alamos National Laboratory, Los Alamos, 2018.

[4] Š. Čerba, *et al.,* "Preliminary Results of the STU Mini Labyrinth Radiation Shielding Experiment," in *Proc. APCOM 2021,* St. Luis: AIP Publishing, 2018.

[5] Thermo Scientific," RadEye SPRD Personal Radiation Detector," 2021, [Online]. Available: https://www.thermofisher.com/order/catalog/product/4250817#/4250817 . Accessed on: April 6, 2021.

[6] R. V. Griffith, *et. al*., "Compendium of Neutron Spectra and Detector Responses for Radiation Protection Purposes, " Technical Report Series No. 318, IAEA, 1990.

[7] M. B. Chadwick, *et. al,* "ENDF/B-VII.0: Next Generation Evaluated Nuclear Data Library for Nuclear Science and Technology," *Nucl. Data Sheets*, vol. 107, pp. 2931-3060, Dec. 2006, DOI. 10.1016/j.nds.2006.11.001.

[8] M. B. Chadwick, *et.al.*, "ENDF/B-VII.1 Nuclear Data for Science and Technology: Cross Sections, Covariances, Fission Product Yields, and Decay Data," *Nucl. Data Sheets*, vol. 112, pp. 2887-2996, Dec. 2011. DOI: 10.1016/j.nds.2011.11.002

[9] International Commission of Radiation Units and Measurements, "ICRU Report 57: Conversion Coefficients for Use in Radiological Protection Against External Radiation," Bethesda, MD, Aug 1998.

[10] O. Sato, *et. al*., "Calculations of Effective Dose and Ambient Dose Equivalent Conversion Coefficients for High Energy Photons," *J. Nucl.* Sci. Technol., vol. 36, no. 11, pp. 977-987, Nov. 1999. DOI. 10.1080/18811248.1999.9726290

[11] W. Wieselquist, *et. al*., "Overview of Nuclear Data Uncertainty in Scale and Application to Light Water Reactor Uncertainty Analysis," NUREG/CR-7249, ORNL, USA, 117 pages., 2017.

[12] M. Herman, "Development of ENDF/B-VII.1 and its covariance component," BNL-93619-2010, BNL, USA, 6 pages, 2010.

Sorting fission from parasitic coincidences of neutrons and gamma rays in plastic scintillators using particle times of flight

V. Bottau[a]*, C. Carasco[a], B. Perot[a], C. Eleon[a], R. De Stefano[a], L. Isnel[a], I. Tsekhanovich[b]
(a) CEA, DES, IRESNE, DTN, SMTA, Nuclear Measurement Laboratory, Cadarache, F-13108 Saint-Paul-Lez-Durance, France
(b) Université de Bordeaux, CNRS, CENBG, UMR 5797, F-33170 Gradignan, France
(*) corresponding author: vincent.bottau@cea.fr

***Abstract*— This work addresses the use of plastic scintillators as an alternative to ^{3}He detectors for radioactive waste drum characterization. The time response of scintillators is three orders of magnitude faster than that of gas proportional counters and they offer similar neutron detection efficiency at lower cost. However, they are sensitive to gamma rays and the commonly used Pulse Shape Discrimination technique is not possible with basic PVT scintillators. This paper reports on an innovative data processing technique allowing to extract spontaneous fission events from parasitic coincidences, such as those from the (α,n) reactions accompanied by correlated gamma rays or from pure gamma-ray sources emitting correlated radiations. The proposed approach makes advantage of differences in the pulse detection times recorded in measurements with the ^{252}Cf, AmBe and ^{60}Co sources. More precisely, a 2D histogram of time delays between the detected 2nd and 1st pulses, on the x-axis, and between the 3rd and 2nd pulses, on the y-axis, is found to allow for selection of a region of interest most relevant to spontaneous fission events.**

***Keywords* — Plastic PVT scintillators, spontaneous fission prompt neutrons and gamma rays, coincidence analysis, radioactive waste drums.**

I. Introduction

In the context of plutonium bearing waste drum characterization by passive neutron coincidence counting, conventional measurement collars based on ^{3}He gas counters permit a temporal discrimination between spontaneous fission neutrons and parasitic neutrons from (α,n) reactions [1]. The ^{3}He proportional counters are well suited for neutron multiplicity counting because of their large capture cross section for thermal neutrons, leading to excellent detection efficiency, and their insensitivity to gamma rays. However, the shortage of ^{3}He gas has greatly increased the cost of these detectors, leading to a worldwide effort to find possible alternatives, especially in the field of homeland security and safeguards [2], [3]. The Nuclear Measurements Laboratory of CEA Cadarache (France) makes studies of low-cost EJ-200 Polyvinyl Toluene (PVT) plastic scintillators as a potential alternative for neutron coincidence collars [4], [5]. Efficient neutron detection with ^{3}He counters requires neutron thermalization, which lasts several tens of microseconds, whereas plastic proton recoil detectors are directly sensitive to fast neutrons, which need just a few tens of nanoseconds to cover the source-detector distance. Therefore, the coincidence window can be three orders of magnitude shorter with PVT scintillators than with ^{3}He counters, which limits dramatically accidental coincidences.

However, plastic scintillators are strongly sensitive to gamma rays, which, in addition, may scatter between detectors and thus produce parasitic coincidences that will be merged with useful (i.e., different-particle) coincidence events expected from fission. For instance, (α,n) reactions accompanied by correlated gamma rays, or pure gamma ray sources emitting correlated gamma radiations can induce undesired high-order coincidences [6] due to the mentioned multi-detector hit events by a single gamma-ray or neutron, that are currently under worldwide investigations [4], [7-10]. Although some plastic scintillators allow via the Pulse Shape Discrimination (PSD) to sort neutron and gamma ray events [11-14], they are not a practical alternative for radioactive waste drum characterization. Indeed, a large detection efficiency is needed for coincidence measurements, but the PSD figure of merit (FOM) decreases for large-size scintillators due to multiple light scattering and lower photoelectron statistics [15-18]. In addition, the FOM also decreases at intermediate and high-count rates due to the pile-up effect-expected to be present for intermediate and high-level radioactive wastes. A possible solution would be to use a large number of small size PSD scintillators (typically less than 5”×5”) but more than 100 detectors would be needed to fully cover the detection solid angle, even for limited-in-size radioactive waste packages like 120 L drums (∅ ≈ 44 cm, h ≈ 80 cm). The cost and complexity of the measurement system, with its electronics, would not be advantageous with respect to ^{3}He detectors.

To overcome the above limitations, we have developed a method exploiting the differences in the time-of-flight of neutrons and gamma rays detected in triple coincidences with an array of PVT detectors. A 2D histogram of the time delays between the detected 2nd and 1st pulses, on the x-axis, and between the 3rd and 2nd pulses, on the y-axis, is found to allow for selection of a region of interest most relevant to spontaneous fission events. This papers reports the very first tests performed with bare radioactive sources measured separately to highlight the region of interest of the useful signal (^{252}Cf spontaneous fission source) with respect to different sources of background noise (^{60}Co and AmBe). The ^{252}Cf source emits correlated fission prompt neutrons and gamma rays, ^{60}Co emits correlated 1.173 and 1.332 MeV gamma rays and AmBe emits a single neutron correlated with a 4.439 MeV gamma ray following the ^{9}Be(α,n)^{12}C* reaction. The multiplicity analysis with the traditional shift register method is also reported to allow a comparison with the new approach.

II. Experimental setup

The experimental setup consists of eight EJ-200 (Eljen Technology) 10 × 10 × 100 cm^3 plastic scintillators coupled to 3M-P-E1-X-N (ET Enterprises Ltd.) photomultipliers. These detectors assembled by Scionix are shown in Fig. 1. Calibrations sources are in central position, at 35 cm from each detector considering the orthogonal direction, and surrounded by 2mm of lead, mainly to cut the 59 keV gamma rays of ^{241}Am in the 3.68 GBq AmBe source. The 8-channel DT5730B digitizer (CAEN S.p.A), used with the built-in DPP-PSD firmware, samples the analog pulse trains with a 500 MHz frequency on 14 bytes [19] and delivers a list of the pulse time stamps calculated by a Constant Fraction Discriminator algorithm (CFD) to the acquisition computer through a maximum 50 MB/s USB2 connector. The v1.0.4 CoMPASS software from CAEN drives acquisition [20], in which the long gate for charge integration and the record length (size of the acquisition window, i.e. number of samples) have been set to 60 ns and 182 ns, respectively. A low-energy threshold of 100 keVee (electron equivalent energy) is also set in order to filter the intense 59 keV line from AmBe. In practice, this threshold will always be useful for plutonium characterization in radioactive wastes because of a plurality of intense gamma and X-ray lines from plutonium isotopes and ^{241}Am [6,21].

Fig. 1. Experimental set-up: 8 EJ-200 scintillators surrounding the ^{252}Cf source placed on a 50 cm height support.

III. Data analysis

A. *Shift Register method*

The data acquired with ^{252}Cf, AmBe and ^{60}Co sources are processed with CERN's ROOT software [22] based on the Rossi-Alpha distribution and on the historical shift register method for the determination of number of fired detectors, which was originally developed for neutron coincidence collars using ^{3}He counters [1], [23]. The Rossi-Alpha distribution (detection times of next pulses after each detected pulse) obtained with the ^{252}Cf source is given in Fig. 2. Two main contributions are visible, first γγ pairs with time intervals below 10 ns, then γn pairs with time delays in the (10-50 ns) range.

For the shift register application, a 100 ns coincidence window is used to integrate real coincidences. The basics of the shift register method are illustrated in Fig. *3*. The "Real + Accidental" (R+A) register accounts for all types of coincidences, while "Accidental" (A) records only random coincidences, (R+A) and (A) windows being separated by 1 µs. The "M0" multiplicity corresponds to the absence of coincidence in the 100 ns window opened by a neutron or gamma pulse (no pulse in the other detectors), M1 records coincidences with another pulse (i.e. a pair), etc.

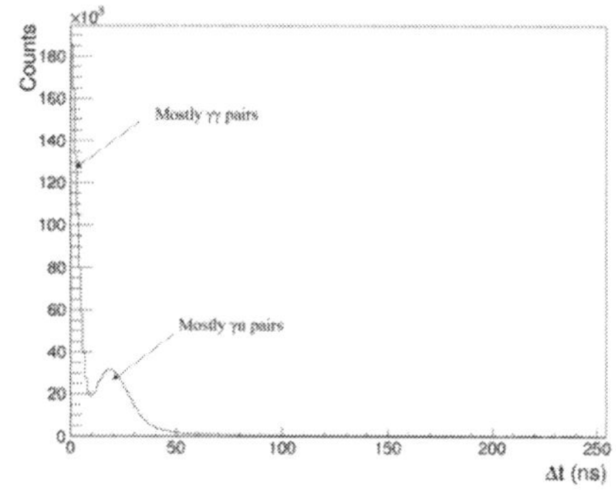

Fig. 2. Rossi-Alpha distribution obtained for the measurement in presence of the ^{252}Cf source

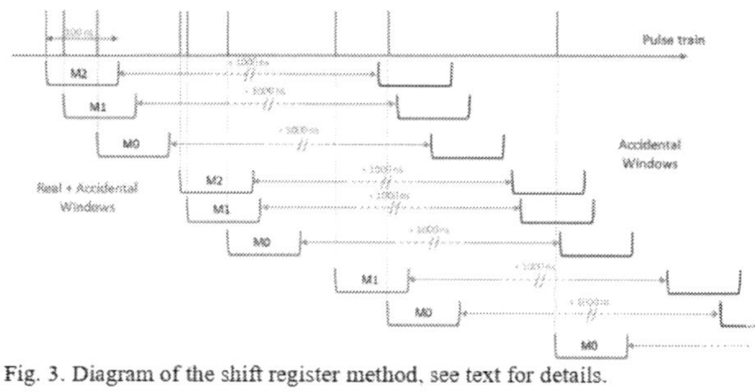

Fig. 3. Diagram of the shift register method, see text for details.

Table I and Table II report 5 min acquisition data of the background noise and of the individual radioactive sources measured separately.

TABLE I
NUMBER OF EVENTS DETECTED FOR DIFFERENT MULTIPLICITIES, ALONG WITH THE COUNT RATES OF SINGLES, DOUBLES, AND TRIPLES (S, D AND T, RESPECTIVELY) DURING BACKGROUND NOISE MEASUREMENT

	Background noise	
	R+A	A
M0	2259505±1503	2349237±1533
M1	89387±299	1860±43
M2	2066±45	64±8
M3	130±11	2±1
M4	46±7	1±0
M5	20±4	0±0
M6	8±3	0±0
M7	2±1	0±0
S	7837 s^{-1}	
D	308 s^{-1}	
T	10 s^{-1}	

TABLE II
NUMBER OF EVENTS DETECTED FOR DIFFERENT MULTIPLICITIES, (AFTER SUBTRACTION OF THE BACKGROUND NOISE) ALONG WITH THE COUNT RATES OF SINGLES, DOUBLES, AND TRIPLES (S, D AND T, RESPECTIVELY)

	^{252}Cf (288 kBq)		AmBe (3.68 GBq)		^{60}Co (882 kBq)	
	R+A	A	R+A	A	R+A	A
M0	3379137 ±2810	4479459 ±3029	15577017 ±4483	16671802 ±4623	53638186 ±7626	57771864 ±7904
M1	857974 ±1018	8310 ±110	1141198 ±1149	102194 ±325	4947762 ±2264	989025 ±996
M2	209237 ±462	1296 ±38	59474 ±252	5790 ±77	251249 ±505	81991 ±287
M3	38098 ±196	257 ±16	2312 ±51	253 ±16	9079 ±97	3516 ±59
M4	4806±70	42±7	67±13	14±4	239±18	97±10
M5	401±21	5±2	-9±6	0±0	-11±5	2±1
M6	18±6	1±1	-3±4	0±0	-7±3	0±0
M7	-1±2	0±0	-2±1	1±0	-2±1	0±0
S		$14966\ s^{-1}$		$55934\ s^{-1}$		$196155\ s^{-1}$
D		$4668\ s^{-1}$		$3843\ s^{-1}$		$14381\ s^{-1}$
T		$1175\ s^{-1}$		$175\ s^{-1}$		$342\ s^{-1}$

The count rates of singles (S, sum of all the triggers), doubles (D, which is S times the difference between the mean values of (R+A) and (A) registers of M1) and triples (T, which is a complex deconvolution of (R+A) and (A) distributions with the registers M2, M3, M4, etc.) are calculated following the equations detailed in [23-25]. As follows from (2-4), the signal-to-noise ratio (SNR) increases significantly from singles to triples because M1 (two fired detectors) is dominated by correlated γn pairs of AmBe and γγ pairs of ^{60}Co, whereas higher-order multiplicities (M2, M3, etc.) are dominated by ^{252}Cf spontaneous fission coincidences (cf. Table II).

$$\mathrm{SNR_S} = \frac{S_{252Cf}}{S_{AmBe} + S_{60Co}} = \frac{14966}{55933 + 196155} = 0.06 \quad (1)$$

$$\mathrm{SNR_D} = \frac{D_{252Cf}}{D_{AmBe} + D_{60Co}} = \frac{4668}{3843 + 14381} = 0.3 \quad (2)$$

$$\mathrm{SNR_T} = \frac{T_{252Cf}}{T_{AmBe} + T_{60Co}} = \frac{1175}{175 + 342} = 2.3 \quad (3)$$

Plastic scintillators are however sensitive to neutron and gamma-ray multi-detector scattering (respectively elastic and Compton scattering), also known as scattering crosstalk. Therefore, compared to ^{3}He neutron counters that are practically insensitive to crosstalk, multiplicity analysis is not straightforward.

B. Useful vs. background coincidence discrimination based on particle time of flight

In an attempt to further improve SNR, we focus coincidence analysis on M2 and higher-order coincidences, and on the differences in times of flight between neutrons and gamma rays. The main idea is to represent all possible combinations of three coincident pulses in a two-dimensional time histogram, as shown in Fig. 4.The axes are the time intervals between the 1st and 2nd pulses ($\Delta_{2\text{-}1}$ axis), and between the 2nd and 3rd pulses ($\Delta_{3\text{-}2}$ axis). The regions of interest (ROI) on these 2D histograms are determined form the times of flight of gamma rays and neutrons between the sources and the detectors. The energy spectrum of ^{252}Cf neutrons typically follows a Watt/Maxwell law that spreads between approximately 0 and 10 MeV [26]. The lowest detectable neutron energy with a 100 keVee threshold is about 700 keV [27] which is the real neutron energy corresponding to the scintillation light produced by 100 keV electrons. This 700 keV energy corresponds to a neutron speed of 1.15 $cm.ns^{-1}$, while 2 MeV (approximately the average energy of ^{252}Cf neutrons) and 10 MeV neutrons (practically their maximum energy, even if higher energies are still possible but for few neutrons) travel at a speed of 2.0 $cm.ns^{-1}$ and 4.3 $cm.ns^{-1}$, respectively. On the other hand, gamma rays travel much faster at a speed of 30 $cm.ns^{-1}$. With the 35 cm source-to-detector radial distance, the time of flight of detected neutrons is thus approximately in the range [8–30] ns, while gamma rays are grouped within 1 ns interval. In reality, the length of the used detectors (cf. Fig. 1) enlarges the span of the source-to-detector distances. Taken together with the time resolution of the detectors and data acquisition electronics (FWHM of about 4 ns), it increases the range between pulse detection times.

Therefore, the γγγ coincidences appear in the red triangles near the origin of the histograms, which extend to about 5 ns on each axis. The γγγ region is densely populated in the case of the ^{252}Cf source because of the high prompt gamma ray multiplicity per spontaneous fission, which is close to 8 [26]. This region is also dense in the case of ^{60}Co because Compton scattering between detectors increases the order of original M1 coincidences between the 1.173 MeV and 1.332 MeV gamma rays. In addition, the ^{60}Co histogram also shows intense bands along the axes corresponding to accidental coincidences, when an uncorrelated pulse is detected with a correlated pair (noted "Acc+γγ" and "γγ+Acc" on Fig. 4).

Another noticeable region populated by γγn coincidences is visible in the ^{252}Cf and AmBe histograms near the $\Delta_{3\text{-}2}$ axis. It corresponds to a short time interval between the two first pulses, i.e. mainly γγ pairs, and to a longer delay between the 2nd and 3rd pulses of the triple coincidence, i.e. a γn pair.

The ^{252}Cf plot differs from the others by the region populated with γnn coincidences due to spontaneous fissions. The latter are mostly triggered by a prompt gamma ray, followed by the detection of two prompt neutrons. The first one is in the [8-30] ns interval on $\Delta_{2\text{-}1}$ axis, and the second arrives up to 30 ns later, in the [0-30] ns range on $\Delta_{3\text{-}2}$ axis. The number of counts in this region of interest (ROI in red in Fig. 4) are given in Table III. The SNR is 5.1 with this new approach, which is significantly better than the SNR of 2.3 obtained in (3) with the shift register traditional method, which illustrates the potential interest of a coincidence selection based on particle times of flight. However, more work is needed to confirm this promising result, especially by performing acquisition with more intense sources measured simultaneously, to highlight the limits in terms of accidental coincidences. Moreover, measurements of these sources placed in mock-up drums filled with organic and metallic matrices (simulating technological wastes) have to be carried out, in order to assess detection sensitivity of high-order coincidences and the impact of gamma and neutron scattering on the contrast between the regions of interest of Fig. 4.

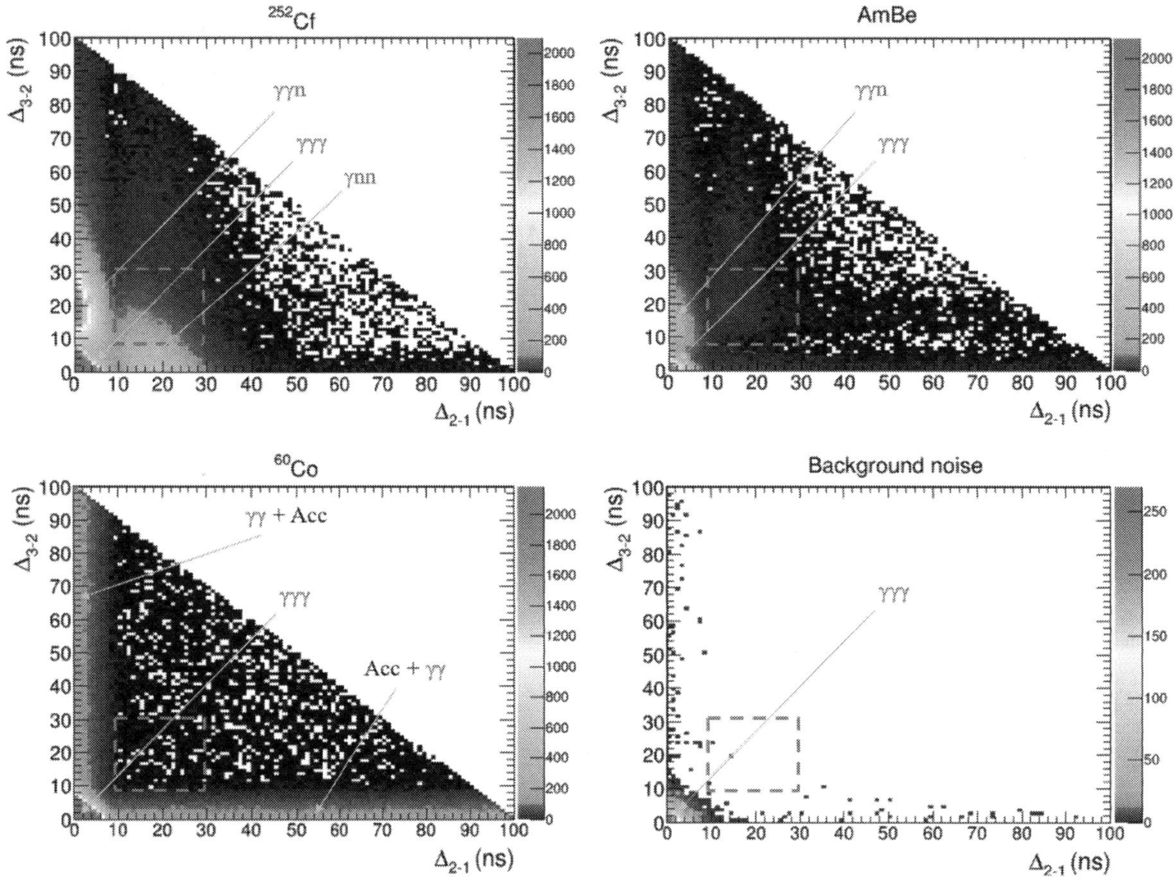

Fig. 4. Experimental two dimensional time histograms of all combinations of three coincident pulses in the measurements with ^{252}Cf (top left), AmBe (top right), ^{60}Co (bottom left) and background noise (bottom right). The color scale of ^{252}Cf and ^{60}Co plots is adjusted to that of AmBe (i.e. with a maximum near 2000 counts/bin). The ROI selected to provide best counting for the ^{252}Cf spontaneous fission coincidences while minimizing the parasitic ones is indicated, on each plot, by the red dashed-line rectangle (10 ns $\leq \Delta_{2\text{-}1} \leq$ 30 ns and 10 ns $\leq \Delta_{3\text{-}2} \leq$ 30 ns).

TABLE III
NUMBER OF COUNTS IN THE REGION OF INTEREST OF FIG. 3 (RED RECTANGLES). SNR IS DEFINED BY THE RATIO OF THE COUNTS OBTAINED WITH ^{252}CF, OVER COUNTS OBTAINED WITH OTHER SOURCES

Radioactive source	Counts in ROI
^{252}Cf (3.6×10^{4} n.s^{-1})	28767
AmBe (2.3×10^{5} n.s^{-1})	4698
^{60}Co (882 kBq)	951
Background noise	3
SNR	5.1

IV. Conclusion

The present work has shown that a discrimination between spontaneous-fission and background noise events of different origins can be achieved even with basic plastic scintillators without PSD capability, based on the time-of-flight differences of the detected neutrons and gamma rays. The preliminary one-by-one measurements performed with correlated neutron and gamma-ray sources (^{252}Cf, AmBe, and ^{60}Co separate acquisitions) have evidenced that the SNR (signal-to-noise ratio) with the classical shift register method can be improved by sorting triple coincidences in a two-dimensional histogram of time intervals between the 2nd and 1st pulses (X axis) and between the 3rd and 2nd pulses (Y axis). Contrary to PSD detectors that allow gamma rejection (after neutron-gamma discrimination), gamma pulses participate here to the useful signal, the ROI (region of interest) of γnn coincidences leading to the best contrast (best SNR) between useful coincidences (spontaneous fissions of ^{252}Cf) and parasitic ones (^{60}Co and AmBe).

Future work will focus on scattering crosstalk rejection to further improve SNR. MCNPX-PoliMi simulations will also be implemented and validated with experimental data, in order to get better insight into the origin of the detected triple coincidences (γγγ, γnn, γγn, but also the difficult-to-observe nnn coincidences) and to optimize scattering crosstalk rejection.

On the other hand, as the present study addressed one-by-one measurements of neutron and gamma-ray sources (separate acquisitions), we will focus on high count rate measurements with all the sources together, and more specifically on the subtraction of accidental coincidences.

The number of detectors in the array will also be doubled, with 16 large PVT scintillators (10 cm × 10 cm × 100 cm) and a new high-count-rate FASTER electronics will be used [28]. The performances of this upgraded setup will be assessed with ≈ 100 L mock-up drums filled with organic or metallic matrices, to simulated realistic technological waste, in which the same sources but also plutonium samples will be placed in different positions. A 5 cm thick lead shield will also be used to limit the influence of plutonium gamma rays, and that of other gamma emitters like ^{137}Cs or ^{60}Co. In this way, the discrimination between useful and background coincidences will be tested in realistic conditions, i.e. with significant neutron and gamma attenuation in the matrices, and with time-of-flight spreads due to neutron and gamma scattering. The assessment of measurement uncertainties due to matrix attenuation and source position will be of first importance in view of future developments towards an industrial application.

Acknowledgment

This work was supported by the CEA R&D program for radioactive waste characterization.

References

[1] N. Ensslin, "Principles of Neutron Coincidence Counting", Passive Nondestructive Analysis of Nuclear Materials, Los Alamos National Laboratory, chapter 16, 457-491.
[2] R.Kouzes et al. "Neutron detection alternatives to 3He for national security applications", Nucl. Instrum. Meth. A 623, 1035-1045 (2010)
[3] D. Henzlova et al. "Current Status of 3He Alternatives Technologies for Nuclear Safeguards" LA-UR-15-21201 Ver. 3, PNNL-24307, July 2015.
[4] B. Simony, C. Deyglun, B. Pérot, C. Carasco, N. Saurel, S. Colas, J. Collot, "Cross-talk characterization in passive neutron coincidence counting of radioactive waste drums with plastic scintillators", IEEE Transactions on Nuclear Science, Vol. 63, No. 3, June 2016, 1513-1519.
[5] B. Simony, B. Pérot, C. Carasco, F. Jallu, N. Saurel, S. Colas, P. Girones, J. Collot, "Passive neutron coincidence counting with plastic scintillators for the characterization of radioactive waste drums", IEEE Transactions on Nuclear Science, Vol. 64, No. 10 (2017) 2719-2724.
[6] V. Bottau, L. Tondut, P.-G. Allinei, B. Pérot, C. Eleon, C. Carasco, R. De Stefano, G. Faussier, "High Resolution Gamma Spectrometry of Plutonium Bearing Waste Drum With High-Energy Reaction-Induced Gamma Rays" IEEE Transactions on Nuclear Science, Vol. 67, No. 4 (2020) 575-584.
[7] S. Li, S. Qiu, Q. Zhang, Y. Huo, H. Lin, "Fast-neutron multiplicity analysis based on liquid scintillation", Applied Radiation and Isotopes 110, 53-58 (2016).
[8] T. H. Shin, M. Y. Yua, M. J. Marcath, D. L. Chichester, I. Pázsit, A. Di Fulvio, S. D. Clarke, S. A. Pozzi, "Neutron Multiplicity Counting Moments for Fissile Mass Estimation in Scatter-Based Neutron Detection Systems", Nuclear Science and Engineering, 188:3, 246-249.
[9] R. Sarwar, V. Astromskas, C. H. Zimmerman, G. Nutter, A. T. Simone, S. Croft, M. J. Joyce, "An event-triggered coincidence algorithm for fast-neutron multiplicity assay corrected for cross-talk and photon breakthrough", Nucl. Instrum. Meth. A 903, 152-161 (2018).
[10] M. J. Joyce, R. Sarwar, V. Astromskas, A. Chebboubi, S. Croft, O. Litaize, R. Vogt, C. H. Zimmerman, "High-order angular correlation of californium-252 fission neutrons and the effect of detector cross-talk", Nucl. Instrum. Meth. A 954, 161866 (2020).
[11] N. Zaitseva et al. "Plastic scintillators with efficient neutron/gamma pulse shape discrimination", Nuclear Instruments and Methods in Physics Research A 668 (2012) 88-93.
[12] P. Roy, K. Banerjee, A. K. Saha, C. Bhattacharya, J. K. Meena, P. Bhaskar, S. Mukhopadhyay, S. Bhattacharya, "Detailed investigation on the possibility of using EJ-299-33A plastic scintillator for fast neutron spectroscopy in large scale experiments", Nucl. Instrum. Meth. A 911, 198-202 (2018).
[13] N. P. Zaitseva, A. M. Glenn, A. N. Mabe, M. L. Carman, C. R. hurlbut, J. W. Inman, S. A. Payne, "Recent developments in plastic scintillators with pulse shape discrimination", Nucl. Instrum. Meth. A 889, 97-104 (2018).
[14] M. P. Taggart, P. J. Sellin, "Comparison of the pulse shape discrimination performance of plastic scintillators coupled to a SiPM", Nucl. Instrum. Meth. A 908, 148-154 (2018).
[15] M. Ellis, C. Tintori, P. Schotanus, K. Duroe, P. A. Kendall, G. Mini, "The Effect of Detector Geometry on EJ-309 Pulse Shape Discrimination Performance", IEEE Nuclear Science Symposium Conference Record, 2013.
[16] C. Payne, P. J. Sellin, M. Ellis, K. Duroe, A. Jones, M. Joyce, G. Randall, R. Speller, "Neutron/gamma pulse shape discrimination in EJ-299-34 at high flux", IEEE Nuclear Science Symposium Conference Record, 2015.
[17] M. G. Paff, S. D. Clarke, S. A. Pozzi, "Organic liquid scintillation detector shape and volume impact on radiation portal monitors", Nucl. Instrum. Meth. A 825, 31-39 (2016).
[18] Banerjee, K. et al. "Variation of neutron detection characteristics with dimension of BC501A neutron detector". Nucl. Instrum. Meth. A 608, 440–446 (2009).
[19] CAEN Electronic Instrumentation, CAEN DT5730 (https://www.caen.it/products/compass/), accessed 28/07/2021
[20] CoMPASS, Multiparametric DAQ Software for Physics Applications (https://www.caen.it/products/dt5730/), accessed 28/07/2021
[21] N. Ensslin, "The Origin of Gamma Rays", Passive Nondestructive Analysis of Nuclear Materials, Los Alamos National Laboratory, chapter 1, pp. 1-25.
[22] ROOT Data Analysis Framework, CERN, https://root.cern.ch, accessed 28/07/2021.
[23] K. Boehnel "Determination of plutonium in nuclear fuels using the neutron coincidence method", March 1978.
[24] N. Ensslin, W. C. Harker, M. S. Krick, D.G. Langner, M. M. Pickrell, J. E. Stewart, "Application Guide to Neutron Multiplicity Counting", LA-13422-M, November 1998.
[25] D. Henzlova, H. O. Menlove, S. Croft, A. Favalli, P. Santi, "The impact of gate width setting and gate utilization factors on plutonium assay in passive correlated neutron counting", Nucl. Instrum. Meth. A 797, 144-152 (2015).
[26] N. Ensslin, "Principles of Neutron Coincidence Counting", Passive Nondestructive Analysis of Nuclear Materials, Los Alamos National Laboratory, chapter 11, pp. 337-354.
[27] S. A. Pozzi, J. A. Mullens, J. T. Mihalczo, "Analysis of neutron and photon detection position for the calibration of plastic (BC-420) and liquid (BD-501) scintillators", Nucl. Instrum. Meth. A 524, 92-101 (2004).
[28] FASTER, Fast Acquisition SysTem for nuclEar Research, https://faster.in2p3.fr, accessed 28/07/2021.

UAV prototype for localization and identification of radioactive contamination and emitters

S.Moretto[1,2], F.E. Pino Andrades [1,3], J. Delgado[1], C.L. Fontana[1,2], D. Fabris[2], G. Nebbia[2], M. Turcato[2], D. Brunelli[4], L. Pancheri[4], A. Quaranta[4]

[1]Università degli Studi di Padova, via Marzolo 8, 35131, Padova ITALY

[2]Istituto Nazionale di Fisica Nucleare, Sez. Padova, via Marzolo 8, 35131, Padova ITALY

[3]Istituto Nazionale di Fisica Nucleare, Laboratori Nazionali di Legnaro, Viale dell'Università, 2, 35020 Legnaro (Padova) ITALY

[4]Università di Trento, Trento,ITALY

Corresponding author: Sandra.moretto@unipd.it

***Abstract*—The Dragon (Drone for RAdiation detection of Gammas and Neutrons) prototype aims at designing and developing an unmanned aerial vehicle (UAV) equipped with a detection system able to identify radioactive materials, spread over an area or located in a specific position. The system is focused on the localization of the unknown emitter and its subsequently identification.**

The proposed prototype is made up of two easily interchangeable detection systems, one will be used as a counter while the second will be aimed to perform good-resolution gamma spectroscopy. Both solutions have neutron - gamma discrimination capability in order to be suitable for special nuclear materials (SNM) detection in gamma contamination background.

The data acquisition module is made up of a compact digitizer board (RedPitaya, sampling rate of 125 MHz and 14 bits of resolution.), a mini computer (Raspberry, for example). This combination allows to install an embedded operating system (e.g. Linux) that can run the necessary software for the Data Acquisition (DAQ), like the ABCD distributed DAQ.

Our contribution will be aimed to show a comprehensive characterization of the two detection systems, a medium size CLLB scintillation detector, and a large plastic scintillator, EJ-276, in order to assess their potential use in a UAV-based radiation monitoring system.

***Keywords* —UAV, neutron-gamma, scintillation detectors**

I. Introduction

Radioactive and nuclear materials may compose a threat to public health and homeland security in the form of terrorism menaces, lost orphan sources, nuclear accidents or radioactive contamination. Since the incident in March 2011 at the Fukushima Daiichi Nuclear Power Plant, a lot of effort has been devoted on the development of new unmanned ground and aerial vehicles (UGV and UAV respectively), equipped with radiation monitoring devices [1] . UAV are mainly used in accident scenarios, in areas of difficult access, and in cases where the radiation dose is expected to be too high for a manned survey. Frequently, the radiation monitoring capabilities of the detectors embedded on a UAV are suitable specifically for gamma-ray detection (dosimetry and spectrometry).

At present, the most common gamma ray detectors used are: Geiger-Müller (GM) tubes [2] , cadmium zinc telluride (CZT) detectors [3] [4] Martin et al. (2015, 2016); [5] Selivanova et al. (2019), and scintillators such as LaBr3:Ce [6] ; NaI:Tl [7] and CsI [8] . The GM tubes are used for dosimetry purposes, while, CZT detectors are adopted for identification of radionuclides (spectrometry). However, both kinds of devices have an inherently very low sensitivity, being lightweight (low electronic density, GM tubes) or small volume (max. ~ 1 cm3, CZT). Small and medium-size scintillation detectors have been employed, they offer the possibility to perform dosimetry and spectrometry at the same time. In case of medium-size scintillators, the sensitivity is significantly increased with respect to the GM tubes and CZT detectors. However, the increase of size and weight of the UAV payload means more power consumption, thus less flight time or more powerful motors.

The Dragon (Drone for RAdiation detection of Gammas and Neutrons) project aims at designing, developing, and characterizing a UAV equipped with a detection system able to identify radioactive materials, spread over an area or located in a specific position. The Dragon project is being funded by the INFN_E commission of the Italian National Institute for Nuclear Physics (Istituto Nazionale di Fisica Nucleare, INFN).

The type of radioactive sources that will be detected are gamma emitters and special nuclear materials (SNM). In particular, the proposed technology incorporates an integrated thermal neutron and gamma ray detector, able to discriminate between the two kinds of particles and to perform good-resolution gamma spectroscopy. Neutron and gamma measurements are complementary: their combined power is expected to improve the system performances. SNMs (Highly Enriched Uranium and Plutonium) are difficult to detect, especially when masked or shielded. Gamma rays and neutrons emitted by SNMs have to be detected at the same time, in order to increase the sensitivity against natural background [9] . The Li-containing elpasolite scintillators (CLYC, CLLB, etc.) seem to be the right choice to fulfil the mentioned requirements: they exhibit excellent energy linearity; high light output, resulting in good energy resolution; have good thermal neutron detection efficiency, through the 6Li(n;)3H capture reaction. One aspect to take into account is that La-containing scintillators such as CLLB, CLLBC, LaBr3, LaCl3, etc. exhibit an intrinsic radioactive background, from naturally occurring 138La and from alpha decays from actinium (227Ac) contamination [10] [11] [12] [13] [14]

In the framework of the Dragon project, we present in this work the characterizations of a large plastic scintillator EJ276 and of a medium size CLLB (Cs2LiLaBr6:Ce) scintillation detector, in order to assess their potential use in a UAV-based radiation monitoring system. The parameters studied are: energy resolution, time resolution, neutron/gamma discrimination capability, high counting rate performance and the neutron in a high gamma flux. Fig. 1a shows a 3D rendering of the prototype of the detection system; it includes: the detector, a compact fast digitizer, a mini computer to analyse the data and a high voltage power supply. An anti-impact structure has been added in the mechanical design in order to protect all mentioned components. Fig. 1b shows a picture of the 3D-printed structure.

II. EXPERIMENTAL OVERVIEW

The detection system of the present study is made up of :

- a 3x5" cylindrical EJ-276 plastic scintillator (from Eljen Technology, Texas)
- a 2" x 2" cylindrical CLLB scintillation detector, purchased from Saint Gobain crystals.

The main physical properties of the plastic EJ-276 scintillator are: density 1.096, light output (% Anthracene) 56, scintillation efficiency 8600 ph/MeV, wavelength of maximum emission 425 nm, mean decay time 460 ns for neutron and 270 ns for gamma.

The plastic scintillator is coupled with a PMT(H6559) Hamamatsu photomultiplier.

The main physical properties of the CLLB scintillation detector are: density 4.2 g/cm3, light output 40000 ph/MeV, hygroscopic, wavelength of maximum emission 420 nm and decay time, according to the manufacturer, 180 ns (50% for gamma and 61% for n) and 1080 ns (50% for gamma and 39% for n).

The scintillator was coupled to a 2" dia. PMT (R6231) Hamamatsu photomultiplier together with a voltage divider (AS20).

The PMT signals were acquired by two digitizers: a CAEN DT5725 fast digitizer (250MSamples/s, 14-bit ADC resolution and Digital Constant Fraction Discriminator, DCFD, embedded in the firmware for precise timing measurements) as reference for the CLLB and the plastic scintillator, and the RedPitaya STEMLAB 125-14 (125Ms/s, 14-bit ADC resolution, power consumption 2 A 5V) for the plastic one. The Red Pitaya is a portable light digitizer with two input channels, digitizer expected to be installed in the DRAGON project, due to its light weight and low power consumption. These digitizers allow to record the PMT current pulses as digitized waveforms, to perform subsequent off-line analysis. The installed firmware in the DT5725 is the Digital Pulse Processing for Charge Integration and Pulse Shape Discrimination, DPP-PSD, which can pre-process data determining the pulses timestamps and integrating them over two gates, for the so-called double integration PSD method Pino et al. (2014a).

Bias voltage was applied using a CAEN V6533 VME HV Power Supply Module. A CAEN V1718 VME-USB2.0 Bridge was used to control the HV power supply unit. A CAEN A3818 PCI Express CONET2 Controller installed on a PC was used to control the digitizer (by means of optical fibre connections using the CAEN CONET2 protocol). All parameters of the nuclear electronics, and also the data acquisition, were managed using a software called ABCD (Acquisition and Broadcast of Collected Data) [15] [16] [17] released as an open-source project.

To study the gamma response of the detectors, different calibration sources are used (137Cs, 22Na, 60Co, 133Ba, 57Co and 88Y) with different activities between 200 kBq and 300 kBq.

To study the neutron and gamma response, a 252Cf source was used (neutron yield at the time of the experiments ~ $5.2\times10^4 \pm 20\%\ s^{-1}$).

III. EXPERIMENTAL RESULTS

One of the objectives of the DRAGoN project is to detect and identify SNM, and this can represent a challenging task taking into account that the SNM can be hidden or masked by a high gamma background, but with the simultaneous detection of neutrons and gamma-rays, it is possible to increase the sensibility. Therefore a very important feature that should be well known is the performance of the detection system in a high counting rate field. In the present work, the determination of the dead time of the detection system and the neutron detection response in a high gamma counting rate will be presented. Moreover, comparative results between different readouts with the plastic scintillator will be shown.

A. High Counting Rate Tests

In this kind of tests, in order to minimize the operators radiation exposure, the detector was fixed to a movable carriage on a guide rail, and its movements (change of position and speed) were controlled from a computer using the Linux CNC platform. The position of the gamma sources was fixed at one edge of the guide rail.

As far as the plastic scintillator system, the gamma sources used were ^{60}Co, ^{22}Na, ^{137}Cs and ^{133}Ba, while for the CLLB the source was a 185 MBq ^{60}Co gamma source.

1) Plastic Scintillator EJ-276

A comparison of the digitizer performance can be seen in Fig. 1, where the red line represents an equivalent behaviour between both digitizers. The dead time of the RedPitaya was determined for each measurement using the CAEN DT5725 digitizer as a reference. The experimental results (blue line in Fig.1) show that when the counting rate is increasing the RedPitaya performance decreases since it can be seen that there is a logarithmic growth of the dead time. We can conclude that for a counting rate equal to 3 kcps the rate of lost events is about 15%.

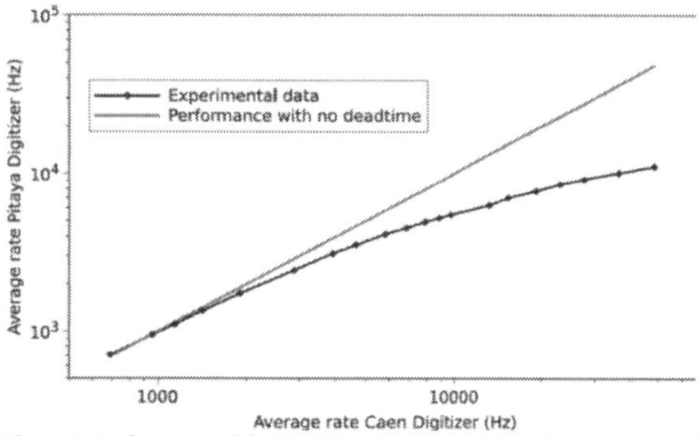

Figure 1: Performance of the RedPitaya digitizer as a function of the counting rate. The average rate of the Red Pitaya digitizer respect to the CAEN digitizers is presented as a function of the counting rate.

2) CLLB detection system

In this case, several gamma energy spectra for a determined counting rate were acquired (each corresponding to a source-detector distance). Only on-line analysis was employed. From the spectra, the relative position of the 1332 keV full energy peak and the energy resolution (at the same energy) were studied as a function of the counting rate. The results are given in Fig. 2. It can be concluded that with counting rates up to about 40kcps the detector gives a response with a gain shift lower than 4.5%, and with an energy resolution @1332keV lower than about 5%.

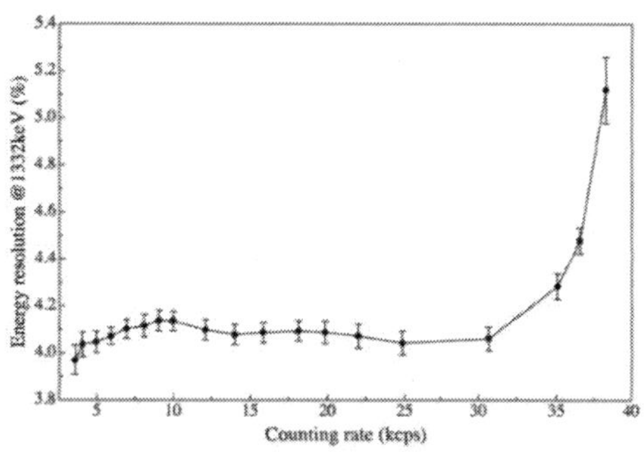

Figure 2: Energy resolution as a function of the counting rate for the CLLB detector.

B. Neutron detection response in a high gamma counting rate

Neutron detection response in a high gamma field is important taking into account the DRAGON project field of application. In the case of high gamma filed gamma events can be detected and determined as fake neutrons counts, giving unwanted false alarms. In these tests a 252Cf source is used together with a 60Co gamma source.

The PSP paramater used in the neutron/gamma discrimination for the plastic scintillator is defined as Q_{tail}/Q_{short}, while for the CLLB detector is defined as $Q_{short}t/Q_{long}$).

1) Plastic Scintillator EJ-276

In these tests, a 1 hour ^{252}Cf source measurement has been performed as a reference for the neutron counts. In particular, the region with the following limits have been considered as "neutron region": E1= 0.1 MeV, E2=0.57 MeV, E3=2.5 MeV, PSDmin=0.2 and PSDmax=0.614, and the slope is a four-degree polynomial function, see (1).

$$f(x) = 1.08756 - 6.74552x + 21.3637x^2 + 31.0788x^3 + 16.9939x^4 \quad (1)$$

The neutron region is clearly indicated by the red line in the Fig.3 of the 2D plot, PSP parameter for the plastic scintillator vs light output.

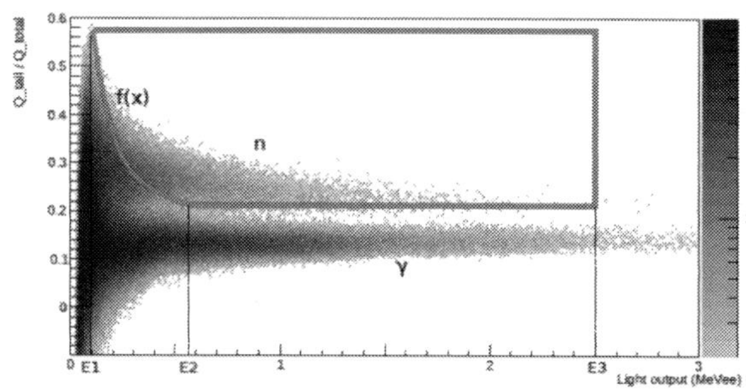

Figure 3: 2D-PSD of a 252Cf source, with the neutron contoured region (red line). Using the Ej-2763"x5", the RedPitaya digitizer and the PMT model H6559.

Once defined the neutron region, the high gamma rate measurements were performed placing a 60Co source next to the 252Cf source where different gamma rates were obtained by approaching the gamma source to the detector, each measurement has a time of around 10 min. The number of events in the contoured region was determined as a function of the mean rate. The results with respect to the 252Cf source measurement can be seen in Fig. 4. Up to 3-4kHz, no increase in the neutron region can be seen since there is not a significant increase in the neutron contoured region.

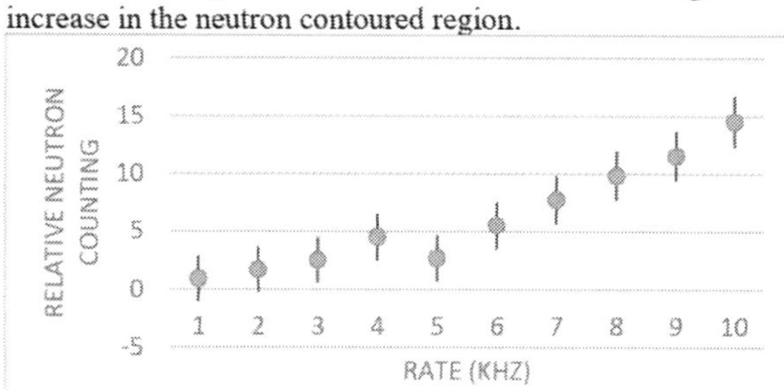

Figure 4: The neutron counting rate in a high gamma field is reported as a function of the counting rate.

2) CLLB Detector

In the CLLB detection system, the results of high gamma intense field is shown in Fig.5, where the 2D plot, PSP parameter for the CLLB versus light output is reported.

From Fig.5 it is clear that the high gamma events are in the region of smallest PSP respect to the neutron region (red line in Fig.5).

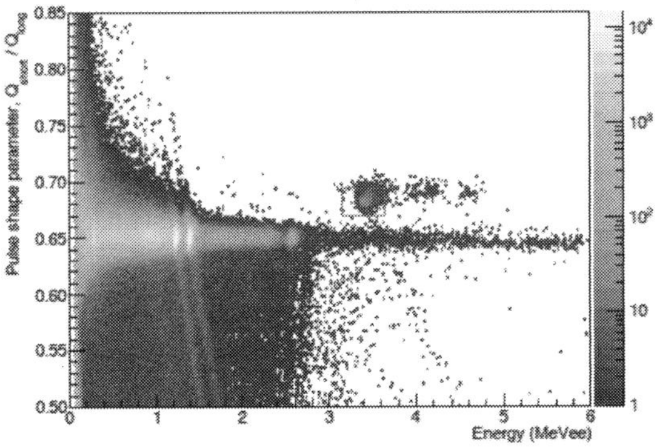

Figure 5: PSP parameter vs light output foe the CCLB detector in a 15kHz gamma rate.

For this reason, a higher gamma flux can be resolved with this detector, in fact up to gamma rate of 30 kcps, the neutron counts lost are lower than 10 %. The losses are mainly due to pile-up and coincidence summing effects, which become more important as gamma rate increases. This is clearly seen in Fig. 6, where the relative neutron counting rate is reported as a function of the count rate.

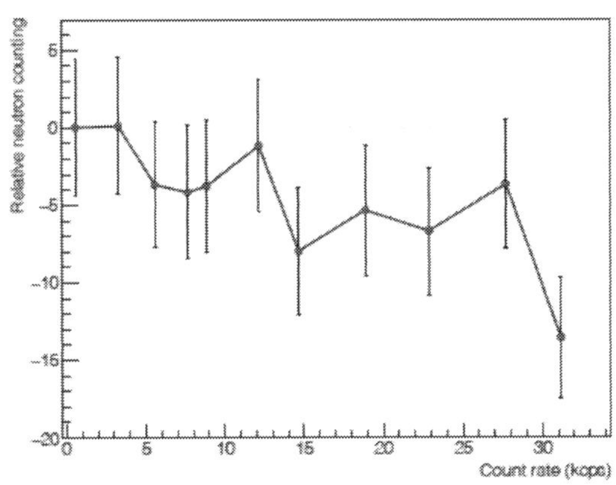

Figure 6: The relative neutron counting as a function of the total counting rate.

3) Comparative results of Red Pitaya and CAEN digitizer with plastic scintillator EJ276.

The Red Pitaya is a portable light digitizer, expected to be installed in the DRAGON project, due to its light weight and low power consumption. An important characteristic to be checked is its intrinsic capabilities to discriminate the neutron and gamma events in the case of the plastic scintillator.

In order to characterize the discrimination capability of the detector, the Figure of Merit (FoM), defined as , $FoM = \frac{\Delta}{(\delta_n+\delta_\gamma)}$ was computed by analyzing the PSP distribution[18] , Δ is the difference between the two mean values of the neutron and gamma distributions and $(\delta_n + \delta_\gamma)$ is the sum of the neutron and gamma Full Width at Half Maximum (FWHM). The results are shown in Fig.7. The resulted FoM is greater with the CAEN digitizer as expected, but, the results with Red Pitaya are still very good, obtaining a FoM of about 1.1 at 1 MeV.

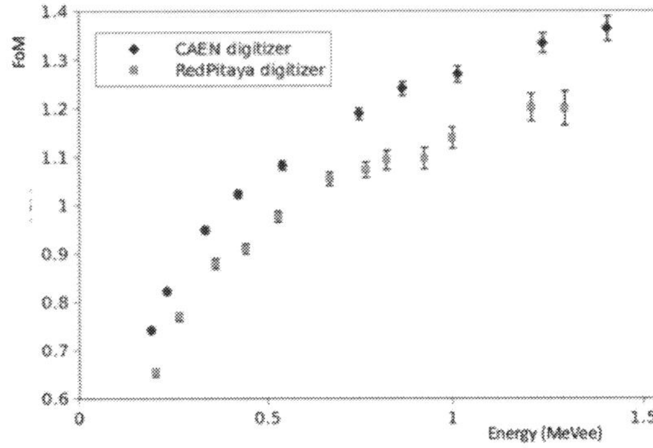

Figure 7: The FoM vs Energy for the plastic scintillator EJ276.

IV. CONCLUSION

In this paper, we presented the DRAGON project and the detection system performances in high gamma rate, in terms of dead time for the plastic scintillator system and energy resolution for the CLLB system. Moreover, the relative neutron counting rate in a high gamma files is presented for both detection solution.

As a result, for the plastic scintillator and the RedPitaya, up to 3 kHz (about 3 time more the natural radioactivity), the rate of lost events is about 15% and there is no a significant increase in the neutron contoured region. As far as the CLLB detector and the CAEN digitizer, higher counting rates can be considered, since, with counting rates up to about 40kcps the detector gives a response with a gain shift lower than 4.5%, and with an energy resolution @1332keV lower than about 5%. Moreover, up to gamma rate of 30 kcps, the neutron counts lost are lower than 10 %. For the CLLB the performances with RedPitaya are on-going, a slightly change in the total counting rate is expected but no increasing in the neutron counting rate is expected.

The use of the Red Pitaya has bene demonstrated to be a very good choice in this kind of application, where the power consumption is a key parameter to be considered, given a very satisfactory results in terms of discrimination capabilities between gamma and neutron.

The next steps will be the study of the detection system performances on the UAV itself with a whole characterization in terms of efficiency, altitude and velocity.

ACKNOWLEDGEMENTS

This work is supported by the INFN_E commission of the Italian National Institute for Nuclear Physics through the Dragon (Drone for RAdiation detection of Gammas and Neutrons) project.

REFERENCES

[1] D. Connor, P.G. Martin, T.B. Scott, "Airborne radiation mapping: Overview and application of current and future aerial systems" International Journal of Remote Sensing 2016;37(24):5953–87, 10.1080/01431161.2016.1252474.

[2] K. Kurvinen, P. Smolander, R. Pöllänen, S. Kuukankorpi, M. Kettunen , J. Lyytinen, "Design of a radiation surveillance unit for an unmanned aerial vehicle" Journal of Environmental Radioactivity 2005;81(1):1–10, 10.1016/j.jenvrad.433 2004.10.009.

[3] P. Martin, O. Payton, J. Fardoulis, D. Richards, T. Scott, "The use of unmanned aerial systems for the mapping of legacy uranium mines" Journal of Environmental Radioactivity 2015;143:135–40, 10.1016/j.jenvrad.2015.437 02.004.

[4] P. Martin, O. Payton, J. Fardoulis, D. Richards, Y. Yamashiki, T. Scott, "Low altitude unmanned aerial vehicle for characterising remediation effectiveness following the FDNPP accident" Journal of Environmental Radioactivity 441 2016;151:58–63, 10.1016/j.jenvrad.2015.09.007.

[5] A. Selivanova, J. Hulka, T. Vrba, I. Češpírová, "Efficiency calibration of a CZT detector and MDA determination for post accidental unmanned aerial vehicle dosimetry" Applied Radiation and Isotopes 2019;154:108879, doi:10.1016/j.apradiso.2019.108879.

[6] Y. Sanada, T. Torii, "Aerial radiation monitoring around the Fukushima Dai-ichi nuclear power plant using an unmanned helicopter", Journal of Environmental Radioactivity 2015;139:294–9, 10.1016/j.jenvrad.2014.06.027.

[7] K. Kurvinen, P. Smolander, R. Pöllänen, S. Kuukankorpi, M. Kettunen, J. Lyytinen, "Design of a radiation surveillance unit for an unmanned aerial vehicle", Journal of Environmental Radioactivity 2005;81(1):1–10, 10.1016/j.jenvrad.2004.10.009.

[8] R. Pöllänen, H. Toivonen, K. Peräjärvi, T. Karhunen, T. Ilander, J. Lehtinen, K. Rintala, T. Katajainen, J. Niemelä, M. Juusela, "Radiation surveillance using an unmanned aerial vehicle", Applied Radiation and Isotopes 2009;67(2):340–4, 10.1016/j.apradiso.2008.10.008.

[9] D. Cester, G. Nebbia, L. Stevanato, G. Viesti, F. Neri, S. Petrucci, *et al.*, "Special nuclear material detection with a mobile multi-detector system", Nuclear Instruments and Methods in Physics Research Section A: Accelerators, Spectrometers, Detectors and Associated Equipment 2012;663(1):55–63, 10.1016/j.nima.2011.10.011.

[10] J. Hartwell, R. Gehrke, "Observations on the background spectra of four LaCl3(Ce) scintillation detectors", Applied Radiation and Isotopes 2005;63(2):223–8, 10.1016/j.apradiso.2005.02.009.

[11] W. Kernan, "Self-activity in lanthanum halides", IEEE Transactions on Nuclear Science 2006;53(1):395–400, 10.1109/TNS.2006.869849.

[12] R.S. Woolf,B.F. Phlips, W.A.Wulf, "Characterization of the internal background for thermal and fast neutron detection with CLLB", Nuclear Instruments and Methods in Physics Research Section A: Accelerators, Spectrometers,Detectors and Associated Equipment 2016;838:147–53, 10.1016/j.nima.2016.09.013.

[13] G. Hull, F. Camera, G. Colombi, M. Josselin, B. Million, N. Blasi, "Detection properties and internal activity of newly developed La-containing scintillator crystals", Nuclear Instruments and Methods in Physics Research Section A: Accelerators, Spectrometers, Detectors and Associated Equipment 2019;925:70–5, doi:10.1016/j.nima.2019.01.094.

[14] R.S. Woolf, E.A. Wulf , B.F. Phlips, P. Chowdhury, E.G. Jackson, "Identification of internal radioactive contaminants in elpasolites (CLYC, CLLB, CLLBC) and other inorganic scintillators", Nuclear Instruments and Methods in Physics Research Section A: Accelerators, Spectrometers, Detectors and Associated Equipment 2020;954:161228, 10.1016/j.nima.2018.09.063.

[15] C. L.,Fontana, M. Lunardon, F. E. Pino, L. Stevanato, A. Carnera, C. Sada, F. Soramel, S.Moretto, (2018). "A distributed data acquisition system for nuclear detectors". International Journal of Modern Physics: Conference Series, 48 , 1860118., 10.1142/S2010194518601187.

[16] C.L. Fontana, M. Lunardon, F.E. Pino, L. Stevanato, S. Moretto, (2019) "Resource sharing in nuclear physics laboratory classes: A distributed data acquisition system for experiments with shared resources and data management." In 25th international conference on the application of accelerators in research and industry (p. 050024). Texas, doi:10.1063/1.5127716.

[17] C. L. Fontana, M. Lunardon, F. E. Pino, L. Stevanato, A. Carnera, C. Sada, F. Soramel, S.Moretto, (2017b). "A distributed data acquisition system for signal digitizers with on-line analysis capabilities". In 2017 IEEE Nuclear Science Symposium and Medical Imagi

[18] D. Cester, G.Nebbia, L. Stevanato, F.E. Pino, G. Viesti, (2014). "Experimental tests of the new plastic scintillator with pulse shape discrimination capabilities EJ-299-33". Nuclear Instruments and Methods in Physics Research Section A: Accelerators, Spectrometers, Detectors and Associated Equipment, 735 , 202–206. doi:10.1016/j.nima.2013.09.031.

The gamma and neutron monitor counters for the MICADO project

Paolo Finocchiaro[1], Luigi Cosentino[1], Quentin Ducasse[2], Sergio Lo Meo[3], Fabio Longhitano[1], Carmelo Marchetta[1], Antonio Massara[1], Alfio Pappalardo[1,#], Giuseppe Passaro[1], Salvatore Russo[1]

[1]INFN Laboratori Nazionali del Sud, 95123 Catania, Italy
[2]CEA, DES, IRESNE, Nuclear Measurement Laboratory, Cadarache, F-13108 Saint-Paul-lez-Durance, France
[3]ENEA Centro Ricerche, 40129 Bologna, Italy
[#] Current address: Extreme Light Infrastructure Nuclear Physics, 077125 Magurele, Romania
finocchiaro@lns.infn.it

***Abstract*—In the framework of the MICADO (Measurement and Instrumentation for Cleaning And Decommissioning Operations) European Union (EU) project, aimed at the full digitization of low and intermediate-level radioactive waste (radwaste) management, a set of 32 solid state thermal neutron detectors, named SiLiF, and 36 gamma-ray counters based on a scintillating fiber readout at each end by a silicon photomultiplier, named SciFi, have been built and characterized. MICADO project encompasses a complete active and passive characterization of the radwaste drums with neutrons and gamma rays, followed by a longer-term monitoring phase. The detectors described are suitable for the monitoring of nuclear materials and can be used around radioactive waste drums possibly containing small quantities of actinides, as well as around spent fuel casks in interim storage or during transportation. Suitable polyethylene moderators, for the SiLiF detectors, can be exploited to better shape the detector response to the expected neutron spectrum, according to Monte Carlo simulations that were performed. The SciFi detectors were thoroughly tested with a ^{22}Na and a ^{137}Cs gamma-ray sources. Both detectors described were tested with an intense AmBe source of neutron and gamma ray. The results are satisfactory and show a quite uniform and reproducible behavior. The next step will be the test in a real environment.**

***Keywords* — gamma-ray counters; neutron counters; radwaste management; radwaste monitoring.**

I. Introduction

The MICADO (Measurement and Instrumentation for Cleaning and Decommissioning Operations) Euratom (EU) project is aimed at the full digitization of low-level and intermediate-level radioactive waste (radwaste) management [1,2]. Following a complete active and passive characterization of the radwaste drums with neutrons and gamma-rays, the project contemplates a longer-term monitoring phase in the Work Package 7 by means of low-cost dedicated detectors for neutrons, named SiLiF [3], and for gamma-rays, named SciFi [4]. The proposed system for the online real-time monitoring consists of an array of many radiation sensors to be deployed all around a number of radioactive waste drums, in order to collect counting-rate data in real time and to make them available to a software platform named DigiWaste. A continuous automatic monitoring of the radwaste drums after their characterization represents an added value in terms of safety and security, and the availability of continuous streams of counting-rate data around each drum would be a comfortable tool toward the transparency, which now more than ever is a relevant topic of the nuclear industry [1,3–8]. As an evolution of the proof-of-principle systems discussed in [5–8], the detectors described here will be part of a prototype system which will be installed in a few real radwaste storage sites to prove its effectiveness. To be suitable for mass deployment these detectors must be small, reasonably inexpensive, robust, easy-to-use and reliable. Furthermore, the sensors must be configurable in a modular and scalable fashion, so that one can tailor the system to small, medium, and large-scale storage configurations. The proposed monitoring system, indeed, is based on detectors which can be easily installed and/or reassembled in different geometrical configurations, as they are mechanically very simple and are based on commercial electronics.

The radiological monitoring of radwaste must be based on the measurement of gamma-rays and neutrons because these are penetrating and thus more easily detectable out of the drums.

The SciFi gamma and SiLiF neutron monitor counters built and tested are described. The SciFi sensors were thoroughly tested with a ^{22}Na and a ^{137}Cs gamma-ray sources. Both detectors were tested with an intense AmBe source of neutron, which also emits a wide spectrum of gamma-rays thus somehow resembling a real radwaste drum. The results of the tests are shown.

II. SiLiF NEUTRON DETECTOR

The operating principle of the SiLiF neutron detector, as a viable alternative to ^{3}He tubes [9], is straightforward: following a thermal neutron capture by ^{6}Li, the ^{7}Li compound nucleus decays into an alpha particle (^{4}He) and a triton (^{3}H), emitted back-to-back with high energy according to the reaction

$$n + {}^6Li \rightarrow {}^3H\,(2.73\ MeV) + {}^4He\,(2.05\ \text{MeV}). \quad (1)$$

Semiconductor detectors, e.g., silicon diodes, can be used in combination with a neutron reactive film, usually made of ^{6}Li or ^{10}B and called neutron converter. Such a film converts thermal neutrons into charged particles which are then detected by the silicon diode. The cross section at thermal neutron energy decreases with the inverse of the neutron velocity. In such a scheme fast neutron can also be detected by surrounding the detector with a suitable moderator box, typically made from polyethylene, which slows neutrons down to thermal energy. The ^{6}Li, following neutron's capture, has a unique decay channel with no gamma rays emitted and with high kinetic energy, and decays into easy-to-detect particles [10]. Since ^{6}Li is chemically very reactive it was decided to employ ^{6}LiF, a stable salt enriched at 95% in ^{6}Li, that can be deposited by evaporation onto a suitable substrate. Instead of the direct deposition onto the silicon diode it was chosen to use independent detector and converter, which allows a better modularity and reconfigurability [11].

In Fig. 1 it is possible to see schematically the detection of neutrons by a SiLiF detector.

The detector configuration, shown in Fig. 2, consists of the double-sided silicon diode MSX09-300 sandwiched between two ^{6}LiF converter layers deposited onto carbon fiber substrates.

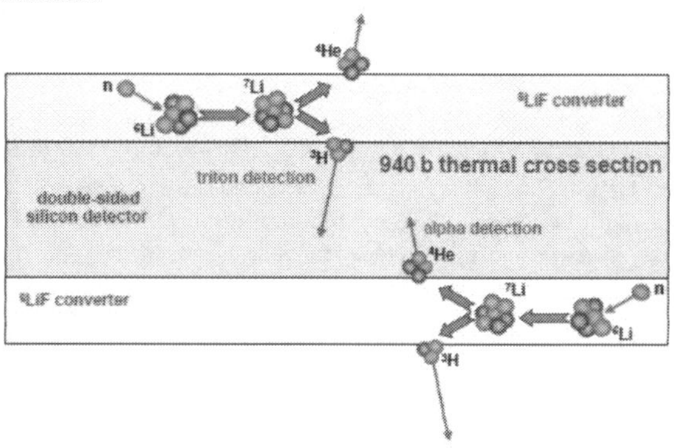

Fig. 1. The SiLiF scheme neutron detection.

To increase the detection efficiency for faster neutrons one makes use of a moderator. For the purposes of the MICADO project, the focus was on optimizing the moderator for neutrons with energy ranging from thermal to a few MeV. The moderator of 4 cm thickness, shown in Fig. 3a, gave the best results and was replicated in 32 units. The detector in its standard configuration, with 10 × 10 × 10 cm^3 size, is shown in Fig. 3c.

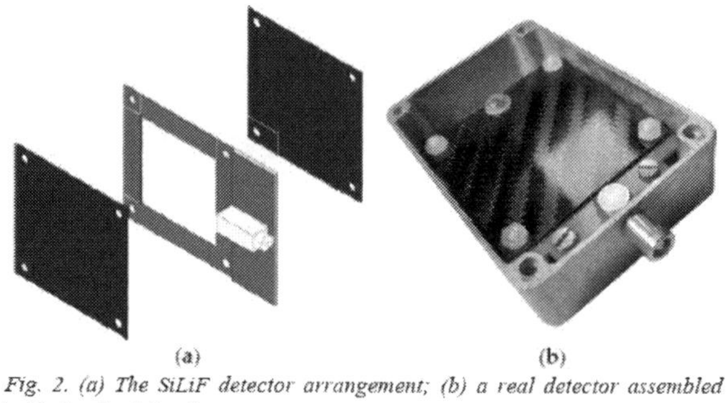

Fig. 2. (a) The SiLiF detector arrangement; (b) a real detector assembled inside its aluminium box.

Fig. 3. (a) 3D sketch of the SiLiF detector; (b) a SiLiF hosted in a half moderator; (c) final assembling of a SiLiF.

A. Test results

To test and characterize the SiLiF detectors, an intense AmBe source (2.2×10^6 neutrons/s) was used, installed in an experimental hall at the INFN Laboratori Nazionali del Sud (LNS). The source is surrounded by a first polyethylene case followed by 30 cm thick paraffin that slow down the high energy neutrons it emits.

The first set of measurements that was performed was a preliminary check that the detectors were operational and behaved as expected. Each detector in turn, inside its moderator, was placed on a small cart and positioned in front of the source just out of the box with the open door (Fig. 4a).

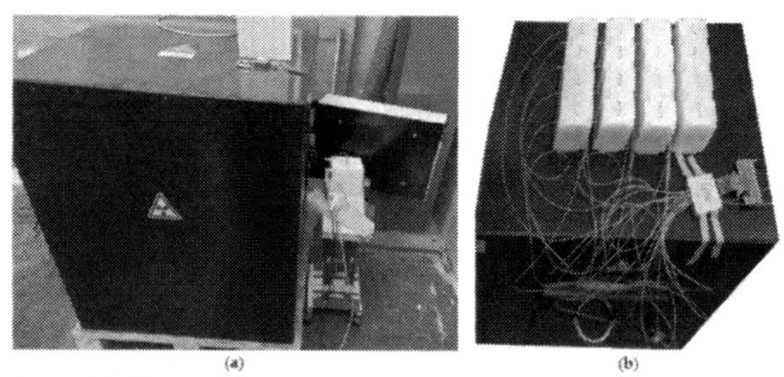

Fig. 4. (a) The AmBe neutron source box with a SiLiF detector in the front position. (b) The 4 × 4 array of SiLiF detectors in the top position.

A final test was performed using simultaneously 16 detectors aimed at measuring a low intensity flux of fast neutrons. To this purpose the detectors were arranged in a square 4 × 4 array in the top position upon the neutron source box (Fig. 4b). The detector was biased at 50 V, in order to operate in full depletion regime and be sensitive on both faces.

 https://doi.org/10.1051/epjconf/202125308002

A charge preamplifier, with a nominal gain of 25 mV/MeV, connected to a spectroscopy amplifier whose output was sent to an Amptek MCA8000A multichannel analyzer was used.

The superposition of the triton and alpha deposited energy, plus the effect of the emission from different depths/angles, produces a characteristic spectrum shape, as can be observed in Fig. 5, that also proves that all the detectors were working properly. The measurements were compared with Fluka [12] simulations giving an excellent agreement.

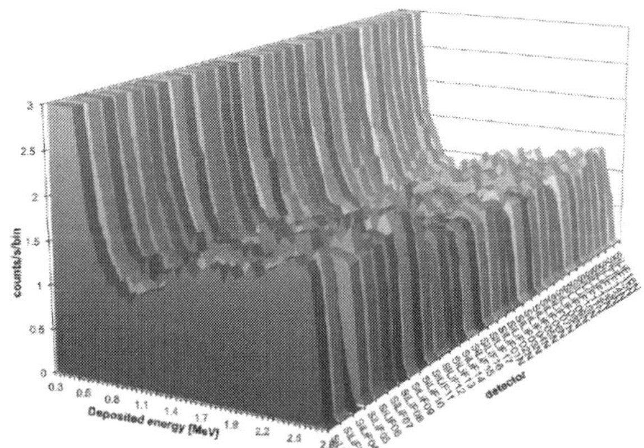

Fig. 5. The characteristic spectrum shape, measured for all the SiLiF detectors in the front position

In Fig. 6 the result of the final test where have been compared the (a) GEANT4 simulated and (b) evaluated fluxes, and still the (c) GEANT4 simulated and (d) measured counting rates. Unfortunately, due to a broken channel in the electronics, a centrally located detector did not provide useful data (white square with 0.00). It is however possible to state that the measurements have a good agreement with the simulations.

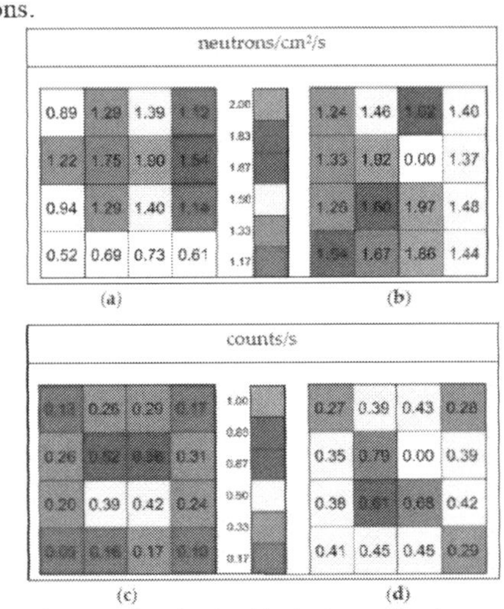

Fig. 6. (a) Total neutron flux simulated in the top position in correspondence of the 4 × 4 SiLiF array; (b) The measured total neutron flux with the 4 × 4 array. (c) The simulated neutron counting rates. (d) The measured neutron counting rates.

III. SciFi gamma-ray detector

To monitor the gamma radiation coming from the radioactive waste drums (height 76 or 88 cm), we opted for a simple and robust design of the detectors. A 3 mm diameter scintillating fiber was allocated inside a 2 cm diameter and 1 mm thick aluminium pipe and held in place by two cylindrical holders designed to host a circular PCB with a SiPM and its support circuitry. The holder has a central hole to allow for fiber-to-SiPM alignment and optical coupling by means of a tiny grease drop, and two side grooves to allow for the passage of cables. Two light-tight rubber caps complete the setup, with three cables coming out of one single side of detector for the common voltage bias and the two output signals. A sketch and three pictures of the SciFi detector components are shown in Fig. 7.

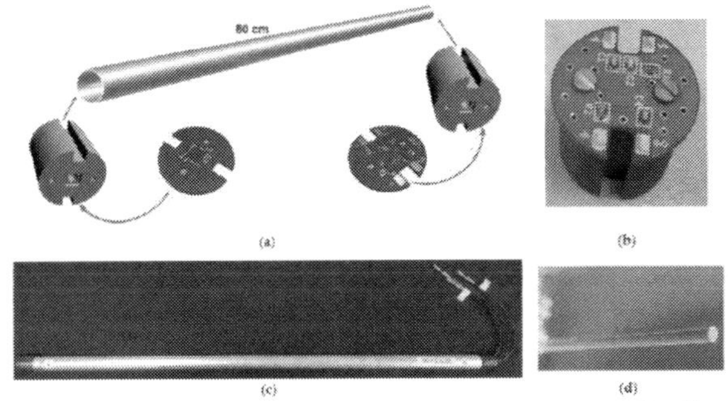

Fig. 7. (a) Sketch of the components of the SciFi detector assembly; (b) A fiber holder with the SiPM PCB in place; (c) A detector fully assembled; (d) A 3mm diameter fiber.

The operating principle is the following: whenever a gamma-ray interacts with the fiber it deposits a variable amount of kinetic energy, mainly due to the Compton effect, which gives rise to a short flash of scintillation light. A fraction of the scintillation photons is trapped into the fiber and propagates toward both ends. If the number of these photons is large enough to produce a signal above a predefined threshold simultaneously on both fiber ends, such a coincidence event is considered as the detection of a gamma-ray and can be counted.

The chosen SiPM for these detectors, after careful evaluation and testing, is the MicroFC-30035-SMT, produced by ON Semiconductor [13]. For all the tests described was used of a homemade voltage amplifier (200x gain and 4 GHz bandwidth). The SiPM bias was set at a 2.5 V overvoltage, i.e., 27 V bias. A threshold at 175 mV was used that corresponds to ≈3.5 photons and reduces the noise rate by three orders of magnitude down to about 350 counts per second. As the signals have a duration of about 30 ns, a duration of 100 ns for the coincidence window between the two SiPMs makes the probability of spurious coincidences negligible.

The chosen scintillating fiber, after GEANT4 simulation, evaluation and testing, is the 3mm diameter BCF-20 produced by Saint Gobain [14].

A. Test results

In order to characterize the detectors several tests were

made with standard laboratory gamma sources and a high activity gamma and neutron source. The electronic setup was very simple: the outputs of the SiPM amplifiers were connected to two discriminators, with threshold set at 175mV. The two logic outputs were used as inputs to a coincidence unit with a 100 ns window, and the final output was sent to a counter.

The validation of the detector behaviour was performed by means of a pointlike ^{22}Na source of activity A = 42±2 kBq which emits gamma-rays of 1274 keV. The source was placed at several distances from the fiber, in the range of few centimeters, and the number of detected gamma-rays was recorded. The observed count rate as function of the distance between the point-like ^{22}Na source and the 3 mm diameter fiber with 1/r fit is shown in Fig. 8.

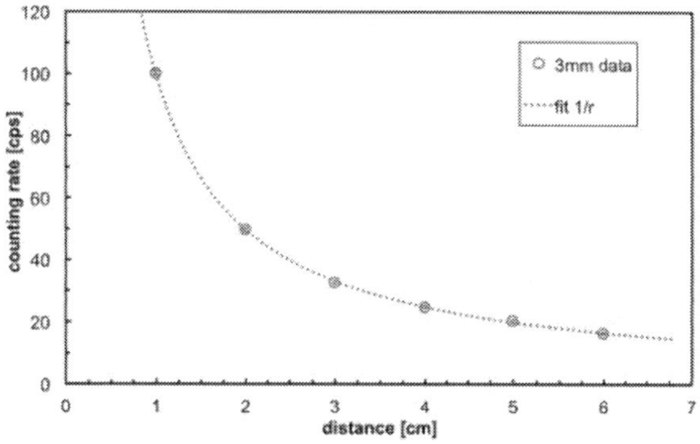

Fig. 8. Measured count rate vs distance between the point like 22 Na source and the 3 mm diameter fiber.

To verify the uniformity of the response within the set of 36 SciFi detectors, a ^{137}Cs gamma source was used, with activity A = 1.49±0.07 Mbq, and a detector holder that kept the mid-point of the fiber at a distance d = 52 cm from the source. The counting rate of each detector was measured in 200 s, as well as the background rate and the signal-to-background ratio, and the results are plotted in Fig. 9.

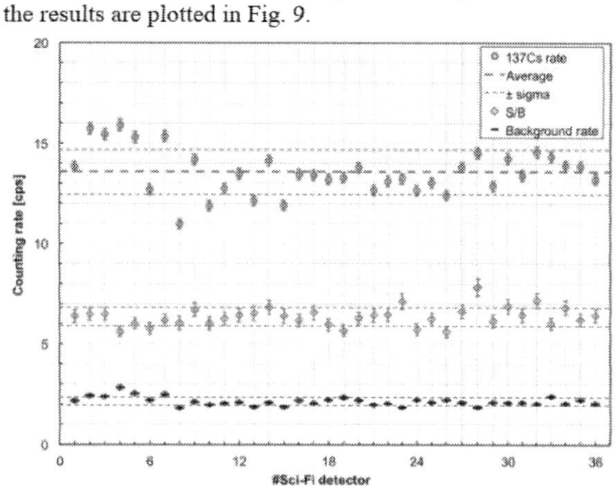

Fig. 9. The counting rate measured with the ^{137}Cs the background rate and the signal to background ratio of the 36 SciFi detectors.

For a series of tests of the SciFi detectors, the same previously mentioned AmBe source was used. As it also emits a wide spectrum of gamma rays it represents a realistic setup recalling the emission from a drum of radioactive waste. All the SciFi detectors were tested on top of the source box, and the measured counting rate, the background and the signal-to-background ratio are reported in Fig. 10.

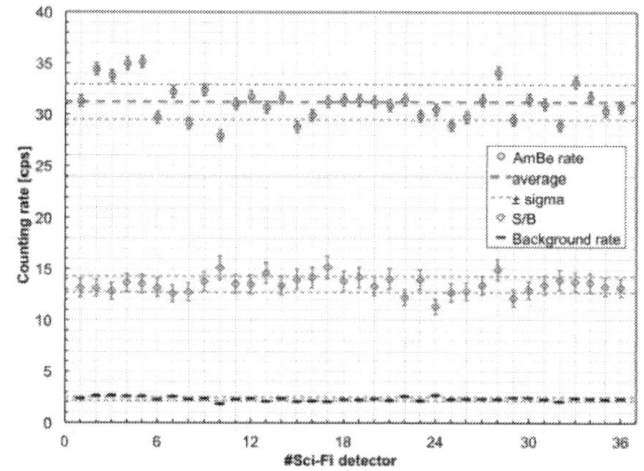

Fig. 10. The counting rate measured on top of the AmBe source box the background rate and the signal to background ratio of the 36 SciFi detectors

For each detector, the ratio between the observed count rates with ^{137}Cs and the AmBe sources, respectively, was also calculated. The results, plotted in Fig. 11, indicate a quite reasonably constant behaviour of this ratio.

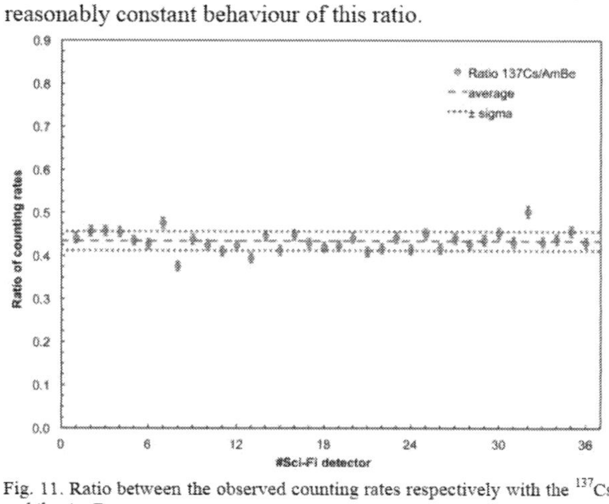

Fig. 11. Ratio between the observed counting rates respectively with the ^{137}Cs and the AmBe sources.

IV. Discussion

The dimension of a standard 220 L radioactive waste drum is 86 cm height and 57 cm diameter with a steel/polyethylene matrix (67/33% in mass). A possible arrangement of the SiLiF and SciFi detectors around such a radwaste drums is shown in Fig. 12. In the MICADO perspective the drums will be first characterized with several methods based on active and passive neutron and gamma ray measurements, then the monitoring detectors will be used to check the longer-term stability of their radiological behavior. Possible counting asymmetries between the detectors would signal an asymmetry in the radioisotopes distribution or, should this occur afterwards, a change in the internal structure of the radwaste package. Obviously, for a more precise evaluation of the sensitivity one should take into account also the background from the neighboring drums.

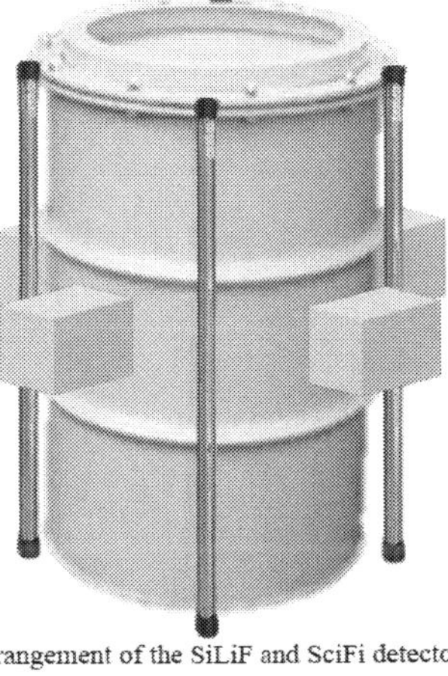

Fig. 12. Possible arrangement of the SiLiF and SciFi detectors on the radwaste drum.

The simulations allowed us to demonstrate the suitability of SiLiF and SciFi technology for the monitoring of radwaste drums that contain typical radioactive elements such as ^{60}Co kBq, ^{133}Ba, ^{134}Cs, ^{137}Cs, ^{152}Eu, ^{154}Eu, ^{241}Am and ^{240}Pu.

The tests carried out have shown that SiLiF detectors can be fruitfully employed to detect neutrons, in a wide energy range and with excellent gamma rejection, and that the SciFi detectors can be successfully employed to detect and count gamma rays in a broad energy range. These results prove that SiLiF and SciFi detectors can successfully be employed in the MICADO project in order to demonstrate the reliability of neutron and gamma-ray monitoring in a medium-long-term storage of radioactive waste drums.

One last point concerns the possible radiation damage of silicon in a radiation field. The radiation hardness claimed by the manufacturer of the silicon is up to 10^{13} n/cm^2. It is known that the main damage comes from fast neutrons which can dislocate silicon atoms thus creating defects in the semiconductor lattice. However, even considering the total neutron flux on the detectors, a life longer than hundreds of years is to be foreseen under these conditions. Indeed, as observed in reference [16], similar detectors were exposed to a fast neutron fluence of $3{\cdot}10^{10}$/cm^2 without appreciable changes in their characteristics.

V. Conclusions

The described tests, measurements and simulations have allowed us to prove the suitability of the SiLiF and SciFi technology for monitoring of radwaste drums.

The 32 SiLiF and 36 SciFi detectors we built have a reasonably uniform behavior, considering their robustness, low cost, and simple construction based on commercial components. We conclude that the SiLiF and SciFi detectors are promising candidates for low and intermediate level radioactive waste monitoring, and they are going to be tested soon in a real environment within the MICADO project.

References

[1] MICADO Project. https://www.micado-project.eu/
[2] Italy in MICADO WP7. https://www.micado-project.eu/news/particles-of-italianinnovation-project/
[3] Cosentino, L.; Ducasse, Q.; Giuffrida, M.; Lo Meo, S.; Longhitano, F.; Marchetta, C.; Massara, A.; Pappalardo, A.; Passaro, G.; Russo, S.; et al. SiLiF Neutron Counters to Monitor Nuclear Materials in the MICADO Project. Sensors 2021, 21, 2630. https://doi.org/10.3390/s21082630
[4] Cosentino, L.; Giuffrida, M.; Lo Meo, S.; Longhitano, F.; Pappalardo, A.; Passaro, G.; Finocchiaro, P. Gamma-Ray Counters to Monitor Radioactive Waste Packages in the MICADO Project. Instruments 2021, 5, 19. https://doi.org/10.3390/
[5] Finocchiaro, P. DMNR: A new concept for real-time online monitoring of short and medium term radioactive waste. In Radioactive Waste: Sources, Types and Management; Nova Science Publishers: New York, NY, USA, 2011; pp. 1–40.
[6] Finocchiaro, P. Radioactive Waste: A System for Online Monitoring and Data Availability. Nucl. Phys. News 2014, 24, 34.
[7] Cosentino, L.; Calì, C.; De Luca, G.; Guardo, G.; Litrico, P.; Pappalardo, A.; Piscopo, M.; Scirè, C.; Scirè, C.; Scirè, S.; et al. Real-Time Online Monitoring of Radwaste Storage: A Proof-of-Principle Test Prototype. IEEE Trans. Nucl. Sci. 2012, 59, 1426–1431.
[8] Finocchiaro, P.: Ripani, M. RadioactiveWaste Monitoring: Opportunities from New Technologies. In Proceedings of the IAEA International Conference on Physical Protection of Nuclear Material and Nuclear Facilities, IAEA-CN-254/117, Vienna, Austria, 13–17 November 2017.
[9] Henzlova, D.; Kouzes, R.; McElroy, R.; Peerani, P.; Aspinall, M.; Baird, K.; Bakel, A.; Borella, M.; Bourne, M.; Bourva, L.; et al. Current status of 3He alternative technologies for nuclear safeguards. NNSA USDOE EURATOM. LA-UR-15-21201 2015 doi:10.2172/1227248.
[10] Barbagallo, M.; Cosentino, L.; Forcina, V.; Marchetta, C.; Pappalardo, A.; Peerani, P.; Scire, C.; Scire, S.; Schillaci, M.; Vaccaro, S.; et al. Thermal neutron detection using a silicon pad detector and 6LiF removable converters. Rev. Sci. Instrum. 2013, 84, 033503.
[11] Pappalardo, A.; Barbagallo, M.; Cosentino, L.; Marchetta, C.; Musumarra, A.; Scirè,C.; Scirè, S.; Vecchio, G.; Finocchiaro, P. Characterization of the silicon+6LiF thermal neutron detection technique. Nucl. Instrum. Methods Phys. Res. Sect. A 2016, 810, 6.
[12] Fasso, A. et al. P.R. FLUKA: A Multi-Particle Transport Code; CERN Technical Report No. INFN/TC_05/11; CERN-2005-10; 2005.
[13] ON Semiconductor https://www.onsemi.com/products/sensors/silicon-photomultipliers-sipm
[14] Saint Gobain Scintillating Fibers. Available online: https://www.crystals.saint-gobain.com/products/scintillating-fiber
[15] Nordlund, K.; Zinkle, S.J.; Sand, A.E.; Granberg, F.; Averback, R.S.; Stoller, R.E.; Suzudo, T.; Malerba, L.; Banhart, F.; Weber, W.J.; et al. Primary radiation damage: A review of current understanding and models. J. Nucl. Mater. 2018, 512, 450.
[16] Amaducci, S.; Cosentino, L.; Barbagallo, M.; Colonna, N.; Mengoni, A.; Massimi, C.; Meo, S.L.; Finocchiaro, P.; Aberle, O.; Andrzejewski, J.; et al. Measurement of the 235U(n,f) cross section relative to the 6Li(n,t) and 10B(n,α) standards from thermal to 170 keV neutron energy range at n TOF. Eur. Phys. J. A 2019, 55, 120.

Localization of nuclear materials in large concrete radioactive waste packages using photofission delayed gamma rays

Manon Delarue[1], Eric Simon[1*], Bertrand Pérot[1], Pierre-Guy Allinei[1], Nicolas Estre[1], Daniel Eck[1], Emmanuel Payan[1], David Tisseur[2], O. Gueton[3], Denise Ricard[4] and Johann Collot[5]
[1]CEA, DES, IRESNE, DTN, SMTA, Nuclear Measurement Laboratory, F-13108 St Paul-lez-Durance, France
[2]CEA, DES, IRESNE, DER, SPESI, Instrumentation Sensors and Dosimetry Laboratory, F-13108 St Paul-lez-Durance, France
[3]CEA, DES, IRESNE, DTN, F-13108 St Paul-lez-Durance, France
[4]ANDRA, French National Radioactive Waste Management Agency, F-92298 Châtenay-Malabry, France
[5]Université Grenoble Alpes, CNRS, Grenoble INP, LPSC-IN2P3, 38000 Grenoble, France
*Corresponding author: eric.simon@cea.fr

***Abstract*— The characterization of radioactive waste packages is mandatory for their transport, interim storage and final disposal. In this framework, the Nuclear Measurement Laboratory of CEA DES IRESNE Institute, at Cadarache, France, uses a high-energy electron linear accelerator (LINAC) to produce an interrogating bremsstrahlung beam with endpoint energies ranging from 9 to 21 MeV to perform X-ray imaging and high-energy photon interrogation on large concrete packages. In particular, high-energy photon beam induces photofission reactions in both fissile (^{235}U, ^{239}Pu, ^{241}Pu) and fertile (^{238}U, ^{240}Pu, ^{232}Th, etc.) actinides possibly present in the radioactive waste. In order to assess their mass, we use delayed gamma rays emitted by their photofission products, which are measured with a 50 % relative efficiency High-Purity Germanium (HPGe) detector. Actinide differentiation, which is important for the fissile mass estimation, is based on the ratios of gamma rays emitted by different photofission products and requires appropriate corrections for the gamma attenuation in concrete. To this aim, we report here a localization method of point-like nuclear materials in the concrete matrix, based on the differential attenuation of several gamma rays emitted by a same photofission product. We use here the 1435.9 and 2639.6 keV lines of ^{138}Cs, with both experimental data and MCNP numerical simulations to determine the (r,θ) coordinates of nuclear materials. Then, the depth inside the concrete matrix, which is determined with a precision of a few percent, mainly depending on counting statistics on 1435.9 and 2639.6 keV net peak areas, is used to correct for the different gamma ratios used in the actinide identification method. Experimental tests with uranium samples have been performed to validate the localization method.**

***Keywords* — Photofission, Uranium, Delayed gamma rays, Bremsstrahlung, MCNP, Concrete matrix, Nuclear material localization**

I. Introduction

THE characterization of radioactive waste packages is a major challenge that has to be addressed to ensure the safety during their management (transport, interim storage, final repository). To that extent, the nuclear material content must be accurately evaluated. Among the active non-destructive methods implemented to obtain characterization information in the case of large and dense packages, such as concrete drums [1] -[5] , Active Photon Interrogation with high-energy X-rays (i.e. photofission) is the only one that could bring sufficient information on the nuclear material content inside the package. This interrogating technique relies on the fact that all actinides undergo fission when using photons of energy higher than 6 MeV. Coupled with the detection of delayed gamma rays emitted by photofission products, this method has the potential to evaluate the actinide mass present in a package, and possibly to separate fertile and fissile isotopes (for example ^{235}U and ^{238}U) using a discrimination signal based on their delayed gamma rays.

The potential of analyzing the delayed gamma ray signal following fission to obtain an actinide identification information has already been demonstrated in the past. Hollas *et al.* [6] and Beddingfield *et al.* [7] have reported the use of delayed gamma-ray ratios for actinide differentiation, respectively for photofission and thermal neutron fission. Also, experimental work conducted by Gmar *et al.* [8] highlighted variations in the delayed gamma-ray emissions for uranium samples of different enrichments. Further investigation made by Carrel *et al.* [1] brought information about the delayed gamma emission following photofission in mixed samples of uranium. Besides, the uranium isotopes differentiation in an 870 L waste drum by using delayed gamma-ray ratios has already been investigated experimentally with a mockup package [1] and by using Monte-Carlo simulations in the work of Simon *et al.* [5] .

We present the first steps in the development of a method

using photofission as a probe to assess the nuclear material content in large volume hydrogenous radioactive waste drums. We propose to exploit gamma-ray ratios of some photofission products emitting multiple gamma rays (such as ^{138}Cs, ^{142}La, ^{135}I, etc.) as attenuation indicators to estimate the depth at which nuclear materials are located inside the radioactive waste package. To that extent, a concrete matrix slice of dimension representative of an 870 L waste package is used to test a localization technique with uranium samples. The localization of actinides is an essential step in the characterization method to apply attenuation corrections. It avoids the need to use only close-in-energy gamma rays for the actinide identification. Once the actinide location is obtained, the attenuation corrections according to the depth of nuclear materials in the package, also known as matrix and localization effect corrections, are needed for the final assessment of their mass. Accordingly, we propose in this work a first overview of the gamma-ray attenuation correction based on the localization results obtained from the relative attenuation signal.

This paper presents the experimental setup deployed at CEA Cadarache (France) using a high-energy electron linear accelerator (LINAC) in the CINPHONIE irradiation cell [9] to interrogate uranium samples placed in a concrete matrix, and measure the photofission delayed gamma rays with a High-Purity Germanium detector (HPGe).

II. MATERIALS AND METHODS

A. Experimental setup

The electron beam accelerated by the LINAC impinges a 5 mm thick tungsten target to produce an interrogating bremsstrahlung photon beam, the endpoint energy of which can be modulated from 9 to 21 MeV. The pulse frequency is up to 200 Hz with a 4 μs pulse width and a peak current of 100 mA in the target. The dose rate at 1 m from the tungsten target in the beam axis can reach up to 60 Gy/min. A 20 cm thick lead collimator with a rectangular aperture allows to focus the beam on a slice of the radioactive waste package. The collimator is surrounded by 20 cm thick polyethylene blocks and a cadmium layer of 2 mm is added at the front face in order to limit the photoneutron flux produced in the tungsten target and in the lead collimator by (γ,n) reactions of the bremsstrahlung high energy photons [10] . The LINAC and the collimator surrounded by the neutron shielding are shown in Fig. 1.

Fig. 1. SATURNE LINAC with its neutron shielding housed in the CINPHONIE casemate.

A remotely controlled mechanical bench, usually dedicated to high-energy photon imaging of waste drums [11] , transfers the concrete mock-up slice (Fig. 2a) from the irradiation position to the counting position, which is distant from the LINAC to limit fast photoneutron damages to the HPGe crystal. Delayed gamma rays are recorded with a 50 % relative efficiency n-type coaxial HPGe detector (ORTEC GMX50-83-A-PL) equipped with a transistor-reset preamplifier. Acquisition is performed with a LYNX Digital Signal Analyzer driven by the Genie2000 software (MIRION Technologies), which also allows gamma spectra analysis. Dead-time is taken into account with the live-time correction method. The energy resolution is 2.0 keV (FWHM) at the 1.33 MeV gamma line of ^{60}Co. In addition to the abovementioned distance to the LINAC, the detector is shielded by lead and polyethylene to limit neutron damages to the germanium crystal.

The localization technique is tested with two uranium samples described in Table I.

TABLE I
URANIUM SAMPLES CHARACTERISTICS

Sample	Depleted Uranium (DU)	Highly Enriched Uranium (HEU)
Mass	> 100 g	> 1 g
Enrichment (^{235}U content)	0.3 %	> 90 %
Composition	Metallic uranium (ρ=18.96 g.cm^{-3})	Metallic uranium core (ρ=18.96 g.cm^{-3}) held between 2 Zircaloy sheets (ρ=6.56 g.cm^{-3})
Dimension	1 cm thickness	< 1 mm thickness

The uranium samples are placed inside a concrete matrix slice of density 2.25 g.cm^{-3}, 1 m diameter and 15 cm height, with an external 4 mm stainless steel strap, shown in Fig. 2a. The matrix contains holes of different diameters where the samples can be inserted, as described in Fig. 2b.

Fig. 2. (a) Overview of the concrete matrix slice of 1 m diameter (b) Uranium sample in its holder inserted in the matrix hole.

B. Localization method

In heterogeneous packages such as technological waste, the localization of nuclear materials is crucial for a precise characterization of actinides. The method presented in this paper considers a unique point-like uranium sample inserted in the concrete matrix, inside a hole whose center is located at 40 cm from the matrix center. The resulting concrete thickness between the sample and the periphery of the concrete slice is at least 7.2 cm. The concrete slice is irradiated with the LINAC photon beam in front of the sample during 2 h. Then, the mechanical bench transfers the concrete matrix from the irradiation position to the counting position in 22 s, while realizing a 45° rotation. This rotation increases the concrete thickness crossed by delayed gamma rays emitted by photofission products (see Fig. 3, counting position 1). After a 45 min acquisition, a backward 45° rotation puts back the uranium sample in front of the HPGe detector (see Fig. 3, counting position 2, which is the same as during irradiation), and delayed gamma rays are counted during a few hours (precise acquisition times are given in Table II).

TABLE II
EXPERIMENTAL IRRADIATION AND COUNTING TIMES

Sample	Irradiation time	Counting time in Position 1	Counting time in Position 2
DU	2 h	45 min	1 h 17 min
HEU	2 h	43 min	20 h 26 min

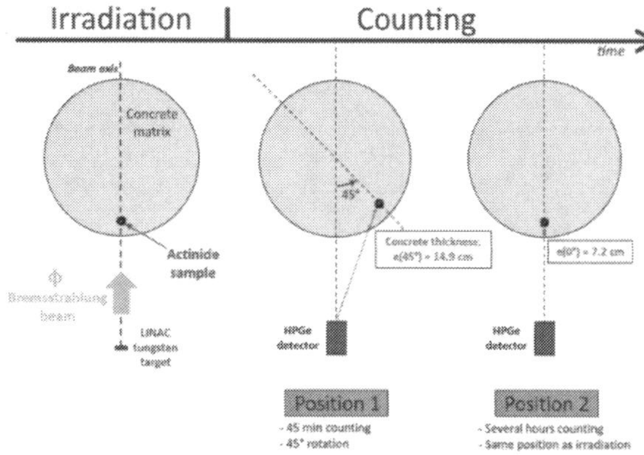

Fig. 3. Experimental steps of the localization method of a point-like uranium sample in a concrete matrix.

The delayed gamma spectra are analyzed with Genie2000 software to extract the net peak areas of the photofission products of interest. For the localization, we are interested in the ratio of two gamma rays emitted by a same photofission product to exploit the differential attenuation of its gamma rays at different energies, when crossing different thicknesses of concrete. The localization combines experimental data with Monte-Carlo simulations performed with MCNP 6.1 computer code [12] . Simulations of a point-like photon source emitting the selected gamma rays are performed for any (r,θ) position inside the concrete matrix. A point detector (F5 tally) located in front of the HPGe detector estimates the flux of gamma rays of each energy reaching the entrance window, as seen in Fig. 4.

The ratio of these fluxes is evaluated for each (r,θ) coordinates so as to estimate the "distortion degree" of gamma attenuation (see the map in further Fig. 6). The "distortion" of the ratio is defined as its normalization to the theoretical gamma-ray ratio without attenuation (i.e. without the concrete matrix, see the text above Fig. 6), which only depends on emission intensity and detection efficiency of the gamma rays of interest. The final step is to compare experimental gamma-ray ratios in positions 1 and 2 with the MCNP distortion map to find the location of the sample in the concrete matrix.

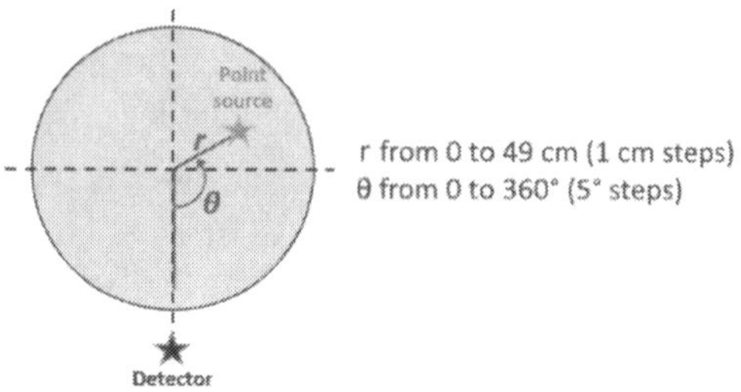

Fig. 4. Principle of the MCNP simulation of the gamma-ray ratio distortion

III. DATA ANALYSIS AND RESULTS DISCUSSION

A. *Uranium samples localization*

Ref. [10] reports delayed gamma-ray spectra obtained with a similar setup on the bare uranium samples (without the concrete matrix), and presents usable peaks of some photofission products in view to localize actinides. Given acquisition times shown in Table II, especially 45 min in the first position, ^{138}Cs is a particularly suitable photofission product because of its half-life $T_{1/2} = 2004.6\ s$. For the localization, we selected the 1435.9 keV and 2639.6 keV gamma rays, with respective intensities of 76.3 % and 7.63 % [13] . These peaks shown in Fig. 5, for the depleted uranium sample, are well above the active background (i.e. concrete matrix irradiated without uranium sample).

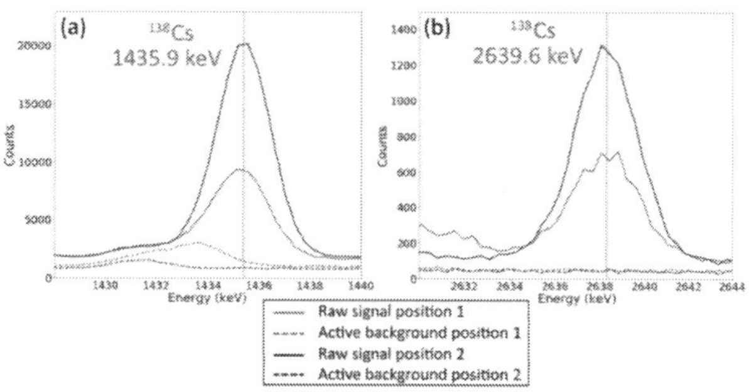

Fig. 5. 1435.9 keV and 2639.6 keV peaks of ^{138}Cs for the DU sample in the concrete matrix and in the two counting positions (and active background).

The net peak areas for the two samples in each position are listed in Table III.

TABLE III
^{138}CS GAMMA-RAY NET PEAK AREAS FOR THE DU AND HEU SAMPLES

Sample	Gamma-ray energy (keV)	Position 1	Position 2
DU	1435.9	53847 ± 549	131812 ± 1471
	2639.6	5972 ± 279	11069 ± 243
HEU	1435.9	8257 ± 341	34528 ± 1833
	2639.6	760 ± 37	2832 ± 280

From the net peak areas, the distortion of the measured ratio in each position, due to gamma-ray attenuation, is defined as:

$$D_i = \frac{N_1}{N_2} \times \frac{I_2}{I_1}\frac{\varepsilon_2}{\varepsilon_1} \qquad (1)$$

Where:

- D_i is the 1435.9 keV/2639.6 keV ratio distortion factor in position i,
- N_1 and N_2 are the net peak areas of the gamma-ray lines at E_1=1435.9 keV and E_2=2639.6 keV,
- I_1 and I_2 are the emission intensities of the gamma rays, respectively 0.763 and 0.0763,
- ε_1 and ε_2 are respectively the intrinsic detection efficiencies of the gamma rays of energies E_1 and E_2. They are determined via simulation and for the two gamma rays of ^{138}Cs, $\frac{\varepsilon_1}{\varepsilon_2} = 1.60$.

Once the experimental distortion factor is obtained in the two positions, it can be compared to the ratio distortion map built from MCNP simulations, covering the full area of the concrete matrix as shown in Fig. 6. The brightest color corresponds to the minimum attenuation and the darkest color to the maximum attenuation and maximum ratio distortion, i.e. the case where the gamma rays cross 1 m of concrete. The scale from 0 to 1 is obtained by rescaling the ratio distortion at every position by considering that the minimum attenuation corresponds to 1 (the real value is 0.905) and the maximum attenuation to 0 (the real value is 0.046).

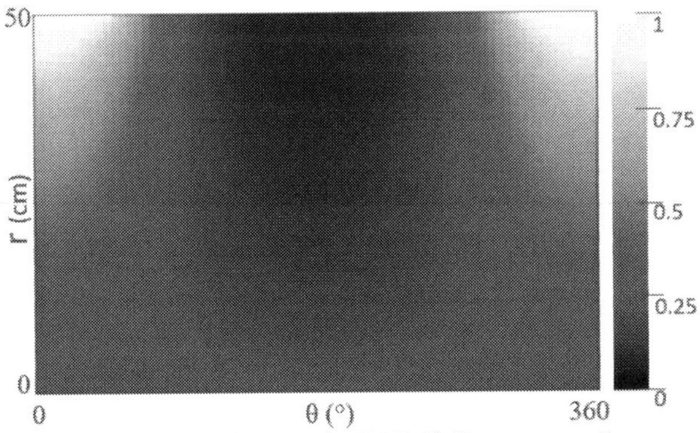

Fig. 6. Distortion map of the 1435.9 keV/2639.6 keV gamma-ray ratio (^{138}Cs), see text for details.

Then, the locations corresponding to the measured ^{138}Cs ratio distortion factor in positions 1 and 2 are determined on the simulated ratio distortion map of Fig. 6. The positions for which the experimental and simulated ratio distortions are in good agreement are represented by the orange and yellow tracks in Fig. 7, respectively for positions 1 and 2, in case of the DU sample. The location of the sample corresponds to the intersection of these tracks and is represented by the blue dot.

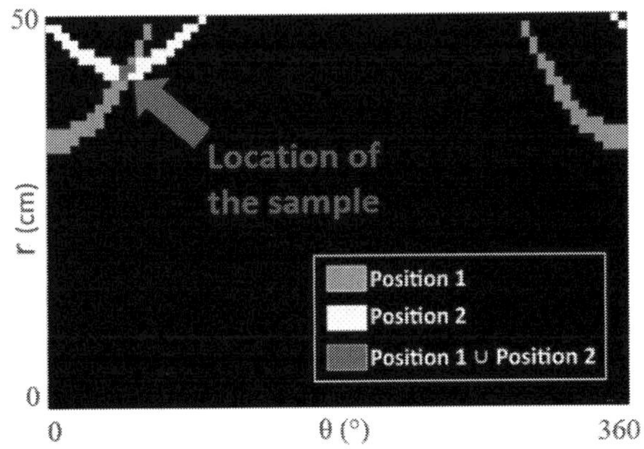

Fig. 7. Sample location from the ratio distortion in the two positions for the DU sample.

Finally, the localization results are compared to the real position of the samples, see Table IV. The uncertainty on the localization, mainly due to counting statistics and photon attenuation in concrete (related to its density), are discussed in section III.B. The estimation of the sample radius is very close to the real position for both DU and HEU samples, within 1 cm. Concerning the angle, it is also very close for the DU sample, but the discrepancy is quite large for the smaller HEU sample due to a poor counting statistics (see Table III).

TABLE IV
LOCALIZATION RESULTS FROM THE ANALYSIS OF THE 1435.9 KEV/2639.6 KEV ^{138}CS RATIO DISTORTION

Sample	Real position (r,θ)	Localization result (r,θ)
DU	43 cm, 45°	(42 ± 1.5) cm, (47 ± 10)°
HEU		(43 ± 3) cm, (33 ± 17)°

B. Localization method sensitivity

In order to test the sensitivity of the localization to the gamma-ray ratios in the two measurement positions, mainly due to counting statistics fluctuations, we varied them independently between -25% and +25%. For instance, the (r,θ) coordinates are calculated from the gamma-ray ratios measured with the DU sample in the two angular positions 1 (R_1) and 2 (R_2). The effect of a ± 25% variation of R_1 and R_2 on r and θ is reported in Fig. 8 and Fig. 9, respectively. Parameters r and θ have very different sensitivities to ^{138}Cs gamma-ray ratio variations: radius r has a low sensitivity with a maximum deviation of -22% (with -25% on R_1 and R_2) but θ can change up to 230% (with +25% on R_1 and -25% on R_2).

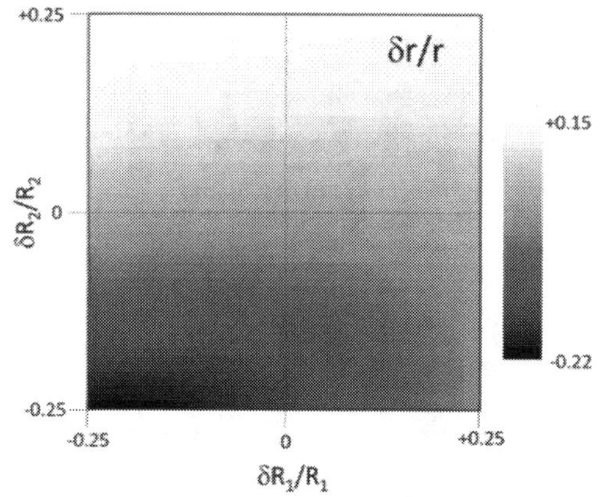

Fig. 8. Effect of the 1435.9/2639.6 keV gamma-ray ratio variations, in positions 1 and 2 (R_1 and R_2, respectively), on the sample radius r.

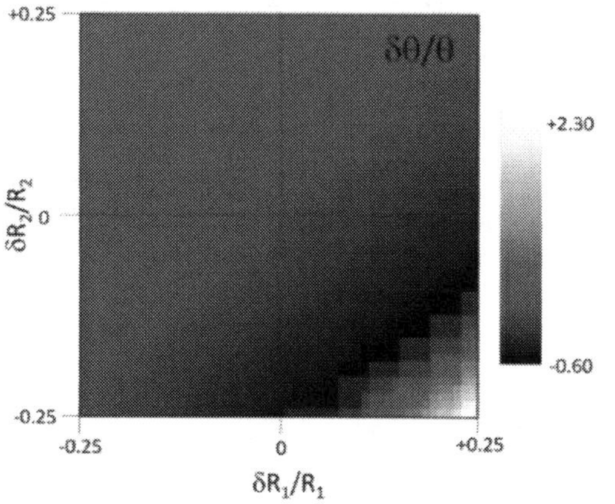

Fig. 9. Angle parameter θ sensitivity to the gamma-ray ratio variations.

On the other hand, we studied the effect of the uncertainty on gamma attenuation, which can be due to the precision on nuclear data (mass attenuation coefficients), concrete composition and density. We grouped these causes of uncertainty in a study with large density variations. The distortion map used for the location determination (Fig. 6) was calculated with a density of 2.25 g.cm^{-3} for the concrete matrix, and we varied it here between -50% and +50% around this value. Fig. 10 shows the relative deviations of r and θ as a function of the variability of concrete density in the model.

Globally, r and θ are not very sensitive to concrete density when using the 1435.9 keV/2639.6 keV ratio of ^{138}Cs. The localization error is the largest when the simulated density is underestimated compared to the real value. An underestimation of 50% of the concrete density (i.e. 1.12 instead of 2.25 g.cm^{-3}) induces a 17% error on r and 40% on θ, while a 50% overestimation (3.37 instead of 2.25 g.cm^{-3}) produces only a 6.2% error on r and 11.2% on θ.

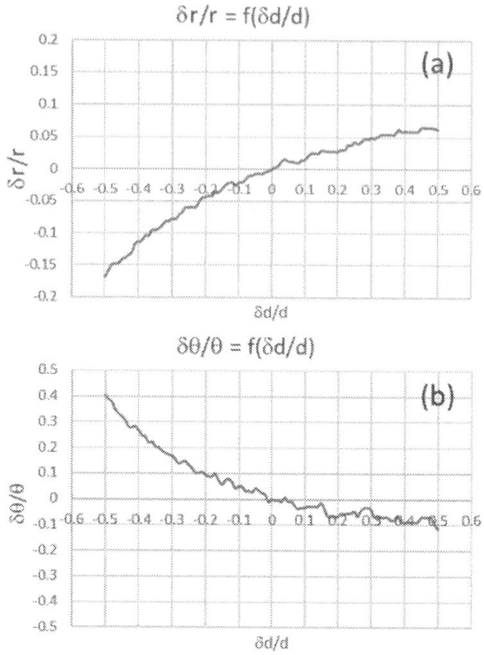

Fig. 10. Sensitivity of (a) r and (b) θ parameters as a function of concrete density variability.

The localization sensitivity is thus primarily driven by the statistical uncertainty of the gamma-ray ratios measured in the two angular positions. In addition, it is worth mentioning that the determination of θ, which is more sensitive than r to statistical fluctuations, does not affect the quantification of the photofission rate in the sample in view of its mass assessment. Indeed, the package will be rotated during irradiation and only the knowledge of r is crucial to calculate the photofission rate in nuclear materials. The knowledge of θ is more important for the delayed gamma-ray detection step, in view to position the point-like nuclear material sample in front of the detector after its localization, and for the gamma detection efficiency calculation necessary to estimate the actinides activity or mass. The identification of actinides (e.g. uranium enrichment) based on gamma-ray ratios of different photofission products also requires a gamma attenuation correction using this localization.

C. Gamma-ray ratio corrections in view of actinides identification

The localization providing the depth at which the actinides are located in the concrete matrix, a correction factor can be calculated for any measured gamma ratio used in the actinide differentiation technique. Originally, this method was only based on gamma rays with close energies to avoid differential gamma attenuation [1] , but this correction significantly opens the possibilities with gamma rays having very different energies.

In the experiments, the concrete matrix is surrounded by a 4 mm layer of stainless steel, which is therefore taken into account in the ratio correction examples detailed hereafter. For gamma rays crossing a concrete thickness e_c and a stainless steel thickness e_s, the ratio can be expressed as:

$$
\begin{aligned}
R_{12} &= \frac{N_1}{N_2} \\
&= \frac{I_1 \varepsilon_1}{I_2 \varepsilon_2} e^{\left(\left[\left(\frac{\mu}{\rho}\right)_{E_2} - \left(\frac{\mu}{\rho}\right)_{E_1}\right]_{concrete} \rho_C . e_C\right)} e^{\left(\left[\left(\frac{\mu}{\rho}\right)_{E_2} - \left(\frac{\mu}{\rho}\right)_{E_1}\right]_{steel} \rho_S . e_S\right)} \qquad (2) \\
&= \frac{I_1 \varepsilon_1}{I_2 \varepsilon_2} e^{\left(\Delta\left(\frac{\mu}{\rho}\right)_{concrete} \rho_C . e_C\right)} e^{\left(\Delta\left(\frac{\mu}{\rho}\right)_{steel} \rho_S . e_S\right)} = \frac{I_1}{I_2} \times C
\end{aligned}
$$

Where:

- R_{12} is the ratio of the net peak areas of gamma rays of energies E_1 and E_2, respectively N_1 and N_2,
- I_1 and I_2 are the emission intensities of these gamma rays,
- ε_1 and ε_2 are the intrinsic detection efficiencies of these gamma rays,
- $\left(\frac{\mu}{\rho}\right)_{E_i}$ is the mass attenuation coefficient (in $cm^2.g^{-1}$) of concrete or stainless steel for photons of energy E_i [14] ,
- ρ_C and ρ_S are the densities of concrete (2.25 $g.cm^{-3}$) and stainless steel (7.87 $g.cm^{-3}$),
- e_C and e_S are the concrete and stainless steel thicknesses crossed by gamma rays,
- C is the correction factor taking into account the relative detection efficiency of the two gamma rays and the matrix attenuation.

The corrected gamma ratio can be obtained by dividing the measured ratio by the correction factor C, calculated with the concrete thickness deduced from actinides localization. We give in Table V an example of ratio correction for different photofission products of the DU sample in the concrete matrix, for which the stainless steel thickness is 4 mm and the concrete thickness is 8 cm (localization result).

TABLE V
RATIO CORRECTION FOR THE DU SAMPLE IN THE CONCRETE MATRIX

FP	Gamma-ray ratio (keV)	Gamma-ray intensity (%)	Real ratio	Measured ratio	Corrected ratio
^{138}Cs	2218.0 2639.6	15.18 7.63	1.99 ± 0.07	1.91 ± 0.08	2.07 ± 0.09
^{135}I	1131.5 1260.4	22.59 28.70	0.79 ± 0.04	0.75 ± 0.02	0.80 ± 0.03
^{142}La	1545.8 2971.0	2.99 3.13	0.96 ± 0.06	0.68 ± 0.10	0.95 ± 0.13
	2542.7 2971.0	10.00 3.13	3.19 ± 0.17	2.95 ± 0.41	3.16 ± 0.44

The real ratio corresponds to the intensity ratio of the two considered gamma rays. Its relative uncertainty is calculated as the quadratic combination of the uncertainties on its two gamma-ray intensities, found in JEFF-3.3 database [13] . The relative uncertainty on the measured ratio corresponds to the quadratic combination of the statistical uncertainties on the net peak areas of the gamma rays and on the detector intrinsic efficiency estimated to 2% with an optimized MCNP model of the HPGe detector. Finally, the relative uncertainty on the corrected ratio is dominated by that of the measured ratio, the uncertainties on the mass attenuation coefficient and concrete thickness being smaller. Finally, the discrepancy between the real and corrected ratios is at most a few percent for the 2218.0/2639.6 keV ratio of ^{138}Cs, by using the localization provided with the 1435.9/2639.6 keV ratio of ^{138}Cs.

IV. Conclusion

The localization of point-like nuclear materials characterized by high-energy photon interrogation in large concrete radioactive waste packages is possible using the differential attenuation of gamma rays emitted by a same photofission product, here the 1435.9 and 2639.6 keV lines of ^{138}Cs. In particular, the radius is determined with a precision of a few cm, thus giving the depth of nuclear materials inside the concrete matrix.

This information is important to correct for gamma attenuation in the method used to differentiate actinides, which is based on delayed gamma-ray ratios from different photofission products like ^{138}Cs, ^{135}I and ^{142}La. Thanks to this correction, it will be possible to use delayed gamma rays that are not close in energy, contrary to the original method, which will open the possibilities to improve actinide identification.

Another important task for the characterization of nuclear materials by active photon interrogation is the accurate measurement of the cumulative yields of photofission products, as reported in [10] for ^{235}U and ^{238}U. Additional experiments have been recently performed to measure the cumulative photofission yields of ^{239}Pu, which will be published soon, as such data are extremely rare so far in the literature.

A series of photofission products liable to allow ^{235}U vs. ^{238}U discrimination in the global photofission signal have been identified in [10] . This work will be continued for uranium vs. plutonium differentiation using our measured uranium and plutonium photofission yields.

Acknowledgment

The authors would like to thank the French National Radioactive Waste Management Agency (ANDRA) contributing for several years to this research and development activity within a cooperation framework under the COV – F32678 DEN4908 – C43R9T5480 contract.

References

[1] F. Carrel, M. Agelou, M. Gmar, F. Lainé, J. Loridon, J.-L. Ma, C. Passard, B. Poumarède, "Identification and Differentiation of Actinides Inside Nuclear Waste Packages by Measurement of Delayed Gammas", IEEE Trans. Nucl. Sci., 57 (5) 2862, 2010.

[2] F. Carrel, M. Agelou, M. Gmar, F. Lainé, "Detection of high-energy delayed gammas for nuclear waste packages characterization", Nucl. Inst. and Meth. A, vol. 652 (1), pp. 137-139, 2011.

[3] T. Nicol, B. Pérot, C. Carasco, F. Brackx, A. Mariani, C. Passard, E. Mauerhofer, J. Collot, "Feasibility study of ^{235}U and ^{239}Pu characterization in radioactive waste drums using neutron-induced fission delayed gamma rays", Nucl. Inst. and Meth. A, vol. 832, pp. 85-94, 2016.

[4] R. De Stefano, C. Carasco, B. Pérot, E. Simon, T. Nicol, E. Mauerhofer, "Feasibility study of fissile mass detection in 870 L radioactive waste drums using delayed gamma rays from neutron-induced fission", Journ. Radioanal. Nucl. Chem, vol. 322, pp. 1185-1194, 2019.

[5] E. Simon, F. Jallu, B. Pérot, S. Plumeri, "Feasibility study of fissile mass quantification by photofission delayed gamma rays in radioactive waste packages using MCNPX", Nucl. Inst. and Meth. A, vol. 840, pp. 28-35, 2016.

[6] C. L. Hollas, D. A. Close, C. E. Moss, "Analysis of fissionable material using delayed gamma rays from photofission", Nucl. Inst. and Meth. B, vol. 24-25 (1), pp. 503-505, 1987.

[7] D. H. Beddingfield, F. E. Cecil, "Identification of fissile materials from fission product gamma-ray spectra", Nucl. Inst. and Meth. A, vol. 417, pp. 405-412, 1998.

[8] M. Gmar, J. M. Capdevila, "Use of delayed gamma spectra for detection of actinides (U, Pu) by photofission", Nucl. Inst. and Meth. A, vol. 422, pp. 841-845, 1999.

[9] B. Pérot, F. Jallu, C. Passard, O. Gueton, P.-G. Allinei, L. Loubet, N. Estre, E. Simon, C. Carasco, C. Roure, L. Boucher, H. Lamotte, J. Comte, M. Bertaux, A. Lyoussi, P. Fichet and F. Carrel, "The characterization of radioactive waste: a critical review of techniques implemented or under development at CEA, France", EPJ Nucl. Sci. and Tech. 4 (3), 2018.

[10] M. Delarue, E. Simon, B. Pérot, P.-G. Allinei, N. Estre, E. Payan, D. Eck, D. Tisseur, I. Espagnon, J. Collot, "Measurement of cumulative photofission yields of ^{235}U and ^{238}U with a 16 MeV Bremsstrahlung photon beam", Nucl. Inst. and Meth. A, vol. 1011, https://doi.org/10.1016/j.nima.2021.165598, 2021.

[11] N. Estre, D. Eck, J.-L. Pettier, E. Payan, C. Roure, E. Simon, "High-Energy X-Ray Imaging Applied to Nondestructive Characterization Characterization of Large Nuclear Waste Drums", IEEE Trans. On Nucl. Sci., vol. 62 (6), pp. 3104-3109, 2015.

[12] Los Alamos National Laboratory, "MCNP 6^{TM} User's Manual", D. B. Pelowitz, 2013.

[13] A. J. M. Plompen, O. Cabellos, C. De Saint Jean et al., "The joint evaluated fission and fusion nuclear data library, JEFF-3.3", Eur. Phys. J. A, vol. 56, 181, 2020.

[14] J. H. Hubbell, S. M. Seltzer, "Tables of X-Ray Mass Attenuation Coefficients and Mass Energy-Absorption coefficients 1 keV to 20 MeV for Elements Z=1 to 92 and 48 Additional Substances of Dosimetric Interest", National Institute of Standards and Technology, 1995.

Study of Neutron Background in order to Improve Radioactive Waste Drum Characterization

Gabrielle Lelaizant[1]
[1]CEA, France
gabrielle.lelaizant@cea.fr

***Abstract*— A usual way of radioactive waste drums characterization combines gamma spectrometry measurements with passive neutron measurements. The CEA facility for waste drum characterization hereby described has been operated for more than 30 years. In this framework, a large variety of waste drums has been characterized in terms of spectra, densities, materials and radioactivity levels. As the facility was first dedicated to measure Intermediate-Level Long-lived Waste, the neutron background was not significant compared to expected neutron emitters from waste packages. These last years, Dismantling and Decommissioning operations have been advanced in this CEA site to the point where they are now associated with mostly Very Low Level Waste production. Therefore, neutron background is becoming significant. Using the large variety of past characterized drums brings the opportunity to study this background. The present study has been led over a sample of almost 1500 drums over a wide waste density range. These drums have been selected over the last 20 years by taking into account only one criterion: without any expected neutron emitters from the waste itself. This work first presents the technical settings of the measurement facility before describing the raw data of the measurements. Next, a statistical study over raw data enables to better acknowledge the neutron spallation background behavior in terms of time, density and materials. Ensues a way of using this new knowledge in order to improve how to take into account neutron spallation background in passive neutron measurements of packages of low actinides activities and high densities.**

***Keywords* —neutron background, coincidence counting, radioactive waste characterization, cosmic-ray induced neutrons**

I. Introduction

The combination of gamma spectrometry with passive neutron measurement has so far been a usual way to characterize radioactive waste [1] . Neutrons are identified and quantified using coincidence counting following thermalization in an assay chamber. The first device enables to quantify radioactivity in a waste package spectrum through the identification of its gamma emitters, whereas the latter not only allows the measurement of nuclear material mass through the neutron count rate associated with spontaneous fission of fissile emitters (e.g. ^{240}Pu, the usual neutron tracer for plutonium assay) but also to confirm the measurement data obtained by gamma spectrometry. These combined measurement devices are well suited in case of alpha spectra, for which both measurements provide different pieces of information and then enable to measure actinides for a large range of waste densities. But some difficulties are encountered when alpha radionuclides activities are so low that they cannot be measured in reasonable measurement time durations in gamma spectrometry. In high density waste which prevents gamma signal from reaching detector cells, characterization with a high confidence level becomes a major issue.

Hence, the role of passive neutron measurement becomes even more important as it is not sensitive to the same parameters. On the contrary to gamma spectrometry, neutron absorption increases with the quantity of light matrices and especially with the proportion of low atomic number materials such as hydrogen. The fast neutrons emitted from spontaneous fission of the fertile isotopes provide good penetrability of bulk materials.

As the facility was first dedicated to measure Intermediate-Level Long-lived Waste (ILW-LL), gamma spectrometry has been paid more attention in waste characterization rather than neutron measurement results which are mainly used to get the value of nuclear mass. In this context, the neutron spallation background was not significant enough compared to expected neutron emitters from waste packages. Thus, the neutron background has been corrected so far by a fixed value which does not depend on the waste characteristics. In recent years, Dismantling and Decommissioning (D&D) operations have been advanced in this CEA site to the point that they are now mostly of Very Low Level Waste (VLLW) category. Therefore, considering a proper correction of neutron background level for each measurement is becoming a major issue in order to investigate low levels of neutron count rates from nuclear material.

Neutron background originates from different contributions [1] :

1. External-area radiation sources (e.g. waste storage area proximity);
2. Radioactive alpha or beta decay in the walls of the detector tubes;
3. (α, n) reactions in α emitters matrices;
4. External-area cosmic-ray sources (e.g. spallation reactions in the room);
5. Spallation reactions from cosmic rays in the sample or in the detector body.

Items 1 to 4 only produce single neutrons, whereas item 5 is accompanied by both single and coincidence neutrons.

Passive neutron coincidence counting is a well-known method for non-destructive assay of nuclear material contaminated waste [3] , [4] . Coincidence counting distinguishes the time-correlated neutrons, produced from fission, from neutrons born randomly in time such as those originating from items 1 to 4 (and mainly from #3's (α, n) reactions). Expressed as the equivalent mass of ^{240}Pu, the nuclear material detection limit is constrained by the cosmic-ray background of spallation neutron events taking place in the vicinity and body of the assay chamber, as well as in the waste itself. This background cannot be removed by a "simple" coincidence counting. Moreover, the cosmic-ray background of spallation neutron events increases with the volume and density of the He-3 tubes, as well as with waste density and is so matrix-dependent. Therefore, the waste itself becomes a spallation neutron source.

The intrinsic cosmic-ray background from the chamber can be minimized at the design stage by avoiding the use of materials of high atomic number in the detector vicinity. During operation, judicious location and shielding of the instrument can also help to reduce cosmic interaction in the chamber's body. Additional means of suppressing and stabilizing the cosmic-ray background are provided by data rejection algorithms, applied on an event-by-event basis by the software, to filter the major part of the cosmic-ray variability from the data, or to eliminate the high-multiplicity cosmic-ray neutron bursts. Nevertheless, no such algorithm can be applied in the current technique whose commissioning took place few decades ago. Besides, the electronics currently used do not allow the definition of a specific background for each measurement, but only a fixed value can be set as correction to calculate the net signal. This fixed value has been obtained from repeated measurements performed in the same configuration in a short period of time during the commissioning stage of the device. Also, any change of configuration leads to a new definition of this correction parameter. One may imagine a proper background correction depending on waste density. Indeed, some works have been carried out in order to define a material component as described in [1] and [5] . However, the results obtained in these studies cannot be considered as universal, but as equipment-dependent. The approach hereby presented deals with a data analysis study aiming at defining this specific correction on the basis of measurement feedback. As the end of D&D operations is coming in this CEA site, the same electronics have been laid out in order to keep the same raw data format and shall remain so until the end of operations.

II. Experimental setup

A. Data selection criteria

Preliminary implicit criterion resides in no expected nuclear material. Some criteria have been taken into account over the whole dataset. First, the current data format has been applied only since 2008. Therefore, the last 11 years' data have been considered (2008–2019). From this dataset, only waste packages without any expected nuclear material (according to historical information provided by waste producers), were selected. 631 drums remain after this selection.

In order to avoid statistical effect on a single measurement, the waste packages (i.e. 200 L drums), are usually measured at least 3 times (and up to 10 times), with a duration of either 1800 seconds (the most common case) or 900 seconds (the former case). Each measurement result is considered in the dataset as an independent result. Nevertheless, a slight difference has to be specified here: the second applied criterion consisted in rejecting raw data which belong to a sample composed of the different measurements of the same drum but which show a standard deviation greater than 10% to the average value. 2050 measurement results remain after the application of the previous condition.

A comparison of each measurement value with its specific decision threshold enables to keep only significant raw data. 2.4% of the dataset was removed through this procedure. Finally, a statistical Grubbs test was applied to identify and discard outliers. 1797 measurement results compose the final dataset.

B. Passive Neutron Measurement Cavity and Variable Dead-Time System Electronics

The subsurface passive neutron measurement device is composed of 24 ^{3}He proportional counters (i.e. detector tubes) embedded in a HDPE layer. The 15 cm thick HDPE layer enables not only to minimize neutron background coming from the outside of the cavity but also to thermalize neutron energy in order to increase the cross-section of their capture into the ^{3}He tubes. Electronic modules are connected to the ^{3}He tubes in order to discriminate neutron events from γ pulses as well as to shape the output signal by summing up input signals from all electronics (one module for 3 ^{3}He tubes).

Passive neutron coincidence counting using shift register electronics is an established technique for non-destructive assay of nuclear material-contaminated waste. The "Variable Dead-Time System" (VDTS) electronics hereby chosen, following historical considerations, are slightly different. Even these modules are less common, they were well described in French standards [6] (in effect until 2020) as an alternative means to shift registers in order to address the neutron coincidence count rate.

https://doi.org/10.1051/epjconf/202125308004

As shown in Fig. 1, two counters are used in these electronics. The fast channel counter, named C_{fast}, increases by increments for each neutron event, whereas each first neutron event triggers the opening of a fixed time window named "dead-time τ_{L}". As the value of applied dead-time is equivalent to the detector die-away time, time-correlated neutrons from spontaneous fission are removed from the slow channel counting, C_{slow}.

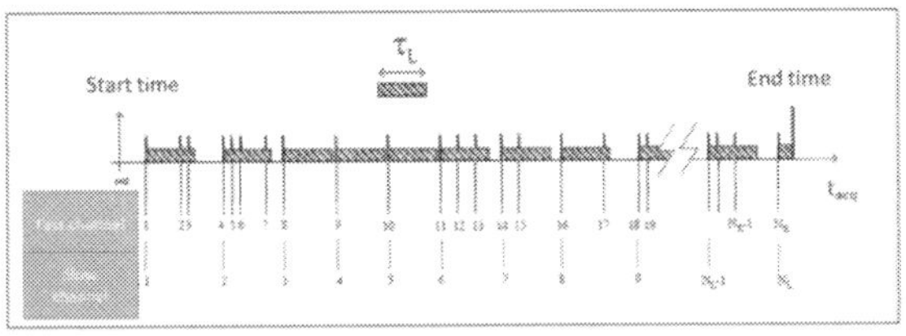

Fig. 1. Description of fast and slow channels' counters

Therefore, the difference between both counters C_i enables to calculate the number of correlated neutron events. Then, by applying the proper correction of times, i.e. the acquisition time t_{acq} and the dead-time τ_{L}, a coincident count rate X_i, equivalent to the spontaneous fission rate, is available when following the equation (1)

$$X_i = \left(\frac{C_{\text{fast}}}{t_{acq}} - \frac{C_i}{t_{acq} - C_i.\tau_L}\right), \qquad (1)$$

The second step consists in subtracting the background coincident count rate X_{bkg} in order to address the equivalent mass of ^{240}Pu directly related to the nuclear mass of the waste package. It is useful to remind here that any density correction is applied in these calculations.

C. Density characteristics of the dataset

However, the density of waste packages varies over a wide range: from 0.1 to 1.4. The histogram shown in Fig. 2 represents the density distribution of the 631 selected drums.

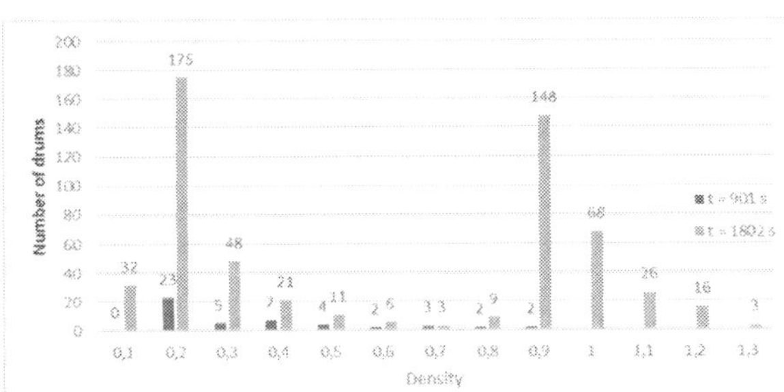

Fig. 2. Density distribution of selected drums

Two density ranges are mostly available in the selected dataset: 0.2 and ≥ 0.9. These two density ranges are targeted in the following parts.

III. Data analysis

A. Background coincident count rate versus density

As previously described, an increase of background coincident count rate is expected with density due to cosmic-ray-induced spallation neutron events. This effect is indeed highlighted in Fig. 3.

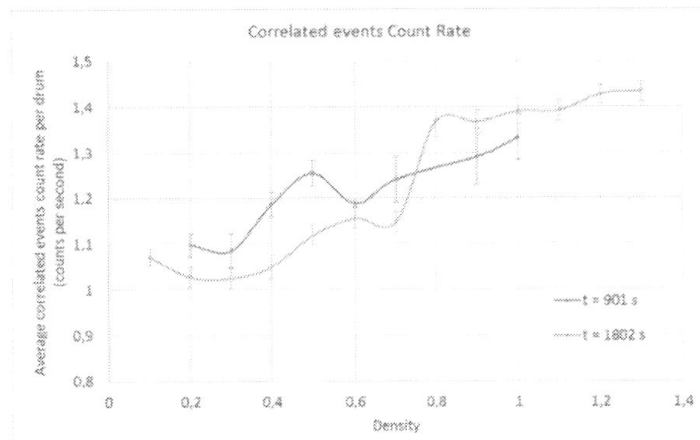

Fig. 3. Average correlated events count rate versus density for two different measurement durations (900 seconds in blue and 1800 seconds in green)

Error bars correspond to standard deviations (1σ) calculated over density samples for each density range. Even if the expected increase with density is visible in both plots, the green (1800 s) plot highlights a clearer cutoff at density ~ 0.8 as compared to the blue (900 s) plot. Smaller statistics available in shorter measurement durations (900 s vs 1800 s) dataset explains the observed flattening of the blue plot compared to the green one.

All background coincident count rates – without any averaging – are plotted in Fig. 4. The same behavior hereby appears in the shape of two ranges of background coincident count rates: one for light matrices (in the bottom part) and the other for heavy matrices (in the top part). Scatter plots in Fig. 4 express deviations to average.

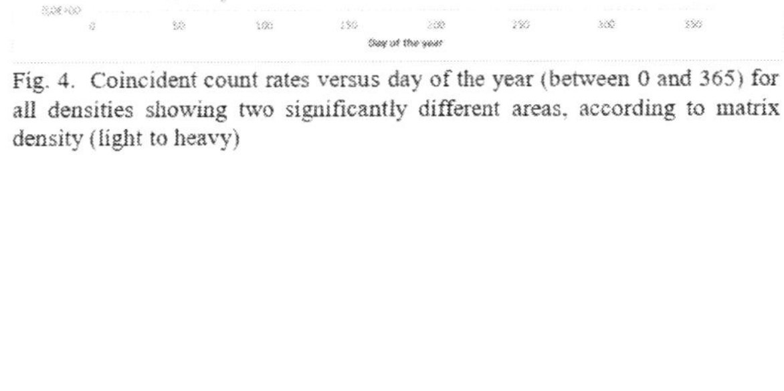

Fig. 4. Coincident count rates versus day of the year (between 0 and 365) for all densities showing two significantly different areas, according to matrix density (light to heavy)

TABLE I
BACKGROUND COINCIDENT COUNT RATE PER DENSITY

Density	Averaged background coincident count rate (counts per second)	Standard deviation
0.2	0.014	0.002
0.9	0.051	0.011
1.0	0.053	0.007
> 1.0	0.057	0.010

Table I highlights a 4-fold factor between high density and low density background coincident count rates.

Following the decision threshold as defined in equation (2),

$$DT = 2\sqrt{\left(X_{\text{bkg}} + \left(\frac{C_{\text{f,bkg}}}{t_{\text{bkg}}}\right)^2 \cdot \tau_L\right) \cdot \left(\frac{1}{t_{\text{acq}}} + \frac{1}{t_{\text{bkg}}}\right)} \quad (2)$$

improving the background coincident count rate consideration enables to better calculate the decision threshold and consequently to better consider significant signals. The 4-fold factor previously highlighted in Table I also stems from this equation.

B. *Background coincident count rate versus time*

The neutron background arises primarily from cosmic origin with negligible contributions from radioisotopes in the local environment [7] . The cosmic-ray-induced neutrons observable at ground level are mainly produced through spallation reactions on nitrogen, oxygen and argon nuclei in air from high-energy particles (mainly protons and helium ions).

Neutrons can originate from miscellaneous sources either diffuse or point-like [8] :

- Solar flares (neutrons, either from direct emission from the Sun or as secondary products in the Earth's atmosphere);
- Galactic neutrons (relativistic neutrons created in active galactic nuclei);
- Cosmic neutrons (from supernova remnants or other high energy cosmic ray point sources);
- Atmospheric neutrons (e.g. secondary neutrons produced by cosmic-ray induced hadronic showers in the Earth atmosphere).

According to [8] , solar neutrons are the most likely source of spallation neutrons.

Therefore, a time-dependent background coincident count rate may be inferred following either solar flares or different solar cycles, ranging from the shortest (day/night cycle), to the longest period (eleven years from the heliospheric and geomagnetic fields modulations [9]). The impact of solar wind could also be investigated as a cosmic ray flux has been shown to be inversely correlated with the Sun's activity-generated solar winds [7] .

Fig. 5 plots the background coincident count rate versus the day of the year since 2008, for the four density ranges introduced in Fig. 4. Both yearly effect and seasonal effect were targeted in these plots. Horizontal lines correspond to respective averaged background coincident count rates.

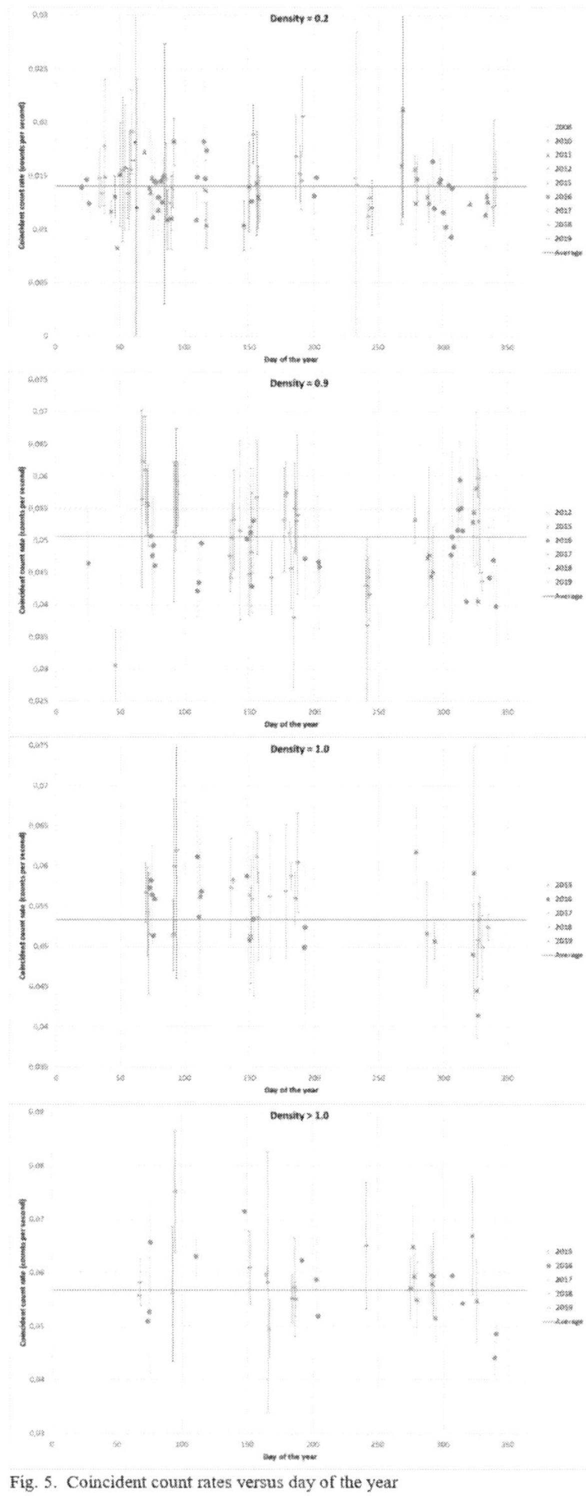

Fig. 5. Coincident count rates versus day of the year

Error bars correspond to standard deviations (1σ) over the sample for each day of the year. By considering error bars as uncertainties over the dataset, a greater deviation to the average value could be characteristic of higher densities. In this case, more discrepancies (i.e. isolated points) could show time effects related to cosmic rays events or solar cycles. Nevertheless, establishing a clear relation is complex, owing to different time scales between measurement times (hundreds of seconds) and solar flares durations (minutes, less than 100 seconds) or cosmic events durations (seconds, up to minutes for the most energetic phenomena).

When searching for an hourly effect during the day in Fig. 6, deviations to average are even less significant.

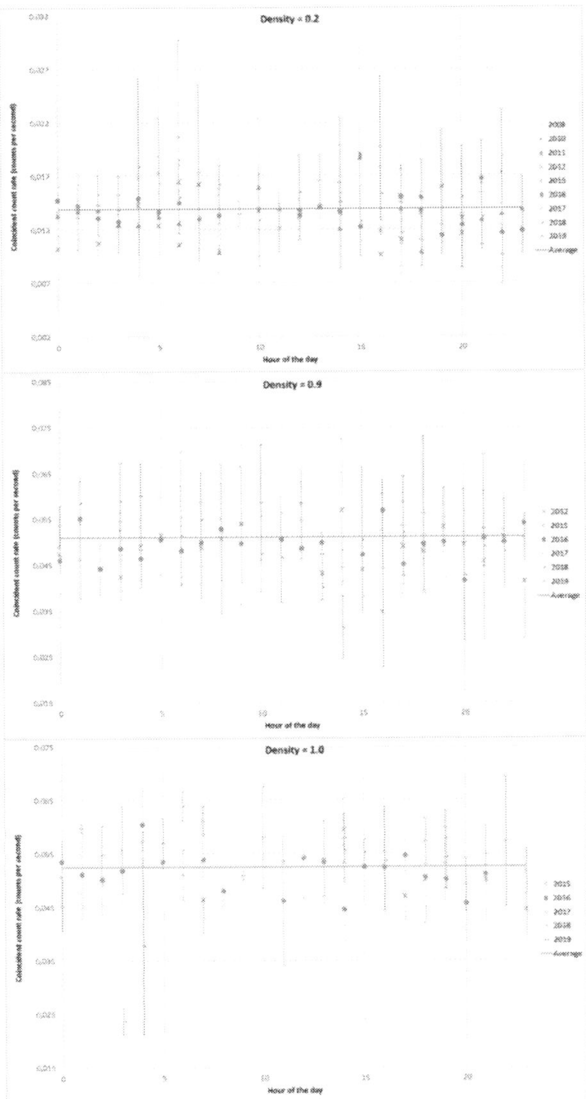

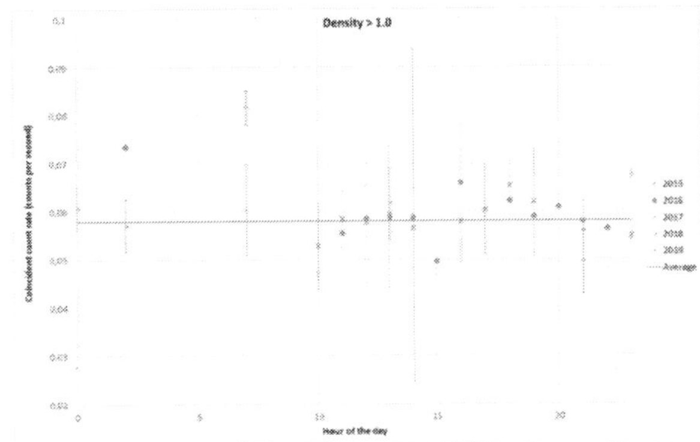

Fig. 6. Coincident count rates versus hour of the day

IV. Conclusion

The data analysis study presented here over the analysis of 2050 measurements undertaken over 11 years has confirmed the expected density effect on background coincident count rate observed in passive neutron measurements, due to stronger spallation effects in higher densities materials. Density-dependent values have been defined from the feedback study in order to improve background consideration not only to better calculate net signal by applying a proper background correction, but also to take into account the right decision threshold in order to discard insignificant measurement data.

The search for time-dependent variations related to galactic or extragalactic cosmic rays events or solar cycles has not been fruitful but could be improved by using a timestamp module in order to investigate neutron flares in shorter time windows.

References

[1] IAEA, Strategy and Methodology for Radioactive Waste Characterization, IAEA-TECDOC-1537, Vienna, Austria, 2007, pp. 66-72

[2] H.O. Menlove, D.H. Beddingfield and M.M. Pickrell, The Design of a High-efficiency Neutron Counter for Waste Drums to Provide Optimized Sensitivity for Plutonium Assay, https://www.canberra.com/fr/literature/waste_special_systems/tech_papers/menlove.pdf

[3] G.F. Knoll, Radiation Detection and Measurement, New York, John Wiley and Sons, 1999

[4] D. Reilly, N. Ensslin and H. Smith Jr, Passive Nondestructive Assay of Nuclear Materials, NuREG/CR-5550,LA-UR-90-732, 1991, pp. 457-528

[5] S. Croft and L.C.A. Bourva, The specific total and coincidence cosmic-ray-induced neutron production rates in materials, Nuclear Inst. And Methods in Physics Research A 505 (2003) 536-539

[6] NF M60-306:1998 Standard, Nuclear energy – Fuel cycle technology – Waste – Determination of radioactive characteristics of solid waste packages by passive neutron counting

[7] M.F. Becchetti, M. Flaska, S.D. Clarke, S.A. Pozzi, Measurements and simulations of the cosmic-ray-induced neutron background, Nuclear Inst. And Methods in Physics Research A 777 (2015) 1-5

[8] D. Casadei, Neutron Astronomy, School of Physics and Astronomy, University of Birmingham and School of Engineering, FHNW, arXiv:astro-ph/1701.02788, 2017

[9] S.E. Forbush, Cosmic-ray intensity variations during two solar cycles, Journal of Geophysical Research Vol. 63, 1958, pp. 651-669

An advanced blind-tube monitoring instrument to improve the characterization of subsurface radioactive plumes

Soraia Elísio[1], Malcolm J. Joyce[1], James Graham[2], Barrie Greenhalgh[3]
[1]Department of Engineering, Lancaster University, United Kingdom
[2]Central Laboratory, National Nuclear Laboratory Ltd., United Kingdom
[3]Radiometric Systems Group, Sellafield Ltd., United Kingdom
s.elisio@lancaster.ac.uk

***Abstract* — This research presents a design of a resilient blind-tube radiometric logging probe, which satisfies key, operational site constraints, for direct in-situ characterization of subsurface radioactivity, particularly caesium-137 and strontium-90. The probe comprises a commercially-available Ø10 mm × 9.5 mm $CeBr_3$ scintillator detector attached to a compact digitizer unit, in a resistant and waterproof housing. The probe is designed to be lowered down into metallic Ø 75 mm blind-tubes by means of a winch system.**

***Keywords* — blind-tube logging, $CeBr_3$ scintillator, gamma-ray spectroscopy, decommissioning**

I. Introduction

THE wet interim storage of nuclear waste in ponds and silos is a common approach at nuclear licensed sites. However, some of these legacy facilities were not designed to modern building standards, and some cases leakage of aqueous radioactive into the subsurface has occurred, dispersing downward and outward in the ground and contaminating the soil and sediment along its path. If unmitigated, then these plumes pose a potential risk to groundwater, surface water receptors and subsequently public health and the environment. nearby streams, presenting an elevated risk to public health and the environment.

Licensed sites have specific regulations associated with the protection and assessment of the soil and groundwater associated with such facilities. Monitoring programs are often focused on the detection and characterization of sources of contamination, and on the evaluation of temporal changes through surveillance measurements at target locations. Additionally, the spread of contamination can be reduced via emplacement of engineered barriers, or remediation schemes ensuring the preservation of the baseline quality of the soil [1] and groundwater [2-5].

The facility on which this research is focused is the Magnox Swarf Storage Silos (MSSS). It was built in the 1960s in Sellafield (UK) and expanded three times during 1970s and 1980s. There are a total of 22 individual vertical concrete silos installed partially below ground level. The aging compartments contain a mixture of intermediate level solid waste underwater from Magnox's reprocessing for approximately forty years. In addition, fuel corrosion sludge residues and swarf stock contain large amounts of various beta gamma residues. The MSSS is a legacy facility currently assessed as the highest risk nuclear facility in the Nuclear Decommissioning Authority estates. MSSS was not built to modern standards and the risk continues to increase significantly with time as the building structure ages. SL has commenced a programme of work to the retrieve solid waste from MSSS and transport it for storage in a more modern storage facility [6,7]. It is known that liquor from the MSSS silos have leaked into the ground contaminated the land below the silo, in the 1970s [6,7], and more recently [7, 8] and this contamination is the focus of routine radiological surveillance [9-10].

The current regulations follow the contamination land best practice risk management system to assess the impact of release, or remobilization, of radioactive contaminants. This involves both a routine groundwater sampling programme and more focused site characterization activities involving borehole drilling and associated soil sampling. The samples are processed and analyzed in the laboratory [11]. Sampling and laboratory-based radiological measurements are well established and provide accurate identification and quantification of gamma-emitting, and most prominent hard-to-measure alpha/beta-emitting radionuclides. However, these approaches are limited by: the datasets being discrete; the practice is expensive, laborious, and time-consuming; the treatment steps generate secondary wastes, and there can be considerable radiological risk to workers.

Alternative methods, such as down-hole radiometric logging systems, have also been applied as the optimal method to characterize underground radioactivity. These solve some practical limitations imposed by the traditional method, as it provides continuous, real-time assessment of vertical soil profiles, is less intrusive approach and allows multiple measurements to investigate trends of contamination distribution over the time. Besides this it also enables investigation of a volume of ground many times larger than the borehole size. Further, if a grid of boreholes is available, this mode of deployment can be extended, and a three-dimensional model of the subsurface could be derived from the data.

Borehole logging simply consists of a continuous recording of a certain physical property as a function of depth, inside boreholes (or drill holes). The recording device (logging probe) comprises a sensing system, and its associated electronic circuity, in a resistant and waterproof case. The logging probe is connected to a cable and lowered downhole by means of a winch system, and information on the probe position in the borehole is provided by a depth encoder. The logging cable provides data transfer, power supply, and mechanical support to lower and raise the probe [12].

In the context of this research, we are interested in direct assessment of radioactive contamination in the ground, this is, to locate, identify and quantify radionuclides. Therefore, a borehole logging probe comprising ionizing radiation detectors as sensor system is highly desirable. Radiometric logging originally evolved from geoscience applications for measure natural gamma-ray radioactivity from rocks in and around oil-bearing formations for example [13-15]. It has also been applied to aid numerous underground assessments at nuclear waste disposal sites, such as the Oak Ridge site (Tennessee, US), the Maxey Flats site (Kentucky, US) [13,15], and the Hanford site (Washington, USA) [17].

The limitations of borehole approach are mostly dominated by the need of construction of ground assets, and the intrinsic limitations of the radiometric detectors. The obstruction by belowground assets on site and the limiting penetration power of radioactivity in matter may limit the ease with which horizontally-resolved measurements are obtained, as well. Moreover, there is a significant challenge to detect principally alpha-/beta-emitting radionuclides given the low penetration power characteristic of such radiations, despite their frequent association with nuclides of significant radiotoxicity and therefore concern.

The challenge monitoring of the ground beneath the MSSS legacy facility is compounded by operational infrastructure and deployment constraints due to the complexity and congestion of the site [9,10]. One existing in-ground monitoring asset local to MSSS building are an array of carbon steel blind-tubes, of around 10 m depth, and inner diameter of 75 mm. The radioactivity fingerprint in the soil monitored by the blind-tube array is believed to be dominated by caesium-137 with strontium-90 and non-isotopic specific dose measurements in some of the blind-tubes reach 1 Gy/h. These dose rates represent a significant limitation in the choice of detectors which may be deployed and their long-term functionality.

The radiometric logging probes used in such scenarios, typically comprise gamma-ray detectors such as gas-filled detector Geiger-Muller (G-M) tubes [17], sodium-iodide (NaI:Tl) scintillators [13-18] and portable high-purity germanium (HPGe) detectors [17-19].

GM tubes have the advantage of being robust, simple, and inexpensive detectors; however, they do not discriminate between different types of radiation, nor radiation energies. As a result, GM tubes are only applied as for basic survey of gross dose rate. Nevertheless, depending upon the specific application, GM tubes perform adequate assessment of radiation levels and trends or changes over time.

NaI:Tl scintillation crystals are the most extensively used sensing material for gamma spectroscopy applications. Sodium iodide is cheap, highly efficient, available in varying sizes and geometries, and very well characterized in the literature. The resolution is sufficient to speciate dominant and well-separated characteristic peaks of some radionuclides, allowing relatively fast surveys at depth. However, the relatively low-resolution characteristic of NaI:Tl might lead to inaccurate analysis when used in some applications. For scenarios comprising a mixture of radioactive materials, shielded and with a high background, a device with higher energy resolution is essential to resolve and clearly distinguish gamma lines in the associated complex spectra. Another limitation of NaI:Tl detectors is their slow scintillation decay time constants. Thus, when deployed in a high radiation level environment the signals tend to pile up. Moreover, NaI:Tl crystals are also susceptible to performance degradation from radiation damage for dose levels above 1 Gy [20].

HPGe semiconductor detectors hold excellent energy resolution, which provide accurate analysis of distinct gamma energies and thus outstanding identification of radionuclides in a wide range of scenarios. The limitation of HPGe detectors is their high cost and the requirement to operate at cryogenic temperatures. Recent advances in HPGe technology have resulted in modules with reduced size and complexity, with electromechanical cooling, and although still expensive are promising in terms of their potential deployment in small spaces.

The selection of an optimal technology for this application is a crucial step to achieve the purpose of this research, with an alternative detection material desirable to ensure quality in underground gamma-ray spectrometry. This will fill the gap between low-resolution scintillators and expensive semiconductor detectors. So, on this basis, the new radiometric logging probe system shall:

- Improve data gathering in existing metallic assets,
- Be operational at high count rates
- Have high intrinsic radiation hardness,
- High energy resolution, and
- Be compatible to be left in harsh environmental conditions.

The proposed solution consists of a scintillator gamma detection system. Among the many available scintillating materials (see Table 1), a small lanthanide halide cerium bromide ($CeBr_3$) crystal was chosen as an optimal compromise with good key characteristics for this application. It has good gamma-ray detection efficiency (due to a high effective atomic mass), good energy resolution, fast decay time and high radiation hardness. Moreover, its reduced intrinsic activity also offers an advantage over $LaBr_3$ for accurate measurements of low-intensity gamma rays.

TABLE I
AVAILABLE CANDIDATE SCINTILLATOR MATERIALS AND KEY PROPERTIES [20-29]

Key Parameters	NaI:Tl	$LaBr_3$	$CeBr_3$	CLYC:Ce	CLLB:Ce	GAGG:Ce
Detection efficiency						
Zeff; ρ	51; 3.7	47; 5.1	46; 5.2	54; 3.3	47; 4.1	54; 6.7
Energy resolution						
FWHM @ 662 keV	6.5 – 10 %	2.5 – 3 %	3.2 – 4 %	4.5 – 5 %	2.9 – 3.5 %	4.5 – 4.8 %
Radiation hardness						
Dose [Gy] @ Gamma-ray	1	10^5	10^5	-	-	10^3
High count-rate capability						
τ [ns]	230	16	17	1 50 1000	120 500	88
Hygroscopic	Yes	Yes	Yes	Yes	Yes	No

Drawing upon strands of research into the desirable properties of a gamma-ray optimal detector, this study also aims to improve in-ground radiological risk assessment of long-lived gamma-emitting fission products, such as caesium-137, and energetic beta emitters, such as strontium-90 (derived via detection of the associated bremsstrahlung radiation).

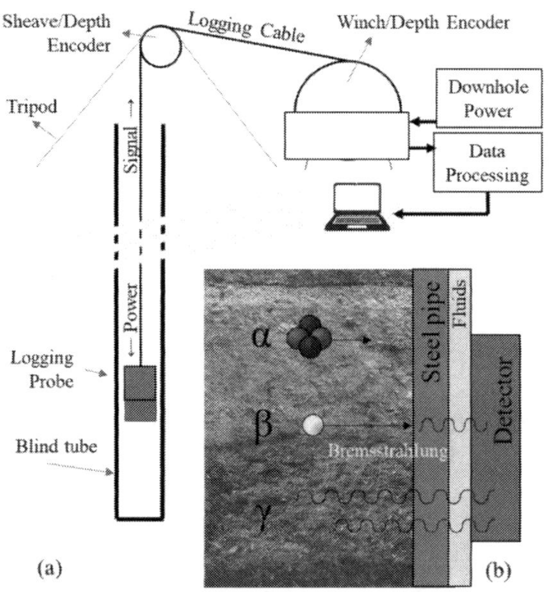

Figure 1 – A schematic diagram of a (a) blind-tube radiation sensing and downhole deployment system to (b) detect gamma and energetic beta emitting radionuclides (via produced bremsstrahlung photons) present in the surrounding soil formation (but close to the blind-tube). Alpha and beta low-energy emitting radionuclides are out of the scope of detection.

II. THE BLIND-TUBE RADIOMETRIC PROBE

The proposed blind-tube radiometric logging probe is comprised of a gamma-ray spectrometer. The sensing material is a Ø10 mm x 9.5 mm cerium bromide ($CeBr_3$) scintillation crystal which is coupled to a silicon photodiode photomultiplier (PM6660-SiPM), and a preamplifier, in a single unit (VS-1402-20) commercialized by Scionix (Netherlands) (figure 2a). This selection satisfies the key characteristics for the application highlighted above. It has good gamma-ray detection efficiency (Z_{eff}=46, ρ=5.2 g/cm^3), energy resolution (3.2-4% @ 662 keV), high count-rate capability (decay time=17 ns) and radiation hardness (10^5 Gy) [22-24]. Note that the SiPM device has practical advantages over traditional photomultiplier tubes, such as low voltage operation, insensitivity to magnetic fields, and superior roughness.

The signal processing chain is accomplished via a digital processor module placed downhole in the probe. The main advantage of digitizing the signal at the probe is that the digital signal can convey information over long distances with less noise, distortion, and environment interference, in contrast with analogue data transmission. Moreover, digitizers are more user-friendly and flexible, and available in a compact unit compatible with the dimensions of typical blind-tubes. Furthermore, processing operations can be altered, and parameters optimized using programmable software to suit a specific application. The proposed digitizer is the compact, stand-alone digital Multi-Channel Analyser (MCA) Topaz-SiPM module commercialized by BrightSpec NV (figure 2b). Note that it is designed to process pulses produced by the silicon photodiode photomultiplier (SiPM) detector, and that it also provides the necessary low-ripple and low-noise 5V voltage supply for the SiPM preamplifier to operate. Therefore, a high-voltage power supply unit is not needed, and the probe can be easily powered from the PC via USB, or Ethernet (PoE) cable. Another important consideration is the applicability of the $CeBr_3$+MCA system to high radiation environment (figure 3).

(a)

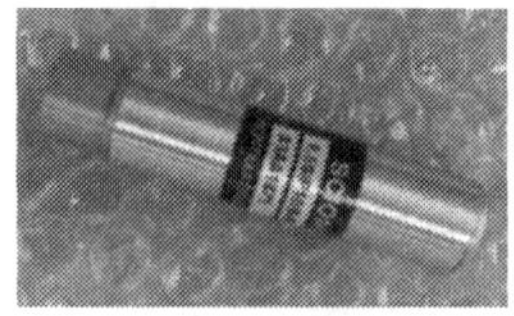

(b)

Figure 2 – (a) Ø10 mm x 9.5 mm $CeBr_3$ scintillator detector, with PM6660-SiPM and built-in preamplifier commercialised by Scionix. (b) Topaz-SiPM digital MCA module commercialised by BrightSpec NV.

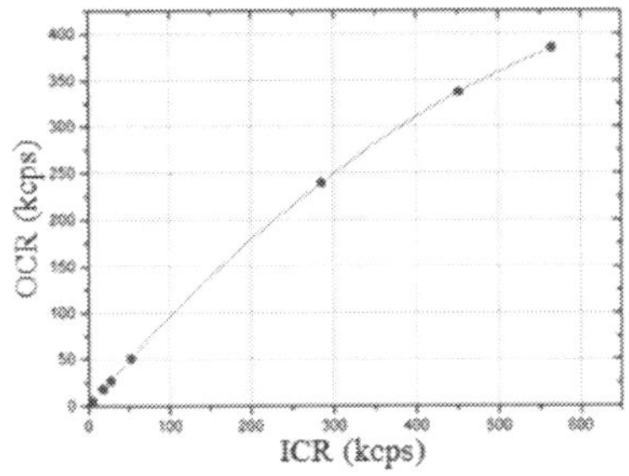

Figure 3 – High count rate capability of system Ø10 mm x 10 mm $CeBr_3$ + SiPM with Topaz-SiPM MCA, applying short shaping times (RT=0.8, FT=1). The radiation source sample was a technetium-99 bottle (figure given by the manufacturer).

III. The blind-tube test pit

Gamma-ray spectrometers must be calibrated to perform quantitative measurements of the emissions from radionuclides, and such calibrations are specific to the sample geometry of the field measurement. In borehole logging measurements, the probe is virtually surrounded by the source, being effectively in a 4π geometry, excluding only a small solid angle above and below it. Traditionally, nuclear probes for geophysical application are calibrated in primary borehole models (or test pits) containing known and uniform concentration of potassium (K model), uranium (U model), or thorium (T model) minerals [14].

Limitations fall to the extent of what is possible and practical to mimic the characteristics of the ground, such as density and composition. The detector response function is defined for a particular geometry, density, and material composition, and correction factors are needed for other conditions [14,30]. Monte Carlo simulations provide an alternative source from which to obtain the detector response functions for a single radionuclide, or decay series, without interference from other radionuclides, and to investigate the effects of heterogeneities in the formation, for example. However, models must be validated with experimental data [30,31].

Figure 4 shows the design and construction of a blind-tube test bed that replicates as close as practical the existing in-ground blind-tubes, such as solid angle coverage of the detector and blind-tube characteristics. Moreover, sand is used as the ground content to facilitate simulations.

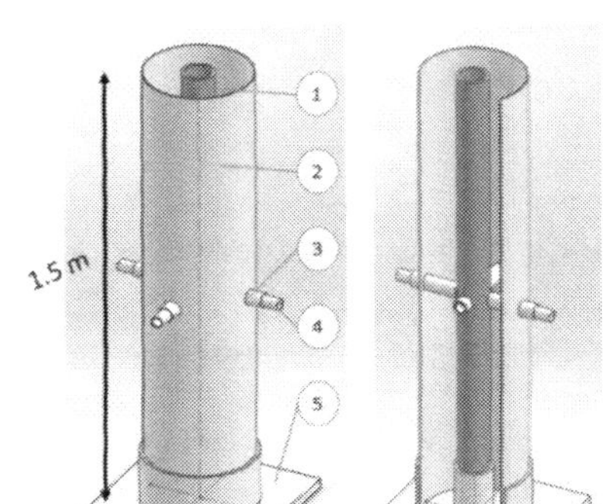

(a)

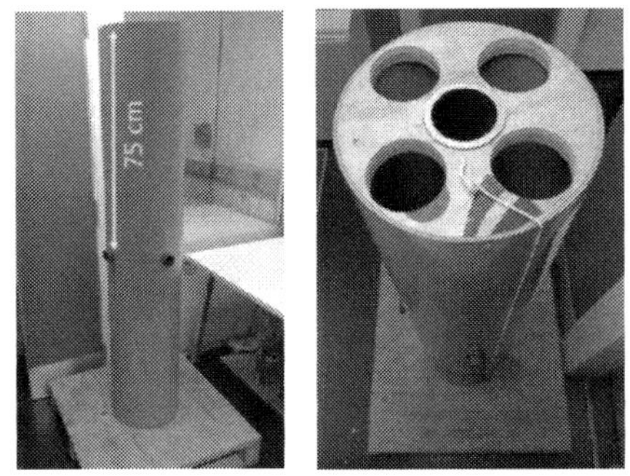

(b)

Figure 4 – (a) CAD design of the blind-tube test bed assembled using Solidworks. It includes (1) a carbon steel tube with inner diameter of 75 mm, wall-thickness of 9.5 mm, and cut to a length of 1.5 m, (2) a plastic sand retaining tube with outer diameter of about 320 mm, wall-thickness of 5 mm, and cut to a length of 1.5 m, (3) four plastic 'void' tubes placed horizontally at 75 cm from the top, (4) four plastic tubes to hold Ø 25 mm disk point sources, and (5) the base support. (b) Picture of the current configuration.

IV. Detector performance

This section presents preliminary desk studies under laboratory conditions to determinate the gamma-ray spectral performance and the angular dependence of response of a cerium bromide scintillator detector, and to test the hypothesis of detecting indirectly beta-emitting nuclide strontium-90, via the bremsstrahlung radiation produced in the materials surrounding its location.

A. Experimental setup

The detection setup comprises a Ø10x10mm $CeBr_3$ scintillation crystal, coupled to a photomultiplier Hamamatsu R9880-110, and an in-built HV-supply and voltage divider, commercialized by Scionix (unit VS-0087-50). The detector readout was connected to digitizer MCA4 commercialised by FAST ComTec (sourced from JCS Ltd., UK). Data were transferred to a computer via USB 2.0 interface and spectra analysed using the MPANT-MCA4 software. The gamma- and beta-emitting disk sources used in this campaign were caesium-137 (304 kBq), europium-152 (370 kBq) and strontium-90 (370 kBq). Measurement times were long enough to achieve good statistical precision for peak evaluation. Figure 5 presents the detector system applied in this campaign of measurements.

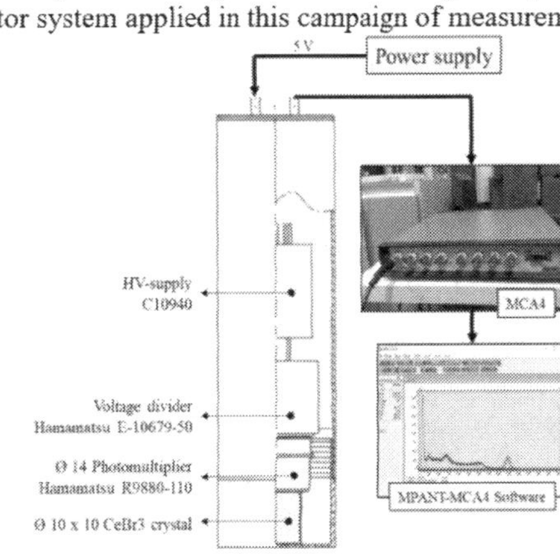

Figure 5 – Simple schematic of detector assembly and experimental detection system used for the preliminary desk studies.

B. Results and discussion

To study the gamma spectroscopy performance of the detector system an Cs-137 and Eu-152 gamma-ray point sources were placed perpendicularly to the central axis of the detector, centered to its sensing volume (figure 6). This geometric configuration is equivalent to a situation in which the detector is positioned vertically in a hole with radioactive material present on the side.

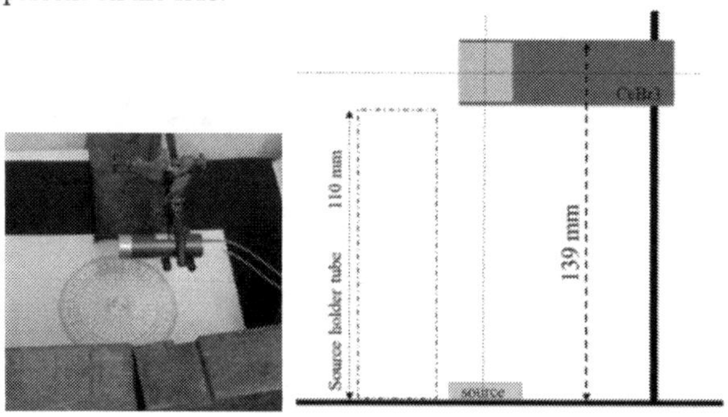

Figure 6 – Experimental setup applied to study the detector gamma-ray spectral performance.

TABLE II
GAMMA-RAY ENERGIES AND EMISSION PROBABILITIES FOR THE RADIONUCLIDES USED IN THE EXPERIMENT [32]

Nuclide	Energy (keV)	Emission probability
Cs-137	661.657 (3)	0.8499 (20)
Eu-152	121.7817 (3)	0.2841 (13)
	244.6974 (8)	0.0755 (4)
	344.2785 (12)	0.2658 (12)
	411.1165 (12)*	0.02237 (10)
	443.965 (3)*	0.03125 (14)
	778.9045 (24)	0.1296 (6)
	867.38 (3)	0.04241 (23)
	964.072 (18)	0.1462 (6)
	1085.837 (10)*	0.1013 (6)
	1089.737 (5)*	0.01731 (10)
	1112.076 (3)*	0.1340 (6)
	1212.948 (11)*	0.01415 (9)
	1299.142 (8)*	0.01632 (9)
	1408.013 (3)	0.2085 (9)

* Not included in spectral analysis calculations.

Regions-of-interest (ROI) of each peak were set manually, and a single Gaussian peak fit with linear background was applied using the MPANT software. The respective centroid-peak and net-peak area values were extracted. Table 2 shows a list of the expected gamma-ray energies and emission probabilities for the radionuclides Cs-137 and Eu-152. Note the selected energies in the calibration procedure.

The gamma-ray spectra are shown in figure 7, as well as the energy and peak width calibration curves. These results suggest a good detector response linearity in the energy range considered 122 to 1408 keV, with the most prominent nuclide peaks well distinguished from background. The detector energy resolution and the photopeak efficiency achieved at 662 keV was 5.2% and 0.002% respectively. This resolution values is analogous to the result presented in [24], nevertheless higher than the value obtained by the manufacturer. Moreover, overlapping peaks can be distinguished by fitting anomalous peak shapes on the basis of inconsistent FWHM values, relative to the calibration curve. This might be useful to indicate the existence of multiple peaks in each region.

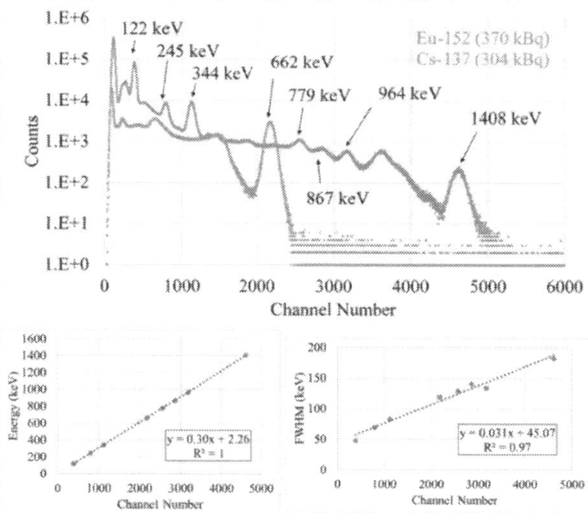

Figure 7 – Ø10x10mm^2 $CeBr_3$ + digitizer MCA4 gamma-ray spectrum for Cs-137 and Eu-152 isotropic point sources. Energy calibration curve and energy resolution response. Acquisition time of about 19h and 25h respectively.

The angular response of the detector to an isotropic point source was obtained. Measurements were performed placing the detector to a turntable with its sensing volume centered on the rotation axis and a Cs-137 gamma-ray source fixed along the axis of the cylindrical detector at 10 cm from it. Thus, the source testing path is a circle around the crystal center with radius of $D = 105 \pm 2$ mm (figure 8).

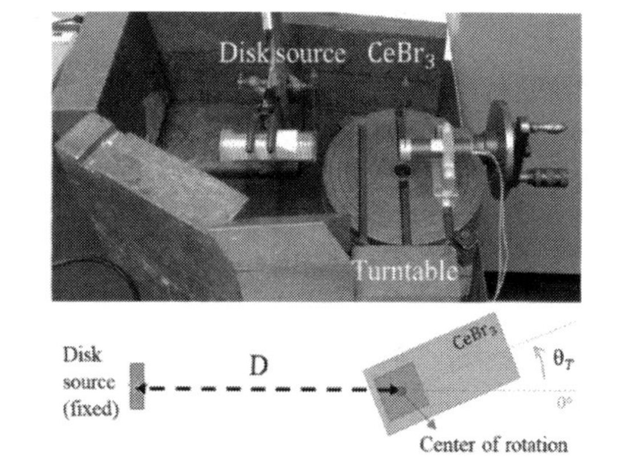

Figure 8 – Experimental setup applied to study the angular dependence of the detector.

Figure 9 shows the detector angular dependence curves for Cs-137 photopeak 662 keV and x-ray 32 keV energies. No significant variation of response was observed over the angle of incidence, and despite the cylindrical shape, the detector may be considered independent of angular directions over the energy range of interest.

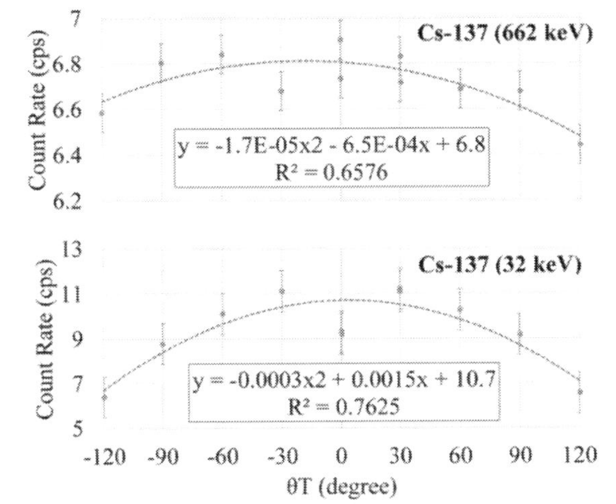

Figure 9 – Angular dependence of a Ø10x10mm^2 $CeBr_3$ detector for Cs-137 photopeak 662 keV and x-ray 32 keV energies. Acquisition times of about 1h each point

Another experiment was the test of the hypothesis of detection Sr-90 (hence Y-90) via produced bremsstrahlung radiation in the presence of gamma-ray emitting sources. The detector was placed centered in a Ø73x88 mm PVC case and inserted in a carbon steel tube (figure 10). Here, Cs-137 and Sr-90 disk sources were fixed to the outside of the metallic tube wall. Spectra were obtained for the situations where only Cs-137 is present next to the wall, only Sr-90, or where both are presented (figure 11).

Figure 10 – Experimental setup applied to study the Sr-90 (via Bremsstrahlung) + Cs-137 assessment. (1) is PVC cylinder with outer size Ø73x88mm, and a cavity to insert the detector centered. (2) is a 25 cm cut of the carbon steel tube presented in the blind-tube test pit section.

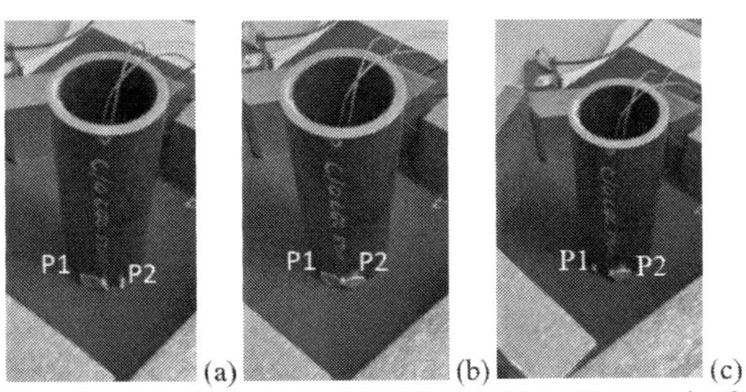

Figure 11 – (a) Cs-137 source placed at position P1. (b) Cs-137 source placed at position P1 + Sr-90 source placed at position P2. (c) Sr-90 source placed at position P2.

Figure 12 shows the detector's response for the three situations mentioned above. The results show that the detection of Sr-90 is feasible applying this method for the activity sources considered. Furthermore, the difference in shape of the Cs-137 and Cs-137+Sr-90 spectrum suggest the possibility of discrimination between the different radiation contributions.

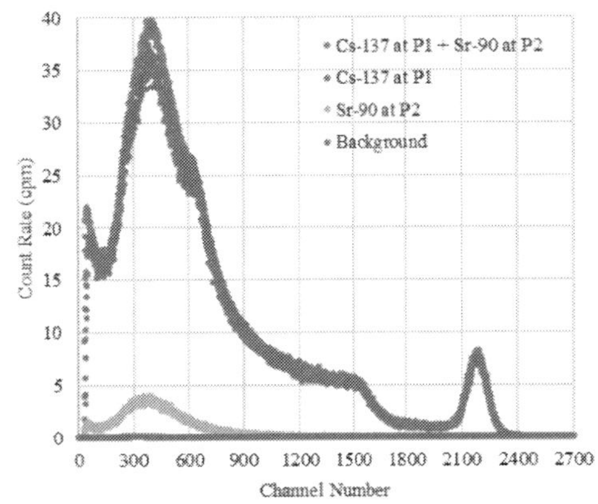

Figure 12 – Measured spectra of bremsstrahlung radiation and gamma-ray radiation of Sr-/Y-90 and Cs-137 respectively. The sources were placed in outer side of a 1cm wall carbon steel tube.

V. Conclusions and Future Work

This paper reports a design of a blind-tube monitoring instrument to improve the characterization of subsurface radioactive plumes beneath the MSSS facility at the Sellafield site (UK). Other prospective uses envisaged are to aid environmental and safety monitoring in and around reactor fuel processing plants, leak detection at interim storage sites for high level waste, and in other worldwide nuclear operating sites which are undergoing land remediation as part of decommissioning.

Overall, the unit comprises a Ø10x9.5mm $CeBr_3$ scintillation detector attached to a compact, stand-alone digitizer Topaz-SiPM module in a resistant and waterproof housing.

Preliminary experimental studies were carried out to study the gamma-ray spectral performance and angular dependence of a small Ø10x10mm $CeBr_3$ scintillation detector. Moreover, a design of a blind-tube test bed for efficiency calibration purposes is also included in this work. Additionally, the hypothesis of detecting Sr-90 via bremsstrahlung radiation has also been assessed, including measurements of this isotope in the presence of Cs-137.

The detector features observed suggest that the Ø10x10mm $CeBr_3$ detector is a promising solution for monitoring underground gamma-emitting radionuclides and high energetic beta-emitting radionuclides, addressing geometric requirements imposed by the existing blind-tubes on site.

Validation of the concept on the blind-tube test pit are underway, alongside with development of a Monte Carlo model in Geant4 software. As well as investigation of a method able to discriminate the contribution of bremsstrahlung photons from the Compton scattering photons at low energy and quantify this discrimination.

Long term future work involves testing of the spectroscopy performance of the system under high count rate conditions, as well as radiation hardness, as well as temperature and shock tests. Finally, the ultimate goal involves extension of a single detector to a blind-tube string network and create a 3-dimensional picture of radiation present on the ground within the MSSS building in the Sellafield site.

ACKNOWLEDGMENT

This research is supported by NNL and Sellafield Ltd. (UK) via the EPSRC programme TRANSCEND (TRANSformative Science & Engineering for Nuclear Decommissioning), code EP/S01019X/1. M. J. J. acknowledges the support of the Royal Society via a Wolfson Research Merit Award.

REFERENCES

[1] ISO, B. "Measurement of radioactivity in the environment – Soil," BS ISO, 18589, Part 2, 2017, Part 3, 2007, Part 5, 2009, Part 6, 2009.
[2] ISO, B. "Water quality – Sampling," BS ISO, 5667, Part 11, 2009, Part 22, 2010.
[3] ISO, B. "Geotechnical investigation, and testing – Sampling methods and groundwater measurements," BS ISO, 22475, Part 1, 2006.
[4] ISO, B. "Water quality – Determination of the activity concentration of radionuclides – Methods by high resolution gamma-ray spectrometry," BS ISO, 10703, 2007.
[5] ISO, B. "Water quality – Strontium 90 and strontium 89 – Test methods using liquid scintillation counting or proportional counting," BS ISO, 13160, 2020.
[6] ONR, "Magnox Swarf Storage Silo (MSSS) – Risk and Hazard Reduction Programme Application for Agreement to Commence Installation and Inactive Commissioning of SEP2 and 1 Mobile Caves in MSSS, " Office for Nuclear Regulation, 2016. [Online]. Available: https://www.onr.org.uk/pars/2015/sellafield-15-008.pdf. Accessed on: 14 July, 2021.
[7] ONR, "Magnox Swarf Storage Silo Retrievals Project Agreement to implement Phase 2 active commissioning trials of the MSSS compartment 10 Miscellaneous Beta Gamma Waste retrieval scheme," Office for Nuclear Regulation, 2019. [Online]. Available: https://www.onr.org.uk/pars/2019/sellafield-19-005.pdf. Accessed on: 14 July, 2021.
[8] ONR, "Quarterly Statement of Civil Incidents reported to ONR," Office for Nuclear Regulation, 2019. [Online]. Available: https://www.onr.org.uk/quarterly-stat/2019-4.htm. Accessed on: 14 July, 2021.
[9] Sellafield Ltd., "Delivering nuclear innovation – Characterisation and monitoring using in-ground assets," Game Changers, 2020. [Online]. Available: https://www.gamechangers.technology/static/u/Characterisation%20and%20Monitoring%20Challenge%20Statement.pdf. Accessed on: July 7, 2021.
[10] Sellafield Ltd., "Game Changers webinar – Characterisation and monitoring using in-ground assets," Game Changers, 2020. [Online]. Available: https://vimeo.com/458899566. Accessed on: July 7, 2021.
[11] Sellafield Ltd., "Groundwater monitoring at Sellafield – Annual data review," Report n. LQTD000758, 2016.
[12] Erdi-Krausz, G., Matolin, M., Minty, B., Nicolet, J. P., Reford, W. S., & Schetselaar, E. M., "Guidelines for radioelement mapping using gamma ray spectrometry data," International Atomic Energy Agency (IAEA), 2003.
[13] Keys, W. S., "A practical guide to borehole geophysics in environmental investigations," CRC Press, 1996.
[14] Schubert, G., "Treatise on geophysics," Elsevier, 2015.
[15] Ellis, D. V., & Singer, J. M., "Well logging for earth scientists", Vol. 692, Dordrecht: Springer, 2007.
[16] Keys, W. S., Eggers, D. E., & Taylor, T. A., "Borehole geophysics as applied to the management of radioactive waste: site selection and monitoring," In Management of low-level radioactive waste, Volume II, 1979.
[17] Johnson, M. E., & Field, J. G., "Hanford SX-Farm Leak Assessments," Report In RPP-ENV-39658, Rev. 0, Washington River Protection Solutions Richland, Washington, 2010.
[18] Keys, W. S., Senftle, F. E., & Tanner, A. B., "Use of NaI(Tl) and germanium detectors for in situ x-ray spectral monitoring of boreholes at nuclear waste-disposal sites," No. 79-1220, US Geological Survey, 1979.
[19] Brodeur, J. R., & Nicaise, W. F., "Subsurface radionuclide assessment at Hanford using HPGe gamma-ray borehole geophysics," Transactions of the American Nuclear Society, 70(CONF-940602-), 1994.
[20] Saint-Gobain Ceramics & Plastics, Inc., "NaI(Tl) and Polyscin NaI(Tl) – Sodium Iodide Scintillation Material," 2020. [Online]. Available: sodium-iodide-material-data-sheet_0.pdf (saint-gobain.com). Accessed on: July 7, 2021
[21] McGregor, D. S., "Materials for gamma-ray spectrometers: Inorganic scintillators," Annual Review of Materials Research, 48, 245-277, 2018.
[22] Quarati, F. G. A., et al., "Scintillation and detection characteristics of high-sensitivity $CeBr_3$ gamma-ray spectrometers," Nuclear Instruments and Methods in Physics Research Section A: Accelerators, Spectrometers, Detectors and Associated Equipment 729: 596-604, 2013.
[23] Shah, K. S., Glodo, J., Higgins, W., Van Loef, E. V., Moses, W. W., Derenzo, S. E., & Weber, M. J., "$CeBr_3$ scintillators for gamma-ray spectroscopy," IEEE Transactions on Nuclear Science, 52(6), 3157-3159, 2005.
[24] Drozdowski, W., Dorenbos, P., Bos, A. J., Owens, A., & Richaud, D., "Gamma radiation hardness of Ø1″×1″$LaBr_3$: Ce, $LaCl_3$: Ce, and $CeBr_3$ scintillators," In 2008 IEEE Nuclear Science Symposium Conference Record (pp. 2856-2858), IEEE, 2008.
[25] Kari Peräjärvi, S. T. U. K., Csome, F. C., & Borg, J., "European Reference Network for Critical Infrastructure Protection: Novel detection technologies for nuclear security", 2018.
[26] John Caunt Scientific Ltd., "Gamma-Neutron Scintillation Detector," 2021. [Online]. Available: (CLYC)CLYC | JCS Nuclear Solutions (johncaunt.com). Accessed on: July 7, 2021.
[27] X-Z LAB, Inc., "GAGG(Ce) – Scintillation Crystal," 2021. [Online]. Available: https://www.x-zlab.com/product/gagg-scintillation-crystal/. Accessed on: July 7, 2021.
[28] Yoneyama, M., Kataoka, J., Arimoto, M., Masuda, T., Yoshino, M., Kamada, K., Usuki, Y., "Evaluation of GAGG:Ce scintillators for future space applications," Journal of Instrumentation, 13(02), P02023, 2018.
[29] Mesick, K. E., Bartlett, K. D., Coupland, D. D. S., Stonehill, L. C., "Effects of proton-induced radiation damage on CLYC and CLLBC performance," Nuclear Instruments and Methods in Physics Research Section A: Accelerators, Spectrometers, Detectors and Associated Equipment, 948, 162774, 2019.
[30] Hendriks, P. H. G. M., Limburg, J., De Meijer, R. J., "Full-spectrum analysis of natural γ-ray spectra," Journal of Environmental Radioactivity, 53(3), 365-380, 2001.
[31] Koomans, R., "Calibration of spectral gamma tools," The Medusa Institute, 2018. [Online]. Available: https://the.medusa.institute/display/GW/Calibration+of+spectral+gamma+tools. Accessed on: 15 February, 2021.
[32] IAEA, "X-ray and gamma-ray decay data: Standards for detector calibration and other applications," International Atomic Energy Agency (IAEA), 2005. [Online]. Available: https://www-nds.iaea.org/xgamma_standards/. Accessed on: 7 July, 2021.

Active Dosimetry with the Ability to Distinguish Pulsed and Non-Pulsed Dose Rate Contributions

Krystsina Makarevich[1], Roland Beyer[2], Jürgen Henniger[1], Yuzhen Ma[1], Sarah Polter[1], Marian Sommer[1], Tobias Teichmann[1], David Weinberger[2], and Thomas Kormoll[1]
[1]Technische Univetsität Dresden, Germany
[2]Helmholz Zentrum Dresden-Rossendorf, Germany
krystsina.makarevich@tu-dresden.de

***Abstract*—This paper presents the concept of an active dosimetry system and its operational regime for pulsed radiation dose rate measurements. The plastic scintillator is suggested to be used for absorbed dose rate measurements. As long as the detector can be considered tissue equivalent, the energy weighting of pile-up events in terms of the dose is achieved. The real-time distinction of pulsed and non-pulsed dose rate contributions is based on the time structure of a single interaction and requires only basic information about the beam time structure (pulses duration and period). The detector connected to a fully digital signal processing board creates an active dosimetry system with adjustable parameters. Such a system was used for absorbed dose rate measurements in pulsed photon field mimicking radiation field outside the bunker of a medical LINAC, but also in the presence of a constant radiation component. The results show a linear dependence of a pulsed radiation contribution on the accelerator current in the investigated range of the total dose rate up to 8 μGy/h.**

***Keywords*—Active dosimetry, Plastic scintillator, Pulsed radiation, Radiation monitoring.**

I. Introduction

Many practical applications (radiation therapy, research accelerators, laser-induced X-rays, etc.) involve pulsed radiation. In such facilities where radiation is generated or used, appropriate dosimetry monitoring has to be accomplished. For continuous radiation one can exploit well-tested area monitoring dosimeters based on ionization chambers, semiconductor detectors or proportional counters. But several studies [1 – 4] revealed that it is not possible to perform accurate dose rate measurements in pulsed radiation fields using the before mentioned dosimeters.

Let us define some criteria that must be satisfied by the dosimeter used for area monitoring in pulsed radiation field. Foremost, such a system has to provide direct readings of the ambient dose or dose rate equivalent [5] and warn users in case of exceeding the established limits. The dosimeter has to be not affected by the pile-up effect, dead-time behavior, energy dependency, or the named effects should be easily corrected. It is also preferable if the dosimeter is fast enough to provide time-resolved measurements necessary, for example, in radiotherapy.

In this regard, plastic scintillators demonstrate several advantages for using them in pulsed radiation dosimetry: they have almost tissue equivalent response to a dose, the results of measurements are not affected by pressure and humidity fluctuations, response time of many of them is in the range of ns, and they can be easily made of a desired shape and size. If appropriate hardware is used, fast scintillators are, in principle, not limited in detecting of pulsed radiation [6].

This paper will give the main considerations regarding the concept of such a detector and its operational regime. The approach of distinguishing pulsed and non-pulsed dose rate contributions will be discussed. The results of dose rate measurements in scattered pulsed radiation field with a time structure imitating the radiotherapeutical beam will be presented and discussed.

II. Materials and Methods

A. Detector

To perform ambient dose equivalent measurements needed for area monitoring, one can exploit a tissue equivalent material for the detector. In this case, simultaneous energy depositions in the detector from multiple photons (pile-up) lead to a proportional value of the absorbed dose as from summing up contributions from separately coming photons in soft tissue. This means that pile-up events in the detector material are energy weighted: higher weighting of pile-up events is achieved by their higher energy values. Thus, tissue equivalency of the detector material allows dealing with the pile-up and consequent energy dependence of the detector.

First of all, let's investigate the response of the plastic scintillator materials to the same dose as in soft tissue. For this, the comparison of mass energy-absorption coefficients of vinyltoluene and polystyrene which are polymer bases of many plastic scintillators and ICRU 4-component soft tissue [5] that is used as the reference material for ambient dose equivalents measurements was performed. The results of the comparison are shown in Fig. 1.

In the energy region from 200 keV up to 4 MeV, the ratio of the mass energy-absorption coefficients is almost constant. This means that the absorbed dose in a plastic scintillator can be proportionally related to the absorbed dose in soft tissue.

Beyond this range, plastic scintillators underestimate the absorbed dose up to 2.5 times depending on energy. If one considers dosimetry behind the treatment room, then this behavior is completely satisfactory as the overwhelming part of the photon energy spectrum is in the energy region higher than 100 keV [7, 8]. Regarding laser-induced X-rays, when the energy of incident photons is in the range of tens keV [9], such a simple correction does not work anymore, and the deconvolution of the spectrum can be required to restore the ratio of mass energy-absorption coefficients. This task is not so trivial as no full absorption peaks are observed in low-*Z* organic scintillators.

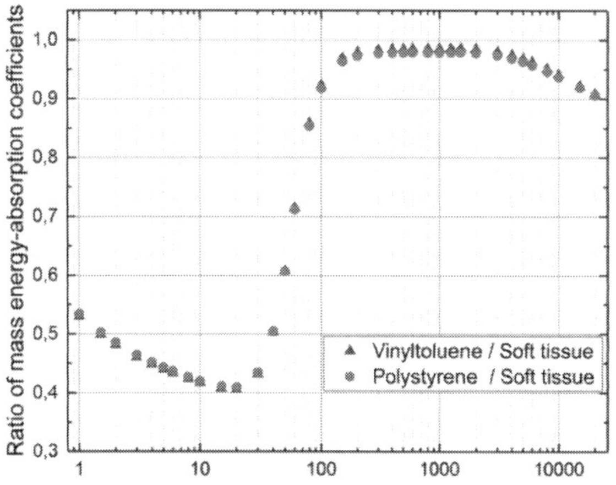

Fig. 1. Comparison of mass energy-absorption coefficients of the main polymer bases of plastic scintillators with the same for ICRU 4-component soft tissue [10].

It should be stressed that the comparison above only refers to the materials' response to the dose and does not consider the relationship between the dose measured with the detector as the whole and ambient dose equivalents which are defined in a point. Additional investigation is needed to find out how to correspond the dose measured with the detector to the value of ambient dose equivalent.

In this work, the detector (SCIONIX Holland B.V.) consists of the cylindrical crystal (1" × 1") of the plastic (polyvinyltoluene) scintillator EJ-200 (Eljien Technology) in an aluminum housing of a 0.4 mm thickness and a Hamamatsu R1924A photomultiplier with a built-in high voltage power supply.

B. Data acquisition and signal processing

The output signal from the detector is shaped, sampled and processed with a fully digital data acquisition board DAQ125 from Serious Dynamics. An event timestamp, an integral over a whole pulse, and a pulse shape are determined and transferred into a listmode file. Continuous integration gate when trigger is ready again in one sample after the last sample of the integration gate means completely non-paralyzable behavior with a known dead-time (the integration gate length plus one sample [11]). This can be implemented in dead-time corrections if the counting regime is of interest.

C. Dosimetry Considerations

In the energy range of 0.2 – 4 MeV, the absorbed dose in a plastic scintillator is directly proportional to the same in soft tissue. To estimate the energy deposition in a detector's material and, subsequently, the absorbed dose in soft tissue, the energy calibration of the detector was performed using the wide-angle Compton-coincidence technique [12].

The absorbed dose D in soft tissue can be derived from the dose D_{plastic} in a plastic scintillator as

$$D = D_{\text{plastic}} / r \tag{1}$$

where r is the ratio of mass energy-absorption coefficients of a plastic scintillator to the same of soft tissue. For the detector's material (polyvinyltoluene) this ratio is $r = (0.981 \pm 0.004)$.

D. Experiment

Initially, the detector was tested in continuous radiation field. Dose rate measurements were carried out with a ^{137}Cs source with a well-known dose rate. The dose rate estimated using (1) was (14.0 ± 0.1) μGy h^{-1} while the ambient dose rate equivalent at the same distance was stated to be (15 ± 2) μSv h^{-1}. The moderate difference between these two values can be explained that the detector is not calibrated to measure the $H^*(10)$.

The actual experiment was performed in pulsed radiation field at the γELBE facility of the HZDR, Germany. The detector was placed in scattered radiation field that keeps the time structure of the initial beam but has an average photons' energy of hundreds keV. For this purpose, a PMMA phantom (10 cm × 10 cm × 20 cm, the density of 1.18 g/cm^3) was used, the distance between the phantom and the detector was of 29 cm. Such geometry of the experiment (see Fig. 2) ensures that the maximum energy of once scattered photons is less than 777 keV.

Fig. 2. Experimental setup.

In some measurements, an additional source (^{22}Na) of continuous radiation was attached to the PMMA phantom opposite to the detector. The main idea to do this was to develop and investigate an algorithm that would allow real-time distinction of pulsed and non-pulsed radiation contributions to the dose rate.

To mimic the time structure of the radiation therapy beam,

the accelerator was working in the macropulse mode with the characteristics indicated in Table I. The DAQ125 board was synchronized to the accelerator frequency and was running with 13×8 MSamples/s.

TABLE I
TIME CHARACTERISTICS OF THE BEAM

accelerator frequency	13 MHz
micropulse period	77 ns
macropulse duration Δt	5 µs
macropulse period T	5 ms

III. Results and Discussion

A. How to Distinguish Pulsed and Non-Pulsed Radiation

The distinguishing of pulsed and non-pulsed radiation is based on the time structure of a single interaction. Fig. 3 shows the time difference histograms for the measurement with the ^{22}Na source only, and also when the accelerator only was operated (without the ^{22}Na source attached).

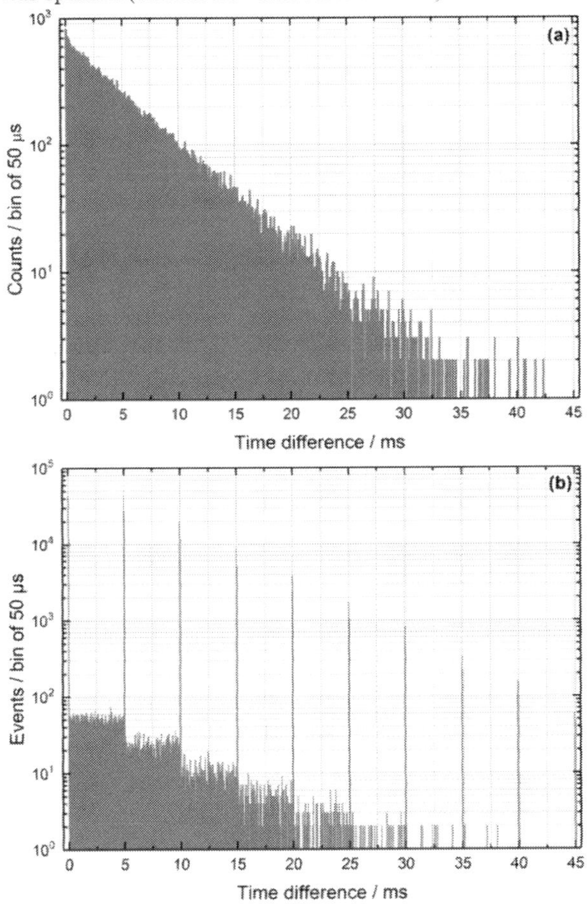

Fig. 3. Time difference histograms for the case when the ^{22}Na source only was used (a) and when the accelerator was switched on and no ^{22}Na (b). The accelerator current in (b) was of 25 µA.

One can see that the radiation coming from the ^{22}Na only (Fig. 3(a)) behaves according to the Poisson process, while prominent peaks with the time difference that is a multiple of the macropulse period are observed for the case when the accelerator only was working (Fig. 3(b)). One can conclude that the pulsed radiation coming from the accelerator has the time difference that belongs to the intervals of

$$[kT-\Delta t; kT+\Delta t] \quad (2)$$

where T is the macropulse period, Δt is the macropulse duration and k is used for positive integers. If one considers the events from the same macropulse, the time difference for them belongs to

$$[\text{integration gate}; \Delta t] \quad (3)$$

If the signal from the detected event crosses a threshold, a leading edge trigger fires, and the integration gate starts. Within the integration gate, no other trigger can occur. This means that all events within the same integration gate are not distinguished from each other and are counted as a single event which is confirmed by Fig. 4.

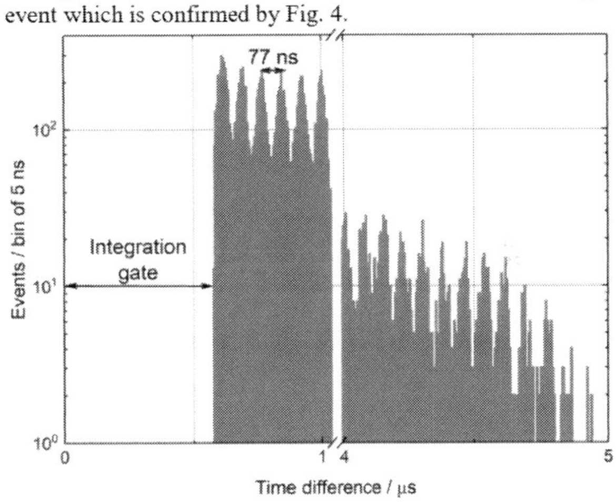

Fig. 4. Enlarged first peak from Fig. 3(b) which corresponds to the time difference histogram for events from the same macropulse. No events have the time difference less that the integration gate length that was of 577 ns for this experiment. The small peaks are spaced by approximately 77 ns, and this corresponds to the period of micropulses.

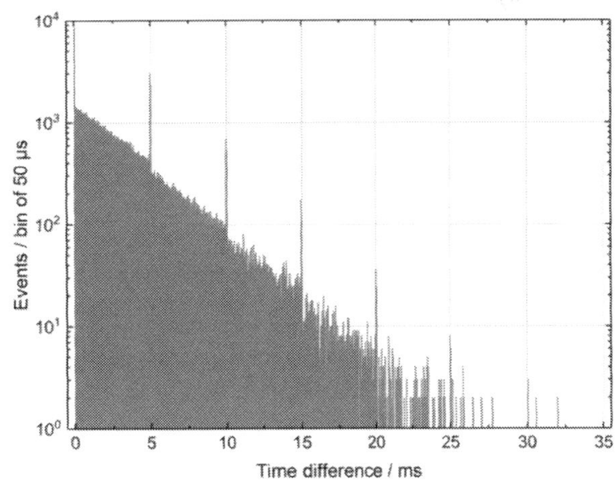

Fig. 5. Time difference histogram for the case of the accelerator switched on and with the ^{22}Na source. The accelerator current was of 25 µA.

Fig. 5 shows the difference histogram for case when the accelerator was operating at the current of 25 µA, and the ^{22}Na source was attached to the PMMA phantom. The interesting thing is that Fig. 5 cannot be obtained by simple summation of normalized Fig. 3(a) and Fig. 3(b). The explanation to this is that photons from the source of constant radiation can be detected almost at any time, and this leads to the decrease of the intensity of peaks with a high time difference and increases the intensity of peaks with a small time difference. In other words, the probability for the detector not to be occupied becomes lower when the source of constant radiation is used.

B. Pulsed and Non-Pulsed Dose Rate Contributions to the Total Dose Rate

The results of the total dose rate, pulsed and non-pulsed radiation dose rates are shown in Fig. 6. It is necessary to mention that Fig. 6 displays an average pulsed radiation dose rate (calculated for the whole measurement time). This means that the dose rate per pulse will be 10^3 times higher.

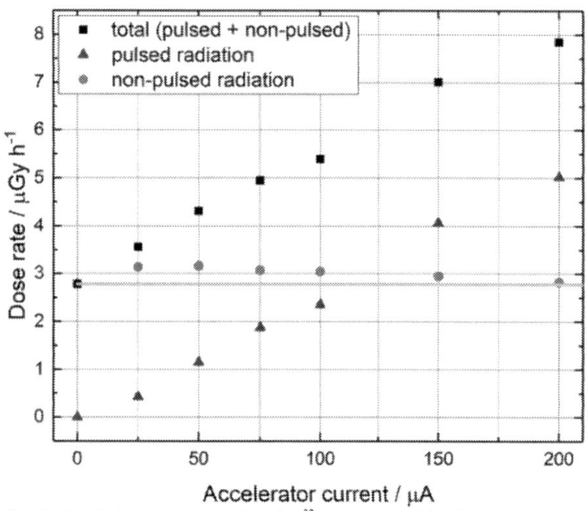

Fig. 6. Total dose rate (assuming the ^{22}Na source and background), pulsed and non-pulsed radiation dose rates versus accelerator current. Uncertainties are enclosed within markers borders. Orange horizontal line indicates the dose rate measured for the ^{22}Na source only.

For these data, the increase of the accelerator current by 1 μA leads to the increase of the total dose rate and the pulsed radiation dose rate by (25±1) nGy/h and (26.2±0.9) nGy/h respectively.

The behavior of the non-pulsed radiation dose rate is not constant and approaches the reference value (measured with the ^{22}Na source only) while increasing the accelerator current. This can be explained by underestimating the pulsed radiation dose rate at low accelerator current values when only one event per macropulse is detected, and this event cannot be unambiguously attributed to pulsed or non-pulsed radiation. It is supposed, that at high values of accelerator current when pulsed radiation significantly prevails over non-pulsed radiation, the influence of non-pulsed radiation contribution that belongs to the "pulsed" time intervals on the dose rate calculations can be neglected. But this question requires further investigation.

IV. Conclusions and Outlook

The presented algorithm of distinguishing pulsed and non-pulsed radiation based on the time difference between events can be easily implemented in the software of any active dosimeter. However, there are some overestimations of pulsed radiation dose rate at low values of accelerator current. This can become significant if an additional source of non-pulsed radiation is present, and the dose rate from the source is comparable to or exceeds the pulsed radiation dose rate.

In the investigated dose rate range up to 8 μGy h^{-1}, the linear dependence of the total and pulsed radiation dose rate on the accelerator current is observed. But this may not work with the further increase of the accelerator current when the photons flux is so high that a time gap between two consequent integration gates leads to losses in detected events and the dose rate is underestimated. Also, the pile-up results in some losses of information about the total absorbed energy in the detector. One of the possible decisions is to expand the integration gate up to the duration of the macropulse (for example, 5 μs in this work). In this case, the detector will integrate all events from the first detected within the macropulse.

One more thing to be studied is adjusting of the current detector for area monitoring when ambient dose rate equivalents ($\dot{H}^*(10)$ and $\dot{H}'(0.07)$) are to be measured. The corrected dose introduced in (1) cannot be treated as the operational dose, even for photon radiation the absorbed dose and the equivalent dose are numerically equal. The difference is hidden in different ways of introducing of the named doses. The corrected dose according to (1) is given for the whole detector, when the ambient dose equivalents are given in a point. A calibration to correspond the operational values is needed or even energy-dependent correction shall be required.

An additional challenge is connected with $\dot{H}'(0.07)$ measurements. The thickness of the housing of the detector is larger than the depth in the ICRU sphere where this quantity is to be measured. This means that for measuring low-penetrating radiation one needs housing for the detector which is equivalent to 0.07 mm thickness of the tissue-equivalent material or even smaller.

Acknowledgment

Parts of this research were carried out at ELBE at the Helmholtz-Zentrum Dresden-Rossendorf e. V., a member of the Helmholtz Association. We would like to thank the experimental facility staff for assistance.

References

[1] Ankerhold, U., Hupe, O. and Ambrosi, P. "Deficiencies of active electronic radiation protection dosemeters in pulsed fields." Radiat. Prot. Dosim, vol. 135, no. 3, pp. 149–153, 2009, DOI:10.1093/rpd/ncp099.

[2] Gotz, M. "Dosimetry of highly pulsed radiation fields." HZDR, Dresden-Rossendorf, Germany. HZDR-90, 2018. Available: https://www.hzdr.de/publications/Publ-27317.

[3] "EPD TruDose for dosimetry in short pulsed X-ray fields using the example of industrial pulsed X-ray source XR-200." Available: https://www.thermofisher.com/order/catalog/product/EPDTRUDOSE#/EPDTRUDOSE.

[4] Friedrich, S. and Hupe, O. "Dose measurements in pulsed radiation fields with commercially available measuring components." Radiat. Prot. Dosim., vol. 168, no.3, pp. 322–329, 2016, DOI:10.1093/rpd/ncv355.

[5] International Commission on Radiation Units and Measurements (ICRU) "Determination of Dose Equivalents Resulting from External Radiation Sources (Report 39)". Journal of the ICRU, vol. os20, no. 2, Feb. 1985, DOI: 10.1093/jicru/os20.2.Report39.

[6] A.R. Beierholm *et al.*, "Investigation of linear accelerator pulse delivery using fast organic scintillator measurements." Radiat. Meas., vol. 45, no. 3-6, pp. 668-670, Mar.-Jul. 2010. DOI: 10.1016/j.radmeas.2009.11.023.

[7] Horst, F., Fehrenbacher, G. and Zink, K. "On the neutron radiation field and air activation around a medical electron LINAC." Radiat. Prot. Dosim., vol. 174, no. 2, pp. 147-158, 2017, DOI:10.1093/rpd/ncw120.

[8] Keehan, S. *et al.* "Dose and gamma-ray spectra from neutron-induced radioactivity in medical linear accelerators following high-energy total body irradiation." Radiat. Prot. Dosim., vol. 172, no. 4, pp. 327-332, 2016, DOI:10.1093/rpd/ncv480.

[9] Behrens, R., Pullner, B. and Reginatto, M. "X-ray emission from materials processing lasers." Radiat. Prot. Dosim., vol. 183, no. 3, pp. 361–374, 2019, DOI:10.1093/rpd/ncy126.

[10] J.H. Hubbell and S.M. Seltzer, "Tables of X-ray mass attenuation coefficients and mass energy-absorption coefficients 1 keV to 20 MeV for elements Z=1 to 92 and 48 additional substances of dosimetric interest." NISTIR, Gaithersburg, MD, USA. Tech. Rep. NISTIR-5632, 1995.

[11] "DAQ125 Firmware (datasheet)." Serious Dynamics.

[12] K. Roemer *et al.*, "A technique for measuring the energy resolution of low-Z scintillators." 2009 IEEE Nucl. Sci. Symp. Conference Record (NSS/MIC), Orlando, FL, 2009, pp. 6-11, DOI: 10.1109/NSSMIC.2009.5401909.

Dose Rate Measurements in Pulsed Radiation Fields by Means of an Organic Scintillator

Theresa Werner[1], Roland Beyer[2], Richard Biedermann[1], Marko Gerber[1], Jürgen Götze[3], Philipp Herzig[1], Vincent Melzer[1], Elena Metzner[1], David Weinberger[2,4] and Thomas Kormoll[1,4]
[1]Instiute of Nuclear and Particle Physics, Technische Universität Dresden, Germany
[2]Helmholtz-Zentrum Dresden-Rossendorf, Germany
[3]Helios Klinikum Aue, Germany
[4]Serious Dynamics, Germany
theresa.werner@tu-dresden.de

***Abstract*—A deficiency in the implementation of current radiation protection is the determination of the ambient dose equivalent *H**(10) and the directional dose equivalent *H'*(0.07) in pulsed radiation fields. Conventional dosimeter systems are not suitable for measurements in photon fields comprising short radiation pulses, which consequently leads to high detector loads in short time periods. Nevertheless, due to the implementation of advanced medical accelerators for cancer therapy, new medical diagnostic devices as well as various laser machining systems, there is an urgent need for suitable dosimeter systems for real time dosimetry. In this paper, a detector concept based on an organic scintillator and a full digital data analysis with the aim of developing a portable, battery powered measurement system is presented.**

***Keywords* — radiation protection, dose rate measurements, pulsed radiation fields, organic scintillation detector**

I. Introduction

Many dosimetric measurement systems are not suitable for the application in pulsed radiation fields regarding radiation protection scenarios. Especially in applications where a low mean dose rate in the range of 1 μSv/h is applied. A main challenge in such fields is to process high detector loads within short time periods, while an appropriate dead time behavior and the suppression of pile up effects must be ensured.

A promising approach for an active dosimetric system fulfilling these requirements is the combination of a fast tissue equivalent scintillation detector coupled to a full digital signal processing unit. Such a system could allow real time dosimetry by measuring the deposited energy in the detector, while a discrimination between pulsed and non-pulsed events is realized by comparing the individual time stamps of the measured events. Additionally, pile up events can be identified by analyzing the pulse shape of the individual events.

Furthermore, due to the completely non-paralyzed dead time behavior of the detection system, it is possible to correct signal losses for the respective measurement. The principle of this method, which is in detail described in [1], was implemented in an appropriate detection system based on an organic scintillator and a digital data acquisition board. The developed prototype was tested in a continuous as well as pulsed photon fields.

In a measurement campaign at the institute of nuclear and particle physics (IKTP) of Technische Universität Dresden, Germany, with a ^{137}Cs source, quantitative dose rate measurements were performed. To determine the dose values an appropriate analyzing algorithm for dose rate measurements was developed.

At the γELBE bremsstrahlung-facility [2] a pulsed photon beam was available, where measurements under various pulse frequencies (up to 10 kHz) and a macro pulse duration of 40 μs were performed. The detection system was placed next to a polymethylmethacrylat (PMMA) phantom, which was irradiated with the bremsstrahlung-beam.

Additionally, dose rate measurements at a clinical TrueBeam therapeutic system by Varian [3] were performed, where the detector was placed outside the treatment room.

For both measurements in pulsed radiation fields, it was possible to reconstruct the characteristic structure of the pulsed beam, which comprises the identification of the pulse length and repetition rate.

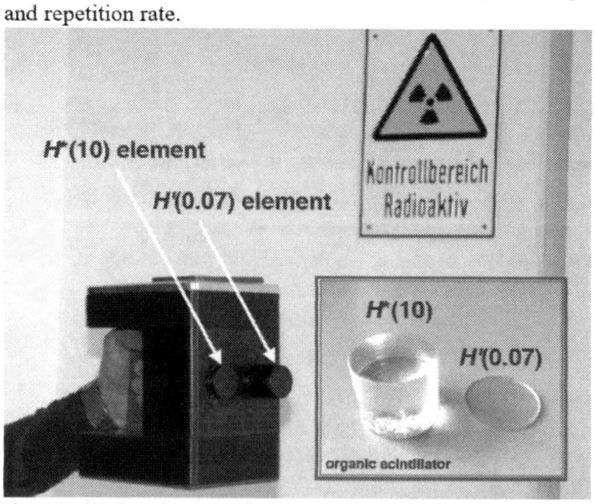

Figure 1. The detection system is a portable, battery powered device with two organic scintillation elements (*H**(10) element, and *H'*(0.07) element) each coupled on an individual photomultiplier tube with a subsequent free running 14 bit analog-to-digital converter with a sampling rate of 125 MHz for data

processing.

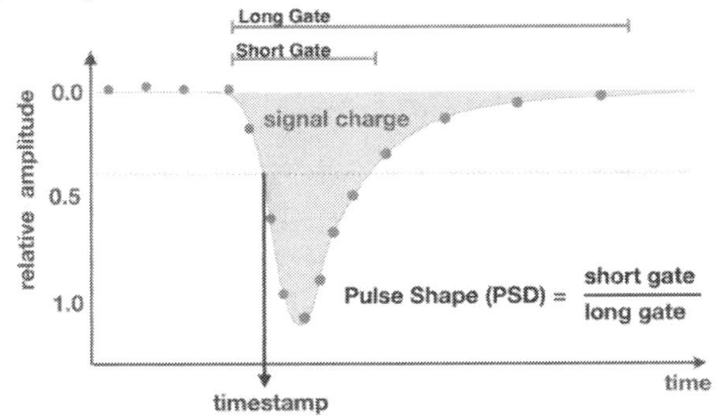

Figure 2. The voltage pulses coming from the two photomultiplier tubes are digitized by an analogue-to-digital converter (125 MHz, 14 Bit) and processed by a field programmable gate array (FPGA). In the sketch the three individual analysis parameter are shown: (1) the timestamp, which determines the arrival time of the detected event, (2) the pulse charge, which is the integral over the signal and proportional to the deposited energy and (3) the pulse shape parameter (PSD), which represents the shape of the signal.

II. Methods and Material

A photograph of the detection system is shown in Figure 1. It consists of two detection elements ($H^*(10)$ element, $H'(0.07)$ element) for the respective radiation protection quantities. The cylindrical organic scintillators have a height of 19 mm and a diameter of 23 mm for measuring $H^*(10)$ and a height of 1 mm and a diameter of 24 mm for measuring $H'(0.07)$. Each scintillator is coupled to an individual photomultiplier tube (PMT: PDM 04-9111) produced by ETEnterprises, which are read out with a free running 14 bit analogue-to-digital converter (ADC). For this, the output signal is sampled with a data acquisition board DAQ125 by Serious Dynamics, Dresden, Germany, which allows a real-time processing of the data via an included field programmable gate array (FPGA).

A. Data processing

As shown in the sketch of Figure 2, for each recorded event signal the FPGA algorithm determines three distinct parameters: (1) The first one is the timestamp, which represents the pulse onset with respect to the number of elapsed clock cycles. (2) The second parameter is the pulse charge, which is defined by the integral over the whole pulse (long gate). (3) The third is the pulse shape parameter (PSD), which is given by the ratio of the partly integrated pulse (front part of the signal / short gate) to the integral of the whole pulse. This parameter is therefore a measure for the shape of the pulse and thus allows to distinguish between different "event scenarios" in the scintillation detector (e.g. single event, pile-up event, particle discrimination).

B. Data analysis

As introduced in section 1, a main challenge for a dose compliance measurement in pulsed radiation fields is to handle the high detector load within short time periods (Hz - kHz) and at the same time short pulse durations (down to ns - fs). In the detected data the dead time behavior of the detection system as well as the probability of pile-up events must be considered.

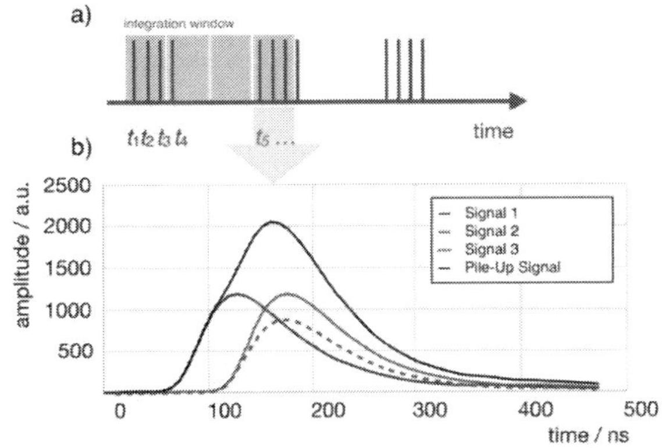

Figure 3.The residual signal of multiple events (Pile-Up Signal), which are detected in the same integration window (event 1: signal 1, event 2: signal 2), is proportional to the deposited energy in the detector (tissue equivalent) and therefore to the ambient and directional dose. This also applied to for different energy depositions (signal 3) due to the equal weighting factor.

Because of the mentioned FPGA real-time processing the detection system shows a non-paralysable dead time behavior due to the subsequent integration gates (Figure 3a).

The system dead time is thereby defined by the length of the integration window (long gate for determining the pulse charge). If multiple events occur in the same integration window the resulting multi-event-signal (Pile-up signal) comprises the deposited energies of each detected single event (Figure 3b). As the organic scintillation material is approximately tissue equivalent, the weighting factor of different energy deposition (Figure 3b: signal 1 and signal 3 (dashed line)) is the same as for equal energy deposition (Figure 3b: signal 1 and signal 2).

C. Experimental Setups

In the following, the experimental setups of the three performed experiments are described. In section a) the measurement in a continuous photon field is characterized, while section b) and c) comprise two experiments in different pulsed photon fields.

a) Radioactive nuclide (continuous): At the IKTP an experiment was performed in a continuous radiation field of a ^{137}Cs source (Figure 4). Within this experiment, it was intended to proof the implemented algorithm by measuring the dose and dose rate under defined conditions. The measured values at different distances to the source were therefore compared to the data from an ionization chamber UNIDOS webline [4] by Physikalisch-Technische Werkstätten (PTW) Freiburg.

b) Research accelerator (pulsed): In a further experiment at the γELBE bremsstrahlung-facility the detector was exposed to a scattered photon field, in which the time structure of the initial beam was received. The micro-frequency of 13 MHz is not regarded in this study and assumed to be quasi continuous at the studied time scales. The accelerator was operated to produce a

 https://doi.org/10.1051/epjconf/202125309002

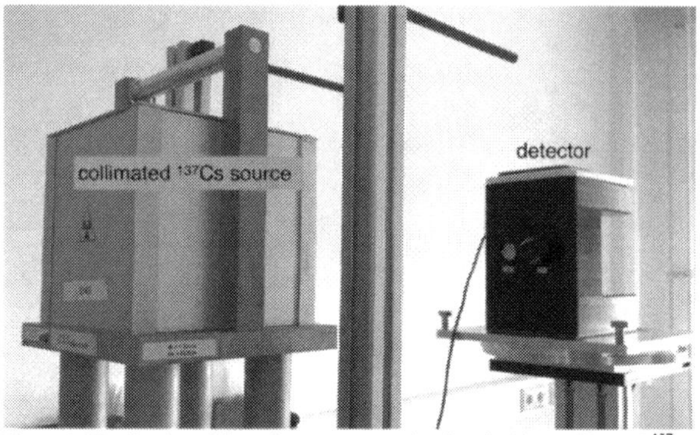

Figure 4. The developed detection system was irradiated with a collimated ^{137}Cs source to determine quantitative dose/dose rate values.

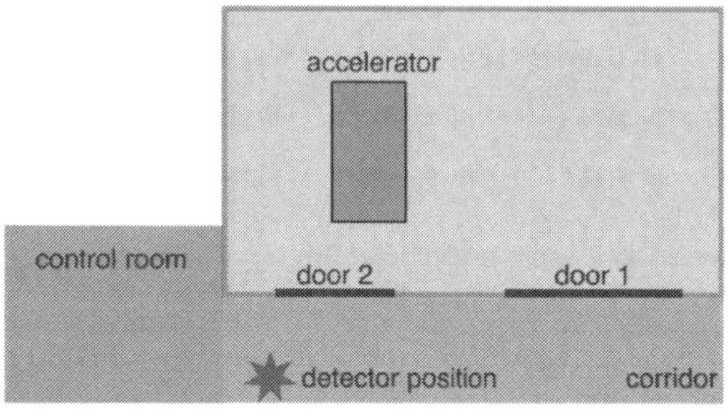

Figure 5. At the Helios Klinikum, Aue, Germany, a measurement next to the control room was performed, while at the same time an experiment was carried out inside the treatment room.

continuous wave (cw) as well as a pulsed bremsstrahlung-beam with a respective pulse frequency of 5 kHz and 10 kHz. The pulse duration was set to 40 µs.

c) Medical accelerator (pulsed): At the Helios Klinikum Aue, Germany, a TrueBeam therapeutic system by Varian produces a pulsed photon beam, where the frequency depends on the requested dose / dose rate in the treatment and can therefore vary. Here, the detector was placed outside the treatment room in the adjacent corridor behind a door to the treatment room (Figure 5). Inside the treatment room a PMMA target was irradiated with a 30 cm x 30 cm field and a 15 MV photon beam. The dose rate in the treatment room was set to 5 Gy/min for the performed experiment.

III. Results and Discussion

A. Radioactive Source: ^{137}Cs

Figure 6 shows the pulse charge spectra for the different detection elements (red: $H^*(10)$ element, gray: $H'(0.07)$ element) measured in the radiation field of a ^{137}Cs source. As the scintillation material is a low Z-scintillator ($Z_{eff} \sim 4.5$ [5]) the characteristic ^{137}Cs peak at 662 keV is not visible in the respective spectra. To calibrate the detectors, the Compton edges (c.f. Figure 6) and dosimetric values from the data of the

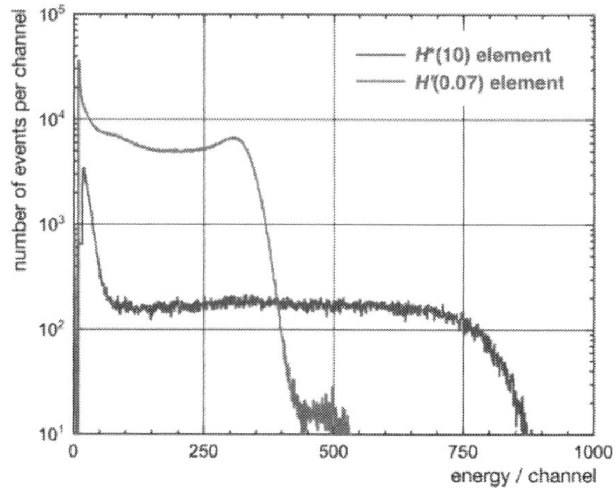

Figure 6. The pulse-charge histograms for the two individual detector elements (gray: $H^*(10)$ element, red: $H'(0.07)$ element) measured during the irradiation with a ^{137}Cs source. The Compton edges are clearly visible.

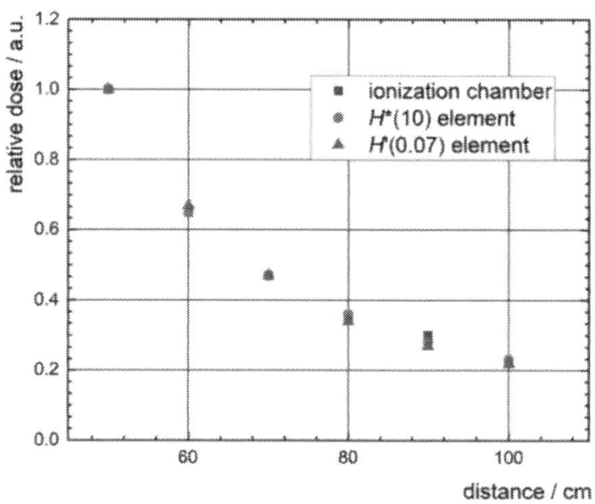

Figure 7. The relative dose decreases with greater distances. The black values belong to the reference measurement with an ionization chamber (UNIDOS webline), the red to the $H^*(10)$ element and the blue to the $H'(0.07)$ element of the detection system. All measured values are normalized to the respective first value recorded at a distance of 50 cm.

UNIDOS webline ionization chamber were used. For this, a reference measurement at a fixed position was performed. This, together with the pulse charge spectra (Compton edge), allowed an energy calibration and therefore a comparison of dose and dose rate values between the detection system and the ionization chamber. The measured relative dose values at the different positions are given in Figure 7.
Here, the values were normalized to the respective first value at a distance of $d = 50$ cm. Consequently, at this position the

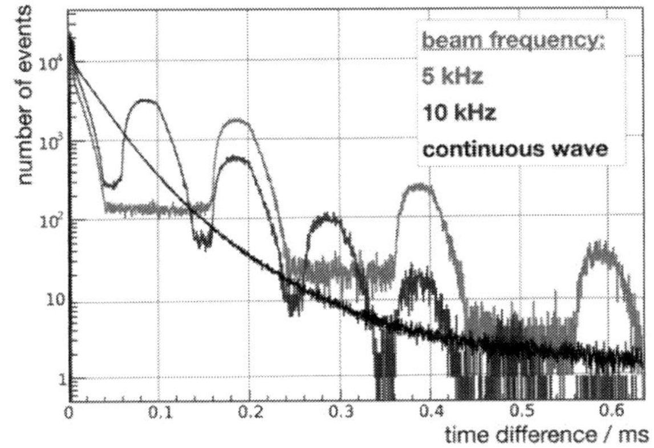

Figure 8. In the time-difference histogram, the timing structure of the initial beam is reproducible. The black curve belongs to the continuous wave beam and the red and blue curve to a pulsed beam with a pulse duration of 40 µs and a frequency of 5 kHz and 10 kHz, respectively.

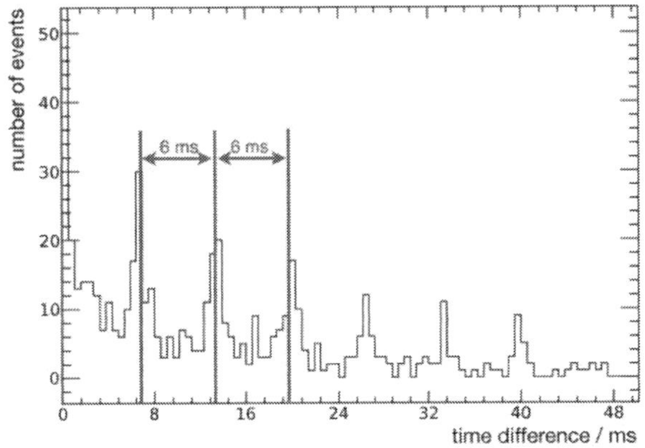

Figure 9. The time-difference histogram measured in front of the treatment room at Helios Klinikum, Aue, Germany, during the operation of a clinical accelerator. In the recorded data, the time structure of the initial beam could be reconstructed with a frequency of around 166 Hz.

relative dose for the different detectors is equal to 1. For the different detector elements the exposed dose decreases, as expected, with greater distances and the single values match with the reference measurement (ionization chamber). This proofs that the developed detection system is suitable to evaluate dose as well as dose rate quantities in continuous radiation photon fields.

B. Research Accelerator: ELBE

To reconstruct the timing structure of the primary beam the difference between two subsequent measured events was calculated. The resulting time-difference histogram is shown in Figure 8. The red curve represents the measurement of a pulsed beam with a frequency of 5 kHz and the blue one with a frequency of 10 kHz. The black curve belongs to the cw measurement. In these histograms the time structure of the initial beam can be clearly identified, where the periodic peak structure gives the respective beam period (0.2 ms and 0.1 ms) (consequently, the above mentioned frequencies) and the peak widths encode the beam pulse duration. In contrast to the pulsed beam, the time differences for events, which are measured during a continuous beam, follow a falling monotonous distribution.

C. Medical Accelerator: TrueBeam

Similar to the above described experiment, the time-difference histogram (Figure 9) was measured outside the treatment room during the operation of a clinical accelerator. The determined frequency of the accelerator was reconstructed from the measured data with around 166 Hz for the specific beam setting.

IV. Summary and Outlook

Based on the performed experiments, it could be shown that the timing structure of various pulsed radiation photon fields can be reconstructed out of the recorded data. This allows, for example, the discrimination between detected events, which are correlated to the pulsed beam, and possible continuous background radiation in future experiments. Furthermore, this enables a correction of the dose and dose rate values for applications under different conditions (mixed photon fields).
The first quantitative measurements in a continuous photon field proof that the system is in principle suitable for the evaluation of dose and dose rate values. In future experiments, this will be additionally tested in pulsed photon fields.
The detections system will be also tested in low energy photon fields (~ keV) to ensure an appropriate calculation of the radiation protection quantities for low doses and dose rates.
For this, an accurate energy calibration is needed, which will be obtained from a further experiment based on a coincidence scattering technique. The detailed method is described in reference [6] and [7].

References

[1] K. Makarevich et al., "Dosimetry with the ability to distinguish pulsed and non-pulsed dose contributions", Radiation Protection Dosimetry, 2020.
[2] R. Schwengner et al., "The photon-scattering facility at the superconducting electron accelerator ELBE", Nucl. Instr. Meth., 2005.
[3] C- Glide-Hurst et al., "Commissioning of the Varian TrueBeam linear accelerator: a multi-insitutional study", Med Phys. 2013.
[4] PTW Freiburg, "Grebrauchsanweisung 30cm^3 Stielkammer Typ 23361", operating instruction, 2001.
[5] G. F. Knoll, „Radiation detection and measurement", Wiley 4th ed. 2010.
[6] Swiderski et al., "Measurement of Compton edge position in low-Z scintillators", Radiation Measurements, 2010.
[7] V. Melzer, "Realisierung eines roboter-unterstützten Weitwinkel-Compton-Aufbaus zur Energiekalibrierung von Niedrig-Z-Szintillatoren", bachelor thesis, 2020.

Deconvolution methods used for the development of a neutron spectrometer

Claude-Alexandre Simonetti[1,2,3], Marc Labalme[2,3], Jean-Lionel Trolet[1] and Patrick Mary[1]
[1]EAMEA Cherbourg, France
[2]ENSICAEN, France
[3]LPC Caen, France
simonetti@lpccaen.in2p3.fr

***Abstract*— To fulfill the requirements of the industry, the feasibility of a transportable neutron spectrometer is under study by our collaboration. Preliminary studies have led to a solution based on a unique-multi-detectors-Bonner sphere, which has the advantage to simultaneously perform many neutrons-measurements through thermal neutron detectors placed at different depths inside a polyethylene sphere. The incident neutron energy is then reconstructed using unfolding methods. The optimization of the layout of the detectors in the sphere and of the diameter of the sphere was performed thanks to GEANT4 simulations, coupled to unfolding methods such as least-squares, maximum of entropy or maximum of likelihood. Different unfolding methods have been tested. For the time being, the best unfolding results were obtained by the computer codes M.A.X.E.D. and G.R.A.V.E.L., formalized by the Nuclear Energy Agency (N.E.A.). A "personal" method based on the maximum of entropy coupled to the maximum of likelihood gives good results as well, but a few convergence parameters have still to be optimized. In the present paper, a solution of a multi-detectors-Bonner sphere is presented as well as results of unfolded neutron energy spectra. The results obtained show a good agreement with unmoderated and moderated Am/Be and ^{252}Cf spectra.**

***Keywords* — Environmental and Medical Sciences, unfolding methods, Monte-Carlo simulations.**

I. Introduction

ACCURATE measurements of neutron dose rates is of great importance for radiation protection applications. To fulfill these objectives, Bonner spheres [1] were developed more than 40 years ago and are still used, due to continuous optimizations of the detectors efficiencies and new patterns of use, from one detector per Bonner sphere at the beginning to multiple detectors at different depths in only one sphere [1,2,3].

This report is about a project in progress, to develop a neutron spectrometer with the following general specifications: it has to be transportable, to return a 0 to 20 MeV neutrons spectrum and be able to measure an ambient dose equivalent (H*(10)) from 1 µSv/h to ~10 mSv/h in less than 5 to 10 minutes.

To fulfill theses specifications, a multi-detectors Bonner sphere setup was chosen. It is composed of a sphere made of a solid polyethylene moderator, with multiple thermal-neutron detectors inside. For this prototype, the ^{6}Li(n,α)^{3}H reaction was chosen for the detectors to optimize the neutron gamma discrimination thanks to the high Q of reaction, equal to 4.78 MeV. An important work of optimization has been done with GEANT4 [4,5,6] Monte-Carlo simulations coupled to unfolding methods. This optimization concerns not only the size of the sphere, but also its thermal-neutron-detectors layout, number and geometry. Numerous simulations were performed to get information on the sensitivity of the diameter of the sphere and of the detectors depth and number in order to get an adequate layout compatible with the previously mentioned specifications. Then, different unfolding methods were tested, leading to results, with a good agreement, for unmoderated and moderated Am/Be and ^{252}Cf neutron energy spectra.

In the following, the geometry of the prototype and the reference spectra will be presented. Then, the unfolding methods used in this project will be exposed. Finally, the results of GEANT4 optimization and of unfolding will be exposed.

II. Geometry optimization

Monte-Carlo simulations with the GEANT4 environment have been processed to optimize the geometry of the multi-detectors-Bonner sphere. The version 4.10.5 of GEANT4 was selected, and the QGSP_BIC_HP physics list was loaded to allow a high precision for neutrons energies below 20 MeV [7].

A. Geometry in GEANT4

GEANT4 simulations were performed with Bonner sphere diameters from 20 to 30 centimeters, in order to determine the most appropriate diameter to the specifications. In these spheres made of polyethylene, one thermal-neutrons detector was placed each centimeter along the 3 axis (x, y and z). The detectors were composed of lithium fluoride, LiF, with an isotopic ratio of 14.3% for ^{6}Li and 85.7% for ^{7}Li, to get 8.81 atoms/cm^3 of ^{6}Li for a LiF density of 2.64 g/cm^3. The dimensions of the detectors were 1.0 cm × 1.0 cm × 0.04 cm. Fig. 1 presents an example of the 25 cm diameter sphere, which makes a total of 73 detectors.

Fig. 1. Unique-multi-detectors Bonner sphere modelized with GEANT4, with a parallel beam in the direction : (-1 0 0). Here, a 25 cm diameter polyethylene sphere is represented in red, with one LiF detector placed each centimeter along x, y and z axis.

B. Neutron source

The first step of the study was to get a maximum of information on the Bonner sphere with GEANT4. Two types of sources were necessary to reach this goal :

- Monoenergetic sources vectors named $\mathbf{X_j}$, where j is the number of the energy bin considered, j∈{0,…,m-1} with m the total number of energy bins, which corresponds to the number of components of the **X** source vector. These sources were used to obtain the response function of the setup.
- Reference sources vectors, corresponding to unmoderated and moderated Am/Be and ^{252}Cf energy spectra, named $\mathbf{X_{AmBe}}$, $\mathbf{X_{Cf}}$, etc. These spectra were used to validate the unfolding.

For the moment, only parallel sources were sent to the sphere, with the direction : (-1 0 0). Other directions, parallel or isotropic, will be studied in the future.

1) Monoenergetic sources

Monoenergetic sources of neutrons were sent to the Bonner sphere to build a response matrix of the setup, named **R**. For monoenergetic-parallel sources, the detector responses will be stored in a rectangular response matrix named $\mathbf{R_{parallel}}$, with n lines and m columns : the R_{ij} scalar represents the response of the detector or group of equidistant from the center detectors i, i∈{0,…,n-1} to a mono-energetic-parallel source of neutron $\mathbf{X_j}$ with the energy bin j, j∈{0,…,m-1}. In our case, 56 monoenergetic parallel beams were simulated with the following energy values:

$$E_j = 10^{-9+0.2j} MeV \tag{1}$$

Where j∈{0,…,55} to cover the energy range from 10^{-9} MeV to 10^2 MeV. For a given mono-energetic source j, an associated logarithmic bin of energy is then defined, whose bin edges are:

$$E_{e(j)} = 7.942 \cdot 10^{-10+0.2j} MeV \tag{2}$$

and

$$E_{e(j+1)} = 7.942 \cdot 10^{-10+0.2(j+1)} MeV \tag{3}$$

So that E_j is the geometrical mean of these two bin edges. A beam surface fluence (Φ) of 1.41×10^5 cm^{-2}, which corresponds to 10^8 neutrons sent into a 30 cm diameter beam, was used.

2) Reference spectra

Four reference-input-parallel spectra were tested, with the same surface fluence :

- Unmoderated and moderated Am/Be ($\mathbf{X_{AmBe}}$ and $\mathbf{X_{AmBe\_mod}}$).
- Unmoderated and moderated ^{252}Cf ($\mathbf{X_{Cf}}$ and $\mathbf{X_{Cf\_mod}}$).

The Am/Be source was defined according to the ISO 8529 standard spectrum [8]. The ^{252}Cf spectrum was defined following a normalized Maxwellian energy distribution [9]:

$$\Phi_M(E) = \frac{2\sqrt{E}}{\sqrt{\pi E_M^3}} e^{-E/E_M} \tag{4}$$

With E the energy in MeV and E_M the temperature parameter equal to 1.42 MeV.

The moderated spectra were obtained by GEANT4 simulations of the previous spectra moderated by 6 cm of polyethylene. Fig. 2 shows an example of the reference spectrum, $\mathbf{X_{AmBe\_mod}}$, corresponding to the moderated Am/Be source.

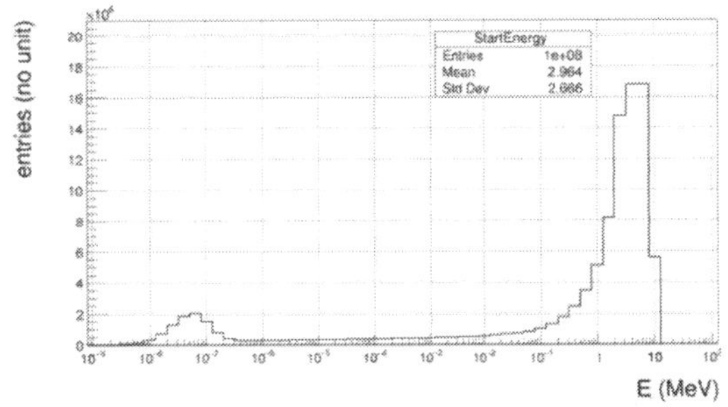

Fig. 2. Energy spectrum (in MeV) of the moderated Am/Be source.

III. UNFOLDING METHODS

Whereas the response-matrix building, the spectra initialization and the prototype-geometry optimization were done with the GEANT4 environment, the aim of the unfolding is to retrieve the neutron source energy spectra **X** by solving the following equation:

$$\boldsymbol{R} \cdot \boldsymbol{X} = \boldsymbol{M} \tag{5}$$

With **X**, the unknown source vector (in cm^{-2}) to be reconstructed, **M** the measured vector (no unit) associated to **X**, and **R** the matrix corresponding to the response function of the setup (in cm^2). In our case, the **X** vector has 56 components which correspond to 56 groups of energies from 10^{-9} MeV to 10^2 MeV. In general, **X** is normalized to 1, image of the total probability, and then multiplied by the calculated surface fluence Φ. If the unfolding perfectly works, **X** has to be equal to the energy spectra of the incident source.

A. General principles

The method is composed of two parts : at first, getting **M**, the measure vector **M** associated to one of the four reference spectra mentioned above, thanks to GEANT4 simulations. Then, getting **X** from **M**, thanks to an iterative unfolding method.

In the first step, the measure vector **M** has n components corresponding to the n detectors or groups of equidistant from the center detectors. $\mathbf{R}=\mathbf{R}_{\mathbf{parallel}}$ is the response matrix obtained thanks to the parallel monoenergetic beams, whose dimensions are n×m, with m=56 (number of groups of energies). **R** is constant, whereas the measure vector **M** depends on the input energy spectrum **X**.

In the second step, the aim is to get **X** from the **M** vector with an iterative deconvolution method. Indeed, as **R** is not square in general, it is not possible to simply invert it to get the unknown spectrum **X**, and even if **R** is invertible, the slightest disturbance in the system implies unstable solutions, even negative ones for **X**.

B. Unfolding methods

Three unfolding methods have been experimented in this project:

- Two methods, M.A.X.E.D. [10] and G.R.A.V.E.L. [11], formalized by the N.E.A. within the U.M.G. package [12]. M.A.X.E.D. is a computer code based on the maximum of entropy. G.R.A.V.E.L. is based on a non-linear least-squares method.
- One "personal method" inspired by [13,14] and based on the maximum of entropy and likelihood.

Concerning the "personal method", the general principles are based on the maximization of the likelihood (L) and of the entropy (H) simultaneously.

The natural logarithm of the likelihood is described in equation (6), considering a Poisson distribution:

$$\ln L = \sum_{i=0}^{n-1} [m_i \cdot (\ln\langle m_i\rangle - \ln(m_i!) - \langle m_i\rangle)] \quad (6)$$

With:

- n the total number of detectors.
- m_i the measure of the detector i.
- $\langle m_i\rangle$ the mathematical expectation of m_i.

Then, according to Shannon, the entropy follows the equation (7):

$$H = -\sum_{j=0}^{m-1} p_j \cdot \ln p_j \quad (7)$$

With:

- m the total number of energy groups.
- p_j the probability for a neutron to belong to the energy group j.

If considering a scalar equation, the differentiation of the function $Q = \ln L + H$ and the use of the Newton's method imply the following iterative equation:

$$x_{k+1} = x_k - \frac{f(x_k)}{f'(x_{k+1})} \quad (8)$$

With k the iteration number. In this case, there are no convergence issues in general, unless the initial value (x_0) is far from the solution.

Though, if considering a matrix equation, the Newton's method leads to the following equation, where x becomes a vector (**X**), f becomes a vector (***f***) and f' becomes a Jacobian matrix (**J**):

$$\boldsymbol{X}_{k+1} = \boldsymbol{X}_k - \mathbf{J}^{-1} \cdot \boldsymbol{f}(\boldsymbol{X}_k) \quad (9)$$

This Jacobian matrix, even invertible, can rapidly become non-invertible with the slightest perturbation of one of its components. This issue can be solved by the Cholesky method [13,15]. Moreover, to help the Newton's method to converge, an adjustment of the size of $\boldsymbol{\delta X_k} = (\boldsymbol{X}_{k+1} - \boldsymbol{X}_k)$ can be necessary, by multiplying $\boldsymbol{\delta X_k}$ by a factor after each iteration, thanks to Armijo's method [13,16]. Despite Cholesky and Armijo's methods, optimizations still remain to be applied to optimize the "personal" method, the results obtained are in good agreement with the source spectra.

The results obtained with these three methods will be discussed in the following part.

IV. Results

The result which have led to the geometry optimization will be presented. Then, the results of unfolding will be exposed, in a first step with high statistics (i.e. a beam of 10^8 neutrons) and then with lower statistics (i.e. a beam of 10^6 neutrons) to validate the performances of the setup.

A. Geometry optimization

The first step of the optimization of the system consists in optimizing both the size of the polyethylene sphere and the number and layout of the detectors. The diameter of the polyethylene sphere has to be a compromise between the capability to moderate the neutrons and the weight of the spectrometer. The mean free path of a 20 MeV neutron in a moderator like polyethylene is about 25 centimeters, by using the Fermi age method [17]. This length should approximately correspond to the most appropriate diameter of the polyethylene sphere. To confirm this rough estimate, GEANT4 simulations were performed, with 56 groups of energies, for 3 different Bonner sphere diameters : 20, 25 and 30 centimeters, with one thermal neutrons detector each centimeter along x, y and z, in accordance with Fig. 1.

Then, the G.R.A.V.E.L. unfolding code was used on the 3 response matrices related to these 3 different diameters. Fig. 3 presents an example of the results obtained, for the three diameters, for a moderated Am/Be source.

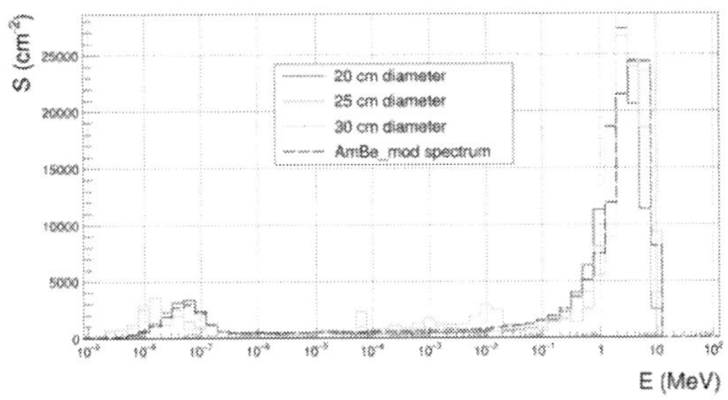

Fig. 3. Unfolding results (cm^{-2}) as function of energy (MeV) obtained with G.R.A.V.E.L., for a moderated Am/Be source. The source spectrum is represented in blue dotted line. The unfolding results obtained with the 3 different Bonner sphere diameters, 20, 25 and 30 cm, are in red, green and yellow respectively. Surface fluence $\Phi = 1.41 \times 10^5$ cm^{-2}.

For a 20 centimeters diameter, the low energies components seem to be correctly returned. But the high energies slightly underestimated, probably due to the small sphere diameter, not sufficient to correctly moderate the highest energy neutrons. For a 25 or a 30 centimeters diameter, high energies are better reconstructed, despite the fact that the energy of the moderated component was underestimated. In any case, for the calculus of the ambient dose equivalent H*(10), to have an accurate estimation of the high energy values is very important because the ambient dose equivalent per unit neutron fluence coefficients h^*_Φ are much higher above 0.1 MeV, according to the I.C.R.U. 66 [18]. Consequently, to minimize the weight of the spectrometer, a 25 centimeters diameter sphere has been retained in the following.

The second step is the optimization of the detectors layout within the sphere. The analysis of the response matrix $\mathbf{R}_{\mathbf{parallel}}$ obtained for the 56 monoenergetic beams with detectors every centimeters is represented in Fig. 4 for a selection of energies.

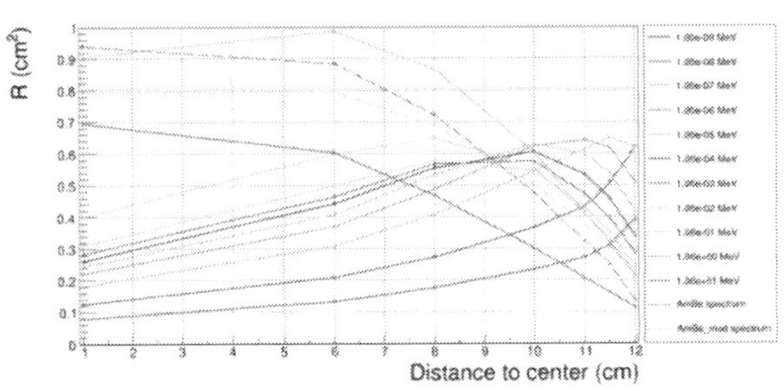

Fig. 4. Response matrix (cm²) of the detectors as function of the distance to the center (cm).

This figure emphasized that close to the surface (~10 to 12 cm to the center), the variation of the curves is more pronounced than next to the center. Consequently, a more important detector granularity is required close to the surface than in the depth of the sphere. An optimization, by removing some detectors, particularly next to the center, leads to an optimized LiF detectors layout based on 6 depths : 1, 6, 8, 10, 11 and 11.5 cm from the center. With this new layout, thermal-neutrons detectors are concentrated next to the surface.

Fig. 5 shows the responses, of the 6 selected groups of detectors, for the 56 groups of energies. These different depths lead to different responses, which contribute to improve the deconvolution.

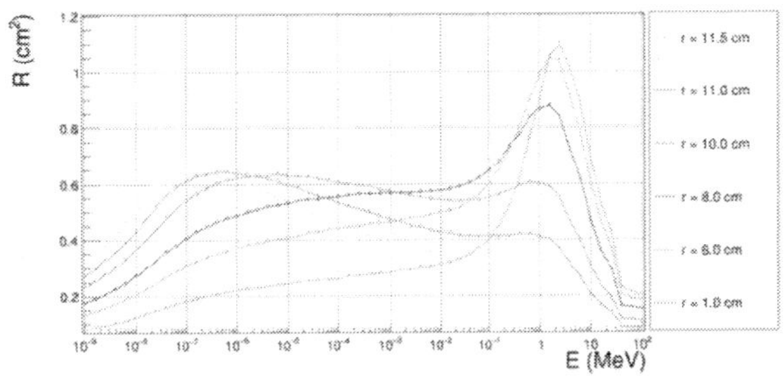

Fig. 5. Response matrix (cm²) as function of energy (MeV) obtained for the 6 groups of detectors.

This layout is the result of a first optimization and will may be subject to more refinements in the future [19].

B. Unfolding results

The aims of these studies are to confirm the optimized geometry and to study the most appropriate unfolding method for different energy spectra (moderated or not) and neutron flux.

1) With high statistics

The first unfolding tests were done with a surface fluence of 1.41×10^5 neutrons per square centimeter, which approximately corresponds to an ambient dose equivalent of about 50 µSv for Am/Be or ^{252}Cf sources.

For each unfolding method, the first iteration has to be initialized with a guess spectrum. Three kinds of guess spectra were tested : a flat one, an Am/Be or ^{252}Cf spectrum and a moderate Am/Be or ^{252}Cf spectrum.

Fig. 6 presents the results obtained with a flat guess spectrum. Fig. 6 (a) shows the unfolding results for an Am/Be source and Fig. 6 (b) for a moderated Am/Be source. The blue dotted line represents the source energy spectrum. An offset in the high energies with the 3 tested methods can be observed for both sources. This offset is much more significant for a moderated source. This offset will lead to an overestimation of the ambient dose equivalent (H*(10)). Thus, a flat guess spectrum does not seem to be adapted to initialize the unfolding.

 https://doi.org/10.1051/epjconf/202125309003

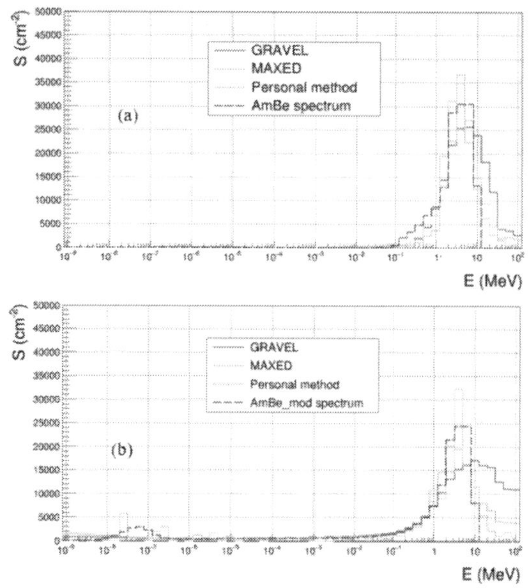

Fig. 6. Unfolding results (cm^{-2}), as function of energy (MeV), with a flat guess spectrum and high statistics (Φ = 1.41×10^{5} cm^{-2}). (a), unmoderated Am/Be source. (b), moderated Am/Be source.

Fig. 7 presents the unfolding results obtained for Am/Be and moderated Am/Be sources with a non-moderated Am/Be guess spectrum. The results are good with M.A.X.E.D. and G.R.A.V.E.L. in Fig. 7 (a), where the unfolded spectra and the source spectrum are quite equivalent. With the personal method, the mean value is correct, but not the shape of the spectrum, showing that further optimizations in this algorithm are to be performed as expected. In Fig. 7 (b), the results totally diverge for M.A.X.E.D. and G.R.A.V.E.L., because the first-iteration-spectrum components are equal to zero for the thermal and epithermal energies. In the personal method, first-iteration components are never equal to zero, they are at least equal to an infinitesimal value, that gives more degrees of freedom to the algorithm and leads to a more appropriate result.

Thus, if a thermal component is present in the source spectrum, which will be usually the case in the industrial environment, a thermal component has probably to be introduced in the guess spectrum to reconstruct thermalized sources.

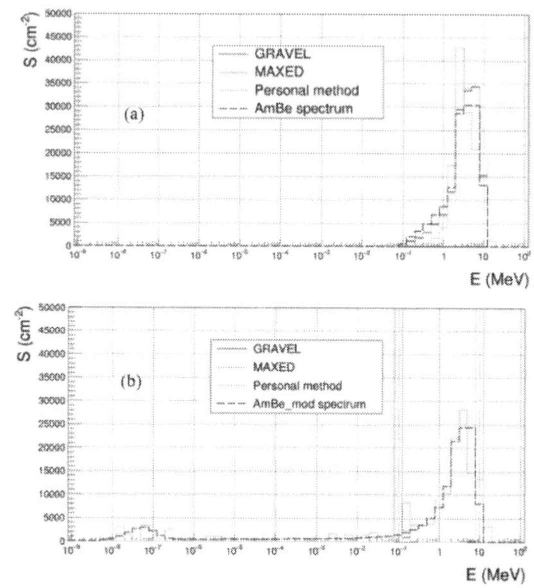

Fig. 7. Unfolding results (cm^{-2}), as function of energy (MeV), with an unmoderated Am/Be guess spectrum and high statistics (Φ = 1.41×10^{5} cm^{-2}). (a), unmoderated Am/Be source. (b), moderated Am/Be source.

Fig. 8 presents the unfolding results obtained for Am/Be sources with a moderated ^{252}Cf guess spectrum. This initialization gives better results for every method than the previous guess spectra. Even if the source is not moderated, the unfolded spectra agreement is good.

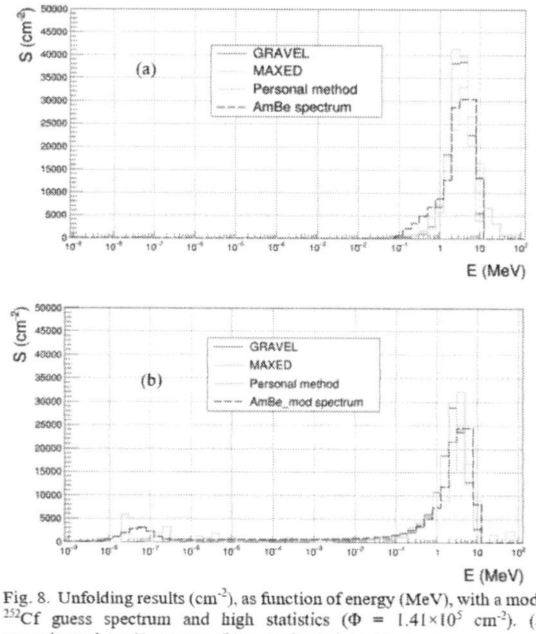

Fig. 8. Unfolding results (cm^{-2}), as function of energy (MeV), with a moderated ^{252}Cf guess spectrum and high statistics (Φ = 1.41×10^{5} cm^{-2}). (a), an unmoderated Am/Be source. (b), a moderated Am/Be source.

As a partial conclusion, when the awaited spectrum is totally unknown, it is better to have a moderated guess spectrum, which covers a wide range of energies and confers to the algorithm more degrees of freedom. Indeed, good results were obtained, regardless of the source spectra. Similar results were obtained for ^{252}Cf and moderated ^{252}Cf sources.

2) With lower statistics

The previous results with high statistics correspond to the following ambient dose equivalent : H*(10) ~ 50 µSv, which can be converted into the following ambient dose equivalent rate : $\dot{H}$*(10)) ~ 300 µSv/h measured in 10 minutes. The aim in this subsection is to test the unfolding methods with one hundred times less statistics (H*(10) ~ 0.5 µSv), in order to validate our setup at lower dose rates ($\dot{H}$*(10) ~ 3 µSv/h in 10 minutes), to approach the specifications discussed in the introduction.

Fig. 9 presents an example of the results obtained for a moderated Am/Be source and an unfolding initialization with a moderated-guess ^{252}Cf spectrum. With a dose divided by one hundred, the results still show a good agreement. Similar results were obtained for moderated or not Am/Be and ^{252}Cf sources.

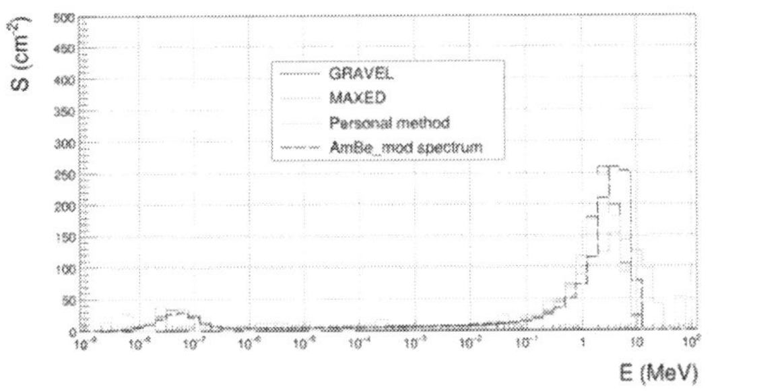

Fig. 9. Unfolding results (cm^{-2}), as function of energy (MeV), with a moderated ^{252}Cf guess spectrum and low statistics ($\Phi = 1.41\times10^3$ cm^{-2}). On this figure, a moderated Am/Be source.

3) Conversion into ambient dose equivalent (H(10))*

One of the main objectives of the ratemeter prototype in development is to be able to measure an ambient dose equivalent rate. To check the accuracy of the results obtained with the different unfolding methods, the neutron energy spectra will now be converted into dose equivalent. This was achieved by applying to the previous spectra the ambient dose equivalent per unit neutron fluence coefficients (h^*_Φ) from the I.C.R.U. 66 [18].

Fig. 10 presents an example of the results obtained when converting the spectra of the Fig. 8 (b) and Fig. 9 into ambient dose equivalent H*(10). The results obtained for a moderated Am/Be source unfolded with a moderated-guess ^{252}Cf spectrum, for 2 different doses are presented : in (a), with high statistics, i.e. 1.41×10^5 cm^{-2} or a dose rate ($\dot{H}$*(10)) of approximately 300 µSv/h measured in 10 minutes, and in (b), with one hundred times lower statistics, i.e. 1.41×10^3 cm^{-2} or a dose rate ($\dot{H}$*(10)) of approximately 3 µSv/h measured in 10 minutes. The doses obtained after deconvolution are in good agreement with the doses calculated for the source spectra. Errors, below 15% from the reference doses are obtained even in the worst case, i.e. for low statistics and an input spectrum which is not the one of the source. This accuracy corresponds to the one obtained usually with Bonner spheres [19], validating the detector setup and the optimized detectors layout.

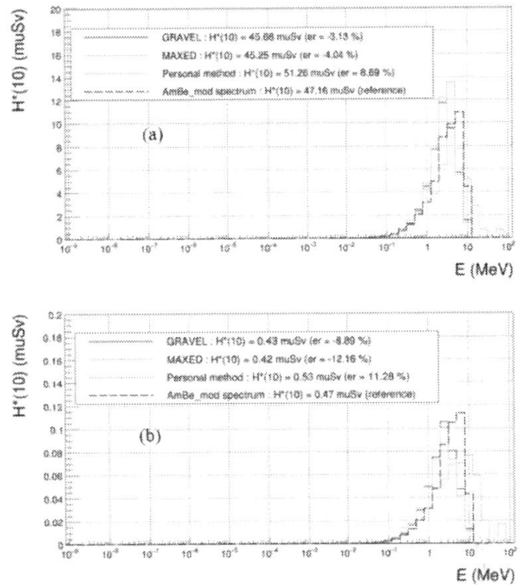

Fig. 10. Unfolding results (µSv), as function of energy (MeV), with a moderated ^{252}Cf guess spectrum. (a), with high statistics ($\Phi = 1.41\times10^5$ cm^{-2}) and a moderated Am/Be source. (b), with low statistics ($\Phi = 1.41\times10^3$ cm^{-2}) and a moderated Am/Be source.

V. Conclusions

Monte-Carlo simulation with GEANT4 associated to unfolding methods of U.M.G. 3.3 and to one "personal" method have been processed, in order to optimize the geometry of a neutron dose ratemeter based on a multi-detectors-Bonner sphere. Results obtained are in good agreement with initial sources energy spectra, thanks to the optimization of the sphere with a diameter of 25 centimeters and the detectors layout composed of 6 groups of detectors.

Concerning the unfolding initialization, guess spectra have to be pre-defined with the following components based on real spectra:

- A rapid component.
- An epithermal component, at least equal to non-zero values.
- A thermal component.

The adequacy of the guess spectra could be determined by calculating the χ^2 per degree of freedom after the first iteration.

In this work, 4 simulated energy spectra of moderated and unmoderated Am/Be and ^{252}Cf sources have been tested, with high and low statistics. Results have shown that even with low statistics, the errors are less than 15% on the dose calculation.

This study was performed with parallel-monoenergetic beams in only one direction (-1 0 0). Thus, two other types of sources will have to be tested:

- Parallel sources with a 45 degrees-shift-angle from the initial beam (-1 -1 0).
- Isotropic sources.

Furthermore, a prototype will be developed and validated with sources in the laboratory and in situ.

REFERENCES

[1] M. Lis, J.M. Gómez-Ros, R. Bedogni and A. Delgado, "Design and feasibility of a multi-detector neutron spectrometer for radiation protection applications based on thermoluminescent 6LiF:Ti,Mg (TLD-600) detectors," in Nuclear Instruments and Methods in Physics Research Section A: Accelerators, Spectrometers, Detectors and Associated Equipment, vol. 584, no. 1, pp. 196–203, Jan. 2008.

[2] M. Lis, J.M. Gómez-Ros, R. Bedogni and A. Delgado, "Preliminary design of a single sphere multi-detector directional neutron spectrometer using TLD-600," in Radiation Measurements, vol. 43, no. 2-6, pp. 1033–1037, Feb. 2008.

[3] J.M. Gómez-Ros, R. Bedogni, A. Delgado, A. Romero and A. Esposito, "A multi-detector neutron spectrometer with nearly isotropic response for environmental and workplace monitoring," in Nuclear Instruments and Methods in Physics Research Section A: Accelerators, Spectrometers, Detectors and Associated Equipment, vol. 613, no. 1, pp. 127–133, Jan. 2010.

[4] S. Agostinelli, J. Allison, K. Amako, J. Apostolakis, H. Araujo, P. Arce, et al., "GEANT4 – a simulation toolkit," in Nuclear Instruments and Methods in Physics Research Section A: Accelerators, Spectrometers, Detectors and Associated Equipment, vol. 506, no. 3, pp. 250-303, Jul. 2003.

[5] J. Allison, K. Amako, J. Apostolakis, H. Araujo, P. Arce Dubois, M. Asai et al., "GEANT4 developments and applications," in IEEE Transactions on Nuclear Science, vol. 53, no. 1, pp. 270-278, Feb. 2006.

[6] J. Allison, K. Amako, J. Apostolakis, P. Arce, M. Asai, T. Aso et al., "Recent developpments in GEANT4," in Nuclear Instruments and Methods in Physics Research Section A: Accelerators, Spectrometers, Detectors and Associated Equipment, vol. 835, pp. 186-225, Nov. 2016.

[7] *GEANT4 – Guide for phyiscs lists – Release 10.6*, rev. 4, CERN, Switzerland, dec. 2019. [Online]. Available: https://geant4.web.cern.ch/support/user_documentation.

[8] International Standard Organization, Reference Neutron Radiations Part I : Characteristics and Methods of Production, ISO 8529-1, ed.1, Feb. 2001.

[9] M. Koçak, H. Ahmadov and G. Dere, "On prompt fission spectrum in spontaneous fission of ^{252}Cf," Annals of Nuclear Energy, vol. 70, pp. 82-86, Aug. 2014.

[10] M. Reginato and P. Goldhagen, "MAXED, a computer code for the deconvolution of multisphere neutron spectrometer data using the maximum entropy method," Environmental Measurements Laboratory, USDOE report EML595, 1998.

[11] W.N. Mc Elroy, S. Berg, T. Corckett and R.G. Hawkings, "A computer-automated iterative method for neutron flux spectra determination by foil activation," U.S. Air Force Weapons Laboratory Report AFWL-TR-67-41, Kirkland AFB, NM, 1967.

[12] M. Reginato, B. Wiegel, A. Zimbal and F. Langner, "UMG 3.3, Analysis of data measured with spectrometers using unfolding techniques," Energy Agency of the OECD (NEA), 2004.

[13] S. Itoh and T. Tsunoda, "Neutron Spectra Unfolding with Maximum Entropy and Maximum Likelihood," in Journal of Nuclear Science and Technology, vol. 26, no. 9, pp. 833-843, Nov. 1989.

[14] S. Maeda and T. Iguchi, "A new unfolding code combining maximum entropy and maximum likelihood for neutron spectrum measurement," in Journal of Nuclear Science and Technology, vol. 50, no. 4, pp. 381-386, Apr. 2013.

[15] P.E. Gill, G.H. Golub, W. Murray and M.A. Saunders, "Methods for modifying matrix factorizations," in Mathematics of Computation, vol. 28, no. 126, pp. 505-535, May. 1974.

[16] L. Armijo, "Minimization of functions having Lipschitz continuous first partial derivatives," in Pacific Journal of Mathematics, vol. 16, no. 1, pp. 1-3, Nov. 1966.

[17] P. Corompt, J. Molinari and P. Calmels, "Theoretical and Experimental Study of Flows of Thermal Neutrons in Hydrocarbons and Application of Results to the Measurement of Their Density and the Search for Their Composition," in EUR Series on Nuclear Energy Resarch, vol. 3624, 1968. [Online]. Available : http://aei.pitt.edu/91917/.

[18] J. Deng, "ICRU Report 66: Determination of Operational Dose Equivalent Quantities for Neutrons," in Physics in Medicine and Biology, vol. 47, no. 14, Jul. 2002.

[19] D.J. Thomas and A.V. Alevra, "Bonner sphere spectrometers, a critical review," in Nuclear Instruments and Methods in Physics Research Section A: Accelerators, Spectrometers, Detectors and Associated Equipment, vol. 476, no. 1-2, pp. 12–20, Jan. 2002.

New probe for the improvement of the Spatial Resolution in total-body PET (PROScRiPT)

A. Ros[1], L. Barrientos[1], M. Borja-Lloret[1], J.V. Casaña[1], E. Muñoz[1], J. Roser[1], J.M. Udías[2], R. Viegas[1] and G. Llosá[1]

[1] Instituto de Física Corpuscular (CSIC / UVEG), Paterna, Spain
[2] Grupo de Física Nuclear, EMFTEL, IPARCOS, Universidad Complutense de Madrid, CEI Moncloa, Madrid, Spain

***Abstract*—In recent decades, PET scanners have been widely used for diagnosis and treatment monitoring in nuclear medicine. The continuous effort of the scientific community has led to improvements in scanner performance. Total-body PET is one of the latest upgrades in PET scanners. These kinds of scanners are able to scan the whole body of the patient with a single bed position, since the scanner tube is long enough for the patient to fit inside. While these scanners show unprecedented efficiency and extended field-of-view, a drawback is their low spatial resolution compared to dedicated scanners. In order to improve the spatial resolution of specific areas when measuring with a total-body PET scanner, the IRIS group at IFIC-Valencia is developing a probe. The proposed setup of the probe contains a monolithic scintillation crystal and a SiPM. The signal of the probe is read out by a TOFPET2 ASIC from PETsys, which has shown good performance for PET in terms of spatial and time resolutions. Furthermore, the PETsys technology generates a trigger signal that will be used to time synchronise the probe and the scanner. The proof-of-concept of the probe will be tested in a Preclinical Super Argus PET/CT scanner for small animals located at IFIC. Preliminary simulations of the scanner and the probe under ideal conditions show a slight improvement in the position reconstruction compared to the data obtained with the scanner, therefore we expect a considerable improvement when using the probe in a total-body PET scanner. Characterisation tests of the probe have been performed with a ^{22}Na point-like source, obtaining an energy resolution of 9.09% for the 511 keV energy peak and a temporal resolution of 619 ps after time walk correction. The next step of the project is to test the probe measuring in temporal coincidence with the scanner.**

***Index Terms*—SiPMs, scintillation crystals, molecular imaging, PET/CT, total-body imaging**

I. Introduction

THE development of new PET (Positron Emission Tomography) scanners with improved performance is still an important line of research in nuclear medicine. One current line of research is the study of a total-body scanner, the development of which would represent an improvement on the current features of PET scanners [1]. Currently, the typical spatial resolution of reconstructed PET images is between 4 and 6 mm [2]. The limitation in spatial resolution is a result of the detector module design, the positron range and the non-collinearity of the detected photons [3]. Using smaller detector modules in PET scanners would improve the spatial resolution but would worsen the sensitivity [4]. For this reason there have been studies suggesting the use of PET probes. Such probes would improve spatial resolution without compromising the scanner sensitivity. The use of a PET probe has been studied in several research groups with a view to improving the spatial resolution: theoretical studies based on Monte Carlo simulations [5], hardware studies using silicon detectors for the PET probe [6] and hardware studies using PsPMT for the PET probe [7]. Furthermore, there are studies of small PET modules with scintillation crystals and SiPM that could be used in PET probes [8]. A PET probe consists of a small detector that operates in temporal correlation with a PET scanner (by means of the scanner electronics or with signal post-processing). In this way, one of the annihilation photons interacts with the PET scanner and the other one with the probe, which has much better spatial resolution than the scanner. The fact that the PET probe can be positioned in the whole field-of-view (FOV) of the PET scanner is of particular interest since the probe increases the spatial resolution of a specific region/lesion of the patient's body [9,10,11]. Advances in recent years in commercial data acquisition electronics for PET with high energetic and temporal resolution capabilities has led to the possibility of a new design and implementation for PET probes. The suitability of a PET probe based on scintillation crystals and silicon photomultipliers (SiPM) is the focus of this study. The functioning of the probe was tested in a Preclinical Super Argus PET/CT scanner for small animals.

This work was supported by grant number SEJIGENT/2019/14. Corresponding author: arosgar@ific.uv.es.

II. Materials and Methods

A. Small Animal PET/CT

The Sedecal Preclinical Super Argus PET/CT is a 4 ring scanner with 24 modules per ring (see Fig. 1a). Each module is composed of a 13×13 pixel array crystal in a phoswich fashion (LYSO and GSO). The scanner has 1.25 ns nominal time resolution, 25 % nominal energy resolution and 1 mm nominal spatial resolution in the centre of the FOV. TTL signals of up to 3 V can be injected via a gating input socket and the data of the injected signal are stored and correlated with the timing of the stored data of the PET scanner. The measured data are singles (raw data) with 40 ns precision. Calibration measurements with a ^{22}Na point-like source were carried out in order to test the reconstruction code and to determine the

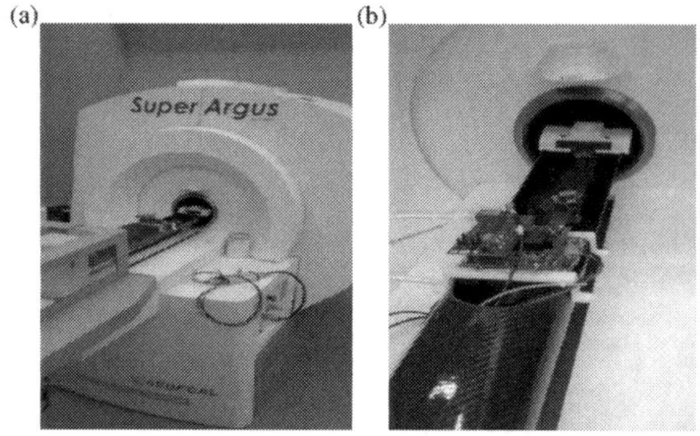

Fig. 1: (a) Super Argus 4R and probe. (b) Detail of the probe.

spatial resolution available with the system. A test with the point-like source in 4 different positions 20 mm apart along the central axis of the tube (z-axis, coronal plane) was carried out. Additionally, measurements with the point-like source in two positions 10 mm apart was made, the first with the source at the centre of the FOV and the second with the source 10 mm above the first (y-axis, transverse plane).

B. Simulations

GATEv8.2 simulations of the probe and the Preclinical Super Argus PET/CT were performed in order to study the feasibility of the PROScRiPT project. The chosen physics list for the simulation was QGSP_BERT_EMV. The simulation was an ideal simulation; optical photons, reflective material between pixelated crystals, photodetectors and electronics were not simulated The simulated setup consisted of a PET with 4 rings and 24 detector modules per ring, with each module containing an array of 13x13 scintillation crystals (GSO and LYSO phoswich). The probe, consisting of a continuous scintillation crystal (LYSO) of 25.8x25.8x5 mm^3, was positioned at 4 mm from a small Derenzo-like phantom (a water cylinder with hot-spots) which was in turn positioned in the centre of the FOV of the scanner (see Fig. 2a). The small Derenzo-like hot-spot phantom, contains an activity concentration of 0.108 mCi/cc and has several inserts measuring 4.8, 4.0, 3.2, 2.4, 1.6 and 1.2 mm in diameter. The coincidence time window for the simulation was set to 2 ns. A reconstructed image of the Derenzo-like phantom was obtained for three cases: events in which there were temporal coincidences between any detector in the setup, events in which there were temporal coincidences between the scanner detectors and events in which at least one of the detectors involved in the temporal coincidence was the probe.

C. The probe

Regarding the hardware, the probe is being read out by the TOFPET2 ASIC kit from PETsys, a commercial time-of-flight PET system. The system provides the digitalised signal of 64 channels per detector and has a high time resolution (time binning 20 ps) [11]. The probe is made of a SiPM and a scintillation crystal. The size of the crystal ($CeBr_3$) is $25.8\times25.8\times5$ mm^3. The SiPMs used for the probe are model S14161-3050HS-08 of Hamamatsu with 64 channels (3×3 mm^2 and pixel pitch 50 μm). All the tests were carried out at a controlled temperature with a maximum variation of three degrees celsius. A set of measurements with two detector modules (DM1 and DM2) were done in order to calibrate the chosen system (see Fig. 3). The measurements were carried out for a distance of 90 mm between modules and a ^{22}Na point-like source positioned equidistantly between them. The TOFPET2 ASIC has three thresholds [12] (time thresholds (vth_t1 and vth_t2) and energy threshold vth_e) that have been characterised for our setup. Energy and time resolution were measured for different combinations of the threshold values. In order to estimate the best temporal resolution achievable with the system, time-walk corrections (TWC) were applied to the data [13] [14]. Furthermore, to estimate the spatial resolution and test the reconstruction code, measurements with the point-like source in nine different positions separated by 6.45 mm were performed. The TOFPET2 ASIC kit has the ability to generate square pulses (with different periods, lengths and total times) that can be configured by the user. The rising times of the generated pulses are stored together with the hits detected by the SiPMs. This signal has been used in order to synchronise the data measured with PETsys and the scanner.

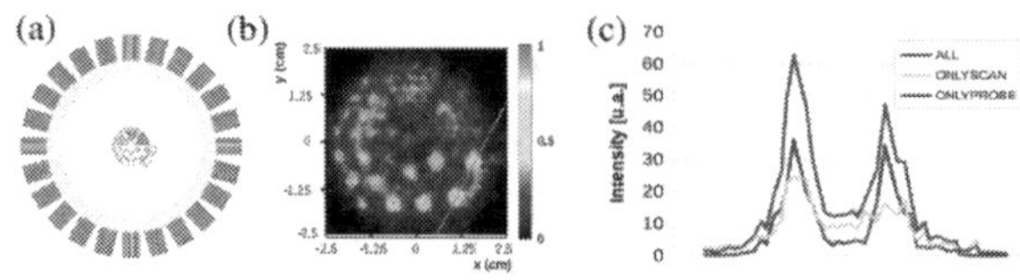

Fig. 2: GATEv8.2 simulations of the probe and the Preclinical Super Argus PET/CT. (a) Gate simulation setup. (b) Source reconstruction. (c) Source reconstruction profiles.

D. Probe and the scanner measurements

In order to be able to measure with the probe and the scanner it is necessary to synchronise them and to determine

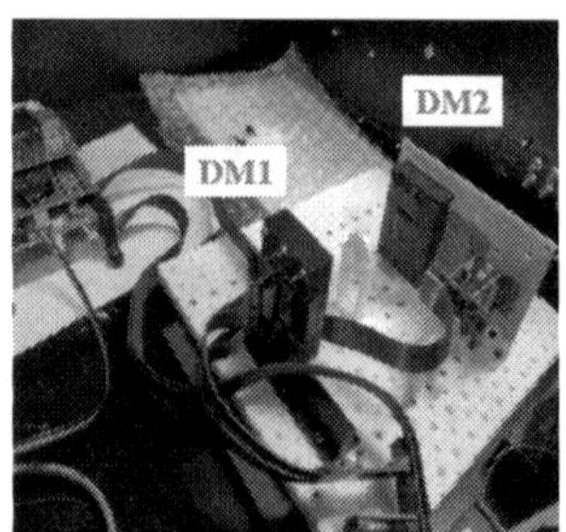

Fig. 3: Probe calibration setup. Detector modules 1 (DM1) and 2 (DM2) are composed of a SiPM (S14161-3050HS-08, Hamamatsu) and a monolithic $CeBr_3$ crystals.

https://doi.org/10.1051/epjconf/202125309004

the position of the probe within the scanner. For the synchronisation the output trigger signal generated by PETsys has been used. A series of four trigger signals were generated every time a measurement was performed. The first, second and forth trigger signals lasted 30 s. The data measured during these triggers were not used for the reconstruction since these triggers were only used to check the integrity of the generated signal. The third trigger, 1800 s long, and the data measured during this time with the probe and the scanner, are used for the reconstruction. A dead-time of around 80 ns has been measured between the generation of one trigger signal and the next. The trigger signal generated by PETsys was injected into a TTL input port of the Super Argus 4R scanner. The scanner reads out the input trigger signal and stores the rising and falling time of the pulses, relating their timing with the internal time of the scanner. The position of the probe in the scanner was determined based on the position of the reconstructed ^{22}Na point-like source and the fact that the probe and the source were mounted in a custom holder of known dimensions.

III. Results

All the images were reconstructed using an ideal MLEM code (no sensitivy matrix or corrections) and all the reconstructions were stopped at the 5th iteration.

A. *Simulation results*

Fig. 2b shows a detail of the superimposed reconstructed Derenzo-like phantom for the three cases. The profiles shown in Fig. 2c correspond to the data of the slice represented by the orange line in Fig. 2b. It can be appreciated that the spatial resolution has slightly improved for the case in which the coincidences occurred between the probe and the scanner with respect to the other two cases.

B. *Probe characterisation*

Table I shows the tested values for the characteristic TOFPET2 ASIC thresholds, the energy resolution (E_R) and the coincidence time resolution (CTR) obtained for each case. The timing was calculated using two different methods: (i) selecting the time of the most energetic hit (CTR_E) and (ii) calculating the average time for the first five hits (CTR_{av}). A clear improvement of the time resolution is obtained when using the averaged time for the five first hits.

The parameters that led to the best results were: threshold vth_t1, vth_t2 and vth_e set to 20, 5, and 5 respectively. The best value for the energy resolution was 9.09% for the 511 keV energy peak. In order to determine the best CTR, the average of the five first hits was calculated, following which the energy was filtered between 450 and 600 keV and, after applying time-walk corrections, a CTR of 619 ps was found. Fig. 4 shows the time difference (ps) versus energy (keV) for the studied detector modules (DM1 and DM2) before and after time-walk correction (Fig. 4a and 4b show the data with no time-walk correction and Fig. 4c and 4d show the data with time-walk correction). This shows good agreement with the value of 660 ps obtained by Lamprou *et al.*, using two monolithic scintillation crystals (50×50×15 mm^3) optically coupled to 8×8 SiPMs arrays (ON-Semi, J-series mode) [15].

TABLE I:
Characterisation of the TOFPET2 ASIC thresholds. Values of energy resolution at 511 keV peak and CTR (calculated by two different ways) are shown.

Vth_t$_1$	20	15	10	10	10	10	10	10	10	10	10	10
Vth_t$_2$	5	5	5	20	18	15	10	10	10	10	10	10
Vth_e	5	5	5	5	5	5	10	9	8	7	6	5
E_R 511keV	9.09	9.43	9.72	9.55	9.79	10.25	10.83	10.64	10.56	10.84	10.45	10.61
CTR_E (ns)	2.61	2.69	2.61	2.40	2.44	2.66	2.37	2.51	2.66	2.40	2.77	3.03
CTR_{av} (ns)	0.88	0.82	0.80	1.07	0.93	0.83	0.80	0.78	0.80	0.81	0.82	0.85

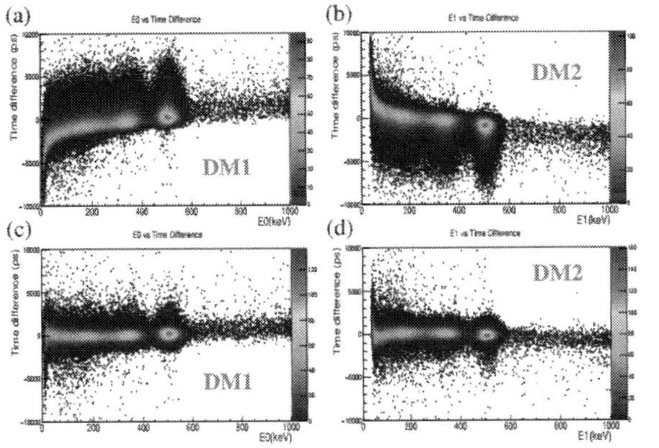

Fig. 4: Time difference (ps) versus energy (keV) for the studied detector modules.

Fig. 5a shows a schematic of the nine different positions in which the ^{22}Na point-like source was positioned between DM1 and DM2. The reconstructed positions can be seen in Fig. 5b. Furthermore, the horizontal profiles for positions [1,2,3], [4,5,6] and [7,8,9] are shown in Fig. 5c, 5d and 5e respectively. A value of 1.29 mm was obtained for the FWHM for the central position (position 5).

C. *Scanner characterisation*

A ^{22}Na point-like source was used to perform measurements with the Sedecal Preclinical Super Argus PET in order to estimate the spatial resolution achievable with the scanner and the imaging reconstruction code used in the present research. Fig. 6 shows the reconstructed position of the point-like source in four different positions (20 mm apart) along the centre of the scanner tube (z-axis). Fig. 6a shows the reconstructed positions in the transverse plane while Fig. 6b shows the same data for the sagittal plane. Fig. 6c corresponds to the profile through the x-axis for all the z-voxels in the FOV and Fig. 6d corresponds to the profile through the y-axis, summing five neighbour voxels in the x and z-axis. The difference in the

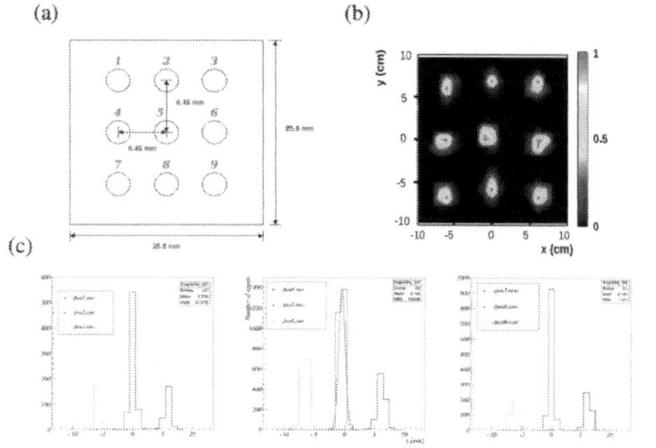

Fig. 5: ^{22}Na point-like source position reconstruction for nine positions. (a) Positioning of the ^{22}Na point-like source. (b) Reconstructed data. (c) Profiles of the reconstructed data.

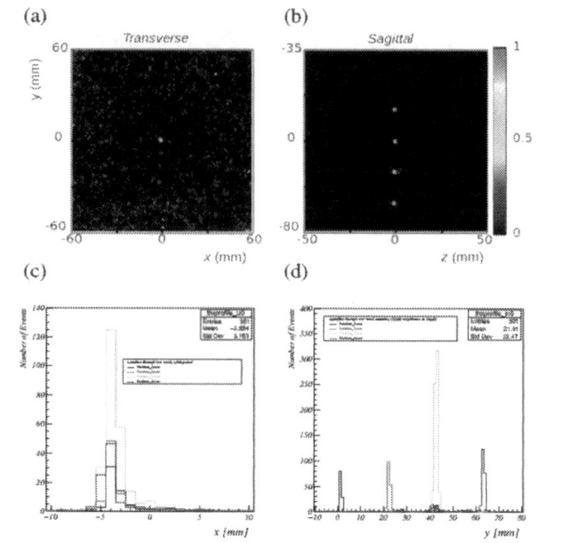

Fig. 6: Reconstructed position (transverse (a) and sagittal (b) planes) of the point-like source in four different positions (20 mm apart) along the centre of the scanner tube (z-axis). Profiles of the reconstructed data through the x-axis (c) and y-axis (d).

numbers of events that is visible in Fig. 6d is due to the lack of a sensitivy matrix in the reconstruction code. A FWHM of 1.94 mm was obtained for the reconstructed distribution for the position at around 40 mm, in the centre of the FOV, as can be seen from the number of events measured in that position.

Fig. 7 shows the reconstructed ^{22}Na point-like source position in the centre of the scanner and 10 mm above. The images for the transverse and sagittal planes are shown in Fig. 7a and 7b, respectively. Fig. 7c shows the profile of the data through the x-axis summing five neighbour voxels in the y and z-axis. Fig. 7d shows the profile of the data through the y-axis summing five neighbour voxels in the x and z-axis. In this case, Fig. 7d also clearly shows the difference in the number of events measured outside the central area of the FOV, but it is still possible to correctly determine the position of the source.

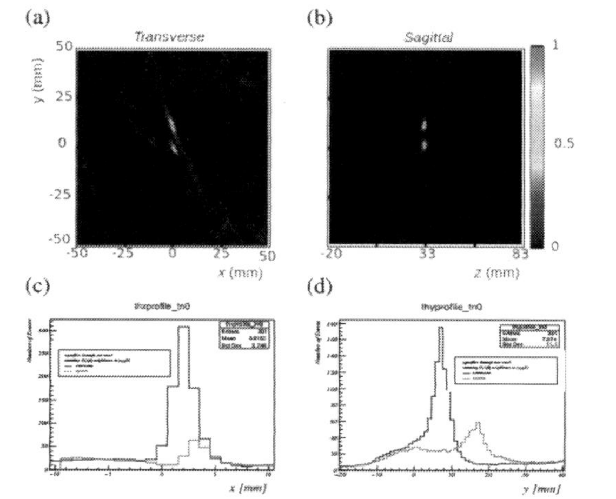

Fig. 7: Reconstructed ^{22}Na point-like source position (transverse (a) and sagittal (b) planes) in the centre of the scanner and 10 mm above. Profiles of the reconstructed data through the x-axis (c) and y-axis (d).

D. Probe and scanner measurement

Fig. 8 shows two reconstructed images for a ^{22}Na point-like source. The first one corresponds to the case in which data was stored when a temporal coincidence occurs between the detectors of the scanner (*scanner-scanner*) and between the scanner detectors and the probe (*scanner-probe*). In the second case, only the temporal coincidences of the type *scanner-probe* for the case in which the coincidences occured between the probe and the detector of the scanner facing the probe were considered. In both cases the coincidence time window was set to 100 ns and an energy filter from 400 keV to 660 keV was applied to the data. In Fig. 8a and 8b the upper point corresponds to the first case while the lower point corresponds to the second case. A mispositioning of the reconstructed point-like source can be appreciated between the two cases. The normalised profiles for the reconstructed source are shown in Fig. 8c and 8d. The black distribution corresponds to the data with coincidences of the type *scanner-scanner* and *scanner-probe* and the red distribution to the data with coincidences of the type *scanner-probe*. Fig. 8 shows that it is possible to reconstruct the ^{22}Na point-like source distribution for both cases.

IV. Discussion and conclusion

The simulation results show a slight improvement in the spatial resolution while measuring with the probe and the scanner. Small animal PETs have high spatial resolution, and

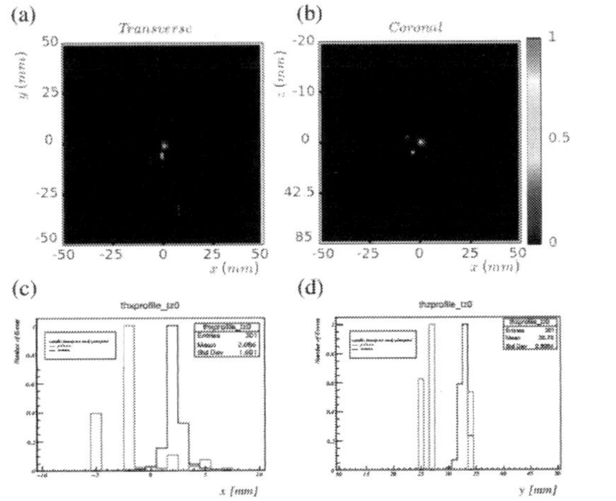

Fig. 8: Two reconstructed images for the ^{22}Na point-like source. The upper point corresponds to the cases *scanner-scanner* and *scanner-probe* and the lower point to the case *scanner-probe*. Transverse and sagittal planes are shown in (a) and (b), respectively. Profiles of the reconstructed data through the x-axis and y-axis are shown in (c) and (d), respectively.

therefore it is expected that using the probe in a total-body PET scanner, in which the intrinsic spatial resolution is worse, would improve the resolution notably. The values obtained in the characterisation of the probe suggest that it is a good candidate for the proof-of-concept.

The measured value of the FWHM measured with the probe setup for the central position is promising although a resolution of approximately 1 mm would be desirable. Tests with different scintillation crystals, both monolithic and pixelated, and with different types of SiPM should be perfomed. Furthermore, the obtained value for the CTR, of the order of ps, suggests that the TOFPET ASIC kit is a good candidate to read out the probe signal. Regarding the trigger signal generated by the TOFPET ASIC kit, the signals generated have been shown to be sufficiently accurate for the probe synchronisation. The reconstructed data using the Super Argus 4R scanner show that the reconstruction code is acceptable for the initial stages of this project, although it should be improved for the final probe prototype.

The results obtained for the measurement of the point-like source with the scanner and the probe show that the synchronisation of the probe and the scanner was succesful using the signal trigger of the TOFPET ASIC kit and the input gating signal of the scanner. The mispositioning of the reconstructed point-like source for the *scanner-probe* case suggests that the methodology used to determine the positioning of the probe within the scanner must be improved. The lack of a sensitivity matrix in the reconstruction code would also have influenced the mispositioning of the probe. Overall, the results obtained so far in the PROScRiPT project are sufficiently promising to suggest that the positioning and synchronisation of a probe within a PET scanner could realistically be used to improve the spatial resolution in TBP scanners in the future.

References

[1] Cherry, Simon R et al. *et al.*, *Journal of Nucl. Med.*, Society of Nuclear Medicine vol. 59, 13-12, 2018.
[2] Zanzonico P.B. *et al.*, *Essentials of in vivo biomedical imaging*, CRC Press, Boca Raton Florida U.S.A., 2015.
[3] Zhou, J., *et al.*, *Phys. Med. Biol.*, vol. 54, no. 17, 5193-5208, 2009.
[4] Park, S-J. *et al.*, *Phys. Med. Biol.*, vol. 52, no. 15, 4653-4677, 2007.
[5] Clinthorne N. *et al.*, *Physics Procedia*, vol. 37, 1488-1496, 2012.
[6] Wu, H. *et al.*, *Journal of Nucl. Med.*, Society of Nuclear Medicine, vol. 49, no. 10, 1668-1676, 2008.
[7] Jiang J. *et al.*, *IEEE transactions on medical imaging*, vol. 39, no. 11, 3268-3277, 2019.
[8] Song, T. Y. *et al.*, *Phys. Med. Biol.*, vol. 55, no. 9, 2573- 2587, 2010.
[9] Grkovski, Milan. *et al.*, *Nucl. Inst. and Meth. A*, vol. 788, 86-94, 2015.
[10] Brzeziński, K. *et al.*, *JINST*, vol. 11, P09016, 2016.
[11] Bugalho R. *et al.*, *JINST*, vol. 14, no. 03, 2019.
[12] Schug D. *et al.*, *IEEE Transactions on Radiation and Plasma Medical Sciences*, vol. 3, no. 04, 444-453, 2018.
[13] Poulson D. *et al.*, *Space Telescopes and Instrumentation 2020: Ultraviolet to Gamma Ray*, vol. 11444, 114446D, 2020.
[14] Hueso-Gonzalez F. *et al.*, *JINST*, vol. 9, no. 05, P05002, 2014.
[15] Lamprou E. *et al.*, *Phys. Med.*, vol. 70, no. 70, 10-18, 2020.

Plutonium and Americium Inventories in Soil Cores from the English Lake District, Cumbria (UK)

A. Madina[1], C. Tighe[1] and M. J. Joyce[1]
[1]Lancaster University, United Kingdom
m.joyce@lancaster.ac.uk

***Abstract*—Environmental radioactivity has been reported in the stationary wetlands closer to the natural water basins in Great Britain, precisely in the English Lake District (Cumbria, UK) often emphasise on the investigation carried out through either high-resolution γ-ray spectroscopy or α-particle spectroscopy. The objective of the study is the evaluation of trace actinide (^{241}Am) through environments radioactivity measurements of two Lakes in the Lake District using a broad-energy, high-resolution γ-ray spectroscopy germanium (BEGe) detector. These appeal to the need to determine the radioactive concentration (Bq/g) in the soil samples with the intent being to discern natural and anthropogenic contributions, and trends in abundances associated with influences of the landscape over time, at trace levels in the environment. These measurements highlight potential benefits for the assessment of ^{241}Am and ^{241}Pu, especially comparisons that might be made with accelerator mass spectrometry assessments (AMS). This study intents to offer an inclusive investigation practise for determining the levels of radioactivity in soil cores, as well as the method for statistical analysis approach to calculate the activity of a soil sample, with a special emphasis on categorising the natural and anthropogenic contributors to trace amounts of plutonium in the environment, that is, less than500 fg/g.**

***Keywords* —Americium; plutonium; soil samples, radioactive concentration**

I. Introduction

The Lake District is situated in the north of England in the United Kingdom; ~ 4 hours from London (see Fig. 1). It comprises an area of approximately 2,292 sq. km, and is a UNESCO world heritage site [1] receiving around 18 million tourists every year under normal circumstances. As its name suggests, the Lake District comprises a significant number of land-locked lakes and water bodies. Many of these are relatively undisturbed and some of them have been subject to prior trace actinide analysis via high-resolution γ-spectroscopy and α-particle assay in the past. The English Lake District offers a unique prospect to investigate the long-term dynamics of the long-term trace radioactivity in a natural environment that is relatively undisturbed, to inform studies of whether the majority of radioactivity arises naturally or, anthropogenically, from fallout from weapons tests and reactor accidents.

Fig. 1. The location of the Lake District in the United Kingdom

II. Description

Environmental radioactivity mensuration are essential to find the level of background radiation arising from natural radioactive sources (that include cosmic radiation originating in outer space and release through the atmosphere), and the terrestrial components arising from the radionuclides decays existing in rocks, water and sediments [2]. Furthermore to natural sources, the background radiation level encompasses man-made contributions such as those released from nuclear facilities such as Sellafield and other industrial sources [3].

The Lake District comprises uplands at elevations significant (maximum 978 m in height) relative to the surroundings and which are likely to have accumulated fallout from the Chernobyl accident in 1986 and atmospheric weapons tests. Sellafield is geographically close (~15 km) and is the source of past radioactive effluent discharges to the environment (air, terrestrial and marine). It was also the site of the Windscale accident in 1957, which resulted in uncontrolled dispersion of radioactivity to the environment, mainly via airborne deposition.

The objective of the research described in this summary was to assess and compare trace quantities of plutonium and americium in soil cores collected from the shores of two specific sites in the Lake District (see: Fig. 2 and 3, Cumbria,

UK): Blelham Tarn and Lake Windermere (at Fell Foot). For instance, this has been attempted by broad-energy germanium (BEGe), high-resolution γ-ray spectroscopy to determine the radioactivity concentration in the soil samples.

However, the established route to americium-241 assay, that is, via γ-ray spectroscopy of the ^{241}Am line at 59.5 keV, requires detailed analysis in a critical part of the γ-ray spectrum in terms of i) detector efficiency and ii) Compton background. This paper reports on the use of a BEGe detector to determine americium activity in soil samples from both sides Blelham Tarn and Fell Foot, Cumbria (UK).

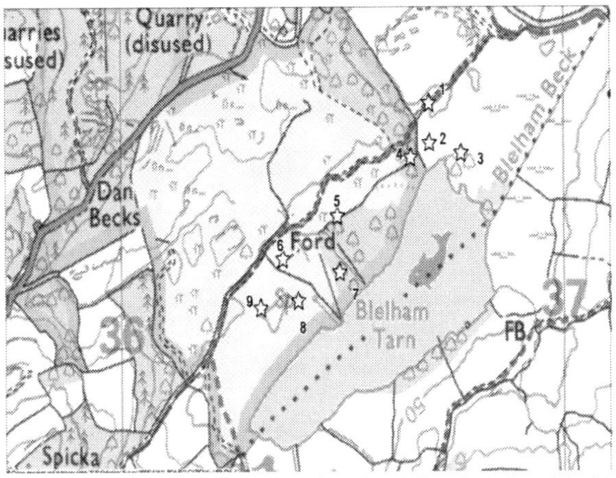

Fig. 2. The location (stars) of samples from the Blelham Tarn, Lake District (Cumbria, UK).

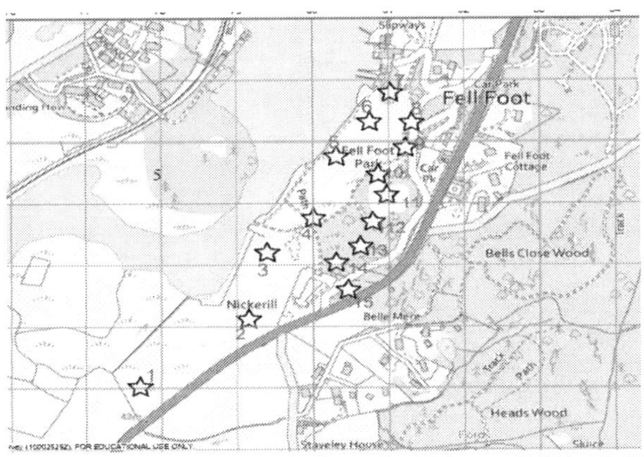

Fig. 3. The location (stars) of samples from Lake Windermere (Fell Foot Park) in the Lake District, (Cumbria UK).

III. Experimental details

The experimental methods used to detect natural and fallout radioactivities from Blelham Tarn and Lake Windermere soil samples were used to estimate radioactivity levels in other environmental samples using γ-ray spectrometers. This research was carried out at Central Teaching Laboratories (CTL Redaction Laboratory) at Liverpool University (UK), using a broad energy germanium (BEGe) counter (MIRON Technologies).

In this research program, a number of samples have been sourced from the environs of two sites in the Lake District to compare hydrogeological migration of trace actinide material: Blelham Tarn (54.3966° N, 2.9766° S) which was studied 20 years ago [3], and constitutes a relatively undisturbed environment (that is, not used by boats or machines) and is also a site of special scientific interest (SSSI) which provides some guarantee to our assumed lack of disturbance [4].

Nine soil samples from Blelham Tarn and fifteen soil samples from Lake Windermere have been sourced, the latter from the northern shore, in this research (see: Figs. 2 and 3) with an aluminum, 50-mm diameter soil core sampler (see: Fig. 4). These samples were dried at 100°C for 24 hrs to remove all impurities (that is, roots and stones) from samples and milled to grind soil in powder form. In addition, their locations were recorded with GPS, selected to correlate with the prior art [5] as closely as practicable and assessed in a BEGe γ-ray spectrometer to estimate the activity of ^{241}Am in each sample; the acquisition live time for each sample were 24 hours.

These data can be used to infer the ^{241}Pu activity based on a number of lifetime scenarios, ahead of accelerator mass spectrometry assessment (AMS) [6, 7] of the neighboring Pu activities and isotope ratios, scheduled later in this research program.

IV. Results

In Fig.5, an example BEGe spectrum for a sample studied in this research with the ^{241}Am peak highlighted and seen adjacent to the more intense lines derived from, for examples, the uranium and thorium decay chains that occur naturally.

In Fig 6, a graph is provided for a γ-ray spectrometry measurements of ^{241}Am from Blelham Tarn samples. Combining results from the soil core with data from the literature, the best estimate of ^{241}Am in-growth from ^{241}Pu decay is 41± 5 (Bq/m^2) for sample 6 in Fig. 4, if we compare that with Michel *et al.* [3], they find just 18±2 (Bq/m^2) at Blelham Tarn. For data for Blelham Tarn, measurements are presented in Table 1.

Fig. 4. The sample preparation strategy is illustrated in pictures adopted in this research, from left to right.

In Fig. 7, a similar graph is derived for a γ-ray spectrometry measurements of ^{241}Am for Lake Windermere, the mean of activity concentration of fifteen soil cores (Bq/g) of ^{241}Am was (1.57±0.03) ×10^{-4}, where the life time was 24 hours and weight of each sample was 250 g for the measurement of mass concentration /g in this lake. For data for Lake Windermere, measurements are present in table 2.

From these results, the activity of ^{241}Pu can be estimated from the abundance of ^{241}Am based on an estimate of 30 years where the global deposition has occurred from, for example, weapons fallout and Chernobyl [8], using the Bateman equation [9]:

$$A_{Pu}(0) = \frac{A_{Am}(t)(\lambda_{Am} - \lambda_{Pu})}{(e^{-\lambda_{Pu}t} - e^{-\lambda_{Am}t})\lambda_{Am}} , \tag{1}$$

where A_{Pu} and A_{Am} are the activities of plutonium-241 and americium-241, respectively, and λ_{Pu} and λ_{Am} are the corresponding decay constants.

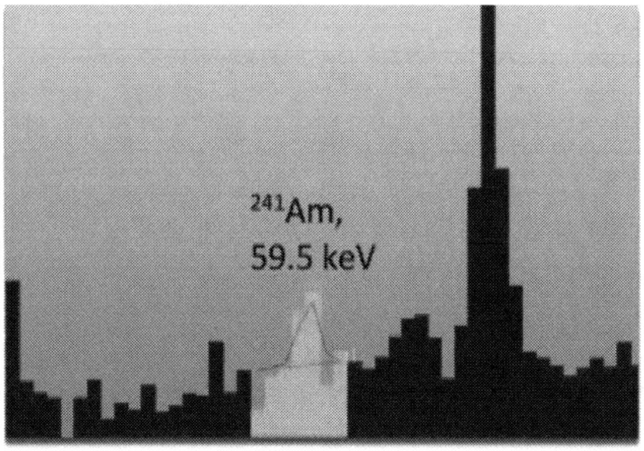

Fig. 5. A section of a BEGe γ-ray spectrum of soil sample #2 in Blelham Tarn site.

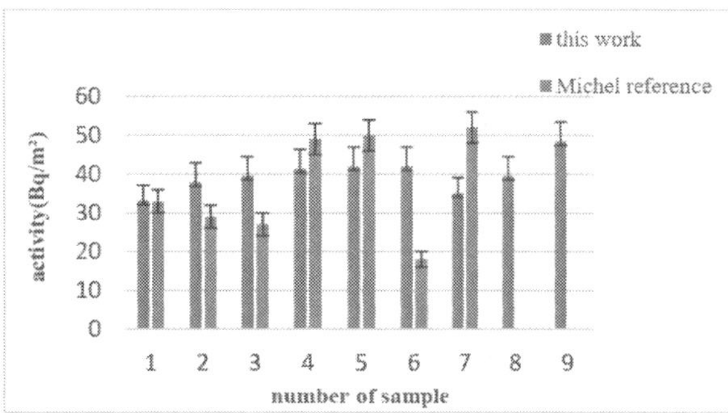

Fig. 6. Comparison of ^{241}Am activity (Bq/m²) ^{241}Am with that from Michel *et al.* [3] from Blelham Tarn, Like District (Cumbria, UK).

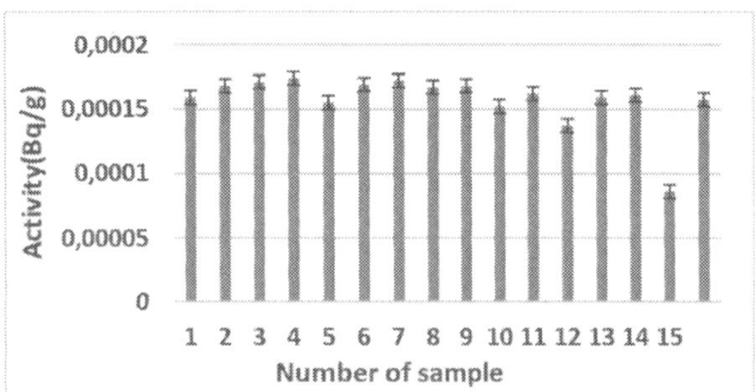

Fig. 7. Activity of ^{241}Am (Bq/g) from Lake Windermere (Fell Foot Park), Lake District (Cumbria, UK).

TABLE I
^{241}AM ACTIVITY IN SOIL CORES FROM BLELHAM TARN

core	Activity (Bq m^{-2})
1	33 ± 4
2	38 ± 5
3	39 ± 5
4	41 ± 5
5	42 ± 5
6	42 ± 5
7	25 ± 4
8	39 ± 5
9	48 ± 5
means	39 ± 5

TABLE II
^{241}AM ACTIVITY IN SOIL CORES FROM LAKE WINDERMERE

core	Activity (× 10^{-4} Bq/g)
1	1.60 ± 0.03
2	1.69 ± 0.03
3	1.71 ± 0.03
4	1.75 ± 0.03
5	1.55 ± 0.03
6	1.69 ± 0.03
7	1.72 ± 0.03
8	1.68 ± 0.03
9	1.68 ± 0.03
10	1.52 ± 0.03
11	1.62 ± 0.03
12	1.37± 0.03
13	1.60 ± 0.03
14	1.61 ± 0.03
15	0.86 ± 0.03
means	1.57 ± 0.03

V. CONCLUSIONS

This research indicates that this approach could benefit related radioanalytical methods for ^{241}Am based on the 59.5 keV γ-ray line, and hence assay of ^{241}Pu, especially since AMS measurements are generally more sophisticated and destructive assessments than BEGe, and can be limited in discerning ^{241}Am and ^{241}Pu from one another because of their closely-lying isobars.

ACKNOWLEDGMENTS

The authors would like to express their gratitude to the Royal Society, for a Wolfson Research Merit Award (M. J. J.), the UK National Trust and Lancaster University for funding A. M.'s PhD studentship.

REFERENCES

[1] BBC News, 9th July 2017, 'UNESCO awards Lake District World Heritage site status', https://www.bbc.co.uk/news/uk-england-cumbria-40547691
[2] Q. M.R. Nizam et al., 'Assessment of activity concentrations of Radionuclides from Upper Level Sediment in Charfassion', Nuclear and Particle Physics 3(3):36-39 (2013).
[3] H. Michel et al., 'Plutonium, americium and cesium records in sediment cores from Blelham Tarn, Cumbria (UK)', J. Radio. Nucl. Chem. 247 (1) pp. 107-110 (2001).
[4] http://www.environmentlaw.org.uk/rte.asp?id=303.

[5] H. Michel et al., 'Plutonium and americium inventories in atmospheric fallout and sediment cores from Blelham Tarn, Cumbria (UK), J. Envir. Radio. 59 pp.127-137 (2002).
[6] M. J. Joyce et al. 'Searching for zero plutonium in the far north of Scotland', Actinide Research Quarterly, (2) pp. 27 (2019).
[7] M. J. Joyce et al. 'A comparison of plutonium abundance in soil from sites in the United Kingdom measured with high-efficiency, high-resolution γ-ray spectroscopy, neutron assay and accelerator mass spectrometry', Plutonium Futures 2018, San Diego, September 2018.
[8] M. J. Joyce et al. 'A Comparison of γ-ray spectroscopy with Accelerator Mass Spectrometry for environmental assay of plutonium', EEE Nuclear Science Symposium and Medical Imaging Conference Proceedings (NSS/MIC), 2018, pp. 1-3, doi: 10.1109/NSSMIC.2018.8824687.
[9] M.J. Joyce 'Nuclear Engineering: A Conceptual Introduction to Nuclear Power'. Netherlands: Elsevier Science, 2017, pp.56-57.

Radionuclides Contamination in Soil: Effects, Sources and Spatial Distribution

Peter Oluwadamilare Olagbaju[1], Olanrewaju Bola Wojuola[1], and Victor Tshivhase[2]
[1] Department of Physics, North-West University, Mafikeng Campus, Mmabatho, South Africa
[2] Centre for Applied Radiation Science and Technology, North-West University, Mafikeng Campus, Mmabatho, South Africa
corresponding author: olagbajupetero@gmail.com

Abstract— **Radionuclide concentrations in the soil depend on the geological and anthropogenic activities of an area. They influence level of gamma radiation in the environment, which can cause significant health risk in humans. Due to the non-uniform distribution of radionuclides in the soil, various measurement methods have been adopted to monitor our environment. The challenges involved in collecting environmental samples, duration, technicality and cost of measurement have led to various models for predicting unmeasured locations. This article presents geostatistical method using kriging techniques, which adopt the theory of regionalized variables, to characterize the spatial distribution of radionuclide in un-sampled locations using data obtained from sampled locations. Among other things, this paper presents results of application of this method to a study area. Spatial distribution of radionuclides reveal the non-uniform distribution in the study area. Though the radionuclides level in the study area are below global average set by United Nations Scientific Committee on the Effects of atomic Radiation (UNSCEAR), the spatial distribution map can be used to provide adequate information needed by regulatory authorities in identifying the contaminated area in need of remediation.**

Keywords: Radionuclides, Kriging interpolation and Spatial mapping

I. Introduction

Radionuclides are unstable isotopes that emit ionizing radiations to attain stability. Emitted radiations have reportedly led to various health effects in exposed humans, ranging from cancer to skin burns, depending on exposure level [1]. They are one of the major concerns to human health, and they exist in various geological formation such as water, soil and rocks.

Potassium-40 and the decay series of uranium and thorium are the major naturally-occurring radionuclides in the environment [2]. They are widespread in the earth's environment and exist in environmental media such as rocks, soils, water, air and plants [3, 4]. Other than the natural sources of radionuclides which account for about 85% of the annual ionizing radiation received by members of the public [5], anthropogenic activities such as nuclear accident, nuclear testing, mineral explorations, industrial activities and agrochemicals used in agricultural practices have been reported to enhance its level in the environment [4, 6-8].

The varying concentration of naturally-occurring radionuclides in the soil is due to the geographic and geological influence on its non-uniform distribution across the earth surface [4, 9]. Several studies [7, 10-12] have reported that regions with granitic rock have high concentrations of radionuclide in soil compared to regions with sedimentary rock. Mining and other anthropogenic activities have also enhanced its level above the recommended threshold in the environment [13].

Due to non-uniform distribution and presence of radionuclide above recommended threshold level in some regions coupled with associated health effects, monitoring and assessment of radionuclides' distribution is essential [2, 4, 9, 12]. The rest of this article gives an overview of sources, effects and spatial distribution of radionuclides. This article also presents results of investigation and modelling of spatial distribution of naturally-occurring radionuclides in a selected study area.

II. Sources of Radionuclides

Soil radioactivity is a result of naturally-occurring radionuclides present in the environment [14]. This is as a result of weathering of bedrock, the main reservoir of the primordial radionuclides, releases uranium (U), thorium (Th), and potassium (K) to the soil and other geological formations. These can migrate to the environment via weathering of rocks and minerals, erosion and dust particles. Naturally-occurring radionuclides are not uniformly distributed in soils and rocks, but vary from one region to another depending on the geological, geographic and anthropogenic activities of each region [3, 15].

Soil is a significant source of radiation exposure as well as a direct source of radionuclides leading to contamination of agricultural lands and products. They play essential role in the human food chain and are employed in human health risk assessments of ionizing radiation exposure in the environment [16, 17]. Soil radioactivity level is influenced by the radionuclide level in the bedrock and soil contamination has been widely associated with industrial and mining activities [12]. Monged, Hassan [6] ascribed enhanced level of radionuclides in soil to both natural and anthropogenic sources, highlighting the contribution from mining and industrial wastes along with agrochemicals to soil radioactivity level in the environment [18].

Radioactive materials from mining, agrochemicals, industrial and chemical waste, deposited in the environment can migrate to non-contaminated fields and water bodies via wind dispersal and erosion [19, 20]. Characterization of environmental contaminants in terms of concentration and distribution is essential in environmental monitoring and protection. Hence, the knowledge of the distribution of soil radionuclides is important in establishing baseline data for radiation impact assessment, radiation protection and exploration.

III. Effect of Radionuclides

Humans are regularly exposed to ionizing radiation as a result of interaction with their environment either by inhalation of radioactive dust, ingestion of radioactive materials or direct exposure to a radioactive source [1, 4]. Emitted radiation from naturally-occurring radionuclides has sufficient energy to cause ionization, resulting in either stochastic or non-stochastic effect of radiation. Stochastic and non-stochastic effects of radiation exposure depend primarily on the dose and duration of exposure. Stochastic effects such as cancer, teratogenic and genetic mutation are associated with long-term low-level radiation exposure. Non-stochastic effects are non-cancerous health effects such as burns and radiation sickness/radiation poisoning, and are associated with acute radiation exposure or short-term, high-level radiation exposure [1].

The dispersal of radioactive dust, radon gas, and its radioactive progeny from anthropogenic sources of radionuclides to the surrounding environment is a global concern because of the human health risk associated with it [14]. The principal radiation risk from exposure to naturally-occurring radionuclides in soil and other environmental media is the gamma radiation from radium decay leading to irradiation of body tissues [15].

Most populace are concerned about medical and occupational exposure risk of non-ionizing radiation, but often neglect risks due to exposure to ironizing radiation resulting from environmental radionuclide. Ironizing radiation are capable of causing genetic damage and cancer to the public, especially in children [16]. Radiological impact from exposure to naturally-occurring radionuclide is due to gamma rays exposure and irradiation of body tissues.

Due to the health risk associated with radiation and heavy metals' exposures, many government and international bodies such as International Commission on Radiation Protection (ICRP), World Health Organization (WHO) have adopted measures to minimize exposures. Such measures include setting threshold limits and implementing precautionary measures whenever the dose is found to be above recommended limits.

IV. Spatial Distribution

Estimation of environmental variable in un-sampled location, with the same anthropogenic loads of contamination is essential in environmental studies due to their non-uniform distribution, time, cost and technicality involve in measurement [17, 18]. Also, the need to determine areas requiring remediation as a result of increasing industrial and agricultural activities, at minimal cost is enough motivation to monitor and investigate the spatial distribution of these contaminants in our environment [17, 18].

Geo-statistical method is an important tool for mapping regional and local soil pollution and are often used to produce contamination maps. It is based on the theory of regionalized variable. They are used to reveal the spatial variability of environmental variables while providing quantitative basis for interpolating and predicting unmeasured location [19, 20].

Kriging techniques, based on the overall spatial arrangement of sampled location (auto-correlation) and the distance between the measured and prediction point, are widely used in environmental science to characterize spatial patterns and variability of pollutants concentration [9, 21, 22]. According to Dindaroğlu [9], Kriging technique is one of the accurate linear estimator and widely known techniques used in spatial predictions due to its reliability in estimating surface variables. Kriging technique is widely used in environmental science to characterize spatial patterns and variability of pollutants concentration [9, 21, 22]. It uses the property of semi-variance to express the degree of relationship between points on a surface by considering both the distance and the degree of variation between known data points when estimating values in unknown areas.

V. Methodology

In this study, Kriging technique was used for generating spatial mapping model for a selected study area in Rustenburg of South Africa. The study area comprises of cultivated as well as uncultivated farmlands. Basic theory underlying the Kriging technique is outline below as follows.

Kriging technique are defined as the optimal weighted average of the neighboring sampled data in an un-sampled location, is given as equation (1) [23].

$$\hat{Z}(s_o) = \sum_{i=1}^{N} \lambda_i Z(s_i) \quad (1),$$

where $Z(s)$ represents value at location s (x, y), $\varepsilon(s)$ is random errors with spatial dependence and μ is constant on which the model is based, λ_i is an unknown weight for the measured value at the i-th location [9, 19, 20].

The statistical prediction is minimized to make an unbiased prediction using $(Z(s_0) - \sum_{i=1}^{N} \lambda_i Z(s_i))^2$ to obtain the kriging equation

$$\Gamma * \lambda = g \quad (2),$$

according to Johnston, Ver Hoef [21]. The matrix Γ, is created using semivariogram

$$\gamma(h) = \frac{1}{2N(h)} \sum_{N(h)} [Z(S) - Z(S+h)]^2 \quad (3),$$

for all possible measured location u, where N(h) is the number of pairs for lag distance h and vector g is estimated using the location to be predicted by applying its Euclidean distance from all measured points. Ordinary kriging equation

$$\lambda = \Gamma^{-1} * g \quad (4),$$

is solved to predict the un-sampled location using equation (1).

In this study, Geostatistical analysis and geographic information system-based spatial mapping was used to study the spatial distribution of measured ^{238}U, ^{232}Th and ^{40}K in the farming area of Rustenburg. Activity concentration of naturally-occurring radionuclides was measured using a well calibrated broad energy hyper pure germanium detector. In the study area, measured activity concentration of naturally-occurring radionuclides is observed to be below the global average reported by United Nations Scientific Committee on the Effects of Atomic Radiation [4].

VI. Results and Discussion

The spatial distribution of the three measured naturally-occurring radionuclides presented in Figure 1 to 3, reveals the uniform distribution of naturally-occurring radionuclides in the study area. This observation is attributed to the geological and lithological homogeneity of the study area [24-26]. Figures 1, 2 and 3 show the activity concentration of naturally occurring radionuclides to vary from (6.0 to 31.0) Bq/kg, (5.0 to 24.0) Bq/kg and (15.0 to 135.0) Bq/kg for ^{238}U, ^{232}Th and ^{40}K. The slight variation observed in distribution of ^{238}U, ^{232}Th and ^{40}K, is due to different land use, which include farming, industries, mining, and residential [27]. Spatial distribution of radionuclides provides easy and comprehensive interpretation of the spread of radionuclides in the environment.

The average activity concentration of ^{238}U, ^{232}Th and ^{40}K measured in the study area is found to be 17.03 ±0.98 Bq/kg, 12.93±1.47 Bq/kg and 82.56 ± 7.80 Bq/kg respectively. They are below the global average reported by United Nations Scientific Committee on the Effects of atomic Radiation (UNSCEAR). This indicate the area is radiological safe for living and agricultural purposes. Spatial distribution map can be employed in the identification of regions in need of remediation. As a result, it is essential in environmental monitoring and protection.

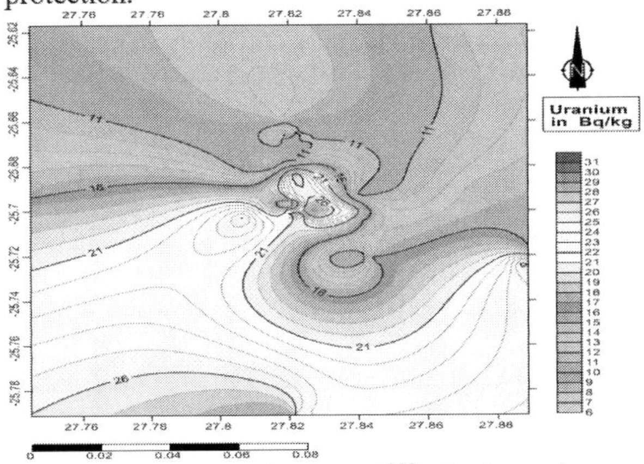

Figure 1: Spatial distribution of ^{238}U in the study area

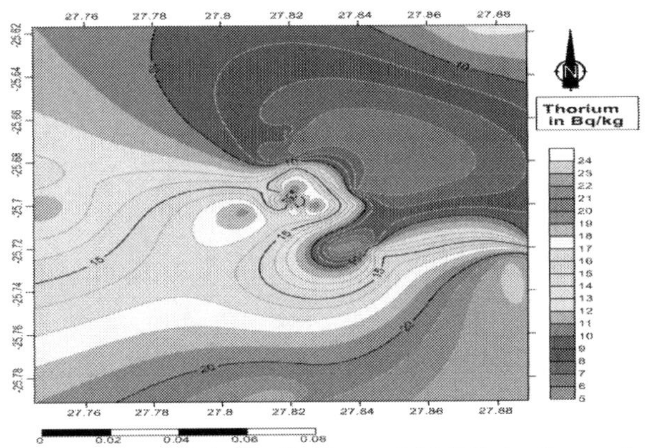

Figure 2: Spatial distribution of 232Th in the study area

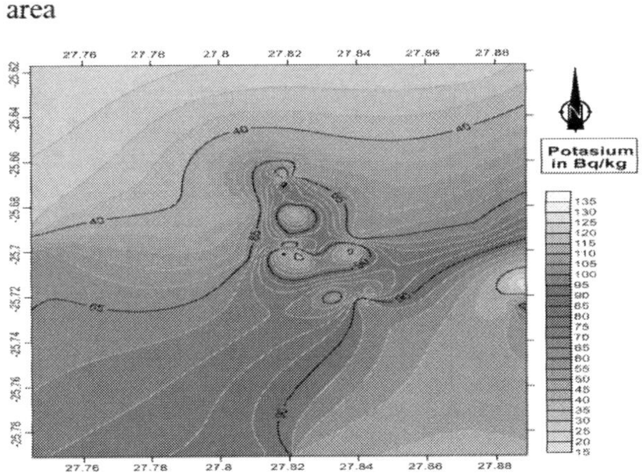

Figure 3: Spatial distribution of ^{40}K in the study area

VII. Conclusions

The potential health impact and regional variation of radionuclides in the earth as a result of geological and anthropogenic influence have resulted in continual investigation of its concentration and distribution in the environment for proper environmental monitoring and protection from its associated health risk globally.

The cost, effects, difficulty and technicality involved in collecting and measuring environmental data, coupled with its non-uniform distribution in the earth, have necessitated the need to develop environmental model for the study of spatial distribution of radionuclides to predict its concentration in unmeasured location using data obtained from measured location.

References

1. Cember, H., T.E. Johnson, and P. Alaei, *Introduction to Health Physics*. MedPh, 2008. **35**(12): p. 5959.
2. Obed, R., I. Farai, and N. Jibiri, *Population Dose Distribution due to Soil Radioactivity Concentration Levels in 18 Cities Across Nigeria*. Journal of Radiological Protection, 2005. **25**(3): p. 305.
3. Ajayi, O.S., K.O. Balogun, and C.G. Dike, *Spatial distributions and dose assessment of natural radionuclides in rocks and soils of some selected sites in southwestern Nigeria*. Human and Ecological Risk Assessment: An International Journal, 2017. **23**(6): p. 1373-1388.
4. UNSCEAR, *Sources and Effects of Ionizing Radiation: Sources*. Vol. 1. 2000: United Nations Publications.
5. WNA, *Naturally-occurring Radioactive Materials (NORM)*. 2014, World Nuclear Association.
6. Monged, M.H., H.B. Hassan, and S.A. El-Sayed, *Spatial Distribution and Ecological Risk Assessment of Natural Radionuclides and Trace Elements in Agricultural Soil of Northeastern Nile Valley, Egypt*. Water, Air, & Soil Pollution, 2020. **231**(7): p. 1-24.
7. Navas, A., J. Soto, and J. Machín, *Edaphic and Physiographic Factors Affecting the Distribution of Natural Gamma-emitting Radionuclides in the Soils of the Arnás Catchment in the Central Spanish Pyrenees*. European Journal of Soil Science, 2002. **53**(4): p. 629-638.
8. Otwoma, D., et al., *Radioactivity and dose assessment of rock and soil samples from Homa Mountain, Homa Bay County, Kenya*. 2012.
9. Dindaroğlu, T., *The use of the GIS Kriging technique to determine the spatial changes of natural radionuclide concentrations in soil and forest cover*. Journal of Environmental Health Science and Engineering, 2014. **12**(1): p. 130.
10. Ramasamy, V., S. Senthil, and V. Meenakshisundaram, *Distribution of Natural Radionuclides and Minerals in Beach Sediments from North East Coast of Tamilnadu, India*. African

Journal of Basic and Applied Sciences, 2009. **1**(1-2): p. 15-20.
11. Ramola, R., et al., *Radionuclide Analysis in the Soil of Kumaun Himalaya, India, using Gamma ray spectrometry*. Current Science, 2011: p. 906-914.
12. Jibiri, N. and I. Okeyode, *Activity concentrations of natural radionuclides in the sediments of Ogun River, Southwestern Nigeria.* Radiation protection dosimetry, 2011. **147**(4): p. 555-564.
13. Azeez, H.H., H.H. Mansour, and S.T. Ahmad, *Effect of Using Chemical Fertilizers on Natural Radioactivity Levels in Agricultural Soil in the Iraqi Kurdistan Region.* Polish Journal of Environmental Studies, 2019.
14. Lottermoser, B. and P. Ashley, *Tailings dam seepage at the rehabilitated Mary Kathleen uranium mine, Australia.* Journal of Geochemical Exploration, 2005. **85**(3): p. 119-137.
15. Abdelouas, A., *Uranium Mill Tailings: Geochemistry, Mineralogy, and Environmental Impact.* Elements, 2006. **2**(6): p. 335-341.
16. Kamiya, K., et al., *Long-term effects of radiation exposure on health.* The lancet, 2015. **386**(9992): p. 469-478.
17. Linnik, V.G., et al., *Spatial Distribution of Heavy Metals in Soils of the Flood Plain of the Seversky Donets River (Russia) based on Geostatistical Methods.* Environmental Geochemistry and Health, 2020: p. 1-15.
18. Chu, H.-J., et al., *Delineating the Hazard Zone of Multiple Soil Pollutants by Multivariate Indicator Kriging and Conditioned Latin Hypercube Sampling.* Geoderma, 2010. **158**(3-4): p. 242-251.
19. Matheron, G., *Principles of geostatistics.* Economic geology, 1963. **58**(8): p. 1246-1266.
20. Dai, L., H. Wei, and L. Wang, *Spatial distribution and risk assessment of radionuclides in soils around a coal-fired power plant: a case study from the city of Baoji, China.* Environmental research, 2007. **104**(2): p. 201-208.
21. Johnston, K., et al., *The principles of geostatistical analysis.* 2003. **4980**.
22. White, J., R. Welch, and W. Norvell, *Soil Zinc Map of the USA using Geostatistics and Geographic Information Systems.* Soil Science Society of America Journal, 1997. **61**(1): p. 185-194.
23. Lin, Y.-P., et al., *Factorial and indicator kriging methods using a geographic information system to delineate spatial variation and pollution sources of soil heavy metals.* Environmental Geology, 2002. **42**(8): p. 900-909.
24. Olagbaju, P., et al., *Background radiation level measurement using hand held dosimeter and gamma spectrometry in Ijebu-Ife, Ogun State Nigeria.* International Journal of Radiation Research, 2021. **19**(3): p. 591-598.
25. Ramasamy, V., S. Senthil, and V. Meenakshisundaram, *Distribution of natural radionuclides and minerals in beach sediments from north east coast of Tamilnadu, India.* Afr J Basic Appl Sci, 2009. **1**(1-2): p. 15-20.
26. Viljoen, M. and R. Hieber. *The Rustenburg Section of Rustenburg Platinum Mines Limited, with Reference to the Merensky Reef.* in *Mineral deposits of southern Africa.* 1986.
27. Nde, S.C., et al., *Modelling the dynamics of the cancer risk due to potentially toxic elements in agricultural soils, in the upper Crocodile River catchment, North-West province, South Africa.* Ecotoxicology and Environmental Safety, 2021. **211**: p. 111961.

Particle Identification and Tracking by the Use of a Pixel-Based Semiconductor Radiation Detector Coupled with Voltage Controlled Oscillators

K. Coulié[1], W. Rahajandraibe[1], L. Ottaviani[1]

[1]Aix Marseille Université, CNRS, Université de Toulon, IM2NP UMR 7334, Marseille, France
Corresponding author: karine.coulie@im2np.fr

***Abstract*— A particle detection chain based on a CMOS-SOI VCO circuit associated to a matrix of detection is presented. The solution is optimized for the recognition and tracking of various particles. Two ions are considered: an alpha and an aluminum. These two ions were chosen because there are very different in terms of energy and LET variations.**

***Keywords* — Sensor, detection, VCO, radiation.**

I. Introduction

SEMICONDUCTOR radiation detectors, also called solid-state detectors, can be classified into several categories, depending on the application field and the particles to be detected. These detectors are based on charge collection and amplification using a semiconductor volume such as a p-n junction [1-2]. They exhibit several advantages in comparison with other types of detectors such as gas-filled counters. For instance, they usually offer faster charge-collection time which provide the ability to process higher counting rates. Another advantage is their compactness which allows the measurement of intensity variations over small distances, as well as their low electron-hole pair energy threshold, improving the measurement resolution. A pixel organization including several sensors could be used when a high detector sensitivity is required [2-8]. This configuration will be investigated in this work.

Particles counting and/or recognition require the conditioning of the signal generated by the sensor. Most of the existing conditioning chains are based on a direct reading of the detected currents created by the crossing of an ionizing particle through the matter. When an embedded reading system is necessary, this solution could not be suitable for the detection of low energy particles where the current signature could be a narrow pulse with a duration of few nanoseconds. Indeed, the readout chain must ensure sufficiently high resolution (optimized Signal-to-Noise-Ratio) for an appropriate post-processing. Moreover, a reading amplifier exhibiting a fast response time is required which means a higher current consumption and complex circuit implementation. Designing high precision, low-power wideband amplifier is a critical issue. As a consequence, most practical detectors are based on direct reading of the detected currents using Charge Sensitive Amplifiers (CSA) [9]. In this case, the circuit delivers a limited amount of information on the actual current shape and no information about the electrical signature of the particles is available. In applications such as gas prospection or medical therapy, the characteristics of the electrical signature of the particles must be known. Moreover, the knowledge of the generated current shape at the output of the detector allows easier post processing of the signal.

In this context, a new approach based on the use of a ring oscillator was developed [10]. This approach is based on the reading of the information related to indirect output parameters of the detection chain signal (i.e. voltage variation), instead of directly measuring the current from the sensor. This solution avoids most of the design problems described upper. Then, the information is extracted by correlating the initial oscillating signal of the system with the oscillating signal after the particle has passed the detector. The particle recognition only requires to link the output information (*i.e.* voltage variation of the oscillating signal) to the input information (current stimuli). The detection system presented here should be enough sensitive to detect low energy particle, to track ions and to provide information about the shape of the detected currents. The

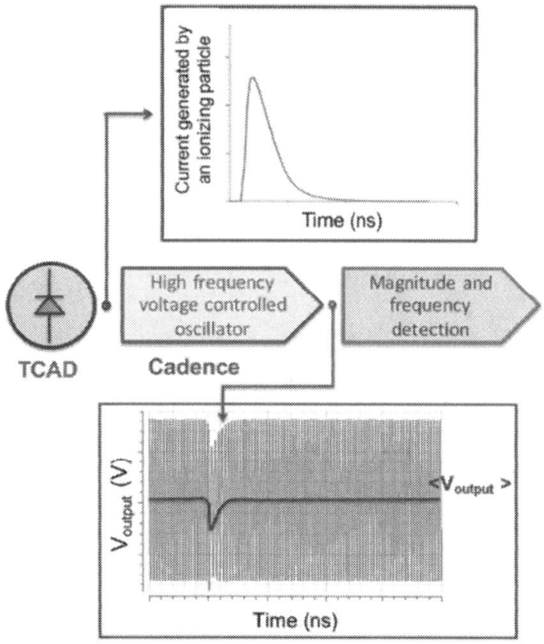

Fig. 1. Simulation chain including the current from TCAD modeling (top) and the variation of the output voltage versus time at the output of the VCO (bottom).

analyses are performed using the device simulator of Synopsys® CAO tool for current generation and the electrical simulator of Cadence® for the VCO simulation.

The full detection system (VCO + matrix) will be described in section II. Section III is dedicated to the system characterization which is mandatory for particle recognition. Part IV deals with two case studies: an alpha particle and an aluminum ion and section V presents a complementary method to allow charge determination. The last section concludes the paper.

II. THE DETECTION SYSTEM

A. The VCO chain

The detection chain has been designed and implemented on 130 nm CMOS SOI technology, then simulated at circuit level using "Spectre" simulator (SPICE-based) under Cadence *Virtuoso* © CAO tool. The concept itself was presented in [10-13]. The VCO chain is composed of three parts: a CMOS based pixel detector, a shaping circuit based on a Voltage Controlled ring Oscillator (VCO) and a system for frequency and magnitude detection (Fig. 1). This system allows the evaluation of the circuit sensitivity to radiation by measuring the oscillator responses. In the very first versions of the system, the operating frequency of 0,35 µm Bulk Silicon substrate CMOS VCO was 1 MHz. Thanks to a new solution based on another CMOS process (130 nm SOI), it reaches now 4.3 GHz. This increase of the operating frequency makes possible the development of new methods for signal recognition. As actual operating frequencies are reaching 4.3 GHz, the shape of the signal is directly reproduced at the output of the VCO. Then, the output parameter is now the average voltage variation (Fig.1). The advantage of this solution is a direct recognition of the signal shape as an amplitude modulation of the VCO output frequency.

B. The detection matrix

In order to improve the efficiency of the detection and the tracking ability, a *3x5* matrix is modeled and analyzed (Fig. 2). The input current comes from the realistic simulation of the matrix using TCAD device simulation tools (Synopsis ©). The effect of the ion strike is simulated using the *Heavy Ion* module of Synopsys [14], considering an electron-hole pair column centered on the ion track axis. The Linear Energy Transfer is defined as the energy lost by the

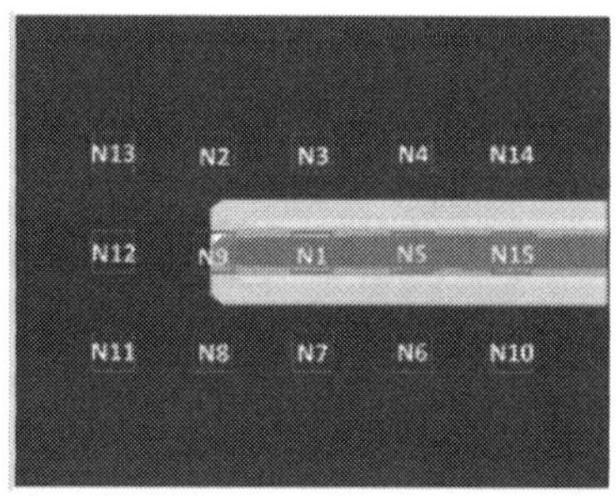

Fig. 2. Top view of the fifteen contacts matrix for the simulation of the alpha particle.

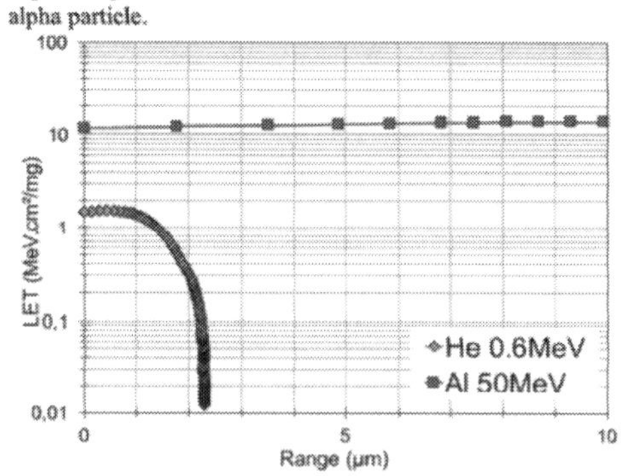

Fig. 3. LET versus range for the aluminum and alpha ions.

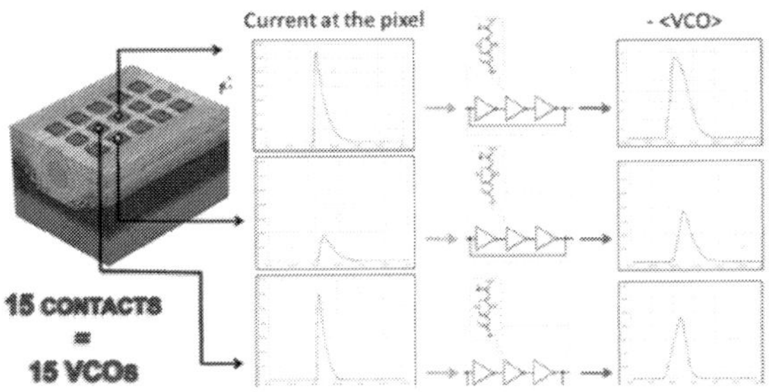

Fig. 4. Schematic presenting the process of simulation for three pixels.

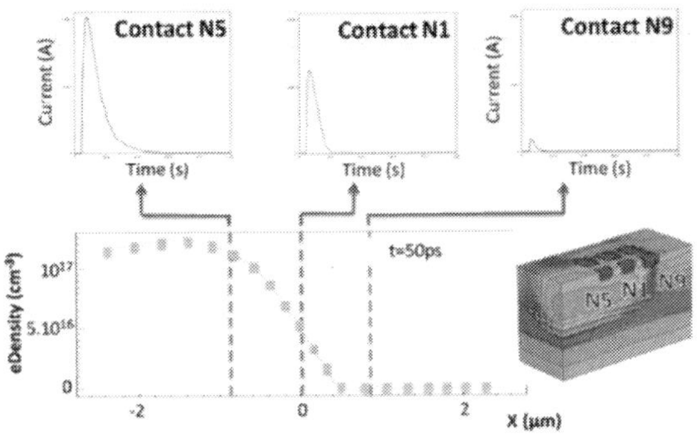

Fig. 5. Example of variation of the current at the output of three pixels related to the electron density, for an alpha particle generated horizontally.

particle, by unit of length and varies along the track depending on the initial energy of the particles [15]. An actual variation of the LET was integrated in our simulations, based on the value given by SRIM tables [16] (Fig. 3).

As this detector should be suitable for low energy particles detection, one of the simulated ion is an alpha particle crossing the device with an initial energy of 0.6 MeV, which corresponds to a maximum range of 2.3 µm in the silicon matrix . The aim is to study the end of the path corresponding to an alpha particle of 1.47 MeV generated by the initial interaction of thermal neutron with boron-10. The case of a more energetic particle is also considered. This is a 50 MeV aluminum ion which could be produced by the interaction of fast protons with silicon [17]. The maximum of the charge generation occurs at 50 ps.

C. The whole system

Thus, the effect of the chosen particle in the detection cell is simulated at device level and the oscillator at circuit level. The current generated at each electrode of the matrix is injected at the corresponding oscillator input. The average voltage is taken from the output oscillating signal of the VCO. Then if the matrix includes 15 contacts, 15 oscillators are needed (Fig. 4).

III. The system characterization

A. Information for tracking

Several information can be provided by the currents at the pixels. First one, is the charge density variation. The electron density variation is plotted versus the particle path on Fig. 5. At the top of this figure, the currents extracted from the contacts just above the charge generation are presented. The variation of the currents clearly follows the one of the charge density. The highest current, on the left, corresponds to the highest electron density and the lowest current, on the right, corresponds to the lowest charge density. These results, coming from the simulation of a 3x3 matrix presented in [19], show that the current variation can provide an information about the charge density in the detection cell.

In [19], it was also shown that the current shape can also give information about the distance between the charge generation and the contact. Indeed, for a contact located just above the ion, the current peak will appear earlier than for contacts located further from the generation point.

Another important result is that the average voltage from the VCO clearly fits the current variation from the pixels. This is clearly visible in the example of

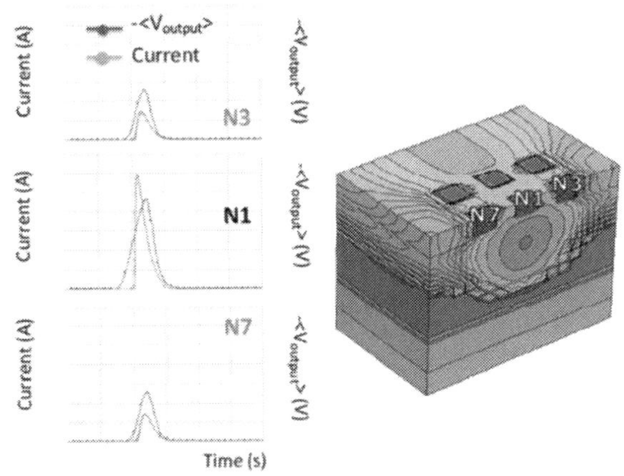

Fig. 6. Example of variation of the average output voltage of the VCO for three pixels related to the input current from the matrix, for an alpha particle generated horizontally.

https://doi.org/10.1051/epjconf/202125309007

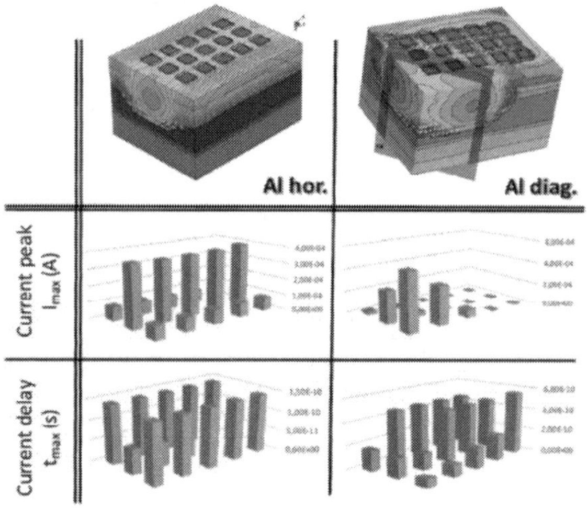

Fig. 7. Variation of the current peak and current delay at the output of the 15 pixels for the aluminum generated horizontally and diagonally.

Fig. 6. This corresponds to the horizontal strike of an alpha particle presented in Fig.5 [19]. The cut is in the direction perpendicular to the particle path. The orange curves correspond to the current from the pixel and the blue curves to the average voltage from the VCO. The scale is the same for the 3 presented contacts. The symmetry of the track can be seen by comparing the green contacts corresponding to top and bottom graphs. Their currents are clearly the same. The current corresponding to the black contact is higher because it is located just above the ion.

B. Calibration of the VCO chain

The key point is how the output parameters of the VCO chain can give information related to the input current. This could be done through the analysis of various characteristics extracted from the average voltage variation. In a previous work, we demonstrated that several metrics can be used but we decide to mainly focus on the variation of the average output signal (ΔVmax) versus the maximum of the input current (Imax). In [19], a linearity curve linking input to output parameters was determined for the 4.3 GHz VCO studied here. The curve which characterizes the VCO response is named « calibration curve ». It is obtained by the injection of various analytical currents at the input of the VCO. As this curve is linear, through calibration curves, the output parameters can be linked to the input currents, which could allow the incoming particle identification.

IV. Cases study

A. Ion tracking

1) Case of an aluminum ion

The first case study deals with the crossing of an aluminum ion in the 15 pixels structure. On the left of Fig. 7, the horizontal case is presented and on the right, the diagonal case. We check here the opportunity to perform particle tracking from the currents from the detection matrix. Two metrics are explored: the current peak and the time corresponding to this current peak. We know that these two parameters can give information about the charge density and the position of the particle strike. For both presented cases, the particle impact is visible on the current peak and the current delay. Indeed, the current peak is clearly higher for the contacts just above the particle and decreases for the most distant contacts. Concerning the current delay, it is lower for the contacts just above and increases for the most distant contacts.

The horizontal configuration has been extensively studied in [19] so we will focus on the diagonal configuration. Fig. 8 presents the electron density at depth=0.8 μm. The local electron density is presented in Fig. 9 for t=50 ps, which corresponds to the maximum time of the generation.

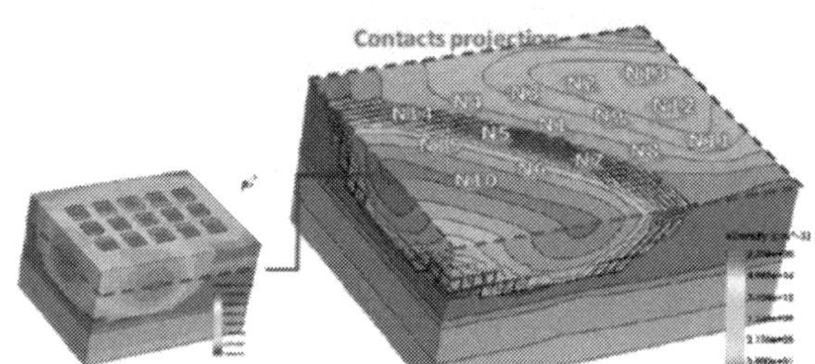

Fig. 8. Aluminum generation in the 15-pixels matrix. The figure on the right is a cut at depth=0.8μm. The represented contacts are a projection of the fifteen top contacts.

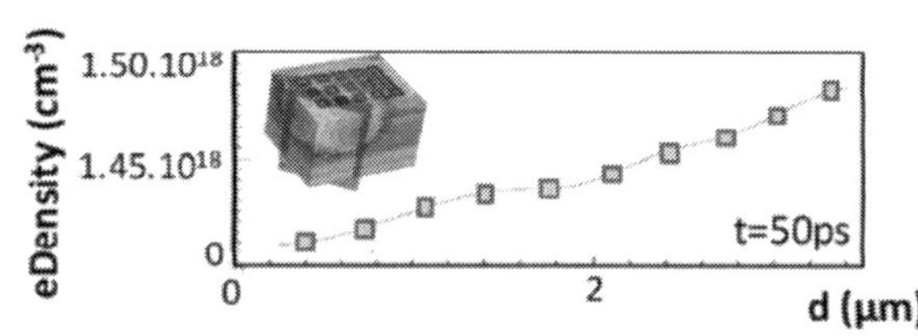

Fig. 9. Electron density along the particle track for the aluminum particle.

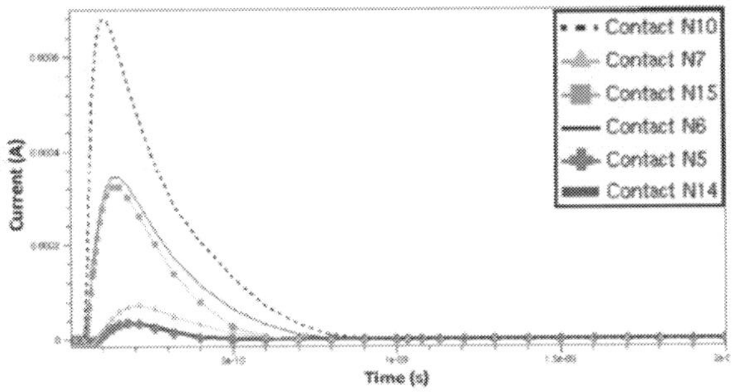

Fig. 10. Comparison of the currents of the six highest currents (N10, N15, N6, N7, N5 and N14 contacts) for the 5x3 matrix, aluminum particle.

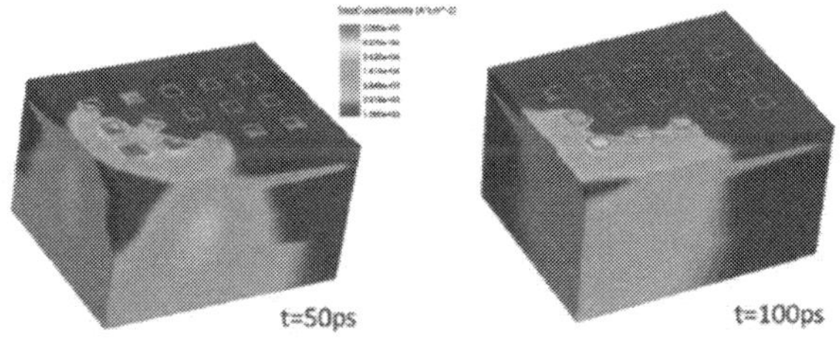

Fig. 11. Total current density at t=50ps (maximum of the generation) and t=100ps for the aluminum particle.

The six highest currents collected during the particle crossing are shown in Fig. 10. The highest current corresponds to N10 because this contact is just under the particle track. The maximum of the generation appears at t=108 ps. Two other currents correspond to the N6 and N15 contacts. The maximum time corresponding to these currents is the same: t=138 ps. The N6 current is higher because the local electron density increases along the track (Fig. 9). The N14 and N5 currents are the same and seem to be at the same distance from the track. However, a highest current would have been expected for N5 because the electron density increases along the track. It could be the effect of the surrounding contacts as it has been highlighted in [19]. The N7 current is higher and its maximum generation time arrives later than for N5 and N14. The N7 contact is probably further from the track. The total current density is shown in Fig. 11. The same variation is observed and visible even at t=100 ps.

The total collected charge for this configuration is 400 fC. This gives an information about the deposited energy.

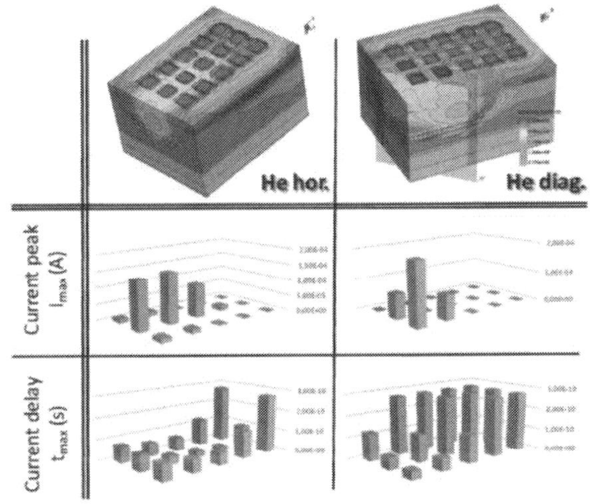

Fig. 12. Variation of the current peak and current delay at the output of the 15 pixels for the alpha generated horizontally and diagonally.

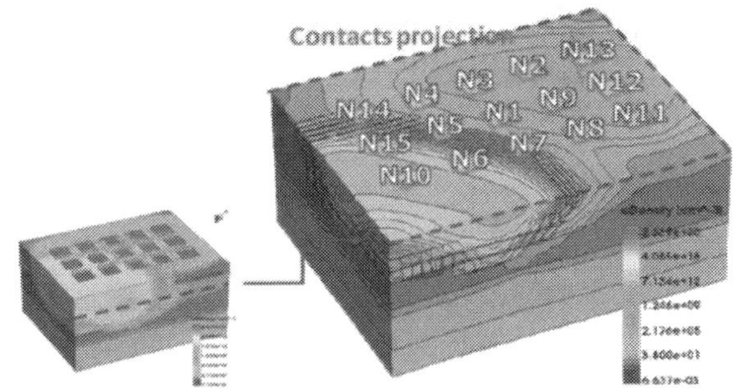

Fig. 13. Alpha generation in the 15-pixels matrix. The figure on the right is a cut at depth=0.8μm. The represented contacts are a projection of the fifteen top contacts.

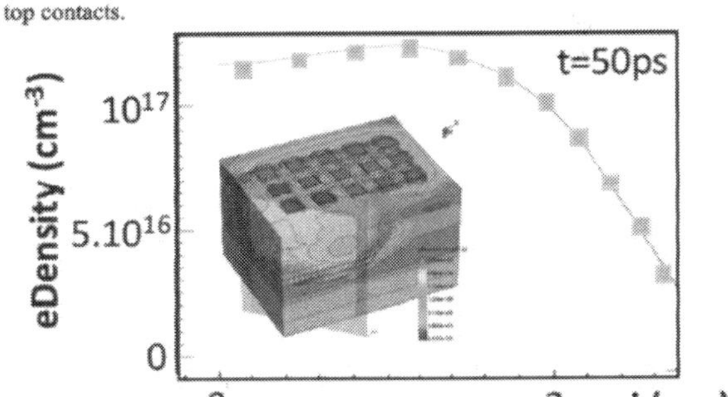

Fig. 14. Electron density along the particle track for the alpha particle.

2) *Case of an alpha particle*

The second case study deals with the crossing of an alpha particle in the same structure. Once again, the current peak and the current delay follow the expected trend and the particle tracking is possible. As for the aluminum particle, Fig. 13 presents the

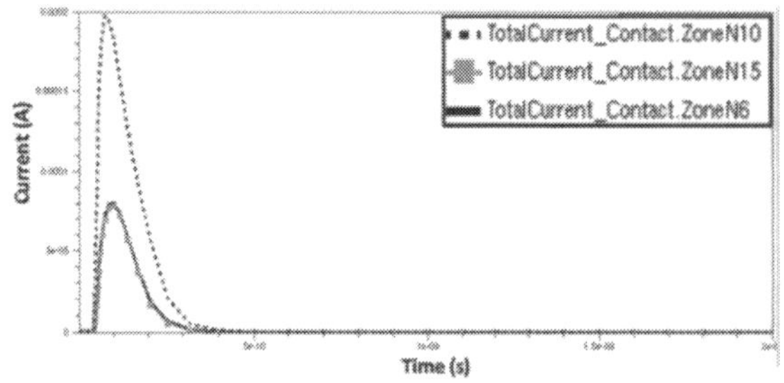

Fig. 15. 5x3 matrix (15 contacts). Comparison of the currents of the three highest currents (N10, N15, N6 contacts) for the alpha particle.

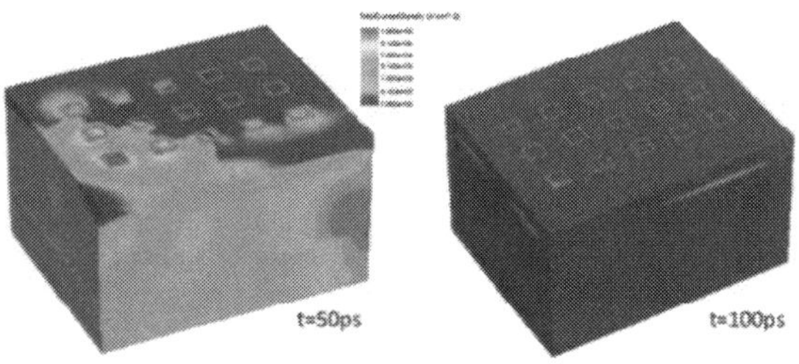

Fig. 16. Total current density at t=50ps (maximum of the generation) and t=100ps for the alpha particle.

electron density at depth=0.8 μm. The local electron density is presented in Fig. 14 for t=50 ps, which corresponds to the maximum time of the generation.

The three highest currents collected during the particle crossing are shown in Fig. 15. The highest one corresponds to N10 contact. This contact is just under the particle track. The maximum of the generation appears at t=78 ps. The two other currents correspond to the N6 and N15 contacts. They are approximatively at the same distance from the trace. This is visible through the maximum generation time which is the same for both contacts: 96 ps. As the electron density follows a bell curve, it is almost the same at both contacts. The total current density is shown in Fig. 16. We can state that the same variation is observed. By analyzing the maximum current and the maximum generation time, the particle crossing can be approximately determined. The maximum generation time gives an idea of the distance between the ion and the contact. Then thanks to Fig. 15, we know that the N10 contact is closer from the particle than the N6 and N15 contacts.

The total collected charge for this configuration is 42,8 fC which is about 10 times lower than for the aluminum.

3) VCO detection

As previously said, the possibility to detect alpha and aluminum horizontal tracks has been checked in a previous study. So we will focus on diagonal configurations. Plotting the corresponding output versus input response, we can see that all the currents are detected, the trend is preserved and even for the shortest currents, the detection is possible (Fig. 17). Only one point on the graph is out of the 50% detection zone what is an unexpected behavior. So we had a look on the average voltages.

All the currents corresponding to both particles in diagonal configuration have been injected sequentially in the VCO. The average voltage is presented in Fig. 18 for the highest currents corresponding to the two particles. All the currents of Fig. 10 and 15 are correctly reproduced except

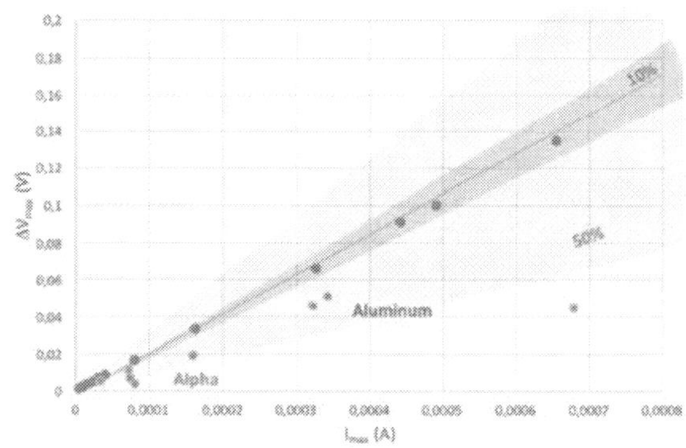

Fig. 17. Calibration curve and detected currents concerning diagonal configurations for the aluminum ion or the alpha particle crossing in a fifteen contacts matrix.

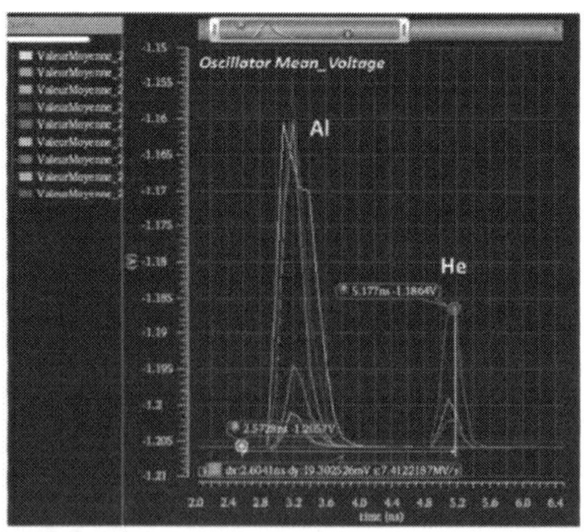

Fig. 18. Mean voltage for the highest currents: aluminum on the left, alpha on the right.

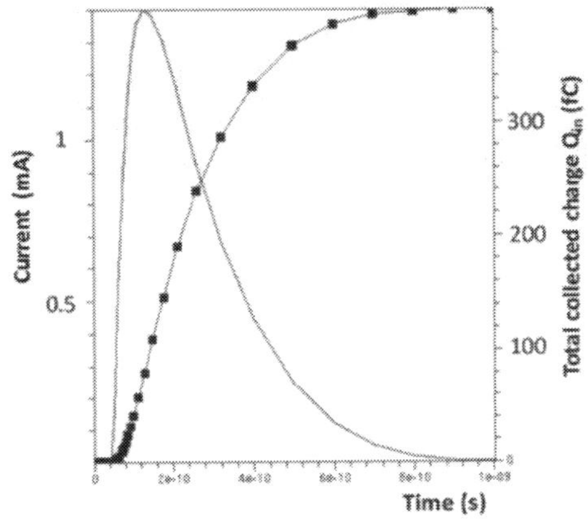

Fig. 19. Current at the bottom contact of the pixel (green curve) and corresponding total collected (black curve) versus time.

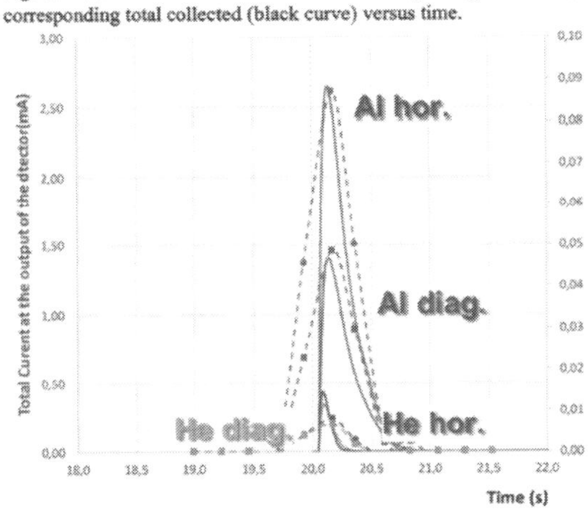

Fig. 20. Current at the bottom contact of the pixel (full line curves) and corresponding average voltage (dotted line curves) versus time.

the N10 current of the aluminum, visible on Fig. 10 and corresponding to the yellow average voltage of Fig. 18. This is surprising because this current is large enough to be correctly detected by the VCO. This current is too high and the oscillator saturates. It is not mandatory for the particle tracking but important for the determination of the total collected charge. Then we decide to use another oscillator, with a lower operating frequency for charge detection. This oscillator is not suitable for particle tracking because it is too slow. But it is able to drive a higher current at the input.

V. Determination of the total collected charge

The addition of all the currents collected at the fifteen pixels of the matrix corresponds to the current collected at the bottom electrode of the matrix. An example of this bottom current is presented on Fig. 19. The total collected charge, in black, corresponds to the integral of this current. This bottom current lasts about 1 ns what is large enough to be detected by the 2.1 GHz VCO we have selected for charge detection. This VCO was presented in a previous paper [18].

Then we injected the 4 bottom currents, corresponding to our 4 configurations (He hor., He diag., Al hor., Al diag.), at the input of a 2.1 GHz oscillator. The current peaks are much higher than the one which saturated the 4.3 GHz VCO (Fig. 20). Dotted curves correspond to the average voltages at the output of the VCO. The variation of these voltages follows the variation of the input currents and the 2.1 GHz oscillator does not saturate, as required.

The linearity curve corresponding to the 2.1 GHz oscillator was presented in a previous paper [18] (Fig. 21). The output charge corresponds to the integral of the average voltage. It is plotted versus the input charge corresponding to the integral of the current. The charges for aluminum and alpha particles are added on this graph. They are in the 20% detection zone which is a quite good result to allow particle identification.

Finally, the full detection system would require one oscillator oscillating at 2.1 GHz for charge identification and 15 VCO oscillating at 4.3 GHz for particle tracking.

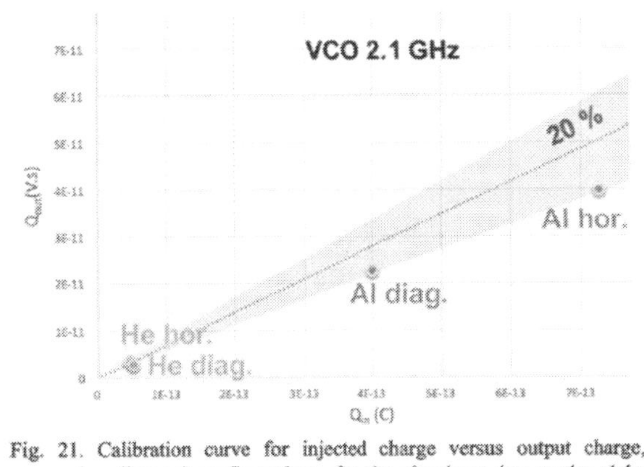

Fig. 21. Calibration curve for injected charge versus output charge, concerning diagonal configurations, for the aluminum ion or the alpha particle crossing in a fifteen contacts matrix. Only bottom currents are injected in the 2.1GHz VCO.

VI. Conclusions

Finally, with a VCO oscillating at 4.3 GHz we demonstrate that tracking various kinds of particles is possible. It is also possible to get information about the total collected charge by using a 2.1 GHz VCO.

Now, we would like to optimize the detector response in order to improve the detection efficiency, especially by exploring the possibility to use a faster oscillator for charge identification. We are also working on the material implementation of the voltage controlled oscillator and its associated electronics.

References

[1] C. Fiorini, A. Pullia, E. Gatti, A. Longoni, and W. Buttler, "A monolitic implementation of a switch-current 'wheel' amplifier for multi-channel silicon drift detectors, *IEEE Trans.* Nucl. Sci., vol. 46 n°3 p. 161, 1999.

[2] D. Hussona,, A. Bozierb, S. Higueretc, T.D. Lec, A. Nourreddinea, "AlphaRad, a new integrated CMOS System-on-Chip for high efficiency alpha particles counting", *Nuclear Instruments and Methods in Physics Research A 569*., pp. 845-852 161, 2006.

[3] I. Perić, C. Kreidl, P. Fischer, "Particle pixel detectors in high-voltage CMOS technology–New achievements", Nuclear Instruments and Methods in Physics Research A, Vol. 650, Issue 1, p. 158-162, 2011.

[4] M. Trocmé, S. Higueret, D. Husson, A. Nourreddine, T.D. Lê, "A new compact device for efficient neutron counting using a CMOS active pixel sensor", Radiation Measurements, pp. 1100-1103, 2008.

[5] D. Passeri, P. Placidi, L. Verducci, P. Ciampolini, G. Matrella, A. Marras, G/M. Bilei, "High-resolution CMOS particle detectors: design and test issues", 331-334 Vol.1, *2003* IEEE Nuclear Science *Symposium Conference Record*.

[6] M. Campbell, V. Havranek, E. HEIjne, T. Holy, J. Idarraga, J. Jakubek, C. Lebel, C. Leray, X. Llopart, J. Novotny, S. Pospisil, L. Tlustos, Z. Vykydal, "Charge collection from proton and alpha particle tracks in silicon pixel detector devices", 2007 IEEE Nuclear Science *Symposium Conference Record.*

[7] M. J. Baudot ; G. Bertolone ; A. Brogna ; G. Claus ; C. Colledani et al., "First test results Of MIMOSA-26, a fast CMOS sensor with integrated zero suppression and digitized output", 2009 IEEE Nuclear Science *Symposium Conference Record,* 2009.

[8] Tomas Baca ; Martin Jilek ; Petr Manek ; Petr Stibinger ; Vladimir Linhart ; Jan Jakubek ; Martin Saska, "Timepix Radiation Detector for Autonomous Radiation Localization and Mapping by Micro Unmanned Vehicles", 2019 IEEE/RSJ International Conference on Intelligent Robots and Systems (IROS), 3-8 Nov. 2019.

[9] Ying Zhang, Christine Hu-Guo, Daniel Husson, The-Duc Le , Stéphane Higueret, Yann Hu, "A High-Sensitivity Low-Power CMOS Sensor for a Future Neutron Personal Dosimeter", *in IEEE Trans. On Nucl. Sc.*, vol.59, no.4, pp. 1465-1471, August 2012.

[10] Rahajandraibe W., Aziza H., Coulié-Castellani K., Micolau, "Device and Method for Detecting Radiation Particles"
G. Patent AMU/CNRS/UAPV – Septembre 17, 2015 WO2015136220, https://patentscope.wipo.int/search/en/detail.jsf?docId=WO2015136220

[11] K. Castellani-Coulié, H. Aziza, W. Rahajandraibe, G. Micolau, J.-M. Portal, "Development of a CMOS Oscillator Concept for Particle Detection and Tracking", *IEEE Trans Nucl. Sci.*, Vol. 60 , Issue 4 , pp. 2450 - 2455 , Aug. 2013. DOI: 10.1109/TNS.2013.2254723

[12] K. Coulié-Castellani, W. Rahajandraibe, G. Micolau, H. Aziza, J.-M. Portal, "Improvement of a detection chain based on a VCO concept for microelectronic reliability under natural radiative environment", *IEEE LATW Conference*, 2015.

[13] S. Ben Krit, K. Coulié-Castellani, W. Rahajandraibe, L. Ottaviani, G. Micolau, H. Aziza, J.-M. Portal, "Comparison of a Readout Chain Dedicated to the Signal Conditioning of a Particle Detector and an Innovative Chain Based on a VCO Concept", *RADECS 2015*, Moscou, Russia, September 2015. DOI: 10.1109/RADECS.2015.7365592

[14] TCAD Sentaurus user's manual (Synopsys).

[15] A. Akkerman, J. Barak, D. Emfietzoglou, "Ion and electron track-structure and its effects in silicon: model and calculation", *Nuclear Instruments and Methods in Physics Research*, pp. 319-336, 2005.

[16] www.srim.org

[17] Vasiliy V. Markelov and Michael G. Tverskoy, "Evaluation of LET Spectra Produced by High Energy Protons in Si", *RADECS 2015*, Moscou, Russia, September 2015. DOI: 10.1109/RADECS.2015.7365592

[18] K. Coulié, W. Rahajandraibe, H. Aziza, G. Micolau and R. Vauché, "Detection limit of a VCO based detection chain dedicated to particles recognition and tracking", EPJ Web Conf., 170 (2018) 09002 , DOI: https://doi.org/10.1051/epjconf/201817009002

[19] K. Coulié, W. Rahajandraibe, H. Aziza, G. Micolau, R. Vauché, "Detection limit of a VCO based detection chain dedicated to particles recognition and tracking", *2005 8th European Conference on Radiation and Its Effects on Components and Systems*, 2005. DOI: 10.1109/RADECS.2005.4365629.

A prototype of pCT scanner: first tests

J. A. Briz[1], I. Posadillo[1], V.G. Távora[1], E. Nácher[2], M.J.G. Borge[1], O. Tengblad[1], A. Perea[1], A. Ortiz[1], J. D. Ovejas[1], and S. Viñals[1,3]
[1]Instituto de Estructura de la Materia, CSIC, Madrid, Spain
[2]Instituto de Física Corpuscular, CSIC-UV, Valencia, Spain
[3]Centro de MicroAnálisis de Materiales (CMAM), UAM, Madrid, Spain
jose.briz@csic.es

Abstract—**Proton therapy technique for cancer treatment offers a high selectivity with respect to conventional radiotherapy with X- and γ- rays due to the properties of the interaction of protons with matter. Very accurate and precise treatment plans and a good control on the dose deposition are required to exploit the full potential of the technique. The substitution of the currently used X-ray Computed Tomography (xCT) by proton Computed Tomography (pCT) in the design of treatment plans would allow for a reduction in proton range uncertainties. This would make possible an important improvement in the accuracy and precision of treatment plans. With this aim, a prototype of pCT scanner is under study. It includes two tracking detectors which provide information on the proton trajectories and a residual energy detector to determine the energy loss while traversing the object scanned. A proof-of-concept experiment has been performed using low-energy protons and a simplified prototype with only the two tracking detectors. The results obtained in the measurement are presented and discussed.**

Keywords — **proton, proton therapy, proton Computed Tomography, particle tracking, silicon detectors, DSSD, scintillators, $LaBr_3$.**

I. Introduction

Cancer is a leading cause of death worldwide and accounted for nearly 10 million deaths in 2020. This motivates the tremendous efforts which are devoted worldwide to improve the diagnostics and treatment techniques and the mitigation of its consequences.

Proton therapy as a tumor treatment technique was first proposed by Robert R. Wilson in 1946 and the first treatments were performed in Berkeley (USA) in 1954 and Uppsala (Sweden) in 1957. Proton therapy offers the great advantage with respect to conventional radiotherapy with X- or γ-rays of being more selective applying a higher dose to the tumor with respect to the surrounding healthy tissues. The reason is that the stopping power of protons increases continuously while slowing down. This causes the deposition of a large fraction of the proton initial kinetic energy in the very last millimeters of its range. The curve describing the behavior of the stopping power vs. depth for heavy-charged particles such as protons shows a narrow peak at the end of its path where most of the energy is lost, which is the so-called Bragg peak.

The fact that the energy deposition occurs in a very reduced region at the end of the proton path requires, at least, two improvements. First, the design of more accurate and precise treatment plans than those of conventional radiotherapy. This is needed to reduce the uncertainties on proton ranges avoiding a high dose to be applied to healthy tissues. Second, a very good control of proton ranges during treatment which can be achieved by means of a reliable and precise system of on-line proton range verification providing information in real time of the location of the deposited dose.

Currently proton therapy treatment plans are designed using images obtained via X-ray Computed Tomography (xCT). The result is a map of Hounsfield Units (HU) which basically provides information on the attenuation coefficient at each location relative to that of water and air. In continuation, to obtain the treatment plan with protons, a conversion to a map of Relative Stopping Powers (RSP), which is the interesting physical quantity in proton therapy, is made. This conversion induces relatively large uncertainties in the range of protons which can be as large as 5% in the abdomen or 11% in the head [1]. Assuming that an optimized imaging system of Computed Tomography directly with protons is developed, these uncertainties could be reduced to below 1% since one directly measures RSP and the conversion from HU would not be needed.

In this context, a prototype of proton Computed Tomography scanner is under study. A typical proton scanner is formed by a tracking system including two detectors placed one before and another after the object to be scanned as it is shown in Fig. 1. Then, a residual energy detector is used to fully stop the protons at the end of the path and be able to determine their energy loss in the object.

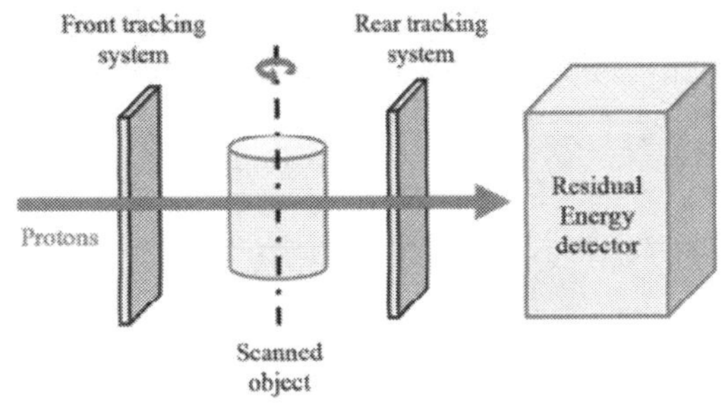

Fig. 1. Schematic structure of a pCT scanner

II. PROTOTYPE OF PCT SCANNER

The prototype to be tested includes two Double-sided Silicon Strip Detectors (DSSDs) placed before and after the object (sample) as tracking detectors. The DSSD detector, described in detail in [2], is formed by 16 horizontal and 16 vertical strips of 3 mm width each which provide an approximate field of view of 50 x 50 mm^2 with 256 pixels of 3x3 mm^2 area each. An array of 2x2 $LaBr_3$(Ce) crystals is used as residual energy detector, being each of them of 27 x 27 mm^2 and 4-cm-long. The $LaBr_3$(Ce) crystals are coupled to 6-cm-long $LaCl_3$(Ce) in "phoswich" configuration as described in [3]. For proton imaging the second layer of $LaCl_3$(Ce) scintillator is not used as protons of up to 120 MeV are fully stopped in the first layer of $LaBr_3$(Ce) crystals. This detector was already tested for proton detection showing successful results published in [4].

III. PROOF-OF-CONCEPT EXPERIMENT

A proof-of-concept measurement was performed in 2019 at Centro de MicroAnálisis de Materiales (CMAM) facility in Madrid. The 5-MV Cockcroft-Walton tandem accelerator of CMAM provides protons of up to 10 MeV kinetic energy. As the energy of protons provided was low, the prototype was simplified to two tracking detectors being DSSDs of 60 and 500 μm thickness. In this measurement, the second tracking detector acted both as tracking and residual energy detector at the same time. The experimental chamber used is shown in Fig. 2 and a closer look on the detectors and sample can be seen in Fig. 3. The proton beam was arriving from the left-hand side then impinged on a 1.2 mg/cm^2 bismuth scattering target. The prototype was placed at an angle of 24 degrees with respect to the primary beam incidence direction to both, reduce the beam intensity up to counting rate values below 10 kHz and completely and continuously illuminate the detectors and sample to be scanned. The experimental chamber was kept under a vacuum of approximately 1x10^{-6} mbar during the whole experiment.

Three different samples, shown in Fig. 4, were analyzed. They were thin samples due to the low beam energy and were

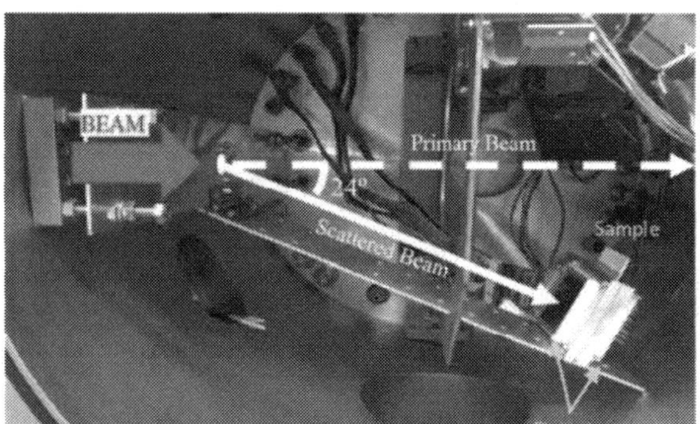

Fig. 2. Experimental setup used at the CMAM facility. The 10-MeV proton beam was scattered at a bismuth target. The detectors and sample were placed at 24° with respect to the primary beam direction. A close-up view on the detectors and sample area is shown in Fig. 3.

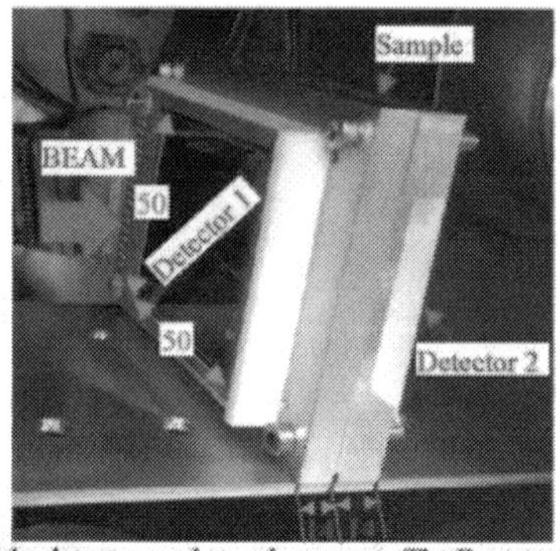

Fig. 3. Image of the detectors and sample support. The Detector 1-sample and sample-Detector 2 distances are indicated in blue. Dimensions are given in mm.

all built on a 500 μm PVC matrix layer. The first sample, shown on the left-hand side, contains a cross pattern made with a 200-μm-thick aluminum layer on top of the PVC layer. The second sample, which can be seen at the center of Fig. 4, contained four different circular regions of 10(1) mm diameter each: one with only a 200-μm-thick aluminum layer (1), another with just a hole (2), a third one containing 200-μm-thick aluminum layer on top of the 500-μm-thick PVC matrix (3) and the last one built with 1-mm-thick aluminum (4). The third sample was built with holes of different diameters and separations emulating a Derenzo pattern. The holes and separations are of 4, 6 and 8 mm organized in three rows as it can be seen in Fig. 4.

The detectors were connected to Mesytec MPR-64 preamplifiers and STM16 amplifiers and their signals were finally digitized by a CAEN ADC V785 VME module integrated in a Multi Instance Data Acquisition System [5]. The data was collected in an event-by-event basis and stored in list mode files.

IV. DATA ANALYSIS

The energy spectra of both Double-sided Silicon Strip detectors were obtained from the list mode data files. The image of the samples was obtained by imposing a time coincidence condition between both DSSD detectors and that only one vertical and one horizontal strip were hit with similar energy. Then, the pixel corresponding to the proton hit in each detector was determined as the crossing between the horizontal and vertical strips activated. The image was built by determining the crossing position of the trajectory of the protons with the sample plane, also known as image plane. The trajectory of protons was assumed to be a line connecting the pixels hit in each detector. Since both detectors were placed exactly at the same distance from the sample, 14 mm as shown in Fig. 3, the XY coordinates at the image plane were obtained just by arithmetic average of the coordinates at each detector.

The statistics accumulated was not the same in every pixel. The reason for that is the Rutherford scattering occurring at the bismuth target. Consequently, the detector hit map as well as the image showed an intensity gradient following the conical symmetry of the scattered-protons distribution. As a result, higher statistics was obtained at pixels closer to the incident

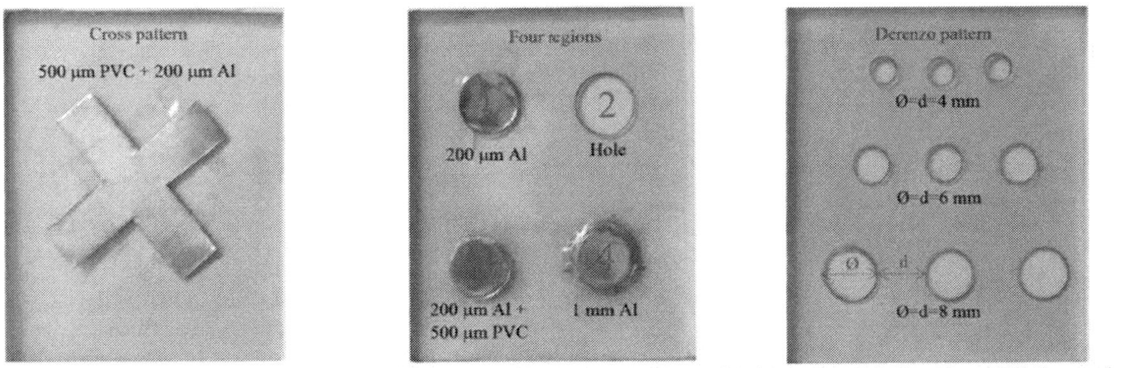

Fig. 4. Set of samples made with a 500-µm PVC matrix layer containing different patterns that were analyzed at the CMAM facility using a 10-MeV-proton beam.

proton beam direction (equivalent to 0° deviation angle) decreasing as one moves towards larger deflection angles.

Having a look at the experimental setup shown in Fig. 2, one can see that the upper part of both detectors is closer to 0 degrees and, therefore, accumulates larger statistics. To correct for this effect, the summed energy in each pixel is normalized by the number of hits detected. In this way, the average energy deposited per detected proton at each pixel is obtained.

V. RESULTS

The images shown in Fig. 5 were obtained from the samples shown in Fig. 4. Remarkable similarities between the images and the real samples are observed. A nice uniformity of the PVC matrix region can be seen in the three images with a value of about 3200 keV. A slight vertical distortion can be seen in the three images corresponding to X = 30 mm caused by a dead strip in the frontal DSSD detector.

A. Sample with cross pattern

A picture of the sample containing the cross pattern is shown on the left-hand side of Fig. 4. It was built by placing a 200 µm aluminum layer on top of the 500-µm-thick layer of PVC. The corresponding image obtained in the experiment is shown at the same side in Fig. 5. The cross pattern can be seen in red including a penumbra region shown in yellow. The uniform PVC background is seen with an energy loss of the protons of about 3200 keV (in green), meanwhile the region of aluminum + PVC of the cross is observed with 6500 keV approximately (in red) and the region of penumbra with about 5500 keV (in yellow). A detail showing the sensitivity of our device can be highlighted by the two more intense red regions indicated with two white circles. They correspond to the thin cellophane tape layers of about 20 µm thickness used to stick the cross pattern to the PVC matrix. The length and width of cross arms were determined in the projection of a slice chosen in the region of interest. Then a fit to a generalized gaussian function, i.e. $f(x) = \frac{A}{\sqrt{2\pi}\sigma} e^{-\frac{1}{2}\left(\frac{|x-\mu|}{\sigma}\right)^n}$ was performed, where n=4 order was chosen for the arm width and n=10 for the arm length. The FWHM of the 1D distributions was obtained using the expression $FWHM = 2 \cdot \sigma \cdot (2 \cdot \log(2))^{1/n}$. The length and width of the two cross arms were determined as the average of the FWHMs. The resulting values are compared to the real dimensions of the sample in Table I. The results agree within the error bars with the real dimensions of the sample.

B. Sample with four regions

The sample with four different regions is shown in the central

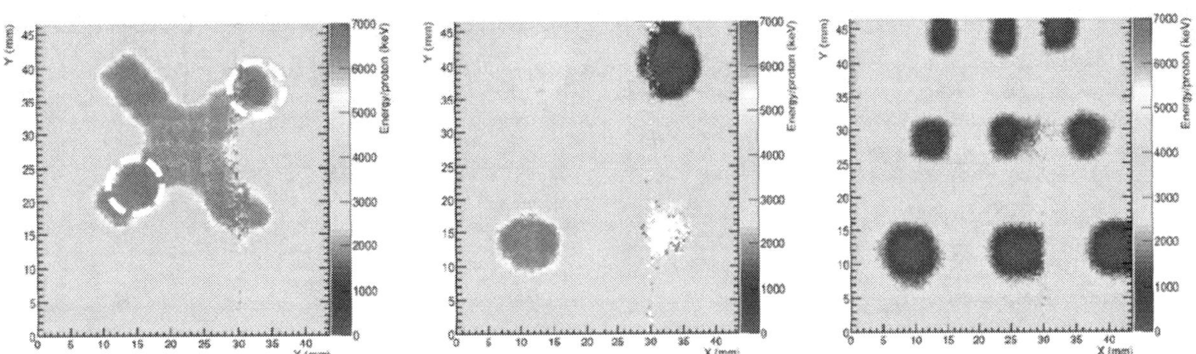

Fig. 5. Set of images obtained at the CMAM facility using a 10-MeV-proton beam of the samples shown in Fig. 4. The colour palette shows the values of deduced energy lost in the sample. The cross-pattern sample is shown on the left, the sample with four regions in the middle and the one with Derenzo pattern on the right side. The sample with cross pattern has indicated two regions which appear in red in the image which correspond to thin cellophane-tape layers of ~20 µm that can be seen in the picture of Fig. 4 and that our system was able to detect.

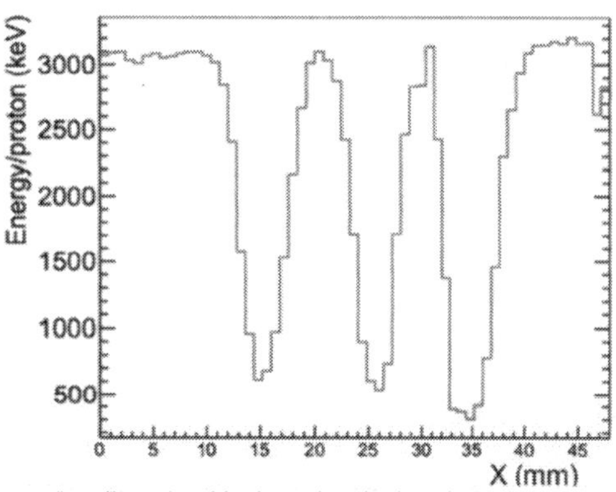

Fig. 6. One-dimensional horizontal projection obtained from the central slice of pixels for the row with 4-mm-diameter holes in the image of the sample with Derenzo pattern shown on the right-hand side of Fig. 5.

part of Fig. 4. The corresponding image is shown in the central part of Fig. 5. The uniform PVC background is found to be of similar value of energy loss than in the previous sample. The region 1 shows a value of proton energy loss of about 2500 keV meanwhile the region 2, as expected, shows a value of 0 keV. The region 3 is obtained with a value of approximately 6500 keV. Protons are stopped in the 1-mm-thick aluminum foil placed in position 4. For this reason, a white color is seen which corresponds to no signal in the back DSSD detector. The values estimated for the diameter of the three regions with data (empty hole, 200-μm Al and region with 200-μm Al + 500 μm PVC) are given in Table I. The real diameter, as already mentioned, was 10(1) mm. It is very encouraging to see that the values obtained are consistent with the real sample.

TABLE I
PROPERTIES OF THE IMAGES OBTAINED

SAMPLE WITH DERENZO PATTERN		SAMPLE WITH CROSS PATTERN			SAMPLE WITH FOUR REGIONS		
Sample hole diameter (mm)	Image hole diameter (mm)		Sample size (mm)	Image size (mm)	Region	Sample region diameter (mm)	Image region diameter (mm)
4(1)	4.1(2)	Arm length	35(1)	35.9(2)	1	10(1)	11(2)
6(1)	6.5(3)	Arm width	8(1)	7.9(2)	2	10(1)	10.1(8)
8(1)	7.5(2)				3	10(1)	10.1(15)

Dimensions of the holes in the sample with Derenzo pattern, arms of the sample with cross pattern and diameter of 3 regions in the sample with four regions obtained from the images as compared to real values. The dimensions are obtained in the images as the Full Width at Half-Maximum (FWHM) from the 1D horizontal projections of slices at the regions of interest.

C. *Sample with Derenzo pattern*

A sample with a Derenzo pattern was used for the evaluation of the spatial resolution. It is shown in the right-hand side picture of Fig. 4. The corresponding image can be seen at the same side of Fig. 5. To study the size of the holes obtained in the image, a one-dimensional spectrum was obtained by selecting a slice formed by the central line of pixels corresponding to each row of holes. This result can be seen in Fig. 6 for the case of the 4-mm diameter holes as an example. Each hole in the sample is seen as an inverted gaussian. A gaussian fit was performed and an average of the obtained FWHM values allowed to estimate the hole dimensions. The resulting values are given on the left-hand side of Table I. A good agreement is found between the real sizes of the holes and the ones determined from the image.

VI. DISCUSSION

The results obtained for the three samples studied are promising. For the sample with cross pattern, the dimensions obtained for the arm length and width agree within the error bars with the real dimensions. In addition, the PVC background shows a good uniformity. The device has sensitivity for thin layers as the presence of cellophane tape was identified in the image, see Fig. 5. The results for the sample with four regions showed that different thicknesses and materials are well identified and that the measured diameters for the three holes evaluated agree nicely with the values of the real sample. The study of the sample with the Derenzo pattern demonstrated that our device has better spatial resolution than 4 mm. In addition, the diameters of the 4-, 6- and 8-mm holes were well determined from the image.

VII. CONCLUSIONS

Proton therapy has an enormous potential for cancer treatment however improvements in treatment plans and dose verification are required to fully exploit its capabilities. With this purpose, our collaboration is working in the development of a proton Computed Tomography scanner.

In the present article, the results of a proof-of-concept experiment performed at the Centro de MicroAnálisis de Materiales facility in Madrid are presented. A simplified version of our prototype is adopted due to the low energy of the proton beam. Direct images were obtained for different thin samples. The properties of those images are presented and discussed in this contribution. In general, satisfactory results were obtained reproducing nicely the samples analyzed.

The results presented here constitute the first step in the development of our prototype that should be tested under more realistic conditions. With this aim, a new measurement was proposed to perform tomography scans of 3D objects with 100 MeV proton beams. The experiment has been performed in June 2021 and the data analysis is currently on-going.

ACKNOWLEDGMENT

Authors want to thank the CMAM (Madrid) staff for their contribution in the performance of the measurement. This work has been supported by the PRONTO-CM B2017/BMD-3888 project funded by Comunidad de Madrid (Spain).

REFERENCES

[1] Robert P Johnson, Rep. Prog. Phys. 81, (2018) 016701.
[2] O. Tengblad *et al.*, Nuclear Instruments and Methods in Physics Research A 525, (2004) 458–464.
[3] O. Tengblad *et al.*, Nuclear Instruments and Methods in Physics Research A 704, (2013) 19–26.
[4] E. Nácher *et al.*, Nuclear Instruments and Methods in Physics Research A 769 (2015) 105–111.
[5] Multi Instance Data Acquisition System. http://npg.dl.ac.uk/MIDAS/

On teaching experimental reactor physics in times of pandemic

Jan Malec[1], Michael Österlund[2], Andreas Solders[2], Ali Al-Adili[2], Anže Jazbec[1], Sebastjan Rupnik[1], Vladimir Radulović[1], Igor Lengar[1] and Luka Snoj[1]
[1]Jožef Stefan Institute, Slovenia
[2]Uppsala university, Sweden
Corresponding author: jan.malec@ijs.si

***Abstract*—The COVID-19 induced restrictions have prevented reactor physics students from attending in-person reactor physics exercises which are a vital part of their education. Jožef Stefan Institute has organized remote exercises with the help of off-the-shelf technology, including multiple videoconferencing setups, remote desktop software, portable cameras, a dome camera, shared spreadsheets, and a common whiteboard. The students were encouraged to actively participate in the exercises by giving instructions to the reactor operator, asking and answering questions, logging data, operating digital acquisition systems, and performing analysis during the exercise. The first remote exercises were organized as a five-day course of experimental reactor physics for students from Uppsala University. The feedback was collected after the course using an anonymous online form and was generally positive but has revealed some problems with sound quality which were resolved later. The Jožef Stefan Institute can also organize a remote course during a full lockdown when the reactor is not able to operate using the in-house developed Research Reactor Simulator based on a point kinetics approximation and a simple thermohydraulic module.**

***Keywords* — education, training, quarantine, reactor physics, nuclear reactor**

I. Introduction

The Jožef Stefan Institute (JSI) regularly organizes training [1], [2] for local and foreign students of nuclear engineering and related sciences. The exercises are very valuable for the students because they give them the opportunity to perform practical experiments on a working nuclear reactor, and in some cases represent the only link between theory and experimental research. Travel restrictions in 2020 due to the COVID-19 pandemic prevented foreign students from traveling to Slovenia, and the closure in spring 2020 meant that practical exercises at the reactor were not possible at all.
This paper describes some innovative solutions that were implemented to allow participants the closest possible approximation of in-person exercises given the restrictions. This paper presents a five-day distance learning course in experimental reactor physics organized for Uppsala University using commercially available equipment. The performance of the course was evaluated by an anonymous online survey in which all students and their mentor participated. The organizers asked both open questions and questions that were answered using a rating scale. Jožef Stefan Institute can also organize remote exercises with a specially developed simulator [3]–[5] when the nuclear reactor is not allowed to operate. In this paper we focus on the experimental reactor physics course organized for Uppsala University and its implementation, followed by an evaluation of the course and an outlook on future improvements.

II. Technical Execution

Instructors used available technology to approximate the face-to-face exercise experience as closely as possible, transforming students from passive spectators to active participants. The primary means of communication was standard videoconferencing software installed on PCs, smartphones that served as portable cameras and a dome camera mounted on the ceiling (Figure 1).

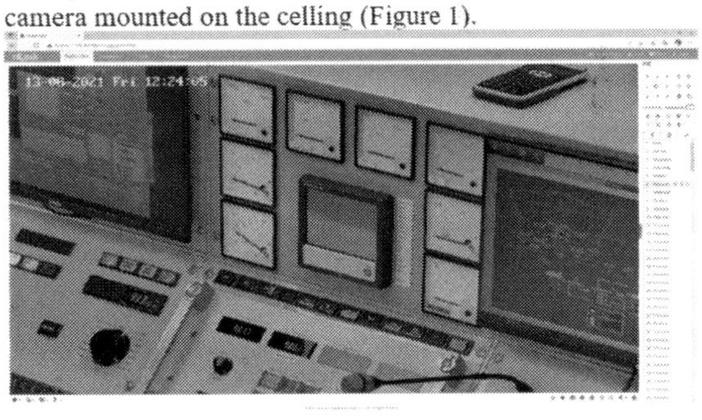

Figure 1 Dome camera pointed to the nuclear instrumentation. It could be remotely moved and zoomed-in to any point in the control room by using controls on the top right corner. Predefined positions (such as instrumentation and rod positions) could be selected in

The personal computers with video conferencing software are marked as "2." on photo Figure 4 and diagram Figure 5. The portable cameras, marked "5." on Figure 5 and Figure 8, could be taken into the reactor hall by the instructors to take a virtual tour, show the experimental setup or other activities at the reactor, such as the removal and re-installation of the neutron

https://doi.org/10.1051/epjconf/202125310001

source during criticality approach. A lecturer using the remote camera can be seen on Figure 8. In addition to the smartphone and PC cameras, a dome camera in the control room and a camera installed above the reactor core could be viewed and controlled by the students via the remote desktop software.

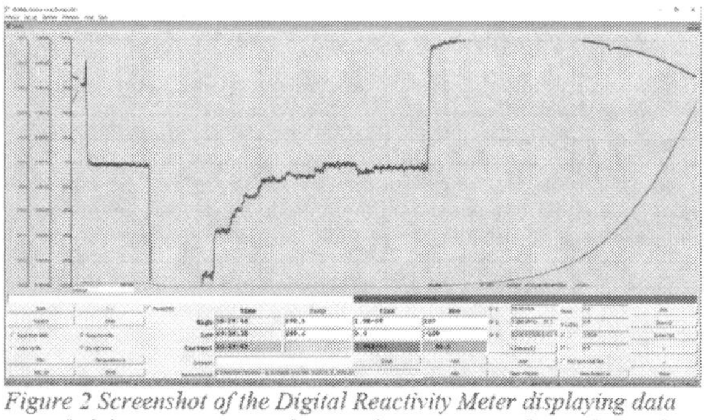

Figure 2 Screenshot of the Digital Reactivity Meter displaying data recorded during an approach criticality experiment by students of Uppsala University. The graph on the top displays the neutron flux in red, fuel temperature in green and the calculated reactivity in blue. The controls on the bottom control the plotting range, data logging and processing of the flux signal.

Students could point the dome camera at any area of the control room and zoom in or out to view the lecturer, the reactor operator's actions, or the nuclear instrumentation. The dome camera location is visible on Figure 6. The camera mounted above the reactor is used to see the bubbles being blown through the reactor core, the neutron source being removed, or the Cherenkov radiation (Figure 7). The dome camera and the camera mounted above the reactors were also both viewable from the television in the control room (marked "1." on Figure 4 and Figure 5). The remote desktop software was also used to control the digital acquisition software, the shared whiteboard, and Digital Reactivity Meter [6], an in-house developed software to monitor the neutron flux and solve inverse kinetic equations to calculate the reactivity of the reactor online. The Digital Reactivity Meter and the digital acquisition software were running on a dedicated computer ("3." on Figure 4 and Figure 5, screenshot on Figure 2).

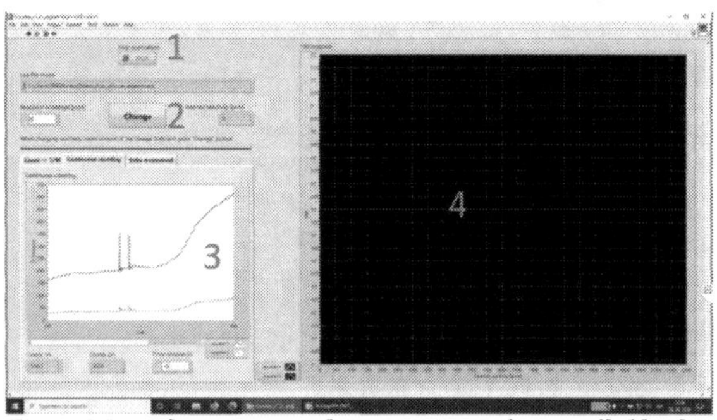

Figure 3 Digital acquisition software was remotely controlled by the students. The purpose of the application on screenshot above was to measure the number of counts in the start-up channel and assist in plotting the 1/M diagram. The students were able to start and stop the application by pressing the Start/Stop button (1), enter the inserted reactivity or differences in inserted reactivity (2) perform timed counts or view count rate as a continuous time plot (3), and view the current 1/M diagram (4), which is not visible on the screenshot.

The exercise program made extensive use of shared online spreadsheet, which was used by multiple students with different roles at the same time. In the control room, the shared spreadsheets (Figure 9) were sometimes viewable on one of the personal computers ("2." on the diagram on Figure 5) or the common whiteboard ("4." on the same diagram).

Figure 4 The students are communicating with the reactor operator and the lecturer, observing the operator movements and controlling a shared spreadsheet

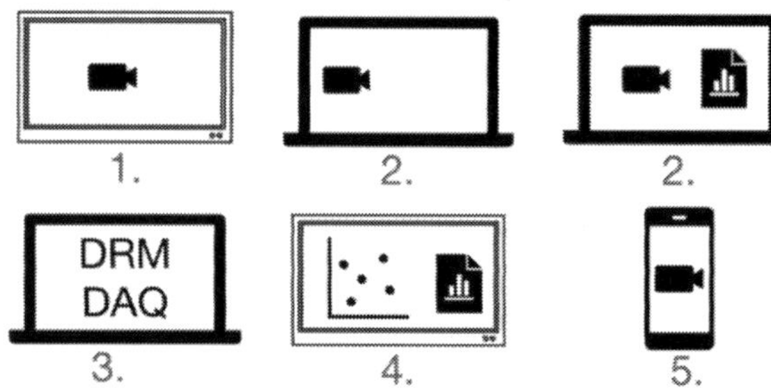

Figure 5 1. TV with dome/reactor camera, 2. Video conferencing/Spreadsheet laptops 3. Digital acquisiton/Digital Reactivity Meter laptop, 4. Common digital whiteboard, 5. Portable remote cameras

 https://doi.org/10.1051/epjconf/202125310001

Figure 6 The dome camera is mounted on the ceiling and positioned so it can see the reactor control panel, nuclear instrumentation readings, and the lecturer in front of the shared whiteboard.

Figure 8 We used smartphones with video conference software as portable cameras.

III. Uppsala University Exercise Scope

Jožef Stefan Institute has offered a standard five-day course in reactor physics for undergraduate students from Uppsala University. The course started with an introductory session where students became familiar with the reactor. After a presentation of the facility and an overview of the course, the instructors took the students on a virtual tour of the reactor hall using portable cameras with zoom capability. The first exercise was the criticality approach, where students were asked to give step-by-step instructions to the operators on rod movements and estimate the current position at each step.

Students were assigned roles to encourage their active participation. The students participating in the exercise can be seen on Figure 10. One of the students entered the neutron count rates measured at each step into a shared online spreadsheet and estimated the current reactivity. Another student used the same spreadsheet to draw an automatic graph of the inverse count rate versus the inserted reactivity. One of the students drew the same chart on a shared whiteboard, one used the dome camera mounted in the control room to monitor the control rod positions, and another communicated with the reactor operator. Students were able to request that the neutron source be removed from the reactor core and observe the process and the reaction of the detectors using remote

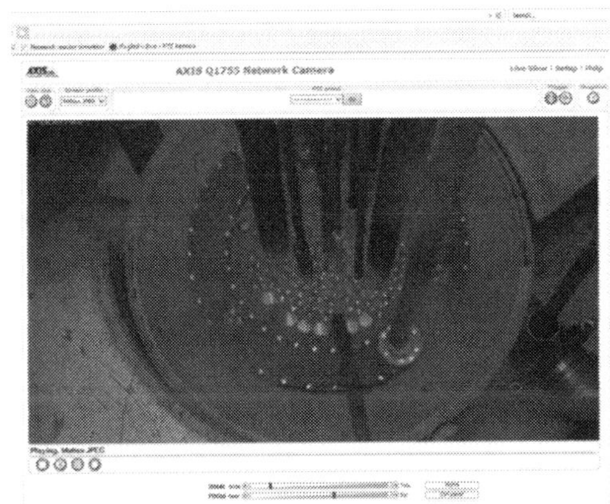

Figure 7 Cherenkov radiation is visible on the remote camera mounted above the reactor core. Students could view this camera at any time and use the zoom controls

cameras.

The second exercise dealt with the reactor's response to step changes in reactivity. As with the previous exercise, students were able to give operating instructions to reactor operators and observe the reactor's response to reactivity changes using remote cameras.

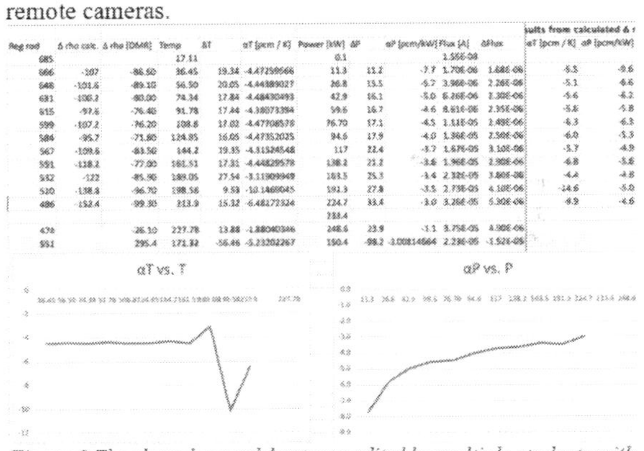

											sults from calculated Δ r	
Reg rod	Δ rho calc.	Δ rho (DMR)	Temp	ΔT	αT [pcm / K]	Power [kW]	ΔP	αP [pcm/kW]	Flux [A]	ΔFlux	αT [pcm / K]	αP [pcm/kW]
685			17.11			0.1			1.55E-08			
666	-107	-86.50	36.45	19.34	-4.47259566	11.3	11.2	-7.7	1.70E-06	1.68E-06	-5.5	-9.6
648	-101.6	-89.10	56.50	20.05	-4.44389027	26.8	15.5	-5.7	3.96E-06	2.26E-06	-5.1	-6.6
631	-100.2	-80.00	74.34	17.84	-4.48430493	42.9	16.1	-5.0	6.26E-06	2.30E-06	-5.6	-6.2
615	-97.6	-76.40	91.78	17.44	-4.38073394	59.6	16.7	-4.6	8.61E-06	2.35E-06	-5.8	-5.8
599	-107.2	-76.20	108.8	17.02	-4.47708578	76.70	17.1	-4.5	1.11E-05	2.49E-06	-6.3	-6.3
584	-95.7	-71.80	124.85	16.05	-4.47352025	94.6	17.9	-4.0	1.36E-05	2.50E-06	-6.0	-5.3
567	-109.6	-83.50	144.2	19.35	-4.31524548	117	22.4	-3.7	1.67E-05	3.10E-06	-5.7	-4.9
551	-118.2	-77.00	161.51	17.31	-4.44829578	138.2	21.2	-3.6	1.96E-05	2.90E-06	-6.8	-5.6
532	-122	-85.90	189.05	27.54	-3.11909949	163.5	25.3	-3.4	2.32E-05	3.60E-06	-4.4	-4.8
520	-138.8	-96.70	198.58	9.53	-10.1469045	191.3	27.8	-3.5	2.73E-05	4.10E-06	-14.6	-5.0
486	-152.4	-99.30	213.9	15.32	-6.48172324	224.7	33.4	-3.0	3.26E-05	5.30E-06	-9.9	-4.6
						233.4						
478		-26.10	227.78	13.88	-1.88040346	248.6	23.9	-1.1	3.75E-05	4.90E-06		
551		295.4	171.32	-56.46	-5.23202267	150.4	-98.2	-3.00814664	2.23E-05	-1.52E-05		

Figure 9 The shared spreadsheet was edited by multiple students with dedicated roles simultaneously. The students were entering the rod positions, changes in reactivity and temperatures and calculating the temperature and power reactivity coefficients (plots on the bottom part).

In addition to the rod positions and neutron flux monitors, students were able to use the remote camera mounted above the reactor core to observe Cherenkov radiation (visible in Figure 7) at higher powers and the remote desktop software

TeamViewer to operate and read out the Digital Reactivity Meter [Figure 3].

Figure 10 The students were using personal computers, larger monitors and a projector to both interact with the experiments and monitor the exercises

During reactor operation, each student starts up the reactor, brings it to full power, and shuts it down under the supervision of a licensed operator. Since direct remote control of the reactor is not possible, students use videoconferencing software to give instructions to the operator and remote cameras to monitor the reactor's response. The void reactivity coefficient is measured by blowing air bubbles into the reactor core and observing the effects on reactivity using the Digital Reactivity Meter [6]. In this exercise, the portable cameras were used extensively to show the details of the experimental setup on the reactor platform. The last two exercises were measurements of the value of control rods, where students measured and reconstructed calibration curves using either the rod swap method or the rod-in method, and a measurement of the temperature-reactivity coefficient, where students instructed operators to move the control rods over the point of active heat and recorded the change in fuel temperature and reactivity at each step.

IV. FEEDBACK INFORMATION

Feedback was obtained anonymously from students and their mentors using an online form. Instructors asked both choice-based questions and open-ended questions about the content of the exercises, staff performance, and technical execution of the remote exercises. Figure 11 is a summary of the results relating to the 'Technical content of the Educational Course", "Quality of lectures", "Quality of printed material and other material" and "Educational Course equipment", all of which were rated as 'Excellent' or 'Good'. The responses to the open-ended questions were also mostly positive. Participants praised the content of the lectures and the opportunity to actively participate in the exercise and experience the role of the nuclear reactor operator. Some participants noted that students were sometimes not given enough time to find the answers to the questions and stated that students sometimes felt lost, especially at the beginning of the exercise. One of the participants recommended that the lecturer should give clear, practical instructions before starting the experiments or give the students more time to solve the task themselves. The most frequently mentioned weakness was the quality of the audio. Several participants complained that they sometimes could not hear the lecturer when they moved away from the microphone.

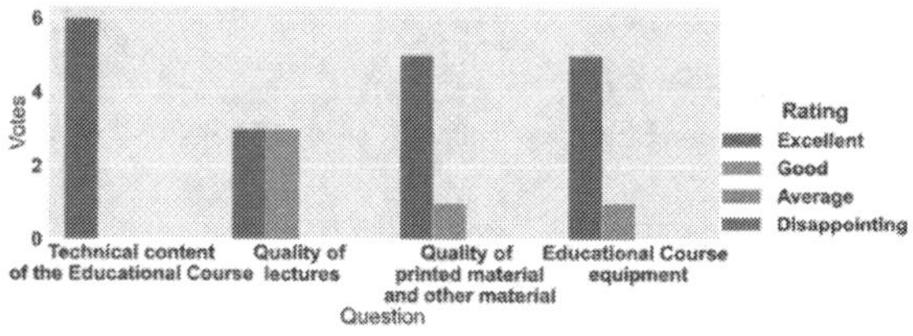

Figure 11 Survey results concerning content and quality of the lectures

The sound difficulties were also reflected in the ratings of the remote session on Figure 12, where the picture quality and latency were rated as 'Excellent' or 'Good' by all participants while the sound quality was also rated 'Average' or 'Disappointing'. Later, correction measures were taken and demonstrators wore wireless headsets.

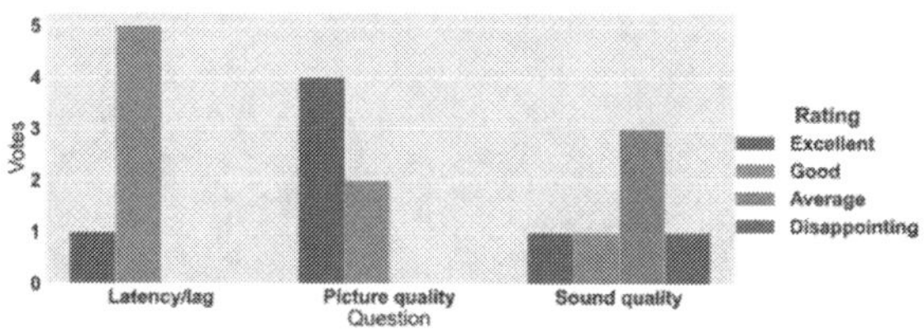

Figure 12 Survey results about the remote session

The participants agreed that such course should be organized routinely and that other nuclear workers and students could benefit from similar exercises. Five participants described the organization and administration of the Education Course as "Excellent" and one as "Good". When asked about the staff and their readiness to help with resolving their problems five participants answered again "Excellent" and one "Good".

V. RESEARCH REACTOR SIMULATOR

Jožef Stefan Institute also organised remote exercises in times of complete lockdown, when the reactor was not allowed to operate. This was done for students of physics at the University of Ljubljana with the help of the Research Reactor Simulator developed [3]–[5] by Jožef Stefan Institute. It solves point kinetic equations and contains a simple thermohydraulic model, which is well suited for simulating a research reactor with a small core and homogeneous neutron flux distribution. Students were able to perform all of the exercises that undergraduate students are expected to perform on a real reactor, i.e.: Approach to criticality, response to reactivity changes below the point of active heat, measurement of the temperature reactivity coefficient, and measurement of the power decay after rod drop.

VI. Conclusions

Students were able to perform all the planned exercises in the full scope in times of restricted travel and even during a full lockdown when the reactor was not allowed to operate. The authors of this article believe that this is incredibly important as the live exercises give students a different insight into reactor physics from the theoretical lectures. The remote exercises even have some advantages over the in-person exercises, especially lower cost because there is no need to travel, and the time agenda can be more flexible. Additionally, the simulator-based exercises guarantee that the measurements will be perfectly reproducible. Despite the advantages of remote exercises and simulator-based exercises, the instructors believe that they should be used exclusively as a backup option when in-person exercises are not possible. The lecturers observed that the students were asking fewer questions and did not participate as actively at the remote exercises compared to in-person exercises. The full predictability of the simulator output also takes away some of the educational value that comes from the exercises. Phenomena such as non-ideal neutron detector response or uncertainty in rod calibration curves are problems that nuclear reactor designers and operators have to recognize and know how to mitigate.

Acknowledgment

The successful execution of the exercises in given conditions is the result of a quick response to circumstances and active collaboration of multiple departments from different institutions. Special thanks go to the staff of the Reactor Infrastructure Center of Jožef Stefan Institute for adapting to the new exercise format and installing the necessary equipment, and to the students and professors from Uppsala University for working closely with the staff from Jožef Stefan Institute and making the exercises possible.

References

[1] L. Snoj, S. Rupnik, and A. Jazbec, 'New practical exercises at the JSI TRIGA Mark II reactor', in *PHYSOR 2014 International conference Proceedings*, vol. 2014, p. 12 str.

[2] L. Snoj, M. Kromar, G. Žerovnik, and M. Ravnik, 'Advanced methods in teaching reactor physics', *Nuclear Engineering and Design*, vol. 241, no. 4, pp. 1008–1012, Apr. 2011, doi: 10.1016/j.nucengdes.2010.02.040.

[3] J. Malec, D. Toškan, and L. Snoj, 'PC-based JSI research reactor simulator', *Annals of Nuclear Energy*, vol. 146, p. 107630, Oct. 2020, doi: 10.1016/j.anucene.2020.107630.

[4] 'Research Reactor Simulator official website', *Research Reactor Simulator official website.* http://reactorsimulator.ijs.si/ (accessed Oct. 01, 2019).

[5] D. Toškan, J. Malec, and L. Snoj, *Research reactor simulator : User manual*, vol. 12332. Ljubljana: Inštitut Jožef Stefan, 2017.

[6] I. Lengar, A. Trkov, M. Kromar, and L. Snoj, 'Digital meter of reactivity for use during zero-power physics tests at the Krško NPP = Uporaba digitalnega merilnika reaktivnosti pri zagonskih testih na ničelni moči v NE Krško', *Journal of energy technology.*, vol. 5, pp. 13–26, 2012.

The European Nuclear Experimental Educational Platform – ENEEP: Progress, Prospects and Remote Education Capabilities

Vladimir Radulović[1], Anže Jazbec[1], Luka Snoj[1], Ján Haščík[2], Branislav Vrban[2], Štefan Čerba[2], Jakub Lüley[2], Filip Osusky[2], Ľubomír Sklenka[3], Marcel Miglierini[3], Ondřej Novák[3], Helmuth Böck[4], Marcella Cagnazzo[4], Mario Villa[4], Szabolcs Czifrus[5], Attila Tormási[5]

[1]Jožef Stefan Institute, Slovenia
[2]Slovak University of Technology in Bratislava, Slovakia
[3]Czech Technical University in Prague, Czech Republic
[4]Technische Universität Wien, Austria
[5]Budapest University of Technology and Economics, Hungary
Corresponding author: vladimir.radulovic@ijs.si

***Abstract* — The European Nuclear Experimental Educational Platform – ENEEP is currently being established by five European educational and research organizations in the framework of a Horizon 2020 project, initiated in 2019. The ENEEP partner institutions are the Jožef Stefan Institute (JSI, Slovenia), the Slovak University of Technology in Bratislava (STU, Slovak Republic), the Czech Technical University in Prague (CTU, Czech Republic), Technische Universität Wien (TU Wien, Austria) and the Budapest University of Technology and Economics (BME, Hungary). ENEEP is intended as an open educational platform, offering experimental hands-on education activities at the ENEEP partner facilities. ENEEP education activities will be offered in different formats (group and individual) and are targeted at university students at all educational levels and young professionals in the nuclear field. This paper gives an overview of the ENEEP project activities and the progress achieved thus far, highlighting the experimental capabilities which will be offered. In the first implementation phase, ENEEP will be based on a comprehensive set of experiments comprising the basics in Reactor Physics and Nuclear Engineering curricula, as well as more specific experiments focusing on particular aspects – investigated phenomena, types and working principles of detectors, etc. Subsequently, novel education activities will be introduced and implemented in ENEEP, following scientific development in nuclear science and technology and nuclear instrumentation detectors, stemming from research activities. Attention will be devoted to the development and optimization of remote education capabilities at the ENEEP partner institutions, of particular relevance during the current Covid-19 pandemic, which is responsible for major changes in education activities worldwide.**

***Keywords* — Experimental Education activities, Nuclear Science and Technology, Remote capabilities.**

I. INTRODUCTION

EDUCATION activities in nuclear science and engineering are mainly based on theoretical lectures and exercises supplemented by modelling of real or simplified reactor systems by various computational codes. Nowadays, computer modelling is slowly replacing real experiments and, at most European universities, teaching of nuclear engineering is supplemented by the performance of only few basic experiments in radiation laboratories. This fact is very natural, as computer modelling is inexpensive compared to real experiments and it can be easily implemented into academic curricula without the need for complex laboratories requiring high operation costs and human resources. This trend is understandable, however, it should be considered that without real experimental education activities and without hands-on experience, the future nuclear engineers will suffer from a significant handicap in their professional careers. Experimental and hands-on experience is necessary to obtain high quality and complete nuclear education.

In nuclear education, research reactors are the most appropriate experimental facilities to achieve these goals. Critical and subcritical assemblies and training reactors are most suitable, as they are specifically designed for education and training. Research reactors in general are suitable for education activities at all academic levels, not only in nuclear engineering, but also in various non-nuclear engineering studies, such as power engineering, electrical engineering, natural sciences, medical sciences, physical sciences, etc. Experimental education at research reactors can significantly increase the quality of theoretical academic education and extend practical hands-on experience of students at real nuclear installations. In addition to research reactors, specific nuclear laboratories can be used to carry out experimental education and hands-on activities. In addition, it is also very valuable to have access to nuclear power plants, fuel cycle or decommissioning facilities where students can acquire experience with real nuclear installations, and obtain a full view on their complexity. Nowadays, for several reasons it is difficult to enable access to research reactors for students and their lecturers in order to provide them with a possibility of dedicated reactor physics experiments or hand-on reactor technology experience.

The mission of the ENEEP project [1-4] is to establish a platform at the European level which will fulfil the needs of European users. ENEEP – the European Nuclear Experimental Educational Platform – will contribute to significantly enhance the experimental education and hands-on activities in nuclear curricula, particularly in the fields of reactor physics, nuclear safety, radiation detection, radiation protection, spent nuclear fuel and radioactive waste management. The ENEEP platform will offer access opportunities to nuclear experimental facilities for university students at all academic levels (bachelors, masters and doctoral), professors, lecturers, experts in nuclear education, etc. It will allow specific nuclear training of professionals, particularly young professionals and post-docs at the beginning of their professional career, staff from governmental and non-commercially oriented companies, such as regulatory bodies, governmental organizations dealing with various aspects of peaceful use of nuclear energy, research institutions, etc. ENEEP is currently being established as an open platform for any European university, European research institute or nuclear training organization that is actively involved in experimental nuclear education, training and competence building. The ENEEP founding members are: the Jožef Stefan Institute (JSI, Slovenia), the Slovak University of Technology in Bratislava (STU, Slovak Republic), the Czech Technical University in Prague (CTU, Czech Republic), Technische Universität Wien (TU Wien, Austria) and the Budapest University of Technology and Economics (BME, Hungary) - institutions which are heavily involved in experimental nuclear education, training and competence building. Four institutions operate research reactors and the fifth one has access to specific laboratories for nuclear education and training. Fig. 1 – Fig. 5 display the facilities operated by the ENEEP project partners.

Fig. 1. *JSI TRIGA reactor, JSI, Ljubljana.*

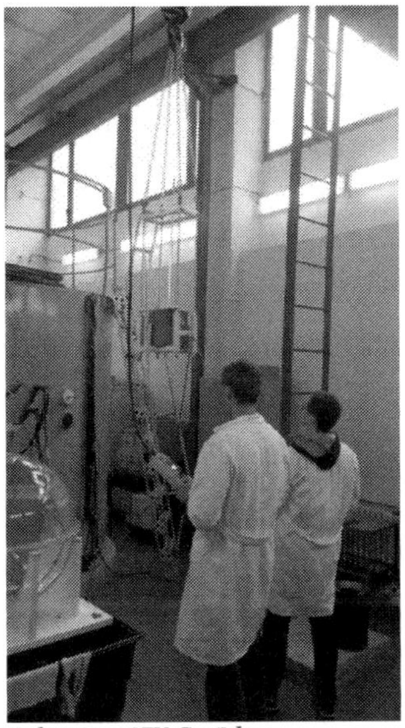

Fig. 2. *Radiation Laboratory, STU, Bratislava.*

Fig. 3. *Training reactor VR-1, CTU, Prague,*

Fig. 4. *TRIGA Mark II reactor, TUWien, Vienna.*

Fig. 5. *BME Training reactor, BME, Budapest.*

In the past, four ENEEP institutions, namely STU, CTU, TU Wien and BME, established in 2003 the Eugene Wigner course which connected three research reactors and offered experimental reactor courses for master students in nuclear engineering. The Eugene Wigner course ran annually for five years. Subsequently, four ENEEP institutions, namely CTU, TU Wien, BME, and JSI established in 2009 a Group Fellowship Training Program on Research Reactors which has been running under the Eastern European Research Reactor Initiative (EERRI). Training courses are performed regularly for International Atomic Energy Agency (IAEA) trainees. Up to now, 15 courses have successfully been accomplished and the 16th will be carried out in the autumn 2021. A six-week EERRI course is organised in two or three countries using two or three research reactors. ENEEP will be established based upon the existing experience of the current project partners originating from the Eugene Wigner courses, the EERRI courses as well as from the experience in experimental nuclear education, training and competence building activities carried out by the project partners. The 3-year ENEEP project was proposed in response to the 2018 EURATOM call for projects on the topic of "Availability and use of research infrastructures for education, training and competence building", and was kicked off in June 2019.

II. ENEEP Project – Progress and Prospects

The activities carried out within the ENEEP project are structured into work packages (WPs) devoted to distinct aspects:

- WP1: Post-grant requirements
- WP2: Consortium coordination and management
- WP3: Capabilities, users and SWOT analysis of ENEEP
- WP4: Establishment of ENEEP
- WP5: Demonstration of ENEEP
- WP6: Dissemination and feedback

WP1 and WP2, led by STU, are focused respectively on the Post-grant requirements set out by the European Commission (EC), and on the coordination and management of the ENEEP consortium and liaison with the EC.

WP3, led by TU Wien, covered the activities performed as a foundation of the ENEEP platform. The ENEEP partners have collected and compiled information in the form of databases covering the ENEEP facilities, the experiments to be offered within the ENEEP platform and the available instruments (e.g. detectors, equipment, etc.). A further important aspect covered in WP3 was the collection of information on potential ENEEP users in order to establish direct contacts, and for the purposes of promotion. Finally, a SWOT (Strengths, Weaknesses, Opportunities, Threats) analysis was performed, which allowed for the identification of key areas for the objectives of the ENEEP platform to be achieved, and thus providing directions for the ENEEP partners.

WP4, led by JSI, focused on the basis for the establishment of the ENEEP platform. The WP4 tasks are described in the following subsections.

A. ENEEP Agreement

The most important task of WP4 was achieving an agreement among the ENEEP partners, defining the platform's structure and legal status, the financial arrangements and responsibilities. The ENEEP partners have taken the decision to establish ENEEP as an association under the laws of the Republic of Slovakia. Work is in progress at present on the formal legal review of the ENEEP Statute and relevant documents. The ENEEP Association is expected to start running in mid-2022, following its formal establishment and the successful completion of the ENEEP demonstration.

B. ENEEP Application procedure

The purpose of the ENEEP application procedure is defining the application process and requirements, defining roles and responsibilities within ENEEP and criteria for application evaluation. The procedure will assist the ENEEP users in submitting applications to ENEEP; part of the content generated within this task will be made available publicly through the ENEEP web portal upon announcement of education and training activities.

C. Access requirement database

Access conditions to nuclear facilities are generally similar, however facility-specific regulations and procedures do exist.

An Access requirement database was created, synthetizing the access requirements and conditions for each partner-operated research facility, including facility-specific forms.

D. Procedures for group education and training activities

This task consisted in the definition of procedures for carrying out group education and training activities at the ENEEP partner institutions. Additionally, a dedicated template was created for the preparation of detailed instructions for each experiment, and presentation materials were created based on the outputs of WP3, providing a synthetic description of each experiment, including information of practical nature, e.g. education level, experiment duration, maximal number of participants, etc. ENEEP currently offers 61 experiments; the presentation materials are available on the ENEEP web portal at http://eneep.org/experiments/. Fig. 6 – Fig. 10 display examples of the presentation materials. It will be possible to prepare and implement additional experiments based on users' requirements.

Fig. 6. Example of experiment presentation materials – JSI, Measurement of the void coefficient of reactivity.

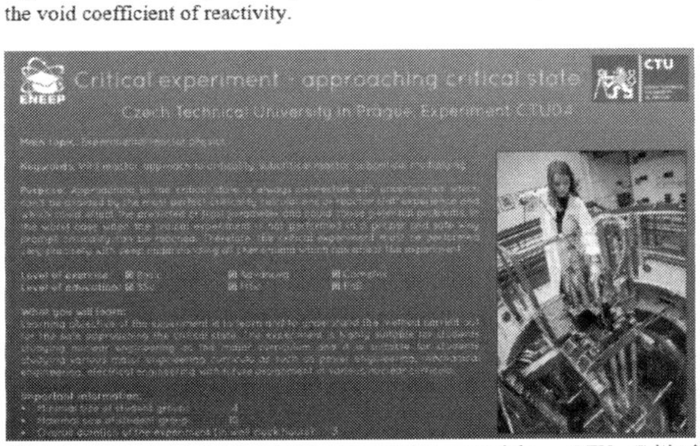

Fig. 7. Example of experiment presentation materials – CTU, Critical experiment.

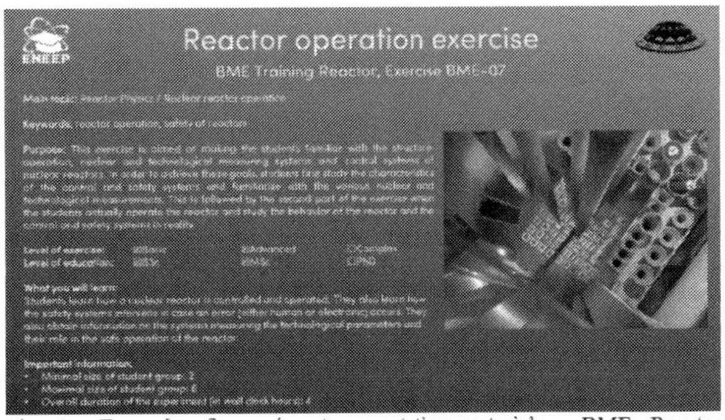

Fig. 8. Example of experiment presentation materials – BME, Reactor operation exercise.

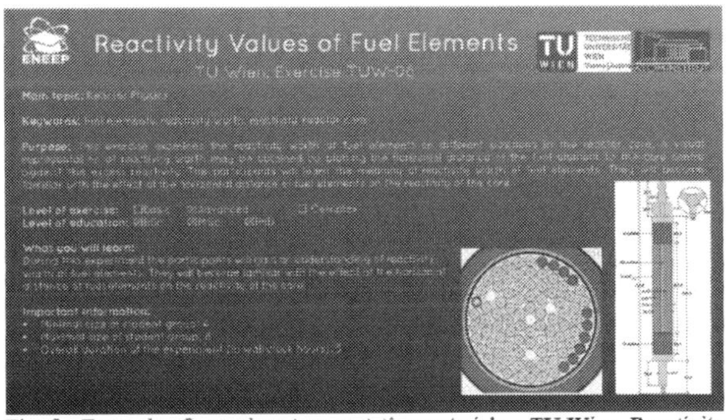

Fig. 9. Example of experiment presentation materials – TU Wien, Reactivity values of fuel elements.

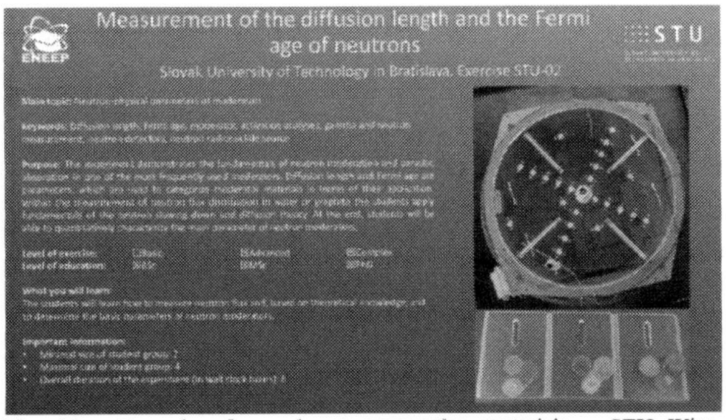

Fig. 10. Example of experiment presentation materials – STU Wien, Measurement of the diffusion length and the Fermi age of neutrons.

E. Shortlist of novel education and training activities

Recognizing the need for updating the ENEEP education and training activities with regard to scientific advancement, a search was made for novel education and training activities with realistic possibilities of implementation at the ENEEP partner facilities. The search involved professors, operating personnel and students; in the resulting shortlist, 17 novel education and training activities were identified. Three novel activities proposed by the JSI were already tested in practice in collaboration with nuclear engineering students of the University of Ljubljana, Faculty of mathematics and Physics:

- Cherenkov light intensity measurements as a means of reactor power monitoring,

- Neutron detection with SiC semiconductor detectors,
- Measurement of self-powered neutron and gamma detector signals.

F. Synthesis of implementation workflow and conditions for novel activities

The implementation of novel, non-routine experimental education activities at nuclear facilities is generally subject to prior fulfilment of technical and administrative requirements. The technical requirements are determined by operational limits and conditions in place at the facilities, the administrative requirements are laid-out in operational procedures / legislation. A synthesis was made of the implementation workflow in place at the ENEEP partner facilities with the aim to support the implementation of novel activities. In addition to the activities identified in the Shortlist of novel education and training activities, the synthesis supports the implementation of experiments at the ENEEP partner institution facilities in the framework of individual education activities, e.g. M.Sc. or Ph.D. theses or other. The synthesis describes the administrative procedures, the operating limits and conditions and presents the relevant technical data, e.g. the neutron flux levels, neutron spectra, irradiation facility dimensions, neutron beam dimensions, gamma dose rates, etc.

G. Facility operation schedule

In order to assist in the time slot selection and the prior preparation activities, the ENEEP project partners agreed to establish a facility operation schedule, published as a subpage on the ENEEP web portal: http://www.eneep.org/schedule/. The subpage provides simple information on the availability of the ENEEP partner facilities in the time period approximately between three months and one year from any given date. The schedule states the periods in which the facilities are either available, available in part, unavailable or unavailable due to long-term outages (> 2 months).

III. ENEEP Demonstration and dissemination activities

The work package WP5 led by CTU concentrates on demonstration of the ENEEP educational and training capabilities and its real educational potential. In early 2022, the ENEEP partners will organize:

- a 2-week course involving two ENEEP facilities for 10 master students which will be held from January 24, 2022 till February 4, 2022 at STU in Bratislava (first week) and CTU in Prague (second week)
- a 1-week course for 10 trainees at JSI (Slovenia) on February 14-18, 2022
- two individual activities involving two ENEEP facilities for 1+1 Ph.D. students or trainees, both during February 7-11, 2022 at TU Wien and BME Budapest.

The experience gained by the ENEEP project partners during WP5 will represent the basis for the organization of education activities in future ENEEP operation. ENEEP will offer:

- **Package courses** with pre-determined content, targeted at Bachelor and Master students, on an annual or semiannual basis. The package courses will consist of selected experimental activities covering a specific topic, determined by the ENEEP project partners, e.g. experiments demonstrating the basic principles of reactor control, nuclear instrumentation, radiation dosimetry or other.
- **Custom courses,** organized on the basis of users' requirements, targeted at closed groups of Master or Doctoral students from one university or institution. Custom courses will be based on experimental activities already established at the ENEEP facilities or on activities implemented specifically for this purpose.
- **Individual activities**, targeted primarily at Doctoral students and intended as a means of deepening the knowledge in one particular field in nuclear science and technology.

WP6 coordinated by BME, covers the dissemination of the results of the ENEEP project and the promotion of ENEEP within the European nuclear community and worldwide. In the framework of WP6 a dedicated ENEEP website was created (www.eneep.org) and is being continuously updated. ENEEP is also being promoted on the social media platforms LinkedIn (https://www.linkedin.com/groups/13834594/) and Facebook (https://www.facebook.com/ENEEP-108437037552996).

IV. Remote Education Capabilities

The relevance of remote education capabilities in the field of nuclear science and technology has been increasing in the last several years, in the context of limited availability of research and training facilities for in-person activities, and with increasing possibilities of content broadcasting. Remote education activities have several advantageous aspects:

- possibility of wider coverage in case in-person attendance is not possible, e.g. due to no or limited possibilities of traveling,
- streamlining and increased flexibility: less time is spent on practical issues related to traveling (e.g. registration, access to facilities), remote activities can effectively be included into educational programs,
- some aspects may benefit from a remote format, e.g. ensuring all students are able to adequately observe the performance of a particular experiment.

However, it must be recognized that remote education activities can never completely replace in-person attendance, as several important aspects are lost, i.e. the actual experience of visiting a nuclear facility, observing the rules and regulations for access and work in a radiation-controlled environment, the possibility to perform oneself the experiment and interact on a more direct and more personal level with instructors or facility staff.

The global Covid-19 pandemic has imposed significant difficulties and restrictions in the performance of in-person education activities at all education levels, including nuclear science and technology. The reality we are currently experiencing has led to significant effort devoted to restructuring education activities and transferring them to a remote format.

At present two ENEEP partners, the JSI and CTU have fully operational remote education capabilities through videoconferencing solutions. Thus far, the JSI performed a dedicated remote course in reactor physics for students of the University of Uppsala in September 2020, practical exercises in the framework of two 1-semester courses for students of the University of Ljubljana and foreign students enrolled in the SARENA programme [5] in winter and spring 2021, and participated with a presentation of the JSI TRIGA reactor and a practical exercise in the MOBIL-APP mobility programme [6], organized for first year master students at the Aix-Marseille University in July 2021. Figs. 11 and 12 display students of the University of Uppsala and demonstrators from the JSI during the performance of remote education activities.

Fig. 11. *Students of the University of Uppsala on location during the performance of a remote experiment – September 28th-October 2nd, 2020.*

Fig. 12. *Reactor operator and demonstrator in the JSI TRIGA control room conductin an remote experiment - September 28th-October 2nd, 2020.*

In collaboration with the IAEA, the Czech Technical University in Prague developed the Internet Reactor Laboratory (IRL) instrumentation for distance education, which connects an operating VR-1 reactor, as a host reactor, with universities in other countries, as a guest institution. This IRL instrumentation offers the opportunity to add real reactor experiments to the academic curriculum, particularly nuclear engineering and reactor physics, where students access an operating research reactor is not feasible. The IRL is in operation in Europe and Latin America, and it is under implementation in Africa and Asia. In Europe, the Training Reactor VR-1 replaced French reactor ISIS as a host reactor this year (2020). The first broadcasting for IAEA to four European and African countries will start in September 2021.

Furthermore, during the first wave of the COVID-19 pandemic in Europe, the IRL instrumentation was upgraded to broadcast outside the IAEA IRL area for groups and individuals. The upgraded IRL system was frequently used during the whole academic year 2020/21 for remote experimental education of Czech students. In 2021 remote educational courses were also carried out for international students from the Middlebury Institute of International Studies at Monterey, California, USA and students from several British universities through NTEC (Nuclear Technology Education Consortium, http://www.ntec.ac.uk/consortium.html). The IRL system also allows carry out distance nuclear training and research.

Fig. 13. *Remote experimental education for students from the VR-1 Reactor to students from the Middlebury Institute of International Studies at Monterey, California, USA in January 2021.*

The remote education capabilities implemented recently primarily due to the Covid-19 pandemic will be maintained by the ENEEP partners in order to support the ENEEP activities.

V. Conclusions

This paper presents an overview of the ENEEP project, aiming at the establishment of a platform at the European level offering experimental nuclear education activities in order to enhance nuclear education curricula. The ENEEP project was initiated in 2019 and achieved significant progress, leading to the formal establishment of the ENEEP Association in 2022, which will be responsible for the organization of education activities. In early 2022 the ENEEP project partners will organize several demonstration activities (group and individual) at the ENEEP partner facilities. The increasingly limited availability in Europe of nuclear facilities suitable also for research and training purposes, in particular research reactors, and more importantly the Covid-19 pandemic, led to increased relevance of remote education capabilities. At present two ENEEP partners, the JSI and CTU have fully operational remote capabilities. ENEEP will offer package and custom courses as well as individual activities to students at all educational levels as well as young professionals from governmental or non-profit organizations dealing with the peaceful use of nuclear energy, starting from mid-2022.

ACKNOWLEDGMENT

The ENEEP project has received funding from the European Union's Horizon 2020 research and innovation programme under grant agreement No. 847555. The JSI would like to acknowledge the support from the University of Uppsala, University of Ljubljana and the Aix-Marseille University in the performance of remote education activities.

REFERENCES

[1] V. Radulović et al, "European Nuclear Experimental Educational Platform – ENEEP", poster presentation at ANIMMA2019, Portorož, Slovenia, 2019.
[2] M. Cagnazzo et al., "The European Nuclear Experimental Educational Platform (ENEEP) for education and training", Atw Internationale Zeitschrift fuer Kernenergie, 65(5), 251-256, 2020.
[3] B. Vrban et al., "Concept of education and training provided by the European nuclear experimental educational platform (ENEEP)", *in Proc.* NENE 2020, September 7-10, Portorož, Slovenia, 2020.
[4] B. Vrban et al., "Introduction of European Nuclear Experimental Educational Platform (ENEEP)", CONTE2021, February 9–11, 2021, Virtual meeting, ANS, USA.
[5] J. Malec et al, "On Teaching Experimental Reactor Physics in Times of Pandemic", presented at ANIMMA2021, Prague, Czech Republic, 2021.
[6] C. Reynard-Carette, "A New Project dedicated to developing International MOBILity of APPrentices in the Master's Degree in Instrumentation, Measurement and Metrology (MOBIL-APP project)", *in Proc.* ANIMMA2019, Portorož, Slovenia, 2019.

Development of Active-Learning Units in Nuclear Engineering

Stephen U. Egarievwe
Alabama A&M University, USA
stephen.egarievwe@aamu.edu

***Abstract*—Active learning engages students in activities that could enhance their ability to analyze, synthesize, and evaluate the material being learned. Evidence-based studies have shown that active learning increases student performance in Science, Technology, Engineering, and Mathematics (STEM) courses. This paper presents the design of active learning units in nuclear engineering. The goal is to enhance students learning and technical skills, thereby improving their preparation for success in pursuing STEM graduate programs and careers in nuclear engineering. Three modes of active learning that are of interest are problem-solving, lab-based hands-on activities, and simulation. The active learning units are aimed at using interactive mode to provide students with the mastering of fundamental principles and concepts, and better understanding of how equations translate and apply to real-life engineering situations. The introductory nuclear engineering topics to be covered include radioactivity and half-life, binding energy, atom density, radiation interactions, radiation dose, radiation shielding, stopping power, and fission. An assessment plan for the effectiveness of the active-learnings units is also presented.**

***Keywords* —Active-Learning, Hands-On Learning, Nuclear Engineering, Problem-Solving, Simulation, STEM Education.**

I. Introduction

NUCLEAR engineers are very important to the technical workforce of the nuclear and radiological industry. One major problem facing STEM education is the production of nuclear engineers that will replace the aging workforce in the nuclear industry. A 2015 survey shows that a high percentage of the nuclear workforce in the United States are nearing retirement [1]. As part of the solution to future shortages in the nuclear workforce, there are efforts and recommendations for government entities and industries to partner with universities to broaden participation by improving diversity enrollments in nuclear pipeline programs [1 – 3]. There is increasing need to grow the nuclear workforce. As of 2017, there are 31 countries having 448 nuclear power plants [4]. Historically, nuclear power generation has been dominated by developed countries in North America, Europe, and Asia (particularly Russia), but increasing power demands has led to increase of the number of developing countries looking towards nuclear power generation, mostly by China, Bangladesh, India, and Pakistan [4]. As part of the solution to nuclear workforce development, 1) STEM programs need to expose students to early hands-on and the mastering of skills and science, engineering and technology principles needed in developing the nuclear workforce; and 2) there should be increased efforts in developing nuclear pipeline education initiatives that will enhance recruitment and retention activities, industry collaborations, and graduate-school/job placement drives. Some logical questions on nuclear workforce development and solution approaches from academic point of view are listed in Table 1.

TABLE I
KEY QUESTIONS AND SOME SOLUTION APPROACHES IN NUCLEAR WORKFORCE DEVELOPMENT

	Question	Solution Approach and Activities
1	How can STEM curriculum enhance nuclear engineering skillset development?	Integrated active learning and hands-on technical training.
2	How can we increase the retention and graduation rates of women and underrepresented minorities in nuclear engineering programs?	Early alert and mentoring based on continuous assessment with improvement feedback.
3	How can STEM programs address the problem of replacing the aging nuclear workforce?	Students-industry bridge activities and the broadening of participation of underrepresented minorities.

One of the successful models used in developing nuclear workforce pipeline is the Vertical Education Enhancement (VEE) model [5] that promotes collaborations between academic institutions, industries, government agencies, and local/regional communities, and has adequate funding and research supports. The VEE model has been used to develop a Nuclear Education and Research Vertical Enhancement (NERVE) program that spans recruitment outreach to secondary schools (middle/high schools) and two-year colleges, multitrack technical training in undergraduate programs, graduate school, and postdoctoral training [5]. The VEE model was also used successfully in faculty development and collaborations among international institutions [6]. This paper focuses on the curriculum development aspect of the VEE model where active-learning approaches are used to support technical training and skillset development. The goal is to

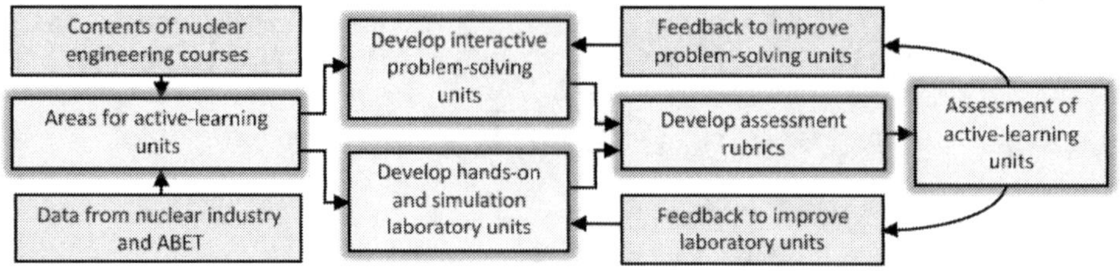

Fig. 1. Approach and methodology for developing and incorporating active-learning units into nuclear engineering courses. ABET is the Accreditation Board for Engineering and Technology.

develop innovative active-learning units to enhance undergraduate STEM education in nuclear engineering curricula.

Some of the recent efforts in nuclear workforce development are focused on the recruitment and undergraduate training stages of the education pipeline [7,8]. According to Karayaka *et al.* [7], one-on-one interaction with undergraduate students is the most effective tool in recruiting into nuclear scholarship programs. A survey of 31 nuclear scholarship students showed the about 80% of then were recruited through one-on-one with faculty advisors and mentors [7]. Other recruitment efforts used include emails, flyers, and peer interactions. In a project-based learning program for nuclear workforce development, students take courses in *Engineering Practices* and *Engineering Capstone* (an *engineering senior design project* class), along with professional training and technical classes that place emphasis on power generation [8].

II. Design Approach for the Active Learning Units

Active learning model is a learning model that engages students in class and/or laboratory activities that promote problem-solving [9], analysis, synthesis and evaluation of the material being learned. In active learning, the teaching is structured for students to participate in doing things instead of just listening. Teaching practices play major role in improving student learning in STEM education [10,11]. Evidence-based data confirms that active learning increases student performance in STEM courses [12]. In a study by Freeman *et al.* [12], Kernel density plots showed a 12% mean improvement in student learning in an active-learning classroom over a lecture classroom.

The three modes of active learning that are of interest in the present work include a) participation of student groups in problem-solving, b) participation of student groups in hand-on skill development, and c) the use of simulation in teaching and learning. Fig. 1 shows the approach and methodology that will be used in this work for developing and incorporating active-learning units into nuclear engineering courses.

A. Problem-Solving Student Groups

The problem-solving interactive units are aimed at mastering fundamental principles and concepts and better understanding of how equations/formulae translate and/or apply to real-life engineering situations. It also enhances the understanding of how different parameters in an equation interact with each other (such as dependency relationships). The practicality aspect is understanding how different components of an engineering systems function together to accomplish the goals for which they are designed. The problem-solving interactive units to be developed at the introductory nuclear engineering level are as follow:

i. Radioactivity and half-life.
ii. Binding energy.
iii. Atom density.
iv. Radiation interactions.
v. Radiation dose.
vi. Radiation shielding.
vii. Stopping power.
viii. Fission.

B. Hands-On Student Groups

This element of the active-learning units will enhance skill-set development in nuclear engineering. Lab experiments will be designed to support nuclear skill-set development. The lab units will cover mostly nuclear instrumentation techniques and methods. See Table II.

TABLE II
TOPICAL AREAS FOR HANDS-ON ACTIVE-LEARNING UNITS

Unit	Hands-On Lab	Knowledge and Skills to be Acquired
1	Nuclear Instruments Electronics	Understand and master basic nuclear instrumentation electronics and techniques in radiation measurements.
2	Geiger-Mueller Counters	Half-life determination, linear absorption coefficient, and counting statistics.
3	Calibration of Gamma Source and Radiation Survey Meters	Skills to calibrate a gamma standard source and use that information to calibrate radiation survey meters.
4	Gamma Spectroscopy Using NaI Scintillation Detectors	Master the use of NaI scintillation detectors. Radio-isotope identification skills: identification and composition of an unknown nuclear material.
5	Measurement of Dose Using TLD and other Dosimeter Systems	Understand the use of thermo-luminescent (TL) material and other dosimeter systems to measure dose. Calibration curve for gamma rays on TL dosimeters.
6	Radiation Survey/Mapping	Skills on surveying an area to locate hidden radioactive materials. Skills to survey areas of contaminations applicable to nuclear power plants.
7	Radiation Shielding	Know how to apply the principles of As Low As Reasonably Achievable (ALARA) in radiation protection as practiced in nuclear plants.
8	Neutron Detection	Skills on neutron activation measurements of trace elements.

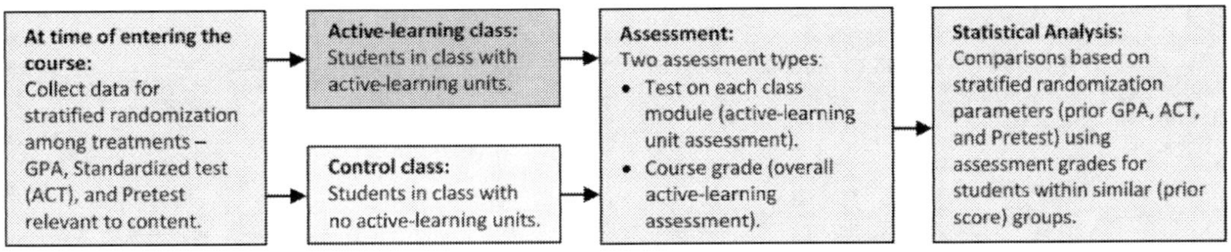

Fig. 2. Approach and experimental design for the research on the effect of active-learning on nuclear skill-set development.

C. *Simulation of Nuclear Systems*

Simulation is particularly very important for concepts and systems that cannot be made available in a student lab setting, such as reactor startup and shutdown. Simulation helps to enhance skill-set development, where students use software packages to design and simulate nuclear systems. Examples of simulation topical areas are:

i. Neutron interaction and flux distribution in a nuclear reactor core.
ii. Nuclear heat generation and heat transports in reactor elements.
iii. Power plant cooling systems.
iv. Power generation system efficiency.

III. Assessment Plan for Active Learning Units

The assessment plan designed for studying the effects of active learning on skillset development in nuclear engineering is shown in Fig. 2. It schematically shows the approach and experimental design for testing the active learning unit. To ensure reliability and consistency, the following criteria could be used to govern the students that will participate in the study groups:

i. Student Equivalence: Stratified randomization will be used to ensure that the comparison groups (active-learning group and control group) are similar as much as possible in terms of characteristics that may influence the outcome of the active-learning units. The parameters that will be used for student equivalence are GPA, standardized test (ACT), and pretest relevant to course module content. Students with equal number of score and GPA levels will be randomly assigned to the comparison groups: active-learning group and control group.
ii. Instructor Equivalence: According to Freeman et al. (2014), instructor equivalence means "whether the instructors in the lecture and active learning treatments were identical, randomly assigned, or consisted of a group of three or more in each treatment". In this study, the same instructor will be in both the active-learning group and control group classes. This meets the identical instructor requirement for instructor-equivalence.
iii. Examination Equivalence: The assessments that will be given to students in the active-learning group and the control group will be identical.

The assessment tools include posttests (for modules in a course), tests and exams. The assessment parameters are 1) post module test scores for assessing the effects of each active-learning unit (formative evaluation), and 2) class grade for assessing the overall effects of the active-learning model (summative evaluation).

IV. Conclusions

This paper presented the design for developing active learning units to enhance students learning and technical skillset development in nuclear engineering. The active learning modes discussed include problem-solving, lab-based hands-on activities, and simulation. These are interactive learning modes that provide students with the mastering of fundamental principles and concepts, and better understanding of how equations translate and apply to real-life engineering situations. The topics to be covered at the introductory nuclear engineering level include radioactivity and half-life, binding energy, atom density, radiation interactions, radiation dose, radiation shielding, stopping power, and fission. In future work, the active learning units will be created, implemented, and tested. The effectiveness of the active-learning units on student learning will also be assessed.

Acknowledgment

This work was supported in part by the National Science Foundation (NSF) HBCU-UP Program through award number 1818732, the U.S. Department of Energy (DOE) National Nuclear Security Administration (NNSA) award number DE-NA0003980, and the U.S. Nuclear Regulatory Commission (NRC) through award 31310018M0035.

References

[1] E. McAndrew-Benavides, "NEI's 2015 Nuclear Workforce Survey and Results," presented at the *Nuclear Energy Institute (NEI) 2015 Nuclear Workforce Survey*, Rockville, MD, USA, Oct. 19, 2015.
[2] American Physical Society, "Readiness of the U.S. Nuclear Workforce for 21st Century Challenges," American Physical Society (APS) Panel on Public Affairs, Committee on Energy and Environment, June, 2008. Available online: https://www.aps.org/policy/reports/popa-reports/upload/Nuclear-Readiness-Report-FINAL-2.pdf. Accessed on: August 14, 2021.
[3] Center for Energy Workforce Development, "Gaps in the Energy Workforce Pipeline 2015 CEWD Survey Results," Available online: https://www.cewd.org/documents/surveyreport/CEWD2015SurveySummary.pdf. Accessed on: August 14, 2021.
[4] F. Alam, R. Sarkar and H. Chowdhury, "Nuclear power plants in emerging economies and human resource development: A review," *Energy Procedia*, vol. 160, pp. 3–10, Feb. 2019, 10.1016/j.egypro.2019.02.111.
[5] S. U. Egarievwe, "Vertical education enhancement–a model for enhancing STEM education and research," *Procedia-Social and Behavioral Sciences*, vol. 177, pp. 336–44, Apr. 2015, 10.1016/j.sbspro.2015.02.354.

[6] S. U. Egarievwe and R. B. James, "Faculty Development and International Collaborations Using Vertical Education Enhancement Model," presented at the *2015 ASEE International Forum*, Seattle, WA, USA, June 14, 2015. Available online: https://peer.asee.org/faculty-development-and-international-collaborations-using-vertical-education-enhancement-model. Accessed on: August 14, 2021.

[7] H. B. Karayaka, C. W. Ferguson and A. C. Thompson, "An Evaluation of Focused Outreach and Recruiting Efforts in a Nuclear-related Workforce Development Program," presented at the *2020 ASEE Virtual Annual Conference Content Access*, Virtual Conference, June 22 – 26, 2020. Available online: https://peer.asee.org/an-evaluation-of-focused-outreach-and-recruiting-efforts-in-a-nuclear-related-workforce-development-program. Accessed on: August 14, 2021.

[8] H. B. Karayaka, C. W. Ferguson and A. C. Thompson, "Project-based Learning Program for Nuclear Workforce Development Phase II: Implementation," presented at the *2021 ASEE Virtual Annual Conference Content Access*, Virtual Conference, July 26 – 29, 2021. Available online: https://peer.asee.org/project-based-learning-program-for-nuclear-workforce-development-phase-ii-implementation. Accessed on: August 14, 2021.

[9] J. R. Savery, "Overview of Problem-Based Learning: Definitions and Distinctions," in Essential Readings in Problem-Based Learning, A. Walker, H. Leary, C. E. Hmelo-Silver and P. A. Ertmer (Editors). West Lafayette, IN, USA: Purdue University Press, 2015, pp. 5–15.

[10] M. K. Smith, F. H. Jones, S. L. Gilbert and C. E. Wieman, "The Classroom Observation Protocol for Undergraduate STEM (COPUS): A New Instrument to Characterize University STEM Classroom Practices," *CBE—Life Sciences Education*, vol. 12, no. 4, pp. 618–27, Dec. 2013, 10.1187/cbe.13-08-0154.

[11] L. M. Cleveland, J. T. Olimpo and S. E. DeChenne-Peters, "Investigating the relationship between instructors' use of active-learning strategies and students' conceptual understanding and affective changes in introductory biology: A comparison of two active-learning environments," *CBE—Life Sciences Education*, vol. 16, no. 2, ar19, Jun. 2017, 10.1187/cbe.16-06-0181.

[12] S. Freeman, S. L. Eddy, M. McDonough, M. K. Smith, N. Okoroafor, H. Jordt and M. P. Wenderoth, "Active Learning Increases Student Performance in Science, Engineering, and Mathematics," Proceedings of the National Academy of Sciences, vol. 111, no. 23, pp. 8410–8415, Jun. 10, 2014, 10.1073/pnas.1319030111.

EASY: Educational Alibava System

Carmen García[1], José Bernabeu[1], Juan Herranz[3], Carlos Lacasta[1], Manuel Lozano[2], Salvador Martí-García[1], Giulio Pellegrini[2] and Miguel Ullán[2]

[1]Instituto de Física Corpuscular-IFIC (CSIC-Universidad de Valencia)
[2]Instituto de Microelectrónica de Barcelona-IMB-CNM (CSIC)
[3]Alibava System
Corresponding author: Carmen.Garcia@ific.uv.es

***Abstract*—**

EASY, a plug-and-play educational system, is portable, compact and a complete system for micro-strip sensor characterization. Ideal for making basic or complex experiments. It is based on the Classic Alibava System [1], largely used within the CERN community to test micro-strip detectors for particle experiments. The system can be configured to work with pulsed laser light or radioactive sources.

The aim of this system is to illustrate students in the operation of a silicon strip detectors

The components of the EASY systems are the Control Unit and the Sensor Unit. The Control Unit is the heart of the system communicating with the Sensor Unit and the Computer software. It contains the Data Acquisition Control and it is also in charge of processing of the sensor data and trigger inputs. In addition, it contains an adjustable Hight Voltage unit for micro-strip sensor bias, with voltage and current display and includes the laser source. The Control Unit communicates with computer software via USB. The Sensor Unit accommodate a p-on-n silicon micro-strip sensor segmented in 128 strips.

EASY comes with an activity book where the students, through 10 exercises, are introduced in the main concepts and functionalities of micro-strip silicon detectors, used in the actual particle physic experiments. The book also provides a full description of the EASY device and the data Acquisition system.

***Keywords* —Education, ionization detector, micro-strip sensor, .data acquisition and processing, pedestal, noise, signal cluster, charge collection, depletion voltage, charge sharing, laser penetration.**

I. Introduction

EASY , the Educational Alibava System (EASY), a plug-and-play educational system, is portable, compact and a complete system for micro-strip sensor characterization, ideal for making basic or complex experiments with silicon micro-strip detectors. The students will be introduced in high-energy physics and particle detectors with hands-on experiences. They will be familiarizing with concepts such as MIP (Minimum Ionization Particles), charge collection, full depletion and charge sharing in strip detectors, among others concepts.

II. The EASY System Components

The components of the EASY systems are the Control Unit (system's main board) and the Sensor Unit (containing the detector board), Fig. 1 shows the EASY components. The Control Unit is the heart of the system communicating with the sensor unit and the computer software. It contains the Data Acquisition Control and it is also in charge of processing of the sensor data and trigger inputs. In addition, it contains an adjustable High Voltage unit for micro-strip sensor bias, with voltage and current display and includes the laser source. This laser source is a pulse laser of 5ns pulse width and a wavelength of 980 nm. It is used to induce charge in the detector. The laser peak power is 0.6m W and its spot around 20 µm. The Control Unit communicates with computer software via USB.

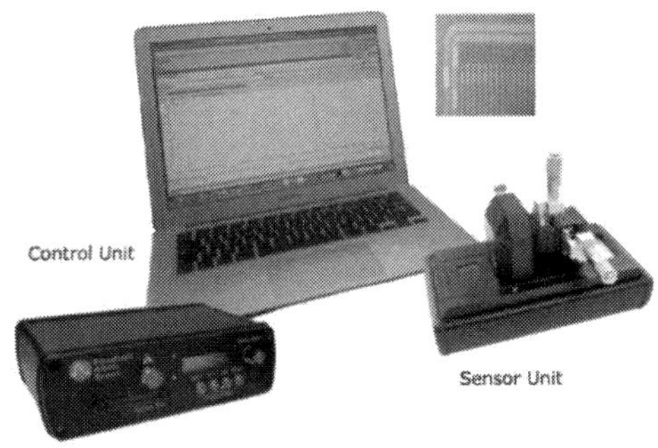

Fig. 1. EASY system with the Control Unit, the Sensor Unit and the acquisition program *Alibava-gui* running in the computer.

The Sensor Unit (Fig. 2) accommodate a p-on-n silicon micro-strip sensor, the same type of sensor used in actual ATLAS detector at LHC. P-on-n strip detectors consist typically of an around 300 µm thick n-type bulk. On one side, a thin layer segments form strips out of p-type material, whereas the back implant is n+. The abrupt junctions are formed between the strips and the bulk silicon. The depletion zone grows from the strips to the backplane so that, the device can operate partially depleted and the movement of holes forms the

strips with a pitch of 160 μm. The Sensor Unit has a mobile part that slides over the box walls with horizontal and vertical micro-positioners screws. It has two positions to use the device with radioactive source or with laser light. In order to use it with radioactive source, the user will place the source on a carbon fiber rectangle aligned with the micro-strip sensor. When the device is use with the laser, the moving part of the Sensor Unit slides to focus the laser on the detector micro-strip sensor. The two micro-positioners allow to the movement of the laser lens in 10μm steps, horizontally and vertically, to focus the laser and to move the laser beam over the sensor, perpendicular to the strips.

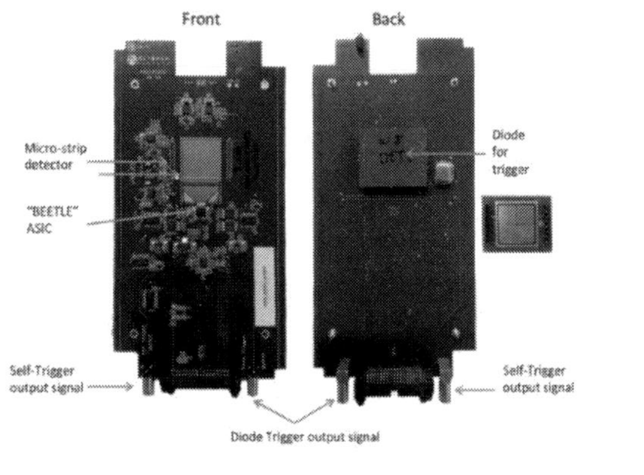

Fig. 2. Sensor Unit holding the micro-strip detector and the BEETLE in the front side and the Diode for trigger proposes in the back. The Board have two connectors for the trigger signal: Diode Trigger and Self-Trigger.

III. Principle of Operation of MIPs

A MIP particle that traverses a silicon detector loses energy mainly by ionization. When the ionizing particle passes through the silicon lattice (see sketch on Fig. 3), this will tend to liberate a loosely bound valence electron generating an electron-hole pair. The free charge carriers formed by ionization in the silicon will move in the detector field in the depleted part of the detector. This is because the ionization in the depleted region is not recombined and the charges drift in the electric field. While in the non-depleted part, there is not electric field, charge will be recombined. Therefore, only the charge generated in the depletion part will contribute to the signal [3]. The width of the depletion zone is given by:

$$d = D\sqrt{\frac{V}{V_{dep}}} \qquad (1)$$

where D is the detector thickness, V de bias voltage applied to de sensor and V_{dep} the voltage to reach full depletion. Thus, the charge collected of a silicon detector will be proportional to:

$$\begin{matrix} \sqrt{V/V_{dep}} & 0 \le V \le V_{dep} \\ 1 & V > V_{dep} \end{matrix} \qquad (2)$$

The signal generated in silicon detectors is generally of small amplitude, around 24000 electrons for full depleted $300\mu m$ sensor, hence amplification and shaping stages are required for its further processing. The read out and amplification of the signal of the detector is done by BEETLE chip [2], an ASIC (Application-Specific Integrated Circuit) developed for CERN LHC experiments.

IV. Trigger Modes

The system incorporates three trigger modes: two asynchronized triggers (an external trigger and a self-trigger) and a synchronized trigger.

The external trigger, consists of discriminating a signal generated by a low noise silicon diode (Fig. 2 and 3). The silicon diode is mounted underneath the micro-strip sensor with a gap of few millimeters. When a particle crosses the micro-strip and reaches the diode, the generated charge in the diode is discriminated to produce the trigger signal that starts the acquisition. The Diode Trigger mode is suitable for MIPs when studying the passage of particles through matter.

The self-trigger signal is generated by the fast comparator incorporated in the BEETLE chip. Each one of the 128 comparators has a different offset and response. Thus, an individual calibration of each comparator is needed in order to calculate the individual threshold value for each channel. With this calibration the a uniform for all channels is obtained. The EASY software provides an automated procedure to generate and calibrate all the thresholds. The self-trigger mode is convenient for acquiring data when the incident particles are completely absorbed, as it can be the case of photons.

Finally, the synchronize trigger consists in a pulse generated to trigger the laser, that should be synchronized before data acquisition.

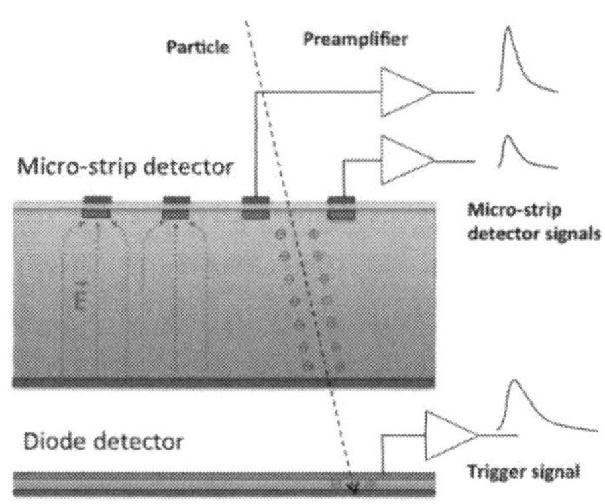

Fig. 3. Sketch of signal formation and the EASY diode trigger.

V. Data Acquisition and Monitoring

The software for data acquisition and monitoring (*Alibava-gui*) is a graphical user interface, written in C++, that controls the Control Unit. It is able to configure the sensor device, receive the data via USB bus and store it in a file for further analysis. *Alibava-gui* can be run in Windows, Linux or MacOs.

This software also monitors the data during data acquisition, displaying parameter as: pedestal (ADC offset), electronic and common noise, gain, pulse shape, collected charge, single events per channel, etcetera. The software allow to demonstrate the basic principles of silicon sensor operation during the data acquisition, or, as in a real experiment, to record the data for a detailed analysis.

The system has five different run modes:

- *Calibration*: *Alibava-gui* programs the BEETLE chips to inject calibration pulses to all channels in order to characterize the electrical behavior of the ASIC. The calibration also allows obtaining the gain of the ASIC and the ADC to number of electron calibration curve.
- *Laser synchronization*: *Alibava-gui* sends a pulse that can be used to trigger the laser. This run mode scans the delay between the pulse sent by the Control Unit to the laser and the data acquisition.
- *Pedestals Run*: *Alibava-gui* generates an internal trigger that will allow computing the baseline for pedestals and its variation (i.e. the electronic noise).
- *Laser Run* makes a run where a laser beam injects the charge; one needs to run the laser synchronization mode beforehand in order to read back the optimal signal produced by the laser.
- *Radioactive Source Run* makes a run in which the acquisition is triggered by signals above the threshold in the input connectors suitable to work with radioactive source.

VI. Some Exercises

The EASY system illustrates the operation of a silicon strip detector with LHC readout electronics in 10 EXERCISES. Teachers can select the exercises according with the student level, from the basic principles of silicon detector operation to high level concepts on detector operation, electronics and data processing. Some of the aspects that can be covered are:

- Observation of electronic noise of a silicon strip detector as a function of bias voltage.
- Study of signal spectra due to MIPs crossing the silicon detector and the demonstration the Landau distribution shape of collected charge.
- Measurement of Charge Collection for different bias voltage and calculation of depletion voltage.
- Illustration of charge sharing between strip and relate this to the position resolution of the detector.
- Verification of the structure of a typical micro-strip detector.
- Comparison of the charge deposition of MIP particles crossing the detector and laser light stopping in the sensor bulk
- And, the calculation the laser penetration in the silicon.

Some of these exercises are described in detail below.

A. Charge collection for a MIP particle

The students can investigate the charge collected by a silicon micro-strip sensor when a MIP goes through it and triggering the diode behind the silicon sensor. The MIPs used are 0.546 MeV electrons irradiated by a ^{90}Sr source.

Before studying the charge formation in a silicon sensor, the student should understand the concept of pedestal, electronic noise (its dependence with the reverse voltage bias) and the common noise. The pedestal value of the ADC output is an effective offset. Namely, the pedestal value is the measured ADC counts when there is no signal at the input. In the EASY system the ADC offset has been set to the mean range of the ADC, around 500 ADC counts, to allow read positive and negative signals. The pedestal fluctuations are due to electronic non-uniformities of each strip. There are two sources of noise. The common mode shift or common noise, resulting for an electronic parasite pick-up which is specific of the chip and varies with time, fluctuating event by event. And, the electronic noise, defined as the fluctuation channel by channel. The common noise, the mean electronic noise and the pedestals are constantly monitored during the data acquisition.

Since the deposited charge can be collected in adjacent strips, a cluster algorithm has been implemented to integrated the charge deposition.

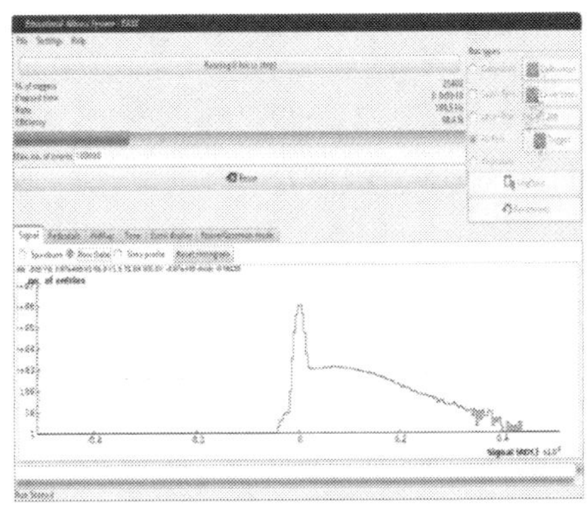

(a)

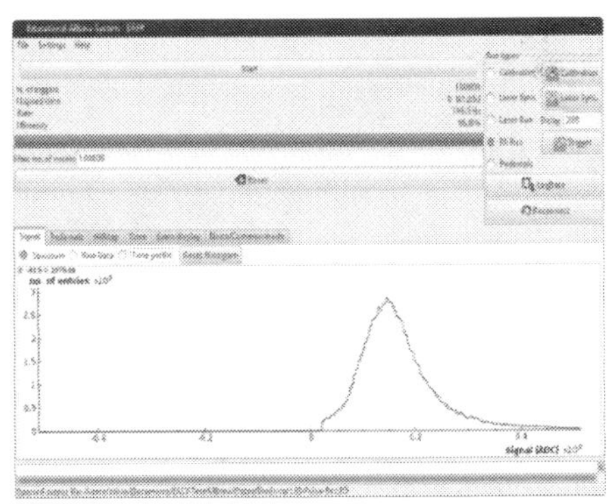

(b)

Fig. 4. Charge spectrum, in ADC count, of a Sr-90 source, (a) raw data in logarithmic scales and (b) after data analysis.

Fig. 4 (a) shows, in logarithmic scale, the accumulated charge (raw data) during the data acquisition. The large pick corresponds to electronic noise of the channel without charge deposition and the tail contain the collected charge. After data analysis (Fig. 4 (b)), one can get the charge spectrum showing

https://doi.org/10.1051/epjconf/202125310004

an asymmetric distribution. As mention in section III, when the ionizing particle passes through the silicon he lattice, this will tend to liberate a loosely bound valence electron generating an electron-hole pair. Occasionally the ionizing particle will interact with a more tightly bound electron than the silicon valence electrons and, as a result, will create an electron with a significant amount of kinetic energy. Due to these rare high-energy events, the energy spectrum of the deposited energy in the silicon detector is a non-symmetric distribution. The distribution could be fit by a Landau function convoluted with a Gaussian distribution, and obtaining the more probably charge from the peak of the distribution. Notice that the more probable charge differs from the mean charge (the more probable charge is 0.7 times the mean charge). After calibration (converting ADC counts in number of elections) and considering the average energy neeed to produce a single electron-hole in silicon, one can obtain that the mean energy deposition of a MIP in 300 µm is around 110 KeV.

Since only the charge generated in the depletion part of the sensor contributes to the signal. Varying bias voltage, the students can observe how the charge spectrum variate proportionally to the deplete area of the sensor. By measuring the more probable charge for different depletion voltages, they can obtain the characteristic curve of charge collection as a function of the bias voltage (equation (2)). Fig. 5 shows how the charge collected grows with bias voltage till a plateau is reached when the sensor is fully depleted. A bias voltage over 80V, is needed for full depletion.

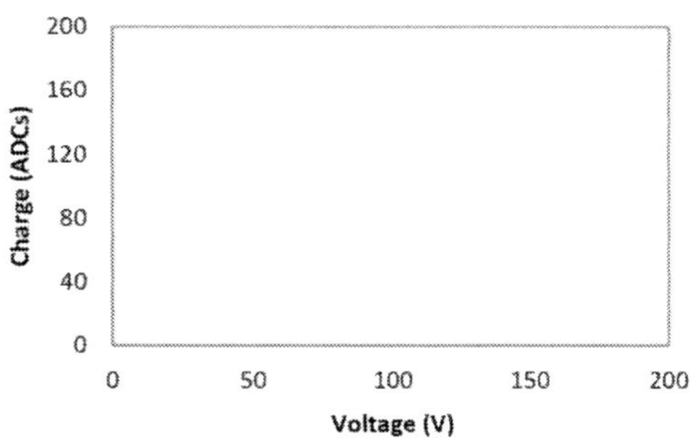

Fig. 5. The plot shows how the charge collected, in ADC counts, increases with bias voltage till a plateau is reached when the sensor is fully depleted.

B. *Strip Structure and Charge Sharing*

EASY system incorporates a 980 nanometers pulsed laser, to induce charge in the micro-strip sensor, mimicking the incidence of a charged particle in the detector. Two micro-positioners screws allow to movement the laser lent in 10 micro-meter steps, horizontally and vertically. The vertical micro-meter screw is used to focusing the laser while the horizontal screw will allow to move the laser bean across the strip. One should be aware that laser beam is reflected in aluminum strips. So, if the laser beam insides over the strip no charge will be collected.

Starting for an initial incident position of the laser beam, for instance over strip 40. One can measure the mean charge in the channel 40 and the adjacent strips, strip 39 and 41. In this initial position since laser beam is reflected in the aluminum strip a charge compatible with the electronic noise should be observe for all strips. When the laser moves away from the center of the

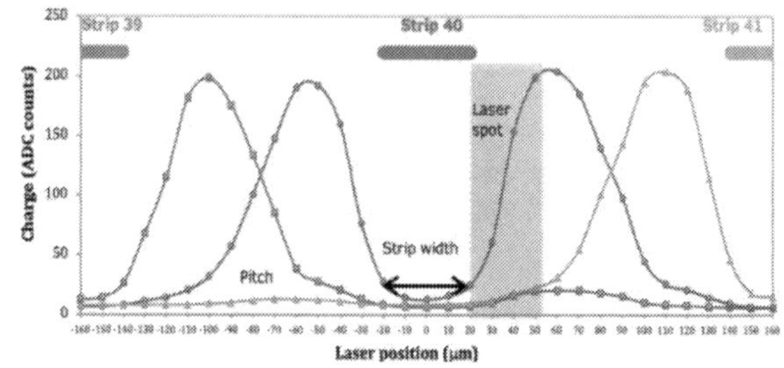

Fig. 6. Signal deposition by the laser bean in three consecutive strips obtained by moving the laser in 10µm steps perpendicular to the strips. The signal recorded by strip 39 is represented in blue, in red the signal in strip 40 and in green the signal in strips 41.

strip 40 to strip 41, a small amount of charge begins to be collected in the strips 40. This is due to the fact that part of the incident light is still reflected on the aluminum. keeping moving the laser beam, the charge on strip 40 will increase. When the beam falls between the strips, the charge will be shared between strips 40 and 41. If we continue moving forward, strip 41 will start collecting more charge till reach its maximum. On the other hand, the charge collected by strip 40 will degrees till became compatible with electronic noise. We will continue this process moving the laser, right and left, for the initial position, towards strip 41 and towards strip 39.

By plotting the mean charge collected by three adjacent strips, showed on Fig. 6, we can illustrate how the charge is shared between strips and prove the strip structure. Also, to determinate the strip width, for the laser path without collected signal collected. Moreover, the beam spot will be measuring from the distance between the position when the laser beam starts generating charge in the detector and the maximum of the charge. And finally, the micro-strip pitch can be measure as the distance between two minima of collected charge as shows in Fig.6.

C. *Laser beam and MIP particles*

Using the laser beam, the growing of the depletion zone with the bias voltage and the laser penetration depth, can be study, by recording the charge collected for various bias voltages. The measured charge will be a convolution of the extension of the depletion zone and the energy deposition within the silicon.

In the case of an incident MIP, the energy deposition is uniformly in the silicon bulk, therefor the measured charge is proportional to depletion voltage till full depletion is reached as expressed in equation (2). While for a laser beam, the energy deposition is exponential and given by:

$$E(y) = E_0\left(1 - e^{-y/a}\right) \quad (3)$$

where y is the penetration distance of the laser and a is the stopping distance (penetration distance). The stopping distance depends on the laser energy or equivalently the laser wavelength. Therefore, for laser light, the charge collection will depend on the depth of the depletion zone but also on the energy lost (i.e. on the laser wavelength). The higher the wavelength of the laser the faster the raise of the charge collection with the bias voltage to reach the plateau [4]).

The penetration of laser in silicon has been measured for some wavelengths, for example, for the 960 nm the penetration is around 74 µm and for 1037 nm is 380 µm [5]. The laser of the EASY system has a wavelength of 980 nm, a priori; its penetration in silicon is unknown.

The Charge Collection Efficiency (CCE) is defined as the fraction between the measured charge for a certain bias voltage and the charge when the detector is fully depleted. For an incent laser light beam the CCE as a function of the bias voltage (V) is given by:

$$CCE(V) = \frac{\left(1 - e^{-d(V)/a}\right)}{\left(1 - e^{-300/a}\right)} \qquad (4)$$

Where $d(V)$ is the width of the depletion area as defined in equation (1) and can be calculated for each bias voltage (V) from the exercise with radioactive source. Fig. 7 represents the CCE for different laser penetrations (a), together with the CCE for MIP as measure in the radioactive source exercise. For laser penetration larger than 1 mm the laser data is a good approximation to a MIP.

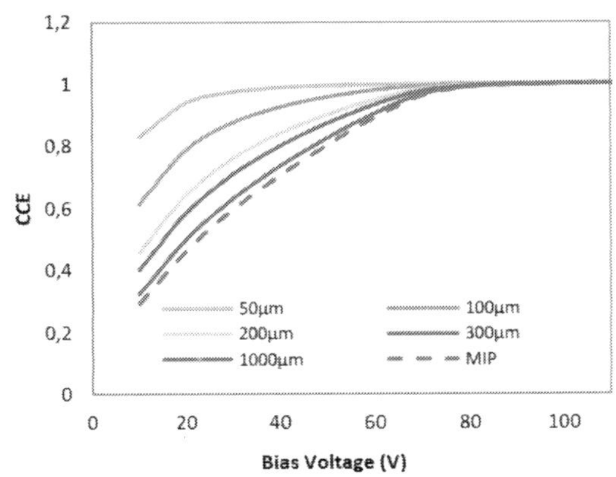

Fig. 7. Theoretical CCE for different laser penetrations and measure CCE for a MIPs.

By measuring CCE with the laser beam for different bias voltages and fitting the obtained curve using equation (4), one can obtain that penetration of a 980nm wavelength laser, the valuer obtain is $167 \mu m$. Fig. 8 shows the fitting curve and the MIP curve for comparation.

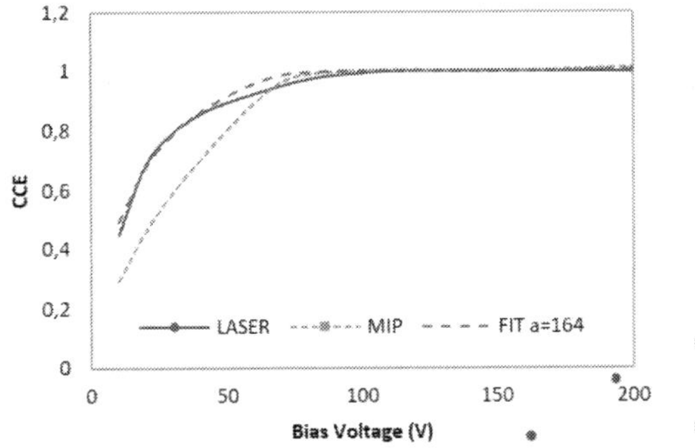

Fig. 8. Charge Collection Efficiency from laser data, fit to a penetration length of 167µm and data for a MIP electron calculated in EXERCISE 6 for comparison.

D. Spectroscopy using the self-trigger

The BEETLE ASIC incorporates a self-trigger signal that will be issued whenever one of its 128 channels registers a signal larger than a preprogramed threshold. To accomplish that, the incoming signal is split into two branches. One goes through a shaper and then to a comparator that will issue the trigger signal for values above the threshold. The signal of the second branch is used to measure the generated charge in the detector.

The self-trigger mode allows measuring the charge induced by particles crossing the detector but also photons or alpha particles absorbed in it. The range of the ADC incorporated in the Control Unit and the dynamic range of the BEETLE define the upper energy limit the system can measure. While the lower energy limit is defined by the noise of the system. Typically, the energy range goes from 20KeV to 300KeV.

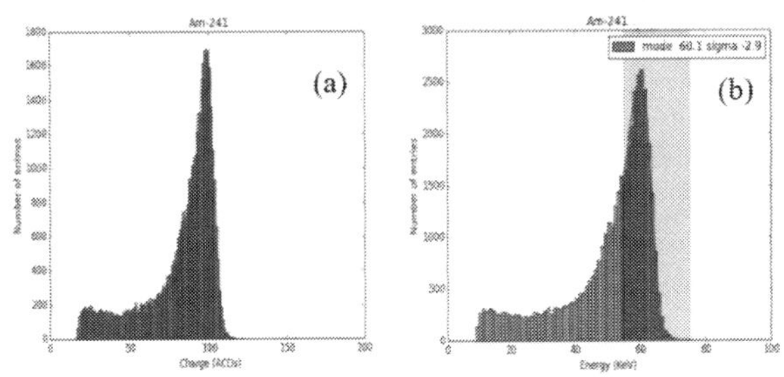

Fig. 9. Charge collected from Am-241 encapsulated source in ADC counts (a) and in energy (b).

As an example, Fig. the collected charge from an ^{241}Am encapsulated source in ADC count (a) and after energy calibration (b). Since the encapsulation of the source absorbers the beta emersion from the ^{241}Am, only the 60KeV gamma rays reach the sensors. Fig. 9 (b) show clearly the gamma emission piek.

VII. Conclusions

EASY is a complete instrumentation system dedicated to Silicon micro-strip Radiation Detectors, representing the state-of-the-art in detector characterization using the same technologies that the actual LHC strip detectors. The aim of this system is to illustrate the operation of a silicon strip detectors, ideal for making basic or complex experiments with silicon micro-strip detectors according with the student level. A complete data acquisition and monitoring software allows to use the system for demonstration of the principles of micro-strip detectors operation. Also, the data can be downloaded for a full analysis as in a real experiment.

The "Activity Book For Students" provides a full description of EASY device, data Acquisition system and 10 exercises to guide the student in the basic principles of micro-strip silicon sensors. The system is commercialized by ALIBAVA SYSTEMS (www.alibavasystems.com), under license of the Spanish National Research Council (CSIC) and the University of Valencia (Spain).

References

[1] ALIBAVA Collaboration. "A portable readout system for microstrip silicon sensors". IEEE Trans.Nucl.Sci. Vol. 56 (2009): 1642-1649.
[2] Löchner, S y Schmelling, M. "The Beetle Reference Manual - chip version 1.3, 1.4 and 1.5". Max Planck Inst.: LHCb-2005-105.
[3] Lutz, G. "Semiconductors Radiation Detectors, volume 2nd Edition". Springer, 1999.
[4] Abt, I. et al. "Characterization of Silicon Microstrip Detectors Using an Infrared Laser System". Nucl. Instr. Meth A423 (1999): 303-319.
[5] Lide, D. R:. "Handbook of Chemistry and Physics". Editor-inChied, 77th Edition, 1996.

Design of a High-Energy and High-Resolution detector for X-ray computed tomography

Maëva Maulin[1], Daniel Eck[1], Nicolas Estre[1], Emmanuel Payan[1], Alix Sardet[1], David Tisseur[2], Grégoire Kessedjian[2]
[1]CEA, DES, IRESNE, DTN, Cadarache F-13108 Saint-Paul-lez-Durance, France.
[2]CEA, DES, IRESNE, DER, Cadarache F-13108 Saint-Paul-lez-Durance, France.
Maeva.maulin@cea.fr

Abstract—**The Nuclear Measurement Laboratory (LMN) at CEA Cadarache in France is developing a high-energy tomograph reaching energies up to 21 MeV with high dose rates. It allows performing tomographies on massive objects (5 tons, 140 cm diameter) with a multimillimetric spatial resolution. For the control of absence of cracks, bubbles or defects in the concrete coating of some CEA waste drums, the laboratory needs a "high-resolution" version of this tomograph. The purpose of this study is to design an imaging detector able to provide a high spatial resolution on large objects with high dose resistance and scanning times of a few hours. To achieve this design, several steps are necessary and will be presented:**

- **A first step consists in characterizing the current detection elements of the tomograph, such as the camera and its lens with experimental measurements. This enables to compare performance of these elements with that available on the market and to consider a replacement.**
- **Then, the scintillator is selected according to experimental measurements on few scintillator types and a state of the art.**
- **Next, the detector configuration is optimized to achieve the higher spatial resolution.**

Throughout these steps, we present the design of a high-resolution detector with a spatial resolution around 300 µm for a 10%contrast.

Keywords — **Linear electron accelerator, High-energy computed tomography, High-resolution computed tomography, X-ray detector**

I. Introduction

THE Nuclear Measurement Laboratory (LMN) at CEA Cadarache has an imaging system currently implemented in the CHICADE experimental platform (Fig. 1) used to perform radiographic and tomographic analyses on large objects (5 tons, 140 cm diameter) with multimillimetric spatial resolution (recently measured around 2.1 mm, obtained with (1)). This system, whose geometry is represented in Fig. 2, is composed of a high-energy photon source generated by the SATURNE accelerator which can reach energies up to 21 MeV with high dose rates (60 Gy/min) [1]. The detector is composed of a scintillator and a sCMOS camera coupled to a lens, kept out of the primary irradiation beam with an angle-deflecting mirror. The objects to be scanned are placed between the source and the detector on a mechanical bench.

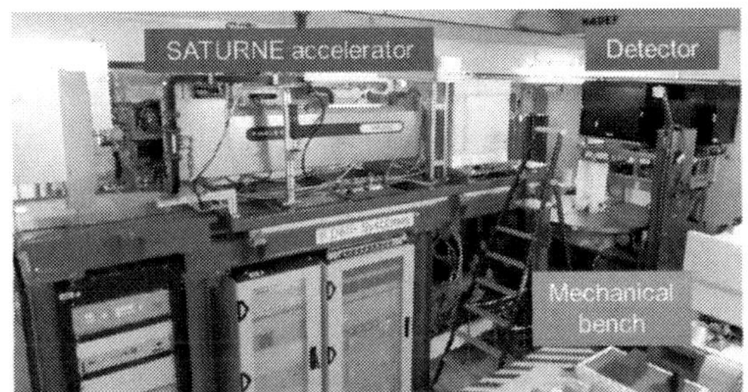

Fig. 1. Panoramic view of the accelerator SATURNE in the CHICADE platform (July 2019)

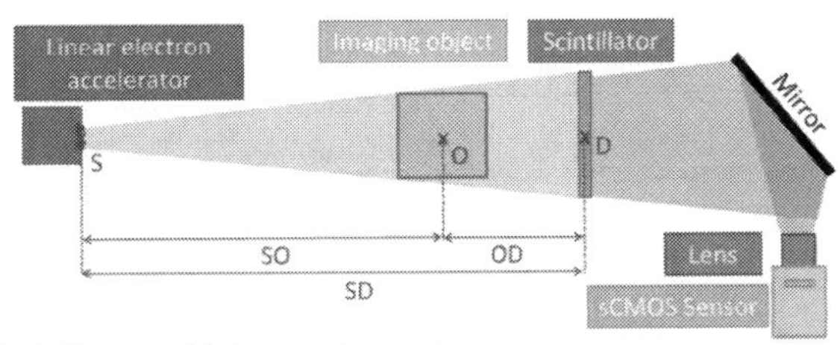

Fig. 2. Diagram of the tomograph geometry

The laboratory needs a "high-resolution" version of this tomograph to check that there are no cracks, bubbles or defects in the concrete coating of some CEA waste drums. Moreover, it can be used for non-destructive testing of metal pieces produced by additive manufacturing, such as specific heat exchangers.

Several types of high energy detectors enabling to obtain high resolution are commercially available:

- Linear detectors producing one-dimensional images [2]. The main advantage of using this type of detector is the significant reduction of the scattered radiation, which allowing to obtain higher contrast and high resolution images. However, the linear technique using a small part of the X-rays field requires high scan times.
- Flat panel detectors (FPD) integrate in the same structure a scintillator that converts X-rays into light (indirect conversion), which is read by a thin film transistor (TFT) [3]. This type of detector is quickly set up and allows short scan times with good spatial resolution, but their compact structure are rapidly damaged by the X-ray beam.

- The X-ray imaging detectors made of a scintillator coupled via a mirror to a sCMOS camera (Fig. 2) have the advantage of being easily adaptable to system expectations and produced with large surface areas. The different elements can be adapted to the high resolution. However, due to their massive geometry and the optimization of the different elements to be realized, they can be difficult to implement compared to a flat panel detector.

The purpose of this study is to design an imaging detector made of a scintillator coupled to a sCMOS camera able to provide a high spatial resolution on large objects with high dose resistance and scanning times of a few hours. This corresponds to a resolution performance increase of one decade compared to the current tomograph.

To achieve this design, five main steps were necessary.

II. Step 1: Optical elements characterization

During the first step, the blur related to the different current detection elements are evaluated using experimental measurements on an optical bench.

The camera currently used on the tomograph is a Zyla camera from Andor [4] coupled with a HyperPrime lens [5]. We also performed tests on the Marana camera from Andor comporting a larger and back-illuminated sensor[6].

In general, the tool used to estimate the spatial resolution is the MTF (Modulation Transfer Function) defined in Fourier space. It is used to evaluate the resolution performance of a device as a signal passes through it. The MTF curves represent the loss of contrast as a function of spatial frequency. Imaging relies largely on the notion of spatial resolution as well as on criteria in terms of contrast at a certain spatial frequency, often the MTF for 10%. In this study, the spatial resolution is defined as the period of a 10% attenuated sinusoidal signal, SR10%, with (1).

$$SR10\% = \frac{1}{f(MTF=10\,\%)} \quad (1)$$

The experimental measurements provided MTF curves for the different detection elements. The corresponding spatial resolution values are listed in Table I. The spatial resolution of the HyperPrime lens was limiting, so a replacement of the lens would improve the total spatial resolution of the optical system.

Several lens were considered and compared based on their MTF curves provided by the manufacturers. The most suitable lens for our study with the lowest spatial resolution is the Sigma 105 mm lens [7]. After experimental measurements in the laboratory, its spatial resolution was measured at 7 μm (Table I).

TABLE I
MTF AND SPATIAL RESOLUTION VALUES OF THE DIFFERENT DETECTION ELEMENTS OF THE TOMOGRAPH

Detection element	MTF(10%) *In sensor plane*	SR10% *In sensor plane*
Zyla camera sensor	111 ± 12 cycle.mm^{-1}	9 ± 1 μm
Lens HyperPrime	83 ± 18 cycle.mm^{-1}	12 ± 3 μm
Optic global (Zyla + HyperPrime)	59 ± 2 cycle.mm^{-1}	17 ± 1 μm
Marana camera sensor	39 ± 4 cycle.mm^{-1}	26 ± 3 μm
Lens Sigma 105 mm	140 ± 30 cycle.mm^{-1}	7 ± 1 μm

III. Step 2: Scintillator selection

The scintillator currently used with the SATURNE accelerator is a Gadox screen (Gadolinium oxysulfide (Gd_2O_2S)). A state of the art of high-energy and high-resolution tomography systems in use around the world has identified a high-energy tomography system with the highest resolution published so far. It is used in the United States to monitor the performance of aging weapons and is called CoLOSSIS (Confined Large Optical Scintillator Screen and Imaging System). This system is developed by the Lawrence Livermore National Laboratory (LLNL) [8] [9]. The real technological achievement of this system is the CoLOSSIS detector. It is composed of a transparent ceramic GLO ($GD_{0.3}Lu_{1.6}Eu_{0.1}O_3$) scintillator [10]. This scintillator is specially designed for high energy and high resolution. Its characteristics are listed in Table II and compared with those of other scintillators typically used in imaging, including the one currently used by the LMN (Gadox).

The GLO is very thin (1.58 mm) but is much denser than typical imaging scintillators, with a density of 9.1 g.cm^{-3}. Therefore, it has a higher photon stopping power than Gadox for an equivalent thickness. In addition, it has a high light output and can be produced in large dimensions, up to 30 cm x 30 cm. However, its most interesting feature is its high spatial resolution. The GLO scintillator has a 10% spatial resolution of 150 μm.

TABLE II
COMPARISON OF THE CHARACTERISTICS OF DIFFERENT SCINTILLATORS TYPICALLY USED IN IMAGING AND THE GLO SCINTILLATOR

Scintillator	Volumic mass ρ (g/cm^3)	Photon yield (ph/MeV)
GLO ($GD_{0.3}Lu_{1.6}Eu_{0.1}O_3$)	9.1	55 00
CsI(Tl)	4.5	65 00
$CdWO_4$	7.9	28 00
Gadox (GOS (Gd_2O_2S:Tb)	4.5	55 000

In view of all its characteristics, the GLO scintillator seems to be the most appropriate for the design of our detector. It offers the best compromise between stopping power, light output and spatial resolution.

IV. STEP 3: OPTIC CONFIGURATION

The aim of this section is to numerically determine which configuration of the detection optics is the best to achieve the smallest possible detector total spatial resolution in the object plane.

Different configurations were tested by changing the cameras, the lenses (using the performance tests from Step 1, see section II) and the camera configuration. In fact, we can use the current configuration of the tomograph used by the LMN with a single out-of-field camera imaging the whole scintillator (Fig. 2), or a coupling of four cameras, each imaging a quarter of the scintillator, which would allow a better spatial resolution. The geometries of these configurations are shown in Fig. 3. The four studied configurations are:

- one Zyla camera with a HyperPrime lens (current tomograph configuration),
- one Zyla camera with a Sigma 105 mm lens,
- a coupling of four Zyla cameras with Sigma 105 mm lenses,
- one Marana camera with a Sigma 105 mm lens.

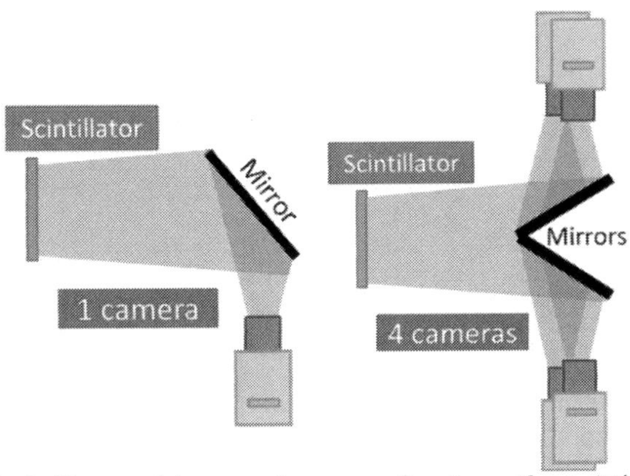

Fig. 3. Diagrams of the proposed camera configurations: either a single camera imaging the whole scintillator or four cameras each imaging a quarter of the scintillator.

To select the best configuration, a calculation of the detector total spatial resolution in the object plane is performed by using the MTFs obtained in the previous steps (MTF Zyla camera, MTF GLO scintillator, MTF Sigma lens, etc.) according to the magnification, M, applied to the system which is defined with (2).

$$M = \frac{Source-detector\ distance}{Source-object\ distance} = \frac{SD}{SO} \quad (2)$$

The use of MTFs is advantageous because the total detector MTF in the object plane, and therefore the associated spatial resolution, can be obtained by multiplying the MTF of the different elements in the object plane.

The total detector spatial resolution values in the object plane for a magnification of 1.3 and 2 are listed in Table III.

TABLE III
VALUES OF THE SPATIAL RESOLUTION TOTAL OF THE DETECTOR IN THE OBJECT PLANE FOR DIFFERENT MAGNIFICATIONS FOR THE FOUR PROPOSED OPTICAL CONFIGURATIONS

Optic configuration	M = 1.3 $SR10\%_{Total\ detector}$	M = 2 $SR10\%_{Total\ detector}$
1 Zyla + HyperPrime	417 ± 109 μm	395 ± 103 μm
1 Zyla + Sigma	341 ± 61 μm	375 ± 68 μm
4 Zyla + Sigma	278 ± 41 μm	300 ± 45 μm
1 Marana + Sigma	358 ± 70 μm	395 ± 77 μm

The configuration with the coupling of four Zyla cameras and Sigma lenses gives the best detector total spatial resolution. It allows a spatial resolution at 10% of approximately 300 μm for a magnification of 1.3.

This study also showed that as the magnification increases, the detector total spatial resolution in the object plane deteriorates, so low magnifications should be chosen. The envisaged magnification will therefore be between 1.1 and 1.3.

V. CONCLUSIONS

All these steps enabled to design a detector to obtain the best total spatial resolution of the system. The final elements composing the dimensioned detector are:

- a GLO scintillator in transparent ceramic of dimension 30 cm x 30 cm,
- a camera configuration with the coupling of four Zyla cameras, coupled with four Sigma 105 mm lenses,
- a magnification between 1.1 and 1.3.

Through all these steps, the detector dimensioned is expected with a spatial resolution about 300 μm for a 10% contrast without taking into account the contribution of the scattered radiation from the scanned object.

Indeed, complementary investigation is in progress to complete this study by taking into account the amount and spread of radiation scattered by the object using Monte-Carlo simulations and Build-Up factor calculations. Depending on the distance between the object and the source, the radiation scattered by the object will not be detected in the same way due to the scattering angles and attenuation. This study will therefore determine an optimum between the influence of scattered radiation and the overall spatial resolution of the system.

One of the perspectives of this study is the implementation of blur and scattered radiation correction methods on the images from deconvolution in order to further reduce the spatial resolution. This correction could be done with the deconvolution of the projections from the global PSF of the system or via the implementation of neural networks.

References

[1] N. Estre, D. Eck, J.-L. Pettier, E. Payan, C. Roure, and E. Simon, "High-energy X-ray imaging applied to non destructive characterization of large nuclear waste drums," in *2013 3rd International Conference on Advancements in Nuclear Instrumentation, Measurement Methods and their Applications (ANIMMA)*, 2013, pp. 1–6. doi: 10.1109/ANIMMA.2013.6727987.

[2] Detection Technology, "X-Ray Line Detectors - Detection Technology." https://www.deetee.com/x-ray-line-detectors/ (accessed Jul. 16, 2021).

[3] PerkinElmer, "Flat Panel X-ray Detectors - PerkinElmer." https://www.perkinelmer.com/PDFs/downloads/SPC-NDT-Detectors-012163A_01.pdf (accessed Jul. 16, 2021).

[4] "ANDOR : Zyla 5.5 camera specifications," 2019. https://andor.oxinst.com/products/scmos-camera-series/zyla-5-5-scmos (accessed Jan. 21, 2019).

[5] "SLR Magic HyperPrime 25 mm lens specifications," 2020. http://www.slrmagic.info/product/HyperPrime+CINE+III+2595MFT (accessed Sep. 30, 2020).

[6] "ANDOR : Marana camera specifications," 2020. https://andor.oxinst.com/products/scmos-camera-series/marana-scmos (accessed Sep. 30, 2020).

[7] "Sigma 105 mm f/1.4 DG HSM Art lens specifications," 2020. https://www.sigma-photo.fr/objectifs-photo/310-105mm-f14-dg-hsm-art.html (accessed Sep. 30, 2020).

[8] A. Heller, "A cat scanner for nuclear weapon," *Sci. Technol. Rev.*, p. 28, Aug. 2009.

[9] Lawrence Livermore National Laboratory (LLNL), "Scientists develop new CT scanner to image nuclear weapon components," Jun. 08, 2020. https://www.llnl.gov/news/scientists-develop-new-ct-scanner-image-nuclear-weapon-components

[10] N. Cherepy, "MeV Radiography with GLO Transparent Ceramic Scintillator," p. 15, 2016.

Pulse shape simulations for organic scintillation detectors using Geant4

Caroline Holroyd[1], Michael Aspinall[1], and Tom Deakin[2]
[1]Lancaster University, United Kingdom
[2]LabLogic Systems Ltd., United Kingdom
c.holroyd2@lancaster.ac.uk

***Abstract*— The accurate simulation of the temporal pulse shapes from organic scintillation detectors capable of pulse shape discrimination (PSD) presents the opportunity to assess the pulse shape discrimination of these detectors prior to fabrication. The aim of this study is the simulation of the temporal pulse shapes from EJ-276, a PSD-capable plastic scintillator developed by Eljen Technologies. PSD plastic scintillators are increasingly replacing organic liquid scintillators for the detection of neutrons in the presence of mixed radiation fields for nuclear security applications. Plastics are inexpensive, robust and can be fabricated in a variety of shapes and sizes. They offer a solid-state alternative to liquid scintillators which can be difficult to transport due to the risk of leakage. However, the PSD performance of plastic scintillators has been observed to decrease due to various factors which combine to influence the overall shape of the pulse. The Monte Carlo toolkit Geant4 has been used to simulate the temporal pulse shapes from an EJ-276 plastic scintillator coupled to a photomultiplier tube (PMT). All three decay time components of EJ-276 have been modelled, utilising new methods available in the latest version of Geant4, for two different scintillator geometries. The simulated n/γ pulse shapes reproduce the features important for PSD. Future work will involve integrating the temporal response of the PMT with existing pulse shape simulations. Simulated data will then be compared with experimental measurements.**

***Keywords* —Monte Carlo simulations, Geant4, plastic scintillators, pulse shape discrimination**

I. INTRODUCTION

THE temporal shape of the output pulse formed following the interaction of ionising radiation with a scintillation detector enables characterisation of the incident radiation. The pulses derived from organic and inorganic scintillators are typically represented by an exponential rise time and multiple decay time components. These time components relate to the scintillation mechanism of the material and can differ with the type of interacting particle. This allows information on the properties of the incident radiation to be extracted from the pulse shape using pulse shape analysis techniques.

Pulse shape discrimination (PSD) is a pulse shape analysis technique used for the separation of neutron and gamma ray induced signals. Signal separation is achieved by exploiting the subtle differences in the shapes of the pulses formed following neutron and gamma ray interactions in a scintillator material. Pulse shape discrimination (PSD) using organic scintillators utilises the fact that organic material with certain molecular properties emit scintillation light with two decay components, referred to as the fast and slow components. The fast component is emitted a few nanoseconds following particle interaction, while the slow component is emitted a few hundred nanoseconds after. Various techniques have been developed for PSD with the aim of achieving the greatest separation of neutron and gamma ray induced signals.

One of the most commonly used methods for PSD with organic scintillators in a mixed radiation field is based on analysis of the decay tail of the pulse [1]. A distinct difference is observed in the decay tail for neutron and gamma ray induced pulses. This is because neutrons experience a greater rate of energy loss in the scintillator compared to gamma rays, resulting in an enhanced slow component for neutron interactions. This is observable in the decay tails of the pulse shapes and signal separation is achieved on this basis.

The aim of this study is the accurate simulation of the temporal pulse shapes from an EJ-276 PSD-capable plastic scintillation detector using the Monte Carlo simulation toolkit Geant4. This has the potential to enable the PSD performance of organic scintillation detectors to be assessed in the initial stages of detector design. Factors which reduce the performance can then be identified and their effects mitigated with the aim of optimising detector performance. This is important for plastic scintillation detectors since the PSD performance of these detectors can be unreliable and is known to fluctuate as a result of numerous factors which combine to influence the shape of the output pulse. One such factor is the geometry of the scintillator, where the ability to separate neutron and gamma ray induced signals is observed to degrade as the size of the scintillator is increased [2]. Other factors include reflector material, photodetector position and location of particle interaction in the scintillator [3].

Geant4 is widely used for the simulation of plastic scintillation detectors. However, simulations of the temporal pulse shapes from PSD-capable plastic scintillators have previously neglected to model all known decay time components associated with the scintillator and their dependency on particle type. This has impacted on the overall accuracy of pulses simulated using Geant4 when compared with experimental measurements. This was identified as a limitation of Geant4 for simulating the pulse shapes of scintillation detectors with more than two decay time components [4]. This includes the PSD-capable plastic scintillator EJ-276 (previously EJ-299-33) which is stated by the manufacturer to have three particle-dependent decay time components. Prior to Geant4 version 10.7 the number of decay time components was limited. For particle-independent yields two decay time components could be specified, while for particle-dependent yields this was reduced to one [5].

Work by Hartwig and Gumplinger [6] used Geant4 to simulate the output pulse shapes from EJ-309, a PSD-capable liquid scintillator developed by Eljen Technologies that has three known decay time components. The first two decay time components were simulated. When compared to experimental data, the simulated pulse shapes did not accurately model the rise times, however the decay tail of the pulses was accurately reproduced. Additionally, the influence of the timing response of the photodetector chosen to convert scintillation photons into an electrical output was not included in the simulation. This is a potential source of inaccuracy when comparing simulated pulse shapes to experimental data.

The pulse shapes derived from inorganic scintillation detectors have been simulated by Ogawara and Ishikawa [7] using a combination of the optical photon tracking capabilities of Geant4 and the experimental response function of a photomultiplier tube (PMT). The arrival time distribution produced by Geant4 to represent the light pulse shape was convoluted with the experimentally measured pulse shape from the PMT to produce the final simulated pulses. While the simulated pulses compared favourably to experimental measurements, the method used to capture the PMT response was identified as a possible source of inaccuracy. This is because the response included the influence of not only the PMT but also the response of a plastic scintillator with fast timing properties that was used in the experimental set-up to obtain the PMT response function.

II. GEANT4 PULSE SHAPE MODELLING

Geant4 is an object-oriented C++ Monte Carlo toolkit for simulating the passage of particles through matter [5]. Particles and their interactions can be tracked through arbitrarily complex geometries, material compositions and electromagnetic fields. Geant4 is widely used for the simulation of organic and inorganic scintillation detectors. This is due to its ability to handle optical physics processes, including scintillation and optical photon transportation. Scintillator materials are characterised by their light yield (photons/MeV), photon emission spectrum and the distribution of emission times. These must be defined by the user when constructing the simulation.

An EJ-276 PSD-capable plastic scintillator coupled to a PMT has been simulated using Geant4 version 10.7. A cube and slab geometry have been constructed to enable a comparison of the pulse shapes from two different geometries and investigate the capabilities of Geant4 to model more than two decay time components. The cube and slab scintillator geometries are shown in Fig. 1a and Fig. 2a, respectively. The location of the photocathode of the PMT is outlined in red. The light yield was set to 8,600 photons/MeV for neutron and gamma radiation. All three decay time components and their relative weightings for neutron and gamma radiation were input into the simulation, utilising the new enhanced time constant method available in Geant4 10.7. This allows for the specification of more than two decay time components for both particle-independent and particle-dependent yields. Experimental values obtained by Iwanowska-Hanke *et al.* [8] have been used for the three decay time components and their relative weightings for neutron and gamma radiation.

The Geant4 physics list was constructed using the QGSP_BIC_HP_EMZ reference physics list. This contains QGSP_BIC_HP for hadronic interactions and G4EmStandardPhysics_option4 for electromagnetic interactions. G4OpticalPhysics was included for the generation and transportation of scintillation photons.

Geant4 has been used for the generation and transportation of individual scintillation photons up to their detection at the photocathode of the PMT. Once detected, their time of arrival was recorded and histogrammed to produce the simulated neutron and gamma ray pulses shown in Fig. 1b and Fig. 2b. These pulses preclude the influence of the timing response of the PMT and any associated data electronics on the shape of the pulse. For each geometry, 10,000 events were simulated using a monoenergetic source of neutron and gamma radiation with an energy of 2 MeV.

III. RESULTS

The average simulated pulse shapes for an EJ-276 plastic scintillation detector with a cube geometry is shown in Fig. 1b for neutron (blue) and gamma (red) radiation. The pulses have been normalised to enable comparison between neutron and gamma pulses.

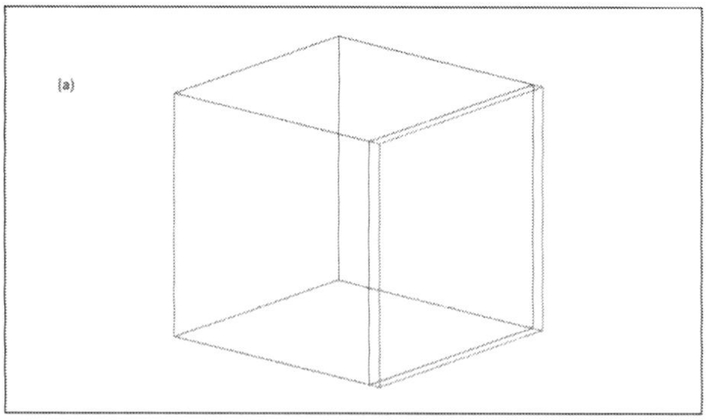

https://doi.org/10.1051/epjconf/202125311002

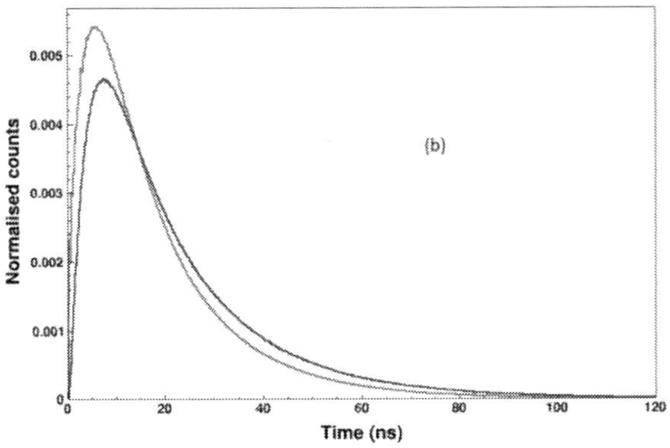

Fig 1. Simulated cube geometry (a) and average output pulse shapes (b) for neutron (blue) and gamma (red) radiation. The dimensions of the cube are 5 cm x 5 cm x 5 cm.

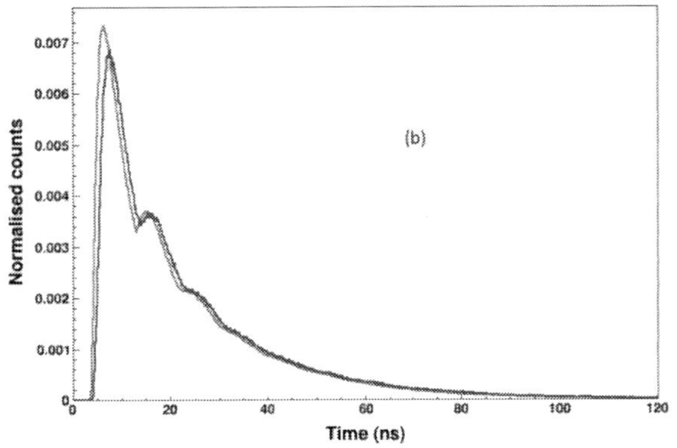

Fig 2. Simulated slab geometry (a) and average output pulse shapes (b) for neutron (blue) and gamma (red) radiation. The dimensions of the slab are 5 cm x 15 cm x 100 cm.

The pulses simulated approximate those that are observed experimentally for an organic scintillation detector; however, an experimental comparison is required to quantify the accuracy of the simulations. Fig. 1b demonstrates the capability of Geant4 to simulate the important features of the pulses derived from an organic scintillation detector; the rising edge and decay tail. The decay tail is the crucial component for distinguishing particle type. Fig. 1b shows the expected differences between neutron and gamma ray induced pulse shapes. The enhanced slow component is observed in the decay tail of the neutron pulse from approximately 20 ns, demonstrating the subtle differences expected between the pulses generated as a result of the interaction of different particles. This forms the basis of PSD and demonstrates the ability of Geant4 to correctly simulate the expected differences in pulse shape for neutron and gamma radiation using the new enhanced time constant method available in Geant4 version 10.7.

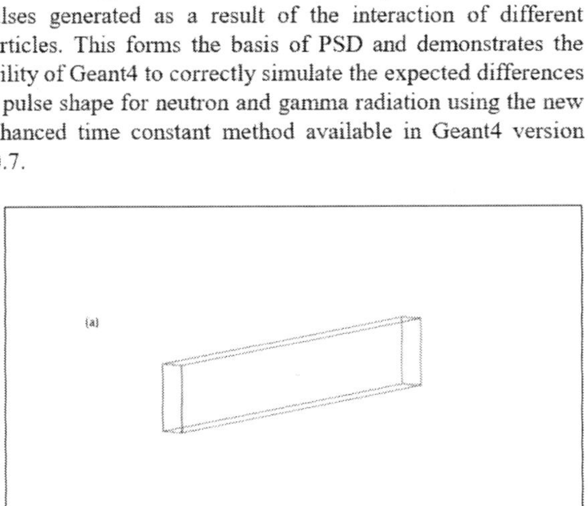

For a comparison between two different geometries, a slab geometry was simulated using the same source and number of events. This is shown in Fig. 2b. Compared with the pulse shapes generated for the cube geometry, the differences between the neutron and gamma pulse shapes are small. There is a peak at 10 ns and a secondary peak at approximately 15 ns for both pulses. This secondary peak is not visible for the cube geometry. This secondary peak is likely the result of optical photons travelling longer distances and experiencing more reflections in the scintillator before being detected at the photocathode. The poorer separation of pulses for the slab geometry, particularly in the decay tail of the pulse, will result in the slab having a reduced PSD performance compared to the cube geometry. This is expected and agrees with previous studies [3] that observed a dependence of PSD performance on geometry.

IV. FUTURE WORK

Future work will focus on improving the pulse shape simulations with the aim of reproducing experimental pulse shapes output from an EJ-276 plastic scintillation detector. While Geant4 can successfully model the temporal pulse shapes from PSD-capable plastic scintillation detectors, they do not provide an accurate representation of that which would be measured experimentally. This is because Geant4 is only capable of outputting the arrival time distribution of scintillation photons at the photodetector and not the final output pulse shape. This is influenced by not only the scintillation emission but also the complex processes occurring in the photodetector.

Experimental work is planned to measure the temporal pulse shape of an ET-Enterprises 9214 PMT, independent of the properties of the scintillator. This will be done using a nanosecond pulsed laser with a wavelength of 425 nm, matching the peak emission spectrum of EJ-276. The PMT response will be integrated with existing simulations to obtain the final simulated pulse shape. This will enable PSD to be performed on the simulated pulse shapes and comparisons to be made with experimental data. The effect of the PMT response on the overall shape of the pulses can also be quantified.

Simplifications have been made to the pulse shape simulations to allow for an investigation into the influence of the various factors which affect the shape of the pulses. In order to replicate experimental pulses, optical surface effects and materials used for coupling the PMT to the scintillator will need to be included.

V. CONCLUSIONS

This paper demonstrates the capability of Geant4 version 10.7 to model the temporal pulse shapes output from a PSD-capable plastic scintillation detector with three known decay time components. The expected differences between neutron and gamma pulse shapes due to the way the different particles lose energy in the scintillator are visible, primarily in the decay tail of the pulses. Two different geometries have been compared to show how the pulse shapes change with geometrical configuration. Future work will integrate an experimentally obtained PMT temporal response with existing pulse shape simulations with the aim of accurately reproducing experimental data. The ability to accurately simulate the pulse shapes from organic scintillation detectors allows for the possibility to assess the PSD performance of such detectors in the early stages of detector development.

REFERENCES

[1] F.D. Brooks "Development of Organic Scintillators", Nuclear Instruments and Methods, VOL. 162 477-505, 1979.

[2] C. Payne, P. L. Sellin, M. Ellis, K. Duroe, et al., Neutron/gamma pulse shape discrimination in EJ-299-34 at high flux. 2015 IEEE Nuclear Science Symposium and Medical Imaging Conference. pp.1-5. 7581994.

[3] Hubbard, Michael W. J, "Light transport modelling of pulse shape discrimination within plastic scintillators. Doctoral thesis, University of Surrey.

[4] Comrie, A., Buffler, A., Smit, F. D. & Wortche, H. J. Tests of pulse shape discrimination with EJ299-33 plastic scintillator for use in portable spectroscopy in Proceedings, 3rd International Conference on Technology and Instrumentation in Particle Physics (TIPP 2014): Amsterdam, Netherlands, June 2-6, 2014 TIPP2014 (2015), 251. doi:10.22323/1.213.0251.

[5] S. C. J. Allison, K. Amako, J. Apostolakis, H. Araujo, P. Arce Dubois, M. Asai, G. Barrand, R. Capra *et al.*, "Geant4 Developments and Applications," *IEEE Trans. Nucl. Sci.*, vol. 53, no. 1, pp. 270–278, 2006.

[6] Z. S. Hartwig and P. Gumplinger, "Simulating response functions and pulse shape discrimination for organic scintillation detectors with Geant4," *Nucl. Instruments Methods Phys. Res. Sect. A Accel. Spectrometers, Detect. Assoc. Equip.*, vol. 737, pp. 155–162, 2014.

[7] R. Ogawara and M. Ishikawa, "Signal Pulse Emulation for Scintillation Detectors using GEANT4 Monte Carlo with light tracking simulation," VOL., 075114, No. July, 2016.

[8] Iwanowska-Hanke, J et al. Comparative study of large samples (2"2") plastic scintillators and EJ309 liquid with pulse shape discrimination (PSD) capabilities. *Journal of Instrumentation* 9, P06014 (2014).

https://doi.org/10.1051/epjconf/202125311003

Performance and Applications of Silicon Carbide Neutron Detectors in Harsh Nuclear Environments

Frank H. Ruddy[1], Laurent Ottaviani[2], Abdallah Lyoussi[3], Christophe Destouches[3], Olivier Palais[2], and Christelle Reynard-Carette[2]

[1]Ruddy Consulting, USA
[2]Aix Marseille Univ, Université de Toulon, CNRS, IM2NP, Marseille, FRANCE
[3]CEA Cadarache, FRANCE
FrankHRuddy@gmail.com

***Abstract*— Silicon carbide (SiC) semiconductor is an ideal material for solid-state nuclear radiation detectors to be used in high-temperature, high-radiation environments. Such harsh environments are typically encountered in nuclear reactor measurement locations as well as high-level radioactive waste and/or "hot" dismantling-decommissioning operations. In the present fleet of commercial nuclear reactors, temperatures in excess of 300 °C are often encountered, and temperatures up to 800 °C are anticipated in advanced reactor designs. The wide bandgap for SiC (3.27 eV) compared to more widely used semiconductors such as silicon (1.12 eV at room temperature) has allowed low-noise measurements to be carried out at temperatures up to 700 °C. The concentration of thermally induced charge carriers in SiC at 700 °C is about four orders of magnitude less than that of silicon at room temperature. Furthermore, SiC radiation detectors have been demonstrated to be much more resistant to the effects of radiation-induced damage than more conventional semiconductors such as silicon, germanium, or cadmium zinc telluride (CZT), and have been demonstrated to be operational after extremely high gamma-ray, neutron, and charged-particle doses. The purpose of the present review is to provide an updated state of the art for SiC neutron detectors and to explore their applications in harsh high-temperature, high-radiation nuclear reactor applications. Conclusions related to the current state-of-the-art of SiC neutron detectors will be presented, and specific ideal applications will be discussed.**
***Keywords* —Neutron detectors, silicon carbide, SiC, semiconductor, radiation damage.**

I. Introduction

SILICON Carbide (SiC) semiconductor neutron detectors have many advantages for measurements in harsh high-temperature, high-radiation environments. The material properties of SiC, particularly the 4H polytype [1], make operation in these environments more feasible than for other conventional semiconductors such as Si, Ge, and CZT (Cadmium Zinc Teluride). The 3.27-eV bandgap for 4H SiC enables low-noise measurements at room temperature and temperatures up to at least 700 °C. SiC has also been shown to be resistant to the cumulative effects of gamma, neutron, and charged-particle irradiation. Other factors that are advantageous for SiC include:

- high thermal conductivity (10-22 W/cm-K)
- a maximum breakdown field that is eight times that of silicon allowing higher biases to be applied resulting in higher drift velocities and more efficient charge collection
- a high saturated drift velocity (nearly twice that of silicon) leading to low charge trapping .

The use of SiC as a radiation detector material was reviewed by Nava, *et al.* in 2008 [1], Strokan, *et al.* in 2009 [2] and Ruddy in 2013 [3]. Designs and properties of SiC neutron detectors were reviewed by Franceschini and Ruddy in 2011 [4]. The purpose of the present review is to provide an updated state of the art for SiC neutron detectors and to explore their applications in harsh high-temperature, high-radiation nuclear applications.

II. Silicon Carbide Detector Designs

By far the most prevalent design for SiC nuclear detectors is the epitaxial layer design shown schematically in Figure 1. [5-9] Although many different polytypes of SiC exist, 4H-SiC is most frequently used for nuclear applications as discussed in detail in reference [1]. The conducting SiC substrate depicted in Fig. 1 typically has a nitrogen dopant concentration of about 10^{18} cm^{-3} and a thickness of about 300-350 µm. An n^{-} layer is grown onto the substrate layer using vapor-phase epitaxy. The dopant concentration, which corresponds to the unintentional residual nitrogen concentration, is typically less than about 10^{15} cm^{-3}, and thicknesses ranging from a few µm up to 250 µm

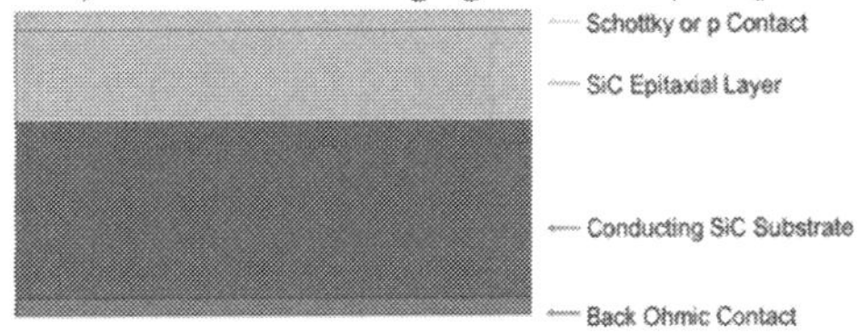

Fig. 1. Schematic representation of a SiC epitaxial diode detector structure.

have been used. A metallic ohmic contact is applied to the back of the detector using a suitable material such as gold, titanium or nickel. The front contact can be either a Schottky contact using a suitable metal such as nickel or a p^+ contact formed by diffusing or implanting materials such as aluminum, phosphorus, or boron [5].

If a reverse bias is applied to the SiC diode (negative Schottky or p contact relative to the back ohmic contact), charge carriers are depleted from the epitaxial layer, and this depleted or intrinsic SiC layer corresponds to the active layer of the detector. The thickness of the depleted region, measured from the Schottky contact into the epitaxial layer, increases with reverse bias, V, according to the following approximation

$$d \cong \left[\frac{2\varepsilon V}{eN}\right]^{1/2}$$

where ε is the dielectric constant for SiC, *e* is the electron charge and N is the nitrogen concentration. It can be seen that the depletion thickness is directly proportional to the square root of the applied voltage and inversely proportional to the square root of the dopant concentration. Therefore, to achieve large active volumes with thick epitaxial layers, one must have low nitrogen concentrations and/or apply high bias voltages. In practice, one might be constrained to voltages less than 1000 V (to protect the counting chain electronics from voltage discharges). Therefore, thick epitaxial layers with the lowest possible dopant concentrations are desirable. As discussed in reference [3], state-of-the-art SiC epitaxy is limited to nitrogen concentrations of about 10^{13} cm^{-3} , but a goal of 10^{12} cm^{-3} is reasonable [10].

III. NEUTRON RESPONSE MEASUREMENTS

Neutrons have no electrical charge and are not capable of producing ionization directly in a detector. Therefore, as with all neutron detectors, SiC neutron detectors require a nuclear reaction to form reaction products that are capable of producing ionization within the detector active volume. For thermal-neutron detection, the $^{10}B(n,\alpha)^7Li$ and $^6Li(n,\alpha)^3H$ reactions have proven useful [4,11], because solid-state compounds of boron and lithium are readily available and can be configured to allow the energetic charged-particle reaction products to enter the detector active volume [4-6,12]. In practice, the thermal-neutron convertor material can be in the form of a thin layer juxtaposed near the active volume or can be incorporated into the detector by ion implantation [13] or diffusion [14].

Fast neutrons can be detected by ionization produced by ions resulting from nuclear reactions with silicon or carbon atoms in or near to the SiC active volume of the detector. A partial list of reactions is shown in Table I. [4] Other more complex reactions such as (n,2n), (n,pn), (n,nα), *etc.* are also possible as well as reactions with less abundant isotopes of carbon and silicon. For neutron spectra derived from down-scattering of fission neutrons, the response of a SiC detector will be dominated by the detection of energetic ^{12}C and ^{28}Si ions from (n,n') elastic and inelastic scattering reactions, because the average energy of fission neutrons is ~2 MeV with most of the neutrons having energies less than 6 MeV. Those threshold reactions that are energetically possible will add to the pulse-height continuum, because the incident neutron spectrum is a distribution of

TABLE I
FAST-NEUTRON INDUCED REACTIONS IN SILICON CARBIDE

Reaction	Threshold Energy (MeV)
$^{12}C(n,n')^{12}C$ (elastic scattering)	0
$^{28}Si(n,n')^{28}Si$ (elastic scattering)	0
$^{12}C(n,n')^{12}C$ (first excited state)	4.809
$^{28}Si(n,n')^{28}Si$ (first excited state)	1.843
$^{12}C(n,\alpha)^9Be$	6.180
$^{28}Si(n,\alpha)^{25}Mg$	2.749
$^{12}C(n,p)^{12}B$	13.643
$^{28}Si(n,p)$	3.999

energies leading to a distribution of recoil-ion energies.

A comparison of the thermal-neutron and down-scattered fission-neutron responses for a SiC detector is shown in Fig. 2, which has been reproduced from Reference 15. When the ^6LiF

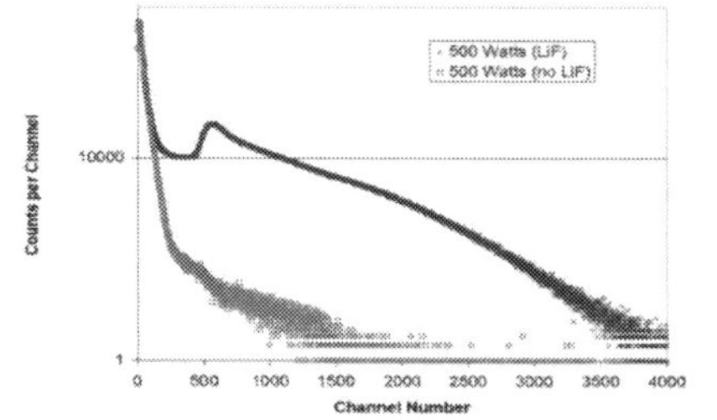

Fig. 2. Comparison of the neutron response of a 500-μm diameter x 3-μm SiC Schottky diode to reactor fission neutrons with and without a 24.2-μm ^6LiF convertor foil (Figure reproduced from Reference 15).

converter foil is present, the detector response is dominated by the ^4He and ^3H ions produced by thermal-neutron reactions, and the events from fast-neutron induced reactions are obscured .

The fast-neutron spectrum (no converter foil) shown in Fig. 2 is a continuum dominated by elastic and inelastic neutron scattering reactions with carbon and silicon atoms in the SiC detector as discussed previously. The events below channel 300 are due primarily to gamma-ray induced secondary-electrons. The ranges of these electrons in SiC are large compared to the dimensions of the detector active volume resulting in only a fraction of the energy from each being detected. Although only a featureless low-energy continuum results, the detector response to gamma rays has been shown to be linear. [16-18] The fast-neutron response is complex and results in no peaks in the spectrum. Nevertheless, incident neutron energy information is present as shown by the data in Fig 3. [19]. The average energy of neutrons from fission of ^{252}Cf is 2.15 MeV, whereas the ^{241}Am-Be source produces a much higher average energy of 4.5 MeV. The ^{241}Am-Be detector response continuum is clearly shifted to higher pulse heights compared to the ^{252}Cf

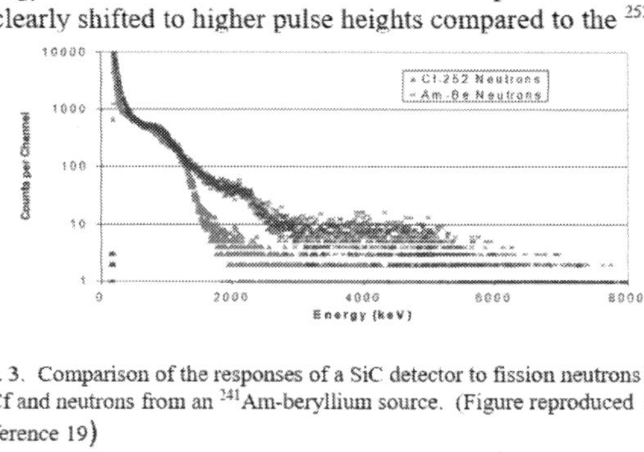

Fig. 3. Comparison of the responses of a SiC detector to fission neutrons from ^{252}Cf and neutrons from an ^{241}Am-beryllium source. (Figure reproduced from Reference 19)

response spectrum. In order to extract neutron energy-spectrum information from the pulse-height response, a complex unfolding procedure using a combination of neutron transport and recoil range calculations is required [20].

In the case of monoenergetic neutrons impinging on a SiC detector, peaks will be present for those reactions that result only in charged-particle products. For example, the $^{12}C(n,\alpha)^{9}Be$ reaction (see Table I) results only in ^{4}He and ^{9}Be ions which share the energy of the incident neutron plus the reaction energy. Therefore, a peak will be observed corresponding to ionization resulting from a total energy of 8298.8 keV being deposited in the detector active volume. The response spectrum for a SiC detector exposed to 14.1 MeV neutrons from a Deuterium-Tritium (DT) neutron generator is shown in Fig. 4, which has been reproduced from Reference 19.

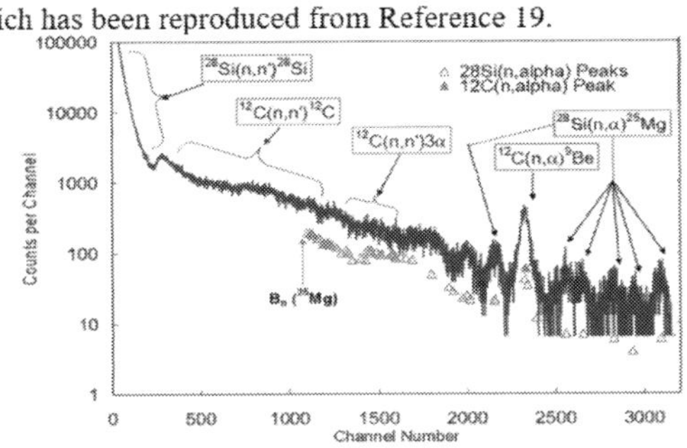

Fig. 4. Response spectrum for a 6 mm^2 x 100 μm SiC detector exposed to 14.1 MeV DT neutrons. (Reproduced from Reference 19)

In addition to the prominent peak from the $^{12}C(n,\alpha)^{9}Be$ reaction, a family of peaks corresponding to the $^{28}Si(n,\alpha)^{25}Mg$ reaction is also present. The highest-energy peak corresponds to the production of the ground state of ^{25}Mg with the deposition of a total energy of 11,346 keV in the SiC detector active volume. The other peaks in this set correspond to excited states of ^{25}Mg where less energy is deposited in the detector. The difference in energy corresponds to gamma rays produced when these excited states of ^{25}Mg decay to the ground state. As the energies of these excited states increases, the energy spacing between them decreases, and the peaks are not resolved above about the tenth excited state. The higher energy states will produce a continuum of events up to the ^{25}Mg neutron binding energy, B_n, above which the $^{28}Si(n,n'\alpha)^{24}Mg$ reaction will dominate. Since the energy of the neutron produced in this reaction is not fixed, no peaks will result from these reactions.

Continua from elastic and inelastic neutron scattering as well as more complex reactions such as $^{12}C(n,n')3\alpha$ are also present. No single peak is observed for these reactions, because the energy of the reaction product neutron is not fixed.

The neutron responses for SiC detectors have been shown to be linear for thermal [12,21], epithermal [16,17], fast [15] and 14-MeV neutrons [22].

IV. NEUTRON-RESPONSE MEASUREMENTS AT HIGH TEMPERATURES

Some applications such as fusion-reactor technologies require stable measurement methods to detect neutrons and charged particles under extreme conditions such as high temperatures and high magnetic fields. To maintain the fusion process, it is mandatory to measure simultaneously the neutron flux and the tritium breeding ratio.

Specific fast neutron detectors, based on 4H-SiC pn diodes, were developed in the framework of the European I_SMART project [23]. Their characteristics were measured at room temperature and compared with a single-crystal chemical vapor deposited (sCVD) diamond-based detector. Measurements were carried out at the DT neutron generator at Technical University of Dresden (TUD) at a fluence rate of about 9.4×10^{6} cm^{-2}-s^{-1} with a neutron energy of 14.12 MeV.

Both detectors produced pulse-height spectra with the well-resolved peak due to the $^{12}C(n,\alpha)^{9}Be$ reaction. The count rate of the diamond-based detector was higher than that of the SiC-based detector as a consequence of the higher thickness of the diamond active detection volume. The energy resolution of the $^{12}C(n,\alpha)^{9}Be$ peak was slightly better for the 4H-SiC-based detector (260 keV at FWHM which corresponds to 3.09%), with respect to 303 keV (3.6%) for the diamond detector. An important phenomenon is the improvement of the energy resolution with increased neutron flux for the SiC detector, in complete opposition to the behavior of diamond [23].

Several prototypes were also tested in industrial conditions at the fast neutron generator at Schlumberger (Clamart, France), at room temperature and at 106 ° C [24]. The spectra showed good stability, preserving features over the whole temperature range. Prototypes with gold metallic contacts were then tested up to 500 °C at the DT neutron generator at the TUD [25]. In the recorded spectra, the different signal structures arising from high-energy deep inelastic reactions can be distinguished, independently of the temperature (see Fig. 5). The most prominent orientation point in the spectrum is the full energy peak of the $^{12}C(n,\alpha)^{9}Be$ reaction, which can be clearly distinguished at all the applied temperatures.

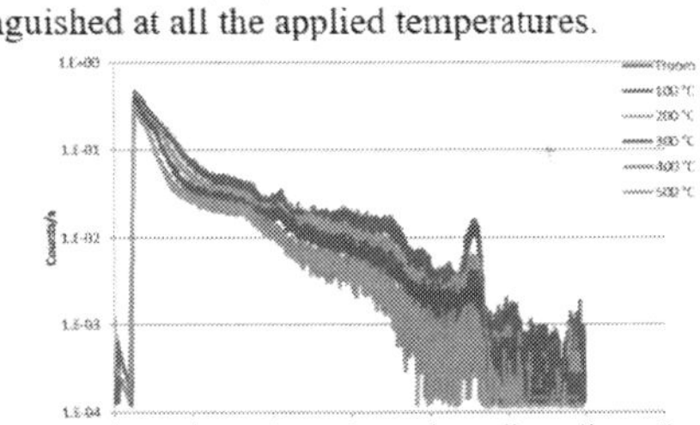

Fig. 5. Resulting energy spectra recorded during irradiation tests of SiC-based detectors with 14 MeV neutrons at different ambient temperatures. (Figure reproduced from reference 25).

Abubakar, *et al.* [26] reported stability of the SiC-based alpha detector signal at 500 K (227 °C) over a period of 18 hours.

No temperature-induced polarization effect has been observed to date with epitaxial Silicon Carbide, in contrast to the case of diamond where this effect appears above 600 K (327 °C) [27].

V. RADIATION DAMAGE EFFECTS IN SILICON CARBIDE DETECTORS

Whereas conventional neutron detectors such as fission chambers and self-powered neutron detectors exhibit excellent resistance to the cumulative damaging effects of radiation, some monitoring applications are potentially better suited to semiconductor detectors. Common semiconductor detectors based on silicon or germanium are not only limited to low-temperature operation, because of low band gaps, but also have low service lifetimes due to radiation damage effects. Therefore, these detectors have been of little use in harsh nuclear environments where large doses of gamma rays, neutrons and charged particles must be tolerated.

For example, silicon detectors show increasing leakage current when exposed to radiation due to the formation of defects that correspond to charge-donor levels in the band gap. Charge-trapping sites are also produced, but the increased leakage current is the primary radiation-damage limitation for the use of silicon detectors and is an indirect consequence of the low band gap for silicon.

On the other hand, wide band-gap semiconductor detector such as SiC are limited more by the accumulation of charge-trapping sites. Increased leakage current due to the formation of charge-donor states is less due to the larger band gap.

SiC has been shown to be highly resistant to the effects of radiation damage. The effects of large gamma-ray dose on the performance of SiC detectors have been investigated by several groups. Kang, *et al.* [28] showed that irradiating 6H SiC diodes with a dose of 120 kGy of ^{60}Co gamma rays lead to a leakage-current decrease. Metzger, *et al.* [29], observed no change in the ^{60}Co gamma-ray detection efficiency of 6H SiC photodiodes after a 1.080 MGy dose of ^{60}Co gamma rays, and Kinoshita, *et al.*[30], observed no change in the 100% charge-collection efficiency of 6H SiC p-n diodes after a ^{60}Co gamma-ray dose of 2.5 MGy.

Ruddy and Seidel [31,32] irradiated 4H SiC Schottky diodes to a cumulative ^{137}Cs gamma-ray dose of 22.7 MGy. ^{137}Cs gamma rays were used to simulate the radiation environment of spent-fuel assemblies five years after discharge from a nuclear power reactor. The irradiated detector was found to perform well as a detector for fast ^{252}Cf fission neutrons and for 14-Mev D-T neutrons [32].

Whereas the radiation effects of gamma rays on the response and operational lifetimes of SiC detectors are minor, the effects of charged particle and fast neutron damage are more formidable. For SiC detectors exposed in neutron environments, the radiation effects of fast neutrons as well as the charged particles produced by neutron convertor reactions (*e.g.*, ^{6}Li[n,α]^{3}H), ^{10}B[n,α]^{7}Li, ^{235}U[n,f]) are both primary concerns.

An early neutron-damage study carried out by Ferber and Hamilton [33] showed that good alpha-particle spectrometry results were obtained with a SiC diode equipped with a ^{235}U convertor foil that had been exposed to a thermal neutron fluence of 6 x 10^{15} cm^{-2}. However, the fast-neutron fluence was not reported, and any radiation damage to the detector would have been caused by fast neutrons and fission fragments.

Tikhomirova, *et al.* [34] noted degradation of Beryllium-doped 6H SiC detectors at neutron fluences greater than 10^{13} cm^{-2} for devices using a uranium convertor foil. However, the degradation was more likely due to the fission product ions from the convertor foil than from the neutrons impinging on the detector [4].

Seshadri, *et al.*, [6] found that the charge collection efficiency for ^{238}Pu decreased systematically, and that self-biased operation was not possible following an accumulated fast-neutron (E > 1 MeV) fluence of 5.7 x 10^{16} cm^{-2}. The reduction in charge-collection efficiency was systematic and corresponded to a carrier removal rate of 9.7±0.7 cm^{-1}, which the authors attributed to the introduction of deep-level traps by fast-neutron interactions.

Dulloo, *et al.*, [12,21] showed that the thermal-neutron response of a SiC Schottky diode with a ^{6}LiF convertor foil was indistinguishable from that of an unirradiated detector after a fast-neutron (E >1 MeV) fluence of 1.3 x 10^{16} cm^{-2}.

Nava, *et al.*, [35] observed only minor (~20%) losses in charge-collection efficiency for 4H SiC Schottky diodes at 1-MeV neutron fluences up to 10^{15} cm^{-2}. At fluences higher than 10^{15} cm^{-2} the charge-collection losses increase monotonically to ~80% loss at 8 x 10^{15} cm^{-2}.

Lo Giudice , *et al.*, [36] irradiated large area 4H-SiC Schottky diodes equipped with a ^{6}LiF convertors with epithermal neutrons and observed less than 0.3% decrease in the count rate after a neutron fluence of 10^{13} cm^{-2}. They attributed this minor decrease to alpha-particle damage from the ^{6}Li(n,α)^{3}H convertor layer. The more energetic ^{3}H ions would be expected to produce less localized damage.

Afanasyev, *et al.*, [37] examined the photosensitivity of SiC UV photodiodes and observed no change in performance after a fast-neutron fluence of 5 x 10^{12} cm^{-2} but observed reduction of the photosensitivity of their devices at higher fluences up to 1 x 10^{14} cm^{-2}.

A study by Wu, *et al.* [38] found that the charge collection efficiency for self-biased 4H SiC Schottky diodes reduced to 1.3% of the unirradiated value after irradiation with a fast-neutron fluence of 8.26 x 10^{14} cm^{-2}.

Systematic investigations were made by Ruddy, *et al.*, [39] of the effects on detector performance of both fast (E > 1 MeV) neutrons and energetic ^{3}H and ^{4}He ions from reactions in a ^{6}LiF convertor layer. Following neutron irradiations, the detectors were tested with a ^{238}Pu alpha-particle source as described in Reference [5] in order to evaluate charge-collection efficiency as a function of accumulated neutron and charged-particle fluence. Irradiations were carried out at two temperatures, 45 °C and 230 °C. The neutrons-only results are shown in Fig. 6, where ^{238}Pu pulse height, which was taken to be a measure of

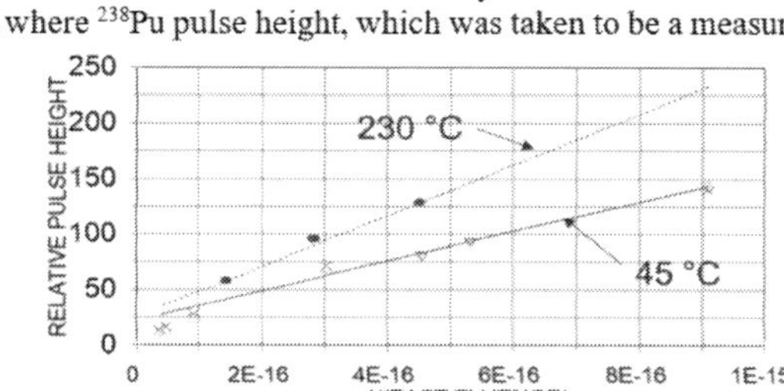

Fig. 6. Effect of elevated temperature during neutron irradiation on the observed pulse heights from a ^{238}Pu alpha test source. Pulse height is plotted as a function of reciprocal fast neutron fluence. The symbols refer to different irradiated SiC detectors. (Adapted from Reference 39)

charge-collection efficiency, is plotted as a function of reciprocal accumulated neutron fluence. The results at both temperatures are fit well by a straight line, consistent with the predictions [6,39,40] of a simple first-order model which assumes that loss of charge carriers is directly proportional to the density of charge traps produced in the SiC active volume by fast neutrons. A similar relationship was observed for the charged-particle results, where ^{238}Pu pulse height was found to be proportional to the reciprocal of the calculated displacements caused by the fast neutrons, ^{3}H ions and ^{4}He ions.

It can be seen from Fig. 6 that the 230 °C irradiations result in significantly higher charge collection in the SiC detectors than the 45 °C irradiations. It was concluded by the authors [39] that elevated temperature mitigates the effects of the radiation damage on charge collection, but that more data are needed at temperatures of 300 °C and higher. An important implication of these results is that operation in an elevated-temperature environment may *extend* the service lifetime of a SiC neutron detector.

A similar conclusion was reached by McGarrity, *et al.*[41,42], who reported results for the performance of 6H SiC JFETs irradiated with neutrons at temperatures up to 300 °C.

VI. APPLICATIONS IN HARSH NUCLEAR ENVIRONMENTS

A. *Spent Fuel Monitoring*

During reactor irradiation, nuclear fuel acquires a neutron specific activity, primarily due to the buildup of higher actinides by successive neutron reactions originating with the ^{238}U component of the fuel. [16-18] These actinides produce neutrons through spontaneous fission of isotopes such as ^{240}Pu, ^{242}Cm and ^{244}Cm. During the first operating cycle, ^{240}Pu will dominate, but for a fully burnt fuel assembly several years after discharge, the primary neutron-producing isotope will be 18.1-year ^{244}Cm. In the case of oxide fuels, neutrons are also produced by the alpha-emitters present through $^{18}O(\alpha,n)$ and $^{17}O(\alpha,n)$ reactions. Gamma rays will also be present primarily from the fission-product inventory present in the fuel. For example, a Pressurized Water Reactor fuel assembly with a burnup of 25,655 MWd/MTU had a total surface neutron fluence rate of 8000 cm^{-2}-s^{-1} at midplane three years after discharge [43], and typical gamma dose rates in spent-fuel environments are in the 10^2 -10^3 Gy/hr range [17-18].

Natsume, *et al.* [18] tested a SiC neutron detector in a spent fuel pool, and over a 2050-hour (85.4-day) period of continuous monitoring the detector neutron and gamma response did not change or deteriorate. The fluence rates were 180 cm^{-2}-s^{-1} and 11 Gy/hr for fast neutrons and gamma rays, respectively. Both the gamma-ray and neutron count rates were stable to precisions of 1.7% and 2.9% for gamma rays and neutrons, respectively.

The high radiation resistance, stability in elevated and changing temperatures and small size and versatility of SiC detectors make them extremely useful for spent-fuel monitoring applications.

B. *Safeguards Monitoring*

A related safeguards-monitoring application is long-term monitoring of spent fuel. The axial distribution of the neutron emission rate at the surface of a spent fuel assembly will depend on the burnup history of the assembly [43]. Deviations from the expected distributions can be used as an indicator of off-normal conditions resulting from diversion of fuel rods, mechanical or corrosive deterioration of the assembly, water incursion, *etc.* SiC neutron detectors can be used to monitor the neutron energy spectrum as a function of axial position either by affixing detectors near the surface of the assembly or inserting a stringer of detectors within an assembly guide tube.

It has been demonstrated [39] that after a fast (E>1MeV) fluence of 1.7 x 10^{17} cm^{-2}-sec^{-1}, SiC neutron detectors are still functional. In a typical spent fuel neutron environment, it would take 5-50 million years to reach this dose. It has also been shown that SiC detectors operate well after a ^{137}Cs dose of 22.7 MGy [39]. If a SiC detector were to be placed adjacent to or within a spent fuel assembly immediately after discharge and left in place indefinitely, this dose would *never* be reached. Therefore, SiC neutron detectors are capable of continuously monitoring a spent fuel assembly following discharge from the reactor, during transportation and following deposition in a spent fuel repository.

C. *Monitoring of Nuclear Fusion Devices*

Monitoring of nuclear fusion devices and reactors is required to obtain energy on important operational parameters such as fusion power and plasma temperature. In a fusion power reactor, such as the International Thermonuclear Experimental Reactor (ITER) currently under construction in Cadarache, France, the neutron intensity and energy distribution can be used to deduce information on the fusion power and ion temperature [44].

Fusion reactions are produced in a dense, high-temperature, magnetically confined plasma. The primary energy-producing reaction is the DT reaction accompanied by 14-MeV neutrons, but DD and other secondary reactions also take place. The plasma and its surroundings are an extremely harsh measurement environment with temperatures exceeding 150,000,000 °C. In Test Blanket Modules in the ITER wall, temperatures can reach up to 920 °C in the breeder material, up to 550 °C in the EUROFER structure, and up to 650 °C in the neutron multiplier. Magnetic fields of ~4T are present and measurements are confined to extremely limited access/space.

SiC neutron detectors are potentially capable of providing detailed neutron spectra under these conditions as discussed in Sections III-IV. Peaks from $^{12}C(n,\alpha)^{9}Be$ and $^{28}Si(n,\alpha)^{25}Mg$ reactions can provide information on the D-T fusion energy as well as energy width. The reaction peaks can be measured to an energy resolution of 2.2-2.3% FWHM, equal to the energy resolution obtainable with Si detectors [19], which have no prospects for operating in a fusion reactor environment.

Concepts are also being developed to demonstrate on-line breeding of tritium to fuel the fusion reaction. Test Blanket Modules integrated inside the tokamak wall will produce tritium via neutron reactions with lithium. SiC neutron detectors can be used to monitor the thermal-neutron fluence rate at key locations in and around the Test Blanket Modules. If a LiF convertor is used, the detector count-rate is a direct measure of the tritium production reaction rate.

In addition to neutron measurements, SiC charged-particle detectors can be deployed to obtain information on yields of charged particles (^{1}H, ^{2}H, ^{3}H, ^{3}He, ^{4}He) from fusion and breeder reactions.

D. *Monitoring of Test and Research Reactors*

New test and research reactors are coming online to address new requirements that can not be met by the current aging fleet. An example is the Jules Horowitz Reactor (JHR), which is currently under construction at CEA Cadarache, France [45]. These reactors will address nuclear materials and fuels irradiation testing needs being generated by advanced commercial nuclear power plant designs that employ innovative high-temperature coolants such as liquid metals and molten salts. Neutron monitoring is required for testing in the high-temperature and high fluence-rate environments typical of reactors such as JHR.

In the framework of the European I_SMART project [46], new 4H-SiC based neutron detectors were developed and tested and have been demonstrated to be able to operate in harsh environments and detect both fast and thermal neutrons. [47-49] Prototypes with various designs were fabricated, some of them optimized for thermal neutron detection via a specific boron converter layer (either deposited on top of the structure or implanted in the anode ohmic contact layer). These detectors have been tested at the MINERVE zero power reactor and in the SCK•CEN BR1 reactor (at fluence rates of the order of 10^9 cm^{-2}-s^{-1}) on prototypes of various sizes, with and without ^{10}B (implanted at room temperature or at 400 °C). These measurements proved the linearity of the detector responses with reactor power. In addition, thermal neutron detection spectra were recorded as a function of bias voltage. Zero-bias operation was demonstrated and shown to give the best gamma-ray discrimination consistent with the thin active volume produced by the zero-bias p-n diode potential.

Tests are planned in the Joseph Stefan Institute TRIGA reactor at fluence rates of the order of 10^{12} cm^{-2}-sec^{-1}, leading to high fluence and fluence -rate testing at the 100-MW JHR.

E. Monitoring of Advanced Nuclear Power Reactors

Many of the advanced nuclear power reactors under development utilize extremely high-temperature coolant concepts. See, for example Reference [50] The Natrium Reactor is a sodium-cooled fast reactor with a molten-salt energy storage system. The KP-FHR reactor employs fluoride-salt cooling, and the Molten Chloride Reactor also requires molten-salt cooling. The newly developed reactors will require neutron monitoring at temperatures up to 850 ° C, which will present challenges to the present gas-filled and self-powered neutron detector technology. SiC neutron detectors have been demonstrated at temperatures up to 700 °C and likely can operate at higher temperatures (4H SiC sublimes at 2857 °C).

VII. Conclusions and Recommendations

SiC neutron detectors are an emerging technology which provides many advantages for monitoring harsh nuclear environments. SiC detectors can operate up to at least 700 °C and have been shown to withstand high doses of gamma-rays, neutrons and charged particles. Additionally, measurements have shown that irradiation of SiC at high temperatures reduces the radiation damage effects of the irradiation.

More elevated-temperature data are needed to understand the effects of temperature on mitigation of radiation damage in SiC. These effects are much more dramatic than those observed from post-irradiation annealing [39], and their understanding is crucially important for evaluating the applicability of SiC detectors in high-temperature, high-fluence nuclear environments.

It has been hypothesized [33] that so-called self-annealing operation may be possible with SiC detectors. If the rate of mitigation of radiation-induced damage at an elevated temperature is greater than or equal to the rate of introduction of radiation-induced damage, the effects of radiation should not limit the lifetimes of SiC detectors in that environment. Although an intriguing possibility, it remains to be demonstrated.

References

[1] F. Nava, G. Bertuccio, A. Cavallini and E. Vittone, "Silicon Carbide and Its Use as a Radiation Detector Material," *Meas. Sci. Technol.*, vol. 19, pp 1-25, Aug. 2008. http://iopscience.iop.org/0957-0233/19/10/102001

[2] N. Strokan, A. Ivanov, and A. Lebedev, "Silicon Carbide Nuclear-Radiation Detectors," in SiC Power Materials: Devices and Applications, (Z. Feng, Ed.), New York, NY, USA, Springer-Verlag, 2009, Chapter 11 pp 411-442.

[3] F. H. Ruddy, "Silicon Carbide Radiation Detectors: Progress, Limitations and Future Directions," *Materials Research Society Online Proceedings* vol. 5761, Nov. 2013 http://journals.cambridge.org/action/displayAbstract?fromPage=online&aid=9071147

[4] F. Franceschini and F. H. Ruddy, "Silicon Carbide Neutron Detectors," in Properties and Applications of Silicon Carbide (R. Gerhardt, Ed.), Chapter 13, Rijeka, Croatia, InTech, 2011, pp 275-296.

[5] F. H. Ruddy, A. R. Dulloo, J. G. Seidel, S. Seshadri and L. B. Rowland, "Development of a Silicon Carbide Radiation Detector," *IEEE Trans. Nucl. Sci.*, vol. 45, pp 536-541, June 1998.

[6] S. Seshadri, A.R. Dulloo, F.H. Ruddy, J. G. Seidel, and L. B. Rowland, "Demonstration of a SiC neutron detector for high-radiation environments," *IEEE Trans. Electron Devices*, vol. 46, pp. 567-571, March 1999.

[7] F. Nava, P. Vanni, C. Lanzieri, and C. Canali, "Epitaxial Silicon Carbide Charge Particle Detectors," *Nucl. Instr. Meth. A*, vol. 437, pp 354-358, Dec. 1999.

[8] G. Bertuccio, R. Casiraghi, and F. Nava, "Epitaxial Silicon Carbide for X-Ray Detectors," *IEEE Trans. Nucl. Sci.*, vol 48, pp 232-233, April 2001.

[9] M. Bruzzi, S. Lagomarsino, S. Nava, and S. Sciortino, "Characteristics of Epitaxial SiC Schottky Barriers as Particle Detectors," *Diamond and Related Materials*, vol. 12, pp 1205-1208, March 2003.

[10] G. Bertuccio, "HiTSiC: High Thickness Epitaxial Silicon Carbide Detectors," presentation given in Rome, September 25, 2006 (unpublished).

[11] G. F. Knoll, Radiation Detection and Measurement, 4th Edition, New York, NY, USA, Wiley, 2019, pp 519-523.

[12] A. R. Dulloo, F. H. Ruddy, J. G. Seidel, J. M. Adams, J. S. Nico, and D. M. Gilliam, "The Neutron Response of Miniature Silicon Carbide Semi-conductor Detectors," *Nucl. Instr Meth. A*, v. 422, pp 47-48, Feb. 1999.

[13] F. Issa, L. Ottaviani, D. Szalkai, L. Vermeeren, V. Vervisch, A. Lyoussi, R. Ferone, A. Kuznetsov, M. Lazar, A. Klix, O. Palais, and A. Hallén, "Improvements in Realizing 4H-SiC Thermal Neutron Detectors," ISRD 15 – International Symposium on Reactor Dosimetry, (A. Lyoussi [Ed.]), EPJ Web of Conferences, vol. 106, 05004, 2016.

[14] K. C. Mandal, T. A. Chowdhury, C. Oner, and F. H. Ruddy, "Design and Response Testing of Boron-Diffused Silicon Carbide Neutron Detectors for Dosimetry and Monitoring Applications," ASTM STP1608, (M. H. Sparks, K. R. Depriest, and D. W. Vehar, [Eds.]), W. Conshohocken, PA, USA, ASTM International, 2018, pp. 353–360.

[15] F. H. Ruddy, J. G. Seidel, T. E. Blue, and D. W. Miller, "Reactor Power Monitoring Using Silicon Carbide Fast Neutron Detectors," in PHYSOR-2006 – American Nuclear Society's Topical Meeting on Reactor Physics, LaGrange Park, Illinois, USA, American Nuclear Society, 2006.

[16] A. R. Dulloo, F. H. Ruddy, J. G. Seidel, C. Davison, T. Flinchbaugh, and T. Daubenspeck, "Simultaneous Measurement of Neutron and Gamma-Ray Radiation Levels from a TRIGA Reactor Core Using

Silicon Carbide Semiconductor Detectors," *IEEE Trans. Nucl. Sci*, vol. 46, pp 275-279, June1999.

[17] A. R. Dulloo, F. H. Ruddy, J. G. Seidel, T. Flinchbaugh, C. Davison, and T. Daubenspeck, "Neutron and Gamma Ray Dosimetry in Spent-Fuel Radiation Environments Using Silicon Carbide Semiconductor Radiation Detectors," in Reactor Dosimetry: Radiation Metrology and Assessment, ASTM STP 1398, (John G. Williams, David W. Vehar, Frank H. Ruddy, and David M. Gilliam, [Eds.]), W. Conshohoken, PA, USA, American Society for Testing and Materials, 2001, pp 683-690.

[18] T. Natsume, H. Doi, F. H. Ruddy, J. G. Seidel, and A. R. Dulloo, "Spent Fuel Monitoring with Silicon Carbide Semiconductor Neutron/Gamma Detectors," *Journal of ASTM International*, Online vol. 3, March 2006.

[19] F. H. Ruddy, J. G. Seidel, and F. Franceschini, "Measurements of the Recoil-Ion Response of Silicon Carbide Detectors to Fast Neutrons," in Reactor Dosimetry State of the Art 2008, (W. Voorbraak, L. Debarberis, P. D'hontd, and J. Wagemans, Eds.), London, England, World Scientific, 2009, pp 77-84.

[20] F. Franceschini and F. H. Ruddy "Simulation of the Response of Silicon Carbide Fast Neutron Detectors," in Reactor Dosimetry State of the Art 2008, (W. Voorbraak, L. Debarberis, P. D'hontd, and J. Wagemans, Eds.), London, England, World Scientific, 2009, pp 128-135.

[21] A. R. Dulloo, F. H. Ruddy, J. G. Seidel, J. M. Adams, J. S. Nico, and D. M. Gilliam, "The Thermal Neutron Response of Miniature Silicon Carbide Semiconductor Detectors," *Nuclear Instr. and Meth. A*, vol. 498, pp 415-423, Feb. 2003.

[22] A. R. Dulloo, F. H. Ruddy, J. G. Seidel, and B. Petrović, "Monitoring of D-T Accelerator Neutron Output in a PGNAA System Using Silicon Carbide Detectors," *Applications of Accelerators in Research and Industry – Sixteenth International Conference*, American Institute of Physics Conference Proceedings, vol. 576, p 499, September 2001.

[23] O. Obraztsova, L. Ottaviani, A. Klix, T. Döring, O. Palais, A. Lyoussi, "Comparing the response of a SiC and a sCVD Diamond detectors to 14 MeV neutron radiation," *IEEE Transactions on Nuclear Science*, vol. 65, pp. 2380-2384, Sept. 2018.

[24] R. Ferone, F. Issa, D. Szalkai, A. Klix, L. Ottaviani, S. Biondo, V. Vervisch, L. Vermeeren, R. Saenger, A. Lyoussi, "SiC-based neutron detector in quasi-realistic working conditions: efficiency and stability at room temperature and high temperatures under fast neutron irradiations," in 4th International Conference on Advancements in Nuclear Instrumentation Measurement Methods and their Applications (ANIMMA), IEEE, pp. 1–4, 2015.

[25] D. Szalkai, R. Ferone, F. Issa, A. Klix, M. Lazar, A. Lyoussi, L. Ottaviani, P. Tutto, V. Vervisch, "Fast Neutron Detection with 4H-SiC Based Diode Detector up to 500 °C Ambient Temperature," *IEEE Trans. Nucl. Sci*, vol. 63, pp 1491-1498, June 2016.

[26] Y.M. Abubakar, A. Lohstroh, and P.J. Sellin, "Stability of Silicon carbide particle detector performance at elevated temperatures," *IEEE Transactions on Nuclear Science*, vol. 62, pp. 2360-2366, October 2015.

[27] M. Angelone, S. Cesaroni, S. Loreti, G. Pagano, and M. Pillon, "High temperature response of a single CVD diamond detector operated in current mode," *Nuclear Inst. And Meth. Phys. Res. A*, vol. 943, Nov. 2019, 162493.

[28] S. M. Kang, J. H. Ha, S. H. Park, H. S. Kim, S. D. Chun, and Y. K. Kim, "Study of the Current-Voltage Characteristics of a SiC Radiation Detector Irradiated by Co-60 Gamma-Rays," *Nuclear Inst. And Meth. Phys. Res. A*, vol. 579, pp 145-147, Aug. 2007.

[29] S. Metzger, H. Henschel, O. Kohn, W. Lennartz, "Silicon Carbide Radiation Detector for Harsh Environments," *IEEE Trans. Nucl Sci.*, vol. 49, pp 1351-1355, June 2002.

[30] A. Kinoshita, M. Iwami, K. Kobayashi, I. Nakano, R. Tanaka, T. Kamiya, A. Ohi, T. Oshima, Y. Fukushima, "Radiation Effect on pn-SiC Diode as a Detector," *Nucl. Instr. and Meth. A*, vol. 541, pp 213-220, April 2005.

[31] F. H. Ruddy and J. G. Seidel, "The Effects of Intense Gamma Irradiation on the Alpha-Particle Response of Silicon Carbide Semiconductor Radiation Detectors," *Nucl. Inst. And Meth. Phys. Res. B*, vol. 263, pp 163-168, Oct. 2007.

[32] F. H. Ruddy and J. G Seidel, "Effects of Gamma Irradiation on Silicon Carbide Semiconductor Radiation Detectors," *IEEE Nuclear Science Symposium Conference Record*, Paper #4179063, pp 583-587, 2006.

[33] R. R. Ferber and G. N. Hamilton, "Silicon Carbide High-Temperature Neutron Detectors for Reactor Instrumentation," *Nuclear Technology* vol. 2, pp 246-251, June 1966.

[34] V. A. Tikhomirova, O. P. Fedoseeva and V. V. Bol'shakov, "Silicon Carbide Detectors as Fission-Fragment Counters in Reactors," *Izmeritel'naya Tekhnika*, vol.6, pp 67-68, June 1973.

[35] F. Nava, A. Castaldini, A. Cavallini, P. Errani and V. Cindro, "Radiation Detection Properties of 4H-SiC Schottky Diodes Irradiated Up to 10^{16} n/cm^2 by 1 MeV Neutrons,", *IEEE Trans. Nucl. Sci*, vol. 53, pp 2977-2982, Oct. 2006.

[36] A. Lo Giudice, F.Fasolo, E. Durisi, C. Manfredotti, E. Vittone, F. Fizzotti, A. Zanini, and G. Rosi, "Performance of 4H-SiC Schottky Diodes as Neutron Detectors," *Nucl. Instr. and Meth. A*, vol. 583, pp 177-180, Dec. 2007.

[37] A.V. Afanasyev, V.A. Ilyin, V.V. Luchinin, S.A. Reshanov, A. Schöner, K.A. Sergushichev and A.A. Smirnov, "Effect of neutron irradiation on epitaxial 4H-SiC PiN UV-photodiodes," *Materials Science Forum*, vol. 897, pp 614-617, May 2017.

[38] J. Wu, Y. Jiang, J. Lei, X. Fan, Y. Chen, M. Li, D. Zou and B. Liu, "Effect of Neutron Irradiation on Charge Collection Efficiency in 4H-SiC Schottky Diode," *Nucl. Instr. and Meth. A*, pp 218-222, Jan. 2014.

[39] F. H. Ruddy, A. R. Dulloo and J. G. Seidel, "Study of the Radiation Resistance of Silicon Carbide Radiation Detectors," *Trans.Am. Nucl. Soc.*, vol. 90, pp 348-349, June 2004.

[40] G. C. Messenger and M. S. Ash, The Effects of Radiation on Electronic Systems, New York, NY, USA, Van Nostrand Reinhold, 1991.

[41] F. B. McLean, J. M. McGarrity, C. J. Scozzie, C. W. Tipton, and W. M. DeLancey, "Analysis of Neutron Damage in HighTemperature Silicon Carbide JFETs," in High-Temperature Electronics, (R. Kirschman,Ed.), Institute of Electrical and Electronics Engineers, 1999, pp.530-540.

[42] C. J. Scozzie, J. M. McGarrity, J. Blackburn, and W. M. DeLancey, Silicon Carbide FETs for High Temperature Nuclear Environments," *IEEE Trans. Nucl. Sci.*, vol. 43, pp 1642-1648, June 1996.

[43] R. Gold, F. H., Ruddy, E. P. Lippincott, W. N. McElroy, and J. H. Roberts, "Spent Thermal Reactor Fuel Assembly Characterization with Solid State Track Recorders," in Dosimetry Methods for Fuels Cladding and Structural Materials - Proceedings of the Third ASTM-Euratom Symposium on Reactor Dosimetry, (H.Röttger, [Ed.]), EUR-6813, Ispra, Italy, Commission of European Communities, pp. 858-867, Oct. 1980.

[44] L. Bertalot, V. Krasilnikov, L. Core, A. Saxena, N. Yukhnov, R. Barnsley, and M. Walsh, "Present Status of ITER Neutron Diagnostic Development," *J. Fusion Energy*, vol. 38, pp 283-290, July 2019.

[45] D. Iracane, and P. Yvon, "The JHR reactor: a multipurpose asset for materials," *Clefs CEA*, vol. 55, pp 102-105, 2007.

[46] A. Lyoussi, "I_SMART a Collaborative Project on Innovative Sensors for Material Ageing and Radiation Testing: European Innovative Project for SiC Applications in Harsh Media," KIC_Innoenergy,SSCV_SMART Proposal/Exhibit 1, 2012.

[47] F. Issa, V. Vervisch, L. Ottaviani, D. Szalkai, L. Vermeeren, A. Lyoussi, A. Kuznetsov, M. Lazar, A. Klix, O. Palais, and A. Hallen, "Radiation Silicon Carbide Detectors based on Ion Implantation of Boron," *IEEE Trans. Nucl. Sci*, Vol. 61, pp 2105-2111, Aug. 2014.

[48] F. Issa, L. Ottaviani, D. Szalkai, L. Vermeeren, V. Vervisch, A. Lyoussi, R. Ferone, A. Kuznetsov, M. Lazar, A. Klix, O.Palais, "4H-SiC Neutron Sensors Based on Ion Implanted ^{10}B Neutron Converter Layer," *IEEE Trans. Nucl. Sci*, vol. 63, pp 1967-1980, June 2016.

[49] V. Radulovic, Y. Yamazaki, Z. Pastuovic, A. Sarbutt, K. Ambrozic, R. Bernat, Z. Eres, J. Coutinho, T. Ohshima, I. Capan, and L. Snoj, "Silicon Carbide neutron detector at the JSI TRIGA reactor for enhanced border and port security," *Nucl. Inst. And Meth. Phys. Res. A*, vol. 972, 164122, Aug. 2020.

[50] United States Department of Energy, "Advanced Reactor Development. Paving a Path to Commercialization" https://www.energy.gov/ne/articles/infographic-advanced-reactor-development .

The effect of the aging of liquid organic scintillators used for gamma-neutron separation

Jiří Janda[1], Jaroslav Jánský[2], Věra Mazánková[2], František Cvachovec[2]
[1]University of Defence, NBC Defence Institute, Vyškov, Czech Republic
[2]University of Defence, Faculty of Military Technology, Department of Mathematics and Physics, Brno, Czech Republic
Corresponding author: jiri.janda@unob.cz

***Abstract*— Since the beginning of using liquid scintillators for gamma-neutron separation, there have been many articles dealing with long-term degradation especially due to oxygen presented during scintillator encapsulation. The effect of aging of liquid organic scintillators namely EJ301, EJ309 (both Eljen Technology), and new custom-made cocktails based on 1-Phenyl-3-(2,4,6-trimethylphenyl)-2-pyrazoline and 2,5-Bis(5-tert-butyl-benzoxazol-2-yl)thiophene fluors were investigated for more than half a year. The research was focused on the Compton edge shifting of gamma particles since the position is proportional to the light yield of the selected scintillator. Furthermore, the gamma-neutron separation was observed and quantified using FOM (Figure Of Merit) for samples prepared and filled under normal and nitrogen atmosphere during the mentioned period. All stated parameters of liquid scintillator NE213 manufactured by Nuclear Enterprises Limited opened more than three decades ago were measured and used for comparison.**

***Keywords* — liquid organic scintillator; γ-n discrimination; long-term stability; scintillator aging.**

I. Introduction

QUENCHING due to dissolved oxygen is an important effect and must be avoided to obtain optimum scintillation efficiency. The explanation is given, that the quenching at aromatic molecules of the singlet state leads to the creation of a triplet state. The ground state of the oxygen molecule is a triplet and the next state is a singlet lying about 0.98 eV over the ground state [1]. Therefore, oxygen molecules surrounding the aromatic molecules can absorb the energy of the singlet state of aromatic molecules and make a spin allowed transition to the triplet state causing a decrease of the fluorescence. Such a transition only occurs in aromatic molecules, which have an energy gap of S1-T1 greater than 0.98 eV [2].

The magnitude of the quenching factor depends on the oxygen solubility in the solvent, which is determined by the external pressure and the temperature, and on the effect of oxygen quenching on the solvent and fluor and possibly other presented compounds (e.g. emulsifiers). At low concentrations, transfer quenching is predominant, but at increased concentrations, it decreases more rapidly than solute quenching. It appears likely that, at practical concentrations, near-ideal concentration for a given system, oxygen quenching of the fluor emission is the main factor.

Recently, few attempts were made to evaluate the basic performances of 18 years old scintillation cocktail. It has been shown that the 18-year storage has less impact on the basic performances (including pulse shape analysis) than batch-to-batch variability in cocktail compositions [3].

This study was focused on the evaluation of the long-term basic characteristics, such as Compton edge stability and g-n PSD, of the EJ301, EJ309, and different custom-made di-iso-propylnaphthalene (DIPN) based liquid scintillation cocktails encapsulated under the nitrogen and air atmosphere.

II. Experimental methods and materials

A. Chemicals and materials

The NE213 cocktail was obtained by Nuclear enterprises limited, UK (1986). Two commercially available scintillation cocktails were obtained by Eljen technology, USA - EJ301 scintillation cocktail based on xylene solution and 2,5-Diphenyloxazole (PPO) and EJ309 scintillation cocktail based on DIPN solution and PPO. Two custom-made scintillators based on DIPN solution were prepared using 1-Phenyl-3-(2,4,6-trimethyl-phenyl)-2-pyrazoline (labeled PYR) and 2,5-Bis(5-tert-butyl-benzoxazol-2-yl)thiophene (labeled THIO) as luminophore. These luminophores were selected based on previous studies [4, 5]. Both custom-made cocktails were prepared in two concentrations of luminophore, 2 $g.L^{-1}$ (labeled 2) and 5 $g.L^{-1}$ (labeled 5), to verify the effect of quenching on luminophore concentration. Furthermore, two batches were prepared due to encapsulation under the nitrogen and ambient atmosphere.

Encapsulation of the liquid scintillators was performed using 20 mL glass vials with a low concentration of K-40 (Wheaton, UK) and the Teflon tape was used as a reflector.

An encapsulated stilbene crystal, with a diameter of 45 mm and a length of 45 mm is used for comparison.

The Cs-137 radionuclide was used for energy calibration in the same way as in [4]. The Cf-252 radionuclide was used to evaluate the neutron-gamma separation. The gamma-neutron discrimination was performed using NGA- 01 [6], which is the two-parameter spectrometric system and was used in earlier works [4, 7, 8].

B. Procedures

Preparation of scintillation cocktails

The volume of 14 mL of the selected liquid scintillator was poured inside the 20 ml glass vials with an upper lid containing a Teflon seal to ensure gas tightness and to better reflect photons to the photomultiplier tube (PMT). The glass vial was wrapped with Teflon tape to reflect photons. Two batches of previously mentioned scintillators were prepared this way; the first batch, labeled "N", was prepared under a nitrogen atmosphere and each liquid scintillator was bubbled with nitrogen for several minutes. After the selected period, the vial was properly closed. The second batch was prepared under an ambient atmosphere with no bubbling and was labeled as "A". The batches during the whole observation period were stored in a dry, clean, and dark place with temperatures varying from 19°C to 24 °C.

Measurement setup

The selected scintillator was placed on an RCA 8575 PMT and enclosed in a light-tight box. The radiation source was placed on the holder which was about 4 cm far from the scintillator horizontally. The PMT's anode output was connected to NGA-01. The high voltage of the PMT was set to 1650 V for each experiment. This should guarantee the fixed position of the Compton edge.

The Cs-137 was used for energy calibration, due to the creation of the Compton edge (CE) at 0.477 MeV. This energy belongs to a certain channel, which was determined using the derivative (differentiation) method. Due to the calibration, the term MeV electron equivalent (MeVee) could be introduced to quantify the light yield. The energy calibration was based on a linear function fit of Cs-137 measured data and zero energy channel and zero energy [4].

Gamma-neutron separation was measured using the same setup as was described in the paragraph dealing with the energy calibration, but instead of Cs-137, the neutron source Cf-252 was used.

To guarantee satisfying results, more than a million pulses were acquired in each experiment. The count rate was in the range of from 150 up to 300 cps for both mentioned radiation sources.

III. Results and discussion

The long-term observation of mentioned parameters started on the 30th of September 2020 and ended on the 26th of May 2021. Every month in this interval, the measurements of these two batches were performed using the same adjustment as described above.

A. Light yield

Since the magnitude of light collected at the photocathode of the PMT is proportional to the height of the electrical pulse, the light yield can be measured indirectly using the position of the Compton edge. As was stated in the introduction, the oxygen prevents excitation of the luminophore, thus causing quenching. Therefore, there should be some difference in the position of CE between samples prepared under the nitrogen and ambient atmosphere.

Since the diameter of the used PMT was 51 mm, the variance in CE position of the used scintillators was performed. This variance of the CE was caused due to differently placed vials on the face of the PMT, different orientations of the vial, and different states of optical contact. It was observed, that the deviation was up to 7% for all scintillators based on 10 repetitions for each cocktail.

As can be seen in Table I, the differences in the position of the CE of samples prepared under nitrogen and ambient atmosphere on the first day corresponded with those previously published [9, 10], and the difference in quenching, i.e. position of CE between samples prepared under nitrogen/ambient atmosphere and lower and higher (optimal concentration C_O) is visible. Furthermore, the values of the standard deviation of the CE means of investigated scintillation cocktails and concentrations expressed in percentage are within the range of variation, when the samples are changed. The obtained data support previously published data, that the dissolved oxygen causes only quenching at some degree (based on the composition of the solvent-solute system), as was stated above.

https://doi.org/10.1051/epjconf/202125311004

TABLE I
THE DEPENDENCE OF THE COMPTON EDGE POSITION ON THE DAY OF MEASUREMENT

	Compton edge position [channel] vs time [day of measurement]											
Day	**EJ 301 N**	**EJ 301 A**	**EJ 309 N**	**EJ 309 A**	**PYR 2 N**	**PYR 2 A**	**PYR 5 N**	**PYR 5 A**	**THIO 2 N**	**THIO 2 A**	**THIO 5 N**	**THIO 5 A**
1.	329	316	310	306	294	268	324	290	267	238	248	247
9.	298	260	323	316	291	283	307	297	254	252	260	255
22.	331	312	325	303	293	279	312	313	268	258	270	275
36.	287	277	315	316	296	286	317	308	266	255	251	258
50.	284	276	321	311	285	279	275	302	262	248	246	248
66.	294	292	310	300	280	286	281	305	263	247	253	252
99.	314	292	320	325	265	260	282	300	230	225	233	231
126.	280	286	319	303	276	261	264	293	262	243	239	249
153.	309	288	330	307	270	279	269	307	246	225	237	226
188.	304	284	331	314	248	269	266	299	259	229	231	228
238.	292	282	314	301	230	238	249	283	234	214	211	213
Mean	**302**	**288**	**322**	**309**	**275**	**272**	**286**	**300**	**255**	**240**	**244**	**244**
σ [%]	**6**	**6**	**4**	**4**	**8**	**5**	**9**	**3**	**5**	**6**	**7**	**7**

HV of PMT was set to 1650 V. N and A stand for nitrogen and air atmosphere respectively.

B. Gamma-neutron discrimination

In addition to measuring the stability of the CE, the efficiency of gamma-neutron discrimination was also observed. To characterize the discrimination of two peaks originated from gamma and neutron radiation, the Figure of Merit (FOM) was used in the same way as in [4]. It was observed (Fig. 1) that the discrimination, especially at low energies, was improving with elapsing time. This improvement was noticeable for all cocktails in the energy range up to 0.4 MeVee. The highest improvement was observed in cocktails marked THIO. Since this luminophore is very problematic to dissolve in DIPN solvent, it is believed, that after the dissolution process, undissolved microscopic particles remained in the cocktail and they were gradually dissolving during the first weeks. On the other hand, the most stable cocktails were based on DIPN solvent and PPO and PYR luminophore prepared under a nitrogen atmosphere. The influence of the nitrogen atmosphere was especially noticeable for the xylene + PPO and DIPN + THIO systems. Systems composed of DIPN + PPO and DIPN + PYR did not show clear differences.

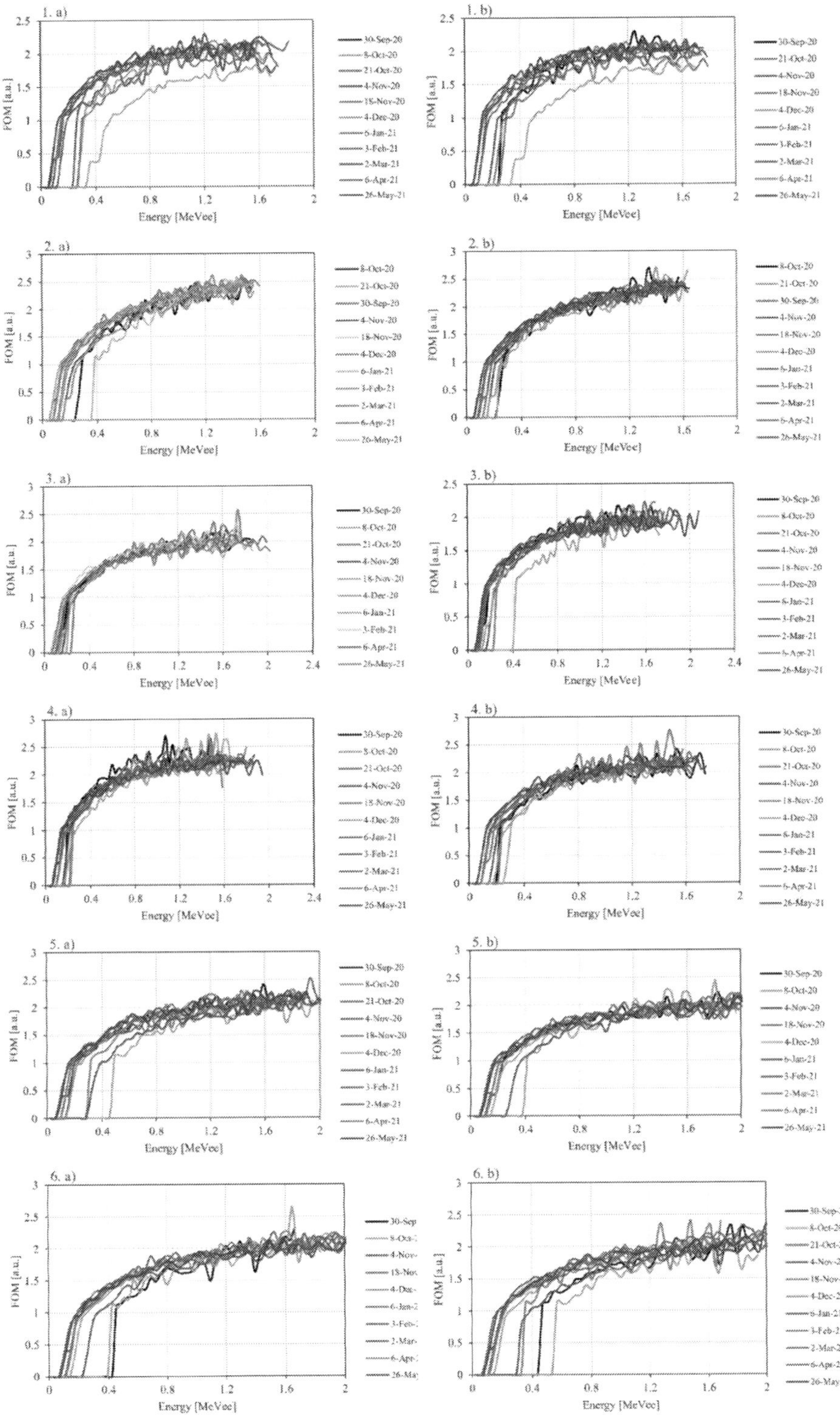

Fig. 1. The gamma-neutron separation of EJ301 (1), EJ309 (2), PYR 2 (3), PYR 5 (4), THIO 2 (5), and THIO 2 (6) scintillation cocktails prepared under nitrogen (a) and ambient (b) atmosphere expressed in FOM.

https://doi.org/10.1051/epjconf/202125311004

C. Comparison with NE213

To verify and compare our results, the NE213 scintillator was measured. This scintillator has a similar composition as EJ301, which means it is based on xylene solvent with PPO as a luminophore. This bottle was opened in 1986 meaning that the effect of oxygen deterioration should be visible after more than 35 years.

To perform energy calibration, the Compton edge of Cs-137 had to be investigated. However, the data obtained when NE213 was measured was completely different and as can be seen in Fig. 2, the shape or even the evidence of CE was almost missing. The position of the CE was estimated to be at the 180th channel at HV=1750 V.

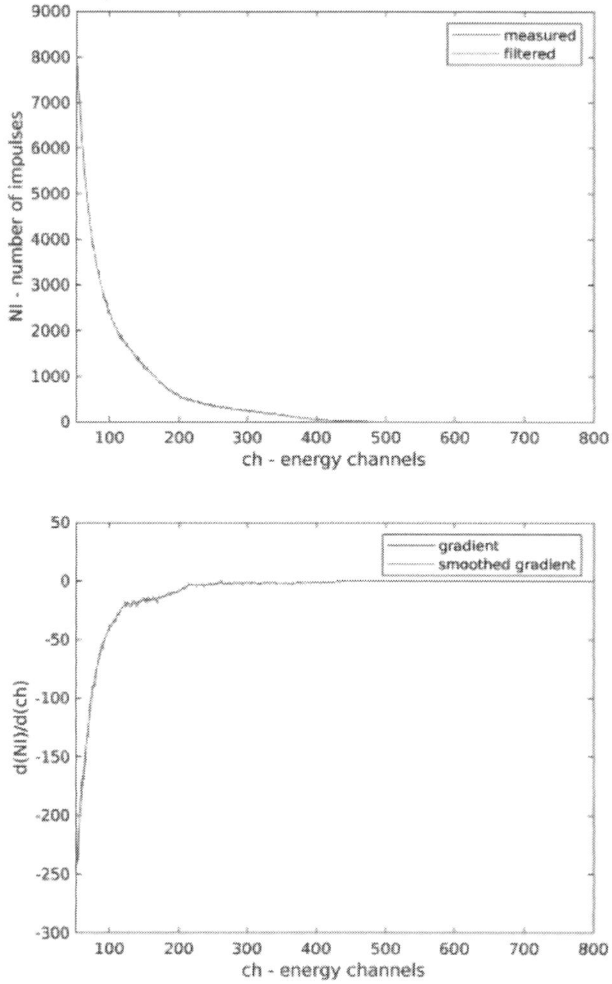

Fig. 2. Energy calibration using CE of Cs-137 at high voltage of PMT 1750 V.

Despite the worse energy threshold for good discrimination, the gamma-neutron separation was still possible even after 35 years from the opening of the bottle and exposure to the cocktail to atmospheric oxygen. As can be seen (Fig. 3) in comparison with other scintillation cocktails (same geometry of measurement) the light output was lower and the energy resolution was bad.

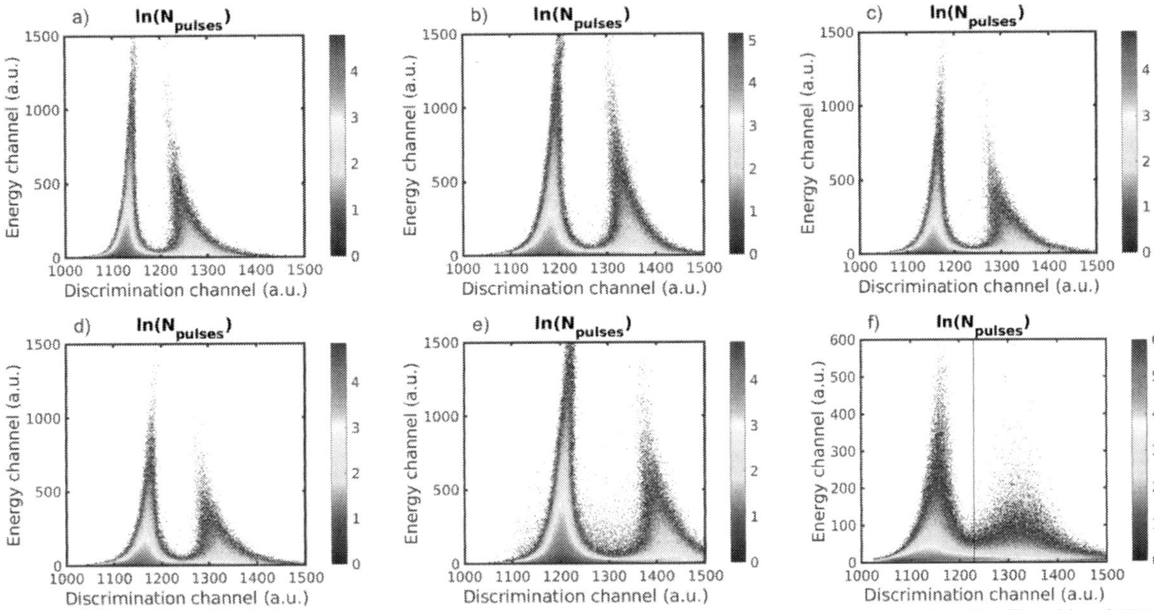

Fig. 3. Two-parameter histogram (left gammas, right neutrons) from the measurement of EJ301 (a), EJ309 (b), PYR 5 (c), THIO 5 (d), stilbene (e), and NE213 (f) using CE of Cs-137; HV of PMT in cases (a-e) was 1650 V, in the case of (f) was 1750 V.

IV. Conclusion

This study aimed to evaluate the long-term performance of different scintillation cocktails prepared under nitrogen and ambient atmosphere. Especially, the focus was aimed to observe the stability of the Compton edge due to the light yield and, ability to discriminate between gamma rays and neutrons.

The effect of nitrogen and ambient atmosphere on CE position was following literature, where the higher shift in the position of the CE was observed at low concentration of luminophore and for higher concentration, the quenching was negligible in comparison of the variance in the position of CE due to sample changing, which was 7 %. It was proved, that the position of the CE in time is not stable with a negative slope at almost all cocktails meaning, that the light yield was constantly decreasing.

To compare observed results, the 35 years old and opened NE213 scintillation cocktail was measured and evaluated. It was demonstrated, that the cocktail didn't exhibit any visible Compton edge and therefore the energy calibration was very difficult to perform. Despite this difficulty, the cocktail was able to separate pulses originated from gamma rays and neutrons. The explanation of this worsening probably lies in chemical changes in the phosphor molecules caused by oxygen (e.g. oxidation) rather than in simple interrupting energy transfer from solvent to luminophore.

Acknowledgment

This research work has been carried out within the ADAR project (Accelerator Driven Advanced Reactor). The authors gratefully acknowledge the financial support from the Ministry of Education, Youth, and Sports of the Czech Republic under the INTER-ACTION research program (project No. LTAUSA 18198).

References

[1] T. L. Brewer, "Oxygen quenching of the aromatic excited singlet state," *J. Am. Chem. Soc.*, vol. 93, no. 3, Feb. 1971.

[2] X. Hua-Lin, D. Jing-Shan, and W. Nai-Yan, "Oxygen quenching in a LAB based liquid scintillator and the nitrogen bubbling model," *Chinese Phys. C*, vol. 34, no. 5, May 2010.

[3] X. gui Feng, Q. ge He, J. chen Wang, and J. Chen, "The long-term stability on basic performances of a diisopropylnaphthalene-based liquid scintillation cocktail," *Appl. Radiat. Isot.*, vol. 70, no. 8, pp. 1536–1540, 2012.

[4] J. Jánský, J. Janda, V. Mazánková, and F. Cvachovec, "Optimization of composition of liquid organic scintillators for fast neutron spectrometry," *Nucl. Instruments Methods Phys. Res. Sect. A Accel. Spectrometers, Detect. Assoc. Equip.*, p. 165523, 2021.

[5] J. Janda and E. Rajchl, "Study of new luminophores for use in modern scintillation cocktails," *J. Lumin.*, vol. 201, pp. 390–396, Sep. 2018.

[6] Z. Matěj, J. Cvachovec, F. Cvachovec, and V. Prenosil, "Digitized two-parameter spectrometer for neutron-gamma mixed field," *Adv. Mil. Technol.*, vol. 7, pp. 51–58, 2012.

[7] M. Košťál *et al.*, "Comparison of fast neutron spectra in graphite and FLINA salt inserted in well-defined core assembled in LR-0 reactor," *Ann. Nucl. Energy*, vol. 83, pp. 216–225, Sep. 2015.

[8] M. Košťál *et al.*, "Measurement and calculation of fast neutron and gamma spectra in well defined cores in LR-0 reactor," *Appl. Radiat. Isot.*, vol. 120, pp. 45–50, Feb. 2017.

[9] J. B. Birks and F. W. K. Firk, "The Theory and Practice of Scintillation Counting," *Phys. Today*, vol. 18, no. 8, pp. 299–300, Aug. 1965.

[10] G. Laustriat and A. Coche, "Influence de la température sur les scintillateurs liquides," *J. Phys. le Radium*, vol. 21, no. 5, pp. 487–489, 1960.

Development of a position-sensitive fast scintillator ($LaBr_3(Ce)$) detector setup for gamma-ray imaging application

Biswajit Das, R. Palit, R. Donthi, A. Kundu, Md. S. R. Laskar, P. Dey, D. Negi, F. S. Babra, S. Jadhav, B. S. Naidu, and A. T. Vazhappilly

Department of Nuclear and Atomic Physics,
Tata Institute of Fundamental Research, Mumbai - 400 005, India
Corresponding author: biswajit.das@tifr.res.in

Abstract—**We have characterized a Cerium doped Lanthanum Bromide ($LaBr_3(Ce)$) crystal coupled with the position-sensitive photo-multiplier system for the gamma-ray imaging application. One can use this detector set-up for the scanning of high purity germanium detectors for pulse shape analysis in gamma-ray spectroscopy experiments and the image formation of an object by Compton back-scattering . The sensor has been tested for energy, timing and position information of the gamma-rays interacting within the detector crystal. The GEANT4 simulation results are consistent with the experimental results. We have reconstructed the image of irradiation spots in different positions throughout the detector crystal. Position resolution is found to be around 3.5 mm with the 2 mm collimated gamma-rays. The 2-d image of hexagonal Bismuth Germanate (BGO) crystal and a cylindrical $LaBr_3(Ce)$ crystal have been reconstructed in coincidence technique. The performance of the detector for imaging application has been investigated by coincidence technique in GEANT4 simulation and compared with the experimental data. We have reconstructed the 2-d images of objects with various geometrical shapes by Compton back-scattered events of the gamma-rays. This position-sensitive detector can be used as an absorber of a Compton camera for the image reconstruction of an extended radioactive source. One can also use this kind of set-up as in radiation imaging and many other applications where the energy and source position of the gamma-ray is the main interest.**

Keywords —**γ-ray instrument; $LaBr_3$ (Ce) crystal coupled with PSPMT; γ-ray imaging**

I. Introduction

Position Sensitive Photo Multiplier Tube (PSPMT) coupled with scintillator crystal converts the associated scintillation light into a current signal in the localized photo-cathode [1, 2]. The current signal distribution contains important information regarding energy, timing and the interaction position of the gamma-ray in the detector crystal [3, 4, 5]. Such position measurements are also often carried out using other detector arrays by combining signal channels in a multichannel analog readout, or by using complex segmented detectors, but potentially, at the expense of other performance parameters, and also not viable in compact arrangements. In the present work, a Ce doped $LaBr_3$ crystal coupled with a position sensitive photo-multiplier assembly has been characterized, where the signal processing unit could be made highly compact by coupling a resistive chain with the PMT. In this way, the electronics for signal processing can be reduced without impacting the image quality [4]. There are some work on position characterization of the similar kind of detector where individual readout has been read and digitized[6, 7]. This setup has a versatile impact, either as a stand-alone assembly or in concurrence with an ancillary detector in various important fields. It is expected to find use in i) scanning of segmented high purity Ge (HPGe) detectors for pulse shape analysis [8-11] in gamma-ray spectroscopy experiments, in order to reconstruct the interaction path within the detector volume, ii) testing the electrical response of granular Si charged-particle detectors/sensors as a function of the interaction position of incident particle [12], iii) medical imaging of large organs/sections of the body, for situations where conventional cameras may be difficult to optimize, iv) small animal imaging for pre-clinical research in the pharmaceutical sector, v) the defense and nuclear waste-disposal sectors for identification of radiation sources as well as for environmental radiation surveys by means of a scatterer-absorber configuration.

In this work, a $LaBr_3(Ce)$ scintillator crystal, with high light output, low self-absorption, and high sensitivity [13, 14] for low energy gamma-rays, has been coupled with a compact readout PSPMT and employed for gamma-ray imaging. This detector has an energy resolution value of about 6% for 511 keV gamma-ray photo peak. The signal produced in the photomultiplier tube by scintillation light output of the crystal has been distributed through a resistive chain circuit. The front, side, and back view of the detector assembly are shown in Fig. 1. The scintillator crystal is chosen on account of its good timing resolution, in order to reject the scattered and background events by coincidence technique for image reconstruction [15]. Spots at different positions on the PSPMT-coupled-crystal have been irradiated, and the position resolution has been determined therefrom. Subsequently, the 2-d images of auxiliary detectors, such as a hexagonal-shaped BGO crystal as well as a cylindrical $LaBr_3(Ce)$ crystal, have been reconstructed in the PSPMT detector using the coincidence technique. The experimental results are found to be in good agreement with simulations carried out using the GEANT4 architecture (version 4.10.05). Subsequently, the applicability of the present system for imaging via the Compton back-scattering technique has been illustrated by simulating the 2-d image of a ring-shaped scatterer. Further more, this kind of setup can be use as a Compton's camera by adding a thin position sensitive detector as scatterer with this $LaBr_3$(Ce)-PSPMT detector.

Fig. 1. $LaBr_3$(Ce)-PSPMT detector coupled with resistive network: a) Front; b) Side; c) Back views, showing four anode (position) signals at four corners along with the cathode signal (lower middle).

II. Materials and Detector Details

The position sensitive detector crystal used in this work has been manufactured using pure $LaBr_3$ with 5% doping of Ce. The crystal was grown by Saint Gobain Crystals and Detectors [16]. It has a high density (5.08 gm/cm 3), high light output of ≈ 63 photons/keV, and a fast timing characteristics with time resolution less than 500 ps. ^{138}La (natural abundance 0.089%) is known to decay by two parallel paths, namely (i) β^- decay to ^{138}Ce (34.4%) and (ii) electron capture decay to ^{138}Ba (65.6%) [17]. These processes are associated with the emission of 789 keV and 1436 keV γ-transitions [18], respectively, which also contribute to the total internal activity of the detector material, measured to be ~ 3 counts sec^{-1} cm^{-3} in the present work. An appropriate thickness of the crystal is chosen that allows seamless detection for a few hundred keV gamma-rays, without much contamination owing to internal activity. The crystal has dimensions measuring 5.08 cm × 5.08 cm × 0.6 cm , and has been coupled to a compact readout PSPMT (Hamamatsu-8500C) [19] through a 5 mm long light guide. The entire assembly is housed inside a 0.5 mm thick aluminium casing. The PSPMT of demountable design is equipped with a mesh of 8 horizontal (X) and 8 vertical (Y) anode-wires, as well as a bi-alkali photo-cathode. The output signal from the photo-cathode is distributed among the anode mesh-wires through an external network of resistors, from which four position signals are derived by means of charge-dividing detection. This allows a compact setup with nominal electronic circuitry.

III. Experimental Details

The cathode signal of the $LaBr_3$(Ce)-PSPMT gives the gamma-ray energy and timing information, and the four anode signals contain X-Y position information of the gamma-ray interaction point in the detector crystal. An operating voltage of -1050 V leads to low internal noise and high signal gain. For characterization of the position sensitive detector, such as estimation of spatial resolution of irradiation spots, as well as for position calibration of the crystal, a ^{57}Co (0.5 μCi) radioactive source (122 keV gamma-ray) [5] has been used. In the course of data acquisition, the performance of the crystal as an imaging detector has been investigated, wherein the events due to internal radioactivity and scattering within the crystal have been suppressed as a result of acquiring data in coincidence mode. In this regard, a ^{22}Na (14 μCi) radioactive source has been used where positron-electron annihilation leads to simultaneous emission of two 511 keV gamma-rays in the diametrically opposite directions [20]. The schematic diagram of the set-up for the coincidence technique used in this work is shown in Fig. 2.

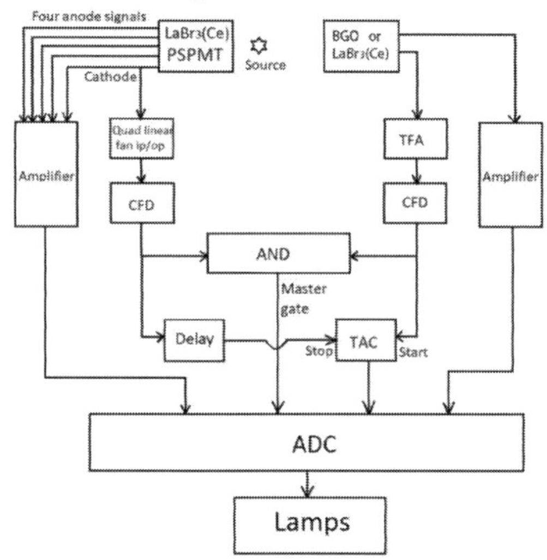

Fig. 2. Schematic diagram of the coincidence set-up, where a hexagonal BGO/ cylindrical $LaBr_3$(Ce) is chosen as the complementary detector along with the position sensitive detector under investigation.

To detect these two 511 keV gamma-rays, the $LaBr_3$(Ce)-PSPMT is used along with an auxiliary detector kept on the opposite side of the radioactive source. The distance of the source from both the detectors, placed face-to-face, has been optimized for count rate handling capacity of the data acquisition system and also to obtain an optimum size of the image of the complementary detector reconstructed by the $LaBr_3$(Ce)-PSPMT. For this purpose, a hexagonal BGO detector and a cylindrical $LaBr_3$(Ce) detector have been used as auxiliary detectors, with dimensions of 5.59 cm × 7.62 cm (height × length) and 5.08 cm × 5.08 cm (diameter × length) respectively. Only those yields recorded by the $LaBr_3$(Ce)-PSPMT assembly have been registered as true events which were in coincidence with the BGO/cylindrical $LaBr_3$(Ce) detector.

All current approaches to the characterization of the

 https://doi.org/10.1051/epjconf/202125311005

position sensitive detector are based on the standard CAMAC acquisition system. To extract energy and position information, the cathode and four anode signals of the $LaBr_3$(Ce)-PSPMT, respectively, have been processed using spectroscopic amplifiers. The coincidence interval is defined using a TAC module, wherein the signal of the complementary BGO/cylindrical $LaBr_3$(Ce) detector, after being processed through a Constant Fraction Discriminator (CFD) module, triggers/starts the coincidence event while the corresponding CFD output of $LaBr_3$(Ce)-PSPMT cathode terminates the window. The common timing reference window for the registered events, i.e. the master gate, has been generated by the AND output of the two aforementioned CFD signals, in order to control the trigger rate of the data acquisition system. A typical time profile of the raw energy signal and position signal are shown in Fig. 3 . Only those events have been accepted during acquisition wherein the four anode signals as well as the TAC output remain well within this window. The analog pulses have been digitized using 8-channel 12-bit ADC. The Linux Advanced Multi-Parameter System (LAMPS) [21, 22] package has been used for the data acquisition system and further analysis has been performed by employing the ROOT package [23].

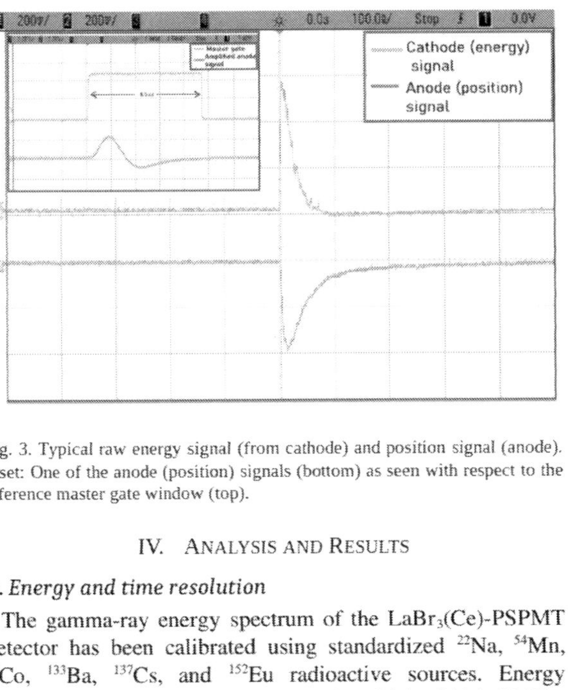

Fig. 3. Typical raw energy signal (from cathode) and position signal (anode). Inset: One of the anode (position) signals (bottom) as seen with respect to the reference master gate window (top).

IV. Analysis and Results

A. Energy and time resolution

The gamma-ray energy spectrum of the $LaBr_3$(Ce)-PSPMT detector has been calibrated using standardized ^{22}Na, ^{54}Mn, ^{57}Co, ^{133}Ba, ^{137}Cs, and ^{152}Eu radioactive sources. Energy resolutions of ~9.5% for 122 keV and ~6% for 511 keV (Fig. 4(a)) have been measured, in good agreement with GEANT4 simulations. In coincidence with the auxiliary cylindrical $LaBr_3$(Ce) crystal, timing resolution of 761(2) ps has been obtained from the TAC spectrum generated between the auxiliary detector and the position sensitive detector, as shown in Fig. 4(b).

B. Position resolution

Nine different spots were selectively irradiated on the $LaBr_3$(Ce)-PSPMT detector with 122 keV gamma-ray source, collimated using a perforated (2 mm) lead block of 20 mm thickness.

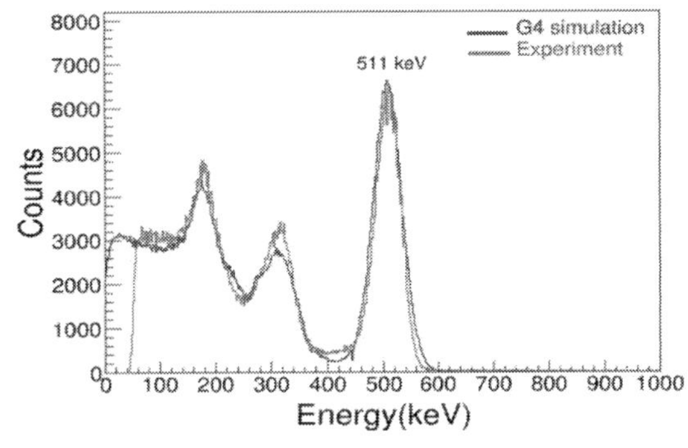

4(a)

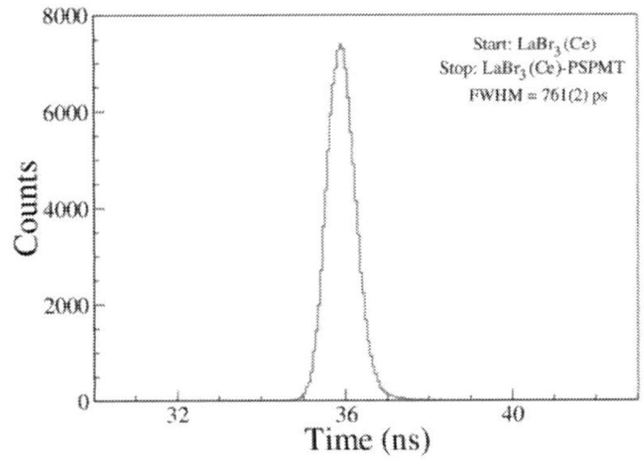

4(b)

Fig. 4. a) Simulated and experimental gamma-ray energy spectra of ^{22}Na source for the $LaBr_3$Ce)-PSPMT detector obtained in coincidence with 511 keV γ-energy in the BGO spectrum. The calculated energy resolution of the 511 keV photo-peak is ~6%. b) TAC spectrum, with the timing from the auxiliary cylindrical $LaBr_3$(Ce) taken as the START signal, and that from the position sensitive detector as the STOP signal.

The arrangement is represented in Fig. 5 . The resultant amplitudes of the four anode signals corresponding to each irradiated spot depend on the voltage drop through the resistive network, and the interaction points (X-position and Y-position) of the incident gamma-rays on the detector crystal have subsequently been determined as follows [3]:

$$X_{position} = \frac{(a_2 B + a_4 D)}{(A+B+C+D)} + b$$

$$Y_{position} = \frac{(a_3 C + a_4 D)}{(A+B+C+D)} + b$$

where, A, B, C and D are the amplitudes of the four anode

signals, a_1 , a_2 , a_3 and a_4 are the gain factors of the amplifiers (adjusted to be identical in the present measurement), with b as a manually-defined offset factor for the position spectra. The spectrum of the nine irradiated spots has been generated with an energy gate on 122 keV gamma-rays recorded from $LaBr_3$ (Ce)-PSPMT cathode, and the X- and Y-positions of the interaction points have been obtained therefrom.

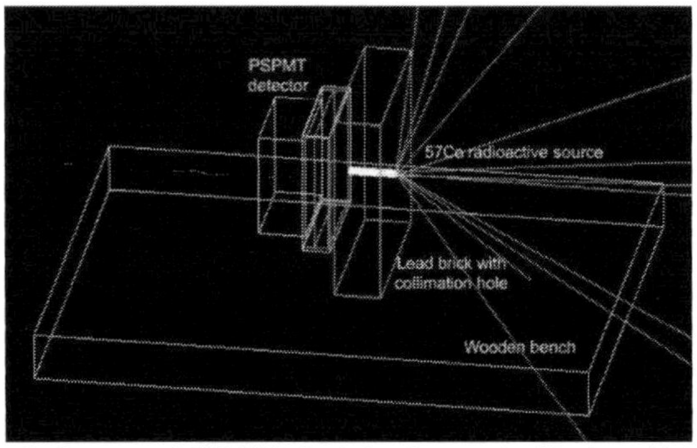

Fig. 5. Geometry of GEANT4 simulation for crystal calibration with lead brick collimator with ^{57}Co source.

The observed spectrum has also been validated by means of GEANT4 simulation, taking all possible physics processes into account. gamma-ray photons, electrons and positrons are expected to be produced during the interaction of the incident gamma-rays. The considered interaction mechanisms for the γ-photons are photoelectric effect, Compton scattering, and pair production. The electrons and positrons so produced can undergo multiple scatterings, ionizations and emit Bremsstrahlung radiation, while in addition, the positron can also undergo annihilation. The image has been reconstructed using those points at which maximum energy is deposited by the electrons generated along the interaction path of the incident gamma-rays in the crystal volume. The FWHM for each spot is ≈ 3.5 mm in the simulated spectrum (Fig. 6(a)), while the measured FWHM for the different spots lies in the range of about 3.5-4.0 mm (Fig. 6(b)). Since the measured spectrum is susceptible to electronic noise, a slight degradation is seen compared to the simulated outcome.

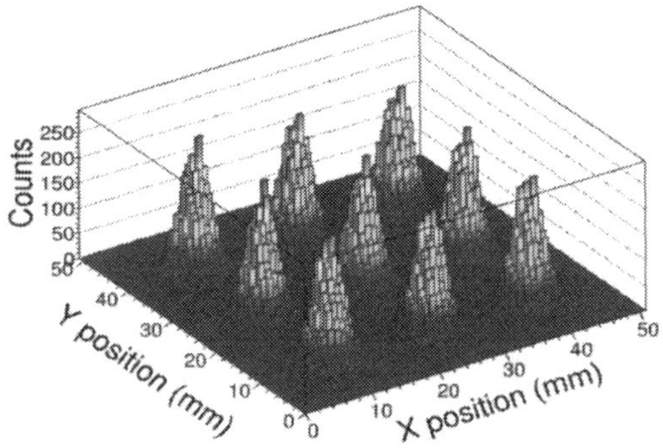

6(a)

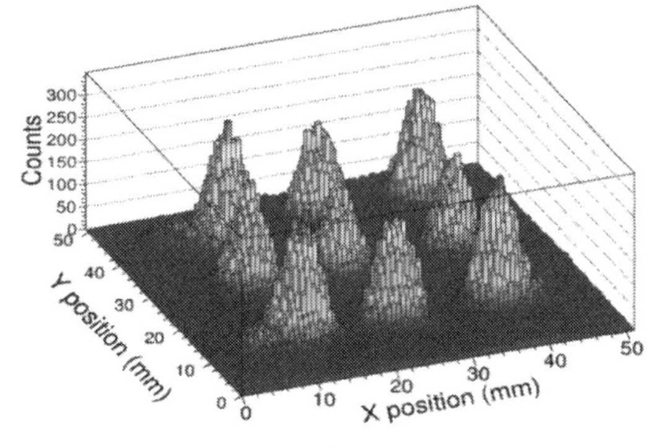

6(b)

Fig. 6: a) GEANT4 simulated and b) experimental result of nine irradiation spots with 2 mm collimated 122 keV gamma-rays on the $LaBr_3$(Ce)-PSPMT crystal.

C. Performance as an imaging detector

The images of a hexagonal BGO detector and a cylindrical $LaBr_3$(Ce) crystal have been generated using the position sensitive detector in a coincidence setup, as illustrated in Fig. 7 for the BGO detector, with energy gate on 511 keV gamma-ray photo-peak of both the detectors.

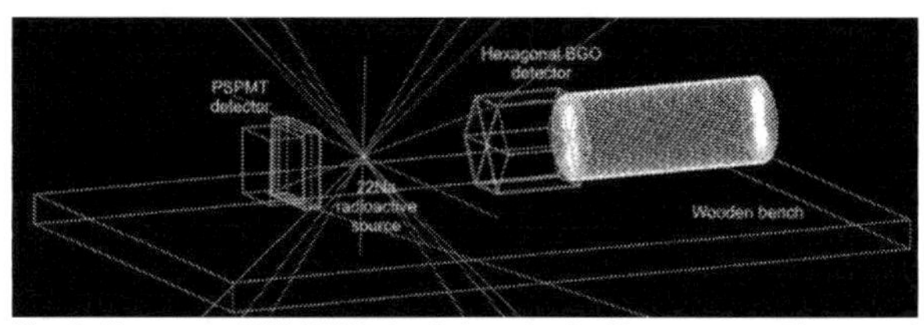

Fig. 7: Geometry of GEANT4 simulation of coincidence set-up of hexagonal BGO and $LaBr_3$(Ce)-PSPMT detectors.

The results compared with simulations are found to be in reasonable agreement in either case. The typical imaging performance is illustrated in Fig. 8. The position resolution, measured from the fit of the 1-d Y-projection of a slice of X-axis (22.5-27.5 mm) of the BGO image (Fig. 9), is 3.77(08) mm.

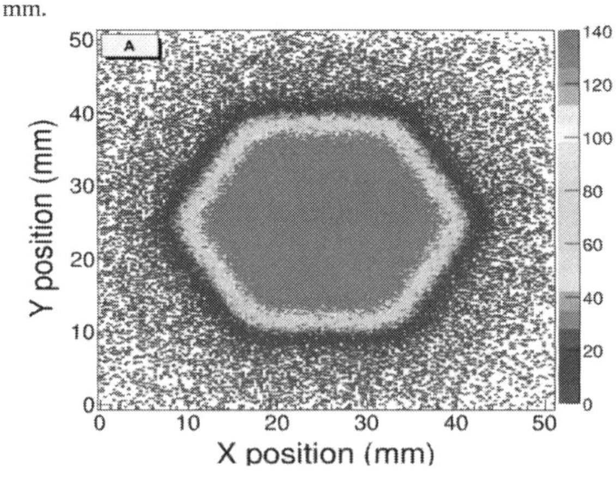

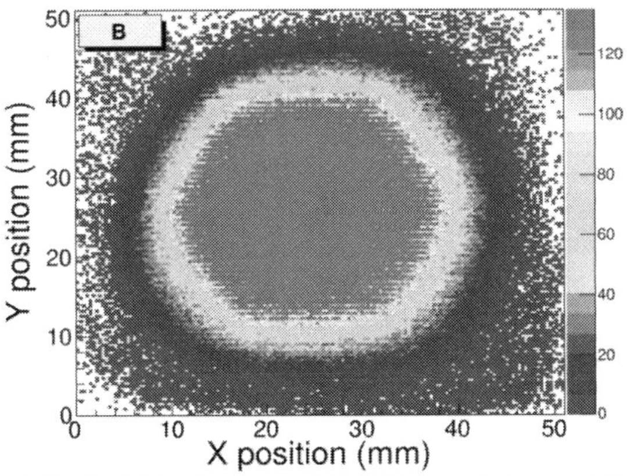

Fig. 8: Simulated (A) and experimental (B) 2-d images of a hexagonal BGO detector reconstructed with the $LaBr_3(Ce)$-PSPMT, in coincidence mode with energy-gating on the 511 keV photo-peak in both detectors.

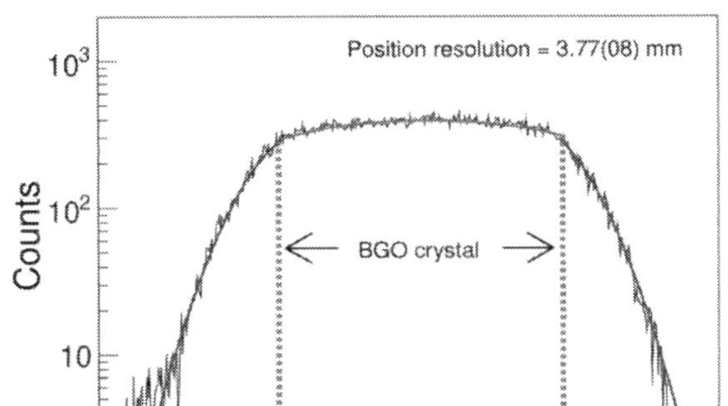

Fig. 9: Y-projection of a slice (X = 22.5 to 27.5 mm) of the experimentally reconstructed BGO image in figure 8(B) showing the variation of the absorbed gamma-ray counts throughout the BGO crystal.

An optimum geometry for image reconstruction by back-scattering technique has been investigated in the GEANT4 framework using the position sensitive $LaBr_3(Ce)$ detector in coincidence with a BGO detector as shown in Fig. 10. A steel ring having inner and outer diameter of 3.6 cm and 5.6 cm respectively with 1.1 cm thickness has been used as a scatterer for the incident photons. The position of the lead brick, scatterer, radioactive source, and detectors are chosen in such, so that the ancillary detector can't see the source directly. The lead brick having appropriate thickness to block the gamma-ray directly coming from the source.

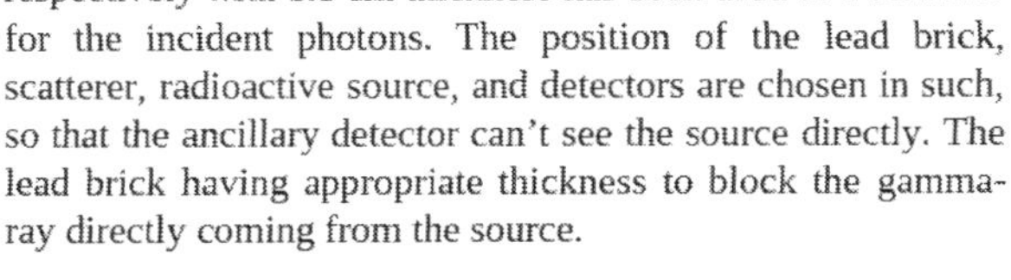

Fig. 10: GEANT4 Geometry of coincidence set-up of BGO and $LaBr_3(Ce)$-PSPMT detectors with a scatterer for Compton back-scattering imaging.

The 150 keV - 230 keV of the gamma-rays recorded in the BGO detector correspond to the back-scattering events from the ring-shaped scatterer. We have reconstructed the image of the ring as shown in Fig. 11, in the simulation with the energy gating condition: 490 keV < E_{PSPMT} <530 keV in the position sensitive $LaBr_3(Ce)$ detector and 150 keV < E_{BGO} < 250 keV in the BGO detector. In future, we are planing to do this back-scattering experiment to verify the simulated result as well as to investigate some applications of back-scattering technique.

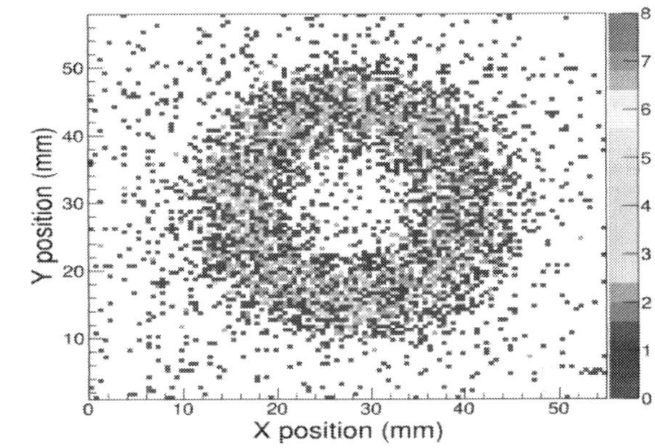

Fig. 11: GEANT4 simulated image of the scatterer (ring) in position sensitive detector with energy gated condition: 490 keV < $E_{LaBr_3(Ce)\text{-}PSPMT}$ < 530 keV and 150 keV < E_{BGO} < 250 keV.

V. Summary and Conclusions

A $LaBr_3(Ce)$-PSPMT detector setup is developed for gamma-ray tracking and imaging applications. The energy and time characterization of the detector has been studied. A spatial resolution of about 3.5 mm has been obtained with a 2 mm collimated source. GEANT4 simulation for this detector reproduces the measured energy response of the detector and the spatial resolution. The reflection of the scintillation light from the periphery of the detector slightly hampers the position resolution. Sharp 2-d images of a hexagonal shaped BGO and a cylindrical $LaBr_3(Ce)$ crystals have been reconstructed using this position-sensitive detector setup, by employing data acquisition in coincidence mode. The 2-d image of a ring shaped scatterer has been reconstructed in GEANT4 by Compton back-scattering technique of the gamma-rays. This setup is expected to find applications in the fields of nuclear and high energy physics for scanning of detectors, as well as for the purpose of imaging in the medical and defense sectors.

Acknowledgment

This work is supported by the Department of Atomic Energy, Government of India, under project No. 12P-R&D-TFR-5.02-0400.

References

[1] S. Majewski, F. Farzanpay, A. Goode, B. Kross, D. Steinbach, A. Weisenberger, M. Williams, and R. Wojcik, (1998) Nuclear Instruments and Methods in Physics Research Section A **409** 520–523

[2] N. Tamda, A. Bakkali, H. Boulahdour, M. Parmentier, A. Pousse, B. Kastler, and J Chavanelle, (2006) Nuclear Instruments and Methods in Physics Research Section A **557** 537 – 543.
[3] A. Banerjee, S. Mandal, P. Roy, S. Mukhopadhyay, G. Mukherjee, M. Kumar, A. Jhingan, and R. Palit, (2019) Nuclear Instruments and Methods in Physics Research Section A **930** 100 – 104.
[4] V. Popov, (2019) Journal of Instrumentation (JINST) **6 C01061**
[5] P. D. Olcott, J. A. Talcott, C. S. Levin, F. Habte, and A. M. K. Foudray, (2005) IEEE Transactions on Nuclear Science **52** 21–27
[6] R. Pani, R. Pellegrini, M.N. Cinti, P. Bennati, M. Betti, F. Vittorini, M. Mattioli, G. Trotta, V. Orsolini Cencelli, R. Scaf, L. Montani, F. Navarria, D. Bollini, G. Baldazzi, G. Moschini, P. Rossi, F. de Notaristefani, Nuclear Instruments and Methods in Physics Research A **572** (2007) 268–269
[7] A. Fabbri,V. Orsolini Cencelli, P. Bennati, M. Nerina Cinti, R. Pellegrini, G. De Vincentis and R. Pani, 2013 JINST **8 C02022**.
[8] M. K. Nguyen, T. T. Truong, M. Morvidone, and Zaidi, International journal of biomedical imaging, Artical id **913893** (2011).
[9] A. Olariu, Member, IEEE, P. Désesquelles, Ch. Diarra, P. Médina, C. Parisel, and C. Santos On behalf of the AGATA Collaboration, IEEE Transactions on nuclear science, VOL. 53, NO. 3, JUNE 2006.
[10] B. Bruyneel, B. Birkenbach, and P. Reiter, The European Physical Journal A **52** (2016) 70.
[11] I. Kojouharov, S. Tashenov, T. Engert, J. Gerl, and H. Schaffner, IEEE Nuclear Science Symposium Conference Record **N46-7** (2007).
[12] B. Alpat and G. Esposito (2003) AIP Conference Proceedings **674** 283.
[13] F. Quarati, A. Bos, S. Brandenburg, C. Dathy, P. Dorenbos, S. Kraft, R. Ostendorf, V. Ouspenski, and A. Owens, Nuclear Instruments and Methods in Physics Research Section A **574** (2007) 115 – 120.
[14] N. Cherepy, S. Payne, S. Asztalos, G. Hull, J. Kuntz, T. Niedermayr, S. Pimputkar, J. Roberts, R. Sanner, T. Tillotson, E. van Loef, C. Wilson, K. Shah, U. Roy, R. Hawrami, A. Burger, L. Boatner, W.-S. Choong, W. Moses, Nuclear Science, IEEE Transactions on **56** (2009) 873 – 880.
[15] A. Kuhn, S. Surti, K. S. Shah, J. S. Karp, in: IEEE Nuclear Science Symposium Conference Record, 2005, volume 4, pp. 5 pp.–2026.
[16] URL http://www.detectors.saint-gobain.com
[17] NNDC URL https://www.nndc.bnl.gov/nudat2/chartNuc.jsp
[18] A. Sonzogni (2003) Nuclear Data Sheets **98** 515 – 664.
[19] Valentino, Francesco and DAbramo E (2007) Nuclear Instruments and Methods in Physics Research Section A **571** 389 – 391.
[20] Robert J.de Meijer, R. Lindsay, M. Tijs, Steven, and H. Limburg, (2019) Radiation Measurements **121** 77 – 85.
[21] A. Chatterjee, A. Kumar, K.Ramachandran, Proceeding of the DAE Symp. on Nucler Physics **55** (2010).
[22] TIFR homepage URL http://www.tifr.res.in/~pell/lamps.html
[23] Root-CERN homepage URL https://root.cern/

Measurement and simulation of the new liquid organic scintillator response to fast neutrons

Jaroslav Jánský [1] , Jiří Janda [2] , Zdeněk Matěj [3] , Filip Mravec [3] , Michal Košťál [4] , František Cvachovec[1]
[1] University of Defence, Faculty of Military Technology, Department of Mathematics and Physics, Brno, Czech Republic
[2]University of Defence, NBC Defence Institute, Vyškov, Czech Republic
[3]Masaryk University, Faculty of Informatics, Brno, Czech Republic
[4]Research Centre Řež, Husinec-Řež, Czech Republic
Corresponding author: jaroslav.jansky@unob.cz

Abstract—**Liquid organic scintillators are important devices for measurements of neutron radiation. This work aims to develop and optimize the composition of liquid organic scintillators so it can be used for fast neutron spectrometry. As the neutron radiation is usually accompanied with γ ray radiation, it is important to have quality γ/n discrimination. The new cocktail for house made liquid organic scintillator is prepared and studied with intention of being able to separate gamma and neutron for neutron energies above 0.5 MeV while keeping lower constraints on practical use (e.g., sealing because of oxygen) than commercial liquid scintillators. In preceding work the composition of liquid scintillators was optimized. Two two-component scintillators were selected for further studies. Solvent DIPN (Di-iso-propyl-naphthalene Mixed Isomers) is selected for both. First is mixed with luminophore PYR (1-Phenyl-3-(2,4,6-trimethyl-phenyl)-2-pyrazoline) of concentration 5 g/l. Second is mixed with luminophore THIO (2,5-Bis(5-tert-butyl-benzoxazol-2-yl)thiophene) of concentration 5 g/l. In this work the response of scintillator to monoenergetic beam of neutrons was measured for multiple neutron energies at PTB in Braunschweig. The two parameter spectrometric system NGA-01 is used to analyze the energy and discrimination characteristics. 137 Cs and 60 Co are used as radiation sources for calibration with pure γ rays. Then the response of scintillator for same neutron energies was simulated using GEANT4. The dissipated energy in the scintillator in response to monoenergetic neutrons is obtained. Both, measured and simulated responses are compared. Functional dependence for yield of recoiled products is estimated. It is seen that main recoil product hydrogen proton is well observed in both. From the edge of proton response one can assume the yield for given neutron energy. The recoiled carbon ion (from elastic collision) is on the other side difficult to observe in measured results but clearly seen in dissipated energy plots. It suggests that yield of carbon ion is very small relatively to proton yield. These results will serve as basis for response function evaluation of scintillator which is necessary for evaluation of unknown neutron spectra from measurements with scintillator.**

I. Introduction

Liquid organic scintillators are important devices for measurements of neutron radiation. Organic liquid scintillators can consist simply of an organic solvent and a luminophore. A third constituent is sometimes added as a wavelength shifter to fit the emission spectrum to better match the spectral response of common photomultiplier tubes. In certain applications, large-volume detectors with dimensions of several meters may be required. In these cases, the liquid scintillator is often the only practical choice because of the cost. Liquid scintillators are expected to be more resistant to radiation damage effects than crystalline or plastic scintillators, because of their lack of a solid structure that could be damaged by an exposure to intense radiation.

This work aims to develop and optimize the composition of liquid organic scintillators so it can be used for fast neutron spectrometry. As the neutron radiation is usually accompanied with γ ray radiation, it is important to have quality γ/n discrimination. The responses of scintillator to monoenergetic neutrons are shown. These results are planned to be used in future for development of response matrix for our scintillators.

II. Experimental Description

The two scintillators are used in this work. Two two-component scintillators were selected for these studies. Solvent DIPN (Di-iso-propyl-naphthalene Mixed Isomers) is selected for both. First is mixed with luminophore PYR (1-Phenyl- 3 - (2 , 4 ,6 - trimethyl – phenyl) - 2 - pyrazoline) of concentration 5 g/l. Second is mixed with luminophore THIO (2 , 5 - Bis (5 - tert - butyl - benzoxazol - 2 - yl) thiophene) of concentration 5 g/l.

In this work the response of scintillator to monoenergetic beam of neutrons was measured for multiple neutron energies at PTB in Braunschweig. The two parameter spectrometric system NGA-01 is used to analyze the energy and discrimination characteristics. 137 Cs and 60 Co are used as radiation sources for calibration with pure γ rays.

For numerical simulations GEANT 4 version 10.0.6 was

used. Physics list Shielding was used as it supports the measured cross section data for neutron collisions below 20 MeV. The simulated geometry consisted of 10 cm cube box with vacuum. The scintillator cylinder with radius 1cm and height 4.5cm is placed inside. The particle source is placed on the axis of symmetry of cylinder 1 cm ahead.

The scinitllator density is considered to be 0.949 g/cm^3. Atom ratio for plain DIPN H:C is 1.25. Including the luminophore PYR shifts the H:C ratio to 1.2494 and nitrogen is present in ratio N:C 0.0005. The results of GEANT4 simulations are presented in form of absorbed (deposited) energy inside the scintillator after each of the source neutrons leaves the simulation domain.

III. RESULTS

In this section we first present the measured responses of our two scintillators to the monoenergetic beams.

First the response of the scintillator with THIO luminophore to neutrons of energy 1.5 MeV is shown in Figure 1. Horizontal axis denotes the energy in electron equivalent units. Vertical axis denotes the pulse height spectra.

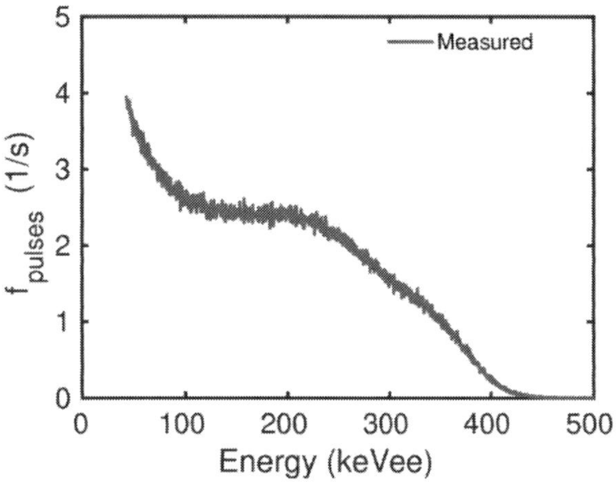

Fig 1: Response of scintillator with THIO luminophore to neutrons of energy 1.5MeV.

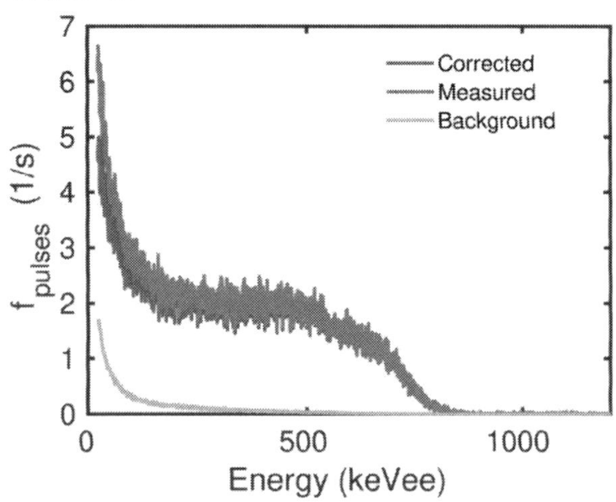

Fig 2: Response of scintillator with THIO luminophore to neutrons of energy 2.5 MeV.

Then the response of the scintillator with THIO luminophore to neutrons of energy 2.5 MeV is shown in Figure 2. The measured pulse height spectra is shown with red line. Then the measurement with shadow cone to subtract the scattered background neutrons is shown with yellow line. And finally the corrected pulse height spectrum is shown with blue line.

Then the response of the scintillator with THIO luminophore to neutrons of energy 19 MeV is shown in Figure 3.

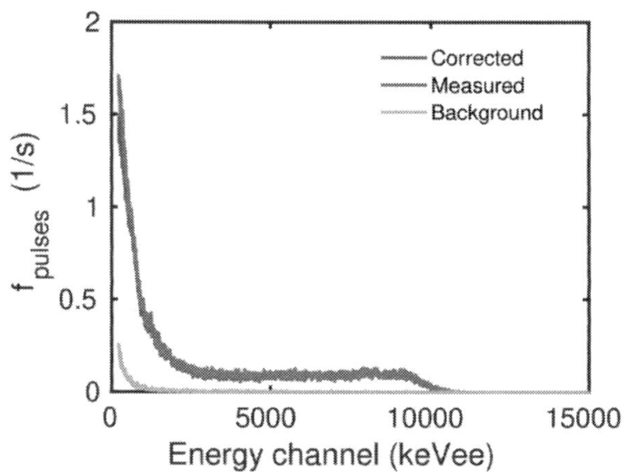

Fig 3: Response of scintillator with THIO luminophore to neutrons of energy 19 MeV.

The response to scintillator with PYR luminophore is measured twice for energy 2.5 MeV and once for energy 19 MeV. The pulse height spectra are shown below in Figure 4,5 and 6.

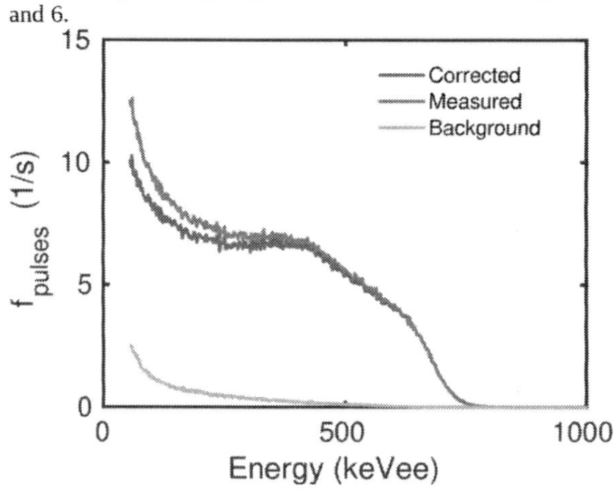

Fig 4: Response of scintillator with PYR luminophore to neutrons of energy 2.5 MeV.

https://doi.org/10.1051/epjconf/202125311006

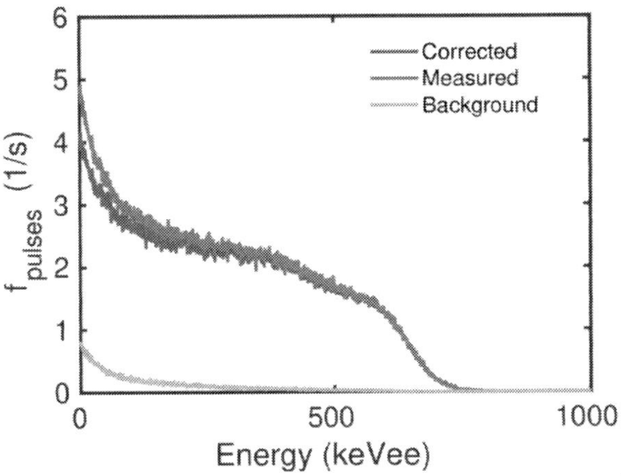

Fig 5: Response of scintillator with PYR luminophore to neutrons of energy 2.5 MeV.

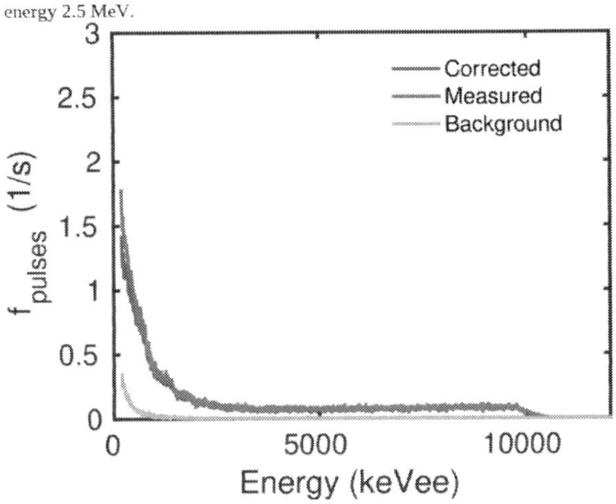

Fig 6: Response of scintillator with PYR luminophore to neutrons of energy 19 MeV.

The simulated deposited energy for neutrons of energy 1.5 and 2.5 MeV using GEANT4 are shown in Figures 7 and 8.

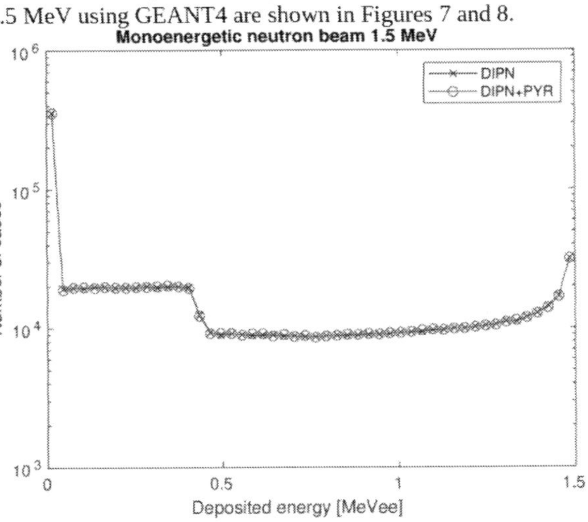

Fig 7: Deposited energy of scintillator with only DIPN and DIPN with luminophore to neutrons of energy 2.5 MeV.

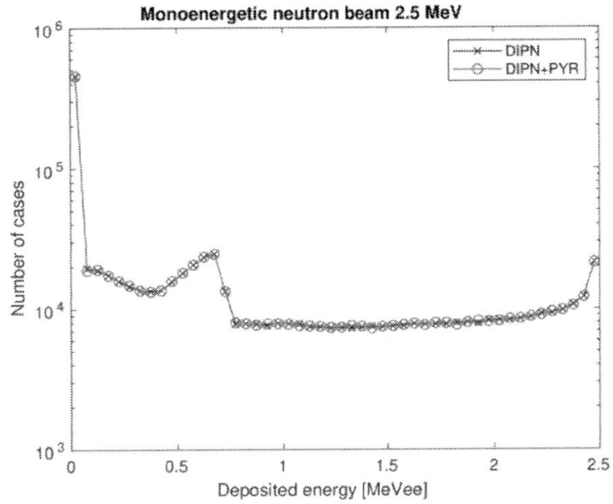

Fig 8: Deposited energy of scintillator with only DIPN and DIPN with luminophore to neutrons of energy 2.5 MeV.

It clear to observe deposited energy due to carbon ions from increased spectra below 0.4 MeV and 0.6 MeV for 1.5 MeV neutron source, respectively 2.5MeV. However this pattern is not observed in experimental results suggesting that light yield of recoiled carbon ion is much smaller than light yield of recoiled proton after neutron elastic collisions.

The summary of light yield function from experimental measurements is shown in Figure 9. The line represents theoretical function for gamma energy which is identical to electron equivalent energy. The points represents results from above measured responses and they all show that neutrons produce several times less light than gamma rays of same energy.

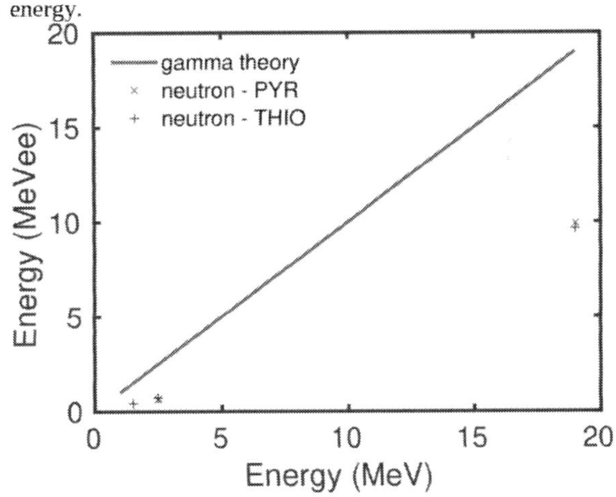

Fig 9: Summary of the light yield function relating the neutron energy to electron equivalent energy.

IV. Conclusions

In this work the response of scintillator to mono-energetic beam of neutrons was measured for multiple neutron energies and compared to the dissipated energy response of scintillator for same neutron energies simulated using GEANT4. Functional dependence for yield of recoiled products is estimated. It is seen that main recoil product hydrogen proton is well observed in both. From the edge of proton response one can assume the yield for given neutron energy. The recoiled carbon ion (from elastic collision) is on the other side difficult to observe in measured results but clearly seen in dissipated energy plots.

ACKNOWLEDGEMENT

This research work has been carried out within the ADAR project (Accelerator Driven Advanced Reactor). The authors gratefully acknowledge the financial support from the Ministry of Education, Youth, and Sports of the Czech Republic under the INTER-ACTION research program (project No. LTAUSA18198). The authors gratefully acknowledge the financial support from the grant VAROPS (DZRO FVT 3) granted by the Ministry of Defence of the Czech Republic. This work has received funding from the Euratom research and training programme 2014-2018 under grant agreement No 847594 (ARIEL).

Spectral Enhancement of a SiPM Array-Based Radiation Detector

R. Harn[1,3], A. Osovizky[1,2], Y. Kadmon[1], S. Rotman [3], N. Kopeika [3], and M. Ghelman[1,3]

[1]Nuclear Research Center – Negev, Israel
[2]Radiation Detection Department, Rotem Industries Ltd, Israel
[3]Ben-Gurion University, School of Electrical and Computer Engineering, Israel
RonH@nrcn.org.il

Abstract— **Silicon Photomultipliers (SiPMs) have many advantages when used in radiation detectors. Low bias voltage, compactness and immunity to electromagnetic interference are among their prominent benefits. However, due to their small size, usually an array of SiPM components is required in order to cover the coupling surface area of a scintillator. Since the SiPM is a semiconductor, biased in a reversed voltage, gain variation and strong temperature dependence are introduced. As a result, SiPM-based detectors, particularly an array of SiPMs, undergo spectral signal to noise ratio reduction. This work studies the effect of the SiPM breakdown voltage variation on the obtained energy spectrum and proposes an electronic approach to overcome this technological drawback. This developed technology provides an adequate temperature-dependent, commonly distributed high bias voltage and an individual offset-voltage fine tuning that enables adjustment of all the SiPM components to their optimum operating points. Power-wise it is beneficial to operate SiPM at lower voltages, where undesirable gain variation is more dominant. The proposed solution enables working at lower bias voltages, which provides lower power consumption and better radiation hardness, while yielding an enhanced spectrum resolution. The proposed electronic approach enhances the obtained spectra, reducing the noise threshold by 16 % when working at 1 V overvoltage. Hence provides an enhanced signal to noise ratio over the traditional biasing methods.**

Keywords — SiPM, Array, Radiation detectors, SNR, Breakdown Voltage.

I. Introduction

RADIATION detectors based on arrays of silicon photomultipliers (SiPMs) coupled to a light scintillator are becoming more and more popular, gradually replacing traditional photomultiplier tube (PMT) based detectors [1]. Among their many uses, SiPM based detectors can be found in medical imaging devices such as positron emission tomography (PET) [2] and single photon emission tomography (SPECT) [3], homeland security applications [4], environmental monitoring with drones [5], and particle astrophysics experiments with radiation detector mounted on a satellite [6,7]. Its compactness, relatively low bias voltage and immunity to electro-magnetic interference (EMI) give the SiPM superiority over the traditional PMT in radiation detector implementations, especially for the two last applications mentioned above.

The gain of the SiPM is voltage dependent. The overvoltage, V_{OV}, is defined as the difference between the supplied bias voltage, V_B, and the SiPMs breakdown voltage, V_{BD}, as shown in (1),

$$V_{OV} = V_B - V_{BD}\,. \quad (1)$$

For low V_{OV}, the SiPM amplification rises linearly with V_{OV}. The avalanche initiation probability (AIP) increases linearly together with V_{OV} as well. Hence, the SiPMs total light detection gain has a parabolic dependence on V_{OV}. Despite the lower obtained gain, using lower V_{OV} can provide advantages such as power consumption reduction and radiation hardness [8,9], both of which are crucial for radiation detectors mounted on satellites and exposed to cosmic ray radiation, and for drones scanning polluted areas with high radiation fields.

Nevertheless, SiPMs have two fundamental drawbacks compared to the PMT. First, the SiPM breakdown voltage has a strong temperature dependence, which leads to gain dependent on the temperature, and second, variation in the breakdown voltage as a result of the limitations of the manufacturing process. The breakdown voltage temperature dependence was widely studied in previous works [10,11] and several solutions were presented to address it [12,13]. However, the breakdown voltage variation, that result from fluctuations in the resistivity of the epitaxial layer that hosts the SiPM active region [11], and its effect on the obtained spectra were less studied. This variation results in gain fluctuation between SiPM components in the same array and a reduction of the total signal to noise ratio (SNR) obtained from the detector at lower voltages. The reduced SNR is an outcome of the photon detection efficiency (PDE) curve shape [14], which rises more moderately as the overvoltage increases. Hence, a reduction of ΔV in the overvoltage, due to the breakdown variation, will decrease the PDE more significantly in respect to the increase of the PDE as a result of a growth of the overvoltage in the same ΔV. The result is a reduction of the average PDE value for an array of SiPM components as compared to the average PDE of an ideal homogeneous SiPM array. This reduction will be expressed in a lower number of fired pixels and therefore lower statistical significance.

Over the years, several approaches for addressing the

breakdown voltage variation have been introduced. One approach is to provide a separate adjustable gain pre-amplifier for each SiPM [15]. Although this approach provides a uniform gain from all SiPM readouts, it does not address the reduction in the obtained SNR. Moreover, due to the SiPM pulse shape, high bandwidth amplifiers would be needed, hence increasing the power consumption and the electronic noise. Another approach is to provide a separate temperature dependent bias voltage to each SiPM separately. Even though this approach compensates for the breakdown voltage variation and its temperature dependence, it is very inefficient in power consumption, electronics size, and cost, especially for large SiPM arrays. In this paper, we propose an electronic approach to overcome this technological drawback.

II. Breakdown voltage variation investigation

Various SiPM manufacturers possess different production technologies, hence providing SiPM components with different temperature and breakdown voltage variation characteristics. **Erreur ! Source du renvoi introuvable.** summarizes the temperature dependence and variation of the breakdown voltages of three leading SiPM manufacturers.

TABLE I
CHARACTERISTICS OF SIPMS OF LEADING MANUFACTURERS

Property	On Semi. [16,17]	Ketek [18]	Hamamatsu [19]
Series	C-30050	PM3347-WL	S14160-3050
AVG gain	6×10^6	7×10^6	2.5×10^6
MAX PDE (%)	47	51	50
AVG V_{BD} (V)	24.4	29.5	38
STD(V_{BD}) (mV)	73	92	100
Vov range (V)	1-5	1-6.5	2-7
STD(V_{BD}) / Vov (%)	2.43	3.33	2.2
ΔV_{BD} /ΔT (mV/ °C)	21.5	22	34
ΔV_{BD} / Vov (%)	0.7	0.58	0.75

AVG = Average, V_{BD} = Breakdown voltage, STD = Standard deviation, Vov = Overvoltage, TMP = Temperature.

As can be seen in **Erreur ! Source du renvoi introuvable.**, SiPM components from different manufacturers vary in their breakdown voltages, their overvoltage range, and their temperature dependence. However, all of them have a variation in the breakdown voltage that is not negligible.

In order to implement a 2 × 2 SiPM array detector, the current curves of 10 C-series 60035 SiPM components [16], were measured as a function of the bias voltage. The curves obtained are presented in **Erreur ! Source du renvoi introuvable.**. The breakdown voltage of the 10 SiPM components was obtained from the current measurement using the inverse logarithmic derivative (ILD) method [20]. The usefulness of this method is that it cancels out constant parameters that affect the current level, such as the number of incoming photons, and yields a linear dependence in V_{OV} in the near breakdown voltages. Hence, by using (2) to calculate the ILD, and finding the intersection of the linear fit for over breakdown voltages, with the voltage axis, we can obtain the measured SiPMs breakdown voltages.

$$ILD(V) = \left(\frac{d}{dV}\ln[I(V)]\right)^{-1}. \quad (2)$$

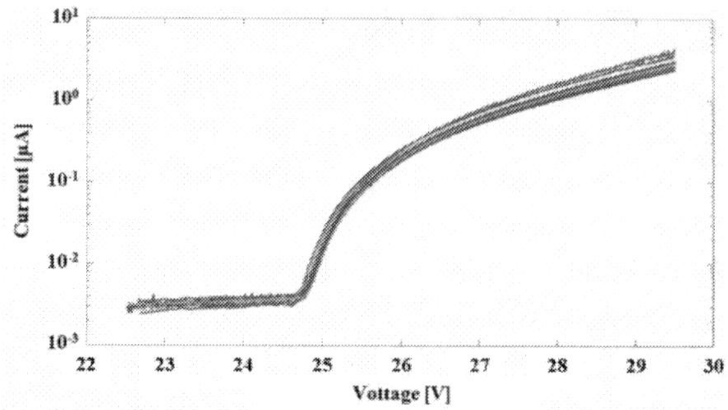

Fig. 1. I-V curve as a function of the bias voltage of ten C-series SiPM components.

The ILD curve and the linear fit of the near breakdown voltages are shown in Fig. 2. The breakdown voltages that were obtained are presented in Fig. 3 over a normal probability density function plotted using the data for the C-30050 series breakdown variation from **Erreur ! Source du renvoi introuvable.**.

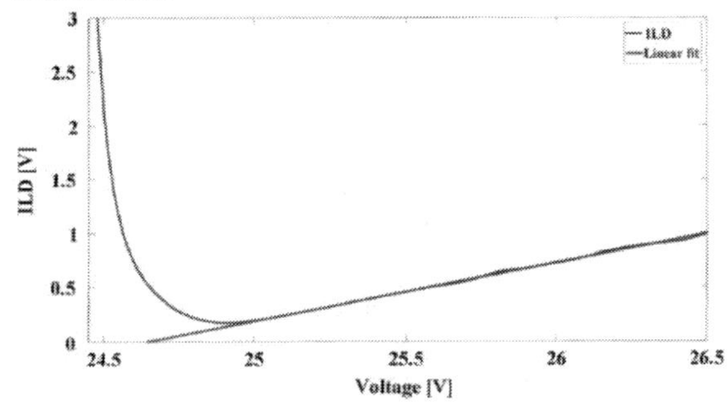

Fig. 2. ILD curve and the linear fit intersect with the voltage axis to yield the breakdown voltage.

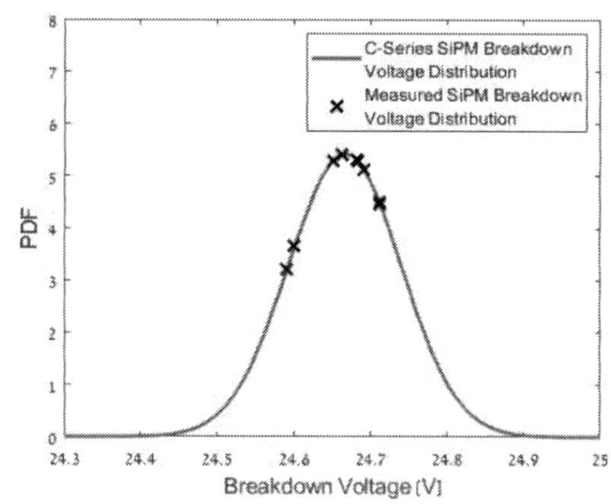

Fig. 3. Ten breakdown voltages that were calculated marked on the normal probability density function plotted using the standard deviation from Table I.

The average voltage that was obtained from the measurements is shifted from the manufacturer average voltage. This is an expected result due to the temperature differences of the measurements and the breakdown voltage dependence of 21½ mV/ºC [16]. All the obtained breakdown voltages are in the single standard deviation (STD) interval.

III. Breakdown Variation Correction Topology

A scintillator detector based on a 2 × 2 SiPM array coupled to a 15 mm × 15 mm × 15 mm Cs(Tl) scintillator was implemented. The breakdown voltages of the SiPMs assemble the array are listed in **Erreur ! Source du renvoi introuvable.**II. Each SiPM output is loaded with a 200 Ω resistor in order to obtain an equivalent 50 Ω load for all the SiPMs in the array.

TABLE II
SIPM ARRAY BREAKDOWN VOLTAGES

Component	Measured Breakdown Voltage (V)	ΔV from AVG (mV)
SiPM 1	24.71	+57
SiPM 2	24.59	-62
SiPM 3	24.60	-52
SiPM 4	24.71	+57

In order to overcome the breakdown voltage variation, an offset voltage topology was implemented and connected to the SiPM load as shown in Fig. 4. Each SiPM in the array was connected to a separate offset topology while all the SiPMs were fed from a single temperature dependent bias voltage. The SiPMs output is fed through a coupling capacitor to a single shaping channel which sums the 4 SiPM outputs.

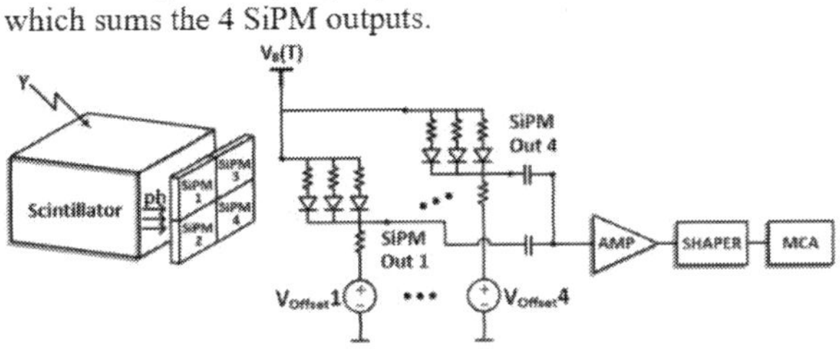

Fig. 4. Simplified block diagram of the proposed detector topology.

A simple approach of implementing the offset voltage would be by using wideband, high power consumption amplifiers that would lead both AC and DC tracks. In order to reduce power consumption, the suggested topology implements an accurate DC path using narrowband, low power amplifiers, and an AC path through a capacitor, C_{Offset}, as shown in Fig. 5. The offset value is set by using a voltage divider or by micro-controller DAC output. Due to the use of narrow bandwidth amplifiers the increase in the electronic noise is minor. The stabilization of the offset amplifier for the capacitive load drive was implemented using the in loop compensation method. In this way the DC feedback comes from the capacitive load itself whereas the AC feedback is fed from the amplifier output through *Cf*, bypassing *Rs* and *Rf*. The values of *Rs*, *Rf*, and *Cf* were calculated according to C_{Offset} value in order to cancel the additional phase shift added to the amplifier's loop gain as a result of the capacitive load. In this way, the capacitive load can be driven without adding a DC error due to *Rs*. The supplied bias voltage to each SiPM is the difference between the bias temperature dependent voltage, $V_B(T)$, and the SiPMs offset voltage.

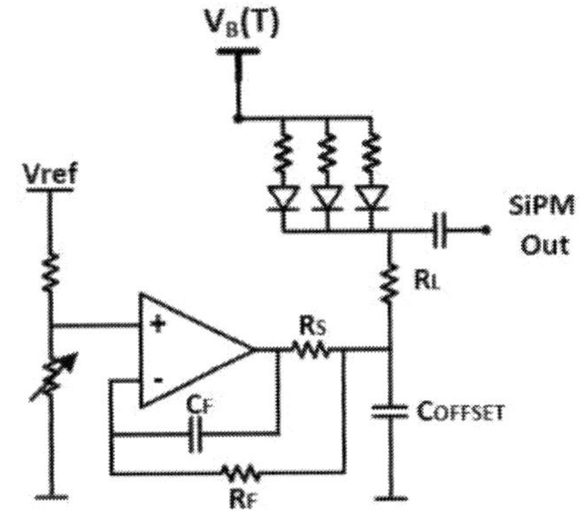

Fig. 5. Offset adjustment topology.

The temperature-dependent bias voltage can be implemented as shown in Fig. 6. The voltage is regulated by adding an accurate temperature sensor to the DC-DC boost converter feedback loop. Hence, the bias voltage is continuously regulated without involving a separate sampling process. The voltage in the boost feedback input is a superposition of the reference voltage and the temperature sensor voltage. By adjusting the values of resistors *Ra* and *Rb* a linear temperature dependence can be implemented.

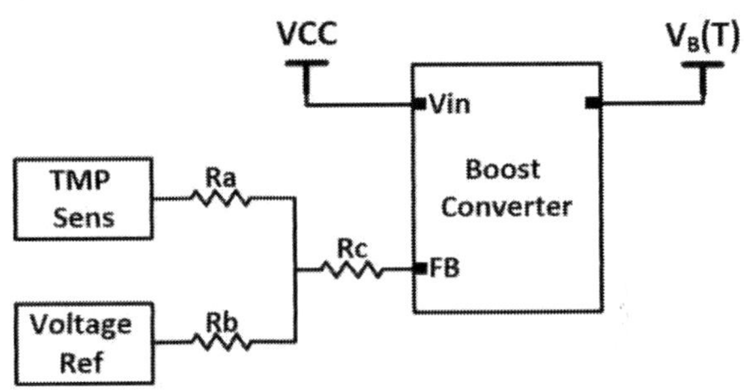

Fig. 6. Temperature dependent bias voltage topology.

IV. Measurement And Analysis

In order to verify that the offset values obtained from the breakdown voltage calculation are the optimal values in SNR perspective, a voltage sweep for each SiPM offset voltage was conducted. At each offset value, the noise threshold was measured while the overall gain remained constant by adjusting the bias voltage of the remaining three SiPMs. This measurement yields a curve of the noise threshold as a function of the offset voltage for each SiPM separately, as shown in Fig. 7. The gap between the average breakdown voltage and the value calculated using the ILD method, ΔV, listed in **Erreur ! Source du renvoi introuvable.**II, is transformed to the correlating offset voltage by taking ΔV negative value. The offset values that were calculated using the ILD method are marked as 'X' on the curves. It can be seen that there is a good agreement between the measured optimal offset and the calculated one. A connection scheme and an image of the measuring setup are shown in Figs. 8 and 9.

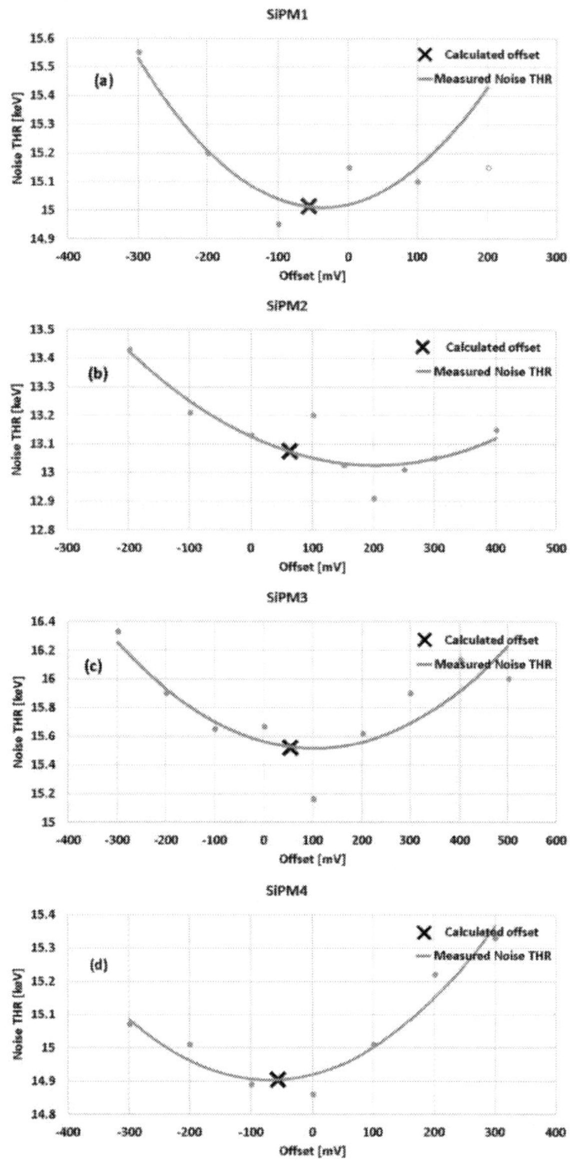

Fig. 7.Noise threshold as a function of the offset voltage for SiPM1 (a), SiPM2 (b), SiPM3 (c), and SiPM4 (d). The 'X' marks the correlating value from Table II.

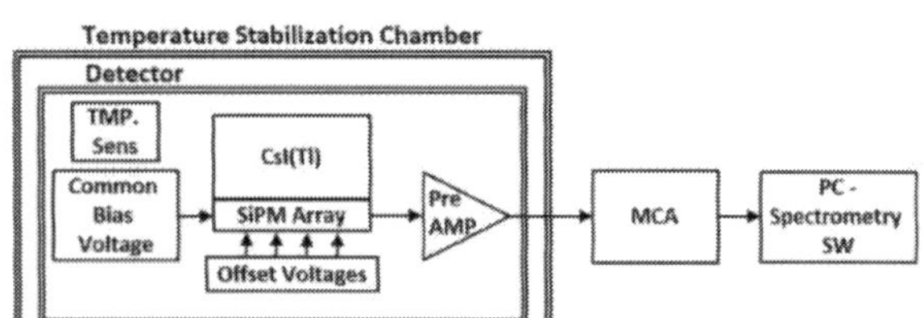

Fig. 8. Connection scheme of the detector and the measuring setup.

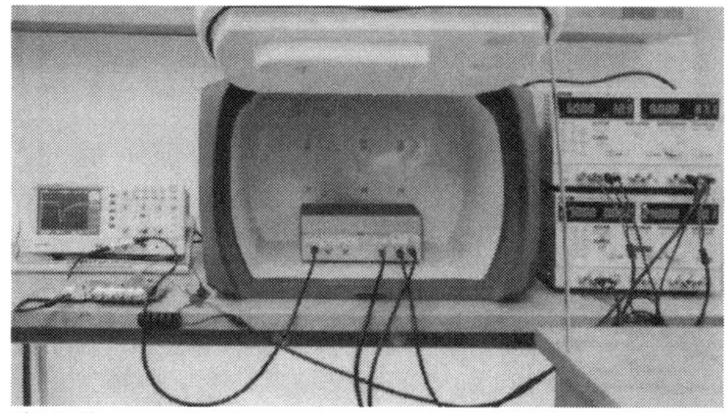

Fig. 9. The measuring setup.

At the second stage, the noise threshold as a function of the V_{OV} was obtained for a V_{OV} interval of 750 mV – 1.75 V. At each voltage, a calibration of the operating point was made followed by a background measurement. The measurement was conducted twice. First, with a uniform voltage offset, hence without the correction of the breakdown voltage variation, and second, with the suggested offset voltage topology based on the values from **Erreur ! Source du renvoi introuvable.**II. In both measurements there was careful attention for the resemblance of the operating point, especially in gain and temperature perspective. The noise threshold was set to obtain 15 counts per second (CPS) above the threshold. Then the noise threshold location on the energy axes was recorded. Assuming normal distribution, the error interval was taken as the square root of the counts number above the threshold. Three STD were taken in each direction and the correlating energy threshold was recorded. The obtained noise threshold curve as a function of V_{OV} without the topology, "Regular", and with the proposed adjustment topology, "ADJ", is shown in Fig. 10.

V. Results

As expected, improvement is seen primarily for lower V_{OV} below 1.5 V, as a result of the moderation in the PDE slope for higher V_{OV}.

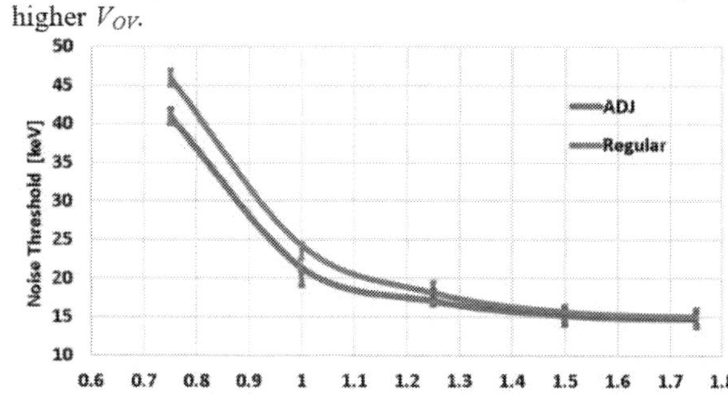

Fig. 10. Noise threshold curve as a function of Vov without the presented topology, "Regular", and with the suggested topology, "ADJ".

As can be seen from Fig. 10, for 1 V overvoltage the noise threshold was reduced from 24 keV to 20 keV, providing 16 % improvement. A ^{137}Cs spectrum was obtained with, and without the adjustment topology is shown in Fig. 11. The spectrum was obtained using 1 V overvoltage. As can be seen, when observing the spectra on a wide scale, the difference of the obtained spectra with and without the suggested topology is not significant. However, when focusing on the low energy interval we note that the noise peak is shifted to the left in the "ADJ" curve (red), as compared to the spectra without the suggested topology in the "Regular" curve (blue).

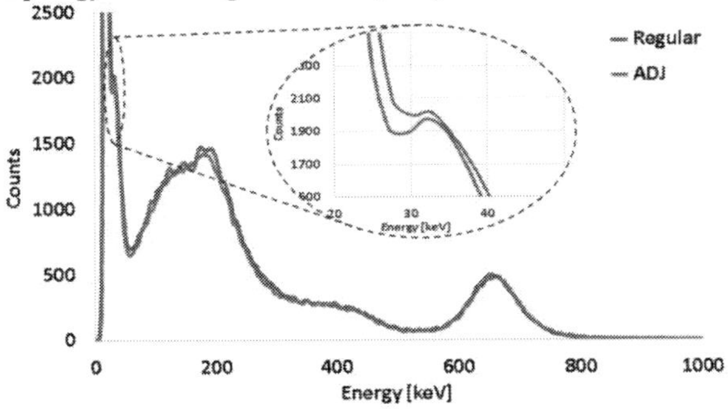

Fig. 11. Spectrum of ^{137}Cs obtained without the use of the breakdown correction topology, "Regular", in blue, and with the topology, "ADJ", in red.

Moreover, while the 32 keV peak is not recognizable in the "Regular" curve, in the "ADJ" curve obtained by the suggested topology the peak is slightly more distinct. Furthermore, the proposed biasing topology is power efficient compared to the straightforward approach of providing a separate bias voltage to each SiPM separately. For comparison, a n × n SiPM array with the proposed topology will have a power consumption as shown in (3) where a separate bias voltage approach will consume as shown in (4).

$$P_{Adj} = V_{In} \cdot I_Q + \frac{1}{\eta_{Adj}} \cdot V_B \cdot I_{SiPM} \cdot n^2 \,, \quad (3)$$

$$P_{Regular} = (V_{In} \cdot I_Q + \frac{1}{\eta_{Reg}} \cdot V_B \cdot I_{SiPM}) \cdot n^2 \,. \quad (4)$$

P_{Adj} and $P_{Regular}$ are the power consumption of the SiPM array when biased with the proposed adjustment topology and in the regular separate bias voltage approach, respectively. The DC-to-DC converter input voltage is V_{In}, I_Q is the converters supply current, V_B is the voltage converters output voltage, and I_{SiPM} is the SiPM current. The DC-to-DC converter efficiency for the proposed topology and for the separate bias voltage are η_{Adj} and η_{Reg}, respectively. From (3) and (4) it can be seen, that when using a separate bias voltage approach the supply current for the DC-to-DC converter is n^2 times bigger. Additionally, the DC-to-DC converter efficiency rises as the output current increases, especially for low output currents such as the SiPM DC current. Hence, the DC-to-DC efficiency in the suggested topology, η_{Adj}, is expected to be significantly bigger than in the separate bias voltage approach.

VI. Conclusions

This work investigated the breakdown variation effect on the spectrum that is obtained from SiPM array-based radiation detectors. An electronic topology that addresses the breakdown variation issue and uses a single temperature dependent bias voltage was presented. An improvement in the noise threshold was observed for overvoltages smaller than 1.5 V. Hence, the suggested topology can be useful in cases where a low working voltage can be advantageous. One example is radiation detectors mounted on satellites and exposed to cosmic rays. The SiPMs at such a detector are exposed to proton particle radiation that causes an additional deviation in the breakdown voltage and an increase of the dark current [7,8]. Working at 1 V overvoltage can reduce the dark current by a factor of 20, compared to standard 3.6 V overvoltage, at a cost of increasing the SiPM sensitivity to the breakdown voltage variation [21]. Hence, the dark current can be reduced while the breakdown voltage variation is addressed using the proposed topology. Moreover, the proposed topology can correct the mentioned deviation in the breakdown voltage, due to the radiation damage, as a part of a periodic calibration process. Further studies showed that there is a correlation between the working voltage of the SiPM and the effect of the radiation on the SNR and resolution obtained by the detector [9]. Hence, working at low overvoltages and combining the suggested topology can yield a reduced influence of the cosmic rays, provide decreased power consumption, and improved SNR, providing an enhanced spectroscopic preformance.

References

[1] Peter R. Menge, "Efficient positioning of silicon photomultipliers on large scintillation crystals", 2016 IEEE Nuclear Science Symposium, Medical Imaging Conference and Room-Temperature Semiconductor Detector Workshop (NSS/MIC/RTSD), Nov. 2016, 10.1109/NSSMIC.2016.8069734.

[2] Dennis R. Schaart, "A novel, SiPM-array-based, monolithic scintillator detector for PET", Physics in Medicine & Biology, Vol. 54, Number 11, May 2009, 10.1088/0031-9155/54/11/015.

[3] Paolo Busca, "Experimental Evaluation of a SiPM-Based Scintillation Detector for MR-Compatible SPECT Systems", IEEE Tran. on Nuclear Science, Vol. 62/5, Oct. 2015, 10.1109/TNS.2015.2481184.

[4] Hye MinPark, "Performance characteristics of a silicon photomultiplier based compact radiation detector for Homeland Security applications", Nuclear Instruments and Methods in Physics Research Section A, Vol. 781, May 2015, 10.1016/j.nima.2015.01.080.

[5] Marco Carminati, "A Lightweight SiPM-Based Gamma-Ray Spectrometer for Environmental Monitoring with Drones", presented at the International Conference on Applications in Electronics Pervading Industry, Environment and Society, Lisbon, Jan. 2021.

[6] P. Lv, "A low-energy sensitive compact gamma-ray detector based on LaBr3 and SiPM for GECAM", Journal of Instrumentation, Vol. 13, Aug. 2018, 10.1088/1748-0221/13/08/P08014.

[7] Jason Link, "SiPM Use in Future Space Instruments", presented at the 35th International Cosmic Ray Conference, Korea, Aug. 2018.

[8] Yu. Musienko, " Radiation damage studies of silicon photomultipliers for the CMS HCAL phase I upgrade", Nuclear Instruments and Methods in Physics Research Section A, Vol. 787, pp. 319-322, July 2015, 10.1016/j.nima.2015.01.012.

[9] S. Mianowski, "SiPM proton irradiation for application in cosmic space", Journal of instrumentation, Vol. 15, Mar. 2020, 10.1088/1748-0221/15/03/P03002.

[10] N.Dinu, "Breakdown voltage and triggering probability of SiPM from IV curves at different temperatures", Nuclear Instruments and Methods in Physics Research Section A, Vol. 845, pp. 64-68, Feb. 2017, 10.1016/j.nima.2016.05.110.

[11] Nicola Serra, "Experimental and TCAD Study of Breakdown Voltage Temperature Behavior in n/p SiPMs", IEEE Tran. on Nuclear Science, Vol. 58/3, Jun. 2011, 10.1109/TNS.2011.2123919.

[12] Zhengwei Li, "A gain control and stabilization technique for Silicon Photomultipliers in low-light-level applications around room temperature", Nuclear Instruments and Methods in Physics Research Section A, Vol. 695, pp. 222-225, Dec. 2012, 10.1016/j.nima.2011.12.037.

[13] Seiichi Yamamoto, "A temperature-dependent gain control system for improving the stability of Si-PM-based PET systems", Physics in Medicine & Biology, Vol. 56, no. 9, pp. 2873–2882, Apr. 2011, 10.1088/0031-9155/56/9/015.

[14] G. Gallina, "Characterization of SiPM Avalanche Triggering Probabilities", IEEE Tran. on Nuclear Science, Vol. 66/10, Oct. 2019, 10.1109/TED.2019.2935690.
[15] Y. C. Shih, "An 8×8 Row-Column Summing Readout Electronics for Preclinical Positron Emission Tomography Scanners", in Proc. IEEE Nucl Sci Symp Conf., Orlando, FL, USA, 2009, pp. 2376–2380, 10.1109/NSSMIC.
[16] ON Semiconductor, "Silicon Photomultipliers (SiPM), Low-Noise, Blue-Sensitive", C-Series SiPM Sensors datasheet, May 2021, www.onsemi.com.
[17] ON Semiconductor, "How to Evaluate and Compare Silicon Photomultiplier Sensors", White paper, Oct. 2015.
[18] Ketek, " SiPM – Silicon Photomultiplier", WL-Series datasheet, 2020, www.ketek.net.
[19] Hamamatsu, "Low breakdown voltage type MPPC for scintillation detector", S14160/S14161 Series datasheet, Jun. 2020, www.hamamatsu.com.
[20] V. Chmill, "Study of the breakdown voltage of SiPMs", Nuclear Instruments and Methods in Physics Research Section A, Vol. 845, pp. 56-59, Feb. 2017, 10.1016/j.nima.2016.04.047.
[21] Alexei Ulyanov, "Radiation damage study of SensL J-series silicon photomultipliers using 101.4 MeV protons", Nuclear Instruments and Methods in Physics Research Section A, Vol. 976 , Oct. 2020, 10.1016/j.nima.2020.164203.

https://doi.org/10.1051/epjconf/202125311008

Analog Pulse Shape Discrimination Based on Time Duration and Pulse Height

R. Harn[1], A. Osovizky[1,2], Y. Kadmon[1], A. Manor [1], and M. Ghelman[1]

[1]Nuclear Research Center – Negev, Israel
[2]Radiation Detection Department, Rotem Industries Ltd, Israel
RonH@nrcn.org.il

Abstract— **Pulse shape discrimination is a name of a group of techniques used to detect and distinguish between different types of radiation interactions. Analog pulse shape discrimination methods can be more suitable than digital methods, for high-speed scintillators both from rate and power consumption perspectives. Common analog discrimination methods are based on pulse-height and pulse-energy discrimination techniques. Other techniques rely on the time difference in the pulse width such as the Zero-Crossing methods. Neither of the above combine both amplitude and time methods. We present a novel analog pulse shape discrimination topology that combines both height and time domain. The topology is based on discrimination according to the pulse duration in time combined with compensation function of the pulse height. Amplitude of the pulse is used as a restraining factor. Subsequently, our topology correctly identifies fast pulses that are prolonged in time due to their high amplitude. The topology yields superior discrimination capabilities, under degraded light collection conditions, with an uncertainty gap smaller than 1 ns in pulse width. The ability to control both the time and the amplitude parameters individually, provides tailored adjustment for various detectors and pulse shape discrimination applications.**

Keywords — PSD, Discrimination, Classification, PHA, Plastic Scintillator.

I. Introduction

Pulse shape discrimination (PSD) is a general name for a group of techniques, aimed to discriminate between different types of radiations, generally used to distinguish between neutrons and gamma radiation, and between alpha and beta radiation. The core of all PSD techniques is the identification of differences in the nuclear pulse shape generated by the detector for different radiations. In order to increase variation in the generated pulse shape, when using scintillator-based detectors, an integration of scintillators is often used. For example, ZnS(Ag) and plastic scintillators are often combined for alpha and beta radiation discrimination. Moreover, different scintillator structures can be used to enhance the pulse shape deviation, such as in the phoswich detector [1].

Different PSD methods vary in their discrimination capabilities, detection efficiency, and power consumption. PSD methods can be categorized into two groups, analog-based techniques, and digital techniques. In the last two decades, the development of high-speed acquisition hardware such as high frequency analog to digital converters (ADC) and field programmable gate arrays (FPGA) gave the digital approach significant advantages over traditional analogue PSD techniques [2]. Digital PSD methods enable simultaneous analysis of a large number of different parameters. Furthermore, digital PSD methods enable implementation of complex algorithms, providing superior discrimination capabilities, as compared to analog techniques [3]. Digital PSD discrimination capabilities improved even further with the development of the artificial intelligence field [4,5]. However, complex algorithms require long processing times, thereby limiting the count rate and increasing the dead time. Sampling of nuclear signals, especially from scintillators with fast decay constants, requires fast samplers, which have significant power consumption. For this reason, an analog PSD topology can be more suitable when working with fast decay scintillators, especially for battery-based applications. The analog methods can be divided into two categories: classification by height (pulse height discrimination - PHD), and classification by time (pulse width discrimination - PWD). PHD is based on classification either by the pulse height or by pulse charge [6]. Classification by time is based on discrimination by the duration of the pulse, using the pulse width over a defined threshold level [7]. Under degraded light collection conditions, such as long and opaque scintillators, a lower pulse-height is obtained, while the noise level is unaffected. Consequently, lower signal to noise ratio is obtained, causing amplitude-based methods to yield a considerable number of misclassifications. Contrary to pulse height and energy that decrease proportionally to the light collection efficiency, the pulse shape is less affected. However, time and pulse width-based methods are not error-proof either. Pulse width at over threshold of fast and high amplitude pulses can have a similar width as slower pulses with lower amplitude.

In this work a novel analog PSD topology that overcomes this issue is presented. The topology is based on discrimination according to the pulse duration in time, combined with a compensation as a function of the pulse height. The amplitude of the pulse is used as a restraining factor. Subsequently, our topology correctly identifies fast pulses that are prolonged in time due to their high amplitude. The topology was realized using high speed printed circuit board and was found to present superior discrimination capabilities with an uncertainty gap smaller than 1 ns in pulse width. The ability to control both time and amplitude parameters separately provides adjustment for various detectors and PSD applications. The topology is introduced in two stages. In sections II and III, a time duration-based PSD is introduced, in section IV and V the combined time and pulse height PSD topology is presented.

II. Analog PSD Based on Time Duration

As mentioned, when the detector's light collection efficiency is reduced, for example when radiation-scintillator interaction occurs in the corner of the detector, the probability of a mis-classification increases. This is due to the strong correlation between the light collection efficiency and the pulse energy. Alpha and beta pulses obtained from a plastic and ZnS(Ag) scintillator coupled to a photomultiplier tube (PMT) are shown in Fig. 1 (a) and (b) respectively. While in the amplitude domain there is an overlap between the alpha and the beta pulse height between 100 mV to 150 mV, their time decay constants intervals are separated.

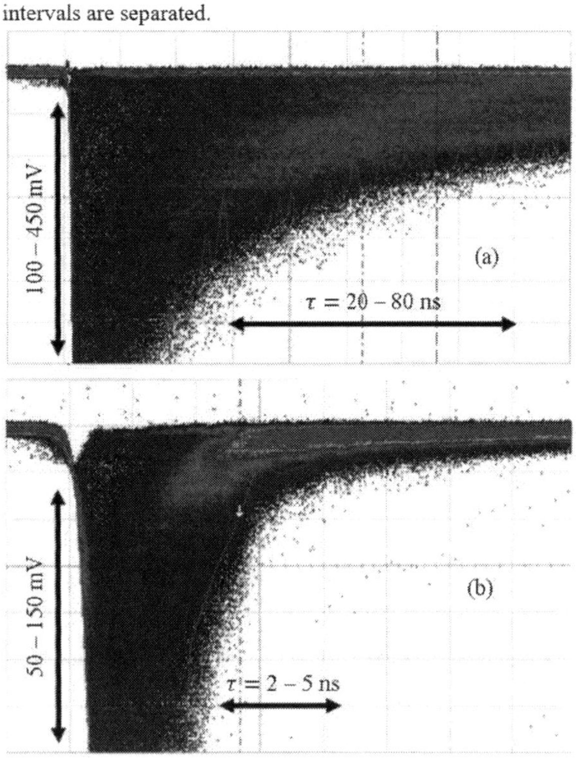

Fig. 1. Alpha pulses (a) and beta pulses (b) obtained from a detector based on plastic and ZnS(Ag).

Hence, for a radiation detector that produces the pulses shown in Fig. 1, separation based on time domain will yield less misclassifications than pulse height discrimination techniques. The distribution of the pulses on the amplitude – decay time span is shown in Fig. 2. As seen, a vertical separation by the time will yield a better separation compared to a horizontal separation, based on amplitude discrimination.

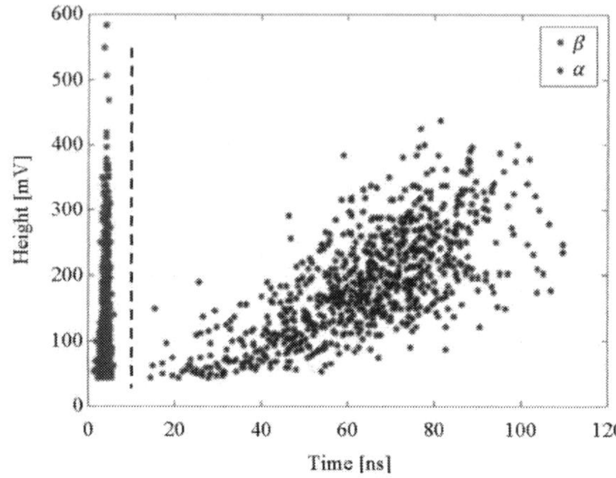

Fig. 2. The distribution of the alpha and beta pulses over the amplitude-decay time span.

A scheme of the proposed time-based discrimination topology is shown in Fig. 3. The nuclear pulse is fed from the detector to a high bandwidth amplifier. The pre-amplifier's bandwidth should be sufficient in order to preserve the detector's pulse shape. An attenuation of the high frequency components at the nuclear signal will increase the resemblance between the fast and slow pulses, hence, increasing mis-classification probability. In addition, the amplifier should have high amplification and saturation recovery capabilities, in order to obtain a full width of the pulse over the threshold detection level. Higher pulses, beyond threshold, generate a more linear connection between the width of the amplified pulse, T_W, and the pulse decay constant, τ_f. On the other hand, wide bandwidth amplifies generate higher noise levels, and hence necessitate higher threshold detection levels. Therefore, the input operational amplifier should be carefully selected, by referring to its gain-bandwidth product and its noise level, compared to the detector's pulse shape and internal noise. The amplifier output is fed into a comparator level that generates a digital pulse. The digital pulse width, T_W, matches the amplifier output pulse width. Hence, the time domain property of the detector pulse is separated. The comparator functions as a low-level discriminator (LLD). Therefore, all the pulses that exceed the threshold detection level is counted at the "ALL" output port, meaning it is either a slow pulse (neutron / alpha), or a fast pulse (beta / gamma). Therefore, the topology detection efficiency depends on the detection threshold level. Hence, the threshold is set as low as possible as a function of the detector noise level. The comparator output controls a single pole double through (SPDT) analog switch. The SPDT switch inputs are connected to the circuit analog power supply, VCC, and to the circuit ground potential. The SPDT switch output is fed into an integrator. Therefore, when comparator output is high the integrator is fed with a regulated stable voltage, VCC, instead of the high output of the comparator that deviate from one

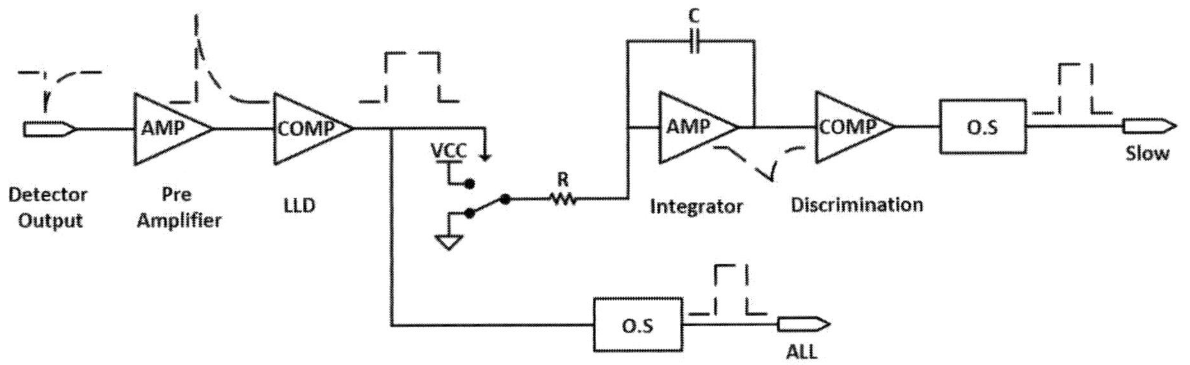

Fig. 3. Scheme of the proposed time based discrimination topology.

component to the other. When the comparator output is low the integrator is fed with the ground potential instead of the comparator's low-level voltage. The output of the integrator peak has a linear dependence in the digital pulse width, T_W, as shown in (1),

$$V_{Integrator} = -\frac{VCC}{RC}(T_W + \tau_{Const}). \quad (1)$$

Where, R and C are the integrator resistor and capacitor respectively. The constant τ_{Const} is a component added due to the switch parasitic junction capacitance discharge decay time. The integrator output is fed into a discrimination comparator with an adjustable threshold. By adjusting the discrimination comparator threshold, the width that classifies between fast and slow pulses is determined. The signal shapes at different nodes of the electronic topology are shown in Fig. 4 (a), for a slow pulse with a decay time constant of 20 ns, and for a fast pulse, Fig. 4 (b), with a decay time constant of 10 ns.

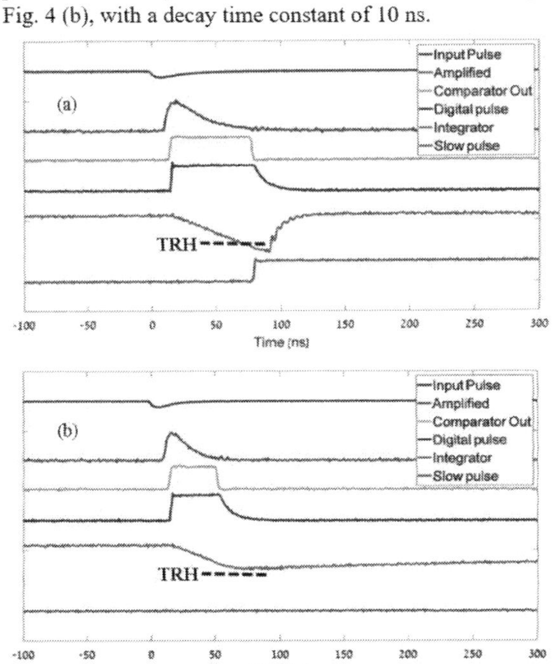

Fig. 4. Signal shape at different nodes, (a) for a slow pulse, (b) for a fast pulse.

Once the integrator curve exceeds the threshold, the signal is detected as a slow pulse, and the "Slow" signal rises. A discharge logic is enabled to implement a fast discharge of the integrator capacitor, in order to enable the process of a new pulse, hence reducing the dead time.

III. Time Duration PSD Results

The topology was implemented on a FR-4 PCB designed for high frequency operation. The topology was evaluated using CAENs DT5810 fast digital emulator. The time width threshold was adjusted to 50 ns. The topology input was fed with pulses with different decay constants at four amplitudes, 100, 200, 400, and 800 mV. At each point, the percentage of slow classifications were recorded. The result is presented in Fig. 5.

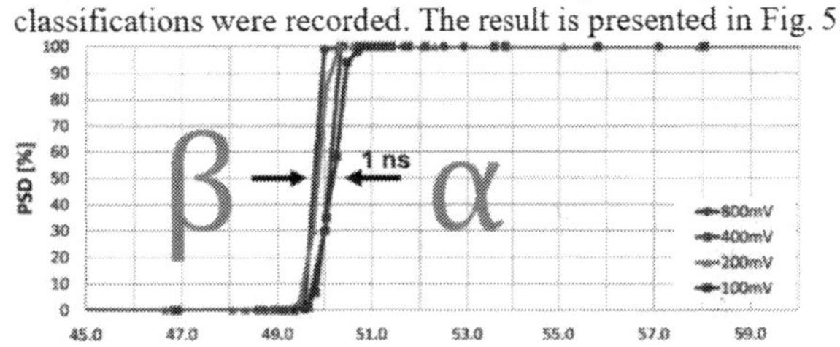

Fig. 5. Percentage of "Slow" classification for different pulses widths for four amplitudes, classification width threshold is set to 50 ns.

As seen in Fig. 5, pulses wider than 50.5 ns were all classified as slow pulses, while all pulses narrower than 49.5 ns were classified as fast pulses, in accordance with the 50 ns threshold. The 1 ns gap where classification is not certain is the uncertainty gap. One can note that the higher the input amplitude, the smaller the uncertainty gap. This is because of the improvement in the linear connection between the pulse width and the pulse decay time with amplitude growth. Furthermore, the congestion of the curves regardless of their amplitude shows that there is a very small amplitude dependence, as desired in a time-based amplitude PSD. Experimental results for the vertical separation line presented in Fig. 2. is shown in Fig 6. The reduction in the linear pulse width – decay constant connection can be observed for small amplitudes.

 https://doi.org/10.1051/epjconf/202125311008

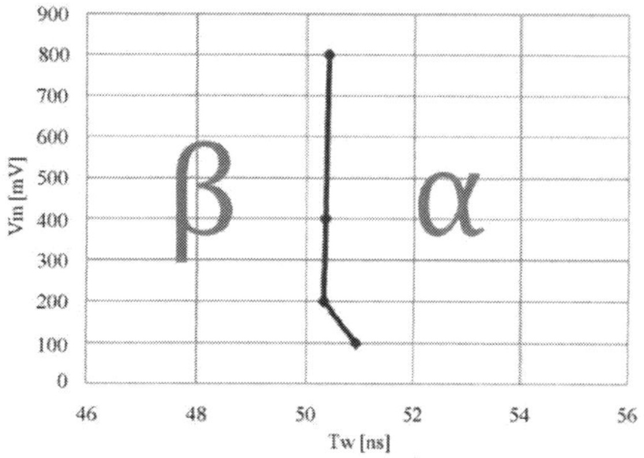

Fig. 6. Measured results for the vertical classification line in the height – time span produced by the time duration PSD topology.

IV. Analog PSD Based on Time Duration and Pulse Height Combination

As mentioned, time duration classification techniques are not error proof. Light collection efficiency reduction can cause misclassifications between fast pulses that are prolonged in time due to their high amplitude, and between slow attenuated pulses. A detector based on a 30 mm X 30 mm X 1 mm plastic scintillator combined with a ZnS(Ag) plating was assembled.

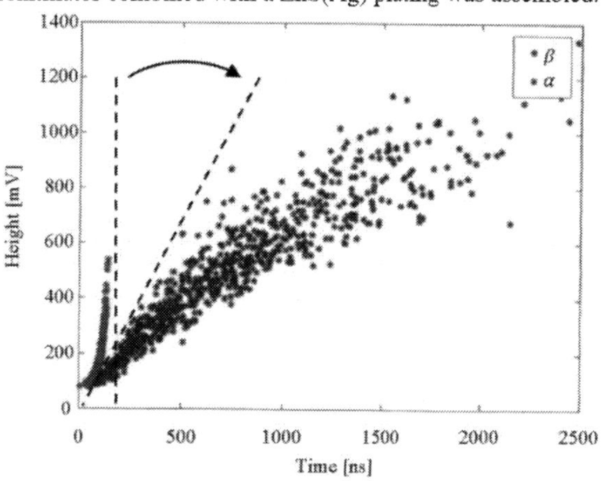

Fig. 7. The distribution of the alpha and beta pulses over the amplitude-decay time span in a single SiPM detector.

The scintillator was coupled to a single 6 mm X 6 mm silicon photodiode (SiPM) with a 10 mm light guide. Pulse distribution on the amplitude time with span is presented.

It can be noticed that in the case presented in Fig. 7, both PHD and PWD techniques will generate a significant number of misclassifications. However, a diagonal classification line, as shown in Fig. 7, can reduce the number of misclassifications. In order to produce such a diagonal separation line, a forward feedthrough was added to the topology from section II, as shown in Fig. 8. The time and amplitude combined topology implement a forward feed of the inverse pulse via a feedforward resistor, R_F. Hence, higher pulses will generate stronger discharge current that discharge the integrator capacitor simultaneously to its charging process. Hence, higher pulses would have to be prolonged in time, compared to lower pulses, in order to be classified as a slow pulse. The angle of the diagonal separation line is a function of the discharge current with respect to the integrator charging current. Hence, the angle is a function of the ratio R_F/R.

V. Results

The discrimination curves of the height-time combination topology for 100, 200, 400, and 800 mV is presented in Fig. 9. Compared to the time-based discrimination presented in Fig. 5, the curves are not grouped together as expected, as the discrimination depends on the amplitude of the pulse. As seen from Fig. 9, for higher pulse amplitudes, larger values of T_W are required in order for the classification to reach the slow region. Furthermore, as opposed to the time-duration topology, where higher amplitude generated smaller uncertainty gaps, in the height-time combination topology the opposite is obtained. This is because in the combined topology the input is fed directly to the integrator input. Hence, injected noise is integrated. Higher pulses require longer integration time, hence more noise is integrated, increasing the uncertainty gap. Measured results for the diagonal classification line yield using the time domain and pulse height combination are shown in Fig. 10. The blue curve is the diagonal classification line obtained from the combine amplitude-time topology while the red curve is the classification line based only on time domain. As can be seen, the results match the desired separation diagonal curve shown in Fig. 7.

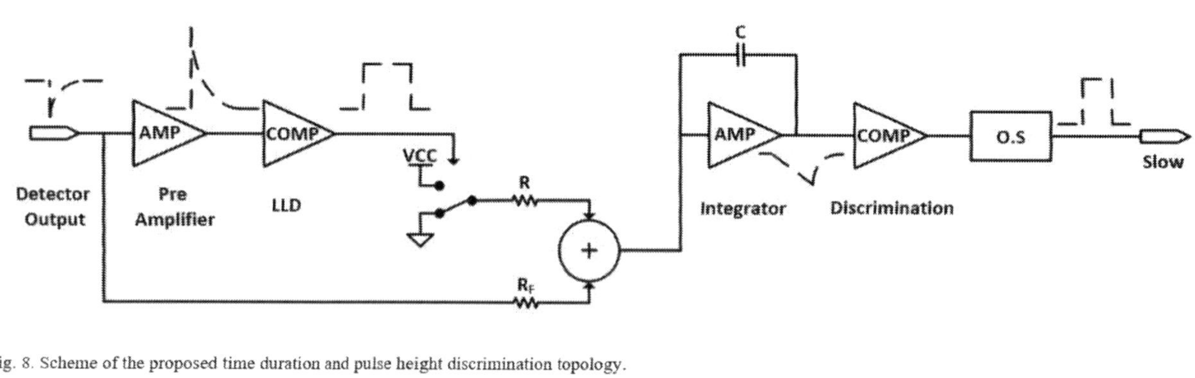

Fig. 8. Scheme of the proposed time duration and pulse height discrimination topology.

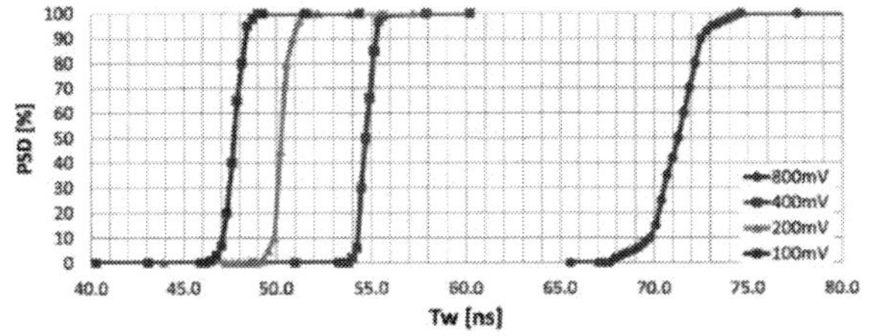

Fig. 9. Percentage of "slow" classification for different pulses widths using time duration and pulse height discrimination topology.

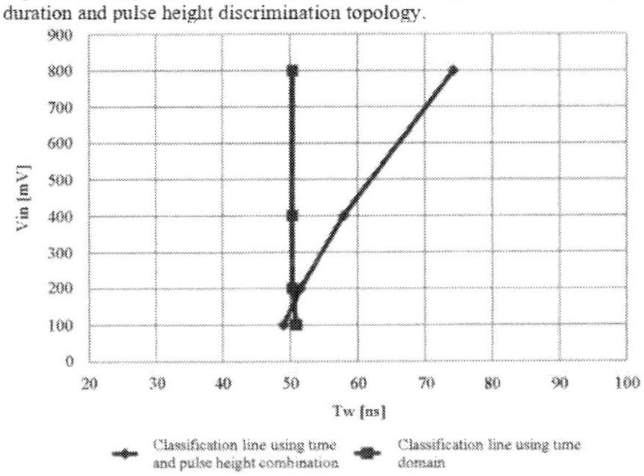

Fig. 10. Measured results of the classification curve in the height – time span, red curve using time duration topology, blue curve using the combined height-time topology.

VI. Conclusions

An analog topology that can discriminate between fast and slow pulses from plastic scintillators was developed. Using the height of the pulse as a restraining factor and combining it with a time domain PSD, the topology provides enhanced discrimination capabilities in inadequate light collection structures. The topology is not based on high frequency samplers or complex algorithms; hence, it has advantages both in power consumption and in processing time. The absence of long processing time ensures shorter dead time. By adjusting a few parameters in the electronics, the topology can easily be adapted for different detectors with different time constants and light yield coefficients.

References

[1] D. H. Wilkinson, "The Phoswich-A Multiple Phosphor", Review of Scientific Instruments, vol. 23, no. 8, Aug. 1952, 10.1063/1.1746324.

[2] Matthew J.I.Balmer, " Comparative analysis of pulse shape discrimination methods in a Li-6 loaded plastic scintillator ", Nuclear Instruments and Methods in Physics Research Section A, Vol. 788, Jul 2015, 10.1016/j.nima.2015.03.089.

[3] C.S. Sosa, " Comparison of analog and digital pulse-shape-discrimination systems", Nuclear Instruments and Methods in Physics Research Section A, Vol. 826, May 2016, 10.1016/j.nima.2016.03.088.

[4] T. S. Sanderson, " Machine learning for digital pulse shape discrimination", 2012 IEEE Nuclear Science Symposium and Medical Imaging Conference Record (NSS/MIC), Oct. 2012, 10.1109/NSSMIC.2012.6551092.

[5] P. Holl, "Deep learning based pulse shape discrimination for germanium detectors", The European Physical Journal C, Vol. 79. no. 450, Jun. 2019, 10.1140/epjc/s10052-019-6869-2.

[6] M.Flaska, " Identification of shielded neutron sources with the liquid scintillator BC-501A using a digital pulse shape discrimination method", Nuclear Instruments and Methods in Physics Research Section A, Vol. 577, Jul 2007, 10.1016/j.nima.2007.04.141.

[7] M.Nakhostin, " Application of digital zero-crossing technique for neutron–gamma discrimination in liquid organic scintillation detectors", Nuclear Instruments and Methods in Physics Research Section A, Vol. 621, Sep. 2010, 10.1016/j.nima.2010.06.252.

Scanning of a Double-Sided Germanium Strip Detector

Arzoo Sharma[1,3], R. Palit[2], I. Kojouharov[3], J. Gerl[3], M. Gorska-Ott[3], H. Schaffner[3], T. Habermann[3], S. Saha[4], Biswajit Das[2], P. Dey[2], R. Donthi[2], B.S. Naidu[2], S. Mandal[5], Pushpendra P. Singh[1]

[1]Department of Physics, Indian Institute of Technology Ropar, Rupnagar – 140 001, Punjab, India
[2]Department of Nuclear and Atomic Physics, Tata Institute of Fundamental Research, Homi Bhabha Road, Mumbai – 400 005, India
[3]GSI Helmholtzzentrum für Schwerionenforschung GmbH, Planckstrasse 1, 64291 Darmstadt, Germany
[4]University of Massachusetts Lowell, Lowell, Massachusetts 01854, USA
[5]Department of Physics, North Campus, University of Delhi, Delhi – 110 007, India
palit@tifr.res.in, 2017phz0005@iitrpr.ac.in

***Abstract*— This work presents the results from the characterization of a Position-Sensitive Planar Germanium (PSPGe) detector. The PSPGe detector is a double-sided orthogonal strip detector consisting of 10x10 electrical segmentation along the horizontal and vertical directions. The characterization was performed using the coincidence setup between the PSPGe detector and the well-characterized scanning system employing the positron annihilation correlation principle. The scanning system consists of a Position Sensitive Detector (PSD) and ^{22}Na positron source. The main objective of this study is to deploy PSPGe detector for future decay experiments at the Facility for Antiproton and Ion Research (FAIR), Germany. The measurements have been performed to find the depth of gamma-ray interaction in the planar segmented detector. The 2-Dimensional image obtained from the PSD has been used to find the depth of gamma-ray interaction in the planar strip detector using pulse shape analysis. In addition, the sensitivity of PSPGe detector has been investigated by calculating the rise-time from pulse shapes for the front and back strips of the detector.**

***Keywords* —Double-Sided Strip Detector; Positron Annihilation Correlation; Position Sensitive Detector; Pulse Shape Analysis.**

I. Introduction

While the nucleus of an atom is a complex object; it has been studied extensively with the aid of characteristic gamma-ray emitted from its decay using gamma spectroscopy tools. The understanding of the structure of nuclei has been obtained through the development of sophisticated solid-state detectors. An advent in this field has been through the development of highly segmented High Purity Germanium (HPGe) detectors like AGATA [1] and GRETA [2]. These detectors are employed to track the path of gamma-ray interaction inside the detector crystal using energy, time, and location information. They are coaxial 4π HPGe detector arrays based on the principle of Pulse Shape Analysis (PSA) [3]. The pulses are stored for each segmented crystal within the defined time window of supporting electronics. Various PSA algorithms have been employed to locate the interaction position that is based on the principle of the Shockley-Ramo theorem [4]. It states that the moving charge, produced by gamma-ray interaction, induces a charge on the electrode, which is detected by the charge-sensitive pre-amplifiers. The correct PSA algorithm helps to reconstruct the complete path of a photon interacting at various locations using the Compton scattering formula. If the algorithm is not reasonably constructed, it may lead to the complete loss of several interacting events. This may further provide wrong energy. Hence, the reasonable choice of PSA algorithm has to be made depending upon the detector geometry, electrode, and segmentation. Therefore, the complexity of signals in these segmented detectors requires complete characterization with known gamma-ray sources.

This manuscript has been put forth to characterize a Planar Position-Sensitive Germanium (PSPGe) detector along the depth. The detector depth is not segmented. Hence, the rise-time studies have been performed to find depth of gamma-ray interaction [5]. Further, the detector has been scanned using a scanning facility developed at GSI Helmholtz Centre for heavy-ion research, Germany [6, 7]. The main aim of performing these tests is to employ the detector as an implantation detector in the Decay SPECtroscopy (DESPEC) experiments in the framework of Facility for Antiproton and Ion Research (FAIR) at GSI [8]. This article is organized as follows. Section II describes the detailed experimental setup. In section III, the analysis procedure is discussed. The Results and conclusions are presented in sections IV and V.

II. Experimental details

The PSPGe detector is a p-type electrically segmented detector. It consists of 10 segmentation along both horizontal as well as vertical directions. The horizontal strips are DC coupled, and the vertical strips are AC coupled. The detector has dimensions of 6 x 6 x 2 cm^3. The energy resolution and efficiency details of the detector are described in the reference [9]. The DC electrodes are etched on the front side of the detector, while AC strips are at the backside. In the present work, DC and AC electrodes are referred to as DC0 to DC9,

 https://doi.org/10.1051/epjconf/202125311009

and AC16 to AC25, respectively, as indicated in figure 1.

Fig. 1: Typical coincidence setup between PSD (left) and PSPGe detector (right) (DC and AC strips are marked in orange and blue color, respectively).

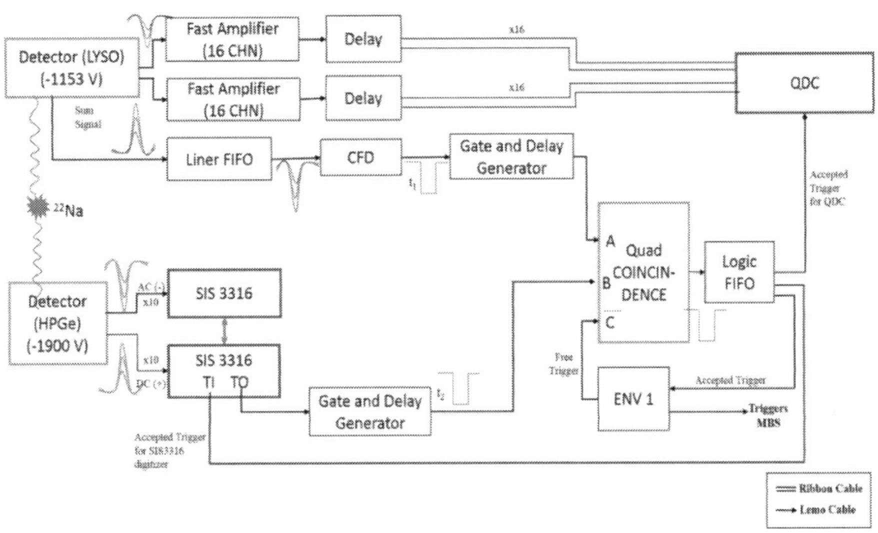

Fig. 2: Typical block diagram of electronics arrangement used for setting up the coincidence between PSD and PSPGe detectors.

The experimental setup is shown in Fig. 1. The PSPGe detector has been set up in coincidence with the PSD using ^{22}Na source. The PSD consists of a LYSO scintillator coupled to Position-Sensitive Photo Multiplier Tube (PSPMT) [10]. After pair annihilation, the two gamma-rays are detected in PSPGe and PSD. The distance between ^{22}Na source and PSPGe detector was kept 8.6 cm. The distance was optimized to get a complete image by looking at the online 2-Dimensional (2-D) image using GSI Object Oriented On-line Off-line (GO4) software. The 2-D image is extracted from PSD, consisting of PSPMT with 16-X and 16-Y anodes mesh. The complete electronics block diagram is shown in Fig. 2. The 32 anode output signals from the PSD were amplified using Fast Amplifier (N979). Afterward, the output was fed to Versa Module Eurocard (VME) based multi-event Charge to Digital Convertor (QDC V792) unit with 12-bit resolution. The cathode output from the PSD was used to generate timing gate, t_1 (depicted in Fig. 2), via Constant Fraction Discriminator (CFD - CF4000). The 10 DC and 10 AC outputs of the PSPGe detector were fed to 100 MHz digitizers from Struck Innovative Systeme (SIS3316) modules. It provides information on pulse shape and energy for each interacting gamma-ray on an event-by-event basis. The digitizer's Trigger-Out (TO) output was a logic pulse that contained information of the OR (OR of the time) of all channels, indicated as time t_2 in Fig. 2. Afterward, t_1 and t_2 were set up in coincidence using the Quad Coincidence unit (CO4001). The data from QDC and SIS3316 was read out using the Multi-Branching Data Acquisition System (MBS). The ENV1 module contains the dead time of the MBS readout, which is put in anti-coincidence with t_1 and t_2 using CO4001. The output was used as an accepted trigger for the QDC and SIS3316 digitizer. The settings of the digitizer were made using desktop Graphical User Interface (GUI). The monitored data rate was ~250 Hz.

III. Data Analysis Procedure

The data analysis has been performed using GO4, C++, and ROOT. The critical points of the analysis procedure are listed here. First, the 2-D image is reconstructed from the charge collected at X and Y anodes in PSD for the coincidence events. The reconstruction has been obtained by the centroid fitting method, as displayed in Fig. 3.

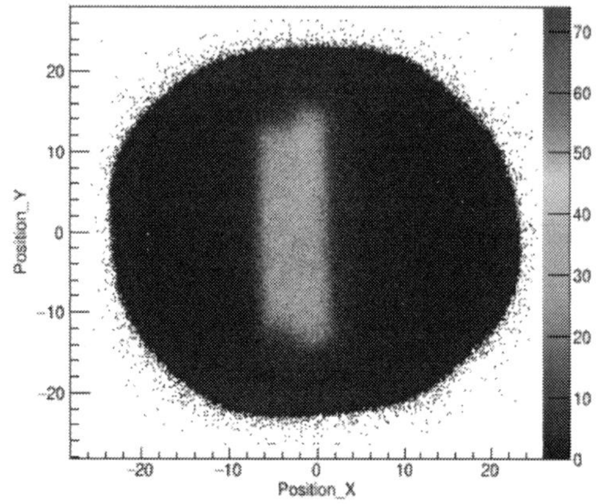

Fig. 3: The 2-D image obtained from PSD.

The middle region with maximum counts represents the active region of detector volume. 2-D image is projected onto X-axis, and it represents the depth of PSPGe. It is used for dividing the depth into segments because of no segmentation along the depth. First, the obtained energy spectra from the digitizer are calibrated. Then, it is used to select an electrode of interest by gating on the Compton edge. The pulse shape and/or trace obtained from the digitizer provides information on the gamma-interaction event for the selected electrode. The analysis has been performed for the obtained traces as described in reference [11], i.e., the baseline subtraction is performed, followed by normalization. Afterward, traces are time-aligned at 10% of the final amplitude.

The work has been presented to understand the analysis for finding the depth of gamma-ray interaction using the setup shown in Fig. 1. The Compton edge of 511 keV photopeak is used to select two middle strips DC4 and AC20. The pulse shape is analyzed for different points along the depth. The rise time is measured for selected traces at a specific depth location. The average of multiple traces is considered to find the rise-time. The sensitivity of the PSPGe detector towards depth would depend on how precisely charge collection could be measured for two charge carriers (electrons and holes) at the two opposite-facing electrodes. It would then lead to different rise-time values at different depth locations. In the scope of the present analysis, T30 rise-times have been calculated for DC4 and AC20 strips. It is defined as the time required to reach 30% of the maximum amplitude minus the time required to reach 10% of the maximum amplitude. The analysis results, for the behavior of both DC and AC strips, have been discussed in detail in the upcoming section.

IV. Results and Discussion

The coincidence technique using the principle of positron annihilation has been used to characterize the electrically segmented PSPGe detector. The increased segmentation is beneficial to track each Compton scattered event in a different segment. The centroid fitting method has been used to reconstruct the image from PSD.

To understand the charge induced by the incoming gamma-rays, the rise-time has been calculated for the pulse shapes obtained from SIS3316 digitizer. The analysis has been performed using energy cut at Compton edge for two middle strips, DC4 and AC20. As a representative case, the behavior of DC4 and AC20 strips is shown in Fig. 4, for one of the middle cuts. In the figure, the amplitude of the trace is plotted as a function of time. The obtained T30 rise-time values for DC4 and AC20 strips are 30 ns and 80 ns, respectively. This shows the appreciable time difference between the two traces. This feature can be further used to find the rise-time as a function of depth.

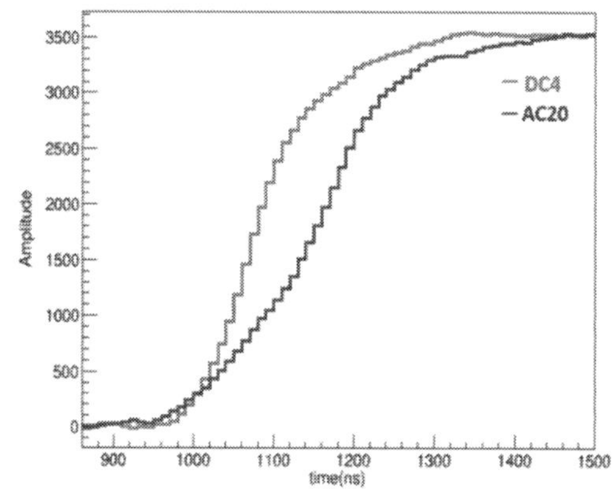

Fig. 4: T10 aligned traces for a middle cut selection (-2.1<Position_X<-1.9) on active region in 2-D image obtained from PSD. Traces of DC4 and AC20 are represented by red and blue color, respectively.

The difference in rise-times of the two strips may be attributed to the movement of charge carriers towards the respective electrodes. The front face of the detector consists of DC segments, and the AC segments are etched on the backside of the detector. The electrons would travel towards the DC electrode while holes would be collected at the AC electrode. The rise-time distribution is plotted for one of the depths in Fig. 5. In this plot, T30 rise-time counts are represented by a black histogram. The red curve is the Gaussian fit to find the centroid of the distribution. Hence, the centroid can be used to find the rise-time at a specific depth. In addition, the spread in the distribution may be used to infer the error in the obtained rise-time values [12].

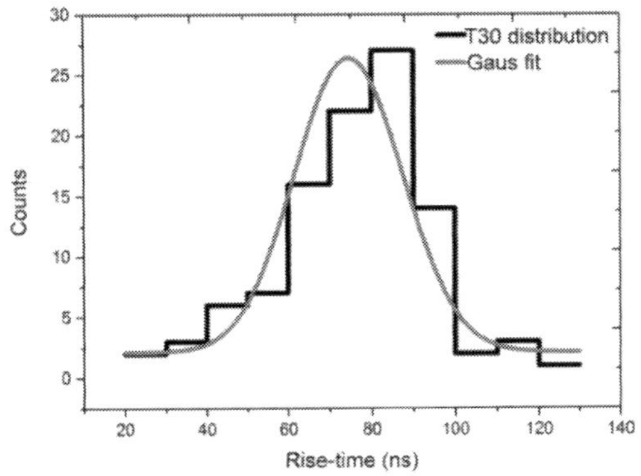

Fig. 5: Plot of counts vs rise-time (T30) at a specific depth. The red curve is the Gaussian fit (used to find centroid of the distribution).

The analysis has been extended to find the rise-time as a function of depth for both the DC4 and AC20 electrodes. It is shown in Fig. 6(a)-(b).

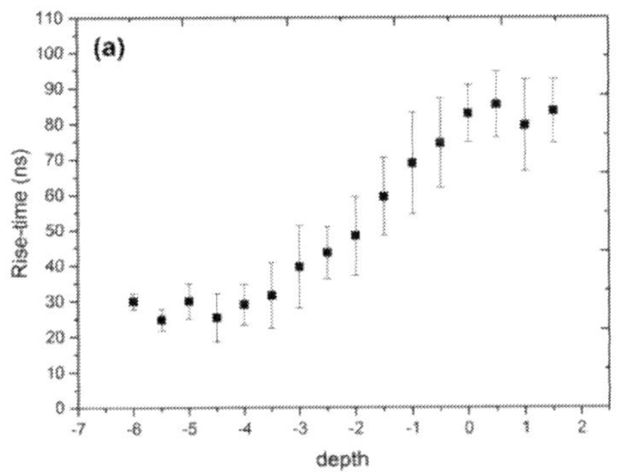

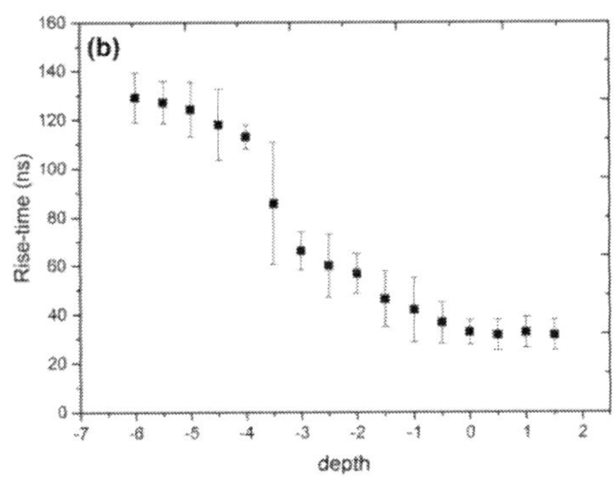

Fig. 6: T30 rise-time values as a function of depth for (a) DC4 and (b) AC20 segments, respectively.

In this figure, X-axis represents depth, i.e., going from the front to the backside of the detector. The values plotted are the centroids of the rise-time distribution obtained for each depth. The error has been calculated for each data point from the rise-time distribution spread, as indicated in Fig. 5. For DC4 (Fig. 6(a)), with an increase in depth, rise-time increases. Fig. 6(b) shows that the rise-time of AC20 decreases with an increase in depth. The rise-time of the DC4 segment starts to increase while traversing from the left to right side, i.e., from DC towards the AC electrodes. It occurs because of the slow collection of electrons at the front surface with an increase in depth.

In comparison, the holes are collected faster, therefore, the faster rise-time response at the AC electrode. This implies that the difference in collection times of the two charge carriers, between the front and backside of the detector, can successfully be applied to find the depth of the gamma-ray interaction. From the above analysis, it may be deduced that the position response of the detector is sensitive towards charge collection at the respective electrodes.

V. Summary and conclusions

In summary, a PSPGe detector has been characterized using a time-efficient scanning facility at GSI Germany. The response of the detector has been studied for gamma-ray interaction along the depth using rise-time values obtained from the pulse shape analysis. The behavior of induced charge along the detector surface has been obtained and analyzed using rise-time calculation as a function of depth. The T30 rise-time distribution has been used to find the position sensitivity of the detector towards a gamma-ray hit. The results obtained in this work have been promising for finding the position sensitivity of the detector as the subsequent analysis steps.

Acknowledgment

This work is partially supported by the Department of Atomic Energy, Government of India, under project No. 12P-R&D-TFR-5.02-0400. The authors would like to acknowledge the Helmholtz International Center for FAIR in Germany and Prof. H. J. Wollersheim for constant support for carrying out this work.

References

[1] S. Akkoyun, A. Algora, B. Alikhani, F. Ameil, G. De Angelis, L. Arnold, A. Astier, A. Ataç, Y. Aubert, and C. Aufranc, *"Agata—advanced gamma tracking array"*. Nuclear Instruments Methods in Physics Research Section A: Accelerators, Spectrometers, Detectors Associated Equipment 2012. **668**: p. 26-58.

[2] K. Vetter, A. Kuhn, I. Lee, R. Clark, M. Cromaz, M. Deleplanque, R. Diamond, P. Fallon, G. Lane, and A. Macchiavelli, *"Performance of the GRETA prototype detectors"*. Nuclear Instruments Methods in Physics Research Section A: Accelerators, Spectrometers, Detectors Associated Equipment 2000. **452**(1-2): p. 105-114.

[3] F. Crespi, F. Camera, B. Million, M. Sassi, O. Wieland, and A. Bracco, *"A novel technique for the characterization of a HPGe detector response based on pulse shape comparison"*. Nuclear Instruments Methods in Physics Research Section A: Accelerators, Spectrometers, Detectors Associated Equipment, 2008. **593**(3): p. 440-447.

[4] Z. He, *"Review of the Shockley–Ramo theorem and its application in semiconductor gamma-ray detectors"*.

Nuclear Instruments Methods in Physics Research Section A: Accelerators, Spectrometers, Detectors Associated Equipment 2001. **463**(1-2): p. 250-267.

[5] M. Momayezi, W. K. Warburton, and R. A. Kroeger. "*Position resolution in a Ge-strip detector*". in *Hard X-Ray, Gamma-Ray, and Neutron Detector Physics*. 1999. International Society for Optics and Photonics.

[6] T. Habermann, F. Didierjean, G. Duchêne, M. Filliger, J. Gerl, I. Kojouharov, G. Li, N. Pietralla, H. Schaffner, and M.-H. Sigward, "*Application of gamma imaging techniques for the characterisation of position sensitive gamma detectors*". Nuclear Instruments Methods in Physics Research Section A: Accelerators, Spectrometers, Detectors Associated Equipment, 2017. **873**: p. 24-26.

[7] C. Domingo-Pardo, N. Goel, T. Engert, J. Gerl, I. Kojouharov, H. Schaffner, F. Didierjean, G. Duchêne, and M. Sigward, "*A novel γ-ray imaging method for the pulse-shape characterization of position sensitive semiconductor radiation detectors*". Nuclear Instruments Methods in Physics Research Section A: Accelerators, Spectrometers, Detectors Associated Equipment, 2011. **643**(1): p. 79-88.

[8] A. Khaplanov, B. Cederwall, and S. Tashenov, "*Position sensitivity of segmented planar HPGe detectors for the DESPEC project at FAIR*". Nuclear Instruments Methods in Physics Research Section A: Accelerators, Spectrometers, Detectors Associated Equipment, 2008. **592**(3): p. 325-333.

[9] J. Sethi, R. Palit, B. Naidu, S. Saha, T. Trivedi, R. Donthi, and S. Jadhav. "*Characterization of a Planar Ge Strip Detector and its Coupling with a Segmented HPGe Clover Detector*". in *Proceedings of the DAE Symp. on Nucl. Phys*. 2011.

[10] N. Goel, C. Domingo-Pardo, T. Engert, J. Gerl, I. Kojouharov, and H. Schaffner, "*Spatial calibration via imaging techniques of a novel scanning system for the pulse shape characterisation of position sensitive HPGe detectors*". Nuclear Instruments Methods in Physics Research Section A: Accelerators, Spectrometers, Detectors Associated Equipment, 2011. **652**(1): p. 591-594.

[11] N. Goel, C. Domingo-Pardo, T. Habermann, F. Ameil, T. Engert, J. Gerl, I. Kojouharov, J. Maruhn, N. Pietralla, and H. Schaffner, "*Characterisation of a symmetric AGATA detector using the γ-ray imaging scanning technique*". Nuclear Instruments Methods in Physics Research Section A: Accelerators, Spectrometers, Detectors Associated Equipment, 2013. **700**: p. 10-21.

[12] M. Descovich, P. Nolan, A. Boston, J. Dobson, S. Gros, J. Cresswell, J. Simpson, I. Lazarus, P. Regan, and J. Valiente-Dobon, "*The position response of a large-volume segmented germanium detector*". Nuclear Instruments Methods in Physics Research Section A: Accelerators, Spectrometers, Detectors Associated Equipment, 2005. **553**(3): p. 512-521.

Experimental and Simulation Investigation of Micro- and Nano-Structured Neutron Detectors

Faruk Logoglu[1], Patrick Albert[2,3], Douglas Wolfe[2,3], and Marek Flaska[1]
1) Ken and Mary Alice Lindquist Department of Nuclear Engineering
2) Department of Materials Science and Engineering
3) Applied Research Laboratory
The Pennsylvania State University
Corresponding author: ful51@psu.edu[1]

***Abstract*— We are investigating different micro- and nano-structure approaches to neutron detection based on inorganic scintillators. Specifically, we have been assessing various neutron converter-scintillator configurations through simulations and experiments. One promising inorganic scintillator is ZnO due to its relatively high light yield[1], reasonable optical transparency in the visible region[2], and relatively low refractive index[3] compared to other Zn-based crystals such as ZnS[4]. Accurate optical data and rigid simulation tools are necessary to optimize the dimensions of the neutron converter/scintillator systems. Accurate optical data are necessary since the optical parameters of a material depend on a variety of factors, including but not limited to its morphology, crystal structure, surface quality (surface roughness), as well as the temperature at which it was manufactured. Therefore, literature data show significant discrepancy when it comes to the optical parameters for the material and it is important to accurately measure these quantities for the specific sample of interest. Neutron detection is a complex process that includes neutron transport, charged particle transport, and light transport in the active detection medium. Hence, a rigid simulation tool is required to handle all these different areas of physics with sufficient accuracy. In this work, Geant4 has been chosen to carry out the simulations of these processes. Geant4 (GEometry ANd Tracking) is a toolkit used in various applications including high energy physics, astrophysics, and radiation detection[5]. The optical simulation capabilities of Geant4 have been validated by comparing the transmission and reflection data from UV-Vis spectroscopy to the Geant4 models for different Zn-based crystals. After validating the optical response of single crystals, simulation models were constructed to model more complex structures of ZnS-based alpha detection sheets (EJ-440) from Eljen Technology. Optical parameters validated with experimental results have been used in radiation simulation in Geant4. This study will serve as a basis for our ongoing effort to optimize and manufacture an efficient and compact fast neutron detection module with micro- and nano-structures.**

***Keywords* — Spectroscopy, Geant4, Optimization, Neutron Detectors, Scintillation**

I. Introduction

Fast neutron detection is achieved through either neutron scattering or neutron capture reactions. Scattering-based fast neutron detection systems typically employ hydrogen atoms due to the similar mass to that of neutrons. On the other hand, capture-based neutron detection systems utilize isotopes that have high neutron capture cross section. Three of the most important neutron capture reactions for neutron detection are given here[6]:

$$^{3}He + n \rightarrow {}^{3}H + {}^{1}H + 0.764MeV$$
$$^{6}Li + n \rightarrow {}^{4}He + {}^{3}H + 4.8MeV$$
$$^{10}B + n \rightarrow \left\{ \begin{matrix} ^{7}Li + {}^{4}He + 2.31MeV + 0.48MeV(\gamma)\ (94\%) \\ ^{7}Li + {}^{4}He + 2.79MeV\ (6\%) \end{matrix} \right\} \quad (1)$$

Although ^{3}He has superior thermal neutron capture cross section (5333 barns) than those of ^{6}Li (940 barns) and ^{10}B (3842 barns)[6], its worldwide shortage prohibits its widespread use in newly developed detectors[7]. This aforementioned shortage of ^{3}He has led researchers to develop neutron detectors based on alternative approaches[8].

In our recent studies, we have investigated several approaches to maximize the neutron detection efficiency of various neutron converter/scintillator configurations[9]. Optimizations of different components of proposed neutron detectors were carried out in Geant4 simulations. Although neutronic data and charged particle models are relatively consistent between Geant4, MCNP6[10], and experiments[7], optical data have to be accurately measured for the material samples of interest. The difficulty arises due to the fact that optical constants are not only dependent on elemental composition of the materials, but also on their morphology, crystal structure, surface roughness, and the temperature at which they were manufactured. Therefore, optical parameters of the specific detector components should be measured, and those same optical parameters must be used in optimization and validation studies.

Moreover, the simulation tool has to be capable of accurately capturing different optical phenomena. These optical phenomena include refraction, surface reflection, and total internal reflection, as well as bulk refraction and bulk reflection for samples with a polymeric-matrix containing particles, such as EJ-440 detection sheets from Eljen Technologies.

In this work, we are investigating how Geant4 models compare against the experimental results of transmission/reflection spectroscopy for samples with thicknesses larger than/comparable to the wavelength of electromagnetic radiation in the UV-Vis region.

II. Experiments

A ZnO sample (99.99% purity) with dimensions of 10 mm x 10 mm x 0.5 mm was obtained from MSE supplies (Tucson, AZ). A ZnS sample with 1-inch diameter and 3.175 mm thickness was obtained from II-VI Incorporated (Saxonburg, PA). Two alpha detection sheets (EJ-440) were obtained from Eljen Technologies (Sweetwater, TX) with two different thicknesses of 0.1 mm and 0.25 mm. The EJ-440 sheets consist of ZnS(Ag) powders bound to a 0.25-mm thick polyester sheets on only one side, which allows Scanning Electron Microscopy (SEM) images to be taken off the surface of the samples. Moreover, several polyester sheets with 0.25-mm thickness were obtained from Eljen Technologies to make separate transmission/reflection measurement to decouple the effect of polyester sheets from the EJ-440 responses.

Micro- and nano-structured ZnO and boron samples with different thicknesses were manufactured in Material Research Institute (MRI) at Pennsylvania State University (PSU). Wavelength dependent refractive indices and extinction coefficients of these samples were measured with Woollam M-2000F Focused Beam Spectroscopic Ellipsometer. Transmission/reflection spectra were measured with Cary-5000 UV-Vis-NIR spectrophotometer with an integrating sphere attachment.

SEM images of EJ-440 samples were taken to calculate the average particle dimensions of ZnS(Ag) powders, as well as the standard deviation of their sizes. Iridium coating was performed on the samples before SEM images were taken in order to eliminate the charging and to make the images clearer. It was observed through several different SEM images that the samples were relatively homogenous in powder distribution. Line-intercept analysis was conducted to find the mean particle size and the standard deviation using SEM images. Additionally, Energy Dispersive Spectroscopy (EDS) was conducted to observe the impurities in the samples.

III. Data analysis

UV-VIS spectroscopy results for transmission and reflection were analyzed, and the optical absorption coefficients of the samples were calculated. Measured transmission and reflection can be related to the bulk absorption coefficient and the surface reflection coefficient as follows for a single layer optical system[11]:

$$T_m = \frac{(1-R)^2 e^{-\alpha d}}{1-R^2 e^{-2\alpha d}} \qquad (2)$$

$$R_m = R + \frac{R(1-R)^2 e^{-2\alpha d}}{1-R^2 e^{-2\alpha d}} \qquad (3)$$

where T_m, R_m, α, R, and d are the measured transmission, measured reflection, bulk absorption coefficient, surface reflection coefficient, and sample thickness, respectively. Reflection coefficient can be related to the refractive index of the material for normal incidence for samples with insignificant absorption through [11]:

$$R = \frac{(n-1)^2}{(n+1)^2} \qquad (4)$$

where n is the refractive index of the crystal. Equations 2 and 3 can be solved numerically to find the bulk absorption coefficient and the surface reflection coefficient simultaneously.

In most of the literature, the absorption coefficient is calculated with the Beer-Lambert Law and the so-called "Modified Beer-Lambert Law"[11]. Although these models can be applied to many samples, it is important to understand their limitations. The Beer-Lambert Law and the Modified Beer-Lambert Law are given as follows:

$$T_m = e^{-\alpha d} \qquad (5)$$

$$T_m = (1-R_m)^2 e^{-\alpha d} \qquad (6)$$

Both of these equations give accurate results in certain regions of the transmission/reflection spectrum; however, they do not give plausible results when their approximations are no longer valid. As an example, Equation 5 is only valid when the measured reflection is negligible compared to the measured transmission in Equations 2 and 3. On the other hand, Equation 6 is only valid when the sample is considerably absorbent[11]. These limitations will be shown through Geant4 simulations in *Results and Discussion*.

For two-layer systems such as those in EJ-440 sheets, none of the equations discussed so far gives accurate results due to the additional polyester backing. Polyester backing introduces an extra bulk region, as well as two surfaces with different refractive index mismatch (ZnS-polyester and polyester-air). Therefore, Transfer Matrix Method (TMM)[12] was used to extract the absorption coefficient of the active detection region for these samples. Expressions for the transmission and reflection of a two-layer system are fairly complicated; as an example, measured transmission of a ZnS-Polyester system when the light is impinged from ZnS side is expressed as follows:

$$T_{ZnS} = \frac{(1-R_1)(1-R_2)(1-R_3)}{\{[(1-2R_2)R_3 e_2^{-1} e_1^{-1} + R_2 e_2 e_1^{-1}](-R_1) - R_2 R_3 e_2^{-1} e_1 + e_2 e_1\}} \qquad (7)$$

where R_1, R_2, R_3 are the reflection coefficients of air-ZnS, ZnS-polyester, and polyester-air surfaces, respectively. e_1 and e_2 are given by:

$$e_1 = e^{\alpha_1 d_1}, \qquad e_2 = e^{\alpha_2 d_2} \qquad (8)$$

where α_1, α_2, d_1, and d_2 are the ZnS absorption coefficient, polyester absorption coefficient, ZnS thickness, and polyester thickness, respectively. Similarly, there are the measured reflection when light is impinged on ZnS side, as well as the measured transmission and reflection data when light is impinged on polyester side, which, in total, amount to four measured values. If the absorption coefficient of polyester was measured separately using Equations 2 and 3, α_1, R_1, R_2, and R_3 can be found by solving four equations simultaneously.

Wave nature of light must be taken into account while analyzing the UV-Vis spectra of thin samples. The wave nature of light gives rise to interference fringes due to multiple reflections of light waves inside the thin sample. A mathematical expression can be used to represent these effects in order to extract the absorption coefficients of the samples. The mathematical expression of transmission for double-layer systems involving a thin film and a thick substrate is shown here[13]:

$$T = \frac{|t_{12}t_{23}t_{34}|^2|e_2|^2|e_3|^2}{|1 + r_{23}r_{34}e_3^2 + r_{12}r_{23}e_2^2 + r_{12}r_{34}e_2^2e_3^2|^2} \tag{9}$$

where t_{nm} and r_{nm} are the transmission and reflection coefficients, respectively. The term e_x can be expressed as

$$e_x = e^{-2\pi\kappa_x k d_2} e^{2\pi i n_x k d_2} \tag{10}$$

where κ_x and n_x are the extinction coefficient and refractive index of the material of interest, respectively, and k is the wavenumber.

Particle size analysis was carried out using line-intercept analysis on SEM micrographs from a Tescan FESEM. The analysis processes used multiple lines drawn both horizontally and vertically on the images. The number of particles crossed by the lines was determined and the average particle size was calculated by dividing the total line length by the number of particles. Standard deviation of particle size was calculated through error propagation using the uncertainty in the number of particles crossed by the intercepting lines.

Mean particle size and the standard deviation are found using

$$\mu = \frac{L}{N} \tag{11}$$

$$\sigma_\mu = \mu \frac{\sigma_N}{N} \tag{12}$$

where μ, L, and N are the mean particle size, total line length, and total number of particles intercepted, respectively. σ_μ and σ_N are the uncertainties in mean particle size and number of particles intercepted, respectively.

IV. Computational Methods

For the simulations of ZnS and ZnO crystals, a single volume was created in Geant4 and a circular polychromatic beam of light was impinged on one side of the target to mimic the experimental setup. No surface definition was used in the simulations; therefore, probabilities of reflection and refraction are calculated directly from the refractive index data found in the literature, which were verified with the experiments using Equation 4. Transmission and reflection are determined by tallying the total number of photons that escaped the crystal through the back surface and the front surface (surface on which light is impinged), respectively, and dividing them by the total number of photons simulated. Similar simulation setup was utilized for thin films with only difference being the addition of quartz substrate into the models. Simulations were conducted using the optical constants measured with ellipsometry and verified with UV-Vis spectroscopy. Moreover, the neutron detection efficiency of a single sandwich layer (ZnO-boron-ZnO) was calculated with Geant4 simulations.

The EJ-440 sheets were simulated by modeling the individual powders, size of which was measured with SEM images. Modeling of individual powders was found necessary to accurately reflect the complex nature of the detection sheets. When EJ-440 was modeled as a single layer, simulations severely underestimated the amount of reflected light due to no reflection inside the sample. However, when powders were modeled, reflection results in Geant4 converged to the experimental data due to the multiple reflections taken place inside the sample.

A Gaussian function was written in Geant4 DetectorConstruction class with the mean particle size and standard deviation information taken from SEM images. For EJ-440, spherical particles (ZnS(Ag) powders) with pseudorandomized radii were placed in a cubical volume of air. Positions of the particles in the cubical volume were also randomized. The cubical volume containing the ZnS(Ag) powders were repeated and the entire active volume of the detectors was filled with spherical ZnS(Ag). Finally, a 0.25mm polyester sheet was created and placed behind the ZnS(Ag) volume.

V. Results and Discussion

We investigated the optical transport capabilities of Geant4 toolkit for single crystal samples, complex powder-based samples, and thin samples. We used measured data from ellipsometry, transmission/reflection spectroscopy and SEM in Geant4 simulations. Results for different samples are shown in the following subsections.

A. ZnS Crystal

Bulk absorption coefficient of the ZnS crystal was calculated using the three methods discussed in Section III and wavelength dependent absorption length ($1/\alpha$) was entered as input into the Geant4 model. Refractive index data were taken from literature[14]. The wavelength dependent refractive index was also calculated through Equations 2-4 and compared to the literature values. Experimental and simulation results for transmission/reflection/absorption spectra are shown in Fig. 1.

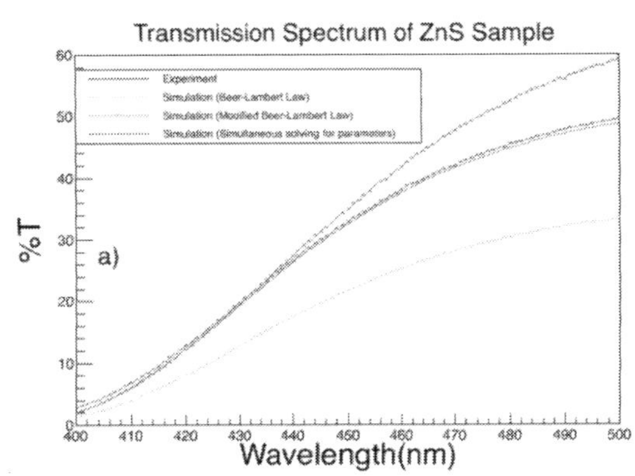

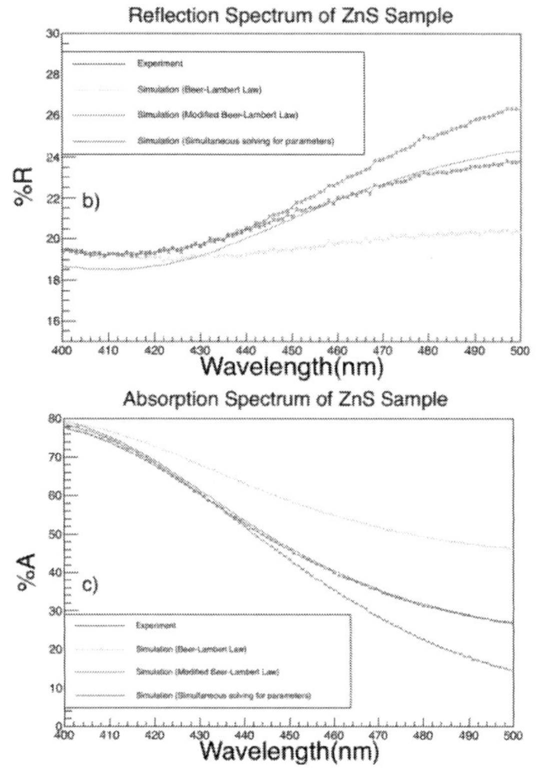

Figure 1. a) Transmission, b) reflection, and c) aAbsorption spectra of the ZnS sample. Both experimental and simulation data are shown.

As shown in Fig.1, transmission increases as the wavelength of the light photon increases (or energy of the photon decreases). Experimental data are consistent with the data sheet provided by the company. As stated in Section III of this paper, both the Beer-Lambert Law and the Modified Beer-Lambert Law capture certain parts of the spectra accurately. However, simultaneous calculation of bulk absorption and reflection coefficients through Equations 2 and 3 give better results in the whole wavelength region. The Beer-Lambert Law fails significantly as reflection increases, whereas the Modified Beer-Lambert Law fails where absorbance is minimal. Since the Beer-Lambert Law completely ignores reflection, it overestimates the absorption in the entirety of the spectrum.

Uncertainties in the simulations are less than 0.2% and they are shown in the figures. Uncertainty estimation in UV-Vis analysis was reported to be 0.15%-1%[15,16].

The largest percentage difference between the experimental and simulation results was found in the reflection spectrum with the largest difference being 4.5%. The reason for this difference was attributed to the inaccurate refractive index data and imprecise surface definitions used in Geant4 models that ignored possible surface roughness. Refractive indices calculated with Equation 4 were on average 5% lower than the literature values. When the simulations were done with the calculated refractive indices, we found that the reflection decreased and matched with the experiments in higher photon energy region; however, it deviated from the experiments in the lower photon energy region.

B. ZnO Crystal

The wavelength dependent bulk absorption coefficient of ZnO crystal was calculated through Equations 2 and 3. The sample cross-sectional area was 10 mm x 10 mm; therefore, a micro-focusing lens was used to collimate the light onto the center of the sample. Refractive indices calculated with Equation 4 were used in the Geant4 models. Spectra calculated from the simulations were compared against the experimental spectra. As an example, the absorption spectrum is shown in Fig. 2.

We observed a very good match between the experimental and simulation results in the absorption spectrum. The highest difference between the experiments and simulations was around 2% in the low energy region of the spectrum. Although not shown here, we observed the largest difference between experiments and simulations in the reflection spectrum with the highest relative error being approximately 9.5%. We assign this discrepancy to the combination of imprecise data of refractive index and non-realistic surface conditions in the models.

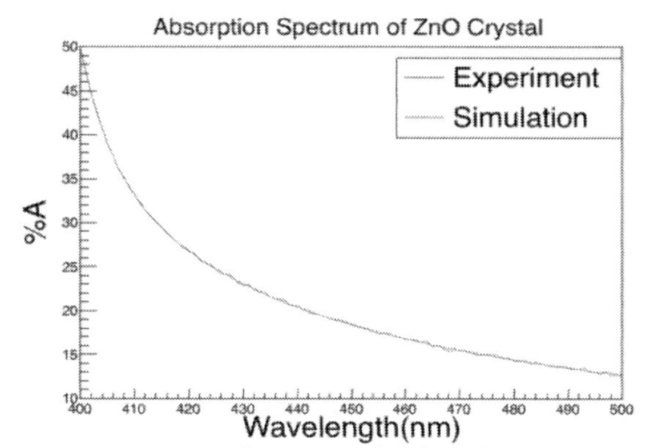

Figure 2. Comparison of experimental and simulation absorption spectra.

C. EJ-440 Sheets

Transmission/reflection spectroscopy measurements were carried out for EJ-440 sheets with two different thicknesses. Since these sheets are made up of two layers (ZnS(Ag) and polyester), a total of four measurements were taken for each sample with two transmission and two reflection measurements with light impinged on both sides of the samples.

Moreover, UV-Vis measurements were carried out for a single polyester layer obtained from Eljen Technologies to find the bulk absorption coefficient of the polyester backing. The polyester bulk absorption coefficient (α_2 in Equation 8) was used to determine the bulk absorption coefficient of ZnS(Ag) layer using Equation 7 and the additional three equations discussed in Section III (not explicitly shown in Section III).

The mean particle size was calculated as 4.4 μm and 4.5 μm for 0.1 mm and 0.25 mm EJ-440 sheets, respectively. The standard deviations associated to those values were calculated to be 0.88 μm and 0.76 μm. The active detector volume was created in Geant4 with spherical ZnS(Ag) powders whose sizes were pseudorandomized through the following probability density function (PDF) with experimentally verified mean and standard deviation:

$$f(x) = \frac{1}{\sqrt{2\pi}\sigma} e^{\frac{-(x-\mu)^2}{2\sigma^2}} \quad (13)$$

where μ and σ are the mean particle size and standard deviation associated to it, respectively.

An SEM image taken with the secondary electron detector is shown in Fig. 3 for a 0.1-mm thick EJ-440 sheet. Highly complex internal structure of the powdery geometry of EJ-440 sheets was approximated by using spheres with pseudorandomized radii calculated with Equation 13. Although the model falls short to mimic the true structure, it nonetheless introduces complicated passage for light and therefore causes light to reflect off multiple boundaries inside the bulk material. A unit-cell created in Geant4 using the data obtained from SEM images is shown in Fig. 4.

Figure 3. SEM image for 0.1-mm thick EJ-440 sheet.

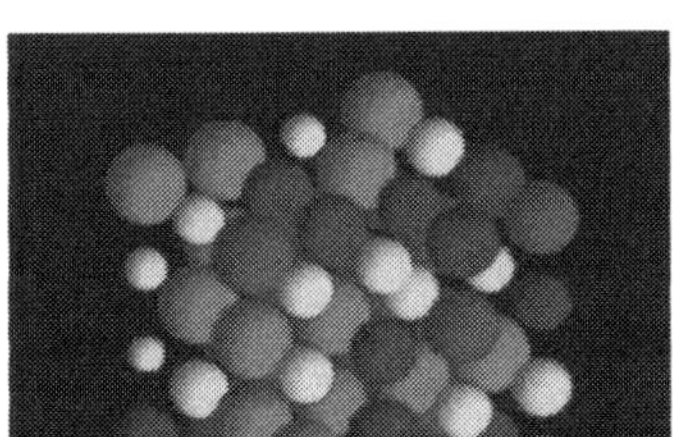

Figure 4. Unit-cell in Geant4 for EJ-440 sheets. Particles are color coded according to their sizes. All spheres are ZnS(Ag) with green representing smaller than μ-σ, blue representing between μ-σ and μ+σ, and red representing larger than μ+σ.

All the volumes shown in Fig. 4 are ZnS(Ag) with red representing particles larger than $\mu + \sigma$, blue representing particles whose sizes vary between $\mu - \sigma$ and $\mu + \sigma$, and green representing particles smaller than $\mu - \sigma$. The unit-cell was copied and repeated to create the active volume that represents the EJ-440 sheet. A polyester sheet was defined as a single rectangular-parallelepiped volume and placed behind the EJ-440 active volume.

Absorption spectra obtained from the simulations for 0.1-mm and 0.25-mm thick EJ-440 samples were compared to the experimental spectra. Moreover, the spectra simulated using the absorption length obtained by Beer-Lambert Law were also compared to the experimental spectra. The comparison is shown in Fig. 5.

The Beer-Lambert Law severely overestimates the absorption in the sample in both cases. As explained in the discussion in section *ZnS Crystal*, the Beer-Lambert Law does not take into account the reflection and assumes that reflection is zero for all wavelengths. Consequently, all the light that is not transmitted is considered absorbed according to Equation 5; hence, the absorption is overestimated.

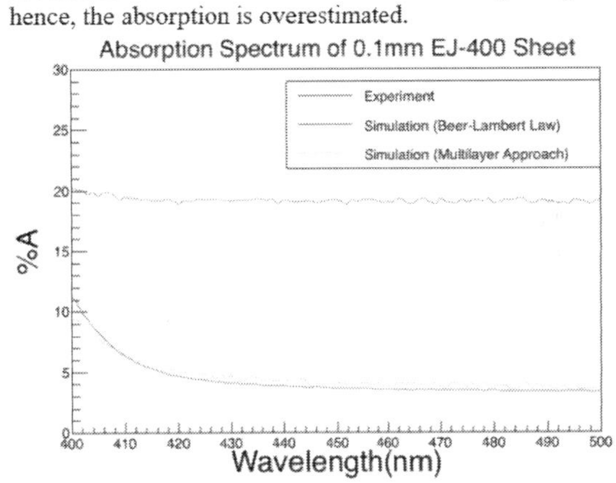

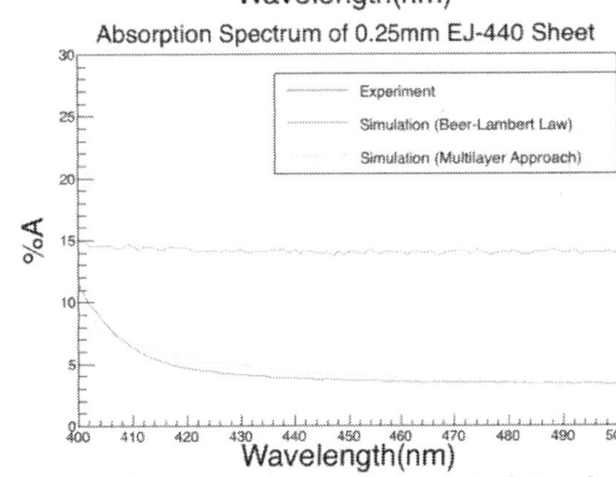

Figure 5. Comparison of experimental and simulation absorption spectra.

On the other hand, simulations with the bulk absorption coefficient calculated through the "Multilayer Approach" gives much better results and they are in agreement with the experiments in terms of both absolute values and the general trend.

Ignoring the internal structure of the samples results in a significantly lower simulated reflection. This is a result of no reflection inside the bulk material and reflection only from the surfaces between air and the sample. Introducing the ZnS(Ag) powders as micro-structures inside the bulk material causes incoming photons to reflect and change their paths multiple times before they leave the sample; hence, increasing the simulated reflection.

Capturing the general trend and the absolute values of absorption is only possible by taking into account the polyester layer and by modeling the internal structure of the sheets.

D. Thin Films of ZnO and Boron

Two ZnO samples with 450-nm and 939-nm thicknesses and three boron samples with 150-nm, 998-nm, and 2027-nm thicknesses were manufactured in MRI at PSU. The wavelength dependent refractive indices and extinction coefficients of these samples were measured with ellipsometry. Measured extinction coefficients and refractive indices were plugged into the model transmission (Equation 9), and the mathematical model for transmission was compared to the experimental spectrum.

SEM images of boron and ZnO films alongside with their transmission spectra are shown in Fig. 6.

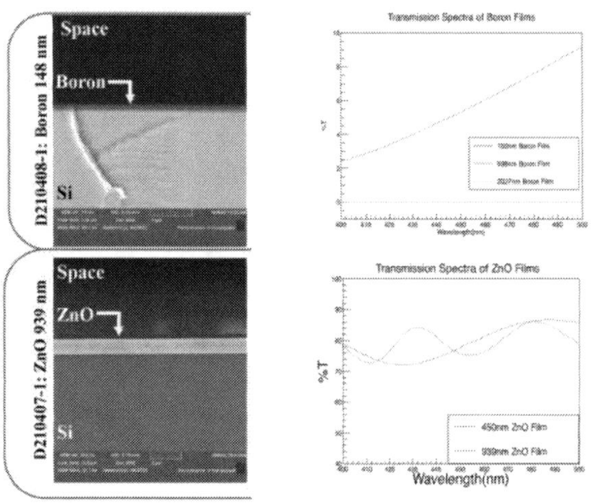

Figure 6. SEM images of ZnO and boron thin films with their transmission spectra.

Sample thickness and refractive indices are the main factors that determine the interference fringe pattern. There was a disagreement between the sample thicknesses measured with SEM and ellipsometry. Although the difference is less than 10% for all the samples, it changes the constructive and destructive interference pattern significantly in the model transmission spectra. These discrepancies between the ellipsometry and SEM data were taken into account while creating the model transmission. The wavelength dependent extinction coefficient and refractive index from the ellipsometry were plugged into the model to finalize the model transmission. Model transmission spectra for 450 nm and 939 nm ZnO sample are shown in Fig. 7.

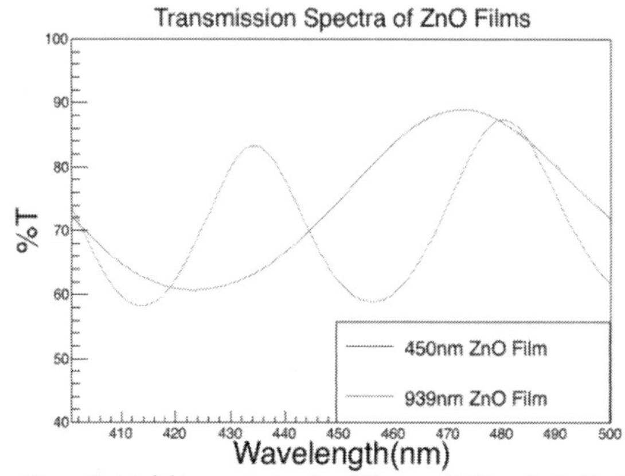

Figure 7. Model transmission for 450nm and 939nm ZnO thin films.

Although the experimental UV-Vis transmission and the model transmission do not match in terms of oscillation amplitudes, wavelength dependent extinction coefficient measured with ellipsometry predicts the dampening of light waves to approximately 80%.

Since Geant4 considers optical photons to be point-like particles, the interference fringes cannot be simulated in Geant4. However, Geant4 predicts a reasonable mean transmission for most of the wavelength region of interest using the ellipsometry data as shown in Fig. 8. At high energies, there is a consistent pattern of ellipsometry overestimating the extinction coefficient of the layers; therefore, Geant4 predicts a significantly lower transmission.

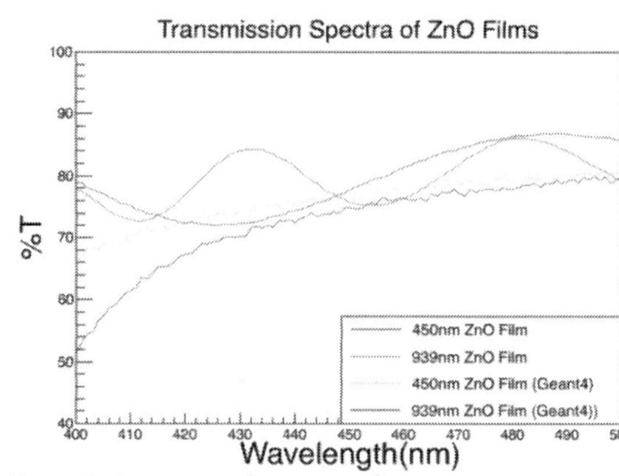

Figure 8. Comparison of experimental transmission spectra to Geant4 transmission spectra for ZnO thin films.

Finally, using the optical constants measured with ellipsometry and verified with UV-Vis transmission and reflection spectrometry, Geant4 models were constructed to simulate the neutron detection efficiency of sandwich-like structures. A circular thermal neutron beam was impinged on the ZnO-boron-ZnO sandwich geometry with a cross-sectional area of 3 cm x 3 cm. A photocathode surface was placed on one end of the geometry and the neutron capture and neutron detection efficiencies were tallied. A 20-photon threshold was applied in order to consider a successful neutron detection. The neutron capture efficiency and the neutron detection efficiency as a function of boron thickness are shown in Fig. 9. The ZnO thickness was held constant at 4 μm, which is the estimated range of 1.78-MeV alpha particle in ZnO.

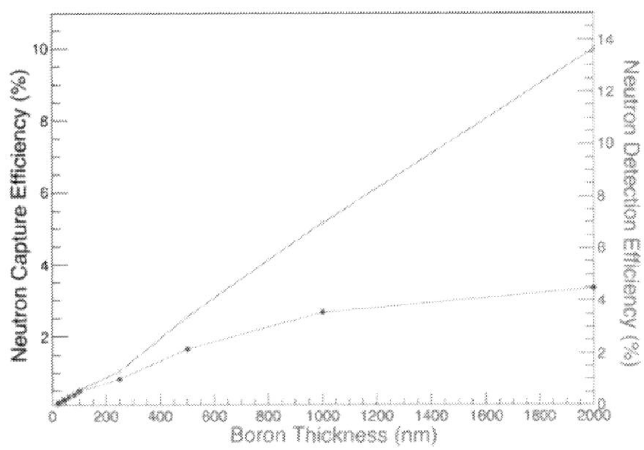

Figure 9. Neutron capture efficiency and neutron detection efficiency of ZnO-B-ZnO sandwich geometry with respect to varying boron thickness.

Although the thermal-neutron capture efficiency reaches approximately 10%, the thermal-neutron detection efficiency saturates at approximately 4.5% due to the relative opaqueness of the boron film and some of the reaction products not reaching the scintillator medium. The neutron detection efficiency can be substantially increased by increasing the number of layers and by arranging multiple photosensors around the neutron detector. The neutron detection efficiency is comparable to that of similar structures for thermal neutrons[17].

VI. Conclusion

It has been shown that Geant4 is capable of modeling the optical response of various structures including single crystal and micro-structures, as well as nano-structures to a certain extent. The transmission and reflection spectra of various materials have been measured and it has been demonstrated that Geant4 accurately predicts the optical behavior of these materials when optical constants such as absorption coefficients and refractive indices are calculated with appropriate mathematical expressions. Optical constants of ZnO and boron thin films have been measured with ellipsometry and verified with UV-Vis spectroscopy. Finally, the neutron detection efficiency of single sandwich geometry of ZnO-B-ZnO was simulated with experimentally measured optical properties. It has been found that a single micro-structure sandwich geometry can achieve approximately 4.5% neutron detection efficiency, which is comparable to those reported previously for similar structures.

Acknowledgement

This research is sponsored by the Defense Threat Reduction Agency (DTRA) as part of the Interaction of Ionizing Radiation with Matter University Research Alliance (IIRM-URA) under contract number HDTRA1-20-2-0002.

References

[1] P. A. Rodnyi *et al*, "Novel Scintillation Material-ZnO Transparent Ceramics," *IEEE Transactions on Nuclear Science*, vol. 59, *(5)*, pp. 2152-2155, 2012;2011;.
[2] E. I. Gorokhova *et al*, "Optical, luminescence, and scintillation properties of ZnO and ZnO:Ga ceramics," 2010.
[3] W. L. Bond, "Measurement of the Refractive Indices of Several Crystals," *Journal of Applied Physics*, vol. 36, *(5)*, pp. 1674-1677, 1965.
[4] M. Debenham, "Refractive indices of zinc sulfide in the 0405–13-µm wavelength range," *Applied Optics*, vol. 23, *(14)*, pp. 2238, 1984.
[5] J. Allison *et al*, "Recent developments in Geant4," *Nuclear Instruments & Methods in Physics Research. Section A, Accelerators, Spectrometers, Detectors and Associated Equipment*, vol. 835, *(C)*, pp. 186-225, 2016.
[6] J E McMillan et al 2015 J. Phys.: Conf. Ser. 620 012011
[7] Y. Yehuda-Zada *et al*, "Optimization of 6LiF:ZnS(Ag) scintillator light yield using GEANT4," *Nuclear Instruments & Methods in Physics Research. Section A, Accelerators, Spectrometers, Detectors and Associated Equipment*, vol. 892, pp. 59-69, 2018.
[8] G. Tsiledakis et al 2017 JINST 12 P09006
[9] F. Logoglu, M. Flaska and D. E. Wolfe, "Development of a Coated-Micro-Particle Neutron Detector Based on LiF/ZnS Scintillator," *EPJ Web of Conferences*, vol. 225, pp. 7009, 2020.
[10] C Guardiola et al 2011 JINST 6 T09001
[11] J. Vac. Sci. Technol. B 34, 04J105 (2016); https://doi.org/10.1116/1.4954211
[12] M. Claudia Troparevsky, Adrian S. Sabau, Andrew R. Lupini, and Zhenyu Zhang, "Transfer-matrix formalism for the calculation of optical response in multilayer systems: from coherent to incoherent interference," Opt. Express 18, 24715-24721 (2010)
[13] J. Appl. Phys. 111, 0.73109 (2012); https://doi.org/10.1063/1.3700178
[14] S. Ozaki and S. Adachi, "Optical constants of cubic ZnS". Jpn. J. Appl. Phys. 32, 5008-5013 (1993)
[15] Rihua Mao, Liyuan Zhang and Ren-Yuan Zhu, "Optical and scintillation properties of inorganic scintillators in high energy physics," *2007 IEEE Nuclear Science Symposium Conference Record*, 2007, pp.2285-2291
[16] Rev. Sci. Instrum. 88, 015105 (2017); https://doi.org/10.1063/1.4973633
[17] M. Roth, E. Mojaev, O. Khakhan, A. Fleider, E. Dul'Kin, M. Schieber Composite boron nitride neutron detectors J. Cryst. Growth (2014)

Design and first tests of the S^3 detector of reactor antineutrinos

Mária Slavíčková[1,2], Vyacheslav Belov[4], Jan Broulím[1], Victor Brudanin[4], Viatcheslav Egorov[4], Lukáš Fajt[1],Maria Fomina[4], Rastislav Hodák[1], Sergei Kazartsev[4], Zdeněk Kruliš[3], Miroslav Macko[1], Petr Mašek[1], Danuše Michálková[3], Maroš Petro[1], Petr Přidal[1], Ekaterina Rukhadze[1], Egor Shevchik[4], Karel Smolek[1], Tomáš Slavíček[1], Ivan Štekl[1], and Igor Zhitnikov[4]
[1]Institute of Experimental and Applied Physics, CTU in Prague, Czech Republic
[2]Faculty of Nuclear Sciences and Physical Engineering, CTU in Prague, Czech Republic
[3]Institute of Macromolecular Chemistry, Czech Academy of Sciences, Czech Republic
[4]Joint Institute for Nuclear Research, Dubna, Russia
maria.slavickova@utef.cvut.cz

***Abstract*—The new experiment S^3 devoted to the study of reactor antineutrinos was designed and constructed as a common activity of IEAP CTU in Prague and JINR (Dubna). The S^3 detector is a compact, highly segmented polystyrene-based scintillating detector composed of 80 detector elements with a gadolinium neutron converter between elements layers. A positron and a neutron are produced in an inverse beta decay initiated with an electron antineutrino in the detector. A modular multi-channel fast ADC was developed for the data acquisition for the whole 80-channel S^3 detector and the 4-channel cosmic veto system. The detector meets very strict safety rules of nuclear power plants and can be installed in a chamber located immediately under the reactor. The close vicinity from the reactor enables to study neutrino properties with a higher efficiency, to investigate neutrino oscillations at short baselines and try to verify the hypothesis of a sterile neutrino. The details of the design and construction of the S^3 detector, as well as properties of the modular multi-channel fast ADC system, and first tests of the device are presented.**

***Keywords* — S^3 experiment, reactor neutrinos, neutrino oscillations.**

I. INTRODUCTION

NUCLEAR power reactor is one the most intense man-controlled source of the pure electron antineutrinos. A typical 3 GW_{th} (e.g. VVER-1000 in Temelin, CZE) nuclear power reactor emits about 10^{21} electron antineutrinos/s/4π. Since the antineutrino flux decreases with a distance as $1/d^2$, it is important to install the detector as close from the reactor as possible or construct a large-volume detector. However, safety rules of a nuclear power plant do not allow any flammable, toxic, caustic or other dangerous materials in the reactor building.

The aim of the S^3 experiment [1], [2], [3] was to construct a compact antineutrino detector, which does not contain dangerous materials and can be installed in the reactor building. Such detector could contribute to the investigation of the neutrino oscillations at very short baselines (~m) and verification or rejection of the sterile neutrino hypothesis. Moreover, such detector could be used in an applied research for the real-time measurement of the reactor power, determination of fuel burnout process and control of the illegal manipulation with ^{239}Pu. Furthermore, antineutrinos penetrate all building materials practically without the interaction and carry direct information about processes taking place in the reactor core. Therefore, the S^3 detector could be an ideal candidate to be a monitoring device of the reactor core.

An idea to develop the S^3 detector is based on the DANSS detector (1 m x 1 m x 1 m) [4], which was constructed in JINR, Dubna. The DANSS detector is a highly-segmented polystyrene based scintillating detector, which consists of 2500 scintillating elements (4 cm x 1 cm x 100 cm). It is installed on a movable platform thus measuring the antineutrino flux at different positions from the reactor. The antineutrino flux is not constant at different positions from the reactor and changes in time with the change of the reactor power and isotopic composition. Therefore, it was decided to develop a reference detector measuring the antineutrino flux at a fixed position from the reactor. As a result, two S^3 detectors were constructed. The first S^3 detector together with a complex shielding and DAQ system is situated at IEAP CTU in Prague and is planned to be installed in Temelin Nuclear Power Plant in Czech Republic. Simultaneously, the second S^3 detector is constructed in JINR, Dubna. It will be used as a reference detector measuring the reactor antineutrinos at a fixed position from the reactor while the position of the DANSS detector will change.

II. DETECTION TECHNIQUE

A detection principle of the S^3 detector is based on the inverse beta decay interaction, where an antineutrino emitted from the reactor core interacts with a proton in the detector volume and a neutron and a positron are created:

$$\bar{\nu}_e + p \longrightarrow n + e^+ \tag{1}$$

The positron losses its energy in a very short range and annihilates with an electron producing two 511 keV gammas at 180° in the Center of the Mass System of the electron and positron pair. The neutron is moderated by the scattering on a hydrogen atom and then captured on a neutron converter, in our case gadolinium. The gadolinium nucleus comes into an excited state and deexcites with the emission of gamma quanta with the total energy of 8 MeV and within 20 cm radius. As a result, the inverse beta decay interaction can be detected as a prompt signal from the positron interaction and annihilation gammas and a delayed signal from the neutron capture. The time interval between the prompt and the delayed signal is expected to be 2 – 50 μs.

The kinetic energy of the positron T_{e^+} is related to the energy of the incoming antineutrino $E_{\overline{\nu}_e}$ [5] as:

$$T_{e^+} \approx E_{\overline{\nu}_e} - 1.806 \text{ MeV}. \quad (2)$$

Therefore, we can reconstruct the energy of the incoming antineutrino by the measurement of the prompt signal.

The neutron produced in the inverse beta decay interaction propagates in the direction of the incoming antineutrino thus providing information about the momentum of the antineutrino.

High segmentation of the detector enables us to suppress the background as well as identify very characteristic time and spatial pattern of the antineutrino interaction in the detector.

III. EXPERIMENTAL SETUP

The S^3 detector (40 cm x 40 cm x 40 cm), see Fig. 1, consists of 80 pieces of polystyrene-based scintillating elements (40 cm x 20 cm x 1 cm each) produced by the NUVIA, a.s. [6]. The polystyrene matrix is doped with 2.0% p-terfenyl (pTP) and 0.025% of 1,4-bis(5- phenyloxazol-2-yl) benzene (POPOP) [7]. This concentration of dopants ensures 1.4 times better light output in comparison with scintillators standardly manufactured by the NUVIA, a.s.

Fig. 1. The S^3 detector consisting of 80 scintillating elements.

Each scintillating element is covered by the 800 μm thick Teflon layer working as a reflective material. The Teflon layer ensures 1.6 times better light yield in comparison with the bare scintillator and 80% homogeneity of the light collection along the scintillating element. Scintillating elements are arranged in the detector so that each layer (2 elements) lies perpendicularly to the previous one thus forming X-Y structure. Between each two layers there is a neutron conversion layer based on Gd_2O_3, which was prepared in cooperation with IMC of Czech Academy of Science.

Light collection along the scintillating element is done using 19 Kuraray Y-11(200) wavelength-shifting fibers [8], which are glued into grooves by the optical epoxy glue. A distance between two neighboring grooves is 1 cm, which ensures an uniform light collection. Light produced in the scintillating element after the particle interaction is guided via wavelength-shifting fibers to the 6 mm x 6 mm SensL SiPM, where the light is converted to an electric signal. The PCB with four pieces of SensL SiPMs, which collect signals from four individual scintillating elements, is shown in Fig. 2a. The location of the PCB on the S^3 detector is shown Fig. 2b.

Fig. 2. The both-sided view on the PCB with four pieces of SensL SiPMs (a) and the position of the PCB with SiPMs on the S^3 detector (b).

A big advantage of the front-end electronics is its negligible power consumption (maximal value of 1.7 W). This eliminates any need of cooling, even if the detector is located in a complex shielding. Electric signal from the SiPM is amplified and shaped by the signal shaping module (shaping constant is 1 μs), which is shown on the left part of Fig. 3. The signal shaping module is mechanically connected with the ADC module, see the right part of Fig. 3. The amplified and shaped signal is processed in a modular multichannel fast ADC, which consists of 84 ADC modules (15 bits + bit for a sign, 100 MS/s @ 84 channels). This multichannel system was designed and constructed at IEAP CTU in Prague and it is used for the data acquisition from the whole 80-channel S^3 detector and the 4-channel cosmic muon veto.

Fig. 3. The signal shaping module (left) attached to the ADC module (right).

The front and back view on the 84-channel digitizer is shown in Fig. 4a and Fig. 4b.

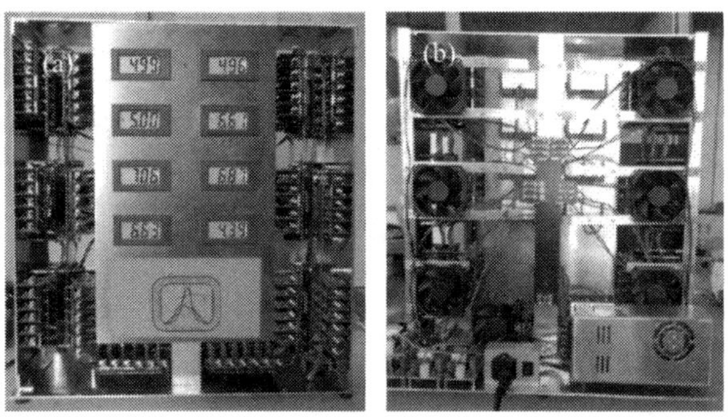

Fig. 4. The Front (a) and back view (b) on the 84-channel digitizer.

After exceeding the threshold, the analog signal is digitized with the sampling rate of 100 MHz and with the required number of samples. Each event has a timestamp counted with a step of 10 ns. The digital signal is then sent to a PC using USB2 (max 480 Mb/s), where it is processed by the DAQ software developed by IEAP CTU in Prague. The DAQ software enables:

- individual configuration of each channel,
- information about the status of each channel,
- real-time visualization of the signal from each channel,
- real-time visualization of energy spectrum from each channel,
- visualization of coincidences,
- data export.

An illustration of the DAQ software is shown in Fig. 5. For clarity, not all functions of the DAQ software are shown.

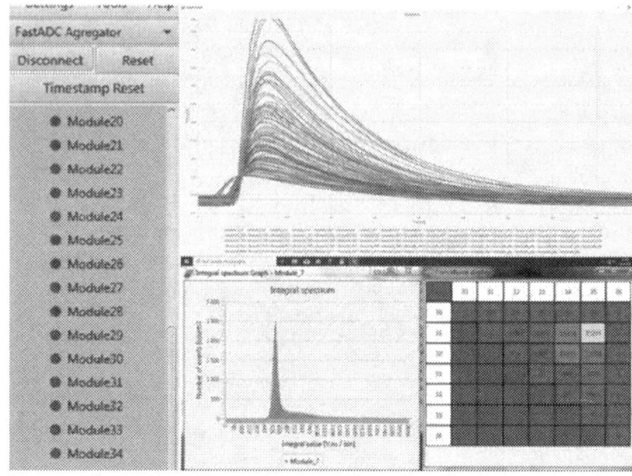

Fig. 5. The DAQ software showing information about the status of each channel (left), signals from 80 channels, energy spectrum from one chosen channel and number of coincidences between two channels.

The whole experimental setup consisting of the detector part and DAQ system is shown in Fig. 6.

Fig. 6. The experimental setup consisting of the detector part and the DAQ system. The S^3 detector is surrounded by a black light-tight box (right).

The S^3 detector is surrounded by a complex shielding consisting of (from inside of the detector):

- 10 cm of 100 years old lead shielding in form of bricks and bars,
- 8 cm of polyethylene with 3.5% admixture of boron,
- 16 cm of standard polyethylene.

Lead shielding was designed to suppress the low energy background, in which the signal from two 511 keV gammas from the positron annihilation can be hidden. The low-energy part was suppressed to 4.5 ± 2.1 events/s using the shielding [1], which provides ideal conditions for the measurement of 511 keV gammas. The lead shielding is shown in Fig. 7a. The polyethylene shielding is used to suppress fast neutrons, which can mimic the antineutrino interaction – the prompt signal from the neutron thermalization and the delayed signal from the neutron capture on gadolinium. The complete shielding is shown in Fig. 7b. The effectivity of the shielding was measured in the civil protection shelter Bezovka (45 m w. e.) located in Prague.

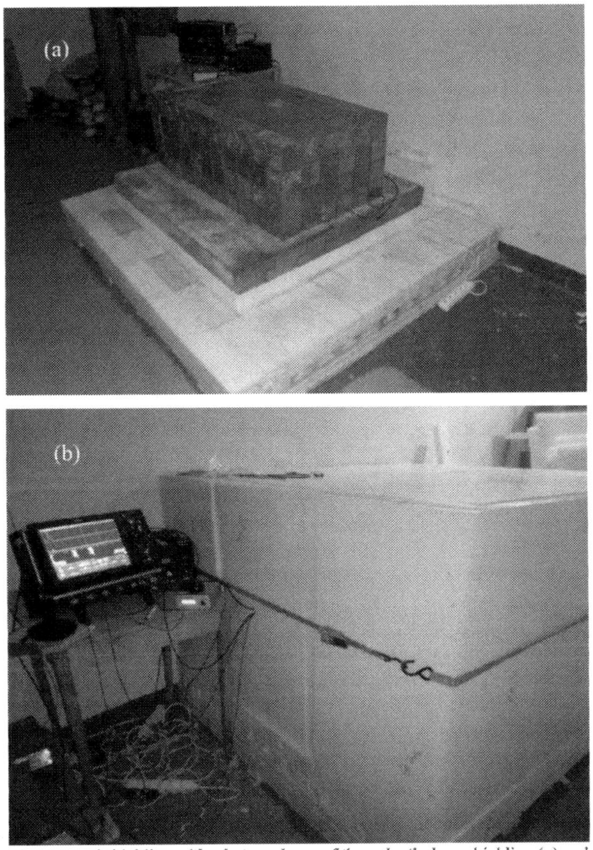

Fig. 7. Lead shielding with a bottom layer of the polyethylene shielding (a) and complete lead and polyethylene shielding (b).

IV. DETERMINATION OF THE TRESHOLD SENSITIVITY OF THE S^3 DETECTOR

For better identification of the inverse beta decay interaction caused by an antineutrino, the detection of two 511 keV gammas is important. However, 511 keV gamma interacts mainly via the Compton scattering in the plastic scintillator depositing only part of its energy in the detector. In some cases, the analysis could be complicated because the low-energy events from 511 keV gammas are hidden in the noise of the SiPM. To distinguish physical events from the electronics noise, analysis of the SiPM self-noise was done. The measurement was performed in a light-tight box while the scintillating element was not connected to the sensitive area of the SiPM. Therefore, only the self-noise of the SiPM was detected. The single photoelectron spectrum of the FC60035 SiPM is shown in Fig. 8. Values obtained during the fitting of the photoelectron peaks are shown in Table I. Based on this measurement, the energy threshold was determined. To demonstrate that the threshold was determined properly the measurement of the physical events was carried out after the connection of the scintillating element to the sensitive area of the FC60035 SiPM. Both measurements were performed under the same conditions, i.e. bias current of the SiPM was 6 μA, the same amplification and shaping were used.

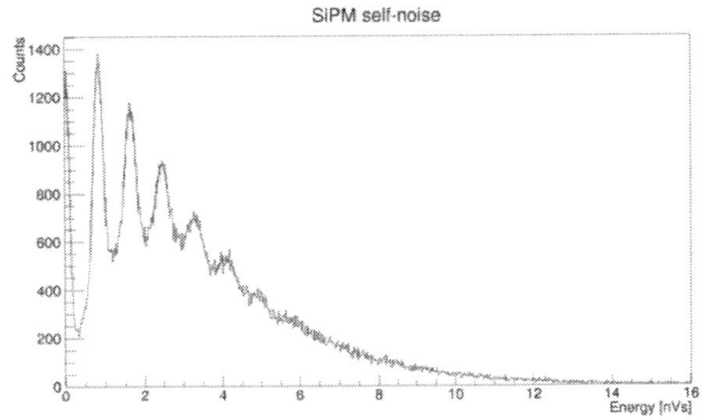

Fig. 8. An example of the photoelectron spectrum of the FC60035 SiPM (blue line) measured at 6 μA and fitted photoelectron peaks (red line).

TABLE I
FIT VALUES OF THE PHOTOELECTRON PEAKS

Number of p.e.	Mean [nVs]	Error [nVs]	Distance between neighboring p.e.
0	0.002	0.001	-
1	0.799	0.002	$\Delta_{10} = 0.797$
2	1.608	0.003	$\Delta_{21} = 0.809$
3	2.431	0.005	$\Delta_{32} = 0.823$
4	3.238	0.010	$\Delta_{43} = 0.807$
			$\overline{\Delta} = 0.820$

Fig. 9 shows the measured pulse-height spectrum which consists of two parts. The red line corresponds to the SiPM self-noise, which is shown in Fig. 8, in detail. The blue line represents physical events induced mainly by the gamma and cosmic muon background.

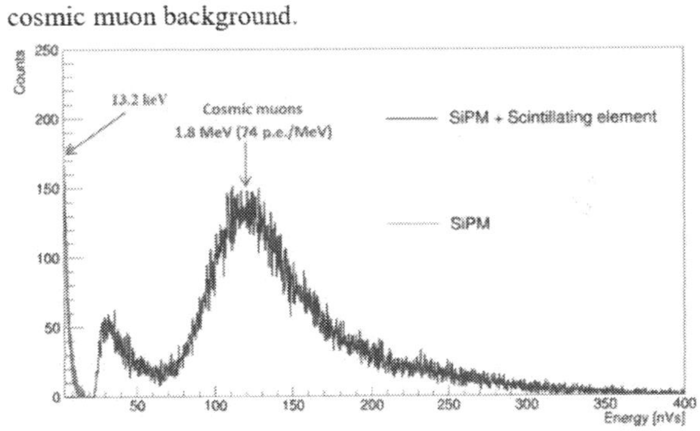

Fig. 9. An example of the spectrum. Red line corresponds to the SiPM self-noise. The blue line represents the pulse-height spectrum of the physical background – the gamma and cosmic muon background.

It is evident that the SiPM self-noise is well separated from the scientific data. Based on the Monte Carlo simulation, the peak in the background spectrum (blue line) corresponds to cosmic muons with the energy of 1.8 MeV and gives the light gain of 148 photoelectrons (p.e.). Therefore, the light gain of one scintillating element is 70-75 p.e./MeV. The maximum of the SiPM self-noise is at 13.2 keV corresponding to the first p.e. and the amount of noise events decreases to 1/2 at the energy of 50 keV. This analysis showed that the threshold sensitivity of the detector is about 50 keV and low-energy events from the antineutrino interaction could be well identified.

V. DETECTOR CALIBRATION

The calibration of each scintillating element of the S^3 detector is done by the use of cosmic muons. When muon penetrates the scintillating element, it behaves like a minimum ionizing particle and deposits 1.8 MeV of its energy per each 1 cm of its track. Fig. 10 presents the amplitude (left) and the energy (right) spectrum from a selected scintillating element. If all collected events are used for the histogram construction, see Fig. 10a, the muon peak is almost invisible. If the events with 5 or more hit scintillating elements are selected for the histogram construction, see Fig. 10b, the noise is suppressed, the muon peak is clear, and it can be used for the energy calibration.

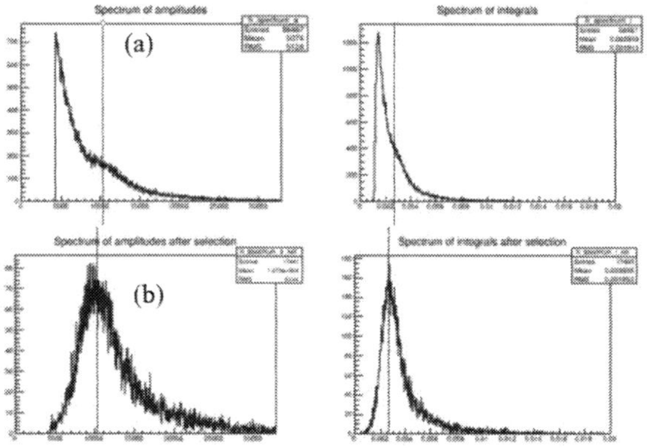

Fig. 10. The amplitude and the energy spectrum from one scintillating element before (a) and after (b) the selection of events with penetrating muons (events with more hitting scintillating elements).

VI. CONCLUSIONS

The S^3 detector represents a compact antineutrino spectrometer, which does not contain any dangerous material and can be installed in the building of the nuclear power plant. The big advantage of the detector is its affordable price, user friendly DAQ software and low demands on number of staff. The multichannel digitizer was designed and constructed to measure in real time and with fast signals from plastic scintillators.

At present, the S^3 detector is tested and calibrated in the IEAP CTU laboratory and is planned to be installed in the Bezovka civil protection shelter to perform a long-term stability tests. Moreover, MC simulations were carried out to optimize the selection criteria for the reactor antineutrino events. All such results are planned to be experimentally verified. Because the S^3 detector is located in the laboratory and is not exposed to the flux of antineutrinos from the reactor, we need to carry out tests using fast neutrons, which can mimic antineutrino interactions.

After tests of long-stability and measurements with radiation sources, the S^3 detector is planned to be installed in the Temelin Nuclear Power Plant, where it will be used for the investigation of the neutrino properties and reactor monitoring in a real time. The second S^3 detector is constructed in JINR, Dubna and is planned to be used as a reference detector for the DANSS experiment.

REFERENCES

[1] M. Spavorova et al., "S^3 - prototype of reactor antineutrino detector", JINST 13, C12013 (2018).
[2] H. Burešová, L. Fajt, R. Hodák, P. Přidal, M. Špavorová, I. Štekl, "Plastic scintillator-based antineutrino detector," CZ patent PV 2015-359. July 13, 2016.
[3] H. Burešová, L. Fajt, R. Hodák, P. Přidal, M. Špavorová, I. Štekl, "Plastic scintillator-based antineutrino detector," CZ patent PV 2015-360. July 13, 2016.
[4] I. Alekseev et al., "DANSS: Detector of the reactor AntiNeutrino based on Solid Scintillator", JINST 11, P11011 (2016).
[5] K. Zuber, Neutrino physics. Florida, USA: CRC Press, 2020.
[6] NUVIA, a.s. Třebíč, Czech Republic [Online]. Available: https://nuvia.cz/en. [Online]
[7] L. Fajt et. al., "Present status of sensitive detector of reactor's antineutrinos using scintillating detectors," in AIP Conference Proceedings, Seattle, USA, 2015.
[8] KURARAY. Tokyo, Japan [Online]. Available: http://kuraraypsf.jp/psf/ws.html. [Online]

A High-Granularity Timing Detector for the Phase-II upgrade of the ATLAS Calorimeter system: detector concept, description and R&D and beam test results

H. Imam on behalf of the ATLAS HGTD Collaboration *
HASSAN II University - Faculty of Sciences Ain Chock
Casablana,Morocco
Corresponding author: hajar.imam@cern.ch

***Abstract*—The particle flux increase (pile-up) at the HL-LHC with luminosities of L = 7.5 × 10^{34} cm^{-2} s^{-1} will have a significant impact on the reconstruction of the ATLAS detector and on the performance of the trigger. The forward region and the end-cap where the internal tracker has poorer longitudinal track impact parameter resolution, and where the liquid argon calorimeter has coarser granularity, will be significantly affected. A High Granularity Time Detector (HGTD) is proposed to be installed in front of the LAr end-cap calorimeter for the mitigation of the pileup effect, as well as measurement of luminosity.**

It will have coverage of 2.4 to 4.0 from the pseudo-rapidity range. Two dual-sided silicon sensor layers will provide accurate timing information for minimum-ionizing particles with a resolution better than 30 ps per track (before irradiation), for assigning each particle to the correct vertex. The readout cells are about 1.3 mm × 1.3 mm in size, which leads to a high granular detector with 3 million channels. The technology of low-gain avalanche detectors (LGAD) with sufficient gain was chosen to achieve the required high signal-to-noise ratio. A dedicated ASIC is under development with some prototypes already submitted and evaluated. The requirements and general specifications of the HGTD will be maintained and discussed. R&D campaigns on the LGAD are carried out to study the sensors, the related ASICs and the radiation hardness. Both laboratory and test beam results will be presented.

***Index Terms*—HL-LHC, ATLAS, HGTD, LGAD, Si sensors, Timing detectors**

I. Introduction

The HL-LHC [1] is an upgrade of the LHC project, which will deliver an integrated luminosity of up to 4 000 fb^{-1} over the subsequent decade, and achieve an instantaneous luminosity of L = 7.5x10^{34} $cm^{-2}s^{-1}$, compared to the current value of L = 2.5x10^{34} $cm^{-2}s^{-1}$. This will increase the pileup interactions to an average of 200 simultaneous proton-proton interaction within the same bunch crossing interval ($<\mu> =$ 200),which is 6 times higher than the current pileup rate. The Inner Tracker upgrade (ITk) [2] will have a better position resolution and pseudo-rapidity extended to $|\eta| = 4.0$ ($\eta <$ 2.5 up to now), but the ITk will have low Z_0 resolution in the forward region. So the timing information can help to further reduce pileup effect in this region. Hence by using only the ITk, multiple tracks nearby in the space can lead to track/vertex association ambiguities. The timing information of the HGTD [3] reduces the density of vertices for a given track, and allows the effective separation of the hard scatter vertex from pile-up vertices surrounding it in the Z direction, hence the improvement of the ATLAS [4] performance in the forward region.

II. HGTD Requirements and Layout

A. Overview & requirements

To ensure the performances needed to deal with the high luminosity expected in the HL-LHC phase ($<\mu> = 200$), a part of the ATLAS [4] detector has to be upgraded, where the electromagnetic calorimeter [5] has larger granularity and the inner tracker resolution degrades. Consequently, The HGTD detector which will provide a high granularity and timing information is proposed to be located in the gap region between the end-cap calorimeters and the barrel. Each HGTD end-cap is the integration of one hermetic vessel, two instrumented double sided layers and two moderators placed inside also outside the airtight vessel. The modules with sensors and on-detector electronics are mounted on the front and back sides of a common cooling disk. Each disk of this last is physically separated in two half circular disks. The HGTD will cover the pseudo-rapidity range between $2.4 < |\eta| < 4.0$, complementing the inner tracker by providing high-precision time measurements for charged particles, and enhancing the performance of physics object reconstruction. Fig. 1 shows the position of the HGTD detector.

Due to space limitations and the high expected radiation level at the HGTD position, a technology of silicon-based time sensor is selected to be used, with thin and configurable array sensors. These requirements are satisfied using the Low Gain Avalanche Detector (LGAD) [6] pads of 1.3mm × 1.3 mm × 50 μm (length × width × thickness). Such pad dimensions ensure under 10% of occupancy at high pileup ranges, limited

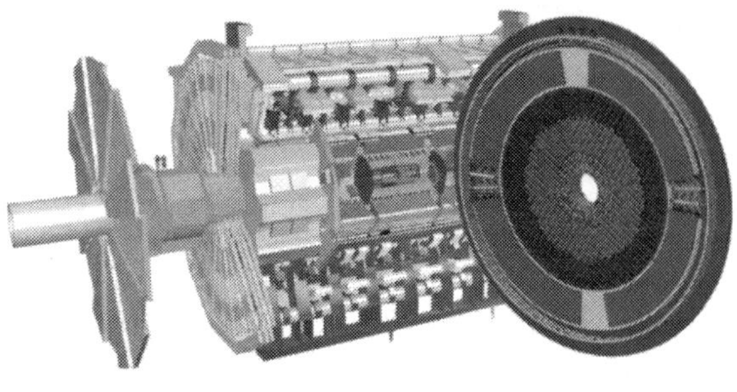

Fig. 1: Position of the HGTD within the ATLAS Detector [3].

Pseudo-rapidity coverage	$2.4 < \lvert\eta\rvert < 4.0$
Thickness in z	75 mm (+50 mm moderator)
Position of active layers in z	∘ 3.5 m
Weight per end-cap	350 kg
Radial extension:	
Total	110 mm < r < 1000 mm
Active area	120 mm < r < 640 mm
Pad size	1.3 mm × 1.3 mm
Active sensor thickness	50 µm
Number of channels	3.6 M
Active area	6.4 m^2
Module size	30 x 15 pads (4 cm × 2 cm)
Modules	8032
Collected charge per hit	> 4.0 fC
Average number of hits per track	
$2.4 < \lvert\eta\rvert < 2.7$ (640 mm > r > 470 mm)	≈2.0
$2.7 < \lvert\eta\rvert < 3.5$ (470 mm > r > 230 mm)	≈2.4
$3.5 < \lvert\eta\rvert < 4.0$ (230 mm > r > 120 mm)	≈2.6
Average time resolution per hit (start and end of operational lifetime)	
$2.4 < \lvert\eta\rvert < 4.0$	≈ 35 ps (start), ≈ 70 ps (end)
Average time resolution per track (start and end of operational lifetime)	≈ 30 ps (start), ≈ 50 ps (end)

Fig. 2: Main parameters of HGTD [3].

inter-pad space, with a low capacitance of the detector, and this is relevant for the temporal resolution. For the read-out a custom low-noise ASIC ALTIROC [7], [8] is developed and bump-bonded to the sensors, to answer the demands for temporal resolution and radiation hardness. Another functionality of the ASIC will be to provide a non-biased bunch by bunch luminosity measurement, and the implementation of a trigger with minimal bias, by counting the amount of hits stored in the detector and deliver this data at 40 MHz. A 35 ps time-resolution at the start of HL-LHC (Run 4) is required, followed by a 70 ps time-resolution by the end of the HL-LHC run period. Fig. 2 summarize the main parameters of the HGTD.

B. Layout

The objective from the design of the detector is to ensure the optimal timing resolution to effectively remove the impact of the pile-up in the forward region. For this reason, each layer of the HGTD is double-sided, and each side is rotated in opposite directions by 15°. On a common cooling disk, modules with on-detector electronics and sensors are installed on both sides, front and back. The modules are arranged in an overlapping manner, with 20% overlap for the radius of the disk r > 470 mm, 54% for r ∈ [230,470] mm and 70% for r < 230 mm. The choice of the optimal overlap for each ring was based on a full simulation, in order to achieve the required timing resolution via the average number of simulated hits, given the expected time resolution of the pads.

C. Radiation hardness

The radiation hardness of the sensors and electronics is one of the most important parameters of the HGTD. To achieve sufficient performance of the sensors and ASICs, the detector layout has been designed considering a replacement scenario during the HL-LHC. Through an intensive R&D campaign, a minimum charge of 4 fC is required, this can be achieved up to a radiation damage of 2.5×10^{15} $n_{eq}cm^2$ and total ionising dose (TID) about 2.0 MGy. As a result, the sensors and electronics within the lowest-radius (r < 230 mm) will be replaced after each $1\,000\,fb^{-1}$ and the sensors and ASICs within 230 mm < r < 470 mm should be replaced at half of the data-taking ($2\,000\,fb^{-1}$) during the HL-LHC program. In the inner ring the total Si 1 MeV n_{eq} fluence has a similar proportion of charged particles and neutrons, whereas in the outer and central rings, the effect is dominated by neutrons. To account for uncertainties in the simulation, a total safety factor of 1.5 is applied for the sensors which are most sensitive to the fluence of the particles, and for the electronics which is more sensitive to TID, 2.25 as a factor. The maximum fluence and total ionising dose as a function of the radial position including the replacement of the rings can be found in Fig. 3.

D. Module components

The hybrid module is made of two parts: two LGAD sensors and two ASICs, called bare module, and the flexible printed circuit board (flex cables). The flex is made of two pieces, a small flex board permanently glued to the bare module and a long flex tail whose length, of up to about 60 cm, depends on the module position in the detector. The sensors and the ASICs are connected through a flip-chip bump bonding process called hybridization. All connections between the ASIC and the peripheral electronics are routed through the flex cable. The bare module is glued on the back side of the sensor to the flex module small piece, and all the signals are wire bonded between the ASIC and the flex cable and for the high voltage between the sensor and the flex. The size of the bare modules is around 2 × 4 cm^2, the pads size is 1.3 × 1.3 mm^2. Each sensor is connected to one ASIC, reading a matrix of 225 channels. A view of the HGTD hybrid module is showed in the Fig. 4.

III. Sensors and Readout Electronics

A. LGAD sensors

LGADs are n-on-p silicon detectors that contain an additional highly doped p-layer under the n-p junction to create a high field that causes internal amplifications referred as the gain of the LGAD as illustrated in Fig. 5. The gain is related to the dose of doping in the multiplier layer and reduces with the fluence of radiation. The HGTD LGADs were initially designed by the Centro Nacional de Microelectrnica (CNM) in Barcelona [6] and developed during the last 5 years within the CERN-RD50 community [9] including collaboration with two other LGAD vendors: Hamamatsu Photonics (HPK, Japan) and Fondazione Bruno Kessler (FBK, Italy).

Three major effects determine the time resolution: time walk from amplitude variations, jitter from electronic noise and

https://doi.org/10.1051/epjconf/202125311012

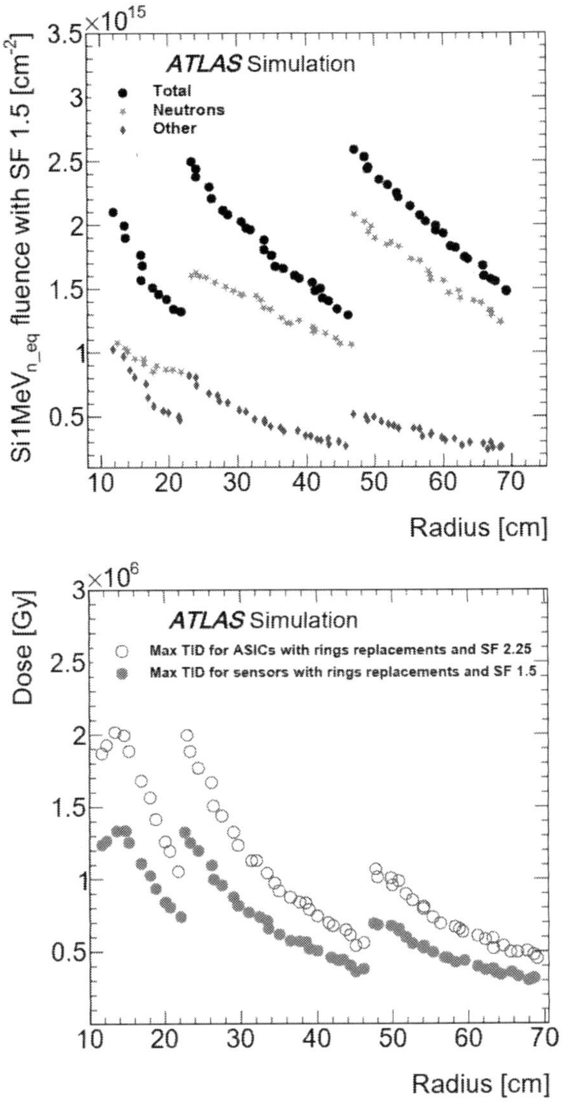

Fig. 3: Expected Si 1 MeV n_{eq} radiation levels in HGTD, using Fluka simulations, as a function of the radius considering a replacement of the inner ring every 1 000 fb^{-1} and the middle ring replaced at 2 000 fb^{-1} [ ].

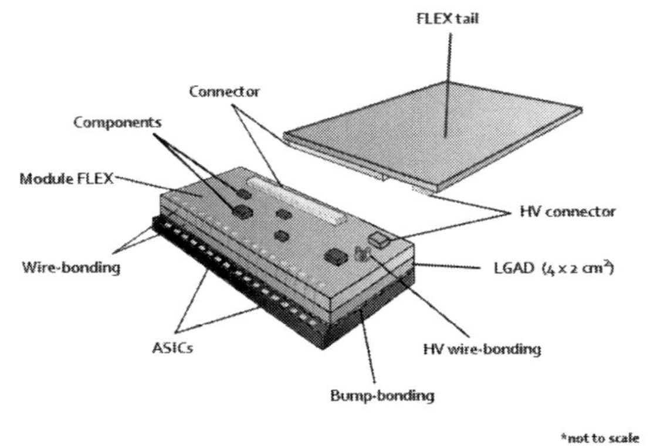

Fig. 4: View of an HGTD hybrid module equipped with its read-out flex cable tail [ ].

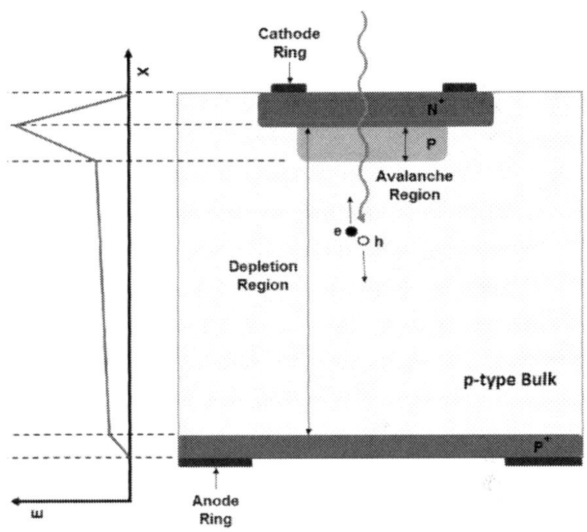

Fig. 5: View of the cross section of an LGAD [ ].

Landau fluctuations from charge deposition non-uniformities along the particle path. Time walk and noise jitter depend on the type of readout electronics chosen. Both depend inversely on the signal slope (voltage slope at the output of the amplifier) dV/dt:

$$\sigma_{TimeWalk} = [\frac{V_{th}}{\frac{S}{t_{rise}}}]_{RMS} \propto [\frac{N}{\frac{dV}{dt}}]_{RMS'},$$

$$\sigma_{Jitter} = \frac{N}{\frac{dV}{dt}} \simeq \frac{t_{rise}}{\frac{S}{N}},$$

where S refers to the signal which is proportional to the gain, V_{th} to the threshold voltage, t_{rise} to the rise time and N to the noise. The third effect, referred to Landau fluctuation is due to the non-uniform charge deposition along the particle path leading to time-of-arrival fluctuations. It is a contribution depending on the thickness of the sensor (thin is beneficial) and the setting of the threshold. Adding the three contributions in quadrature yields the overall time resolution. After time walk correction, the noise jitter is the dominating contribution for low S/N and the Landau term takes over for high S/N.

B. Electronics readout

Each LGAD will be read out using the ASIC chip, ALTIROC that will have 225 readout channels. Two iterations of this chip have been produced and tested so far: the first, ALTIROC0, integrated four pads in a 2×2 array, with the analog part of the single-channel readout: the preamplifier and the

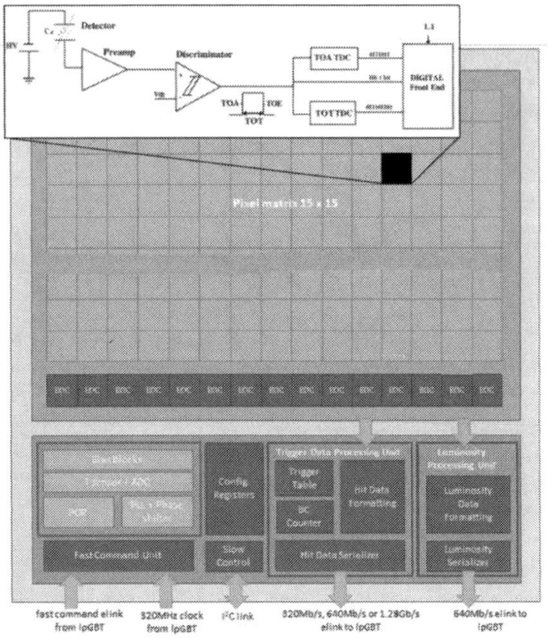

Fig. 6: Global architecture of the ALTIROC ASIC. The schematic of one Front End electronics channel is displayed on top of the channels matrix, with the preamplifier followed by a discriminator, two TDCs, and a digital front end block. The ASIC common digital part is shown in the bottom [3].

discriminator. The second iteration, ALTIROC1, consists of a 5×5 pad matrix, in which the digital components have been added to the single-channel readout. The next major ASIC iterations, ALTIROC2, will integrate all the functionality of the final ASIC and will have its final size, and the ALTIROC3 which will be the radiation hard version of the ALTIROC2. The ASIC chip will provide the Time Of Arrival (TOA), with measurements restricted to a 2.5 ns window centered on the bunch crossing, and the Time Over Threshold (TOT) used offline to correct the TOA for time walk effect in order to achieve the target time resolution with an ASIC contribution smaller than 25 ps, and its discriminator threshold expected to be about 2 fC. The design of the whole ASIC of the HGTD with 225 channels is shown in the Fig. 6.

IV. Sensors measurements at laboratory

The dynamic characteristics of LGADs, like the temporal resolutions, gain and charge collection have been determined in the laboratory with Sr^{90} β particles in response to ionizing particles. Fig. 7 illustrates the charge collected as a function of the biasing voltage for different fluences for HPK-3.2 (top) and at maximum fluence for all vendors (bottom). The horizontal lines indicate the HGTD lower charge limit of 4 fC at all fluences. Solid markers indicate neutron irradiation (n), open markers p irradiation at CYRIC (pCy). Measurements were performed at -30 °C except for the pre-rad measurement that was done at 20 °C. The HGTD requirement to collect up to 4 fC for efficient HGTD operation until the end of the HL-LHC phase has been met by these sensors.

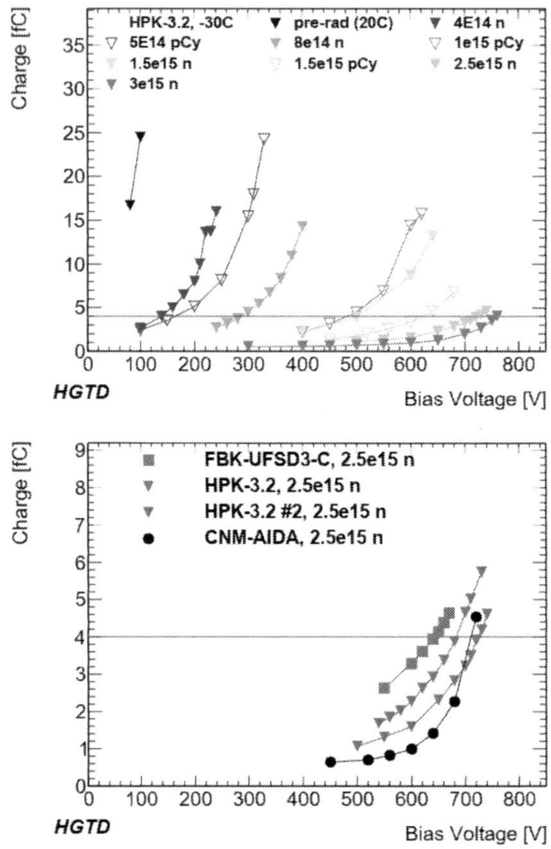

Fig. 7: Collected charge as a function of bias voltage for different fluences for HPK-3.2 (top) and at maximum fluence for all vendors (bottom) [10].

V. Test beam campaigns

To study the LGAD performance before and after irradiation, and the HGTD modules, the sensors have been exposed to charged-particle beams. The HGTD community performed the test beam in more than fifteen periods between 2016 and 2020 at the H6 beam line of the CERN SPS [11] with 40 to 120 GeV pions, at Fermilab with 120 GeV protons, at SLAC with 15 GeV electrons, and at DESY with 5 GeV electrons [12], [13]. In the next sections, we will present the experimental setup using by the DESY laboratory as a model, and the test beam results for different laboratories.

A. *Experimental setup*

An EUDET-type [14] telescope based on MIMOSA pixel planes was used to provide the position-dependent measurements. Also for the triggering purposes, an additional pixel

 https://doi.org/10.1051/epjconf/202125311012

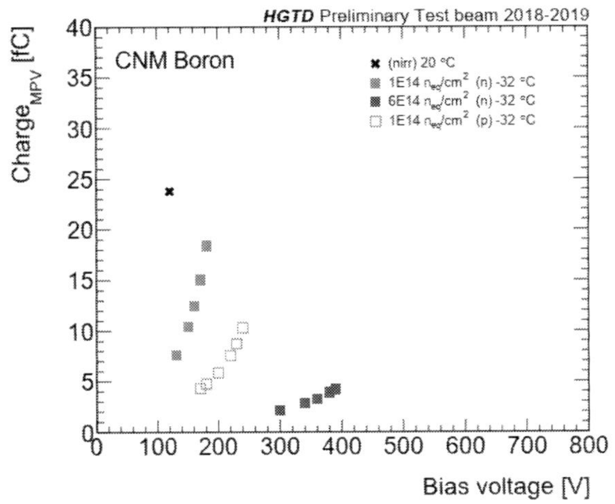

Fig. 8: Collected charge as a function of the the bias voltage for different single pad sensors built by CNM [10].

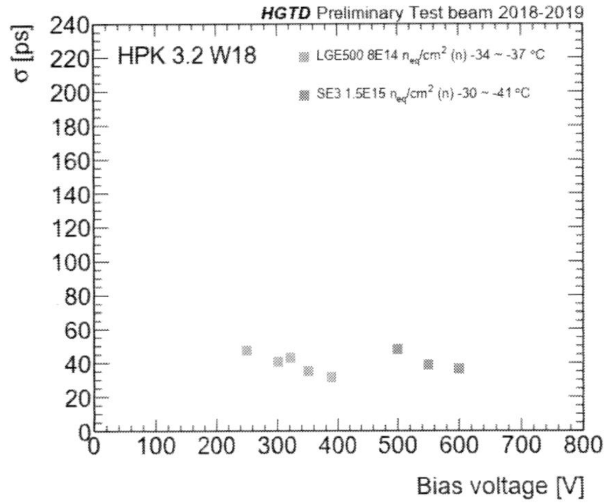

Fig. 9: Time resolution as a function of the bias voltage for different single pad sensors built by HPK [10].

reference plane was employed, using the FE-I4 [15] readout chip. The signals of the trigger were combined in the Trigger Logic Unit (TLU), whose output was used by the telescope and connected to the oscilloscope, ensuring a perfect match between the events recorded by the telescope and the oscilloscope(s). In addition Cherenkov light emitting quartz bars coupled to Silicon Photo-multipliers (SiPM) were used as timing references. The installation was assembled on a plate base linked to a remote controlled stage motors that moved in both vertical and horizontal axes orthogonal to the direction of the beam with a micrometer accuracy. The polystyrene container that covers the devices under test has been used to ensure both the light-tightness and to stabilize the temperature to about 25 °C with a water cooled system. To prevent condensation of moisture that may harm the sensors, a locally available nitrogen outlet was added to the box.

B. Results

Fig. 8 shows the collected charge as a function of the the bias voltage for different single pad sensors built by CNM from wafer W4 produced in the run 10 478, doped with Boron. Measurements were performed at -32 °C for sensors irradiated with neutrons (at $1\text{x}10^{14}$ n_{eq}/cm^2 and $6\text{x}10^{14}$ n_{eq}/cm^2) or with protons (empty markers) at $1\text{x}10^{14}$ n_{eq}/cm^2 and at 20 °C for the unirradiated one. The fluences are provided with a precision of 10 %. At every event, the collected waveform charge is computed as the area integral of the signal. Using the Landau-Gauss convoluted function, the distribution of the charge is fitted, where the charge collected is determined to be the likeliest value of the fitted function. After irradiating this sensor to $6\text{x}10^{14}$ n_{eq}/cm^2, it achieves the 4 fC value that is demanded for an ALTIROC optimal performance, and specifically 4.2 fC at 390 V.

The Constant Fraction Discriminator (CFD) have been chosen in order to optimise time resolution. Using this algorithm, for every single event in a run, the waveform TOA is calculated as the time of crossing a given threshold. To determine this threshold we use the CFD fraction from the signal amplitude, defined using the time resolution dominant factor. For unirradiated sensors, the Landau effect is the dominant, whereas is the noise in irradiated sensor. In order to provide time reference signal we used a non-irradiated LGAD. The time resolution was calculated by fitting the distribution of the difference between time at a CFD = 50% for the device under test (DUT) and the time at a CFD = 20% for non-irradiated sensor. As it is shown in Fig. 9 the HPK-3.2 sensors doped with Boron irradiated to $1.5\text{x}10^{15}$ n_{eq}/cm^2 reaches the HGTD timing requirement with 36 ps, at 600 V for a collected charge of 22.8 fC.

The hit reconstruction efficiency is determined by calculating the ratio between the number of tracks that trigger the sensor response (with a collected charge greater than 2 fC) and the total number of reconstructed tracks traversing the DUT at a given position. Fig. 10 (top) manifest the efficiency as a function of bias voltage for different single pad sensors built by CNM from wafer W6 produced in the run 10 924, doped with Gallium. Sensors were irradiated with neutrons or protons with different fluences. The efficiency is computed for a charge threshold of 2 fC which corresponds to the threshold of the ALTIROC and from hits in the pad center region (in 0.5x0.5 mm^2 area). For CNM sensors doped with gallium irradiated to $3\text{x}10^{15}$ n_{eq}/cm^2 the efficiency is about 99.7 % for the bias voltage of 740 V and with 5.3 fC as collected charge. The bottom plot of Fig. 10 shows the efficiency 2D map for the same sensor with same voltage. This map confirms that even after irradiation the sensor still very efficient in the center, while the borders are less efficient.

VI. Conclusion

The addition of the timing information is a powerful new tool to improve the performance of the HL-LHC detectors. Therefore, the HGTD detector has been proposed to meet the

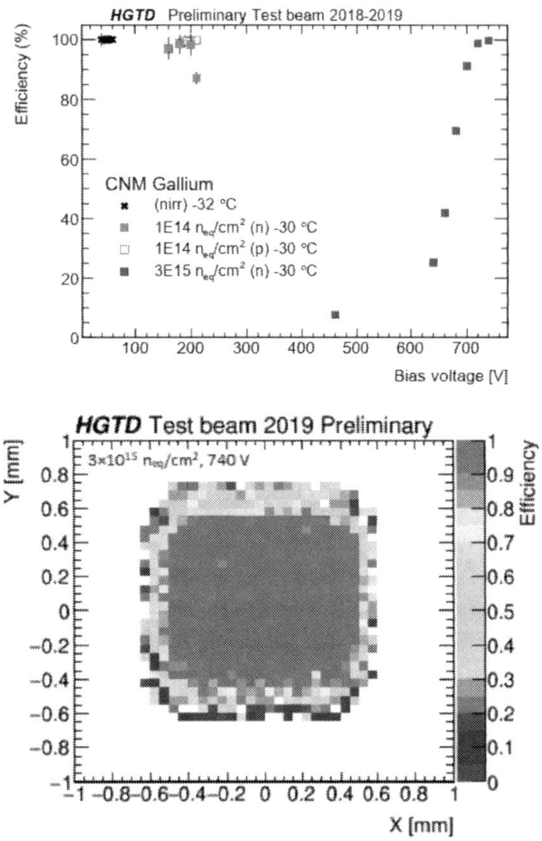

Fig. 10: Efficiency as a function of bias voltage for different single pad sensors built by CNM (top), Signal efficiency in a 1 x 1 mm2 CNM Gallium-doped LGAD irradiated to 3e15 neq/cm2, as a function of the reconstructed position X and Y of particles (in mm) (bottom) [10].

requirements of the forward region. This detector is planned to be built using LGAD sensor technology, which will provide a time resolution of 30-50 ps per track until the end of the HL-LHC lifetime. The R&D carried out shows that HGTD will improve the physical performance of the ATLAS detector in the dense environment of the HL-LHC. In addition, the technology development opens up new possibilities for the application of LGAD detectors in fields other than particle physics.

References

[1] ATLAS Collaboration, *High-Luminosity Large Hadron Collider (HL-LHC): Technical Design Report V. 0.1*, ser. CERN Yellow Reports: Monographs. Geneva: CERN, 2017. [Online]. Available: http://cds.cern.ch/record/2284929

[2] ATLAS Collaboration, "Technical Design Report for the ATLAS Inner Tracker Pixel Detector," CERN, Geneva, Tech. Rep., Sep 2017. [Online]. Available: https://cds.cern.ch/record/2285585

[3] ATLAS Collaboration, "Technical Design Report: A High-Granularity Timing Detector for the ATLAS Phase-II Upgrade," CERN, Geneva, Tech. Rep., Jun 2020. [Online]. Available: https://cds.cern.ch/record/2719855

[4] ATLAS Collaboration, "The ATLAS Experiment at the CERN Large Hadron Collider," *JINST*, vol. 3, p. S08003. 437 p, 2008, also published by CERN Geneva in 2010. [Online]. Available: https://cds.cern.ch/record/1129811

[5] ATLAS Collaboration, "ATLAS Liquid Argon Calorimeter Phase-I Upgrade: Technical Design Report," Tech. Rep., Sep 2013, final version presented to December 2013 LHCC. [Online]. Available: https://cds.cern.ch/record/1602230

[6] G. Pellegrini, P. Fernández-Martínez, M. Baselga, C. Fleta, D. Flores, V. Greco, S. Hidalgo, I. Mandić, G. Kramberger, D. Quirion, and M. Ullan, "Technology developments and first measurements of Low Gain Avalanche Detectors (LGAD) for high energy physics applications," *Nuclear Instruments and Methods in Physics Research Section A: Accelerators, Spectrometers, Detectors and Associated Equipment*, vol. 765, pp. 12–16, 2014. [Online]. Available: https://www.sciencedirect.com/science/article/pii/S0168900214007128

[7] C. Agapopoulou, S. Blin, A. Blot, L. Castillo García, M. Chmeissani, S. C. Di Lorenzo, C. de La Taille, P. Dinaucourt, A. Fallou, J. G. Rodriguez *et al.*, "Performance of a front end prototype asic for picosecond precision time measurements with lgad sensors," *Journal of Instrumentation*, vol. 15, no. 07, p. P07007, 2020.

[8] C. de la Taille, S. Callier, S. Conforti, P. Dinaucourt, G. Martin-Chassard, N. Seguin-Moreau, C. Agapopoulou, N. Makovec, L. Serin, and S. Simion, "ALTIROC0, a 20 pico-second time resolution ASIC for the ATLAS High Granularity Timing Detector (HGTD)," *PoS*, vol. TWEPP-17, p. 006, Sep 2017. [Online]. Available: https://cds.cern.ch/record/2319292

[9] RD50 Collaboration. [Online]. Available: https://rd50.web.cern.ch/rd50

[10] ATLAS HGTD Public Plots. [Online]. Available: https://twiki.cern.ch/twiki/bin/view/AtlasPublic/HGTDPublicPlots

[11] C. Allaire, J. Benitez, M. Bomben, G. Calderini, M. Carulla, E. Cavallaro, A. Falou, D. Flores, P. Freeman, Z. Galloway, E. Gkougkousis, H. Grabas, S. Grinstein, B. Gruey, S. Guindon, A. Henriques Correia, S. Hidalgo, A. Kastanas, C. Labitan, D. Lacour, J. Lange, F. Lanni, B. Lenzi, Z. Luce, N. Makovec, G. Marchiori, L. Masetti, A. Merlos, F. McKinney-Martinez, I. Nikolic-Audit, G. Pellegrini, R. Polifka, D. Quirion, A. Rummler, H. Sadrozinski, A. Seiden, L. Serin, S. Simion, E. Spencer, S. Trincaz-Duvoid, M. Wilder, A. Zatserklyaniy, D. Zerwas, and Y. Zhao, "Beam test measurements of Low Gain Avalanche Detector single pads and arrays for the ATLAS High Granularity Timing Detector," *JINST*, vol. 13, p. P06017. 28 p, Apr 2018. [Online]. Available: https://cds.cern.ch/record/2313951

[12] R. Diener, J. Dreyling-Eschweiler, H. Ehrlichmann, I. M. Gregor, U. Kötz, U. Krämer, N. Meyners, N. Potylitsina-Kube, A. Schütz, P. Schütze, and M. Stanitzki, "The DESY II test beam facility," *Nuclear Instruments and Methods in Physics Research Section A: Accelerators, Spectrometers, Detectors and Associated Equipment*, vol. 922, pp. 265–286, 2019. [Online]. Available: https://www.sciencedirect.com/science/article/pii/S0168900218317868

[13] L. Castillo Garcia, "A High-Granularity Timing Detector for the Phase-II upgrade of the ATLAS Calorimeter system: detector concept, description and R&D and beam test results," CERN, Geneva, Tech. Rep., May 2020. [Online]. Available: https://cds.cern.ch/record/2717972

[14] H. Jansen, S. Spannagel, J. Behr, A. Bulgheroni, G. Claus, E. Corrin, D. Cussans, J. Dreyling-Eschweiler, D. Eckstein, T. Eichhorn *et al.*, "Performance of the eudet-type beam telescopes," *EPJ Techniques and Instrumentation*, vol. 3, pp. 1–20, 2016. [Online]. Available: https://link.springer.com/content/pdf/10.1140/epjti/s40485-016-0033-2.pdf

[15] U. Koetz and I. Rubinskiy, "User manual: Atlas fe-i4a pixel module as a trigger plane for the beam telescope," 2013. [Online]. Available: https://telescopes.desy.de/images/0/08/131014_koetz-rubinsky_FEI4_as_trigger_plane.pdf

Towards the experimental validation of a small Time-Projection-Chamber for the quasi-absolute measurement of the fission cross section

Carole Chatel[1,2], Ludovic Mathieu[1], Mourad Aïche[1], Maria Diakaki[2], Gilles Noguere[2] and Olivier Bouland[2]

[1]Centre d'Etude Nucléaire de Bordeaux-Gradignan, FRANCE
[2]CEA Cadarache, DES/IRESNE/DER/SPRC/LEPh, FRANCE
Corresponding author: carole.chatel@cea.fr

Abstract—**To accurately measure neutron-induced fission cross sections, to characterize neutron-beam lines or to make dosimetric investigations, it is necessary to have high accuracy measurements of neutron fluence. It is possible to perform independent and precise neutron flux measurements with respect to the $^{1}H(n,n)p$ elastic scattering cross section. The use of a silicon detector is recommended from 1 to 70 MeV neutron energy. However, it has been observed that a high electrons background forbids its use below 1 MeV. Hence, a new gaseous proton-recoil telescope is developed and characterized to overcome this limit. It should provide quasi-absolute neutron flux measurements with an accuracy around 3% and is not sensible to gamma and electrons background. It consists in two ionization chambers read by a segmented micromegas technology detection plane. The gas pressure inside is adjustable to the proton range in the detector and therefore to the neutron energy. This detector is described in details below and the newest results of its characterization are presented. A special attention is paid to detection efficiency measurements.**

***Keywords* — Gaseous proton recoil telescope, low energy neutron flux measurements.**

I. Introduction

Absolute measurements of neutron fluence is a prerequisite for neutron-induced cross section measurements, neutron-beam lines characterization or dosimetric investigations. The development of the Generation IV nuclear systems (fast nuclear reactors) [1] or the accuracy improvement of the current neutronics code calculation require precise cross sections of neutron-induced reactions, mainly on heavy nuclei. However, despite the great general effort made by the nuclear community, discrepancies still exist between different measurements, especially in the actinides region.

To measure neutron-induced reaction cross sections, secondary standards ($^{235}U(n,f)$, $^{238}U(n,f)$, $^{237}Np(n,f)$) are usually used for the neutron flux measurements. These standards are measured with respect to a primary standard, so their accuracy ranges from 0.5 to 10% between 0 to 20 MeV. Moreover, the use of the same secondary creates correlations between the different measurements. An independent and precise technique of neutron-flux measurements becomes desirable. It is then possible to short-circuit secondary standards by using directly a primary standard. This can be performed with respect to the $^{1}H(n,n)p$ elastic scattering cross section, known with an accuracy between 0.2 % to 0.5% in a large energy range (0 to 20 MeV) [2, 3] and reproducible with ab initio calculations. This cross section has also the advantage of being monotonous. Hence, observed resonances are not correlated to standards' one. Silicon junction to make silicon-based proton-recoil telescope is the best choice to perform mono-energetic neutron flux measurements between 1 to 70 MeV neutron energy [4, 5, 6, 7]. However, the maximum flux is between 100 keV to few MeV in fast nuclear reactors and measurements at these low energies prevent the use of silicon detectors. Indeed, even if adapting silicon's thickness to neutrons energies is recommended, a high gamma and electrons background due to the neutrons' source prevent the use of a silicon below 1 MeV [4]. This is because the energy of the electron deposited in the whole detector thickness in the silicon detectors becomes comparable to the energy of the proton which prohibits a clear discrimination between the protons' signal and the background.

Of course, other proton-recoil detectors have been developed. In this neutron energy range, plastic scintillators [8] are commonly used. Yet, the light produced by these detectors is not linear with neutron energy. Moreover, neutron reactions on the ^{12}C must be discriminated [9-12]. Finally, their detection efficiency cannot be accurately measured. Thus, these detectors are not suited to our needs. Proportional counter detectors also exist [13-15]. However, their efficiency must be rebuilt by simulation, along with other constraints (gas ionization potential, ionizing particle energy, edge-effects, interferences from photons and carbon recoils [13, 16 17]) which prohibit again their use for our needs.

Thus, it seems that no detector already exists to answer our needs. That is why the ACEN team of the CENBG decided to develop a gaseous proton-recoil telescope able to accurately measure neutron fluxes in the neutron energy range of 0.2 to 2 MeV and not sensitive to high gamma and electrons background.

II. Detector description

The gaseous proton-recoil telescope (GPRT) is composed of two ionization chambers (ΔE and E) separated by a collimator. The electric field is made by the presence of a cathode and a micromegas technology detector plane. This plane is composed by two electrodes: an anode segmented in 64 pads and a biased mesh 120 μm above. A field cage homogenizes the electric field [18]. A macor disk to carry samples stands upstream the detector. A first collimator stands after and represents the entrance of the detector. A disposition of the detector is shown in figure 1. A zoom on the detection plane is made in figure 2.

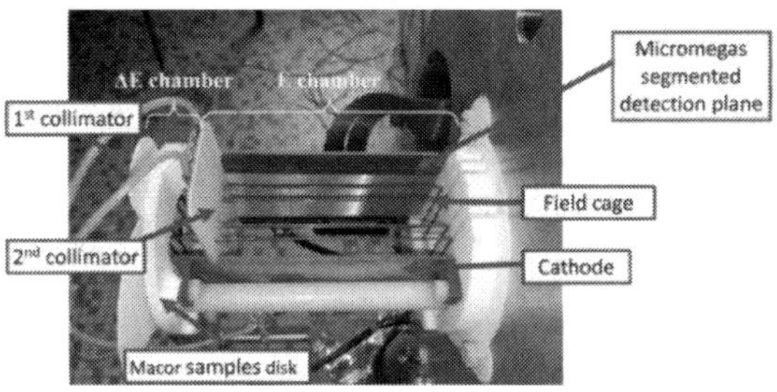

Figure 1: Picture of the Gaseous Proton-Recoil Telescope

Figure 2: Micromegas detector plane. Red thick lines represent the places of both collimators (ΔE chamber)

A. *Choices made for the GPRT optimal functioning*

The macor disk can carry up to eight H foils. The aim is to adapt the foil thickness to the neutron energy so the energy loss of the recoil-proton is less than 10% in the foil. This prohibits a too big energy spread, which is essential to make accurate measurements.

Neutrons imping an H foil on the samples disk and recoil protons are collimated a first time before entering the detector to clearly define the sample active area. They are collimated a second time before being detected in the second chamber (E chamber). The aim is to limit the protons' angle so that their energy is similar to that of the impinging neutrons. The presence of these two collimators also well-defines the geometry, and thus the geometrical efficiency, and allows a physical separation between the two chambers. The detector plane is segmented in 64 pads that will each collect an individual signal. Thus, it is possible to check the proton's track. Correct proton tracks have to activate pads in the ΔE chamber and in the E chamber and to have passed through both collimators. Parasitic tracks, for instance those induced by scattered neutrons, have to be discarded.

To reduce as much as possible the number of these unwanted protons, the collimators, as well as the structures, the samples disk and the detector chamber, are made of macor, a H-free material and with as little material as possible to reduce neutrons scattering.

The chamber is filled with an H-free gas. The one chosen is made of 70% N_2, 30% CO_2. We will discuss later this choice. The gas pressure is chosen so that protons are stopped in the detector. Thus, it is possible to see their Bragg's peak. Once the gas pressure is decided, the polarization voltage have to be applied. It has to be as high as possible to maximize the amplification gain. The breakdown voltages as a function of pressure for this gas have been investigated. To avoid micro-breakdown occurring when operating near the breakdown voltage, it has then been decided to work 20V below breakdown voltages.

B. *Acquisition*

The signal collected by the micromegas pads is treated by three successive electronics cards. The first one: the Zap card, is useful to protect the two other cards in case of breakdowns. The second card, called SAM for Single AGET Module, pre-amplifies the signal with four possible gains chosen by the user. A discriminator also receives the signal. The discriminator compares the signal level with the amplitude threshold and with the multiplicity threshold, both chosen by the user. This means that the signal will be registered only if the number of activated pads by the incoming particle is equal or higher than the number chosen by the user as triggers and that their amplitude is higher than the threshold. The last card, named ZedBoard, reduces the data size, sends events and timing to the computer. All the signal treatment last few nanoseconds per point and the dead time is of 20.5 μs per channel, corresponding to 1.4 ms for the 64 pads channels.

Once treated, the signal is sent to the acquisition system. The software, named Cobo, has been developed by CEA Saclay. It allows the user to determine adjustable parameters as threshold, multiplicity, pads triggers and gain; to start and stop the acquisition and to visualize detected events.

Another software (CoboFrameViewer, developed at CEA Saclay) is used to read recorded data once stopped and saved. A Frame is composed of 64 curves, each corresponding to the response of one pad. An example is given in figure 3.

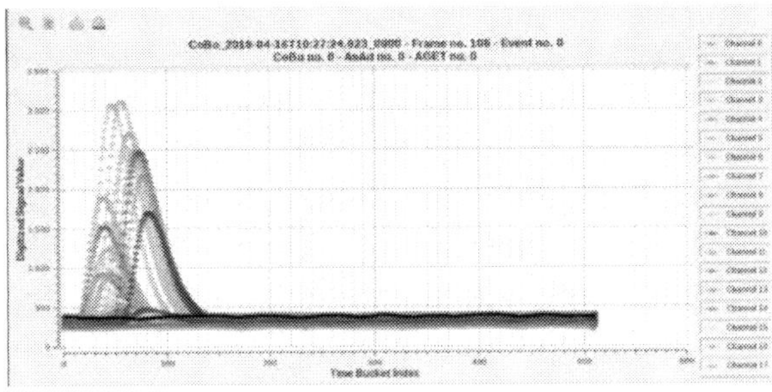

Figure 3: Frame obtained during an acquisition

C. Tracks reconstruction and discrimination

As each activated pad can be spotted by its curve, it is then possible to reconstruct the track accordingly to the activated pads. An example of a reconstructed track of an event is given in figure 4.

Figure 4: Example of a proton track reconstruction.

Reconstructing tracks is very useful to check the detected particle origin. Indeed, protons coming from scattered neutrons may not have a straight track passing through both collimators. Checking the ΔE-E coincidence is thus essential to discriminate and exclude from the analysis protons from scattered neutrons. Another possible but less convenient way to discriminate them is to compare the signal amplitude in frames. Indeed, a proton coming from a scattered neutron have a lower energy than a recoil proton and so has a shorter range. Its Bragg's pic is then before the one of a recoil proton. The track of a proton coming from a scattered neutron is presented in figure 5 (left). Sometimes, the GPRT detects other events coming from cosmics. Their tracks are easily recognized due to their vertical trajectory. The track of a cosmic ray is presented in figure 5 (right).

Figure 5: Reconstruction of the track of a proton coming from a scattered neutron (left) and of a cosmic ray (right)

Furthermore, it is possible to reconstruct the third dimension of tracks thanks to electrons drift velocity. Indeed, the chosen gas can be considered as a slow gas, namely electrons move slowly to join the anode. The drift velocity is quite constant at about 4 cm/µs for the different conditions used. Considering that a particle ionizes all the gas at the same time, electrons produced the closest of the detection plane arrive the first. The position shift of the different peaks (see Figure 3) allows to infer the arrival time via the sampling frequency (10 ns per point). For example, in the frame showed in Figure 3, the curves' maximum of the last pads in the segmented plane arrive in last so the particle got further from the detection plane as represented in figure 6.

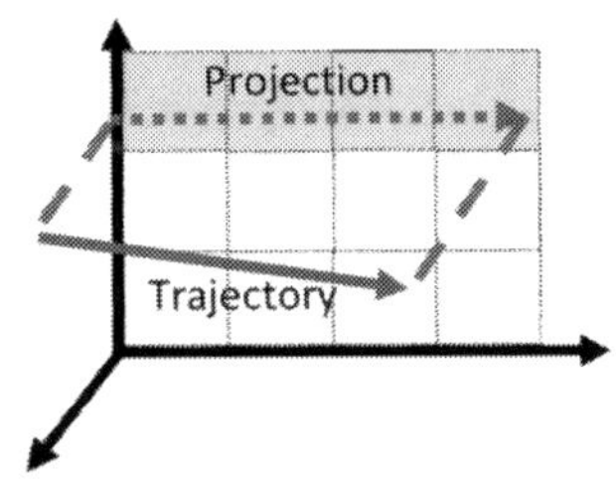

Figure 6: reconstruction of the 3rd dimension thanks to the electron drift velocity.

Background tracks from γ-rays or electrons has also been investigated. As long as the threshold was suitable for proton detection, no unwanted-even frame was detected. One can deduce that the GPRT is non-sensitive neither to γ-rays nor to electrons, once proper trigger and threshold are set for protons.

Once the recoil-protons tracks can be discriminated from parasitic events, it is necessary to count them to make accurate measurements. In fact, any recoil-proton entering the detector must be detected to reduce as much as possible the uncertainties. This means that the detector's intrinsic efficiency must be equal to 100%. Now that the device, acquisition and tracks reconstruction has been described, the determination of the GPRT's efficiency is described in the following section.

III. Detection efficiency measurements

First, the detection efficiency had been measured with an alpha source. Then, a mono-energetic proton micro-beam has been used to verify the results and extend them to other energies and other counting rates.

A. Intrinsic efficiency with an alpha source

The intrinsic efficiency depends of two other terms: the geometric efficiency and the detection efficiency as shown in equation (1):

$$\varepsilon_{intrinsic} = \frac{\varepsilon_{detection}}{\varepsilon_{geometric}} \quad (1)$$

In the case of the use of a source, the detection efficiency is equation (2)

$$\varepsilon_{detection} = \frac{N}{A_{source} * t} \quad (2)$$

With N, the number of events detected during an acquisition, t the time of this acquisition and A_{source} the activity of the alpha source. Hence, the geometric efficiency has to be calculated and the source activity has to be accurately measured before running an acquisition of a precise time.

The geometric efficiency was calculated by means of a Monte Carlo simulations. The parameters to calculate the GPRT geometric efficiency with the error are presented in Table I. The sensitivity of the geometric efficiency to each parameter has been investigated to infer the final uncertainties and is also presented in Table I. The same effort has been made to define the alpha activity uncertainty.

TABLE I
PARAMETERS TO CALCULATE THE GPRT GEOMETRIC EFFICIENCY

Parameters	Values (mm)	Absolute uncertainties (mm)	Sensitivity (%/mm)
Collimator radius	5.00	0.01	38.91
Source radius	4.10	0.25	0.82
Distance source-collimator	26.2	0.5	7.33

The geometric efficiency of the GPRT device is (0.87 ± 0.03) %.

The alpha activity A_{source} was measured by a dedicated alpha spectrometry. A simple device with the source and a silicon detector was used to reach this goal. Silicon detectors are known to have an intrinsic efficiency of 100%. The detector was placed as far as possible to minimize the geometric efficiency error. This reduces the statistics that was compensated by a longer run. As before, Monte-Carlo simulations were used to estimate the geometric efficiency of this setup.

The geometric efficiency of the device to measure the alpha source activity is (0.0947 ± 0.0008)% for a measured activity of $(517.9 \pm 4.4^{sys} \pm 1.9^{stat})$ Bq.

Knowing the alpha source activity and the GPRT geometric efficiency, acquisitions were performed to infer the intrinsic efficiency.

To run acquisitions, parameters introduced in part I.A. must be defined. Even if the alpha energies are much higher than foreseen recoil protons energies, we still want to be representative of the intended use of the GPRT. Thus, the chosen pressure is 100 mbar and the voltage is -2000 V for the cathode and -430 V for the electrified mesh. Intrinsic parameters of the firmware must also be defined. To stay in consistency with the intended use, the chosen triggers are the three columns after the second collimator. To be in optimal conditions, the multiplicity is one so a frame is registered if at least one of the pads of these three columns is activated. A multiplicity of two has also been tested. Tests have been done with different gains and different thresholds in order to optimize the intrinsic efficiency and the rate of parasitic events to reject.

Optimal conditions appeared to be with the highest gain and a multiplicity of one. The intrinsic efficiency is then $(99.2 \pm 3.7^{sys} \pm 1.1^{stat})$ % for 4.5 alphas per second passing through the second collimator. This result is compatible with the expected 100%. However, the uncertainties are large. We can see that the limiting factor is the detector geometry that will have to be improved in the next version of the prototype. Tests proved that the threshold does not influence the efficiency for such high-energy particles. However, a higher multiplicity or a smaller gain decrease significantly the efficiency. The hardware, firmware and software response to these parameters must further be investigated.

Now that one knows the intrinsic efficiency with optimal conditions and with an alpha source, the use of a proton beam is required to check the results.

B. Intrinsic efficiency with a proton beam

The proton beam was performed by the AIFIRA facility (Bordeaux, France). The typical current at such facilities is few nA (i.e. 10^{10} protons/s) [19, 20] for classic analysis but can be tuned from 50 µA to few ions/s depending on the purpose of the experiment and the chosen beamline. For this experiment, we need only few protons/s. Thus, the beam was defocused and passed through a micro collimator with a diameter of 20 µm.

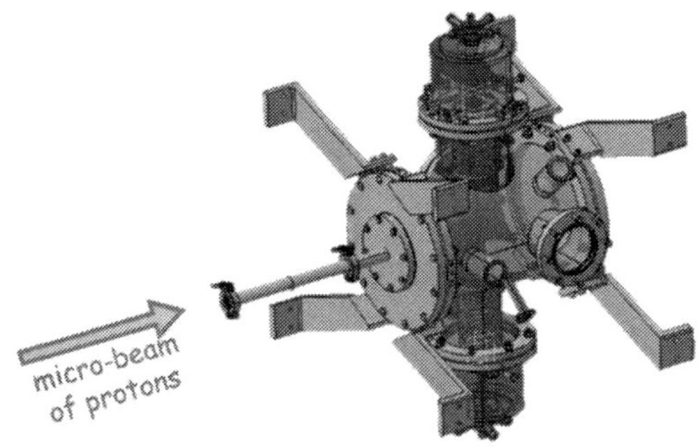

Figure 7: Scheme of the designed chamber for this experiment.

For this experiment, a dedicated chamber has been designed. This chamber is schematically shown in Figure 7. It has been designed to answer many goals. First of all, the chamber must be connected to the AIFIRA beam line. A window separates the accelerator vacuum from the gas filled chamber. The GPRT fits in the chamber, is connected to the electronics cards and aligned with the beam. As the number of protons entering the detector must be low, a collimator is placed at the entrance of the chamber and a second is before the entrance of the detector (to stop scattered protons from the first collimator). A set-up has also been designed so the GPRT can be moved inside the chamber to send the beam in different places of the detector (for instance, to test its spatial resolution). Finally, at the end of this chamber, a silicon detector can be fixed and connected so it will count the number of protons passing through the GPRT. All this device allows a much more precise analysis of the behavior and of the reproducibility of events than with an alpha source thanks to a mono-directional and mono-energetic protons beam.

The main test was to check the detector efficiency. The protons were delivered with an energy of 3 MeV. A silicon detector was placed downstream as explained above. Acquisitions were run with the GPRT optimal conditions mentioned above and the number of particles per second ranged from 0.6 protons/s to 366 protons/s. The gas pressure used was 100 mbar. At this pressure, the energy deposition of 3 MeV protons is similar to that of low energy recoil protons. The efficiency resulted to be 100% for less than 3 protons/second. Above this value, the efficiency decreased to reach 7% for a beam of 366 protons/s. The 50% efficiency corresponded to 38 protons/s. This means that the GPRT seems to have an effective dead time of 26 ms, much higher than the theoretical 1.4 ms presented in section I.B. However, these results are still coherent with the efficiency measured with the low activity alpha source.

A second set of tests was then performed. The silicon detector was there taken as a trigger for the GPRT. With these conditions, one can expect to overcome every GPRT issue and to register as many events as the silicon detector. However, the results were very similar to the first test. An investigation is currently carried out but the problem seems to come from the hardware or the software.

IV. Conclusions

The Gaseous Proton Recoil Telescope is a miniaturized Time Projection Chamber developed at the CENBG (Bordeaux, France). This detector is intended to accurately measure a neutrons flux for fission cross sections measurements.

The GPRT characterization has been investigated with an alpha source and a direct protons beam. It appeared that the GPRT has an intrinsic efficiency equal to 100% with optimal initial conditions and with less than few particles per second. However, this efficiency falls when more restrictive conditions are used or when the number of particles becomes too high.

Further investigations are currently carried out to solve the efficiency issue. A special attention will be paid to the hardware and the software. Finally, once these problems are solved, another experiment will be planned on the AIFIRA facility to check the response of the detector with many different conditions.

Acknowledgment

The authors would like to acknowledge the AIFIRA staff for their help to realize and manage the experiment as well as their ideas to make better tests and to get better results.

References

[1] NEA, "Nuclear data high priority request list", 2018
[2] J.C. Hopkins, G. Breit, "The $^1H(n,n)^1H$ scattering observables required for high-precision fast-neutron measuremenst", Nuclear Data Tables A 9 137, 1971
[3] G.M. Hale, *et al.*, Proc. 17th Int. IUPAP, conf. on few-body problems in physics, Elsevier, Amsterdam pp S120-S122, 2004
[4] P. Marini, *et al.*, "Systematic investigation of background sources in neutron flux measurements with proton-recoil silicon detector", Nuclear Instruments and Methods in physics A 841 56-64, 2016
[5] P. Marini, *et al.*, "^{242}Pu neutron-induced fission cross section measurement from 1 to 2 Mev", Physical Review C 96, 054604 (2017)
[6] S.J.Bame Jr, *et al.*, "Absolute determination of monoenergetic neutron flux in the energy range 1 to 30 MeV", Review of Scientific Instruments 28, 997, 1957
[7] T. Ryves, Nuclear Instruments and Methods in physics 135, 455, 1976
[8] R. Beyer, *et al.*, "Proton-recoil detectors for time-of-flight measurements of neutrons with kinetic energies from some tens of keV to a few MeV," Nuclear Instruments and Methods in physics A 575, 449, 2007
[9] M.W. Wu, *et al.*, "Absolute neutron fluence measurements between 0.5 and 3 MeV and their intercomparisons", Nuclear Instruments and Methods in physics A 422, 79, 1999
[10] B. Daub, *et al.*, "Response of BC-418 plastic scintillator to low energy protons", Nuclear Instruments and Methods in physics A 701, 171, 2013
[11] M. Kovash, *et al.*, ANIMMA 2011, PP 1-3, 2011
[12] D. Schmidt, *et al.*, "Characterization of liquid scintillation detectors", Nuclear Instruments and Methods in physics A 476, 186, 2002
[13] T.H.R. Skyrme, *et al.*, "A proportional counter for neutron flux measurements in the energy range 0.1 to 1 MeV", Review of Scientific Instruments 23, 204, 1952
[14] E.F. Bennett, "Proportional counter proton-recoil spectrometer with gamma discrimination", Review of Scientific Instruments 33, 1153, 1962
[15] E.F. Bennett, T.J. Yule, "A neutron spectrum map of the JANUS irradiation facility using proton-recoil proportional counters", Radiation Research society 50, 219, 1972
[16] R. Babut, V Gressier, "(n,p) reactions contribution in spherical proton recoil proportional counter responses", Journal of Instrumentation 2, P01005, 2007
[17] R. Nolte, "Detection of neutrons", Ecole Joliot Curie 2014, 2014
[18] P. Marini, *et al.*, "Development of a Gaseous Proton-Recoil Detector", WONDER-2018, 2018
[19] P. Barberet, *et al.*, "AIFIRA : a light ion beam facility for ion beam analysis and irradiation", the European Physical Journal Plus, 2020
[20] S. Sorieul, P. Alfaurt, L. Daudin, L. Serani, P. Moretto, Nuclear Instruments and Methods in Physics Research Section B: Beam Interactions with Materials and Atoms 332, 68 (2014)

AUTHOR INDEX

AUTHOR INDEX

AUTHOR INDEX

AUTHOR INDEX

AUTHOR INDEX

EDP Sciences – Web of Conferences
17, Avenue du Hoggar
Parc d'Activité de Courtabouf
BP 112
F-91944 Les Ulis Cedex A
France

ISBN 978-1-7138-3952-1

9 781713 839521

CURRAN ASSOCIATES INC.
proceedings.com